PARTICLES
AND
FIELDS

Previous Proceedings in the Series of Mexican Workshops on Particles and Fields

	Year	Held in	Publisher	ISBN
8th	2002	Zacatecas, Zacatecas	AIP Conf. Proceedings vol. 623	0-7354-0072-5
7th	1999	Mérida, Yucatán	AIP Conf. Proceedings vol. 531	1-56396-954-8
6th	1997	Morelia, Michoacán	AIP Conf. Proceedings vol. 445	1-56396-791-X
5th	1995	Puebla, Puebla	AIP Conf. Proceedings vol. 359	1-56396-548-8

Previous Proceedings in the Series of Mexican Schools on Particles and Fields

	Year	Held in	Publisher	ISBN
X	2003	Playa del Carmen, Quintana Roo	AIP Conf. Proceedings vol. 670	0-7354-0135-7
IX	2000	Metepec, Puebla	AIP Conf. Proceedings vol. 562	1-56396-998-X
VIII	1998	Oaxaca, México	AIP Conf. Proceedings vol. 490	1-56396-895-9
VII	1996	Mérida, Yucatán	AIP Conf. Proceedings vol. 400	1-56396-686-7

To learn more about these titles, or the AIP Conference Proceedings Series, please visit the webpage **http://proceedings.aip.org/proceedings**

PARTICLES AND FIELDS

X Mexican Workshop on Particles and Fields

Morelia, Michoacán, México   6 –12 November 2005

■ PART A

EDITORS
Adnan Bashir
Victor Villanueva
Luis Villaseñor

*University of Michoacán
Morelia, Michoacán, México*

SPONSORING ORGANIZATIONS
University of Michoacán (UMSNH)
Division of Particles and Fields of the Mexican Physical Society (DPyC-SMF)
Mexican Council on Science and Technology (CONACyT)

Melville, New York, 2006
AIP CONFERENCE PROCEEDINGS ■ VOLUME 857

Editors

Adnan Bashir
Victor Villanueva
Luis Villaseñor

Instituto de Física y Matemáticas
UMSNH
Edificio C3, Ciudad Universitaria
58040, Morelia, Michoacán, México

E-mail: adnan@ifm.umich.mx
vvillanu@ifm.umich.mx
villasen@ifm.umich.mx

Authorization to photocopy items for internal or personal use, beyond the free copying permitted under the 1978 U.S. Copyright Law (see statement below), is granted by the American Institute of Physics for users registered with the Copyright Clearance Center (CCC) Transactional Reporting Service, provided that the base fee of $23.00 per copy is paid directly to CCC, 222 Rosewood Drive, Danvers, MA 01923, USA. For those organizations that have been granted a photocopy license by CCC, a separate system of payment has been arranged. The fee code for users of the Transactional Reporting Services is: 978-0-7354-0354-3/06/$23.00

© 2006 American Institute of Physics

Permission is granted to quote from the AIP Conference Proceedings with the customary acknowledgment of the source. Republication of an article or portions thereof (e.g., extensive excerpts, figures, tables, etc.) in original form or in translation, as well as other types of reuse (e.g., in course packs) require formal permission from AIP and may be subject to fees. As a courtesy, the author of the original proceedings article should be informed of any request for republication/reuse. Permission may be obtained online using Rightslink. Locate the article online at http://proceedings.aip.org, then simply click on the Rightslink icon/"Permission for Reuse" link found in the article abstract. You may also address requests to: AIP Office of Rights and Permissions, Suite 1NO1, 2 Huntington Quadrangle, Melville, NY 11747-4502, USA; Fax: 516-576-2450; Tel.: 516-576-2268; E-mail: rights@aip.org.

L.C. Catalog Card No. 2006932275
ISBN 978-0-7354-0354-3
ISSN 0094-243X

Printed in the United States of America

CONTENTS

Preface ix
Acknowledgments x

INVITED LECTURES

The Brave ν World 3
    A. de Gouvêa
Experimental Status of B Physics 18
    S. Stone
Higgsless Models: Lessons from Deconstruction 34
    E. H. Simmons, R. S. Chivukula, H.-J. He, M. Kurachi, and M. Tanabashi
A Perspective on Hadron Physics 46
    A. Höll, C. D. Roberts, and S. V. Wright
Experimental Highlights of the RHIC Program 62
    P. Fachini
Theoretical Status of the RHIC Program 76
    J. Jalilian-Marian
Precision Electroweak Physics 85
    J. Erler
The Pierre Auger Observatory—Status and First Results 100
    L. Nellen *(On behalf of the Pierre Auger Collaboration)*

INVITED LABORATORY COURSES

Detection of Cherenkov Photons with Multi-anode Photomultipliers 113
    H. Salazar, E. Moreno, T. Murrieta, and L. Villaseñor
X-ray Spectroscopy with PIN Diodes 121
    F. J. Ramírez-Jiménez
A Pedagogical Multiwire Proportional Chamber 134
    R. Alfaro
Silicon Microstrip Detectors 143
    L. M. Montaño
Laboratory Course on Drift Chambers 152
    Ix-B. García-Ferreira, J. García-Herrera, and L. Villaseñor

CONTRIBUTED TALKS

The Schwinger Model on a Nonperiodic Lattice 170
    R. G. Campos and E. S. Tututi
Proton/Pion Ratios and Radial Flow in pp and Peripheral Heavy
Ion Collisions 175
    E. Cuautle and G. Paić

Rare Top Quark Decays in Extended Models. 179
  R. Gaitán, O. G. Miranda, and L. G. Cabral-Rosetti
High-accuracy Critical Exponents for O(N) Hierarchical 3D Sigma Models 186
  J. J. Godina, L. Li, Y. Meurice, and M. B. Oktay
Towards the Establishment of Nonlinear Hidden Symmetries of the Skyrme Model 194
  A. Herrera-Aguilar, K. Kanakoglou, and J. E. Paschalis
Predictions for the Higgs Mass from the Stability and Triviality Conditions 202
  H. G. Solís R., S. R. Juárez W., and P. Kielanowski
Recent Results from the H1 Experiment of HERA 210
  R. Lopez-Fernandez
Radiation Hardness Tests of a Scintillation Detector with Wavelength Shifting Fiber Readout 218
  R. Alfaro, E. Cruz, M. I. Martinez, L. M. Montaño, G. Paić, and A. Sandoval
On Gauge Independent Dynamical Chiral Symmetry Breaking 226
  A. Bashir and A. Raya
Growth of Magnetic Field in a Two-component Plasma 234
  J. F. Nieves and S. Sahu
Pair Creation in Inhomogeneous Fields from Worldline Instantons 240
  G. V. Dunne and C. Schubert
Two Time Physics and Hamiltonian Noether Theorem for Gauge Systems 249
  J. A. Nieto, L. Ruiz, J. Silvas, and V. M. Villanueva
Search for Gamma Ray Bursts at Sierra Negra, México 259
  H. Salazar, C. Alvarez, O. Martínez, and L. Villaseñor

POSTER SESSIONS

Fermions in $d=1+2$ Dimensions from First Principles 271
  M. G. Carrillo-Ruiz and M. Napsuciale
The Trigonometric Rosen-Morse Potential as a Prime Candidate for an Effective QCD Potential 275
  C. B. Compeán Jasso and M. Kirchbach
Scalar Quantum Electrodynamics: Perturbation Theory and Beyond 279
  A. Bashir, L. X. Gutierrez-Guerrero, and Y. Concha-Sánchez
Azimuthal Correlations in p-p Collisions 283
  E. Cuautle, I. Domínguez, and G. Paić
On Quark-Lepton Complementarity 287
  F. González Canales and A. Mondragón
Description of Charge Conjugation from First Principles 293
  C. Luján-Peschard and M. Napsuciale
Sterile Neutrinos and the Solar Mixing 297
  J. C. Gómez-Izquierdo and A. Pérez-Lorenzana

Tritium-Beta Spectrum in a Left-Right Symmetric Model..................302
    A. Gutiérrez-Rodríguez, M. A. Hernández-Ruíz, and F. Ramírez-Sánchez
Optical Simulation for V0A..306
    C. Pérez Lara, A. Gago Medina, and G. Herrera Corral
Dynamical Mass Generation in QED with Weak Magnetic Fields............311
    A. Ayala, A. Bashir, A. Raya, and E. Rojas
Empirical Testing of Tsallis' Thermodynamics as a Model for Dark Matter Halos..316
    D. Nunez, R. A. Sussman, J. Zavala, L. G. Cabral-Rosetti, and T. Matos
Entropy Considerations in Constraining the mSUGRA Parameter Space..321
    D. Nunez, R. A. Sussman, J. Zavala, L. Nellen, L. G. Cabral-Rosetti, and M. Mondragón
Probing New Physics with Neutrino Interactions at the Pierre Auger Observatory..326
    L. Díaz-Cruz, M. I. Pedraza-Morales, A. Rosado, and H. Salazar
Neutrino Telescopes and Non Standard Interactions......................330
    R. Pérez Martínez, A. Zepeda, and O. G. Miranda
Fermions Living in a Flat World..334
    M. de Jesús Anguiano-Galicia and A. Bashir
Electronics for the Extensive Air Shower Detector Array at the University of Puebla..338
    E. Pérez, R. Conde, O. Martínez, T. Murrieta, H. Salazar, and L. Villaseñor

Proceedings Published by the Division of Particles and Fields of the Mexican Physical Society...343
Author Index..345

PREFACE

The X Mexican Workshop on Particles and Fields took place in Morelia, Michocán, México, from November 7-12, 2005. It belongs in a series of workshops organized every two years, along with the Mexican Schools on Particle and Fields (organized also every two years on alternating years), by the Division of Particles and Fields of the Mexican Physical Society. They have been successfully organized without interruption since 1984. The purpose of these workshops is to bring together research leaders from around the world with scientists and students from Mexico to review recent advances in theoretical, phenomenological and experimental high energy physics.

On this occasion we had a very rich program of invited and contributed lectures, as well as poster sessions, covering the most active and fascinating research topics on high energy physics, including the status and first results on the detection of ultra high energy cosmic rays by the Pierre Auger Observatory.

This year we decided to include laboratory sessions as a part of the program. Like the fist time that we did so at the IX Mexican School on Particles and Fields in 2004, this experience was highly successful according to the students interested in experimental high energy physics who took these laboratory courses. Likewise, on this occasion we organized discussion sessions on theory and phenomenology for the benefit of students and young researchers with these orientations.

As can be inferred from the articles in these proceedings, there is an increasing number of Mexican groups involved in experimental collaborations in high energy and cosmic ray physics, leading to an ever increasing maturity of these research areas within our community.

Around 100 participants from around 10 countries attended this Workshop in attractive historical buildings in the colonial-style city of Morelia. This volume contains almost the totality of the lectures and posters presented at the Workshop. We hope it will remain as a useful and up-to-date reference to students and researchers in the covered areas.

Luis Villaseñor
President
Division of Particles and Fields
Mexican Physical Society

ACKNOWLEDGEMENTS

I would like to thank CONACyT and the University of Michoacan for providing financial support to celebrate the X Mexican Workshop of Particles and Fields. I also thank the members of the **National Organizing Committee**:

- Alfredo Aranda, Universidad de Colima
- Alejandro Ayala, Instituto de Ciencias Nucleares, UNAM
- Heriberto Castilla Valdez, Departamento de Física, Cinvestav
- Lorenzo Díaz-Cruz, FCFM, BUAP
- Jurgen Engelfried, Instituto de Física, UASLP
- Gabriel Lopez-Castro, Departamento de Física, Cinvestav
- Miguel Angel Perez Angón, Departamento de Física, Cinvestav
- Humberto Salazar, FCFM, BUAP
- Maria Elena Tejeda, Universidad de Sonora
- Luis Villaseñor, Instituto de Física y Matemáticas, UMSNH
- Arnulfo Zepeda, Departamento de Física, Cinvestav

and the members of the **Local Organizing Committee**:

- Adnan Bashir, Instituto de Física y Matemáticas, UMSNH
- Ulises Nucamendi, Instituto de Física y Matemáticas, UMSNH
- Eduardo Tututi, Facultad de Ciencias Físico-Matemáticas, UMSNH
- Victor Villanueva, Instituto de Física y Matemáticas, UMSNH
- Axel Weber, Instituto de Física y Matemáticas, UMSNH.

The help provided by Francisco Alarcón, Jaime Nieto, José Vega, Gabriel Arroyo and Laura Godinez was greatly appreciated.

Finally, I thank the speakers of invited and contributed talks as well as the participants who presented posters for their efforts to make this Workshop a success.

Luis Villaseñor
President
Division of Particles and Fields
Mexican Physical Society

INVITED LECTURES

The Brave ν World

André de Gouvêa

Northwestern University, Department of Physics & Astronomy, 2145 Sheridan Road, Evanston, IL 60208, USA

Abstract. I briefly summarize our current understanding of neutrino physics, which was severely "shaken up" by the discovery that neutrinos have nonzero, albeit tiny, masses. I discuss the evidence for neutrino masses – neutrino oscillations, list what have been identified as the "known unknowns" of neutrino physics, and try to argue why non-zero neutrino masses provide the first unambiguous evidence that the Standard Model is incomplete. I conclude with a summary of different approaches to interpreting neutrino masses and lepton mixing and some of their consequences.

Keywords: neutrino masses, lepton mixing
PACS: 14.60.Pq,12.15.Ff,14.60.St

BRIEF INTRODUCTION

Over the past eight years our understanding of neutrino physics has changes dramatically. A suite of measurements involving solar, atmospheric, accelerator, and reactor neutrinos has revealed, beyond any doubt, that neutrinos have non-zero – but tiny – masses, and that leptons, just like the quarks, mix [1]. On the other hand, unlike quarks, we have discovered that lepton mixing is large. All known fermion masses are schematically depicted in Fig. 1. The purpose of this figure is three-fold. First, it illustrates how disparate supposedly fundamental fermion masses are. Second, it illustrates how small neutrino masses are compared to all other charged fermion masses. Finally, it is curious to note that the "gap between the lightest and heaviest charged fermion masses – which is nicely populated by all other charged fermions – is larger than the "gap" between the largest possible neutrino mass and the lightest charged fermion. Unlike the former, the latter gap is, as far as we can tell, "empty."

Neutrino masses provide the only "conditionless" evidence that the Standard Model (SM) of electroweak interactions is, at least incomplete. It turns out, however, that we don't have enough information to determine what the "new" Standard Model – the νSM – shoud be. More experimental information is required.

Here, I briefly overview what we know about neutrino masses and mixings, and present the questions that must be addressed by next-generation neutrino experiments. I then briefly discuss different possibilities for the νSM, and spend some time presenting the LSND anomaly – the only currently unsolved neutrino puzzle.

Before proceeding, I would like to point that there are several excellent reviews of neutrino physics [1, 2]. I refer readers to these for all the proper references and more detailed discussions.

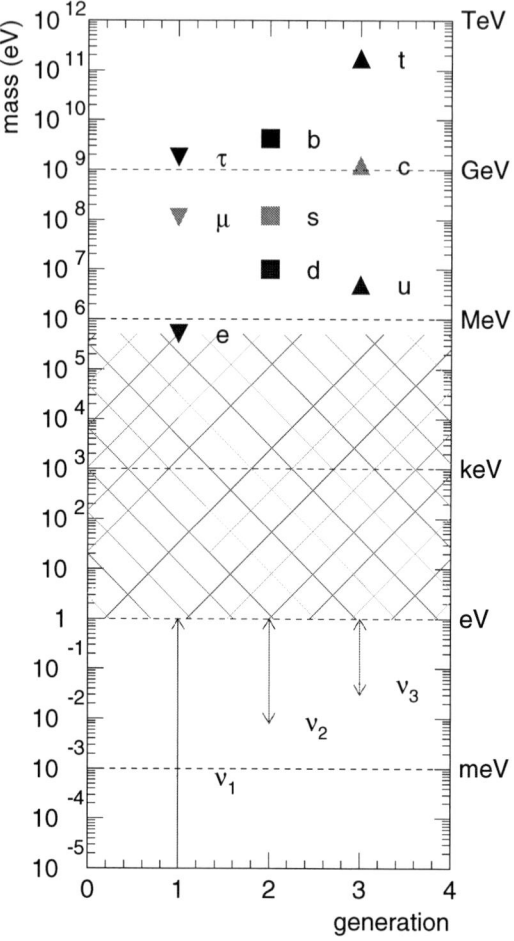

FIGURE 1. Masses of all known fundamental fermions. The arrows are meant to represent, qualitatively, our current knowledge of the neutrino masses.

WHAT WE KNOW AND DON'T KNOW ABOUT NEUTRINOS

Neutrinos change flavor. This means that a neutrino produced in a well-defined weak eigenstate ν_α can be detected in a distinct weak eigenstate ν_β[1] after propagating a

---

[1] Weak eigenstates are labeled by the flavor of the charged lepton involved in the charged current weak process responsible for either producing or detecting neutrinos, *i.e.*, $\alpha, \beta = e, \mu, \tau, \ldots$.

macroscopic distance. Furthermore, it has been established that the rate of flavor change $P_{\alpha\beta}$ depends on the neutrino energy E_ν and on the propagation distance L.

The simplest (and only completely satisfactory) way to explain the observed neutrino flavor changing phenomena is to postulate that neutrinos have distinct, nonzero masses, and that the neutrino mass eigenstates[2] are different from the neutrino weak eigenstates. This being the case, neutrinos will undergo neutrino oscillations as they propagate, and $P_{\alpha\beta}$ is the oscillation probability. Besides L and E_ν, $P_{\alpha\beta}$ is a function of the neutrino mass-squared differences $\Delta m_{ij}^2 \equiv m_j^2 - m_i^2$ and the elements of the lepton mixing matrix U. In the weak basis where the charged current and the charged lepton masses are diagonal, the lepton mixing matrix is the unitary matrix that relates the neutrino weak eigenstates to the neutrino mass eigenstates: $v_\alpha = U_{\alpha i} v_i$.

All neutrino data, with the exception of the LSND anomaly [3] are explained by three flavor neutrino oscillations. I will comment on the LSND anomaly later, and, for simplicity and the sake of the presentation, will ignore its existence henceforth.

For three neutrino flavors, the elements of the lepton mixing matrix are defined by

$$\begin{pmatrix} v_e \\ v_\mu \\ v_\tau \end{pmatrix} = \begin{pmatrix} U_{e1} & U_{e2} & U_{e3} \\ U_{\mu 1} & U_{\mu 2} & U_{\mu 3} \\ U_{\tau 1} & U_{e\tau 2} & U_{\tau 3} \end{pmatrix} \begin{pmatrix} v_1 \\ v_2 \\ v_3 \end{pmatrix}, \qquad (1)$$

and are, of course, not all independent. It is customary [4] to parameterize U in Eq. (1) with three mixing angles $\theta_{12}, \theta_{13}, \theta_{23}$ and three complex phases, δ, ξ, ζ, defined by

$$\frac{|U_{e2}|^2}{|U_{e1}|^2} \equiv \tan^2 \theta_{12}; \quad \frac{|U_{\mu 3}|^2}{|U_{\tau 3}|^2} \equiv \tan^2 \theta_{23}; \quad U_{e3} \equiv \sin \theta_{13} e^{-i\delta}, \qquad (2)$$

with the exception of ξ and ζ, the so-called Majorana CP-odd phases. These are only physical if the neutrinos are Majorana fermions, and have, unfortunately, virtually no effect in flavor-changing phenomena. We have no idea what their values are or whether they are physical observables, and I will ignore them henceforth, unless otherwise noted.

In order to relate the mixing elements to experimental observables, it is necessary to define the neutrino mass eigenstates, *i.e.*, to "order" the neutrino masses. This will be done in the following way: $m_2^2 > m_1^2$ and $\Delta m_{12}^2 < |\Delta m_{13}^2|$. In this case, there are three mass-related observables: Δm_{12}^2 (positive definite), $|\Delta m_{13}^2|$, and the sign of Δm_{13}^2. A positive sign for Δm_{13}^2 implies $m_3^2 > m_1^2$ – a so-called normal mass-hierarchy – while a negative sign for Δm_{13}^2 implies $m_3^2 < m_1^2$ – a so-called inverted mass-hierarchy. These are depicted in Fig. 2

Detailed combined analyses of all neutrino data are consistent, at the two sigma confidence level, with [5]

1. $\sin^2 \theta_{12} = 0.31 \pm 0.06$, mostly from solar and KamLAND data;
2. $\sin^2 \theta_{23} = 0.50 \pm 0.15$, mostly from atmospheric neutrino data;

---

[2] Mass eigenstates are eigenstates of the free particle Hamiltonian in vacuum, *i.e.*, $H|v_i\rangle = m_i |v_i\rangle$ in the neutrino rest frame. These are referred to as v_i, $i = 1, 2, 3, \ldots$.

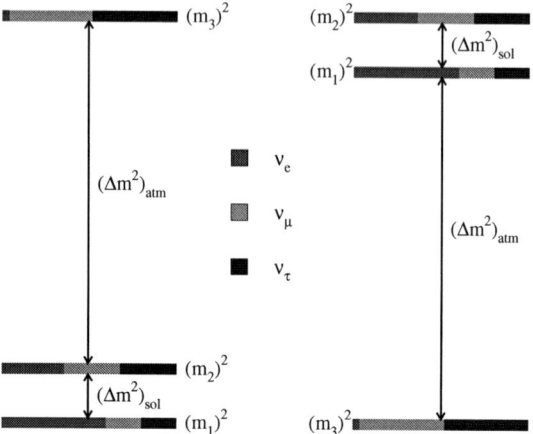

FIGURE 2. Cartoon of the two distinct neutrino-mass hierarchies that fit all of the current neutrino data, for fixed values of all mixing angles and mass-squared differences. The color coding (shading) indices the fraction $|U_{\alpha i}|^2$ of each distinct flavor ν_α, $\alpha = e, \mu, \tau$ contained in each mass eigenstate ν_i, $i = 1,2,3$. For example, $|U_{e2}|^2$ is equal to the fraction of the $(m_2)^2$ "bar" that is painted red (shading labeled as 'ν_e').

3. $\sin^2 \theta_{13} \leq 0032$, mostly from atmospheric and Chooz [6] data;
4. $\Delta m_{12}^2 = (7.9 \pm 0.7) \times 10^{-5}$ eV^2, mostly from solar and KamLAND data;
5. $|\Delta m_{13}^2| = (2.4 \pm 0.6) \times 10^{-3}$ eV^2, mostly from atmospheric neutrino data.

I refer readers to [5] for details of the analyses and the results, including correlations, etc.

We still need to find out

1. the sign of Δm_{13}^2 or what is the neutrino mass hierarchy?;
2. the value of θ_{13} or is $|U_{e3}| \neq 0$?;
3. the value of δ or is there CP-invariance violation in neutrino oscillations?;
4. the sign of $1/2 - \sin^2 \theta_{23}$ or is atmospheric mixing really maximal?

Finally, it is important to realize that massive neutrinos can be either Dirac or Majorana fermions. The reason for this is simple: neutrinos are the only electrically neutral fermions. Dirac fermions and Majorana fermions are very distinct. If the neutrino is a Dirac fermion, it is described by four degrees of freedom – a left-handed neutrino, a right-handed antineutrino, a right-handed neutrino and a left-handed antineutrino. If, on the other hand, the neutrino is a Majorana fermion, it is only described by two degrees of freedom – a left-handed neutrino and a right-handed antineutrino. Furthermore, massive

Majorana fermions cannot transform under any continuous symmetries. This leads to the very important consequence that, if neutrinos are Majorana fermions, lepton number (or, its non-anomalous "cousin" $U(1)_{B-L}$) is *not* a good symmetry of Nature.

THE ν STANDARD MODEL

The SM is a Lorentz invariant quantum field theory, and its renormalizable Lagrangian is uniquely determined once one specifies its internal symmetries (gauged $SU(3)_c \times SU(2)_L \times U(1)_Y$ invariance) and particle content (Q, u, d, L, e, the matter fields, plus H, the Higgs doublet scalar field). The fact that the Lagrangian is renormalizable implies that, in principle, the SM is valid up to arbitrarily high energy scales.[3] It is easy to check that given the SM as defined above, neutrinos are strictly massless. We need to qualitatively change the SM in order to accommodate neutrino masses. The way to modify the standard model depends significantly on whether neutrinos are Dirac or Majorana fermions.

Majorana Neutrinos

Arguably, the simplest way to add neutrino masses to the SM is to give up on the renormalizability of the Lagrangian. This allows one to add (an infinite number of) irrelevant operators consistent with the symmetries:

$$\mathscr{L}_{\text{new}} = \mathscr{L}_{\text{SM}} - \lambda^{\alpha\beta} \frac{L_\alpha H L_\beta H}{2M} + \mathscr{O}\left(\frac{1}{M^2}\right). \tag{3}$$

L is the left-handed lepton double field. Note that, unless otherwise noted, all fermion fields are understood to be Weyl fermions, such that, for example, the charged lepton Yukawa operator is written as LHe, where e is the (positively) charged lepton singlet field.

Two facts are remarkable. One is that $(LH)^2$ is the *only* dimension-five operator allowed by the SM gauge invariance and particle content [7]. The other is that, as long as M is much larger than $\langle H \rangle$, the Higgs vacuum expectation value, the only observable consequence of Eq. (3)[4] is that neutrinos get a nonzero mass after electroweak symmetry breaking: $m_\nu = \lambda \langle H \rangle^2 / M$. An extra "bonus" is that neutrino masses are naturally much smaller than all other fermion masses $m_f \propto \langle H \rangle$ by a factor $\langle H \rangle / M$.

---

[3] This is not the case. Gravitational interactions are not accounted for in the SM. This is usually a very good approximation, which breaks down, naively, if one tries to describe processes that occur around the Planck scale $M_{\text{Pl}} \sim 10^{18}$ GeV. I'll argue that neutrino masses may be interpreted as evidence that the SM is not an appropriate description for physical processes at significantly lower energy scales, and will ignore "quantum gravity" concerns.

[4] This is not necessarily correct. One also has to worry about dimension six operators that lead to baryon number violation and a finite lifetime for the proton.

Another important consequence of Eq. (3) is that lepton number is not a good symmetry ($(LH)^2$ breaks lepton number by two units). Lepton number violation is "encoded" in the fact that the neutrinos are Majorana fermions.

It remains to discuss the "new physics" scale M. It can be described, roughly, as the energy scale above which Eq. (3) is no longer valid. There is very little information regarding the magnitude of M, but one can set an upper bound for M by assuming that $\lambda \sim 4\pi$ [8], i.e., by assuming that the physics replacing Eq. (3) at the scale M is strongly coupled. In this case

$$M < 4\pi \frac{\langle H \rangle^2}{m_\nu} \sim 10 \frac{100^2 \text{ GeV}^2}{10^{-10} \text{ GeV}} \left(\frac{100 \text{ meV}}{m_\nu}\right) = 10^{15} \text{ GeV} \left(\frac{100 \text{ meV}}{m_\nu}\right). \quad (4)$$

If Eq. (3) is indeed the correct low-energy description of Nature, neutrino masses represent the first direct evidence that the SM is an effective field theory, valid up to an energy scale less that 10^{15} GeV (or so), which is, in turn, much less than the Planck scale [8, 9]. It is also impressive that the upper bound above coincides, qualitatively, with the energy scale where all three running gauge coupling constants of the SM seem to meet, i.e., the grand unified scale $M_{\text{GUT}} \sim 10^{15-16}$ GeV.

The SeeSaw Mechanism

There are several different proposals for the physics that replaces Eq. (3). The most celebrated one is the seesaw mechanism [10], described by the following Lagrangian

$$\mathscr{L}_{\text{seesaw}} = \mathscr{L}_{\text{SM}} - y_{\alpha i} L_\alpha H N_i - \frac{M_N^{ij}}{2} N_i N_j + H.c.. \quad (5)$$

N_i are SM singlet fermion fields, and Eq. (5) is the most general renormalizable Lagrangian consistent with gauge invariance and the addition of these new fields to the SM particle content. Note that the presence of N_i breaks lepton number since these have Majorana masses M_N. In the limit $M_N \gg \langle H \rangle$, one can integrate out these "right-handed neutrino" fields in order to describe phenomena at and below the weak scale. The effective Lagrangian turns out to be

$$\mathscr{L}_{\text{new}} = \mathscr{L}_{\text{SM}} - \left(y^t M_N^{-1} y\right)^{\alpha\beta} L_\alpha H L_\beta H + \mathscr{O}\left(\frac{1}{M_N^2}\right). \quad (6)$$

Eq. (6) is the same as Eq. (3) after one equates $y^t M_N^{-1} y = \lambda/2M$. Eq. (5) is, therefore, an ultraviolet completion of Eq. (3), where M is associated with the mass of the right-handed neutrino fields.

In the limit $M_N \gg \langle H \rangle$, Majorana neutrino masses are the only observable physical consequence of Eq. (5), with the possible important exception of the decay of right-handed neutrino fields in the very early universe. These may have lead to the observed baryon asymmetry of the universe, generated via the lepton asymmetry which naturally arises from CP-violating and lepton-number violating decays of the N_i [11]. I refer readers to [12] for recent, detailed studies of this so-called leptogenesis mechanism.

Note, however, that there is very little *experimental* information regarding the magnitude of M. Indeed, it could be as low as a few eV [13], or even zero! Theoretically, any value of M is technically natural [13]. If M is indeed small, the right-handed neutrinos are potentially observable thanks to active–sterile neutrino mixing. Indeed, there are several "uses" for right-handed neutrinos, including the LSND anomaly [13] (discussed below), evidence for warm dark matter [14], etc.

Dirac Neutrinos

A different option is to assume that neutrinos are, similar to all charged matter fields, Dirac fermions. In this case, in order to render the neutrinos massive, it suffices to add extra SM gauge singlet Weyl fermions N_i ("right-handed neutrinos"), and Yukawa couplings between H, L_α, N_i:

$$\mathscr{L}_{\text{new}} = \mathscr{L}_{\text{SM}} - y_{\alpha i} L_\alpha H N_i + H.c.. \tag{7}$$

This Lagrangian is, of course, identical to Eq. (5) in the limit $M \to 0$.

After electroweak symmetry breaking, the neutrino mass matrix is given by $m_\nu = y\langle H \rangle$, similar to the up-type and down-type quark mass matrices and the charged lepton mass matrix. The magnitude of the neutrino masses requires $y < 10^{-11}$, at least six orders of magnitude smaller than the electron Yukawa coupling. It is clear that a natural explanation for the smallness of the neutrino mass is *not* contained in Eq. (7). On the positive side, Eq. (7) is renormalizable, meaning that this new SM version is, naively, valid up to arbitrarily high energy scales.

Modulo a natural explanation for the size of the neutrino mass, one could argue that Eq. (7) is a rather innocuous addition to the SM Lagrangian. This is not correct, and he reason for it is somewhat subtle. Eq. (7) is not the most general, renormalizable Lagrangian consistent with the symmetries of the SM. Once the fields N_i are introduced, the dimension-three Majorana mass operators $\frac{1}{2}M_N^{ij} N_i N_j$ should also have been introduced, given the fact that N_i's are gauge singlets, *cf.* Eq. (5). One needs, therefore, to modify the *symmetry structure* of the SM in order to forbid a Majorana mass term for the right-handed neutrinos. One simple way of doing this is to add to the SM an internal global symmetry, *e.g.* $U(1)_{B-L}$, where B stands for baryon number, and L for lepton number. Note that, in the massless-neutrino SM, $U(1)_{B-L}$ is an *accidental* global symmetry, *i.e.*, it arises as a consequence of the imposed gauge symmetries and the particle content. Among other things, this means that there was, *a priori*, no reason to believe that it needed to be conserved by allowed SM extensions, including SUSY, quantum gravitational effects, etc. If Eq. (7) is indeed the correct description of neutrino masses, $U(1)_{B-L}$ needs to be "upgraded" to an imposed fundamental global symmetry, and it is expected to be preserved by, say, quantum gravity, etc.

Understanding Neutrino Mixing

The other theoretical question that has received a significant amount of attention is the fact that the lepton mixing matrix is qualitatively different from the quark one.

The quark mixing matrix, or the CKM matrix V_{CKM} is, at zeroth order, equal to the identity matrix – quark mass (or strong) eigenstates are almost identical to weak (or flavor) ones. Furthermore, the off-diagonal entries of V_{CKM} are hierarchical. Schematically,

$$|V_{CKM}| \sim \begin{pmatrix} 1 & \varepsilon & \varepsilon^3 \\ \varepsilon & 1 & \varepsilon^2 \\ \varepsilon^3 & \varepsilon^2 & 1 \end{pmatrix}, \quad \varepsilon \sim 0.2. \tag{8}$$

For more quantitative details, see, for example, citepdg.

The lepton mixing matrix, sometimes refereed to as the MNS matrix V_{MNS} is characterized by two large mixing angles, and most of its entries are of the same order of magnitude. Schematically,

$$|V_{MNS}| \sim \begin{pmatrix} 1 & 1 & U_{e3} \\ 1 & 1 & 1 \\ 1 & 1 & 1 \end{pmatrix}, \quad U_{e3} < 0.2. \tag{9}$$

It is quite apparent that that Eq. (9) does not resemble Eq. (8) at all, with the possible exception that ε is approximately the same as the upper bound for U_{e3}.

The suspiciously ordered form of Eq. (8), together with the fact that the quark masses are very hierarchical, has received a huge amount of attention over the past thirty years or so. We interpret quark masses and mixing as evidence of some new organizing principle (symmetry) of Nature, which is perhaps broken at inaccessibly high energy scales. This means that the small mixing angles in the CKM matrix may be indicative of the existence of, say, new "flavor" quantum numbers capable of distinguishing a top-quark from a charm-quark from an up-quark.

The question one would like to address is whether the same should apply in the lepton sector, *i.e.*, is there evidence that Nature distinguishes, at a more fundamental level, between electrons and muons and taus? There are several indirect reasons to naively believe that this should be the case. First of all, charged lepton masses are hierarchical. Second of all, there are several indications that quarks and leptons may turn out to be different "low-energy" manifestations of the same fundamental matter field, as is predicted by most grand unified theories (GUTs).

GUT relations have lead to the first prediction for the lepton mixing matrix: $V_{MNS} \simeq V_{CKM}$, which has clearly been falsified by the neutrino oscillation data. More recently, several GUT neutrino mass models have succeeded in properly accommodating all neutrino oscillation data [2]. The challenge has now shifted from trying to understand whether GUTs can explain large neutrino mixing to sorting out the different possibilities.

Even in the absence of any guidance from the quark sector, symmetry-explanations to neutrino masses and lepton mixing have been heavily undertaken. These are driven by peculiar features of neutrino mixing, such as

- some (or all) neutrino masses could be quasi-degenerate in absolute value;

- $|U_{e3}|$ could be much smaller than all other entries in the MNS matrix;
- $|U_{\mu 3}|$ could be very close to $|U_{\tau 3}|$ (maximal atmospheric mixing),

and are summarized in the literature [2]. Perhaps the most relevant feature of all lepton flavor models (with or without grand unification) is that they can be falsified by precision neutrino oscillation experiments. Indeed, the values of the currently unknown neutrino oscillation observables ($|U_{e3}|$, the mass-hierarchy, etc), are predicted by the different lepton flavor models (see, for example, table 1 in [15]).

It may, however, turn out that we have been asking the wrong question when it comes to neutrino mixing. Upon closer inspection, while Eq. (8) looks very structured, Eq. (9) looks quite structureless and "ordinary." It has been pointed out [16], and studied in detailed [17, 18], that Eq. (9) is a good representative of a random three-by-three unitary matrix distribution, indicating, perhaps, that, unlike the quark sector, mixing in the lepton sector is "anarchical" – all 'peculiar features' mentioned above could turn out to be accidental!

The anarchical hypothesis, spelled out in detail in [18], is, contrary to naive expectations, quite consistent with hierarchical charged leptons and hierarchical neutrino masses [17, 9]. More interestingly, the anarchical hypothesis is falsifiable – anarchy prefers values of $|U_{e3}|^2$ not too far from the current upper bound. As a matter of fact, at the 95% confidence level, anarchy predicts $|U_{e3}|^2 > 0.01$, which will be probed definitively by next-generation neutrino experiments (see, for example, [15, 19]). Hence, if $|U_{e3}|^2 \ll 0.01$, we will be lead to strongly believe that there is indeed structure in lepton mixing.

The biggest potential criticism against the anarchical hypothesis today is the fact that the atmospheric mixing seems to be maximal: $\sin^2 2\theta_{23} = 1$ is preferred by the current atmospheric and accelerator neutrino data. Maximal mixing implies that the ν_3 mass eigenstate is "composed" of equal amounts of ν_μ and ν_τ: $|U_{\mu 3}|^2 = |U_{\tau 3}|^2$. It would be rather surprising if this turned out to be the case without a "fundamental reason" (symmetry), in spite of the fact that a statistical test of anarchy "prefers" maximal atmospheric mixing[18].

Currently, in spite of the fact that $\sin^2 2\theta_{23} = 1$ is preferred, the atmospheric mixing angle need not be particularly close to maximal at all. The uncertainty on $\cos 2\theta_{23}$, a good measure of maximal atmospheric mixing, is very large. The key question that can be addressed in practice is how close to zero should $\cos 2\theta_{23}$ be if $\cos 2\theta_{23} = 0$ is required by some approximate symmetry.[5] An attempt at studying this issue can be found in [20]. For other related studies concerning the deviation of θ_{23} from $\pi/4$, see [21].

Another popular approach to neutrino mixing is to explore special forms – "textures" – for the neutrino mass matrix (usually in the basis where charged-fermion masses are diagonal) which may be interpreted as evidence of some new organizing principle [2]. Results of "generic" textures are presented in table 1. See [20] for all the details. Note the interesting correlations among the several currently unknown neutrino oscillation

---

[5] It is quite unlikely that any flavor symmetry related to maximal atmospheric mixing is exact. Indeed, flavor symmetries usually need to be broken in order to properly fit the charged lepton and the neutrino observables [2].

TABLE 1. Different leading-order neutrino mass textures and their "predictions" for various observables. The fifth column indicates the "prediction" for $|\cos 2\theta_{23}|$ when there is no symmetry relating the different order one entries of the leading-order texture ('n.s.' stands for 'no structure', meaning that the entries of the matrices in the second column should all be multiplied by and order one coefficient), while the sixth column indicates the "prediction" for $|\cos 2\theta_{23}|$ when the coefficients of the leading order texture are indeed related as prescribed by the matrix contained in the second column. One may argue that the anarchical texture prefers but does not require a normal mass hierarchy.

Case	Texture	Hierarchy	$\|U_{e3}\|$	$\|\cos 2\theta_{23}\|$ (n.s.)	$\|\cos 2\theta_{23}\|$	Solar Angle
A	$\frac{\sqrt{\Delta m_{13}^2}}{2}\begin{pmatrix} 0 & 0 & 0 \\ 0 & 1 & 1 \\ 0 & 1 & 1 \end{pmatrix}$	Normal	$\sqrt{\frac{\Delta m_{12}^2}{\Delta m_{13}^2}}$	$\mathcal{O}(1)$	$\sqrt{\frac{\Delta m_{12}^2}{\Delta m_{13}^2}}$	$\mathcal{O}(1)$
B	$\sqrt{\Delta m_{13}^2}\begin{pmatrix} 1 & 0 & 0 \\ 0 & \frac{1}{2} & -\frac{1}{2} \\ 0 & -\frac{1}{2} & \frac{1}{2} \end{pmatrix}$	Inverted	$\frac{\Delta m_{12}^2}{\|\Delta m_{13}^2\|}$	—	$\frac{\Delta m_{12}^2}{\|\Delta m_{13}^2\|}$	$\mathcal{O}(1)$
C	$\frac{\sqrt{\Delta m_{13}^2}}{\sqrt{2}}\begin{pmatrix} 0 & 1 & 1 \\ 1 & 0 & 0 \\ 1 & 0 & 0 \end{pmatrix}$	Inverted	$\frac{\Delta m_{12}^2}{\|\Delta m_{13}^2\|}$	$\mathcal{O}(1)$	$\frac{\Delta m_{12}^2}{\|\Delta m_{13}^2\|}$	$\|\cos 2\theta_{12}\| \sim \frac{\Delta m_{12}^2}{\|\Delta m_{13}^2\|}$
Anarchy	$\sqrt{\Delta m_{13}^2}\begin{pmatrix} 1 & 1 & 1 \\ 1 & 1 & 1 \\ 1 & 1 & 1 \end{pmatrix}$	Normal	> 0.1	$\mathcal{O}(1)$	—	$\mathcal{O}(1)$

parameters.

MORE SURPRISES? – THE LSND ANOMALY

The LSND experiment [3] measured the neutrino flux produced by pion decay in flight ($\pi^+ \to \mu^+ \nu_\mu$) and antimuon decay at rest ($\mu^+ \to e^+ \nu_e \bar{\nu}_\mu$). It observed a small *electron-type antineutrino* flux some 30 meters away from the production region [3]. The originally absent $\bar{\nu}_e$-flux can be interpret as evidence that $\bar{\nu}_\mu$ is transforming into $\bar{\nu}_e$ with $P_{\bar{\mu}\bar{e}}$ of the order a fraction of a percent. The data also contain a weak hint of a ν_e excess, which is both consistent with the $\bar{\nu}_\mu \to \bar{\nu}_e$ hypothesis and consistent with zero. If interpreted in terms of two-flavor neutrino oscillations, the LSND anomaly, combined with constraints imposed by several other experiments, points to a mass-squared difference $\Delta m^2_{\text{LSND}} \sim 0.1 - 10$ eV^2, as depicted in Fig. 3.

The LSND result will be confirmed or refuted by the on-going MiniBooNE experiment, perhaps as early as the end of Summer 2006 [23]. MiniBooNE consists of a standard ν_μ-beam experiment (from π^+ decay), with neutrino energies in the sub-GeV to GeV range and a baseline of around 500 m. Its beam (neutrino versus antineutrino) and systematics are quite distinct from the LSND setup.

It is easy to understand why the LSND anomaly does not "fit" in the three flavor mixing scheme described in the previous section. With three neutrinos, one can define only two independent mass-squared differences, and these are completely determined by the solar, atmospheric, reactor, and accelerator data. As discussed earlier, both mass-squared differences are much smaller than Δm^2_{LSND}. Given that the LSND results are

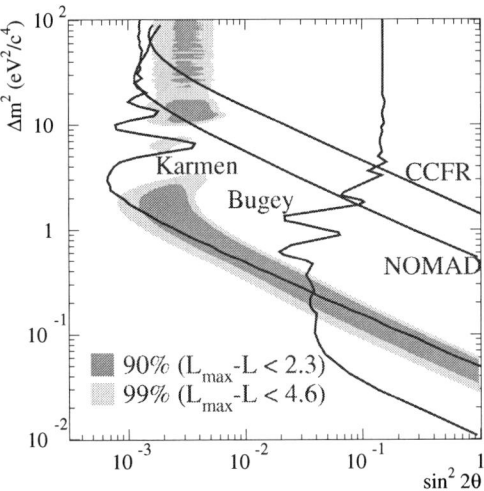

FIGURE 3. A $(\sin^2 2\theta, \Delta m^2)$ oscillation parameter fit for the entire data sample, $20 < E_e < 200$ MeV. The fit includes primary $\bar{\nu}_\mu \to \bar{\nu}_e$ oscillations and secondary $\nu_\mu \to \nu_e$ oscillations, as well as all known neutrino backgrounds. The inner and outer regions correspond to 90% and 99% CL allowed regions, while the curves are 90% CL limits from the Bugey reactor experiment, the CCFR experiment at Fermilab, the NOMAD experiment at CERN, and the KARMEN experiment at ISIS. From A. Aguilar *et al.* [LSND Collaboration], Phys. Rev. D **64**, 112007 (2001).

yet to be confirmed by another experiment — indeed, the Karmen 2 experiment [24] using a similar setup but with a shorter baseline ($L = 18$ m), could have confirmed the LSND anomaly but did not observe any excesses, ruling out a significant portion of the LSND allowed parameter space (Fig. 3) — it is widely believe that "ordinary" three-flavor oscillations are responsible for "all-but-LSND-data," while the LSND anomaly could be due to more exotic new physics. Reinforcing this bias is the fact that, if indeed present, the LSND anomaly requires a very small transition probability.

One possible solution to the LSND anomaly is to add extra, standard model gauge singlet neutrinos, capable of mixing with the ordinary, or active, neutrinos. While this allows one to define at least three mass-squared differences, it is not guaranteed that one is capable of fitting all neutrino data with four (or more) neutrino mixing. Indeed, detailed analyses [25] suggest that four neutrino mixing schemes are either very poor or at best mediocre fits to all neutrino data.

The reason for this can be qualitatively understood in the following way. There are two general neutrino mass-patterns capable of describing one large and two small mass-squared differences. These are referred to as the "2+2" and "3+1" scheme, and are depicted in Fig. 4.

The 2+2 schemes are ruled out by combined fits to solar and atmospheric data (see, for example, [25]). The 3+1 schemes fit the atmospheric and solar data just fine, given

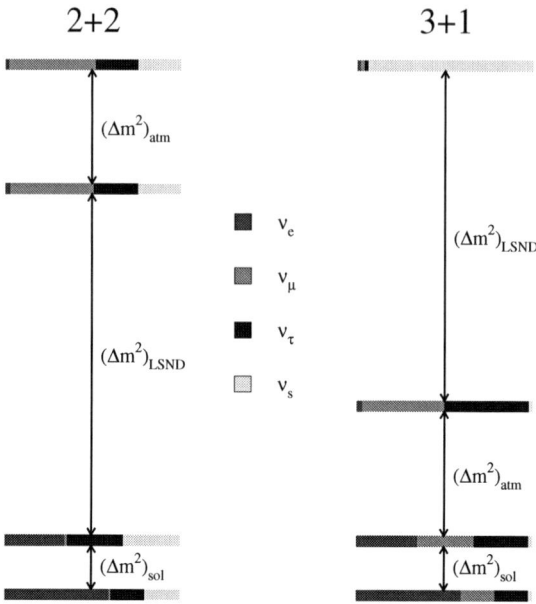

FIGURE 4. Two possible mass-patterns potentially capable of addressing all neutrino data, including those from LSND. The one on the left (right) is characteristic of a "2+2" ("3+1") mass-scheme.

that sterile neutrino effects are just a small perturbation to the ordinary three neutrino fit to all-but-LSND data. They run into some trouble when it comes to short-baseline searches for v_e and v_μ disappearance driven by the LSND frequency (see Fig. 3). LSND $v_\mu \leftrightarrow v_e$ oscillations are given by[6]

$$P_{e\mu} \simeq 4|U_{e4}|^2|U_{e4}|^2 \sin^2\left(\frac{\Delta m_{\text{LSND}}^2 L}{4E}\right), \qquad (10)$$

while the survival probability of a species $\alpha = e, \mu$, for "short" L, is given by

$$P_{\alpha\alpha} \simeq 1 - 4|U_{\alpha 4}|^2(1 - |U_{\alpha 4}|^2) \sin^2\left(\frac{\Delta m_{\text{LSND}}^2 L}{4E}\right). \qquad (11)$$

The absence of electron-type and muon-type neutrino disappearance constrains $|U_{e4}|^2, |U_{\mu 4}|^2$ to be small, while the LSND data require $|U_{e4}|^2 \times |U_{\mu 4}|^2$ to be larger than a fraction of a percent [26]. The tension in the current data is not enough to rule out the 3+1 schemes, but it does lead to a mediocre fit [25].

Five neutrino mixing schemes have also been explored (see, for example, Ref. [27]). These look like "3+1+1" schemes ("2+2+1" schemes do not fare much better than 2+2

---

[6] Here, $\Delta m_{\text{LSND}}^2 \simeq |\Delta m_{i4}^2|$, $\forall i = 1, 2, 3$.

schemes) and are designed in a such a way that the tension between the short-baseline and the LSND data is alleviated. With five neutrinos, it is possible to fit all the neutrino data properly, but some worry that the choices for mixing parameters and mass-squared differences are rather "finely tuned."

More exotic solutions to the LSND anomaly have been proposed, and none of them seem to fit all data particularly well. Here I list some of them.

The possibility that there are rare lepton-flavor violating $\mu^+ \to e^+ \nu_\alpha \bar{\nu}_e$ decays [28] could explain the LSND data as long as the branching ratio for the flavor-violating decay was of order a fraction of a percent. Such decays, however, should also have been observed by the Karmen experiment, which disfavored this hypothesis at around the 90% confidence level [29]. Recently available precision data of the Michel electron energy spectrum [30] seem to safely rule out flavor changing muon decays as a solution to the LSND anomaly.

Postulating that neutrinos and antineutrinos have different masses and mixing angles [31] received a significant amount of attention in the recent pass. The original idea was inspired by the fact that solar data required the disappearance of electron-type *neutrinos*, while those from LSND required the appearance of electron-type *antineutrinos*. If neutrinos and antineutrinos oscillated at different frequencies (different Δm^2), all data could be rendered compatible. Aside from all sorts of theoretical issues, the original CPT-violating setup was ruled out when KamLAND published the first evidence for antineutrino oscillations at solar frequencies. A second manifestation of CPT-violating solutions the the LSND data consisted of postulating that atmospheric oscillations in the antineutrino sector were driven by Δm^2_{LSND}. This possibility is strongly disfavored (at the three sigma level) by the atmospheric data [32].

Finally, different authors have explored violations of Lorentz invariance as the physics behind neutrino masses and mixing [33, 34]. None appear to be satisfactory. The analysis performed in [34], for example, properly explained all solar, atmospheric and LSND data, but ran into trouble with short baseline experiments.

Given the fact that none of the solutions to the LSND anomaly proposed to date seem completely satisfactory, it is fair to say that if MiniBooNE confirms the observations made by LSND, there is a good chance we have uncovered a novel physical phenomenon, *i.e.*, we are yet to figure out what the LSND result is teaching us. This being the case, if the LSND anomaly is confirmed, all the necessary experimental and theoretical efforts will most likely concentrate, first, on uncovering the mechanism responsible for the LSND flavor change. This will likely require (i) detailed analysis of all available data, (ii) new, compelling ideas, and (iii) a series of other experimental neutrino efforts, capable of mapping out the LSND/MiniBooNE potential parameter spaces. We are still not sure what these should be!

CONCLUDING REMARKS

The venerable Standard Model of electromagnetic, weak, and strong interactions has, finally, been falsified (or, at least, proven to be incomplete) – neutrinos have mass. Presently, we have a very successful parameterization of the newly uncovered neutrino masses and mixings. This parameterization tells us that there are several neutrino properties we do not yet know – the neutrino mass hierarchy, whether CP-invariance is violated in neutrino oscillations, whether the v_3 mass eigenstate contains a non-zero v_e component, whether atmospheric mixing is maximal, and, perhaps most important, whether neutrinos are Majorana fermions. All of these questions will need to be addressed *experimentally*.

Regardless of the answer to the questions above, we know that neutrino masses are non-zero but very small (when compared to all charged fermion masses). We have not figured out why that is, but there is consensus in the high energy community that this (small neutrino masses) means something important. Similarly, the neutrino oscillation data have revealed that the lepton mixing matrix is very different from the quark mixing matrix. Again, we have no satisfactory for this, but we feel it means something important.

Phenomenological approaches to neutrino masses and lepton mixing depend strongly on whether the neutrinos are Dirac or Majorana fermions. This issue can be resolved by determining whether $U(1)_{B-L}$ is indeed a good global symmetry of the Standard Model. Experimentally, the deepest probe of the Majorana nature of the neutrino is the search for neutrinoless double-beta decay. For details, please see [35].

Finally, while we have learned a lot about neutrino properties, there is plenty of room for surprises. If the MiniBooNE experiments confirms the LSND anomaly beyond any reasonable doubt (which may have happen by the time these proceedings are published!) we will be faced with a new puzzle. We do not have a good solution for yet (in spite of several years of very creative approaches), but it is safe to say it will improve and qualitative modify our understanding of fundamental physics.

ACKNOWLEDGMENTS

I am happy to thank Lorenzo Dias-Cruz for the invitation to present this brief lecture at the "X Mexican Workshop on Particle and Fields" in Morelia.

REFERENCES

1. For a historic introduction and lots of references see, for example, A. de Gouvêa, "2004 TASI lectures on neutrino physics," hep-ph/0411274.
2. Very recent reviews on the physics of neutrinos include A. Strumia and F. Vissani, hep-ph/0606054; R.N. Mohapatra and A.Yu. Smirnov, hep-ph/0603118; R.N. Mohapatra et al., hep-ph/0510213; A. de Gouvêa, Mod. Phys. Lett. A **19**, 2799 (2004).
3. A. Aguilar *et al.* [LSND Collaboration], Phys. Rev. D **64**, 112007 (2001).
4. S. Eidelman *et al.* [Particle Data Group Collaboration], Phys. Lett. B **592**, 1 (2004).
5. see, for example, G.L. Fogli, E. Lisi, A. Marrone and A. Palazzo, hep-ph/0506083.
6. M. Apollonio *et al.* [CHOOZ Collaboration], Phys. Lett. B **466**, 415 (1999).
7. S. Weinberg, Phys. Rev. Lett. **43**, 1566 (1979).
8. F. Maltoni, J.M. Niczyporuk and S. Willenbrock, Phys. Rev. Lett. **86**, 212 (2001).
9. A. de Gouvêa and J.W.F. Valle, Phys. Lett. B **501**, 115 (2001).
10. P. Minkowski, Phys. Lett. B **67**, 421 (1977); M. Gell-Mann, P. Ramond and R. Slansky in *Supergravity*, eds. D. Freedman and P. Van Niuenhuizen (North Holland, Amsterdam, 1979), p. 315; T. Yanagida in *Proceedings of the Workshop on Unified Theory and Baryon Number in the Universe*, eds. O. Sawada and A. Sugamoto (KEK, Tsukuba, Japan, 1979); S.L. Glashow, *1979 Cargèse Lectures in*

Physics – Quarks and Leptons, eds. M. Lévy *et al.* (Plenum, New York, 1980), p. 707; R.N. Mohapatra and G. Senjanović, Phys. Rev. Lett. **44**, 912 (1980). The first general $SU(2) \times U(1)$ description of the seesaw mechanism was performed by J. Schechter and J.W.F. Valle, Phys. Rev. D **22**, 2227 (1980).
11. M. Fukugita and T. Yanagida, Phys. Lett. B **174**, 45 (1986).
12. For example, G.F. Giudice, A. Notari, M. Raidal, A. Riotto and A. Strumia, Nucl. Phys. B **685**, 89 (2004); W. Buchmuller, P. Di Bari and M. Plumacher, Annals Phys. **315**, 305 (2005); Nucl. Phys. B **665**, 445 (2003).
13. A. de Gouvêa, Phys. Rev. D **72**, 033005 (2005).
14. T. Asaka, S. Blanchet and M. Shaposhnikov, Phys. Lett. B **631**, 151 (2005).
15. K. Anderson *et al.*, hep-ex/0402041.
16. L.J. Hall, H. Murayama and N. Weiner, Phys. Rev. Lett. **84**, 2572 (2000).
17. N. Haba and H. Murayama, Phys. Rev. D **63**, 053010 (2001).
18. A. de Gouvêa and H. Murayama, Phys. Lett. B **573**, 94 (2003).
19. D. Ayres *et al.* [Nova Collaboration], hep-ex/0210005; Y. Itow *et al.*, hep-ex/0106019. See also http://neutrino.kek.jp/jhfnu/
20. A. de Gouvêa, Phys. Rev. D **69**, 093007 (2004).
21. For other recent approaches to this issue see, for example, P.H. Frampton, S.T. Petcov and W. Rodejohann, Nucl. Phys. B **687**, 31 (2004). W. Rodejohann, Phys. Rev. D **70**, 073010 (2004); C.A. de S. Pires, J. Phys. G **30**, B29 (2004); W. Grimus, *et al.*, Nucl. Phys. B **713**, 151 (2005); R.N. Mohapatra, JHEP **0410**, 027 (2004); S.T. Petcov and W. Rodejohann, Phys. Rev. D **71**, 073002 (2005).
22. S. Antusch, P. Huber, J. Kersten, T. Schwetz and W. Winter, Phys. Rev. D **70**, 097302 (2004).
23. A.A. Aguilar-Arevalo [MiniBooNE Collaboration], hep-ex/0408074, and references therein. See also http://www-boone.fnal.gov/.
24. B. Armbruster *et al.* [KARMEN Collaboration], Phys. Rev. D **65**, 112001 (2002).
25. A. Strumia, Phys. Lett. B **539**, 91 (2002).
26. O.L.G. Peres and A.Yu. Smirnov, Nucl. Phys. B **599**, 3 (2001); C. Giunti and M. Laveder, JHEP **0102**, 001 (2001). V.D. Barger, B. Kayser, J. Learned, T.J. Weiler and K. Whisnant, Phys. Lett. B **489**, 345 (2000).
27. M. Sorel, J.M. Conrad and M. Shaevitz, Phys. Rev. D **70**, 073004 (2004).
28. For a recent discussion see K.S. Babu and S. Pakvasa, hep-ph/0204236
29. B. Armbruster *et al.*, Phys. Rev. Lett. **90**, 181804 (2003).
30. J. R. Musser [TWIST Collaboration], Phys. Rev. Lett. **94**, 101805 (2005).
31. H. Murayama and T. Yanagida, Phys. Lett. B **520**, 263 (2001). See also G. Barenboim, L. Borissov, J. Lykken and A. Y. Smirnov, JHEP **0210**, 001 (2002); G. Barenboim, L. Borissov and J. Lykken, Phys. Lett. B **534**, 106 (2002); Ref. [25].
32. M.C. Gonzalez-Garcia, M. Maltoni and T. Schwetz, Phys. Rev. D **68**, 053007 (2003).
33. See, for example, T. Katori, A. Kostelecky and R. Tayloe, hep-ph/0606154; V.A. Kostelecky and M. Mewes, Phys. Rev. D **70**, 031902 (2004); A. de Gouvêa, Phys. Rev. D **66**, 076005 (2002).
34. A. de Gouvêa and Y. Grossman, hep-ph/0602237.
35. S.R. Elliott and P. Vogel, Ann. Rev. Nucl. Part. Sci. **52**, 115 (2002).

Experimental Status of B Physics

Sheldon Stone

Physics Department, Syracuse University, Syracuse, N. Y., USA 13244-1130

Abstract. A short summary is given of the current status of B physics. Reasons for physics beyond the Standard Model are discussed. Constraints on New Physics are given using measurements of B mixing, B_S mixing, and CP violation, along with $|V_{ub}|$. Future goals, and upcoming new experiments are also mentioned.

Keywords: Weak decays, B physics
PACS: 13.20.He, 14.65.Fy

INTRODUCTION

"New Physics" (NP) refers to physics beyond the "Standard Model," the paradigm that we have constructed to explain our current high energy physics data [1]. We know, however, that NP is required to explain certain global phenomena including the Baryon asymmetry of the Universe, without which we could not exist, or the "Dark Matter," found first by Zwicky studying rotation curves of galaxies [2]. An even more mysterious phenomena called "Dark Energy" may also have a connection to particle physics experiments [3], perhaps via "Extra Dimensions" [4]. The fundamental goals of B decay studies are to discover, or help interpret, NP found elsewhere. Additional goals include measuring "fundamental constants" revealed to us by studying Weak interactions and understand the theory of strong interactions, QCD, necessary to interpret our measurements.

Baryogenesis

When the Universe began with the Big Bang, there was an equal amount of matter and antimatter. Now we have mostly matter. How did it happen? A. Sakharov gave three necessary conditions: Baryon (B) number violation, departure from thermal equilibrium, and C and CP violation [5]. (The operation of Charge Conjugation (C) takes particle to anti-particle and Parity (P) takes a vector $\vec{r}$ to $-\vec{r}$.)

These criteria are all satisfied by the Standard Model. B is violated in Electroweak theory at high temperature, though baryon minus lepton number is conserved; in addition we need quantum tunneling, which is powerfully suppressed at the low temperatures that we now have. Non-thermal equilibrium is provided by the electroweak phase transition. C and CP are violated by weak interactions. However the violation is too small. The ratio of the number of baryons to the number of photons in the Universe needs to be $\sim 6 \times 10^{-10}$, while the SM can provide only $\sim 10^{-20}$. Therefore, there must be new physics.

The Hierarchy Problem

Definition from the WIKIPEDIA encyclopedia [6]: "In theoretical physics, a hierarchy problem occurs when the fundamental parameters (couplings or masses) of some Lagrangian are vastly different (usually larger) than the parameters measured by experiment. This can happen because measured parameters are related to the fundamental parameters by a prescription known as renormalization. Typically the renormalized parameters are closely related to the fundamental parameters, but in some cases, it appears that there has been a delicate cancellation between the fundamental quantity and the quantum corrections to it."

Our worry is why the Planck scale at $\sim 10^{19}$ GeV is so much higher than the scale at which we expect to find the Higgs Boson, ~ 100 GeV. We expect the explanation lies in physics beyond the Standard Model.

THE BASICS: QUARK MIXING AND THE CKM MATRIX

The CKM matrix parameterizes the mixing between the mass eigenstates and weak eigenstates as couplings between the charge +2/3 and -1/3 quarks. I use here the Wolfenstein approximation [7] good to order λ^3 in the real part and λ^4 in the imaginary part:

$$V_{CKM} = \begin{pmatrix} 1-\lambda^2/2 & \lambda & A\lambda^3(\rho - i\eta(1-\lambda^2/2)) \\ -\lambda & 1-\lambda^2/2 - i\eta A^2\lambda^4 & A\lambda^2(1+i\eta\lambda^2) \\ A\lambda^3(1-\rho-i\eta) & -A\lambda^2 & 1 \end{pmatrix}. \quad (1)$$

In the Standard Model A, λ, ρ and η are fundamental constants of nature like G, or α_{EM}; η multiplies i and is responsible for all Standard Model CP violation. We know $\lambda=0.226$, $A \sim 0.8$ and we have constraints on ρ and η.

Applying unitarity constraints allows us to construct the six independent triangles shown in Fig. 1. Another basis for the CKM matrix are four angles labelled as χ (sometimes called β_S), χ' and any two of α, β and γ since $\alpha + \beta + \gamma = \pi$ [8]. (These angles are also shown in Fig. 1.)

B meson decays can occur through various processes. Some decay diagrams are shown in Fig. 2. The simple spectator diagram is dominant. Semileptonic decays, discussed next, proceed through this diagram.

SEMILEPTONIC DECAYS AND LIFETIMES

These are the simplest decays to describe theoretically. The transformation of the virtual W^- to a lepton-antineutrino pair proceeds through the axial-vector current just as in pion decay. Because of their relative simplicity, semileptonic decays are used to probe the $b \to c$ and $b \to u$ transitions. The overall semileptonic branching ratio, $\mathscr{B}_{SL}$ is defined as $\mathscr{B}(\to Xe^-\bar{\nu})$ equal to $\mathscr{B}(\to X\mu^-\bar{\nu})$ and has a measured value of (10.2 ± 0.9)% and (10.5 ± 0.8)%, for B^- and $\overline{B}^0$ mesons, respectively. The average for these mesons is much better measured as (10.87 ± 0.17)% [9].

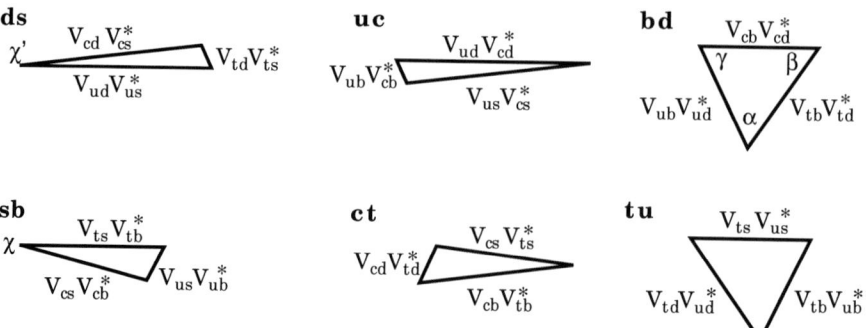

FIGURE 1. The 6 CKM triangles resulting from applying unitarity constraints to the indicated row and column. The CP violating angles are also shown.

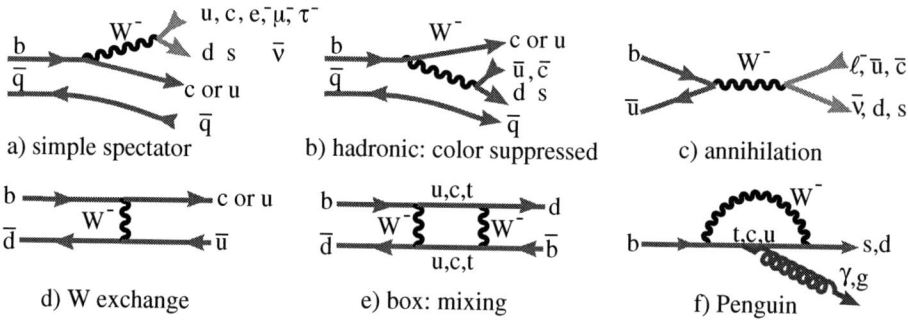

FIGURE 2. Some B decay diagrams.

The rather long average B lifetime, ~ 1.5 ps is an important aspect of B decays and is a crucial property allowing for more precise measurements of CP violation and other properties. The lifetime ratio $\tau_{B^-}/\tau_{\bar{B}^0} = 1.071 \pm 0.0009$ clearly demonstrates a longer, but not much longer lifetime for charged versus neutral B mesons.

Measurements of the CKM matrix elements $|V_{cb}|$ and $|V_{ub}|$ have been made using both exclusive decays to specific final states, such as $B \to D^* \ell^- \bar{\nu}$ and inclusive final states. Values have been compiled by the Heavy Flavor Averaging Group [10]. $|V_{cb}|$ is measured to be 0.038 ± 0.001 from exclusive decays using Heavy Quark Effective Theory (HQET) [9]. Inclusive decays have also been used and good precision has been achieved, although the accuracy depends critically on whether or not the assumption of "duality" is indeed correct. Measurements of $|V_{ub}|$ have also been made also using exclusive and inclusive decays. It is in the range of $3 - 4 \times 10^{-3}$. The main uncertainties are theoretical since there isn't a firm theoretical basis similar to HQET that can be used. The combination $|V_{ub}/V_{cb}| \approx \lambda^2 \sqrt{\rho^2 + \eta^2}$.

CURRENT B DECAY EXPERIMENTS

Current e^+e^- Experiments

The BaBar and Belle collaborations both work at e^+e^- colliders, with asymmetric energies, at a center-of-mass energy equal to the mass of the $\Upsilon(4S)$ resonance. Here there is almost equal production of both B^-B^+ and $\overline{B}^0B^0$ pairs, totaling 1 nb of cross section on top of 3 nb of background quark-antiquark production. The asymmetric energies are necessary to boost the B^0 mesons so that CP violation measurements can be made; the time integrated asymmetries would otherwise vanish as they are in $J^{PC} = 1^{--}$ states [11]. The boost, however, is small so the decay time resolution is only $\sim$900 fs r.m.s.

CLEO and ARGUS collected data on the Υ resonances using symmetric e^+e^- machines. CLEO is now concentrating in studying charm meson decays at lower energies. It is also worth noting that many e^+e^- experiments have provided a wealth of interesting data including the aforementioned ones and experiments at LEP (operating at the Z^0 resonance) and the PEP and PETRA machines (operating near 30 GeV).

Both CLEO and Belle have taken data at the $\Upsilon(5S)$ resonance. CLEO has determined the B_S fraction $\sim$16% of the 0.3 nb $b\bar{b}$ cross-section, about 1/20 the production rate at the $\Upsilon(4S)$ [12]. Not only is the yield small but the proper time resolution is not sufficient to allow time dependent CP violation measurements.

Current Hadron Collider Experiments

The CDF and D0 experiments at the Fermilab Tevatron are designed to study high energy phenomena, such as finding the top-quark and Higgs boson. However, they have some b capabilities and are well suited to study the B_S meson, which cannot be studied with e^+e^- colliders. The most important measurement that may be within reach of these experiments is that of B_S mixing. Production of b-flavored hadrons is a large 100 μb at the 2 TeV center-of-mass energy of the Tevatron. Unfortunately the detectors are as not fully equipped as dedicated heavy flavor experiments. They lack the excellent particle identification and crystal based electromagnetic calorimetry of the state-of-the-art e^+e^- experiments. They do, however, have good $\sim$100 fs decay time resolution [13].

B_D AND B_S MIXING

A diagram for B_d mixing is shown in Fig. 2(e). For B_S mixing just replace the d quarks with s quarks. The flavor eigenstates, degenerate in pure QCD mix under the weak interactions. Designating the base states as $\{|1>, |2>\} \equiv \{|B^0>, |\overline{B}^0>\}$, the Hamiltonian is

$$H = M - \frac{i}{2}\Gamma = \begin{pmatrix} M_{11} & M_{12} \\ M_{12}^* & M_{22} \end{pmatrix} - \frac{i}{2}\begin{pmatrix} \Gamma_{11} & \Gamma_{12} \\ \Gamma_{12}^* & \Gamma_{22} \end{pmatrix}. \quad (2)$$

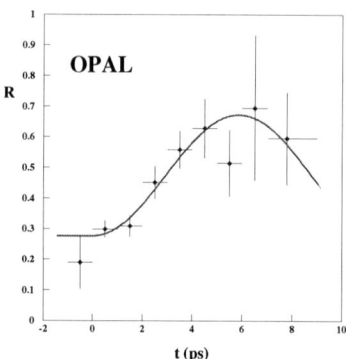

FIGURE 3. The probability $R = \left(B^0 \to \overline{B}^0\right) / \left(B^0 \to B^0\right)$ as a function of time from the OPAL experiment.

Diagonalizing the matrix we find the mass difference $\Delta m = m_{B_H} - m_{B_L} = 2|M_{12}|$. For B_d we predict $\Delta \Gamma \sim 0$.[1] The probability for a B^0 meson to appear as a $\overline{B}^0$ as a function of time is given by $0.5\Gamma e^{-\Gamma t}[1 + \cos(\Delta mt)]$. R is often defined as the ratio $\left(B^0 \to \overline{B}^0\right) / \left(B^0 \to B^0\right)$. B_d mixing was first discovered by the ARGUS experiment [14]. (There was a previous measurement by UA1 indicating mixing for a mixture of B_d^0 and B_s^0 [15].) At the time it was quite a surprise, since m_t was thought to be in the 30 GeV range. It is usual to define R as probability for a B^0 to materialize as a $\overline{B}^0$ divided by the probability it decays as a B^0. An early mixing result from OPAL is shown in Fig. 3 [16]. The world average value for Δm_d is a very precise 0.509 ± 0.005 ps^{-1} [9]. The measurement is dominated by the BaBar and Belle experiments.

The probability of mixing is given by [17] as

$$x \equiv \frac{\Delta m}{\Gamma} = \frac{G_F^2}{6\pi^2} B_B f_B^2 m_B \tau_B |V_{tb}^* V_{td}|^2 m_t^2 F\left(\frac{m_t^2}{M_W^2}\right) \eta_{QCD}, \qquad (3)$$

where B_B is a parameter related to the probability of the d and $\bar{b}$ quarks forming a hadron and must be estimated theoretically, F is a known function which increases approximately as m_t^2, and η_{QCD} is a QCD correction, with value about 0.8. By far the largest uncertainty arises from the unknown decay constant, f_B. In principle f_B can be measured. The decay rate of the annihilation process $B^- \to \ell^- \overline{\nu}$ is proportional to the product of $f_B^2 |V_{ub}|^2$. This is a very difficult process to measure, and even if this were done, the uncertainty on V_{ub} will lead to an imprecise result. Our current best hope is to rely on unquenched lattice QCD which can use the measurements of the analogous $D^+ \to \mu^+ \nu$ decay as check. These checks are currently in progress at CLEO-c [18].

---

[1] This is because the fraction of final states that of the same CP parity that both B^0 and $\overline{B}^0$ can decay into is very small. This is not the case for B_S.

Since
$$|V_{tb}^*V_{td}|^2 \propto |(1-\rho-i\eta)|^2 = (\rho-1)^2 + \eta^2, \tag{4}$$

measuring mixing gives a circle centered at (1,0) in the $\rho-\eta$ plane. This could in principle be a very powerful constraint. Unfortunately, the parameter B_B is not experimentally accessible and f_B must be calculated; the errors on the calculations are quite large.

B_s^0 mesons can mix in a similar fashion to B_d^0 mesons. The diagram in Fig. 2(e) is modified by substituting s quarks for d quarks, thereby changing the relevant CKM matrix element from V_{td} to V_{ts}. Measuring x_s allows us to use ratio of x_d/x_s to provide constraints on the CKM parameters ρ and η. We still obtain a circle in the (ρ,η) plane centered at (1,0):

$$|V_{td}|^2 = A^2\lambda^4\left[(1-\rho)^2 + \eta^2\right] \tag{5}$$
$$\frac{|V_{td}|^2}{|V_{ts}|^2} = (1-\rho)^2 + \eta^2 \ .$$

Now however we must calculate only the SU(3) broken ratios B_{B_d}/B_{B_s} and f_{B_d}/f_{B_s}.

B_s^0 mixing has been searched for at LEP, the Tevatron, and the SLC. A combined analysis has been performed. The probability, $\mathscr{P}(t)$ for a B_s to oscillate into a $\overline{B}_s$ is given as

$$\mathscr{P}(t)\left(B_s \to \overline{B}_s\right) = \frac{1}{2}\Gamma_s e^{-\Gamma_s t}\left[1+\cos(\Delta m_s t)\right] \ , \tag{6}$$

where t is the proper time.

To combine different experiments a framework has been established where each experiment finds a amplitude A for each test frequency ω, defined as

$$\mathscr{P}(t) = \frac{1}{2}\Gamma_s e^{-\Gamma_s t}\left[1+A\cos(\omega t)\right] \ . \tag{7}$$

Fig. 4 shows the world average measured amplitude A as a function of the test frequency $\omega = \Delta m_s$ [10]. For each frequency the expected result is either zero for no mixing or one for mixing. No other value is physical, although measurement errors admit other values. The data do indeed cross one at a Δm_s of 16 ps^{-1}, however here the error on A is about 0.6, precluding a statistically significant discovery. The quoted upper limit at 95% confidence level is 16.6 ps^{-1}. This is the point where the value of A plus 1.645 times the error on A reach one. Also, one should be aware that all the points are strongly correlated.

As this work was being completed the D0 experiment announced that they had limited the Δm_s between 17 ps^{-1} and 21 ps^{-1} at 90% confidence level [19]. Fig. 5 shows their amplitude analysis results. Clearly the significance of the result, although limited, relies on seeing an amplitude in excess of the expected value of 1, in fact, nearly at 3. Further data will be needed to confirm this result. The inferred values of ρ and η are within the range expected by fits to other parameters (see Fig. 8), and are consistent with Standard Model expectations.

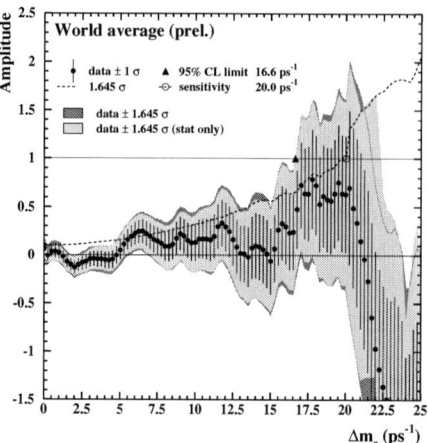

FIGURE 4. Combined experimental values of the amplitude A versus the test frequency $\omega = \Delta m_s$ as defined in equation 7. The inner (outer) envelopes give the 95% confidence levels using statistical (statistical and systematic) errors. The "sensitivity" shown at 20.0 ps^{-1} is the likely place a 95% c.l. upper limit could be set.

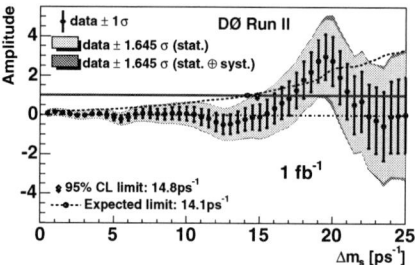

FIGURE 5. B_S^0 oscillation amplitude as a function of oscillation frequency, Δm_s from D0. The red (solid) line shows the $\mathscr{A} = 1$ axis for reference. The dashed line shows the expected limit including both statistical and systematic uncertainties.

CP VIOLATION MEASUREMENTS

Introduction

CP violation can occur because of the imaginary term in the CKM matrix, proportional to η in the Wolfenstein representation [11].

Decays of neutral K mesons were the first to show CP violating effects. In this decade the BaBar and Belle experiments provided precision measurement of one of the four CP violating angles (β) and gave first measurements of two other angles (α and γ).

Consider the case of a process $B \to f$ that goes via two amplitudes A and B each of which has a strong part e. g. $s_{\mathscr{A}}$ and a weak part $w_{\mathscr{A}}$. Then we have

$$\Gamma(B \to f) = \left(|\mathscr{A}|e^{i(s_{\mathscr{A}}+w_{\mathscr{A}})} + |\mathscr{B}|e^{i(s_{\mathscr{B}}+w_{\mathscr{B}})}\right)^2 \tag{8}$$

$$\Gamma(\overline{B} \to \overline{f}) = \left(|\mathscr{A}|e^{i(s_{\mathscr{A}}-w_{\mathscr{A}})} + |\mathscr{B}|e^{i(s_{\mathscr{B}}-w_{\mathscr{B}})}\right)^2 \tag{9}$$

$$\Gamma(B \to f) - \Gamma(\overline{B} \to \overline{f}) = 2|\mathscr{A}\mathscr{B}|\sin(s_{\mathscr{A}} - s_{\mathscr{B}})\sin(w_{\mathscr{A}} - w_{\mathscr{B}}) \ . \tag{10}$$

Any two amplitudes will do, though its better that they be of approximately equal size. Thus charged B decays can exhibit CP violation as well as neutral B decays. In some cases, we will see that it is possible to guarantee that $|\sin(s_{\mathscr{A}} - s_{\mathscr{B}})|$ is unity, so we can get information on the weak phases. In the case of neutral B decays, mixing can be the second amplitude.

Formalism of CP Violation in Neutral B Decays

For neutral mesons we can construct the CP eigenstates

$$|B_1^0\rangle = \frac{1}{\sqrt{2}}\left(|B^0\rangle - |\overline{B}^0\rangle\right), \quad |B_2^0\rangle = \frac{1}{\sqrt{2}}\left(|B^0\rangle + |\overline{B}^0\rangle\right), \text{ where} \tag{11}$$

$$CP|B_1^0\rangle = |B_1^0\rangle, \quad CP|B_2^0\rangle = -|B_2^0\rangle \ . \tag{12}$$

Since B^0 and $\overline{B}^0$ can mix, the mass eigenstates are a superposition of $a|B^0\rangle + b|\overline{B}^0\rangle$ which obey the Schrodinger equation

$$i\frac{d}{dt}\begin{pmatrix}a\\b\end{pmatrix} = \mathscr{H}\begin{pmatrix}a\\b\end{pmatrix} = \left(M - \frac{i}{2}\Gamma\right)\begin{pmatrix}a\\b\end{pmatrix}. \tag{13}$$

If CP is not conserved then the eigenvectors, the mass eigenstates $|B_L\rangle$ and $|B_H\rangle$, are not the CP eigenstates but are

$$|B_L\rangle = p|B^0\rangle + q|\overline{B}^0\rangle, \quad |B_H\rangle = p|B^0\rangle - q|\overline{B}^0\rangle, \text{ where} \tag{14}$$

$$p = \frac{1}{\sqrt{2}}\frac{1+\varepsilon_B}{\sqrt{1+|\varepsilon_B|^2}}, \quad q = \frac{1}{\sqrt{2}}\frac{1-\varepsilon_B}{\sqrt{1+|\varepsilon_B|^2}}. \tag{15}$$

CP is violated if $\varepsilon_B \neq 0$, which occurs if $|q/p| \neq 1$.

CP violation for B via interference of mixing and decays

Here we choose a final state f which is accessible to both B^0 and $\overline{B}^0$ decays. The second amplitude necessary for interference is provided by mixing. It is necessary only that f be accessible directly from either state; however if f is a CP eigenstate the situation is far simpler. For CP eigenstates $CP|f_{CP}\rangle = \pm|f_{CP}\rangle$. It is useful to define the amplitudes $A = \langle f_{CP}|\mathcal{H}|B^0\rangle$, $\bar{A} = \langle f_{CP}|\mathcal{H}|\bar{B}^0\rangle$. If $\left|\frac{\bar{A}}{A}\right| \neq 1$, then we have "direct" CP violation in the decay amplitude, which we will discuss in detail later. Here CP can be violated by having

$$\lambda = \frac{q}{p} \cdot \frac{\bar{A}}{A} \neq 1, \tag{16}$$

which requires only that λ acquire a non-zero phase, i.e. $|\lambda|$ could be unity and CP violation can occur.

A comment on neutral B production at e^+e^- colliders is in order. At the $\Upsilon(4S)$ resonance there is coherent production of $B^0\bar{B}^0$ pairs. This puts the B's in a $C = -1$ state. In hadron colliders, or at e^+e^- machines operating at the Z^0, the B's are produced incoherently. The asymmetry is defined as

$$a_{f_{CP}} = \frac{\Gamma\left(B^0(t) \to f_{CP}\right) - \Gamma\left(\bar{B}^0(t) \to f_{CP}\right)}{\Gamma\left(B^0(t) \to f_{CP}\right) + \Gamma\left(\bar{B}^0(t) \to f_{CP}\right)}, \tag{17}$$

which for $|q/p| = 1$ gives

$$a_{f_{CP}} = \frac{\left(1 - |\lambda|^2\right)\cos(\Delta mt) - 2\mathrm{Im}\lambda \sin(\Delta mt)}{1 + |\lambda|^2}. \tag{18}$$

For the cases where there is only one decay amplitude A, $|\lambda|$ equals 1, and we have

$$a_{f_{CP}} = -\mathrm{Im}\lambda \sin(\Delta mt). \tag{19}$$

Only the amplitude, $-\mathrm{Im}\lambda$ contains information about the level of CP violation, the sine term is determined only by B_d mixing; the time integrated asymmetry is given by

$$a_{f_{CP}} = -\frac{x}{1+x^2}\mathrm{Im}\lambda = -0.48\mathrm{Im}\lambda. \tag{20}$$

This is quite lucky as the maximum size of the coefficient for any x is -0.5.

Let us now find out how $\mathrm{Im}\lambda$ relates to the CKM parameters. Recall $\lambda = \frac{q}{p} \cdot \frac{\bar{A}}{A}$. The first term is the part that comes from mixing:

$$\frac{q}{p} = \frac{(V_{tb}^* V_{td})^2}{|V_{tb}V_{td}|^2} = \frac{(1-\rho-i\eta)^2}{(1-\rho+i\eta)(1-\rho-i\eta)} = e^{-2i\beta} \text{ and} \tag{21}$$

$$\mathrm{Im}\frac{q}{p} = -\frac{2(1-\rho)\eta}{(1-\rho)^2 + \eta^2} = \sin(2\beta). \tag{22}$$

To evaluate the decay part we need to consider specific final states. Let's consider the final state $J/\psi K_s$. The decay process proceeds via the diagram in Fig. 2(b), where the $c\bar{c}$

forms a J/ψ. Here we do not get a phase from the decay part because

$$\frac{\bar{A}}{A} = \frac{(V_{cb}V_{cs}^*)^2}{|V_{cb}V_{cs}|^2} \qquad (23)$$

is real to order $1/\lambda^4$.

In this case the final state is a state of negative CP, i.e. $CP|J/\psi K_s\rangle = -|J/\psi K_s\rangle$. This introduces an additional minus sign in the result for Imλ. Before finishing discussion of this final state we need to consider in more detail the presence of the K_s in the final state. Since neutral kaons can mix, we pick up another mixing phase (similar diagrams as for B^0, see Fig. 2(e)). This term creates a phase given by

$$\left(\frac{q}{p}\right)_K = \frac{(V_{cd}^*V_{cs})^2}{|V_{cd}V_{cs}|^2}, \qquad (24)$$

which is real to order λ^4. It is necessary to include this term, however, since there are other formulations of the CKM matrix than Wolfenstein, which have the phase in a different location. It is important that the physics predictions not depend on the CKM convention.[2]

CP Violation Measurements

The CP asymmetry $\sin(2\beta)$ has been measured by both Belle and BaBar using both CP+ and CP- final states. Most of the latter are $J/\psi K_S$, while most of the former are $J/\psi K_L$. Fig. 6 shows the raw asymmetries and the fit results for $(c\bar{c})K_S$ (top) and $J/\psi K_L$ (bottom) [20]. The world average value of $\sin(2\beta)$ is 0.685±0.032 [10].

The Belle collaboration pioneered the measurement of γ using the charged decays $B^\mp \to D^0 K^\mp$, where the $D^0 \to K_S\pi^+\pi^-$. Here D^0 decays cannot be distinguished from $\bar{D^0}$ decays, and they interfere. Measurements from BaBar and Belle have been reported. BaBar averages in additional information from $D^{*0}K^\mp$ and $D^0K^{0\mp}$, finding $\gamma = (67 \pm 28 \pm 13 \pm 11)°$ and Belle, omitting the last mode, obtains $\gamma = (67^{+14}_{-15} \pm 13 \pm 11)°$, where the last error is due the parametrization of the D^0 decay Dalitz plot, and could be helped greatly by CLEO-c measurements of the CP+ and CP- Dalitz plots [21, 22].

The angle α can be probed by measuring processes such as $B^0 \to \pi^+\pi^-$ or $\rho^+\rho^-$ as shown in Fig. 7(a), because the combination of weak phases in the mixing amplitude and the $b \to u$ decay amplitude are $\sin(2(\beta+\gamma)) = \sin(2(180-\alpha)) = -\sin(2\alpha)$. Unfortunately, the Penguin diagram in Fig. 7(b) has no weak phase and can be significant in these processes. Thus the Penguin process can "pollute" the measurement of α in these modes, but it can be limited by using the upper limit on the branching ratio for $B^0 \to \rho^0\rho^0$ as shown by Grossman and Quinn [23]. BaBar first used this final state for

---

[2] Here we don't include CP violation in the neutral kaon since it is much smaller than what is expected in the B decay. The term of order λ^4 in V_{cs} is necessary to explain K^0 CP violation.

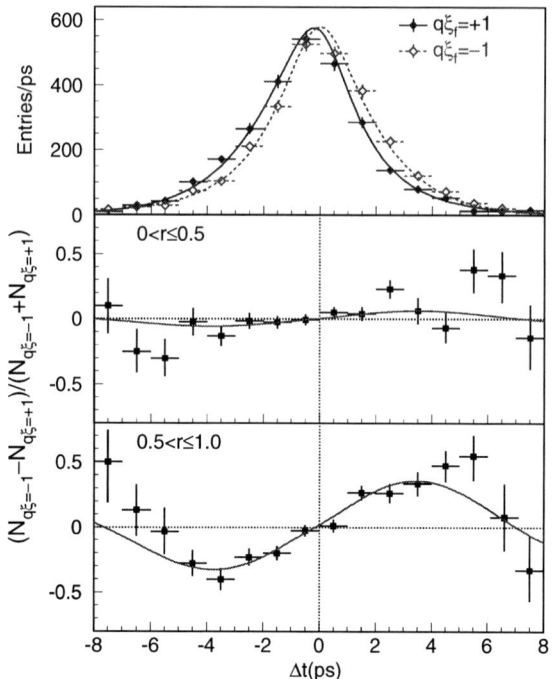

FIGURE 6. Δt distributions from Belle for the events with $q\xi_f = -1$ (open points) and $q\xi_f = +1$ (solid points) with all modes combined (top), asymmetry between $q\xi_f = -1$ and $q\xi_f = +1$ samples with $0 < r \leq 0.5$ (middle), and with $0.5 < r \leq 1$ (bottom). The variable r refers to the probability of the correctness of the flavor tag, where $r=1$ is almost assuredly correct, while $r=0$ conveys no information. The results of the global unbinned maximum-likelihood fit ($\sin(2\beta)=0.728$) are also shown.

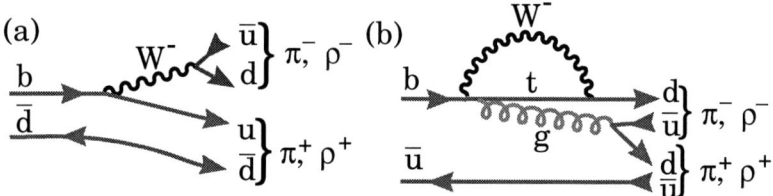

FIGURE 7. Tree (a) and Penguin (b) processes for neutral B decay into either $\pi^+\pi^-$ or $\rho^+\rho^-$.

CP violation measurement by showing that it is almost fully polarized and that the Penguin term could be usefully limited. Currently we have $\alpha = (96 \pm 13 \pm 11)°$, where the last error is due to the possible Penguin contribution [21].

LIMITS ON NEW PHYSICS

Constraints on the Wolfenstein ρ and η parameters are given by many measurements and summarized in Fig. 8 [24]. (For alternative fits see Ref. [25].) This plot is based

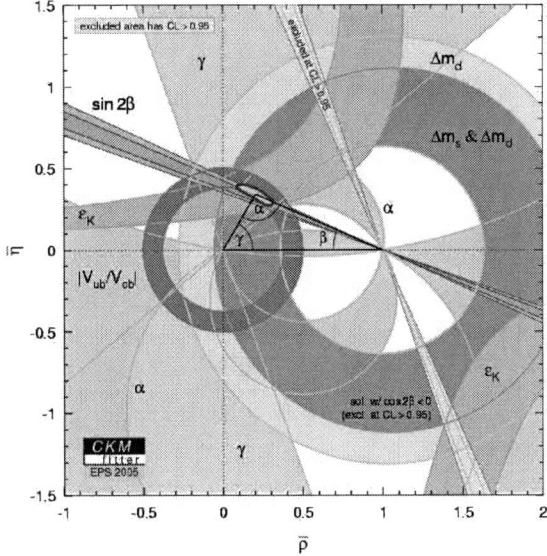

FIGURE 8. Constraints on the ρ and η Wolfenstein parameters after summer 2005, $\overline{\rho} = \rho(1-\lambda^2/2)$ and $\overline{\eta} = \eta(1-\lambda^2/2)$.

on measurements of $|V_{ub}/V_{cb}|$, $B^0 - \overline{B}^0$ mixing, upper limits on B_S mixing and the CP violation measurements discussed here of α, β and γ as well as CP violation in the K_L^0 system.

Agashe et al. have established limits on New Physics (NP) arising via B^0 mixing [26], using a method that was modified from that first used by Grossman et al. and Ligeti [27]. They assume that NP in tree level processes, such as those used to measure $|V_{ub}|$ is negligible. They then parameterize NP in terms of an amplitude h and a phase σ as

$$\Delta m_d = \left|1 + h_d e^{2i\sigma_d}\right| \Delta m_d^{SM}, \qquad S_{\psi K} = \sin[2\beta + 2\theta_d], \qquad (25)$$

where $\theta_d = \arg\left(1 + h_d e^{2i\sigma_d}\right)$. The CP asymmetry in $B^\mp \to DK^\mp$, A_{DK}, is also a SM tree level transition and therefore is unaffected by NP; $A_{DK} \sim \tan\gamma = \frac{\eta}{\rho}$. Note that A_{DK} depends only on ρ, η in a combination different than V_{ub}. The CP asymmetry in $B^0 \to \rho^+\rho^-$, $S_{\rho\rho}$, is given by $S_{\rho\rho} \propto \sin(2\gamma + 2\beta + 2\theta_d)$. Thus, $S_{\rho\rho}$ also depends only on ρ, η *after* subtracting the phase of B_d mixing (including the NP phase) using $S_{\psi K_S}^{exp}$. Thus ρ, η can be determined even in presence of NP. The allowed size of NP admits a range for h_d of $h_d = 0 - 0.4$, for $2\sigma_d = \pi - 2\pi$. This is demonstrated in Fig. 9 where the $h_d - \sigma_d$ allowed regions are shown. Thus, the data do not yet exclude substantial contributions to NP via B_d mixing. In the case of NP via B_S mixing, there are almost no restrictions.

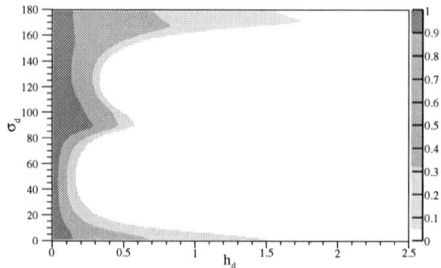

FIGURE 9. The allowed range for h_d and σ_d from [26]. The shaded scale on the right side indicates the confidence level.

A hint of NP may be showing up in measurements of CP violation in Penguin decays. A data summary is shown in Fig. 10. The trend is for these modes to have asymmetries below that in $J/\psi K_S$ related modes. These modes may have additional amplitudes, but calculations tend to show that these would result in positive asymmetries, opposite to the observed effect [28]. Each mode must be considered individually so averaging them is not a reasonable approach.

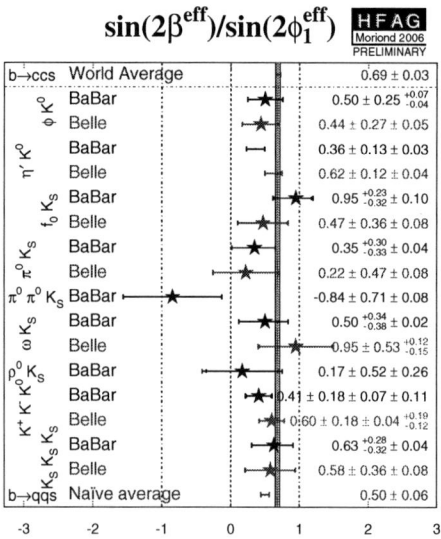

FIGURE 10. Measurement of $\sin(2\beta)$ in Penguin dominated modes versus that in $(c\bar{c})s$ modes. Note that $\sin(2\beta)$ is sometimes called $\sin(2\phi_1)$. The superscript "eff" indicates that no attempt has been made to correct for the possible presence of a $\cos(\Delta mt)$ term, see equation 18.

FUTURE B DECAY EXPERIMENTS

The future of heavy physics may well be the provenance of experiments at CERN starting in ~2008 when significant data will be taken by experiments at the LHC, a proton-proton collider with 14 TeV of energy in the center-of-mass.

Three experiments are equipped to study B decays. The LHCb experiment is the only one specifically designed for this purpose. The ATLAS and CMS experiments can, however, make some useful measurements; they are intended to run a very high luminosity, 10^{34}cm^{-2}/s, while LHCb will run around 2×10^{32}cm^{-2}/s. While CMS and ATLAS are designed to measure new high mass particles in the central region, LHCb will detect b-flavored hadrons produced in the forward direction along one of the beams. The production mechanism tends to put both particles in the detector acceptance, crucial for flavor tagging, i. e. distinguish the flavor of the b's at birth.

A sketch of the LHCb detector is shown in Fig. 11. A silicon strip detector called "VELO" is used to measure decay vertices. The detectors are segmented along the radial and azimuthal directions. The layout is shown Fig. 12, the sensor geometry in (b) and a photograph in (c). There are two ring imaging Cherenkov counters used to distinguish pions from kaons, required because of the large range of momenta (1-100 GeV/c) that occur. An electromagnetic calorimeter constructed from scintillating fibers and lead detects γ's π^0's and η's; it also identifies electrons. The iron filter after the hadron calorimeter (HCAL) interspersed with the chambers M2-M5 is used to identify muons. The calorimetry, both electromagnetic and hadronic provide real time information used in the first trigger level (called Level 0) for charged particles or neutral energy at transverse momenta that are likely to come from b decays. A "pile-up" device is also used to identify beam crossings with more than one interaction. More details about the detector can be found in [29].

The KEK accelerator has produced very impressive luminosities and there are plans to improve it. This concept is called "Super-Belle." There would be both machine and detector improvements allowing running up to an instantaneous luminosity of ~ 5×10^{35}cm^{-2}/s. Currently, this is a proposal that has yet to be acted on. Another similar proposal was also formulated by the "Super-BaBar" group. It however has not been supported by SLAC or the U. S. Dept. of Energy.

A group in Frascati, Italy has been exploring the possibility of using recirculating electron linacs as the basis of a novel e^+e^- collider in the Upsilon region [30]. This machine would not have appreciable synchrotron radiation, so current detector technologies would work just fine. However, the number of interactions per crossing could be large.

CONCLUSIONS

The study of the decays of b-flavored hadrons has advanced greatly from its early beginnings. We have one precision measurement, namely that of $\sin(2\beta)$ and initial measurements of α and γ. Yet much more needs to be done. $|V_{ub}|$ needs to be made more precise by improvements in QCD calculations and comparisons with charmless semileptonic decays in the appropriate kinematic regions suitable for reliable theoretical predictions. Measurements of CP violation in B_S decays are of prime importance. After

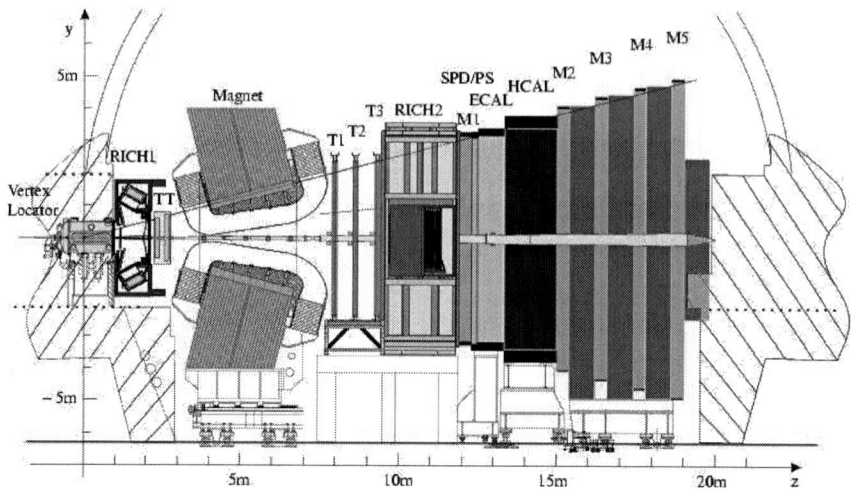

FIGURE 11. A sketch of the LHCb detector showing the Vertex Locator (VELO), the two RICH subsystems, the tracking trigger stations (TT) before the magnet, the tracking stations after the magnet (T1-T3), the Scintillating Pad detector (SPD), the Prewshower (PS), the Electromagnetic Calorimeter (ECAL), the Hadronic Calorimeter (HCAL) and the Muon Stations (M1-M5).

the termination of current experimental efforts in flavor physics in the U. S. at the end of this decade, experimental progress will depend on experiments at the LHC, in particular LHCb, and at Belle or a possible Super-Belle in Japan. These experiments will be essential in interpreting the New Physics we expect to find at the LHC.

ACKNOWLEDGMENTS

This work was supported by the U. S. National Science Foundation under grant #0353860. I thank A. Alejandro and A. Bashir for making the X Mexican Workshop on Particles and Fields a pleasant and useful meeting. I also had useful discussions with K. Agashe and M. Artuso.

REFERENCES

1. G, Altarelli, "The Standard Electroweak Theory and Beyond," in *Proceedings of the Summer School on Phenomenology of Gauge Interactions*, Zuoz, Switzerland, [hep-ph/0011078].
2. M. A. Srednicki, "Particle Physics and Cosmology: Dark Matter," Amsterdam, Netherlands, North-Holland (1990).
3. D. Hunter, G. D. Starkman and M. Trodden, Phys. Rev. **D66** 043511 (2002).
4. J. Hewett and M. Spiropulu, Ann. Rev. Nucl. Part. Sci. **52**, 397 (2002).
5. A. Sakharov, JETP, **5**, 32 (1967).
6. http://en.wikipedia.org.
7. L. Wolfenstein, Phys. Rev. Lett. **51**, (1983) 1945.

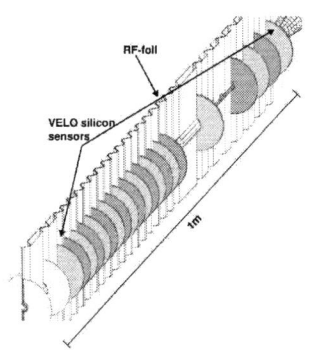 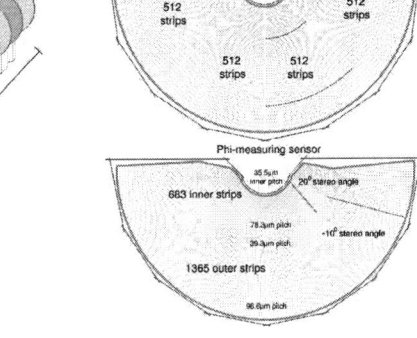 

(a) VELO with RF-foil, 21 $r - \phi$ detector stations, and two upstream r stations for the Pile Up system.

(b) r and ϕ sensors. For each sensor, 2 readout strips are indicated by dotted lines, for illustration.

(c) Prototype Si sensor with readout electronics

FIGURE 12. The LHCb Vertex Locator (VELO)

8. R. Aleksan et al., Phys. Rev. Lett. **73**, (1994) 18; J. P. Silva and L. Wolfenstein, Phys. Rev. **D55** (1997) 5331.
9. S. Eidelman et al., Phys. Lett. B **592**, 1 (2004).
10. E. Baberio et al. (HFAG), [hep-ex/0603003].
11. I. I. Bigi and A. I. Sanda, *CP Violation*, Cambridge Univ. Press, Cambridge, U. K. (1999).
12. M. Artuso et al. CLEO, Phys. Rev. Lett. **95**, 261801 (2005).
13. G. Giurgiu, talk at Frontiers in Contemporary Physics III, Vanderbilt Univ. (2005) http://www.hep.vanderbilt.edu/~wjohns/fcp05/recent/.
14. H. Albrecht et al. ARGUS, Phys. Lett. B**192**, 245 (1987).
15. C. Albajar et al. UA1, 1987, Phys. Lett. B **186**, 247, erratum-ibid B**197** 565 (1987).
16. R. Akers et al. OPAL 1995, Z. Phys. C **66**, 555 (1995).
17. M. Gaillard and B. Lee, 1974, Phys. Rev. D **10**, 897 (1974).
18. M. Artuso, "Charm Decays Within the Standard Model and Beyond," Submitted to the Proceedings of Lepton Photon 2005, XXII International Symposium on Lepton-Photon Interactions at High Energy July 2005, Uppsala, Sweden [hep-ex/0510052].
19. V. Abazov et al., (D0), [hep-ex/0603029].
20. K. Abe et al. (Belle), Phys. Rev. **D71**, 072003 (2005); Erratum-ibid. **D71**, 079903 (2005).
21. M. H. Schune, "CP violation and Heavy Flavours," plenary talk at HEP-EPS 2005, July 21-27, Lisboa, Portugal.
22. D. M. Asner, and W. M. Sun, Phys. Rev. **D73**, 034024 (2006).
23. Y. Grossman and H. Quinn, Phys. Rev. D58, 017504 (1998).
24. J. Charles et al., Eur. Phys. J. C **41**, 1 (2005) [hep-ph/0406184]; http://ckmfitter.in2p3.fr/.
25. M. Bona et al., [hep-ph/0501199]; M. Bona et al., [hep-ph/0509219]; http://www.utfit.org.
26. K. Agashe et al. [hep-ph/05091171].
27. Y. Grossman, Y. Nir and M. P. Worah, Phys. Lett. **B407**, 307 (1997); Z. Ligeti, Int. J. Mod. Phys. **A20** 5105 (2005), [hep-ph/0408267].
28. M. Benecke, Phys. Lett. **B620**, 143 (2005).
29. http://lhcb.web.cern.ch/lhcb/
30. J. Albert et al., [physics/0512235].

Higgsless Models: Lessons from Deconstruction

E.H. Simmons*, R.S. Chivukula*, H.-J. He[†],
M. Kurachi** and M. Tanabashi[‡]

*Department of Physics and Astronomy, Michigan State University, East Lansing, MI, USA
[†]Department of Physics, Tsinghua University, Beijing, P.R. China
**Department of Physics and Astronomy, SUNY Stony Brook, NY, USA
[‡]Department of Physics, Tohoku University, Sendai, Japan

Abstract. This talk reviews recent progress in Higgsless models of electroweak symmetry breaking, and summarizes relevant points of model-building and phenomenology.

Keywords: Higgsless Theories, Electroweak Symmetry Breaking Dimensional Deconstruction, Delocalization, Multi-gauge-boson vertices, Chiral Lagrangian, Extended Electroweak Groups
PACS: 12.60.-i, 11.10.K.k, 12.60.Cn

INTRODUCTION

Higgsless models [1] do just what their name suggests: they break the electroweak symmetry and unitarize the scattering of longitudinal W and Z bosons without employing a scalar Higgs [2] boson. In a class of well-studied models [3, 4] based on a five-dimensional $SU(2) \times SU(2) \times U(1)$ gauge theory in a slice of Anti-deSitter space, electroweak symmetry breaking is encoded in the boundary conditions of the gauge fields. In addition to a massless photon and near-standard W and Z bosons, the spectrum includes an infinite tower of additional massive vector bosons (the higher Kaluza-Klein or KK excitations), whose exchange is responsible for unitarizing longitudinal W and Z boson scattering [5, 6, 7, 8]. The electroweak properties and collider phenomenology of many such models have been discussed in the literature [9, 10, 11, 12, 13, 14, 15, 16].

An alternative approach to analyzing the properties of Higgsless models [17, 18, 19, 20, 21, 22, 23, 24] is to use deconstruction [25, 26] and to compute the electroweak parameters [27, 28, 29] in a related linear moose model [30]. We have shown [24] how to compute all four of the leading zero-momentum electroweak parameters defined by Barbieri et. al. [15] in a very general class of linear moose models. Using these techniques, we showed [24] that a Higgsless model whose fermions are localized (i.e., derive their electroweak properties from a single site on the deconstructed lattice) cannot simultaneously satisfy unitarity bounds and precision electroweak constraints unless the model includes extra light vector bosons with masses comparable to those of the W or Z.

It has recently been proposed [31, 32, 33] that the size of corrections to electroweak processes may be reduced by including delocalized fermions. In deconstruction, a delocalized fermion is realized as a fermion whose $SU(2)$ properties arise from several sites on the deconstructed lattice [34, 35]. We have studied [36] the properties of deconstructed Higgsless models with fermions whose $SU(2)$ properties arise from delocaliza-

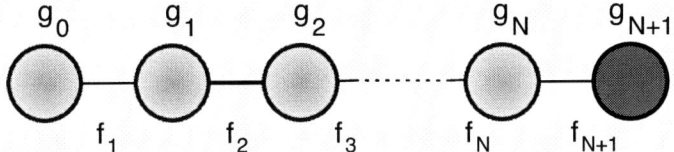

FIGURE 1. Moose diagram of the type of model analyzed in this talk. Sites 0 to N are $SU(2)$ gauge groups, site $N+1$ is a $U(1)$ gauge group. The gauge couplings and f-constants indicated are left arbitrary in the analysis. In part of the analysis, the fermions are delocalized in the sense that the $SU(2)$ couplings of the fermions arise (potentially) from the gauge groups at all sites from 0 to N. The $U(1)$ coupling comes from the gauge group at site $N+1$.

tion over many sites of the deconstructed lattice. In an arbitrary Higgsless model we showed that if the probability distribution of the delocalized fermions is appropriately related to the W wavefunction (a condition we call "ideal" delocalization) then deviations in precision electroweak parameters are minimized. In particular, three ($\hat{S}$, $\hat{T}$, W) of the four leading zero-momentum precision electroweak parameters defined by Barbieri, et. al. [15] vanish at tree-level.

Because Higgsless models with ideally delocalized fermions have vanishing precision electroweak observables, it is necessary to look elsewhere for experimental signatures of these models [37][38]. We have computed [37] the form of the triple and quartic gauge boson vertices in these models. These constraints were shown to provide lower bounds of order a few hundred GeV on the masses of the lightest KK resonances above the W and Z bosons. We have also computed [37] the leading Appelquist-Longhitano coefficients in the electroweak chiral Lagrangian [39, 40, 41, 42, 43] for these models. We also studied the collider phenomenology of the KK resonances in models with ideal delocalization; because these resonances are fermiophobic, traditional direct collider searches are not sensitive to them and measurements of gauge-boson scattering [44] will be needed to find them.

This talk summarizes some of the recent results for Higgsless models, especially those related to model-building and phenomenology. Each topic can only be touched on briefly here, and the reader is encouraged to visit the cited literature for more extensive discussion and derivations.

A GENERAL HIGGSLESS MODEL

This discussion focuses on a deconstructed Higgsless model of the general type shown diagrammatically (using "moose notation" [30]) in fig. 1. The model incorporates an $SU(2)^{N+1} \times U(1)$ gauge group, and $N+1$ nonlinear $(SU(2) \times SU(2))/SU(2)$ sigma models in which the global symmetry groups in adjacent sigma models are identified with the corresponding factors of the gauge group. The Lagrangian for this model at leading order is given by [36]

$$\mathcal{L}_2 = \frac{1}{4}\sum_{j=1}^{N+1} f_j^2 \, \text{tr}\left((D_\mu U_j)^\dagger (D^\mu U_j)\right) - \sum_{j=0}^{N+1} \frac{1}{2g_j^2} \, \text{tr}\left(F_{\mu\nu}^j F^{j\mu\nu}\right), \tag{1}$$

with

$$D_\mu U_j = \partial_\mu U_j - iA_\mu^{j-1} U_j + iU_j A_\mu^j, \tag{2}$$

where all gauge fields A_μ^j ($j = 0, 1, 2, \cdots, N+1$) are dynamical. The first $N+1$ gauge fields ($j = 0, 1, \ldots, N$) correspond to $SU(2)$ gauge groups; the last gauge field ($j = N+1$) corresponds to the $U(1)$ gauge group. The symmetry breaking between the A_μ^N and A_μ^{N+1} follows an $SU(2)_L \times SU(2)_R/SU(2)_V$ symmetry breaking pattern with the $U(1)$ embedded as the T_3-generator of $SU(2)_R$. Our analysis proceeds for arbitrary values of the gauge couplings and f-constants. In the continuum limit, therefore, this allows for arbitrary background 5-D geometry, spatially dependent gauge-couplings, and brane kinetic energy terms for the gauge-bosons.

The neutral vector meson mass-squared matrix is of dimension $(N+2) \times (N+2)$

$$M_Z^2 = \frac{1}{4}\begin{pmatrix} g_0^2 f_1^2 & -g_0 g_1 f_1^2 & & & \\ -g_0 g_1 f_1^2 & g_1^2(f_1^2+f_2^2) & -g_1 g_2 f_2^2 & & \\ & \ddots & \ddots & \ddots & \\ & & -g_{N-1} g_N f_N^2 & g_N^2(f_N^2+f_{N+1}^2) & -g_N g_{N+1} f_{N+1}^2 \\ & & & -g_N g_{N+1} f_{N+1}^2 & g_{N+1}^2 f_{N+1}^2 \end{pmatrix}. \tag{3}$$

and the charged current vector bosons' mass-squared matrix is the upper-left $(N+1) \times (N+1)$ dimensional block of the M_Z^2 matrix. The neutral mass matrix (3) is of a familiar form that has a vanishing determinant, due to a zero eigenvalue. Physically, this corresponds to a massless neutral gauge field – the photon. The non-zero eigenvalues of M_Z^2 are labeled by m_{Zz}^2 ($z = 0, 1, 2, \cdots, N$), while those of M_W^2 are labeled by m_{Ww}^2 ($w = 0, 1, 2, \cdots, N$).

The lowest massive eigenstates corresponding to eigenvalues m_{Z0}^2 and m_{W0}^2 are, respectively, identified as the usual Z and W bosons. We will refer to these last eigenvalues by their conventional symbols M_Z^2, M_W^2; the distinction between these and the corresponding mass matrices should be clear from context. We will denote the eigenvectors corresponding to the photon, Z, and W by v_i^γ, v_i^Z, and v_j^W. These eigenvectors are normalized as

$$\sum_{i=0}^{N+1} (v_i^\gamma)^2 = \sum_{i=0}^{N+1} (v_i^Z)^2 = \sum_{j=0}^{N} (v_j^W)^2 = 1. \tag{4}$$

Inspection of the matrix M_Z^2 reveals that each component of the photon eigenvector is inversely related to the gauge coupling at the corresponding site

$$v_i^\gamma = \frac{e}{g_i}, \quad \text{where} \quad \frac{1}{e^2} = \sum_{i=0}^{N+1} \frac{1}{g_i^2}. \tag{5}$$

In the continuum limit, the eigenstates with masses m_{Ww}^2 and m_{Zz}^2 correspond to the higher Kaluza-Klein ("KK") excitations of the five-dimensional W and Z gauge fields.

αS AND UNITARITY

One of the central roles of the Higgs boson in the Standard Model is to unitarize the scattering of electroweak gauge bosons at high energies. If the Higgs boson were removed from the Standard Model and no new physics were added, $W_L W_L$ spin-0 isospin-0 scattering would violate unitarity at an energy scale of $\sqrt{8\pi}\, v$, where $v = 246$ GeV is the electroweak scale. In Higgsless models, the Higgs boson is absent, but new physics coupling to the electroweak gauge bosons is present in the form of the higher KK modes of the W and Z. It has been shown that low-energy unitarity of longitudinal electroweak boson scattering is maintained in Higgsless models through exchange of various KK modes [5, 6, 7, 8].

In [24], we studied a slightly more general deconstructed model than the one shown in Figure 1, a model in which there could be a whole chain of $U(1)$ groups at the right-hand end of the moose, rather than just a single one. The gauge couplings and f-constants were, again, left arbitrary to make our results as broadly applicable as possible. We studied the unitarization of $W_L W_L$ scattering in this model and concluded that KK mode exchange can maintain low-energy unitarity only if the mass-squared M_{W1}^2, of the next lightest state after the W-boson, is bounded from above by $8\pi v^2$.

We also calculated the form of the corrections to the electroweak interactions for the case in which fermions are localized in the extra dimension; in deconstructed language, the fermions couple to only a single $SU(2)$ group and to a single $U(1)$ group along the moose. We found [24], that the precision observable $\hat{S}$ (defined below, under Precision Electroweak Corrections) is inversely related to the unitarity-imposed upper bound on M_{W1}^2. In other words, the upper bound on the mass-squared of the first KK mode places a lower bound on $\hat{S}$. Specifically, we computed

$$\hat{S} = \frac{1}{4s^2}\left(\alpha S + 4c^2(\Delta\rho - \alpha T) + \frac{\alpha\delta}{c^2}\right) \geq \frac{M_W^2}{8\pi v^2} \simeq 4\times 10^{-3}, \qquad (6)$$

However, this precision observable is constrained by experiment [15] to be less than of order 10^{-3} in magnitude. The value predicted by our Higgsless model with localized fermions is significantly disfavored by experiment.

We concluded that Higgsless models with localized fermions cannot simultaneously satisfy the constraints imposed by precision electroweak corrections and by the need to ensure that scattering of electroweak gauge bosons is unitary.

IDEAL DELOCALIZATION OF FERMIONS

Given the problems with localized fermions, it is logical to consider the possibility that the standard model fermions have wavefunctions with finite extent in the fifth dimension. In practice, this means that the observed fermions are the lightest eigenstates of five-dimensional fermions, just as the W and Z gauge-bosons are the lightest in a tower of "KK" excitations. Refs. [31, 32] show that by adjusting the five-dimensional wavefunction of the light fermions, one can modify (and potentially eliminate) the dangerously large corrections to precision electroweak measurements. Ref. [36] took this a step further by introducing ideal fermion delocalization to guarantee that the precision corrections will be small.

Let us review the basics of fermion delocalization in the language of deconstruction [36]. Since a five-dimensional spinor is equivalent to a four-dimensional Dirac fermion, one introduces a separate Dirac fermion at each site (*i.e.* one left-handed and one right-handed Weyl spinor per site, ψ_L^j and ψ_R^j) on the interior of the moose diagram of fig. 1. The chirality of the standard model fermions is introduced by adjusting the boundary conditions for the fermion fields at the ends of the moose. A convenient choice [45] (consistent with the weak interactions) that we will adopt corresponds to

$$\psi_L^{N+1} = 0, \quad \psi_R^0 = 0. \tag{7}$$

Discretizing the Dirac action for a five-dimensional fermion in an arbitrary background metric then corresponds to introducing site-dependent masses (m_j) for the Dirac fermions at each interior site and postition-dependent Yukawa interactions (y_j) which couple the left-handed modes at site j to the right-handed modes at site $j+1$

$$\mathscr{L}_{5f} = -\sum_{j=1}^{N-1} m_j \bar{\psi}_L^j \psi_R^j - \sum_{j=0}^{N-1} f_{j+1} y_{j+1} \left(\overline{\psi_L^j} U_{j+1} \psi_R^{j+1} \right) + h.c., \tag{8}$$

where gauge-invariance dictates that each such interaction include a factor of the link field U_{j+1}, and we therefore write the corresponding interaction proportional to f_{j+1}.

Note that in eqn. (8) we have not included a Yukawa coupling corresponding to link $N+1$. We will analyze the model in the limit where the lightest fermion eigenstates (which we identify with the standard model fermions) are massless. The absence of the Yukawa couplings at site $N+1$ insures that the right-handed components of these massless modes are localized entirely at site $N+1$. For simplicity, in what follows we will also assume flavor universality, *i.e.* that the same five-dimensional fermion mass matrix applies to all flavors of fermions.

In this limit, only the left-handed components of the massless fermions are delocalized, and their behavior is characterized by a wavefuntion $|\psi_L\rangle = (\alpha_0, \alpha_1, \cdots, \alpha_N)$ where the α_j are complex parameters. Denoting $|\alpha_i|^2 \equiv x_i$ and recognizing that $\sum_{i=0}^N x_i = 1$, we find that the couplings of the ordinary (zero-mode) fermions in this model may be written

$$\mathscr{L}_f = \vec{J}_L^\mu \cdot \left(\sum_{i=0}^N x_i \vec{A}_\mu^i \right) + J_Y^\mu A_\mu^{N+1}. \tag{9}$$

As usual, $\vec{J}_L^\mu$ denotes the isotriplet of left-handed weak fermion currents and J_Y^μ is the fermion hypercharge current.

In the presence of fermion delocalization, the coupling of a gauge boson mass eigenstate to a fermion current is the sum of the contributions from each site

$$g_W = \sum_{j=0}^{N} x_j g_j v_j^W \qquad g_Z^W = \sum_{j=0}^{N} x_j g_j v_j^Z \qquad g_Z^Y = g_{N+1} v_{N+1}^Z , \qquad (10)$$

and therefore reflects the fermion (x_i) and gauge boson (v_i^W, v_i^Z) wave-functions and the site-dependent couplings (g_i).

In [36], we introduced a scheme of "ideal fermion delocalization" which guarantees that the precision corrections can be made small. This scheme exploits the fact that the eigenvector for the lightest massive W and those for each of the W KK modes are mutually orthogonal

$$\sum_i v_i^W v_i^{W_w} = 0 . \qquad (11)$$

Suppose we choose our "ideally delocalized" fermion wavefunction x_i to be related to the form of the W wavefunction

$$g_i x_i = g_W v_i^W , \qquad (12)$$

where the site-independent normalization factor g_W is fixed by the constraint $\sum_i x_i = 1$. Then the coupling of the fermion to the KK modes vanishes

$$g_{W'} = \sum_{j=0}^{N} x_j g_j v_j^{W'} = g^W \sum_{j=0}^{N} v_j^W v_j^{W'} = 0 \qquad (13)$$

and the KK mode is fermiophobic.

PRECISION ELECTROWEAK OBSERVABLES

To see the implications of ideal delocalization for precision electroweak observables, recall [22] that the most general amplitude for low-energy four-fermion neutral weak current processes in any "universal" model [15] may be written as

$$-\mathcal{M}_{NC} = e^2 \frac{\mathcal{Q}\mathcal{Q}'}{Q^2} + \frac{(I_3 - s^2 \mathcal{Q})(I_3' - s^2 \mathcal{Q}')}{\left(\frac{s^2 c^2}{e^2} - \frac{S}{16\pi}\right) Q^2 + \frac{1}{4\sqrt{2}G_F}\left(1 - \alpha T + \frac{\alpha\delta}{4s^2c^2}\right)} \qquad (14)$$

$$+ \sqrt{2} G_F \frac{\alpha\delta}{s^2 c^2} I_3 I_3' + 4\sqrt{2} G_F (\Delta\rho - \alpha T)(\mathcal{Q} - I_3)(\mathcal{Q}' - I_3') ,$$

and the matrix element for charged current process may be written

$$-\mathcal{M}_{CC} = \frac{(I_+ I_-' + I_- I_+')/2}{\left(\frac{s^2}{e^2} - \frac{S}{16\pi}\right) Q^2 + \frac{1}{4\sqrt{2}G_F}\left(1 + \frac{\alpha\delta}{4s^2c^2}\right)} + \sqrt{2} G_F \frac{\alpha\delta}{s^2 c^2} \frac{(I_+ I_-' + I_- I_+')}{2} . \qquad (15)$$

The parameter s^2 is defined implicitly in these expressions as the ratio of the $\mathcal{Q}$ and I_3 couplings of the Z boson. $\Delta\rho$ corresponds to the deviation from unity of the ratio of the strengths of low-energy isotriplet weak neutral-current scattering and charged-current scattering. S and T are the familiar oblique electroweak parameters [27, 28, 29], as determined by examining the *on-shell* properties of the Z and W bosons. The contact interactions proportional to $\alpha\delta$ and $(\Delta\rho - \alpha T)$ correspond to "universal non-oblique" corrections arising from the exchange of heavy KK modes [22]. Hence, if the heavy KK modes are fermiophobic, these electroweak corrections will vanish at leading order.

These "on-shell" paramters may be recast [22] as the the zero-momentum electroweak parameters defined in [15]

$$\hat{S} = \frac{1}{4s^2}\left(\alpha S + 4c^2(\Delta\rho - \alpha T) + \frac{\alpha\delta}{c^2}\right), \quad (16)$$

$$\hat{T} = \Delta\rho, \quad (17)$$

$$W = \frac{\alpha\delta}{4s^2c^2}, \quad (18)$$

$$Y = \frac{c^2}{s^2}(\Delta\rho - \alpha T), \quad (19)$$

In this language, it can be shown that the electroweak corrections take on a very compact form for ideally delocalized fermions [36]:

$$\hat{S} = \hat{T} = W = 0, \qquad Y = M_W^2(\Sigma_W - \Sigma_Z) \quad (20)$$

where Σ_Z and Σ_W are the sums over inverse-square masses of the higher neutral- and charged-current KK modes. While Y is not precisely zero, its value is small (of order $M_W^4/M_{W'}^4$)

DIRECT SEARCHES

Because the electroweak gauge bosons' KK modes are fermiophobic in Higgsless models with ideal fermion delocalization, it will be difficult for direct collider searches to detect the W' and Z' states.

Existing searches for W' bosons [46] assume that the W' bosons couple to ordinary quarks and leptons (generally with SM strength). All assume that the W' is produced via these couplings; all but one also assume that the W' decays only to fermions and that the $W' \to WZ$ decay channel is unavailable. Likewise, existing direct searches for Z' bosons [46] rely on $Z'f\bar{f}$ couplings and assume that the decay channel $Z' \to W^+W^-$ is not available. None of these searches can constrain the $W_{(n\geq 1)}$ and $Z_{(n\geq 1)}$ states of a Higgsless model with ideal fermion delocalization. Proposed future searches that rely on the W' and Z' couplings to fermions for either production or decay of the KK modes (e.g. [47] , [48]) will not apply either.

The only way to perform a direct search for the $W_{(n\geq 1)}$ and $Z_{(n\geq 1)}$ states will be to study WW or WZ elastic scattering. If no resonances are seen, these processes will also afford the opportunity to constrain the values of the chiral Lagrangian parameters α_4 and α_5 (see discussion below).

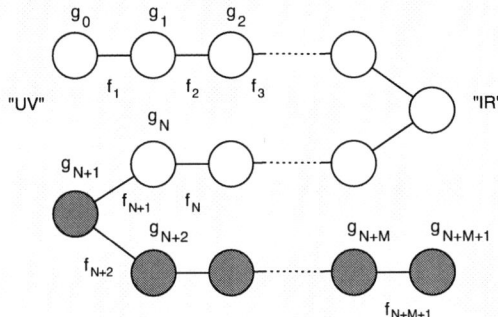

FIGURE 2. Moose diagram of the deconstructed version of the $SU(2)_A \times SU(2)_B \times U(1)$ 5-dimensional gauge theory in flat space discussed in the sections on Triple Gauge Vertices and Chiral Lagrangian Parameters. The upper (lower) string of $SU(2)$ groups corresponds to $SU(2)_A$ ($SU(2)_B$) in the continuum.

TRIPLE GAUGE VERTICES

Experimental signatures of Higgsless models with ideally-delocalized fermions will need to rely on couplings among the many electroweak gauge bosons in the theories, since more traditional signatures involving on fermion couplings to the W' and Z' are not available. One of the best tests of Higgsless models comes from the WWZ vertex.

To leading order, the CP-conserving triple gauge boson vertices may be written in the Hagiwara-Peccei-Zeppenfeld-Hikasa triple-gauge-vertex notation [49]

$$\begin{aligned}\mathscr{L}_{TGV} &= -ie\frac{c_Z}{s_Z}[1+\Delta\kappa_Z]W^+_\mu W^-_\nu Z^{\mu\nu} - ie[1+\Delta\kappa_\gamma]W^+_\mu W^-_\nu A^{\mu\nu} \\ &\quad - ie\frac{c_Z}{s_Z}[1+\Delta g^Z_1](W^{+\mu\nu}W^-_\mu - W^{-\mu\nu}W^+_\mu)Z_\nu \\ &\quad - ie(W^{+\mu\nu}W^-_\mu - W^{-\mu\nu}W^+_\mu)A_\nu, \end{aligned} \qquad (21)$$

where the two-index tensors denote the Lorentz field-strength tensor of the corresponding field. In the standard model, $\Delta\kappa_Z = \Delta\kappa_\gamma = \Delta g^Z_1 \equiv 0$. The Higgsless models predict $\Delta\kappa_\gamma = 0$ but generally have non-zero values of Δg^Z_1 and $\Delta\kappa_Z$.

In models with ideally-delocalized fermions, experimental constraints from LEP II measurements of the triple-gauge-boson vertices can provide valuable bounds on the KK masses [37]. The 95% c.l. upper limit (recalling that Δg^Z_1 is positive in our models) is $\Delta g^Z_1 \leq 0.028$ [50]. We can estimate the degree to which this constrains Higgsless models with ideal delocalization by considering how Δg^Z_1 is related to the mass of the lightest KK resonance.

For a deconstructed version of an $SU(2)_A \times SU(2)_B \times U(1)$ gauge theory in 5-dimensional flat space (see Figure 2), the form of Δg^Z_1 is found in [37] to be

$$\Delta g^Z_1 = \Delta\kappa_Z = \frac{\pi^2}{12c^2}\left(\frac{M_W}{M_{W_1}}\right)^2\left[\frac{1}{4}\cdot\frac{7+\kappa}{1+\kappa}\right] \qquad (22)$$

where $\kappa = g_{5WB}^2/g_{5WA}^2$. Inserting numerical values for M_W, and c and denoting the 95% c.l. experimental upper bound on Δg_1^Z as Δg_{max}, we find the bound

$$M_{W_1} \geq 500 \text{GeV} \sqrt{\frac{0.028}{\Delta g_{max}} \left[\frac{1}{4} \cdot \frac{7+\kappa}{1+\kappa}\right]} \quad (23)$$

The LEP II data therefore implies a 95% c.l. lower bound of 500 GeV on the first KK resonance in flat space models for $\kappa = 1$ and lower bounds of 250 - 650 GeV as κ varies from ∞ to 0.

CHIRAL LAGRANGIAN PARAMETERS

The language of the effective electroweak chiral Lagrangian is useful for phenomenological studies. As discussed in [37], the operators [39, 40, 41, 42, 43] present in our Higgsless models are:

$$\mathcal{L}_1 \equiv \frac{1}{2}\alpha_1 g_W g_Y B_{\mu\nu} Tr(TW^{\mu\nu}) \quad (24)$$

$$\mathcal{L}_2 \equiv \frac{1}{2}i\alpha_2 g_Y B_{\mu\nu} Tr(T[V^\mu, V^\nu]) \quad (25)$$

$$\mathcal{L}_3 \equiv i\alpha_3 g_W Tr(W_{\mu\nu}[V^\mu, V^\nu]) \quad (26)$$

$$\mathcal{L}_4 \equiv \alpha_4 [Tr(V^\mu V^\nu)]^2 \quad (27)$$

$$\mathcal{L}_5 \equiv \alpha_5 [Tr(V_\mu V^\mu)]^2 . \quad (28)$$

Here $W_{\mu\nu}$, $B_{\mu\nu}$, $T \equiv U\tau_3 U^\dagger$ and $V_\mu \equiv (D_\mu U)U^\dagger$, with U being the nonlinear sigma-model field arising from $SU(2)_L \otimes SU(2)_R \to SU(2)_V$, are the $SU(2)_W$-covariant and $U(1)_Y$-invariant building blocks of the expansion. The relationships of the α_i to several alternative parametrizations are given in [37]; we note here that $S = -16\pi\alpha_1$ and that the leading corrections to WW and WZ elastic scattering arise from $\alpha_{4,5}$. The following table shows the values of the Appelquist-Longitano parameters in the deconstructed version of a 5-dimensional $SU(2)_A \times SU(2)_B \times U(1)$ gauge theory in flat space

TABLE 1. Longhitano's parameters in $SU(2)_A \otimes SU(2)_B$ flat Higgsless models for brane localized and ideally delocalized fermions.

flat $SU(2) \times SU(2)$ Longhitano parameters	brane localized	ideally delocalized
$e^2\alpha_1$	$-\frac{2}{3}\lambda s^2$	0
$e^2\alpha_2$	$-\frac{1}{12}\left(\frac{7+\kappa}{1+\kappa}\right)\lambda s^2$	$-\frac{1}{12}\left(\frac{7+\kappa}{1+\kappa}\right)\lambda s^2$
$e^2\alpha_3$	$-\frac{1}{12}\left(\frac{1+7\kappa}{1+\kappa}\right)\lambda s^2$	$\frac{1}{12}\left(\frac{7+\kappa}{1+\kappa}\right)\lambda s^2$
$e^2\alpha_4$	$\frac{1}{30}\frac{(1+14\kappa+\kappa^2)}{(1+\kappa)^2}\lambda s^2$	$\frac{1}{30}\frac{(1+14\kappa+\kappa^2)}{(1+\kappa)^2}\lambda s^2$
$e^2\alpha_5$	$-\frac{1}{30}\frac{(1+14\kappa+\kappa^2)}{(1+\kappa)^2}\lambda s^2$	$-\frac{1}{30}\frac{(1+14\kappa+\kappa^2)}{(1+\kappa)^2}\lambda s^2$

These values are consistent with several symmetry considerations. First, α_2 is the coefficient of an operator that is not related to the $SU(2)_W$ properties of the model; as such, this coefficient should be unaffected by the degree of delocalization of the $SU(2)_W$ properties of the fermions. Indeed, we see that α_2 is the same for both the brane-localized and ideal fermions. Conversely, we expect the values of α_1 and α_3 to be sensitive to the $SU(2)_W$ properties of the fermions and this is observed in our results, yielding an example of theories in which $\alpha_2 \neq \alpha_3$. Third, in the limit where $\kappa \to 1$, the models with brane-localized fermions should display an $A \leftrightarrow B$ parity; this is consistent with the fact that $\alpha_2 = \alpha_3$ for $\kappa = 1$. Finally, since $\Delta \kappa_\gamma \equiv 0$, we find $\alpha_2 = -\alpha_3$ for the case of ideal delocalization, in which $\alpha_1 = 0$.

CONCLUSIONS

In this talk, we have seen that Higgsless models are intriguing candidate solutions to the puzzle of Electroweak Symmetry Breaking. While these models inherently arise from extra-dimensional gauge theories, we find it convenient to employ the technique of deconstruction in order to study the 5-dimensional gauge theories as consistent effective field theories in four dimensions. We have shown that Higgsless with localized fermions are not phenomenologically viable, because those able to properly unitarize longitudinal electroweak gauge boson scattering have values of αS that are larger than experiment allows. However, delocalizing the fermions along the 5th dimension can yield viable models; models with ideal delocalization have vanishingly small precision electroweak observables. The best experimental limits on these models presently come from LEP II bounds on triple-gauge-boson vertices; the mass of the lightest extra W' boson must be at least 500 GeV. Moreover, the electroweak chiral Lagrangian parameters are calculable in Higgsless models and are of a size that should be accessible to future experiments.

ACKNOWLEDGMENTS

EHS gratefully acknowledges the kind hospitality of the conference organizers and the support of NSF grant PHY-0354226.

REFERENCES

1. C. Csaki, C. Grojean, H. Murayama, L. Pilo and J. Terning, *Gauge theories on an interval: Unitarity without a Higgs*, Phys. Rev. D **69**, 055006 (2004) [arXiv:hep-ph/0305237].
2. P. W. Higgs, *Broken symmetries, massless particles and gauge fields*, Phys. Lett. **12** (1964) 132–133.
3. K. Agashe, A. Delgado, M. J. May and R. Sundrum, *RS1, Custodial Isospin and Precision Tests*, JHEP **0308**, 050 (2003) [arXiv:hep-ph/0308036].
4. C. Csaki, C. Grojean, L. Pilo, and J. Terning, *Towards a realistic model of higgsless electroweak symmetry breaking*, Phys. Rev. Lett. **92** (2004) 101802, [arXiv:hep-ph/0308038].
5. R. Sekhar Chivukula, D. A. Dicus, and H.-J. He, *Unitarity of compactified five dimensional yang-mills theory*, Phys. Lett. **B525** (2002) 175–182, [arXiv:hep-ph/0111016].
6. R. S. Chivukula and H.-J. He, *Unitarity of deconstructed five-dimensional yang-mills theory*, Phys. Lett. **B532** (2002) 121–128, [arXiv:hep-ph/0201164].
7. R. S. Chivukula, D. A. Dicus, H.-J. He, and S. Nandi, *Unitarity of the higher dimensional standard model*, Phys. Lett. **B562** (2003) 109–117, [arXiv:hep-ph/0302263].
8. H.-J. He, *Higgsless deconstruction without boundary condition*, arXiv:hep-ph/0412113.

9. G. Cacciapaglia, C. Csaki, C. Grojean and J. Terning, *Oblique corrections from Higgsless models in warped space*, Phys. Rev. D **70**, 075014 (2004) [arXiv:hep-ph/0401160].
10. Y. Nomura, *Higgsless theory of electroweak symmetry breaking from warped space*, JHEP **11** (2003) 050, [arXiv:hep-ph/0309189].
11. R. Barbieri, A. Pomarol and R. Rattazzi, *Weakly coupled Higgsless theories and precision electroweak tests*, Phys. Lett. B **591** (2004) 141 [arXiv:hep-ph/0310285].
12. H. Davoudiasl, J. L. Hewett, B. Lillie and T. G. Rizzo, *Higgsless electroweak symmetry breaking in warped backgrounds: constraints and signatures*, Phys. Rev. D **70** (2004) 015006 [arXiv:hep-ph/0312193].
13. G. Burdman and Y. Nomura, *Holographic theories of electroweak symmetry breaking without a Higgs boson*, Phys. Rev. D **69** (2004) 115013 [arXiv:hep-ph/0312247].
14. H. Davoudiasl, J. L. Hewett, B. Lillie, and T. G. Rizzo, *Warped higgsless models with ir-brane kinetic terms*, JHEP **05** (2004) 015, [arXiv:hep-ph/0403300].
15. R. Barbieri, A. Pomarol, R. Rattazzi and A. Strumia, *Electroweak symmetry breaking after LEP1 and LEP2*, Nucl. Phys. B **703** (2004) 127 [arXiv:hep-ph/0405040].
16. J. L. Hewett, B. Lillie, and T. G. Rizzo, *Monte carlo exploration of warped higgsless models*, JHEP **10** (2004) 014, [arXiv:hep-ph/0407059].
17. R. Foadi, S. Gopalakrishna, and C. Schmidt, *Higgsless electroweak symmetry breaking from theory space*, JHEP **03** (2004) 042, [arXiv:hep-ph/0312324].
18. J. Hirn and J. Stern, *The role of spurions in Higgs-less electroweak effective theories*, Eur. Phys. J. C **34**, 447 (2004) [arXiv:hep-ph/0401032].
19. R. Casalbuoni, S. De Curtis and D. Dominici, *Moose models with vanishing S parameter*, Phys. Rev. D **70** (2004) 055010 [arXiv:hep-ph/0405188].
20. R. S. Chivukula, E. H. Simmons, H. J. He, M. Kurachi and M. Tanabashi, *The structure of corrections to electroweak interactions in Higgsless models*, Phys. Rev. D **70** (2004) 075008 [arXiv:hep-ph/0406077].
21. M. Perelstein, *Gauge-assisted technicolor?*, JHEP **10** (2004) 010, [arXiv:hep-ph/0408072].
22. R. S. Chivukula, E. H. Simmons, H.-J. He, M. Kurachi, and M. Tanabashi, *Universal non-oblique corrections in higgsless models and beyond*, Phys. Lett. **B603** (2004) 210–218, [arXiv:hep-ph/0408262].
23. H. Georgi, *Fun with Higgsless theories*, Phys. Rev. D **71**, 015016 (2005) [arXiv:hep-ph/0408067].
24. R. Sekhar Chivukula, E. H. Simmons, H. J. He, M. Kurachi and M. Tanabashi, *Electroweak corrections and unitarity in linear moose models*, Phys. Rev. D **71** (2005) 035007 [arXiv:hep-ph/0410154].
25. N. Arkani-Hamed, A. G. Cohen, and H. Georgi, *(de)constructing dimensions*, Phys. Rev. Lett. **86** (2001) 4757–4761, [arXiv:hep-th/0104005].
26. C. T. Hill, S. Pokorski, and J. Wang, *Gauge invariant effective lagrangian for kaluza-klein modes*, Phys. Rev. **D64** (2001) 105005, [arXiv:hep-th/0104035].
27. M. E. Peskin and T. Takeuchi, *Estimation of oblique electroweak corrections*, Phys. Rev. **D46** (1992) 381–409.
28. G. Altarelli and R. Barbieri, *Vacuum polarization effects of new physics on electroweak processes*, Phys. Lett. **B253** (1991) 161–167.
29. G. Altarelli, R. Barbieri, and S. Jadach, *Toward a model independent analysis of electroweak data*, Nucl. Phys. **B369** (1992) 3–32.
30. H. Georgi, *A tool kit for builders of composite models*, Nucl. Phys. **B266** (1986) 274.
31. G. Cacciapaglia, C. Csaki, C. Grojean and J. Terning, *Curing the ills of Higgsless models: The S parameter and unitarity*, Phys. Rev. D **71** (2005) 035015 [arXiv:hep-ph/0409126].
32. R. Foadi, S. Gopalakrishna and C. Schmidt, *Effects of fermion localization in Higgsless theories and electroweak constraints*, Phys. Lett. B **606** (2005) 157 [arXiv:hep-ph/0409266].
33. G. Cacciapaglia, C. Csaki, C. Grojean, M. Reece and J. Terning, *Top and bottom: A brane of their own*, arXiv:hep-ph/0505001.
34. R. S. Chivukula, E. H. Simmons, H. J. He, M. Kurachi and M. Tanabashi, *Deconstructed Higgsless models with one-site delocalization*, Phys. Rev. D **71**, 115001 (2005) [arXiv:hep-ph/0502162].
35. R. Casalbuoni, S. De Curtis, D. Dolce and D. Dominici, *Playing with fermion couplings in Higgsless models*, Phys. Rev. D **71**, 075015 (2005) [arXiv:hep-ph/0502209].
36. R. Sekhar Chivukula, E. H. Simmons, H. J. He, M. Kurachi and M. Tanabashi, *Ideal fermion delocalization in Higgsless models*, Phys. Rev. D **72**, 015008 (2005) [arXiv:hep-ph/0504114].

37. R. S. Chivukula, E. H. Simmons, H. J. He, M. Kurachi and M. Tanabashi, Phys. Rev. D **72**, 075012 (2005) [arXiv:hep-ph/0508147].
38. C. Grojean, W. Skiba and J. Terning, Phys. Rev. D **73**, 075008 (2006) [arXiv:hep-ph/0602154].
39. T. Appelquist and C. W. Bernard, *Strongly Interacting Higgs Bosons*, Phys. Rev. D **22**, 200 (1980).
40. A. C. Longhitano, *Low-Energy Impact Of A Heavy Higgs Boson Sector*, Nucl. Phys. B **188**, 118 (1981).
41. A. C. Longhitano, *Heavy Higgs Bosons In The Weinberg-Salam Model*, Phys. Rev. D **22**, 1166 (1980).
42. T. Appelquist, *Broken Gauge Theories And Effective Lagrangians*, Print-80-0832 (YALE) *Based on lectures presented at the 21st Scottish Universities Summer School in Physics, St. Andrews, Scotland, Aug 10-30, 1980*
43. T. Appelquist and G. H. Wu, *The Electroweak chiral Lagrangian and new precision measurements*, Phys. Rev. D **48**, 3235 (1993) [arXiv:hep-ph/9304240].
44. A. Birkedal, K. Matchev and M. Perelstein, *Collider phenomenology of the Higgsless models*, Phys. Rev. Lett. **94**, 191803 (2005) [arXiv:hep-ph/0412278].
45. H.-C. Cheng, C. T. Hill, and J. Wang, *Dynamical electroweak breaking and latticized extra dimensions*, Phys. Rev. **D64** (2001) 095003, [arXiv:hep-ph/0105323].
46. S. Eidelman et al. (Particle Data Group), Physics Letters B592, 1 (2004) and 2005 partial updates for edition 2006 (URL: http://pdg.lbl.gov).
47. S. Godfrey, P. Kalyniak, B. Kamal and A. Leike, *Discovery and identification of extra gauge bosons in e+ e- → nu anti-nu gamma*, Phys. Rev. D **61**, 113009 (2000) [arXiv:hep-ph/0001074].
48. S. Godfrey, P. Kalyniak, B. Kamal, M. A. Doncheski and A. Leike, *Discovery and identification of W' bosons in e gamma → nu q + X*, Phys. Rev. D **63**, 053005 (2001) [arXiv:hep-ph/0008157].
49. K. Hagiwara, R. D. Peccei, D. Zeppenfeld and K. Hikasa, *Probing The Weak Boson Sector In E+ E- → W+ W-*, Nucl. Phys. B **282**, 253 (1987).
50. The LEP Collaborations ALEPH, DELPHI, L3, OPAL and the LEP TGC Working Group. LEPEWWG/TC/2005-01; June 8, 2005.

A Perspective on Hadron Physics

Arne Höll[*], Craig D. Roberts[*,†] and Stewart V. Wright[†]

[*]*Institut für Physik, Universität Rostock, D-18051 Rostock, Germany*
[†]*Physics Division, Argonne National Laboratory, Argonne IL 60439, USA*

Abstract. The phenomena of confinement and dynamical chiral symmetry breaking are basic to understanding hadron observables. They can be explored using Dyson-Schwinger equations. The existence of a systematic, nonperturbative and symmetry preserving truncation of these equations enables the proof of exact results in QCD, and their illustration using simple but accurate models. We provide a sketch of the material qualitative and quantitative success that has been achieved in the study of pseudoscalar and vector mesons. Efforts are now turning to the study of baryons, which we exemplify via a calculation of nucleon weak and pionic form factors.

Keywords: Bethe-Salpeter equation; Confinement; Covariant Faddeev equation; Dynamical chiral symmetry breaking; Dyson-Schwinger equation; Electroweak properties of baryons and mesons; pion-nucleon interaction
PACS: 11.10.St, 12.38.Aw, 13.20.-v, 13.40.Gp, 13.75.Gx,

1. INTRODUCTION

We begin with a key question: how does one make an almost massless bound state from two massive constituents? Naturally, the bound state is the pion and the massive components are constituent-quarks. It has long been known [1] that

$$m_\pi^2 \propto m_q, \qquad (1)$$

where m_q is the light-quark current-mass that appears in QCD's Lagrangian. While it is possible to construct a quantum mechanical model with a potential finely tuned to give a massless pseudoscalar bound state composed of heavy constituents, in such a framework $m_\pi \propto M_q$, where M_q is the constituents' mass. This is plainly not the way to a veracious understanding of strong interaction physics.

True comprehension of the visible universe requires that we learn just what it is about QCD which enables the formation of an unexpectedly light pseudoscalar meson from two rather massive constituents. The correct understanding of hadron observables must explain why the pion is light but the ρ-meson and the nucleon are heavy. The keys to this puzzle are QCD's **emergent phenomena**: *confinement* and *dynamical chiral symmetry breaking* (DCSB). Confinement is the feature that no matter how hard one strikes a hadron, it never breaks apart into quarks and/or gluons that reach a detector. DCSB is signalled by an apparently unnatural pattern of bound state masses in the strong interaction spectrum, and can only be fathomed once one grasps the nature of a well-defined and valid chiral limit. Thereafter can follow an understanding of the connection between a current-quark and a constituent-quark, and subsequently Eq. (1). QCD's emergent phenomena are not apparent in the action. Yet they are the dominant determining characteristics of hadron properties. Attaining an understanding of these phenomena is one of the greatest intellectual challenges in physics.

A nonperturbative method for solving quantum field theory is necessary in order to answer the question we have posed, and those which shall follow. The Dyson-Schwinger equations (DSEs) are one such tool. At the simplest level the DSEs provide a generating tool for perturbation theory and, because QCD is asymptotically free, this means that any model-dependence in their application can be restricted to the infrared (long-range) domain. The solutions of the DSEs are Schwinger functions (Euclidean space Green functions) and because all cross-sections can be constructed from such n-point functions the DSEs can be used to make predictions for real-world experiments. In this mode they provide a means by which to use nonperturbative strong interaction phenomena to map out, e.g., the behaviour at long-range of the interaction between light-quarks. A nonperturbative solution of the DSEs enables the study of: hadrons as composites of dressed-quarks and -gluons; the phenomena of confinement and DCSB; and therefrom an articulation of any connection between them. One of the merits of this is that any assumptions employed, or guesses made, can be tested, verified and improved, or rejected in favour of more promising alternatives. The modern application of these methods is described in Refs. [2, 3, 4, 5], while Ref. [6] provides a pedagogical overview.

The DSEs are a countable infinity of equations, which are vitally important in proving the renormalisability of quantum field theories. However, the coupling between equations is at the heart of a persistent challenge to their application. This relationship means that in order to arrive at a tractable problem one must employ a truncation. Perturbation theory is ever popular. However, it is not useful in connection with the nonperturbative phenomena that provide the keystones of hadron physics. Fortunately, at least one systematic, nonperturbative and symmetry preserving truncation of the DSEs exists [7, 8]. This enables the proof of exact results using the DSEs. Moreover, that the truncation scheme is also tractable provides a method by which the exact results may be illustrated and, furthermore, a practical tool for the prediction of observables that are accessible at contemporary experimental facilities. The consequent opportunities for rapid feedback between experiment and theory brings within reach an intuitive understanding of nonperturbative strong interaction phenomena.

2. GAP EQUATION

The renormalised gap equation in QCD may be written

$$S(p)^{-1} = Z_2(i\gamma \cdot p + m^{\text{bm}}) + \Sigma(p), \qquad (2)$$

$$\Sigma(p) = Z_1 \int_q^\Lambda g^2 D_{\mu\nu}(p-q) \frac{\lambda^a}{2} \gamma_\mu S(q) \Gamma_\nu^a(q,p), \qquad (3)$$

where $\int_q^\Lambda$ represents a Poincaré invariant regularisation of the integral, with Λ the regularisation mass-scale [9, 10], $D_{\mu\nu}(k)$ is the dressed-gluon propagator, $\Gamma_\nu(q,p)$ is the dressed-quark-gluon vertex, and m^{bm} is the Λ-dependent current-quark bare mass. The quark-gluon-vertex and quark wave function renormalisation constants, $Z_{1,2}(\zeta^2, \Lambda^2)$, depend on the renormalisation point, ζ, the regularisation mass-scale and the gauge parameter. The gap equation's solution has the form

$$S(p) = \frac{1}{i\gamma \cdot p A(p^2,\zeta^2) + B(p^2,\zeta^2)} = \frac{Z(p^2,\zeta^2)}{i\gamma \cdot p + M(p^2)}. \quad (4)$$

It is obtained from Eq. (2) augmented by the renormalisation condition

$$S(p)^{-1}\big|_{p^2=\zeta^2} = i\gamma \cdot p + m(\zeta), \quad (5)$$

where $m(\zeta)$ is the renormalised mass:

$$Z_2(\zeta^2,\Lambda^2)m^{\text{bm}}(\Lambda) = Z_4(\zeta^2,\Lambda^2)m(\zeta), \quad (6)$$

with Z_4 the Lagrangian mass renormalisation constant. In QCD the chiral limit is strictly and unambiguously defined by

$$Z_2(\zeta^2,\Lambda^2)m^{\text{bm}}(\Lambda) \equiv 0, \forall \Lambda \gg \zeta, \quad (7)$$

which states that the renormalisation-point-invariant current-quark mass $\hat{m} = 0$.

In the absence of interactions $Z(p^2) = 1$ and $M(p^2) = m_q$ in Eq. (4). On the other hand, the behaviour of these functions in QCD is a longstanding prediction of DSE studies [11], which could have been anticipated from Refs. [12, 13]: the functions receive strong momentum-dependent corrections at infrared momenta so that $Z(p^2)$ is suppressed and $M(p^2)$ enhanced. These DSE predictions are confirmed in numerical simulations of lattice-QCD [14], and the conditions have been explored under which pointwise agreement between DSE results and lattice simulations may be obtained [15, 16].

The gap equation's kernel, Eq. (3), is constructed from the contraction of the dressed-gluon two-point function and the dressed-quark-gluon vertex. In Landau gauge

$$D_{\mu\nu}(p) = \left(\delta_{\mu\nu} - \frac{p_\mu p_\nu}{p^2}\right)\frac{F(p^2,\zeta^2)}{p^2}. \quad (8)$$

The modern DSE perspective on $F(p^2,\zeta^2)$ is reviewed in Ref. [4]: these studies predicted that $F(p^2)$ is suppressed at small p^2; i.e., in the infrared, with the deviation from expectations based on perturbation theory becoming apparent at $p^2 \simeq 1\,\text{GeV}^2$. A mass-scale of this magnitude has long been anticipated as characteristic of nonperturbative gauge-sector dynamics and its origin is fundamentally the same as that of Λ_{QCD}, which appears in perturbation theory. These DSE predictions, too, have been verified in contemporary simulations of lattice-regularised QCD [17].

The remaining piece of the gap equation's kernel is the dressed-quark-gluon vertex, whose form is the subject of contemporary research. In correlating lattice-QCD results on the dressed-quark and -gluon propagators via the gap equation it was found [15] that the vertex must exhibit an infrared enhancement. This was anticipated in Ref. [18] and confirmed in Ref. [16]. The exact nature of this enhancement and its origin in QCD is currently being explored; e.g., Refs. [19, 20, 21].

3. MESONS

Dyson-Schwinger equation studies have established a reliable picture of key propagators and vertices in QCD. It is now natural to ask: what about bound states? Without them, of course, a direct comparison with experiment is impossible. Bound states appear as pole contributions to colour-singlet Schwinger functions and this observation may be viewed as the origin of the Bethe-Salpeter equation (BSE).

The DSE for the dressed-quark-gluon vertex can be viewed as a BSE. So can that for the dressed-quark-photon vertex. The latter is a colour singlet vertex and its lowest mass pole-contribution is the ρ-meson [22]. This fact underlies the success of ρ-meson dominance phenomenology.

The axial-vector vertex is of primary interest to hadron physics. It may be obtained as the solution of the inhomogeneous Bethe-Salpeter equation

$$[\Gamma_{5\mu}(k;P)]_{tu} = Z_2 [\gamma_5 \gamma_\mu]_{tu} + \int_q^\Lambda [\chi_{5\mu}(q;P)]_{sr} K_{tu}^{rs}(q,k;P), \tag{9}$$

where $\chi_{5\mu}(q;P) = S(q_+)\Gamma_{5\mu}(q;P)S(q_-)$, $q_\pm = q \pm P/2$, and the colour-, Dirac- and flavour-matrix structure of the elements in the equation is denoted by the indices r,s,t,u. In Eq. (9), $K(q,k;P)$ is the fully-amputated quark-antiquark scattering kernel. It is one-particle-irreducible and hence does not contain quark-antiquark to single gauge-boson annihilation diagrams, such as would describe the leptonic decay of the pion, nor diagrams that become disconnected by cutting one quark and one antiquark line. If one knows the form of K then one *completely* understands the nature of the interaction between quarks in QCD.

Model-independent results

In quantum field theory, chiral symmetry and the pattern by which it is broken are expressed via the chiral Ward-Takahashi identity ($k_\pm = k \pm P/2$):

$$P_\mu \Gamma_{5\mu}^H(k;P) = \check{S}(k_+)^{-1} i\gamma_5 \frac{T^H}{2} + i\gamma_5 \frac{T^H}{2} \check{S}(k_-)^{-1} - i\{M^\zeta, \Gamma_5^H(k;P)\}, \tag{10}$$

where the pseudoscalar vertex satisfies

$$[\Gamma_5^H(k;P)]_{tu} = Z_4 \left[\gamma_5 \frac{T^H}{2}\right]_{tu} + \int_q^\Lambda [\chi_5^H(q;P)]_{sr} K_{tu}^{rs}(q,k;P), \tag{11}$$

with $\check{S} = \text{diag}[S_u, S_d, S_s, \ldots]$ and $M^\zeta = \text{diag}[m_u(\zeta), m_d(\zeta), m_s(\zeta), \ldots]$. We have written Eqs. (10), (11) for the case of a flavour-nonsinglet vertex in a theory with N_f quark flavours. The matrices T^H are constructed from the generators of $SU(N_f)$ with, e.g., $T^{\pi^+} = \frac{1}{2}(\lambda^1 + i\lambda^2)$ providing for the flavour content of a positively charged pion.

The axial-vector Ward-Takahashi identity relates the solution of a BSE to that of the gap equation. If the identity is always to be satisfied and in a model-independent manner, as necessary in order to preserve an essential symmetry of the strong interaction and its breaking pattern, then the kernels of the gap and Bethe-Salpeter equations must be intimately related. Any truncation or approximation of these equations must preserve that relation. This is an extremely tight constraint. Perturbation theory is one truncation that, order by order, guarantees Eq. (10). However, perturbation theory is inadequate in the face of QCD's emergent phenomena. Something else is needed.

That need is satisfied by the systematic, nonperturbative and symmetry preserving truncation of the DSEs explained in Refs. [7, 8, 20, 23]. It enables a proof of Goldstone's theorem in QCD [9]. Namely, in the chiral limit, Eq. (7), and with chiral symmetry dynamically broken: the axial-vector vertex, Eq. (9), is dominated by the pion pole for $P^2 \sim 0$ and the homogeneous, isovector, pseudoscalar BSE has a massless ($P^2 = 0$) solution. The converse is also true, so that DCSB is a sufficient and necessary condition for the appearance of a massless pseudoscalar bound state of dynamically-massive constituents, which dominates the axial-vector vertex for infrared total momenta.

Furthermore, from the axial-vector Ward-Takahashi identity and the existence of a systematic, nonperturbative symmetry-preserving truncation, one can prove the following identity involving the mass-squared of a pseudoscalar meson [9]:

$$f_H m_H^2 = \rho_H(\zeta) M_H^\zeta, \qquad (12)$$

where $M_H^\zeta = m_{q_1}(\zeta) + m_{q_2}(\zeta)$ is the sum of the current-quark masses of the meson's constituents;

$$f_H P_\mu = Z_2 \mathrm{tr} \int_q^\Lambda \frac{1}{2}(T^H)^T \gamma_5 \gamma_\mu \check{S}(q_+) \Gamma^H(q;P) \check{S}(q_-), \qquad (13)$$

where $(\cdot)^T$ indicates matrix transpose, the trace is over all matrix indices; and

$$\rho_H(\zeta) = Z_4 \mathrm{tr} \int_q^\Lambda \frac{1}{2}(T^H)^T \gamma_5 \check{S}(q_+) \Gamma^H(q;P) \check{S}(q_-) =: \frac{-\langle \bar{q}q \rangle_\zeta^H}{f_H}. \qquad (14)$$

The renormalisation constants in Eqs. (13), (14) play a pivotal role because the expressions would be meaningless without them. They serve to guarantee that the quantities described are gauge invariant, and finite as the regularisation scale is removed to infinity. Moreover, Z_2 in Eq. (13) and Z_4 in Eq. (14) ensure that both f_H and the product $\rho_H(\zeta) M_H^\zeta$ are renormalisation point independent, which is an absolute necessity for any observable quantity.

Taking note that in a Poincaré invariant theory a pseudoscalar meson Bethe-Salpeter amplitude assumes the form

$$\Gamma_H^j(k;P) = T^H \gamma_5 \left[i E_H(k;P) + \gamma \cdot P F_H(k;P) + \gamma \cdot k\, k \cdot P G_H(k;P) + \sigma_{\mu\nu} k_\mu P_\nu H_H(k;P) \right], \qquad (15)$$

then, in the chiral limit, one can also prove

$$\begin{array}{ll} f_H^0 E_H(k;0) = B(k^2), & F_R(k;0) + 2 f_H^0 F_H(k;0) = A(k^2), \\ H_R(k;0) + 2 f_H^0 H_H(k;0) = 0, & G_R(k;0) + 2 f_H^0 G_H(k;0) = 2A'(k^2), \end{array} \qquad (16)$$

where f_H^0 is the chiral limit value from Eq. (13), which is nonzero when chiral symmetry is dynamically broken. The functions F_R, G_R, H_R are associated with terms in the axial-vector vertex that are regular in the neighbourhood of $P^2 + m_H^2 = 0$ and do not vanish at $P_\mu = 0$. These four identities are quark-level Goldberger-Treiman relations for the pion. They are: exact in QCD; and a pointwise expression of Goldstone's theorem. These identities relate the pseudoscalar meson Bethe-Salpeter amplitude directly to the dressed-quark propagator, Eq. (4). The first explains why DCSB and the appearance of a Goldstone mode are so intimately connected, and the remaining three entail that in general a pseudoscalar meson Bethe-Salpeter amplitude has what might be called pseudovector components; namely: F_H, G_H, H_H. It is the latter which, in a covariant treatment, guarantee that the electromagnetic pion form factor behaves as $1/Q^2$ at large spacelike momentum transfer [24].

Equation (12) and its corollaries are fundamental in QCD. To exemplify we'll focus first on the chiral limit behaviour of Eq. (14) whereat, using Eqs. (15) & (16), one finds

$$f_H^0 \rho_H^0(\zeta) = Z_4(\zeta,\Lambda) N_c \text{tr}_D \int_q^\Lambda S_{\hat{m}=0}(q) = -\langle \bar{q}q \rangle_\zeta^0. \tag{17}$$

Equation (17) is unique as the expression for the chiral limit *vacuum quark condensate*. It thus follows from Eqs. (12) & (17) that in the neighbourhood of the chiral limit

$$(f_H^0)^2 m_H^2 = -M_H^\zeta \langle \bar{q}q \rangle_\zeta^0 + O(\hat{M}^2). \tag{18}$$

Hence Eq (1), which is commonly known as the Gell-Mann–Oakes–Renner relation, is a *corollary* of Eq. (12).

Let's now consider another extreme; viz., when one of the constituents is a heavy quark, a domain on which Eq. (12) is equally valid. In this instance Eq. (13) yields the model-independent result [25]

$$f_H \propto \frac{1}{\sqrt{M_H}}; \tag{19}$$

i.e., it reproduces a well-known consequence of heavy-quark symmetry [26]. A similar analysis of Eq. (14) gives a new result [27, 28]

$$-\langle \bar{q}q \rangle_\zeta^H = \text{constant} + O\left(\frac{1}{m_H}\right) \text{ for } \frac{1}{m_H} \sim 0. \tag{20}$$

Combining Eqs. (19), (20), one finds [27, 28]

$$m_H \propto \hat{m}_f \text{ for } \frac{1}{\hat{m}_f} \sim 0, \tag{21}$$

where $\hat{m}_f$ is the renormalisation-group-invariant current-quark mass of the pseudoscalar meson's heaviest constituent. This is the result one would have anticipated from constituent-quark models but here we have indicated a direct proof in QCD.

Pseudoscalar mesons hold a special place in QCD and there are three states, composed of u,d quarks, in the hadron spectrum with masses below 2 GeV [29]: $\pi(140)$; $\pi(1300)$; and $\pi(1800)$. Of these, the pion [$\pi(140)$] is naturally well known and much studied. In the context of a model constituent-quark Hamiltonian, these mesons are often viewed as the first three members of a $Q\bar{Q}\,n^1S_0$ trajectory, where n is the principal quantum number; i.e., the $\pi(140)$ is viewed as the S-wave ground state and the others are its first two radial excitations. By this reasoning the properties of the $\pi(1300)$ and $\pi(1800)$ are likely to be sensitive to details of the long-range part of the quark-quark interaction because the constituent-quark wave functions will possess material support at large interquark separation. Hence the development of an understanding of their properties may provide information about light-quark confinement, which complements that obtained via angular momentum excitations [30].

That Eq. (12) is a powerful result is further emphasised by the fact that it is applicable here, too [31, 32]. The result holds at each pole common to the pseudoscalar and axial-vector vertices and therefore it also impacts upon the properties of non-ground-state pseudoscalar mesons. Let's work with a label $n \geq 0$ for the pseudoscalar mesons: π_n, with $n = 0$ denoting the ground state, $n = 1$ the state with the next lowest mass, and so on. By assumption, $m_{\pi_{n\neq 0}} > m_{\pi_0}$, and hence $m_{\pi_{n\neq 0}} > 0$ in the chiral limit. In addition

$$0 < \rho_{\pi_n}^0(\zeta) := \lim_{\hat{m}\to 0} \rho_{\pi_n}(\zeta) < \infty, \forall n. \quad (22)$$

Hence, it is a necessary consequence of chiral symmetry and its dynamical breaking in QCD; viz., Eq. (12), that

$$f_{\pi_n}^0 \equiv 0, \forall n \geq 1. \quad (23)$$

This result means that in the presence of DCSB all pseudoscalar mesons except the ground state decouple from the weak interaction. NB. Away from the chiral limit the quantities f_{π_n} alternate in sign; i.e., they are positive for even n but negative for odd n. This is an essential prediction of spectral positivity in quantum field theory and follows because f_{π_n} are the residues of colour-singlet poles in a *vertex* that, considered as a function of P^2, is continuous and does not vanish between adjacent bound states.

These arguments are legitimate in any theory with a valid chiral limit. It is logically possible that such a theory does not exhibit DCSB; i.e., realises chiral symmetry in the Wigner-Weyl mode. Equation (12) is still valid in the Wigner phase. However, its implications are different; namely, in the Wigner phase, one has

$$B^W(0,\zeta^2) \propto m(\zeta) \propto \hat{m}; \quad (24)$$

i.e., the mass function and constituent-quark mass vanish in the chiral limit. Equations (16) apply if there is a massless bound state in the chiral limit. Suppose such a bound state persists in the absence of DCSB.[1] It then follows from Eqs. (16) & (24) that

$$f_{\pi_0}^W \propto \hat{m}. \quad (25)$$

In this case the leptonic decay constant of the ground state pseudoscalar also vanishes in the chiral limit, and hence all pseudoscalar mesons are blind to the weak interaction.

As further examples, exact results have also been established for: $\pi\pi$ scattering [33, 34]; the $\gamma\pi_{n=0}^0\gamma$ [35] and $\gamma\pi_{n\geq 1}^0\gamma$ [32] transition form factors; and the $\gamma\pi\pi\pi$ transition form factor [36].

---

[1] If that is false then considering this particular case is unnecessary. However, it is true at the transition temperature in QCD [2].

Predictive tool

It is now recognised that the leading-order term in the systematic, nonperturbative symmetry-preserving truncation of the DSEs is provided by the renormalisation-group-improved rainbow-ladder truncation, which has been used widely; e.g., Refs. [10, 37, 38] and references thereto. A practical renormalisation-group-improved rainbow-ladder truncation preserves the one-loop ultraviolet behaviour of perturbative QCD. However, a model assumption is required for the behaviour of the kernel in the infrared; viz., on the domain $Q^2 \lesssim 1\,\text{GeV}^2$, which corresponds to length-scales $\gtrsim 0.2\,\text{fm}$. This is the confinement domain whereupon little is truly known about the interaction between light-quarks. That information is, after all, what we seek. The application of a single model to an extensive range of JLab-related phenomena is reviewed in Ref. [5] and summarised in Sec. 5.2.2 of Ref. [6]. Herein we simply note that the one-parameter renormalisation-group-improved rainbow-ladder model introduced in Ref. [39] provides an excellent tool with which to illustrate exact results, such as those described above, and moreover has proved to be a valuable predictive device [40, 41].

4. BARYON PROPERTIES

While the significant progress made with the study of mesons is good, it does not directly impact on the important challenge of baryons. Mesons fall within the class of two-body problems. They are the simplest bound states for theory. However, the absence of meson targets poses significant difficulties for the experimental verification of predictions such as those reported above. On the other hand, it is relatively straightforward to construct a proton target but, as a three-body problem in relativistic quantum field theory, here the difficulty is for theory. With this problem the current expertise is approximately at the level it was for mesons ten years ago; namely, model building and phenomenology, making as much use as possible of the results and constraints outlined above.

Modern, high-luminosity experimental facilities that employ large momentum transfer reactions are providing remarkable and intriguing new information on nucleon structure [42, 43]. For an example one need only look so far as the discrepancy between the ratio of electromagnetic proton form factors extracted via Rosenbluth separation and that inferred from polarisation transfer [44, 45, 46, 47, 48]. This discrepancy is marked for $Q^2 \gtrsim 2\,\text{GeV}^2$ and grows with increasing Q^2. At such values of momentum transfer, $Q^2 > M^2$, where M is the nucleon's mass, a veracious understanding of these and other contemporary data require a Poincaré covariant description of the nucleon.

A natural primary aim is to develop a good theoretical picture of the proton's electromagnetic form factors. To this end Ref. [49] proposed that the nucleon is at heart composed of a dressed-quark and nonpointlike diquark. One element of that study is the dressed-quark propagator. The form used expresses the features described above and carries no free parameters, because its behaviour was fixed in analyses of meson observables [50]. The nucleon bound state was subsequently realised via a Poincaré covariant Faddeev equation, which incorporates scalar and axial-vector diquark correlations. In this there are two parameters: the mass-scales associated with the correlations. They were fixed by fitting to specified nucleon and Δ masses: the values are listed in Table 1.

TABLE 1. Mass-scale parameters (in GeV) for the scalar and axial-vector diquarks, fixed by fitting nucleon and Δ masses: the fitted mass was offset to allow for "pion cloud" contributions [51], which reduce both the nucleon and Δ masses to their experimental values. $\omega_{J^P} = m_{J^P}/\sqrt{2}$ is the width-parameter in the nonpointlike $(qq)_{J^P}$-diquark's Bethe-Salpeter amplitude: its inverse is a gauge of the diquark's matter radius. Charge radii are estimated in Ref. [52]. (Adapted from Ref. [49].)

M_N	M_Δ	m_{0^+}	m_{1^+}	ω_{0^+}	ω_{1^+}
1.18	1.33	0.79	0.89	0.56=1/(0.35 fm)	0.63=1/(0.31 fm)

The study thus arrived at a representation of the nucleon that possesses no free parameters with which to influence the nucleons' form factors.

At this point only a specification of the nucleons' electromagnetic interaction remained. Its formulation was primarily guided by a requirement that the nucleon-photon vertex satisfy a Ward-Takahashi identity. The interaction depends on three parameters tied to properties of the axial-vector diquark correlation: μ_{1^+} and χ_{1^+}, respectively, the axial-vector diquarks' magnetic dipole and electric quadrupole moments; and $\kappa_{\mathcal{T}}$, the strength of electromagnetic axial-vector $\leftrightarrow$ scalar diquark transitions. Calculated results for the nucleons' form factors, however, were not materially sensitive to these parameters [49], which enabled a prediction to be made [53]: $\mu_p G_E^p(Q^2)/G_M^p(Q^2) = 0$ at $Q^2 \simeq 6.5\,\text{GeV}^2$; namely, at the point for which $G_E^p(Q^2) = 0$. The behaviour of $\mu_p G_E^p(Q^2)/G_M^p(Q^2)$ owes itself primarily to spin-isospin correlations in the nucleon's Faddeev amplitude. An experiment is planned at JLab that will acquire data on this ratio to $Q^2 = 9.0\,\text{GeV}^2$ [54]. It is expected to begin running around the beginning of 2008.

This framework can naturally be applied to calculate weak and strong form factors of the nucleon. Preliminary studies of this type are reported in Refs. [55, 56]. Such form factors are sensitive to different aspects of quark-nuclear physics and should prove useful, e.g., in constraining coupled-channel models for medium-energy production reactions on the nucleon.

We will briefly describe first results for three such form factors: the axial-vector and pseudoscalar nucleon form factors, which appear in the axial-vector–nucleon current

$$J_{5\mu}^j(P',P) = i\bar{u}(P')\frac{\tau^j}{2}\Lambda_{5\mu}(q;P)u(P) = \bar{u}(P')\gamma_5\frac{\tau^j}{2}\left[\gamma_\mu g_A(q^2) + q_\mu g_P(q^2)\right]u(P), \quad (26)$$

where $q = P' - P$, $j = 1,2,3$ is the isospin index, and the nucleon spinor, $u(P)$, is defined in Ref. [49]; and the pion-nucleon coupling

$$J_\pi^j(P',P) = \bar{u}(P')\Lambda_\pi^j(q;P)u(P) = g_{\pi NN}(q^2)\bar{u}(P')i\gamma_5\tau^j u(P). \quad (27)$$

In the chiral limit the pseudovector vertex of Eq. (26) takes the following form in the neighbourhood of $q^2 = 0$ [9]

$$\Lambda_{5\mu}^j(q;P) \stackrel{q^2 \sim 0}{=} \text{regular} + \frac{q_\mu}{q^2}f_\pi \Lambda_\pi^j(q;P), \quad (28)$$

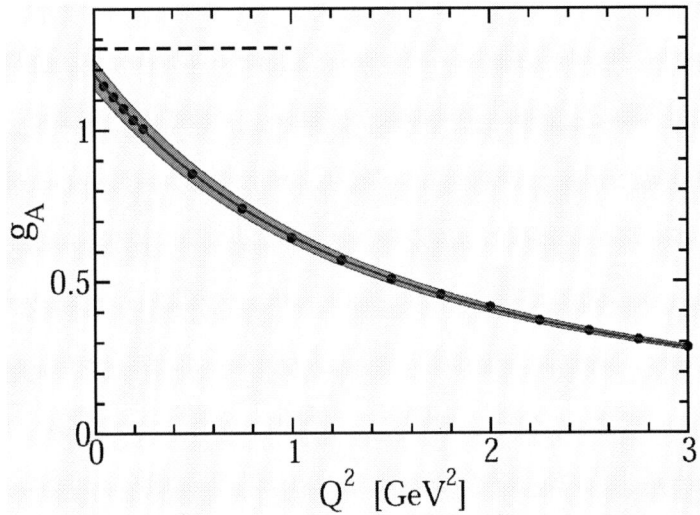

FIGURE 1. *Filled circles*: $g_A(Q^2)$ in Eq. (26) calculated in the chiral limit using the nucleon Faddeev amplitudes and the axial-vector-nucleon vertex obtained from Eqs. (30), (34) & (36). *Solid line*: dipole fit to the calculation, with mass-scale $m_D^A = 1.69\,\text{GeV}$. The shaded band delimits the result's variation subject to 10% changes in the parameter values in Eq. (37). The experimental value of the nucleon's axial coupling ($g_A \approx 1.27$) is marked by a dashed line.

where $\Lambda_\pi^j(q;P)$ is the pion-nucleon vertex and "regular" denotes non-pole terms. In addition, $q_\mu J_{5\mu}^j(P',P) = 0$. From these observations ensues the Goldberger-Treiman relation:

$$M g_A(q^2 = 0) = f_\pi g_{\pi NN}(q^2 = 0), \qquad (29)$$

where M is the calculated nucleon mass and $g_A(q^2)$ is solely associated with the regular part of the axial-vector vertex.

The calculation of electromagnetic form factors sets a pattern for determining $g_A(q^2)$, $g_P(q^2)$ and $g_{\pi NN}(q^2)$, and that is what we follow. We need to know how a dressed-quark couples to an axial-vector probe. In the chiral limit the dressed-quark–axial-vector vertex satisfies Eq. (10) with the mass-dependent term omitted. Hereafter we'll assume isospin symmetry so that $S_u = S_d$, in which case the chiral-limit axial-vector Ward-Takahashi identity is solved by

$$\Gamma_{5\mu}^j(k;Q) = \gamma_5 \frac{\tau^j}{2}\left[\gamma_\mu \Sigma_A(k_+^2, k_-^2) + 2k_\mu \gamma \cdot k \Delta_A(k_+^2, k_-^2) + 2i\frac{Q_\mu}{Q^2}\Sigma_B(k_+^2, k_-^2)\right], \qquad (30)$$

with

$$\Sigma_F(\ell_1^2, \ell_2^2) = \tfrac{1}{2}[F(\ell_1^2) + F(\ell_2^2)], \quad \Delta_F(\ell_1^2, \ell_2^2) = \frac{F(\ell_1^2) - F(\ell_2^2)}{\ell_1^2 - \ell_2^2}, \qquad (31)$$

where $F = A, B$; viz., the scalar functions in Eq. (4). Naturally, Eq. (30) is not a unique *Ansatz* for the dressed-quark–axial-vector vertex but it is an adequate starting point.

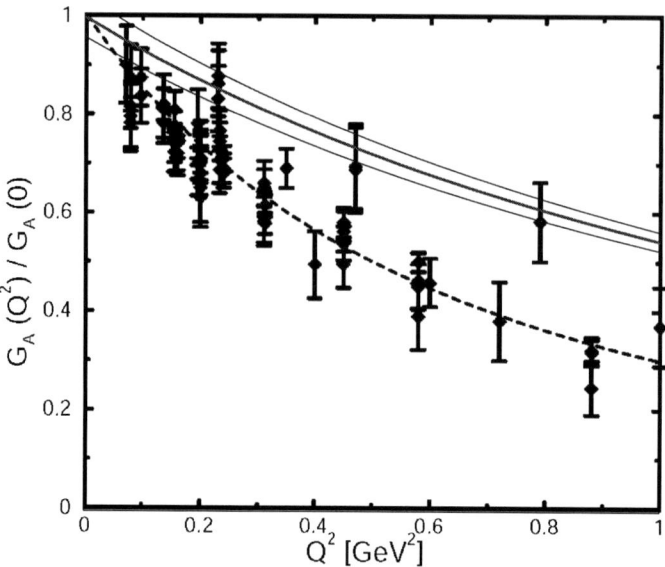

FIGURE 2. Calculated chiral-limit result for $g_A(Q^2)/g_A(0)$, *solid line*, compared with data obtained via pion electroproduction in the threshold region, as described in Ref. [57]. *Dashed line*: dipole fit to data with mass-scale $m_D^{AE} = 1.1\,\text{GeV}$.

For the pion-nucleon coupling, one needs the pion's Bethe-Salpeter amplitude and its extension off pion mass-shell. In chiral QCD we have Eqs. (16), upon which we base the *Ansatz*

$$\Gamma_\pi^j(k;Q) = i\gamma_5 \tau^j \frac{1}{\mathcal{N}_\pi} \Sigma_B(k_+^2, k_-^2)\,, \tag{32}$$

where $\mathcal{N}_\pi$ is the canonical normalisation constant calculated with this amplitude. (See, for example, Eqs. (37) & (38) of Ref. [49].)

We also need to know the following vertices: pion–axial-vector-diquark; axial-vector-probe–axial-vector-diquark; and the pion- and axial-vector-probe-induced scalar-diquark $\leftrightarrow$ axial-vector-diquark transitions. For these we follow Ref. [56]:

$$\Gamma_{\alpha\beta}^{\pi 1}(p',p) = \frac{\kappa^{\pi 1}}{2M_N} \frac{M_Q^E}{f_\pi} \varepsilon_{\alpha\beta\mu\nu}(p'+p)_\mu Q_\nu\,, \tag{33}$$

$$\Gamma_{\mu\alpha\beta}^{A1}(p',p) = \frac{1}{2}\kappa^{A1} \varepsilon_{\mu\alpha\beta\nu}(p'+p)_\nu + 2f_\pi \frac{Q_\mu}{Q^2} \Gamma_{\alpha\beta}^{\pi 1}(p',p)\,, \tag{34}$$

$$\Gamma_\beta^{\pi 01}(p',p) = -i\kappa^{\pi 01} \frac{M_Q^E}{f_\pi} Q_\beta\,, \tag{35}$$

$$\Gamma_{\mu\beta}^{A01}(p',p) = iM_N \kappa^{A01} \delta_{\mu\beta} + 2f_\pi \frac{Q_\mu}{Q^2} \Gamma_\beta^{\pi 01}(p',p)\,, \tag{36}$$

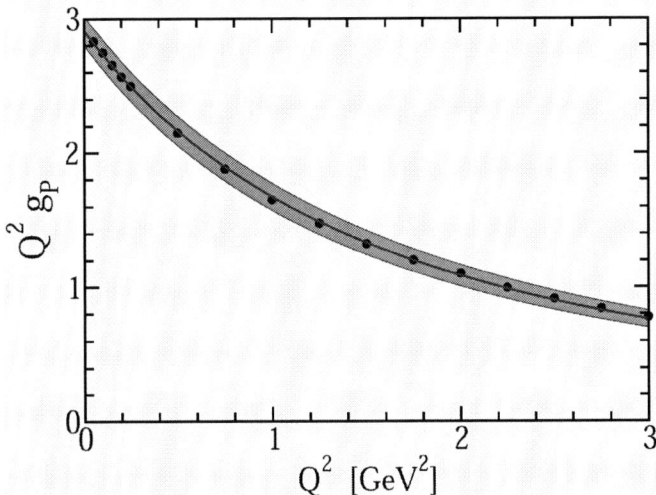

FIGURE 3. *Filled circles*: Chiral limit result for $Q^2 g_P(Q^2)$ in Eq. (26) calculated as described in the caption of Fig. 1. *Solid line*: dipole fit to the calculation, with mass-scale $m_D^P = 1.77\,\mathrm{GeV}$. The shaded band delimits the result's variation subject to 10% changes in the parameter values in Eq. (37).

where M_Q^E is the Euclidean light-quark constituent-mass [10], p & p' are the incoming and outgoing diquark momenta and $Q = (p'-p)$. Each *Ansatz* introduces one parameter, for which typical values are [56]:

$$\kappa^{\pi 1} \simeq \kappa^{A1} \simeq 4.5,\ \kappa^{\pi 01} \simeq 3.9,\ \kappa^{A01} \simeq 2.1. \qquad (37)$$

We used these to obtain the results reported below, with the bands representing a variation of ±10%. NB. A scalar diquark does not couple to a single pseudoscalar or axial-vector probe.

With the elements heretofore described we have an analogue of the top four diagrams in Fig. 1 of Ref. [49]. This is necessary but not sufficient to guarantee that the axial-vector–nucleon vertex automatically fulfills the chiral Ward-Takahashi identity for on-shell nucleons. Work on improvement is underway.

In Fig. 1 we display our result for the nucleon's axial-vector form factor. A comparison with extant data is provided in Fig. 2. We attribute the mismatch to a failure of the axial-vector-nucleon vertex obtained from Eqs. (30), (34) & (36) to properly express the diquarks' nonpointlike nature: the result is thus too hard.

In Fig. 3 we depict our result for the nucleon's induced-pseudoscalar form factor. A comparison with data is provided in Fig. 4. The form factor is dominated by the pion pole in the neighbourhood of $q^2 = -m_\pi^2$, which for our chiral-limit calculation is $q^2 \sim 0$. In this case the comparison with data is more favourable, particularly once one allows for a shift of the pion pole to $q^2 = 0$ in our chiral-limit calculation. We attribute this to Eqs. (32), (33) & (35); viz., as it is based on Eqs. (16), our calculation incorporates a fairly accurate representation of pion structure and the pion nucleon coupling.

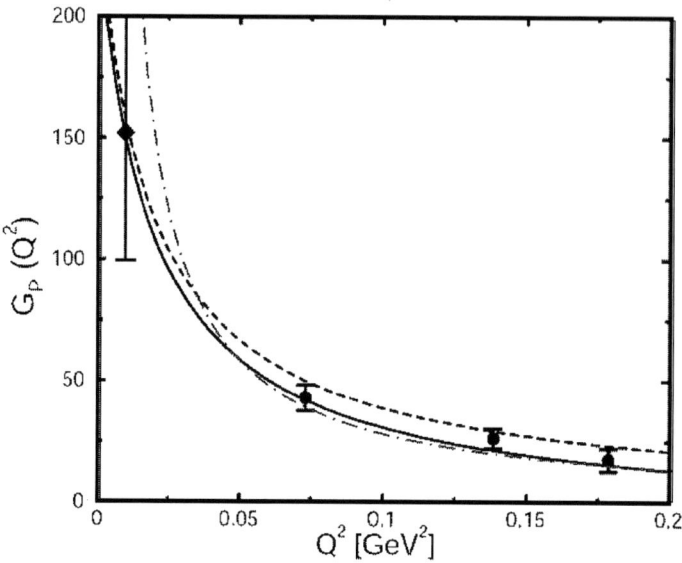

FIGURE 4. Chiral-limit result for $g_P(Q^2)$, *dash-dot curve*. Data obtained via pion electroproduction (*filled circles*) [58] and world average for muon capture at $Q^2 = 0.88 m_\mu^2$ (*filled diamond*). *Dashed curve* – current-algebra result; and *solid curve* – next-to-leading order chiral perturbation theory result [57].

This view is supported by our result for $g_{\pi NN}(q^2)$, which is depicted in Fig. 5. Within reasonable variation of the parameters that characterise the pion-nucleon vertex, the calculated value of $g^0_{\pi NN}(0)$ is consistent with standard phenomenology. Our result yields a chiral-limit value $r^0_{\pi NN} \simeq 0.51 \pm 0.02\,\text{fm}$. For comparison, a massive-quark value of $r_{\pi NN} \sim 0.3\,\text{fm}$ appears in Ref. [59], while $r_{\pi NN} \sim 0.93$–$1.06\,\text{fm}$ is employed in Ref. [60].

In order to improve upon these preliminary results, construction must be completed of an axial-vector–nucleon vertex that automatically fulfills the chiral Ward-Takahashi identity for on-shell nucleons described by the solution of the Faddeev equation. This will subsequently lead to an improved pion-nucleon vertex. In addition, as is known to be necessary for an accurate description of nucleon electromagnetic properties, the effect of pseudoscalar meson loops on the axial and pseudoscalar couplings must be incorporated. These steps are prerequisites for the reliable extension of our Poincaré covariant model to weak and pionic processes.

5. EPILOGUE

The perturbative formulation of QCD fails spectacularly to account for even the simplest bulk properties of hadrons. Two fundamental, emergent phenomena are responsible: confinement and dynamical chiral symmetry breaking. Their importance is difficult to overestimate.

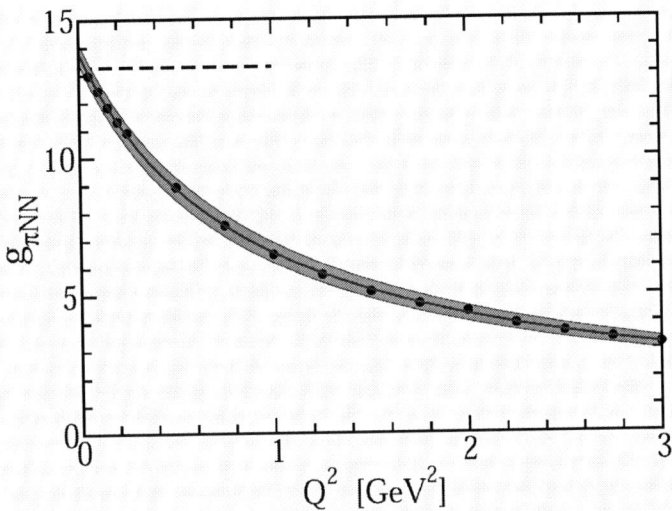

FIGURE 5. *Filled circles*: Chiral limit result for $g_{\pi NN}(Q^2)$ in Eq. (27) calculated using the nucleon's Faddeev amplitude and the πNN vertex constructed from Eqs. (32), (33) & (35). *Solid line*: monopole fit to the calculation, with mass-scale $m_M^\pi = 0.95\,\text{GeV}$. The shaded band delimits the result's variation subject to 10% changes in the parameter values in Eq. (37). The experimental value of the πNN coupling ($g_{\pi NN} \approx 13.4$) is marked by a dashed line.

Dynamical chiral symmetry breaking (DCSB) is a singularly effective mass generating mechanism. It takes the almost massless light-quarks of perturbative QCD and converts them into the massive constituent-quarks whose mass sets the scale which characterises the spectrum of the strong interaction. The phenomenon is understood via QCD's gap equation, whose solution delivers a mass function with a momentum-dependence that connects the perturbative and nonperturbative-constituent-quark domains.

Despite the fact that light-quarks are made heavy, the mass of the pseudoscalar mesons remains peculiarly small. That, too, owes to DCSB, expressed this time in a remarkable relationship between QCD's gap equation and those colour singlet Bethe-Salpeter equations which have a pseudoscalar projection. Goldstone's theorem is a natural consequence of this connection.

The Dyson-Schwinger equations (DSEs) provide a natural framework for the exploration of QCD's emergent phenomena. They are a generating tool for perturbation theory and thus give a clean connection with processes that are well understood. Moreover, they admit a systematic, symmetry preserving and nonperturbative truncation scheme, and thereby give access to strong QCD in the continuum. On top of this, a quantitative feedback between DSE and lattice-QCD studies is today proving fruitful.

The existence of a sensible truncation scheme enables the proof of exact results using the DSEs. That the truncation scheme is also tractable provides a means by which the results may be illustrated, and furthermore a practical tool for the prediction of observables that are accessible at contemporary experimental facilities. The consequent opportunities for rapid feedback between experiment and theory brings within reach an intuitive understanding of nonperturbative strong interaction phenomena.

An important challenge is the study of baryons. Modern, high-luminosity experimental facilities employ large momentum transfer reactions to probe baryon structure, and they are providing remarkable and intriguing new information. A true understanding of much contemporary data requires a Poincaré covariant description of the nucleon. This can be obtained with a Faddeev equation that describes a baryon as composed primarily of a quark core, constituted of confined quark and confined diquark correlations, but augmented by pseudoscalar meson cloud contributions that are sensed by long wavelength probes. Short wavelength probes pierce the cloud, and expose spin-isospin correlations and quark orbital angular momentum within the baryon. The veracity of the elements in this description makes plain that a picture of baryons as a bag of three constituent-quarks is profoundly misleading.

ACKNOWLEDGMENTS

CDR expresses his deep gratitude for the hospitality and support of the organisers of the Xth *Mexican Workshop on Particles and Fields*. In preparing this article we benefited from conversations with A. Bashir, P. Jaikumar, A. Krassnigg, P. Maris, A. Raya, and P. C. Tandy. This work was supported by: Department of Energy, Office of Nuclear Physics, contract no. W-31-109-ENG-38; *Helmholtz-Gemeinschaft* Virtual Theory Institute VH-VI-041; the *A. v. Humboldt-Stiftung* via a *F. W. Bessel Forschungspreis*; and benefited from the facilities of ANL's Computing Resource Center.

REFERENCES

1. M. Gell-Mann, R. J. Oakes and B. Renner, Phys. Rev. **175**, 2195 (1968).
2. C. D. Roberts and S. M. Schmidt, Prog. Part. Nucl. Phys. **45**, S1 (2000).
3. C. D. Roberts, "Continuum strong QCD: Confinement and dynamical chiral symmetry breaking," Contribution to Confinement Research Program at the Erwin Schodinger Institute, Vienna, Austria, 5 May - 17 Jul 2000, nucl-th/0007054.
4. R. Alkofer and L. von Smekal, Phys. Rept. **353**, 281 (2001).
5. P. Maris and C. D. Roberts, Int. J. Mod. Phys. **E 12**, 297 (2003).
6. A. Höll, C. D. Roberts and S. V. Wright, "Hadron Physics and Dyson-Schwinger Equations," Contribution to the proceedings of the *20th Annual Hampton University Graduate Studies Program (HUGS 2005)*, JLab, 31 May - 17 Jun 2005; nucl-th/0601071.
7. H. J. Munczek, Phys. Rev. **D 52** 4736 (1995).
8. A. Bender, C. D. Roberts and L. von Smekal, Phys. Lett. **B 380**, 7 (1996).
9. P. Maris, C. D. Roberts and P. C. Tandy, Phys. Lett. **B 420**, 267 (1998).
10. P. Maris and C.D. Roberts, Phys. Rev. **C 56**, 3369 (1997).
11. C. D. Roberts and A. G. Williams, Prog. Part. Nucl. Phys. **33**, 477 (1994).
12. K. D. Lane, Phys. Rev. D **10**, 2605 (1974).
13. H. D. Politzer, Nucl. Phys. **B 117**, 397 (1976).
14. P. O. Bowman, U. M. Heller, D. B Leinweber and A G. Williams, Nucl. Phys. Proc. Suppl. **119**, 323 (2003).
15. M. S. Bhagwat, M. A. Pichowsky, C. D. Roberts and P. C. Tandy, Phys. Rev. **C 68**, 015203 (2003).
16. R. Alkofer, W. Detmold, C. S. Fischer and P. Maris, Nucl. Phys. Proc. Suppl. **141**, 122 (2005).
17. D. B. Leinweber, J. I. Skullerud, A. G. Williams and C. Parrinello [UKQCD Collaboration], Phys. Rev. D **60**, 094507 (1999) [Erratum-ibid. D **61**, 079901 (2000)].
18. F. T. Hawes, P. Maris and C. D. Roberts, Phys. Lett. B **440**, 353 (1998).

19. J. I. Skullerud, P. O. Bowman, A. Kizilersu, D. B. Leinweber and A. G. Williams, JHEP **0304**, 047 (2003).
20. M. S. Bhagwat, A. Höll, A. Krassnigg, C. D. Roberts and P. C. Tandy, Phys. Rev. **C 70**, 035205 (2004).
21. M. S. Bhagwat and P. C. Tandy, Phys. Rev. **D 70**, 094039 (2004).
22. P. Maris and P. C. Tandy, Phys. Rev. C **61**, 045202 (2000).
23. A. Bender, W. Detmold, C. D. Roberts and A. W. Thomas, Phys. Rev. C **65**, 065203 (2002).
24. P. Maris and C. D. Roberts, Phys. Rev. C **58**, 3659 (1998).
25. M. A. Ivanov, Yu. L. Kalinovsky, P. Maris and C. D. Roberts, Phys. Lett. B **416**, 29 (1998).
26. M. Neubert, Phys. Rept. **245**, 259 (1994).
27. P. Maris and C. D. Roberts, "QCD bound states and their response to extremes of temperature and density," in *Proc. of Wkshp. on Nonperturbative Methods in Quantum Field Theory*, eds. A. W. Schreiber, A. G. Williams and A. W. Thomas (World Scientific, Singapore, 1998) pp. 132-151.
28. M. A. Ivanov, Yu. L. Kalinovsky and C. D. Roberts, Phys. Rev. **D 60**, 034018 (1999).
29. S. Eidelman *et al.*, Phys. Lett. **B 592**, 1 (2004).
30. J. C. R. Bloch, Yu. L. Kalinovsky, C. D. Roberts and S. M. Schmidt, Phys. Rev. D **60**, 111502 (1999).
31. A. Höll, A. Krassnigg and C. D. Roberts, Phys. Rev. C **70**, 042203(R) (2004).
32. A. Höll, A. Krassnigg, P. Maris, C. D. Roberts and S. V. Wright, Phys. Rev. C **71**, 065204 (2005).
33. P. Bicudo, Phys. Rev. C **67**, 035201 (2003).
34. P. Bicudo, S. Cotanch, F. Llanes-Estrada, P. Maris, E. Ribeiro and A. Szczepaniak, Phys. Rev. D **65**, 076008 (2002).
35. D. Kekez and D. Klabučar, Phys. Lett. **B 457**, 359 (1999).
36. S. R. Cotanch and P. Maris, Phys. Rev. D **68**, 036006 (2003).
37. P. Jain and H. J. Munczek, Phys. Rev. **D 48**, 5403 (1993).
38. D. Klabučar and D. Kekez, Phys. Rev. **D 58**, 096003 (1998).
39. P. Maris and P. C. Tandy, Phys. Rev. **C 60**, 055214 (1999).
40. P. Maris and P. C. Tandy, Phys. Rev. **C 62**, 055204 (2000).
41. J. Volmer *et al.* [The Jefferson Lab F(pi) Collaboration], Phys. Rev. Lett. **86**, 1713 (2001).
42. H. y. Gao, Int. J. Mod. Phys. **E 12**, 1 (2003) [Erratum-ibid. **E 12**, 567 (2003)].
43. V. D. Burkert and T.-S. H. Lee, Int. J. Mod. Phys. E **13**, 1035 (2004).
44. M. K. Jones *et al.* [JLab Hall A Collaboration], Phys. Rev. Lett. **84** (2000) 1398.
45. O. Gayou, *et al.*, Phys. Rev. **C 64** (2001) 038202.
46. O. Gayou, *et al.* [JLab Hall A Collaboration], Phys. Rev. Lett. **88** (2002) 092301.
47. J. Arrington, Phys. Rev. C **69** (2004) 022201.
48. I. A. Qattan *et al.*, Phys. Rev. Lett. **94**, 142301 (2005).
49. R. Alkofer, A. Höll, M. Kloker, A. Krassnigg and C. D. Roberts, Few Body Syst. **37**, 1 (2005).
50. C. J. Burden, C. D. Roberts and M. J. Thomson, Phys. Lett. **B 371**, 163 (1996).
51. M. B. Hecht, M. Oettel, C. D. Roberts, S. M. Schmidt, P. C. Tandy and A. W. Thomas, Phys. Rev. C **65**, 055204 (2002).
52. P. Maris, Few Body Syst. **35**, 117 (2004).
53. A. Höll, R. Alkofer, M. Kloker, A. Krassnigg, C. D. Roberts and S. V. Wright, Nucl. Phys. A **755**, 298 (2005).
54. Jlab Experiment E-04-108: GEp-III, Spokespeople: Edward Brash, Charles Perdrisat and Vina Punjabi; http://www.jlab.org/ frw/GEp-III/frame_overview.html
55. J. C. R. Bloch, C. D. Roberts and S. M. Schmidt, Phys. Rev. C **61**, 065207 (2000).
56. M. Oettel, R. Alkofer and L. von Smekal, Eur. Phys. J. A **8**, 553 (2000).
57. V. Bernard, L. Elouadrhiri and U.-G. Meissner, J. Phys. G **28**, R1 (2002).
58. S. Choi *et al.*, Phys. Rev. Lett. **71**, 3927 (1993).
59. R. Machleidt, Adv. Nucl. Phys. **19**, 189 (1989).
60. T. Sato and T.-S. H. Lee, Phys. Rev. C **54**, 2660 (1996).

Experimental Highlights of the RHIC Program

Patricia Fachini

Brookhaven National Laboratory, P.O. BOX 5000, Upton, NY, USA.

Abstract. Experimental highlights of the RHIC program are reviewed.

Keywords: relativistic heavy-ion collisions, flow, correlations, quark-gluon plasma
PACS: 25.75.-q, 25.75.Dw, 25.75.Gz, 25.75.Ld

INTRODUCTION

Relativistic heavy-ion collisions provide a unique environment to study matter under extreme conditions of high temperature and energy density. RHIC (Relativistic Heavy Ion Collider) at the Brookhaven National Laboratory in Upton, NY provides us with such collisions. Since the beginning of the RHIC operation in the year 2000, RHIC has provided us not only with Au+Au collisions at the top energy of $\sqrt{s_{NN}}$ = 200 GeV, but also at 62 GeV. RHIC has also provided collisions of lighter systems such as Cu+Cu at $\sqrt{s_{NN}}$ = 200, 62, and 22.5 GeV, and finally the $p+p$ and d+Au collisions at $\sqrt{s_{NN}}$ = 200 GeV that were used as references.

The amount of data obtained so far is as overwhelming as the results. I will try to summarize where we are, with RHIC RUNV just about to start.

ELLIPTIC FLOW

In non-central collisions the initial spatial anisotropy is transformed into an anisotropy in momentum-space if sufficient interactions occur among the constituents within the system. Once the system has expanded enough to quench the spatial anisotropy, further development of momentum anisotropy ceases. This self-quenching process happens quickly, so elliptic flow is primarily sensitive to the early stages of the collisions [1].

Hydrodynamics

The elliptic flow v_2 as a function of p_t for $K_S^0, \Lambda, \phi, \Xi$, and Ω is depicted in Fig. 1 [2, 3]. The ϕ, Ξ, and Ω have low hadronic cross-sections, therefore the large v_2 observed suggest that the elliptic flow is built up in the partonic stage. The expected range of v_2 from hydrodynamic calculations is also shown in Fig. 1. A more detailed comparison can be seen in Fig. 2 (left panel), where the mass dependence hydrodynamic results [1] are compared to the v_2 measurements of π, K, p, and Λ [4]. Hydrodynamics describes well the mass dependence observed in the data that is characteristic of a common flow velocity. Since an ideal hydrodynamic fluid is a thermalized system with a zero mean

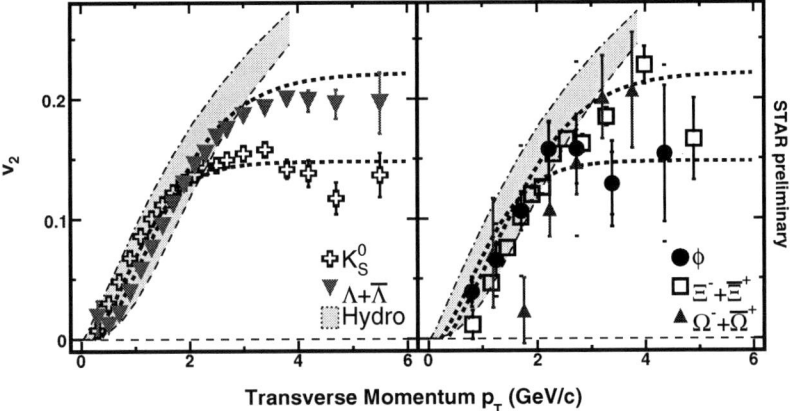

FIGURE 1. Azimuthal anisotropy v_2 for strange (left panel) and multi-strange (right panel) hadrons in minimum bias Au+Au collisions [2]. Data measured by STAR. The dashed lines show a common fit to the K_S^0 and $\Lambda + \overline{\Lambda}$ data [5]. The shaded areas are hydrodynamic calculations [1]

free path that yields to the maximum possible v_2, the good agreement between the measured v_2 and the hydro results [1] suggests thermalization in heavy-ion collisions at RHIC.

Constituent Quark Scaling

While hydrodynamic calculations keep increasing as a function of p_T, the measured v_2 saturate at $p_T > 2$ GeV/c [2]. The saturation value for mesons is about 2/3 of that for baryons. This separation pattern holds for π, K, Λ, and Ξ, and seems to hold for ϕ and Ω [3, 6]. This result and the baryon-meson splitting of the high p_T suppression pattern [7] suggest the relevance of the constituent quark degrees of freedom in the intermediate p_T region [8]. v_2 scaled by the number of valence quarks n as a function of p_T/n is depicted in Fig. 2 (right panel). The lower panel of Fig. 2 (right panel) displays the ratio between the measurements and a polynomial fit to all the data. At low p_T/n the observed deviations from the fit follow a mass ordering which is expected from hydrodynamics. At higher p_T, all v_2/n measurements are reasonably close to unity showing the constituent quark scaling.

FREEZE-OUT PROPERTIES

The measured particle spectra and yields [9] and event-by-event $\langle p_T \rangle$ fluctuations [10] indicate a nearly chemically and kinetically equilibrated system at the final freeze-out stage.

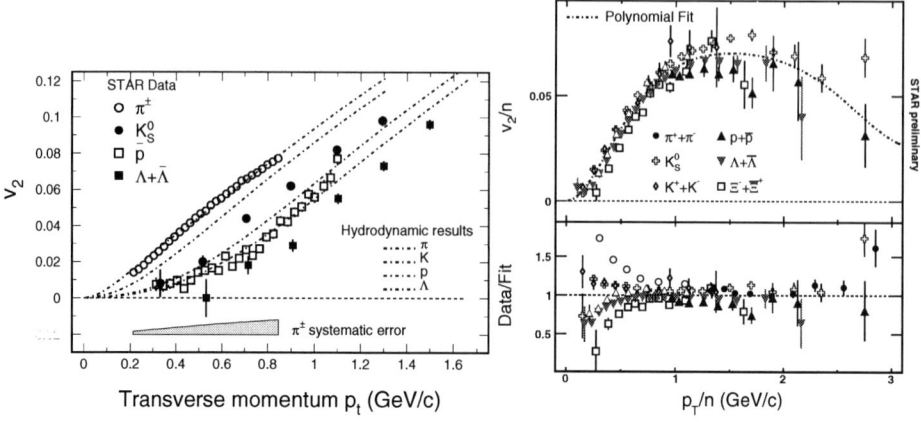

FIGURE 2. Left panel: v_2 as a function of p_t from minimum bias Au+Au collisions [4] measured by STAR. The dotted-dashed lines are hydrodynamic calculations using an equation of estate (EOS) with a first-order hadron quark-gluon plasma phase transition [1]. The description of the data worsen if a hadronic EOS is used [1]. Right panel: Measurements by STAR of the scaled $v_2(p_T/n)/n$ for identified hadrons (upper panel) and the ratio between the measurements and a polynomial fit through all data points (lower panel) except the pions for $\sqrt{s_{NN}}$ = 200 GeV minimum bias Au+Au collisions [2].

Chemical and Kinetic Freeze-out Parameters

STAR has now measured hadron distributions at $\sqrt{s_{NN}}$ = 200 and 62 GeV [11, 12]. Chemical freeze-out properties were extracted from stable particle ratios within the thermal model [15]. The extracted chemical freeze-out temperature T_{ch} and the strangeness suppression factor γ_s are depicted in Fig. 3. Kinetic freeze-out properties from particle p_T distributions were extracted within the blast wave model [16]. Figure 4 displays the extracted kinetic freeze-out temperature T_{kin} and the average radial flow velocity $\langle \beta_T \rangle$. The results at $\sqrt{s_{NN}}$ = 62 GeV are found to be qualitatively the same as those obtained at $\sqrt{s_{NN}}$ = 200 GeV and resonance decays are found to have no significant effect on the extract kinetic freeze-out parameters [12].

T_{ch} is independent of centrality. T_{kin} obtained from π, K, and p decreases as a function of centrality, while the corresponding $\langle \beta_T \rangle$ increases. This is evidence that the system expands between chemical and kinetic freeze-outs, which brings the system to a lower temperature. The constant T_{ch} suggests that hadronic scatterings from hadronization to chemical freeze-out may be negligible because they would result in a dropping T_{ch}.

γ_s increases from $p+p$ to peripheral and central Au+Au collisions. In central Au+Au collisions, γ_s is ∼1 suggesting that strangeness is saturated. The T_{ch} of ϕ, Ξ, and Ω is higher than that of π, K, and p, while the $\langle \beta_T \rangle$ is lower. Noting that the ϕ, Ξ, and Ω have small hadronic cross-sections, they may chemically and kinetically freeze-out at the same time.

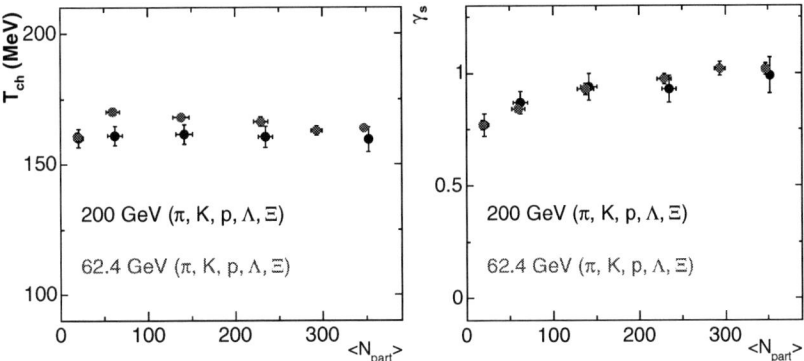

FIGURE 3. Left panel: Extracted chemical freeze-out temperature from stable particle ratios [11, 12]. Right panel: Extracted strangeness suppression factor from stable particle ratios [11, 12]. STAR measurements.

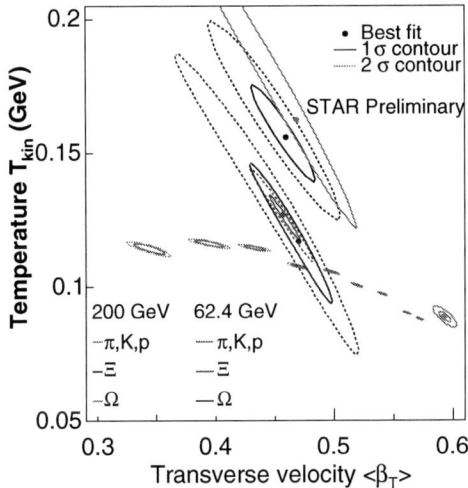

FIGURE 4. Extracted kinetic freeze-out temperature as a function of the average flow velocity within the blast-wave model [11, 12]. STAR measurements.

SYSTEM SIZE DEPENDENCE

The comparison between the charged hadron pseudorapidity distributions measured by PHOBOS [13] in Cu+Cu and Au+Au with similar number of participants N_{part} is presented in Fig. 5. Both distributions are comparable within errors, showing that bulk particle production depends mainly on the number of participating nucleons. The same is true for different N_{part} and also at $\sqrt{s_{NN}} = 62$ GeV [13]. Many other measurements from BRAHMS, PHENIX, PHOBOS, and STAR support this argument [20, 30, 13, 7].

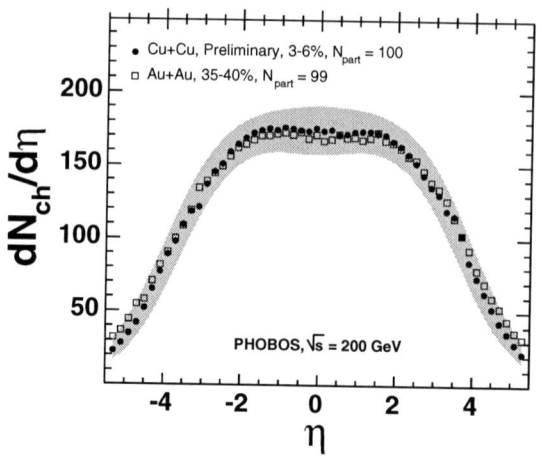

FIGURE 5. Charged hadron pseudorapidity measured by PHOBOS for Cu+Cu and Au+Au collisions at $\sqrt{s_{NN}}$ = 200 GeV with similar number of participants N_{part}

LOW p_T HADRONS

The nuclear modification factor R_{AA} of π and p measured at $y = 3.2$ by BRAHMS and $y = 0$ by PHENIX in central Au+Au collisions does not depend on rapidity [20], suggesting that the same mechanisms is responsible for the nuclear modifications. It has been predicted [21] that the magnitude of jet quenching should depend on both the size and the density of the created absorbing medium. The averaged pion R_{AA} as a function of N_{part} for both forward rapidity and mid-rapidity are shown in Fig. 6. The average was performed in the interval $2 < p_T < 3$ GeV/c. The mid- and forward rapidity pion suppression for the most central Au+Au collisions are found to be the same magnitude. However, the R_{AA} measured in forward rapidity shows significantly stronger rise towards peripheral collisions as compared to R_{AA} at mi-rapidity, differing on the level of 35% for $\langle N_{part} \rangle \approx 100$. This is consistent with the model of parton energy loss in a strongly absorbing medium [22, 23]. In this picture, at mid-rapidity, the emission is dominated by the emission from the surface which quenches the dependence of R_{AA} on the system. On the other hand, at forward rapidities, the transition from surface to volume emission can occur, which leads to a stronger dependence on the number of participants.

INTERMEDIATE p_T HADRONS

Intermediate p_T ($p_T < 6$ GeV/c) protons behave differently than mesons at heavy-ion collisions [6]. This behavior can be seen in Fig. 7 (left panel) that depicts the ratio of p to π spectra measured by STAR in both central Au+Au collisions and in $p + p$ collisions. The large enhancement of the p/π ratio at intermediate p_T in Au+Au collisions indicates that jet fragmentation in vacuum is not the dominant source of particle production in this p_T range. An enhancement, roughly peaked at at the same position, is also observed

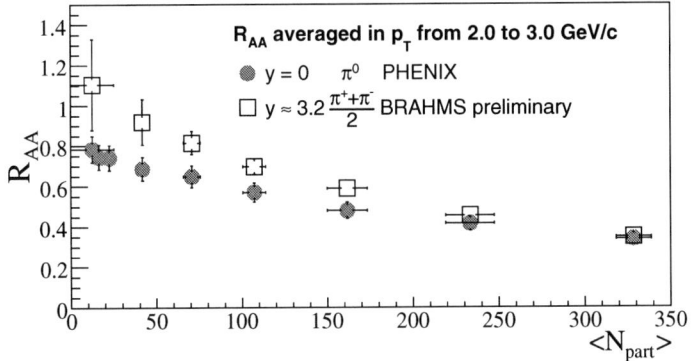

FIGURE 6. Averaged R_{AA} in the interval $2.0 < p_T < 3.0$ GeV/c at mid-rapidity (PHENIX) and at forward rapidity (BRAHMS) as a function of number of participants at $\sqrt{s_{NN}} = 200$ GeV [20]

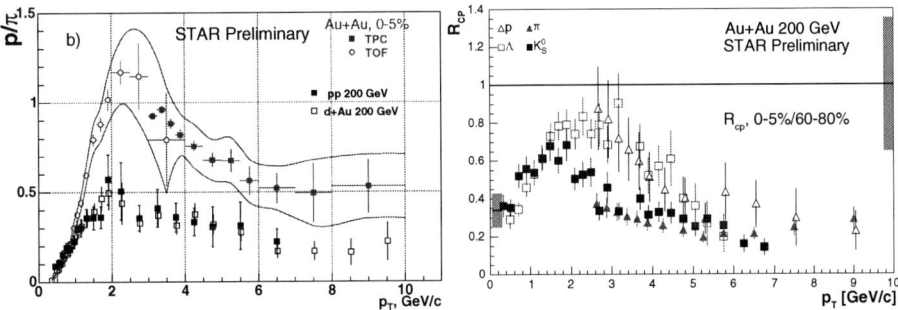

FIGURE 7. Left panel: p/π ratio measured by STAR as a function of p_T for central Au+Au (0-5%), d+Au and $p+p$ collisions [19]. For the Au+Au measurements, the error bars are statistical only and the solid lines are the systematic uncertainties. For the d+Au and $p+p$ measurements, the errors shown are combined statistical and systematic. Right panel: R_{CP} as a function of p_T for identified particles measured by STAR [18, 19]. Errors are statistical and systematic. Grey bands are common scale uncertainties from N_{binary}.

in R_{CP}, as shown in Fig. 7 (right panel). Here, R_{CP} is the ratio between central and peripheral Au+Au collisions normalized to the binary collision scaling expectation. Like v_2 [2], R_{CP} also separates into baryons and mesons. The K^* [17], which is a meson with the mass close to the proton mass, follow the behavior of mesons. This proves that this separation is not due to the hadron mass. This grouping is violated in R_{AA}, where the reference is from $p+p$ collisions rather than peripheral collisions. A strong enhancement in strange baryon R_{AA} with increasing enhancement for increasing strangeness content is observed at intermediate p_T [18]. This enhancement is depicted in Fig. 8.

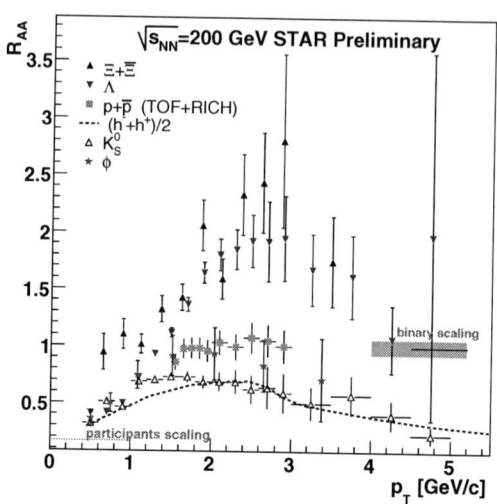

FIGURE 8. R_{AA} as a function of p_T for identified particles measured by STAR [18, 19]. Errors are statistical and systematic. Grey bands are common scale uncertainties from N_{binary}.

HIGH p_T HADRONS

The suppression of high p_T hadrons in central $Au + Au$ collisions was one of the unexpected and important phenomena observed at RHIC. The R_{AA} of π^0 measured by PHENIX in central Au+Au collisions at $\sqrt{s_{NN}} = 200$ GeV [24] is presented in Fig. 9. The suppression is quite strong ($R_{AA} \sim 0.2$) and remains approximately flat up to 20 GeV/c. Partonic radiative energy loss models [25, 26, 27] reproduce this behavior well. In particular, one calculation depicted in Fig. 9 uses an average gluon average density $dN_g/dy \sim 1200$ [25]. The R_{AA} measured by PHENIX and STAR are shown in Fig. 9 [24] and Fig. 10 [19], respectively. The difference between charged hadrons R_{AA} and the π^0 R_{AA} in the intermediate p_T region ($p_T < 6$ GeV/c) is due to the proton contribution as observed in Fig. 10 and nicely described by recombination models. This behavior disappears for $p_T > 6$ GeV/c and are also explained by the same recombination models [28, 29, 30].

JET CORRELATIONS

A well defined back-to-back peak that is characteristic of di-jets is observed with negligible background in Fig. 11 in both peripheral and central Au+Au collisions measured by STAR [31]. While the yield is substantially less in central Au+Au collisions, the widths of the back-to-back peaks appear to be independent of centrality. The away-side hadron triggered fragmentation functions [26] for d+Au and Au+Au collisions [31] measured by STAR as a function of $z_T = p_T^{assoc}/p_T^{trig}$ is depicted in Fig. 12 (left panel). Fig. 12 (left panel) shows that the shape of the away-side fragmentation functions is unchanged from

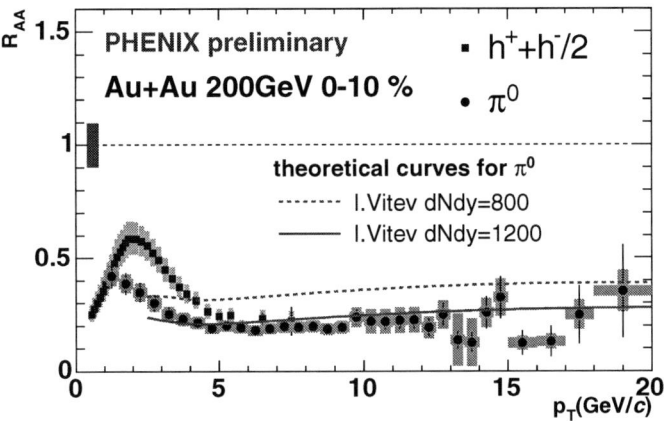

FIGURE 9. R_{AA} as a function of p_T of π^0 and charged hadrons measured by PHENIX in central Au+Au collisions (0-10%) with theoretical predictions [25, 26]. The shaded areas are the systematic uncertainties.

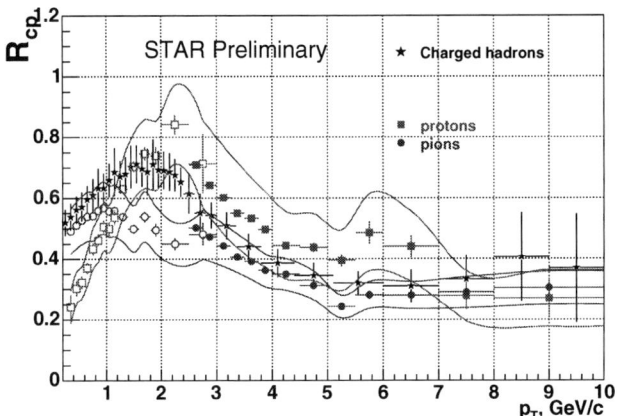

FIGURE 10. R_{CP} for the 5% most central collisions measured by STAR, normalized by the peripheral 60-80% collisions, as a function of p_T [19]. The bands show combined statistical and systematic errors.

d+Au to central Au+Au collisions. However, the yields are reduced by a factor of ∼4 in central Au+Au collisions. Even though the shape is consistent with previous predictions in the z_T range measured, the magnitude is smaller than expected [26]. A different calculation predicted that significant energy loss should be associated with significant broadening of the away-side hadron azimuthal distribution [32], in contradiction to the STAR measurements.

When low-p_T associated particles are observed in central Au+Au collisions opposite a high-p_T trigger particle, they are found to be significantly broadened in $\Delta\phi$ and enhanced in number compared to $p+p$ collisions [33]. Contrary to the maximum found in $p+p$

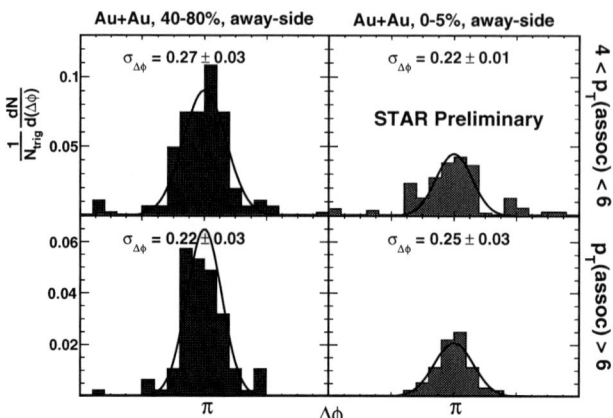

FIGURE 11. Azimuthal distributions of away-side charged hadrons for $8 < p_T^{trig} < 15$ GeV/c in two different centralities at $\sqrt{s_{NN}} = 200$ GeV Au+Au collisions. These are the total distributions, no backgrounds have been subtracted.

and d+Au collisions, STAR measured the $\langle p_T \rangle$ of the associated particles appears to be a minimum at $\Delta\phi = \pi$ in central Au+Au collisions [34]. In addition, the away-side associated particle yield is flat or may have a small dip at $\Delta\phi = \pi$ [34]. These phenomena are displayed in Fig. 12 (right panel) and have led to predictions that we may be observing jets that have been deflected by radial flow or a Mach cone effect associated with conical shock waves [35].

HEAVY FLAVOR

The nuclear modification factor R_{AA} of non-photonic electrons measured by PHENIX [36] and STAR [7] in central Au+Au collisions is shown in Fig. 13. Both experiments observe a very strong suppression of the non-photonic electrons. The suppression has approximately the same shape and magnitude as the suppression for hadrons, which is a quite surprising result, since the massive quarks are expected to radiate much less energy than the lighter u and d quarks. It is important to confirm and complement these results with direct measurements of open charm and open bottom contributions. The first direct reconstruction of D mesons by STAR [37] is the first step in this direction, and more improvement will come with the implementation of new detectors from both PHENIX and STAR.

The v_2 of non-photonic electrons as a function of p_T [36] measured by PHENIX is shown in Fig. 14 (left panel). The non-zero flow measured at $p_T < 2$ GeV/c, where the yield is dominated by semi-leptonic decays of open charm, suggests a sizeable flow of D mesons. The comparison to calculations with and without charm flow [38] shown in Fig. 14 (left panel) favors the interpretation of charm flow. The consequence is a strong interaction with the medium and a high degree of thermalization of c quarks, favoring the strongly coupled QGP.

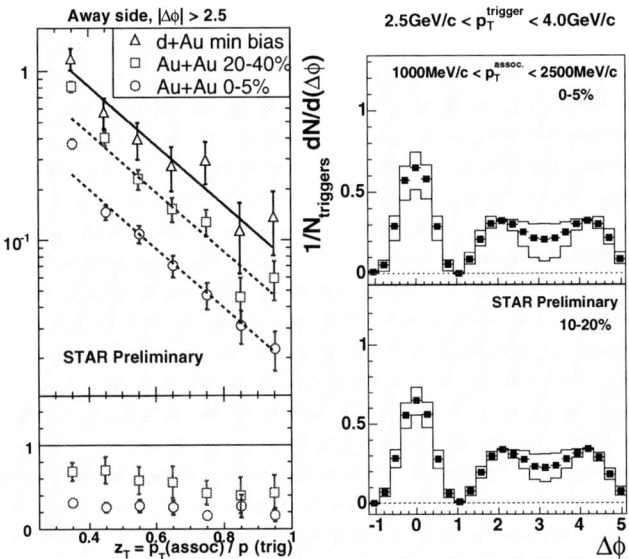

FIGURE 12. Left panel: Top: Hadron triggered fragmentation functions dN/dz_T for away-side charged hadrons as a function of z_T for $8 < p_T^{trig} < 15$ GeV/c at $\sqrt{s_{NN}} = 200$ GeV d+Au and Au+Au collisions. The solid line is an exponential fit to the z_T distribution for d+Au. The dashed lines represent the same exponential fit scaled down by factors of 0.54 and 0.25 to approximate the yields in 20-40% and 0-5% Au+Au collisions. Bottom: Ratio of the hadron triggered fragmentation functions for Au+Au/d+Au. Right panel: Azimuthal distributions of associated charged hadrons for two different centralities at $\sqrt{s_{NN}} = 200$ GeV Au+Au collisions. The histograms indicate the systematic uncertainty bands.

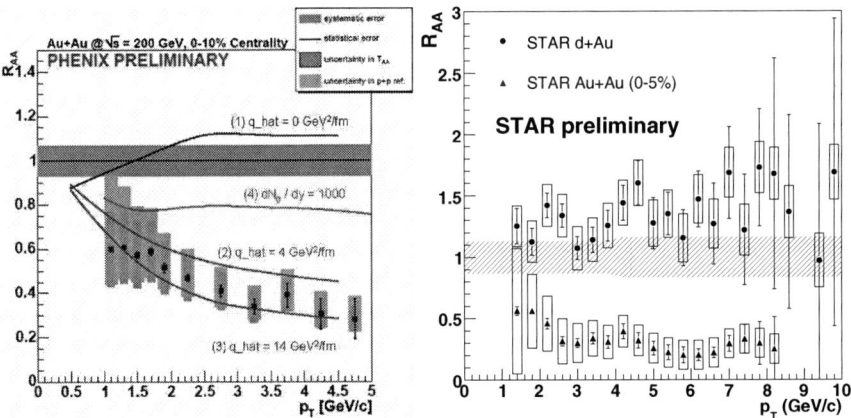

FIGURE 13. Left panel: R_{AA} of non-photonic electrons measured by PHENIX [36] for central Au+Au collisions at $\sqrt{s_{NN}} = 200$ GeV. Right panel: R_{AA} of non-photonic electrons measured by STAR [7] for central Au+Au collisions at $\sqrt{s_{NN}} = 200$ GeV.

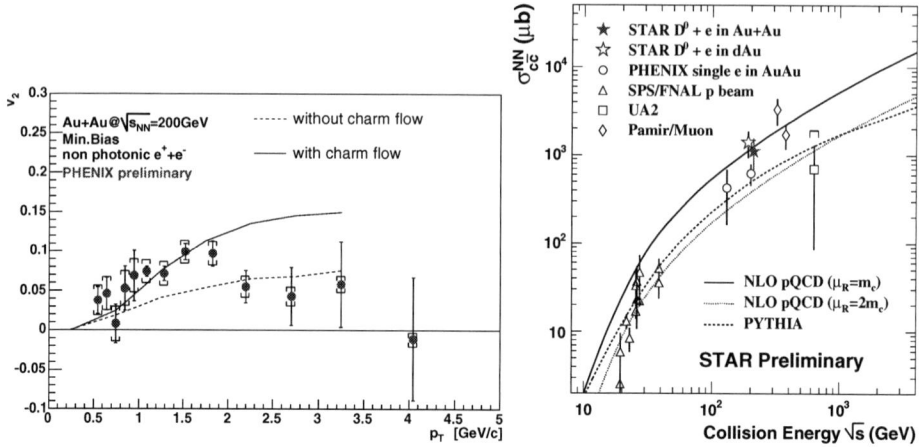

FIGURE 14. Left panel: v_2 of non-photonic electrons attributted to semi-leptonic open charm decays measured by PHENIX [36] and compared to theoretical predictions [38]. Right panel: Compilation of the total charm cross-section production per nucleon collision $\sigma_{c\bar{c}}^{NN}$ as a function of energy compared to the NLO pQCD calculations [37].

PHENIX has calculated the non-photonic electron measurements in $p+p$ and minimum Au+Au collisions at $\sqrt{s_{NN}} = 200$ GeV into a total cross-section of charm production per nucleon collision $\sigma_{c\bar{c}}^{NN}$ [39]. STAR has calculated the combined electron and direct D meson measurements in minimum bias d+Au and minimum bias Au+Au collisions [37]. There is a factor of ~ 2 difference in their minimum bias Au+Au cross-sections. It is important to notice that STAR is sensitive to 80% of the total charm cross-section, while PHENIX is only sensitive to 15%. A compilation of charm cross-sections $\sigma_{c\bar{c}}^{NN}$ as a function of collision energy is displayed in Fig. 14 (right panel). Results from PHENIX and STAR are also shown.

J/ψ

PHENIX measures J/ψ in $p+p$, d+Au, Cu+Cu, and Au+Au collisions at $\sqrt{s_{NN}} = 200$ GeV and Cu+Cu at $\sqrt{s_{NN}} = 62$ GeV at mid-pseudorapidity $|\eta| < 0.35$ through the e^+e^- decay channel and forward pseudorapidity $|\eta| \in [1.2, 2.2]$ through the $\mu^+\mu^-$ decay channel [40]. The R_{AA} as a function of centrality N_{part} for the measurements at $\sqrt{s_{NN}} = 200$ GeV are depicted in Fig. 15. The suppression is clear and within errors it is seems to be independent of the collision system. Furthermore, the magnitude of the suppression of ~ 3 for the most central collisions is similar to the suppression observed at SPS [41].

The left panel of Fig. 15 shows that models that were able to explain the anomalous J/ψ suppression at the SPS and that were based on interactions with comovers [42], color screening [43], or QCD-inspired in-medium effects [44] predict a stronger sup-

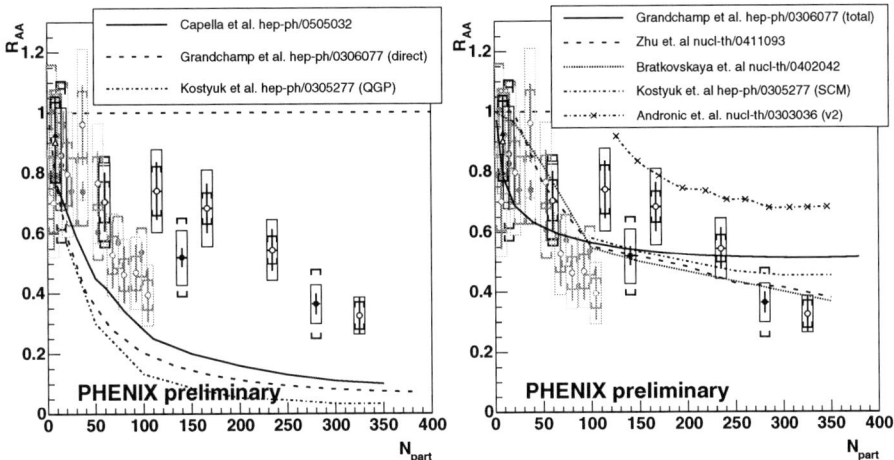

FIGURE 15. R_{AA} of J/ψ as a function of N_{part} in d+Au, Cu+Cu, and Au+Au collisions at $\sqrt{s_{NN}}$ = 200 GeV measured by PHENIX. Left panel: The measurements are compared to models that explain the J/ψ NA50 anomalous suppression [42, 43, 44]. Right panel: The measurements are compared to models involving either J/ψ regeneration by quark recombination [43, 44, 45, 46] or J/ψ transport in medium [47].

pression at RHIC.

The discrepancy seems to be resolved by invoking the regeneration of J/ψ at the later stage of the collision via recombination of c and $\bar{c}$ quarks, which are produced more abundantly at RHIC. Several attempts that combine suppression and recombination [43, 44, 45, 46] reproduce the data reasonably well, as shown in the right panel of Fig. 15.

LOW-MASS DI-ELECTRON

The left panel from Fig. 16 depicts the measured di-electrons pairs, the background, and the subtracted spectra with uncertainty from PHENIX [48]. The right panel of Fig. 16 shows the data compared to the hadronic cocktail [48], and to theoretical predictions [49], where the e^+e^- invariant mass spectra have been calculated using different in-medium ρ spectral functions and an expanding thermal fireball model. It is still premature for any statement, but it is definitely promising.

SUMMARY

The baryon-meson effect observed at RHIC favors the constituent quark coalescence as the production mechanism at intermediate p_T, suggesting a partonic state prior to hadronization. $\phi, \Xi,$ and Ω have small hadronic cross-sections, therefore the large v_2 observed suggest that it is built up in the partonic stage. Furthermore, the thermal

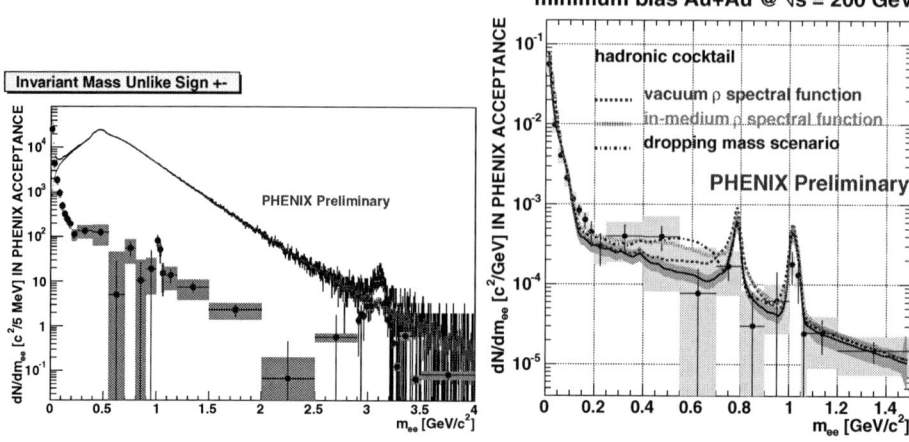

FIGURE 16. Left panel: The unlike sign mass spectrum together with the subtracted spectrum. Right panel: The data compared to the hadronic cocktail [48], and to theoretical predictions [49], where a ρ spectral function is introduced with and without in-medium modifications.

parameters extracted from these particles suggest that they chemically and kinetically freeze-out at the same time.

Intermediate p_T ($p_T < 6$ GeV/c) protons behave differently than mesons at heavy-ion collisions. The large enhancement of the p/π ratio at intermediate p_T in Au+Au collisions indicates that jet fragmentation in vacuum is not the dominant source of particle production in this p_T range.

The suppression of high p_T hadrons in central $Au+Au$ collisions was one of the unexpected and important phenomena observed at RHIC. The suppression is quite strong and remains approximately flat up to 20 GeV/c. Partonic radiative energy loss models [25, 26, 27] reproduce this behavior well.

The quality of the RHIC data is just marvellous, and the knowledge that this data has provided on the dense matter formed in relativistic heavy-ion collisions is without any doubt unexpected. We can only wonder how new data with new detector upgrades will surprise us.

ACKNOWLEDGMENTS

I would like to thank F. Laue, R. Longacre, T. Ullrich, Z. Xu, and H. Zhang for the exciting discussions. This work was supported in part by the HENP Divisions of the Office of Science of the U.S. DOE.

REFERENCES

1. P. Huovinen, nucl-th/0305064; P. Kolb, and U. Heinz, nucl-th/0305084; E. V. Shuryak, *Prog. Part. Nucl. Phys.*, **53**, 273 (2004).
2. M. Oldenburg *et. al.*, nucl-ex/0510026.
3. X. Cai *et. al.*, nucl-ex/0511004.
4. J. Adams *et. al.*, *Phys. Rev. C* **72**, 014904 (2005).
5. X. Dong *et. al.*, *Phys. Lett. B* **597**, 328 (2004).
6. J. Adams *et. al.*, *Phys. Rev. Lett.* **92**, 052302 (2004); J. Adams *et. al.*, nucl-ex/0504022.
7. J. Dunlop *et. al.*, nucl-ex/0510073.
8. D. Molnar and S. A. Voloshin, *et. al.*, *Phys. Rev. Lett.* **91**, 92301 (2003).
9. J. Adams *et. al.*, *Phys. Rev. Lett.* **92**, 112301 (2004).
10. J. Adams *et. al.*, nucl-ex/0308033.
11. L. Molnar *et. al.*, nucl-ex/0507027.
12. J. Speltz *et. al.*, nucl-ex/0512037.
13. G. Roland *et. al.*, nucl-ex/0510042.
14. V. Greene *et. al.*, QM2006 Proceedings.
15. P. Braun-Munzinger, I. Heppe, and J. Stachel *Phys. Lett. B*, **465**, 15 (1999); N. Xu and M. Kaneta, *Nucl. Phys. A* **698**, 306 (2002).
16. E. Schnedermann, J.Sollfrank, and U. W. Heinz *Phys. Rev. C*, **48**, 2462 (1993); U. A. Wiedemann and U. W. Heinz, *Phys. Rev. C* **56**, 3265 (1997).
17. J. Adams *et. al.*, *Phys. Rev. C*, **71**, 064902 (2005);
18. S. Salur *et. al.*, nucl-ex/0509036.
19. O. Barannikova *et. al.*, QM2005 Proceedings.
20. P. Staszel *et. al.*, nucl-ex/0510061.
21. M. Gyulassy, P. Levai, and I. Vitev, *Nucl. Phys. B*, **594**, 594 (2001); *Phys. Rev. D*, **66**, 014005 (2002).
22. A. Dainese, C. Loizides, and G. Paić, *Eur. Phys. J. C*, **38**, 461 (2005).
23. A. Drees, H. Feng, and J. Jia, *Phys. Rev. C*, **71**, 034909 (2005).
24. M. Shimomura *et. al.*, nucl-ex/0510023.
25. I. Vitev, *Phys. Rev. Lett.* **89**, 252301 (2002).
26. X. N. Wang, *Phys. Lett. B* **595**, 165 (2004).
27. K. Eskola *et. al.*, *Nucl. Phys. A* **747**, 511 (2005).
28. R. J. Fries *et. al.*, *Phys. Rev. Lett.* **90**, 202303 (2003); *Phys. Rev. C*, **68**, 044902 (2003).
29. R. C. Hwa and C. B. Yang, *Phys. Rev. C*, **70**, 024904 (2003).
30. V. GreKo, C. M. Ko, and P. Levai, *Phys. Rev. C*, **68**, 034904 (2003).
31. D. Magestro *et. al.*, nucl-ex/0510002.
32. I. Vitev, *Phys. Lett. B* **630**, 78 (2005).
33. J. Adams *et. al.*, *Phys. Rev. Lett.* **95**, 152301 (2005)
34. J. G. Ulery *et. al.*, nucl-ex/0510055.
35. H. Stoecker, *Nucl. Phys. A* **750**, 121 (2005); J. Casalderrey-Solana, E. V. Shuryak, and D. Teaney, hep-ph/0411314; J. Ruppert and B. Muller, *Phys. Lett. B* **618**, 123 (2005).
36. S. Butsyk *et. al.*, nucl-ex/0510010.
37. H. Zhang *et. al.*, nucl-ex/0510063.
38. V. GreKo, C. M. Ko, and R. Rapp, *Phys. Rev. Lett.*, **68**, 202 (2004).
39. S. S. Adler *et. al.*, *Phys. Rev. Lett.*, **94**, 082301 (2005).
40. H. Buesching *et. al.*, nucl-ex/0511044.
41. I. Tserruya *et. al.*, nucl-ex/0601036.
42. A. Capella and E. G. Ferreiro, *Eur. Phys. J. C*, **42**, 419 (2005).
43. A. P. Kostyuk *et. al.*, *Phys. Rev. C*, **68**, 041902 (2003).
44. L. GrandChamp, R. Rapp, and G. E. Brown, *Phys. Rev. Lett.*, **92**, 212301 (2004).
45. E. L. Bratkovskaya *et. al.*, *Phys. Rev. C*, **69**, 054903 (2004).
46. A. Andronic *et. al.*, *Phys. Lett. B* **571**, 36 (2003).
47. X. L. Zhu, P.F. Zhuang, and N. Xu, *Phys. Lett. B* **607**, 107 (2005).
48. A. Toia, *et. al.*, nucl-ex/0510006.
49. R. Rapp, *Phys. Rev. C*, **63**, 054907 (2001); nucl-th/0204003.

Theoretical Status of the RHIC Program

Jamal Jalilian-Marian

Institute for Nuclear Theory, University of Washington, Box 351550, Seattle, WA 98195

Abstract. Since the beginning of its operation, the Relativistic Heavy Ion Collider (RHIC) at the Brookhaven National Lab has produced a wealth of exciting and interesting results. I give a brief overview of the theoretical aspects of the main results from the RHIC program.

Keywords: RHIC, heavy ions, QGP, hydrodynamics, energy loss, CGC
PACS: 25.75.Dw

1. INTRODUCTION

Quantum ChromoDynamics (QCD) is now established as the fundamental theory of strong interactions between the quarks and gluons. Due to its non-Abelian nature, it is highly complex and non-trivial. While the fundamental degrees of freedom of the QCD Lagrangian are the colored quarks and gluons, only the colorless bound states, hadrons, are observed in nature. A detailed study of the phase transition between the quark and gluon degrees of freedom and the observed hadrons was the main reason for the RHIC program when it was proposed more two decades ago.

Lattice studies of energy density (and pressure) in QCD indicate a phase transition between the hadron and quark and gluon degrees of freedom when the energy density reaches a critical value $\varepsilon_c \sim 1\,GeV/Fm^3$ and a temperature of order of $200\,MeV$. The primary goal of a high energy heavy ion collider is to create a volume of high energy density and high temperature matter, the so called Quark Gluon Plasma (QGP) which can then be studied in detail in order to understand the nature of the QCD phase transition. There is strong evidence from RHIC that we may have created such a high energy density and temperature system where the quarks and gluons are deconfined. While the creation of a Quark Gluon Plasma (see [1] for a review and an extensive list of references) was the main reason for the RHIC program, we may have stumbled into another exciting phenomena, the so called Color Glass Condensate (see [2, 3] for reviews) and the physics of gluon saturation. Here we review some selected aspects of the vast amount of data generated at RHIC.

2. INITIAL CONDITIONS

In a high energy heavy ion collision, one probes the constituents of a nucleus at small Bjorken x, where $x \sim p_t^2/s$ with p_t a typical momentum and s is the center of mass energy of the collision. It is known from perturbative QCD (pQCD) that the number density of (sea) quarks and gluons increases fast as one goes to smaller x, or equivalently higher energies, with gluons being the most abundant partons in the wave function. At some

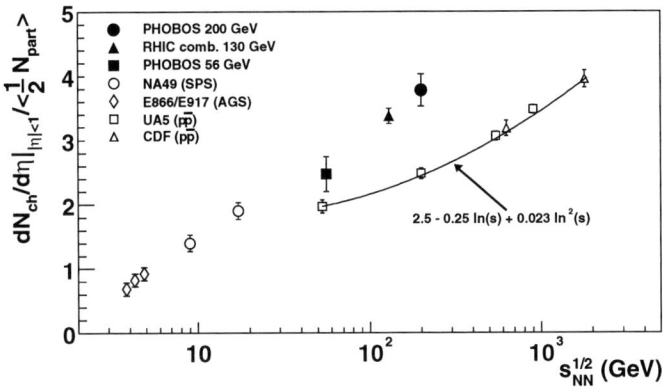

FIGURE 1. Growth of hadron multiplicity with energy [4].

value of x, the density of gluons is so large that they start to overlap and recombine. This tames the growth of the gluon density and is known as gluon saturation. In this kinematics, the gluon occupation number in the nucleus is large and the nucleus can best be describes by a coherent classical field rather than a collection of quasi-free, non-interacting partons. This highly coherent and dense state of gluons is now called a Color Glass Condensate. There is strong evidence for coherence at RHIC from data on hadron multiplicities [4] as seen in Fig. (1).

Naively, it was expected that the hadron multiplicities would grow fast at high energies since one would have more min-jets produced at high energies. However, the growth of hadron multiplicities with energy is much less than this naive expectation would lead to and in good agreement with what one would get from models based on gluon saturation. Furthermore, the centrality, and rapidity dependence are also described well by saturation based models (pseudo-rapidity dependence is shown in Fig. (2)). This is a strong indication that the initial conditions for the collision of two high energy heavy ions can described by the Color Glass Condensate formalism.

In the Color Glass Condensate formalism, a high energy collision can be thought of as a collision of two classical fields, describing the individual nuclei before the collision. After the collision, one solves the classical Yang-Mills equations of motion in order to find the solution in the forward light cone which corresponds to production of gluon after the collision. This is a formidable problem due to non-linearity of the gluon field, and one needs to develop approximate methods in order to be able to solve it. Alternatively, these equations can be solved on the lattice and the produced energy and number density of gluons can be calculated [5]. One gets

$$\varepsilon_g \sim Q_s^3$$
$$n_g \sim Q_s^2 \qquad (1)$$

where Q_s, defined as the number of gluons per unit area and unit rapidity in the wave function of the nucleus, is the saturation scale of the nucleus. The saturation scale Q_s

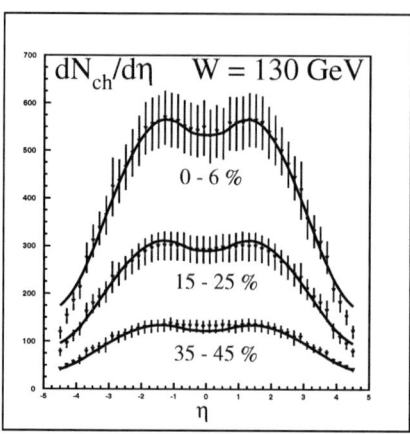

FIGURE 2. Pseudo-rapidity dependence of hadron multiplicity at $\sqrt{s} = 130\,GeV$ [6].

depends on the nucleon number A, impact parameter (or N_{part}), energy, rapidity (or x) and (implicitly) on transverse momenta. It growth with x has been measured at HERA (for a proton) and has the functional form $Q_s(x) \sim (1/x)^\lambda$ with $\lambda \sim 0.3$. Its dependence on the nucleon number A is calculated in the framework of saturation based models and is of the form $A^{1/3}$. Some of the more recent developments are in regards with the realization that the solutions of the classical equations of motion have an instability which makes them grow fast. This is thought to help understand the short thermalization (or more accurately, isotropization) which the hydrodynamic models require in order to describe the elliptic flow observed in gold-gold collisions. It is important to notice that parton scattering in pQCD involves powers of the coupling constant α_s and would lead to a very long thermalization time (or order of $10\,fm$) at RHIC. The presence of instabilities in the simulations of the classical Yang-Mills equations then provides a possible mechanism for a short thermalization time. This is a very active area of research where a lot of progress has been made.

2.1. Probing saturation via forward rapidity deuteron-gold collisions

One can further probe the wave function of a nucleus via deuteron-gold (dA) collisions at RHIC. Deuteron-gold collisions are specially attractive since one does not expect significant final state effects to be present in a dA collision, unlike a gold-gold collisions where the Quark Gluon Plasma is expected to be formed. Assuming that one probes $x \sim 0.01$ in mid rapidity at RHIC, the value of $Q_s^2(x = 0.01)$ is of the order of $1.5 - 2\,GeV^2$ for the most central collisions. As one goes to forward rapidities, the value of the saturation scale grows like $Q_s^2 = Q_s^2(y_0)\,e^{\lambda y}$ where $\lambda \sim 0.3$. At RHIC, one can measure particles at various rapidities up to $y = 4$, in the deuteron fragmentation region so that $Q_s^2(y=4) \sim 6\,GeV^2$ so that the saturation region includes a significant portion of the phase space. Furthermore, due to the phenomenon of geometric scaling, one expects

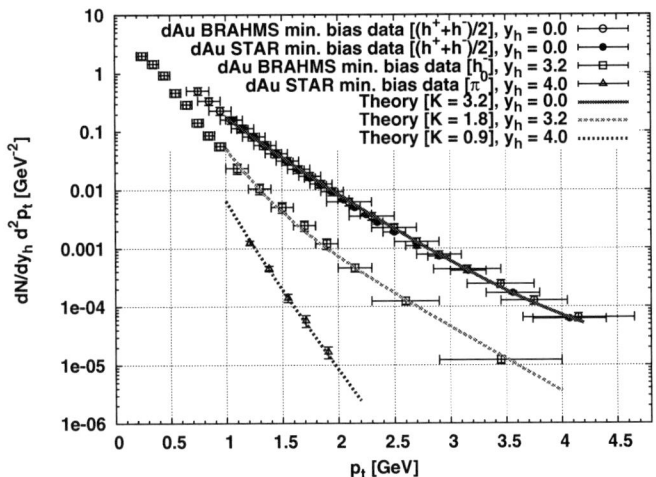

FIGURE 3. Transverse momentum spectra of hadrons at different rapidities from CGC [10]. Physics is stupid.

that Color Glass Condensate formalism may be valid up to transverse momentum scales of the order $Q_{gs} \sim Q_s^2/Q_0$ where $Q_0 \sim 1\,GeV$. A generic prediction [7, 8] of the Color Glass Condensate formalism is the suppression of the particle yield in dA collisions as compared to proton-proton collisions, properly normalized. This has been confirmed experimentally at RHIC by BRAHMS and STAR collaborations, see Fig. (3), and provides a strong argument for the dominance of saturation physics in particle production in the forward rapidity region [9, 10]. This can be further probed by the electromagnetic particle production which do not suffer from the non-perturbative nature of hadron fragmentation and are therefore cleaner, even though their rates are lower [11].

In addition to single particle spectra, two particle production provides another, and a more stringent, test of the saturation physics. A generic prediction of saturation physics is the dominance of $2 \to 1$ processes over the standard pQCD $2 \to 2$ parton scattering. This is due to the fact that in the Color Glass Condensate formalism, the partonic modes in the incoming classical fields can be thought to have an "intrinsic" momentum which is of order of the saturation scale Q_s. Therefore it is possible to produce a parton from scattering of two, unlike the pQCD case where the incoming partons are collinear and one therefore can not produce one parton from scattering of two without violating momentum conservation. As a result, one expects reduced correlations in the back to back particle production. While there is preliminary evidence for this $2 \to 1$ process at RHIC, one needs further detailed studies of two particle correlations before drawing a firm conclusion [12, 13].

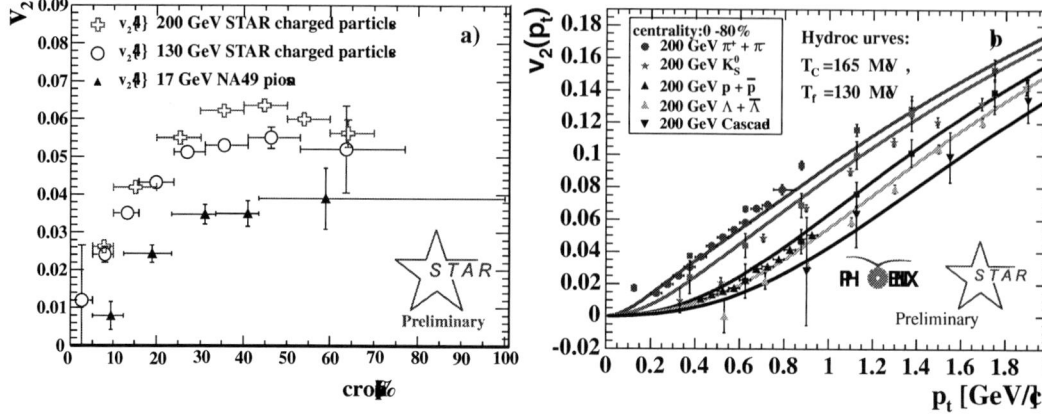

FIGURE 4. Flow at RHIC.

3. HYDRODYNAMICS

While the issue of thermalization in high energy heavy ion collisions needs a better understanding from a theoretical point of view, hydrodynamic approaches to description of soft particle spectra has been quite successful. All hydro models need initial conditions which in principle can be provided by the saturation physics. An essential ingredient is the earliest time at which hydrodynamics can be applied. Hydrodynamics requires local isotropization of the system it is applied to, which in the context of the Quark Gluon Plasma, means that the produced quarks and gluons in a heavy ion collision have come to an (approximate) equilibrium due to multiple elastic and inelastic scatterings. Furthermore, all the hydrodynamics models which are used at RHIC have no viscosity effects included. This would mean that the viscosity of the produced system of quarks and gluons is very low, if non-zero which would imply that we have created a state of strongly interacting matter, more like a liquid than a gas (which was the naive expectation) which would lead to large viscous effects. Sometimes the plasma state created at RHIC is called sQGP in order to signify the nature of the strongly coupled color degrees of freedom.

One of the successes of the hydrodynamics models at RHIC [14] has been understanding flow effect. Starting with an anisotropy in coordinate space, multiple scatterings will always lead to an anisotropy in momentum space. An example is the elliptic flow in non-central heavy ion collisions. At non-zero impact parameter, the overlap region between nuclei at the earliest moments after the collision is asymmetric in $x-y$ plane (where z is the beam axis) and leads to elliptic flow which is measured at RHIC. This large flow (or v_2) effect at RHIC is a clear indication that the system of produced particles does not free flow but undergoes many interactions and flows as a whole, i.e. exhibits large collectivity which was one of the requirements of the system being a QGP. This is shown in Fig. (4).

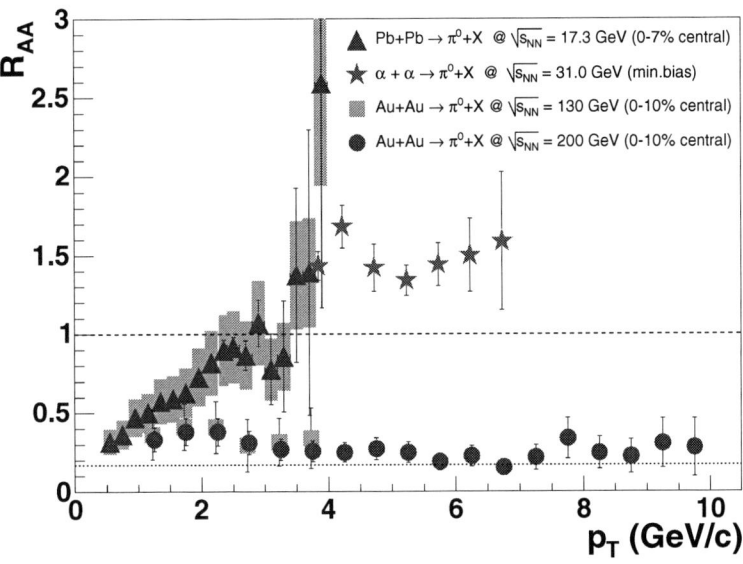

FIGURE 5. Suppression of the neutral pion spectra in gold-gold collisions at RHIC.

4. HARD PROBES

Long before RHIC started operation, it had been realized that a colored parton, a quark or gluon, traversing a colored medium will interact and lose its energy via radiation. This would lead to a depletion of the number of jets at a given transverse momentum. RHIC has measured the spectra of single hadrons at transverse momenta of up to around 20 *Gev*. A strong suppression of the hadron yield in central gold-gold collisions, as compared to proton-proton collisions, in mid rapidity is observed. This suppression factor is of order of 4 − 5 for pions and is shown in Fig. (5) where lower energy data are also shown. While the overall features of the data can be understood by QCD inspired models of energy loss, the quantitative details of the models (for an extensive review of energy loss models, see [15]) are somewhat different. Nevertheless, they all require partonic degrees of freedom and large energy densities which are consistent with the hypothesis of the QGP formed at RHIC. It is interesting to note that the preliminary data at RHIC on the suppression factor for heavy quarks also show a similar pattern which needs to be better understood.

4.1. Back to Back correlations

While one expects a strong suppression of single particle spectra in a heavy ion collision from the Quark Gluon Plasma effects, a depletion of the incoming gluon number density in the incoming nuclei could also lead to a reduced hadron yield. The latter scenario is what one would expect in the saturation physics. In order to clarify

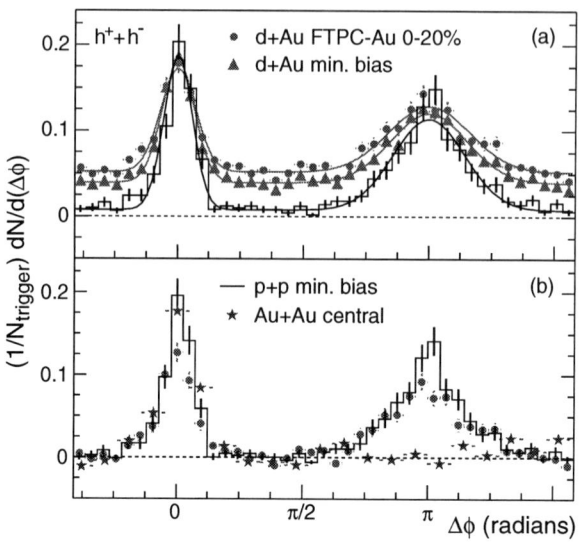

FIGURE 6. Suppression factor for AA and dA at RHIC.

the origin of the suppression, back to back correlations of hadrons were measured at mid rapidity in gold-gold collisions. If the Quark Gluon Plasma, or more precisely, the final state effects are responsible for this suppression, one would not expect to observe a similar suppression in deuteron-gold collisions due to the absence of QGP effects in dA collisions. On the contrary, if the suppression is due to the depletion of parton densities in the initial state of the incoming nuclei due to (leading twist) CGC effects [16], one would expect a suppression of the particle production spectra in dA as well, albeit with a smaller magnitude. The measured spectra are shown in Fig. (6) where, unlike the AA spectra, the dA spectra exhibit an enhancement, known as the Cronin effect which clearly rules out quantum evolution in x as the reason for the suppression factor in gold-gold collisions in the kinematics shown and is a clear indication that the final state, i.e. Quark Gluon Plasma, effects are responsible for the disappearance of the second jet.

5. BARYONS

One of the most interesting and somewhat unexpected results from RHIC is the enhancement of the baryon production spectra as compared to pions. In pQCD, hadronization is described by fragmentation functions of partons into hadrons. While the pion and baryon fragmentation functions are different, this difference is not able to describe the access of baryons (charged hadrons is actually shown) seen in Fig. (7) and there is a clear indication of the role of the final state medium in the process, for example, from the strong centrality dependence of the enhancement factor. One of the more successful phenomenological models which try to explain this effect is based on the idea of valence quark recombination and has been able to describe the baryon enhancement and the flow of baryons albeit at the expense of rigorous theory.

What is clear is that the Quark Gluon Plasma profoundly affects hadronization of partons in a way that can not be described by the standard pQCD methods in terms of fragmentation functions [17]. Further studies are clearly desirable in order to put the recombination models on a more solid theoretical footing. For example, the role played by gluons in recombination models specially since there is clear evidence for importance of gluon production in gold-gold collisions from studies of global aspects of particle production at RHIC. From a more theoretical point of view, assumption on spatial dependence of partons just before recombination and the effect of entropy violation needs to be improved.

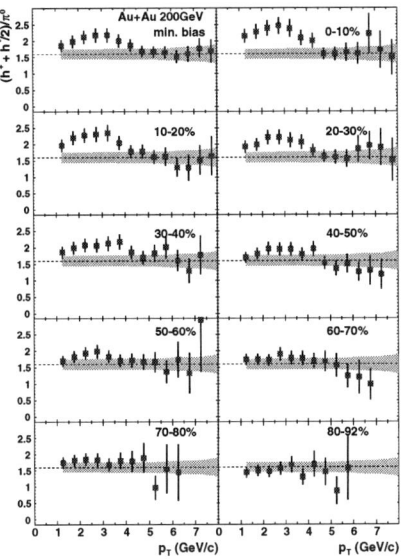

FIGURE 7. Charged hadrons vs. pions.

ACKNOWLEDGMENTS

I would like to thank the organizers of the X Mexican Workshop on Particles and Fields for this very interesting meeting and special thanks to Prof. A. Ayala and Prof. A. Bashir for their invitation to give this talk and their kind hospitality.

REFERENCES

1. R. C. . Hwa and X. N. . Wang, eds. "Quark-gluon plasma. Vol. 3".
2. J. Jalilian-Marian and Y. V. Kovchegov, Prog. Part. Nucl. Phys. **56**, 104 (2006) [arXiv:hep-ph/0505052].
3. E. Iancu and R. Venugopalan, arXiv:hep-ph/0303204.

4. B. B. Back et al. [PHOBOS Collaboration], Phys. Rev. Lett. **85**, 3100 (2000) [arXiv:hep-ex/0007036].
5. A. Krasnitz and R. Venugopalan, Phys. Rev. Lett. **84**, 4309 (2000) [arXiv:hep-ph/9909203].
6. D. Kharzeev, E. Levin and M. Nardi, Phys. Rev. C **71**, 054903 (2005) [arXiv:hep-ph/0111315].
7. D. Kharzeev, Y. V. Kovchegov and K. Tuchin, Phys. Rev. D **68**, 094013 (2003) [arXiv:hep-ph/0307037].
8. J. L. Albacete, N. Armesto, A. Kovner, C. A. Salgado and U. A. Wiedemann, Phys. Rev. Lett. **92**, 082001 (2004) [arXiv:hep-ph/0307179].
9. A. Dumitru and J. Jalilian-Marian, Phys. Rev. Lett. **89**, 022301 (2002) [arXiv:hep-ph/0204028], Phys. Lett. B **547**, 15 (2002) [arXiv:hep-ph/0111357].
10. A. Dumitru, A. Hayashigaki and J. Jalilian-Marian, Nucl. Phys. A **770**, 57 (2006) [arXiv:hep-ph/0512129], Nucl. Phys. A **765**, 464 (2006) [arXiv:hep-ph/0506308].
11. F. Gelis and J. Jalilian-Marian, Phys. Rev. D **67**, 074019 (2003) [arXiv:hep-ph/0211363], Phys. Rev. D **66**, 094014 (2002) [arXiv:hep-ph/0208141], Phys. Rev. D **66**, 014021 (2002) [arXiv:hep-ph/0205037].
12. J. Jalilian-Marian and Y. V. Kovchegov, Phys. Rev. D **70**, 114017 (2004) [Erratum-ibid. D **71**, 079901 (2005)] [arXiv:hep-ph/0405266].
13. R. Baier, A. Kovner, M. Nardi and U. A. Wiedemann, Phys. Rev. D **72**, 094013 (2005) [arXiv:hep-ph/0506126].
14. D. Teaney, Phys. Rev. C **68**, 034913 (2003) [arXiv:nucl-th/0301099].
15. A. Kovner and U. A. Wiedemann, arXiv:hep-ph/0304151.
16. D. Kharzeev, E. Levin and L. McLerran, Nucl. Phys. A **748**, 627 (2005) [arXiv:hep-ph/0403271].
17. R. C. Hwa and C. B. Yang, Phys. Rev. C **70**, 024905 (2004) [arXiv:nucl-th/0401001]; R. J. Fries, B. Muller, C. Nonaka and S. A. Bass, Phys. Rev. Lett. **90**, 202303 (2003) [arXiv:nucl-th/0301087].

Precision Electroweak Physics

Jens Erler

Instituto de Física, Universidad Nacional Autónoma de México, 04510 México D.F., México

Abstract. The status in electroweak precision physics is reviewed. I present a brief summary of the latest data, global fit results, a few implications for new physics, and an outlook.

Keywords: Electroweak interaction; standard model; higgs; strong interaction coupling constant.
PACS: 12.15.-y,12.38.Qk,13.60.-r,13.15.+g,13.66.Jn

OBSERVABLES

Z pole

The Z factories, LEP and SLC, have performed benchmark precision measurements for the electroweak Standard Model (SM) [1]. LEP scanned the Z lineshape yielding the Z boson mass, M_Z, with 2×10^{-5} relative precision, as well as its total width, Γ_Z, and hadronic peak cross section, $\sigma_{had}^0 \equiv 12\pi \Gamma(e^+e^-)\Gamma(had)/M_Z^2 \Gamma_Z^2$, both to better than one per mille accuracy. $\Gamma(\bar{f}f)$ is the Z partial decay width into fermion f and $\Gamma(had)$ is the hadronic Z decay width. Results on the three leptonic ($\ell = e, \mu, \tau$) branching ratios, $R_\ell \equiv \Gamma(had)/\Gamma(\ell^+\ell^-)$, are also at the per mille level. Γ_Z, σ_{had}^0, and the R_ℓ are unique in their sensitivity to the strong coupling constant, α_s, which can be extracted with very small theoretical uncertainty. The SLC was able to compensate its lower luminosity by its electron beam polarization. The left-right polarization asymmetry, A_{LR}, for hadronic final states provides the currently most precise value of the weak mixing angle,

$$\sin^2 \theta_W = \frac{g'^2}{g^2 + g'^2}, \qquad (1)$$

where g and g' are the $SU(2)_L$ and $U(1)_Y$ gauge couplings, respectively. Very high precision could also be achieved in the heavy flavor sector consisting of branching ratios and various asymmetries for b and c quarks. More specifically, the forward-backward (FB) asymmetry into b quarks, A_{FB}^b, amounts to the most precise measurement of $\sin^2 \theta_W$ at LEP, while the few per mille measurement of $R_b = \Gamma(\bar{b}b)/\Gamma(had)$ yields independent information on the top quark mass, m_t, and constraints on new physics affecting the third generation in a non-universal way. The heavy flavor results have been finalized very recently. Analogous results are also available for s quarks, albeit with larger uncertainties. Other Z pole observables include the three leptonic FB asymmetries, A_{FB}^ℓ, the final state τ polarization and its FB asymmetry, and charge asymmetry measurements.

TABLE 1. Z pole observables compared with the SM best fit predictions. $\bar{s}_\ell^2$ is an effective mixing angle which absorbs all radiative corrections and corresponds most closely to what enters the Z pole asymmetries. The first is extracted from the hadronic charge asymmetry, a weighted average over light-quark FB asymmetries. The second is from the final state electron FB asymmetry, A_{FB}, from CDF [2]. The three values of A_e are (i) from A_{LR} for hadronic final states [3]; (ii) from A_{LR} for leptonic final states and from polarized Bhabba scattering [4]; and (iii) from the angular distribution of the τ polarization (LEP) [1]. The two A_τ values are from SLD and the total τ polarization, respectively. The uncertainties in the SM predictions are induced by the errors in the SM parameters, and their correlations have been accounted for.

observable	experimental value	SM prediction	pull
M_Z [GeV]	91.1876 ± 0.0021	91.1874 ± 0.0021	0.1
Γ_Z [GeV]	2.4952 ± 0.0023	2.4968 ± 0.0011	-0.7
σ_{had}^0 [nb]	41.541 ± 0.037	41.467 ± 0.009	2.0
R_e	20.804 ± 0.050	20.756 ± 0.011	1.0
R_μ	20.785 ± 0.033	20.756 ± 0.011	0.9
R_τ	20.764 ± 0.045	20.801 ± 0.011	-0.8
R_b	0.21629 ± 0.00066	0.21578 ± 0.00010	0.8
R_c	0.1721 ± 0.0030	0.17230 ± 0.00004	-0.1
A_{FB}^e	0.0145 ± 0.0025	0.01622 ± 0.00025	-0.7
A_{FB}^μ	0.0169 ± 0.0013		0.5
A_{FB}^τ	0.0188 ± 0.0017		1.5
A_{FB}^b	0.0992 ± 0.0016	0.1031 ± 0.0008	-2.4
A_{FB}^c	0.0707 ± 0.0035	0.0737 ± 0.0006	-0.8
A_{FB}^s	0.0976 ± 0.0114	0.1032 ± 0.0008	-0.5
$\bar{s}_\ell^2$	0.2324 ± 0.0012	0.23152 ± 0.00014	0.7
	0.2238 ± 0.0050		-1.5
A_e	0.15138 ± 0.00216	0.1471 ± 0.0011	2.0
	0.1544 ± 0.0060		1.2
	0.1498 ± 0.0049		0.6
A_μ	0.142 ± 0.015		-0.3
A_τ	0.136 ± 0.015		-0.7
	0.1439 ± 0.0043		-0.7
A_b	0.923 ± 0.020	0.9347 ± 0.0001	-0.6
A_c	0.670 ± 0.027	0.6678 ± 0.0005	0.1
A_s	0.895 ± 0.091	0.9356 ± 0.0001	-0.4

These Z pole measurements are summarized in Table 1. Some results are quoted in terms of asymmetry parameters,

$$A_f \equiv \frac{2v_f a_f}{v_f^2 + a_f^2}, \qquad (2)$$

where at tree level the $Zf\bar{f}$ vector (axial-vector) coupling is given by $v_f = T_3^f - 2Q^f \sin^2\theta_W$ ($a_f = T_3^f$), and where Q^f and T^f denote, respectively, the electric charge and third component of weak isospin of fermion f. The FB asymmetries can also be expressed in terms of these, $A_{FB}^f = 3/4 A_e A_f$.

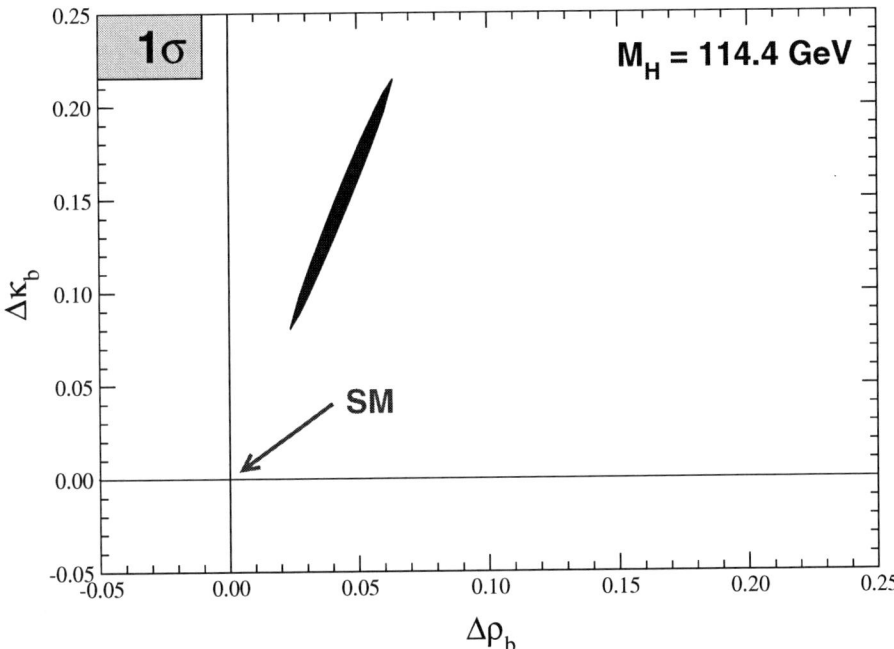

FIGURE 1. New Physics contributions to form factors for the $Zb\bar{b}$ vector and axial-vector couplings.

The pull of an observable gives its deviation from the SM and is defined as,

$$\text{pull}(O_i) = \frac{O_i^{\text{exp.}} - O_i^{\text{SM}}}{\Delta O_i^{\text{total}}}, \tag{3}$$

where $O_i^{\text{exp.}}$ is the experimental central value of observable, O_i, O_i^{SM} is the central value of its SM prediction, and $\Delta O_i^{\text{total}}$ is the sum in quadrature of the contributing statistical, systematical, and theoretical uncertainties, but excludes the parametric uncertainty in the SM prediction. As can be seen from Table 1, there are only three Z pole observables which deviate by two standard deviations (σ) or more, but interestingly these are all among the most precise. In particular, A_{LR} and A_{FB}^b provide valuable information on the mass of the Higgs boson, M_H, and deviate by 3.1 σ from each other. Since A_{FB}^b involves b quarks, and the third fermion generation is often suspected to be affected differently by physics beyond the SM, one can interpret it alternatively as a measurement of $Zb\bar{b}$ couplings. Defining,

$$\begin{aligned} v_b &= (1+\Delta\hat{\rho}_b+\Delta\rho_b)[T_3^f - 2Q^f(1+\Delta\hat{\kappa}_b+\Delta\kappa_b)\sin^2\theta_W], \\ a_b &= (1+\Delta\hat{\rho}_b+\Delta\rho_b)T_3^f, \end{aligned} \tag{4}$$

one can fit to $\Delta\rho_b$ and $\Delta\kappa_b$, which are due to new physics only when all SM contributions are subsumed in $\Delta\hat{\rho}_b$ and $\Delta\hat{\kappa}_b$. R_b and A_b provide additional constraints. The result is shown in Fig. 1, from where it becomes clear that a correction of 10–20% to $\Delta\kappa_b$ would be necessary to account for the data. This would be a very large radiative correction, given that the quadratically enhanced top quark contribution in the SM is less than 1%. It is thus very unlikely that the deviation in A_{FB}^b is due to a loop effect, but it is conceivably of tree-level type affecting preferentially the third generation. Examples include the decay of a scalar neutrino resonance [5], mixing of the b quark with heavy exotics [6], and a heavy Z' with family non-universal couplings [7]. It is difficult, however, to simultaneously account for R_b, which has been measured on the Z peak and off-peak [1] at LEP 1. In this context it is interesting that an average of R_b measurements at LEP 2 at energies between 133 and 207 GeV is 2.1 σ below the SM prediction, and A_{FB}^b is 1.6 σ low [8].

The measurement of σ_{had}^0 is 2 σ higher than the SM prediction. As a consequence, when one fits to the number, N_ν, of active neutrinos[1] one obtains a 2 σ deficit, $N_\nu = 2.986 \pm 0.007$, compared to the SM prediction, $N_\nu = 3$. Amusingly, LEP 2 [8] also sees a 1.7 σ excess in the averaged hadronic cross section.

Other data

Table 2 lists non-Z pole observables. Precise values for the W boson mass, M_W, have been obtained at the high energy frontier at LEP 2 [8] (e^+e^-) and the Tevatron [10, 11] ($p\bar{p}$). The world average, $M_W = 80.410 \pm 0.032$ GeV, has reached a relative precision of 4×10^{-4} with further improvements expected in the near future. The direct measurements [23] of $m_t = 172.7 \pm 2.9 \pm 0.6$ GeV from the Tevatron[2] can be compared to an indirect determination, $m_t = 172.3^{+10.2}_{-7.6}$ GeV, from the other precision data. The agreement is spectacular. As shown in Fig. 2, this comparison can even be carried out for the two parameters, m_t and M_W, simultaneously. In the indirect determination, m_t is mostly constrained by R_b, but Γ_Z and low energy measurements also contribute significantly. M_W is then mostly implied by the asymmetries. The agreement is again remarkable and it should be stressed that there are now two theoretically and experimentally independent indications for a relatively light Higgs boson with a mass of $\mathcal{O}(100\text{ GeV})$. The implications of various sets of observables for M_H and m_t are shown in Fig. 3.

Other important measurements are from comparatively lower energies or momentum transfers [24]. The most precise are determinations of anomalous magnetic moments in leptons, a_ℓ. The measurement [25] of a_μ stands out because of its unique sensitivity to high energy scales. If the new physics [26] couples, respectively, through tree or one-loop effects, a simple dimensional estimate of the scales that can be probed by a_μ at the 1 σ level (Δa_μ denotes the total error) gives,

$$\Lambda_{new} \sim \frac{m_\mu}{\sqrt{\Delta a_\mu}} \sim 3.7 \text{ TeV}, \qquad \frac{\Lambda_{new}}{g} \sim \frac{1}{2\pi}\frac{m_\mu}{\sqrt{\Delta a_\mu}} \sim 590 \text{ GeV}. \qquad (5)$$

---

[1] By definition, an active neutrino is one that couples to the Z boson like a standard neutrino.
[2] The first error is experimental [23] and the second is theoretical from the conversion from the top pole mass to the $\overline{\text{MS}}$ mass, the quantity which actually enters the radiative corrections.

TABLE 2. Non-Z pole observables compared with the SM best fit predictions. The first M_W value is from UA2 [9], CDF [10], and DØ [11], and the second is from LEP 2 [8]. g_L^2 and g_R^2 (see text) are from NuTeV [12], while the older neutrino deep-inelastic scattering (ν-DIS) results from CDHS [13], CHARM [14], and CCFR [15] are included in the fits, but not shown. $g_{V,A}^{\nu e}$ are world averaged effective four-Fermi couplings in νe scattering and dominated by the CHARM II results [16]. A_{PV} is the parity violating asymmetry in Møller scattering [17]. Q_W(Cs) [18, 19] and Q_W(Tl) [20, 21] are the so-called weak charges of Cs and Tl and have been determined in atomic parity violation (APV) experiments. The APV errors shown contain significant theory uncertainties from atomic structure calculations [22]. In the case of τ_τ (see text) the theory uncertainty is included in the SM prediction. In all other SM predictions, the uncertainty is from the SM parameters.

observable	experimental value	SM prediction	pull
m_t [GeV]	172.7 ± 3.0	172.7 ± 2.8	0.0
M_W [GeV]	80.450 ± 0.058	80.376 ± 0.017	1.3
	80.392 ± 0.039		0.4
g_L^2	0.30005 ± 0.00137	0.30378 ± 0.00021	-2.7
g_R^2	0.03076 ± 0.00110	0.03006 ± 0.00003	0.6
$g_V^{\nu e}$	-0.040 ± 0.015	-0.0396 ± 0.0003	0.0
$g_A^{\nu e}$	-0.507 ± 0.014	-0.5064 ± 0.0001	0.0
A_{PV}	-1.31 ± 0.17	-1.53 ± 0.02	1.3
Q_W(Cs)	-72.62 ± 0.46	-73.17 ± 0.03	1.2
Q_W(Tl)	-116.6 ± 3.7	-116.78 ± 0.05	0.1
$a_\mu - \frac{\alpha}{2\pi}$	4511.07 ± 0.82	4509.82 ± 0.10	1.5
τ_τ [fs]	290.89 ± 0.58	291.87 ± 1.76	-0.4

The interpretation of a_μ is complicated by hadronic contributions which first arise at the two-loop level. One can use experimental $e^+e^- \to$ hadrons cross section data to estimate [27] the two-loop effect, which is due to a vacuum polarization (VP) insertion into a one-loop graph, $a_\mu^{(2,VP)} = (69.54 \pm 0.64) \times 10^{-9}$. This value suggests a 2.3 σ discrepancy between the SM and experiment. If one assumes isospin symmetry (which is not exact and appropriate corrections [28] have to be applied) one can also make use of τ decay spectral functions [29] and one obtains [30], $a_\mu^{(2,VP)} = (71.10 \pm 0.58) \times 10^{-9}$. This result implies no conflict (0.7 σ) between data and prediction. It is important to understand the origin of this difference, but the following observations point to the conclusion that at least some of it is experimental: (i) The latest e^+e^- data by the SND Collaboration [31] are consistent with the implications of the τ decay data, and in conflict with other e^+e^- data. (ii) The $\tau^- \to \nu_\tau 2\pi^-\pi^+\pi^0$ spectral function disagrees with the corresponding e^+e^- data at the 4 σ level, which translates to a 23% effect [27] and seems too large to arise from isospin violation. (iii) Isospin violating corrections have been studied in detail in Ref. [28] and found to be largely under control. The largest effect is due to higher-order electroweak corrections [32] but introduces a negligible uncertainty [33]. (iv) Ref. [34] shows on the basis of a QCD sum rule that the spectral functions derived from τ decay data are consistent with values of $\alpha_s(M_Z) \gtrsim 0.120$ (in agreement with the global fit result described in the next section), while the spectral

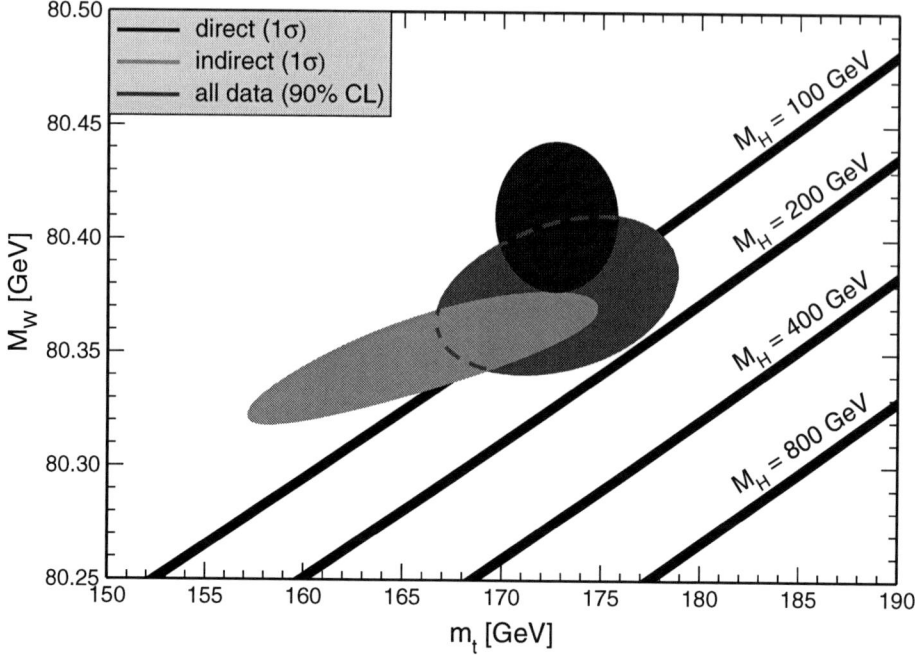

FIGURE 2. One-standard-deviation (39.35%) region in M_W as a function of m_t for the direct and indirect data, and the 90% CL region ($\Delta\chi^2 = 4.605$) allowed by all data. The SM prediction as a function of M_H is also indicated. The width of the M_H bands reflects the theoretical uncertainty in the prediction.

functions from e^+e^- annihilation are consistent only for somewhat lower (disfavored) values. In any case, due to the suppression at large momentum transfer (from where the conflicts originate) these problems are less pronounced as far as $a_\mu^{(2,\mathrm{VP})}$ is concerned, so that it seems justified to view these differences as fluctuations and to average the results. An additional uncertainty is induced by the hadronic three-loop light-by-light scattering contribution. For this the most recent value, $a_\mu^{\mathrm{LBLS}} = (1.36 \pm 0.25) \times 10^{-9}$, of Ref. [35] is employed, which is higher than previous evaluations [36, 37].

The τ is the only lepton which can decay hadronically, offering a luxurious arena to study the strong interaction and to extract α_s. Its mass, m_τ, is large enough that the operator product expansion, OPE (QCD perturbation theory plus almost negligible power corrections in an expansion in the inverse τ mass), can be applied, yet small enough that QCD effects are sizable with great sensitivity to α_s. Upon renormalization group evolution from m_τ to M_Z (where α_s can be compared to the values from Γ_Z, σ_{had}^0, and R_ℓ), the uncertainty scales roughly[3] like $\alpha_s(M_Z)^2/\alpha_s(m_\tau)^2 \sim 0.12$. Furthermore, because the

---

[3] This order of magnitude decrease is sometimes called the "incredibly shrinking error".

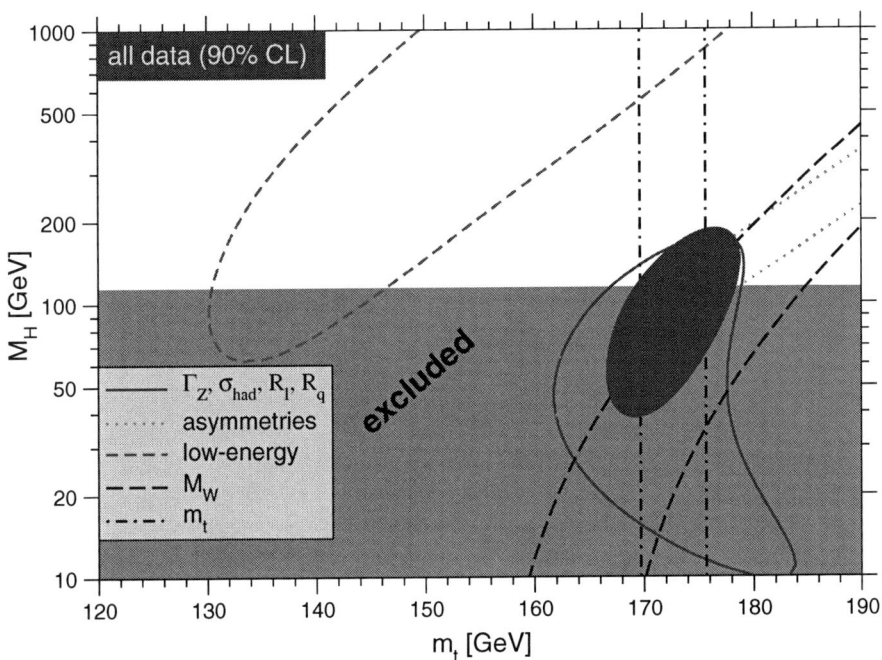

FIGURE 3. One-standard-deviation (39.35%) uncertainties in M_H as a function of m_t for various inputs, and the 90% CL region allowed by all data. $\alpha_s(M_Z) = 0.120$ is assumed except for the fits including the Z lineshape data. The 95% direct lower limit from LEP 2 is also shown.

τ lifetime, τ_τ, and leptonic branching ratios[4] are fully inclusive, there are no uncertainties from hadronization, fragmentation, parton distribution functions, or other modeling of the strong interaction. The only potential theoretical uncertainties are from the truncation of the perturbative series and from non-perturbative OPE breaking effects. The perturbative series is known up to $\mathcal{O}(\alpha_s^3)$ (the same order as the QCD correction to Γ(had)) and should therefore not be combined with only next-to-leading order determinations of α_s. The coefficients that enter the α_s expansion of τ_τ are relatively large, but dominated by terms that arise from analytical continuation and are thus proportional to QCD β-function coefficients. Since the latter are known to $\mathcal{O}(\alpha_s^4)$ and first enter at $\mathcal{O}(\alpha_s^2)$, it is advantageous to treat these effects separately and re-sum them to all orders. This amounts to a re-organization of the perturbative series (also referred to as "contour improvement") with smaller expansion coefficients[5] and where α_s^n is replaced by more complicated functions, $A_n(\alpha_s)$. The dominant uncertainty is from the lack of knowledge of the four-loop coefficient, d_3. One is still exposed to OPE breaking non-perturbative

---

[4] The τ lifetime world average in Table 2 is computed by combining the direct measurements with values derived from the leptonic branching ratios.
[5] These coefficients are given by those of the Adler D-function, d_i.

effects because at one kinematic point one needs to change from quark degrees of freedom (QCD) to hadrons (data), but fortunately this point is suppressed by a double zero. Very precise data on τ spectral functions (mainly from ALEPH [29]) constrain such effects to a sub-dominant level.

Currently the largest discrepancy is from ν-DIS scattering. The NuTeV Collaboration finds for the on-shell definition of the weak mixing angle, $s_W^2 = 0.2277 \pm 0.0016$, which is 3.0 σ higher than the SM prediction. The discrepancy is in the left-handed effective four-Fermi coupling, $g_L^2 = 0.3000 \pm 0.0014$, which is 2.7 σ low, while $g_R^2 = 0.0308 \pm 0.0011$ is 0.6 σ high. At tree level, these are given by,

$$g_L^2 \approx \frac{1}{2} - \sin^2\theta_W + \frac{5}{9}\sin^4\theta_W, \qquad g_R^2 \approx \frac{5}{9}\sin^4\theta_W. \tag{6}$$

Within the SM, one can identify five categories of effects that could cause or contribute to this effect [38]: (i) an asymmetric strange quark sea, although this possibility is constrained by dimuon data [39]; (ii) isospin symmetry violating parton distribution functions at levels much stronger than generally expected [40]; (iii) nuclear physics effects [41, 42]; (iv) QED and electroweak radiative corrections [43, 44]; and (v) QCD corrections to the structure functions [45]. The NuTeV result and the other ν-DIS data should therefore be considered as preliminary until a re-analysis using PDFs including all experimental and theoretical information has been completed. It is well conceivable that various effects add up to bring the NuTeV result in line with the SM prediction. It is likely that the overall uncertainties in g_L^2 and g_R^2 will increase, but at the same time the older ν-DIS results may become more precise when analyzed with better PDFs than were available at the time.

GLOBAL FIT

With these inputs a simultaneous fit to various SM parameters can be performed,

$$\begin{aligned}
M_Z &= 91.1874 \pm 0.0021 \text{ GeV}, \\
M_H &= 89^{+38}_{-28} \text{ GeV}, \\
m_t &= 172.7 \pm 2.8 \text{ GeV}, \\
\alpha_s(M_Z) &= 0.1216 \pm 0.0017, \\
\hat{\alpha}(M_Z)^{-1} &= 127.904 \pm 0.019, \\
\sin^2\hat{\theta}_W &= 0.23122 \pm 0.00015, \\
s_W^2 \equiv 1 - \frac{M_W^2}{M_Z^2} &= 0.22306 \pm 0.00033,
\end{aligned} \tag{7}$$

where the last two lines show the weak mixing angle in the $\overline{\text{MS}}$-scheme (coupling based) and the on-shell scheme (vector meson mass based), respectively. $\hat{\alpha}(M_Z)$ is the $\overline{\text{MS}}$ electromagnetic coupling as it enters at the Z pole. Despite the small discrepancies discussed in the previous section, the goodness of the fit to all data is very good with a $\chi^2/\text{d.o.f.} = 47.5/42$. The probability of a larger χ^2 is 26%. Experimental correlations have been taken into account. Theoretical correlations, e.g. between $\hat{\alpha}(M_Z)$ and $g_\mu - 2$

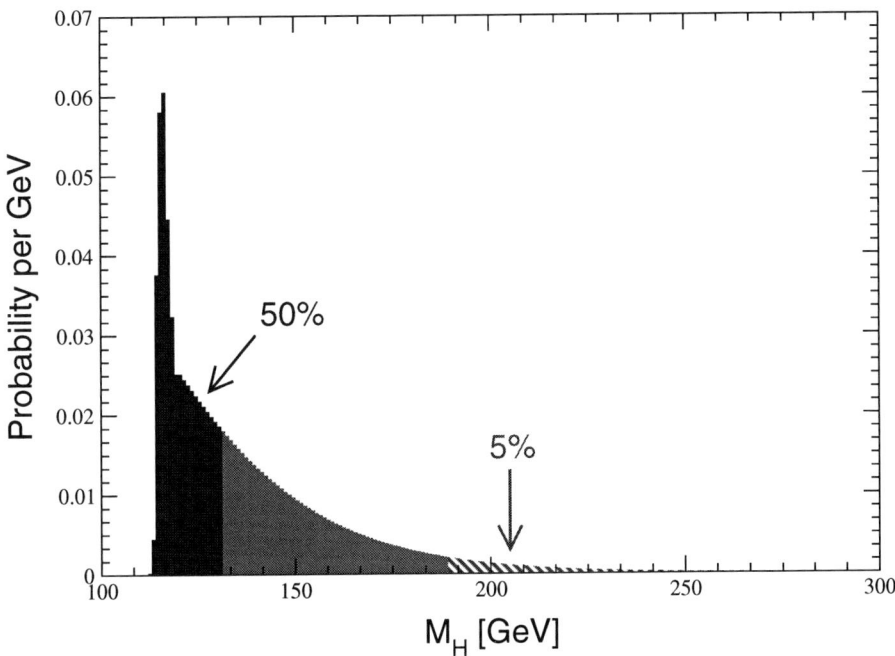

FIGURE 4. Probability distribution function of M_H including direct search results.

are also addressed[6]. The measurement of the latter is higher than the SM prediction, and its inclusion in the fits favors a larger $\hat{\alpha}(M_Z)$ and a lower M_H by about 3 GeV.

The extracted Z pole value of $\alpha_s(M_Z)$ is based on a formula with almost negligible theoretical uncertainty (± 0.0005 in $\alpha_s(M_Z)$) if one assumes the exact validity of the SM. One should keep in mind, however, that this value[7], $\alpha_s = 0.1198 \pm 0.0028$, is very sensitive to such types of new physics as non-universal vertex corrections. In contrast, the value derived from τ decays, $\alpha_s(M_Z) = 0.1225^{+0.0025}_{-0.0022}$, is theory dominated but less sensitive to new physics. The two values are in remarkable agreement with each other. They are also in good agreement with other recent values, such as from a 4-jet analysis at OPAL [46] (0.1182 ± 0.0026) and from jet production at HERA [47] (0.1186 ± 0.0051), but the τ decay result is somewhat higher than the value, 0.1170 ± 0.0012, from the most recent unquenched lattice calculation [48].

There is a strong correlation between the quadratic m_t and logarithmic M_H terms in the radiative corrections except for the $Z \to b\bar{b}$-vertex. M_W has additional M_H dependence

---

[6] This is due to the common use of the experimental $e^+e^- \to$ hadrons cross section and τ decay data. The weak mixing angle for momentum transfers at or below hadronic scales and various hadronic three-loop contributions to a_μ need these inputs, as well, implying additional correlations.

[7] If one adds non-Z pole observables (other than τ_τ or τ leptonic branching ratios), one obtains the slightly higher value, $\alpha_s = 0.1202 \pm 0.0027$.

which is not coupled to m_t^2 effects. The strongest individual pulls toward smaller M_H are from M_W and A_{LR} (SLD), while A_{FB}^b and the NuTeV results favor higher values. The difference in χ^2 for the global fit is, $\Delta\chi^2 = \chi^2(M_H = 1000 \text{ GeV}) - \chi^2_{\min} = 60$. Hence, the data clearly favor a small value of M_H, as in supersymmetric extensions of the SM. The 90% central confidence range from all precision data is 46 GeV $\leq M_H \leq$ 154 GeV. The central value of the global fit result, $M_H = 89^{+38}_{-28}$ GeV, is below the direct LEP 2 lower bound, $M_H \geq 114.4$ GeV (95% CL) [49]. Including the results of these direct searches as an extra contribution to the likelihood function drives the 95% upper limit to $M_H \leq$ 189 GeV. As two further refinements, the theoretical uncertainties from uncalculated higher order contributions and the M_H-dependence of the correlation matrix which gives slightly more weight to lower Higgs masses [50] are accounted for. The resulting limits at 95 (90, 99)% CL are

$$M_H \leq 194 \ (176, 235) \text{ GeV}, \tag{8}$$

respectively. The probability distribution function of M_H is shown in Fig. 4.

NEW PHYSICS AND OUTLOOK

The good agreement between SM predictions and experiments implies strong constraints on new physics scenarios beyond the SM. The Z pole measurements are particularly suitable to study possible new physics effects on the Z couplings to quarks and leptons. The per mille precision which has been achieved at LEP and SLC allows, for example, only very small mixing between the Z and a hypothetical extra Z' boson [51]. On the other hand, a Z' with no or little mixing, or other types of new physics contributing to e^+e^- amplitudes without affecting Z boson properties, could have easily gone unnoticed, since such effects may hide under the Z resonance. It is then expedient to examine precision observables away from the Z pole. High quality and high energy data are provided by LEP 2 [8], although these come with comparatively low rates. An interesting alternative is to utilize low energy observables probing directly the weak interaction. This includes processes which exploit the parity violating character of the weak interaction, as well as neutrino scattering.

For example, the E158 Collaboration at SLAC [17] has extracted the weak charge of the electron, $Q_W(e)$, from the parity violating asymmetry, A_{PV}, in polarized electron scattering, $\vec{e}e$. A 13% error in the Møller asymmetry suffices to access the TeV scale. A similar experiment, Qweak at JLab [52], will determine the analogous proton weak charge, $Q_W(p)$, in $\vec{e}p$-scattering. The new physics scales probed by $Q_W(e)$ [53] and $Q_W(p)$ [54] reach (at the 1 σ level),

$$\frac{\Lambda_{\text{NEW}}}{g} \approx \frac{1}{\sqrt{\sqrt{2}G_F |\Delta Q_W|}} \approx \begin{cases} 4.6 \text{ TeV} \ [Q_W(p)], \\ 3.2 \text{ TeV} \ [Q_W(e)], \end{cases} \tag{9}$$

where g is the coupling strength of the new physics, G_F is the Fermi constant, and $|\Delta Q_W|$ is the total uncertainty (a 4% determination of $Q_W(p)$ is assumed). The reason for the high reach in these experiments is a suppression of the tree-level SM contribution which is proportional to $1 - 4\sin^2\theta_W$. The numerical value of $\sin^2\theta_W$ which enters at very low

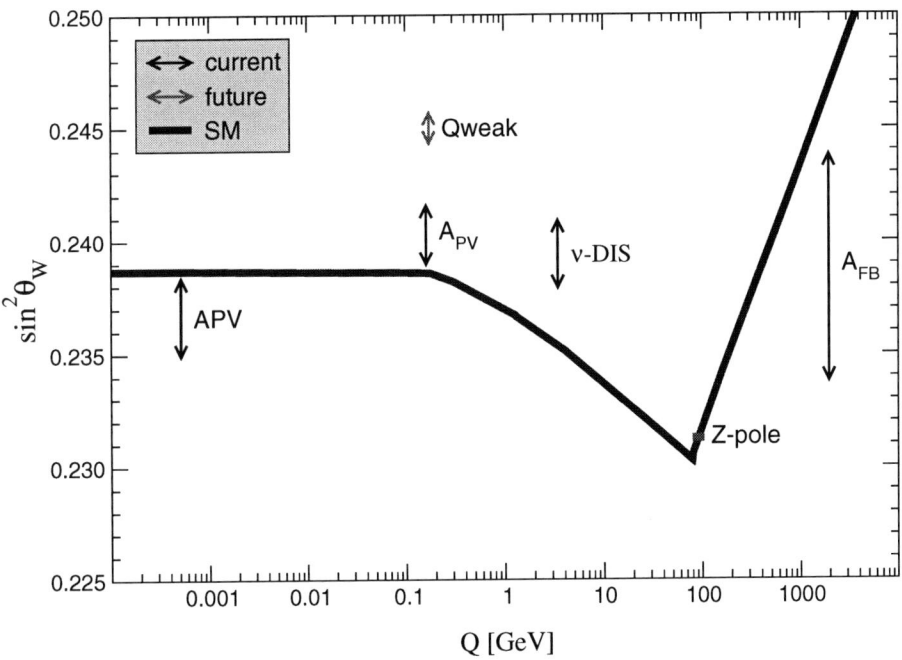

FIGURE 5. The weak mixing angle in the $\overline{\text{MS}}$-scheme as a function of energy, $\sqrt{Q^2}$. The width of the line indicates the uncertainty in the SM prediction. The mixing angle can be determined from a variety of neutral-current processes spanning a very wide Q^2 range. The largest discrepancy is the measurement from ν-DIS which is 2.7 σ above the prediction. This is mostly due to the NuTeV result [12]. The figure is updated from Ref. [55].

momentum transfer ($Q^2 \approx 0.03$ GeV^2 in these experiments) is even closer to the ideal value of 1/4 than the one entering Z pole physics ($Q^2 = M_Z^2$). This is illustrated in Fig. 5.

This kind of low energy, very high statistics measurement may even compete with the Z factories. For example, a factor of 4 improvement in A_{PV} relative to the E158 result (which can conceivably be achieved at an upgraded 12 GeV CEBAF at JLab) would yield a measurement of the low energy mixing angle to about ±0.00035. Fig. 6 shows a breakdown of our current knowledge of the weak mixing angle. A possible projection into the intermediate future is also shown, where a 3.25% A_{PV} and a 4% $Q_W(p)$ measurement are assumed, along with some expected improvements [56] at the Tevatron Run IIA (corresponding approximately to an accumulated luminosity of 2 fb^{-1} of data). It is entertaining to also display what such an outcome would mean for the various laboratories. Fig. 7 shows that JLab with a dedicated asymmetry physics program could contribute almost as much to $\sin^2 \theta_W$ as the high-energy laboratories, SLAC and FNAL.

One can also consider the general effects on neutral-current and Z and W boson observables of various types of heavy new physics which contribute to the W and Z self-energies but which do not have any (or only small) direct coupling to the ordi-

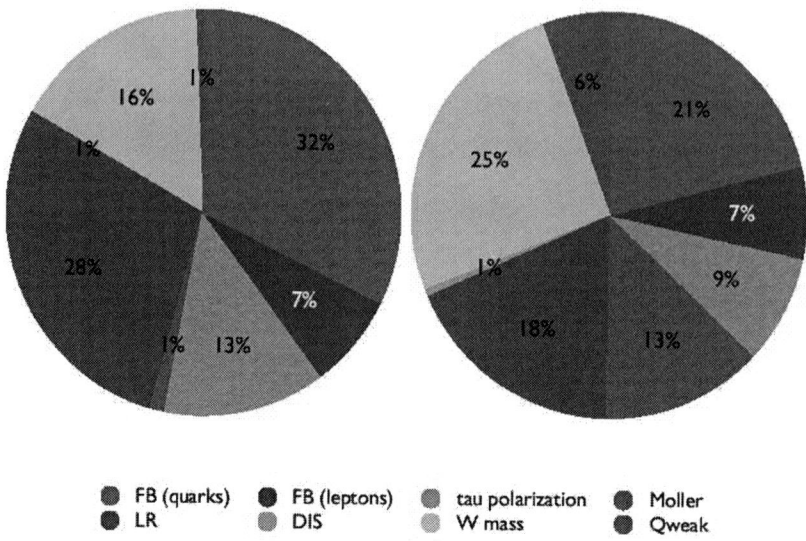

FIGURE 6. Precision weighted contributors to our knowledge of the weak mixing angle, $\sin^2\theta_W$, by type of observable. The left-hand side is the status, while the right-hand side is a projection into the intermediate future, assuming 3.5% A_{PV} and 4% $Q_W(p)$ determinations, as well as 2 fb^{-1} of Tevatron Run IIA data. The W mass can be regarded as a measurement of the on-shell mixing angle, s_W^2. Qweak refers to the weak charges of the proton (Qweak) and heavy nuclei (APV).

nary fermions. In addition to non-degenerate multiplets, which break the vector part of $SU(2)_L$, these include heavy degenerate multiplets of chiral fermions which break the axial generators. Such effects can be described by just three parameters, S, T, and U [57]. T is equivalent to the electroweak ρ-parameter [58] and proportional to the difference between the W and Z self-energies at $Q^2 = 0$ (vector $SU(2)_L$-breaking). S and $S+U$ are associated, respectively, with the difference between the Z and W self-energies at $Q^2 = M_{Z,W}^2$ and $Q^2 = 0$ (axial $SU(2)_L$-breaking). S, T, and U are defined with a factor proportional to $\hat{\alpha}$ removed, so that they are expected to be of order unity in the presence of new physics. A heavy non-degenerate multiplet of fermions or scalars contributes positively to T, while a multiplet of heavy degenerate chiral fermions increases S. For example, a heavy degenerate ordinary or mirror family would contribute $2/(3\pi)$ to S.

The data allow a simultaneous determination of S, T, U, and all SM parameters except for M_H,

$$\begin{aligned} S &= -0.13 \pm 0.10\ (-0.08), \\ T &= -0.13 \pm 0.11\ (+0.09), \\ U &= 0.20 \pm 0.12\ (+0.01), \\ \alpha_s(M_Z) &= 0.1223 \pm 0.0018, \end{aligned} \qquad (10)$$

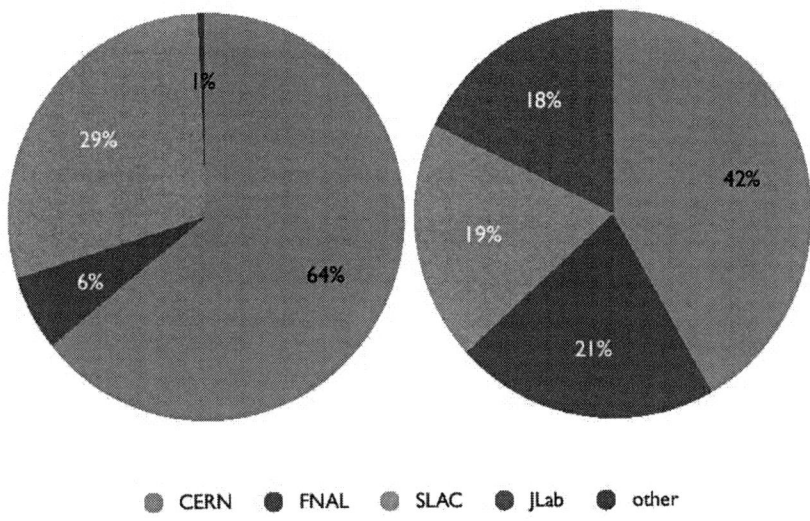

FIGURE 7. Contributors to our knowledge of the weak mixing angle, $\sin^2\theta_W$, by laboratory. The left-hand side is the status, while the right-hand side is the projection using the same assumptions as in Fig 6.

where the uncertainties are from the inputs. The central values assume $M_H = 117$ GeV, and in parentheses the change for $M_H = 300$ GeV is shown. As can be seen, α_s (U) can be determined with no (little) M_H dependence. On the other hand, S, T, and M_H cannot be obtained simultaneously, because the Higgs boson loops themselves are resembled approximately by oblique effects. Eqs. (10) show that negative (positive) contributions to the S (T) parameter can weaken or entirely remove the strong constraints on M_H from the SM fits. The parameters in Eqs. (10), which by definition are due to new physics only, all deviate by more than one standard deviation from the SM values of zero. However, these deviations are correlated. Fixing $U = 0$ (as is done in Fig. 8) will also move S and T to values compatible with zero within errors. Note the strong correlation (84%) between the S and T parameters.

An extra generation of ordinary fermions is excluded at the 99.999% CL on the basis of the S parameter alone, corresponding to $N_F = 2.81 \pm 0.24$ for the number of families. This result assumes that there are no new contributions to T or U and therefore that any new families are degenerate. In principle this restriction can be relaxed by allowing T to vary as well, since $T > 0$ is expected from a non-degenerate extra family. However, the data currently favor $T < 0$, thus strengthening the exclusion limits. A more detailed analysis is required if the extra neutrino (or the extra down-type quark) is close to its direct mass limit [59, 60]. This can drive S to small or even negative values but at the expense of too-large contributions to T.

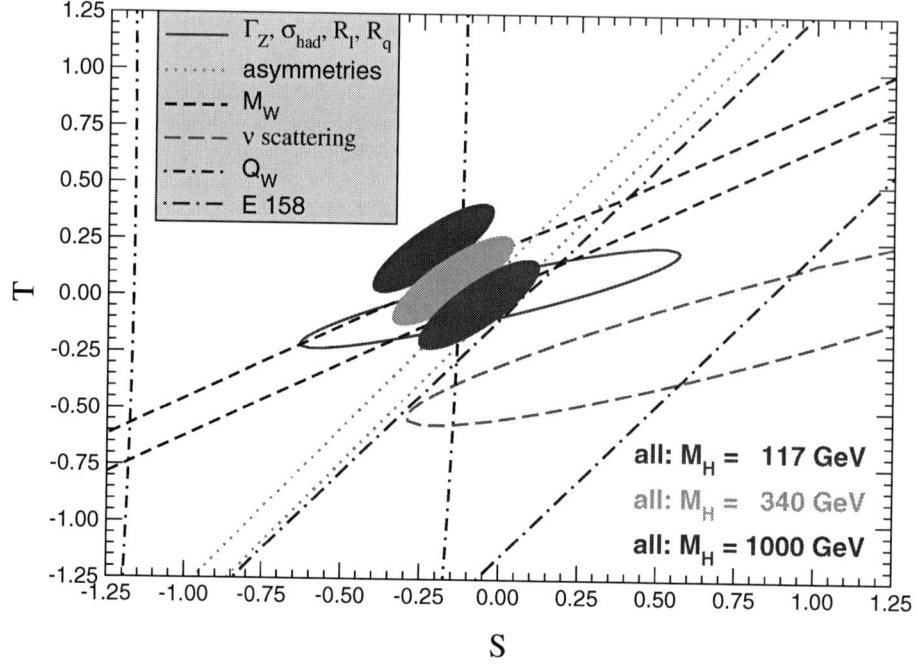

FIGURE 8. 1 σ constraints on S and T from various inputs combined with M_Z. S and T represent the contributions of new physics only. The contours assume $M_H = 117$ GeV except for the central and upper 90% CL contours allowed by all data which are for $M_H = 340$ GeV and 1000 GeV, respectively. α_s is constrained using the τ lifetime as additional input in all fits.

ACKNOWLEDGMENTS

It is a pleasure to thank Paul Langacker and Michael Ramsey-Musolf for fruitful collaborations. This work was supported by CONACyT (México) contract 42026–F and by DGAPA–UNAM contract PAPIIT IN112902.

REFERENCES

1. ALEPH, DELPHI, L3, OPAL, and SLD Collaborations, LEP and SLD Electroweak Working Groups and SLD Heavy Flavour Group: S. Schael *et al.*, eprint `hep-ex/0509008`.
2. CDF Collaboration: D. Acosta *et al.*, *Phys. Rev.* **D71**, 052002 (2005).
3. SLD Collaboration; K. Abe *et al.*, *Phys. Rev. Lett.* **84**, 5945–5949 (2000).
4. SLD Collaboration; K. Abe *et al.*, *Phys. Rev. Lett.* **86**, 1162–1166 (2001).
5. J. Erler, J. L. Feng and N. Polonsky, *Phys. Rev. Lett.* **78**, 3063–3066 (1997).
6. D. Choudhury, T. M. P. Tait and C. E. M. Wagner, *Phys. Rev.* **D65**, 053002 (2002).
7. J. Erler and P. Langacker, *Phys. Rev. Lett.* **84**, 212–215 (2000).
8. ALEPH, DELPHI, L3, and OPAL Collaborations and LEP Electroweak Working Group: J. Alcaraz *et al.*, eprint `hep-ex/0511027`.
9. UA2 Collaboration: J. Alitti *et al.*, *Phys. Lett.* **B276**, 354–364 (1992).
10. CDF Collaboration: T. Affolder *et al.*, *Phys. Rev.* **D64**, 052001 (2001).

11. DØ Collaboration: V. M. Abazov et al., *Phys. Rev.* **D66**, 012001 (2002).
12. NuTeV Collaboration: G. P. Zeller et al., *Phys. Rev. Lett.* **88**, 091802 (2002).
13. CDHS Collaboration: H. Abramowicz et al., *Phys. Rev. Lett.* **57**, 298–301 (1986).
14. CHARM Collaboration: J. V. Allaby et al. *Phys. Lett.* **B177**, 446–452 (1986).
15. CCFR Collaboration: C. Arroyo et al., *Phys. Rev. Lett.* **72**, 3452–3455 (1994).
16. CHARM-II Collaboration: P. Vilain et al., *Phys. Lett.* **B335**, 246–252 (1994).
17. SLAC E158 Collaboration: P. L. Anthony et al., *Phys. Rev. Lett.* **95**, 081601 (2005).
18. C. S. Wood et al., *Science* **275**, 1759–1763 (1997).
19. J. Guena, M. Lintz and M. A. Bouchiat, eprint `physics/0412017`.
20. N. H. Edwards, S. J. Phipp, P. E. G. Baird and S. Nakayama, *Phys. Rev. Lett.* **74**, 2654–2657 (1995).
21. P. A. Vetter et al., *Phys. Rev. Lett.* **74**, 2658–2661 (1995).
22. J. S. M. Ginges and V. V. Flambaum, *Phys. Rep.* **397**, 63–154 (2004).
23. CDF and DØ Collaborations and Tevatron Electroweak Working Group: J. F. Arguin et al., eprint `hep-ex/0507091`.
24. For a review, see J. Erler and M. J. Ramsey-Musolf, *Prog. Part. Nucl. Phys.* **54**, 351–442 (2005).
25. Muon g-2 Collaboration: G. W. Bennett et al., *Phys. Rev. Lett.* **92**, 161802 (2004).
26. For a review, see A. Czarnecki and W. J. Marciano, *Phys. Rev.* **D64**, 013014 (2001).
27. M. Davier, A. Höcker and Z. Zhang, eprint `hep-ph/0507078`.
28. V. Cirigliano, G. Ecker and H. Neufeld, *JHEP* **0208**, 002 (2002).
29. ALEPH Collaboration: S. Schael et al., *Phys. Rep.* **421**, 191–284 (2005).
30. M. Davier, S. Eidelman, A. Höcker and Z. Zhang, *Eur. Phys. J.* **C31**, 503–510 (2003).
31. SND Collaboration: M. N. Achasov et al., *J. Exp. Theor. Phys.* **101**, 1053–1070 (2005).
32. W. J. Marciano and A. Sirlin, *Phys. Rev. Lett.* **61**, 1815–1818 (1988).
33. J. Erler, *Rev. Mex. Fis.* **50**, 200–202 (2004).
34. K. Maltman, *Phys. Lett.* **B633**, 512–518 (2006).
35. K. Melnikov and A. Vainshtein, *Phys. Rev.* **D70**, 113006 (2004).
36. M. Knecht and A. Nyffeler, *Phys. Rev.* **D65**, 073034 (2002).
37. M. Hayakawa and T. Kinoshita, eprint `hep-ph/0112102`.
38. S. Davidson et al., *JHEP* **0202**, 037 (2002).
39. NuTeV Collaboration: M. Goncharov et al., *Phys. Rev.* **D64**, 112006 (2001).
40. A. D. Martin, R. G. Roberts, W. J. Stirling and R. S. Thorne, *Eur. Phys. J.* **C35**, 325–348 (2004).
41. G. A. Miller and A. W. Thomas, *Int. J. Mod. Phys.* **A20**, 95–98 (2005).
42. S. Kumano, *Phys. Rev.* **D66**, 111301 (2002).
43. A. B. Arbuzov, D. Y. Bardin and L. V. Kalinovskaya, *JHEP* **0506**, 078 (2005).
44. K. P. Diener, S. Dittmaier and W. Hollik, *Phys. Rev.* **D72**, 093002 (2005).
45. B. A. Dobrescu and R. K. Ellis, *Phys. Rev.* **D69**, 114014 (2004).
46. OPAL Collaboration: G. Abbiendi et al., eprint `hep-ex/0601048`.
47. H1 Collaboration: A. Specka et al., eprint `hep-ex/0602007`.
48. HPQCD and UKQCD Collaborations: Q. Mason et al., *Phys. Rev. Lett.* **95**, 052002 (2005).
49. ALEPH, DELPHI, L3, and OPAL Collaborations and the LEP Working Group for Higgs Boson Searches: R. Barate et al., *Phys. Lett.* **B565**, 61–75 (2003).
50. J. Erler, *Phys. Rev.* **D63**, 071301 (2001).
51. J. Erler and P. Langacker, *Phys. Lett.* **B456**, 68–76 (1999).
52. Qweak Collaboration: D. S. Armstrong et al., *Eur. Phys. J.* **A24S2**, 155–158 (2005).
53. A. Czarnecki and W. J. Marciano, *Int. J. Mod. Phys.* **A15**, 2365–2376 (2000).
54. J. Erler, A. Kurylov and M. J. Ramsey-Musolf, *Phys. Rev.* **D68**, 016006 (2003).
55. J. Erler and M. J. Ramsey-Musolf, *Phys. Rev.* **D72**, 073003 (2005).
56. The Snowmass Working Group on Precision Electroweak Measurements: U. Baur et al., eprint `hep-ph/0202001`.
57. M. E. Peskin and T. Takeuchi, *Phys. Rev.* **D46**, 381–409 (1992).
58. M. J. G. Veltman, *Nucl. Phys.* **B123**, 89–99 (1977).
59. H. J. He, N. Polonsky and S. Su, *Phys. Rev.* **D64**, 053004 (2001).
60. V. A. Novikov, L. B. Okun, A. N. Rozanov and M. I. Vysotsky, *JETP Lett.* **76**, 127–130 (2002).

The Pierre Auger Observatory
— Status and first results —

Lukas Nellen, for the Pierre Auger Collaboration

Instituto de Ciencias Nucleares, UNAM
04510 México D.F.
MEXICO

Abstract. The Pierre Auger Observatory, is designed to study cosmic rays at energies above 10^{19} eV. To achieve a uniform, full sky coverage, the Observatory is designed with two sites, one in the southern hemisphere, another one in the northern hemisphere. The southern observatory, currently under construction in Malargüe, Argentina, has started data taking with what amounts to about 1/4 of the full installation. We present the current status of the installation and first results from the dataset, collected from January 2004 to June 2005.

Keywords: Cosmic rays, Auger Observatory
PACS: 96.50.S-, 95.55.Vj

INTRODUCTION

The observation of cosmic rays extends to energies of more then 10^{20} eV or 16 J. It is still a mystery by which mechanism nature manages to accelerate subatomic particles to macroscopic energies. The exact nature of these ultra-high energy cosmic rays (UHECR) is also unknown. We don't know if they are protons, heavier atomic nuclei, nor are we sure if there are photons in the primary cosmic ray beam.

To cast light on the mystery of the UHECRs, the Pierre Auger Collaboration decided to build the world's largest cosmic ray observatory [1]. The observatory is split into two sites, one in the northern hemisphere and one in the southern hemisphere. The southern site is currently under construction in the southern hemisphere, near Malargüe, in Mendoza, Argentina. The northern site will be located near Lamar, in Colorado, USA.

UHECRs are detected indirectly, by the observation of the extended air shower (EAS) which they create in the atmosphere. There are two established techniques for the detection of EAS: the sampling of the front of the shower using an array of detectors on the ground and the detection of the fluorescence of atmospheric nitrogen, induced by the charged particles of the air shower. The Pierre Auger Observatory combines the two techniques in its unique hybrid design. The southern site will cover an area of roughly 3000 km² with 1600 surfaced detector stations. A set of 24 fluorescence telescopes is installed in 4 buildings, located along the border of the surface array.

The hybrid design of the Pierre Auger Observatory makes it possible to measure a larger set of observables of EAS, since the surface array measures the lateral distribution of the shower, whereas the fluorescence detector observes the longitudinal development of the shower. A hybrid design also permits to overcome the weaknesses of either of the techniques alone. For example, for the measurement of the spectrum of UHECRs, one combines the superior energy determination of the fluorescence technique with the larger statistics and geometrical aperture of the surface array.

A QUICK REVIEW OF THE PHYSICS OF UHECRS

Our knowledge of the highest energy cosmic rays is still very limited, even 40 years after the detection of the first cosmic ray with an energy above 10^{20} eV, see for example [2, 3] and references therein. Different models for the acceleration of UHECRs and for the distribution of their sources predict different observables. The best know prediction, made independently by K. Greisen [4] and by G.T. Zatsepin and V.A. Kuz'min [5], is probably the so-called GZK cut-off. They observed that protons above $\approx 5 \times 10^{19}$ eV loose most of their energy over distances of 50 to 100 Mpc due to interactions with the cosmic microwave background. Other cosmic ray primaries suffer similar losses and, as a consequence, one expects a rapid fall-off of the cosmic ray spectrum above the GZK energy. Observational indications that the spectrum of UHECR might extend beyond the GZK cut-off motivated the development of a large number of models that avoid one or several of the assumptions behind the prediction of the GZK cut-off. One class, the top-down models, predicts the dominance of photons at the highest energies. Different models, combined with assumptions about the type of the primary cosmic ray and about galactic and inter-galactic the magnetic fields, predict point sources or anisotropy patterns for the arrival of UHECR.

The most straight-forward, but not necessarily easiest, measurement is that of the spectrum of UHECR. The two predecessor experiments to the Pierre Auger Observatory, AGASA [6] and HiRes [7], did this measurement, with contradicting conclusions. HiRes reports the observation of the GZK cut-off [8, 9], whereas AGASA does not see this fall-off [10, 11]. The AGASA experiment uses a surface array to measure cosmic rays, whereas the HiRes experiment is an air fluorescence detector. The Pierre Auger Observatory, with its increased statistics and with its possibility to compare the two techniques, will resolve this puzzle and more of the open questions around the highest energy cosmic rays.

STATUS AND PERFORMANCE OF THE OBSERVATORY

The construction of the southern site of the Pierre Auger Observatory is still on-going. In the following, we present the current status of the installation an the performance of the instrument.

The surface detector

The surface array of the Pierre Auger Observatory consists of a triangular grid of surface stations with a spacing of 1.5 km. Each station consists of a rotomolded tank of $10\,m^2$ area, filled with $12\,m^3$ of purified water. The detection of the particles of the front of the air shower is via the Cherenkov light they emit in the water. This light is

FIGURE 1. A surface detector tank. The mast with the communications and GPS antennae, the solar panel, and electronics dome are clearly visible. The battery enclosure can be seen on the left hand side of the tank.

captured by three 9" photo-multiplier tubes (PMTs). The signal is taken both from the anode and from the last dynode. All signals are sampled continuously by 10 bit, 40 MHz FADCs. The relative amplification of the anode and dynode channels is chosen to create a dynamic range which allows for the detection of signals of a few photo electrons up to 10^5 photo electrons. A front-end board with a PLD-based 1st level trigger, connected to the local station controller, reads the data. The station controller implements a second level trigger. It also handles the communication with the central data acquisition system. The clock of all the local stations are synchronised using the signals from the global positioning system (GPS). The time resolution achieved with this system is better than 20 ns. The local station is powered from a battery which is recharged by a solar panel, mounted on top of the tank. More details of the surface detector can be found in [1, 12].

The conditions, under which the detector operates are hostile, and continuous monitoring is of utmost importance in order to guarantee the quality of the data. Temperatures, battery status and trigger rates as well as general weather data like pressure, humidity, and wind speed, are continuously recorded.

At the time of writing, nearly 1160 stations are in the field, of which about 970 are operational and taking data.

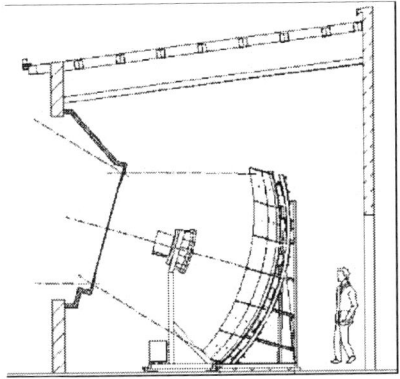 

FIGURE 2. Schematic view of a fluorescence telescope (left) and photo (right). From left to right, one can see the aperture with filter and corrector ring, the PMT camera, and the mirror.

The fluorescence detector

The fluorescence detection system consists of 24 telescopes, installed in 4 sites along the perimeter of the surface array, overlooking the sky above the array. Three of the four sites are fully equipped and operational. The fourth building is currently under construction.

Each telescope has a field of view of $30 \times 30°$. It consists of a Schmidt camera with a corrector ring to enlarge the aperture. A $3.5 \times 3.5 \, m^2$ spherical mirror reflects the image onto the camera, mounted in the focal plane. The camera consists of 440 PMTs, each with a field of view of $1.5°$ diameter. The corrector ring allows to enlarge the aperture of the camera from 1.7 m diameter to 2.2 m, increasing the effective light collecting area by a factor two. A filter, matched to the nitrogen fluorescence spectrum (approximately 300 nm to 400 nm), reduces the background light from the sky and doubles as a window, protecting the delicate optics from the exterior.

The telescopes have to be calibrated carefully, using an absolute and a relative procedure [13, 14]. The absolute calibration is performed three to four times a year, using a mobile light source in a drum diffuser, which is mounted on the outside of the aperture of a telescope [13]. For the relative calibration, optical fibres illuminate the camera and mirrors in three different positions [14]. Tracking the relative calibration information for each pixel fixes the gain information of each of them.

The calibration of the fluorescence detector is verified by detecting and reconstructing vertical and inclined laser shots from the central laser facility (CLF) [15]. By firing a laser beam with a well measured energy and geometry, on can verify the accuracy of the geometrical reconstruction, of the energy determination, and measure the atmospheric scattering and absorption.

During the reconstruction of an event detected by the fluorescence detector, one has to estimate the number of photons emitted by the shower from the number of photons observed. To be able to do this, one has to know the atmospheric conditions. The collaboration operates a sophisticated atmospheric monitoring system for this purpose.

FIGURE 3. The central laser facility. An optical fibre allows to feed light into the *Celeste* surface detector station for hybrid timing studies.

Details can be found in [16, 17].

Hybrid detection

In a hybrid event, an air shower is detected simultaneously but the fluorescence and surface detectors of the observatory. Only about 10% of all events are hybrid, reflecting the duty cycle of the fluorescence detector, which can operate only on clear, moonless nights.

Fluorescence events detected from a single FD site are difficult to reconstruct. The plane containing the shower and the detector site can be determined straightforwardly and with good precision. The determination of the shower inside that plane in difficult though. One needs additional information, either from a second FD site, or from the surface array. It turns out that the precision of the reconstruction is similar for either of the two techniques. The first technique, stereo detection, is used by the HiRes collaboration. The Pierre Auger Collaboration decided to use the second technique, the hybrid detection, as the main technique. For higher energy events, the Pierre Auger Observatory is also capable of performing stereo observation of EAS.

To maximise the number of hybrid events, a cross-triggering mechanism is built into the data acquisition of the observatory. This is necessary, since the trigger threshold of the FD is lower for nearby showers than that of the SD. To overcome this problem, the fluorescence detector notifies the surface detector of any event trigger it generates. Thereby, a trigger can be generated for a sub-threshold SD event.

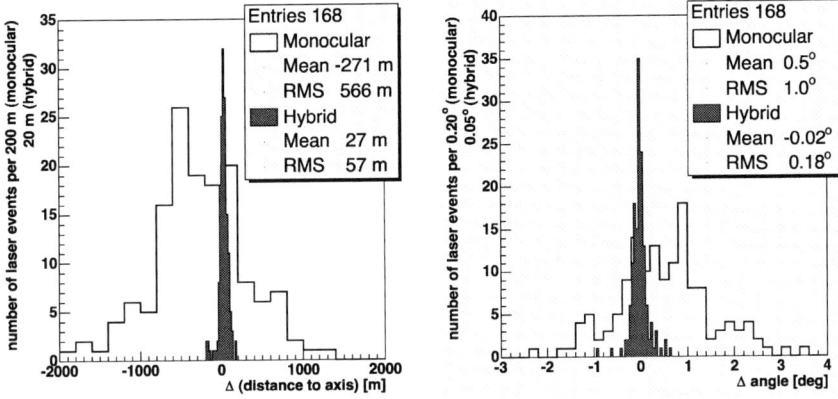

FIGURE 4. Resolution of reconstructed position and direction for vertical laser shots from the CLF.

During the geometrical reconstruction of a hybrid event, one fits simultaneously timing information of the pixel and SD timing. The resulting accuracy is 50 m for the core position and 0.6° for the arrival direction [18]. This resolution is verified analysing reconstructed laser shots from the CLF (fig. 4).

For more details about the hybrid detection technique and performance, see [1, 19] and references therein.

FIRST RESULTS

The data taken during the 18 month period from January 2004 to June 2005 has been analysed by the collaboration. The integrated aperture over that period is 1750 km^2 sr y. This aperture is similar to the integrated aperture over the lifetime of the AGASA experiments and to that of the HiRes experiment. The statistics of the data set is therefore similar.

Photon limit

Some theoretical models predict the presence of a significant amount of photons in the primary cosmic radiation at the highest energies. A limit on the number of photons can therefore be used to eliminate some of those models.

The analysed data set of the Pierre Auger Observatory has been used to derive a limit on the photon fraction at energies above 10^{19} eV [20, 21]. A set of cuts has been applied to this set, requiring a good hybrid geometry and fit, an energy $E > 10^{19}$ eV, at least 6 triggered PMTs, a well-fitted shower profile (see, e.g., fig. 5), and a visible $X_{\max}$. Using a fluorescence based energy estimate leads to conservative estimates of the number of events accepted, since electromagnetic showers, as induced by photons, have

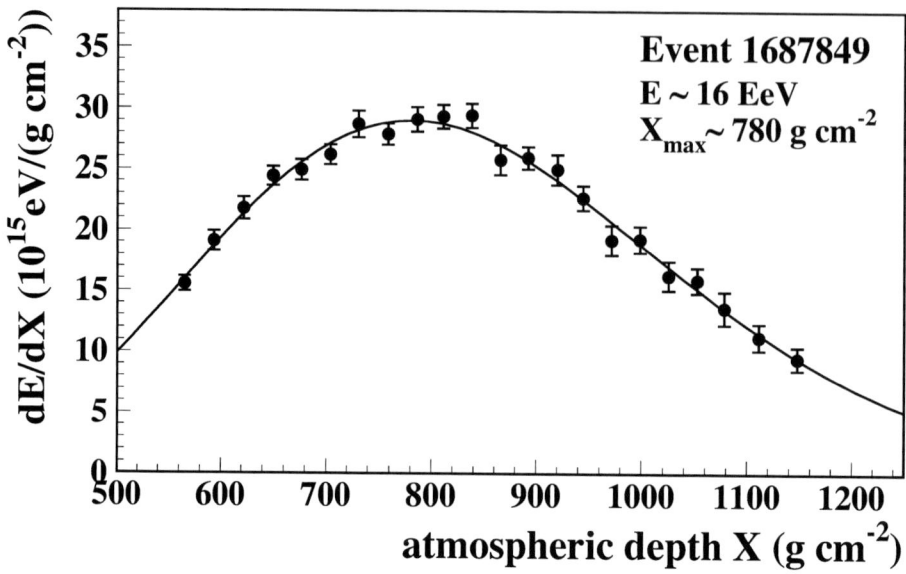

FIGURE 5. Event 1687849: profile fit.

less invisible energy. One is therefore more likely to loose nuclei rather than photons. After all cuts, a set of 29 events remains. Analysing each of these events, comparing them with events simulated under equivalent conditions, one notes that all the events develop higher in the atmosphere than a typical photon event (e.g., fig. 6). After a careful analysis [20], we conclude that the upper limit of the photon fraction is 16% at 95% confidence level.

UHECR spectrum

A first spectrum for UHECRs above 3 EeV has been determined using the first data set analysed. At this energy, the surface detector is fully efficient and the aperture is given by the geometry of the surface array [22, 23].

The energy converter used to assign an energy to the SD events is calibrated using hybrid events, using the SD observable S_{38} [24]. In a first step, on measures $S(1000)$, the integrated signal a 1 km from the shower core, interpolating a fitted lateral distribution function (LDF). The value of $S(1000)$ varies with the slant depth, and therefore with the incident zenith angle. Using the near isotropy of the incident cosmic rays, one can assume that showers at a fixed flux I_0 are equivalent, independent of the incidence direction. Selecting events such that $I(> S(1000)) = I_0$, one obtains curves of constant intensity in the signal-zenith angle plane. We use a particular curve $CIC(\theta)$ of that family, anchored at a median zenith angle of $\theta = 30°$ to relate the signal $S(1000)$ at

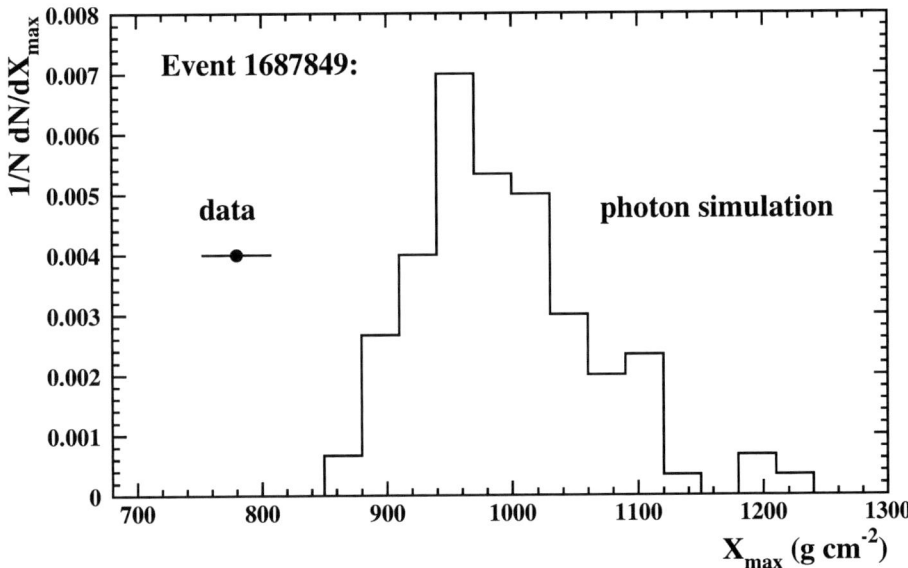

FIGURE 6. Event 1687849: comparison of the X_{max} with that of simulated photon showers.

zenith angle θ to the signal S_{38} at $38°$ via $S_{38} = S(1000)/CIC(\theta)$. Using hybrid events which can be well reconstructed both by the FD and by the SD, one can fit the relation between S_{38} and the energy to be

$$\frac{E}{\text{EeV}} = 0.16 \times \left(\frac{S_{38}}{\text{VEM}}\right)^{1.06}.$$

The resulting spectrum is shown in figure 7. Systematic uncertainties in the energy estimate result in part from uncertainties in the fluorescence yield for EAS and in part in uncertainties in the correlation of FD and SD observables. At the highest energies, the spectrum terminates due to limited statistics. In the period until the middle of 2007, the data set of the Pierre Auger Observatory will increase by a factor 7, which will allow us to make measurements in the region of the GZK prediction.

Anisotropy measurements

Both AGASA [25] and SUGAR [26] report an excess of cosmic rays around 1 EeV from the galactic centre region.

From the analysed data set of the Pierre Auger Observatory, SD events satisfying the quality trigger T5 where selected. The angular resolution is $2.2°$ for events with three SD stations and better than $1.7°$ for events with four or more stations [18]. The zenith angle of the events was limited to not more than $60°$.

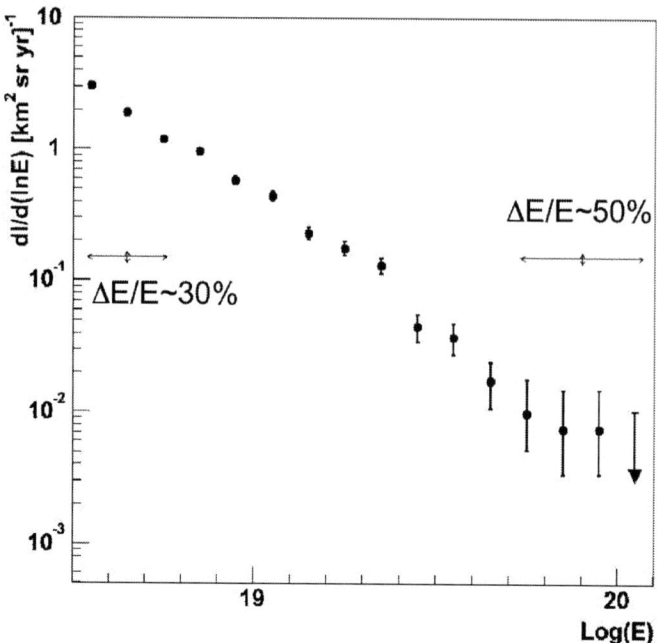

FIGURE 7. The estimated spectrum for $E > 3\,\text{EeV}$. The vertical error bars indicate the statistical errors only. Estimated systematic uncertainties are indicated by the horizontal bars at the left and right of the graph.

The coverage map (fig. 8A) was computed using a shuffling technique. Searches for a possible excess in the galactic centre region were done using the full Auger resolution of 1.5° (fig. 8B), for a SUGAR resolution of 3.7° (fig. 8C), and for an AGASA resolution of 13.3° (fig. 8D), the last with a dataset cut to 1.0–2.5 EeV. In all cases, the observed fluctuations are consistent with chance fluctuations of an isotropic distribution. This observations do not support the excesses claimed by AGASA and by SUGAR.

For a more detailed discussion of anisotropy studies, see [27–29].

OUTLOOK

The construction of the southern site of the Pierre Auger Observatory is well on its way and about to be finished in 2007. The first data set has been analysed and first scientific results are presented.

Once the observatory is completed, the data set will increase rapidly. We expect to have seven times more data to analyse by mid 2007, reducing our statistical errors significantly. This will allow us to perform studies of anisotropies and of spectral features, like the GZK prediction, with unprecedented detail.

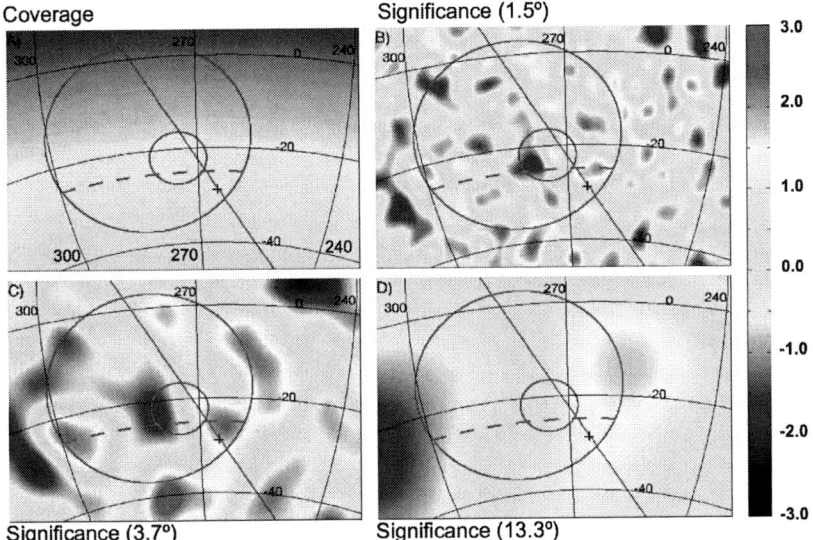

FIGURE 8. Coverage map (A) and significance maps (B–D) for the galactic centre region. The straight line marks the galactic plane, the cross the galactic centre, the dashed line the field of view of AGASA. The circles indicate the regions of excess of AGASA and of SUGAR. Map B is smoothed for a 1.5° degree resolution, map C for 3.7° (SUGAR), and map D for 13.3° (AGASA).

Using our enlarged data set and independent detector studies, systematic and statistical uncertainties of the results will also be reduced. In particular, we expect to see the power of the hybrid technique demonstrated more clearly in the future.

ACKNOWLEDGMENTS

We would like to thank our colleagues in the Pierre Auger Collaboration, who all contributed to the results presented here. We also acknowledge support received by the Universidad Nacional Autonoma de México (PAPIIT IN116503) and from CONACyT (project 46999).

REFERENCES

1. J. Abraham, et al., *Nucl. Instrum. Meth.* **A523**, 50–95 (2004).
2. M. Nagano, and A. A. Watson, *Rev. Mod. Phys.* **72**, 689–732 (2000).
3. D. F. Torres, and L. A. Anchordoqui, *Rept. Prog. Phys.* **67**, 1663–1730 (2004), astro-ph/0402371.
4. K. Greisen, *Phys. Rev. Lett.* **16**, 748–750 (1966).
5. G. T. Zatsepin, and V. A. Kuzmin, *JETP Lett.* **4**, 78–80 (1966).
6. N. Chiba, et al., *Nucl. Instrum. Meth.* **A311**, 338–349 (1992).
7. T. Abu-Zayyad, et al., *Nucl. Instrum. Meth.* **A450**, 253–269 (2000).
8. R. U. Abbasi, et al., *Phys. Rev. Lett.* **92**, 151101 (2004), astro-ph/0208243.

9. T. Abu-Zayyad, et al., *Astropart. Phys.* **23**, 157–174 (2005), astro-ph/0208301.
10. M. Takeda, et al., *Phys. Rev. Lett.* **81**, 1163–1166 (1998), astro-ph/9807193.
11. N. Hayashida, et al. (2000), astro-ph/0008102.
12. X. Bertou, "Performance of the Pierre Auger Observatory surface array," 2005, for the Auger Collaboration, Proceedings of the 29th Cosmic Ray Conference, Pune, India, astro-ph/0508466.
13. P. Bauleo, et al., "Absolute calibration of the Auger fluorescence detectors," 2005, for the Auger Collaboration, Proceedings of the 29th Cosmic Ray Conference, Pune, India, astro-ph/0507347.
14. C. Aramo, et al., "Optical relative calibration and stability monitoring for the Auger fluorescence detector," 2005, for the Auger Collaboration, Proceedings of the 29th Cosmic Ray Conference, Pune, India, astro-ph/0507577.
15. F. Arqueros, et al., "The central laser facility at the Pierre Auger observatory," 2005, for the Auger Collaboration, Proceedings of the 29th Cosmic Ray Conference, Pune, India, astro-ph/0507334.
16. R. Mussa, et al., *Nucl. Instrum. Meth.* **A518**, 183–185 (2004).
17. B. Keilhauer, et al., "Atmospheric profiles at the southern Pierre Auger observatory and their relevance to air shower measurement," 2005, for the Auger Collaboration, Proceedings of the 29th Cosmic Ray Conference, Pune, India, astro-ph/0507275.
18. C. Bonifazi (2005), presented at 29th International Cosmic Ray Conference (ICRC 2005), Pune, India, 3-11 Aug 2005.
19. A. Mostafa, Miguel (2005), for the Pierre Auger Collaboration, Presented at 29th International Cosmic Ray Conference (ICRC 2005), Pune, India, 3-11 Aug 2005.
20. J. Abraham, et al. (2006), astro-ph/0606619.
21. M. Risse, "Upper limit on the primary photon fraction from the Pierre Auger observatory," 2005, for the Auger Collaboration, Proceedings of the 29th Cosmic Ray Conference, Pune, India, astro-ph/0507402.
22. D. Allard, et al., "The trigger system of the Pierre Auger surface detector: Operation, efficiency and stability," 2005, for the Auger Collaboration, Proceedings of the 29th Cosmic Ray Conference, Pune, India, astro-ph/0510320.
23. D. Allard, et al., "Aperture calculation of the Pierre Auger Observatory surface detector," 2005, for the Auger Collaboration, Proceedings of the 29th Cosmic Ray Conference, Pune, India, astro-ph/0511104.
24. P. Sommers, "First estimate of the primary cosmic ray energy spectrum above 3-EeV from the Pierre Auger observatory," 2005, for the Auger Collaboration, Proceedings of the 29th Cosmic Ray Conference, Pune, India, astro-ph/0507150.
25. M. Teshima, et al. (2001), prepared for 27th International Cosmic Ray Conference (ICRC 2001), Hamburg, Germany, 7-15 Aug 2001.
26. J. A. Bellido, R. W. Clay, B. R. Dawson, and M. Johnston-Hollitt, *Astropart. Phys.* **15**, 167–175 (2001), astro-ph/0009039.
27. A. Letessier-Selvon, "Anisotropy studies around the galactic center at EeV energies with Auger data," 2005, for the Auger Collaboration, Proceedings of the 29th Cosmic Ray Conference, Pune, India, astro-ph/0507331.
28. B. Revenu, "Search for localized excess fluxes in Auger sky maps and prescription results," 2005, for the Auger Collaboration, Proceedings of the 29th Cosmic Ray Conference, Pune, India, astro-ph/0507600.
29. J.-C. Hamilton, "Coverage and large scale anisotropies estimation methods for the Pierre Auger observatory," 2005, for the Auger Collaboration, Proceedings of the 29th Cosmic Ray Conference, Pune, India, astro-ph/0507517.

INVITED LABORATORY COURSES

Detection of Cherenkov Photons with Multi-Anode Photomultipliers

H. Salazar*, E. Moreno*, T. Murrieta* and L. Villaseñor[†]

*Facultad de Ciencias Fisico-Matematicas, BUAP, Puebla Pue., 72570, Mexico.
[†]Institute of Physics and Mathematics, University of Michoacan, Edificio C3, Ciudad Universitaria, Morelia, Mich., 58040, Mexico.

Abstract. The present paper describes the laboratory course given at the X Mexican Workshop on Particles and Fields. We describe the setup and procedure used to measure the Cherenkov circles produced by cosmic muons upon traversal of a simple glass radiator system. The main purpose of this exercise is to introduce the students to work with multi-anode photomultipliers such as the one used for this experiment (Hamamatsu R5900-M64), with which measurements requiring position sensitive detection of single photons can be successfully performed. We present a short introduction to multi-anode photomultipliers (MAPMT) and describe the setup and the procedure used to measure the response of a MAPMT to a uniform source of light. Finally, we describe the setup and procedure used to measure the Cherenkov circles produced by cosmic muons upon traversal of a simple glass radiator system.

Keywords: Cosmic ray muons; Cherenkov light; Multi-Anode Photomultiplier.

INTRODUCTION

Detection and identification of ionizing particles constitutes the basis of experimental high energy and cosmic ray physics. In particular, ring imaging Cherenkov detectors [1] are used to determine the velocities of charged particles by measuring the angle of their Cherenkov light cone; this light is an electromagnetic shock wave emitted when the velocity $v = \beta c$ of a charged particle in a medium exceeds the speed of light c/n in that medium (c is the speed of light in vacuum and n is the refractive index of the medium). The angle of emission of this light with respect to the particle's direction of motion is called Cherenkov angle, θ_c; the latter is determined by the formula

$$cos\theta_c = \frac{1}{\beta n} \quad (1)$$

Although the idea of determining these velocities by measuring the Cherenkov angle was proposed in 1960 [2], the first successful laboratory tests had to wait until the 1970Šs and the first generation of operational Ring Imaging Cherenkov (RICH) detectors were built in in the 1980's; RICH detectors play an important role in several high energy physics experiments that are either in their data-taking stage or under construction [3].

If in addition to measuring the particle velocity one can measure its linear momenta, i.e., via a magnetic spectrometer, then the particle mass can be determined, leading to the identification of the type of particle.

In this paper we describe a laboratory course on the basic detection of Cherenkov circles produced when fast cosmic ray muons traverse a 1 cm thick glass radiator.

FIGURE 1. Multi-anode photomultiplier Hamamatsu R5900-M64 used for this laboratory course.

This course is based on a course given at a past ICFA School on Instrumentation [4]. The muons used are produced by decays of pions and kaons at tens of kilometers in the atmosphere; in turn, these pions and kaons are produced by primary cosmic rays interacting with the nitrogen and oxygen nuclei of the atmosphere. These secondary muons arrive at our detector from above with a mean energy of 2 GeV; their arrival rate at ground level in Morelia (2000 m.a.s.l.) is approximately 1.1 $cm^{-2}min^{-1}$. We use a 64-channel multi-anode photomultiplier (Hamamatsu R5900-M64) to detect the positions of the Cherenkov UV photons. We reconstruct the Cherenkov circles on-line by using a simple LabView-based data acquisition system on a PC.

EXPERIMENTAL SETUP

The laboratory session was divided into two parts. The first consists of measuring the uniformity of the response of the M64 MAPMT by using a UV-light emitting diode (LED) which illuminates the 64 MAPMT channels in a uniform way. The second part consists of measuring Cherenkov circles with the M64 MAPMT. Before describing the experimental setups for this two exercises, we give a description of the MAPMT used.

FIGURE 2. Experimental setup used to measure the response of the MAPMT to a uniform source of light.

Description of the Multi-Anode Photomultiplier

The position-sensitive multi-anode PMT used for this laboratory course is the Hamamatsu R5900-M64 shown in Fig. 1.

The Hamamatsu R5900-M64 multi-anode photomultiplier tube has 64 separate anodes arranged in an 8x8 array with each anode having a 2mm x 2mm active area. The tube has an 0.8mm thick borosilicate glass window over a bialkali photocathode. The MAPMT is capable of resolving single photoelectron signals. The dark count for the MAPMT is 5 counts/anode/second at room temperature; it produces an accidental trigger rate which is negligible. The quantum efficiency at 420nm, defined as the ratio of the number of photoelectrons ejected from the photocathode to the number of photons incident on the MAPMT, is around 25% [6].

The dynode system in this multi-anode PMT is very different from conventional single-anode PMTs. It consists of foils with specially shaped channels where multiplication of the number of electrons occurs. This MAPMT has 12 dynode-stages allowing multiplication gains typically above 10^6 when operated at 900 volts. This multi-anode PMT possess the special feature that the 12th dynode is common to the 64 anodes an is connected by a lead through the glass to the exterior case of the MAPMT. This common signal from the 12th dynode is used to trigger the DAQ system. The anode dark current is mainly below 200 nA [6]. Attention must be paid not to exceed the maximum allowed voltage of 1000 V for this MAPMT and the maximum allowed current of 0.01 mA [6].

Of special relevance for Cherenkov ring imaging is the position resolution of this type of MAPMTs, which is determined by the anode pad size, i.e., 2 x 2 mm^2 in the case of the M64 MAPMT used. Cross-talk among adjacent channels is small and the response across the photocathode surface is uniform within a 10% level [6]. According

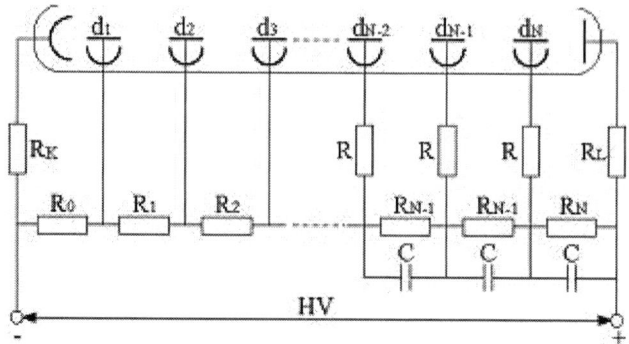

FIGURE 3. Voltage divider used in the base of the MAPMT used for this experiment.

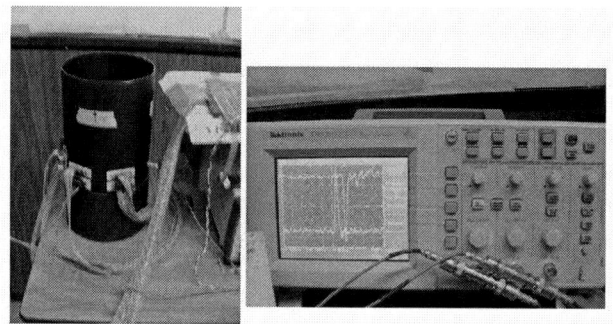

FIGURE 4. DAQ system used to read out the signals from the 64 anodes of the MAPMT.

to specifications [6], the pulse rise time is $\cong 1$ ns, allowing them to be used for timing purposes. This MAPMT's high segmentation (64 channels) jointly with its low cross-talk and fast response were the main criteria that we used to select this MAPMT for this experiment.

Experimental setup used to measure the response of a multi-anode photomultiplier to a uniform source of light

The experimental setup for measuring the response of the MAPMT to a uniform source of light is shown in Fig. 2.

Light from the UV-light emitting diode is reflected by tyvek sheet located at the top of the light-tight cylindrical enclosure containing the light source, the tyvek reflector and the MAPMT; this ensures a uniform illumination of all the 64 channels of the multi-anode PMT. The latter is plugged into a MAPMT base. This base consists of a resistor chain used as a voltage divider as shown in Fig. 3. In some cases, potentiometers are provided for adjusting the voltage on the electrodes for focusing the photoelectrons to the

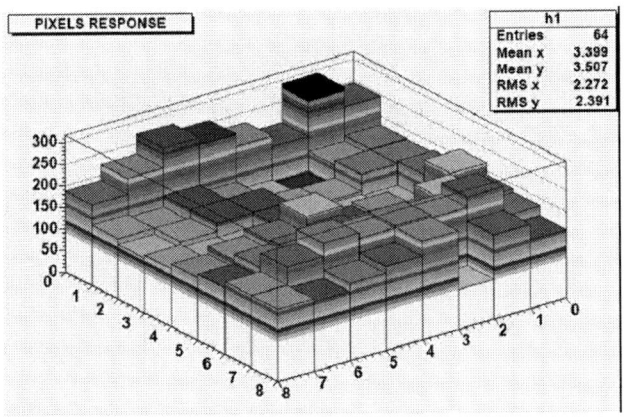

FIGURE 5. Average amplitude recorded as a function of position with the setup shown in Fig. 2.

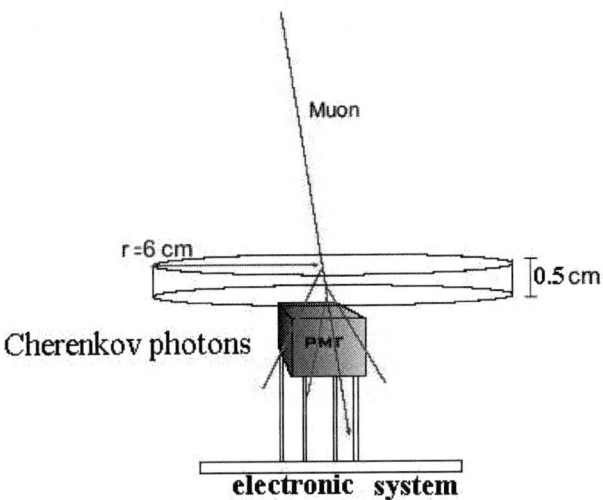

FIGURE 6. Schematic diagram of the radiator borosilicate glass and MAPMT used to detect Cherenkov circles.

first dynode and capacitors or Zener diodes to stabilize the voltage on the last dynodes in case of high rate and high gain operation. High voltage is provided by connecting a HV power supply by using a cable to the MAPMT base. Cables are also used to connect the 64 anode channels to the electronics which consists of amplifiers, discriminators and an ADC system located outside of the detector enclosure.

The signals from the 64 anodes of the MAPMT are read out with a 64-channel home-made FADC multiplexed electronics, see Fig. 4. This DAQ system is read out into the

FIGURE 7. Photograph of the radiator glass of 0.5cm thickness and the MAPMT used to detect Cherenkov circles.

PC by using one of its parallel ports. The data acquisition process is controlled with a LabView-based program. The average amplitude is recorded as a function of position, as shown in Fig. 5. From these results we may obtain, in addition to the uniformity of pad response, also the magnitude of cross-talk signals between adjacent pads.

Experimental setup used to detect Cherenkov circles

The DAQ system used to read out the signals from the 64 anodes of the MAPMT is shown in Fig. 4. Display of Cherenkov circles is done by using appropriate thresholds to guarantee that the signal is above noise on the individual channels.

In this laboratory course we detected the Cherenkov photons radiated by high energy muons in a radiator borosilicate glass of .5 cm thickness and refractive index of 1.47. In contrast with normal glass, borosilicate glass is transparent to UV light. The schematic diagram of the setup used is shown in Fig. 6 and Fig. 7 is a photograph of this setup. The PMT borosilicate glass window is in good optical contact with the radiator glass by means of optical liquid; this is very important to avoid total internal reflection of the Cherenkov light inside the radiator glass.

According to formula (1), muons with $\beta \simeq 1$ traversing a glass with an index of

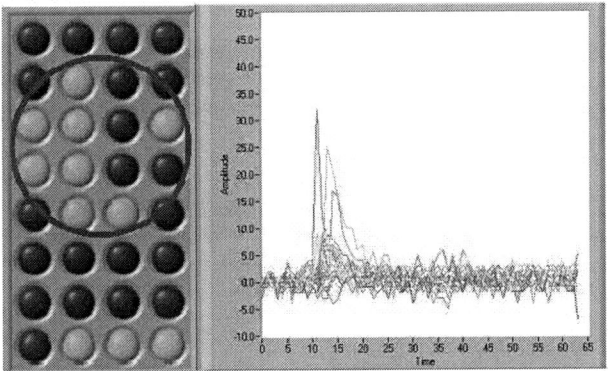

FIGURE 8. Reconstructed Cherenkov circles on the left and digitized signals for 32 out of the 64 channels of the MAPMT for one event.

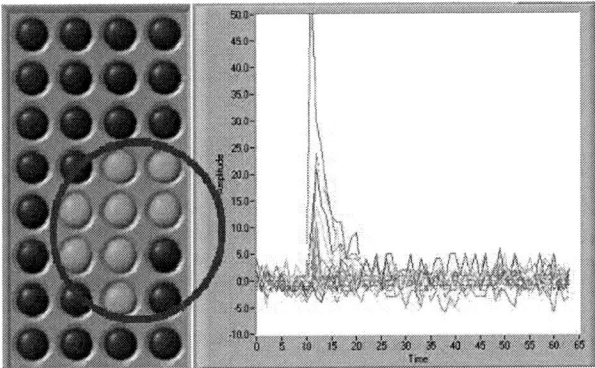

FIGURE 9. Same as Fig. 8 for another event.

refraction $n = 1.47$ and a thickness L, radiate Cherenkov photons in cones with $\theta_c = arccos(1/n) = 822$ mrad. As shown in Fig. 6, these cones intersect the PMT window in circles (assuming vertical muons) of radius

$$R = Ltan(\theta_c) = Ltan(.822) = 1.08L \qquad (2)$$

Therefore we expect a radius R = 0.54 cm for a radiator thickness L = 0.5 cm. Fig. 8 and Fig. 9 show the reconstructed Cherenkov circles on the left and the digitized MAPMT traces for 32 out of the 64 channels of the MAPMT for two events taken with the setup shown schematically in Fig. 6. The radius of the Cherenkov circle from these figures is consistent with the value R = 0.54 cm that we expect for (vertical) muons with $\beta = 1$.

CONCLUSION

We have described the laboratory course given at the X Mexican Workshop on Particles and Fields. The main purpose of this exercise is to introduce the students to work with position-sensitive multi-anode PMTs which can be used for detection of single photons, such as in the case of this experiment. We used a MAPMT of 64 anode channels. We have given a short introduction to MAPMTs. We have shown the setup and the procedure used to measure the response of the MAPMT to a uniform source of light. We have also described the setup and procedure used to measure the Cherenkov rings produced by cosmic muons upon traversal of a simple glass radiator system.

ACKNOWLEDGEMENTS

We are grateful to students and technicians who have helped in the present laboratory course, specially Enrique Varela, Alejandra Palma, Enrique Patiño and Saul Aguilar.

REFERENCES

1. W.R.Leo, *Techniques for Nuclear and Particle Physics Experiments*, Springer Verlag, Berlin, Heidelberg (1987).
2. A. Roberts: A new type of Cherenkov detector for the accurate measurement of particle velocity and direction. Nucl. Instr. and Meth. 9 (1960) 55.
3. J. Engelfried, Lecture Course on Instrumentation given at the 2nd Latin American School of High Energy Physics, San Miguel Regla, Mexico, June 1-14, 2003, CERN Yellow Report 2006-001 (2006) 261-286.
4. S. Korpar, P. Krizan, A. Gorisek, A. Stanovnik, Tests of a position sensitive photomultiplier and measurement of diffraction pattern by counting single photons, ICFA'99 Instrumentation School, Istanbul, Turkey, AIP Conference Proceedings, Vol. 536 (2000) 340-348.
5. http://sales.hamamatsu.com/en/products/electron-tube-division/detectors/photomultiplier-tubes-pmts/r5900-03-m64.php
6. Hamamatsu Photonics K.K., Data Sheet of R5900-L16 and Data Sheet of R5900-M16

X-Ray Spectroscopy with PIN diodes

F. J. Ramírez-Jiménez

Instituto Nacional de Investigaciones Nucleares
Carretera México-Toluca S/N, La Marquesa, Ocoyoacac, 57150, MEXICO
e-mail: fjrj@nuclear.inin.mx

Abstract. A PIN diode and a low noise preamplifier are included in a nuclear spectroscopy chain for X-ray measurements. This is a laboratory session designed to review the main concepts needed to set up the detector-preamplifier array and to make measurements of X-ray energy spectra with a room temperature PIN diode. The results obtained are compared with those obtained from radioactive sources with a high resolution cooled Si-Li detector.

Keywords: PIN diodes, X-rays, spectroscopy.
PACS: 29.40.-n, 29.30.Kv

INTRODUCTION

An experiment is proposed as a laboratory session, in which a PIN diode and a charge sensitive preamplifier specially designed to match the PIN diode characteristics are included in a nuclear spectroscopy chain to measure the X-ray energy spectra from radioactive sources, the PIN diode is operated at room temperature. The main concepts needed to set up the detector-preamplifier array are reviewed. The energy resolution of the system is measured and the results obtained are compared with those obtained with a high resolution cooled Silicon Lithium-Drifted, Si-Li, detector.

X–ray spectroscopy is useful to identify radioactive sources by the energy of the peaks encountered in the spectrum; also it is utilized to determine the elementary composition of samples by X-ray fluorescence, in this technique an excitation source is employed. The excitation source could be: an X-or gamma-ray radioactive source or an X-ray tube.

The main characteristics of X-rays are the energy and the intensity. The energy is in the order of some keV's and depends of the energy of the accelerated electrons in case of generation of X-rays through a collision process. In the case of X-ray fluorescence or nuclear decay, the energy depends of the chemical element. Radioisotope X-ray sources generate characteristic X-rays with well defined energy peaks. These peaks are very useful as absolute energy reference points to make the energy calibration of a spectroscopy system. The number of counts registered in the spectrum depends on the activity of the source.

In the nuclear decay process of an atom, electron capture and internal conversion, lead to the generation of X-rays which are characteristic of the final element of the decay. Radioisotopes that follow this process are sources of characteristic X-rays.

X-RAY SPECTROSCOPY

A basic energy spectroscopy system is shown in the Figure 1. The signal of charge generated in the detector by the radiation is conditioned in the preamplifier, the amplifier gives a further amplification and sets the optimal bandwidth of the system in order to get the best signal to noise ratio, the output from the amplifier is a series of analog pulses with a height directly proportional to the energy of the detected radiation, these pulses are analyzed considering its pulse height, classified an the result of this classification is shown in the screen of the multichannel analyzer system.

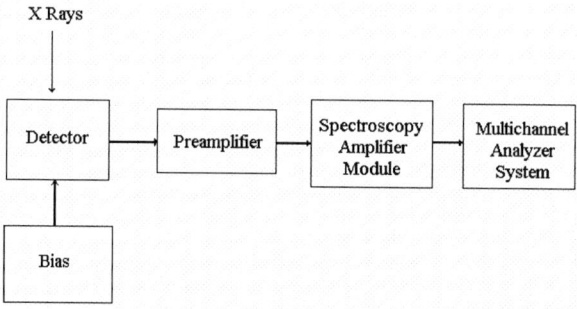

FIGURE 1. Basic Energy Spectroscopy System.

The energy spectrum of an Am-241 radioactive source is shown in the Figure 2, a cooled Si-Li detector was employed to realize the measurement. The characteristic peaks of the radioactive element are clearly defined.

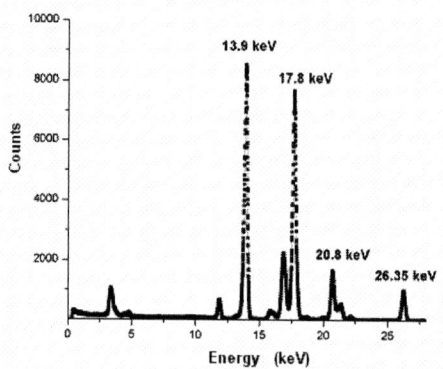

FIGURE 2. X-ray energy spectrum of an Am-241 source, obtained with a Si-Li detector.

The total noise is one of the more important characteristics of the spectroscopy system and is reflected on the spectrum in the width of the peaks, the energy resolution of the system is defined [1] as the Full Width at Half Maximum of a peak, FWHM, and can be measured directly from the spectra, see Figure 3. The resolution for X-ray detectors is measured in the peak, H_o, of 5.89 keV of a Fe-55 radiation source.

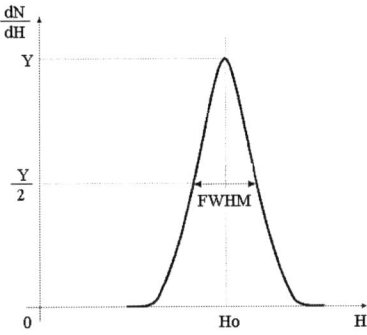

FIGURE 3. Definition of the energy resolution (FWHM) in a spectroscopy system.

The total noise $FWHM(eV)$ is the resultant of two uncorrelated components: the electronic noise $FWHM_{NOISE}$ associated mainly with the preamplifier and the electrical characteristics of the detector, and the statistical component due to the detection process inside the detector.

$$FWHM(eV) = \sqrt{Electronic\ Noise^2 + Detection\ Noise^2} \qquad (1)$$

$$FWHM(eV) = \sqrt{(FWHM_{NOISE})^2 + (2.35\sqrt{F\ E\ w})^2} \qquad (2)$$

where: $F=0.12$ is the Fano factor for Silicon, E is the energy of the incident photons in eV and w is the energy needed to create an electron-hole pair; $w = 3.6$ eV for Silicon detectors [2].

PIN DIODE DETECTORS

Si-Li detectors are normally used to get the best results in X-ray analysis due to its good energy resolution, typically 180 eV, and good efficiency in the energy range from 1 kev to 100 keV, nevertheless there are other kind of detectors commercially available and suitable for X-ray detection, such as: semiconductor drift detectors [3][4] and special PIN diodes [5] [6], these detectors have some nice advantages over the Si-Li detectors mainly with respect to the size.

We demonstrate the detection capability of an OPF420 PIN photo diode [7], originally intended for optical fiber applications, in the proposed experiment. The basic structure of a PIN diode is seen in Figure 4(a), the three regions, n^+, Intrinsic and p^+, can be distinguished clearly. Figure 4(b) shows the real active area of the PIN diode. The mechanical characteristics of this device are: 300 μm of active thickness, 1 mm^2 active area, and a glass window of 1 mm thick.

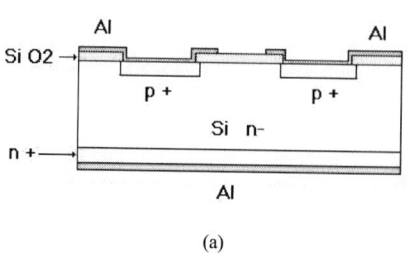 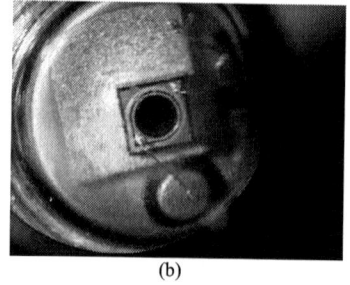

FIGURE 4. a) Basic structure of a PIN diode, b) Picture of an OPF420 PIN diode as seen through a microscope.

Semiconductor diode detectors need to be operated with enough reverse voltage in order to reach the maximum depletion zone and therefore to get the best detection efficiency because in the depletion zone occurs the generation of carriers by effect of radiation in the best conditions. The Figure 5 shows the evolution of the radiation effect in the detector, an electric charge Q_i is induced first by the movement of the generated electrons and then by the movement of holes, both of them generate a pulse of current.

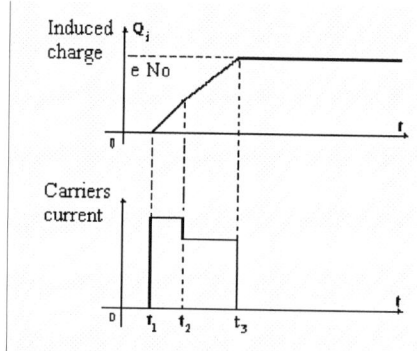

FIGURE 5. Signals generated inside the detector. Electric charge induced by the radiation in the PIN diode and the corresponding generated current.

PIN diodes can be represented by an electrical model as shown in Figure 6, in this model the main components are the equivalent charge source Q_i, the leakage current I_o and the diode capacitance C_j.

The PIN detector has been characterized as X-ray detector for this application [8]; the experimentally obtained electrical characteristics are shown in Figure 7. The graph of variation of the detector capacitance C_j and leakage current I_o, as a function of reverse voltage is used to determine the optimal operating voltage of the PIN diode. The selected point is where the reverse current does not increase abruptly and the diode capacitance gets a minimum, which implies that full depletion is reached. In this case, the operating point is 65 V, where the leakage current is less than 700 pA and the capacitance is less than 2 pF. From these values, we concluded that the PIN diode can be used at room temperature with good performance for X-ray spectroscopy because its leakage current is small, but it requires a low input capacitance and low noise preamplifier as read-out circuit.

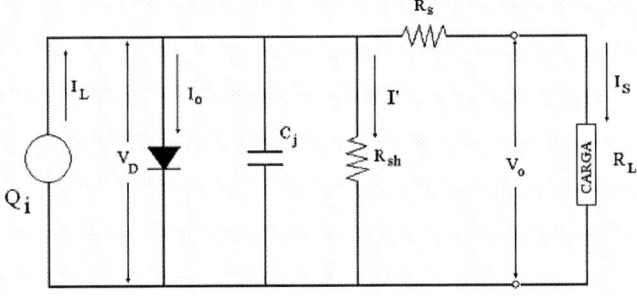

FIGURE 6. Electrical model of a PIN diode in reverse bias mode.

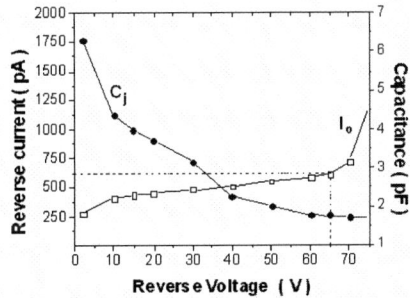

FIGURE 7. Dependence of diode capacitance and leakage current with the reverse voltage for a PIN diode.

The detection efficiency of radiation detectors needs to be specified. The intrinsic efficiency is defined as the ratio between the detected and the total number of events reaching the detector; it was experimentally obtained for the PIN diode by using the standard radioactive sources of the Table 1, with its characteristic energies [1].

Table 1. Characteristics of radioactive sources used for calibration.

RADIOISOTOPE	Radiation	Energy [keV] (yield)	Half-life
^{109}Cd (Cadmium)	X- Rays	22.16 (86 %)	462.6 days
		24.94 (17 %)	
	Gamma Rays	88.03 (3.61 %)	
^{55}Fe (Iron)	X-Rays	5.89 (24.9 %)	2.74 years
		6.49 (3.4 %)	
^{241}Am (Americium)	X- Rays	11.9 (0.86 %)	458 years
		13.9 (13.2 %)	
	Gamma Rays	17.8 (19.25 %)	
		20.8 (4.85 %)	
		26.35 (2.4 %)	
		59.54 (35.9 %)	

Results of these measurements are shown in Figure 8, where it can be seen that the range of operation with good efficiency is from a few keV, limited by the window, up to 60 keV when the relative efficiency is still around 2 %. To use the PIN diode for lower energy X-ray measurements, the original glass window can be taken out or replaced by a 25 µm beryllium foil.

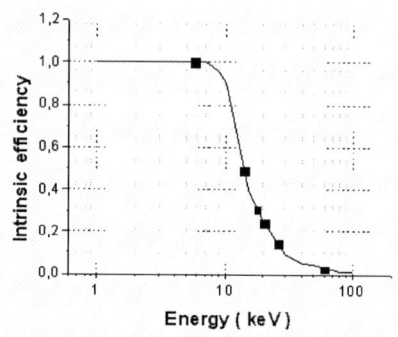

FIGURE 8. Intrinsic efficiency of a PIN diode with an effective thickness of 300 µm.

Principle of Operation

Measurement of the radiation is based in the production of electron-hole pairs in the interaction with the detector material and the further collection of charges (see Figure 5). The number N of electron-hole pairs generated [2], is related to the incident energy E, by:

$$N \equiv \frac{E}{w} \quad (3)$$

Then the total charge Q_i, generated in the detector by the interaction is:

$$Q_i = N\,e \quad (4)$$

e is the electric charge of the electron. Substituting eq. (3) in eq. (4) we have:

$$Q_i \equiv \frac{e\,E}{w} \quad (5)$$

An important conclusion from this equation is that the generated charge in the detector is directly proportional to the energy of the radiation. Due to the very small signal generated in the detector for X-rays, the noise of the measuring system has to be considered with great care.

CHARGE SENSITIVE PREAMPLIFIERS

The interface between detector and preamplifier defines the low noise of the spectroscopy system, therefore for these applications, the best noise performance of the associated preamplifier is required, generally the preamplifiers have to be done with discrete components selected for low noise. The matching between the detector and preamplifier also defines the noise characteristics of the system, an optimal matching is desired for low noise performance. Generally, a field effect transistor, FET, is used as the input device.

Special low noise preamplifiers have been developed to fulfill this requirement like the charge sensitive preamplifier in which, the radiation is measured as an individual event, a charge is generated due to the interaction of radiation with the detector and the preamplifier converts the input charge to voltage at the output. The classical charge sensitive preamplifier in one with a feedback resistor, Figure 9, in this case, the feedback capacitor C_f is charged by the injected signal Q_i from the detector and discharged immediately through the feedback resistor R_f. The feedback resistor has the disadvantage that it is an additional and undesired source of noise.

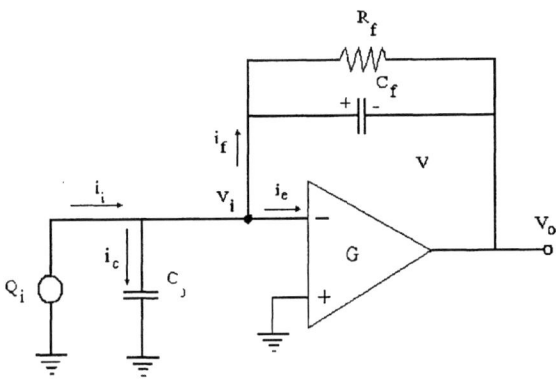

FIGURE 9. Charge sensitive preamplifier with resistive feedback.

The charge sensitive preamplifier with optical feedback is another possibility that do not include an additional source of noise by a feedback resistor, Figure 10. The feedback capacitor C_f is discharged through the input FET, when an optically coupled LED sends a reset light pulse to the FET every time the output voltage, V_o, reaches a defined level.

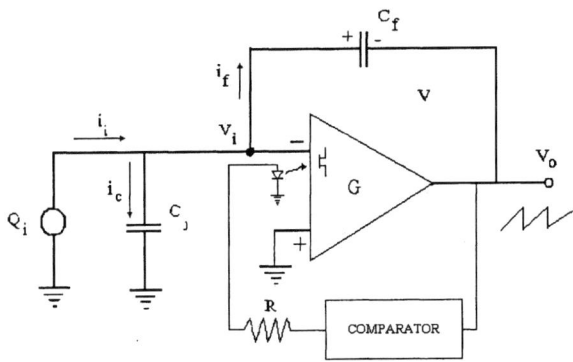

FIGURE 10. Charge sensitive preamplifier with optical feedback reset.

Forward Biased FET Charge Amplifier

We selected a novel preamplifier configuration called forward biased FET charge amplifier, FBFA [9], for the present experiment, in order to get the minimum noise with PIN diodes because this configuration has proven to give a good noise performance when used with silicon diode detectors. The FBFA configuration without feedback resistor but well defined operating point and continuously discharging feedback capacitor C_f is shown in the Figure 11. C_f is discharged through the input field effect transistor biased with the gate in forward mode.

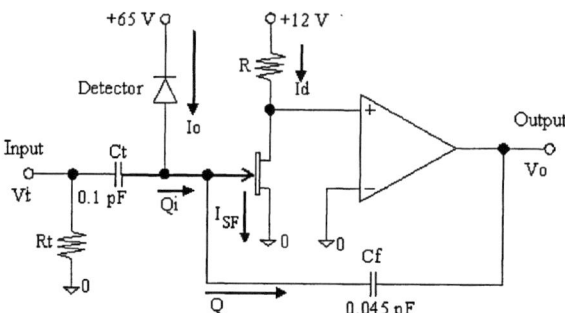

FIGURE 11. Basic idea of the preamplifier with input FET in the condition of forward biased gate.

The output voltage V_o of the preamplifier is:

$$Vo\,(t) = \left(-\frac{Q_i}{C_f}\right)\left\{A_1\,\exp^{-\frac{t}{\tau_s}} + B_1\,\exp^{-\frac{t}{\tau_1}}\right\} \qquad (6)$$

where: Q_i is the charge generated in the PIN diode by effect of the radiation; A_1, B_1, τ_s and τ_1 are parameters depending on the values of the circuit components. The value at the peak of the signal, see Figure 12, is:

$$V_o = -\frac{Q_i}{C_f} \qquad (7)$$

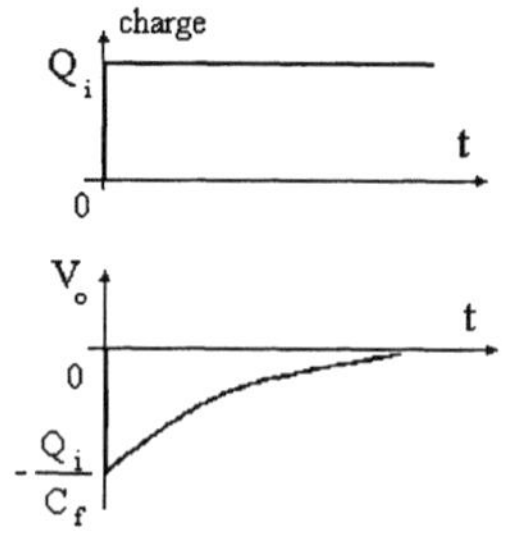

FIGURE 12. Charge generated in the detector and the corresponding output signal from the preamplifier.

In the FBFA, the gate of the front-end JFET is kept at a constant voltage by slightly forward biasing the gate-source junction, by the effect of the flow of the reverse current of the detector (see Figure 11). The reverse current of the detector plus the current from the gate junction of the JFET will flow through the source to ground. As the charge signal from the detector accumulates in the feedback capacitor C_f, the FBFA will find an equilibrium, through the effect of the AC negative feedback, returning to the quiescent condition.

EXPERIMENT

Equipment Required

- NIM (Nuclear Instrumentation Modules) standard bin
- Digital Oscilloscope
- Electrometer
- Curve tracer for semiconductor devices
- NIM nuclear pulse generator
- NIM Spectroscopy Amplifier
- PIN diode detector with special low noise preamplifier
- Multichannel Analyzer (MCA) inside a personal computer
- Calibration sources (Cd-109, Am-241, Fe-55)

Experimental procedure

1.- Measure the V-I characteristics of the PIN diode in the curve tracer. A graph similar to the one shown in the Figure 13 should be obtained in the forward bias condition.

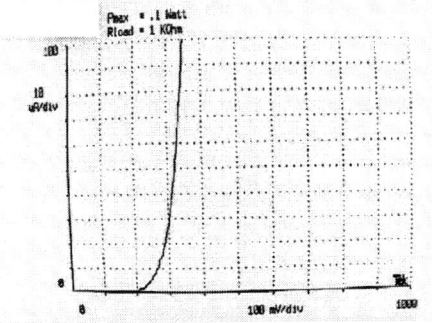

FIGURE 13. V-I curve for a Si PIN diode obtained in a curve tracer.

2.- Measure the V-I characteristics of the input FET of the preamplifier in the curve tracer. A graph similar to the one shown in the Figure 14 should be obtained. Register the values of current and voltage of quiescent points shown to calculate the transconductance g_m, of the device. Notice the special situation in the forward bias condition at the gate.

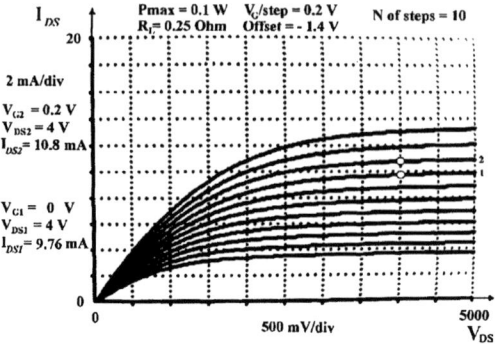

FIGURE 14. V-I characteristics of a FET, showing the forward biased condition of the gate.

$$g_m = \frac{\Delta I_{DS}}{\Delta V_G} \tag{8}$$

3.- The circuit diagram of the preamplifier to be used is shown in Figure 15, it has the following characteristics: Charge sensitive with a feedback capacitance of 0.045 pF and conversion gain of 22 mV/fC, the equivalent noise referred to the input is 20 electrons (rms).

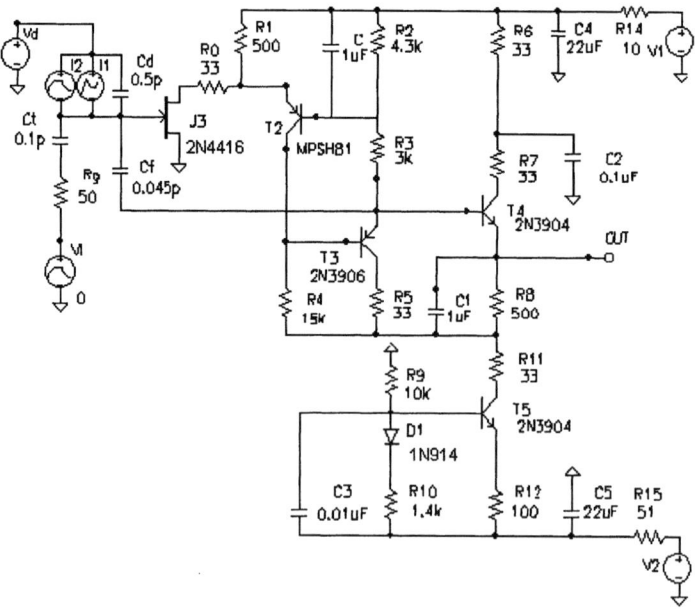

FIGURE 15. Detailed circuit diagram of the FBFA employed in the experiment.

The preamplifier is assembled in a printed circuit board as seen in Figure 16.

FIGURE 16. Printed circuit board of the charge sensitive preamplifier.

In the FBFA, the gate of the front-end JFET is kept at a constant voltage by slightly forward biasing the gate-source junction, see Figure 17. Bias the preamplifier and measure the voltage in the gate with an electrometer, the value should be around 0.24 V.

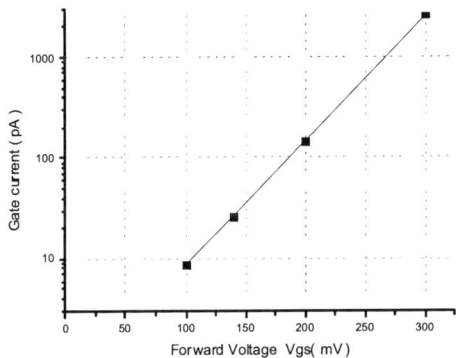

FIGURE 17. Forward bias condition in the gate of the input FET of the preamplifier.

4.- Measure the conversion gain G_V of the preamplifier and compare the result with the theoretical value, according to equation (7):

$$G_V = \frac{output\ voltage}{input\ charge} \qquad (9)$$

the input charge is injected with a pulse generator, see Figure 11, trough the test capacitor C_t. Then, the input charge Q_i is:

$$Q_i = C_t V_t \qquad (10)$$

5.- Assemble the spectroscopy system as indicated in Figure 1. Standard NIM modules are used.

6.- Adjust the controls of the amplifier to have a time constant between 1 and 2 μs, a gain enough to have the 59.54 keV peak of Am-241 in channel 1700 of the MCA used with 2048 channels.

7.- Measure the signals in the outputs of the preamplifier and amplifier. Adjust the pole-zero condition by seeing the signal in the output of the amplifier.

8.- Get a spectrum at least with two known energies (one or two calibration sources). Accumulate at least 10 000 counts in the region of interest (ROI) under the peaks. Register the data in Table 2:

Table 2. Data for energy calibration.

Source	Energy (keV)	Channel Number

With this data, calculate the calibration parameters according to the straight line equation:

$$Energy\,[keV] = slope\left[\frac{keV}{channel}\right] \bullet channel\ number\ +\ intersection\,[keV] \quad (11)$$

9.- Calibrate the system by using the software of the MCA, compare the results.

10.- Obtain the resolution of the system with a Fe-55 source.

11.- Get a spectrum for an Am-241 source. Save the spectrum in ASCII format. The obtained spectrum should be similar to the one shown in the Figure 18(a).

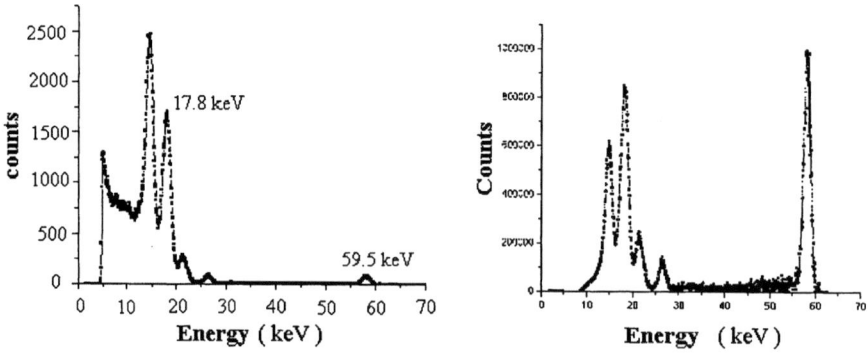

FIGURE 18. a) Raw spectrum obtained with the PIN diode. b) Corrected spectrum considering the intrinsic efficiency of the PIN diode.

Considering the intrinsic efficiency, shown in the Figure 8, the spectrum can be corrected to get the real energy distribution of the radiation as shown in the Figure 18(b).

12.- The spectrum obtained for the Am-241 source with the PIN diode can be compared with the one obtained with a Si-Li detector, as shown in Figure 19. In this figure a good correspondence of the peaks position in both detectors is observed, which guaranties the good linearity of the PIN diode experimental results.

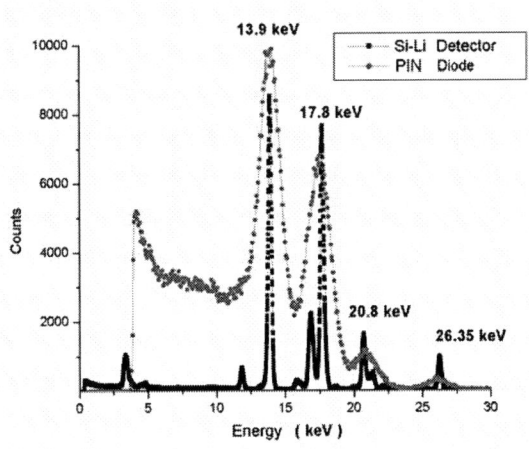

FIGURE 19. Comparison of the spectra obtained with the PIN diode and a Si-Li cooled detector, energy spectra for an Am-241 source.

REFERENCES

1. G. F. Knoll "Radiation Detection and Measurement", John Willey and S. New York, 2000.
2. G. Bertolini, A. Coche, "Semiconductor Detectors", Edited by John Wiley and Sons, Amsterdam, Netherlands, (1968).
3. E. Gatti, P. Rehak "Semiconductor Drift chamber- an application of a novel charge transport scheme" Nucl. Inst. and Meth. in Phys. Res. 225 (1984) 608-614.
4. KETEK, available in the web site: http://www.ketek.com/
5. Hamamatsu, available on web site: http://usa.hamamatsu.com/cmp-detectors/xray.htm
6. AMPTEK Inc, available in the web site: http://www.amptek .com/
7. Optek Technology, Inc. Optek's Product Catalog , Carrollton, TX, USA, 1997.
8. F. J. Ramírez-Jiménez, R. López-Callejas, A. Cerdeira-Altuzarra, J. S. Benítez-Read, M. Estrada-Cueto, J. O. Pacheco-Sotelo, "PIN diode-preamplifier set for the measurement of low energy γ and X-rays", Nucl. Inst. and Meth. A497 (2003) 557-583.
9. G. Bertuccio, P. Rehak and D. Xi, "A novel charge sensitive preamplifier without the feedback resistor", Nucl. Inst. and Meth., A326 (1993) 71-76.

A pedagogical Multiwire Proportional Chamber

R. Alfaro

Instituto de Física, UNAM, A.P.20-364, 01000, México D.F.,México.

Abstract. The purpose of this laboratory session is to provide the basic ingredients for understanding the construction and operation of Multiwire Proportional Chambers (MWPC). During this session the students constructed and tested a simple position sensitive MWPC. Only measurements requiring rather simple hardware (amplifiers, digital oscilloscope) were made and some of them are presented.

INTRODUCTION

Principles of Operation

The principles underlying modern multiwire chambers were already shown around 1920 (Geiger-Müller counter); the first wire chamber used in high-energy physics was, in fact, a spark chamber, whose electrode plates were replaced by grids of parallel wires in order to reduce multiple scattering, energy loss and secondary interactions, and to allow the localization of particle impact points without using photographic methods. Later, the idea of the Geiger-Müller counter was taken up again and developed into modern MWPC. In 1968 Charpak was able to operate such detector and the first large-size MWPCs were successfully used in the early 1970s by Jack Steinberger and collaborators in an experiment on CP violation.

A MWPC in its simplest form is made of a set of thin parallel anode wires stretched and sandwiched between two cathode plates (figure 1). The chamber is filled with an appropriate mixture of gases depending on the desire mode of operation. On application of a high voltage (HV) between anodes and cathodes, the electric field takes the form shown in figure 2.

If an ionizing process occurs in the gas, the produced electrons (primaries) will drift toward an anode wire. Far away from those wires (~20 times the wire-diameter) the electric field is basically constant, however, near them the electric field becomes inversely proportional to the square of distance (r) to the wire, therefore the primary electrons can gain enough kinetic energy so that inelastic collisions with the gas molecules can lead to new ionizations, with the generation of secondary electrons (figure 3).

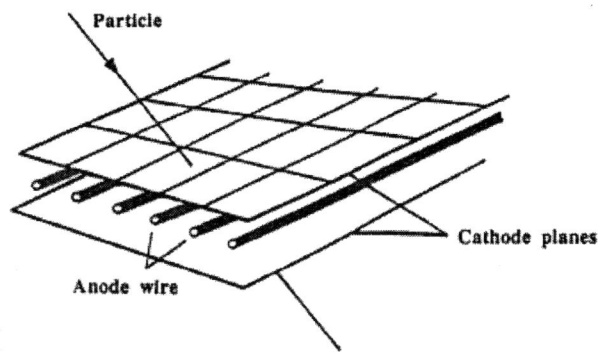

FIGURE 1. Schematics of a typical MWPC.

The latter can undergo further inelastic collisions, eventually resulting in what is called "an electron avalanche" or "charge multiplication". If the total collected charge is proportional to the number of primary electrons, then the chamber is said to operate in the "proportional mode". The proportionality constant is called "multiplication factor" and depends exponential on the applied HV.

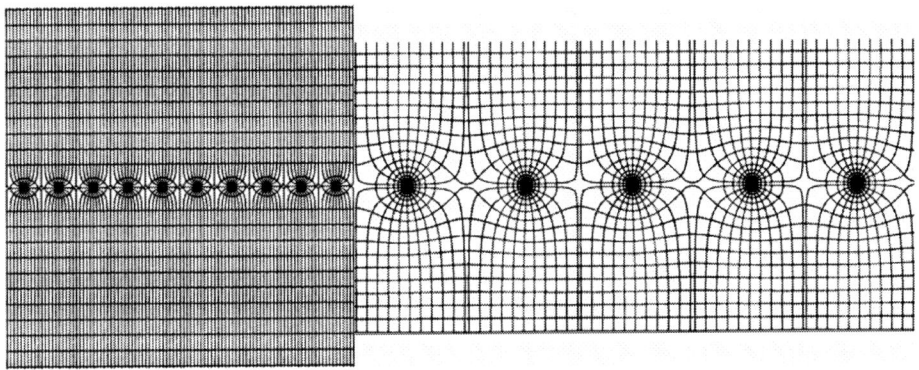

FIGURE 2. Typical Electric field lines and potentials lines in a MWPC.

The electron avalanche is rapidly (~nsec) collected by the wires, the positive ions leftover (figure 3) in the trail of multiplying electrons move in opposite direction toward the cathode. In their motion, they induce image charges in all surrounding electrodes, and these results in a negative signal on the wire where the avalanche originated. So, in principle, each wire could act as an individual detector.

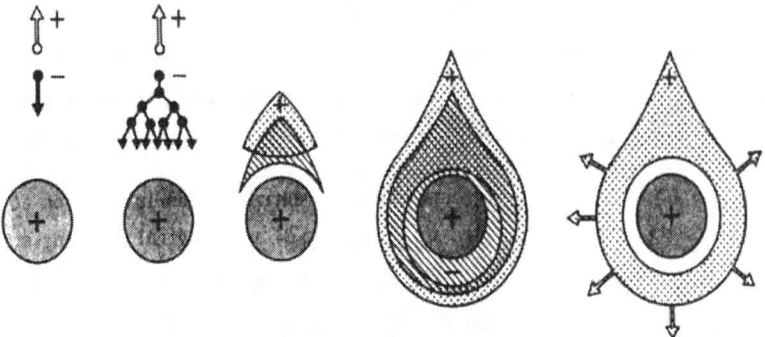

FIGURE 3. Typical Electric field lines and potentials lines in a MWPC.

If it is possible to decode on which of the wires the signal originated, then the MWPC is said to be "position sensitive". The easiest, but most expensive, way to read those electric signals is to connect each wire to a circuit which includes an amplifier, a discriminator and digitizer (ADC or QDC). However, handling a high number of electronic channels is often not affordable. In those cases, cheaper (interpolation) methods such as delay line, or charge division, are recommended (Ref 1).

EXPERIMETAL PROCEDURE

Mechanical Construction

The multiplication factor, efficiency and other operating characteristics of a MWPC depend on both, the mechanical parameters (wires diameter and distances, electrode's gap, etc), and on the gas used. Since our aim here is that the students could build their own position sensitive MWPC, the chambers to be described here have a design and an electronic read out that is easy to assemble, and is sufficient to demonstrate position sensing capabilities. The mechanical parameters here can be considered as typical for a small size MWPC (Ref. 2,3).

Our chamber consists of a central anode wire plane placed between two cathodes. The anode was formed by a 17 Tungsten Au-plated wires (25 microns diameter and 3 cm long) with a 2.5 mm separation between them (the "pitch"). They were soldered to a 1/16"-thick fiberglass printed circuit board (PCB). The cathodes were made from a single 4x4 cm Cu strip laid down on a PCB.

The electrodes were enclosed in a 10x12x2.5 cm^3 gas-thigh, high-density, polyurethane container (see fig 4). For pedagogical reason the box had a transparent

Lucite® cover to show the inner parts of the chamber. A 5x3 cm² entrance window made of 1.5 microns thick Mylar was also necessary to allow a calibration using alpha particles from a radioactive source.

In order to obtain the position information, the charge division readout method was implemented in these chambers. In this method the wires are electrically connected by resistors forming a chain. The ratio of the charge collected at one end of the chain to the total charge determines the position (fig 5).

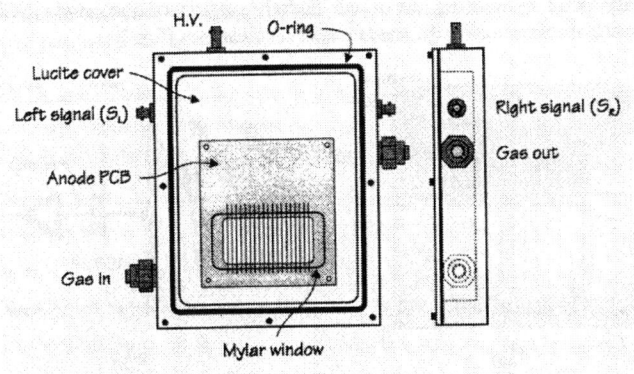

FIGURE 4. Schematic diagram of the mechanical assembling of the MWPC.

Assembling

Each chamber were designed like a ready-made kit, which includes housing, anode and cathode PCBs, anode wires, spacers, plastic screws, resistors, capacitors an soldering equipment. The detector housing was fitted with the appropriate feed-troughs for the HV, signal outputs and gas connectors (fig 2). In order to simplify

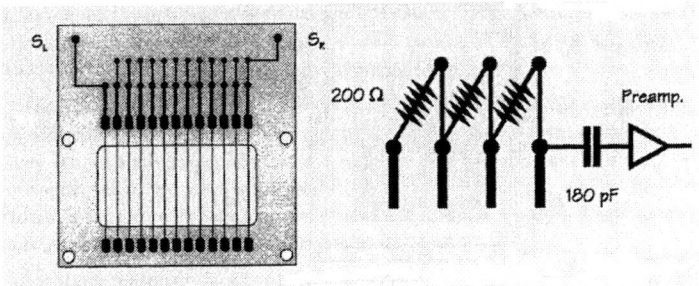

FIGURE 5. Schematic of the anode plane and resistive network implemented in the charge division readout method.

the construction, the students were provided with PCB frames on which the tensed (100 g) Tungsten wires had already been soldered. In this way, the anode plane could be quickly built, by just placing the wired frame directly on top of the PCB anode, and soldering the wires using the anode tracks as guidelines (fig 5).

After soldering the anode wires, it was necessary to cut (very carefully) the wires from the PCB frame. A sharp cuter or single edge knife was found to work just fine. In order to make the resistive chain, a series of resistor (200 ohm each) is placed in the back-side of the PCB anode. Once in place, they were soldered by the students.

The next step was to assemble the anode, and the cathode, planes using the plastic screws and spacers. At this point, it was possible to try different spacers and work with different cathode-anode distances (D), just keeping in mind that for large D's the chamber needed higher voltages to produce the same charge. The chamber was assembled using only one cathode plane to allow a see-through environment. Finally, the chamber was mounted on the plastic box, and each end of the resistive chain was connected through a 180 pF high voltage capacitor to their respective signal outputs (labeled SL and SR in fig 4). The anode wires were connected to the HV through a 10 Mega-ohm resistor to decrease parasite currents. The cathode was connected to a common electric ground. Final testing of electrical connections was made before placing the Lucite® cover of the detector box, with the corresponding O'ring acting as a gas-tight seal. After this procedure, the MWCP was ready to be filled with gas.

INITIAL OPERATION

Gas Flow

As with all gaseous detectors, the choice of gas depends on the multiplication factor, among other experimental requirements. It was found that the avalanche process occurs in noble gases at much lower voltages than in complex molecules. Thus, the gas used in a MWPC usually has noble gas as main component. As an example of how much the operational voltage and the avalanche size can vary, in figure 6 we present the pulse-height measured with a chamber (identical to the one built in the School) but filled with two different gases 80%Ar+20%CO_2 and 90%Ar+10%CH_4. During the laboratory session the gas used was Ar+CO_2 mixture which is non explosive.

Turning on the voltage

For simplicity, the chamber was operated under a steady gas flow at atmospheric pressure. An ^{241}Am radioactive source, producing 5.48 MeV alpha-particles, was

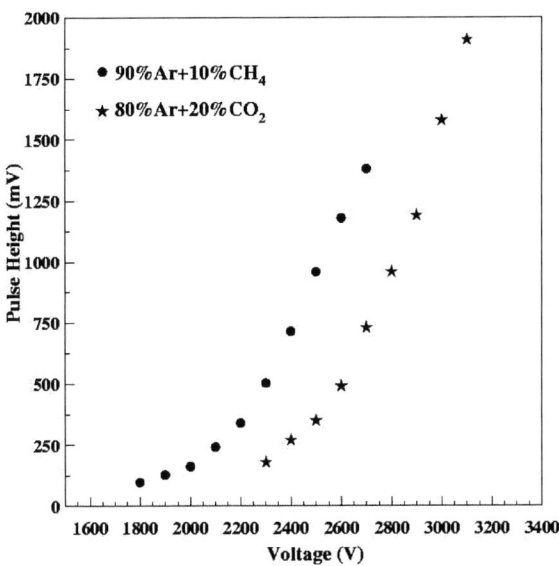

FIGURE 6. Pulse height variation as function of the voltage for two different gasses.

used for testing purposes. This source was mounted on a holder with a collimator of 0.5 mm and placed directly on front of the Mylar entrance window of the chamber. The collimator was used to improve the position resolution as well as to control the counting rate. The MWPC output was connected to an ORTEC 142C charge-sensitive preamplifier, followed by a shaping amplifier ORTEC 855 operated with a shaping time of 0.5 microseconds. After flushing the chamber for several minutes with the gas, the HV power supply was switched on, and the bias slowly increased making sure to start from 0 V, and not to exceed 3.5 kV. After reaching 2.3 KV, the signals were clearly visible in the oscilloscope (1 µs/div, 50 mV/div). Figure 7 shows a typical pulse. It is possible to use either the unipolar or bipolar outputs of the amplifier.

MEASUREMENTS

Pulse height

One of the first things to measure in a gas detector is the relationship between operational HV and pulse height (PH). This measurement was carried out by placing the ^{241}Am source near the center of the chamber and setting the HV to 100 Volts less that the minimum voltage ($\approx 2,200$ V) at which the signal is observed for

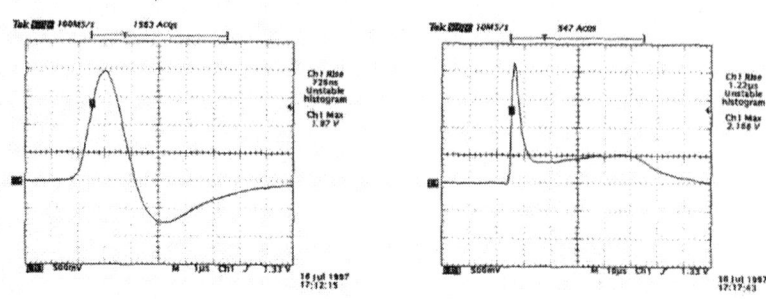

FIGURE 7. Typical signal outputs (256 sample mean) as seen on a digital oscilloscope, bipolar (a), and unipolar (b).

first time. In such conditions, the scope's trigger level is set just above the electronic noise, and kept at the same level from then on. In order to measure the PH, the HV was increased, and our digital oscilloscope was set in an "averaging mode" (64 samples, or higher). Under this condition, a very clean and stable pulse can be observed and measured. The PH was registered for different HV values. Figure 8 shows a log-linear plot of the results obtained while operating one of the chambers. The straight line observed in this plot is the prediction of a simple avalanche growth theory (Ref 4). This exponential growth is also a signature that the chamber is working in the proportional mode. The deviation from linearity at highest HV value indicates the beginning of the Geiger regime, where chamber starts to discharge (sparks).

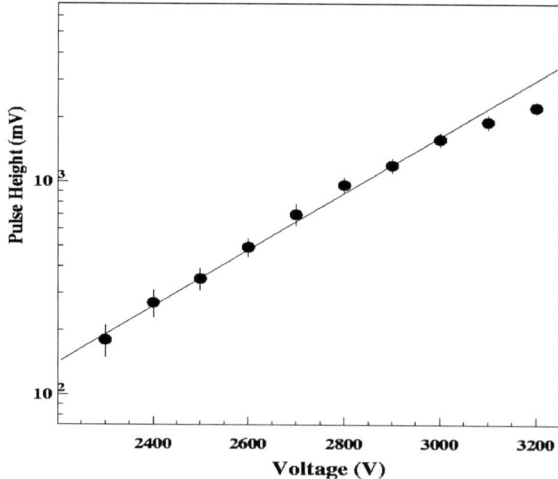

FIGURE 8. Pulse height vs. HV applied between the anode and the cathode. The straight line represents an exponential ($Ae^{(bx)}$) fit.

Position Sensing Capability

The use of digital oscilloscope also helped to illustrated, at least qualitatively, the position sensing capability of the MWPC. Since the charge division method is used, it is expected that the output signals SL and SR (fig 4 and 5) should show a correlation with the position of the source. Placing the source at the middle of the chamber both signals should have approximated the same height. Therefore, when the source is moved closer to the right (left), the SR (SL) increases while SL (SR) decreases, as shown in Fig. 9. This reveals charge conservation, i.e., the SR, SL are proportional to the charge collected at either the end of the resistive chain, so the total signal for a single event has to be proportional to (SL+SR). Thus, SR/(SR+SL) = (1-SL)/(SR+SL), therefore whenever SR (SL) increases, SL (SR) has to decrease. In fact this ratio is also proportional to the position X of the traversing particle (Ref 5), as measured at the left-end of the chamber, and it is given by

$$X = L \ SR/(SR+SL)$$

where **L** is the total active length of the chamber along X (in our case, 4 cm).

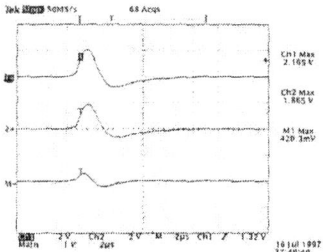 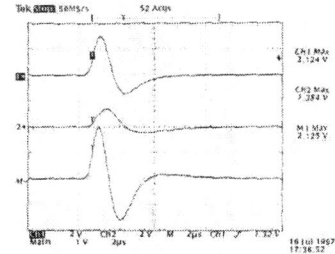

FIGURE 9. Position determination using a digital oscilloscope. In (a) the radioactive source is near the center of the MWPC. In (b) the source is closer to the right of the chamber. The signal output order is: right (top), left (middle), ratio right/left (bottom).

SUMMARY

In this laboratory session the students built from scratch a MWPC. This MWPC was operated in the proportional mode, and shown to have position-sensing capabilities. Of course, time permitting, once the chamber is ready and working, it is possible to demonstrate to the students other features such as rise time, energy and position resolution etc.

ACKNOWLEDGEMENTS

We would like to thank Roberto Núñez for their help in constructing of the MWPC kits.

REFERENCES

1. Lopez-Robles J.M., Alfaro R, Belmont-Moreno E., Grabski V., Martinez-Davalos A., Menchaca-Rocha A., *IEEE Trans. on Nuc. Sc.* **52**, 2841 (2005).

2. Erskine G.A., *Appl. Nuc. Inst. and Meth* **105**, 565 (1972).

3. Charpak,G and Sauli F., *Nuc. Inst. and Meth* **162**, 405 (1979).

4. Rose M. A. and Korff, S.A. *Phys .Rev.* **59**, 850 (1941).

5. Albrei J.L. and Radenka V., *AIEEE Trans. Nuc. Sci.* **23-1**, 251 (1976).

Silicon microstrip detectors

Luis M. Montaño

Physics Department, Cinvestav, Mexico City, Mexico

Abstract. The main scope of this laboratory is to give the students an introduction of some special characteristics of silicon microstrip detectors. The students will perform some exercises using different instruments to appreciate the properties of these detectors, especially its great position resolution. An overview of different instruments such as an oscilloscope, wave function generator as others will be also given as important devices in any experimental laboratory.

INTRODUCTION

Semiconductor detectors were mainly developed to identify the path of charged particles in high energy experiments. Due to the great energy and spatial resolution and the small amount of radiation energy to generate an interaction in them, these detectors have been successfully applied in tracking and identification systems. Therefore, they are generally situated in the closest position of the particle interaction point in any experiment on particle physics.

In general the principle of particle detectors is the identification of the tracks left by the passage of a charged particle by ionization. Also, detectors are sensible to photons through the well known processes: photoelectric effect, pair production, compton scattering, bremsstrahlung, as others. When photons pass through matter, it knocks out electrons from the atoms disturbing the structure of the material and creating loose electrons. Thus this interaction leaves a trace of disturbed matter and move the electrons from the original atoms to some terminals (wires, anodes) where they can be collected. This collected charge will be the signal that the rest of the electronic chain in the system will identify as an electronic pulse.

The spatial resolution of the silicon detector depends on the design. There are pixel or strip detectors, two and one dimensional respectively. There are others which also identify the kind of particle through the energy deposit in it. Their energy resolution depends on the number of interactions on the detector. So. in order to have a good energy resolution it is necessary to increase the number of information carriers creating during the interaction. This is another vantage of semiconductor detectors. Other detectors use light as the carriers of information, instead silicon detectors use electron-hole pairs.

More than two decades ago silicon detectors have been used in high energy physics experiments mainly in the identification of charged particles and tracking position[1]. Due to their versatility they also have been applied in other fields as Astrophysics,

Medical Physics and others.

The great amount of applications of silicon detectors is due to parallel development in electronics of low noise and very large system integration (VLSI). This kind of electronic is necessary for the small amount of charge created in the detector when radiation passes through. This charge is hidden by the electronic noise. So it is necessary to have an electronic system with very low levels of noise. For a good application of these devices it is needed to get at least a signal-to-noise ratio of one order of magnitude. The amount of charge deposited in the typical 300 μm of thickness of a silicon detector is around 25,000 electrons which means a charge of 4fC.

In this laboratory the students will develop some exercises to appreciate the properties of silicon detectors as its great spatial resolution. This lab is based in a previous one[2] given in other course in Mexico. This is an improved version of it.

SILICON DETECTORS

When an electric field is created in the semiconductor, the electron-hole pairs present in the material undergo a net migration. This effect is called the drift velocity of the carriers. This migration or velocity depends on the electric field. The time required to collect the charge could be of nanoseconds, which makes them one of the fastest-responding of all radiation detector types. One can see in fig.1 a charged particle or photon hitting the detector.

The dominant advantage of semiconductor detectors lies in the smallness of the ionization energy. The main elements that semiconductors are made are silicon and germanium. For them it is required only an energy of 3.6 eV to create an electron-hole pair while for a gas-filled detectors a 30 eV is needed. As it was pointed before this increment has two beneficial effect on the energy resolution; the statistical fluctuations diminishes and the signal to noise ratio improves considerable.

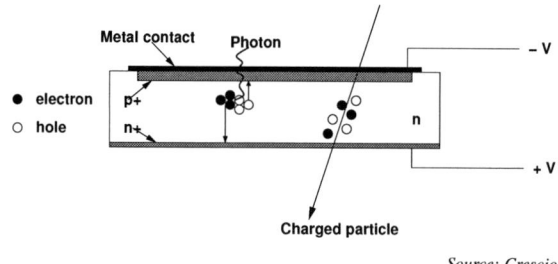

Source: Crescio

FIGURE 1. Charge particle and a photon hitting a silicon detector.

The main characteristics of the silicon detectors that make them useful devices are:

- Speed of reaction when radiation cross the surface of 10 ns.
- Spatial resolution $\sim 10 \mu m$.
- Flexibility of design.
- Small amount of material (0.003 X_0 for 300 μm thick detector).
- Linearity of the response vs. the deposited energy.
- Good resolution in the deposited energy.
- Tolerance to high radiation doses.

A silicon detector can be visualized as a diode with a junction of p^+ and n material. In this junction a depletion zone is created. Applying an inverse voltage this depletion region expands, decreasing the production of charges (leakage current), letting the region prepared for detecting radiation that will originate charges when crossing the wafer surface. The electric field created guides the generated charge to the cathodes. These cathodes are the p^+ material which collect the charge that will be transmitted to the electronics. For our detector these cathodes are the microstrips. Above each cathode there is a metallic cover to permit the connection between the detector and readout electronics via a microbounding.

The detector used in this laboratory is known as microstrip silicon detector. Our detector has 256 strips, 50 microns pitch, 300 microns wide and an area of $2 \times 1.25 cm^2$. The read out of the signal is done by an electronic chip called Viking. It is connected only to 128 strips so it covers an active area of $2 \times 0.625 cm^2$.

Microstrip detectors provide therefore the measurement of one coordinate of the particle's crossing point with high precision. Using very low noise readout electronics, the measurement of the centroid of the signal over more than one strip further improves the precision. Clearly the precision of this procedure depends on the noise of the readout chain (including the quantization error introduced by the analog-to-digital converter, which is important when using small signals). If digital readout is used (strip hit or not hit), the resolution is simply $\sigma = \frac{pitch}{\sqrt{12}}$ [3].

Viking chip

For the application of silicon detectors in identifying charged radiation and X-ray, a low noise electronic system is required. VIKING, a low noise silicon strip readout VLSI chip has been designed for this purpose. It was constructed in $1.5 \mu m$ CMOS technology[4].

The chip contains 128 low power (1.5 mW/channel) charge sensitive preamplifiers followed by CR-RC shapers and sample and hold circuit, input and output multiplexing and one output buffer. Use of time continuous shaping facilitates triggered applications and enables optimum signal to noise ratio.

Two signals activate the start/stop unit in the chip, which creates an internal clock, a "start" signal for the shift registers and an activate signal for the output buffer. This signal is necessary to allow daisy chaining of chips and hence only one output buffer at any time should be switched on. The "start" goes into the output multiplexer and the internal clock shifts it through the 128 channels. For one clock cycle each channel is connected to the output buffer to read the channels out. After 128 clock cycles the outcoming signal from channel 128 stops the internal clock, disables the output buffer and creates a shift-out which can be used as a shift-in for the next readout chip. The peaking time is $1.5\,\mu s$ and the noise is typically $70\,e^- + 12\,e^- C$.

For low energy X-ray applications the noise has to be as low as possible. A 10 keV X-ray will only produce 2800 electron-hole pairs in silicon. So, if a signal to noise ratio of one order of magnitude, as mentioned before, has to be reached it is necessary to reduce all the sources of unwanted electric signals inside the electronic chain. The contributions of the silicon detector to the total noise of the assembly come from:

- The load capacitance of the silicon detector.
- The leakage current in the silicon detector.
- Possible resistance between the active element of the detector and ground, or the bias supply.

Therefore it is important for these kind of detectors to have a very low noise electronics. Generally the development of these detectors implies a similar development of their electronics.

THE LABORATORY SETUP

In this experiment we use the following equipment:

1. Timing unit for readout circuit, VIKING TIMING;
2. Wave function generator for trigger and laser;
3. NIM crate, discriminator and scaler;
4. Oscilloscope;
5. Laser diode;

The silicon microstrip detector is wire bonded to the Viking readout circuit, which has been placed on a readout PCB (Printed Circuit Board). For this system there is a metallic box which contains the microstrip. This box is necessary to avoid the increment of noise in the silicon detector produced by visible light and it will permit us to have a signal as clean as possible.

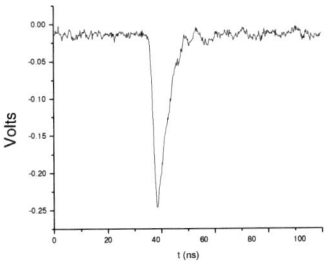

Source: Montano

FIGURE 2. Common pulse of a cosmic ray detected by a scintillator detector coupled to a PMT.

Trigger system

The development of this system includes a start for detecting an event (trigger) in one of the 128 channels. Triggering is a common tool in particle detectors to be able to synchronize them with others and the whole system.

The trigger sends a signal in order to let the channels transmit their information or voltages to the electronic system. When a particle is detected and impinges in the active zone one or some strips show a voltage peak of different values going from 50 to 200 mV, see fig. 2.

LABORATORY COURSE ON SILICON DETECTORS

This laboratory course consists of two different mini sessions, in order to give the student some hands-on experience on various aspects of silicon detectors and related integrated electronics. The exercises to carry out are:

- Familiarization with the different instruments in any experimental lab as oscilloscope, wave function generator, as others.
- Measurement of the position resolution of a microstrip detector with a laser.

The first exercise consists in identifying the important parameters of the oscilloscope to find a wave function selected in the instrument and set it on the screen. In this simple excercise the main scope is to let the student to handle the oscilloscope to set the best parameters of voltage and time scale and also to set the trigger.

The second exercise consists in using a laser to make some studies of the spatial resolution. In this case it is required to modify a little the set-up. A special assembly, a micro manipulator has been mounted on the metallic box containing the silicon detector, which can be precisely moved such that the translation direction is orthogonal to the

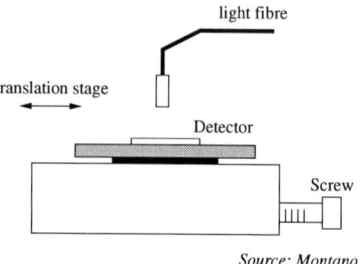

FIGURE 3. Device to move the laser perpendicular to the strips.

strips. An optical fiber has been mounted there to move the light shining the strips and register it in the oscilloscope. There is about 100 μm distance from the laser to the silicon surface.

Also in this lab session the students will acquire some knowledge and skills on diodes, in managing electronic devices such as voltages supplies, NIM technology, etc.

As we have seen during this report, silicon microstrip detectors are the most commonly used device for high resolution tracking in particle physics. The strip design allows a large sensitive area with relatively few readout channels. The basic strip detector is read out on one side giving information of the track position only in one dimension.

To trigger the viking timing module and at the same time to send a laser light pulse to the detector, we will use a laser diode. In order to enable the laser as a light pulse mode, we used a wave generator device. The circuit of the laser diode gives also an electric pulse which will be used as a trigger. So a cable with this electric signal goes to the module and the laser light goes to shine the detector transmitted by an optic fiber. This fiber is placed in a micro manipulator which can be moved with good precision (see fig.3). The transfer direction is orthogonal to the strips so we can see in the oscilloscope the strips hit by the light.

Measurements

Once the students worked out with the instruments we can develop the second exercise which consists in determining the position resolution of the silicon microstrip detector. A little lense was placed in the output of the optic fiber in order to focalize the light, see Fig. 4. In this way we were able to obtain the laser spot a few strips wide. We require the information from a number of strips in order to accurately determine the peak position. This is done by determining the center of mass of the pulse. Moving the laser some known distance, we can calculate the spatial resolution comparing this distance vs. the interval between two center of mass of each pulse. It is important to say

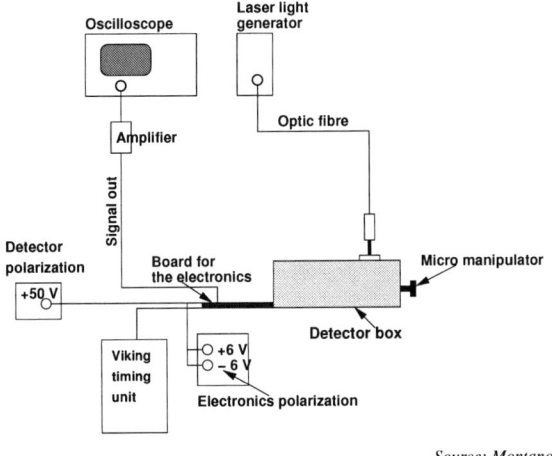

FIGURE 4. Set-up with laser and silicon.

that one complete turn move the laser 0.9 mm. So we can proceed as follows:

1. Move the micro manipulator by turning the micrometer placed on the box. Using an oscilloscope you will now see the signal from the light moving from one strip to another.
2. Place the laser in a region with a nicely distributed signal.
3. Turn the screw one seventh of a complete turn and repeat step 2.

Results

As we can see, the light should hit some strips and that is shown in fig.5 . Sometimes there are more strips hit in one zone than in another. This is due to the not perfect straight movement of the laser mechanism. There were one where one could see 6 strip and other where 7 were seen.

Moving the laser along the detector and being its light spot perpendicular to the strips we register the amplitudes. Table 1 shows the measured amplitudes (in mV) of the signal for different strips in some positions of the laser. We decided to move the laser an equivalent turn of 20 μm which in our case it is a 1/50 of a complete turn. In the screw there are some marks to indicate those intervals.

As it is shown in fig. 5 we write down the amplitude of each strip. These strips are manifested as little steps in the distribution. As the system is build, the signal of the different strips can be seen in the oscilloscope where the signal appears as amplitude vs. time. In this case the amplitude of each strip corresponds to $1\mu s$.

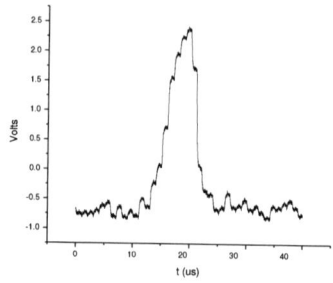

Source: Montano

FIGURE 5. Laser signal of the microstrip in the oscilloscope.

TABLE 1. Raw data from the second setup.

Strip n	0 μm	20 μm	40 μm
1	750	540	400
2	960	720	560
3	1240	950	780
4	1050	1210	1150
5	900	980	1270
6	660	830	1080
7	420	640	910

Source: Experimental results

The peak position can be determined by calculating the centroid :

$$C = \frac{\sum A_i \cdot x_i}{\sum A_i} \qquad (1)$$

A_i represents the amplitude and x_i the strip number multiplied by the pitch (50 μm). Performing the operation of calculating the peak position one can demonstrate that the measured difference between the peak positions is very close to the expected one of 20 μm.

Since we are averaging the signals from the Viking in order to minimize the electronics noise, and taking 7 points to calculate the centroid, it can be seen that position accuracy down to a few μm can be achieved. To determine in a more precise way the resolution of the detector one should take several measurements for "single events" (i.e. not averaging the signal, but taking only one signal for each measurement) and fill an histogram with the distribution of the errors on the measured peak differences. The RMS of this distribution would represent the detector resolution.

Acknowledgments

I want to thank L. M. Villaseñor for having invited me to give this lab and for his warm hospitality.

REFERENCES

1. ALICE Collaboration 1999, *CERN/LHCC, 99/12*, (Swi/CERN)
2. Montano L. 2005, *Proc. XI Mexican School on Particles and Fields*, (USA/Journal of Physics) p368-379.
3. Duerdoth I. 1982, *Nucl. Instr. and Meth.* **203** 291.
4. Toker O, Masciocchi S, Nigard E, Rudge A, Weilhammer P 1994, *Nucl. Instr. and Meth.* **A340** 572.

Laboratory Course on Drift Chambers

Ix-B. García-Ferreira, J. García-Herrera and L. Villaseñor

Institute of Physics and Mathematics, University of Michoacan, Bdg. C3-Ciudad Universitaria, Morelia, Michoacan, 58040, México

Abstract. Drift chambers play an important role in particle physics experiments as tracking detectors. We started this laboratory course [1] with a brief review of the theoretical background and then moved on to the the experimental setup which consisted of a single-sided, single-cell drift chamber. We also used a plastic scintillator paddle, standard P-10 gas mixture (90% Ar, 10% CH_4) and a collimated ^{90}Sr source. During the laboratory session the students performend measurements of the following quantities: a) drift velocities and their variations as function of the drift field; b) gas gains and c) diffusion of electrons as they drifted in the gas.

Introduction

When a charged particle passes through a gas, it will interact electromagnetically with nearby atomic electrons, resulting in the creation of electron/ion pairs along the path of the particle. The number of such pairs created depends on the energy of the particle and the type of gas, but for a typical gas at STP and a particle with unit charge, the mean number of electron/ion pairs formed will be on the order of 100 cm^{-1} atm^{-1}.

If an electric field is applied, the electrons will start to drift through the gas towards the positive electrode, undergoing repeated collisions with the gas molecules. (The ions also drift in the opposite direction. However, their drift speed is much less than that of the electrons, so they can be ignored in this discussion.). If the electric field near the anode is strong enough, an electron can acquire enough energy between collisions to knock an additional electron free from a gas molecule. This additional electron can then go on to ionize more gas molecules; in this way, an avalanche is formed in which the number of electrons increases exponentially. When this avalanche reaches the positive electrode, it gives rise to a measurable current, the size of which is proportional to the original number of ions created. The ratio between the final number of electrons collected and the initial number deposited is called the *gas gain*, and for practical detectors is typically on the order of 10^4–10^6.

The large electric field needed to obtain gas amplification is usually obtained by forming the anode from a very thin (20-100μm) wire. An electron sitting in the gas far away from the anode will see a much smaller electric field, and will drift towards the anode with a velocity roughly proportional to the field. When it gets close to the anode, the electric field will start to rapidly increase, and the electron will initiate an avalanche.

The fact that an electron drifts with a predictable speed over most of the distance to the anode implies that one can turn a measurement of the time an electron took to drift to the anode into a measurement of the distance of the original source particle from the anode. A device designed for this type of measurement is called a drift chamber.

In this way, the drift chambers are electronic detectors providing accurate measurement of the position of a charged particle. The theory behind the drift chamber was first studied by Bressani, Charpak, Rahm and Zupancic in 1969. The first fully operational drift chamber was used to study particles by Prof. A.H. Walenta, Heintze and Schürlein in 1971 [2].

The drift chamber works because of ionization. As a charged particle passes through the gas, it ionizes the gas atoms. These atoms then feel the force of an electric potential and drift towards wires that register a signal. Knowledge of the drift time (time from ionization to signal) and the drift velocity in the gas allows one to determine the position of the incoming particle. Stacking several drift chambers on top of one another allows one to resolve a track left by the particle, which can give information such as incoming angle. If there is an applied magnetic field, the track will curve and one can then determine the momentum of the particle. Drift chambers can achieve spatial resolution of 50 μm or better.

The basic principle of operation makes use of the fact that the timing of the signal in a proportional counter shows a time lag with respect to the moment of particle passage (and consequently the moment of creation of the ionization along the particle track in the gas of the detector volume). This time lag is related to the time the electrons from the ionization take to travel (drift) to the amplifying anode wire (Fig.1). In proportional counters this time lag is considered as a nuisance limiting the time resolution of the detector. If, however, the timing of the particle passage is determined with higher precision by scintillating counters (hodoscopes) or by the timing of the accelerator bunch, the time lag can be used to determine the exact position of the ionization with respect to the anode wire.

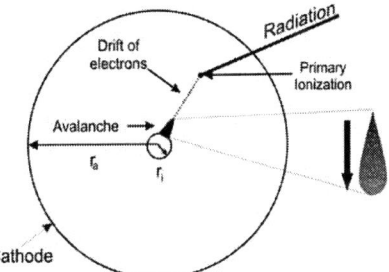

Figure 1. Principle of proportional counter (Adapted from A.H. Walenta) [3].

Mathematically the measured drift time t_{drift} is related to the drift path (for zero magnetic field along the electrical field lines) from the location of ionization creation along the track to the anode by:

$$t_{drift} = \int_{track}^{anode} \frac{ds}{v_{drift}(x)} \qquad (1)$$

In order to cover larger surfaces these cells have to be repeated which requires in addition a solution for the right-left ambiguity. This can be achieved by the "double wire method" (Fig.2a) or by the "staggering method" (Fig.2b).

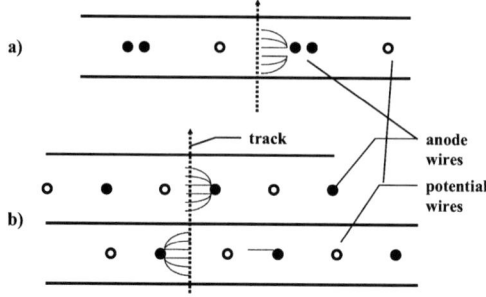

Figure 2. Right-left ambiguity in drift chambers. a) Double wire type, b) Double chamber type [1,3].

Theoretical Background

- Drift and diffusion in gases

The motion of an electron in the gas is governed by two physical phenomena: drift and diffusion. Diffusion is the random motion caused by the thermal energy of the electron. The result is that after a time t, a collection of electrons originally distributed in a straight line at $x = 0$ will be distributed in a Gaussian distribution of the form:

$$\frac{dN}{dt} = \frac{N_0}{(4\pi Dt)^{1/2}} \exp(-\frac{x^2}{4Dt}) \qquad (2)$$

Where N_0 is the total number of charges, and D is the diffusion coefficient given by:

$$D = \frac{2}{3\sqrt{\pi}} \frac{1}{p\sigma_0} \left(\frac{(kT)^3}{m}\right)^{1/2} \qquad (3)$$

Where T is the temperature, k is the Boltzmann's constant, p is the pressure, m is the mass of the electron and σ_0 is the cross section for an electron to collide with a gas molecule. The velocity due to drift can be determined from the mobility of a charge in a gas, which is given by:

$$v_{drift} = \mu E \qquad (4)$$

Where v_{drift} is the drift velocity, μ is the mobility and E is the electric field strength. The mobility is related to the diffusion coefficient by the Einstein relation:

$$D = \frac{\mu kT}{e} \qquad (5)$$

Where e is the charge of electron. For a further explanation of these expressions, see [11].

The drift velocity v_{drift} versus field E is pressure dependent since the collision time depends on the pressure, so that v_{drift} versus E/p is the correct variable to plot. Where E/p is the reduced electric field. The first ion that a charged particle creates is appropriately called the primary ionization. As the electron from this ionization is accelerated by the electric field, it will eventually gain enough energy to ionize other atoms. Then these secondary ionizations can gain enough energy to cause further ionizations, and so on. This effect is called avalanche multiplication, and is important for the gain of the chamber. The electric field is constant through most of the chamber, except very near of the anode wire, where it goes as $E \propto 1/r$, where r is the radius of the wire. Thus most avalanche multiplication takes place within a few radii of the wire.

- Ionization by charged particles.

As a charged particle passes through a gas, it actually loses energy via 2 mechanisms: excitation and ionization. In excitation, the particle passes a specific amount of energy to a gas atom. The cross section for that is around $10^{-17} cm^2$. In ionization, the particle knocks an electron off the gas atom, and leaves a positively charged ion. The cross section for ionization is about $10^{-16} cm^2$. It is only ionization that will lead to a signal in the drift chamber, so excitation isn't directly useful. However, excited atoms can still participate in reactions that lead to ionization, so the energy is not lost. An example of this is the Penning Effect where an atom stays excited long enough to collide with another atom and ionize it.

The ionization clusters are produced in the following way: the time dependent electrical field of the passing particle couples to the electrons of the gas atoms and may have enough impact to kick out the electron of an atomic shell, say the L-shell, causing an energy transfer E'. The excess energy $E_{Kin} = E' - E_L$, represents the kinetic energy of this liberated electron which in turn may ionize again. In this case it is considered as a new charged particle and is called δ-ray. If the excess energy is low it will not ionize and thermalize quickly. The ionization left behind by these individual encounters, including the rearrangement of the L-shell, mediated by Auger-electrons or x-ray emission, is called a cluster and the sum of the clusters represents the footprint of the charged particle, which will be detected in the drift chamber.

Therefore it is important to know some features:
1. The number of the clusters per track length
2. The spatial distribution along the track
3. Their size, measured in terms of the number of free electrons belonging to them
4. Their spatial extension

Ionization energy loss of charged particles

Tracking detectors designed to make position measurements along the trajectories of charged particles can simultaneously be used to identify particle species by measuring the mean ionization loss along the tracks. This applies to any detector in which particles lose energy to ionization, e.g., gas-filled wire chambers or silicon strip detectors. Since the pulse height on each wire or strip must be read out, this technique requires the use of analog readout electronics. Because the ionization energy loss is a statistical process with large fluctuations, many measurements are needed along each track to get a precise mean. For gas-filled detectors, the statistics are improved by operating at higher gas pressure, since this result in more primary ionizations along the particle path. The precision of the mean is improved by using the method of "truncated mean" whereby a certain fraction of the signals of largest size are removed in taking the average. The energy loss mechanism for a charged particle passing through a gaseous medium is mainly by interactions with atomic electrons i.e. ionization loss. The ionization energy loss of charged particles is fundamental to most particle detectors.

If $\Phi(E', E)\, dE'\, dx$ is the probability for an energy loss in the interval between E' and $E'+ dE'$ of the charged particle with energy E along a track segment dx, then the mean number (N) of such encounters per track length is given by:

$$\langle \frac{dN}{dx} \rangle = \int_{E'_{min}}^{E'_{max}} \Phi(E', E) dE' \qquad (6)$$

Where the integration extends from the smallest possible energy transfer E'_{min} to the largest E'_{max}. Since $\Phi(E', E)$ also contains encounters leading to excitation of the atom, the number of ionization clusters is obtained if only those encounters are considered leading to ionization. A good approximation can be obtained from a simple collision model (Bohr's model):

$$\Phi(E', E) dE' = \frac{\tilde{A}\rho}{\beta^2} \frac{1}{E'^2} dE' \qquad (7)$$

With ρ the gas density, $\beta = v/c$, and $\tilde{A} = 0.1536$ Z/A MeV cm^2 g^{-1}; again Z is the atomic number and A is the atomic weight respectively.
Inserting 7) into 6) yields:

$$\langle \frac{dN}{dx} \rangle = \frac{\tilde{A}\rho}{\beta^2} (\frac{1}{E'_{min}} - \frac{1}{E'_{max}}) \approx \frac{\tilde{A}\rho}{\beta^2} \frac{1}{E'_{min}} \qquad (8)$$

Which takes into account that $E'_{max} \gg E'_{min}$. For the calculation of E'_{min} the quantum mechanical effects cannot be neglected and therefore a simple formula has not been developed. Since the generation of an individual ionization cluster is completely random the actual number in a chamber m_p varies according to a Poisson distribution. This is responsible as well for the random distribution along the track resulting in an exponential probability distribution of the gaps between two clusters.

The cluster size distribution is governed by the energy loss distribution described by $\Phi(E', E)$ such that the number of electrons is typically E'/W with W the energy necessary to create an ion pair. Following the steep decrease of $\Phi(E', E)$ most of the clusters will be single electron clusters and larger clusters will occur only occasionally. These large clusters can be considered as small tracks on their own (δ-rays) and their range is responsible for the spatial extension of the clusters, which may reach a few hundred microns in a counting gas at NTP. In a similar way as the mean number of clusters in equation (6) the mean energy loss for a track segment is obtained by

$$\langle \frac{dE}{dx} \rangle = \int_{E'_{min}}^{E'_{max}} E' \Phi(E', E) dE' \qquad (9)$$

This is called the mean energy loss.

Taking into account relativistic effects and relation (7) the well-known Bethe-Bloch formula is obtained:

$$\langle \frac{dE}{dx} \rangle = \frac{\tilde{A}\rho}{\beta^2}(2\ln\frac{2m_e c^2}{\bar{I}^2} + 2\ln\beta\gamma - 2\beta^2 - 2\delta) \qquad (10)$$

Where $\bar{I}$ is the mean ionization potential and δ the density correction.

The fluctuations of the energy loss are given by an asymmetric rather broad Landau distribution with a long tail towards higher energy. A good description of this distribution is obtained using the most probable energy loss E_{mp} and the full width at half maximum ΔE. The properties of this distribution become important in the light of a possible application for particle identification using either the $1/\beta^2$ dependence in the non relativistic case or the logarithmic rise in the relativistic part of the energy loss formula 10). Since the fluctuations of the Landau distribution are rather large ($\Delta E/E_{mp}$ = 0,40 ... 1,0) only a method of multiple sampling yields the necessary resolution of about 6%.

The proportional gas gain

Originally the proportional counter is used in cylindrical geometry, i.e. a wire with a radius r_i of less than 50 μm is stretched in the center of a tube (outer cylinder) with a radius r_a of typically a centimeter (see figure 2). The inner wire is held at positive potential U_0 with respect to the outer cylinder resulting in the known potential U(r) and field distribution E(r) of a cylindrical capacitor:

$$E(r) = \frac{U_0}{r} \frac{1}{\ln\frac{r_i}{r_a}} \qquad (11)$$

and

$$U(r) = \frac{U_0}{\ln\frac{r_i}{r_a}} \ln\frac{r}{r_a} \qquad (12)$$

Each primary electron contributes to the total charge in the avalanche G ion pairs, where G is called gas amplification. Using the assumption that the gas multiplication is only due to the accelerated electrons ionizing the gas atoms, then the increase, dn in the number of electrons, n, (the growth of the avalanche) on the path length dr is described by the Townsend process dn = α·n·dr, where the probability of ionization is defined by the *first Townsend Coefficient* α and consequently for the total avalanche the gain G is calculated:

$$\ln G = -\int_{r_1}^{r_2} \alpha(r)dr = \int_{r_i}^{r_a} \alpha(r)dr \qquad (13)$$

The integration extends from the threshold radius r_1 to the surface of the anode wire $r_2 = r_i$. For simplicity the integration can be carried out from the outer (cathode) radius $r_a = r_k$ since the additional contribution will vanish. The Townsend coefficient α is a function of the electrical field strength and therefore it becomes a function of r as well. In practice the evaluation of the integral fails because of the limited knowledge of α. Mostly simple (linearized) approximations are used resulting in formulas for the gas gain of limited use. A more satisfactory parametrization of α is obtained in the form:

$$\frac{\alpha}{p} = A\exp(-B(\frac{p}{E})^k) \qquad (14)$$

with A, B and k experimental parameters.

The parameter A defines the plateau, which is related to the max cross section for ionization by electron impact. The parameter B defines the onset of the rise and is related to the energy distribution of the electrons in the swarm and therefore is influenced by the inelastic cross section of electron scattering. Small admixtures of molecular gases have a strong influence. The parameter k changes the slope in the steeply rising section. Since the gas gain takes place in the field region E/p = 10^{-3} to 10^{-2} kV/mmTorr; the most important parameter is B which is responsible for the high voltage needed for sufficient gas gain.

Drift and Diffusion in a "mean electron model"

As relation (1) indicates the drift velocity of electrons in gases as function of the electric field is of foremost importance for the proper operation of a drift chamber. The basic process of the motion of free electrons in gases is described by the diffusion of thermalized electrons superimposed by a directional motion of drift under the force created by the electrical field acting on the charge of the electrons [5].

In a "mean electron model" (where the Maxwell-like energy distribution of the electrons is replaced by an appropriate mean energy ε or the mean velocity $c = \sqrt{\frac{8}{3\pi}}\sqrt{\frac{2\varepsilon}{m}}$ the diffusion is described by a broadening of an initial delta function for the spatial distribution into a gaussian distribution with

$$\sigma = \sqrt{2Dt} \qquad (15)$$

where the diffusion coefficient is given by

$$D = \frac{1}{3}c\lambda \qquad (16)$$

with λ the mean free path.
The drift velocity in this model is given by

$$v_{drift} = \frac{8}{3\pi} \frac{e}{m} \frac{E}{c} \lambda = \mu \cdot E \qquad (17)$$

with μ the mobility and E the electrical field strength.

It is seen that the mean free path λ (and consequently the cross section for electron scattering at the atoms) as function of the mean electron energy and the mean electron energy as function of the reduced electric field E/p determine the motion parameters v_{drift} and D. In the usual drift gases a mixture of a noble gas (mostly Ar) with a molecular admixture (mostly hydro-carbons like $C_n H_m$) are used. In order to emphasize the atomic and molecular physics, the mobility μ is considered: with increasing field E the electrons are accelerated but the velocity c is increasing only to a limit since the onset of inelastic cross sections (rotational and excitational processes) absorbs most of the energy from the field. However, the small change in c reduces the total cross section considerably due to the Ramsauer effect, where atoms become transparent to the electrons. Therefore the mobility rises sharply for small fields. At somewhat higher fields and energies the rotational cross sections become constant, therefore c rises quickly reducing the mobility. In concert with the increase of the cross section behind the Ramsauer minimum, the mobility may even drop to an extent that the drift velocity is dropping as well. This effect can be seen very well in the standard P-10 mixture (90% Ar, 10% CH_4). In order to obtain some insight to the diffusion process it is useful to consider the following relation (Einstein relation):

$$e\frac{D}{\mu} = kT_e \qquad (18)$$

Where e is the electron charge and T_e is the electron temperature which coincides with the gas temperature if the "heating" of the electrons by the electric field is negligible. For gases with large inelastic cross section for electron collisions this will be the case even for large electric field strength (cool gases). For noble gases or noble gas mixtures with a small admixture of molecular gases (e.g. P-10) the deviation starts already at small drift field. This is best seen by the relation obtained for the experimentally accessible quantity σ as function of E and l (the drift length) by combining equations. 1,6,8,9:

$$\frac{\sigma^2}{l} = \frac{2}{e}\frac{kT}{E} \qquad (19)$$

The measured quantity σ^2/l representing the quality of the drift process depending inversely on E while the minimum is governed by the relation kT/E, the thermal limit. If measurements are found above this limit, the electrons are not in equilibrium with the gas. The cooling effect of the molecular gases is clearly seen [6-10].

Experimental Setup

A schematic view of the drift chamber is shown in Fig.3 indicating the drift electrodes connected to a resistive divider chain producing a constant drift field. The electrons are amplified at three anode wires of which the two outer ones are connected together.

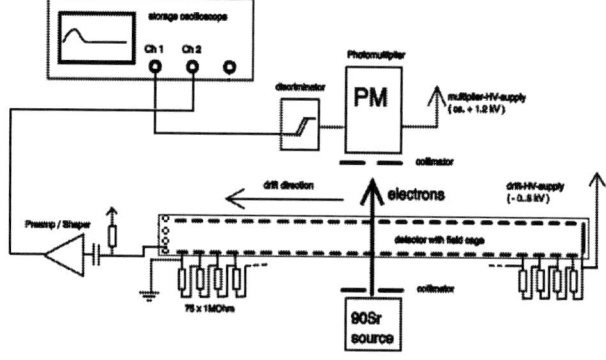

Figure 3. Functional setup of experiment with drift chamber and electronics [1].

The side view (Fig.4) shows some construction details: the drift electrodes are fabricated in hybrid (thick film) technology with structured conducting electrodes (silver-palladium) and resistors deposited in a baking process on ceramic ($Al_2 O_3$) substrate.

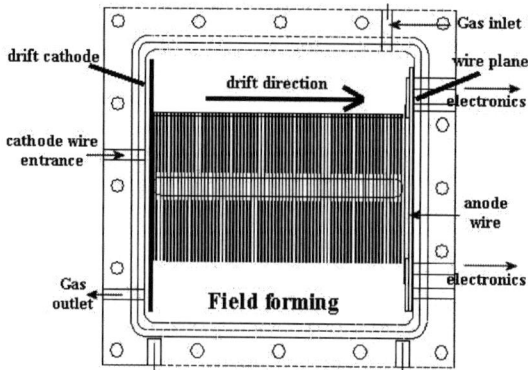

Figure 4. Side view of drift chamber [1].

In principle the ceramic is thin enough to allow x-rays to penetrate. In order perform measurements with minimum ionizing β -rays a slot at half height is cut into the ceramic. The electrodes are bridged by wires soldered to the conducting strips on both sides. The amplifying structure at the end of the drift cage consists of three anode wires enclosed by 4 potential wires in order to generate a controlled transition from the drift field to the almost cylindrical amplification field around the anode wires. The operation of the drift chamber is controlled by two independent power supplies: one for the drift field (negative) and one for the gas gain (positive).

The preamplifier and shaper of the anode signal is optimized for low noise and a shaping allowing at the same time to record the phase of the signal (drift time) with good precision and the shape due to broadening from diffusion. This is achieved by an integrating input stage and the following pole-zero cancelation. A second pole-zero cancelation removes the tails from the signal generation process in the proportional gas gain. Two more integration time constants (not shown) define an approximately gaussian width of the output pulse of ca. 150 ns. The shaped signals are recorded in a digital storage oscilloscope with respect to the trigger from the photomultiplier signal. The averaging mode of the scope allows the recording of fluctuating individual signals from the detector with good precision.

For the timing precision of the trigger signal from the scintillator-photomultiplier combination a value of about 1 ns may be easily reached, good enough for the drift time measurement. But the trigger signal also defines a spatial selection of an ionizing particle crossing the drift space at a well-defined location. This is achieved by collimators placed at the source and the entrance of the scintillator. Beyond the geometrical limitation also bremsstrahlung and multiple scattering broaden the accepted beam. Therefore the chamber thickness has been minimized and the collimator consists of low Z material (Lucite). The gas supply consists of a gas mixing system where the gas is supplied from a pressurized bottle. The gas mixture is P-10 and controlled by calibrated flow meter [1, 3] .

Measurements and experimental results

1. Preparation of the detector

After regulation of the gas flow allow about 5 minutes for full exchange of the gas in the detector. Control the flow at the outlet of the chamber. Check the operation of the amplifier by observing the noise in the single shot mode of the scope or using an analogue scope. Observe the photomultiplier signals on the scope and adjust the voltage such that in the mean a few 100 mV are obtained. Control the output of the discriminator and adjust the threshold.

2. Drift velocity experiment

The drift time is recorded at an appropriate number of detector positions. In principle a straight line should be obtained for measurements not too close to the anode. This measurement is repeated for a number of drift voltage settings and gas mixtures. It is important to control each time the gas gain.

Radiation, gas and high voltage specifications.

Collimated Source ^{90}Sr: 1milliCi
Gas mixture: P-10 (0.9Ar + 0.1CH$_4$)
PMT Voltage: +900 Volts.

Gas pressure: 748 Torrs.
Anode Voltage: +950 Volts.
Field Voltage [Drift Field]: from –300 up to –4000 Volts.

Experimental results:

d (cm)	Drift Field -300 V Drift Time (ns)	Drift Field -900 V Drift Time (ns)	Drift Field -1500V Drift Time (ns)	Drift Field -2000 V Drift Time (ns)	Drift Field -3000 V Drift Time (ns)	Drift Field -4000 V Drift Time (ns)
3.0	172	180	186	192	200	200
3.5	280	268	288	310	320	340
4.0	556	372	386	368	450	440
4.5	1000	532	510	530	570	600
5.0		652	590	580	690	740
5.5		790	710	760	820	880
6.0		950	820	800	930	1020
6.5			910	980	1080	1200
7.0					1180	1330
7.5						1550

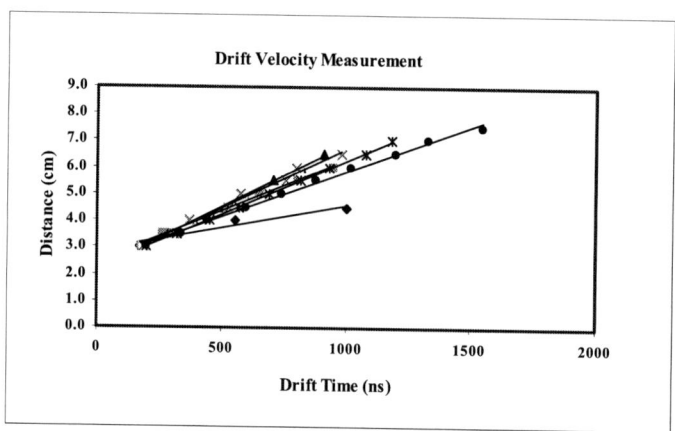

Drift Velocity (from -300 to -4000 V)

Drift Field [Volts]	Linear Fit Equations	Drift Velocity [cm/μsec]	Typical Error
-300	y$_{-300}$ = 0.0017x + 2.9033	1.70	± 0.2076
-900	y$_{-900}$ = 0.0038x + 2.4533	3.80	± 0.0969
-1500	y$_{-1500}$ = 0.0048x + 2.1217	4.80	± 0.0426
-2000	y$_{-2000}$ = 0.0045x + 2.2036	4.50	± 0.1544
-3000	y$_{-3000}$ = 0.004x + 2.1944	4.00	± 0.0329
-4000	y$_{-4000}$ = 0.0034x + 2.4417	3.40	± 0.1043

3. Determination of the reduced field

The determination of the reduced field strength E/p results in values to be compared to literature. The plots of drift velocity vs. E/p for different gas mixtures allow the discussion of the mean electron energy and the cross sections for elastic encounters (Ramsauer effect) and inelastic encounters (rotational and excitational levels in molecules).

Radiation, gas and high voltage specifications

Collimated Source ^{90}Sr: 1milliCi
Gas mixture: P-10 (0.9Ar + 0.1CH$_4$)
PMT Voltage: +900 Volts.

Gas pressure: 748 Torrs.
Anode Voltage: +950 Volts.
Field Voltage: from –300 up to –4000 Volts.

Experimental results:

Drift Field Voltage [Volts]	E/p [Volts/(cm*Torr)] (Reduced Field)	Drift Velocity [cm/µsec]
-300	0.06	1.70
-900	0.17	3.80
-1500	0.29	4.80
-2000	0.38	4.50
-3000	0.57	4.00
-4000	0.76	3.40

E/p vs. Drift Velocity

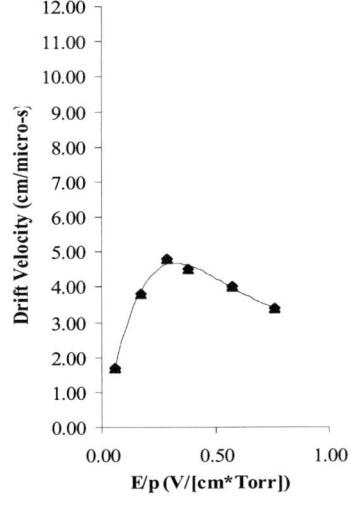

Measured Drift velocity vs. reduced field for the standard P-10 mixture (90% Ar, 10% CH4). See [10] for several gases (CH4, C2H6, C3H, Ar, Ar/CH4).

4. Gas gain

For a medium setting of the drift field (ca. 2 kV) the anode voltage is increased until signals appear occasionally. For each gas mixture a maximum allowable voltage is given (by the assistant) which never should be exceeded, otherwise the detector will be destroyed. If this voltage is reached without observing signals, check the system again. In the averaging mode the proper signal height is adjusted by fine-tuning the gain voltage.

Radiation, gas and high voltage specifications.

Collimated Source ^{90}Sr: 1milliC*i*
Field Voltage: -2000 Volts.
Anode Voltage: from +650 up to +950 Volts.

Gas pressure: 748 Torrs bar.
PMT Voltage: +900 Volts.
Fixed Distance: 40 mm.

Experimental results:

Anode Voltage [Volts]	Amplitude [mV] (Drift Field : -900V)	Amplitude [mV] (Drift Field : -1200V)	Amplitude [mV] (Drift Field : -1500V)
950.0	25.0	40.4	48.8
900.0	16.4	25.6	23.6
850.0	7.4	12.8	10.3
800.0	4.4	5.8	6.9
750.0	2.5	3.5	4.2
700.0	1.4	2.2	2.5
650.0	0.7	1.1	1.2

Anode Voltage [kV]	Log Amplitude [mV] (Drift Field : -900V)	Log Amplitude [mV] (Drift Field : -1200V)	Log Amplitude [mV] (Drift Field : -1500V)
0.95	1.39794	1.60638	1.68842
0.90	1.21484	1.40824	1.37291
0.85	0.87157	1.10721	1.01284
0.80	0.64345	0.76641	0.83885
0.75	0.39794	0.54407	0.62325
0.70	0.14613	0.34242	0.39794
0.65	-0.15490	0.04139	0.07918
Typical Error:	0.03268	0.04267	0.05960

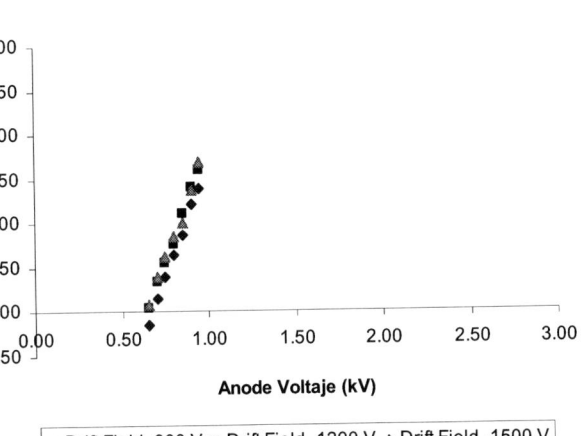

Gas Gain Measurements

♦ Drift Field -900 V ■ Drift Field -1200 V ▲ Drift Field -1500 V

Gas gain measured, see [3] for calculated and measured values for other gases.

5. Diffusion

For a low drift field and small drift velocity (ca. 1 cm/µs in P-10) the signals are averaged and recorded at different positions. Although the amplitude decreases the broadening of the width at half maximum is well visible. Even if the single signal seems to disappear in the noise the averaging will sum up the signals linearly but the noise only by the square root law such that the signals finally will emerge from the noise.

Radiation, gas and high voltage specifications.

Collimated Source ^{90}Sr: 1milliCi
Field Voltage: -300 Volts.
Anode Voltage: +950 Volts.

Gas pressure: 748 Torrs.
PMT Voltage: +900 Volts.
Variable Distance: from 30 mm up to 42.5 mm

Experimental results:

d(cm)	Δt (ns)	Δt^2 (ns^2)
3.00	120	14400
3.50	158	24964
3.75	245	60025
4.00	270	72900
4.25	290	84100

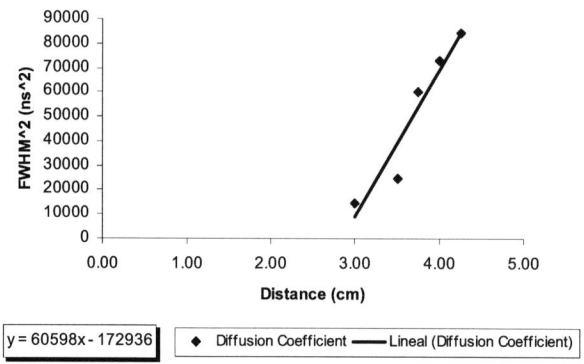

The plot of the width Δt of the signals vs. the position l should follow the general expression: $(\Delta t)^2 = a + bl$

A fit to the data allows the determination of **a** and **b**. In this particular case we have $(\Delta t)^2 = -178936 + (60598) \cdot l$.

Comparing our results with 2 other different measurements under the same conditions we have:

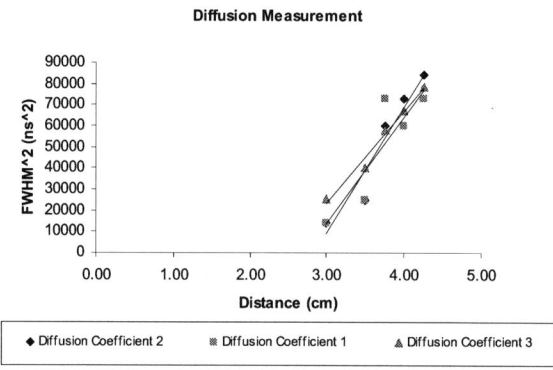

Linear Fit Equation	Typical Error
$y_1 = 50459x - 137661$	15096.22229
$y_2 = 60598x - 172936$	9608.553727
$y_3 = 43632x - 107600$	3401.324596

The first coefficient (a) contains the collimation and the inherent width of the shaper while the second (b) contains the diffusion constant D. Using eq. (18) and (19) D and T_e should be calculated and compared to the literature value. Possible deviations should be discussed. The values for measured diffusion vs. reduced electric field for several gases can be found in [10].

ACKNOWLEDGEMENTS

The authors wish to thank Professor Albert Walenta, from University of Siegen and President of the ICFA Panel on Instrumentation, Innovation and Development, for having donated the drift chamber used in this laboratory course to the ICFA Instrumentation Center in Morelia. In addition, this laboratory course is based on a similar course he gave at the First ICFA Instrumentation School at the ICFA Instrumentation Center in Morelia held in Morelia, Michoacan, Mexico, in November 2002

REFERENCES

1. Ix-B García Ferreira, J García Herrera and L Villaseñor, *The Drift Chambers Handbook, Introductory Laboratory Course* (346), in Proceedings of the XI Mexican School on Particles and Fields. J. Phys.: Conference Series. Volume 18, 2005, 346-361.
2. W. Blum and L. Rolandi, *Particle Detection with Drift Chambers*, Springer-Verlag, 1993.
3. A.H. Walenta et al, *Instrumentation in Elementary Particle Physics: Drift Chamber Experiment* (246), , in Proceedings of the First ICFA Instrumentation School at the ICFA Instrumentation Center in Morelia, AIP Conference Proceedings, Vol. 674, 2002.
4. W.R. Leo, *Techniques for Nuclear and Particle Physics Experiments: A How to Approach*, Springer Verlag, 2nd revised edition, 1994.
5. L.G.H. Huxley, R.W. Crompton, *The Diffusion and Drift of Electrons in Gases*, John Wiley & Sons, 0-471-42590-7.
6. A.H. Walenta, *The time expansion chamber and single ionization cluster measurement*, IEEE Trans. Nucl. Sci. NS - 26.1 (1979) 73.
7. A.H. Walenta et al., *The time expansion chamber as a high precision drift chamber*, in: Proceedings, International Conference on Instrumentation for Colliding Beam Physics, SLAC, February 1982.
8. A.H. Walenta, *Review of the Physics and Technology of Charged Particle Detectors*, 1983 Proceedings of the SLAC Summer Institute on Particle Physics: Dynamics and Spectroscopy at High Energy, SLAC-R-267.
9. F. Sauli, *Principles of Operation of Multiwire Proportional and Drift Chambers,* CERN 77-09, also in Experimental Techniques in High Energy Physics, ed. T. Ferbel, Addison-Wesley, 1987
10. Detlef Mattern, *Bestimmung des Einflusses von Driftparametern auf die Signalform einer TEC(Time Expansion Chamber)*, PhD Thesis, University of Siegen, 1988.
11. C. Grupen, Particle Detectors, Cambridge Monographs on Particle Physic, Nuclear Physics and Cosmology, 1996

CONTRIBUTED TALKS

The Schwinger model on a nonperiodic lattice

[1]Rafael G. Campos and [2]Eduardo S. Tututi

Facultad de Ciencias Físico–Matemáticas
Universidad Michoacana
58060 Morelia, Michoacán, México

Abstract. We study the Schwinger model on a lattice constructed from zeros of the Hermite polynomials that incorporates a lattice derivative and a discrete Fourier transform with many properties. Such a lattice produces a Klein-Gordon equation for the boson field and the correct value of the mass in the asymptotic limit.

Keywords: Schwinger model, lattice theory
PACS: 11.15.Ha, 02.60.Jh

INTRODUCTION

The Schwinger model has proved to be an excellent theoretical laboratory to study physical properties of fermions on the lattice [1, 2, 3, 4, 5]. Several approaches and derivatives have been used to study the mass spectrum, chirality, charge screening and the chiral condensate, among others. In fact, in [1] it has been shown that different types of fermions derivatives, namely the naive, the Wilson's and the so called modified SLAC derivative work well for the Schwinger model, yielding the correct boson mass in the continuum limit, but also producing a infinitely heavy boson mode. In the past years, a nonlocal lattice derivative with a well-defined discrete Fourier transform and useful properties, has been used to circumvent the restrictions of the no-go Nielsen-Ninomiya theorem [6]. Such a derivative, constructed from zeros of the Hermite polynomials, is used in this work to show how this lattice framework yields the correct boson mass for the Schwinger model with no extra modes, when the number of lattice nodes tends to infinity.

The $N \times N$ Hermite differentiation matrix D is given by $D = S\tilde{D}S^{-1}$ where

$$\tilde{D}_{jk} = \begin{cases} 0, & i = j, \\ \dfrac{1}{x_j - x_k}, & i \neq j, \end{cases} \quad (1)$$

S is a diagonal matrix [6] which nonzero elements are given by $\exp[-x_k^2/2]H'_N(x_k)$, $k = 1, 2, \ldots, N$, and x_k is the kth zero of the Hermite polynomial $H_N(x)$. The prime means differentiation with respect to the argument.

---

[1] E-mail: rcampos@umich.mx
[2] E-mail: tututi@umich.mx

The differentiation matrix D is diagonalized by a discrete Fourier transform F whose elements has the asymptotic form

$$F_{jk} = \frac{\Delta \xi}{\sqrt{2\pi}} e^{ix_j p_k} + (1/N), \qquad (2)$$

where $\Delta x = x_{k+1} - x_k = \Delta p = p_{k+1} - p_k = \pi/\sqrt{2N}$ are dimensionless variables.

THE SCHWINGER MODEL ON THE LATTICE

In the temporal gauge $A_0 = 0$, the Hamiltonian of the Schwinger model reads

$$H = \int dx \left[\frac{1}{2} E^2(x) + i \psi^\dagger \sigma_z (\partial_x + ieA(x)) \psi(x) \right], \qquad (3)$$

where $A(x)$ stands for the component $A_1(x)$ of the gauge field. To get a discretized Hamiltonian we need to consider the discrete version of the covariant derivative $\partial_x + ieA(x)$ which is given by

$$\nabla = U^\dagger \circ D \circ U, \qquad (4)$$

where U is the unitary matrix with elements $U_{jk} = \delta_{jk} e^{-ieb_j}$ and $\circ$ denotes the element-wise matrix product.
Thus, our lattice version of (3) is given by

$$H_L = \frac{1}{2} E^t E + i \psi^\dagger (\sigma_z \otimes \nabla) \psi, \qquad (5)$$

where E and ψ are the vectors constructed with the values of the fields $E(x)$ and $\psi(x)$ at the lattice points and the superscript t means transpose.
Taking into account the similarity transformation (4), H_L can be written in the form of a free Hamiltonian

$$H_L = H_\rho + H_\Psi = \frac{1}{2} \rho^t D^{-2} \rho + i \Psi^\dagger (\sigma_z \otimes D) \Psi, \qquad (6)$$

where $\Psi^\alpha = U \psi^\alpha$ and the discretized form of Gauss' law

$$\rho = DE = e \psi^{\alpha \dagger} \circ \psi^\alpha = e \Psi^{\alpha \dagger} \circ \Psi^\alpha$$

has been used. In order to have a nonsingular matrix D, the number of nodes N must be an even integer. The rth value of the axial current j is given by

$$j_r = -e \psi_r^{\dagger \alpha} \sigma_{\alpha\beta} \psi_r^\beta = -e \Psi_r^{\dagger \alpha} \sigma_{\alpha\beta} \Psi_r^\beta,$$

and the fermion fields $\psi, \psi^\dagger$ satisfy the equal-time anticommutation relations

$$\{\psi_r^{\dagger \alpha}, \psi_s^\beta\} = \delta_{rs} \delta_{\alpha\beta}. \qquad (7)$$

Note that in order to make the connection to the continuum case, H_L should be multiplied by $\Delta\xi$ to obtain the Riemann sum of (3), whereas (7) should be divided by $\Delta\xi$ to obtain a delta function.

To find an equation for the bosonic variable ρ on the lattice we begin by considering the lattice version of $\partial_0\rho = [\rho,H]/i$, which can be written as $(\partial_0\rho)_r = [\rho_r,H_L]/i$. By using (7) and the skew-symmetry of D we obtain the continuity equation

$$\begin{aligned}(\partial_0\rho)_r &= [\rho_r,H_L]/i = e\left(\Psi_r^{\dagger\alpha}\sigma_{\alpha\beta}D_{rs}\Psi_s^\beta - \Psi_s^{\dagger\alpha}\sigma_{\alpha\beta}D_{sr}\Psi_r^\beta\right) \\ &= e\left(\Psi_r^{\dagger\alpha}\sigma_{\alpha\beta}(D\Psi^\beta)_r + (D\Psi^{\dagger\alpha})_r\sigma_{\alpha\beta}\Psi_r^\beta\right) \\ &= e[D(\Psi^{\dagger\alpha}\sigma_{\alpha\beta}\Psi^\beta)]_r = -(Dj)_r.\end{aligned} \quad (8)$$

Now, let us compute

$$(\partial_0^2\rho)_r = -[(Dj)_r,H_L]/i = -[(Dj)_r,H_\rho]/i - [(Dj)_r,H_\Psi]/i. \quad (9)$$

Proceeding as before, we find

$$\begin{aligned}[(Dj)_r,H_\Psi]/i &= -e\left(\Psi_r^\dagger[(1_2\otimes D^2)\Psi]_r + 2[(1_2\otimes D)\Psi]_r^\dagger[(1_2\otimes D)\Psi]_r + [(1_2\otimes D^2)\Psi]_r^\dagger\Psi_r\right) \\ &= -(D^2\rho)_r,\end{aligned} \quad (10)$$

where 1_2 is the identity matrix of dimension 2. Therefore, (9) becomes

$$(\partial_0^2\rho)_r - (D^2\rho)_r = -\frac{1}{2i}[(Dj)_r,\rho'D^{-2}\rho]. \quad (11)$$

To obtain the asymptotic form of the mass term it is necessary to compute the Schwinger commutator $[j_j,\rho_{j'}]$. To this end we compute the commutator of Fourier transforms $[\tilde{j}_k,\tilde{\rho}_{k'}]$. First, we expand Ψ_q^α in its Fourier components

$$\Psi_q^\alpha = \sum_k[a_k u_k^\alpha F_{qk} + b_k^\dagger v_k^{\dagger\alpha}F_{kq}^\dagger],$$

where F stands for the discrete Fourier transform (2). Thus, the Fourier representation of $\rho_q = e\Psi_q^\dagger\Psi_q$ is

$$\rho_q = e\sum_{k,k'}[a_k^\dagger a_{k'}F_{kq}^\dagger F_{qk'} + b_k b_{k'}^\dagger F_{qk}F_{k'q}^\dagger],$$

where we have used the orthogonal property of the spinors. Therefore, the Fourier transform $\tilde{\rho}_k = \sum_q F_{kq}^\dagger \rho_q$ becomes

$$\tilde{\rho}_k = e\sum_{q,k',k''}[a_{k''}^\dagger a_{k'} F_{k''q}^\dagger F_{qk'}F_{kq}^\dagger + b_{k''}b_{k'}^\dagger F_{qk''}F_{k'q}^\dagger F_{kq}^\dagger]. \quad (12)$$

Note that the sum of products of elements of the discrete Fourier transform F involved in this equation are proportional to the generic asymptotic form

$$\frac{1}{2\pi}\sum_q e^{ix_q P}\Delta\xi, \quad (13)$$

where P stands for a sum of three lattice momenta, for instance, $P = p_{k'} - p_k - p_{k''}$. It is worth to notice that a sum like (13) is exactly a Kronecker delta if P is the sum of *two* lattice momenta (since F is unitary). Therefore, (13) becomes an exact delta whenever the sum of two of the three momenta summing up P is again an allowed lattice momentum. But this is not always the case, even for uniformly spaced momenta. However, we can consider an infinite lattice, i.e., a lattice with a very great number of nodes with no boundaries and extending to infinity. Since there is a momentum for each lattice site, the lattice momenta do not have neither a maximum nor a minimum and therefore, the sum of two evenly spaced momenta is again an allowed momentum, yielding that

$$\frac{\Delta \xi}{2\pi} \sum_q e^{ix_q P} \Delta \xi \to \delta(P) \Delta \xi. \tag{14}$$

In this case, (12) becomes

$$\tilde{\rho}_k = e \frac{\Delta \xi}{\sqrt{2\pi}} \sum_{k'} [a^\dagger_{k'} a_{k'+k} + b_{k'} b^\dagger_{k'-k}]. \tag{15}$$

A similar computation can be done for $\tilde{j}_k$. To obtain the value of $[\tilde{j}_k, \tilde{\rho}_{k'}]$, it is necessary to write this commutator in terms of operators that are normal-ordered with respect to the vacuum state filled with particles of any negative energy and antiparticles of any positive energy, defined as

$$|\text{vac}\rangle = \prod_{k<0} a^\dagger_k \prod_{k>0} b^\dagger_k |0\rangle.$$

Thus, after the use of the anticommutation relations

$$\{a^\dagger_k, a_{k'}\} = \delta_{kk'}, \quad \{b^\dagger_k, b_{k'}\} = \delta_{kk'}, \tag{16}$$

the normal-ordered commutator becomes[3]

$$[\tilde{j}_k, \tilde{\rho}_{k'}] = e^2 \frac{\Delta \xi}{2\pi} \sum_m (a^\dagger_m a_{m+k+k'} - a^\dagger_{m-k} a_{m+k'} + b^\dagger_{m-k-k'} b_m - b^\dagger_{m-k'} b_{m+k}),$$

and, as it can be seen, it annihilates the vacuum state unless $k' = -k$. Therefore,

$$[\tilde{j}_k, \tilde{\rho}_{k'}] = e^2 \delta_{k',-k} \frac{\Delta \xi}{2\pi} \sum_m (a^\dagger_m a_m - a^\dagger_{m-k} a_{m-k} + b^\dagger_m b_m - b^\dagger_{m+k} b_{m+k})$$

and the action of this commutator on $|\text{vac}\rangle$ is given by

$$[\tilde{j}_k, \tilde{\rho}_{k'}]|\text{vac}\rangle = -\frac{e^2}{\pi} \Delta \xi \, k \, \delta_{k',-k} |\text{vac}\rangle.$$

---

[3] Again, in order to make the connection to the continuum case, (16) should be divided by $\Delta \xi$ to obtain a delta function.

Thus, the inverse transform of $[\tilde{j}_k, \tilde{\rho}_{k'}]$, is

$$[j_j, \rho_{j'}] = -\frac{e^2}{\pi} \Delta\xi \sum_k k F_{jk} F_{j',-k} = -\frac{e^2}{\pi} \Delta\xi \sum_k k F_{jk} F^\dagger_{kj'}. \tag{17}$$

The product $\Delta\xi\, kF_{jk}$ is the asymptotic form of $p_k F_{jk} = -i\sum_l D_{jl} F_{lk}$. By using the fact that $FF^\dagger = 1$, we obtain finally the Schwinger commutator on an infinite lattice

$$[j_j, \rho_{j'}] = i\frac{e^2}{\pi} D_{jj'}. \tag{18}$$

Therefore, the right-hand side of (11) becomes the correct value of the mass term

$$m^2 \rho_r = \frac{1}{2i}[(Dj)_r, \rho^t D^{-2}\rho] = \frac{e^2}{\pi} \rho_r. \tag{19}$$

Thus, we have introduced a nonperiodic lattice that incorporates a lattice derivative and a discrete Fourier transform which yields the correct bosonic mass for the Schwinger model. This result shows that such a lattice can be useful in the study of quantum-field problems.

REFERENCES

1. K. Melnikov and M. Weinstein, Phys. Rev. D**62** (2000) 094504.
2. G.T. Bodwin, E.V. Kovacs, Phys. Rev. D**35** (1987) 3198.
3. S. Durr and C. Hoelbling, Phys. Rev. D**69** (2004) 034503.
4. S. Durr and C. Hoelbling, Phys. Rev. D**71** (2005) 054501.
5. S. Durr and C. Hoelbling, hep-lat/0604005.
6. R.G. Campos and E. S. Tututi, Phys. Lett A**297** (2002) 20-28.
7. R.G. Campos and E.S. Tututi, hep-lat/0603009

Proton/pion ratios and radial flow in pp and peripheral heavy ion collisions

E. Cuautle and G. Paić

Instituto de Ciencias Nucleares, Universidad Nacional Autónoma de México,
Apartado Postal 70-543, México, Distrito Federal 04510 México

Abstract. The production of baryon and mesons in the RHIC heavy-ion experiments has received a lot of attention lately. Although not widely known, the pp data measured concurrently with heavy ion collisions do not find a convincing explanation in terms of simple models. We present the results of an afterburner to Pythia and Hijing event generators, simulating radial flow which seems to qualitatively explain the experimental results when applied to the pp collision data from RHIC at 200 GeV center-of-mass energy.

Keywords: pp and ion-ion collision, radial flow
PACS: 25.75.Dw:14.40.-n,14.20.-c

INTRODUCTION

In heavy ion collisions at RHIC energies some phenomena remain up to now without valid explanation, one of them is the p/π ratios which exhibits a departure from the same ratio, measured in pp collisions at transverse momenta between 2.0 and ~4.5 GeV/c[1]. Although the coalescence model [2] has been widely accepted, it does not provide a satisfactory response to many questions. Beyond the details of the hadronization process which is still a debated question; it is well known that the p_t spectra of particles of different masses cannot be fit with a unique temperature as would predict a naive thermal model. In the nineties the NA44 collaboration has put in evidence the so called *radial flow*[3] in heavy ion collisions and it was interpreted as consequence of multiple interaction among the partons after the collisions of ions and before the freeze out of the created system. This interpretation also was confirmed by recent experimental results from RHIC [4].

The flow is understood as reflecting collective aspects of the interacting medium, depending on the collision energy. In central collisions between spherical nuclei, the initial state is symmetric in azimuth implying an isotropic azimuthal distribution of the final state particles. Consequently, any pressure gradient will cause an azimuthally symmetric collective flow of the outgoing particles. This is what we call radial flow.
The relevant observable to study the radial flow is the transverse momenta of the particles. For each particle, the random thermal motion is superimposed onto the collective radial flow velocity, correspondingly, the invariant p_t distribution depends of the temperature at freeze out, the particle mass, and the velocity profile. The experimental data

on radial flow at RHIC indicate two things:
- the temperature lowers with centrality while the flow velocity increases [5].
- even for pp collisions the analysis of the data from STAR yields a flow value of ≈ 0.2 c

This latter result coupled to the fact that in pp collisions the p/π ratio could not be reproduced by Pythia has encouraged us to investigate what would be the result of the p/π ratio if a flow component would be incorporated.

Consequently, in a first step we have generated *flow-free* pion and proton spectra using: 1) Pythia [6] with the Popcorn baryon production mechanism (note that from earlier work [7] we know that the differences in the ratio at 200 GeV are not very large when other baryon production mechanisms are used). 2) Hijing 1.36.
In a second step a toy model of the radial flow is incorporated in the p_t spectra and finally our results are compared with the baryon to meson ratio from STAR [8] in pp.

THE RADIAL FLOW AFTER BURNER

We are assuming that a fireball, thermalized, and expanding was created in a collision from an event generator. The expansion produces an additional momentum to the one created in the collisions. This contribution we call momentum of the flow $p_{t,f}$ given by $p_{t,f} = \gamma m \beta$, where γ is the Lorentz factor, β is the profile velocity and m is the mass of the particle under consideration. In order to add this radial component to the transverse momenta produced by the generators, it is necessary to attribute to the momentum p_t of each particle a randomized position in the transverse plane. Once this is done, the radial flow component is added vectorially.

RESULTS

In the left side of Fig. 1 we show the midrapidity ($|\eta| < 1$) transverse momentum spectra obtained with Pythia and Hijing. It is apparent that the two generators do not yield for pp collisions the same result. Hijing predicts a slightly steeper p_t dependence than Pythia. In the right side of the figure 1 we show the effect of the afterburner onto the flow free spectra for the proton case. We have used here a flow velocity of 0.6c to be able to clearly demonstrate the effect. As expected the flow free and afterburner spectra coincide at higher p_t.

The ratios p/π are shown in Fig. 2. In the left we show the results obtained for a flow velocity β=.2 and .3, applied to the HIJING events, compared withe STAR pp data the right side of the same figure shows the same data fitted with Pythia applying the same flow parameter and an additional one of 0.6c. It is obvious that the resulting fit is different, Pythia requiring a much larger flow velocity than Hijing to fit the same spectra.

The model works well also for peripheral heavy ion collisions as shown in Fig. 3 comparing with PHENIX [9] data.

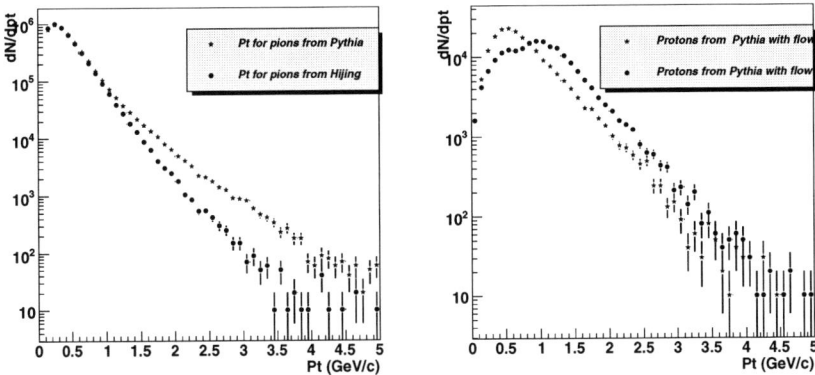

FIGURE 1. Spectra of pions from Pythia and Hijing event generators. The flow effects are show in the right side, for the protons cases.

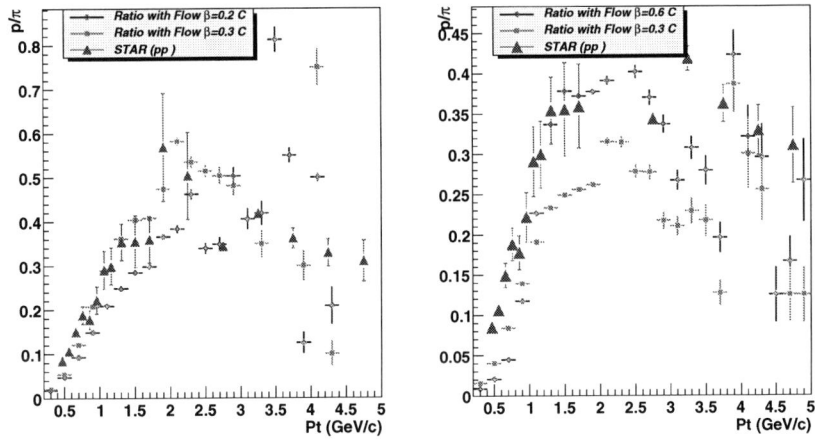

FIGURE 2. Proton to pion ratio as function of p_t. The left correspond to experimental results on pp collisions from STAR, comparing with our model using Pythia. The right part shows our model using Hijing and comparing it with the experimental results from pp collisions.

SUMMARY

The data of pp collisions at RHIC suggest that even in pp collisions the transverse momentum spectra of different particles are not completely parallel to each other. The analysis suggests that a *flow* velocity albeit small has to be added. Without entering in the foundation of the existence or not of flow we have constructed a toy model where we apply a given quantity of flow to flow free transverse momentum spectra generated by

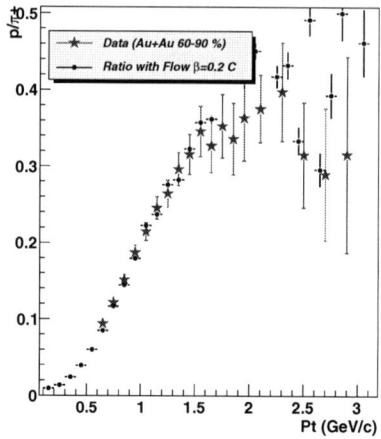

FIGURE 3. Ratio proton to pion from Hijing included radial flow with $\beta = 0.2c$ and compared with PHENIX $Au + Au$ peripheral results.

either Pythia with the popcorn baryon production mechanism or Hijing. The obtained results demonstrate that both for pp collisions and peripheral heavy ion collisions we obtain a remarkably good fit to the data using Hijing spectra and the velocities extracted in the experiments. The differences in the quality of the fit observed using Pythia indicate that the initial shape of the spectrum is key to a good reproduction of the data. We conclude that the flow may be one of the key ingredients in the so called Baryon puzzle at RHIC.

ACKNOWLEDGMENTS

We would like to thanks A. Morsch for his comments and suggestions. This work was supported in part by project IN107105 and Conacyt under grant number 40025-F.

REFERENCES

1. K. Adcox *et. al.* (PHENIX Collaboration), *Phys. Rev.* **C69** (2004) 024904,
2. R. C. Hwa, and C.B. Yang, *Phys. Rev.* **C70**,(2004) 024905. RJ.
3. I.G. Bearden, *et. al.* (NA44 Collaboration), *Phys. Rev. Lett.* 78 (1997) 2080.
4. J. Adams, *et. al.* (STAR collaboration) Phys.Rev.**C 70** (2004) 0141901(R)
5. J. Adams et al (STAR collaboration) *Nucl. Phys.* **A757** (2005) 102, and references therein.
6. T. Sjöstrand *et. al.* Pythia 6.3 Physics and Manual (hep-ph/0308153).
7. E. Cuautle and G. Paic, *Study of the pion and proton production in pp collisions at 14 TeV*, ALICE-INT-2005-027
8. J. Adams *et. al.* (STAR Collaboration), nucl-ex/0601033
9. S.S. Adler *et. al.* (PHENIX collaboration), *Phys.Rev.* **C** 69 (2004)034909

Rare top quark decays in extended models

R. Gaitán*, O. G. Miranda† and L. G. Cabral-Rosetti**

*Centro de Investigaciones Teóricas,
Facultad de Estudios Superiores – Cuautitlán,
Universidad Nacional Autónoma de México, (FESC-UNAM).
A. Postal 142,Cuautitlán-Izcalli, Estado de México, C. P. 54700, México.
†Departamento de Física,
Centro de Investigación y de Estudios Avanzados del IPN
A. Postal 14-740, México D. F. 07000, México.
** Departamento de Posgrado,
Centro Interdisciplinario de Investigación y Docencia en Educación Técnica (CIIDET),
Av. Universidad 282 Pte., Col. Centro, A. Postal 752,
C. P. 76000, Santiago de Queretaro, Qro., México.

Abstract. Flavor changing neutral currents (FCNC) decays $t \to H^0 + c$, $t \to Z + c$, and $H^0 \to t + \bar{c}$ are discussed in the context of Alternative Left-Right symmetric Models (ALRM) with extra isosinglet heavy fermions where FCNC decays may take place at tree-level and are only suppressed by the mixing between ordinary top and charm quarks, which is poorly constraint by current experimental values. The non-manifest case is also briefly discussed.

INTRODUCTION

Flavor-changing neutral currents (FCNC) are absent in the Standard Model (SM) at the tree-level due to the Glashow-Iliopoulos-Maiani (GIM) mechanism. However, new FCNC states can appear in top decays if there is physics beyond the Standard Model. In this context, rare top quark decays are interesting because they might be a source of possible new physics effects. In some particular models beyond the SM, rare top decays may be significantly enhanced to reach detectable levels [1].

Rare top decays have been studied in the context of the SM and beyond [2, 3, 4, 5]. The top quark decays into gauge bosons ($t \to c + V$; $V \equiv \gamma, Z, g$) are extremely rare events in the SM. However, by considering physics beyond the SM, for example, the Minimal Supersymmetric Standard Model (MSSM) or the two-Higgs-doublet model (2HDM) or extra quark singlets, new possibilities open up [1, 6], enhancing this branching ratios to the order of $\sim 10^{-6}$ for the $t \to c + Z$ [7] channel and $\sim 10^{-4}$ for the $t \to c + H$ [8] case.

In the future CERN Large Hadron Collider (LHC), about 10^7 top quark pairs will be produced per year [9]. An eventual signal of FCNC in the top quark decay will have to be ascribed to new physics. Furthermore, since the Higgs boson could also be produced at significant rates in future colliders, it is also important to search for all the relevant FCNC Higgs decays.

On the other hand, while the electroweak SM has been successful in the description of low-energy phenomena, it leaves many questions unanswered. One of them has to do with the understanding of the origin of parity violation in low-energy weak interaction processes. Within the framework of left-right symmetric models, based on the gauge group $SU(2)_L \otimes SU(2)_R \otimes U(1)_{B-L}$, this problem finds a natural answer[10, 11]. Moreover, new formulations of this model have been considered in which the fermion sector has been enlarged to include isosinglet vectorlike heavy fermions in order to explain the mass hierarchy [12, 13]. Most of these models includes two Higgs doublets.

We consider the rare top decay into a Higgs boson and the FCNC decay of the Higgs boson with the presence of a top quark in the final state, within the context of these alternative left-right models (ALRM) with extra isosinglet heavy fermions. Due to the presence of extra quarks the Cabibbo-Kobayashi-Maskawa matrix is not unitary and FCNC may exist at tree-level.

THE MODEL

The ALRM formulation is based on the gauge group $SU(2)_L \otimes SU(2)_R \otimes U(1)_{B-L}$. In order to solve different problems such as the hierarchy of quark and lepton masses or the strong CP problem, different authors have enlarged the fermion content to be of the form

$$l^0_{iL} = \begin{pmatrix} v^0_i \\ e^0_i \end{pmatrix}_L, e^0_{iR} \quad ; \quad \widehat{l}^0_{iR} = \begin{pmatrix} \widehat{v}^0_i \\ \widehat{e}^0_i \end{pmatrix}_R, \widehat{e}^0_{iL}$$

$$Q^0_{iL} = \begin{pmatrix} u^0_i \\ d^0_i \end{pmatrix}_L, u^0_{iR}, d^0_{iR}, \quad ; \quad \widehat{Q}^0_{iR} = \begin{pmatrix} \widehat{u}^0_i \\ \widehat{d}^0_i \end{pmatrix}_R, \widehat{u}^0_{iL}, \widehat{d}^0_{iL}, \quad (1)$$

where the index i ranges over the three fermion families. The superscript 0 denote weak eigenstates. In many of these models, extra neutral leptons also appears in order to explain the neutrino mass pattern, however we will focus in this work only on the quark sector.

In order to break $SU(2)_L \otimes SU(2)_R \otimes U(1)_{B-L}$ down to $U(1)_{em}$ the ALRM introduces two Higgs doublets, the SM one (ϕ) and its partner ($\widehat{\phi}$). Ref. [14] shows that from the eight scalar degrees of freedom, six become the Goldstone bosons required to give mass to the $W^\pm$, $\widehat{W}^\pm$, Z and $\widehat{Z}$; thus two neutral Higgs bosons, H and $\widehat{H}$, remain in the physical spectrum.

The renormalizable and gauge invariant interactions of the scalar doublets ϕ and $\widehat{\phi}$ with the fermions are described by the Yukawa Lagrangian. For the quark fields, the corresponding Yukawa terms are written as

$$\mathscr{L}^q_Y = \lambda^d_{ij}\overline{Q^0_{iL}}\phi d^0_{jR} + \lambda^u_{ij}\overline{Q^0_{iL}}\widetilde{\phi} u^0_{jR} + \widehat{\lambda}^d_{ij}\overline{\widehat{Q}^0_{iR}}\widehat{\phi} \widehat{d}^0_{jL}$$

$$+\widehat{\lambda}^u_{ij}\overline{\widehat{Q}^0_{iR}}\widetilde{\widehat{\phi}} \widehat{u}^0_{jL} + \mu^d_{ij}\overline{\widehat{d}^0_{iL}}d^0_{jR} + \mu^u_{ij}\overline{\widehat{u}^0_{iL}}u^0_{jR} + h.c. \quad (2)$$

where $i,j = 1,2,3$ and $\lambda^{d(u)}_{ij}$, $\widehat{\lambda}^{d(u)}_{ij}$, and $\mu^{d(u)}_{ij}$ are (unknown) matrices. The conjugate fields $\widetilde{\phi}$ $\left(\widetilde{\widehat{\phi}}\right)$ are $\widetilde{\phi} = i\tau_2\phi^*$ and $\widetilde{\widehat{\phi}} = i\tau_2\widehat{\phi}^*$, with τ_2 the Pauli matrix.

We can introduce the generic vectors ψ_L^0 and ψ_R^0 [15], for representing left and right electroweak states with the same charge. These vectors can be decomposed into the ordinary ψ_{OL}^0 and the exotic ψ_{EL}^0 sectors

$$\psi_L^0 = \begin{pmatrix} \psi_{OL}^0 \\ \psi_{EL}^0 \end{pmatrix}, \quad \psi_R^0 = \begin{pmatrix} \psi_{OR}^0 \\ \psi_{ER}^0 \end{pmatrix}, \qquad (3)$$

In the same way we can define the vectors for the mass eigenstates in terms of 'light' ψ_{lL} and 'heavy' ψ_{hL} states. The relation between weak eigenstates and mass eigenstates will be given through the matrices U_L and U_R by $\psi_L^0 = U_L \psi_L$, $\psi_R^0 = U_R \psi_R$ where

$$U_a = \begin{pmatrix} A_a & E_a \\ F_a & G_a \end{pmatrix}, \qquad a = L, R \qquad (4)$$

Here, A_a is the 3×3 matrix relating the ordinary weak states with the light-mass eigenstates, G_a is a 3×3 matrix relating the exotic states with the heavy ones, while E_a and F_a describe the mixing between the two sectors.

In this model, thanks to the extra heavy quarks, it is possible to have a relatively big mixing between ordinary quarks. This is not a particular characteristic of the model but a general feature when considering models with extra heavy singlets [16].

The tree-level interaction of the neutral Higgs bosons H and $\widehat{H}$ with the light fermions are given by

$$\mathcal{L}_Y^f = \frac{g}{2\sqrt{2}} \overline{\psi}_L A_L^\dagger A_L \frac{m_f}{M_W} \psi_R \left(H \cos\alpha - \widehat{H} \sin\alpha \right) \\ + \frac{\widehat{g}}{\sqrt{2}} \overline{\psi}_L \frac{m_f}{M_{\widehat{W}}} F_R^\dagger F_R \psi_R \left(H \sin\alpha + \widehat{H} \cos\alpha \right) + h.c. \qquad (5)$$

The neutral current in terms of the mass eigenstates, including the contribution of the neutral gauge boson mixing, can be written directly from this Lagrangian.

From the last equation we can see that, thanks to the non-unitarity of the A_a matrices we can have FCNC at tree-level. This characteristic appears due to the extra quark content of the model, which is not present in the usual left-right symmetric model.

FCNC TOP AND HIGGS DECAYS IN THE ALRM

Once we have introduced the model in which we are interested, we compute the expected branching ratio for a FCNC top or Higgs decay with a charm quark in the final state. We perform this analysis in this section.

Constraining the top-charm mixing angle

In order to have an expectation on the branching ratio for the FCNC top decay in the ALRM we need first an estimate on the mixing between the top and charm quarks in the model. One may think that the best constrain could come from the flavor-changing coupling of the neutral Z boson to the top and charm quarks, which can be written as:

$$\mathcal{L}_Z^{ct} = \frac{e}{s_{\theta_W} c_{\theta_W}} \bar{c}(g_V + g_A \gamma^5) \gamma^\mu Z_\mu t \tag{6}$$

where

$$g_{V,A} = \tfrac{1}{4}\left(c_\Theta - \tfrac{g}{\hat{g}} \tfrac{s_{\theta_W}^2}{r_{\theta_W}} s_\Theta\right) \eta_{32}^L \pm \tfrac{1}{4} \tfrac{\hat{g}}{g} \tfrac{c_{\theta_W}^2}{r_{\theta_W}} s_\Theta \, \eta_{32}^R \tag{7}$$

and s_{θ_W}, c_{θ_W} and r_{θ_W} are, respectively, $\sin\theta_W$, $\cos\theta_W$ and $\sqrt{\cos^2\theta_W - (g^2/\hat{g}^2)\sin^2\theta_W}$; θ_W is the weak mixing angle, Θ is the mixing between the Z and $\hat{Z}$ neutral gauge bosons. Here, η_{32}^L and η_{32}^R represent the mixing between the ordinary top and charm quarks and are given by

$$\eta_{32}^L = (A_L^+ A_L)_{32} \qquad \eta_{32}^R = (A_R^+ A_R)_{32}. \tag{8}$$

The mixing between the Z and the $\hat{Z}$ neutral gauge bosons, Θ, is expected to be small [17] if the ratio $r_g = g/\hat{g} = 1$. However, one might think that for different values of r_g these bounds are not longer valid. This is actually true, however, for most of the values of r_g, the expected freedom for the mixing angle Θ is still limited.

We can see from the definition of r_{θ_W} that the value of this ratio can not be bigger than $\sqrt{\cos^2\theta_W/\sin^2\theta_W} \approx 1.82$. We can recalculate the constraint obtained in [17] taking into account the freedom of this parameter. In order to do this analysis we simply need to consider the appropriate range for the parameter r_g that will affect the coupling constants for the $Z \to e^+ e^-$ that are needed for such computation:

$$g_{V,A}^e = -\tfrac{1}{2}\left(c_\Theta - \tfrac{g}{\hat{g}} \tfrac{s_{\theta_W}^2}{r_{\theta_W}} s_\Theta\right) \pm \tfrac{1}{2} \tfrac{\hat{g}}{g} \tfrac{c_{\theta_W}^2}{r_{\theta_W}} s_\Theta \, . \tag{9}$$

Note that in this case we are not taking into account the lepton flavor violation that has been discussed in ref [14].

With this formula and the limit for g_A obtained from the experiment: $g_A^{\exp} = -0.4998 \pm 0.00014$ [18] we can obtain a constraint for the Θ mixing depending on the value of r_g. The result of such analysis is shown in Fig. (1) were it is possible to see that $|\sin\Theta| \leq 0.03$, if the value of the ratio r_g is smaller than 1.6. As r_g approaches the critical value of 1.8, the constraint will disappear.

Therefore, the mixing angle Θ can be safely neglected for most of the values of g and $\hat{g}$. In this case the expression in Eq. (7) will not depend on the parameter η_{32}^R. For the present analysis we will consider only the case with $g = \hat{g}$, and therefore from now on we will denote $\eta_{32} = \eta_{32}^L$.

From Eq. (6) we can compute the branching ratio for the decay $t \to Z + c$ and compare it to the experimental limit $B(t \to Z+c) \leq 0.137$ [19] at 95 % C. L. We will get the maximum value for $\eta_{32} \leq 0.53$.

Although we have found a direct constrain to η_{32}, it is possible to get a stronger limit if we use the unitarity properties of the mixing matrix and the constrain on η_{22} that comes from the branching ratio $\Gamma(Z \to c + \bar{c})$. The experimental value for the branching

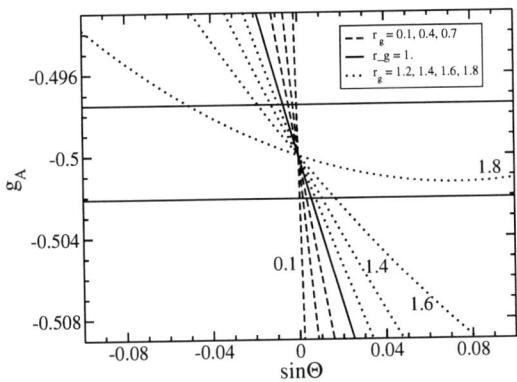

FIGURE 1. Expected value of g_A for the model in dependence on the mixing angle between Z and $\widehat{Z}$, and the ratio $r_g = g/\hat{g}$. The horizontal lines shows the 90 % C. L. allowed by the experiment.

ratio of this process is given by $B(Z \to c\bar{c}) = \Gamma(Z \to c\bar{c})/\Gamma_{total} = 0.1181 \pm 0.0033$ (see [20]). Using this experimental value, the minimum value for η_{22} at 95 % C. L. will be $\eta_{22} \geq 0.99$.

This information is of great help for constraining η_{32} since the unitarity of the mixing matrix has already been analyzed in the general case [21] and leads to the relation $|\eta_{32}|^2 \leq (1-\eta_{33})(1-\eta_{22})$.

Although we don't know the value for η_{33}, the boundary on η_{22} is enough to see that the mixing parameter $\eta_{32} \leq 0.1$. The higher value $\eta_{23} = 0.1$ is obtained when we take the extreme case $\eta_{33} = 0$, as can be seen from the equation in the previous paragraph.

It is possible to obtain more stringent constraints if low-energy data are considered. For the case of two extra quark singlets, this analysis was done in a very general framework in Ref. [6]. After a very complete analysis of all the observables, the author of this article obtained $|\eta_{32}| \leq 0.036$. This relatively large value is allowed for the case of a exotic top mass similar to that of the SM top-quark (There are not stringent lower bounds on the mass of a exotic top quark, being 220 GeV the current direct limit [22]). In the case of a very heavy mass for the exotic top-quark the constraint is more stringent: $|\eta_{32}| \leq 0.009$. In what follows we will use these two values in order to illustrate the expected signals from rare Higgs and top decays.

The decays $t \to H^0 + c$ and $H^0 \to t + \bar{c}$

Now that we have an estimate for the value of η_{32}, we compute the branching ratio for $t \to H^0 + c$ in the framework of ALRM. We take the charged-current two-body decay $t \to b + W$ to be the dominant t-quark decay mode. The neutral Higgs boson H^0 will be assumed to be the lightest neutral mass eigenstate.

Assuming $M_{\hat{H}} \gg M_H$ the vertex tcH^0 is written as $\frac{gm_t\eta_{32}}{2M_W}\cos\alpha\, P_L$. The partial width for this tree-level process can be obtained in the usual way and it is given by:

$$\frac{G_F\eta_{32}^2\cos^2\alpha}{16\sqrt{2}\pi m_t}\left(m_t^2+m_c^2-M_H^2\right)\left[\left(m_t^2-(M_H+m_c)^2\right)\left(m_t^2-(M_H-m_c)^2\right)\right]^{\frac{1}{2}} \quad (10)$$

where G_F is the Fermi's constant, m_t denotes the top mass, m_c is the charm mass, and M_H is the mass of the neutral Higgs boson. We can see from this formula that the branching ratio will be proportional to the product $\eta_{32}\cos\alpha$, of the top-quark mixing with the SM Higgs boson mixing with the extra Higgs boson.

The branching ratio for this decay is obtained as the ratio of Eq. (10) to the total width for the top quark, namely $B(t\to H^0+c)=\frac{\Gamma(t\to H^0+c)}{\Gamma(t\to b+W)}$.

Thanks to the possible combined effect of a big $\cos\alpha$ (null mixing between the SM Higgs boson and the additional Higgs bosons) and a big value of η_{32} this branching ratio could be as high as $\approx 3\times 10^{-4}$, for a Higgs mass of 117 GeV. Perhaps is more realistic to consider the more stringent constraint $\eta_{32}=0.009$, but even in this case, for $\cos\alpha\approx 1$ there is still sensitivity for detecting a positive signal of order 10^{-5}.

Finally we also consider the case of a Standard Higgs with a large mass. The best-fit value of the expected Higgs mass, including the new average for the mass of the top quark, is 117 GeV [23] and the upper bound is $M_H \leq 251$ GeV at 95 % C L. However, the error for the Higgs boson mass from this global fit is asymmetric, and a Higgs mass of 400 GeV is well inside the 3σ region as can be seen in Ref [23].

We estimate the branching ratio for the decay $H^0\to t+\bar{c}$, where H^0 is the light neutral Higgs boson of the ALRM. The expression for the partial width is

$$\frac{3G_F m_t^2 \eta_{32}^2\cos^2\alpha}{8\sqrt{2}\pi M_H^3}\left(M_H^2-m_t^2-m_c^2\right)\left[\left(M_H^2-(m_t+m_c)^2\right)\left(M_H^2-(m_c-m_t)^2\right)\right]^{\frac{1}{2}}. \quad (11)$$

The branching ratio for this decay is obtained as the ratio of Eq. (11) to the total width of the Higgs boson, which will include the dominant modes $H^0\to b+\bar{b}$, $H^0\to c+\bar{c}$, $H^0\to \tau+\bar{\tau}$, $H^0\to W+W$, and $H^0\to Z+Z$. The expressions for these decay widths in the ALRM also includes corrections due to the new parameters introduced in the model, and they are taken into account [24].

We computed the branching ratios for different decay modes, both for the Standard Model case ($\eta_{32}=0$ and $\eta_{ii}=1$) and for the FCNC case. We found that, also for a heavy Higgs, there are chances to either detect or to constrain the mixing angle parameter η_{32}. In this case, since all the partial widths have the same dependence on $\cos^2\alpha$, the branching ratios will depend only on η_{32}.

RESULTS AND CONCLUSIONS

The ALRM allows relatively big values of η_{32}. The $t\to H+c$ branching ratio could be of order of 10^{-4}, which is at the reach of LHC. It has been estimated that the LHC sensitivity (at 95 % C. L.) for this decay is $Br(t\to Hc)\leq 4.5\times 10^{-5}$ [25]; this branching ratio would be obtained in this model for a top-charm mixing $\eta_{32} = 0.015$ and a diagonal ordinary top coupling $\eta_{22} \simeq 0.98$. On the other hand, the FCNC mode $H \to t + \bar{c}$ may reach a branching ratio of order 10^{-3} and can also be a useful channel to look for signals of physics beyond the SM in the LHC.

ACKNOWLEDGMENTS

This work has been supported by CONACyT, SNI and also by PAPIIT project no. IN113206.

REFERENCES

1. S. Béjar, J. Guasch and J. Solá, *Nucl. Phys.* **B600**, 21 (2001).
2. J. L. Diaz-Cruz, R. Martinez, M. A. Perez and A. Rosado, Phys. Rev. **D41** 891 (1990);
3. E. Jenkins, *Phys. Rev.* **D56**, 458 (1997).
4. B. Mele, S. Petrarca, A. Soddu, *Phys. Lett.* **B435**, 401 (1998).
   [5]
5. G. Eilam, J. L. Hewett and A. Soni, Phys. Rev. D **44**, 1473 (1991) [Erratum-ibid. D **59**, 039901 (1999)].
6. J. A. Aguilar-Saavedra, *Phys. Rev.* **D67** 035003 (2003) [Erratum *Phys. Rev.* **D69** 099901 (2004)]
7. J. A. Aguilar-Saavedra and B. M. Nobre, *Phys. Lett.* **B553** 251 (2003).
8. A. M. Curiel, M. J. Herrero and D. Temes, *Phys. Rev.* **D67**, 075008 (2003).
9. M. Beneke, I. Efthymiopoulos, M. L. Mangano, J. Womerslwy et. al., hep-ph/0003033.
10. J. C. Pati and A. Salam *Phys. Rev.* **D10** 275 (1974); R. N. Mohapatra and J. C. Pati *Phys. Rev.* **D11** 566 (1975) and *Phys. Rev.* **D11** 2558 (1975); G. Senjanovic and R. N. Mohapatra *Phys. Rev.* **D12** 1502 (1975);
11. Rabindra N. Mohapatra *Unification and Supersymmetry* Springer 2003 and references therein.
12. A. Davidson and K. C. Wali, *Phys. Rev. Lett.* **59** 393 (1987); S. Rajpoot *Phys. Lett.* **B191** 122 (1987).
13. K. Kiers, J. Kolb, J. Lee, A. Soni and G. H. Wu, *Phys. Rev.* **D66** 095002 (2002); [arXiv:hep-ph/0205082].
14. V. E. Cerón, U. Cotti, J. L. Díaz-Cruz and M. Maya, *Phys. Rev.* **D57** 1934 (1998); U. Cotti et. al. *Phys. Rev.* **D66** 015004 (2002).
15. P. Langacker and D. London, *Phys. Rev.* **D38** 886 (1988).
16. J. A. Aguilar-Saavedra, Acta Phys. Pol. B **35** 2695 (2004) arXiv:hep-ph/0409342.
17. O. Adriani et. al., L3 Coll, *Phys. Lett.* **B306** 187 (1993); J. Polak and M. Zralek, *Phys. Rev.* **D46** 3871 (1992); M. Maya and O. G. Miranda Z. *Phys.* **C68** 481 (1995).
18. M. Acciarri *et al.* [L3 Collaboration], Z. Phys. C **62**, 551 (1994).
19. G. Abbiendi *et al.* [OPAL Collaboration], *Phys. Lett.* **B521** 181 (2001).
20. S. Eidelman *et al.* (Particle Data Group), *Phys. Lett. B* **592** 1 (2004).
21. F. del Aguila, J. A. Aguilar-Saavedra, *Phys. Rev. Lett.* **82** 1602 (1999).
22. D. Acosta et al, Phys. Rev. Lett. **90** 131801 (2003).
23. V. M. Abazov *et al.* [D0 Collaboration], Nature **429**, 638 (2004) [arXiv:hep-ex/0406031].
24. R. Gaitan, O. G. Miranda and L. G. Cabral-Rosetti, Phys. Rev. D **72**, 034018 (2005) [arXiv:hep-ph/0410268].
25. J. A. Aguilar-Saavedra and G. C. Branco, *Phys. Lett.* **B495** 347 (2000). [arXiv:hep-ph/0004190].

High-accuracy critical exponents for O(N) hierarchical 3D sigma models

J. J. Godina*, L. Li and Y. Meurice† and M. B. Oktay**

*Depto. de Física
CINVESTAV-IPN
Ap. Post. 14-740, México, D.F. 07000
†Department of Physics and Astronomy
The University of Iowa
Iowa City, Iowa 52242, USA
**School of Mathematics
Trinity College
Dublin, Ireland

Abstract. The critical exponent γ and its subleading exponent Δ in the $3D$ $O(N)$ Dyson's hierarchical model for N up to 20 are calculated with high accuracy. We calculate the critical temperatures for the measure $\delta(\vec{\phi}.\vec{\phi}-1)$. We extract the first coefficients of the $1/N$ expansion from our numerical data. We show that the leading and subleading exponents agree with Polchinski equation and the equivalent Litim equation, in the local potential approximation, with at least 4 significant digits.

Keywords: Non linear sigma model, Dyson's hierarchical model, critical exponents, $1/N$ expansion
PACS: 11.15.Pg, 11.10.Hi, 64.60.Fr

MOTIVATION

The Berlin-Kac spherical model [1] has been studied for a lot of time, particularly because this is exactly soluble [2], and because its solution has been given various ideas concerning the "Mathematical Mechanism" of the phase transitions [3], in the large limit of the dimensionality N. Actually large N limit and the $1/N$ expansion appear prominently in recent developments in particle physics, condensed matter and string theory, in this way we can face problems like: (1) Numerical solution of QCD in the planar limit using the fermionic sector of large N [4]. (2) The problem of the fermion pair condensation [5]. (3) The problem of the confinement, the stable string at larger N, the Pomeron, the deconfinement, topology, 't Hooft string tensions, chiral physics, etc. [6]. The basic equation for sigma models is obtained using the steepest descent method for the functional integral [8]. To use the expansion $1/N$ as a quantitative tool, we need satisfied several conditions: a) We need understand the large order behavior of the series that we are using to describe the problem. b) We need locate the singularities of its Borel transform. c) We need compare the accuracy of various procedures with the numerical results for various N. This schedule is obviously very difficult, for instance in D=3 we only can calculate up to the $1/N^2$ order the critical exponents [7]. Are known some results concerning the Borel summability of particular $1/N$ expansions [9]; but not there are any model for which the three points of the program outlined above has been

completed.
We are planning to complete this program using Dyson's hierarchical model (HM) [10].

PROGRAM

We provide high-accuracy numerical values for the critical exponent γ, the subleading exponent Δ and the critical parameter β_c for the 3D $O(N)$ hierarchical nonlinear sigma models. Quantities that appear in the parametrization of the magnetic susceptibility of a spin model near its critical temperature:

$$\chi \simeq (\beta_c - \beta)^{-\gamma}(A_0 + A_1(\beta_c - \beta)^{\Delta} + ...). \tag{1}$$

The critical exponents γ and Δ are universal and can be obtained from the calculation of the eigenvalues of the linearized renormalization group (RG) transformation. $A_0, A_1, ...$ are critical amplitudes, which are functions of the microscopic details of the theory. The method of calculation of the critical exponents used here is an extension of one of the methods described at length for the $N = 1$ case [11].

RENORMALIZATION GROUP EQUATION

The fundamentals and applications of the "exact renormalization group" (exact RG) discovered and so christened independently by Wilson and Wegner [12] have been studied intensively since the beginning of the nineties. The central reason for this recrudescence is the general acceptance that, far from being merely formal exact realizations of Wilson's RG ideas, these ideas form the basis for powerful and flexible approximations in non-perturbative quantum field theory. The two most widely used realizations of such exact RG's, are Polchinski's version [13] equivalent to Wilson's by a change of variables, and the version for the Legendre effective action [14]. We will be interested in the case where these are applied to $O(N)$ invariant N-component real scalar field theory in D Euclidean dimensions. The RG transformation can be constructed as a blockspin transformation followed by a rescaling of the field ($k \rightarrow k/s$). For HM, the blockspin transformation affects only the local measure and can be expressed conveniently in terms of the Fourier transform (R) of the local measure. Using $O(N)$ symmetry, this Fourier transform will depend only on $\vec{k} \cdot \vec{k} \simeq u$, a simple variable which is a source conjugate to the local field variable ϕ. For $N = 1$ case:

$$R_{n+1}(k) \propto \exp(-\frac{1}{2}\beta \frac{\partial^2}{\partial k^2})(R_n(\frac{k\sqrt{c}}{2}))^2, \tag{2}$$

replacing k by u and the second derivative by the N-dimensional Laplacian, we obtain the RG transformation for the Fourier transform of the local measure:

$$R_{n+1,N}(u) \propto \exp(-\frac{1}{2}\beta(4u\frac{\partial^2}{\partial k^2} + 2N\frac{\partial}{\partial u}))(R_{n,N}(\frac{cu}{4}))^2, \tag{3}$$

where $c = 2^{1-2/D}$ in order to reproduce the scaling of a Gaussian massless field in D dimensions. The normalization constant is fixed requiring that $R_{n,N}(0) = 1$. $R_{n,N}$ has a direct probabilistic interpretation:

$$R_{n,N}(k) = \sum_{q=0}^{\infty} \frac{(-ik)^{2q}}{2q!} \frac{\langle (M_n)^{2q} \rangle_n}{s^{2qn}}, \tag{4}$$

since $\chi_n(\beta) = \frac{\langle (M_n)^2 \rangle_n}{2^n}$ is the magnetic susceptibility.

POLYNOMIAL TRUNCATION

We are going to use polynomial truncation of degree $lmax$ in the calculations:

$$R_{n,N}(u) = 1 + a_{n,1}u + a_{n,2}u^2 + \ldots + a_{n,l_{max}}u^{l_{max}}, \tag{5}$$

the magnetic susceptibility (zero-momentum two point function) to finite volume is related to the first coefficient by the relation

$$\chi_n = -2a_{n,1}\left(\frac{2}{c}\right)^n. \tag{6}$$

The truncated recursion formula for the $a_{n,m}$ reads

$$a_{n+1,m} = \frac{\sum_{l=m}^{2lmax}(\sum_{p+q=l} a_{n,p}a_{n,q})B_{m,l}}{\sum_{l=0}^{2lmax}(\sum_{p+q=l} a_{n,p}a_{n,q})B_{0,l}}, \tag{7}$$

$$B_{m,l} = \frac{\Gamma(l+1)\Gamma(l+N/2)}{\Gamma(m+1)\Gamma(m+N/2)} \frac{1}{(l-m)!} \left(\frac{c}{4}\right)^l (-2\beta)^{l-m}. \tag{8}$$

In our numerical calculations, no truncation is applied after squaring in the above equations, then the sum does extend up to $2lmax$.

GOAL

Provide the critical exponents calculating the eigenvalues of the matrix $\frac{\partial a_{n+1,l}}{\partial a_{n,m}}$ at the nontrivial fixed point:

$$\gamma = \frac{\ln(2/c)}{\ln(\lambda_1)}, \quad \Delta = \left|\frac{\ln(\lambda_2)}{\ln(\lambda_1)}\right|. \tag{9}$$

They are within numerical errors, independent of the manner in which we approach to the nontrivial fixed point. These will be compared with other RG transformations particularly with the references [15], [16], [17]. This critical exponents will be calculated with the local measure of the nonlinear sigma model

$$\delta(\vec{\phi} \cdot \vec{\phi} - 1). \tag{10}$$

RG TRANSFORMATIONS

The block-spin transformation of HM is an integral formula which transforms the *local* measure $W(\phi)$ according to the rule:

$$W_{n+1}(\phi) \propto e^{\frac{\beta}{2}(\frac{c}{4})^{n+1}\phi^2} \int d\phi' W_n(\frac{(\phi-\phi')}{2}) W_n(\frac{(\phi+\phi')}{2}) . \tag{11}$$

For Eq. (10) the corresponding Fourier transform is

$$R_{0,N}(u) = \sum_{l=0}^{\infty} \frac{(-1)^l u^l \Gamma(\frac{N}{2})}{2^{2l} l! \Gamma(\frac{N}{2}+l)} . \tag{12}$$

This particular choice allows us to calculate the β_c in the large N limit. Other measures have also been used in order to check the universal values of this two exponents. The asymptotic behavior of the ratio $\frac{a_{n+1,1}}{a_{n,1}}$ allows us to decide unambiguously if we are in the symmetric phase (where the ratio approaches $c/2 \approx 0.63$) or in the broken phase (ratio approaches to c) and by successive bifurcations, one can determine the critical value $\beta_c(lmax)$:

$$lmax \to \infty \Longrightarrow \beta_c(lmax) \to \beta_c .$$

The rate at which this limit is reached depends of N as we can see in fig 1. To obtain

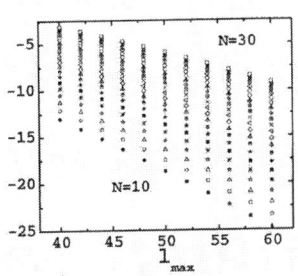

FIGURE 1. $\log_{10} \frac{|\beta(l_{max})-\beta_c|}{\beta_c}$ vs l_{max} for $N = 10$ (filled circles), $N = 11$ (empty circles), $N = 12$ (empty triangles) ...... up to $N = 30$ (empty squares).

20 significant digits for β_c in fig 2 we give the minimum $lmax$ necessary. The nontrivial fixed point for a given value of $lmax$ can be constructed by iterating sufficiently many times the RG map at values sufficiently close to $\beta_c(lmax)$. For instance, to obtain some accuracy ε for the fixed point, we need to iterate n times the map with

$$\lambda_2^n \sim \varepsilon , \tag{13}$$

in order to get rid of the irrelevant directions. At the same time, we want the growth in the relevant direction to be limited, in other words:

$$|\beta - \beta_c(lmax)| \lambda_1^n < \varepsilon . \tag{14}$$

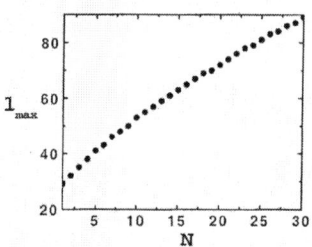

FIGURE 2. Approximately the minimal value for $lmax \simeq 22 + 6.2 N^{0.7}$ in the range of N considered here to obtain $\log_{10} \frac{|\beta_c(l_{\max}) - \beta_c(\infty)|}{\beta_c(\infty)} = -20$.

TABLE 1. β_c and the first two eigenvalues for $N = 1...20$

N	β_c	λ_1	λ_2
1	1.1790301704462697325	1.427172478	0.8594116492
2	2.4735265752919854000	1.38574349	0.8563409066
3	3.8273820333573397671	1.354668326	0.8506945150
4	5.2111615635533656165	1.332749865	0.8440522954
5	6.6104153462855068435	1.317578283	0.8376436746
6	8.0181114053706725941	1.306955396	0.8320345022
7	9.4307096447427796882	1.299321025	0.8273378172
8	10.846330737925124699	1.293666393	0.8234676785
9	12.263918029354988652	1.289354227	0.8202833449
10	13.682844072802585664	1.285978489	0.8176485461
11	15.102717572108367579	1.283274741	0.8154492652
12	16.523283812777939366	1.281066141	0.8135953137
13	17.944370719047342283	1.279231192	0.8120168555
14	19.365858255947423937	1.277684252	0.8106600963
15	20.787660334686062513	1.276363511	0.8094834857
16	22.209713705054412233	1.275223389	0.8084547150
17	23.631970906283518487	1.274229622	0.8075484440
18	25.054395659078177206	1.273356000	0.8067446107
19	26.476959772907788848	1.272582158	0.8060271793
20	27.899641020779716433	1.271892050	0.8053832116

Combining these two requirements we obtain

$$|\beta - \beta_c(lmax)| \lambda_1^n \simeq \varepsilon^{1+1/\Delta}, \tag{15}$$

where was used the definition of the equation 9. The numerical results for $\varepsilon = 10^{-10}$ and N up to 20 are given in Tables 1 and 2.

TABLE 2. γ, Δ and β_c for $N = 1...20$.

N	γ	Δ	β_c/N
1	1.29914073	0.425946859	1.179030170
2	1.41644997	0.475380832	1.236763288
3	1.52227970	0.532691965	1.275794011
4	1.60872817	0.590232008	1.302790391
5	1.67551051	0.642369187	1.322083069
6	1.72617703	0.686892637	1.336351901
7	1.76479863	0.723880426	1.347244235
8	1.79469275	0.754352622	1.355791342
9	1.81827105	0.779508505	1.362657559
10	1.83722291	0.800424484	1.368284407
11	1.85272637	0.817977695	1.372974325
12	1.86561092	0.832855522	1.376940318
13	1.87646998	0.845589221	1.380336209
14	1.88573562	0.856588705	1.383275590
15	1.89372813	0.866171682	1.385844022
16	1.900689026	0.874586271	1.388107107
17	1.906803377	0.882027998	1.390115936
18	1.912215071	0.888652409	1.391910870
19	1.917037518	0.894584429	1.393524199
20	1.921361210	0.899925324	1.394982051

ASYMPTOTIC BEHAVIOR

As N increases, the values of table 2 slowly approach to asymptotic values as is expected. Using the corresponding propagator for the model one finds the asymptotic expansions:

$$\gamma \simeq 2 + a_1/N + ... \quad (16)$$

$$\Delta \simeq 1 + b_1/N + ... \quad (17)$$

$$\beta_c/N \simeq (2-c)/(2(c-1)) + c_1/N + ... \quad (18)$$

The values of the leading coefficients a_1, b_1, c_1 are estimated subtracting the asymptotic values and multiplying by N. Results are in table 3, according to them:

$$a_1 \simeq -1.6, \, b_1 \simeq -2.0, \, c_1 \simeq -0.57 \, .$$

It seems possible to improve these estimates by estimating the next to leading order corrections.

RESULTS

The exponents should be the same for the four models. The change of coordinates that relates the RG transformation considered here and the one studied in Ref. [17] is given in Ref. [18]. Under this formulation in the limit $L \to 1$ it is obtained the Polchinski equation in the local potential approximation by Ref. [16]. In Table [5], we display $v = \gamma/2$ (since $\eta = 0$ here) and $\omega = \Delta/v$ from those references:

TABLE 3. $N(2-\gamma)$, $N(1-\Delta)$ and $N(\frac{2-c}{2(c-1)} - \frac{\beta_c}{N})$ for $N = 1\ldots 20$.

N	$N(2-\gamma)$	$N(1-\Delta)$	$N(\frac{2-c}{2(c-1)} - \frac{\beta_c}{N})$
1	0.7009	0.5741	0.2446
2	1.167	1.049	0.3738
3	1.433	1.402	0.4436
4	1.565	1.639	0.4835
5	1.622	1.788	0.5079
6	1.643	1.879	0.5239
7	1.646	1.933	0.5349
8	1.642	1.965	0.5430
9	1.636	1.984	0.5490
10	1.628	1.996	0.5538
11	1.620	2.002	0.5576
12	1.613	2.006	0.5606
13	1.606	2.007	0.5632
14	1.600	2.008	0.5654
15	1.594	2.007	0.5673
16	1.589	2.007	0.5689
17	1.584	2.006	0.5703
18	1.580	2.004	0.5715
19	1.576	2.003	0.5726
20	1.573	2.001	0.5736

TABLE 4. ν, ω and α for $N = 1\ldots 10$.

N	$\nu = \gamma/2$	$\omega = \Delta/\nu$	$\alpha = 2 - 3\nu$
1	0.649570	0.655736	0.051289
2	0.708225	0.671229	-0.124675
3	0.761140	0.699861	-0.283420
4	0.804364	0.733787	-0.413092
5	0.837755	0.766774	-0.513266
6	0.863089	0.795854	-0.589266
7	0.882399	0.820355	-0.647198
8	0.897346	0.840648	-0.692039
9	0.909136	0.857417	-0.727407
10	0.918611	0.871342	-0.755834

- Our results coincide with 4 digits (for ν) in table 3, with Ref. [16], $\nu(pol)$, $\omega(pol)$.
- They coincide with six digits for ν given in table 8 for $N = 1, 2, 3, 5$ and 10 of Ref. [17], $\nu(sch)$, $\omega(sch)$.
- We found discrepancies of order 1 in the fifth digit of ν and slightly larger for ω with the values found in table 1 of Ref. [15], $\nu(lit)$, $\omega(lit)$.

In Summary, we have provided high-accuracy data for:

- Critical temperatures, and critical exponents for the measure $\delta(\vec{\phi} \cdot \vec{\phi} - 1)$.
- We extracting the critical coefficients of the $1/N$ expansion from this data.

TABLE 5. Results of the comparision in the four approaches, pol= Polchinski, sch=Schnetmann, lit=Litim.

N	$\nu = \gamma/2$	$\omega = \Delta/\nu$	$\nu(pol)$	$\omega(pol)$	$\nu(sch)$	$\nu(lit)$	$\omega(lit)$
1	0.649570	0.655736	0.6496	0.6557	0.649570	0.649562	0.655746
2	0.708225	0.671229	0.7082	0.6712	0.708225	0.708211	0.671221
3	0.761140	0.699861	0.7611	0.6998	0.761140	0.761123	0.699837
4	0.804364	0.733787	0.8043	0.7338	0	0.804348	0.733753
5	0.837755	0.766774	0	0	0.837755	0.837741	0.766735
6	0.863089	0.795854	0	0	0	0.863076	0.795815
7	0.882399	0.820355	0	0	0	0.882389	0.820316
8	0.897346	0.840648	0	0	0	0.897338	0.840612
9	0.909136	0.857417	0	0	0	0.909128	0.857384
10	0.918611	0.871342	0	0	0.918611	0.918605	0.871311

- We show that the leading and subleading exponents agree with the Polchinski equation and the equivalent Litim equation with 4 significant digits.

ACKNOWLEDGMENTS

One of the authors J.J. Godina give thanks to the Physics Department at CINVESTAV IPN for the economical support to attend the workshop.

REFERENCES

1. M. Kac, Phys. Today **17**, 40 (1964)
2. T. H. Berlin and M. Kac, Phys. Rev. **86**, 821 (1952)
3. M. Kac and C. J. Thomson, Proc. Nat. Acad. Sci. **55**, 676 (1966)
4. R. Narayanan and H. Neuber, hep-lat/ 0501031
5. M. Moshe and J. Zinn-Justin hep-th/0306133
6. J. M. Maldacena et.al., hep-th/9905111, M. Teper, hep-lat/0509019
7. Y. Okabe, M. Oku, and R. Abe, Prog. Theor. Phys. **59**, 1825 (1978); J. A. Gracey, J. Phys. **A24**, L197 (1991); A. Pelisseto and E. Vicari, Phys. Rept. **368**, 549 (2002), cond-mat/0012164
8. H. E. Stanley, Phys. Rev. **176**, 718 (1968); F. David, D. A. Kessler, and H. Neuberger, Nucl. Phys. **B257**, 695 (1985)
9. B. de Wit and G. 't Hooft, Phys. Lett. **B69**, 61 (1977); J. Avan and H. J. de Vega, Phys. Rev. **D29**, 2904 (1984); J. L. Kneur and D. Reynaud, JHEP **01**, 014 (2003), hep-th/0111120
10. F. Dyson, Comm. Math. Phys. **12**, 91 (1969), Phys. Rev. **B5**, 2622 (1972); G. Baker, Phys. Rev. **B5**, 2622 (1972)
11. J. Godina, Y. Meurice, M.B. Oktay, Phys. Rev. **D59**, 096002 (1999)
12. K. Wilson, J. Kogut, Phys. Rep. **12C**, 75 (1974); F. J. Wegner, A. Houghton, Phys. Rev. **A8**, 401 (1973)
13. J. Polchinski Nucl. Phys. **B231**, 269 (1984)
14. C. Wetterich, Phys. Lett. **B301**, 90 (1993)
15. D. F. Litim, Nucl. Phys.**B631**, 128 (2002)
16. J. Comellas and A. Travesset, Nucl. Phys. **B498**, 539 (1997)
17. J. Gottker-Schnetmann (1999), cond-mat/9909418.
18. H. Koch and P. Wittwer, Commun. Math. Phys. **138**, 537 (991)

Towards the establishment of nonlinear hidden symmetries of the Skyrme model

A. Herrera–Aguilar[*], K. Kanakoglou[†] and J.E. Paschalis[†]

[*]*Instituto de Física y Matemáticas, Universidad Michoacana de San Nicolás de Hidalgo. Edificio C–3, Cd. Universitaria, CP 58040, Morelia, Michoacán, México.*
[†]*Physics Department, Aristotle University of Thessaloniki, 54124 Thessaloniki, Greece.*

Abstract. We present a preliminary attempt to establish the existence of hidden nonlinear symmetries of the SU(N) Skyrme model which could, in principle, lead to the further integration of the system. An explicit illustration is given for the SU(2) symmetry group.

Keywords: Nonlinear hidden symmetries, Skyrme model.
PACS: 11.10.Lm, 02.30.Ik, 12.39.Dc, 21.60.Fw, 21.60.-n, 21.30.Fe

INTRODUCTION

The theory which describes strong interactions of particles is Quantum Chromodynamics (QCD). This theory possesses two limits which describe different physics. In its high energy limit, the theory describes the interaction between weakly coupled quarks and gluons, and can be solved perturbatively due to the so–called asymptotic freedom. In its low energy limit, QCD is supposed to describe the interactions of mesons, baryons and probably glueballs. However, the theory becomes strongly coupled and, till now, there is no way to afford it analytically since perturbation theory fails in this limit.

Nevertheless, the QCD theory possesses nontrivial chiral symmetry properties in the zero quark mass limit. This remarkable fact allow us to have an effective lagrangian for the description of strong interactions at low energies. As time goes on, the Skyrme model [1] turned out to be one the most successful effective lagrangians of this type thanks to the fact that it contains terms which are compatible with the symmetries of QCD and are quadratic in time derivatives at most [2, 3] (for excellent reviews on the fundamentals of the Skyrme model see [4]–[5] and references therein).

Thus, the main progress which the community has reached in understanding the physics of the low energy limit of QCD, e.g. the description of the dynamics of mesons and baryons, relies basically on the Skyrme model.

In the present report we begin by briefly reviewing some basic aspects of the $SU(N)$ Skyrme model ($N \geq 2$) in order to further point out a path towards the establishment of hidden nonlinear symmetries that could in principle lead to the integration of the system.

Thus, the Skyrme model represents a nonlinear lagrangian written in terms of the unitary and unimodular Skyrme field $U(x)$. In other words, the Skyrme field $U(x)$ constitutes a $N \times N$ matrix which belongs to the $SU(N)$ symmetry group. As proposed by Skyrme, the $SU(N)$ invariant lagrangian compatible with the chiral symmetries of

QCD reads

$$\mathscr{L} = \frac{F_\pi^2}{16} \text{Tr}\left(\partial_\mu U^\dagger \partial_\mu U\right) + \frac{1}{32e^2} \text{Tr}\left(\left[U^\dagger \partial_\mu U, U^\dagger \partial_\nu U\right]^2\right), \qquad (1)$$

where e is a dimensionless constant parameter and F_π denotes the pion decay constant. In fact, a mass term together with a Wess–Zumino–Witten term and chiral symmetry breaking terms can be added to the lagrangian density (1), but we shall omit them in our consideration. The first term in (1) is the standard nonlinear chiral lagrangian, whereas the second one represents the so–called Skyrme term. It turns out that the classical topological solutions described by the first chiral term alone are non stable, moreover, it has been proved that the introduction of the Skyrme term avoids the instability of such field configurations. Thus, the lagrangian density (1) supports the existence of stable topological solitonic configurations.

Let us come back to the Skyrme field $U(x) \in SU(N)$. It can be represented in terms of the generators of $SU(N)$ in the fundamental representation T_α for $\alpha = 1, \ldots, N^2 - 1$ in the standard form

$$U(x) = e^{iL^\alpha T_\alpha}. \qquad (2)$$

By taking the logarithm in both sides of this relation we get the following expression

$$-i \ln U(x) = L^\alpha T_\alpha. \qquad (3)$$

Further, by formally differentiating this equation with respect to the x^μ coordinate we introduce the left–invariant Maurer–Cartan covariant vector

$$-iU^\dagger \partial_\mu U = L_\mu^\alpha T_\alpha \equiv \mathbf{L}_\mu, \qquad (4)$$

where $L_\mu^\alpha \equiv \partial_\mu L^\alpha$.

The covariant vector $\mathbf{L}_\mu(x)$ represents an $SU(N)$ Lie–algebra valued current and, by construction, it acts as a pure gauge connection. This quantity satisfies the Maurer–Cartan identity (in fact this relationship can be viewed as well as the zero curvature condition)

$$\partial_\mu \mathbf{L}_\nu - \partial_\nu \mathbf{L}_\mu = -i\left[\mathbf{L}_\mu, \mathbf{L}_\nu\right]. \qquad (5)$$

Thus, written in terms of the Maurer–Cartan covariant vector (4), the Skyrme lagrangian density (1) reads

$$\mathscr{L} = \frac{F_\pi^2}{16} \text{Tr}\left(\mathbf{L}_\mu \mathbf{L}_\mu\right) + \frac{1}{32e^2} \text{Tr}\left(\left[\mathbf{L}_\mu, \mathbf{L}_\nu\right]^2\right). \qquad (6)$$

Skyrme proved that this model supports the existence of stable solitonic solutions, that later on were called skyrmions in his honor. These field configurations can be characterized by an integer–valued topological number that it is identified with the baryon number of the nucleon [1]. Moreover, the conservation of the baryon number is associated to the conservation of the topological charge

$$\mathbf{Q} = \frac{i}{48\pi^2} \varepsilon_{ijk} \int d^3 x \, \text{Tr}\left(\mathbf{L}_i \left[\mathbf{L}_j, \mathbf{L}_k\right]\right), \qquad (7)$$

where the charge **Q** adopts integer values on each topologically distinct class of field configurations $(U : R^3 \longrightarrow S^3)$ or, equivalently, on each nontrivial homotopy class $\pi_3(S^3) = Z$.

It is straightforward to check that the expression for the topological charge **Q** in (7) constitutes a conserved quantity. In order to prove this fact, we first express the corresponding conserved topological current

$$\mathbf{J}_\mu = \frac{i}{48\pi^2} \varepsilon_{\mu\nu\sigma\rho} \operatorname{Tr}\left(\mathbf{L}_\nu \left[\mathbf{L}_\sigma, \mathbf{L}_\rho\right]\right); \tag{8}$$

then, the conservation of the topological charge **Q**, in which the current component $\mathbf{J}_0$ plays the role of the charge density, is just a direct consequence of the conservation law of the topological current

$$\partial_\mu \mathbf{J}_\mu = 0. \tag{9}$$

It turns out that the Skyrme model possesses a very rich soliton spectrum. For instance, it was recently shown by Cho [6] that, with an appropriate choice of the Skyrme field $U(x)$, the Skyrme model can be interpreted as a theory of self–interacting non–Abelian monopoles and allows as well the existence of knot–like solitonic solutions. These solitons are very similar to the topological knots previously obtained in the Faddeev–Skyrme model [7], but they possess a different behaviour from the well–known skyrmions.

By making use of the standard variational principle, from the Skyrme model (6) one can derive the Euler-Lagrange's equation satisfied by each topological field configuration. This equation reads

$$\partial_\mu \left(F_\pi^2 \mathbf{L}_\mu + \frac{1}{e^2} \left[\mathbf{L}_\nu, \left[\mathbf{L}_\mu, \mathbf{L}_\nu\right]\right] \right) = 0 \tag{10}$$

and it has the structure of a local conservation law with a chiral current which possesses a linear term plus a highly nonlinear component.

Because of the intrinsic nonlinear nature of the Skyrme model, it is quite difficult to solve this equation and only a few analytic solutions are known till now.

In [1] Skyrme pointed out that if, for instance, one considers an $SU(2)$ Skyrme field of the following form

$$U(x) = U(k \cdot x), \tag{11}$$

it constitutes a solution of equation (10) provided that the k_μ is a light–like four–momentum, that is, $k^2 = 0$. Moreover, he showed as well that static Skyrme field configuration with the spherically symmetric ansatz

$$U(\vec{r}) = \cos\theta(r) + i\hat{r}_a \tau_a \sin\theta(r), \tag{12}$$

with unitary vector $\hat{r} = \vec{r}/r$, yields a nontrivial solution of the same field equation (10) which, indeed, realizes the absolute minimum of the energy in the first homotopy class, i.e., among fields with $|\mathbf{Q}| = 1$. In this picture, it is precisely the profile function (in fact, the chiral angle) $\theta(r)$ that minimizes the static energy functional of the model [8]. The corresponding basic equation for the chiral angle reads

$$\left(r^2 + 4\sin^2\theta\right)\theta'' + 2r\theta' = \sin(2\theta)\left[1 - 2(\theta')^2 + \frac{2\sin^2\theta}{r^2}\right] \tag{13}$$

where the prime denotes derivatives with respect to the r coordinate. Until now, no one has managed to obtain an exact solution for this equation.

When considering analytic solutions of the $SU(N)$ (with $N > 2$) Skyrme model, the harmonic map ansatz of S^2 to CP^{N-1} (the direct generalization of tha ansatz (12)), allows us to construct radially symmetric solitonic configurations with nontrivial topological charge [9], which, in the general case, depend on $N - 1$ profile functions that must be determined numerically. On the other hand, the so–called axially symmetric field configurations were proved to realize the energy minimum in higher homotopy classes, that is for $|Q| > 1$. Recently, a class of exact solutions for both the $SU(2)$ and the $SU(3)$ Skyrme model were reported in [10]–[11]. These solutions, in contrast to the previously found field configurations, depend on both couplings constants F_π and e, however, they do not constitute solitonic objects, but possess wave characteristics in the Minkowski space. In some particular cases, these solutions provide an explicit example that renders a nonlinear superposition of wave solutions in the framework of nonlinear field theories.

IN SEARCH OF NONLINEAR SYMMETRIES

By looking at the field equation (10), one realizes that the first term corresponds to a standard chiral field equation, whose properties and exact solutions are well studied. In particular, under the assumption of some symmetry conditions, this chiral equation can be completely integrable. The second term in (10) corresponds to the nonlinear Skyrme term.

Thus, it is natural to think that if there exists some kind of relationship between the first and second terms of the field equation (10), then the solutions corresponding to the former term could be mapped onto solutions of the latter, leading to the integration of the field equation (10). Of course, if such a symmetry does exist, it will be a nonlinear symmetry which involves derivatives of the functions that parameterize the left–invariant Maurer–Cartan covariant vector $\mathbf{L}_\mu(x)$ due to the structure of the nonlinear field equation (10).

One of the reasons for thinking that such a nonlinear symmetry could exist in principle relies in the fact that the covariant vectors $\mathbf{L}_\mu(x)$ represent an $SU(N)$ Lie–algebra valued chiral current. Thus, the algebraic structure of the generators of these chiral currents seems to be the main ingredient in the establishment of such hidden symmetries.

Since the $SU(N)$ matrix–valued chiral currents, with $N > 2$, involve a large number of generators ($N^2 - 1$), the establishment of these hidden symmetries (if they exist) represents a rather complicated task. For this reason, we shall consider just the simplest $SU(2)$ case.

In this case, the Skyrme field $U(x) = e^{iL^\alpha T_\alpha}$ will possess three generators corresponding to the three Pauli matrices: $T_\alpha = \sigma_\alpha$, where

$$\sigma_1 = \begin{pmatrix} 0 & 1 \\ 1 & 0 \end{pmatrix}, \quad \sigma_2 = \begin{pmatrix} 0 & -i \\ i & 0 \end{pmatrix}, \quad \sigma_3 = \begin{pmatrix} 1 & 0 \\ 0 & -1 \end{pmatrix}. \quad (14)$$

By choosing the following complex parametrization for the Skyrme field

$$U(x) = \begin{pmatrix} e^{i\xi} \cos\eta & e^{i\theta} \sin\eta \\ -e^{-i\theta} \sin\eta & e^{-i\xi} \cos\eta \end{pmatrix}, \quad (15)$$

where the real functions $\xi(x)$, $\theta(x)$ and $\eta(x)$ depend on the coordinates x_μ.

By computing the matrix-valued chiral currents $\mathbf{L}_\mu(x)$ we obtain the following expressions for L_μ^α:

$$\begin{aligned} L_\mu^1 &= \sin(\theta - \xi)\, \eta_\mu + \cos(\theta - \xi) \sin\eta \cos\eta\, (\theta + \xi)_\mu \\ L_\mu^2 &= \cos(\theta - \xi)\, \eta_\mu - \sin(\theta - \xi) \sin\eta \cos\eta\, (\theta + \xi)_\mu \\ L_\mu^3 &= \cos^2\eta\, \xi_\mu - \sin^2\eta\, \theta_\mu \end{aligned} \quad (16)$$

where, for instance, we have denoted $\eta_\mu \equiv \partial_\mu \eta$.

Here it is useful to denote the second term involved in the "conserved chiral current" of the field equation (10), which describes the dynamics of the field system corresponding to the nonlinear Skyrme term, in the following form

$$\lambda \left[\mathbf{L}_\nu, [\mathbf{L}_\mu, \mathbf{L}_\nu]\right] \equiv \mathbf{P}_\mu = P_\mu^\alpha T_\alpha, \quad (17)$$

where $\lambda = 1/(F_\pi^2 e^2)$.

Similarly, we can straightforwardly compute the expressions for the components P_μ^α of the chiral vector $\mathbf{L}_\mu(x)$:

$$\begin{aligned} P_\mu^1 &= \sin(\theta - \xi) \left\{ \cos^2\eta\, \xi_\nu\, [\xi_\mu, \eta_\nu] + \sin^2\eta\, \theta_\nu\, [\theta_\mu, \eta_\nu] \right\} \\ &+ \cos(\theta - \xi) \sin\eta \cos\eta \left\{ (\cos^2\eta\, \xi_\nu - \sin^2\eta\, \theta_\nu)\, [\xi_\mu, \theta_\nu] \right. \\ &\left. + \eta_\nu\, [\eta_\mu, (\theta + \xi)_\nu] \right\} \\ P_\mu^2 &= \cos(\theta - \xi) \left\{ \cos^2\eta\, \xi_\nu\, [\xi_\mu, \eta_\nu] + \sin^2\eta\, \theta_\nu\, [\theta_\mu, \eta_\nu] \right\} \\ &+ \sin(\theta - \xi) \sin\eta \cos\eta \left\{ (\cos^2\eta\, \xi_\nu - \sin^2\eta\, \theta_\nu)\, [\theta_\mu, \xi_\nu] \right. \\ &\left. + \eta_\nu\, [(\theta + \xi)_\mu, \eta_\nu] \right\} \\ P_\mu^3 &= \sin^2\eta \cos^2\eta\, (\theta + \xi)_\rho\, [\theta_\mu, \xi_\rho] + \cos^2\eta\, \eta_\nu\, [\eta_\mu, \xi_\nu] \\ &+ \sin^2\eta\, \eta_\nu\, [\theta_\mu, \eta_\nu]. \end{aligned} \quad (18)$$

Here it will be useful to introduce the following notation in order to make things more clear

$$\begin{array}{ll} L_\mu^1 = \chi_\mu & P_\mu^1 = G_\mu \\ L_\mu^2 = \psi_\mu & P_\mu^2 = H_\mu \\ L_\mu^3 = \varphi_\mu & P_\mu^3 = F_\mu \end{array} \quad (19)$$

where χ_μ, ψ_μ and φ_μ are defined through the relations given by (16) and

$$\begin{aligned}
G_\mu &= \{-\cos(\theta-\xi)\sin\eta\cos\eta\ [\eta^2+\cos^2\eta\ \xi^2-\sin^2\eta\ (\theta\xi)] \\
&\quad + \sin(\theta-\xi)\sin^2\eta\ (\theta\eta)\}\theta_\mu \\
&\quad + \{-\cos(\theta-\xi)\sin\eta\cos\eta\ [\eta^2-\cos^2\eta\ (\theta\xi)+\sin^2\eta\ \theta^2] \\
&\quad + \sin(\theta-\xi)\cos^2\eta\ (\xi\eta)\}\xi_\mu \\
&\quad + \{\cos(\theta-\xi)\sin\eta\cos\eta\ [(\theta\eta)+(\xi\eta)] \\
&\quad - \sin(\theta-\xi)\ [\cos^2\eta\ \xi^2+\sin^2\eta\ \theta^2]\}\eta_\mu \\
H_\mu &= \{-\sin(\theta-\xi)\sin\eta\cos\eta\ [\eta^2+\cos^2\eta\ \xi^2-\sin^2\eta\ (\theta\xi)] \\
&\quad + \cos(\theta-\xi)\sin^2\eta\ (\theta\eta)\}\theta_\mu \\
&\quad + \{\sin(\theta-\xi)\sin\eta\cos\eta\ [\eta^2-\cos^2\eta\ (\theta\xi)+\sin^2\eta\ \theta^2] \\
&\quad + \cos(\theta-\xi)\cos^2\eta\ (\xi\eta)\}\xi_\mu \qquad (20) \\
&\quad - \{\sin(\theta-\xi)\sin\eta\cos\eta\ [(\theta\eta)+(\xi\eta)] \\
&\quad + \cos(\theta-\xi)\ [\cos^2\eta\ \xi^2+\sin^2\eta\ \theta^2]\}\eta_\mu \\
F_\mu &= \{\sin^2\eta\cos^2\eta\ [(\theta\xi)+\xi^2\ ]+\sin^2\eta\ \eta^2\}\theta_\mu \\
&\quad - \{\sin^2\eta\cos^2\eta\ [(\theta\xi)+\theta^2\ ]+\cos^2\eta\ \eta^2\}\xi_\mu \\
&\quad + [\cos^2\eta\ (\xi\eta)-\sin^2\eta\ (\theta\eta)]\eta_\mu.
\end{aligned}$$

where, for instance, $\xi^2 = (\xi_\nu \cdot \xi_\nu)$, $(\xi\eta) = (\xi_\nu \cdot \eta_\nu)$, etc.

Thus, we are looking for a symmetry between the left–invariant Maurer–Cartan covariant vectors $\mathbf{L}_\mu$ and $\mathbf{P}_\mu$

$$\mathbf{L}_\mu(x) \longleftrightarrow \mathbf{P}_\mu(x), \qquad (21)$$

where now

$$\mathbf{L}_\mu(x) = \begin{pmatrix} \xi_\mu & \varphi_\mu - i\psi_\mu \\ \varphi_\mu + i\psi_\mu & -\xi_\mu \end{pmatrix} \qquad (22)$$

and

$$\mathbf{P}_\mu(x) = \lambda \begin{pmatrix} F_\mu & G_\mu - iH_\mu \\ G_\mu + iH_\mu & -F_\mu \end{pmatrix}. \qquad (23)$$

At this stage it looks quite difficult to establish a nonlinear symmetry between these chiral vectors. However, in the particular case in which $\theta(x) = \xi(x)$, the components of the chiral vectors $\mathbf{L}_\mu$ and $\mathbf{P}_\mu$ read

$$\begin{array}{lll}
\varphi_\mu = \sin(2\eta)\ \xi_\mu & \longleftrightarrow & G_\mu = \lambda\sin(2\eta)\ [(\xi\eta)\eta_\mu - \eta^2\ \xi_\mu], \\
\psi_\mu = \eta_\mu & \longleftrightarrow & H_\mu = \lambda\ [(\xi\eta)\xi_\mu - \xi^2\ \eta_\mu], \qquad (24) \\
\chi_\mu = \cos(2\eta)\ \xi_\mu & \longleftrightarrow & F_\mu = \lambda\cos(2\eta)\ [(\xi\eta)\eta_\mu - \eta^2\ \xi_\mu].
\end{array}$$

Remarkably, if we choose a set of initial functions for the chiral vectors $\mathbf{L}_\mu$ in the following form

$$\varphi_{0\mu} = \frac{\sin(2\eta)\left[(\xi\eta)\eta_\mu - \eta^2\xi_\mu\right]}{\sqrt[3]{\lambda\left[\xi^2\eta^2 - (\xi\eta)^2\right]^2}} \tag{25}$$

$$\psi_{0\mu} = \frac{(\xi\eta)\xi_\mu - \xi^2\eta_\mu}{\sqrt[3]{\lambda\left[\xi^2\eta^2 - (\xi\eta)^2\right]^2}} \tag{26}$$

$$\chi_{0\mu} = \frac{\cos(2\eta)\left[(\xi\eta)\eta_\mu - \eta^2\xi_\mu\right]}{\sqrt[3]{\lambda\left[\xi^2\eta^2 - (\xi\eta)^2\right]^2}}, \tag{27}$$

and compute the corresponding components of the chiral current $\mathbf{P}_\mu$, we obtain the following expressions

$$\begin{array}{l} G_\mu = \sin(2\eta)\,\xi_\mu \equiv \varphi_\mu \\ H_\mu = \eta_\mu \equiv \psi_\mu \\ F_\mu = \cos(2\eta)\,\xi_\mu \equiv \chi_\mu \end{array} \tag{28}$$

which precisely coincide with the components of the chiral vector $\mathbf{L}_\mu$ in (24).

Thus, the initial functions (25)–(27) map the components of the chiral current $\mathbf{P}_\mu$ onto the components of the chiral vector $\mathbf{L}_\mu$, establishing a nontrivial symmetry between them.

CONCLUDING REMARKS

Obviously, the generalization of this relationship to the general case in which the restriction $\theta(x) = \xi(x)$ does not hold, constitutes a nontrivial task. We are searching for such a generalization at the moment and hope to report further advances in the near future.

It is interesting to try to apply the obtained nonlinear symmetry in order to generate exact solutions to the field equation (10) of the Skyrme model. Even if this constitutes as well a rather difficult task in the general case, it could be analytically treatable for some simple nontrivial initial solutions.

It is worth noticing that a first attempt to obtain the spherically symmetric ansatz (12) or axially symmetric field configurations compatible with the special condition $\theta(x) = \xi(x)$ (or some equivalent restrictions) was unsuccessful. It seems that it is necessary to consider the three nontrivial different functions $\xi(x)$, $\theta(x)$ and $\eta(x)$ in order to achieve this aim. This topic is currently under investigation as well.

ACKNOWLEDGMENTS

AHA and KK are grateful to V. Oikonomou for useful discussions while this investigation was carried out. AHA would like to express his gratitude to the whole Theoretical Physics Department of the Aristotle University of Thessaloniki for providing a stimulating atmosphere while this work was in progress. He acknowledges as well a grant for postdoctoral studies provided by the Greek Government. This research was supported by grants CIC-UMSNH-4.16 and CONACYT-F42064.

REFERENCES

1. T. H. R. Skyrme, *Nucl. Phys.* **31**, 556 (1961).
2. G. 't Hooft, Nucl. Phys. **72**, 461 (1974).
3. E. Witten, *Nucl. Phys.* **160**, 57 (1979); *ibid* **223**, 433 (1983).
4. V.G. Makhan'kov, Yu.P. Rybakov and V.I. Sanuyk, *The Skyrme model: Fundamentals, methods, applications*, Springer, Berlin, 1993; *Sov. Phys. Usp.* **35**, 55 (1992), *Usp. Fiz. Nauk* **162**, 1 (1992).
5. A.P. Balachandran, "The Skyrme soliton: a review", in *Solitons in nuclear and elementary particle physics*, Ed. A. Chodos, E. Hadjimichael and C. Tze, World Scientific, Singapore, 1984.
6. Y. M. Cho, *Phys. Rev. Lett.* **87**, 252001 (2001).
7. L. D. Faddeev and A. J. Niemi, *Nature (London)* **387**, 58 (1997); L. D. Faddeev and A. J. Niemi, *Phys. Rev. Lett.* **82**, 1624 (1999).
8. G. S. Adkins, C. R. Nappi and E. Witten, *Nucl. Phys.* **B 228**, 552 (1983).
9. T. Ioannidou, B. Piette and W. J. Zakrzewski, *J. Math. Phys.* **40**, 6233 (1999); *ibid* **40**, 6353 (1999).
10. M. Hirayama and J. Yamashita, *Phys. Rev.* **D66**, 105019 (2002); M. Hirayama, C. G. Shi and J. Yamashita, *Phys. Rev.* **D67**, 105009 (2003).
11. W.-C. Su, *Phys. Lett.* **B568**, 167 (2003).

Predictions for the Higgs Mass from the Stability and Triviality Conditions

H. Gabriel Solís R.*, S. Rebeca Juárez W.[1] * and P. Kielanowski[†]

*Dpto. de Física, Esc. Sup. de Fís. y Mat. – Instituto Politécnico Nacional,
UP "Adolfo López Mateos" Edif. 9, CP 07738, México, DF
[†]Dpto. de Física, CINVESTAV, Av. IPN 2508, CP 07360, México, DF

Abstract. In the context of the Standard Model (SM), we use the one-loop and two-loop Renormalization Group Equations (RGE) in order to analyze the evolution of the Higgs quartic coupling λ_H in the interval $[m_t, E_{GU}]$, where m_t is the mass of the top quark and $E_{GU} = 10^{14}$ GeV. The analytical solution for the one-loop differential equation (Riccati type) is obtained and analyzed and in the two-loop case we obtain a numerical solution which takes into account all the parameters (couplings) at the same order of approximation. In both cases, we restrict the possible initial values for λ_H by means of imposing the triviality and stability conditions which determine the range of energies where the SM is valid. We obtain the following bounds: $0.387 < \lambda_H < 0.623$ for the one-loop case and $0.360 < \lambda_H < 0.628$ for the two-loop case. These results determine the interval of the possible Higgs mass values: $151.9 < M_H < 192.3$ GeV, $143.8 < M_H < 190.3$ GeV for the one-loop and two-loop cases, respectively.

GENERAL IDEAS

Nowadays, the SM is the main tool that high-energy physicists have to describe the elementary particle phenomena. The Higgs boson is the only experimentally missing ingredient needed to give a physical reality to the model. Therefore, the knowledge of the boundary values for the mass of the Higgs particle is of prime importance.

Let us start with the presentation of the main features of the subject and the notation we use. A Higgs doublet Φ is introduced in order to generate the masses of the particles in the Lagrangian with the Higgs potential of the following form:

$$\mathscr{L}_{int} = -\left(\overline{\mathbf{Q}}_L \Phi \mathbb{Y}_D^\dagger \mathbf{D}_R + \overline{\mathbf{Q}}_L \widetilde{\Phi} \mathbb{Y}_U^\dagger \mathbf{U}_R + \overline{\mathbf{L}}_L \Phi \mathbb{Y}_\ell^\dagger \mathbf{L}_R + hc\right) + m^2|\Phi|^2 - \frac{\lambda_H}{2}|\Phi|^4, \quad (1)$$

where $\widetilde{\Phi} =: i\,\hat{\tau}_2 \Phi^*$. $\mathbb{Y}_U$, $\mathbb{Y}_D$ and $\mathbb{Y}_\ell$ denote the quark and charged lepton Yukawa couplings (which are 3×3 complex matrices); m^2 and λ_H are the quadratic and quartic Higgs couplings, respectively. We do not include any neutrino terms, because their contributions are negligible at the order of approximation we are interested in. Furthermore, after the spontaneous symmetry breaking, only a scalar neutral particle ϕ

---

[1] Becaria EDD and COFAA-IPN.

remains and the Higgs potential becomes

$$V(\phi) = -\frac{1}{2}m^2\phi^2 + \frac{1}{8}\lambda_H \phi^4. \quad (2)$$

The SM Lagrangian includes also the kinetic terms and determines, through the gauge invariance, the possible interactions between particles. The physical results for different processes are obtained through a perturbative expansion. The main problem that arises in the precise evaluation of the parameters of the theory is the appearance of divergencies caused by the inclusion of loops in the Feynman graphs. The renormalization scheme deals with these divergencies. It furnishes a way for subtracting the infinite contributions, and absorbing them into the redefinitions of the physical parameters, which yield numerical results that agree with the experimental data. The renormalized parameters do depend on the energy scale through the renormalization point E. A shift of the renormalization point is a scale transformation that induces a multiplicative transformation. The group of such scale transformations is called the renormalization group (RG).

In order to analyze the evolution of the relevant parameters in the SM according to an energy scale $t = \ln(E/E_0)$, the (generic) RGE reads as follows:

$$\beta_x = E\frac{\partial x}{\partial E} = \frac{dx}{dt} = \frac{1}{16\pi^2}\beta_x^{(1)} + \frac{1}{256\pi^4}\beta_x^{(2)}, \quad (3)$$

denoting by the upper indices (1) and (2) the corrections to one and two loops, respectively. These equations have been analyzed by several authors without considering the knowledge about the top quark mass [1] and using some simplifications [2]. The goal was also to establish the limits on the Higgs Mass [2, 3] and to generalize these ideas in supersymmetric models [1, 4].

Since the Higgs mass is mainly related to λ_H [2, 5], but also to $\mathbb{Y}_U, g_1, g_2$ and Z (the renormalizing factor of the Higgs field), it is important to know precisely the running of all these parameters with energy.

SOLUTION OF THE RGE

The complete and revised form of the equations (3) is shown explicitly in [6]. In solving the equations, we take as our reference point for initial values $E_0 = m_t \approx 172.5\,\text{GeV} \Rightarrow t_{\min} = 0$ and the interval of interest ends at $E_{GU} = 10^{14}\,\text{GeV} \Rightarrow t_{\max} \approx 27.08$.

Since we neglect the neutrino's contribution, there is no mixing in the leptonic sector, so $\mathbb{Y}_\ell$ is already diagonal. With respect to $\mathbb{Y}_D$ and $\mathbb{Y}_U$, they can be diagonalized for each t as proved in [7]. Considering $\mathbb{Y}_U, \mathbb{Y}_D$ and $\mathbb{Y}_\ell$ as diagonalized, we only take into account the heaviest element from each family; this is to say, we retain the contributions of the top, bottom and tau. We also "scalarize" the equations by taking traces in all of them:

$$\text{Tr}\{\mathbb{Y}_U^\dagger \mathbb{Y}_U\} \to Y_U^2 = Y_t^2, \quad \text{Tr}\{\mathbb{Y}_D^\dagger \mathbb{Y}_D\} \to Y_D^2 = Y_b^2 \quad \text{and} \quad \text{Tr}\{\mathbb{Y}_\ell^\dagger \mathbb{Y}_\ell\} \to Y_\ell^2 = Y_\tau^2; \quad (4)$$

with respective initial values 1, m_b^2/m_t^2 and m_τ^2/m_t^2.

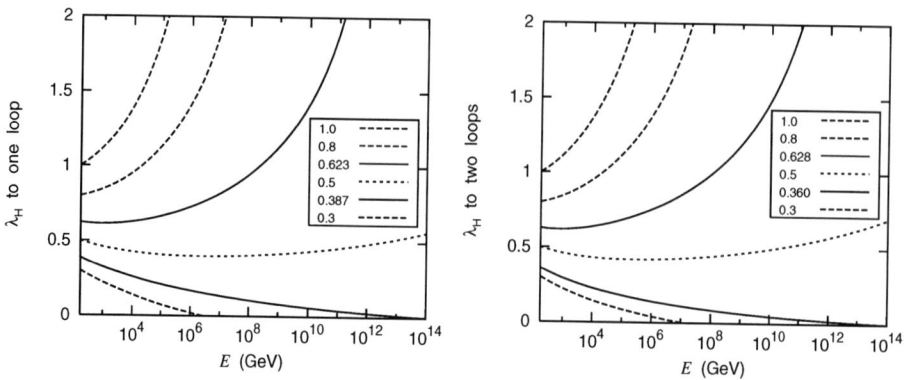

FIGURE 1. The evolution of λ_H with energy for different values of $\lambda_H^0 = \lambda_H(m_t)$. For $\lambda_H^0 > 0.5$ in the one-loop case λ_H has a pole whereas for two loops it tends to the asymptotic value of ≈ 24 (see Ref. [9]). In each case, the third and fifth values are those referred in Eqs. (7) y (10), respectively.

In the case of the gauge couplings g_i's, the values at t_0 are established after performing the one-loop evolution from $t(M_Z)$ to t_0, then the values we use are $g_1^0 = 0.463$, $g_2^0 = 0.648$ and $g_3^0 = 1.173$. The initial value for the Higgs quartic coupling λ_H^0 is unknown but it is restricted through the SM validity conditions. These bounds will turn out to be the boundary values on the Higgs mass.

To solve the one-loop RGE, we retain only the up sector contribution (consistently) in the right hand side of Eq. (3), so the equations decouple and we can obtain the analytical solution:

$$\frac{g_i}{g_i^0} = \left[1 - (g_i^0)^2 \frac{2b_i}{16\pi^2}(t-t_0)\right]^{-\frac{1}{2}}, \quad \frac{Y_U}{Y_U^0} = \sqrt{r(t)}\left[1 - \frac{9(Y_U^0)^2}{16\pi^2}\int_{t_0}^{t} r(\tau)d\tau\right]^{-\frac{1}{2}}, \quad (5)$$

where $b_i = (\frac{41}{10}, -\frac{19}{6}, -7)$, $c_i = (\frac{17}{20}, \frac{9}{4}, 8)$, and we define $r =: \prod_{i=1}^{3}\left[g_i^0/g_i\right]^{2c_i/b_i}$. In the case of Y_D and Y_ℓ, the solutions are generated by direct integration.

For the Higgs quartic coupling λ_H one obtains a Riccati type equation, whose solution [8, 9] is

$$\lambda_H = -\frac{4\pi^2}{3}\frac{3\lambda_H^0 \frac{d}{dt}W_2 - 4\pi^2 \frac{d}{dt}W_1}{3\lambda_H^0 W_2 - 4\pi^2 W_1}, \quad (6)$$

where W_1 and W_2 are solutions of the associated second order linear equation and they are chosen to satisfy $W_1^0 = 1$, $W_2^0 = 0$, $\left(\frac{d}{dt}W_1\right)^0 = 0$, $\left(\frac{d}{dt}W_2\right)^0 = 1$.

To analyze the two-loop case, we also include the down and leptonic contributions in the right hand side of the β's. Due to the mathematical complexity, we proceed numerically (in Maple X). In Fig. 1, we illustrate the λ_H one- and two-loop evolution for various initial values, whereas in Fig. 2 the evolution of gauge and Yukawa couplings is shown. Y_U, Y_D and λ_H are quantities that present an important difference when calculated to one and two loops, see Ref. [10].

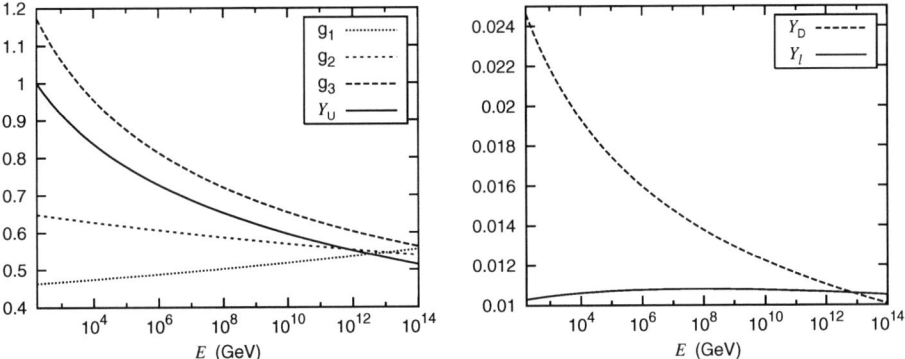

FIGURE 2. The two-loop energy dependence of gauge and Yukawa couplings for the initial value $\lambda_H(m_t) = 0.5$. Except for g_1 and Y_ℓ, the remaining variables are monotonically decreasing. Even though the range is limited to E_{GU}, a "unification" can be appreciated between g_1, g_2 and Y_U on one side and of Y_D and Y_ℓ on the other. The most relevant parameter that affects λ_H is Y_U.

STABILITY AND TRIVIALITY

To assure that Eq. (2) makes physical sense throughout the interval $m_t \leq E \leq E_{GU}$, we require that λ_H *must be positive and finite*. These requirements are also called the stability and triviality conditions, respectively. In the one-loop case (i.e. from Eq. (6)), we observe that the zeros of λ_H are determined by the numerator, whereas the singularities (Landau poles) are obtained from the zeros of the denominator. Thus, both conditions constrain the acceptable values of $\lambda_H^0 = \lambda_H(m_t)$:

$$0.387 < \lambda_H^0 < 0.623. \tag{7}$$

The first value comes from $\lambda_H(\lambda_H^0, t_{max}) = 0$, the second one from $\lambda_H(\lambda_H^0, t_{max}) \to \infty$. Incidentally, there exists an interesting relation between the two conditions in equation (6), the numerator is the derivative of the denominator.

The Two-Loop Asymptotic Condition

In the two-loop approximation, $\lambda_H(t)$ does not diverge for the λ_H^0's that produce an increasing behavior, but it reaches an asymptotic value of ≈ 24. Even though, in this case the stability condition is reliable, the triviality one is not useful anymore and needs to be reconsidered.

In this context, the analysis of Ref. [2] (the most cited article about the Higgs mass) deserves some discussion. At first, the λ used in [2] is equivalent to our $\lambda_H/2$. A "fix point" value of $\lambda_{FP} = 12$ is proposed and interpreted as a local maximum, but it corresponds to an asymptotic value instead. Further, the Y_U evolution is neglected and without justification $Y_U(t)$ is assumed to reveal an increasing behavior, but actually it

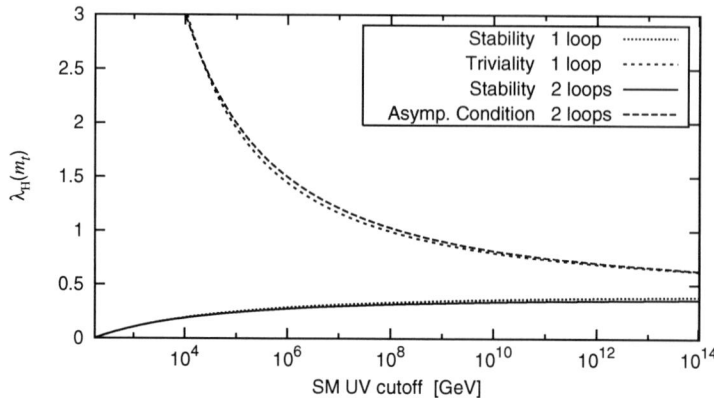

FIGURE 3. The generalization of bounds for λ_H^0 in terms of Standard Model UV cutoff for energies below E_{GU}. In each approximation and for a given cut off, the possible values of λ_H^0 are those that lay between the upper and lower graphs.

decreases. At last, based on the perturbative expansion validity, a UV bound for the SM is proposed as the t_{UV} given by the condition $\lambda(t_{UV}) = \lambda_{FP}/2$ or $\lambda(t_{UV}) = \lambda_{FP}/4$.

On the other hand, from the evolution graphs (Figure 2) we note that Y_U is a strictly decreasing function and $Y_U(t) < 1 = Y_U^0$ in all the energy range. Y_D and Y_ℓ are negligible, in the sense that they do not affect the λ_H evolution. In the RGE for λ_H [6], g_1, g_2 and Y_U appear only in powers of two or greater, so their contributions are small at high energies.

Hence for the asymptotic behavior of λ_H it is valid to retain only the leading terms of the RGE, and to determine the λ_H^{as} from the condition:

$$\beta_{\lambda_H} = 0 \Rightarrow 12\lambda_H^2 - \frac{78}{4\pi^2}\lambda_H^3 = 0 \Rightarrow \lambda_H^{as} \approx 24.294. \tag{8}$$

Inspired by Ref. [2], we propose the following condition for the SM UV cutoff:

$$\lambda_H\left(\lambda_H^0, t_{max}\right) = \lambda_H^{as}/2. \tag{9}$$

Finally, from the stability and Eq. (9), respectively, we obtain limits on λ_H^0 in the two-loop approximation:

$$0.360 < \lambda_H^0 < 0.628. \tag{10}$$

The bounds in Eq. (7) and Eq. (10) arise from our choice of the maximum energy E_{GU}. When this energy is increased, the bounds for λ_H^0 become more restricted. Beyond E_{GU} it is usual to consider the extensions of the SM. At energies below E_{GU} the bounds are less constrained and the corresponding results are shown in Fig. 3.

BOUNDS ON THE HIGGS MASS

According to [11], the potential $V(\phi)$, once renormalized, preserves its tree level form (2), but with the renormalized field and parameters[2]:

$$V(\phi) = -\frac{1}{2}m^2(t)Z^2(t)\phi^2 + \frac{1}{8}\lambda_H(t)Z^4(t)\phi^4. \tag{11}$$

The evolution of the renormalizing factor $Z(t)$ is given through the Higgs anomalous dimension γ_H [1]:

$$\frac{d\ln Z}{dt} = -\gamma_H = -\frac{1}{16\pi^2}\gamma_H^{(1)} - \frac{1}{256\pi^4}\gamma_H^{(2)}. \tag{12}$$

Thus, the potential minimum also evolves according to $Z(t)\mathscr{V}$, and the vacuum expectation value $\mathscr{V} = 246.218\,\text{GeV}$ corresponds to the scale $E = M_Z$. The analytical solution for $Z(t)$ in the one-loop case has the form:

$$Z(t) = \exp\left[-\frac{3}{16\pi^2}\int_{t_0}^{t} Y_U^2(\tau)d\tau\right]\left[\frac{g_1(t)}{g_1^0}\right]^{9/20b_1}\left[\frac{g_2(t)}{g_2^0}\right]^{9/4b_2}. \tag{13}$$

For two loops we proceed again numerically.

Following [12, 9], the running Higgs mass m_H is calculated from the potential in the Lagrangian:

$$m_H^2 = \left.\frac{\partial^2 V(\phi)}{\partial\phi^2}\right|_{\phi_{\min}=\mathscr{V}} = \lambda_H(t)Z^2(t)\mathscr{V}^2. \tag{14}$$

To one loop, the equations (6) and (13) contain all the information needed to determine the running mass. Then, at this level the result is analytic and m_H possesses the same zero and Landau pole as λ_H does. For two loops, Eq. (14) remains unchanged [11], but one has to consider the two-loop evolution.

The Physical Higgs Mass

The definition of mass we take is [12]: *the physical mass is the real part of the pole in the renormalized propagator*. We then have two contributions to M_H, the running mass and the auto-energy factor [9]:

$$M_H^2 =: m_H^2 + \text{Re}\left\{\Pi_R\big|_{p^2=M_H^2} - \Pi_R\big|_{p^2=0}\right\}. \tag{15}$$

The explicit calculation of the last term in the former equation was carried out in Ref. [4, Appendix B].

---

[2] This point of view is different from [4, 5, 9], where an effective potential is considered.

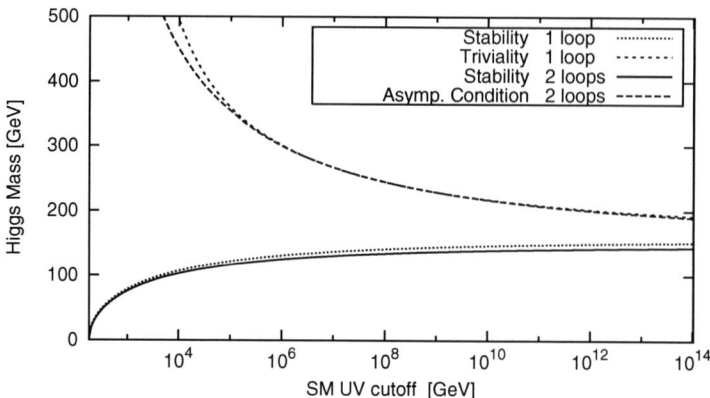

FIGURE 4. The plot of bounds for M_H as a function of E_{UV}. In each approximation and for a given cut off, the range of allowed values of M_H is enclosed by the upper and lower graphs.

All the quantities in the right hand side of Eq. (15) do not depend on the choice of gauge or scale, only when all the perturbative terms are included. Nonetheless, if we restrict the analysis to order n, the result for M_H will depend on the specific value of the scale t_H. Following [2], the error is minimized if we choose

$$M_H^2 \approx \left[M_H^{(n)}\right]^2_{t=t_H} \quad \text{with} \quad t_H = \max\left\{0, \ln\left[\frac{M_H}{m_t}\right]\right\}. \tag{16}$$

At last, we mention that the auto-energy factor is only included in the two-loop case [4, 13], so $M_H^{(1)} = m_H^{(1)}(t_H)$. Since M_H depends on the initial values which are all fixed except for λ_H^0, the limits (7) and (10) become bounds for the physical Higgs mass:

$$\begin{aligned} \text{one-loop} &\longrightarrow \quad 151.9 < M_H < 192.3 \text{ GeV,} \\ \text{two-loop} &\longrightarrow \quad 143.8 < M_H < 190.3 \text{ GeV.} \end{aligned} \tag{17}$$

As before, the bounds are less constrained for $E_{UV} < E_{GU}$, see Fig. 4.

Comparison with other Results

Experimentally, the direct detection [14] and the high precision measurements of electroweak processes [15], give the bounds for the Higgs mass: $114.4 < M_H < 186$ GeV (both at 95% C.L.), respectively. Therefore, these experimental bounds are compatible with a UV cutoff in the range $10^4 - 10^{14}$ GeV.

Most of the previous theoretical works established bounds less restrictive, but compatible with ours. On the other hand, the bound of Ref. [2] is a little bit more restrictive ($m_H \lesssim 180$ GeV) but the $Y_U \approx Y_t$ evolution is neglected and the maximum energy is at the Planck scale 10^{19} GeV. In Ref. [9] we used the effective potential method developed by Casas et al. [4] to obtain the Higgs mass. The small differences with respect to the values presented here stem from the inclusion of the leptonic sector.

FINAL REMARKS

The lower bounds on λ_H^0 are always obtained from the stability condition. The upper limit for λ_H^0 at the one loop level is established from the position of the pole in Eq. (6). For two loops λ_H does not diverge and reaches an asymptotic value, so we propose the condition in Eq. (9) to obtain the upper bound.

In dealing with the Higgs mass, we follow Ref. [11] with respect to the renormalized potential Eq. (11). We calculate the Higgs mass and establish the limits for its acceptable values by a correspondence with the ones for λ_H^0.

In both cases, Fig. 3 and Fig. 4 represent the generalization of bounds for energies below E_{GU}. Once the Higgs boson existence is confirmed, the precise value of its mass will determine, in accordance to the graph, the maximum energy at which the SM can be applied. Hence this also indicates at which minimum energy it is necessary to start thinking about more complex theories.

REFERENCES

1. H. Arason, D. J. Castaño, B. Kesthelyi, S. Mikaelian, E. J. Piard, P. Ramond, B. D. Wright, Phys. Rev. D **46**, 3945 (1992).
2. T. Hambye, K. Riesselmann, Phys. Rev. D **55**, 7255 (1997).
3. J. A. Casas, J. R. Espinosa and M. Quirós, Phys. Lett. **B382**, 374 (1996).
4. J. A. Casas, J. R. Espinosa, M. Quirós and A. Riotto, Nuc. Phys. **B436**, 3 (1995).
5. J. A. Casas, J. R. Espinosa and M. Quirós, Phys. Lett. **B342**, 171 (1995).
6. M. Luo and Y. Xiao, Phys. Rev. Lett. **90**, 011601 (2003).
7. P. Kielanowski, S.R. Juárez and G. Mora, Phys. Lett. **B479**, 181 (2000).
8. S. R. Juárez, A. Morales and P. Kielanowski, Rev. Mex. de Fís. **50**, 401 (2004).
9. P. Kielanowski, S. R. Juárez and H. G. Solís-Rodríguez, Phys. Rev. D **72**, 096003 (2005).
10. S. R. Juárez, H. G. Solís-Rodríguez y P. Kielanowski, AIP Conference Proceedings 809, 196-198 (2005), ISBN 0-7354-0300-7.
11. Jegerlehner, M. Yu. Kalmikov and O. Veretin, Nucl. Phys. **B658**, 49 (2003).
12. S. Willenbrok, G. Valencia, Phys. Lett. **B160**, 267 (1985);
    A. Sirlin, Phys. Rev. Lett. **67**, 2127 (1991); Phys. Lett. **B267**, 240 (1991);
    M. Carena *et al.*, Nucl. Phys. **B580**, 29 (2000).
13. M. Sher, Phys. Rep. **179**, 273 (1989); Phys. Lett. **B317**, 159 (1993);
    C. Ford, I. Jack, D. R. T. Jones, Nucl. Phys **B387**, 373 (1992); **B504**, 551 (1997).
14. A. Heister *et al.*, Phys. Lett. **B565**, 61 (2003).
15. URL: http://lepewwg.web.cern.ch/LEPEWWG/

Recent results from the H1 experiment of HERA

Ricardo Lopez-Fernandez

Departamento de Física, CINVESTAV, AP 14-740, 07000 México DF, Mexico

Abstract. A summary on some of the recent measurements made by the H1 Collaboration in ep collisions at HERA is given. Those include determination of electroweak parameters, charm and beauty structure functions of the proton and heavy quark production. First results on charged current cross section with polarised lepton beam are also presented.

Keywords: Electroweak parameters, Heavy flavours, Deep Inelasic Scattering, Polarisation.

DETERMINATION OF ELECTROWEAK PARAMETERS AT HERA

In the first phase of HERA operation (HERA-I) with the unpolarised $e^{\pm}$ beam colliding with the proton beam, the H1 experiment has collected three major data samples of e^+p in years from 1994 to 1997 at a center-of-mass energy of 301 GeV, e^-p in 1998-1999 and e^+p in 1999-2000 at 319 GeV. The corresponding integrated luminosities are 35.6 pb^{-1}, 16.4 pb^{-1} and 65.2 pb^{-1}, respectively. These data have been used to measure neutral current (NC) and charged current (CC) cross sections covering more than four orders of magnitud in both Q^2, the negative four-moment transfer squared, and Bjorken x. The large kinematic coverage and the different flavour sensitivity of the $e^{\pm}$ NC and CC cross section data have enabled 5 sets of parton distribution functions (PDF) to be determined simultaneouly in a previous QCD analysis [1]. These five PDF sets are the gluon, up-type and dwon-type quarks and their anti-quarks distributions.

The inclusive NC and CC cross sections are not only sensitive to PDFs but also to electroweak (EW) parameters. Indeed, the NC cross sections at high Q^2 depend on up- and down-typ quark couplings to the Z^0 boson, a_q and $v_q (q=u,d)$, via structure functions, whereas the shape of the CC cross sections as a function of Q^2 is controlled by the propagator mass (M_{prop}) of the W boson. It is this natural to extend the QCD analysis of [1] into a combined EW-PDF analysis so that EW parameters can be determined together with the PDFs taking properly into account the small but non-negligible correlation between them.

This is precisely the strategy chosen in [2, 3], namely using the same parametrisation forms for the five PDF sets for the QCD part. The QCD analysis is performed using the DGLAP evolution equations[4] at next-to-leading order in the modified minimal substraction renormalization scheme. All quarks are taken as massless. Several combined EW-PDF fits are performed either in a model independent way (fits $a_u - v_u - a_d - v_d$-PDF and $G - M_{prop}$-PDF) or within the Standard Model(SM, fits M_W-PDF and m_t-PDF).

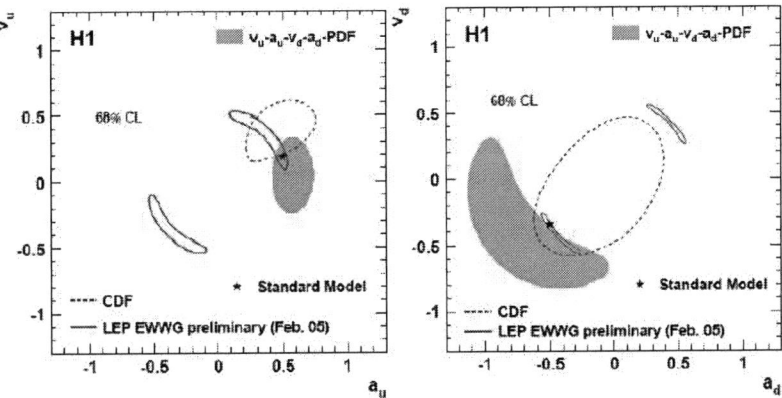

FIGURE 1. H1 results (shaded area) at 68% confidence level (CL) on the couplings of u quark (left) and d quark (right) to the Z^0 in comparison with similar results from LEP (full curves).

First results on light quark couplings to the Z^0 boson at HERA

The sensitivity on the quark couplings at HERA stems from the γZ interference and Z^0 exchange contributions in NC interactions at high Q^2. The results of the combined $a_u - v_u - a_d - v_d$-PDF fit are shown in Fig.1 and compared with similar results obtained by the combined LEP experiments [5]. The HERA determination has comparable precision. These determinations are sensitive to u and d quarks separately, contrary to other measurements of the light quark-Z^0 couplings in νN scattering[6] and atomic parity violation[7] on heavy nuclei. They also resolve any sign ambiguity and the ambiguities between v_q and $a_q (q = u, d)$ of the determinations based on the observables measured at the Z^0 resonance [5].

Improved W propagator mass measured at HERA

The cross section data allow a simultaneous determination of the Fermi coupling constant G_F and the W boson mass, and of the PDFs ($G - M_{prop}$-PDF fit). When treating G and M_{prop} as independent parameters, the sensitivity on G and M_{prop} originates respectively from the normalization and Q^2 dependence of the CC cross sections. The result of the fit is shown in Fig.2 as the shaded area.

Fixing G to the measured G_F value[8], one gets a determination of M_{prop}, also shown in Fig.2, $M_{prop} = 82.87 \pm 1.82_{exp}{}^{+0.3}_{-0.16}|_{model}$ GeV where the first error is experimental and the second corresponds to uncertainties due to input parameters and model assumptions as introduced in Table 1 in [1] (e.g., the variation of $\alpha_s = 0.1185 \pm 0.0020$). The determination differs from all previous ones in the treatment of the correlation between M_{prop} and PDFs and represents the most accurate measurement so far of the CC propagator mass at HERA.

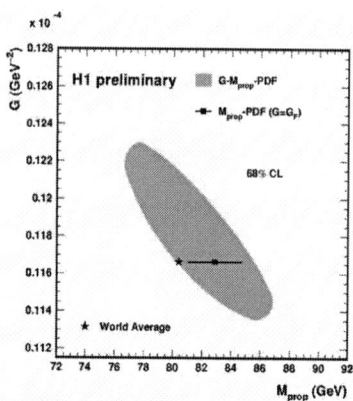

FIGURE 2. The result of the fit to G and M_{prop} at 68% confidence level (CL) shown as the shaded area. The world average values are indicated with the star symbol. Fixing G to G_F, the fit results in a measurement of the propagator mass M_{prop} shown as the circle with the horizontal error bar.

Within the SM, the Fermi coupling constant G_F is connected with the W boson mass M_W through a relation which contains EW radiative corrections including quadratic (logarithmic) dependence on the top quark mass m_t (the Higgs mass M_H). A combined EW-PDF fit in the SM gives

$$M_W = 90.786 \pm 0.205_{exp} \pm 0.025_{\delta m_t} - 0.084_{\delta M_H} \pm 0.033_{\delta(\Delta r)} GeV, \quad (1)$$

where the measured central value corresponds to using the world averaged values of $M_Z = 91.1876 \pm 0.0021$ GeV, $m_t = 178 \pm 4.3$ GeV and a Higgs mass of 120 GeV. The uncertainty on M_Z has a negligible error on M_W whereas the uncertainty on m_t gives rise to the third quoted error on M_W. Varying M_H from 120 GeV to 300 GeV results in the fourth error. The last error is due to higher order radiative correction uncertainties.

Together with the world average value of M_Z given above, the result obtained on M_W from Eq.1 represents an indirect determination of $sin^2\theta_W$ in the on-mass shell scheme: $sin^2\theta_W = 0.2152 \pm 0.0040^{+0.0019}_{-0.0011}$ where the first error is experimental and the second is theoretical covering all remaining uncertainties in Eq.1. The uncertainty due to δM_Z is negligible.

Fixing M_W to the world average value assuming M_H=120 GeV, the fit m_t-PDF gives $m_t = 108 \pm 44$ GeV where the uncertainty is experimental. The result represents the first determination of the top quark mass through loop effects in the ep data at HERA.

Again the precision of these determinations will be improved by a large amount as the best sensitivity comes from the CC e^-p cross section which was measured from a very limited data sample at HERA-I. Polarised e^-p data corresponding to an increase of one order magnitude in the integrated luminosity from HERA-II are being taken.

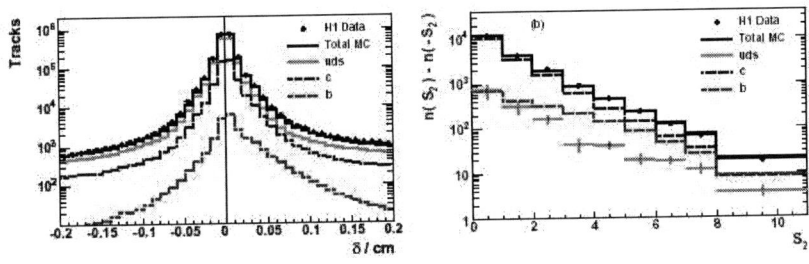

FIGURE 3. The signed impact parameter δ of a track (left) and the substracted significance distribution $S_2 = \delta/\sigma(\delta)$ (right)

CONTRIBUTION OF THE HEAVY QUARKS TO THE PROTON STRUCTURE FUNCTION

Measurements of the charm (c) and beauty (b) contributions to the inclusive proton structure functions F_2 have been made recently in Deep Inelastic Scattering (DIS) at HERA. Measurements of the open charm cross section in DIS at HERA have mainly been of exclusive D or D^* meson production [9, 10]. From the $D*$ measurements the contribution of charm to the proton structure function, $F_2^{c\bar{c}}$, has been derived by correcting from the fragmentation fraction $f(c \to D^*)$ and the unmeasured phase space (mainly at low values of transverse momentum of the meson). The result are found to be in good agreement with perturbative QCD(pQCD) predictions at next-to-leading order (NLO). Events containing heavy quarks can also be distinguished by the long lifetimes of c and b flavoured hadrons, which lead to displacements of tracks from the primary vertex. The impact parameter of a track, which is the transverse distance to the closest approach (DCA) of the track to the primary vertex point, is reconstructed using precise spatial information from silicon vertex detectors. Measurements using this method have been made at low [11] and high [12] values of Q^2.

Impact parameter method

In order to determine a signed impact parameter (δ) for a track, the azimuhal angle of the struck quark ϕ_{quark} must be determined for each event. The angle ϕ_{quark} is reconstructed from the ϕ of the jet with the highest transverse momentum or, if there is no jet reconstructed in the event, from the properties of the hadronic final state or scattered electron. The direction defined in the transverse plane by ϕ_{quark} and the primary vertex is called the quark axis. If the angle between the quark axis and the line joining the primary vertex to the point of DCA of a track is less than $90°$, δ is defined as positive, and is defined as negative otherwise. The δ distribution, shown in Fig.3, is seen to be asymmetric with positive values in excess of negative values indicating the presence of long lived particles.

To further distinguish between the c, and b and light quark flavours the quantities

S_1 and S_2 are defined as the significance $(\delta/\sigma(\delta))$ of the track with the highest and second highest absolute significance, respectively, where $\sigma(\delta)$ is the error on δ. In order to substantially reduce the uncertainty due to the resolution of δ and the light quark normalisation, the contents of the negative bins in the significance distributions are substracted from the contents of the corresponding positive bins. The substracted distributions for S_2 is shown in Fig.3.

The fractions of c, b and light quarks of the data are extracted in each $x - Q^2$ interval using a fit to the substracted S_1 and S_2 distributions and the total number of inclusive events. The results of the fit are converted to a measurement of the "reduced c cross section" defined from the differential cross section as

$$\tilde{\sigma}^{c\bar{c}}(x,Q^2) = \frac{d^2\sigma^{c\bar{c}}}{dxdQ^2}\frac{xQ^4}{2\pi\alpha^2(1+(1-y)^2)}. \tag{2}$$

Results

The measurement of $\tilde{\sigma}_{c\bar{c}}$ are shown in Fig.4 as a function of x for fixed values of Q^2. The H1 impact parameter data[11, 12] are compared with the results extracted from D^* meson measurements by H1[9]. The measurements from the impact parameter analysis and the D^* extraction methods are in good agreement. The σ^{cc} data are compared with two predictions from NLO QCD from MRST[13] and CTEQ[14], and with predictions based on CCFM[15] parton evolution. The predictions provide a reasonable description of the present data, although the data are more precise than the spread in QCD predictions. The measurements of $\tilde{\sigma}_{b\bar{b}}$ are also shown in Fig.4 and the beauty data are reasonably well described by the QCD predictions. The measurements $F_2^{c\bar{c}}$ and $F_2^{b\bar{b}}$ are shown as a function of Q^2 for various x values in Fig.5. The measurements of $F_2^{c\bar{c}}$ $F_2^{b\bar{b}}$ show positive scaling violations which increase with decreasing x.

POLARISATION DEPENDENCE OF THE TOTAL CC ep CROSS SECTION

Data taking of the second, high luminosity phase of the HERA program (HERA-II) started in october 2003. A major success at HERA-II is a longitudinal polarisation of the electron beam which collides with protons inside the H1 detector.

Both NC and CC interactions can be measured at HERA and provide complementary information. In the Standard Model, there is clear prediction on polarisation dependance of these cross sections. Specifically, the e^+p CC cross section for a left handed positron and e^-p CC cross section for right handed electron should be zero and should have full strength for opposite helicities. This follows from absence of right handed weak charged currents in the Standard Model. Thus, for unpolarised electrons(positrons), the cross section is a half of the one for left handed electron(right handed positron).

The incident electron(positron) beam energy is 27.6 GeV, whilst the unpolarised proton beam energy is 920 GeV. The e^+p data collected in 2003 and 2004 consist of two periods with positive and negative polarisations. The first period has a mean

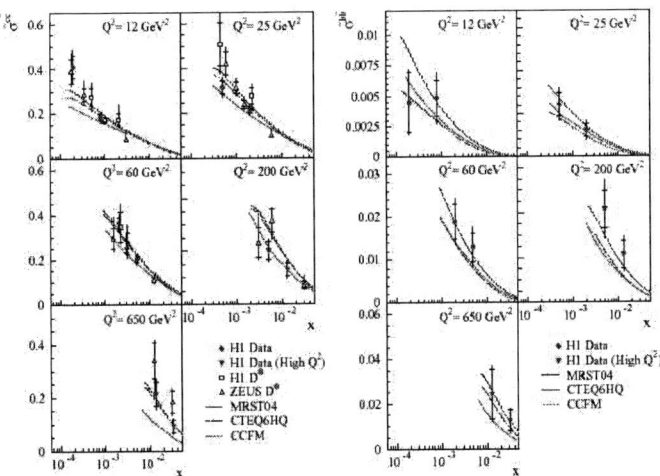

FIGURE 4. The measured reduced cross section $\tilde{\sigma}^{c\bar{c}}$ (left) and $\tilde{\sigma}^{b\bar{b}}$ (right) shown as a function of x for 5 different Q^2 values

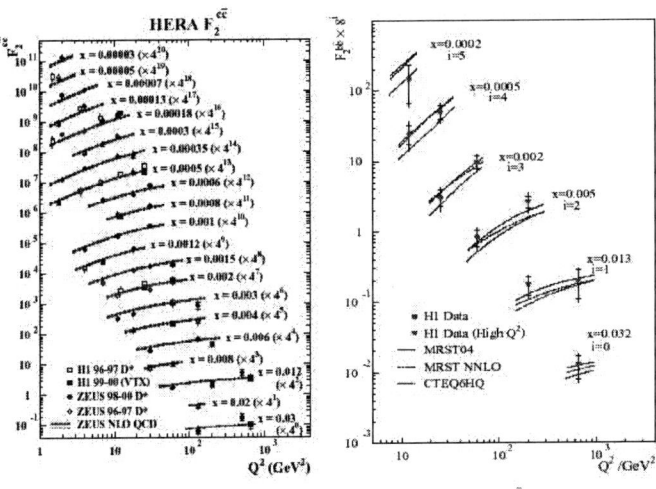

FIGURE 5. The measured structure function $F_2^{c\bar{c}}$ (left) and $F_2^{b\bar{b}}$ (right)

polarisation of (33.0 ± 0.7) and integrated luminosity of $(15.3 \pm 0.4) pb^{-1}$. The second period, with negative polarisation, has (-40.2 ± 0.6) and $(21.7 \pm 0.6) pb^{-1}$. The 2005 $e^- p$ data set has a mean polarisation of (-25.4 ± 0.4) and and integrated luminosity of $(17.8 \pm 0.2) pb^{-1}$.

The double differential CC cross section for collisions of polarised leptons with unpolarised protons corrected for QED radiative effects can be written as

$$\frac{d^2\sigma_{CC}^{e^{\pm}p}}{dxdQ^2} = [1 \pm P_e]\frac{G_F^2}{2\pi x}[\frac{M_W^2}{Q^2+M_W^2}]^2\Phi_{CC}^{\pm}, \qquad (3)$$

where G_F is the Fermi coupling constant, M_W is the mass of the W boson. $\Phi_{CC^{\pm}}$ is the term depending on the partonic content of the proton.

P_e is the degree of longitudinal polarisation defined as $P_e = (N_R - N_L)/(N_R + N_L)$ with $N_R(N_L)$ being the number of right(left) handed polarised leptons in the beam. The cross section has a linear dependence on the polarisation of the lepton beam. For positrons(electrons) with positive(negative) polarisation it is enlarged, whilst the cross section for negative(positive) polarised positrons(electrons) is diminished. For $P_e = -1 (P_e = +1)$ the $e^+p(e^-p)$ CC cross section is identically equal to zero in the Standard Model.

Measurement procedure

Candidates for CC interactions are selected by requiring $P_{T,h} > 12$ GeV. In order to ensure high efficiency of the trigger and kinematic resolution, the analysis is further restricted to the domain of $Q_h^2 > 200$ GeV^2 and $0.03 < y_h < 0.85$. The ep background is dominantly due to photoproduction events in which the scattered electron(positron) escapes undetected in the beam pipe and large missing transverse momentum is faked by fluctuations in the detector response or undetected particles. These and some additional background from cosmic ray and beam-induced interactions is rejected by identifying event topologies characteristics.

Results

The total polarised e^+p and e^-p CC cross section are measured in the range $Q^2 > 400$ GeV^2 and $y < 0.9$. The results are:

$$\sigma_{CC}^{e^+p}(P_e = +0.33) = 34.67 \pm 1.94_{stat} \pm 1.66_{sys} pb \qquad (4)$$

$$\sigma_{CC}^{e^+p}(P_e = -0.40) = 13.80 \pm 1.04_{stat} \pm 0.94_{sys} pb \qquad (5)$$

$$\sigma_{CC}^{e^-p}(P_e = -0.25) = 66.42 \pm 2.39_{stat} \pm 2.99_{sys} pb \qquad (6)$$

The polarisation dependence of the CC cross section is shown in Fig.6. The measurements of the unpolarised CC cross section are also shown in Fig.6. They are based o HERA-I data with a luminosity of 65.2 pb^{-1} for e^+p and 16.4 pb^{-1} for e^-p:

$$\sigma_{CC}^{e^+p}(P_e = 0) = 28.44 \pm 0.77_{stat} \pm 1.22_{sys} pb \qquad (7)$$

$$\sigma_{CC}^{e^-p}(P_e = 0) = 57.03 \pm 2.21_{stat} \pm 1.38_{sys} pb \qquad (8)$$

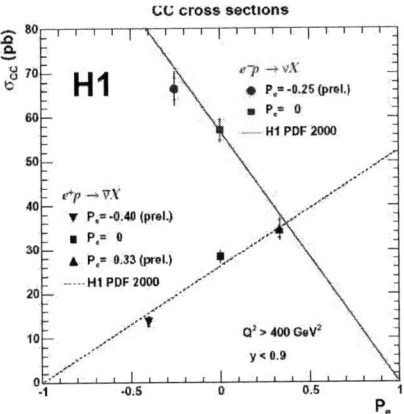

FIGURE 6. The dependence of the e^+p and the e^-p cross section with the lepton beam polarisatin P_e is shown. The data are compared to the Standard Model prediction based on the H1 PDF 2000 fit.

The measured cross sections agree well the Standard Model predictions based on the H1 PDF 200 fit. A linear fit to the polarisation dependence of the three measured e^+p CC cross sections is performed. The fit takes into account a correlation of systematic errors of the measurements. The fit provides a reasonable description of the data with a $\chi^2=2.45$. The cross section extrapolation to $P_e = -1.0$ yields a value:

$$\sigma_{CC}(P_e = -1.00) = -3.7 \pm 2.4_{stat} \pm 2.7_{sys} pb. \tag{9}$$

This value is consistent with zero and indicates the absence of right handed weak charged currents, in agreement with the Standard Model.

REFERENCES

1. C. Adloff et al.,[H1 Collaboration], *Eur.Phys.J.*, **C30** (2003) 1,[hep-ex/0304003].
2. A. Aktas et al.,[H1 Collaboration], accepted for publication in *Phys.Lett.***B**,[hep-ex/0507080].
3. B. Portheault, Ph.D. thesis(Mar. 2005) LAL 05-05 (IN2P3/CNRS, Université de Paris-Sud XI, Orsay).
4. G. Altarelli and G. Parisi, *Nucl.Phys.* **B126** (1977) 298 and references therein.
5. http://lepewwg.web.cern.ch/LEPEWWG/plots/winter005/
6. G.P. Zeller et al.,[NuTeV Collaboration], *Phys.Rev.Lett.***88** (2002) 091802.,[hep-ex/0110059]; Erratum-ibid **90** (2003) 239902.
7. S.C. Bennett and C.E. Wieman, *Phys.Rev.Lett.***82** (1999) 2484,[hep-ex/9903022].
8. Particle Data Group (S. Eidelman et al.), *Phys.Lett.***B592** (2004) 1.
9. C. Adloff et al.,[H1 Collaboration], *Phys.Lett.***B528** (2002) 199.
10. S. Chekanov et al.,[ZEUS Collaboration], *Phys.Rev.***D69** (2004) 012004.
11. A. Aktas et al.,[H1 Collaboration], [hep-ex/0507081].
12. A. Aktas et al.,[H1 Collaboration], *Eur.Phys.J.***C40** (2005) 349.
13. Martin, Roberts, Stirling, R.S Thorne, *Eur.Phys.J.***C39** (2005) 155.
14. Kretzer, Lai, Olness, Tung, *Phys.Rev.***D69** (2004) 114005.
15. M. Ciafaloni, *Nucl.Phys.***B296** (1988) 49.

Radiation Hardness Tests of a Scintillation Detector with Wavelength Shifting Fiber Readout

R. Alfaro*, E. Cruz[†], M. I. Martinez[†], L. M. Montaño**, G. Paic[†] and A. Sandoval*

*Instituto de Física - UNAM.
[†]Instituto de Ciencias Nucleares - UNAM.
**CINVESTAV.

Abstract. We have performed radiation tolerance tests on the BCF-99-29MC wavelength shifting fibers and the BC404 plastic scintillator from Bicron as well as on silicon rubber optical couplers. We used the ^{60}Co gamma source at the Instituto de Ciencias Nucleares facility to irradiate 30-cm fiber samples with doses from 50 Krad to 1 Mrad. We also irradiated a 10×10 cm^2 scintillator detector with the WLS fibers embedded on it with a 200 krad dose and the optical conectors between the scintillator and the PMT with doses from 100 to 300 krad. We measured the radiation damage on the materials by comparing the pre- and post-irradiation optical transparency as a function of time.

Keywords: Radiation hardness, WLS fibers, scintillation detectors
PACS: 29.40.Mc, 78.70.-g, 61.80.Ed, 42.88.+h

INTRODUCTION

Wavelength shifting (WLS) fibers are widely used in high energy physics experiments [1]-[7] mainly for reading out scintillation detectors. Both the detectors and the WLS fibers are, in general, exposed to different levels of ionizing radiation during the lifetime of the experiments. However only a few results on the WLS fibers or scintillating materials radiation tolerance are found in the literature [8, 9]. We have studied the effects of ionizing radiation on WLS fibers and on a typical detector configuration where these type of fibers are used to read a piece of organic plastic scintllator.

We tested the Bicron's BC9929-MC round WLS fibers alone and a piece of Bicron's BC404 plastic scintillator with the fibers embedded on it. We also irradiated a couple of rhodorsil RTV 141 (elastomer silicon bicomposant) optical coupler discs.

We used the ^{60}Co gamma source at the irradiation facility of *Instituto de Ciencias Nucleares* to irradiate the scintillating tiles, the WLS fibers and the optical couplers.

SCINTILLATION DETECTOR

The BC404 plastic scintillator with WLS fibers (BC9929) embedded on it was irradiated with 200 krad at a dose rate of 4.4 krad/min. We used a $12 \times 12 \times 1$ cm^3 tile of the scintillator and 25 cm samples of the fibers to absorb and transmit the light to a Photonics

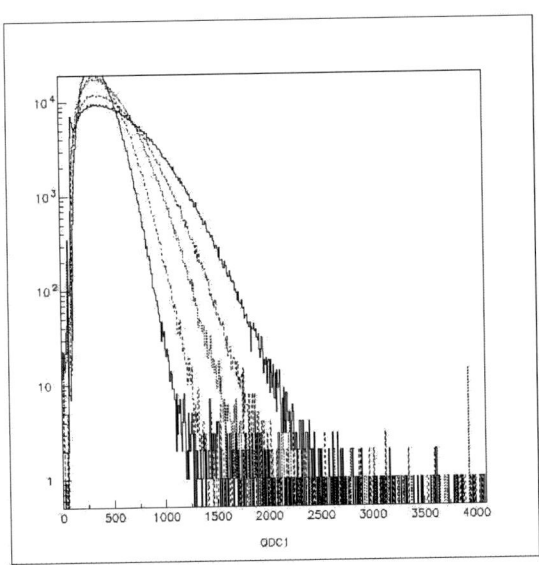

FIGURE 1. Pre-irradiation amplitude spectra of the scintillator for different values of the PMT high voltage. They correspond, from the right, to 1550 V (full line), 1500 V (dashed line), 1450 V (dotted line), 1400 V (dashed-dotted line) and 1350 V (full line). Every 50 V decrease corresponds to a signal decrease of 13%.

xp13 PMT.

Energy spectra of a ^{90}Sr β source were recorded before and after the irradiation. We took several spectra before irradiation for different (decreasing) high voltage levels on the PMT in an attempt to reproduce the expected damage due to the irradiation process. We obtained spectra from the nominal value of 1550 V down to 1350 V in 50 V steps resulting in the curves of figure 1. The relative amplitudes of these spectra indicate that a decrease of 13% is obtained for each 50 V step.

By decreasing the high voltage step we were able to estimate the minimum change in the spectrum that our system can unambiguously detect. Figure 2 shows a clear difference between the spectra corresponding to 1550 V and 1525 V, that is a 6.5% change. This means that we can resolve any change of at least 6%.

Results

We obtained the amplitude spectrum of the irradiated detector (scintillator and fibers) at different times. Figure 3 (*left*) shows these spectra as compared with the non-irradiated one. By comparing the former ones with the latter we can estimate the damage on the scintillator-fibers system. Figure 3 (*right*) shows the degradation of the signal due to the irradiation and the partial recovery of the detector.

It can be seen that it took quite long ($\sim$ 300 hours) to the detector to reach the 80%

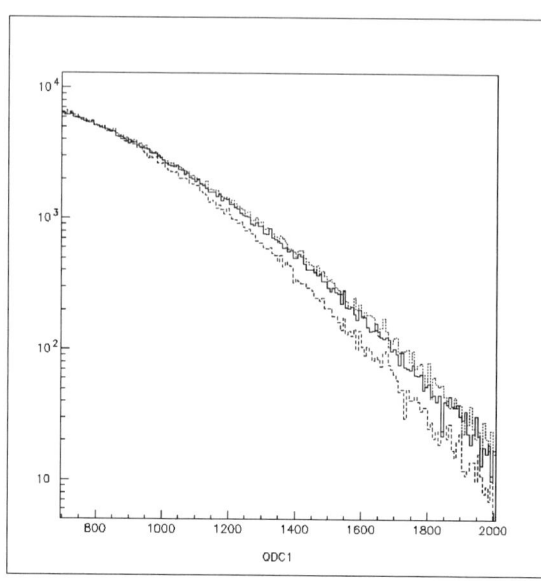

FIGURE 2. Spectra corresponding to a change of 6.5% can clearly be discriminated by our testing setup. Upper spectra (full and dotted lines) correspond to the nominal value (1500 V) of the PMT voltage. Lower (dashed line) spectrum corresponds to 1525 V.

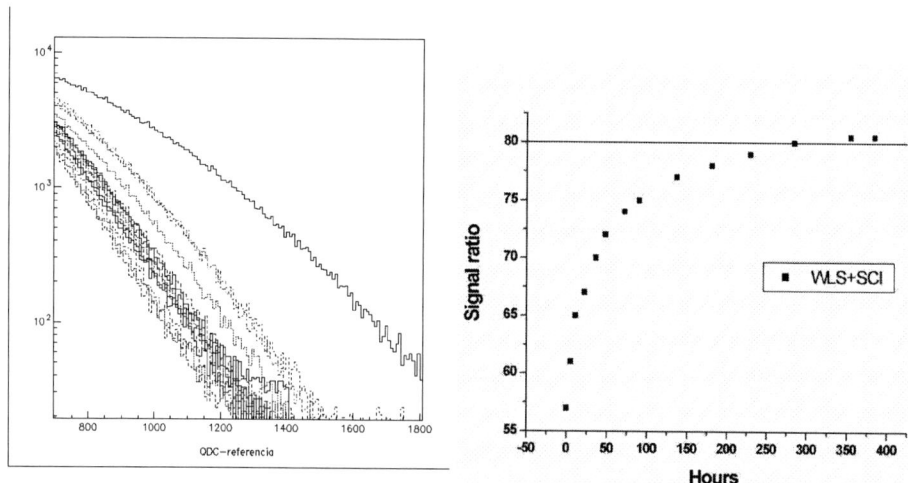

FIGURE 3. Spectra of irradiated scintillator (lower) compared to the non-irradiated spectrum (top, solid line) (*left*). Relative PMT signal amplitude of post-irradiated to pre-irradiated scintillator (*right*).

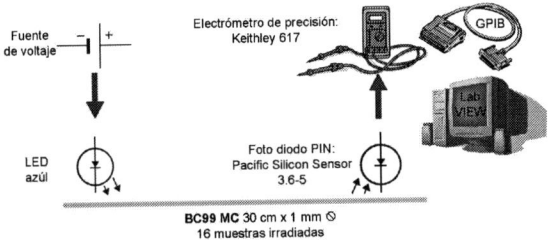

FIGURE 4. Experimental setup for the fibers tests. We measured the voltage across the photodiode 50 times (taking the sample out and back in each time) before irradiating the samples. After the irradiation we measured periodically the photodiode voltage.

of the original signal which we can say it's the permanent effect for the 200 krad dose. The results of the tests with the fibers alone show that the main contribution to the recovery time is given by the scintillator as the recovery time for the fibers is two orders of magnitude shorter (see figure 5 left).

WLS FIBERS

We irradiated six 30 cm samples of BC9929 WLS fibers with 50, 100, 200, 300 and 650 krad, and with 1 Mrad using a quite high dose rate in all cases of 7 krad/min. We measured the transmitted light on 30 cm long samples immediately after irradiation and compared with the mean value measured before the fibers were irradiated. The average of fifty measurements was calculated before irradiation removing and putting back the fiber in its position each time. This takes into account the uncertainty in the position of the sample when it is measured again after irradiation since then all the measurements are automatically taken with the fiber in a fixed position.

Experimental setup

To establish the light loss in the WLS fibers we built the test setup shown in Figure 4. It consists of a blue LED used to excite the fiber on one end and a PIN photodiode to capture the light at the opposite end. A precision voltmeter was used to measure the voltage across the photodiode; and a GPIB interface allowed to read this value into a PC. We wrote a LabVIEW program that drives the GPIB, writes the values in an ASCII file and presents the data in a graphical interface.

We excited the fibers on a lateral spot near one of its ends covering the cross section surface at this end with black tape. The opposite cross section surface was put in direct contact (no optical grease applied) with the photodiode window.

Each fiber sample was manually polished at both ends using a planar glass surface to support the polish paper and a metallic box where the fiber was holed to maintain it in veritical position. Each sample was labeled for identifying purposes and to ensure they were excited at the same end in every measurement. The only arbitrary position of the

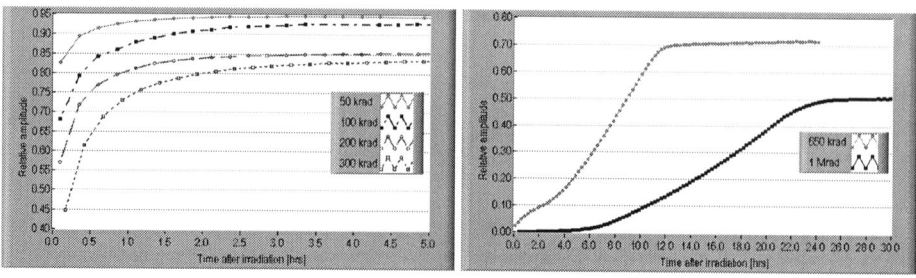

FIGURE 5. Post-irradiation response of WLS fibers for different doses. *Left:* full circles, 50 krad; full squares, 100 krad; hollow circles, 200 krad and hollow squares, 300 krad. *Right:* circles, 650 krad and squares, 1 Mrad. The horizontal axis shows the time elapsed from the end of the irradiation while the vertical one is the ratio of the values measured to the pre-irradiation average.

fibers on the setup was the orientation around their own axis which determines a better or worse optical coupling between the not-perfectly polished surface of the fiber and the photodiode window.

Results

We measured the voltage across the photodiode at regular time intervals (every 15 minutes) for a continuous period of 24 hours. The first value was taken within 5–10 minutes from the end of the irradiation. Then each value was compared with the average obtained before the irradiation.

The curves on the left in figure 5 show the recovering process for the fibers irradiated with 50, 100, 200 and 300 krad. A slightly faster recovery is observed for the 200 krad case. It is to be noticed the fast recovery of the fibers compared to the scintillator+fibers system which recovered a hundred times more slowly. On the other hand the damage only increased from 15 to 20% when irradiating both the scintillator and the fibers.

The plot on the right in figure 5 shows the damage corresponding to 650 krad and 1 Mrad. A large permanent effect can be seen for the case of 1 Mrad where the recovery reaches only half of the pre-irradiation level. The less critical, but still almost 15% larger than the 300 krad case, is the damage corresponding to 650 krad. In both cases the recovery time increases significantly.

The mid and long term response was also investigated. Figure 6 shows these effects. We measured up to 20 hours after the irradiation and we see no further recovery for the 50 and 100 krad cases. The 200 and 300 krad samples, on the other hand, show a marginal further recovery after 20 hours.

We measured the samples a second time for a 5-hour period several weeks after the irradiation. The 50 and 100 krad samples show the expected behavior; the former recovering totally and the latter remaining at the same level. However an extra 9% recovery can be seen for the 200 krad sample and, on the contrary, the 300 krad sample shows an unexpected decrease in its output value.

It should be noticed however that for the long term measurement the samples had

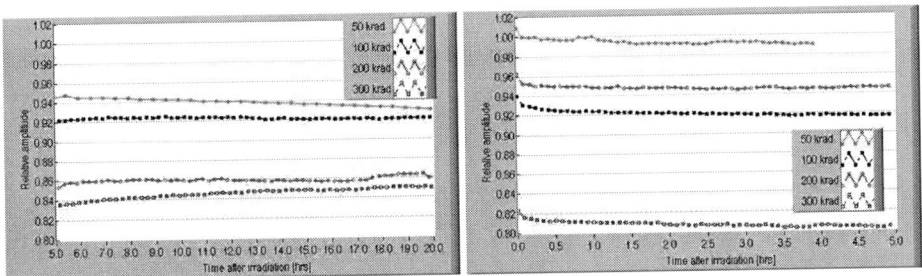

FIGURE 6. Mid- (left) and long- (right) term relative response of WLS fibers. The plot on the left shows the range from 5 to 20 hours after irradiation. The one on the right was taken six weeks after.

to be replaced in the setup so each sample was in a different orientation than for the short and mid term cases where continuous measurements were taken over the 20-hour period. The ~4% decrease in the output of the 300 krad sample can then be understood as it corresponds to a $< 2\sigma$ deviation. Finaly the large recovery of the 200 krad sample in figure 6 (right) is not compatible with the rest of the measurements. Some more measurements should be made to address this problem.

OPTICAL COUPLERS

We analyzed the radiation effects on optical *cookies* which are used to couple light guides to PMT's. They are elastomer silicon bicomponant (technical name rhodorsil RTV 141) discs of 5 cm in diameter and 0.5 cm wide. The advantage of using these optical cookies is to have mechanical flexibility while maintaining a good optical coupling that assures an efficient transmission of light.

We also used the gamma beam facility at Instituto de Ciencias Nucleares for these tests. The dose given to the samples depends on the position in three coordinates from the center of the beam. In order to measure the radiation damage on the couplers we used the VARIAN Cary 100 spectrophotometer to obtain the absorption spectra in the visible and UV wavelength regions.

The absorption measured A is defined as
$$A = -\log_{10} T$$
where the transmitance $T = I_s/I_b$; with I_s the intensity of energy through the sample and I_b the intensity of energy through the calibration reference.

In these tests the spectra were taken from 190 to 500 nm. Some calibration (with air or a non irradiated sample) had to be done beforehand. We irradiated three samples, labeled m_1, m_2 and m_3, with the following doses

m_1 non irradiated (used as the reference)

m_2 irradiated with 200 Gy (20 krad)

m_3 irradiated with 600 Gy (60 krad)

m_2 and m_3 were placed at 40cm far from the center of the beam and at 93cm from the floor. The dose received in this position is ≈ 9.5 Gy/min so that the irradiation took 21

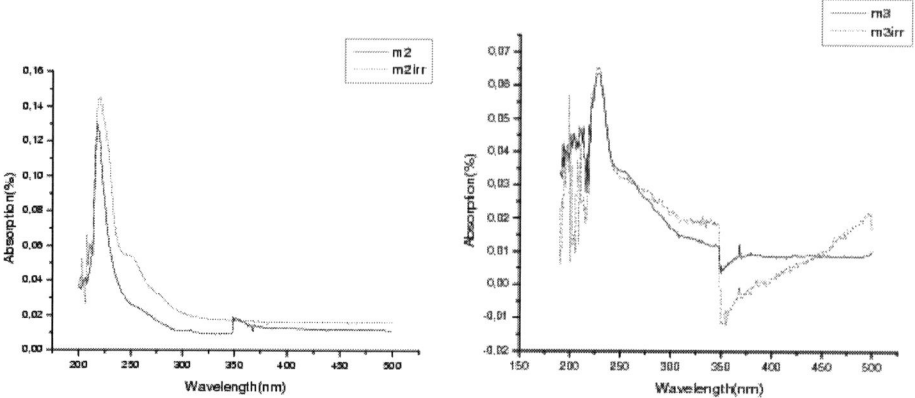

FIGURE 7. Absorption spectra for the samples before (black curve) and after (red curve) irradiation. m_2 was irradiated with 200 Gy (20 krad) and m_3 with 600 Gy (60 krad).

and 63 minutes respectively.

Results

We used the m_1 sample as the calibration reference that must be provided to the spectrophotometer. In this way, we looked for any change in the irradiated samples, m_2 and m_3, as compared to m_1.

In figure 7 the spectra of m_2 vs m_1 and m_3 vs m_1 obtained after the irradiation are compared to the corresponding ones before the irradiation. The light absorption (in percent) in the sample is shown as a function of the wavelength.

In both cases, the changes are too small to be taken as the effect of the radiation on the optical cookies. They are actually of comparable magnitude for both samples while the dose given to sample m_3 was 3 times higher than that given to sample m_2. This shows that this optical couplers are not affected by irradiation up to doses of 60 krad. The differences found might be due to other variables connected to the instrument or the cookies themselves as the dust or dirt accumulated on them during manipulation.

The sharp change in the spectrum of sample m_3 around 350 nm is due to the change of the light source (from the UV source to the visible light source) inside the spectrophotometer.

ACKNOWLEDGMENTS

We wish to thank Dra. Guadalupe Albarrán and Dra. Pilar Carreón from Instituto de Ciencias Nucleares for their valuable help in the realization of this work.

REFERENCES

1. S. Katsanevas, G. Largeron, J. Marteau and G. Moret, *"Auto-triggerable HPD sensors fully readout on ethernet: applications for high-energy physics and medical imaging"*, Nucl. Inst. and Meth. in Phys. Res. A **504** (may 2003) 103-108.
2. Huston J., Iori M., Kuhlmann S., Lami S., Miller R., Paoletti R., Turini N. and Ukegawa F., *"The CDF calorimeter upgrade for RunIIb"* in *21st IEEE Instrumentation and Measurement Technology Conference* IMTC04, Como, Italy, 18 - 20 May 2004, pp. 1879-1880 (Vol. 3).
3. Toh K., Katagiri M., Sakasai K., Matsubayashi M., Birumachi A., Takahashi H. and Nakazawa M. *"High counting rate two-dimensional neutron-imaging method using rectangular scintillators with WLS fibers"*, Appl. Phys. A **74** (2002) pp. s1601-s1603.
4. Mineev O. V., Garber E., Frank J., Ivashkin A. P., Kettell S. H., Khabibullin M. M., Kudenko Y. G., Li K., Littenberg L. S., Mayatski V. and Yershov N. V., *Photon sandwich detectors with WLS fiber readout*, in *8th International Conference on Instrumentation for Colliding Beam Physics*, Novosibirsk, Russia, 28 Feb - 6 Mar 2002. Published in *Nucl. Inst. and Meth. in Phys. Res. A* **494** (2002) 362-368.
5. Tilecal/ATLAS Collaboration, *"Hadron Calorimetry using scintillator Tiles and WLS fibers: the Tilecal/ATLAS"* in *SCIFI97: Conference on Scintillating and Fiber Detectors*, South Bend, Indiana, 2-6 Nov 1997. Published in *AIP Conf. Proc.* **450** (nov. 1998) 448-455.
6. Hagopian V., *"The Hadron Calorimeter of the Compact Muon Solenoid (CMS)"*, CERN's CMS notes, CMS-CR-1997-001.
7. Adinolfi M. et. al., *"The QCAL tile calorimeter of KLOE"*, Nucl. Inst. and Meth. in Phys. Res. A **483** (2002) 649-659.
8. M. J. Varanda, M. David, A. Gomes and A. Maio, *"Recent results on radiation hardness tests of WLS fibers for the ATLAS Tilecal hadronic calorimeter"*, Nucl. Inst. and Meth. in Phys. Res. A **453** (oct. 2003) 255-258.
9. M. David, A. Gomes and A. Maio, *"Dose rate effects in WLS fibers"*, Nucl. Phys. B Proc. Suppl. **54** (march 1997) 222-228.

On Gauge Independent Dynamical Chiral Symmetry Breaking

A. Bashir*,† and A. Raya*

Instituto de Física y Matemáticas, Universidad Michoacana de San Nicolás de Hidalgo, Apartado Postal 2-82, Morelia, Michoacán 58040, México
Instituto de Ciencias Nucleares, Universidad Nacional Autónoma de México, Ciurcuito Exterior, C. U., Apartado Postal 70-543, 04510, México, D. F.
†*Institute for Particle Physics Phenomenology, University of Durham, Durham DH1 3LE, U.K.*

Abstract.
Schwinger-Dyson equations (SDEs) are an ideal framework to study nonperturbative phenomena such as dynamical chiral symmetry breaking (DCSB). Loss of gauge invariance is an obstacle to achieve fully reliable predictions from these equations. In addition to Ward-Green-Takahashi identity (WGTI), Landau-Khalatnikov-Fradkin transformations (LKFT) also play an important role in restoring the said invariance at the level of physical observables. On one hand, they impose useful constraints on the transverse part of the fermion-boson vertex and on the other, they govern the change in dynamically generated fermion propagator with a variation of gauge. We consider the latter in this article and study the gauge (in)dependence of chiral condensate in quantum electrodynamics in (2+1) space-time dimensions (QED3).

Keywords: Schwinger-Dyson equations, Landau-Khalatnikov-Fradkin transformations, Nonperturbative phenomena, Dynamical chiral symmetry breaking, QED3.
PACS: 11.15.Tk,12.20.-m

Gauge theories of fundamental interactions have been highly successful in collating experimental results in perturbative regime. However, not all interesting phenomena can be understood in this approximation scheme. Confinement of quarks and gluons and hadronic structures are two examples. There are also proposals to go beyond the standard model of particle physics which invoke strong dynamics such as technicolor models, top mode standard model and its modern variants, including those with extra dimensions. In view of these interesting possibilities, it becomes all the more important that corresponding studies of Schwinger-Dyson equations (SDEs) be independent of the ambiguities stemming from the approximations employed.

SDEs for QCD are cumbersome to solve due to its non-Abelian nature. QED provides a simpler and popular platform to study the merits of various truncation schemes in a quantitative fashion. However, it has ultraviolet divergences. Most of the numerical investigations of SDEs employ the more convenient cut-off regularization method. Owing to the fact that it can potentially introduce spurious gauge dependence, planar quantum electrodynamics (QED3), which is super renormalizable, can serve as a neater testing ground of the validity of the approximations made. [1] QED3 is an attractive physical the-

---

[1] One should keep in mind that planar field theories display unusual properties such as having arbitrary spin and statistics and peculiar parity and chiral transformations, see for example [1].

ory in its own right in the field of high T_c superconductivity, [2, 3], the recently studied unconventional Quantum Hall Effect in Graphene [4], and in the realm of dynamical generation of fundamental fermion masses, where numerical findings on the lattice and the results obtained by employing SDEs have yet to arrive at a final consensus and continue to provide a popular battle ground [5].

In the absence of computational ambiguities, the problem of gauge invariance can be traced back to not employing, or doing so incorrectly, the gauge identities such as Ward-Green-Takahashi identities (WGTI) [6], the Nielsen identities [7] and the Landau-Khalatnikov-Fradkin transformations (LKFT) [8]. This contribution is based on the work [9], where we address this issue in the light of the LKFT. The LKFT of Green functions describe the specific manner in which these functions transform under a variation of gauge. These transformations are nonperturbative in nature, and they are better described in coordinate space. Momentum space calculations are however more tedious, because of the complications induced by Fourier transforming. These difficulties are reflected in references [10, 11, 12] where a nonperturbative fermion propagator is obtained starting from a perturbative one in the Landau gauge in quenched QED in various dimensions. As LKFT are nonperturbative, we expect them not only to be satisfied at every order in perturbation theory but also in phenomena which are realized only nonperturbatively, such as dynamical chiral symmetry breaking (DCSB). Initial steps were taken in [13] to apply these transformations directly to the dynamically generated mass function. As a continuation of this effort, we carry out an exact numerical exercise to study the gauge dependence of the chiral condensate in the light of the LKFT.

The traditional way to study gauge dependence of the fermion propagator is to make an ansatz for the fermion-boson vertex and then solve the corresponding SDE in different covariant gauges. In practice, one needs to proceed in small steps in varying the gauge parameter away from the Landau gauge and it is prohibitively difficult to be able to compute the result for an arbitrarily large value of this parameter, particularly if a sophisticated form of the 3-point interaction is taken into account [14, 15]. Moreover, this line of attack does not guarantee that the LKFT for the dynamically generated fermion propagator will be satisfied gauge by gauge. We suggest an alternative approach to achieve the same objective. What we have to do is to start from the solution for the fermion propagator in the Landau gauge and simply perform a LKFT to find the result in any other gauge. As LKFT preserve WGTI, the advantage of our proposal is to conserve both the WGTI and LKFT as we move from one value of the gauge parameter to another. We draw the chiral condensate as a function of gauge both for the quenched and the unquenched case and find the results to be practically gauge independent.

This work is organized as follows: We start by introducing the notation and conventions we shall be using to study the gauge covariance of the fermion propagator. Then we review the traditional approach to study DCSB in the context of SDEs and present some truncation schemes widely implemented. Next we make a comparison of the results obtained by solving the SDE gauge by gauge against those arrived at by employing LKFT of the solution in the Landau gauge. Finally we present discussion and conclusions.

GAUGE COVARIANCE OF THE FERMION PROPAGATOR

We start by putting forward the definitions and notations we shall use to study the LKFT of the fermion propagator. We write out the Euclidean space fermion propagator in momentum and coordinate spaces, respectively, in their most general forms as :

$$S(p;\xi) \equiv \frac{F(p;\xi)}{i\not{p} - \mathcal{M}(p;\xi)}, \qquad S(x;\xi) \equiv \not{x}X(x;\xi) + Y(x;\xi). \qquad (1)$$

F is often referred to as the fermion wavefunction renormalization and $\mathcal{M}$ as the mass function. Expressions in Eq. (1) are related through a Fourier transformation. The LKFT relating the coordinate space fermion propagator in the Landau gauge to the one in an arbitrary covariant gauge reads

$$S(x;\xi) = S(x;0) e^{-ax}, \qquad (2)$$

where $a = \alpha\xi/2$ with $\alpha = e^2/(4\pi)$, e^2 being the dimensionful electromagnetic coupling of QED3 and ξ the covariant gauge parameter. The line of action for applying the LKFT is as follows: Start with a a dynamically generated solution of the SDE for the fermion propagator in the Landau gauge and Fourier transform it to coordinate space. Apply the LKFT transformation law and Fourier transform this result back to the momentum space to obtain the fermion propagator in an arbitrary covariant gauge. Fortunately in QED3, the related integrals involved are sufficiently simple to carry out exact numerical calculations. We straightforwardly find [9]

$$\begin{aligned}\frac{F(p;\xi)}{p^2 + \mathcal{M}^2(p;\xi)} &= \frac{a}{\pi p^2} \int_0^\infty dk\, k^2 \frac{F(k;0)}{k^2 + \mathcal{M}^2(k;0)} \left[\frac{1}{\lambda^-} + \frac{1}{\lambda^+} + \frac{1}{2kp}\ln\left|\frac{\lambda^-}{\lambda^+}\right|\right], \\ \frac{F(p;\xi)\mathcal{M}(p;\xi)}{p^2 + \mathcal{M}^2(p;\xi)} &= \frac{a}{\pi p} \int_0^\infty dk\, k \frac{F(k;0)\mathcal{M}(k;0)}{k^2 + \mathcal{M}^2(k;0)} \left[\frac{1}{\lambda^-} - \frac{1}{\lambda^+}\right],\end{aligned} \qquad (3)$$

where $\lambda^\pm = a^2 + (k \pm p)^2$. Thus the knowledge of $S(p;0)$ is the input required to obtain the same in an arbitrary covariant gauge.

DYNAMICAL CHIRAL SYMMETRY BREAKING

We know that perturbation theory in QED does not generate fermion masses dynamically. Nonperturbative methods have to be employed to look for such solutions for the fermion propagator and SDEs are a natural tool for calculation in continuum. SDE for the fermion propagator can be written as follows for QED3 :

$$S^{-1}(p;\xi) = S_0^{-1}(p) + e^2 \int \frac{d^3k}{(2\pi)^3} \Gamma_\nu(k,p) S(k;\xi) \gamma_\mu \Delta_{\mu\nu}(q), \qquad (4)$$

where $q = k - p$ and the photon propagator is

$$\Delta_{\mu\nu}(q) = \frac{\mathcal{G}(q^2)}{q^2}\left(\delta_{\mu\nu} - \frac{q_\mu q_\nu}{q^2}\right) + \xi \frac{q_\mu q_\nu}{q^4}, \qquad (5)$$

$\mathscr{G}$ being the photon wavefunction renormalization and $\Gamma^\mu(k,p)$ the full 3-point vertex.

The photon propagator and the fermion-boson vertex each have their own SDE, which in turn are coupled to the rest of the infinite tower containing higher point Green functions. A practical choice to truncate the infinite tower of these equations is to make an ansatz for $\Gamma^\mu(k,p)$ with the essential ingredients to ensure the correct gauge covariance of the fermion propagator. One requirement is that the choice of vertex satisfies the WGTI

$$q_\mu \Gamma_\mu(k,p) = S^{-1}(k;\xi) - S^{-1}(p;\xi), \tag{6}$$

which allows a decomposition of the vertex into its longitudinal and transverse parts: $\Gamma_\mu(k,p) = \Gamma_\mu^L(k,p) + \Gamma_\mu^T(k,p)$, such that $q_\mu \Gamma_\mu^T(k,p) = 0$. $\Gamma_\mu^T(k,p)$ remains undetermined by the WGTI. Although it is not uniquely fixed by the WGTI, a popular choice of $\Gamma_\mu^L(k,p)$ is the so called Ball-Chiu vertex (BC-vertex) [16]. An ansatz is generally made for the transverse part of the vertex guided by the requirements that (i) under the operations C, P and T, it should transform in the same way as the bare vertex, (ii) it should have not kinematic singularities, (iii) it should reduce to the perturbative Feynman expansion for the vertex when the coupling is small, (iv) it should lead to a fermion propagator with a correct LKFT property and (v) it itself should obey its LKFT law (see Ref. [17] for a comprehensive study of these requirements).

Instead of adopting the hit and trial method, we can choose another path. *We can start from the solution of the SDE in one gauge ensuring that the WGTI is satisfied in that gauge. We can then LKFT the result to other gauges and the WGTI will continue to be satisfied. The hope is then to obtain gauge independence of physical observables.* We shall be interested in the chiral condensate $\langle \bar{\psi}\psi \rangle = -\text{Tr} S(x=0;\xi)$ which acquires a nonzero value when chiral symmetry is dynamically broken.

TRUNCATION SCHEMES

We now present some truncation schemes of the SDE for the fermion propagator. We consider quenched ($\mathscr{G}=1$) and unquenched ($\mathscr{G} \neq 1$) versions of QED3.

- **Rainbow-Ladder approximation.** This is the simplest approximation, which consists is setting $\mathscr{G}=1$ and $\Gamma^\mu = \gamma^\mu$. Although in this approximation WGTI is not exactly verified, in the Landau gauge it might be considered a not too unreliable truncation scheme, as long as this identity is concerned (see Ref. [17]). Its simplicity has allowed to study the gauge dependence of the chiral condensate in a close vicinity of the Landau gauge, [14], as well as in a relatively broader range of values of the gauge parameter, [15].
- **BC-vertex in quenched approximation.** By construction, this vertex ansatz fulfills the WGTI. As compared to the rainbow-ladder approximation, solving the resulting SDE with this vertex is a more demanding exercise, and in literature there exists solutions only close to the Landau gauge, [14, 18].
- **Hybrid vertex in unquenched approximation.** In this truncation scheme the SDE for the fermion propagator is coupled to the corresponding equation for the photon

TABLE 1. $-\langle \bar{\psi}\psi \rangle$ from Ref. [19] in units of $10^{-5}e^4$.

ξ	$N_f = 0$	$N_f = 1$	$N_f = 2$	$N_f = 3$	$N_f = 4$	$N_f = 5$	$N_f = 6$
0	333	121	13	0.026	??	0	0
0.5	340	165	79	39	23	15	11
1	351	202	108	74	55	37	29
2	356	259	189	143	107	92	77

propagator with N_f flavors of degenerate fermions :

$$\Delta_{\mu\nu}^{-1}(q) = \Delta_{\mu\nu}^{0}{}^{-1}(q) - N_f e^2 \int \frac{d^3 k}{(2\pi)^3} \text{Tr}\left[\gamma_\mu S(k;\xi)\Gamma_\nu(k,q)S(q;\xi)\right]. \quad (7)$$

This exercise has been carried out in [19] employing a hybrid choice of the 3-point interaction. For the fermion propagator, the well-tested Curtis-Pennington vertex (CP-vertex) [20] was used, whereas the BC-vertex was implemented in the SDE of the photon propagator to avoid unwanted divergences. Although this is a more ambitious truncation scheme in QED3, it reveals the essence of the problem (see Table 1): The gauge dependence of the chiral condensate is such that in one gauge there is no DCSB, but there is in another!

NUMERICAL FINDINGS

We now compare the gauge dependence of the chiral condensate coming from the solution to the SDE for the fermion propagator in various gauges and that coming from the LKFT of the solution to the SDE in the Landau gauge. In order to carry out the latter exercise, we start from the solution found in all the three cases mentioned above for the Landau gauge and then employ Eq. (3) to find the solution in other gauges. Results are summarized in Figs. 1, 2, and 3. On can immediately see that the LKFT generated solutions are practically gauge invariant in contrast with the ones obtained by solving SDEs in various gauges. In the next section, we present a discussion of our findings.

DISCUSSION AND CONCLUSIONS

If we take the solution obtained in the Landau gauge and LKF transform it to other gauges we ensure that WGTI remains intact. Furthermore, the correct gauge covariance properties of the fermion propagator are ensured in passing from gauge to gauge, and all this can be seen from the virtually gauge independent value of the chiral condensate obtained in this form. Moreover, the simplicity of LKFT, Eq. (3), allows to carry out this exercise for arbitrarily large values of ξ, which cannot be achieved in practice solving the SDE gauge by gauge.

Using the BC-vertex, still in unquenched QED3, ensures that in every gauge the WGTI is fulfilled. Nevertheless, neglecting the transverse part of the vertex leads to wrong gauge covariance properties of the fermion propagator, and this yields (a smaller)

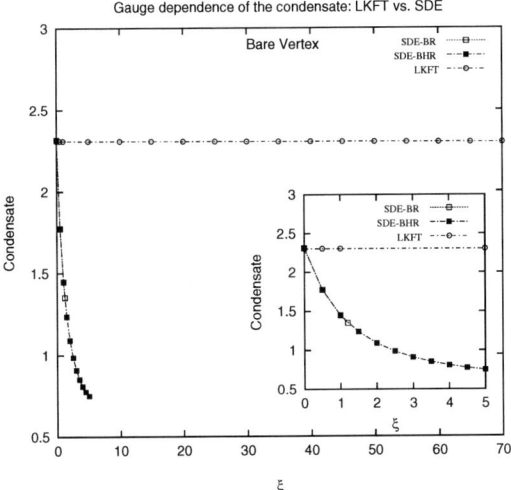

FIGURE 1. $-\langle \bar{\psi}\psi \rangle$ in various gauges for the rainbow-ladder approximation in units of $10^{-3}e^4$.

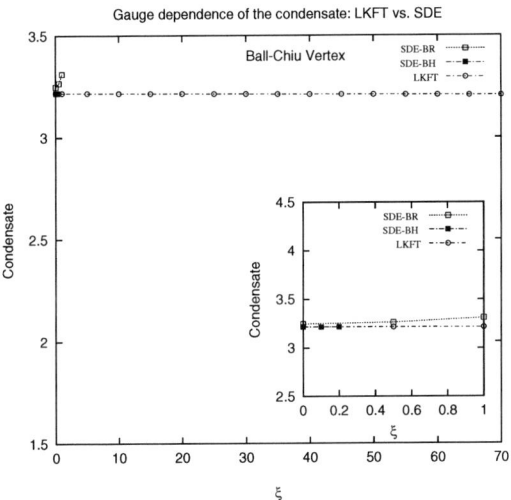

FIGURE 2. $-\langle \bar{\psi}\psi \rangle$ in quenched QED3 in various gauges for the BC-vertex in units of $10^{-3}e^4$.

gauge dependence of the chiral condensate as compared with the rainbow-ladder approximation. The complexity of the resulting SDE in this truncation scheme makes it difficult to study the solutions for arbitrarily large values of ξ. However, in the light of the LKFT, the corresponding transformation to the solution in the Landau gauge forces it to transform gauge covariantly to other gauges, and this can be done for arbitrarily large values of the gauge parameter (Fig. 2). Improved truncation schemes like the Hy-

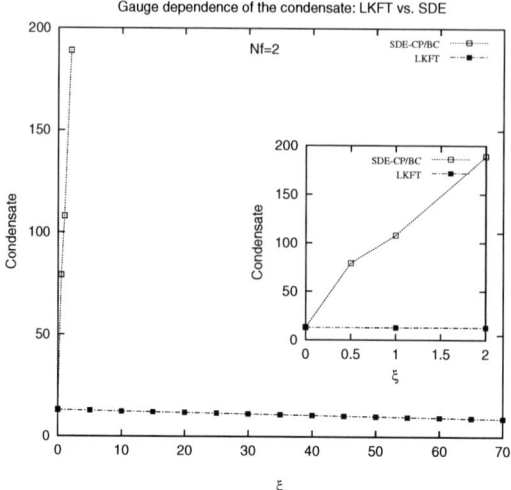

FIGURE 3. $-\langle \bar{\psi}\psi \rangle$ in unquenched QED3 for $N_f = 2$ in various gauges for the hybrid CP-BC vertex. Note that for a direct comparison with the result of [19], the condensate in this graph has been plotted in units of $10^{-5} e^4$. Solving the SDE the condensate changes rapidly from 13 in the Landau gauge with $\xi = 0$ to over 200 by $\xi = 2$. In contrast the LKFT gives a value that is almost constant in comparison falling from a value of 13 in the Landau gauge to 9 when $\xi = 70$.

brid CP-BC vertex in unquenched studies, which is indeed a reliable truncation scheme, does not do any better. Again the LKFT come to rescue as shown in Fig. 3. Looking at the entries of Table 1, one can make a gauge independent statement concluding that with the BC-CP truncation scheme, the critical number of flavors would lie in the range $3 < (N_f)_c < 5$.

To conclude, the inclusion of the WGTI alone is not sufficient to ensure the gauge independence of the physical observable associated with the fermion propagator. It is essential to apply LKFT to the dynamically generated mass function as advocated in [13]. This is what we achieve in an exact numerical fashion. We study the gauge dependence of the chiral condensate for quenched as well as unquenched QED3. The truncations employed correspond to approximating the fermion-boson interaction by the bare vertex, the BC-vertex and a hybrid choice of the CP-BC vertex. Numerically, we obtain an almost gauge independent value of the condensate for a very broad range of values of the covariant gauge parameter for all the above-mentioned cases. One may wonder why we worry about the WGTI if the LKFT are apparently sufficient to obtain the objective. The WGTI is crucial in the hunt for a more reliable result in one gauge, here the Landau gauge. This is where the choice of the full vertex plays an important role and the closer we are to the *true* vertex, the better it is. Once we know the result in the Landau gauge, LKFT will guide us along the path of varying gauge. We hope to apply this method in more realistic theories such as QED and QCD in the future.

ACKNOWLEDGMENTS

We thank M. R. Pennington for academic support at all stages of the work, V.P. Gusynin for valuable comments and C.S. Fischer for providing us with some numerical results. AB wishes to acknowledge a short term visitor grant by a joint scheme of The Royal Society, U.K, and The Mexican Academy of Sciences, Mexico. Support has also been provided by CIC and CONACyT (grants 4.10 and 46614-I).

REFERENCES

1. Yong-Shi Wu, Phys. Rev. Lett. **52** 2103 (1984); B. Binegar, J. Math. Phys. **23** 1511 (1982); Ma. de Jesús Anguiano and A. Bashir, Few Body Systems, **37** 71 (2005).
2. N. Dorey and N.E. Mavromatos, Nucl. Phys. **B386** 614 (1992); K. Farakos and N.E. Mavromatos, Mod. Phys. Lett. **A13** 1019 (1998); M. Franz and Z. Tesanovic, Phys. Rev. Lett. **87** 257003 (2001); I.F. Herbut, Phys. Rev. **B66** 094504 (2002), M. Franz, Z Tesanovic and O. Vafek, Phys. Rev. **B66** 054535 (2002).
3. M. Sutherland et. al., Phys. Rev. Lett. **94** 147004 (2005).
4. V.P. Gusynin and S.G. Sharapov, Phys. Rev. Lett. **95** 146801 (2005); K.S. Novoselov et. al., Nature **428** 197 (2005); Y. Zhang et. al Nature **438** 201 (2005).
5. S.J. Hands, J.B. Kogut, L. Scorzato and C.G. Strouthos, Phys. Rev. **B70** 104501 (2004); S.J. Hands, J.B. Kogut and C.G. Strouthos, Nucl. Phys. **B645** 321 (2002); S.J. Hands, J.B. Kogut, L. Scorzato and C.G. Strouthos, Nucl. Phys. Proc. Suppl. **119** 974 (2003); V.P. Gusynin and M. Reenders, Phys. Rev. **D68** 025017 (2003); M.R. Pennington and D. Walsh, Phys. Lett. **B253** 246 (1991).
6. J.C. Ward, Phys. Rev. **78** (1950); H.S. Green, Proc. Phys. Soc. (London) **A66** 873 (1953); Y. Takahashi, Nuovo Cimento **6** 371 (1957).
7. N.K. Nielsen, Nucl. Phys. **B101** 173 (1975); O. Piguet and K. Sibold, Nucl. Phys. **B253** 517 (1985).
8. L.D. Landau and I.M. Khalatnikov, Zh. Eksp. Teor. Fiz. **29** 89 (1956); L.D. Landau and I.M. Khalatnikov, Sov. Phys. JETP **2** 69 (1956); E.S. Fradkin, Sov. Phys. JETP **2** 361 (1956); K. Johnson and B. Zumino, Phys. Rev. Lett. **3** 351 (1959); B. Zumino, J. Math. Phys. **1** 1 (1960). S. Okubo, Nuovo Cim. **15** 949 (1960). I Bialynicki-Birula. Nuovo Cim. **17** 951 (1960). H. Sonoda, Phys. Lett. **B499** 253 (2001).
9. A. Bashir and A. Raya, *"Gauge Independent Chiral Condensate in QED3"* hep-ph/0511291.
10. A. Bashir, Phys. Lett. **B491** 280 (2000).
11. A. Bashir and A. Raya, Phys. Rev. **D66** 105005 (2002);
12. A. Bashir and R. Delbourgo, J. Phys. **A37** 6587 (2004).
13. A. Bashir and A. Raya, Nucl. Phys. **B709** 307 (2005); Nucl. Phys. Proc. Suppl. **B141** 259 (2005); proceedings of "2004 International Workshop on Dynamical Symmetry Breaking", Nagoya University, Nagoya, Japan, Dec. 21-22, 257-261 (2004).
14. C.J. Burden and C.D. Roberts, Phys. Rev. **D44** 540 (1991).
15. A. Bashir, A. Huet and A. Raya. Phys. Rev. **D66** 025029 (2002).
16. J.S. Ball and T.-W. Chiu, Phys. Rev. **D22** 2542 (1980).
17. A Bashir and A. Raya, *"Gauge Symmetry and its implications for the Schwinger-Dyson equations"* In "Trends in Boson Research". Nova Science Publishers Inc. N. Y., ISBN 1-59454-521-9, (2006). Preprint hep-ph/0411310.
18. A. Huet. *"Numerical Analysis of the Schwinger-Dyson equations"* M. Sc. Thesis, Universidad Michoacana de San Nicolás de Hidalgo, México (2003).
19. C.S. Fischer, R. Alkofer, T. Dahm and P. Maris, Phys. Rev. **D70** 073007 (2004).
20. D.C. Curtis and M.R. Pennington, Phys. Rev. **D42** 4165 (1990).

Growth of magnetic field in a two-component plasma

J. F. Nieves* and Sarira Sahu[†]

*Laboratory of Theoretical Physics, Department of Physics, P.O. Box 23343, University of Puerto Rico Río Piedras, Puerto Rico 00931-3343
[†]Instituto de Ciencias Nucleares, Universidad Nacional Autónoma de México, Circuito Exterior, C. U., A. Postal 70-543, 04510 Mexico DF, Mexico

Abstract. We study the electromagnetic properties of a system that consists of an electron background and a neutrino gas that may be moving or at rest, as a whole, relative to the background. The photon self-energy for this system is characterized by the usual transverse and longitudinal polarization functions, and two additional ones which are the focus of our calculations, that give rise to birefringence and anisotropic effects in the photon dispersion relations. We consider the macroscopic electrodynamic equations for the above system, and discuss the evolution of a magnetic field perturbation in such a medium. This particular phenomena has also been considered by others as a mechanism for the generation of large-scale magnetic fields in the Early Universe as a consequence of the neutrino-plasma interactions.

Keywords: neutrino, magnetic field

INTRODUCTION

This work is concerned with the electromagnetic properties of a medium that consists of a matter background, such as an electron plasma, and a neutrino gas that moves, as a whole, relative to the matter background. Technically, the quantity of interest to us is the photon self-energy, from which the dispersion relations of the photon modes that propagate in the medium can be obtained, and from which other macroscopic quantities of physical interest can be determined.

Some aspects of this composite system were studied in [1] using the methods of real-time finite temperature field theory (FTFT) which, from a modern point of view, provides a natural setting for studying the problems related to the propagation of photons in a medium. A convenient technique employed in FTFT is to carry out the calculations in a manifestly covariant form and is implemented by introducing the velocity four-vector u^μ of the medium, in terms of which the thermal propagators are written in a covariant form. In this way, covariance is maintained, but quantities such as the photon self-energy depend on the vector u^μ in addition to the kinematic momentum variables of the problem. Generally, for practical purposes the vector u^μ is set to $(1, \vec{0})$ in the end, which is equivalent to having carried out the calculation in the rest frame of the medium from the beginning, and this is usually the relevant physical situation.

However, as noted in [1], the system we are considering provides an example for a novel application of FTFT. A distinctive feature of this system is that the matter background on one hand, and the neutrino gas on the other, each is characterized by its own velocity four-vector. Thus, if we denote by u^μ the velocity four-vector of the matter background, and by v^μ the corresponding one for the neutrino gas, then we can take the matter background to be at rest, so that $u^\mu = (1,\vec{0})$ but we must keep $v^\mu = (v^0, \vec{V})$. Therefore, the physical quantities such as the photon self-energy depend on the momentum variables and u^μ as usual, and in addition on the vector v^μ. This additional dependence can have physical effects that cannot be produced by the stationary background alone.

Here we compute the contribution to the photon self-energy due to both electron and neutrino/anti-neutrino background which is proportional to both the neutrino-antineutrino asymmetry and the electron density, and consider the consequences for the propagation of photons and for the electromagnetic properties of the medium. The calculation is based on the application of FTFT. For the calculation of photon self-energy we use the dressed electron propagator due to the neutrino gas. The implicit assumption is that, in its own rest frame, the neutrino gas has a momentum distribution function that is parametrized in the usual way. The presence of the neutrinos gives rise to two additional polarization functions that we denote by π_P and π'_P besides the ordinary longitudinal and transverse polarization functions. A non-zero value of π_P by itself leads to optical activity effects[2], while π'_P induces anisotropic effects. Aside from the implications for the photon dispersion relations, the results for the photon self-energy can be interpreted in terms of the macroscopic electromagnetic properties of the system. In particular, the effects due to π_P and π'_P remain finite in the limit $q \to 0$, and therefore they can manifest themselves in macroscopic effects in the long wavelength and static regimes. As a specific application, we consider the evolution of a magnetic field perturbation in the system, and we arrive at an equation for the dynamics of the magnetic field that has been suggested by Semikoz and Sokoloff[3], in the context of a mechanism for the generation of large-scale magnetic fields in the Early Universe as a consequence of the neutrino-plasma interactions. Our results are applicable in a variety of situations where the neutrino interactions with the other background particles are important, and the method we employ could also be useful in the study of similar problems that may arise in other contexts.

PHOTON SELF-ENERGY

The photon self-energy in the background of the static electron and moving neutrino media can be expressed in the form[4, 5]

$$\pi_{\mu\nu} = \pi_T R_{\mu\nu} + \pi_L Q_{\mu\nu} + \pi_P P_{\mu\nu} + \pi'_P P'_{\mu\nu}, \tag{1}$$

where $R_{\mu\nu}$ and $Q_{\mu\nu}$ are transverse and longitudinal tensors defined as

$$R_{\mu\nu} = \tilde{g}_{\mu\nu} - Q_{\mu\nu}, \quad Q_{\mu\nu} = \frac{\tilde{u}_\mu \tilde{u}_\nu}{\tilde{u}^2}, \tag{2}$$

and

$$P_{\mu\nu} = \frac{i}{Q}\varepsilon_{\mu\nu\alpha\beta}q^\alpha u^\beta, \quad P'_{\mu\nu} = \frac{i}{Q}\varepsilon_{\mu\nu\alpha\beta}q^\alpha v'^\beta, \tag{3}$$

with

$$v'^\mu = v^\mu - u^\mu (u \cdot v). \tag{4}$$

We have defined $\tilde{g}_{\mu\nu} = g_{\mu\nu} - q_\mu q_\nu/q^2$ and $\tilde{u}_\mu = \tilde{g}_{\mu\nu} u^\nu$. Also the quantities $\omega = q \cdot u$ and $Q = \sqrt{\omega^2 - q^2}$. The scalar functions $\pi_{T,L,P}$s are calculated from the photon self-energy diagram where, the internal electron/position propagators are modified due to its interaction with the neutrino background and are given by,

$$\begin{aligned}
\pi_T &= \pi_T^{(e)} - 2e^2 \lambda_V \left(A'_e + \frac{q^2}{Q^2} B'_e \right), \\
\pi_L &= \pi_L^{(e)} + 4e^2 \lambda_V \left(\frac{q^2}{Q^2} \right) B'_e, \\
\pi_P &= (u \cdot v) \pi_P^{(v)} + \pi_P^{(ev)} \\
\pi'_P &= \pi_P^{(v)} + \pi_P^{'(ev)}
\end{aligned} \quad (5)$$

where

$$\begin{aligned}
\pi_P^{(ev)} &= -4e^2 \lambda_A Q[I_1 + (u \cdot v) I_2], \\
\pi_P^{'(ev)} &= -4e^2 \lambda_A Q I_2,
\end{aligned} \quad (6)$$

with $I_{1,2}$ given by

$$I_1 = \frac{v \cdot u}{Q^2} \int \frac{d^3 p}{(2\pi)^3 2E} \frac{\partial (f_e + f_{\bar{e}})}{\partial E} \left[\frac{q^2 (p \cdot u) - (q \cdot u)(p \cdot q)}{q^2 + 2p \cdot q} \right] + (q \to -q), \quad (7)$$

and

$$I_2 = 2m_e^2 \int \frac{d^3 p}{(2\pi)^3 2E} \frac{\partial}{\partial E} \left[\frac{1}{2E} \left(\frac{f_e + f_{\bar{e}}}{q^2 + 2p \cdot q} \right) \right] + (q \to -q). \quad (8)$$

These two integrals can be calculated in different physical situations. The quantity λ_A is given by

$$\lambda_A = \frac{G_F}{\sqrt{2}} \left[(n_{v_\tau} - n_{\bar{v}_\tau}) + (n_{v_\mu} - n_{\bar{v}_\mu}) - (n_{v_e} - n_{\bar{v}_e}) \right], \quad (9)$$

and the functions A'_e and B'_e are defined in Ref.[4].

MACROSCOPIC ELECTRODYNAMICS

Besides modifying the dispersion relations of the propagating modes, the presence of the neutrino gas influence the electromagnetic properties of the system in the static ($\omega = 0$, or the $\vec{Q} = 0$) and long wavelength regime ($\omega \gg \bar{v} Q$), with $\bar{v}$ being the average velocity of the background particles. Here for simplicity we assume that the neutrino gas is at rest with respect to the electron gas, that is $v^\mu = u^\mu$. In this case, $P'_{\mu\nu} = 0$, and the photon self-energy takes the form

$$\pi_{\mu\nu} = \pi_T R_{\mu\nu} + \pi_L Q_{\mu\nu} + \pi_P P_{\mu\nu}. \quad (10)$$

In the presence of an external current $j_\mu^{(ext)}$, the electromagnetic potential in the medium is determined from the field equation, which in momentum space is

$$\left[-q^2 \tilde{g}_{\mu\nu} + \pi_{\mu\nu}\right] A^\nu = j_\mu^{(ext)}. \tag{11}$$

The photon dispersion relations are determined by finding the solutions of the homogeneous equation, i.e., with $j_\mu^{(ext)} = 0$. The above equation can be written in the form,

$$-iq^\mu F_{\mu\nu} = j_\nu^{(ext)} + j_\nu^{(ind)}, \quad \text{with} \quad j_\mu^{(ind)} = -\pi_{\mu\nu} A^\nu. \tag{12}$$

In fact, Eq.(12) is equivalent to the Maxwell equations, with $j_\mu^{(ind)}$ interpreted as the induced current. After some algebra, in terms of dielectric and magnetic permeabilities (ε and μ respectively) we obtain

$$\vec{j}^{(ind)} = i\left[\omega(1-\varepsilon)\vec{E} + \left(1 - \frac{1}{\mu}\right)\vec{Q} \times \vec{B} + i\frac{\omega^2}{Q}\varepsilon_p \vec{B}\right], \tag{13}$$

where $\varepsilon_p = \pi_p/\omega^2$. These equations can of course be used to discuss the dispersion relations of the propagating modes and related effects.

EVOLUTION OF MAGNETIC FIELDS

Here we consider the evolution of an initial magnetic perturbation in the above medium. In the absence of any external sources, the equation for the magnetic field, Eq.(12), is

$$i\vec{Q} \times \vec{B} + i\omega\vec{E} = \vec{j}^{(ind)}, \tag{14}$$

which using Eq.(13) we can write in the form

$$\varepsilon\omega\vec{E} + \frac{1}{\mu}\vec{Q} \times \vec{B} + i\gamma\vec{B} = 0, \tag{15}$$

with

$$\gamma = -\frac{\omega^2}{Q}\varepsilon_p = -\frac{\pi_P}{Q}. \tag{16}$$

By taking the long wavelength limit, $\omega \gg \bar{v}Q$, and making the quasi-static approximation, $\omega \to 0$, we obtain from Eq.(15)

$$\sigma\frac{\partial \vec{B}}{\partial t} = \frac{1}{\mu}\nabla^2 \vec{B} + \gamma \vec{\nabla} \times \vec{B}. \tag{17}$$

where we have used $\varepsilon \to 1 + i\sigma/\omega$, with σ the conductivity of the medium. Eq.(17) describes the self-excitation of a magnetic perturbation. To see how that comes about, consider a plane wave magnetic field of definite helicity

$$\vec{B}_\lambda = B e^{i(\vec{Q}\cdot\vec{x} - \omega t)} \hat{e}_\lambda, \tag{18}$$

where

$$\hat{e}_\lambda = \frac{1}{\sqrt{2}}(\hat{e}_1 + i\lambda \hat{e}_2), \qquad (19)$$

with $\lambda = \pm 1$ and the vectors $\hat{e}_{1,2}$ are the unit vectors. From Eq.(17), the dispersion relation is given by

$$\omega_\lambda = \frac{i}{\sigma}\left(\lambda Q\gamma - \frac{Q^2}{\mu}\right). \qquad (20)$$

The sign of γ depends on the sign of the neutrino-antineutrino asymmetry. But it is clear that for either sign, one helicity amplitude is damped according to

$$B \sim e^{-\frac{Q}{\sigma}\left(|\gamma| + \frac{Q}{\mu}\right)t}, \qquad (21)$$

while for the opposite helicity

$$B \sim e^{\frac{Q}{\sigma}\left(|\gamma| - \frac{Q}{\mu}\right)t}, \qquad (22)$$

which grows for those wavelenghts satisfying $Q < \mu|\gamma|$. The mode with the maximum growth rate corresponds to the value of Q given by $Q_0 = \frac{\mu|\gamma|}{2}$ and the corresponding growth rate is $\Gamma = \frac{\mu\gamma^2}{4\sigma}$. Thus, for example, for the case of the relativistic and non-degenerate electron gas, together with the formula[3] $\sigma^{-1} \simeq e^2/((4\pi)^2 T)$ for the conductivity, we obtain

$$\Gamma = 2.2 \times 10^{-20} \xi^2 \left(\frac{T}{GeV}\right)^5 GeV. \qquad (23)$$

Analogous formulas can be obtained for the other physical situations as well, which are useful for numerical estimates of the possible effects in different physical environments. An equation of the same form as Eq.(17) was obtained by Semikoz and Sokoloff[3] by very different means, in their work suggesting a new mechanism for the generation of large-scale magnetic fields in the Early Universe as a consequence of the neutrino-plasma interactions. As emphasized in that reference, the mechanism can result in the self-excitation of an almost constant magnetic perturbation, as we have illustrated.

CONCLUSIONS

We have studied the electromagnetic properties of an electron background, that contains a neutrino gas which is at rest or moving as a whole relative to the background. Apart from the well known longitudinal and transverse polarization functions of the photon in a medium, the presence of the neutrinos gives rise to two additional polarization functions, that we denote by π_P and π'_P. We have computed that particular contribution to these two functions, $\pi_P^{(ev)}$ and $\pi_P^{\prime(ev)}$, that depends on the neutrino-antineutrino asymmetry in the medium as well as the momentum integral of the electron (and positron) distribution function. One of the consequences of a non-zero value of π_P and π'_P is to give rise to birefringence and anisotropic effects in the propagation of a photon through that medium.

A non-zero value of π_P and π'_P also has consequences related to the electromagnetic properties of the system at a macroscopic level, and we considered specifically the evolution of a macroscopic magnetic field in this system. We arrived at an equation for the dynamics of the magnetic field, that had been suggested in Ref.[3] as a mechanism for the generation of large-scale magnetic fields in the Early Universe as a consequence of the neutrino-plasma interactions.

ACKNOWLEDGMENTS

This material is based upon work supported by the US National Science Foundation under Grant No. 0139538 (JFN); and by DGAPA-UNAM under PAPIIT grant number IN119405 (SS).

REFERENCES

1. J. F. Nieves, *Phys. Rev. D* **61**, 113008(1-13) (2000).
2. S. Mohanty, J. F. Nieves and P. B. Pal, *Phys. Rev. D* **58**, 093007(1-9) (1998).
3. V. B. Semikoz and D. D. Sokoloff, *Phys. Rev. Lett.* **92**, 131301(1-4) (2004).
4. J. F. Nieves and Sarira Sahu, *Phys. Rev. D* **71**, 073006(1-14) (2005).
5. J. C. D'Olivo, J. F. Nieves and Sarira Sahu, *Phys. Rev. D* **67**, 025018(1-20) (2003).

Pair creation in inhomogeneous fields from worldline instantons

Gerald V. Dunne* and Christian Schubert†

*Department of Physics, University of Connecticut, Storrs, Connecticut 06269-3046, USA
†Instituto de Física y Matemáticas, Universidad Michoacana de San Nicolás de Hidalgo, Ed. C-3, C.U., C.P. 58040, Morelia, Michoacán, México

Abstract. We show how to do semiclassical nonperturbative computations within the worldline approach to quantum field theory using "worldline instantons". These worldline instantons are classical solutions to the Euclidean worldline loop equations of motion, and are closed spacetime loops parametrized by the proper-time. Specifically, we compute the imaginary part of the one loop effective action in scalar and spinor QED using worldline instantons, for a wide class of inhomogeneous electric field backgrounds.

Keywords: Pair production; semiclassical approximation.
PACS: 11.27.+d, 12.20.Ds

INTRODUCTION

As is well-known, quantum field theory allows the spontaneous creation of electron-positron pairs from vacuum in external electric fields. This effect has been considered already in the early days of quantum electrodynamics [1, 2], and Schwinger used effective action methods to obtain a simple closed-form expression for the production rate in the constant field case [3]. Although spontaneous pair creation is of potential interest for many branches of physics the chances for its direct experimental verification hitherto seemed very remote. This is due to the exponential smallness of the production rate for field strengths below the critical value $E_c = \frac{m^2 c^3}{e\hbar}(1.3 \times 10^{18} V/m)$, which is far above the electric fields which can be produced in the laboratory macroscopically. However, given the rapid progress of laser technology it seems not any more impossible that in the near future pair production might be observable in laser fields. Both the optical laser system POLARIS [4] under construction at the Jena high-intensity laser facility and the X-ray free electron lasers to be constructed at SLAC [5] and DESY [6] are expected to reach laser field strengths missing E_c only by a few orders of magnitude. Moreover, it has been argued that for focused laser pulses substantial pair creation should set in already somewhat below critical field strength [7].

Laser fields cannot usually be treated in the constant field approximation, and this is particularly true for the pair creation process due to its nonperturbative nature. Much effort has gone into developing methods for the calculation of pair creation rates in inhomogeneous fields, mostly based on WKB [8, 9, 10, 11, 12, 13]. In this talk I will present a substantially different approach [14] based on Feynman's worldline path integral formalism [15, 16] and work done by Affleck et al. in 1982 for the constant field case in scalar QED [17]. The path integral representing the imaginary part of the one loop effective action is calculated in a semiclassical approximation around a closed classical "instanton" trajectory. The worldline action evaluated on this solution directly gives the Schwinger exponent of the imaginary part of the effective action in the weak field approximation.

PAIR CREATION IN ELECTRIC FIELDS

Let us start with Schwinger's well-known formula for the imaginary parts of the scalar and spinor QED effective Lagrangians in a constant electric field (at one loop) [3]:

$$\begin{aligned} \mathrm{Im}\mathscr{L}_{\text{scalar}}^{(1)}(E) &= \frac{m^4}{16\pi^3}\beta^2 \sum_{n=1}^{\infty} \frac{(-1)^{n-1}}{n^2} \exp\left[-\frac{\pi n}{\beta}\right] \\ \mathrm{Im}\mathscr{L}_{\text{spinor}}^{(1)}(E) &= \frac{m^4}{8\pi^3}\beta^2 \sum_{n=1}^{\infty} \frac{1}{n^2} \exp\left[-\frac{\pi n}{\beta}\right] \end{aligned} \quad (1)$$

where $\beta = \frac{eE}{m^2}$. The first term in each series gives (up to a factor of 2) directly the total pair-production rates per volume per time [18]. The higher order ($n \geq 2$) terms are statistics dependent and contain the information on the coherent production of n pairs by the field. All terms are exponentially suppressed for $E < E_{\text{cr}}$. Higher loop corrections have also been considered [19, 20] but can be neglected for subcritical fields as far as the total pair production rate is concerned [21].

The formulas (1) are commonly derived from the standard propertime representation for the one-loop effective Lagrangian. For spinor QED, this is the Euler-Heisenberg Lagrangian [2]:

$$\mathscr{L}_{\text{spinor}}^{(1)}(E) = -\frac{1}{8\pi^2} \int_0^\infty \frac{dT}{T^3} e^{-m^2 T} \left[\frac{eET}{\tan(eET)} + \frac{1}{3}(eET)^2 - 1 \right] \quad (2)$$

The nth term in the series for $\mathrm{Im}\mathscr{L}_{\text{spinor}}$ in (1) is generated y the pole at $T = \frac{n\pi}{eE}$ of the integrand in (2).

SCHWINGER'S FORMULA FROM WORLDLINE INSTANTONS

Feynman in 1950 [16] presented, "as an alternative to the formulation of second quantization", a formula representing the scalar QED effective action $\Gamma_{\text{scalar}}[A]$ in terms of first-quantized worldline path integrals. In the quenched approximation (i.e. with only one scalar loop but any number of photons) it reads

$$\Gamma_{\text{scalar}}^{(\text{quenched})}[A] = \int d^4x \, \mathscr{L}_{\text{scalar}}^{(\text{quenched})}[A] = \int_0^\infty \frac{dT}{T} e^{-m^2 T} \int_{x(T)=x(0)} \mathscr{D}x(\tau) \, e^{-S[x(\tau)]} \quad (3)$$

Here the path integral is over all closed loops in spacetime with a given period T in the proper-time of the loop scalar. The worldline action $S = S_0 + S_e + S_i$ has three parts,

$$S_0 = \int_0^T d\tau \frac{\dot{x}^2}{4}$$
$$S_e = ie \int_0^T \dot{x}^\mu A_\mu(x(\tau))$$
$$S_i = -\frac{e^2}{8\pi^2} \int_0^T d\tau_1 \int_0^T d\tau_2 \frac{\dot{x}(\tau_1) \cdot \dot{x}(\tau_2)}{(x(\tau_1) - x(\tau_2))^2}$$

(4)

Of these S_e incorporates the interaction with the external field, while S_i takes all internal photon exchanges in the loop into account.

Feynman generalized this "worldline representation" to spinor QED in [16]. Although it has always been considered an interesting alternative to standard quantum field theory, only in recent years it has gained some popularity as a calculational tool. Presently there exist, among others, the following approaches to the calculation of this type of path integral:

- The "string-inspired" approach [22, 23, 24, 20, 25] (see [26] for a review) which aims at an analytical calculation using appropriate wordline Green's functions.
- Semiclassical calculation using a stationary phase approximation [17].
- Variational methods [27].
- Numerical calculation using Monte Carlo methods [28, 29]. This approach in principle applies to effective actions in arbitrary backgrounds, and has been applied to the pair creation process in [30].

We will expand here on the first approach, which is inspired by instanton methods in field theory. The idea is to calculate Im$\mathscr{L}_{\text{scalar}}$ for weak fields using an extremal trajectory of the worldline path integral for a stationary phase approximation. In [17] it was shown that for the case of a constant electric field in the z direction this extremal action trajectory ("worldline instanton") is given by a circle in the (euclidean) $t - z$ plane:

$$x_{\text{extremal}}(\tau) = \frac{m}{eE}\big(0, 0, \cos(2\pi\tau), \sin(2\pi\tau)\big) \qquad (T=1) \qquad (5)$$

At leading order in the stationary phase approximation, the exponent of the imaginary part of the effective Lagrangian is given by the worldline action of this trajectory,

$$\text{Im}\mathscr{L}_{\text{scalar}}^{(\text{quenched})}(E) \sim e^{-S[x_{\text{extremal}}]} \qquad (6)$$

This is easily evaluated to be

$$(S_0 + S_e)[x_{\text{extremal}}] = \pi \frac{m^2}{eE}, \qquad S_i[x_{\text{extremal}}] = -\alpha\pi. \qquad (7)$$

The contribution of $S_0 + S_e$ just reproduces the first of the exponentials in Schwinger's one-loop formula (1), while the higher order ones are generated by the "multi-instantons" where the same circle is traversed n times. More surprising is the simplicity of the S_i term, which represents the contribution of all the higher loop corrections involving arbitrary photon exchanges in the loop. According to [17] this is the exact all-orders result in the weak field limit, including renormalization effects:

$$\text{Im}\,\mathcal{L}_{\text{scalar}}^{(\text{quenched})}(E) \stackrel{\beta \to 0}{\simeq} \frac{m^4}{8\pi^3}\beta^2 \exp\left(-\frac{\pi}{\beta} + \alpha\pi\right) \tag{8}$$

At the two-loop level this remarkable formula has been independently confirmed [31].

INHOMOGENEOUS BACKGROUND FIELDS

Despite of the simplicity and elegance of this worldline instanton approach it appears that the work of [17] has never been extended either to spinor QED or to more general backgrounds. As we will now show, at least at the one-loop level the method generalizes to a large class of inhomogeneous backgrounds straightforwardly. Let us return to Feynman's formula (3), omitting the photon exchange term S_i:

$$\Gamma_{\text{scalar}}^{(1-\text{loop})}[A] = \int_0^\infty \frac{dT}{T} e^{-m^2 T} \int \mathcal{D}x\, e^{-\int_0^T d\tau \left(\frac{\dot{x}^2}{4} + ieA\cdot\dot{x}\right)} \tag{9}$$

Rescaling $\tau = Tu$, this becomes

$$\Gamma_{\text{scalar}}^{(1-\text{loop})}[A] = \int_0^\infty \frac{dT}{T} e^{-m^2 T} \int \mathcal{D}x\, e^{-\left(\frac{1}{T}\int_0^1 du\,\dot{x}^2 + ie\int_0^1 du A\cdot\dot{x}\right)} \tag{10}$$

The T integral has a stationary point at

$$T_0^2 = \frac{\int du\,\dot{x}^2}{m^2} \tag{11}$$

leading to

$$\text{Im}\,\Gamma_{\text{scalar}}^{(1-\text{loop})} = \frac{1}{m}\sqrt{\frac{2\pi}{T_0}}\,\text{Im}\int \mathcal{D}x\, e^{-\left(m\sqrt{\int du\,\dot{x}^2} + ie\int_0^1 du A\cdot\dot{x}\right)} \tag{12}$$

The new worldline action,

$$S = m\sqrt{\int du\,\dot{x}^2} + ie\int_0^1 du A\cdot\dot{x} \tag{13}$$

is stationary if

$$m\frac{\ddot{x}_\mu}{\sqrt{\int du \dot{x}^2}} = ieF_{\mu\nu}\dot{x}_\nu \qquad (14)$$

Contracting with $\dot{x}^\mu$ yields $\dot{x}^2 = \text{constant} \equiv a^2$ so that

$$m\ddot{x}_\mu = ieaF_{\mu\nu}\dot{x}_\nu \qquad (15)$$

Now specialize to a *time-dependent electric field directed in the x_3 direction*. Choose a gauge where

$$A_3 = A_3(x_4); \quad A_\mu = 0 \text{ for } \mu \neq 3. \qquad (16)$$

Since $F_{\mu 1} = F_{\mu 2} = 0$, the stationarity conditions together with the periodicity imply that x_1 and x_2 must be constant. Hence one is down to an equation for x_3 and x_4:

$$\ddot{x}_3 = \frac{iea}{m}F_{34}\dot{x}_4, \quad \ddot{x}_4 = -\frac{iea}{m}F_{34}\dot{x}_3. \qquad (17)$$

In terms of A_3 this can be further reduced to

$$\dot{x}_3 = -\frac{iea}{m}A_3(x_4), \quad |\dot{x}_4| = a\sqrt{1+\left(\frac{eA_3(x_4)}{m}\right)^2}. \qquad (18)$$

As an example, let us consider the following single-pulse electric background [11, 12],

$$E(t) = E\,\text{sech}^2(\omega t) \qquad (19)$$

For this background the solution of (17) turns out to be very simple:

$$x_3(u) = -\frac{1}{\omega}\frac{1}{\sqrt{1+\gamma^2}}\text{arcsinh}\left[\gamma\cos(2n\pi u)\right]$$

$$x_4(u) = \frac{1}{\omega}\arcsin\left[\frac{\gamma}{\sqrt{1+\gamma^2}}\sin(2n\pi u)\right] \qquad (20)$$

Here $\gamma \equiv \frac{m\omega}{eE}$ denotes the "adiabaticity parameter" [8] and the integer $n \in \mathbf{Z}^+$ counts the number of times the closed path is traversed. The worldline action (13) evaluated on this instanton is

$$S_0 = n\frac{m^2\pi}{eE}\left(\frac{2}{1+\sqrt{1+\gamma^2}}\right) \quad (21)$$

In fig. 1 we plot the the instanton trajectories for various values of the parameter γ. In the static limit $\gamma \to 0$ we recover the circular paths (5) of the constant field case. In the short-pulse limit $\gamma \to \infty$ the instantons shrink in size and become elongated. Thus the instanton action decreases with increasing γ, leading to a local enhancement of the pair creation rate as compared to the case of a constant field with magnitude E.

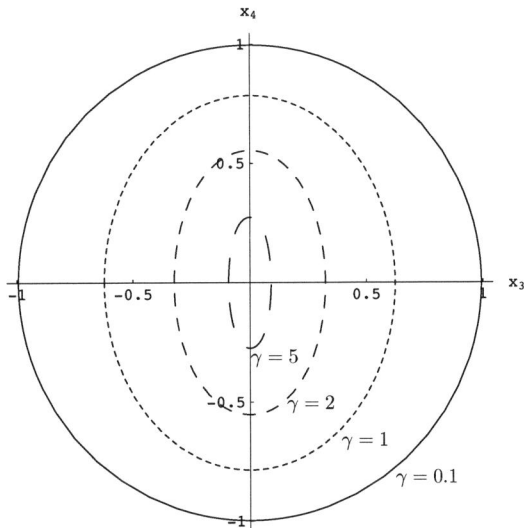

FIGURE 1. Plot of the worldline instanton paths (20) in the (x_3, x_4) plane for the case of a time-dependent electric field $E(t) = E\,\text{sech}^2(\omega t)$. The paths are shown for various values of the adiabaticity parameter γ. $x_{3,4}$ have been expressed in units of $\frac{m}{eE}$.

The case of a *spatially* inhomogeneous electric field in the z direction can be treated completely analogously. The (euclidean) gauge potential can be chosen as

$$A_4 = A_4(x_3); \qquad A_\mu = 0 \quad \text{for } \mu \neq 4. \quad (22)$$

This leads to instanton equations differing from eqs. (18) just by interchanging $3 \leftrightarrow 4$. As an example, let us consider the spatial analogue of (19), i.e. a single-bump electric field depending only on x_3:

$$E(x_3) = E\,\text{sech}^2(kx_3) \quad (23)$$

The instanton solutions are obtained from the single-pulse ones (20) *mutatis mutandis*:

$$x_3(u) = \frac{m}{eE}\frac{1}{\tilde{\gamma}}\text{arcsinh}\left(\frac{\tilde{\gamma}}{\sqrt{1-\tilde{\gamma}^2}}\sin(2\pi nu)\right)$$

$$x_4(u) = \frac{m}{eE}\frac{1}{\tilde{\gamma}\sqrt{1-\tilde{\gamma}^2}}\arcsin\left(\tilde{\gamma}\cos(2\pi nu)\right)$$

(24)

with $\tilde{\gamma} = \frac{mk}{eE}$. The stationary action is

$$S_0 = n\frac{m^2\pi}{eE}\left(\frac{2}{1+\sqrt{1-\tilde{\gamma}^2}}\right) \quad (25)$$

These solutions are plotted in fig. 2 for various values of $\tilde{\gamma}$. Note that again they reduce to the constant field circles for $\tilde{\gamma} \to 0$, but they grow in size with increasing $\tilde{\gamma}$ and become infinitely large when $\tilde{\gamma} \to 1$. Physically, this is where the width of the electric field becomes too small for a virtual pair to extract from it the energy necessary to turn real.

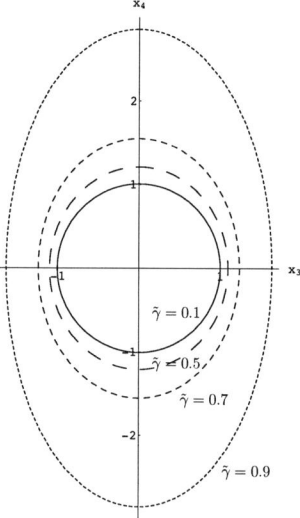

FIGURE 2. Plot of the instanton paths (24) for the case of a space-dependent field $E(x) = E\text{sech}^2(kx)$ for various values of the adiabaticity parameter $\tilde{\gamma}$.

The instanton action increases with increasing $\tilde{\gamma}$, leading to a lower local pair production rate as compared to the constant field case. Thus we see a rule emerging here which we believe holds quite generally:

- *Temporal inhomogeneity increases the pair production rate*
- *Spatial inhomogeneity decreases the pair production rate*

THE SPINOR LOOP CASE

A path integral representation for the one loop effective action in spinor QED [16] can be obtained from the scalar QED one (9) by multiplication by a global factor of $-\frac{1}{2}$ and the insertion of the following "spin factor" $S[x,A]$,

$$S[x,A] = \mathrm{tr}_\Gamma \mathscr{P} e^{\frac{i}{2} e \sigma^{\mu\nu} \int_0^T d\tau F_{\mu\nu}(x(\tau))} \tag{26}$$

For the two-dimensional background fields considered above the path ordering has no effect, since the $F_{\mu\nu}(x(\tau))$'s at different proper-times commute. The spin factor then reduces to

$$S[x,A] = 4\cos\left[eT \int_0^1 du\, E(x(u))\right] \tag{27}$$

Since the exponent of the spin factor (26) is purely imaginary it affects neither the determination of the stationary point T_0 nor the instanton equations (14). Therefore it remains only to evaluate the spin factor on the instanton solutions for the scalar loop. For the class of backgrounds of the form $A_3(x_4)$ or $A_4(x_3)$ considered above the result turns out to be [14]

$$S[x,A] = 4(-1)^n \tag{28}$$

Thus at least for this class of inhomogeneous backgrounds we find that $\mathrm{Im}\mathscr{L}^{(1)}_{\mathrm{scalar}}$ and $\mathrm{Im}\mathscr{L}^{(1)}_{\mathrm{spinor}}$ differ by the the same simple sign changes as in the constant field case, eq.(1).

SUMMARY, WORK IN PROGRESS

We have shown that the worldline instanton approach holds considerable promise as a tool for calculating pair creation rates in inhomogeneous backgrounds in scalar and spinor QED. Although the class of backgrounds which we have considered here is also amenable to a treatment by WKB methods [10, 11, 12, 13] the worldline approach offers a number of distinct advantages: (i) it bypasses the momentum space integrals which usually would have to be done in this type of calculation (ii) it uses proper-time instead of time, which seems more natural in the treatment of an intrinsically relativistic effect such as pair creation (iii) it provides a framework for the inclusion of radiative corrections.

Clearly we have not fully explored here the potential of this method. The instanton equations in the form (14) generalize immediately to the case of an arbitrary electromagnetic background field. While closed-form solutions can be expected only in special cases, the numerical integration of these equations poses no problems in principle. However, the instanton provides only the exponents of the Schwinger exponentials. We have not discussed here the prefactors of the Schwinger exponentials (1), which in the present approach involve the determinant of fluctuations around the worldline instanton. In the constant field case the determinant computation is straightforward, since the fluctuation problem is Gaussian [17]. As will be shown in a forthcoming paper [32] for inhomogeneous background fields the determinant can be computed using the Gelfand-Yaglom technique, since the fluctuation operator is an ordinary differential operator, depending only on the proper-time.

ACKNOWLEDGMENTS

We are very grateful to Don Page for helpful correspondence. We acknowledge the support of the NSF US-Mexico Collaborative Research Grant 0122615.

REFERENCES

1. F. Sauter, *Z. Phys.* **69**, 742 (1931).
2. W. Heisenberg and H. Euler, *Z. Phys.* **98**, 714 (1936).
3. J. Schwinger, *Phys. Rev.* **82**, 664 (1951).
4. J. Hein et. al., *Appl. Phys. B* **79**, 419 (2004).
5. Linac Coherent Light Source, http://www-ssrl.slac.stanford.edu/lcls/
6. European X-Ray Laser Project XFEL, http://xfel.desy.de/
7. S.S. Bulanov, N.B. Narozhny, V.D. Mur, and V.S. Popov, *Phys. Lett. A* **330**, 1 (2004).
8. L.V. Keldysh, *Sov. Phys. JETP* **20**, 1307 (1965).
9. A.I. Nikishov and V.I. Ritus, *Sov. Phys. JETP* **25**, 1135 (1967).
10. E. Brezin and C. Itzykson, *Phys. Rev. D* **2**, 1191 (1970).
11. N.B. Narozhnyi and A.I. Nikishov, *Yad. Fiz.* **11**, 1072 (1970) [*Sov. J. Nucl. Phys.* **11**, 596 (1970)].
12. V. S. Popov, *Sov. Phys. JETP* **34**, 709 (1972); *Sov. Phys. JETP* **35**, 659 (1972); V. S. Popov and M. S. Marinov, *Yad. Fiz.* **16**, 809 (1972) [*Sov. J. Nucl. Phys.* **16**, 449 (1973)].
13. S.P. Kim and D.N. Page, *Phys. Rev. D* **65**, 105002 (2002); *Phys. Rev. D* **63**, 065020 (2006).
14. G.V. Dunne and C. Schubert, *Phys. Rev. D* **72**, 105004 (2005).
15. R.P. Feynman, *Phys. Rev.* **80**, 440 (1950).
16. R.P. Feynman, *Phys. Rev.* **84**, 108 (1951).
17. I.K. Affleck, O. Alvarez, and N.S. Manton, *Nucl. Phys. B* **197**, 509 (1982).
18. A.I. Nikishov, *Sov. Phys. JETP* **30**, 660 (1970).
19. S.L. Lebedev and V.I. Ritus, *Sov. Phys. JETP* **59**, 237 (1984) [*Zh. Eksp. Teor. Fiz.* **86**, 408 (1984)].
20. M. Reuter, M.G. Schmidt, and C. Schubert, *Ann. Phys. (N.Y.)* **259**, 313 (1997).
21. G.V. Dunne and C. Schubert, *Fifth workshop on quantum field theory under the influence of external conditions, Leipzig 2001*, edited by K. Bordag, *Int. J. Mod. Phys. A* **17**, 956 (2002).
22. A. M. Polyakov, *Gauge Fields and Strings*, Harwood Publ., Chur, 1987.
23. Z. Bern and D.A. Kosower, *Nucl. Phys. B* **379**, 451 (1992).
24. M. J. Strassler, *Nucl. Phys. B* **385**, 145 (1992).
25. D. Fliegner, P. Haberl, M.G. Schmidt, and C. Schubert, *Ann. Phys. (N.Y.)* **264**, 51 (1998).
26. C. Schubert, *Phys. Rept.* **355**, 73 (2001).
27. C. Alexandrou, R. Rosenfelder, and W. Schreiber, *Phys. Rev. D* **62** 085009 (2000).
28. H. Gies and K. Langfeld, *Nucl. Phys. B* **613**, 353 (2001).
29. M.G. Schmidt and I.O Stamatescu, *Mod. Phys. Lett.* **18**, 1499 (2003).
30. H. Gies and K. Klingmüller, *Phys. Rev. D* **72**, 065001 (2005).
31. G.V. Dunne and C. Schubert, *IX Mexican Workshop on Particles and Fields, Colima 2003*, arXiv:hep-th/0409021.
32. G.V. Dunne, H. Gies, C. Schubert, and Q.-H. Wang, *Phys. Rev. D* **73**, 065028 (2006) (completed after the workshop).

Two time physics and Hamiltonian Noether theorem for gauge systems

J. A. Nieto[†,][1] L. Ruiz[†], J. Silvas[†*] and V. M. Villanueva[‡,2]

*[†]Escuela de Ciencias Fisico-Matemáticas
Universidad Autónoma de Sinaloa
Culiacán, Sinaloa, México
[†‡]Instituto de Física y Matemáticas
Universidad Michoacana de San Nicolás de Hidalgo
P.O. Box 2-82, Morelia, Michoacán, México

Abstract. Motivated by two time physics theory we revisited the Noether theorem for Hamiltonian constrained systems. Our review presents a novel method to show that the gauge transformations are generated by the conserved quantities associated with the first class constraints.

Keywords: Noether Theorem; Gauge Systems, Two Time Physics
PACS: 04.60.-m, 04.65.+e, 11.15.-q, 11.30.Ly

1.- INTRODUCTION

The purpose of this brief note is to review the main results of Ref. [1]. The key idea in such a reference is to obtain Noether's first and second theorems [2] for gauge systems in the Hamiltonian sector (see Refs. [3]-[6]). It is proved that when the Noether theorem is applied, the conserved quantities can be identified precisely with the first class constraints. It is important to emphasize that these results are derived avoiding the Lagrangian sector in the sense of Gràcia and Pons [7]-[9] construction and focusing completely on the Hamiltonian sector [10]-[14].

In order to motivate our approach we first discuss two examples: two time physics and the Friedberg *et al.* model [15] (see also Refs. [16]-[17]). And then we proceed to generalize our formalism putting special emphasis in the connection between the first class constraints and the conserved quantities.

Two time physics [18]-[23] offers an interesting example for discussing our formalism because in this case the variables q^i and p^j are unified in just one object x_a^i, with $a = 1, 2$, where $x_1^i \equiv q^i$ and $x_2^i \equiv p^i$, and consequently in the corresponding action the hidden symmetry $Sp(2,R)$ or $SL(2,R)$ becomes manifest (see Refs. [24]). Thus, we show that our formalism shed some new light on this hidden symmetry.

---

[1] E-mail: nieto@uas.uasnet.mx
[2] E-mail: vvillanu@ifm.umich.mx

This work is organized as follows. In Section 2, we discuss two times physics. In Sections 3, we describe the helix model of Friedberg *et al.* In Section 4, we generalize our procedure Finally, in Section 5 we make some final remarks.

2.- TWO TIME PHYSICS

Two time physics is a proposal that has been reconsidered by many theoretical physicist in last years. The main motivation for considering this theory is that it can be used to obtain various dynamical systems in one-time physics from the same action, through gauge fixing, providing a more unified point of view of one-time dynamics in a higher dimensional theory which is achieved by introducing new gauge symmetries in order to insure unitarity, causality and absence of ghosts [18]-[23].

The action for two time physics is given by [18] (see also Refs. [19]-[23])

$$S = \int_{\tau_i}^{\tau_f} d\tau \left(\frac{1}{2} \varepsilon^{ab} \dot{x}_a^\mu x_b^\nu \eta_{\mu\nu} - H_T(x_a^\mu) \right), \quad (1)$$

where the overdot means total derivative with respect to the parameter τ, $\mu = 0,1,2,3,4$, H_T denotes the total Hamiltonian and the symbol x_a^μ ($a = 1,2$) is defined as

$$x_1^\mu = q^\mu, \quad x_2^\mu = p^\mu. \quad (2)$$

It can be seen that the first term in the action (1), which up to a total derivative is equivalent to $\dot{q}^\mu p_\mu$, has the manifest $Sp(2,R)$ (or $SL(2,R)$) invariance. It turns out that the simplest possible choice for H_T which maintains the symmetry $Sp(2,R)$ in the massless case is

$$H_T = \frac{1}{2} \lambda^{ab} x_a^\mu x_b^\nu \eta_{\mu\nu}, \quad (3)$$

where $\lambda^{ab} = \lambda^{ba}$ is a Lagrange multiplier. One could think about an extension to consider the "massive" case as

$$H_T = \frac{1}{2} \lambda^{ab} (x_a^\mu x_b^\nu \eta_{\mu\nu} + m_{ab}^2), \quad (4)$$

with $m_{11}^2 = -R^2$, $m_{22}^2 = m_0^2$ and $m_{12}^2 = 0$, but it turns out that the mass term m_{ab}^2 breaks the $Sp(2,R)$-symmetry as shown by considering (4) in the action (1) from where arbitrary variations on λ^{ab} throw out the constraint

$$\Omega_{ab} = x_a^\mu x_b^\nu \eta_{\mu\nu} + m_{ab}^2 = 0, \quad (5)$$

which turns out to be first class.

In terms of notation (2), expression (5) is equivalent to the following set of equations (see Ref. [24])

$$q^\mu q_\mu - R^2 = 0,$$

$$q^\mu p_\mu = 0, \qquad (6)$$

$$p^\mu p_\mu + m_0^2 = 0,$$

but this set of equations does not hold if q^μ and p^μ are interchanged. So, if we are interested in maintaining the $Sp(2,R)$-symmetry for the entire action one must consider $m_{ab} = 0$, but in this case one can observe that if $\eta_{\mu\nu}$ corresponds to just one time, that is, if $\eta_{\mu\nu}$ has the signature $\eta_{\mu\nu} = diag(-1, 1, ..., 1)$ then from (7) it follows that p^μ is parallel to q^μ and this implies that the angular momentum

$$L^{\mu\nu} = q^\mu p^\nu - q^\nu p^\mu \qquad (7)$$

associated with the Lorentz symmetry of (2) should vanish, which is, of course, an unlikely result. This observation is one of the reasons that support two time physics theory. Thus, if we want to consider a system for which $L^{\mu\nu} \neq 0$, keep the constraints (7) and solve the problem without ghosts it will be consistent only in two times physics (see Refs. [18]-[23]).

With these observations at hand, we proceed to consider the total Hamiltonian H_T of the form

$$H_T = H(x_a^\mu; \tau) - \lambda^{bc}(\tau)\phi_{bc}(x_a^\mu; \tau), \qquad (8)$$

where $\phi_{bc}(=\phi_{cb})$ denotes a generalization of the first class constraint Ω_{ab} (see expression (5)).

Now consider the following coordinate transformations

$$\delta\tau = \tau'(\tau) - \tau,$$
$$\delta_\star x_a^\mu = x_a^{\prime\mu}(\tau') - x_a^\mu(\tau) = \delta x_a^\mu + \dot{x}_a^\mu \delta\tau, \qquad (9)$$
$$\delta_\star \lambda_{ab} = \lambda'_{ab}(\tau') - \lambda_{ab}(\tau) = \delta\lambda_{ab} + \dot{\lambda}_{ab}\delta\tau,$$

where $\delta x_a^\mu = x_a^{\prime\mu}(\tau) - x_a^\mu(\tau)$, $\delta\lambda_{ab} = \lambda'_a(\tau) - \lambda_a(\tau)$. Note that $\delta_\star \dot{x}_a^\mu = \delta \dot{x}_a^\mu + \ddot{x}_a^\mu \delta\tau$.

Under transformations (9) the total variation of the action (2) with H_T given by (8) is

$$\delta_\star S = \int_{\tau_i}^{\tau_f} d\tau \left\{ \frac{d}{d\tau}\left[\frac{1}{2}\varepsilon^{ab}\delta_\star x_a^\mu x_b^\nu \eta_{\mu\nu} - \delta\tau H_T\right] + \varepsilon^{ab}\dot{x}_a^\mu \delta x_b^\nu \eta_{\mu\nu} \right. \qquad (10)$$
$$\left. + \delta\tau \dot{H}_T - \delta_\star H_T \right\}$$

but due to action invariance under transformations then (10) can be written as

$$\delta_\star S = \int_{\tau_i}^{\tau_f} d\tau \frac{d}{d\tau}\delta_\star \Lambda(x_a^\mu; \tau), \qquad (11)$$

where $\Lambda(x_a^\mu; \tau)$ is an arbitrary function. Now, if one define the quantity $Q = Q(x_a^\mu; \tau)$ as

$$Q = \frac{1}{2}\varepsilon^{ab}\delta_\star x_a^\mu x_b^\nu \eta_{\mu\nu} - \delta\tau H_T - \delta_\star \Lambda \qquad (12)$$

and observing that $\delta\tau\dot{H}_T - \delta_*H_T = \delta H_T$, then (10) and (12) implies

$$\int_{\tau_i}^{\tau_f} d\tau \left\{ \frac{d}{d\tau}Q + \varepsilon^{ab}\dot{x}_a^\mu \delta x_b^\nu \eta_{\mu\nu} - \delta H_T \right\} = 0. \tag{13}$$

From this expression is not difficult to show that Q is a conserved quantity when equations of motion hold, that is when

$$\dot{x}_a^\mu = \varepsilon_{ab} \frac{\partial}{\partial x_{b\mu}} H_T. \tag{14}$$

Conversely, when equations of motion (14) are not satisfied Q becomes to be the generator of canonical transformations, namely

$$\delta x_a^\mu = \{x_a^\mu, Q\} \tag{15}$$

and

$$\delta_* H_T = \frac{\partial}{\partial t} Q. \tag{16}$$

In order to proceed further, observe that in terms of the symbol x_a^μ the Poisson brackets for any canonical functions $f(x_a^\mu)$ and $g(x_a^\mu)$ can be written as

$$\{f, g\} = \varepsilon_{ab} \frac{\partial f}{\partial x_a^\mu} \frac{\partial g}{\partial x_{b\mu}}. \tag{17}$$

From this expression it follows that

$$\{x_a^\mu, x_b^\nu\} = \varepsilon_{ab}\eta^{\mu\nu}, \tag{18}$$

and for the constraint Ω_{ab} given in (5) it is straightforward to check that

$$\{\Omega_{ab}, \Omega_{cd}\} = C_{abcd}^{ef} \Omega_{ef}, \tag{19}$$

which establishes that Ω_{ab} is in fact a first class constraint. The coefficients C_{abcd}^{ef} are called structure constants and are given by

$$C_{abcd}^{ef} = \tfrac{1}{2}[\varepsilon_{ac}(\delta_b^e \delta_d^f + \delta_d^e \delta_b^f) + \varepsilon_{ad}(\delta_b^e \delta_c^f + \delta_c^e \delta_b^f) \\ + \varepsilon_{bc}(\delta_a^e \delta_d^f + \delta_d^e \delta_a^f) + \varepsilon_{bd}(\delta_a^e \delta_c^f + \delta_c^e \delta_a^f)]. \tag{20}$$

Now consider the quantity Q as

$$Q = \xi^{ab}(\tau)\Omega_{ab}, \tag{21}$$

where $\xi^{ab} = \xi^{ba}$ are infinitesimal parameters. When equations of motion are not satisfied we can use the formulas (15) and (16) to obtain that the transformations generated by the constraint Ω_{ab} are

$$\delta x_a^\mu = \varepsilon_{ab} \xi^{bc} x_c^\mu \tag{22}$$

and

$$\delta\lambda^{ab} = \dot{\xi}^{ab} - \xi^{ef}\lambda^{cd}C^{ab}_{efcd}. \qquad (23)$$

We recognize in the expression (22) the infinitesimal transformation associated with the group $Sp(2,R) \cong SL(2,R)$ with infinitesimal parameter $\varsigma^c_a = \varepsilon_{ab}\xi^{bc}$. Thus, we have proved that if the Lagrange multipliers variation $\delta\lambda^{ab}$ is given by (23) then the action (1) is invariant under the $Sp(2,R)$ gauge transformation (22). The remarkable fact is that $Sp(2,R)$ invariance of the action (1) is generated by the conserved quantity (21) corresponding to the first class constraint Ω_{ab}.

3.- THE FRIEDBERG ET AL. MODEL.

The helix model of Friedberg *et al.* [15] (see also Refs. [16] and [17]) can be described in terms of the fundamental Hamiltonian first order action:

$$S = \int_{t_i}^{t_f} dt \left[\dot{x}p_x + \dot{y}p_y + \dot{z}p_z - \tfrac{1}{2}\left(p_x^2 + p_y^2 + p_z^2\right) \right.$$

$$\left. -U(x,y) - [\lambda p_z + gxp_y - yp_x] \right], \qquad (24)$$

where (x,y,z) and (p_x,p_y,p_z) stand for the three dimensional coordinates and canonical momenta respectively. In (24) $U(x,y) = U\left(x^2+y^2\right)$, and λ is a Lagrange multiplier associated with the first class constraint $\phi = p_z + g(xp_y - yp_x)$, where g denotes a coupling constant.

It is easy to verify that this action is invariant under the infinitesimal gauge transformations

$$\delta x = -\alpha y, \quad \delta y = +\alpha x, \quad \delta z = +\tfrac{1}{g}\alpha,$$
$$\delta p_x = -\alpha p_y, \quad \delta p_y = +\alpha p_x, \quad \delta p_z = 0, \qquad (25)$$

and

$$\delta\lambda = +\frac{1}{g}\dot{\alpha}, \qquad (26)$$

where $\alpha = \alpha(t)$ is a transformation parameter. Note that the variation of the action is exactly zero so there is no need for the surface term. Let us introduce the quantity

$$Q = \delta x^i p_i,$$
$$= p_z + g(xp_y - yp_x). \qquad (27)$$

As we can see Q corresponds to the first class constraint of the physical system whose motion is governed by the action (24).

4.- GENERALIZATION

The above two examples provide a motivation for a generalization. For this purpose let us first rewrite the action (1) in the form

$$S[q,p;\lambda^\alpha] = \int_{t_i}^{t_f} dt \left[\dot{q}^i p_i - H_T\right], \qquad (28)$$

where

$$H_T = H(q,p;t) + \lambda^\alpha(t)\phi_\alpha(q,p;t) \qquad (29)$$

denotes de total Hamiltonian. Our aim is to see the consequences of applying to the action (28) the total variations:

$$\delta t = t'(t) - t,$$

$$\delta_\star q^i = q'^i(t') - q^i(t) = \delta q^i + \dot{q}^i \delta t,$$

$$\delta_\star p_i = p'_i(t') - p_i(t) = \delta p_i + \dot{p}_i \delta t, \qquad (30)$$

$$\delta_\star \lambda^\alpha = \lambda'^\alpha(t') - \lambda^\alpha(t) = \delta\lambda^\alpha + \dot{\lambda}^\alpha \delta t,$$

where $\delta q^i = q'^i(t) - q^i(t)$ and similar expressions hold for δp_i and $\delta \lambda^\alpha$. Also $\delta_\star \dot{q}^i = \delta \dot{q}^i + \ddot{q}^i \delta t$.

It is important to remark that $\delta \dot{q}^i = \frac{d}{dt}\delta q^i$ but $\delta_\star \dot{q}^i \neq \frac{d}{dt}\delta_\star q^i$.
Invariance of the action (28) under total variations implies

$$\delta_\star S = \int_{t_i}^{t_f} dt\, \delta_\star[\dot{q}^i p_i - H_T] + \int_{t_i}^{t_f} dt\, \frac{d\delta t}{dt}[\dot{q}^i p_i - H_T]$$

$$= \int_{t_i}^{t_f} dt\, \frac{d}{dt} \delta_\star \Lambda(q,p). \qquad (31)$$

where $\Lambda(q,p)$ is an arbitrary function.

It is straightforward to show that, in virtue of definitions of the total variations (30) and defining the quantity $Q = Q(q,p;t)$ as

$$Q = \delta_\star q^i p_i - \delta t H_T - \delta_\star \Lambda, \qquad (32)$$

relation (31) leads to

$$\int_{t_i}^{t_f} dt \left\{ \frac{d}{dt} Q + \dot{q}^i \delta p_i - \dot{p}_i \delta q^i - \delta H_T \right\} = 0. \qquad (33)$$

Let us define the quantities

$$A = \int_{t_i}^{t_f} dt \frac{d}{dt} Q \tag{34}$$

and

$$B = \int_{t_i}^{t_f} dt \{\dot{q}^i \delta p_i - \dot{p}_i \delta q^i - \delta H_T\}. \tag{35}$$

Expression (33), offers three different possibilities, namely
  (i) If $A = 0$ then (33) implies that $B = 0$.
  (ii) If $B = 0$ then (33) implies that $A = 0$.
  (iii) If neither A nor B are zero then (33) establishes that they are related by $A + B = 0$.
The first two cases are well known, and it can be easily seen that both are equivalent, but the third one seems to have passed unnoticed. In order to clarify these observations let us briefly discuss each one of these cases. In the first case, we assume that the quantity Q satisfies the expression

$$Q \big|_{t_i}^{t_f} = 0, \tag{36}$$

which is equivalent to say that $A = \int_{t_i}^{t_f} dt \frac{d}{dt} Q = 0$, from where it follows that

$$\int_{t_i}^{t_f} dt \{(\dot{q}^i - \frac{\partial}{\partial p_i} H_T)\delta p_i + (-\dot{p}_i - \frac{\partial}{\partial q^i} H_T)\delta q^i - \delta \lambda^\alpha \phi_\alpha\} = 0 \tag{37}$$

and since variations δq^i, δp_i and $\delta \lambda^\alpha$ are arbitrary, then (37) implies the following equations of motion:

$$\begin{aligned} \dot{q}^i &= \frac{\partial}{\partial p_i} H_T = \{q^i, H_T\}, \\ \dot{p}_i &= -\frac{\partial}{\partial q^i} H_T = \{p_i, H_T\} \end{aligned} \tag{38}$$

and

$$\phi_\alpha = 0. \tag{39}$$

Here the symbol $\{f, g\}$, for any functions f and g of the canonical variables q^i and p_i, stands for the usual Poisson bracket, that is

$$\{f, g\} = \frac{\partial f}{\partial q^i} \frac{\partial g}{\partial p_i} - \frac{\partial g}{\partial q^i} \frac{\partial f}{\partial p_i}. \tag{40}$$

In the second case, we assume that the dynamical system satisfies equations of motion (39) and (40). This means that (37) follows, which says that $B = 0$. Therefore from (33) we see that

$$\int_{t_i}^{t_f} dt \frac{d}{dt} Q = 0. \tag{41}$$

Since the interval $t_f - t_i$ does not have been defined, from (41) we have $\frac{d}{dt} Q = 0$ and therefore we find that Q is a conserved quantity.

The last possibility arises if we assume that neither (37) nor (41) hold, that is, we assume that A and B are different from zero. We shall show that in this case expression (33) implies that Q is the generator of canonical transformations. For this purpose let us first compute $\frac{d}{dt}Q$. Since $Q = Q(q,p;t)$ hence

$$\frac{d}{dt}Q = \frac{\partial Q}{\partial q^i}\dot{q}^i + \frac{\partial Q}{\partial p_i}\dot{p}_i + \frac{\partial Q}{\partial t}. \tag{42}$$

Thus, for an undefined interval $t_f - t_i$, (33) can be rewritten as

$$\int \omega = 0. \tag{43}$$

with the quantity ω defined as

$$\omega = \left(\frac{\partial Q}{\partial q^i} + \delta p_i\right)dq^i + \left(\frac{\partial Q}{\partial p_i} - \delta q^i\right)dp_i + \left(\frac{\partial Q}{\partial t} - \delta H_T\right)dt. \tag{44}$$

From (44) we observe that ω may admit an interpretation of 1-form. Thus, under usual assumptions (43) implies that ω is an exact form which means that

$$\omega = df, \tag{45}$$

where f is an arbitrary zero-form.

We shall assume that $f = f(q,p)$. From (44) and (43) we see that

$$\frac{\partial Q}{\partial t} - \delta H_T = 0. \tag{46}$$

Considering (46), the expressions (43) and (45) yield

$$\left(\frac{\partial Q'}{\partial q^i} + \delta p_i\right)dq^i + \left(\frac{\partial Q'}{\partial p_i} - \delta q^i\right)dp_i = 0, \tag{47}$$

where

$$Q' = Q + f. \tag{48}$$

Since, dq^i and dp_i are 1-form bases then (47) implies

$$\delta q^i = \frac{\partial Q'}{\partial p_i} = \{q^i, Q'\} \tag{49}$$

and

$$\delta p_i = -\frac{\partial Q'}{\partial q^i} = \{p_i, Q'\}. \tag{50}$$

Thus, we have shown that up to an arbitrary function f the quantity Q, which is a conserved quantity when the equations of motion are satisfied, is the generator of canonical transformations.

In order to clarify the meaning of expression (46), we investigate the consequences of invariances under gauge transformations, *i.e.*, we consider the particular case

$$Q' = \xi^\alpha(t)\phi_\alpha(q,p;t), \tag{51}$$

where the quantities $\xi^\alpha(t)$ are infinitesimal parameters associated with the first class constraints $\phi_\alpha(q,p;t)$. Moreover; since we are dealing (by assumption) with only first class constraints, we can write (see Refs. [4] and [25])

$$\{H, \phi_\alpha\} = V_\alpha^\beta \phi_\beta \tag{52}$$

and

$$\{\phi_\alpha, \phi_\beta\} = C_{\alpha\beta}^\gamma \phi_\gamma, \tag{53}$$

where V_α^β and $C_{\alpha\beta}^\gamma$ are structure "constants". Then, (46), (50)-(52) lead to

$$\delta\lambda^\alpha \phi_\alpha = \left(\dot{\xi}^\alpha - \xi^\beta V_\beta^\alpha - \xi^\beta \lambda^\gamma C_{\beta\gamma}^\alpha\right)\phi_\alpha, \tag{54}$$

and by considering the constraints $\phi_\alpha(q,p;t)$ as independent functions this expression implies the result

$$\delta\lambda^\alpha = \dot{\xi}^\alpha - \xi^\beta V_\beta^\alpha - \xi^\beta \lambda^\gamma C_{\beta\gamma}^\alpha, \tag{55}$$

which describes the usual transformations of the Lagrange multipliers λ^α under gauge transformations generated by first class constraints (see Refs. [4], [5] and [25]).

5.- FINAL REMARKS

In this work we revisited two time physics and the Friedberg *et al.* model using fundamental constrained Hamiltonian formalism. We proved that our method may reveal hidden symmetries in specific cases. Since in two time physics the phase space has a unified character in the sense that the spacetime and the momentum space are put together at the same level, we found that an application of our formalism in this context requires a generalization of the usual Noether's procedure. Using this generalization we showed that the gauge transformations for the coordinates and momenta also exhibit a unified character. Further, we also proved that the gauge transformations are generated by the conserved quantities associated with the first class constraints.

ACKNOWLEDGMENTS

We would like to thank I. Bars, J. M. Pons, L. Lusanna, J. L. Lucio-Martinez and H. Villegas for helpful comments. V. M. V. whishes to thank financial support from Universidad Michoacana through Coordinación de Investigación Científica CIC 4.14 and CONACyT project 38293-E.

REFERENCES

1. V. M. Villanueva, J. A. Nieto, L. Ruiz, J. Silvas, J. Phys. A **38**, 7183 (2005); hep-th/0503093.
2. A. E. Noether, Nachr. König. Gesell. Wissen. Göttingen, Math. Phys. Kl, 235 (1918).
3. P. A. M. Dirac, *Lectures on Quantum Mechanics* (New York: Yeshiva UP, 1964).
4. J. Govaerts, *Hamiltonian Quantisation and Constrained Dynamics* (Leuven University Press, Leuven, 1991).
5. M. Henneaux and C. Teitelboim, *Quantization of Gauge Systems* (Princeton University Press, Princeton, New Jersey, 1992).
6. A. Hanson, T. Regge and C. Teitelboim, *Constrained Hamiltonian Systems* (Accademia Nazionale dei Lincei, Roma, 1976).
7. X. Gràcia and J. M. Pons, J. Phys. A: Math. Gen. **25**, 6357 (1992).
8. C. Batlle, J. Gomis, J. M. Pons and N. Roman, J. Math. Phys **27**, 2953 (1986).
9. J. A. García and J. M. Pons, Int. J. Mod. Phys. A**16**, 3897 (2001); hep-th/0012094
10. L. Lusanna, Int. J. Mod. Phys. A8, 4193 (1993).
11. L. Lusanna, Riv. Nuovo Cim. 14, n3, 1 (1991).
12. L. D. Faddeev, Phys. Rev. Lett. **60**, 1692 (1988).
13. L. Castellani, Ann. Phys. **143**, 357 (1982).
14. N. Mukunda Phys. Scripta. **21**, 783 (1980).
15. R. Friedberg, T. D. Lee, Y. Pang and H. C. Ren, Ann. Phys. **246** (1996) 381.
16. V. M. Villanueva, J. Govaerts and J. L. Lucio-Martinez, J. Phys. A**33**, 4183 (2000).
17. V. M. Villanueva, J. Govaerts and J. L. Lucio-Martinez, *Quantizing gauge theories without gauge fixing: the physical projector*. Proceedings of the "*International Conference on Quantization, Gauge Fields and Strings: In memory of Profr. Efim Fradkin*", Moscow, Rep. Fed. Russia, June 2000. World Scientific.
18. I. Bars, Class. Quant. Grav. **18** (2001) 3113; hep-th/0008164.
19. I. Bars, Phys. Rev. D **64**, 126001 (2001); hep-th/0106013.
20. I. Bars and C. Deliduman, Phys. Rev. D **64**, 045004 (2001); hep-th/0103042.
21. I. Bars, Phys. Rev. D **62**, 046007 (2000); hep-th/0003100.
22. I. Bars, AIP Conf.Proc.767:3-27,2005 Also in *Wroclaw 2004, Fundamental interactions and twistor-like methods* 3-27; hep-th/0502065.
23. I. Bars and Soo-Jong Rey, Phys. Rev. D **64**, 046005 (2001); hep-th/0104135.
24. J. M. Romero and A. Zamora, Phys. Rev. D **70**, 105006 (2004); hep-th/0408193.
25. M. Henneaux, Phys. Rep. **126**, (1985).

Search for Gamma Ray Bursts at Sierra Negra, México

H. Salazar*, C. Alvarez*, O. Martinez* and L. Villaseñor[†]

Facultad de Ciencias Fisico-Matematicas, BUAP, Puebla Pue., 72570, México.
[†]*Institute of Physics and Mathematics, University of Michoacan, Edificio C3, Ciudad Universitaria, Morelia, Mich., 58040, México.*

Abstract. We present results from a search for GRBs in the energy range from tens of GeVs to one TeV with an array of 4 water Cherenkov detectors located at 4550 m a.s.l. as part of the high mountain observatory of Sierra Negra (N18°59.1, W97°18.76) near Puebla city in México. The detectors consist of light-tight cylindrical containers of 1 m² and 4 m² cross section filled with purified water; they are spaced 25 m and have a 5" photomultiplier (EMI model 9030A) facing down along the cylindrical axis. We report the measured rates of the electromagnetic and mounic components of the background as the photon estimated flux.

INTRODUCTION

Discovered by military satellites in the 60's and more properly studied until 1991, when NASA launched the *Compton Gamma-Ray Observatory* (*CGRO*), Gamma Ray Bursts (GRBs) are probably the most energetic phenomena in the Universe lasting from few milliseconds to some hundreds of seconds. The origin of GRBs has been associated with the collapse of two neutron stars if they last less than 2 seconds (short-GRBs) or with the explosion of special type of supernovae called hipernova (M>15 solar masses) if they last more than 2 seconds (long-GRBs). GRBs are gamma ray explosions that can liberate up to 10^{53} ergs in about one second. In order to detect and study GRBs, *CGRO* carried onboard 4 instruments: *BATSE, EGRET, COMPTEL* and *OSSE*. These instruments were able to detect gamma-ray photons at different energy ranges; in particular, *BATSE* detected more than 2700 GRBs with photon energies in the range from 20 KeV to 1 MeV, showing a bimodal duration distribution that clearly indicates the existence of two populations of GRBs. The spacial distribution of GRBs events in the sky showed that they do not have a prefer direction. They are observed coming from any direction at a rate of about 1 per day, indicating its cosmological origin as the red-shift measurements proved.

On the other hand, 7 GRB events were observed with photon energies greater than 30 MeV by *EGRET*, with 6 of them with photon energies greater than 1 GeV. The event named GRB940217 had the highest energy photon, 18 GeV [8].

It is important to mention that so far *BATSE* and *EGRET* have not observed a cut-off in the GRB energy spectrum; this suggests that the spectrum may extend up to high energy components, with TeV photons, or even greater as some models predict [5, 14]. GRBs have been very well studied in the range from KeV to MeV by the *CGRO* and *BEPPO-Sax* missions, however the high energy component from GeV to TeV is still unknown.

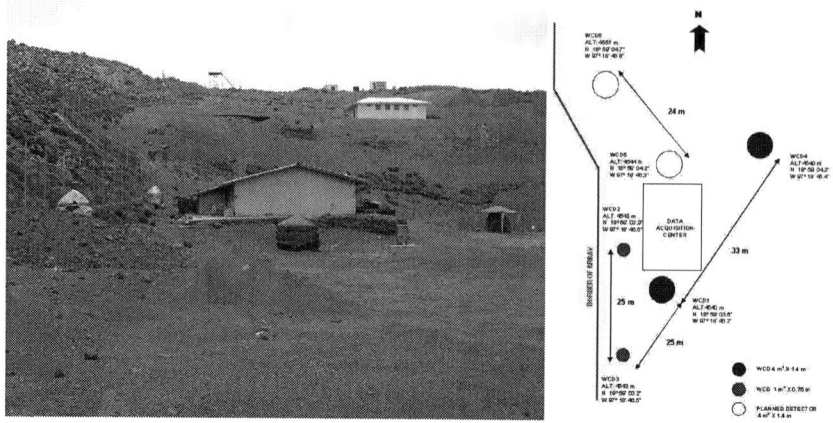

FIGURE 1. Sierra Negra Array site, showing 4 light-tight cylindrical water Cherenk. The tanks are 4 *m* and 1 *m* cross section. The empty circles represent two 4 m cross sections tanks to be deploy in future in order to measure the coincidence events.

The *GLAST* mission will be launched with this purpose next year.

In contrast, we are interested in detecting GRBs with energies in the 10 GeV to 10 TeV range using a ground-based detector array. This array is operating in a single-particle counting and coincidences modes. We describe this water Cherenkov detector array located at the high mountain observatory of Sierra Negra and we also describe the array's capabilities in comparison with other ground-based observatories.

GROUND-BASE EXPERIMENTS

Since gamma rays coming from outside the Earth cannot penetrate the atmosphere, it is necessary to use detectors on balloons or satellites to detect them directly. In addition, as photon energies increase, the photon fluxes decrease as a power law. Therefore, in order to detect small fluxes of gamma radiation or high energy photons in the range of GeV to TeV is necessary to construct more sensitive detectors with larger areas. Satellite- borne detectors with large collecting areas become impractical due to their cost. However, with inexpensive ground-based experiments of large area, it is possible to detect the relativistic secondary particles produced by the interaction of GeV or TeV gamma-ray photons with the nuclei of the upper atmosphere.

Currently, there exists a handful of ground-based experiments around the world searching GRBs: Chacaltaya at 5200 m a.s.l. in Bolivia [6]; Argo at 4300 m a.s.l. in Tibet [4], China; Milagro at 2630 m a.s.l. in New Mexico [11], USA; the Pierre Auger Observatory at 1400 m a.s.l. in Malargüe, Argentina [1] and Sierra Negra at 4550 m a.s.l. in México. Of all these experiments only the prototype of Milagro called Milagrito has reported the possible detection of signals associated to a GRB, GRB 970417 [2]. Milagro is the largest area (60 m×80 m) water Cherenkov detector capable of continuously monitoring the sky at energies between 10 GeV and 100 TeV. Although the Pierre Auger

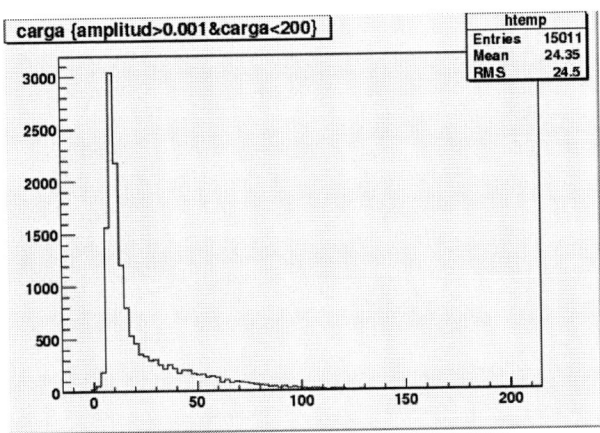

FIGURE 2. Notice that the peak is due to the electromagnetic component of the air shower which is the dominant component. This result is in agreement with Vernetto's prediction [13]. The data were taken at Sierra Negra site.

Observatory was designed to study ultra high energy cosmic rays, it is also a competitive high energy GRB ground-based detector due to its large area and the good sensitivity to photons of its water Cherenkov detectors [1].

Sierra Negra Experiment

The high mountain array prototype of Sierra Negra is located near Puebla city, México, at 4550 m a.s.l. At present, the array consists of 2 cylindrical light-tight water Cherenkov detectors of 1 m cross section separated by 25 m and another two detectors of 4 m cross section added 15 m and 33 m away as it is shown in the Fig. 1. The two latter tanks were added to allow the possibility of detecting secondary particles in coincidence between them. The tanks are covered internally with tyvek as UV-reflector and filled with ultra-pure water (5 μsiemens resistivity) and all of them contain a PMT (EMI 9030A) to collect the Cherenkov light produced in the water. The PMT signals are read out by a DAQ system located in a central shelter, that measures the rates of secondary particles each tenth of second and send it to a PC to their storage and further analysis. The weather conditions are taken every 10 seconds by an Ventage Pro2 whether station that is property of Instinstuto Nacional de Astrofisica, Optica y Electronica (INAOE).

Sensitivity of Sierra Negra to GRBs

GRBs can be detected with ground-based detectors if the secondary particles produced by their interactions with the atmosphere give rise to an excess in the counting rate significantly larger than the statistical fluctuations of the background rate. The method of counting every single particle that hits the tank is known as single particle technique [13]. It is important to mention that any observed counting excess due to GRBs should be temporally coincident with the detection by another cherenkov detector located at high altitude, hundreds or thousands of kilometros away from it as the Large Aperture Gamma Ray Bursts Observatory (LAGO) will work [9]. It is also needed to confirm the GRB event by one of the satellite experiments that observe a common part of the sky, for example *SWIFT* [12]. In this way any other background processes that give rise to particle counting excesses are discarded as for example electric storms.

A GRBs is detected with a statistical significance of n standard deviations if [13]

$$N_s/\sigma_b > n, \qquad (1)$$

where N_s is the signal detected by the array. This signal is proportional to the area and to the flux of secondary particles; σ_b is the background noise and it is proportional to the square root of all the secondary particles produced by cosmic rays. In general, n is taken as 4. The background consisting of all the secondary particles produced by cosmic rays entering into the terrestrial atmosphere varies with altitude as shown in [13]. Then, for the altitude of Sierra Negra, the background rate of charge particles and photons is ≈ 1600 part m^{-2} s^{-1} and ≈ 4000 photons m^{-2} s^{-1} respectively. These components represent the dominant component of the shower. This prediction is in agreement with the data taken at Sierra Negra, where the electromagnetic component is seen as a large peak while the muonic component is lost in the left wing of the charge distribution histogram (Fig. 2). In order to see the muonic component it is necessary to make a cut around 30 mV to discard the electromagnetic component (the peak) as is shown later is this work. It is shown in [13] that knowing the background, we can calculate the minimum flux of particles detectable by an array of a given area, located at a given altitude. It is assumed that the shower is originated by a GRB with a total energy L= 10^{53} ergs and photons of $E > 1$ GeV arriving vertically to the detector array during 1 second. For Sierra Negra, we expect a minimum detectable flux of secondary particles around 50 part m^{-2} s^{-1} for a detector area of 10 m^2. For Chacaltaya in Bolivia, which is the highest altitude array at 5.2 Km a.s.l, the minimum detectable flux is ≈ 26 particles m^{-2} s^{-1} considering an effective area of 48 m^2 and a background of ≈ 2100 particles m^{-2} s^{-1}.

The sensitivity increases strongly with the altitude of observation. Showers generated by primary photons of the same energy increase the size (number of secondary particles) with altitude. As an example, the mean number of particles generated by a photon of 16 GeV at 5200 m is 1 while at 2000 m (altitude of the EAS-TOP experiment) is only 0.03, i.e., the sensitivity at Chacaltaya is better even though its detecting area (48 m^2) is considerable smaller than that of EAS-TOP (350 m^2). The Sierra Negra array is sensible to GRBs of energies E< 160 GeV with the present detectors using the single-particle technique which is represents our first approach to detecting GRBs.

On the other hand, due to the geographic coordinates of Sierra Negra (N 18°59.1, W 97°18.76), we have the advantage of the almost zenithal transit of the Crab Nebula everyday. The Crab Nebula is a constant source of gamma rays that can be used as a standard candle for calibration [7]. Therefore, in Sierra Negra we have the possibility to detect it by using a simple method. The method is based on measured the rate of

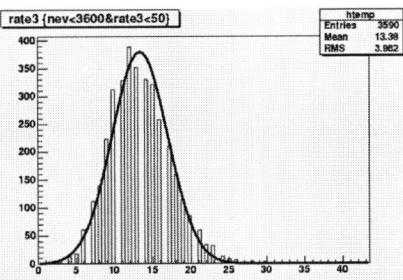

FIGURE 3. Coincidence rate/sec for the 4 water Cherenkov detectors shown in Figure 1. There are about 780 showers per minute.

FIGURE 4. These Figures show the hodoscope with the two layers of iron placed over and under one of the tanks of 4 m cross section.

coincidences in the 4 tanks that forms a symmetric array of triangle shape. The tank placed in the center of this triangle is expected to measure the higher signal while the tanks placed at the vertices are expected to have no signal according to a previous define trigger condition (30 mV or 80 mV). It is assumed that the signal is produced by the second particles induced by a gamma ray impact at the upper layers of the atmosphere. Currently, a prototype array is working a Sierra Negra as it is shown in Fig. 1. With this prototype we have made an estimation of the time it would take to detect the gamma rays coming from the Crab Nebula as follows: First of all, we need to know the photon flux expected from the Crab Nebula when it is located at the zenith of Sierra Negra. Assuming that the source is observed during 4 hours everyday ($\pm 30^o$ from the zenith) with 4 detectors that covers ~ 500 m^2 of area. And knowing the Crab Nebula flux, f=2.68 x $10^{-7} E^{-2.59}$ photons m^{-2} s^{-1} TeV^{-1} [2, 3], we obtain a flux of two photons per day. In other words, we expect to detect two showers per day coming from the Crab Nebula. On the other hand, from the coincidence data of the Sierra Negra array, we are presently detecting around 780 showers per minute (Fig. 3). This means that if we are able to discriminate the 95 % of the muonic component in the showers, and to detect 50% of the photons and reduce the uncertainty in the field of view to $\pm 2^o$ (corresponding to 16 minutes of Crab Nebula observation), we expect to detect the Crab Nebula with a significance of 4 in less than 6 months! Table 1 compare our results with

TABLE 1. Crab Nebula detection by different detectors

	Area m	Altitude meters a.s.l.	(σ)	Detection time
Sierra Negra	~ 500	4550	4	4-6 months
Milagro	4800	2600	5	0.5 years
miniHAWC	22500	>4000	5	2 days

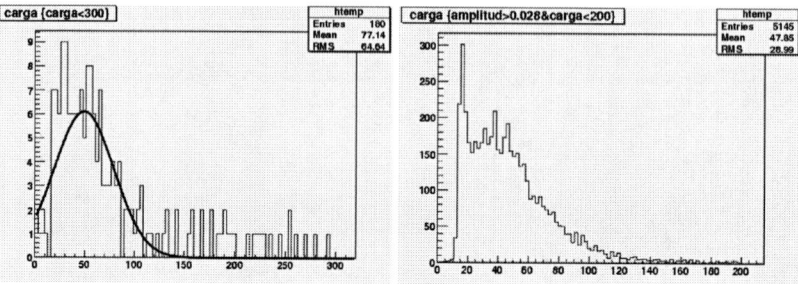

FIGURE 5. Water Cerenkov detector response to muons, electrons and photons. Notice that the vertical muon signal is lost in the Fig. 2.

other experiments.

However, in order to achieve this goal, we need to be able to discriminate out the muonic component. The plot of Fig. 2 and the Fig. 5 show that with a water Cherenkov detector and a fast digitization system we can indeed achieve a separation of the muon component from the electromagnetic one. Further steps planned will optimize the single-particle technique of detection with higher levels of coincidence triggers and a better shower reconstruction. By the moment, one of 4 m tanks deployed on Sierra Negra mountain has been settled with an hodoscope with two layers of iron over and under the tank in order to trigger events containing the vertical muonic component of the extensive air showers (EAS, Fig. 4). A Lab-View program traces the oscilloscope trigger signals corresponding to single muons or isolated electrons, in Fig. 6 it is shown two examples of the trace of a vertical muons triggered at different threshold: 80 mV and 30 mV. The oscilloscope traces provides information about the amplitude, rise time and deposited charge of the signals.

It is worth mentioning that the Crab Nebula has already been detected at high energies by ground-based experiments such as Milagro, located in New Mexico, USA (2630 m a.s.l.) and ARGO, located in the Tibet, China (4300 m a.s.l.). A future Cherenkov water detector called mimiHAWC (a pond of 150 m times 150 m of area) placed above 4000 m a.s.l. of altitude will be able to detect the Crab Nebula gamma ray emission (E>700 GeV) in 2 days with 5σ of significance.

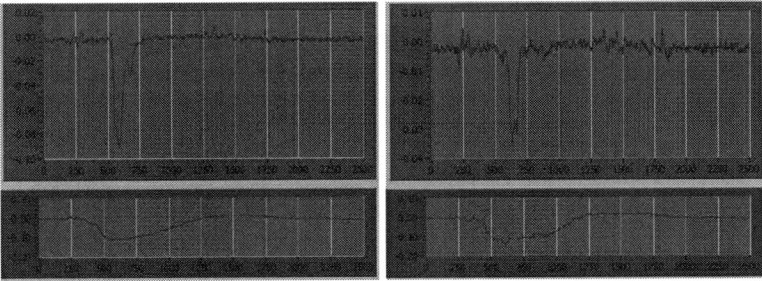

FIGURE 6. Examples of well defined signals of vertical muons at different thresholds. The vertical and horizontal scales are Volts/bin and sec/bin respectively. The Figures at the bottom shown the average pulse.

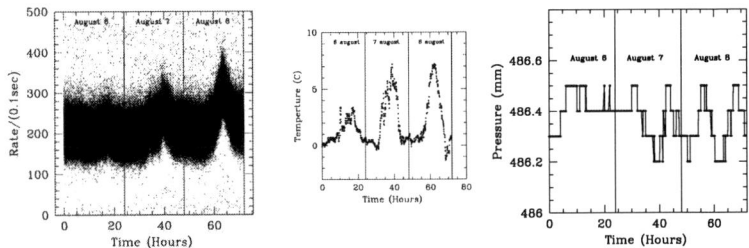

FIGURE 7. Comparison or the background rate for one of the tanks in the array with the atmospheric conditions of temperature and pressure. It is observed that at night time the background rate is stable for all the detectors. The slight increase in the background rate is well correlated with measured changes in temperature and pressure.

Data Analysis and Results

The data were taken from 4 water Cherenkov detectors operating in Sierra Negra, Puebla, México. The background rate measured in 0.1 s intervals versus time for each tank is not constant. The variations in the background rate are mainly due to two factors: the outdoor temperature and the atmospheric pressure, eventually solar activity may also be detectable [10]. Although, we have few atmospheric data at the site of Sierra Negra, we found that the background rates measured with each detector are correlated with temperature and pressure (Fig. 7). However, the time scale of these modulations is much larger than the typical time duration of GRBs and therefore they do not affect our GRB search. In addition, during night time the background rate of detectors is much more stable.

Figure 8 shows the particle rate distribution for the tanks of 4 m^2 and 1 m^2 respectively. The mean rate of the fitted Gaussian for these tanks are 12150 part m^{-2}/s and 2960 part m^{-2} /s. The mean rates of these tanks are in a 4:1 ratio, showing that the expected rate increases in proportion to the physical area of the tanks. The measured

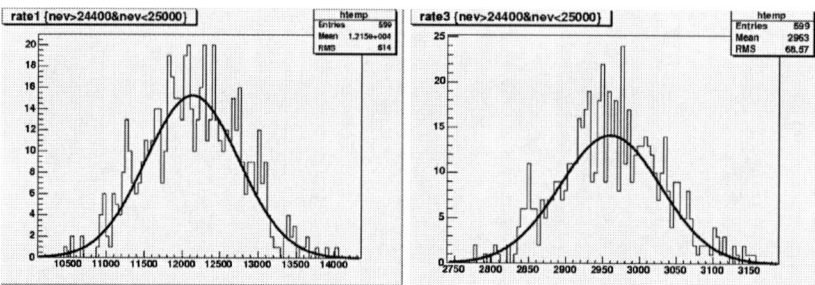

FIGURE 8. Rates of the 4 m^2 (left) and 1 m^2 detectors. The 4:1 ratio holds in a average of 600 seconds period. The rate of 3000 events is due to photons, electromagnetic particles and muons.

background rate in tanks 1 and 3 is about a factor of 0.75 lower than that predicted by Vernetto [13] for the Sierra Negra altitude (about 4000 part m^{-2} s^{-1}). Out of the total background (see [13]), approximately 41 % corresponds to the electromagnetic component (electrons, positrons and photons), 33 % corresponds to the muon component and all the rest is due to the hadronic component. It is also observed that at low altitude places, < 3.5 Km, the muon component is dominant while at higher altitudes the electromagnetic component is dominant. According to our data, 4 standard deviations corresponds to 256 part m^{-2} s^{-1}, then a GRB will be detected if it shows a counting rate excess of at least this number of single particles. Notice that this flux of secondary particles is above the theoretical minimum flux detected by Sierra Negra array (50 part m^{-2} s^{-1}).

Conclusions

From the analysis of data and theoretical calculation, we expect that with a simple array at high altitude and a good method to discriminate the muonic component in the extended air showers we will be able to detect the Crab Nebula in less than 6 months. In addition, GRBs that produce a total energy above 10^{53} erg with photon energies E>160 GeV are also within the detection reach capabilities of Sierra Negra.

Acknowledgments

We thank INAOE, LMT and especially Eduardo Mendoza for all the technical facilities at Sierra Negra that allow us to carry out this work. We thank also Tirso Murrieta, Saúl Aguilar, Gonzalo Pérez and Rubén Conde for helping in the development of electronic devices and deployment of the detectors.

REFERENCES

1. D. Allard et al. *Proceedings of ICRC* (2005)
2. R. Atkins et al. ApJ, **533**, L119, (2000)
3. R. Atkins et al. ApJ, **595**, 803, (2003)
4. C. Bacci et al, A&ASS, **138**, 597 (1999)
5. M. G. Baring, "Towards a Major Atmospheric Cerenkov Detector," Proc. Kruger National Park TeV Workshop, (1997) ed. O. C. de Jager (Wesprint, Pochefstroom)– astro-ph/9711256
6. R. Cabrera et al. A&A **138**, 599, (1999)
7. A. M. Hillas et al., ApJ, **503**, 744, (1998)
8. K. Hurley et al., Nature **372**, 652-654, (1994)
9. Actas del Workshop Astronomia Observacional en Argentina: Problemas y Perspectivas, 2006, P. Benaglia, and S. Cellone, eds. (in spanish)
10. A. Mahrous et al.*Proceedings of ICRC*,3477,(2001)
11. P.M. Saz Parkinson, astro-ph/0505335
12. http://swift.gsfc.nasa.gov/docs/swift/swiftsc.html
13. S. Vernetto Astropart.Phys **13**, 75-86 (2000)
14. M. Vietri Phys.Rev.Lett. **78**, 4328, (1997)

POSTER SESSIONS

Fermions in $d = 1+2$ dimensions from first principles

Ma. Georgina Carrillo-Ruiz and Mauro Napsuciale

Instituto de Física, Universidad de Guanajuato, Lomas del Bosque 103, Fracc. Lomas del Campestre, C.P. 37150, León Gto., México

Abstract. In this work we construct states describing planar electrons ("spin" $\frac{1}{2}$ particles with well defined parity) in $d = 1+2$ from first principles and show that they satisfy Dirac equation, which turns out to be the covariant form of the eigenvalue equation for spatial inversion (parity) just like in $d = 1+3$.

Keywords: Fermion, parity, spin.
PACS: 11.10.Kk, 11.30.Er, 03.65.Pm.

INTRODUCTION

We are interested in studying the implications of space inversion in the definition of spin ½ fermions in the case (1+2)-dimensional, understanding *fermions* as the states of the lowest dimensional irreducible representation (irrep) of the proper orthochronous piece of the homogeneous Lorentz group, $L_+^\uparrow(2)$.

The $L_+^\uparrow(2)$ group is locally isomorphic to $SO(1,2)$. The commutation relations for the generators of $L_+^\uparrow(2)$ are

$$[J, K_1] = iK_2, \quad [J, K_2] = -iK_1, \quad [K_1, K_2] = -iJ, \tag{1}$$

where J is the generator of rotations and $\vec{K} = (K_1, K_2)$ are the boost generators. Besides, the Casimir operator is $C \equiv J^2 - K_1^2 - K_2^2$. This can directly inferred from the mapping $J_1 \to iK_1$, $J_2 \to iK_2$ of the $SO(3)$ algebra. Moreover, this mapping allows us to determine the irreps of $L_+^\uparrow(2)$. Defining the ladder operators as [1]

$$J_\pm = iK_1 \mp K_2, \tag{2}$$

the finite dimensional irreps of the $L_+^\uparrow(2)$ are labelled by the eigenvalues of the Casimir operator, $c(c+1)$, where $c=n/2$, with $n \in Z^+$. These are subspaces of dimension $d=2c+1$ spanned by the states $|c,m\rangle$, where $m = -c, -c+1, \ldots, c-1, c$ are the eigenvalues of J.

The lowest dimensional irrep (beyond the trivial one) is obtained setting $c=\frac{1}{2}$, in whose case we obtain $C=\frac{3}{4}\mathbf{1}$. This irrep has dimension $d=2$ and the states $\{|\frac{1}{2},\pm\frac{1}{2}\rangle\}$ constitute a basis for this subspace. In this basis the operators $\{J, K_i\}$ have the following matrix representations

$$J = \tfrac{1}{2}\sigma^3, \quad K_1 = -\tfrac{i}{2}\sigma^1, \quad K_2 = -\tfrac{i}{2}\sigma^2, \qquad (3)$$

where σ^i are the Pauli matrices, whereas the states themselves are represented as

$$|\tfrac{1}{2},\tfrac{1}{2}\rangle \rightarrow \psi^+(0) = \begin{pmatrix} 1 \\ 0 \end{pmatrix}, \quad |\tfrac{1}{2},-\tfrac{1}{2}\rangle \rightarrow \psi^-(0) = \begin{pmatrix} 0 \\ 1 \end{pmatrix}, \qquad (4)$$

and the superindex denotes the eigenvalue of the J ("spin" projection).

The boost and rotation operators for this representation read

$$B(\hat{n},\varphi) = \cosh\frac{\varphi}{2} - (\vec{\sigma}\cdot\hat{n})\sinh\frac{\varphi}{2}, \qquad D(\theta) = \cos\frac{\theta}{2} - i\sigma^3\sin\frac{\theta}{2}, \qquad (5)$$

where $\hat{n} = (n_1, n_2)$ is the vector in the plane along the direction of the boost and $\vec{\sigma} \equiv (\sigma^1, \sigma^2)$. In the case of boosts acting on rest frame states, the rapidity φ is related to p^μ through the relations $\cosh\varphi = \frac{E}{m}$ and $\sinh\varphi = \frac{|\vec{p}|}{m}$. Thus, the boost operator and its square can be written as

$$B(\hat{n},\varphi) = \frac{E+m-\vec{\sigma}\cdot\vec{p}}{\sqrt{2m(E+m)}}, \qquad B^2(\hat{n},\varphi) = \frac{E-\vec{\sigma}\cdot\vec{p}}{m}. \qquad (6)$$

Acting with the boost operator on the rest frame states in Eq.(4) we obtain

$$\psi^+(p) = \frac{1}{\sqrt{2m(E+m)}}\begin{pmatrix} E+m \\ -p_1 - ip_2 \end{pmatrix}, \quad \psi^-(p) = \frac{1}{\sqrt{2m(E+m)}}\begin{pmatrix} -p_1 + ip_2 \\ E+m \end{pmatrix}. \qquad (7)$$

These are the explicit form of the states in the lowest dimensional representation, i.e. *spinors*, for a planar Minkowski world, constructed from first principles.

SPATIAL INVERSION AND DIRAC EQUATION FOR $d=1+2$

In sharp contrast with the physics in (1+3)-dimensions, the spatial inversion (*parity*) transformation is not an improper transformation in $d=1+2$. At the Minkowski space level this transformation is

$$(t', x', y') = \mathrm{diag}(1, -1, -1)(t, x, y) \qquad (8)$$

and the corresponding matrix $P = diag(1,-1,-1)$ satisfies $\det(P) = 1$. It is easy to see this transformation is just a rotation in an angle π.

The transformation properties for the generators of rotations and boosts in Hilbert space under the parity operator Π can be inferred from Eq.(8) according to [2,3]

$$\Pi(iJ^{\mu\nu})\Pi^{-1} = iP^\mu_\rho P^\nu_\sigma J^{\rho\sigma}, \quad \Pi(iP^\mu)\Pi^{-1} = iP^\mu_\rho P^\rho \qquad (9)$$

or explicitly

$$\Pi J \Pi^{-1} = J, \quad \Pi K_i \Pi^{-1} = -K_i \quad \rightarrow \quad \Pi C \Pi^{-1} = C, \quad \Pi J_\pm \Pi^{-1} = -J_\mp. \qquad (10)$$

Hence, *the irreps $\{c,m\}$ are also invariant subspaces of spatial inversion*. This comes as not surprise since as we noticed above, spatial inversion belongs to $L_+^\uparrow(2)$. From Eq.(10) it is straightforward to show that for a general irrep

$$\Pi|c,m\rangle = \eta(c,m)|c,m\rangle, \qquad (11)$$

where $\eta(c,m) = (-1)^m \eta^{(c)}$ and $\eta^{(c)}$ is a global phase. For the specific case $c = \frac{1}{2}$, we get the following representation for the parity operator in the basis $\{\frac{1}{2}, \pm\frac{1}{2}\}$

$$\Pi = i\eta^{(\frac{1}{2})}\sigma^3. \qquad (12)$$

Notice that modulo a phase this is the very same operator for a rotation in an angle π [see Eq.(5)] as expected. Hereafter we adopt the convention $\eta^{(\frac{1}{2})} = -i$.

A general transformation in $L_+^\uparrow(2)$, $U_\Lambda(\theta,\vec{\varphi}) = \exp[-i(J\theta + \vec{K}\cdot\vec{\varphi})]$, satisfies

$$\Pi[U_\Lambda(\theta,\vec{\varphi})]^\dagger \Pi^{-1} = [U_\Lambda(\theta,\vec{\varphi})]^{-1}, \qquad (13)$$

which can de used to show that the product $(\psi,\varphi) \equiv \langle\psi|\Pi|\varphi\rangle$ is invariant under these transformations.

Let us now consider the parity eigenvalue equation for spinors in the $\{\frac{1}{2}, \pm\frac{1}{2}\}$ basis and in the rest frame

$$\Pi \psi(0) = \pi \psi(0) \qquad (\pi = \pm 1). \qquad (14)$$

Boosting this equation and using Eq.(13) we get

$$[B^2(\vec{p})\Pi - \pi]\psi(\vec{p}) = 0, \qquad (15)$$

which upon using the explicit form of $B^2(\vec{n},\varphi)$ in Eq.(6) can be rewritten to

$$[(E-\vec{\sigma}\cdot\vec{p})\sigma^3 - \pi m]\psi(\vec{p}) = 0. \tag{16}$$

If now we define the matrices $\gamma^0 = \sigma^3$, $\gamma^1 = \sigma^1\sigma^3 = -i\sigma^2$, $\gamma^2 = \sigma^2\sigma^3 = i\sigma^1$ this equation can be written in the standard Dirac form as

$$[\gamma^\mu p_\mu - \pi m]\psi(\vec{p}) = 0. \tag{17}$$

Clearly, the solutions to this equation are the states in Eq.(7). This derivation shows that in $d=1+2$ Dirac equation is just the invariant form of spatial inversion eigenvalue equation, as it is in $d=1+3$.

Despite this formal analogy between $d=1+3$ and $d=1+2$ cases, there are substantial differences in the physical content of Dirac equation. To start with, the states $\psi^\pm(\vec{p})$ in the strict non-relativistic limit $(E \to m, \vec{p} \to \vec{0})$ could be identified as the conventional non-relativistic spin states of an electron. However, it is easy to show that the states $\psi^\pm(\vec{p})$ are parity eigenstates, i.e.

$$P\psi^\pm(\vec{p}) \equiv \Pi\psi^\pm(-\vec{p}) = \pm\psi^\pm(\vec{p}), \tag{18}$$

which require to identify ψ^+ as the state describing the particle and ψ^- as the anti-particle. This requires that a planar electron has fixed "spin" projection $m = +\frac{1}{2}$ and the planar positron has also a fixed "spin" projection $m = -\frac{1}{2}$.

The explicit construction of charge conjugation operator shows that ψ^+ is the state describing the particle and ψ^- is the anti-particle. Moreover, the non-relativistic limit of the gauged Dirac equation shows that in a planar world the spin of a particle in the lowest dimensional representation has only one degree of freedom [3,4].

ACKNOWLEDGMENTS

We wish to thank Adnan Bashir for stimulating discussions on this topic.

REFERENCES

1. B. G. Wybourne, *Classical Groups for Physicists*, John Wiley & Sons, 1974.
2. S. Weinberg, The Quantum Theory of Fields: Vol I Foundations, Cambridge, Cambridge University Press, 1995, pp. 225-226.
3. M. G. Carrillo-Ruiz, "Fermiones en (1+2)-dimensiones de Primeros Principios", M.Sc. Thesis, University of Guanajuato, 2005.
4. R. C. Paschoal ans J. A. Helayël-Neto,, *Phys. Lett.* A313, 412-417 (2003).

The Trigonometric Rosen-Morse Potential as a Prime Candidate for an Effective QCD Potential

Cliffor Benjamín Compeán Jasso and Mariana Kirchbach

*Instituto de Física, Universidad Autónoma de San Luis Potosí.
Av. Manuel Nava No. 6, S.L.P. 78290, México*

Abstract. We make the point that the trigonometric Rosen-Morse (tRM) potential is of possible interest to quark physics in so far as it captures the essentials of the QCD quark-gluon dynamics. This potential (i) interpolates between a Coulomb-like potential (associated with one-gluon exchange) and the infinite wall potential (associated with asymptotic freedom), (ii) reproduces in the intermediary region the linear confinement potential (associated with multi-gluon self-interactions) as established by lattice QCD calculations of hadron properties. Moreover, its exact real solutions given here display a new class of real orthogonal polynomials and thereby interesting mathematical entities in their own.

Keywords: Trigonometric Rosen-Morse potential, orthogonal polynomials, effective QCD potential.
PACS: 02.30.Em, 02.30.Gp, 02.30.Hq, 03.65.Ge, 12.60.Jv

THE SHAPE OF THE TRM POTENTIAL

We adopt the following form of the trigonometric Rosen-Morse potential [1]

$$v(z) = -2b\cot z + a(a+1)\csc^2 z, \qquad (1)$$

with $a > -1/2$ and displayed in Fig. 1. Here, $z = y/d$, $v(z) = V(zd)/(\hbar^2/2md^2)$ and $\varepsilon_n = E_n/(\hbar^2/2md^2)$, with y being the one-dimensional variable, d a properly chosen length scale, $V(y)$ the potential in ordinary coordinate space, and E_n the energy level. Our point here is that $v(z)$ interpolates between the Coulomb-and the infinite wall potential [2] going through an intermediary region of linear-z–and harmonic-oscillator-z^2 dependences. To see this (besides inspection of the figure) it is quite instructive to expand the potential in a Taylor series which for appropriately small z takes the form of a Coulomb-like potential with a centrifugal-barrier like term, provided by the $\csc^2 z$ part,

$$v(z) \approx -\frac{2b}{z} + \frac{a(a+1)}{z^2}, \qquad z \ll 1. \qquad (2)$$

For $z < \pi/2$ we can the take Eq. (2) plus a linear like pertubation

$$\Delta v(z) = a(a+1)/3 + 2bz/3, \qquad (3)$$

as an aproximation of tRM potential. Finally, the potential obviously evolves to an infinite wall as z approaches the limits of the definition interval $0 < z < \pi$, due to the behavior of the $\cot z$ and $\csc z$ for $a > 0$.

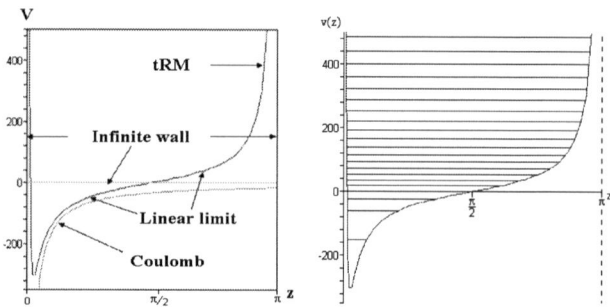

FIGURE 1. tRM potential (left) and its spectra (right).

Above shape captures surprisingly well the essentials of the QCD quark-gluon dynamics where the one gluon exchange gives rise to an effective Coulomb-like potential, while the self-gluon interactions produce a linear potential as established by lattice QCD calculations of hadron properties. Finally, the infinite wall piece of the tRM potential provides the regime suited for the asymptotical freedom of the quarks.

The above considerations suggest that the potential of interest might be a prime candidate for an effective QCD potential.

SPECTRUM AND WAVE FUNCTIONS OF THE TRM POTENTIAL

The one-dimensional Schrödinger equation with the trigonometric Rosen-Morse potential (tRM) reads:

$$\frac{d^2}{dz^2}R_m(z) + (2b\cot z - a(a+1)\csc^2 z + \varepsilon_n)R_m(z) = 0 \ . \tag{4}$$

Our pursued strategy [2, 3, 4] in solving it will be to first reshape it to the particular case of a Sturm-Liouville equation [5, 6] of the form

$$s(x)\frac{d^2}{dx^2}F_m(x) + \frac{1}{w(x)}\left(\frac{d}{dx}s(x)w(x)\right)\frac{d}{dx}F_m(x) + \lambda_m F_m(x) = 0 \ , \tag{5}$$

and then try to solve it by means of the so called Rodrigues representation

$$F_m(x) = \frac{1}{K_m w(x)} \frac{d^m}{dx^m}(w(x)s(x)^m) \ , \tag{6}$$

where K_m is the normalization constant of the $F_m(x)$ polynomials. The constant λ_m in Eq. (5) is supposed to satisfy the following condition [5]:

$$\lambda_m = -m\left(K_1 \frac{d}{dx}F_1(x) + \frac{1}{2}(m-1)\frac{d^2}{dx^2}s(x)\right) \ . \tag{7}$$

Sturm-Liouville equations of the type given in Eq. (5) are called hypergeometric equations. The chosen strategy is inspired by the observation that all the classical polynomials have been obtained precisely from those very Eqs. (5) to (7), appear orthogonal with respect to the weight function $w(x)$ and obey the following restrictions (see Chpt. 10 in Ref. [5] for more details): (i) $F_1(x)$ is a polynomial of first order, (ii) $s(x)$ is a polynomial of at most second order and real roots, (iii) $w(x)$ is real, positive and integrable within a given interval $[a, b]$, and satisfies the boundary conditions $w(a)s(a)a^m = w(b)s(b)b^n = 0$ for all n and m integer.

We here draw attention to the fact that the exact solutions of Eq. (4) can be expressed in terms of real orthogonal polynomials that solve a hypergeometric differential equation of a new class.

Back to Eq. (4) we factorize the wave function as

$$R_m(z) = e^{-\beta z/2}(1+\cot^2 z)^{\alpha/2} C_m^{(\alpha,\beta)}(\cot z) , \qquad (8)$$

with β and α being constants. Upon introducing the new variable $x = \cot z$ and substituting the above factorization ansatz into Eq. (8) and a subsequent division by $(1+x^2)^{\alpha/2}$, Eq. (4) can be cast into the form

$$(1+x^2)\frac{d^2}{dx^2}C_m^{(\alpha,\beta)}(x) + (\beta + 2(\alpha+1)x)\frac{d}{dx}C_m^{(\alpha,\beta)}(x) - m(2\alpha + m + 1))C_m^{(\alpha,\beta)}(x) = 0 , \qquad (9)$$

to the cost that the constants became m depending according to

$$\alpha_m = -(m+1+a), \quad \beta_m = \frac{2b}{m+1+a}, \quad \varepsilon_m = (m+1+a)^2 - \frac{b^2}{(m+1+a)^2}, \quad m \geq 0. \qquad (10)$$

Comparison to Eq. (5) allows to identify $s(x)$ and $w^{(\alpha,\beta)}(x)$ as

$$s(x) = 1+x^2, \quad w^{(\alpha,\beta)}(x) = (1+x^2)^{\alpha} e^{-\beta \cot^{-1} x} . \qquad (11)$$

Notice that Eq. (9) generalizes the hypergeometric differential equation. The orthogonally condition for the $C_m^{(\alpha_m,\beta_m)}(x)$ polynomials is now obtained as

$$\int_{-\infty}^{\infty} dx \frac{\sqrt{w^{(\alpha_n,\beta_n)}(x)} C_n^{(\alpha_n,\beta_n)}(x) \sqrt{w^{(\alpha_m,\beta_m)}(x)} C_m^{(\alpha_m,\beta_m)}(x)}{s(x)} = 0, \quad m \neq n . \qquad (12)$$

In this way, we obtain the exact tRM spectra, displayed in Fig. 1 and the wave functions are given by Eq. (8) with β_m and α_m in place of β and α respectively.

THE $C_m^{(\alpha,\beta)}(x)$ POLYNOMIALS

We obtain the explicit form of solutions of Eq. (9) which are based on a new family of real orthogonal polynomials. Now we going to show some of the properties of this new

family of real polynomials.
Explicitly, the $C_m^{(\alpha,\beta)}(x)$ are obtained from theRodrigues formula as

$$C_m^{(\alpha,\beta)}(x) = \frac{1}{w^{(\alpha,\beta)}(x)} \frac{d^m}{dx^m}(w^{(\alpha,\beta)}(x)s(x)^m). \tag{13}$$

Actually, it easily verifies that the $C_m^{(\alpha,\beta)}(x)$ polynomials relate to the Jacobi polynomials via

$$P_m^{(\eta,\eta^*)}(z) = i^m \frac{1}{2^m m!} C_m^{(Re\,\eta, 2Im\,\eta)}(x), \quad z = ix. \tag{14}$$

In a certain sense, Eq. (14) resembles the relation between the trigonometric and the hyperbolic functions, i.e. $\sin ix = i\sinh x$. Similarly to the real $\sinh x$, the real C polynomials are defined on the $[-\infty, +\infty]$ interval, a property which is not obvious from Eq. (14). The new definition interval is obtained upon substituting η, η^*, and z into the weight-function of the Jacobi polynomials, in which case one ends up with Eq. (11). In the same way, we obtain the $C_n^{(\alpha,\beta)}$ generating function, $G^{(\alpha,\beta)}(x,t) = \sum_{m=0}^{\infty} C_m^{(\alpha,\beta)}(x) t^m / 2^m m!$, and the recurrence relation from the respective generating function and recurrence relations of the Jacobi polynomials as

$$G^{(\alpha,\beta)}(x,t) = \frac{\left(1+\sqrt{1-2xt-t^2}-tx\right)^{-\alpha}}{2^{-\alpha}\sqrt{1-2xt-t^2}} e^{\beta \tan^{-1}(t/(\sqrt{1-2xt-t^2}+1))}, \tag{15}$$

and

$$(m+2\alpha+1)(m+\beta)C_{m+1}^{(\alpha,\beta)}(x) = (2m+2\alpha+1)(2(m+\alpha+1)(m+\beta)x+\alpha\beta)C_m^{(\alpha,\beta)}(x)$$
$$+m(4(m+\beta)^2+\beta^2)(m+\alpha+1)C_{m-1}^{(\alpha,\beta)}(x). \tag{16}$$

Although the properties of the C polynomials can be built up from those of the Jacobi polynomilas with purely imaginary arguments and parameters that are complex conjugate to each-other, nonetheless, similarly to $\sinh x$, the C polynomilas represent autonomous real functions that are different but all the established real classical polynomials.

REFERENCES

1. F. Cooper, A. Khare, U. P. Sukhatme, *Supersymmetry in Quantum Mechanics*, Singapore: World Scientific, 2001.
2. Cliffor Benjamín Compeán Jasso, *Baryon spectra in a quark-diquark model with the trigonometric Rosen-Morse potential*, MS thesis (in Spanish), Institute of Physics, Autonomous University of San Luis Potosí, México, 2005.
3. C. B. Compean, M. Kirchbach, J. Phys. A: Math.Gen. **39**, 547 (2006).
   C. B. Compean, M. Kirchbach, *The Quantum Mechanic Problem of the Schrödinger Equation with the Trigonometric Rosen Morse Potential*, arXiv:quant-ph/**0603232** (2006).
4. H. J. Weber, M. Kirchbach, C. B. Compean, Preprint (Jan. 2006); H. J. Weber, Preprint (Feb. 2006)
5. Phylippe Dennery, André Krzywicki, *Mathematics for Physicists*, New York: Dover, 1996.
6. M. Abramowitz, I. A. Stegun, *Handbook of Mathematical Functions with Formulas, Graphs and Mathematical Tables*, New York: Dover, 1972.

Scalar Quantum Electrodynamics: Perturbation Theory and Beyond

A. Bashir[**,*,†], L.X. Gutierrez-Guerrero[**] and Y. Concha-Sánchez[**]

[*]*Institute for Particle Physics Phenomenology (IPPP), University of Durham, Durham DH1 3LE, United Kingdom*
[†]*Instituto de Ciencias Nucleares, Universidad Nacional Autónoma de México, Circuito Exterior, C.U., Apartado Postal 70-543, 04510, México*
[**]*Instituto de Física y Matemáticas, Universidad Michoacana de San Nicolás de Hidalgo, Apartado Postal 2-82, Morelia, Michoacán 58040, Mexico*

Abstract.
In this article, we calculate scalar propagator in arbitrary dimensions and gauge and the three-point scalar-photon vertex in arbitrary dimensions and Feynman gauge, both at the one loop level. We also discuss constraints on their non perturbative structure imposed by requirements of gauge invariance and perturbation theory.

Keywords: Scalar QED, Scalar Propagator, Three point vertex
PACS: 11.15.Tk, 11.30.Rd, 12.20.-m

INTRODUCTION

Knowledge of the non-perturbative structure of Green functions in gauge field theories has been a challenging problem to deal with. Let alone the complicated, though physically relevant, scenario of Quantum Chromodymaics (QCD), even simpler examples such as Quantum Electrodynamics (QED) have proved to be a hard nut to crack. One might hope that Scalar Quantum Electrodynamics (SQED) could offer a simpler platform in the absence of Dirac matrices to look for non perturbative solutions [1]. Perturbation theory can be a reliable guide to which all acceptable non perturbative structures should reduce in the weak coupling regime. In addition to perturbation theory, constraints imposed by gauge covariance relations such as the Ward-Green-Takahashi Identity (WGTI) and the Landau-Khalatnikov-Fradkin tranformations (LKFT) also contain vital information of nonperturbative structure of propagators and vertices. This is the subject of study of this article.

MESON PROPAGATOR

One loop meson propagator in arbitrary dimensions and gauge is given by

$$S^{-1}(p) = p^2 - m^2 - \frac{e^2}{m^2}\left(\frac{m^2}{4\pi}\right)^{\frac{d}{2}}\Gamma\left(1-\frac{d}{2}\right)\left\{1 - 2\frac{(m^2+p^2)}{m^2} {}_2F_1\left(2-\frac{d}{2},1;\frac{d}{2};\frac{p^2}{m^2}\right)\right.$$

$$+(1-\xi)\frac{(m^2-p^2)^2}{m^4}{}_2F_1\left(3-\frac{d}{2},2;\frac{d}{2};\frac{p^2}{m^2}\right)\Big\},\qquad(1)$$

where e is the electromagnetic coupling, m is the bare meson mass, d is the space-time dimension and ξ is the usual covariant gauge parameter. On the other hand, we can derive a non perturbative expression for the meson propagator through employing LKFT and starting from the bare propagator in the Landau gauge. Unfortunately, we were not able to obtain a closed expression in arbitrary dimensions. However, in four dimensions, we have

$$S(p;\xi) = -\frac{1}{m^2}\left(\frac{m^2}{\Lambda^2}\right)^v \Gamma(1-v)\Gamma(2-v)\,{}_2F_1(1-v,2-v;2;-\frac{p^2}{m^2}),\qquad(2)$$

where $v = \alpha\xi/4\pi$ and Λ the ultraviolet cut-off. This expression contains the correct coefficient of $(\alpha\xi)^n$ term at every order in perturbation theory. For $n = 1$, we can verify this statement by comparing the perturbative expansion of Eq. (2) against Eq. (1), and setting $d = 4 - 2\varepsilon$ and $\ln\Lambda^2 \Longrightarrow \gamma + 2\ln 2 + \ln[\pi] + 1/\varepsilon$. We now turn our attention to the 3-point meson-photon vertex.

MESON-PHOTON VERTEX

The three-point vertex at the one-loop level in arbitrary dimensions and gauge is:

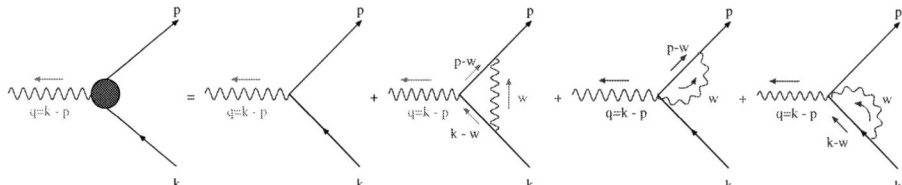

Figure 1: 3-point vertex correction.

The complete one-loop vertex is the sum of these three contributions :

$$\Lambda^\mu = \Lambda_1^\mu + \Lambda_2^\mu + \Lambda_3^\mu,\qquad(3)$$

where the results for individual contributions Λ_1^μ, Λ_2^μ and Λ_3^μ are given below.

$$\Lambda_\mu^1 = -\frac{ie^2}{(2\pi)^d}\Big\{4(k\cdot p)(k+p)_\mu J^{(0)} + [-8(k\cdot p)g_\mu{}^v - 2(k+p)_\mu(k+p)^v]J_v^{(1)}$$
$$+4(k+p)^v J_{\mu v}^{(2)} + (k+p)_\mu K^{(0)} - 2K_\mu^{(1)}\Big\},\qquad(4)$$

where

$$K^{(0)} = \int d^d w \frac{1}{[(p-w)^2 - m^2][(k-w)^2 - m^2]}$$

$$= i\pi^{d/2}\,\Gamma(2-d/2)\,(m^2)^{\frac{d}{2}-2}\,{}_2F_1\left(1,\,2-\frac{d}{2}\,;\,\frac{3}{2}\,;\,\frac{q^2}{4m^2}\right),$$

$$K_\mu^{(1)} = \int d^d w \frac{w_\mu}{[(p-w)^2 - m^2][(k-w)^2 - m^2]} = \frac{1}{2}(p_\mu + k_\mu)\,K^{(0)},$$

$$J_\mu^{(1)} = \int d^d w \frac{w_\mu}{w^2[(p-w)^2 - m^2][(k-w)^2 - m^2]} = \frac{i\pi^2}{2}\left[\,k_\mu\,J_A(k,p) + p_\mu\,J_B(k,p)\,\right],$$

$$J_{\mu\nu}^{(2)} = \int d^d w \frac{w_\mu w_\nu}{w^2[(p-w)^2 - m^2][(k-w)^2 - m^2]}$$

$$= \frac{i\pi^2}{2}\left[\frac{g_{\mu\nu}}{d}K_0 + \left(k_\mu k_\nu - g_{\mu\nu}\frac{k^2}{d}\right)J_C + \left(p_\mu k_\nu + k_\mu p_\nu - g_{\mu\nu}\frac{2(k\cdot p)}{d}\right)J_D\right.$$

$$\left. + \left(p_\mu p_\nu - g_{\mu\nu}\frac{p^2}{d}\right)J_E\right]. \tag{5}$$

The coefficients J_A, J_B, J_C, J_D and J_E in the above expressions are :

$$J_A(k,p) = -\frac{1}{\Delta^2}\Bigg\{\Big[p^2 - (k\cdot p)\Big]\frac{K_0}{2} + \Big[p^2(k^2 - m^2) - (k\cdot p)(p^2 - m^2)\Big]\frac{J_0}{2}$$

$$+\pi^{d/2-2}\Gamma(1-d/2)\Bigg[p^2(m^2 - p^2)^{d/2-2}{}_2F_1\left(2-\frac{d}{2},\frac{d}{2}-1;\frac{d}{2};\frac{p^2}{p^2-m^2}\right)$$

$$-k\cdot p(m^2 - k^2)^{d/2-2}{}_2F_1\left(2-\frac{d}{2},\frac{d}{2}-1;\frac{d}{2};\frac{k^2}{k^2-m^2}\right)\Bigg]\Bigg\}, \tag{6}$$

$$J_B(k,p) = J_A(p,k), \tag{7}$$

$$J_C(k,p) = \frac{d}{(d-2)\Delta^2}\Bigg\{\Bigg[\left(\frac{1}{2}-\frac{1}{d}\right)(p^2-m^2)(k\cdot p) - \left(\frac{1}{2}-\frac{1}{2d}\right)(k^2-m^2)p^2\Bigg]J_A$$

$$-\frac{1}{2d}(p^2-m^2)p^2 J_B + \Bigg[\left(\frac{2}{d}-\frac{1}{2}\right)p^2 + \left(\frac{1}{2}-\frac{1}{d}\right)(k\cdot p)\Bigg]\frac{K_0}{2}$$

$$-\left(1-\frac{2}{d}\right)(k\cdot p)\pi^{d/2-2}\frac{\Gamma(2-\frac{d}{2})\Gamma(\frac{d}{2})}{\Gamma(\frac{d}{2}+1)}(m^2-k^2)^{d/2-2}\times$$

$${}_2F_1\left(2-\frac{d}{2},\frac{d}{2};\frac{d}{2}+1;\frac{k^2}{k^2-m^2}\right)\Bigg\}, \tag{8}$$

$$J_E(k,p) = J_C(p,k), \tag{9}$$

$$J_D(k,p) = \frac{d}{2(d-2)\Delta^2}\Bigg\{\frac{1}{2}\Big[(k\cdot p)(k^2-m^2) + \left(\frac{2}{d}-1\right)k^2(p^2-m^2)\Big]J_A$$

$$+\frac{1}{2}\left[(p^2-m^2)(k\cdot p)+\left(\frac{2}{d}-1\right)p^2(k^2-m^2)\right]J_B$$
$$-\frac{1}{4}\left[\left(1-\frac{2}{d}\right)(p-k)^2+\frac{4}{d}(k\cdot p)\right]K_0$$
$$+\left(1-\frac{2}{d}\right)\left[k^2\pi^{d/2-2}\frac{\Gamma(2-\frac{d}{2})\Gamma(\frac{d}{2})}{\Gamma(\frac{d}{2}+1)}(m^2-k^2)^{d/2-2}\times\right.$$
$$\left.{}_2F_1\left(2-\frac{d}{2},\frac{d}{2};\frac{d}{2}+1;\frac{k^2}{k^2-m^2}\right)+p^2\pi^{d/2-2}\frac{\Gamma(2-\frac{d}{2})\Gamma(\frac{d}{2})}{\Gamma(\frac{d}{2}+1)}\times\right.$$
$$\left.(m^2-p^2)^{d/2-2}{}_2F_1\left(2-\frac{d}{2},\frac{d}{2};\frac{d}{2}+1;\frac{p^2}{p^2-m^2}\right)\right]\right\}. \quad (10)$$

The contributions Λ_2^μ and Λ_3^μ are much simpler as they contain only propagator type loop. Thus :

$$\Lambda_2^\mu = \frac{e^2}{(4\pi)^{d/2}}\Gamma(1-d/2)\,p^\mu\frac{(m^2)^{d/2-1}}{p^2}\left(1+3\frac{p^2}{m^2}\right){}_2F_1\left(2-\frac{d}{2},1;\frac{d}{2}+1;\frac{p^2}{m^2}\right),$$
$$\Lambda_3^\mu = \Lambda_2^\mu(k\leftrightarrow p)\,. \quad (11)$$

Eqs. (3–11) form the complete one-loop meson-photon vertex in arbitrary dimensions at the one-loop level for massive scalars. This is a generalization to arbitrary dimensions and massive scalars of the work of Ball and Chiu, [2]. They calculated the same vertex in 4 dimensions using the cut-off regularization procedure. Following the above mentioned work, we write out the full vertex in its most general form in the following manner independent of any unwanted kinematic singularities:

$$\Gamma^\mu(k,p) = (k^\mu+p^\mu)\frac{S^{-1}(k)-S^{-1}(p)}{k^2-p^2}+\tau(k^2,p^2,q^2)[k\cdot q p^\mu - p\cdot q k^\mu]. \quad (12)$$

Function $\tau(k^2,p^2,q^2)$ is the only unknown to fully fix the meson-photon vertex. Perturbatively, it gets fixed through Eqs. (1,3–11) to the one loop order. Every non perturbative *ansatz* for this function should reduce to this expression in the weak coupling regime.

ACKNOWLEDGMENTS

We thank Maria Elena Tejeda-Yeomans for several useful discussions. AB wishes to acknowledge a short term visitor grant by a joint scheme of The Royal Society, U.K, and The Mexican Academy of Sciences, Mexico. Support has also been provided by CIC and CONACyT (grants 4.10 and 46614-I).

REFERENCES

1. A. Salam and R. Delbourgo, *Phys. Rev.* **135** B1389, (1964); R. Delbourgo, *J. Phys* **A10** 1369, (1977).
2. J.S. Ball and T-W. Chiu, *Phys. Rev.* **D22** 2542, (1980).

Azimuthal Correlations in p-p collisions

Eleazar Cuautle, Isabel Domínguez and Guy Paić

Instituto de Ciencias Nucleares, UNAM, A. P. 70543, 04510 Mexico City, Mexico.

Abstract.
We report the analysis of experimental azimuthal correlations measured by STAR in p-p collisions at $\sqrt{s_{NN}}$ = 200 GeV. We conclude that for a fit of data using Pythia event generator we need to include two values of k_T.

Keywords: kt in p-p collisions
PACS: 13.75.Cs 13.87.Fh 25.75.-q

INTRODUCTION

Jets are produced by the hard scattering of two partons. Two scattered partons propagate nearly back-to-back in azimuth from the collision point and fragment into jet-like spray of final state particles (The schematic view is in Figure 1). These particles have a transverse momentum j_T with respect to the parent partons, with component j_{Tz} projected onto the azimuthal plane. The magnitude of j_{Tz} measured at lower energies has been found to be $\sqrt{s_{NN}}$ and p_T independent.

In collinear partonic collisions, the two partons emerge with the same magnitude of transverse momentum in opposite directions. However, the partons carry the "intrinsic" transverse momentum k_T before the collision. This momentum affects the outgoing transverse momentum p_T, resulting in a momentum imbalance (i.e. transverse momentum of one jet does not lie in the plane determined by the transverse momentum of the second jet and the beam axes) and consequently affects the back-to-back correlations of final high p_T hadrons [1].

The back-to-back azimuthal correlations of high p_T hadrons is written as

$$C(\Delta\Phi) = \frac{1}{N_{trigger}} \int d\Delta\eta \frac{dN}{d\Delta\Phi d\Delta\eta} \quad (1)$$

where $\Delta\Phi$ and $\Delta\eta$ are, respectively, the azimuthal angle and pseudorapidity between a trigger and their associated particles. The azimuthal correlation function displays a two-peak structure, where the width of the near-side peak is denoted by σ_N and the width of the away-side peak is σ_A. The value of σ_N carries information on the fragmentation process only i.e. j_T. For particles with average transverse momenta $< p_{T,trigg} >$ and $< p_{T,associate} >$ from the same jet, the width of the near-side correlation, σ_N, can be related to $< j_{Tz} >$ as [2]:

$$< j_{Tz} > = \frac{< p_{T,trigger} >< p_{T,associate} >}{\sqrt{< p_{T,trigger} >^2 + < p_{T,associate} >^2}} \sigma_N \quad (2)$$

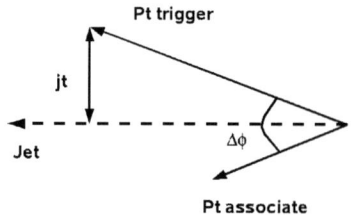

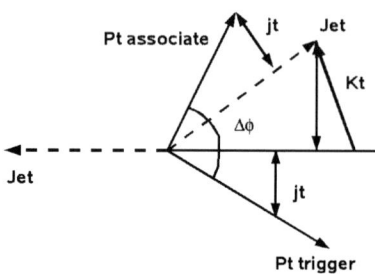

FIGURE 1. Schematic view of a jet fragmentation, near-side jet (upper) and away-side jet (lower)

The width of the away-side peak σ_A contain the contribution of the intrinsic transverse momentum k_T. It has been characterized by a Gaussian distribution [3]

$$g(k_T) = \frac{1}{2\pi\sigma^2} exp(-\frac{k_T^2}{2\sigma^2}) \qquad (3)$$

The azimuthal correlations are used extensively in heavy ions collisions to understand the parton suppression mechanisms. We have concentrated in this work to the simplest case i.e. p-p to understand the size and details of the peaks in azimuthal correlations.

k_T CONTRIBUTION IN THE AZIMUTHAL CORRELATIONS

Correlation function was calculated choosing $< k_T^2 >$=0, 1, 4 GeV^2/c^2 at 200 GeV in a mid-rapidity region ($|\eta| < 0.7$). Charged hadrons in $4 < p_{T,trigg} < 6$ GeV/c and in 2 GeV $< p_{T,assoc} < 4$ GeV are defined to be trigger and associated particles respectively. In the actual calculation, we use PYTHIA 6.325 [4] in AliRoot [5] to simulate each hard scattering where a Gaussian distribution is assuming for k_T. The correlations functions were fitted by the sum of two Gaussians, one for the near-side component (around $\Delta\Phi = 0$ radians) and one for the away-side component (around $\Delta\Phi = \pi$ radians) and a constant for the uncorrelated pairs.

Figure 2 compares experimental data [6] with different $< k_T^2 >$ simulations. In the four cases is observed that can not reproduce experimental data. In order to reproduce the experimental data, we characterize the intrinsic momentum by two Gaussians distributions

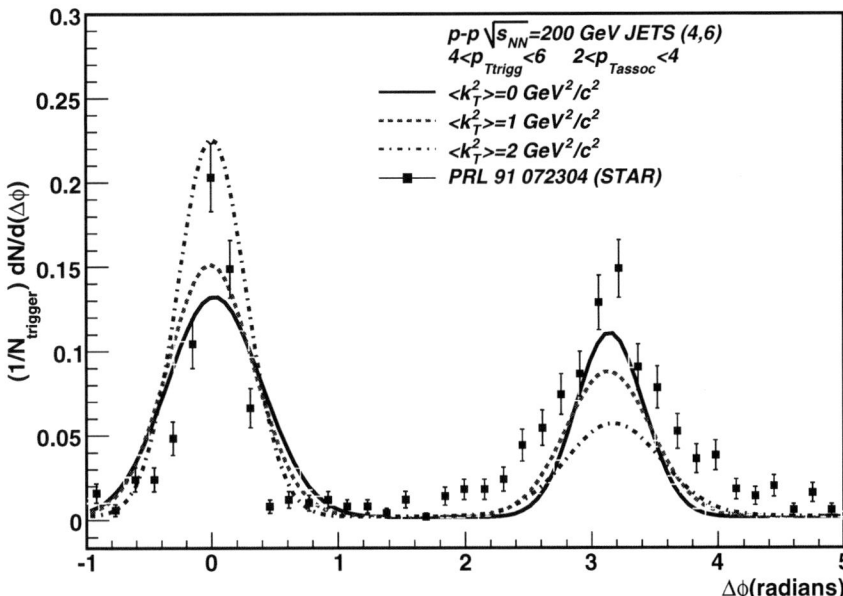

FIGURE 2. Azimuthal distributions for p+p collisions at $\sqrt{s_{NN}}$ = 200 GeV, experimental data [6] and simulations with different $<k_T^2>$

$$g(k_{T1},k_{T2}) = \frac{1}{2\pi\sigma_1^2}exp(-\frac{k_{T1}^2}{2\sigma_1^2}) + \frac{1}{2\pi\sigma_2^2}exp(-\frac{k_{T2}^2}{2\sigma_2^2}) \qquad (4)$$

This distribution was adding in PYTHIA code and calculated the azimuthal correlations. The Figure 3 show the experimental data and the simulation. The simulation is in good agreement with the experimental data. The values of $<k_{T1}>$ and $<k_{T2}>$ are 0.558 ± 0.042 and $<k_{T2}>$ = 0.099 ± 0.050 respectively. In addition the magnitude of the partonic transverse momentum $<j_{Tz}>$ was calculated. The values obtained of $<j_{Tz}>$ = 0.397 ± 0.091 GeV/c are in agreement with the average value $<j_{Tz}>$ = 0.324 ± 0.06 GeV/c obtained experimentally [7].

SUMMARY

We report the analysis of experimental azimuthal correlations measured by STAR in p-p collisions at $\sqrt{s_{NN}}$ = 200 GeV. Comparisons between experimental data and simulation with different $<k_T^2>$ show that the $<k_T^2>$ characterized by a Gaussian distribution can not reproduce experimental data.

Assuming two Gaussians distributions for k_T the simulation is in agreement with the experimental data, as far as, we understand the use of two Gaussians. It has never been

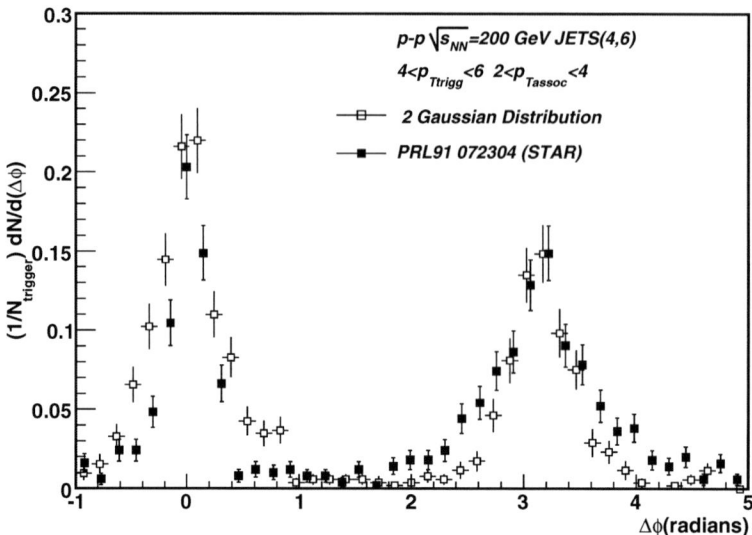

FIGURE 3. Azimuthal distributions for p+p collisions, experimental data (line) [6] and simulations with two Gaussian distributions for the partons intrinsic momentum (circle)

used before to explain the peaks observed in azimuthal correlations.

In addition the magnitude of the partonic transverse momentum $<j_{Tz}>$ was calculated. The values of $<j_{Tz}> = 0.397 \pm 0.091$ GeV/c are in agreement with the average value $<j_{Tz}> = 0.324 \pm 0.06$ GeV/c obtained experimentally.

ACKNOWLEDGMENTS

The authors thanks A. Morsch for his valuable comments and suggestions. Support for this work has been received by PAPIT-UNAM under grant number IN107105.

REFERENCES

1. R. P. Feynman, R. D. Field and G. C. Fox, Nucl. Phys. B. **128**:1, (1977).
2. J. Rak [PHENIX Collaboration] J. Phys G. Nucl. Part. Phys. **31**, S541 (2005).
3. X. N. Wang, Phys. Rev. C **61**, 064910 (2000).
4. T. Söstrand *et al.*, Comp. Phy. Commun. **135**, 238 (2001).
5. http://aliweb.cern.ch/offline/
6. J. Adams *et al.* [STAR Collaboration], Phys. Rev. Lett. **91**, 072304 (2003).
7. J. Rak, J. Phys. G **30**, S1309 (2004).

On quark-lepton complementarity

F. González Canales and A. Mondragón

Instituto de Física, UNAM, 04510, México D.F., MEXICO

Abstract. Recent measurements of the neutrino solar mixing angle and the Cabibbo angle satisfy the empirical relation $\theta_{sol} + \theta_C \simeq \frac{\pi}{4}$. This relation suggests the existence of a correlation between the mixing matrices of leptons and quarks, the so called quark-lepton complementarity. Here, we examine the possibility that this correlation originates in the strong hierarchy in the mass spectra of quarks and charged leptons, and the seesaw mechanism that gives mass to the Majorana neutrinos. In a unified treatment of quarks and leptons in which the mass matrices of all fermions have a similar Fritzsch texture, we calculate the mixing matrices V_{CKM} and U_{MNSP} as functions of quark and lepton masses and only two free parameters, in very good agreement with the latest experimental values on masses and mixings. Three essential ingredients to explain the quark-lepton complementarity relation are identified: the strong hierarchy in the mass spectra of quarks and charged leptons, the normal seesaw mechanism and the assumption of maximal CP violation in the lepton sector.

Keywords: Quark and lepton masses and mixings, Neutrino masses and mixings,CKM matrix
PACS: 12.15.Ff, 14.60.Pq, 12.15.Hh, 14.60.St

INTRODUCTION

In the last few years, the neutrino oscillations between different flavour states were measured in a series of experiments with atmospheric neutrinos[1], solar neutrinos[2][3], neutrinos produced in nuclear reactors [4] and accelerators [5]. As a result, the difference of the squared neutrino masses and the mixing angles in the lepton mixing matrix, U_{MNSP}, were determined:

$$0.34 \leq \sin^2\theta_{23} \leq 0.68, \quad 1.4 \times 10^{-3}(\text{eV})^2 \leq \Delta m_{23}^2 \leq 3.0 \times 10^{-3}(\text{eV})^2, \quad (1)$$

$$0.29 \leq \sin^2\theta_{12} \leq 0.40, \quad 7.1 \times 10^{-5}(\text{eV})^2 \leq \Delta m_{12}^2 \leq 8.9 \times 10^{-5}(\text{eV})^2, \quad (2)$$

$$\sin^2\theta_{13} \leq 0.046, \quad (3)$$

at 90% confidence level [6]-[7]. The CHOOZ experiment [8] determined an upper bound for the θ_{13} mixing angle. It was soon realized [9] that the solar mixing angle θ_{12}^{MNSP} and the Cabibbo angle θ_{12}^{CKM}, which is the corresponding angle in the quark sector, satisfy an interesting and intriguing numerical relation,

$$\theta_{12}^{MNSP} + \theta_{12}^{CKM} = 45^o + 1^o \pm 2.4^o, \quad (4)$$

with $\theta_{12}^{MNSP} = 33.9^o \pm 2.4^o$ (1σ) and $\theta_{12}^{CKM} = 12.8^o \pm 0.15^o$. Equation (4) relates the 1-2 mixing angles in the quark and lepton sectors, it is commonly called the quark-lepton complementarity relation (QLC) and, if not accidental, it could imply a quark-lepton symmetry (for a recent review see [10]) or a quark-lepton unification [11]-[14].

A second QLC relation, $\theta_{23}^{MNSP} + \theta_{23}^{CKM} \approx \frac{\pi}{4}$, is also satisfied. However, this is not as interesting as (4) because θ_{23}^{CKM} is only about two degrees, and the corresponding QLC relation would be satisfied, within the errors, even if the angle θ_{23}^{CKM} had been zero, as long as θ_{23}^{MNSP} is close to the maximal value $\pi/4$. A third possible QLC relation is not realized at all, or at least not realized in the same way, since $\theta_{13}^{CKM} + \theta_{13}^{MNSP}$ is less than ten degrees. In this short note we will focus our attention on understanding the nature of the QLC relation shown in equation (4).

UNIVERSAL FRITZSCH TEXTURE OF QUARKS AND LEPTONS

The quark and lepton flavor mixing matrices, U_{MNSP} and V_{CKM}, arise from the mismatch between diagonalization of the mass matrices of u and d type quarks and the diagonalization of the mass matrices of charged leptons and neutrinos,

$$U_{MNSP} = U_l^\dagger U_\nu, \quad V_{CKM} = U_u^\dagger U_d. \tag{5}$$

Therefore, to get predictions for the flavor mixing angles and CP violating phases, we should specify the mass matrices.

In this work, we propose a unified treatment of quarks and leptons. Lepton and quark mass matrices could have the same mass texture from a universal flavor symmetry (exact at a certain energy scale). Imposing a flavor symmetry has been successful in reducing the number of parameters of the Standard Model. In particular, a permutational S_3 flavor symmetry and its sequential explicit breaking, allows us to represent the mass matrices as a modified Fritzsch texture:

$$\mathbf{M}_i^{(F)} = \begin{pmatrix} 0 & A_i & 0 \\ A_i^* & B_i & C_i \\ 0 & C_i & D_i \end{pmatrix} \quad i = u, d, l, \nu. \tag{6}$$

Some reasons to propose the validity of the modified Fritzsch texture as a universal mass texture for all fermions in the theory are the following:

1. The idea of S_3 flavor symmetry and its explicit breaking has been realized as a modified Fritzsch texture in the quark sector to interpret the strong mass hierarchy of up and down type quarks [15].
2. The quark mixing angles and the CP violating phase appearing in the V_{CKM} mixing matrix were computed as explicit, exact functions of the four quark mass ratios $(m_u/m_t, m_c/m_t, m_d/m_b, m_s/m_b)$, one symmetry breaking parameter $Z^{1/2} = \left(\frac{81}{32}\right)^{1/2}$ and one CP violating phase $\phi_{u-d} = 90^o$, in very good agreement with experiment [16].
3. Since the mass spectrum of the charged leptons exhibits a similar hierarchy to the quark's one, it would be natural to consider the same S_3 symmetry and its explicit breaking for the charged lepton mass matrix.
4. As for the Dirac neutrinos, we have no direct information about the absolute values or the relative values of the neutrino masses, but the Fritzsch texture can

be incorporated in a $SO(10)$ neutrino model [17]. Therefore it would be sensible to assume that the Dirac neutrinos have a mass hierarchy similar to that of the u-quarks and it would be natural to take for the Dirac neutrino mass matrix also a modified Fritzsch texture.

5. The left handed Majorana neutrinos naturally acquire their mass through an effective seesaw mechanism of the form

$$M_{\nu_L} = M_{\nu_D} M_R^{-1} M_{\nu_D}^T, \qquad (7)$$

where M_{ν_D} and M_R denote the Dirac and right handed Majorana neutrino mass matrices. From our conjecture of a universal S_3 flavor symmetry it follows that M_R could have the same texture as that of M_{ν_D} and M_l. Then, it is straightforward to show that M_{ν_L} has the same modified Fritzsch texture [18].

MIXING MATRICES AS FUNCTIONS OF THE FERMION MASSES

When the unitary matrices that diagonalize the mass matrices $M_i^{(F)}$ are written in polar form, $U_i = P_i O_i$ and $M_i^{(F)} = P_i \bar{M} P_i^\dagger$, the expressions (5) for the mixing matrices take the form

$$U_{MNSP} = O_l^T P^{(l-\nu)} O_\nu K, \quad V_{CKM} = O_d^T P^{(u-d)} O_u, \qquad (8)$$

where O_i, $i = u, d, \nu, l$, are the orthogonal matrices that diagonalize the real symmetric mass matrices $\bar{M}_i^{(F)}$ and $P^{(u-d)} = \mathrm{diag}\left[1, e^{i\phi}, e^{i\phi}\right]$, $P^{(l-\nu)} = \mathrm{diag}\left[1, e^{i\Phi}, e^{i\Phi}\right]$, where $\phi = \phi_u - \phi_d$, and $\Phi = \Phi_l - \Phi_\nu$, are the matrices of the Dirac phases and K is the diagonal matrix of the Majorana phases.

We reparametrized the matrices $\bar{M}_i$ in terms of their eigenvalues. The orthogonal matrices are then expressed in terms of the mass eigenvalues of M_i:

$$\mathbf{O}_i = \begin{pmatrix} \left[\frac{\tilde{m}_{i2} f_{i1}}{D_{i1}}\right]^{\frac{1}{2}} & -\left[\frac{\tilde{m}_{i1} f_{i2}}{D_{i2}}\right]^{\frac{1}{2}} & \left[\frac{\tilde{m}_{i1}\tilde{m}_{i2} f_{i3}}{D_{i3}}\right]^{\frac{1}{2}} \\ \left[\frac{\tilde{m}_{i1}(1-\delta_i)f_{i1}}{D_{i1}}\right]^{\frac{1}{2}} & \left[\frac{\tilde{m}_{i2}(1-\delta_i)f_{i2}}{D_{i2}}\right]^{\frac{1}{2}} & \left[\frac{(1-\delta_i)f_{i3}}{D_{i3}}\right]^{\frac{1}{2}} \\ -\left[\frac{\tilde{m}_{i1} f_{i2} f_{i3}}{D_{i1}}\right]^{\frac{1}{2}} & -\left[\frac{\tilde{m}_{i2} f_{i1} f_{i3}}{D_{i2}}\right]^{\frac{1}{2}} & \left[\frac{f_{i1} f_{i2}}{D_{i3}}\right]^{\frac{1}{2}} \end{pmatrix}, \qquad (9)$$

$$f_{i1} = 1 - \tilde{m}_{i1} - \delta_i, \quad f_{i2} = 1 + \tilde{m}_{i2} - \delta_i, \quad f_{i3} = \delta_i, \qquad (10)$$

$$D_{i1} = (1-\delta_i)(\tilde{m}_{i1} + \tilde{m}_{i2})(1-\tilde{m}_{i1}), \qquad (11)$$

$$D_{i2} = (1-\delta_i)(\tilde{m}_{i1} + \tilde{m}_{i2})(1+\tilde{m}_{i2}), \qquad (12)$$

$$D_{i3} = (1-\delta_i)(1-\tilde{m}_{i1})(1+\tilde{m}_{i2}), \qquad (13)$$

the small parameters δ_i are also functions of the mass ratios and the symmetry breaking parameter $Z^{1/2} = (81/32)^{1/2}$.

Substitution of the expressions (9) and (10)-(13) in (8) allows us to express the mixing matrices U_{MNSP} and V_{CKM} as explicit functions of the quark and lepton masses.

QUARK-LEPTON COMPLEMENTARITY

The resulting theoretical expression for the Cabibbo angle written to first order in m_u/m_c and m_d/m_s, is

$$\sin^2\theta_C^{th} = |V_{us}^{th}|^2 \approx \frac{\frac{\tilde{m}_d}{\tilde{m}_s} + \frac{\tilde{m}_u}{\tilde{m}_c} - 2\sqrt{\frac{\tilde{m}_u}{\tilde{m}_c}\frac{\tilde{m}_d}{\tilde{m}_s}}\cos\phi}{\left(1+\frac{\tilde{m}_u}{\tilde{m}_c}\right)\left(1+\frac{\tilde{m}_d}{\tilde{m}_s}\right)}. \tag{14}$$

Taking for the quark masses the values $m_u = 2.75\text{MeV}$, $m_c = 1310\text{MeV}$, $m_d = 6.0\text{MeV}$, $m_s = 120\text{MeV}$ and maximal CP violation, $\phi = 90^o$ [16], we reproduce the numerical value of the Cabibbo angle

$$\sin\theta_C^{th} = 0.225 \quad \text{or} \quad \theta_c = 12.8^o, \tag{15}$$

in very good agreement with the latest analysis of the experimental data [19].

The theoretical expression for the solar mixing angle is derived in a similar way. From $|(U_{MNSP})_{12}|^2 / |(U_{MNSP})_{11}|^2 = \tan^2\theta_{12}^{th}$, we obtain

$$\tan^2\theta_{12}^{th} = \frac{\frac{\tilde{m}_{\nu_1}}{\tilde{m}_{\nu_2}} + \frac{\tilde{m}_e}{\tilde{m}_\mu} - 2\sqrt{\frac{\tilde{m}_{\nu_1}}{\tilde{m}_{\nu_2}}\frac{\tilde{m}_e}{\tilde{m}_\mu}}\cos\Phi}{1 + \frac{\tilde{m}_{\nu_1}}{\tilde{m}_{\nu_2}}\frac{\tilde{m}_e}{\tilde{m}_\mu} + 2\sqrt{\frac{\tilde{m}_{\nu_1}}{\tilde{m}_{\nu_2}}\frac{\tilde{m}_e}{\tilde{m}_\mu}}\cos\Phi}. \tag{16}$$

In the absence of experimental information, we assumed that CP violation is also maximal in the lepton sector *i.e.* $\Phi = 90^o$. Taking for the masses of the left handed Majorana neutrinos a normal hierarchy with the numerical values $m_{\nu_1} = 4.4 \times 10^{-3}\text{eV}$ and $m_{\nu_2} = 1.0 \times 10^{-2}\text{eV}$, and for the charged lepton masses the values $m_e = 0.5109\text{MeV}$, $m_\mu = 105.685\text{MeV}$ and $m_\tau = 1776.99\text{MeV}$, we obtain the following numerical value for the solar mixing angle

$$\tan^2\theta_{12}^{th} = 0.45 \quad \text{or} \quad \theta_{12}^{th} = 33.9^o. \tag{17}$$

We may now address the question of the meaning of the quark-lepton complementarity relation as expressed in eq(4). The previous theoretical analysed allows us to calculate,

$$\tan\left(\theta_C^{th} + \theta_{12}^{th}\right) = 1 + \Delta^{th}, \tag{18}$$

$$\Delta^{th} = \frac{\left(\frac{\tilde{m}_d}{\tilde{m}_s}+\frac{\tilde{m}_u}{\tilde{m}_c}\right)^{\frac{1}{2}}\left[\left(\frac{\tilde{m}_{\nu_1}}{\tilde{m}_{\nu_2}}+\frac{\tilde{m}_e}{\tilde{m}_\mu}\right)^{\frac{1}{2}}+\left(1+\frac{\tilde{m}_{\nu_1}}{\tilde{m}_{\nu_2}}\frac{\tilde{m}_e}{\tilde{m}_\mu}\right)^{\frac{1}{2}}\right]+\left(1+\frac{\tilde{m}_d}{\tilde{m}_s}\frac{\tilde{m}_u}{\tilde{m}_c}\right)^{\frac{1}{2}}\left[\left(\frac{\tilde{m}_{\nu_1}}{\tilde{m}_{\nu_2}}+\frac{\tilde{m}_e}{\tilde{m}_\mu}\right)^{\frac{1}{2}}-\left(1+\frac{\tilde{m}_{\nu_1}}{\tilde{m}_{\nu_2}}\frac{\tilde{m}_e}{\tilde{m}_\mu}\right)^{\frac{1}{2}}\right]}{\left(1+\frac{\tilde{m}_{\nu_1}}{\tilde{m}_{\nu_2}}\frac{\tilde{m}_e}{\tilde{m}_\mu}\right)^{\frac{1}{2}}\left(1+\frac{\tilde{m}_d}{\tilde{m}_s}\frac{\tilde{m}_u}{\tilde{m}_c}\right)^{\frac{1}{2}}-\left(\frac{\tilde{m}_d}{\tilde{m}_s}+\frac{\tilde{m}_u}{\tilde{m}_c}\right)^{\frac{1}{2}}\left(\frac{\tilde{m}_{\nu_1}}{\tilde{m}_{\nu_2}}+\frac{\tilde{m}_e}{\tilde{m}_\mu}\right)^{\frac{1}{2}}}. \tag{19}$$

After substitution of the numerical value of the mass ratios of quarks and leptons in (19), we obtain,

$$\Delta^{th} = 0.061 \quad \text{and} \quad \theta_C^{th} + \theta_{12}^{th} = 45^o + 1.7^o. \tag{20}$$

in very good agreement with the experimental value.

CONCLUSIONS

In this short communication, we outlined a unified treatment of masses and mixing of quarks and leptons in which the left handed Majorana neutrinos acquire their masses via the seesaw mechanism, and the mass matrices of all fermions have a similar Fritzsch texture and a normal hierarchy. In this scheme, we derived exact, explicit expressions for the Cabibbo and solar mixing angles as functions of the quark and lepton masses. The quark-lepton complementarity relation takes the form,

$$\theta_{12}^{CKM} + \theta_{12}^{MNSP} = 45^o + \delta_{12}. \quad (21)$$

The correction term, δ_{12}, is an explicit function of the ratios of quark and lepton masses, given in eq.(19), which reproduces the experimentally determined value, $\delta_{12} \approx 1.7^o$, when the numerical values of the quark and lepton masses are substituted in (19) and maximal violation of CP in the lepton sector is assumed.

Three essential ingredients are needed to explain the correlations implicit in the small numerical value of δ_{12}:

1. The strong hierarchy in the mass spectra of the quarks and charged leptons, realized in our scheme through the explicit breaking of the S_3 flavor symmetry in the Fritzsch mass texture, explains the resulting small or very small quark mixing angles, the very small charged lepton mass ratios explain the very small θ_{13}^{MNSP} which, in our scheme, is independent of the neutrino masses.
2. The normal seesaw mechanism that gives very small masses to the left handed Majorana neutrinos with relatively large values of the neutrino mass ratio m_{v_1}/m_{v_2} and allows for large θ_{12}^{MNSP} and θ_{23}^{MNSP} mixing angles.
3. The assumption of maximal CP violation in the lepton sector.

A more complete and detailed version of this work will be presented in a forthcoming publication [20]

ACKNOWLEDGEMENTS

We thank Dr. M. Mondragón for many inspiring discussions on this exciting problem and for a critical reading of the manuscript.

This work was partially supported by CONACyT Mexico under Contract No. 42026F, and DGAPA-UNAM Contract No. PAPIIT IN116202.

REFERENCES

1. Y. Fukuda et al. (Super-Kamiokande Collaboration), Phys. Rev. Lett. **81**, 1562. (1998)
   Y. Fukuda et al. (Super-Kamiokande Collaboration), Phys. lett. B **539**, 179. (2002)
   M. Sanchez et al., Phys. Rev. D **68**, 11 3004, (2003)
   Y. Ashie et al.,(Super-Kamiokande Collaboration), Phys. Rev. D **71**, 112005 (2005),arXiv/hep-ex/0501064

2. B. T. Cleveland et al., Astrophys. **J. 496**, 505 (2003)
   S.N. Ahmad et al. (SNO Collaboration), Phys .Rev. Lett. **92**, 181301 (2004); arXiv:nucl-ex/0309004,
   M. Altmann et al. (Super-Kamiokande Collaboration), Phys. lett. B **616**, 174 (2005)
3. B. Aharmim et al., (SNO Collaboration), Phys. Rev. C **72**, 055502 (2005); arXiv: nucl-ex/0502021
4. K. Eguchi et al. (KamLAND Collaboration), Phys. Rev. Lett. **90**, 021802,(2003)
   T. Ashie et al. (KamLAND Collaboration), Phys. Rev. Lett. **94**, 081801,(2005)
5. M. H. Ahn et al. Phys. Rev. Lett. **90**, 041801, (2003)
6. M. Maltoni et al. New J. Phys. **6**,122 (2004)
7. T. Schwetz, Acta Phys. Polon. B **36**, 3203-3214 (2005); arXiv:hep-ph/0510331
8. M. Apollonio et al.,(CHOOZ Collaboration), Eur. Phys. J. C **27**; 331-374, 2003, arXiv/hep-ex/0301017 hep-ex/9907037
9. A. Y. Smirnov arXiv:hep-ph/0402264, Published in "Venice 2003, Neutrino oscillations" 1-21
10. H. Minakata, arXiv:hep-ph/0505262, Published in "Venice 2005, Neutrino telescopes" 83-97
11. M. Raidal, Phys. Rev. Lett. **93** (2004) 161801, arXiv:hep-ph/0404046.
12. H. Minakata and A. Y. Smirnov, Phys. Rev. D **70** 073009 (2004), arXiv:hep-ph/0405088
13. P.H. Frampton and R. N. Mohapatra, JHEP 0501:025 (2005), arXiv:hep-ph/0407139
14. P.H. Frampton, S.T. Petcov and W. Rodejohan, Nucl. Phys. B **687**:31-54 (2004)
15. H. Fritzsch, Phys. Lett. B **70**, 436 (1977),
    H. Fritzsch, Nucl. Phys. B **155**, 189 (1979),
    S. Pakvasa and H. Sugawara, Phys. Lett. B **73**, 61 (1978)
    H. Harari, H. Haut and J. Weyers, Phys. Lett. B. **78**, 459 (1979)
16. A. Mondragón and E. Rodríguez Jáuregui, Phys. Rev. D **61**, 113002 (2000),arXiv:hep-ph/0003104
    A. Mondragón and E. Rodríguez Jáuregui Phys. Rev. D, **59** 093009 (1999)
17. W. Buchmüller and D. Wyler, Phys. Lett B **521**, 291 (2001)
    M. Bando and M. Obara, Prog. Theor. Phys. **109**, 995 (2003)
    M. Bando, S. Kaneko, M. Obara and M. Tanimoto, arXiv:hep-ph/0405071
    G.G. Ross and L. Velasco-Sevilla Nucl. Phys. B **623** (2003) 3-26
18. H. Fritzsch and Z. Xing, "Mass and Flavor Mixing Schemes of Quarks and Leptons.", Prog. Part. Nucl. Phys. **45** :1-81 (2000), arXiv:hep-ph/9912358
19. V. Mateu and A.Pich, JHEP **0510**:041 (2005), arXiv:hep-ph/0509045.
20. F. González Canales and A. Mondragón, Work in progress.

Description of charge conjugation from first principles

C. Luján-Peschard and M. Napsuciale

Instituto de Física de la Universidad de Guanajuato
Loma del Bosque 103, Col. Lomas del Campestre, CP 37150, León, Gto.

Abstract. We construct the charge conjugation operator as a *unitary* automorphism in the spinor space $(\frac{1}{2},0) \oplus (0,\frac{1}{2})$ from first principles. We calculate its eigenspinors and derive the equation of motion they satisfy. The mapping associated to charge conjugation is constructed from parity eigenstates which are considered as particle and antiparticle.

Keywords: discrete symmetries, charge conjugation, unitary operators, neutral particle states
PACS: 11.30.Er

Totally neutral particles either fundamental or composite are necessarily eigenstates of charge conjugation [1]. This discrete symmetry was historically obtained as an *anti-unitary* automorphism in the spinor space, $(\frac{1}{2},0) \oplus (0,\frac{1}{2})$, based on the Dirac equation coupled to an external field A_μ. It is precisely the anti-unitary operator connecting spinors describing a particle to spinors describing an antiparticle. In [2] it was shown that Dirac equation is just the boosted parity eigenvalue equation in the spinor space. In this sense, Dirac equation is actually associated to parity eigenstates and in principle has nothing to do with neutral particles. In this work we construct charge conjugation operation as the *linear and unitary* automorphism mapping spinors describing particles (positive parity eigenstates), to spinors describing antiparticles (negative parity eigenstates). We calculate eigenstates of this operator which describe totally neutral particles. In order to construct this mapping we first construct parity eigenstates form first principles, namely: i) all of our particle states live in a Hilbert space $\mathscr{H}$, ii) states belong to the irreducible representations (irreps) of the symmetry group, iii) the symmetry group of a free particle is the Poincaré group.

The most general transformation of the Poincaré Group is given by

$$x'^\mu = \Lambda^\nu{}_\mu x_\nu + a^\mu. \qquad (1)$$

Every transformation belonging to this group follows the composition law

$$T(\Lambda_2,a_2)T(\Lambda_1,a_1) = T(\Lambda_2\Lambda_1, \Lambda_2 a_1 + a_2). \qquad (2)$$

Transformations with $a^\mu = 0$ constitute the Homogeneous Lorentz Group (HLG), which is a subgroup of the Poincaré group. The global structure of the HLG can be characterized by

$$(\det \Lambda)^2 = 1, \qquad (\Lambda^0_0)^2 \geq 1, \qquad (3)$$

hence the HLG is composed of four disjoint sets, which are connected by two discrete symmetries: parity ($\mathscr{P}$) and time reversal ($\mathscr{T}$). Furthermore, since $\mathscr{P}$ and $\mathscr{T}$ belong to the HLG, at the quatum level, there must exist operators implementing these transformations in Hilbert space [3]

$$\Pi \equiv U(\mathscr{P},0) \qquad \mathbf{T} \equiv U(\mathscr{T},0) \qquad (4)$$

which satisfy the group composition law given in (2). Studying infinitesimal transformations it is possible to show that Π must be unitary and $\mathbf{T}$ is necessarily anti-unitary. The transformation properties of the generators of the group are

$$\Pi \mathbf{J} \Pi^{-1} = \mathbf{J}, \qquad \Pi \mathbf{K} \Pi^{-1} = -\mathbf{K}, \qquad \Pi H \Pi^{-1} = H. \qquad (5)$$

$$\mathbf{T} \mathbf{J} \mathbf{T}^{-1} = -\mathbf{J}, \qquad \mathbf{T} \mathbf{K} \mathbf{T}^{-1} = \mathbf{K}, \qquad \mathbf{T} H \mathbf{T}^{-1} = H. \qquad (6)$$

The proper orthochronous Lorentz group is isomorphic to $SU(2)_A \otimes SU(2)_B$, each $SU(2)$ being generated independently by the complex linear combinations of the Lorentz generators

$$\mathbf{A} = \frac{1}{2}(\mathbf{J}+i\mathbf{K}) \qquad \mathbf{B} = \frac{1}{2}(\mathbf{J}-i\mathbf{K}) \qquad (7)$$

The irreducible representations (irreps) of this group are characterized by the eigenvalues of $\mathbf{A}^2(a(a+1))$ and $\mathbf{B}^2(b(b+1))$, hence we will label them as (a,b). This provide us with a specific representation of the generators. In particular for

1. $b=0$ we have $(a,0) \to$ right state (denoted ϕ_R) and $\mathbf{J}_R = i\mathbf{K}_R = \mathbf{A}$.
2. $a=0$ we have $(0,b) \to$ left state (denoted ϕ_L) and $\mathbf{J}_L = -i\mathbf{K}_L = \mathbf{B}$.

These two kinds of states are distinguished by the way they transform under boosts and are indistinguishable under rotations. Under parity $(a,b) \longleftrightarrow (b,a)$ hence if we require parity to be a symmetry it is necessary to enlarge our space to $(a,0) \oplus (0,a)$. We focus here in the latter case with $a = \frac{1}{2}$. In this space, the specific representation of a given operator can be obtained using the $|j,m\rangle_{a,b}$ basis (Weyl basis). In particular, the most general form for parity and boost operators read

$$\Pi = \begin{pmatrix} 0 & \pi\sigma_0 \\ \pi^*\sigma_0 & 0 \end{pmatrix}, \qquad B(\mathbf{p}) = \begin{pmatrix} \frac{E+m+\sigma\cdot\mathbf{p}}{\sqrt{2m(E+m)}} & 0 \\ 0 & \frac{E+m-\sigma\cdot\mathbf{p}}{\sqrt{2m(E+m)}} \end{pmatrix}, \qquad (8)$$

where π is a phase and $\sigma_0 = \mathbf{1}_{2\times 2}$. Now, we consider the eigenvalue equation for parity in the rest frame

$$\Pi\Psi(\mathbf{0}) = \eta\Psi(\mathbf{0}). \qquad (9)$$

Boosting this equation and upon the usage of Eq.(5) we get

$$(B^2(\mathbf{p})\Pi - \eta)\Psi(\mathbf{p}) = 0. \qquad (10)$$

Using the explicit representation of the parity and boost operators we obtain

$$[\gamma^\mu p_\mu - \eta m]\Psi(\mathbf{p}) = 0, \qquad (11)$$

where

$$\gamma^0 = \begin{pmatrix} 0 & \pi \\ \pi^* & 0 \end{pmatrix} \qquad \gamma^j = \begin{pmatrix} 0 & -\pi\sigma_i \\ \pi^*\sigma_i & 0 \end{pmatrix}$$

It can be shown that the full structure of Dirac theory follows from here. The solutions to this equation

$$\hat{v}_1(\mathbf{p}) = N_\pi \begin{pmatrix} \pi(E+m+p_z) \\ \pi p_+ \\ (E+m-p_z) \\ -p_+ \end{pmatrix}, \qquad \hat{v}_2(\mathbf{p}) = N_\pi \begin{pmatrix} \pi p_- \\ \pi(E+m-p_z) \\ -p_- \\ (E+m+p_z) \end{pmatrix} \quad (12)$$

$$\hat{v}_3(\mathbf{p}) = N_\pi \begin{pmatrix} \pi(E+m+p_z) \\ \pi p_+ \\ -(E+m-p_z) \\ p_+ \end{pmatrix}, \qquad \hat{v}_4(\mathbf{p}) = N_\pi \begin{pmatrix} \pi p_- \\ \pi(E+m-p_z) \\ -p_- \\ -(E+m+p_z) \end{pmatrix} \quad (13)$$

with $N_\pi = \frac{1}{\sqrt{4m(E+m)}}$, $p_\pm = p_x \pm i p_y$, are eigenstates of parity with eigenvalue $\eta = +1$ for $\hat{v}_{1,2}$ and $\eta = -1$ for $\hat{v}_{3,4}$.

We proceed with the construction of **C**. We define it as a *unitary* operator transforming a particle state into an anti-particle one and vice versa. In other words charge conjugation operator satisfy

$$\mathbf{C}\left|e^\uparrow\right\rangle = \varepsilon_{e^\uparrow}\left|\bar{e}^\uparrow\right\rangle, \quad \mathbf{C}\left|e^\downarrow\right\rangle = \varepsilon_{e^\downarrow}\left|\bar{e}^\downarrow\right\rangle, \quad \mathbf{C}\left|\bar{e}^\uparrow\right\rangle = \varepsilon_{\bar{e}^\uparrow}\left|e^\uparrow\right\rangle, \quad \mathbf{C}\left|\bar{e}^\downarrow\right\rangle = \varepsilon_{\bar{e}^\downarrow}\left|e^\downarrow\right\rangle \quad (14)$$

where ε_a are phases and we identified

$$\left|e^\uparrow\right\rangle \equiv \hat{v}_1, \quad \left|e^\downarrow\right\rangle \equiv \hat{v}_2, \quad \left|\bar{e}^\uparrow\right\rangle \equiv \hat{v}_3, \quad \left|\bar{e}^\downarrow\right\rangle \equiv \hat{v}_4. \quad (15)$$

Assuming the phases ε_a do not depend on spin, the unitarity of charge conjugation operator yields $\varepsilon_{\bar{e}} = \varepsilon_e^* \equiv \varepsilon$ and the most general form of this operator in the rest frame and in Weyl basis is

$$\mathbf{C} = \begin{pmatrix} \cos\lambda & \pi i \sin\lambda \\ -\pi^* i \sin\lambda & -\cos\lambda \end{pmatrix} \quad (16)$$

where we defined $\varepsilon = e^{i\lambda}$. Using this specific representation we boost the eigenvalue equation fo charge conjugation eigenvalue equation in the rest frame

$$\mathbf{C}\Psi(0) = \kappa\Psi(0) \quad (17)$$

to obtain

$$B(\varphi)(\mathbf{C} - \kappa)B^{-1}(\varphi)B(\varphi)\Psi(0) = 0 \quad (18)$$

which can be rewritten to

$$\left(\Gamma^\mu p_\mu \sin\lambda + \Gamma^5 m \cos\lambda - \kappa m\right)\Psi(\mathbf{p}) = 0 \quad (19)$$

where

$$\Gamma^0 = \begin{pmatrix} 0 & \pi i \\ -\pi^* i & 0 \end{pmatrix} \quad \Gamma^i = \begin{pmatrix} 0 & -\pi i \sigma_i \\ -\pi^* i \sigma_i & 0 \end{pmatrix} \quad (20)$$

These matrices satisfy the same anticommutation relations as the conventional Dirac γ matrices. Current conservation requires $\cos \lambda = 0$, thus $\mathbf{C}$ takes the simple form $\mathbf{C} = \Gamma^0$ and equation (19) becomes

$$\left(\Gamma^\mu p_\mu - \kappa m\right) \Psi(\mathbf{p}) = 0 \quad (21)$$

States satisfying this equation (charge conjugation eigenstates) describe totally neutral particles and are given as

$$\hat{w}_1(\mathbf{p}) = N \begin{pmatrix} i\pi(E+m+p_z) \\ i\pi p_+ \\ (E+m-p_z) \\ -p_+ \end{pmatrix}, \quad \hat{w}_2(\mathbf{p}) = N \begin{pmatrix} i\pi p_- \\ i\pi(E+m-p_z) \\ -p_- \\ (E+m+p_z) \end{pmatrix}, \quad (22)$$

$$\hat{w}_3(\mathbf{p}) = N \begin{pmatrix} -i\pi(E+m+p_z) \\ -i\pi p_+ \\ (E+m-p_z) \\ -p_+ \end{pmatrix}, \quad \hat{w}_4(\mathbf{p}) = N \begin{pmatrix} -i\pi p_- \\ -i\pi(E+m-p_z) \\ p_- \\ (E+m+p_z) \end{pmatrix}, (23)$$

where $N = \frac{1}{\sqrt{4m(E+m)}}$.

Summarizing, in this work we construct the *unitary* unitary automorphism mapping particle into anti-particle states in spinor space. We obtain the eigenspinors of this operator, which are the appropriate spinors to describe neutral particles. We obtain also the equation of motion for charge conjugation eigenstates. It has a similar form to the parity equation but with a different representation for the γ matrices. This points to a different interpretation of the representations of the Clifford algebra. In addition to a simple change of basis of the representation space, different representations of the γ matrices can also be associated to the description of different discrete properties of fermions. As can be easily verified, our unitary charge conjugation operator satisfy $\{C, \Pi\} = 0$, as expected.

ACKNOWLEDGMENTS

This work was supported by CONACyT and U. Gto. under projects 37234-E and DINPO-000085 respectively.

REFERENCES

1. G.C. Branco, L.Lavoura, J. P. Silva, *CP violation*, Oxford Univ. Press Inc. New York, (1999).
2. M. Napsuciale, C. A. Vaquera-Araujo, *Equations of motion as projectors and the gyromagnetic factor $g_s = 1/s$ from first principles*, hep-ph/0310106.
3. S. Weinberg, *The Quantum Theory of Fields Vol. 1*, Cambridge University Press (1995).

Sterile Neutrinos and the Solar Mixing

J.C. Gómez-Izquierdo and A. Pérez-Lorenzana

Departamento de Física, Cinvestav. Apdo. Post. 14-740, 07000, México, D.F., México

Abstract. Neutrino mixing matrix appears to be close to bimaximal mixing, but for the solar angle which is definitively smaller than 45 degrees. Whereas it seems quite easy to understand bimaximal mixings, as in models using $L_e - L_\mu - L_\tau$ global symmetries, understanding the 12 degrees of deviation in the observed solar angle seems less simple. We suggest that such a deviation could be due to a light sterile neutrino that mixes with the active sector. The mass scale needed to produce the effect is shown to be smaller than atmospheric scale, and thus, irrelevant for LSND.

Keywords: Neutrino oscillations. Sterile neutrinos
PACS: 14.60.Pq, 14.60.St, 11.30.Hv

INTRODUCTION

Convincing evidence that neutrinos have mass and oscillate has been provided along recent years by Kamiokande, Super-Kamiokande, MACRO and Soudan results on atmospheric neutrinos; by Chlorine, Kamiokande, Super-Kamiokande, SAGE, GALLEX and most recently the SNO experiment on solar neutrinos; as well as by KamLAND, K2K and CHOOZ, PALO-Verde, long and short base-line neutrino experiments [1]. KamLAND independent confirmation of oscillation parameters observed by SNO indicates that the solar deficit is due to MSW matter effect [2] and large mixing angle oscillations.

In the standard framework, only three weak neutrino species are needed to consistently describe all experimental results, with the sole addition of neutrino masses and mixings for neutrino oscillations. Point is that weak and mass eigenstates are different, but they can be expressed in terms of each other through a unitary rotation: $\nu_\alpha = \sum_i U_{\alpha i} \nu_i$, for $\alpha = e, \mu, \tau$ and $i = 1, 2, 3$. A common parameterization of the mixing matrix, U, is given in terms of three angles and three CP phases, such that $U = U_{PMNS} K$, where $K = \text{diag}\{1, e^{i\phi_1}, e^{i\phi_2}\}$, with ϕ_1, ϕ_2 the physical CP-odd Majorana phases, and with the Pontecorvo-Maki-Nakagawa-Sakata (PMNS) matrix [3]

$$U_{PMNS} = \begin{pmatrix} c_{12}c_{13} & s_{12}c_{13} & s_{13}e^{-i\varphi} \\ -s_{12}c_{23} - c_{12}s_{23}s_{13}e^{i\varphi} & c_{12}c_{23} - s_{12}s_{23}s_{13}e^{i\varphi} & s_{23}c_{13} \\ s_{12}s_{23} - c_{12}c_{23}s_{13}e^{i\varphi} & -c_{12}s_{23} - s_{12}c_{23}s_{13}e^{i\varphi} & c_{23}c_{13} \end{pmatrix} ; \quad (1)$$

where c_{ij} and s_{ij} stand for $\cos\theta_{ij}$ and $\sin\theta_{ij}$ respectively. φ has not yet been measured, but most analysis suggest a small phase. In the CP conservation limit a combined analysis of all data indicates that at two sigma level [1]

$$\sin^2\theta_{12} = 0.314 \, ^{+0.057}_{-0.047} ; \quad \sin^2\theta_{23} = 0.44 \, ^{+0.18}_{-0.096} . \quad (2)$$

The kinematical scales for the oscillation are given by two mass squared differences: (i) the solar/KamLAND scale $\Delta m_{sol}^2 = \Delta m_{12}^2$; and (ii) the atmospheric scale $\Delta m_{ATM}^2 = \Delta m_{23}^2$; with the values [1]

$$\Delta m_{sol}^2 = (7.92 \pm 0.71) \times 10^{-5} \text{ eV}^2 \; ; \quad \Delta m_{ATM}^2 = (2.4 ^{+0.5}_{-0.62}) \times 10^{-3} \text{ eV}^2 \; . \quad (3)$$

Thus, whereas atmospheric angle, $\theta_{ATM} = \theta_{23} \approx 41.55° \pm 6.5°$; is consistent with maximal mixing, solar is smaller: $\theta_{sol} = \theta_{12} \approx 34.08° \pm 1.6°$. Since $\theta_{13} = \theta_{CHOOZ} \approx 0° \pm 7.4°$, it is suggestive to think that U_{PMNS} arises from a theory for neutrino masses and mixings that contains some global (flavor) symmetry G, such that at the exact symmetric limit the mixing is bimaximal, thus of the form:

$$U_{BM} = \begin{pmatrix} \frac{1}{\sqrt{2}} & \frac{1}{\sqrt{2}} & 0 \\ \frac{1}{2} & -\frac{1}{2} & \frac{1}{\sqrt{2}} \\ \frac{1}{2} & -\frac{1}{2} & -\frac{1}{\sqrt{2}} \end{pmatrix} \; ; \quad (4)$$

for which two angles are exactly maximal and third is null. The breaking of G would provide the eleven degrees of deviation in solar angle. One would then write $U_{PMNS} = U_{BM} \times U_A$ where U_A parameterize the additional rotations induced by the breaking of G. Here we work in a class of simple models where the flavor symmetry is chosen to be $L' = L_e - L_\mu - L_\tau$ [4, 5], and suggest that U_A comes due to corrections that involve at least one light sterile neutrino. The reason for this would appear clear when one realizes that the sole active corrections may be unlikely to provide correct values for both the solar parameters (see next section). Our hypothesis may be constrained by solar data, however, with the currently allowed range of active-sterile mixing at one sigma level, $\sin^2 \eta < 0.09$ [6], there is still enough room as to provide the desired corrections.

THE SOLAR MIXING PROBLEM IN L' MODELS

We start by assuming that on the flavor basis where charged lepton masses are diagonal, the Majorana neutrino mass matrix has the form dictated by $L_e - L_\mu - L_\tau$:

$$M_{active} = M_0 + M_\varepsilon = m \begin{pmatrix} 0 & \cos\theta & \sin\theta \\ \cos\theta & \varepsilon & 0 \\ \sin\theta & 0 & \varepsilon \end{pmatrix} \; . \quad (5)$$

where ε is a small parameter representing a typical leading corrections introduced by the breaking of L'. At zero order, taking $\varepsilon = 0$, the above mass matrix exactly produces, for $\tan\theta = 1$, bimaximal mixing. There are two massive neutrinos, with masses: $\pm m$, and a massless one. So the spectrum is inverted, but solar scale is exactly zero at this level. Atmospheric oscillations require $m^2 = \Delta m_{ATM}^2$.

After bimaximal rotation, the ε contributions introduce a mixing only for the heavier states in the 1-2 sector, whereas the lightest state gets a mass of order εm. After diagonalizing M_{active}, we get for the solar mixing

$$\sin^2 2\theta_{sol} = \cos^2 2\alpha = \left[\frac{1+\sqrt{1+\varepsilon^2/4}}{1+\varepsilon^2/4+\sqrt{1+\varepsilon^2/4}}\right]^2 \approx 1 - \frac{\varepsilon^2}{4} . \quad (6)$$

where α is the additional rotation angle. It is the clear that $\theta_{sol} = \pi/4 - \alpha$, which explicitly shows that the ε corrections work in the right direction. On the other hand, the solar scale would be $\Delta m_{sol}^2 = \varepsilon\, m^2\, \sqrt{4+\varepsilon^2}$. Fixing ε from this equation, one gets

$$\varepsilon \approx \frac{1}{2} \frac{\Delta m_{sol}^2}{\Delta m_{ATM}^2} \approx 1.65 \times 10^{-3} ; \quad (7)$$

However, this means a too large $sin^2 2\theta_{sol} = 0.9999$, equivalent to $\sin^2 \theta_{sol} = 0.496$.

STERILE CONTRIBUTIONS

From the previous analysis one sees that the simple corrections we considered are unable to provide both the solar mass scale and mixing angle. A complete understanding along this line of thought for these parameters may need to consider more complicated corrections to the original texture. Another possibility is that the desired corrections may come from another source. We suggest that this extra source could be the coupling to a fourth (sterile) neutrino [7, 8]. In the hypothesis that the sterile mixes preferentially to a single active state after a bimaximal rotation, we can choose, for instance, the active to sterile coupling along the direction $(1, c_\theta, s_\theta)$. Moreover, for simplicity we also assume $m_s = 0$. After bimaximal rotation we then get

$$M = m \begin{pmatrix} 1+\frac{\varepsilon}{2} & -\frac{\varepsilon}{2} & 0 & \delta \\ -\frac{\varepsilon}{2} & -1+\frac{\varepsilon}{2} & 0 & 0 \\ 0 & 0 & \varepsilon & 0 \\ \delta & 0 & 0 & 0 \end{pmatrix} . \quad (8)$$

Since the third state decouples, $\theta_{13} = 0$ still. A semi-perturbative analysis, using only ε as a perturbation, shows that

$$\Delta m_{sol}^2 \approx \Delta m_{ATM}^2 \left[\delta^2 - \lambda_- + \varepsilon\left(1+\lambda_+ \cos^2\beta\right)\right] . \quad (9)$$

for $\lambda_\pm = \frac{1}{2}(1 \pm \sqrt{1+4\delta^2})$ and $\tan\beta = \delta/\lambda_+$, whereas for the solar mixing we get

$$\sin^2 2\theta_{sol} \approx \cos^2\beta \left[1 + \frac{3}{2}\varepsilon \sin^2\beta \left(\frac{1+\delta^2}{(2-\delta^2)\sqrt{1+4\delta^2}} - \frac{2\varepsilon^2}{(3+\lambda_+ - \lambda_-)^2}\right)\right]^2 . \quad (10)$$

Now, both equations can be solved numerically for central values of the solar parameters and one gets $\varepsilon \approx -0.1412$ and $\delta \approx 0.4122$. These values validate our approximations. The fact that ε is negative indicates that the observable values come from a compensation among both the corrections.

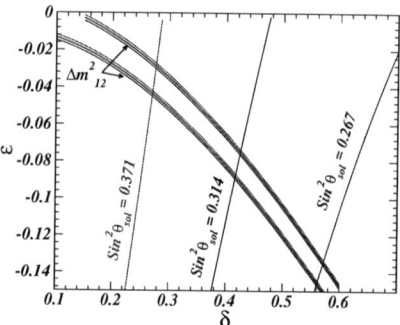

FIGURE 1. Allowed values of ε and δ which give the solar parameters within two sigma deviations. The two narrow regions for the squared mass difference are depicted.

A precise calculation can be done by numerically diagonalizing the mass matrix in Eq. (8) for given values of the pair (δ, ε). This shows that there is actually a narrow range of parameters that would give acceptable results for our model (see Fig. 1). Notice that whereas the mixing angle, $\sin^2 \theta_{sol}$, is little sensitive to ε, and corresponds to a wide region on the parameter space, the solar scale, Δm_{12}^2, is clearly sensitive to both ε and δ parameters. This indicates that some level of fine tuning may be required to get the proper scale.

CONCLUDING REMARKS

We have suggested the possibility that light sterile neutrinos may be the missing ingredient that transforms a maximal mixing into the observed large mixing angle in solar neutrino oscillations. The idea has been realized in a simple toy model with inverted hierarchy and broken global $L_e - L_\mu - L_\tau$ symmetry. With the introduction of appropriate sterile couplings the model gain enough freedom to arrange both solar parameters to the proper order of magnitude. We note that the light sterile neutrino comes with a new mass scale, Δm_{34}^2, which is lighter than solar scale and does not contribute to P_{ee} in solar oscillations, since it is attached to a null U_{e3}. The effective fraction of sterile neutrinos is found to roughly be just below the current limits: $\eta_s \approx 7.1 \times 10^{-2}$. Our results suggest that a light sterile neutrino is an interesting possibility that deserves further exploration.

Acknowledgments.- Work partially supported by CONACyT, México, grant J44596.

REFERENCES

1. For recent reviews and references see for instance: G.L. Fogli, *et al.*, hep-ph/0506083; M.Maltoni, T. Schwetz, M.A. Tortola and J.W.F. Valle, New. J. Phys. **6** (2004) 122 [hep-ph/0405172].
2. L. Wolfenstein, Phys. Rev. **D17** (1978) 2369; S.P. Mikheev and A. Yu Smirnov, Sov. J. Nucl. Phys. **42** (1985) 913.
3. Z. Maki, M. Nakagawa, and S. Sakata, Prog. Theor. Phys. **28** (1962) 870; B. Pontecorvo, Zh. Eksp. Teor. Fiz. **53** (1968) 1717 [Sov. Phys. JETP **26** (1968) 984].

4. For some examples see: S. T. Petcov, Phys. Lett. B110 (1982) 245; R. Barbieri, L. J. Hall, D. R. Smith, A. Strumia and N. Weiner, JHEP 9812, 017 (1998); A. S. Joshipura and S. D. Rindani, Eur. Phys. J. C14, 85 (2000); R. N. Mohapatra, A. Pérez-Lorenzana and C. A. de S. Pires, Phys. Lett. B474, 355 (2000); T. Kitabayashi and M. Yasue, Phys. Lett. B524, 308 (2002); R. N. Mohapatra, Phys. Rev. D 64, 091301 (2001); K. S. Babu and R. N. Mohapatra, Phys. Lett. B 532, 77 (2002).
5. A. Pérez-Lorenzana, C.A. De S. Pires, Phys. Lett. **B522** (2001) 297.
6. See for instance M. Cireli, G. Marandella, A. Strumia, F. Vissani, Nucl. Phys. **B708** (2005) 215.
7. For a complete discussion see: J.C. Gómez-Izquierdo, A. Pérez-Lorenzana, hep-ph/0601223; J.C. Gómez-Izquierdo, .M.Sc. Thesis, CINVESTAV (2005).
8. For early ideas see: P.C. de Holanda, A. Yu. Smirnov, Phys. Rev. **D 69** (2004) 113002; K.R.S. Balaji, A. Pérez-Lorenzana, A.Yu. Smirnov, Phys. Lett. **B509** (2001) 111.

Tritium-Beta Spectrum in a Left-Right Symmetric Model

A. Gutiérrez-Rodríguez*, M. A. Hernández-Ruíz[†] and F. Ramírez-Sánchez*

*Facultad de Física, Universidad Autónoma de Zacatecas, Apartado Postal C-580, 98060 Zacatecas, Zacatecas México.
[†]Facultad de Ciencias Químicas, Universidad Autónoma de Zacatecas, Código Postal 98600 Zacatecas, Zacatecas México.

Abstract. We start with a Left-Right Symmetric Model and we analyze the endpoint of the beta decay of the tritium $^3H \to {}^3He + e^- + \bar{\nu}_e$. We applied this model to incorporate the right currents, whereby we propose an amplitude whose leptonic part contains the parameter λ defined as a left-right asymmetry parameter which measures the parity violation. We realized the numerical computation to the sensibility of the experiments Mainz and Troitsk of $m_{\nu_e} = 2.2\ eV$. We find that the spectrum of energy of the electrons for this experiments it is not affected by the left-right asymmetry parameter.

Keywords: Models beyond the standard model, Neutrino mass and mixing, Beta decay.
PACS: 12.60.-i, 14.60.Pq, 23.40.-s

INTRODUCTION

In modern particle physics, one of the most intriguing and most challenging tasks is to discover the rest mass of neutrinos which bear fundamental implications for particle physics, astrophysics and cosmology. Until recently, the Standard Model (SM) [1] of particle physics assumed neutrinos to be massless. However, actual investigations of neutrinos from the sun and of neutrinos created in the atmosphere by cosmic rays, in particular the recent results of the Super-Kamiokande experiment on the neutrino oscillations [2] as well as on the GALLEX, SAGE, GNO, HOMESTAKE and Liquid Scintillator Neutrino Detector (LSND) [3] experiments, have given strong evidence for massive neutrinos indicated by neutrino oscillations.

The best neutrino mass limits have been extracted from measurements of the tritium β-decay spectrum close to its endpoint. Since neutrinos are very light particles, a mass measurement can best be performed in this region of the spectrum as in other parts the nonlinear dependencies caused by the relativistic nature of the kinematic problem cause a significant loss of accuracy. This by far overwhelms the possible gain in statistics one could hope for. Two groups in Mainz and Troitsk used spectrometers based on Magnetic Adiabatic Collimation combined with an Electrostatic filter (MAC-E technique), which obtained the same value $m_{\nu_e} < 2.2\ eV$ [4, 5], 95% C.L..

A new experiment, KATRIN [6], is presently prepared in Karlsrube, Germany, which is planned to exploit the same technique. It aims for an improvement by about one order of magnitude. The physical dimensions of a MAC-E device scale inversely with the possible sensitivity to a finite neutrino mass. This may ultimately limit an approach with this principle. The new experiment will be sensitive to the mass range where a finite effective neutrino mass value of between 0.1 and 0.9 eV was extracted from a signal in neutrinoless double β-decay in ^{76}Ge [7]. The Heidelberg-Moskow collaboration performing this experiment in the Grand Sasso laboratory reports a 4.2 standard deviation effect for the existence of this decay.

Determination of the absolute scale of neutrinos masses is one of the most important and, at the same time, challengin problems in neutrino physics. Currently, the study of the electron energy spectrum near the endpoint of the Tritium beta decay

$$^3H \rightarrow {}^3He + e^- + \bar{\nu}_e, \quad (1)$$

is the most sensitive direct method of determining the scale of masses. We obtain in the absence of mixing, the energy spectrum of the emitted e^-:

$$\frac{d\Gamma}{dE_e} = \frac{G_F^2}{2\pi^3}(1+\lambda^2)(1+3\rho^2)F(Z,R,E_e)p_e E_e(E_0-E_e)\sqrt{(E_0-E_e)^2 - m_{\bar{\nu}_e}^2}, \quad (2)$$

where G_F is the Fermi constant, λ is the left-right asymmetry parameter, p_e, E_e and E_0 are the momentum, energy, and maximum endpoint energy, respectively, of the electron. The Fermi function, $F(Z, R_e, E_e)$, captures the correction due to the Coulomb interactions of the electron with the charge Ze of the daughter nucleus [8].

The purpose of this paper is to carry out an analysis of the tritium beta decay in the context of a model with left-right symmetry [9]. We start from an extension of the electroweak model applied to the baryons decay [10]. This model contains the parameter λ defined as the parameter of left-right asymmetry which measures the parity violation. We apply this theory to incorporate the right currents, for which we propose an amplitude whose leptonic part is $V + \lambda A$, with $\lambda = -1$ for left currents and $\lambda = 1$ for right currents. The analysis consists of seeing if the endpoint of the tritium beta decay for $m_{\nu_e} = 2.2$ eV (Mainz and Troitsk) is affected by the left-right asymmetry parameter.

This paper is organized as follows: In Sec. II we make the numerical computations. Finally, we summarize our results in Sec. III.

RESULTS

In this section we present our results and conclusions of the beta decay of the tritium. We realized the numerical computation to the sensibility of the experiments Mainz and Troitsk of $m_{\nu_e} = 2.2 \, eV$. An important point here is to see the influence of the parameter of left-right asymmetry λ in the spectrum or distribution of energy of the electrons.

If the lower limit on the mass of heavy vector-boson ($M_{W_R} = 715 GeV$) [11] and the upper limit on the mixing angle ζ of left-and right-handed bosons ($\zeta < 3 \times 10^{-3}$) [11] are taken into account, one finds that the asymmetry parameter lambda (λ) to be very close to minus unity ($-1 < \lambda < -0.98$).

We have calculated differential electron energy spectra, $d\Gamma/dE_e$, in tritium decay using the following values for the mass of ^3H and ^3He: $M_{^3H} = 2809.4319 \, MeV$ [12]

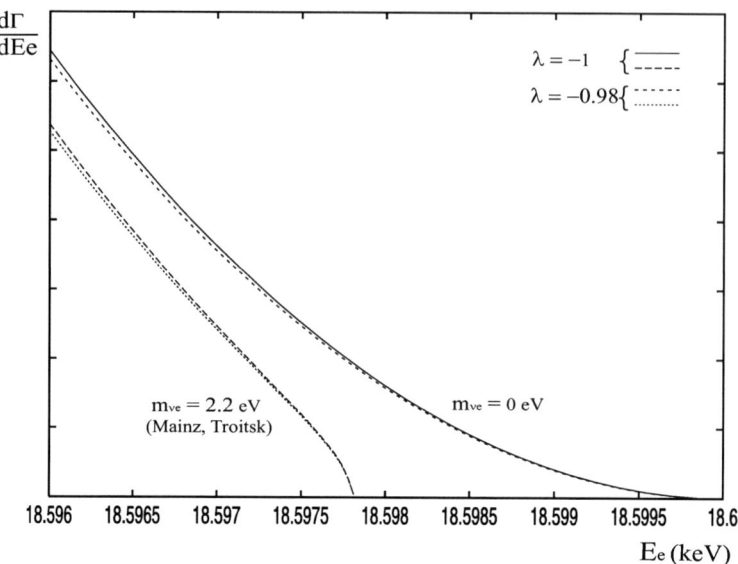

FIGURE 1. The electron energy spectrum of tritium β decay. The β spectrum is shown for neutrino masses of 0 and 2.2 eV (Mainz and Troitsk), with $\lambda = -1, -0.98$.

and $M_{^3_1He} = 2808.9023\ MeV$ [12], respectively. The corresponding value for the electron endpoint kinetic energy is $T_{max} = 18587.56\ eV$. Differential electron energy spectra, $d\Gamma/dE_e$, corresponding to decays with a massive and massless neutrino in the vicinity of the endpoint, are shown in Fig. 1. This figure corresponds to $\lambda = -1, -0.98$, that is, for left currents and mixing of currents.

If $m_{v_e} = 0$, Eq. (2) immediately shows that $d\Gamma/dE_e \propto (E_0 - E_e)^2$. Thus, if we plot the quantity $d\Gamma/dE_e$ vs E_e, we would obtain an almost straight line. Such a plot is called the Kurie plot, as shown in Fig. 1 by the solid line. Since the quantity $d\Gamma/dE_e$ must be non-negative, the maximum value of the electron kinetic energy in this case is E_0.

However, the almost linearity of the Kurie plot is lost if the neutrino has a non-zero mass (Mainz and Troitsk). The effect of this mass becomes appreciable only near the endpoint of the plot where $(E_0 - E_e)$ is comparable to m_{v_e}. Notice that in this case, the maximun kinetic energy of the electron is not given by E_0, but rather by the vanishing of the quantity inside the square root sign in Eq. (2), i.e., $E_e^{max} = E_0 - m_{v_e}$.

CONCLUSIONS

We have done an analysis on the electron energy spectrum. The analysis is for $m_{v_e} = 0$ eV, $m_{v_e} = 2.2$ eV (Mainz and Troitsk). The difference between an electron neutrino with mass and the undistorted β spectrum of a massless v_e is clearly observed. The spectral distortion is only significant in a region close to the β endpoint.

In summary, we conclude that the parameter of asymmetry λ does not modify the curve of the tritium β spectrum, but only raises or drops it depending on the value of λ. In the case of $\lambda = -1$, we reproduce the curve of the spectrum of energy previously reported in the literature.

ACKNOWLEDGMENTS

This work was supported in part by *SEP-CONACyT* (Projet: 2003-01-32-001-057) and SNI (México).

REFERENCES

1. S. L. Glashow, Nucl. Phys. **22**, 579 (1961); S. Weinberg, Phys. Rev. Lett. **19**, 1264 (1967); A. Salam, in *Elementary Particle Theory: Relativistic Groups and Analyticity* (Nobel Symposium No. 8), edited by N. Svartholm (Almqvist and Wiksell, Stockholm, 1968), p. 367.
2. Super-Kamiokande Collaboration, Y. Fukuda *et al.* Phys. Rev. Lett. **81**, 1158 (1998); **81**, 4279 (1998).
3. SAGE Collaboration , J.N. Abdurashitov *et al.*, Phys. Rev. Lett. **83**, 4686 (1999); Phys. Rev. C **60**, 055801 (1999); GALLEX Collaboration, W. Hampel *et al.* Phys. Lett. B**447**, 127 (1997); GNO Collaboration , M. Altmann *et al., ibid.* **490**, 16 (2000); B.T. Cleveland *et al.*, Astrophys. J. **496**, 505 (1998); LSND Collaboration, C. Anthanassopoulos *et al.*, Phys. Rev. Lett. **81**, 1774 (1998).
4. V. M. Lobashev *et al.*, Phys. Lett. **B460**, (1999) 227; Ch. Kraus *et al.*, Eur. Phys. J. **40**, (2005) 447.
5. Ch. Weinheimer, hep-ex/0306057 (2003); Nucl. Phys. **B118**, (2003) 279.
6. A. Osipowicz *et al.*, hep-ex/0109033 (2001).
7. H. V. Klapdor-Kleingrothaus, Phys. Lett. **B586**, (2004) 198.
8. S.M. Berman, Phys. Rev. **112**, (1958) 267; T. Kinoshita and A. Sirlin, Phys. Rev. **113**, (1959) 1652.
9. R. Huerta, Nucl. Phys. **B204**, (1982) 413.
10. A. García, R. Huerta, M. Maya, R. Pérez Marcial, Phys. Lett. **B157**, (1985) 317.
11. Particle Date Group, S. Eidelman *et al.* Phys. Lett. **B592**, 1 (2004).
12. http://www.tunl.duke.edu/NuclData/.

Optical Simulation for V0A

Carlos Pérez Lara*, Alberto Gago Medina* and Gerardo Herrera Corral**

Pontificia Universidad Catolica del Perú: Departamento de Ciencias, Sección Física
**Centro de Investigación y de Estudios Avanzados: Departamento de Física*

Abstract. The V0A detector is one of the forward detectors that will be used for trigger in the ALICE experiment at CERN. Simulation results of the optical response of the V0A elements are presented in this work. The simulations are based on the LITRANI package. The simulation results guarantee a flat response of the whole detector as well as within each cell of the array.

Keywords: simulation, scintillator, V0A, fluorescence
PACS: 29.40.Mc, 29.40.-n, 29.90.+r

INTRODUCTION

The V0A detector is part of a system which consists of two arrays of scintillator cells forming discs and installed on both sides of the ALICE interaction point. The system will provide online signals for the trigger in both proton-proton and ion-ion collisions.

The description of the ALICE V0A detector can be found in the Technical Design Report [1].

Simulation of the detector response is important because it allow us to not only check the resolution and homogeneity of the detector but also to have an estimate of the number of photons reaching the photomultiplier tube.

The package we have used is LITRANI [2], a general purpose Monte-Carlo for the simulation of propagation of optical photons. It is written in C++ and was originally developed for the calibration of the CMS calorimeter.

We have developed a code using LITRANI v3.0 in a Root [3] v4.08f framework for this purpose. Our aims are the following:

- Analysis of the homogeneity in each cell of the detector
- Analysis of the homogeneity in the entire detector
- Propagation of photons through the wavelength shifter fibers

SETUP

V0A array is subdivided in 32 cells as show in figure 1. Due to LITRANI limitations, we have considered each cell as the composition of many prisms whose bases are trapezoids.

Each cell is made of plastic scintillator wrapped with teflon and with fibers embedded in, see figure 1. The response of the materials is shown at figure 2. These information was taken from materials datasheets (see [4] and [5]).

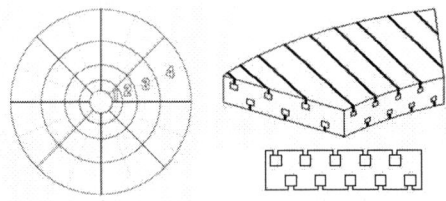

FIGURE 1. Segmentation of V0A disc

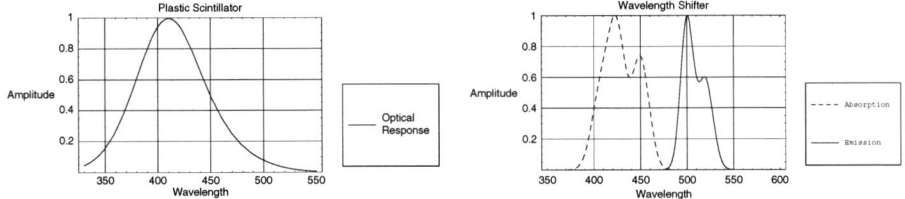

FIGURE 2. Optical response for plastic scintillator and wavelength shifter bars.

Geometrical and Physical Parameters

Plastic scintillator

- Radius: 4.2cm, 7.6cm, 13.8cm, 22.7cm, 41.3cm.
- Width: 2.5cm
- Absorption length: 140cm
- Refraction index: 1.58
- Decay time: 2.5 nanoseconds.

Wavelength shifter fibers

- Square prisms ($1mm^2 x length$).
- Embedding deep: 4mm.
- Channel width: 0.6 mm.
- Absorption length: 400cm

ANALYSIS OF GEOMETRICAL EFFICIENCY

In order to study the efficiency, we define two important quantities:

$$Efficiency = \frac{Number\ of\ photons\ detected}{Number\ of\ photons\ generated\ due\ to\ minimum\ ionization\ particles}$$

and,

$$Normalized\ area\ of\ detection = K_i\ (Number\ of\ fibers)$$

where

$$K_i = \frac{1}{Number\ of\ fibers\ to\ fill\ the\ cell\ i}$$

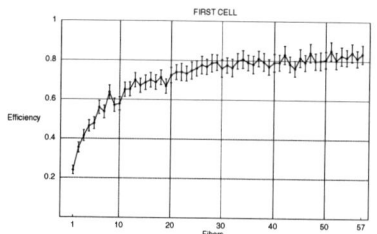
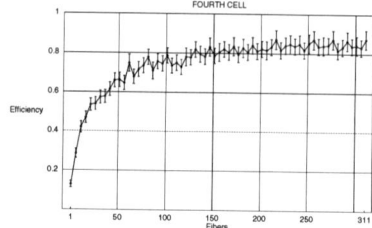

FIGURE 3. Efficiency for two cells.

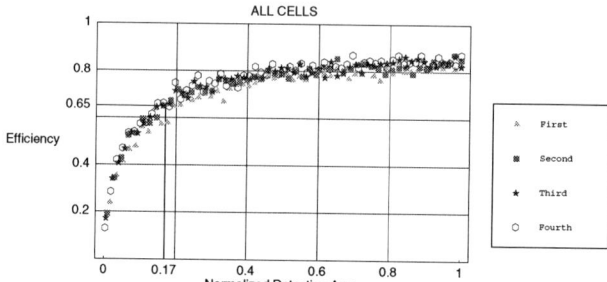

FIGURE 4. Efficiency for photons reaching at fibers.

The values of K_i according to the geometry of each cell are: $K_1 = 0.0175$, $K_2 = 0.0097$, $K_3 = 0.0058$ and $K_4 = 0.0032$

The simulation was performed with a beam of pions with 6.5 GeV/c momentum. The beam scans the whole cell area generating light that is propagated through the material.

The efficiency dependence upon the number of fibers

To study the efficiency dependende upon the number of fibers, we consider that every photon that reachs a fiber is detected. Figure 3 shows the efficiency in two cells of the V0A array. From figure 4, we have the behavior of the efficiency for all cells, which is, at the begining, growing rapidly with the normalized area of detection, being that gradually the efficiency tends to be a constant. An efficiency of 65% represents a good point to characterize this last tendency, which corresponds to a value of 0.17 as show in figure 4.

It means that, in order to have a flat efficiency over the whole detector, cell 1 should have $N = \frac{0.17}{K_1} \approx 10$ fibers. In a similar way, cell 2 should have $N \approx 18$; cell 3, $N \approx 29$ and cell 4, $N \approx 53$ fibers.

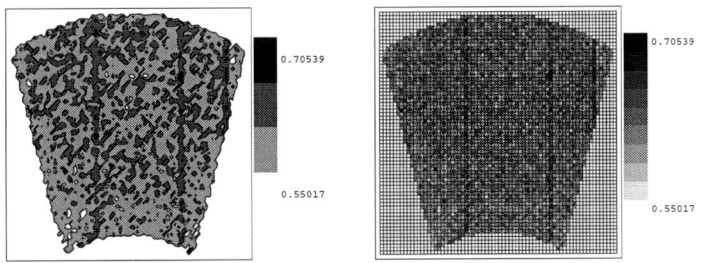

FIGURE 5. Geometric efficiency of the innermost cell.

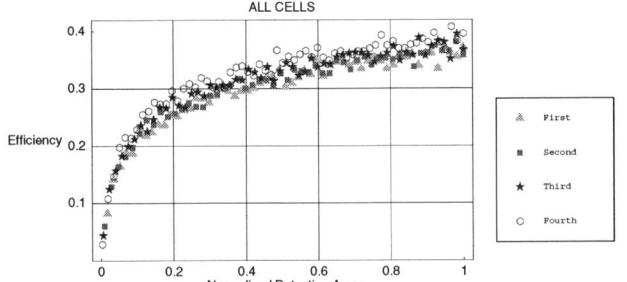

FIGURE 6. Efficiency for photons reaching the PMT.

Analysis of the homogeneity of each cell

It is also important to know the resolution of each cell of the V0A in order to guarantee an even efficiency within the cell. We have tested each cell with a sequential scanning of beams and checked the efficiency at each point. Figure 5 shows the first cell response with 10 fibers. As we can check, the efficiency varies very little within the cell.

Transmission toward PMT

We have used the LITRANI's code to simulate the propagation of photons inside the fibers. For this part, we consider as detected photons only those that impact at the PMT, which is located at the end of the fibers. In order to do so, the LITRANI source code was modified to take into account the wavelength shifting properties of the fibers [5].

After this improvement the efficiency turn out to be as shown in figure 6. Even though the efficiency has decrease, the rate still guarantee a clear signal at the PMT due to the material light yield.

CONCLUSION

The simulation results guarantee two things:

- An even efficiency over the whole detector, which avoids big scaling factors in using the signals coming from the actual measurements, and
- A clear signal at the PMT, even if we assume a pessimistic 1% quantum efficiency, due to the amount of photons produced.

REFERENCES

1. ALICE Collaboration, *Technical Design Report on Forward Detectors: FMD, T0 and V0*, CERN/LHCC/2004-025
2. Francois - Xavier Gentit, *LITRANI, Light Transmission for Anisotropic Media*, http://gentit.home.cern.ch/gentit
3. The ROOT Team, *ROOT, An Object-Oriented Data Analysis Framework*, http://root.cern.ch
4. Saint-Gobain Crystals. *Premium Plastic Scintillators*, http://www.detectors.saint-gobain.com
5. ELJEN Technology. *EJ-280 Wavelength shifting plastics*, http://www.apace-science.com/ljen/ej-280-4.htm

Dynamical mass generation in QED with weak magnetic fields

A. Ayala*, A. Bashir*,†,**, A. Raya†,‡ and E. Rojas*

*Instituto de Ciencias Nucleares, Universidad Nacional Autónoma de México, Circuito Exterior, C. U., Apartado Postal 70-543, 04510, México, D. F.
†Instituto de Física y Matemáticas, Universidad Michoacana de San Nicolás de Hidalgo, Apartado Postal 2-82, Morelia, Michoacán 58040, México
** Institute for Particle Physics Phenomenology, University of Durham, Durham DH1 3LE, U.K.
‡Instituto de Ciencias Nucleares, Universidad Nacional Autónoma de México, Ciurcuito Exterior, C. U., Apartado Postal 70-543, 04510, México, D. F.

Abstract. We study the dynamical generation of masses for fundamental fermions in quenched quantum electrodynamics in the presence of magnetic fields using Schwinger-Dyson equations. We show that, contrary to the case where the magnetic field is strong, in the weak field limit $eB \ll m(0)^2$, where $m(0)$ is the value of the dynamically generated mass in the absence of the magnetic field, masses are generated above a critical value of the coupling and that this value is the same as in the case with no magnetic field. We carry out a numerical analysis to study the magnetic field dependence of the mass function above critical coupling and show that in this regime the dynamically generated mass and the chiral condensate for the lowest Landau level increase proportionally to $(eB)^2$.

Keywords: Schwinger-Dyson equations, Nonperturbative phenomena, Dynamical chiral symmetry breaking.
PACS: 11.15.Tk, 12.20.-m

It is well known that in quantum electrodynamics (QED), fermions can acquire masses through self interactions without the need of a non zero bare mass. This phenomenon, known as dynamical mass generation (DMG), happens above a certain critical value of the coupling and its description can only be carried out in terms of non-perturbative treatments. Schwinger-Dyson equations (SDEs) provide a natural platform to study DMG. In quenched quantum electrodynamics (qQED), a favorite starting point is to make an ansatz for the fermion-photon vertex and then study the fermion propagator equation in its decoupled form. It is also well known that in the presence of strong magnetic fields, it is possible to generate fermion masses for any value of the coupling. This phenomenon has been given the name of *magnetic catalysis* [1, 2, 3, 4]. In this work, we undertake the study of QED in the rainbow approximation in the presence of a weak magnetic field.

When the magnetic field is strong, Landau levels are separated from each other by an amount $\sim \sqrt{eB}$ in such a way that for any value of the coupling α, only the lowest Landau level (LLL) contributes to the DMG [1, 2, 3, 4]. However, in the case of weak external magnetic fields, Landau levels are close to each other and hence all contributions should be taken into account, which adds considerably to the complexity of the problem.

It has been shown [5] that the mass operator in the presence of an electromagnetic field can be written as a combination of the structures

$$\gamma^\mu \Pi_\mu, \; \sigma^{\mu\nu} F_{\mu\nu}, \; (F_{\mu\nu}\Pi^\nu)^2, \; \gamma_5 F_{\mu\nu}\tilde{F}^{\mu\nu} \tag{1}$$

which commute with the operator $(\gamma \cdot \Pi)^2$, where $\Pi_\mu = i\partial_\mu - eA_\mu^{\text{ext}}$, $F_{\mu\nu} = \partial_\mu A_\nu^{\text{ext}} - \partial_\nu A_\mu^{\text{ext}}$, $\tilde{F}^{\mu\nu} = \frac{1}{2}\varepsilon^{\mu\nu\lambda\tau}F_{\lambda\tau}$, $\sigma_{\mu\nu} = \frac{i}{2}[\gamma_\mu, \gamma_\nu]$ and A^{ext} is the external vector potential. We take $A_\mu^{\text{ext}} = B(0, -y/2, x/2, 0)$ in such a way that it gives rise to a constant magnetic field $\mathbf{B} = B\hat{z}$.

The equation relating the two-point fermion Green's function $G(x,y)$ and the mass operator $M(x,y)$ in coordinate space reads

$$\gamma \cdot \Pi(x) G(x,y) - \int d^4 x' M(x,x') G(x',y) = \delta^4(x-y) \tag{2}$$

where the mass function in the rainbow approximation in coordinate space is

$$M(x,x') = -ie^2 \gamma^\mu G(x,x') \gamma^\nu D^{(0)}_{\mu\nu}(x-x'). \tag{3}$$

Weakness of the magnetic field implies that the bare vertex is a reasonable choice in the sense that the Ward Identity is satisfied in the Landau gauge upto a correction connected with the mass function in momentum space $\mathcal{M}$.

Schwinger [6] was the first to obtain an exact analytical expression for the fermion Green's function in the presence of a constant electromagnetic field of arbitrary strength. However, we find the alternative representation of $G(x,y)$ proposed by Ritus [5] more convenient for our purposes. Since, in the presence of a constant external field, the fermion asymptotic states are no longer free particle states represented by plane waves, but are described by wave functions consistent with the particular external field configuration, namely eigenfuncions of $(\gamma^\mu \Pi_\mu)^2$, it is convenient to work in the representation spanned by these eingenfunctions, where the mass operator is diagonal. A matrix is used to *rotate* the SDE to momentum space, yielding [7]

$$\begin{aligned}
\mathcal{M}(p_\|, n_p) &= \frac{-ie^2}{2} \sum_{\sigma_k, \sigma_p = \pm 1} \sum_{n_k, s_k} \frac{s_p! s_k!}{(n_p - \frac{(\sigma_p+1)}{2})!(n_k - \frac{(\sigma_k+1)}{2})!} \\
&\times \int \frac{d^4 q}{(2\pi)^4} \frac{e^{-q_\perp^2/2\gamma}}{q^2 + i\varepsilon} \frac{\mathcal{M}((p-q)_\|, n_k)}{(p-q)_\|^2 - 2eBn_k - \mathcal{M}^2((p-q)_\|, n_k)} \\
&\times \left(\frac{q_\perp^2}{4\gamma}\right)^{l_k - l_p - \frac{(\sigma_k - \sigma_p)}{2}} \left[2 + \frac{1}{q^2}\left(q_\perp^2(1 - \delta_{\sigma_p \sigma_k}) - q_\|^2 \delta_{\sigma_p \sigma_k}\right)\right] \\
&\times \left[L^{n_k - n_p - \frac{(\sigma_k - \sigma_p)}{2}}_{n_p - \frac{(\sigma_p+1)}{2}}(q_\perp^2/4\gamma)\right]\left[L^{s_p - s_k}_{s_k}(q_\perp^2/4\gamma)\right]^2,
\end{aligned} \tag{4}$$

where $q_\perp = (0, q_1, q_2, 0)$, $q_\| = (q_0, 0, 0, q_3)$, $\gamma = eB/2$, $p^2 = E_p^2 - p_z^2 - 2eBn$ and L_n^m are Laguerre functions. We work in the Landau gauge ($\xi = 0$) where we know that for vacuum the wave function renormalization equals one. Since we aim at a description for small magnetic field strengths, we naturally expect that wave function renormalization $\mathscr{F} \sim 1$ in this gauge. Furthermore, let us work with the ansatz that $\mathscr{M}(k)$ is proportional to the unit matrix.

We expect that $\mathscr{M}((p-q)_\parallel, n_k)$ should be independent of s_k since the energy only depends on the principal quantum number n_k. Furthermore we assume that $\mathscr{M}((p-q)_\parallel, n_k)$ is a slowly varying function of n_k and thus make the approximation $\mathscr{M}((p-q)_\parallel, n_k) \sim \mathscr{M}((p-q)_\parallel, n_k = 0)$. For consistency we consider the case $n_p = 0$. Hereafter, we employ the more convenient notation $\mathscr{M}(k_\parallel, n_k = 0) \equiv \mathscr{M}(k_\parallel)$ for generic arguments of the mass function. With these considerations the sum over s_k can be computed by means of the result in Ref. [8]. It is worth mentioning that after summing over s_k, the resulting equation is the same as Eq. (50) in Ref. [2] when considering the case $n_k = 0$, which corresponds to the strong field limit.

In the situation where the magnetic field is weak, we expand $[(p-q)_\parallel^2 - 2eBn_k - \mathscr{M}^2((p-q)_\parallel)]^{-1}$ as a geometric series in powers of eB. The remaining sum over n_k can be performed also by resorting to Ref. [8] yielding, after a Wick rotation

$$\mathscr{M}(p_\parallel) \simeq$$
$$\frac{\alpha}{4\pi^3} \int d^4q \frac{\mathscr{M}((p-q)_\parallel)}{q^2[(p-q)_\parallel^2 + q_\perp^2 + \mathscr{M}^2((p-q)_\parallel)]}$$
$$\left\{ 3 + \left[\frac{4}{[(p-q)_\parallel^2 + q_\perp^2 + \mathscr{M}^2((p-q)_\parallel)]^2} - \frac{6(6 - q_\perp^2/q^2)q_\perp^2}{[(p-q)_\parallel^2 + q_\perp^2 + \mathscr{M}^2((p-q)_\parallel)]^3} \right.\right.$$
$$\left.\left. + \frac{36q_\perp^4}{[(p-q)_\parallel^2 + q_\perp^2 + \mathscr{M}^2((p-q)_\parallel)]^4} \right] (eB)^2 \right\}, \tag{5}$$

keeping only the lowest order contribution in eB. Notice that, as expected, the s_p dependence of the mass function disappears on carrying out the sum over s_k.

Solving the above equation numerically is still not trivial, owing to the fact that the unknown function $\mathscr{M}((p-q)_\parallel)$ within the integral is Lorentz non-invariant. However, we can always expand it out in powers of $(eB)^2$. Therefore we write, $\mathscr{M}((p-q)_\parallel) = \mathscr{M}_0(p-q) + (eB)^2 \mathscr{M}_1$, where $\mathscr{M}_1$ is responsible for breaking the Lorentz invariance of $\mathscr{M}_0(p-q)$. Consistently, we carry out the same expansion on the left hand side of Eq. (5). As the Lorentz invariance should be restored for the leading terms, we justifiably complete the momenta to achieve the same. This filters out the vacuum result. To calculate the magnetic field effect, we solve the integral equation for $\mathscr{M}_1$ obtained by comparing powers of $(eB)^2$. The results for $\mathscr{M}(p_\parallel/\Lambda)$ in the LLL are depicted in Fig. 1, scaled by the ultraviolet cut-off Λ. The dot-dashed line corresponds to the vacuum. The effect of the external field is to increase the dynamically generated mass, preserving the qualitative features of the mass function profile. To see the magnetic contribution to the dynamically generated mass, we show in Fig. 2 the difference $m(eB) - m(0)$, as a function of $(eB)^2$, where m is the dynamical fermion mass, namely, $m \equiv \mathscr{M}(0)$. Notice that this difference grows linearly with $(eB)^2$. We also evaluate the condensate defined as $\langle \bar{\psi}\psi \rangle = i \operatorname{Tr} G(x,x)$. In the weak field limit, spacing between Landau levels becomes

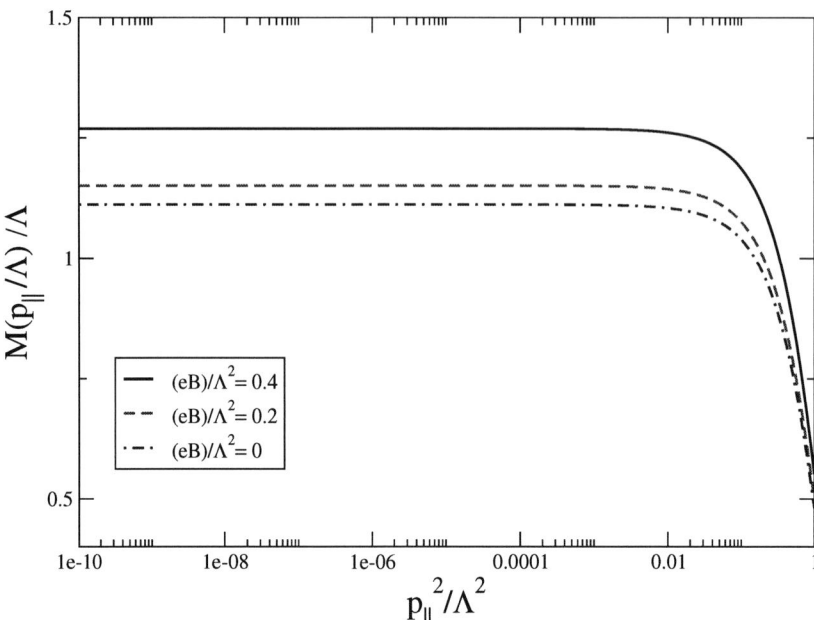

FIGURE 1. Mass function in the LLL for different values of the weak external magnetic field for $\alpha = 6.5\alpha_c$.

small. To compute the condensate, the sum over these levels, can be carried out by replacing $\sum_n$ with the integral $\int d^2k_\perp/(2\pi\, eB)$, along with the substitution $2eBn \to k_\perp^2$. The weak field contribution to the condensate also turns out to be quadratic and with the above mentioned substitutions, it owes itself entirely to the non Lorentz invariant piece of the mass function in our computational set up. Its behavior as a function of $(eB)^2$ is also shown in Fig. 2.

In summary, we have shown that for $\alpha > \alpha_c$ the dynamically generated mass increases quadratically with the magnetic field strength. As compared to the strong field case, this is a four-fold dependence on the magnetic field [1, 3, 2, 4]. Contrary to the widely studied case when the field is strong and the LLL dominates, in the weak field limit, all the Landau levels should be taken into account. This feature has made the problem prohibitively difficult and hence has been scarcely discussed in literature. Here we have shown that under plausible assumptions about the behavior of the mass function, the sum over Landau levels can be performed.

ACKNOWLEDGMENTS

We acknowledge the valuable discussions with V. de la Incera, E. Ferrer, V. Gusynin, C.N. Leung, V.A. Miransky and A. Sánchez. Support has been received in part by PAPIIT under grant number IN107105 and CONACyT under grant numbers 40025-F

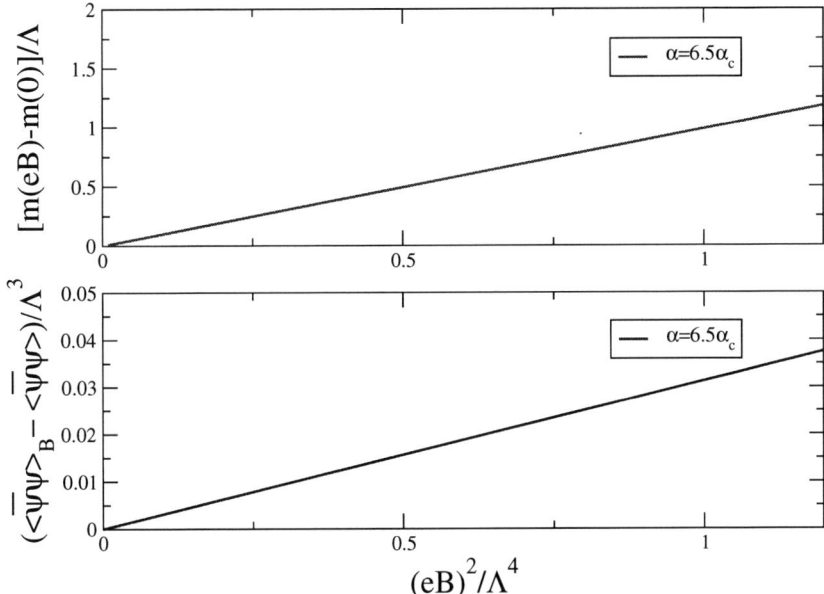

FIGURE 2. Magnetic contribution to the dynamically generated mass (upper graph) and condensate (lower graph) in the LLL as a function of $(eB)^2$ for $\alpha = 6.5\alpha_c$.

and 46614-I. AB wishes to acknowledge a short term visitor grant by a joint scheme of The Royal Society, U.K, and The Mexican Academy of Sciences, Mexico.

REFERENCES

1. V.P. Gusynin, V.A. Miransky and I.A. Shovkovy, Phys. Lett. B **349** 477 (1995); Phys. Rev. D **52** 4747 (1995); Nucl. Phys. B **462** 249 (1996); Nucl. Phys. B **563**, 361 (1999).
2. D.-S. Lee, C.N. Leung and Y.J. Ng, Phys. Rev. D **55**, 6504 (1997).
3. D.K. Hong, Phys. Rev. D **57**, 3759 (1998).
4. E.J. Ferrer and V. de la Incera, Phys. Lett. B **481**, 287 (2000).
5. V.I. Ritus in *"Issues in Intense-Field Quantum Electrodynamics"*, Ed. V.L. Ginzburg Nova Science, Commack, New York (1987).
6. J. Schwinger, Phys. Rev. D **82**, 664 (1951).
7. A. Ayala, A. Bashir, A. Raya and E. Rojas *"Dynamical mass generation in strongly coupled Quantum Electrodynamics with weak magnetic fields"*, hep-ph/0602209.
8. G. Gangopadhyay, J. Phys. A **32**, L433 (1999).

Empirical testing of Tsallis' Thermodynamics as a model for dark matter halos

Dario Nunez, Roberto A. Sussman, Jesus Zavala*, Luis G. Cabral-Rosetti[†] and Tonatiuh Matos**

*Instituto de Ciencias Nucleares,
Departamento de Gravitación y Campos,
Universidad Nacional Autónoma de México (ICN-UNAM).
A. Postal 70-543, 04510 México, D.F., México.
[†]Departamento de Postgrado,
Centro Interdisciplinario de Investigación y Docencia en Educación Técnica (CIIDET),
Av. Universidad 282 Pte., Col. Centro, A. Postal 752, C. P. 76000,
Santiago de Querétaro, Qro., México.
**Departamento de Física, Centro de Investigación y de Estudios Avanzados del IPN,
A.P. 14-740, 07000 México D.F., México.

Abstract. We study a dark matter halo model from two points of view: the "stellar polytrope" (SP) model coming from Tsallis' thermodynamics, and the one coming from the Navarro-Frenk-White (NFW) paradigm. We make an appropriate comparison between both halo models and analyzing the relations between the global physical parameters of observed galactic disks, coming from a sample of actual galaxies, with the ones of the unobserved dark matter halos, we conclude that the SP model is favored over the NFW model in such a comparison.

Keywords: dark matter, galaxy dynamics

INTRODUCTION

An alternative formalism to the micro-canonical ensemble treatment, that allows non-extensive forms for entropy and energy under simplified assumptions, has been developed by Tsallis [1] and applied to self–gravitating systems [2, 3, 4] under the assumption of a kinetic theory treatment and a mean field approximation. As opposed to the Maxwell–Boltzmann distribution that follows as the equilibrium state associated with the usual Boltzmann–Gibbs entropy functional, the Tsallis' functional yields as equilibrium state the "stellar polytrope" (SP), characterized by a polytropic equation of state with index n.

On the other hand, high precision N–body numerical simulations based on Cold Dark Matter (CDM) models, perhaps the most powerful method available for understanding gravitational clustering, lead to the famous results of Navarro, Frenk and White (NFW model) [5] that predicts density and velocity profiles which are roughly consistent with observations, however, it also predicts a cuspy behavior at the center of galaxies that is not observed in most of the rotation curves of dwarf and LSB galaxies [6, 7, 8, 9, 10, 11]. The significance of this discrepancy with observations is still under dispute, nevertheless, the NFW model of collision–less WIMPs remains as a viable model to account for DM in galactic halos, provided there is a mechanism to explain the discrepancies of this model with observations in the center of galaxies.

Since gravity is a long–range interaction and virialized (i.e in virial equilibrium) self–gravitating systems are characterized by non-extensive forms of entropy and energy, it is reasonable to expect that the final configurations of halo structure predicted by N–body simulations must be, somehow, related with states of relaxation associated with non-extensive formulations of Statistical Mechanics; therefore, a comparison between the SP and NFW models is both, possible and interesting. The main purpose of our analysis is to make a dynamical analysis of two halo models, one based on the NFW paradigm, and other based on the SPs derived from Tsallis' non-extensive thermodynamics, compare them and test both with observational results coming from a sample of disk galaxies.

SP AND NFW HALO MODELS

For a face space given by $(\mathbf{r}, \mathbf{p})$, the kinetic theory entropy functional associated with Tsallis' formalism is [2, 3], and [4]

$$S_q = -\frac{1}{q-1} \int (f^q - f) d^3\mathbf{r} d^3\mathbf{p}, \quad (1)$$

where f is the distribution function and $q > 1$ is a real number. In the limit $q \to 1$, the functional (1) leads to the usual Boltzmann–Gibbs functional, corresponding to the isothermal sphere. The condition $\delta S_q = 0$ leads to the distribution function that corresponds to the SP model characterized by the equation of state $p = K_n \rho^{1+1/n}$, where K_n is a function of the polytropic index n, and can be expressed in terms of the central parameters: $K_n = \frac{\sigma_c^2}{\rho_c^{1/n}}$. The polytropic index, n, is related to the Tsallis' parameter $q > 1$ by: $n = \frac{3}{2} + \frac{1}{q-1}$. Inserting the equation of state into Poisson's equation, the Lane-Emden equation [12] is obtained, based on it we can obtain density, mass and velocity profiles for the SP model.

NFW numerical simulations yield the following expression for the density profile of virialized galactic halo structures [5, 13]:

$$\rho_{\text{NFW}} = \frac{\delta_0 \rho_0}{y(1+y)^2}, \quad (2)$$

where: $\delta_0 = \frac{\Delta c_0^3}{3[\ln(1+c_0) - c_0/(1+c_0)]}$, $\rho_0 = \rho_{\text{crit}} \Omega_0 h^2 = 253.8 h^2 \Omega_0 \frac{M_\odot}{\text{kpc}^3}$, $y = c_0 \frac{r}{r_v}$, and Ω_0 is the ratio of the total density to the critical density of the Universe. Using equation (2) its easy to obtain mass and velocity profiles.

COMPARISON OF SP AND NFW MODELS

In order to compare both halo models, it is important to make various physically motivated assumptions. First, we want both models to describe a halo of the same scale, which means same virial mass M_{vir}. Secondly, both models must have the same maximal value for the rotation velocity. This is a plausible assumption, as it is based on the

TABLE 1. Parameters characterizing the polytropes while being compared to NFW halos

$\log_{10}(M_{vir}/M_\odot)$	$\rho_c [M_\odot/\text{pc}^3]$	$\sigma_c [\text{Km/s}]$	n	q	K_n	$v_{\max} [\text{Km/s}]$	$r_{vir} [\text{kpc}]$
15	3.7×10^{-4}	982	4.93	1.29	4873.4	1504	2606.2
12	7.5×10^{-4}	108	4.87	1.30	478.94	164	260.6
11	9.0×10^{-4}	52	4.83	1.30	221.82	79.1	120.9
10	1.2×10^{-3}	25	4.82	1.30	100.68	38.2	56.1

Tully–Fisher relation [14], a very well established result that has been tested successfully for galactic systems, showing a strong correlation between the total luminosity of a galaxy and its maximal rotation velocity. Our third assumption is that the polytropic and NFW halos, complying with the previous requirements, also have the same total energy evaluated at the cut–off scale $r = r_{vir}$. The main justification for this assumption follows from the fact that the total energy is a fixed quantity in the collapse and subsequent virialized equilibrium of dark matter halos [15]. Since the SP model has three free parameters (contrary to only one parameter of the NFW model), and we have selected three comparison criteria, we have a mathematically closed problem once M_{vir} is specified.

Following the guidelines described above, we proceed to compare NFW and polytropic halos for M_{vir} ranging from 10^{10} up to 10^{15} solar masses. From this comparison we find the "best-fit" values for the free parameters of the SP model. The results are displayed explicitly in table 1. The comparison between both models in velocity profiles is shown in the left panel of figure 1. SP and NFW models have both the same virial mass, $M_{vir} = 10^{12} M_\odot$. For other values of M_{vir} the velocity profiles are qualitatively similar to the one displayed in figure 1 (left panel). The detailed description and results of the method of comparison presented above will appear in an article that is being prepared [16].

So far we have analyzed only the global structure of the dark matter halo without considering the effects of the luminous galaxy within. If one wishes to test a given model with observational results, it is necessary to add the galactic baryonic disk as a dynamical component of the model. In order to do so we followed the method described in [13, 17]. Then to compare both models with observational results we used the prescription presented in [18].

Using such a prescription, we may define the ratio of maximum disk velocity to maximum total velocity, $V_{d,m}/V_{c,m}$, which is a global quantity that can be directly compared with theoretical predictions. This ratio is not defined at a given radius, but it can be related to the total mass to disk mass ratio M_t/M_d, defined at an specific radius. In particular at radius r_m where the total rotational curve has its maximum we have [18]:

$$\left(\frac{M_t}{M_d}\right)_{r_m} \propto \left(\frac{V_{d,m}}{V_{c,m}}\right)^{-2} \qquad (3)$$

The use of $V_{d,m}/V_{c,m}$ instead of M_t/M_d is suitable because it can be obtained directly from observational parameters, without the assumptions needed to calculate M_t/M_d.

One of the principal results obtained in the work [18] is that the ratio $V_{d,m}/V_{c,m}$ correlates principally with the disk surface density Σ_d of galaxies. Therefore we will

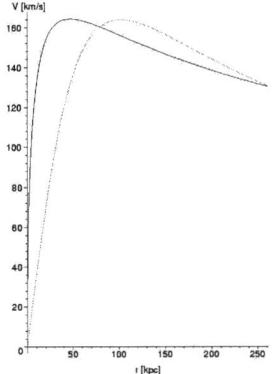 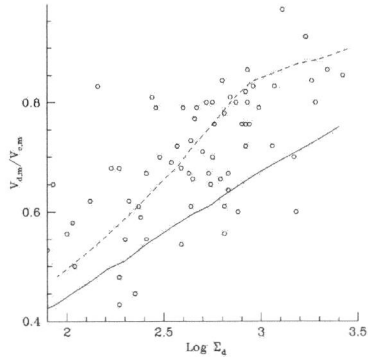

FIGURE 1. Left panel: velocity profiles for the NFW halo (solid line) and its compared SP one (dashed curve). Right panel: luminous to total dark matter content vs central surface density; open circles correspond to observational data, the dashed line represents the NFW model and the solid one to SP model. Both models have a virial mass of $M_v = 10^{12} M_\odot$.

use this result to compare the NFW and the SP models with the observational results coming from the sample. We will take $M_{vir} = 1 \times 10^{12} M_\odot$, which is a characteristic value for the mass of dark matter halos. For simplicity, we will assume that the baryonic mass fraction of the disk, f_d, has the same value for all disks.

In the right panel of figure 1 it is shown the ratio $V_{d,m}/V_{c,m}$ vs $Log\Sigma_d$ for the sample of galaxies used (open circles). An almost linear trend can be seen between this quantities. HSB (high surface brightness) galaxies (corresponding approximately to values of $Log\Sigma_d$ greater than 2.5) have greater values of $V_{d,m}/V_{c,m}$ than LSB (low surface brightness) galaxies. This means that the luminous matter content is greater for HSB than for LSB galaxies. The shown picture is consistent with a well known result: LSB galaxies are dark matter dominated systems within the optical radius.

The value of the graphic presented is that it allow us to bound statistically the possible values of the $V_{d,m}/V_{c,m}$ ratio that galaxies with a given surface density can have. As was proved by [18], the size of this range of values (associated with dispersion on the graphic) is mainly due different virial mass that galaxies with the same Σ_d have.

Right panel of figure 1 also show clearly that NFW models can not reproduce at satisfaction the results obtained for the compiled sample without introducing unrealistic values for the virial mass. This is one of the results that lead us to the possibility of seeking an alternative to the NFW paradigm. The curves shown in figure 1 (right panel) represent both models with average values for their respective parameter; it's clear from the figure that the SP model follows in better agreement the average behavior of the observational sample than the NFW model.

CONCLUSIONS

Motivated by the fact that SPs are the equilibrium state in Tsallis' non-extensive entropy formalism, we have found the structural parameters of those SPs that allows us to compare them with NFW halos of virial masses in the range $10^{10} < M_{vir}/M_\odot < 10^{15}$; the results are displayed in Table 1. It is shown in the left panel of figure 1 that the velocity profile of the SP model is much less steep in the same region than that of the NFW halo. These features are consistent with the fact that NFW profiles predict more dark matter mass concentration than what is actually observed in a large sample of galaxies [19, 8, 18].

We have also shown that the SP model is favored over the NFW model regarding the dark matter content in disk galaxies (within the optical radius) which is shown by the average behavior of the observational sample in figure 1 (right panel). These results show that the NFW halo model can be enhanced with the use of alternative paradigms in Statistical Mechanics, which seems to solve a recurrent item which throws a shadow in such an excellent description as the NFW model is.

We are grateful to Vladimir Avila-Reese for his comments and suggestions to the manuscript of the present work. We acknowledge partial support by CONACyT México, under grants 32138-E 32138-E, 42748, 45713-F and 34407-E, and DGAPA-UNAM IN-122002 and IN117803 grants. TM wants to thank Matt Choptuik for his kind hospitality at the UBC, JZ acknowledges support from DGEP-UNAM and CONACyT scholarships.

REFERENCES

1. Tsallis C, 1999 *Braz J Phys* **29** 1
2. Plastino A R and Plastino A, 1993 *Physics Letters A* **174** 384
3. Taruya A and Sakagami M, 2002 *Physica A* **307** 185 [cond-mat/0107494]
4. Taruya A and Sakagami M, 2003 *Phys. Rev. Lett.* **90** 181101; See also Taruya A and Sakagami M, 2004 *Continuum Mechanics and Thermodynamics* **16** 279-292 [cond-mat/0310082]
5. Navarro J F, Frenk C S and White S D M, 1997 *ApJ* **490** 493 [astro-ph/9611107]
6. de Blok W J G, McGaugh S S, Rubin V C, 2001 *ApJ* **122** 2396 [astro-ph/0107366]
7. de Blok W J G, McGaugh S S, Bosma A and Rubin V C, 2001 *ApJ* **552** L23 [astro-ph/0103102]
8. Binney J J and Evans N W, 2001 *MNRAS* **327** L27 [astro-ph/0108505]
9. Borriello A and Salucci P, 2001 *MNRAS* **323** 285 [astro-ph/0001082]
10. Blais-Ouellette, Carignan C and Amram P, 2002 *Preprint* astro-ph/0203146
11. Bosma A, Invited review at IAU Symposium 220: *Dark Matter in Galaxies*, Sydney Australia, 21-25 Jul 2003, *Preprint* astro-ph/0312154
12. Binney J and Tremaine S, *Galactic dynamics*, 1987, ed. Princeton University
13. Mo H J, Mao S and White S D M, 1998 *MNRAS* **295** 319 [astro-ph/9707093]
14. Tully R B and Fisher J R, 1977 *A&A* **54** 661, see also Colin P, Avila-Reese V and Valenzuela O, 2000 *ApJ* **542** 622 [astro-ph/0004115]
15. Padmanabhan T, *Structure formation in the universe*, 1993, Cambridge University Press.
16. Cabral-Rosetti L G, Matos T, Nunez D, Sussman R and Zavala J, *Empirical testing of Tsallis entropy in density profiles of galactic dark matter halos*, 2006, in preparation.
17. Cabral-Rosetti L G, Matos T, Nunez D, Sussman R and Zavala J,*Stellar polytropes and Navarro-Frenk-White halo models: Comparison with observations*, 2006, in preparation.
18. Zavala J, Avila-Reese V, Hernández-Toledo H and Firmani C, 2003 *A&A* **412** 633 [astro-ph/0305516]
19. McGaugh S S, Rubin V C and de Blok W J G, 2001 *ApJ* **122** 2381 [astro-ph/0107326]

Entropy considerations in constraining the mSUGRA parameter space

Dario Nunez, Roberto A. Sussman, Jesus Zavala, Lukas Nellen*, Luis G. Cabral-Rosetti[†] and Myriam Mondragón**

*Instituto de Ciencias Nucleares,
Departamento de Gravitación y Teoría de Campos,
Universidad Nacional Autónoma de México (ICN-UNAM).
A. Postal 70-543, 04510 México, D.F., México.
[†]Departamento de Postgrado,
Centro Interdisciplinario de Investigación y Docencia en Educación Técnica (CIIDET),
Av. Universidad 282 Pte., Col. Centro, A. Postal 752, C. P. 76000,
Santiago de Querétaro, Qro., México.
**Instituto de Física,
Universidad Nacional Autónoma de México (IF-UNAM).
A. Postal 20-364, 01000 México D.F., México.

Abstract. We explore the use of two criteria to constraint the allowed parameter space in mSUGRA models. Both criteria are based in the calculation of the present density of neutralinos as dark matter in the Universe. The first one is the usual "abundance" criterion which is used to calculate the relic density after the "freeze-out" era. To compute the relic density we used the numerical public code micrOMEGAs. The second criterion applies the microcanonical definition of entropy to a weakly interacting and self-gravitating gas evaluating then the change in the entropy per particle of this gas between the "freeze-out" era and present day virialized structures (i.e systems in virial equilibrium). An "entropy-consistency" criterion emerges by comparing theoretical and empirical estimates of this entropy. The main objective of our work is to determine for which regions of the parameter space in the mSUGRA model are both criteria consistent with the 2σ bounds according to WMAP for the relic density: $0.0945 < \Omega_{CDM}h^2 < 0.1287$. As a first result, we found that for $A_0 = 0$, sgn$\mu = +$, small values of $\tan\beta$ are not favored; only for $\tan\beta \simeq 50$ are both criteria significantly consistent.

Keywords: dark matter, supersymmetry

INTRODUCTION

One of the most accepted candidates to be the major component of dark matter (DM) is the neutralino as an LSP (Lightest Supersymmetric Particle). Supersymmetric models with R-parity conservations predict this type of particles (for an excellent introduction to Supersymmetry see [1]). This type of models have several parameters that can be constrained in its values using observational constraints of the actual density of DM, according with WMAP: $0.0945 \leq \Omega_{CDM}h^2 \leq 0.1287$ [2, 3]. In particular for mSUGRA models this has been done using the standard approach [2, 4] which is based in the Boltzmann equation considering that after the "freeze-out" era, neutralinos cease to annihilate keeping its number constant. In such an approach, the relic density of neutralinos is approximately: $\Omega_\chi \approx 1/\langle\sigma v\rangle$, where $\langle\sigma v\rangle$ is the thermally averaged cross section times the relative velocity of the LSP annihilation pair. Within the mSUGRA model five parameters (m_0, $m_{1/2}$, A_0, $\tan\beta$ and the sign of μ) are needed to specify the supersymmetric spectrum of particles and the final relic density. We will use the numerical code micrOMEGAs [5] to compute the relic density following the past scheme which will be called the "abundance criterion" (AC).

Just after "freeze-out", we can consider neutralinos then as forming a Maxwell-Boltzmann (MB) gas in thermal equilibrium with other components of the primordial cosmic structures. In the present time, such a gas is almost colisionless and either constitutes galactic halos and larger structures or it is in the process of its formation. In this context, we can conceive two equilibrium states for the neutralino gas, the decoupling (or "freeze-out") epoch and its present state as a virialized system. Computing the entropy per particle for each one of this states we can use an "entropy consistency" criterion (EC) using theoretical and empirical estimates for this entropy to obtain the relic density of neutralinos (Ω_χ).

Our objective is then to use AC and EC criteria, to obtain constraints for the parameters of the mSUGRA model by demanding that both criteria must be consistent within them and within the observational constraints required by WMAP.

ABUNDANCE CRITERION

Relic abundance of some stable species χ is defined as $\Omega_\chi = \rho_\chi/\rho_{crit}$, where $\rho_\chi = m_\chi n_\chi$ is the relic's mass density (n_χ is the number density), ρ_{crit} is the critical density of the Universe (see [6] for a review on the standard method to compute the relic density). The time evolution of n_χ is given by the Boltzmann equation:

$$\frac{dn_\chi}{dt} = -3Hn_\chi - \langle\sigma v\rangle(n_\chi^2 - (n_\chi^{eq})^2) \tag{1}$$

where H is the Hubble expansion rate, $\langle\sigma v\rangle$ is the thermally averaged cross section times the relative velocity of the LSP annihilation pair and n_χ^{eq} is the number density that species would have in thermal equilibrium. In the early Universe, the neutralinos (χ) were initially in thermal equilibrium, $n_\chi = n_\chi^{eq}$. As the Universe expanded, their typical interaction rate started to diminish an the process of annihilation froze out. Since then, the number density of neutralinos has remained basically constant.

There are several ways to solve equation (1), one of the more used is based on the "freeze-out" approximation (see for example [7]). However in order to have more precision, we will use the exact solution to Boltzmann equation using the public numerical code micrOMEGAs 1.3.6 [5] which calculates the relic density of the LSP in the Minimal Supersymmetric Standard Model (MSSM). We will take and mSUGRA model and its five parameters (m_0, $m_{1/2}$, A_0, $\tan\beta$ and the sign of μ) as input parameters for micrOMEGAs and use *Suspect* [8], which comes as an interface to micrOMEGAs, to calculate the supersymmetric spectrum of masses of particles. Details about how we used micrOMEGAs for making the calculation will be described in a future paper that is currently in preparation [9].

Using micrOMEGAs, we can obtain the relic density for any region of the parameter space to discriminate regions that are consistent with the WMAP constraints in this abundance criterion.

ENTROPY CONSISTENCY CRITERION

Since the usual MB statistics that can be formally applied to the neutralino gas at the "freeze-out" era can not be used to describe present day neutralinos subject to a long range gravitational interaction making up non-extensive systems, it is necessary to use the appropriate approach that follows from the microcanonical ensemble in the "mean field" approximation which yields an entropy definition that is well defined for a self-gravitating gas in an intermediate state. Such an approach is valid at both the initial ("freeze-out" era, f) and final (virialized halo structures, h) states that we wish to compare. Under these conditions, the change in the entropy per particle (s) between these two states is given by [10]:

$$s^h - s^f = ln\left[\frac{n_\chi^f}{n_\chi^h} \frac{x^f}{x^h}^{3/2}\right] \qquad (2)$$

where $x = m_\chi/T$, T is the temperature of the gas. A region that fits with the conditions associated with the intermediate scale is the central region of halos ($10pc^3$ within the halo core); evaluating the thermodynamical quantities at this region, using equation (2) and some assumptions more, it is possible to construct a theoretical estimate for s^h that depends on the nature of neutralinos (m_χ and $\langle\sigma v\rangle$), initial conditions (given by x^f), cosmological parameters (Ω_χ, the Hubble parameter, h) and structural parameters of the virialized halo (central values for temperature and density); for details of these and the following, see section IV of [10].

An alternative estimate for s^h can be made based on empirical quantities for observed structures in the present Universe using the microcanonical entropy definition in terms of phase space volume, but restricting this volume to the actual range of velocities accessible to the central particles, that is, up to a maximal escape velocity $v_e(0)$ which is related to the central velocity dispersion of the halo (σ_h) by an intrinsic parameter α: $v_e^2(0) \sim \alpha\sigma_h^2(0)$. The authors in [10] give an uncertainty range for the value of α for actual galaxies: $11.2 \leq \alpha \leq 24.8$. The range of values allowed for this parameter is of the highest importance to determine the allowed region of the parameter space in the mSUGRA model as will be clear in the results presented on next section.

Equating the theoretical an empirical estimates for the entropy per particle it is obtained a relation for the relic abundance of neutralinos using the EC criterion[1]:

$$ln(\Omega_\chi h^2) = 10.853 - x^f + ln\left[\frac{(x^f \alpha)^{3/2} m_\chi}{f_g^*(x^f)}\right] \qquad (3)$$

where $f_g^*(x^f)$ is a function related to the degrees of freedom at the "freeze-out" time (see for example [7]) that will be described elsewhere [9].

Modifying the program micrOMEGAs, we can obtain the value for x^f for any region of the parameter space and then Ω_χ using (3), therefore we will be able to discriminate regions that are consistent with the WMAP constraints for the EC criterion.

---

[1] This formula is a small modification to the one presented in [10]

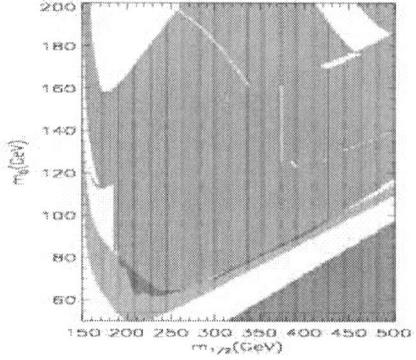 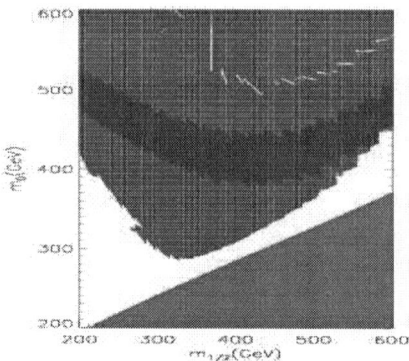

FIGURE 1. Allowed regions in the parameter space for AC (red) and EC (blue) criteria for the mSUGRA model with $A_0 = 0$ and $\mathrm{sgn}\mu = +$. The left panel shows the results for $\tan\beta = 10$ and the right one for $\tan\beta = 50$.

RESULTS AND CONCLUSIONS

Using both AC and EC criteria, that have been briefly described above, we can compute the relic abundance of neutralinos and constrain the region in the mSUGRA parameter space where both criteria are fullfilled. In order to obtain a first result we will fix the value of two parameters in the mSUGRA model: $A_0 = 0$ and $\mathrm{sgn}\mu = +$. In the left panel of figure (1), we present a region of the parameter space with these two values fixed and with $\tan\beta = 10$. The green region is where the $\tilde{\tau}$ is the LSP, the red and blue areas determine the WMAP allowed regions for the AC and EC criteria respectively. As we can see from figure (1) (left panel), the region where both criteria are fullfilled is very small, in fact only for the highest values of α there is an intersection between both criteria. The right panel of figure (1) shows the same regions as in the left panel but for $\tan\beta = 50$; contrary to the case of $\tan\beta = 10$ we see now a complete intersection of both AC and EC criteria.

We have followed the novel idea of [10] to introduce a new criterion to constrain the mSUGRA parameter space using the assumption of entropy consistency for the initial and final states of a neutralino gas. Using the program micrOMEGAs, we explored with precision which regions then satisfy this criterion and the usual AC criteria previously used several times. We found that for the regions so far explored, values with small $\tan\beta$ are not favored, leading to an insignificant allowed region satisfying both criteria. Values with $\tan\beta \gtrsim 50$ fullfill the requirement of both criteria and the WMAP constraints. Further analysis, which is currently being done, is required to give more precise conclusions about this new method to constrain the parameter space of the mSUGRA model.

We acknowledge partial support by CONACyT México, under grants 32138-E, 34407-E and 42026-F, and PAPIIT-UNAM IN-122002, IN117803 and IN116202 grants. JZ acknowledges support from DGEP-UNAM and CONACyT scholarships.

REFERENCES

1. Martin S P, 1999 *Preprint* heph-ph/9709356
2. Bélanger G, Kraml S and Pukhov A, 2005 *Preprint* hep-ph/0502079
3. *The WMAP collaboration*, Bennet et al., 2003 *Preprint* astro-ph/0302207 and Spergel et al., 2003 *Preprint* astro-ph/0302209
4. Bélanger G, Boudjema A, Pukhov A and Semenov A, 2005, *Nuclear Physics B* **706** 411-454 [hep-ph/0407218]
5. Bélanger G, Boudjema A, Pukhov A and Semenov A, *Preprint* hep-ph/0405253
6. Jungman G, Kamionkowski M and Griest K, 1996 *Phys Rept.* **267** 195 [hep-ph/9506380]
7. Gondolo P and Gelmini G, 1991 *Nuclear Physics B* **360** 145-179
8. Djouadi A, Kneur G and Moultaka G *Preprint* hep-ph/0211331
9. Nunez D, Sussman R A, Zavala J, Nellen L, Cabral-Rosetti L G and Mondragón M, 2006, *in preparation*
10. Cabral-Rosetti L G, Hernández X and Sussman R A, 2004 *Phys. Rev. D* **69** 123006

Probing New Physics With Neutrino Interactions At The Pierre Auger Observatory

L. Díaz-Cruz[1], M. I. Pedraza-Morales[2], A. Rosado[2], H. Salazar[1]

[1]*Fac. de Ciencias Físico-Matemáticas,BUAP. Apdo. Postal 1364, C.P. 72000 Puebla,Pue., México.*
[2]*Instituto de Física, BUAP. Apdo. Postal J-48 Col. San Manuel, C.P. 72570 Puebla, Pue., México.*

Abstract. The energy of the shower produced through neutrino-nucleon interactions depends on the neutrino flavor and the type of interaction. Electron neutrinos undergoing CC interactions produce a shower with both an electromagnetic and hadronic component: $E_{sh,em}=(1-y)E_\nu$, $E_{sh,had}=yE_\nu$. Muon neutrinos undergoing CC interactions, along with all neutrino flavors undergoing NC interactions, produce a hadronic shower with an energy, $E_{sh,had}=E_\nu$. Charged current interactions of tau neutrino are somewhat more complicated. The tau lepton produced in the initial CC neutrino interaction with sufficiently high energy generate a second hadronic shower from the tau decay spatially separated from the initial one, identifiable as a "double bang" event. The Auger capability for seeing these events allows to restrict models where non-perturbative effects are present, which induce an enhancement in the neutrino-nucleon cross-section (*e.g.* low-scale Quantum Gravity, EW instantons). We are interested to study EW-instantons, both the Standard Model as well as in an extension of the type $SU(2)_1 \times SU(2)_2 \times U(1)_Y$, where the influence of new physics is only in the third family.

Keywords: Neutrino interactions, Ultra-high energy cosmic rays, Cosmic rays detection.
PACS: 13.15.+g, 13.85.Tp, 95.85.Ry.

INTRODUCTION

The Pierre Auger Observatory (PAO) [1], the largest cosmic ray detector in the world, is a hybrid detector, with a ground array of water Cherenkov and air fluorescence detectors.

The PAO is expected to detect neutrinos with energies above ~108 GeV, detecting the neutrino flux from interactions of extragalactic ultra-high energy cosmic rays with the cosmic rays background, getting information about neutrino interactions at very high energies [2,3].

In this short note we shall discuss save of the characteristics of neutrino induced cosmic rays; although these results are known from the literature, we feel it is worth to summarize them in order to compare the expected standard model reactions with those coming from the new physics we are interested in.

The next figure shows the angular distribution of neutrino-induced showers expected to be observed by the PAO whit different selection criteria.

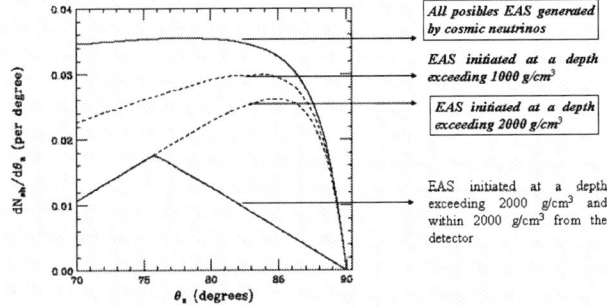

FIGURE 1. Angular distribution of neutrino-induced showers

DETECTION

The selection criteria to consider an Auger event as neutrino-induced showers are the following:
- The zenith angle, $\theta_z > 70°$.
- The neutrino first interaction takes place after 2000 g/cm2 into the atmosphere and within 2000 g/cm2 of the detector.

The rate of neutrino-induced showers expected to be observed in an experiment such as Auger can be written as:

$$\frac{N_{events}}{\Delta T_{obs}} = 2\pi N_A \int dE_\nu \int_0^1 dy \frac{d\sigma_{\nu N}}{dy}(E_\nu) \int d\cos\theta_z\, A_\perp(\cos\theta_z)$$
$$\times \int_{X_{min}}^{X_{ground}} dX\, P[E_{sh}, \cos\theta_z, X] \frac{dN_\nu}{dE_\nu}(E_\nu) , \quad (1)$$

INTERACTIONS

Quasi-Horizontal Showers

The neutrino-nucleon cross-section in Eq.(1) describes both charged current (CC) neutrino-quark scattering and neutral current (NC) neutrino-quark scattering [4].

The energy of the shower produced depends on the neutrino flavor and the type of interaction. Electron neutrinos undergoing CC interactions produce a shower with both an electromagnetic and hadronic component: $E_{sh,em}=(1-y)E_\nu$, $E_{sh,had}=yE_\nu$. Muon neutrinos undergoing CC interactions, along with all neutrino flavors undergoing NC interactions, produce a hadronic shower with an energy, $E_{sh,had}=E_\nu$.

Charged current interactions of tau neutrino are somewhat more complicated. The tau lepton produced in the initial CC neutrino interaction with sufficiently high energy generate a second hadronic shower from the tau decay spatially separated from the initial one, identifiable as a "double bang" event.

Earth Skimming Tau Neutrinos

A second class of neutrino events potentially observable at the PAO is generated by tau neutrinos which interact while skimming the Earth's surface. Such interactions generate tau leptons which escape the Earth and produce a slightly up going hadronic.

The effect of interaction in the Earth on the tau neutrino spectrum when the (CC + NC) interaction cross-section is suppressed.

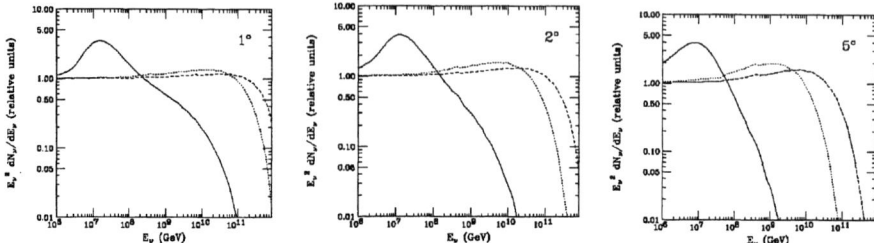

Figure 2. Adopting a spectrum $\approx E_\nu^{-2}$. In each frame, results are shown assuming the SM cross-section (solid line) and a cross-section smaller by a factor of 2 (dotted-line) and a factor of 5 (dashed-line). The three frames are for incoming angles of 1°, 2°, 5° degrees below the horizon

The effect of interaction in the Earth on the tau neutrino spectrum when the (NC) interaction cross-section is enhanced.

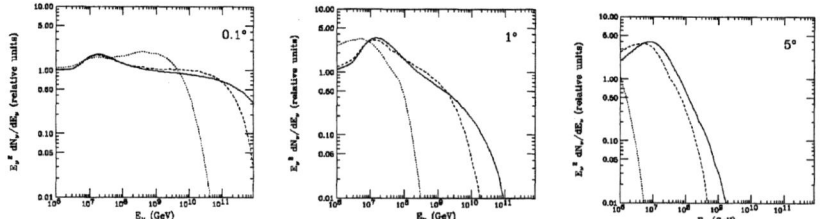

Figure 3. Adopting a spectrum $\approx E_\nu^{-2}$. In each frame, results are shown assuming the SM cross-section (solid line) and a cross-section smaller by a factor of 2 (dotted-line) and a factor of 5 (dashed-line). The three frames are for incoming angles of 1°, 2°, 5° degrees below the horizon

INTERACTION MODELS

Non-perturbative Electroweak Interactions.

The spectrum of quasi-horizontal showers mediated by electroweak sphalerons [5] as would be seen by the PAO; as studied in [6] for the SM is shown in Fig. 4.

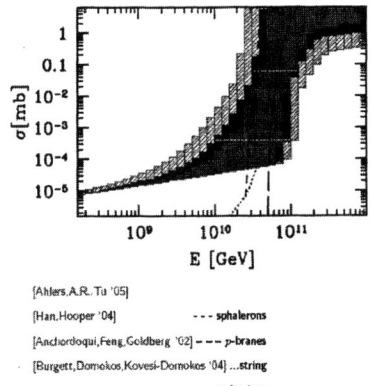
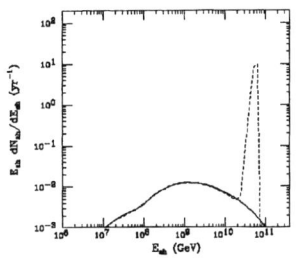

Figure 4. Cross section dependence on the model

Figure 5. Cosmogenic neutrino flux (left). The solid line is the SM prediction

PROSPECTS

The Auger capability for seeing these events let restrict models where the non-perturbative effects are present, which induce an enhancement in the neutrino-nucleon cross-section (low-scales, Quantum Gravity, EW instantons) [7].

We are interested to study EW- instantons, in the Standard Model of the electroweak and strong interactions as well as in some kind of extension $SU(2)_1 \times SU(2)_2 \times U(1)_Y$, which influences only the third family.

ACKNOWLEDGMENTS

J.L.D.-C., A.R. and H.S. would like to thank Sistema Nacional de Investigadores (Mexico) for financial support, and the Huejotzingo Seminar for inspiration. M.I.P.M. acknowledges financial support from CONACYT (México). This research was supported in part by CONACYT (México).

REFERENCES

[1] M. Giller and A.Smialkowski [Auger Collaboration], ``The Pierre Auger Observatory, "AIP Conf. Proc. Vol.801, 82 (2006).
[2] M. H. Reno and C. Quigg, ``On The Detection Of Ultrahigh-Energy Neutrinos", Phys. Rev. Vol.D37, 657 (1988).
[3] R. Gandhi, C. Quigg, M.H. Reno and I. Sarcevic, ``Neutrino interactions at ultrahigh energies," Phys. Rev. Vol. D58, 093009 (1998) [arXiv:hep-ph/9807264].
[4] Acceptance and Flux Limit for the tau neutrino with the Pierre Auger Observatory Surface Detector", O. Blanch and P. Billoir, Preprint LPNHE, Paris (France), September 26, 2005.
[5] J. Ambjorn and A. Krasnitz, ``Sphaleron transition rate: Recent numerical results", Nucl. Phys. Proc. Suppl. Vol. 63, 575 (1998).
[6] A. Ringwald, ``Strongly interacting Neutrinos as The Highest Energy Cosmic Rays", 29th Johns Hopkins Workshop on Current Problems in Particle Theory "Strong Matter in the Heavens", (2005).
[7] Luis Anchordoqui, Tao Han, Dan Hooper, and Subir Sakar, ``Exotic neutrino interactions at the Pierre Auger observatory," Astropart. Phys.\ Vol.25, 14 (2006) [arXiv:hep-ph/0508312].

Neutrino Telescopes and Non Standard Interactions

Ricardo Pérez Martínez, Arnulfo Zepeda and Omar Miranda

Departamento de Física, Centro de Investigación y de Estudios Avanzados del IPN, P.O. Box 14-740, 07000, México D.F., México

Abstract. We study the expected number of atmospheric upward muon neutrino events in a neutrino telescope of ultra high energy. For this computation we introduce the atmospheric neutrino flux, the survival probability, the deep inelastic neutrino-nucleon cross section and the range of muon in a homogeneous medium. As a first application we analyze the sensitivity that these detectors could have to test the effects of Non Standard Interactions of neutrinos with matter.

INTRODUCTION

Neutrino physics is one of the most active areas of investigation in high energy physics. The strong evidence that neutrinos have nonzero masses is a signal of physics beyond the Standard Model (SM). It is a fact that neutrinos offer a new window to explore several phenomena in particle physics and astrophysics. In this line of investigation, neutrino telescopes are one of the most promising projects to study neutrinos coming from astrophysical and terrestrial sources. In this work we compute the expected number of events of atmospheric upward muon neutrinos which could be detected in a neutrino telescope with an effective area of 1km^2. Afterwards, we incorporate the effects of Non Standard Interactions (NSI) and oscillation. Our aim is to study the sensitivity that these experiments could have to NSI.

This work is organized as follows: in section 2 we introduce the expressions which give us the expected number of events and we study the elements needed to achieve its numerical computation. We also describe the NSI in the context of atmospheric neutrinos. Section 3 is devoted to introduce our results showing the sensitivity that these neutrino telescopes could have to NSI. Our conclusions are finally given in section 4.

EVENT RATES AND NON STANDARD INTERACTIONS

In a high energy neutrino telescope muon neutrinos are detected via their charged current (CC) interactions in the matter surrounding the detector. Such interactions produce muons that can be detected in a neutrino telescope. High energy muons have very large average range, therefore, the effective volume is significantly larger than the detector one. In a semi-analytical calculation we can obtain the expected number of upward ν_μ

induced events from [1]

$$N_{ev}^{\nu_\mu} = T \int_{-1}^{0} d\cos\theta \int_{E_\mu^{min}}^{\infty} dE_\mu^0 \int_{E_\mu^0}^{\infty} dE_\nu \qquad (1)$$

$$\frac{d\phi_{\nu_\mu}}{dE_\nu d\cos\theta}(E_\nu,\cos\theta) \frac{d\sigma_{CC}^\mu}{dE_\mu^0}(E_\nu,E_\mu^0) n_T R(E_\mu^0,E_\mu^{min}) A_{eff}^0,$$

where $\frac{d\phi_{\nu_\mu}}{dE_\nu d\cos\theta}$ is the differential muon neutrino flux in the vicinity of the detector after evolution in the Earth matter. The attenuation effect was taken into account [2]. For the atmospheric neutrino flux we use as input the neutrino energy spectrum computed by Volkova [3] and the average flux from Gandhi [2] which is given by

$$\frac{dN_{\nu_\mu+\bar{\nu}_\mu}}{dE_\nu} = 7.8 \times 10^{-11} \left(\frac{E_\nu}{1\text{TeV}}\right)^{-3.6} \text{cm}^{-2}\text{s}^{-1}\text{sr}^{-1}\text{GeV}^{-1}. \qquad (2)$$

Another important element of Eq. (1) is the differential CC cross section for a deep inelastic neutrino nucleon process, producing in the final state a muon of energy E_μ^0. This is given by $\frac{d\sigma_{CC}^\mu}{dE_\mu^0}(E_\nu,E_\mu^0)$. To calculate this cross section we consider [2]

$$\frac{d^2\sigma_{\nu N}^{CC}}{dxdy} = \frac{2G_F^2 M_N E_\nu}{\pi}\left(\frac{M_W^2}{Q^2+M_W^2}\right)^2 \left[xQ(x,Q^2)+x\bar{Q}(x,Q^2)(1-y)^2\right], \qquad (3)$$

where $-Q^2$ is the invariant momentum transfer between the incident neutrino and the outgoing muon, M_N and M_W are the nucleon and intermediate-boson masses, and $G_F = 1.16632 \times 10^{-6}$ GeV^{-2} is the Fermi constant. We use CTEQ6-L1 for evaluating the parton distribution functions [4]. For energies greater than 10^{10} GeV we use the DLA approximation [2]. Our results are in good agreement with those reported in literature [2]. After been produced with an energy E_μ^0, the muon travels in the rock surrounding the detector and looses energy. The expression $R(E_\mu^0,E_\mu^{min})$ gives the distance or range that a muon with initial energy E_μ^0 travels until it reaches a final energy greater or equal than E_μ^{min}. In this work we use the analytical expression [5]

$$R(E_\mu^0,E_\mu^{min}) = \frac{1}{b}ln\left(\frac{a+bE_\mu^0}{a+bE_\mu^{min}}\right), \qquad (4)$$

where $a = 2.0 \times 10^{-3}$ GeV cm^2/g y $b = 3.9 \times 10^{-6} cm^2/g$ are the values of the parameters for standard rock.

Non Standard Interactions

When neutrinos propagate in a medium, an interaction term has to be added to the Hamiltonian to account for neutrino refraction in matter. For this, we consider the

TABLE 1. Upward $\mu^+ + \mu^-$ event rates per year arising from $\nu_\mu N$ and $\bar{\nu}_\mu N$ interactions in rock, for a detector with an effective area $A = 1$ km^2 and different energies thresholds E_μ^{min}. Our rates are compared with other results [5, 7].

Flux	E_μ^{min}/GeV				
	10^2	10^3	10^4	10^5	10^6
Our estimation with ATM (average)	260700	14340	241	1.8	0.0076
Other authors					
K. Giesel[5] 2003		17600	318	2.4	0.0012
L. Moscoso[7] 1998		12000	280	3.6	0.04

Standard Model (SM) weak interactions, and possible NSI, both flavor changing (FC) and flavor preserving (FP). We can write the NSI Lagrangian in the form of effective four-fermion terms:

$$L^{NSI} = -2\sqrt{2}G_F(\bar{\nu}_\alpha\gamma_\rho\nu_\beta)(\varepsilon_{\alpha\beta}^{f\tilde{f}L}\bar{f}_L\gamma^\rho\tilde{f}_L + \varepsilon_{\alpha\beta}^{f\tilde{f}R}\bar{f}_R\gamma^\rho\tilde{f}_R) + h.c., \quad (5)$$

where $\varepsilon_{\alpha\beta}^{f\tilde{f}L}$ ($\varepsilon_{\alpha\beta}^{f\tilde{f}R}$) denotes the strength of the NSI between the neutrinos ν of flavors α and β and the left-handed (right-handed) effective components of the fermions f and $\tilde{f}$. Here we are interested in the oscillation probability in medium with constant density which is given by [6]

$$P(\nu_\mu \to \nu\tau') = \sin^2 2\theta_m \sin^2(\Delta_m L), \quad (6)$$

where the effective mixing θ_m and mass splitting in matter Δ_m depend on the parameters $\varepsilon_{ee}, \varepsilon_{e\tau}$ and $\varepsilon_{\tau\tau}$ which describe the NSI [6].

RESULTS

We present the results of our computations for the total number of events in a 1 km^2 neutrino telescope in table 1. We can see that there is a good agreement with other results previously reported. Now we incorporate the survival probability shown in Eq. (6) and compute the number of events for different allowed values of the NSI parameters according to [6]. In table 2 the relative variation of this number of events is shown for neutrino energies greater than 10^2 GeV.

CONCLUSIONS

We calculated the total atmospheric upward muon event rates in a neutrino telescope with an effective area of 1 km^2 in order to make estimations of the sensitivity to new physics. As a first application we studied the case of NSI and we found that the variation is minor, however we should take into account that for energies greater than 10^2 GeV a neutrino telescope could collect 7×10^5 events in ten years giving us an statistical

TABLE 2. Relative variation $\Delta N/N$ of the total upward event rates N for a neutrino telescope in presence of NSI. We consider an energy threshold of 10^2 GeV.

ε_{ee}	$\varepsilon_{e\tau}$	$\varepsilon_{\tau\tau}$	$\eta = \frac{\Delta N}{N}$ (%)
1.5	2.0	1.5	$\eta = 0.3$
1.5	3.0	4.0	$\eta = 0.41$
0.0	1.0	1.0	$\eta = 0.37$
-0.5	0.2	0.0	$\eta = 0.1$

resolution of 0.12 percent. For this reason a more detailed calculation in the survival probability with NSI could be interesting.

ACKNOWLEDGMENTS

I am grateful with my advisors Arnulfo Zepeda and Omar Miranda for many helpful discussions. I also thank Jose Valle who proposed the subject of this interesting work and Juan Carlos Arteaga and Celio Moura for their valuable suggestions.

REFERENCES

1. M. C. Gonzalez-Garcia, F. Halzen and M. Maltoni, "Physics reach of high-energy and high-statistics Icecube atmospheric neutrino data," Phys. Rev. D **71**, 093010 (2005)
2. R. Gandhi, C. Quigg, M. H. Reno and I. Sarcevic, "Neutrino interactions at ultrahigh energies," Phys. Rev. D **58**, 093009 (1998)
3. L. V. Volkova, "Energy Spectra And Angular Distributions Of Atmospheric Neutrinos," Sov. J. Nucl. Phys. **31**, 784 (1980)
4. J. Pumplin, D. R. Stump, J. Huston, H. L. Lai, P. Nadolsky and W. K. Tung, "New generation of parton distributions with uncertainties from global QCD analysis," JHEP **0207**, 012 (2002)
5. K. Giesel, J. H. Jureit and E. Reya, "Cosmic UHE neutrino signatures," Astropart. Phys. **20**, 335 (2003)
6. A. Friedland and C. Lunardini, "A test of tau neutrino interactions with atmospheric neutrinos and K2K," Phys. Rev. D **72**, 053009 (2005)
7. L. Moscoso, "Neutrino astronomy," DAPNIA-SPP-99-02 *Talk given at 5th ICTP School on Nonaccelerator Particle Astrophysics, Trieste, Italy, 29 Jun - 10 Jul 1998*

Fermions Living in a Flat World

Ma. de Jesús Anguiano-Galicia* and A. Bashir*,†,**

*Instituto de Física y Matemáticas, Universidad Michoacana de San Nicolás de Hidalgo, Apartado Postal 2-82, Morelia, Michoacán 58040, Mexico
†Instituto de Ciencias Nucleares, Universidad Nacional Autónoma de México, Circuito Exterior, C. U., Apartado Postal 70-543, 04510, México, D. F.
** Institute for Particle Physics Phenomenology, University of Durham, Durham DH1 3LE, U.K.

Abstract.
  In a plane, parity transformation, which changes the sign of only one spatial coordinate, swaps the fermion fields living in two inequivalent representations. A parity invariant Lagrangian thus contains fields corresponding to both the representations. For such a Lagrangian, we show that we can also define a chiral symmetry.

Keywords: chiral symmetry, parity symmetry, 3 space-time dimensions
PACS: 11.10.Kk, 11.30.Rd, 03.65.Pm

INTRODUCTION

In this paper, we study the problem of chiral symmetry in odd space-time dimensions. An interesting and somewhat uncomfortable observation is that in the fundamental 2×2 representation of the gamma matrices, chiral symmetry cannot be defined owing to the fact there does not exist any matrix which would anti-commute with all the three Pauli matrices. It is for this reason that the dynamical chiral symmetry breaking in a plane is generally studied in the 4×4 representation of the gammas, e.g., [1, 2]. In the present paper, we argue that a chiral symmetry could formally be de defined within the 2×2 formalism provided we take into account both the inequivalent representations of the gamma matrices, [3].

FUNDAMENTALS

Starting from the Lagrangian $\mathscr{L} = \bar{\psi}(i\hbar \slashed{\partial} - mc)\psi$, we can choose the following representation of the gamma matrices in a plane : $\gamma^0 = \sigma_3, \gamma^1 = i\sigma_1$, and $\gamma^2 = i\sigma_2$ (the A representation hereafter), where σ_i are the Pauli matrices. We then readily obtain the solutions of the free Dirac equation :

$$\psi_A^P(x) = \begin{pmatrix} 1 \\ \frac{c(p_y - ip_x)}{E+mc^2} \end{pmatrix} e^{-\frac{i}{\hbar}x\cdot p} \equiv u_A(p) e^{-\frac{i}{\hbar}x\cdot p} ,$$

$$\psi_A^N(x) = \begin{pmatrix} \frac{c(p_y + ip_x)}{E+mc^2} \\ 1 \end{pmatrix} e^{\frac{i}{\hbar}x\cdot p} \equiv v_A(p) e^{\frac{i}{\hbar}x\cdot p} . \quad (1)$$

Choosing the normalization of the spinors to be such that there are $2E$ number of particles per unit volume, we have $u_A \bar{u}_A = \slashed{p}c + mc^2$ and $v_A \bar{v}_A = \slashed{p}c - mc^2$. This is to say that the completeness relations are not hampered by the fact there is just one particle spinor and one anti-particle spinor. The above relation permits us to define the projection operators $\Lambda_\pm = (\pm \slashed{p} + mc)/2mc$ which project out the particle and ant-particle spinors respectively. Therefore, everything is apparently in order. However, there are reasons to believe that the above Lagrangian fails to incorporate various symmetries and their consequences :

- **Parity Invariance:** There are two independent solutions, one corresponding to a particle (P) and the other to an anti-particle (N). In a plane, there is just one orbital angular momentum which we can define as $L = r_x p_y - r_y p_x$. It does not commute with the Hamiltonian $H = \gamma^0(\vec{\gamma} \cdot \vec{p} + mc^2)$. However, if we define the spin operator as $\Sigma = (\hbar/2)\gamma^0$, the total angular momentum $J = L + \Sigma$ is a conserved quantity. It is easy to see that for the particle at rest, i.e., for $p_x = p_y = 0$, u_A and v_A are eigenfunctions of Σ with eigenvalues $\hbar/2$ and $-\hbar/2$ respectively. It implies a natural interpretation of the solution u_A as that of a particle with spin say clockwise and of v_A as that of an anti-particle with spin anti-clockwise. Let us define parity operation as what reverses the sign of one coordinate. Let us suppose that under this transformation, $r_x \to -r_x$ and $r_y \to r_y$. Consequently, spin, being an angular momentum, changes sign. Therefore, particle with clockwise spin and anti-clockwise spin are related through the parity transformation. But one of these is not a solution of the Dirac equation. The Lagrangian and thus the particle spectrum are not parity invariant.
- **Chiral Symmetry:** In a plane, all the anti-commuting matrices in the fundamental representation are consumed by the Dirac gamma matrices. We are left with no extra γ_5 which will anti-commute with all the gamma matrices. Therefore, chiral transformations (and chiral symmetry) cannot be defined. Apparently, the massless Lagrangian has no more symmetry than the massive one.

If we wish to incorporate these symmetries within the framework of two-component spinors, we can do this thanks to the well-known fact that there exist two inequivalent representations, [3]. We can choose the B representation to be $\gamma^0 = \sigma_3, \gamma^1 = i\sigma_1, \gamma^{2'} = -\gamma^2 = -i\sigma_2$ We transform the corresponding solutions ϕ_B^P and ϕ_B^N of the Dirac equation to ψ_B^P and ψ_B^N for obvious particle identification as follows :

$$\psi_B^P(x) = i\gamma^2 \phi_B^P = \begin{pmatrix} \frac{c(p_y+ip_x)}{E+mc^2} \\ 1 \end{pmatrix} e^{-\frac{i}{\hbar}p \cdot x} = u_B(p) e^{-\frac{i}{\hbar}p \cdot x},$$

$$\psi_B^N(x) = i\gamma^2 \phi_B^N = \begin{pmatrix} 1 \\ \frac{c(p_y-ip_x)}{E+mc^2} \end{pmatrix} e^{\frac{i}{\hbar}p \cdot x} = v_B(p) e^{\frac{i}{\hbar}p \cdot x}. \tag{2}$$

Looking at the stationary case, $p_x = p_y = 0$, by applying the spin operator, we can see that ψ_A^P and ψ_B^P correspond to particles with opposite spins. Similarly, ψ_A^N and ψ_B^N correspond to anti-particles with opposite spins. The parity transformation $\mathscr{P}$:

$(\psi_A)^{\mathcal{P}} = -i\gamma^1 \psi_B\, e^{i\phi_1}$ and $(\psi_B)^{\mathcal{P}} = -i\gamma^1 \psi_A\, e^{i\phi_2}$ swaps the spinors in the two inequivalent representations. It converts particle of one spin to the particle of opposite spin, and the same is true for the anti-particle. The Lagrangian which takes into account both the representations is, [3]:

$$\mathcal{L} = \bar{\psi}_A(i\hbar\,\slashed{\partial} - mc)\psi_A + \bar{\psi}_B(i\hbar\,\slashed{\partial} + mc)\psi_B. \tag{3}$$

It is parity invariant, [3]. Most importantly, we shall show in the next section that this Lagrangian also permits us to define chiral symmetry.

CHIRAL SYMMETRY

We now pose ourselves the question whether the Lagrangian of Eq. (3) has any more symmetries for the massless case. An immediate look reveals there is an exchange symmetry $\psi_A \leftrightarrow \psi_B$ which leaves the massless Lagrangian invariant. As it is only a discrete symmetry, it cannot be considered a serious candidate for the chiral symmetry. However, one can define following sets of simultaneous continuous transformations:
Set 1

$$\begin{aligned}\psi_A \to \psi'_A &= \psi_A + \alpha\psi_B, \\ \psi_B \to \psi'_B &= \psi_B - \alpha\psi_A.\end{aligned} \tag{4}$$

Set 2

$$\begin{aligned}\psi_A \to \psi'_A &= \psi_A + i\alpha\psi_B, \\ \psi_B \to \psi'_B &= \psi_B + i\alpha\psi_A.\end{aligned} \tag{5}$$

Correspondingly, the Lagrangian (3) transforms in the following manner respectively:

$$\begin{aligned}\mathcal{L}'_1 &= \mathcal{L} - 2mc\alpha(\bar{\psi}_A\psi_B + \bar{\psi}_B\psi_A), \tag{6} \\ \mathcal{L}'_2 &= \mathcal{L} - 2imc\alpha(\bar{\psi}_A\psi_B - \bar{\psi}_B\psi_A). \tag{7}\end{aligned}$$

Therefore, we conclude that under the continuous transformations (4,5), the massless Lagrangian is invariant and the corresponding conserved currents are:

$$\begin{aligned}j_1^\mu &= c(\bar{\psi}_A\gamma^\mu\psi_B - \bar{\psi}_B\gamma^\mu\psi_A), \tag{8} \\ j_2^\mu &= c(\bar{\psi}_A\gamma^\mu\psi_B + \bar{\psi}_B\gamma^\mu\psi_A). \tag{9}\end{aligned}$$

To be able to identify continuous transformations (4) as the chiral transformation, we resort to address the issue of chiral symmetry in the often studied 4-dimensional representation of the gamma matrices in the Lagrangian $\mathcal{L} = \bar{\psi}(i\hbar\,\slashed{\partial} - mc)\psi$:

$$\gamma^0 = \begin{pmatrix} \sigma_3 & 0 \\ 0 & -\sigma_3 \end{pmatrix}, \quad \vec{\gamma} = \begin{pmatrix} i\vec{\sigma} & 0 \\ 0 & -i\vec{\sigma} \end{pmatrix},$$

where the vector $\vec{\sigma}$ has only two components, namely, σ_1 and σ_2. In the 4-dimensional representation, we have sufficient freedom to define two matrices which anti-commute with the Dirac gamma matrices :

$$\gamma^5 = i \begin{pmatrix} 0 & I \\ -I & 0 \end{pmatrix}, \qquad \gamma^3 = \begin{pmatrix} 0 & I \\ I & 0 \end{pmatrix},$$

and hence we have two types of chiral transformations which yield the following chiral current in the massless limit :

$$j_5^\mu \equiv c\bar{\psi}\gamma^\mu \gamma^5 \psi, \qquad j_3^\mu \equiv c\bar{\psi}\gamma^\mu \gamma^3 \psi. \qquad (10)$$

Interestingly, if we write out $\psi = (\psi_A, \psi_B)$, the Lagrangian $\mathscr{L} = \bar{\psi}(i\hbar\,\partial\!\!\!/ - mc)\psi$ reduces to the one in Eq. (3). Therefore, the components ψ_A and ψ_B themselves satisfy the Dirac equation in the 2-dimensional representation of the gamma matrices justifying the use of this notation. The chiral currents j_5^μ and j_3^μ of Eq. (10) can then be written as :

$$\begin{aligned} j_5^\mu &= c(\bar{\psi}_A \gamma^\mu \psi_B - \bar{\psi}_B \gamma^\mu \psi_A), \\ j_3^\mu &= c(\bar{\psi}_A \gamma^\mu \psi_B + \bar{\psi}_B \gamma^\mu \psi_A). \end{aligned}$$

Comparing Eqs. (8,11), we see that the currents obtained from the conventional definition of chiral symmetry in the 4-dimensional representation and the ones we propose in the 2-dimensional representation, Eq. (4,5), are identical. *Therefore, Eq. (4,5) are indeed chiral transformations.*

CONCLUSIONS

We demonstrate that by taking into account both the inequivalent fundamental representations of the gamma matrices in 3-dimensional space-time, the resulting Lagrangian is not only parity invariant but also that *we can write out chiral transformations within the two-component description of the fermion spinors which mix the fields belonging to the different inequivalent representations.*

ACKNOWLEDGEMENTS

Support for this work has been received in part by CIC under grant number 4.10 and CONACyT under grant number 32395-E. AB wishes to acknowledge a short term visitor grant by a joint scheme of The Royal Society, U.K, and The Mexican Academy of Sciences, Mexico.

REFERENCES

1. R.D. Pisarski, Phys. Rev. **D29** 2423 (1984).
2. T.W. Appelquist, M. Bowick, D. Karabali and L.C.R. Wijewardhana, Phys. Rev. **D33** 3704 (1986); T.W. Appelquist, M. Bowick, D. Karabali and L.C.R. Wijewardhana, Phys. Rev. **D33** 3774 (1986).
3. K. Shimizu, Prog. Theor. Phys. **74** 610 (1985).

Electronics for the Extensive Air Shower Detector Array at the University of Puebla

E. Pérez*, R. Conde*, O. Martínez*, T. Murrieta*, H. Salazar* and L. Villaseñor[†]

*Facultad de Ciencias Fisico-Matematicas, BUAP, Puebla Pue., 72570, Mexico, hsalazar@fcfm.buap.mx
[†]Institute of Physics and Mathematics, Universidad Michoacana, Edificio C3, Ciudad Universitaria, Morelia, Mich., 58040, Mexico, villasen@ifm.umich.mx

Abstract. In this paper we describe in detail the electronics cards that were designed to be the basis of the data acquisition system (DAS) of the extensive air shower detector array built in the Campus of the University of Puebla. The purpose of this observatory is to measure the energy and arrival direction of primary cosmic rays with energies around 10^{15} eV. The array consists of 18 liquid scintillator detectors (12 in the first stage) and 6 water Cherenkov detectors (one of 10 m^2 cross section and five smaller ones of 1.86 m^2 cross section), distributed in a square grid with a detector spacing of 20 m over an area of 4000 m^2. The electronics described here uses analog to digital converters of 10 bits working at a sampling speed of 40 MS/s and field-programmable gate array (FPGA).

Keywords: Cosmic rays; ADC; FPGA.

INTRODUCTION

The extensive air shower detector array at University of Puebla (EAS-UAP) was designed to measure the lateral distribution and arrival direction of secondary particles for EAS in the energy region of $10^{14} - 10^{16}$ eV. The special location of the EAS-UAP array; 2200 m above sea level; and all the facilities coming from the Campus of the University of Puebla make it a valuable apparatus for the long term study of cosmic rays and at the same time an important training center for new physics students interested in getting a first class education in the field of cosmic rays in Mexico.

The energy spectrum of primary cosmic rays is well described by a power law, i.e., $dE/dx \sim E^{-\gamma}$, over many decades of energy with the spectral index γ approximately equal to 2.7, and steepening to $\gamma = 3$ at $E = 3 \times 10^{15}$ eV. This structural feature is known as the "knee" of the cosmic ray spectrum.

The nature of the knee is still a puzzle despite the fact that it was discovered more than 46 years ago. The best handle to study the composition of primary cosmic rays by using ground detector arrays is the measurement of the ratio of the muonic to the electromagnetic component of EAS; Monte Carlo simulations show that heavier primaries give rise to a bigger muon/EM ratio compared to lighter primaries of the same energy.

The EAS-UAP array is expected to contribute to unveiling the mystery of the origin of the knee by detecting thousands of cosmic rays with energies around $10^{14} - 10^{16}$ eV. In this paper we describe in detail the electronics cards that were designed to be the basis

FIGURE 1. Temporary data acquisition system; it is based on digital oscilloscopes to digitize the signals from the PMTs and commercial NIM modules to discriminate the PMT signals and to generate the coincidence trigger signal.

of the data acquisition system of the EAS-UAP detector array.

EXPERIMENTAL SETUP

The EAS-UAP array has been described in detail elsewhere [1, 2, 3], it is located in the campus of the University of Puebla in Mexico (UAP) at 19° N, 89° W and 800 gcm^{-2}; it consists of 18 liquid scintillator detectors distributed uniformly on a square grid with spacing of 20 m, and six water Cherenkov detectors (one of 10 m^2 cross section and five smaller ones of 1.86 m^2). It uses both Cherenkov and liquid scintillator detectors to attempt a measurement of the muonic component of EAS [4, 5].

Data Acquisition System

At the moment the EAS-UAP array is using a temporary DAS based on a set of digital oscilloscopes that digitize the signals from the PMTs of the liquid scintillator and water Cherenkov detectors. All the digital oscilloscopes are connected to the GPIB port of a PC in a daisy chain configuration.

The system is controlled by the PC running a custom-made acquisition program written in a graphical language called LabView. We used commercial NIM modules to discriminate the PMT signals at a threshold of -30 mV and to generate the coincidence trigger signal Fig. 1. The DAS acquires all the PMT traces for each triggered event. The acquired traces are used by the PC to perform on-line measurements of the integrated charges, arrival times, amplitudes and widths of all signals the PMTs, these data are saved into a hard-disk file for further off-line analysis.

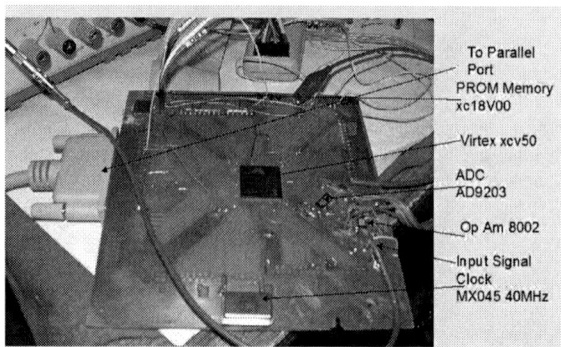

FIGURE 2. Photograph of the new electronics card that will replace the digital oscilloscopes and the commercial NIM and CAMAC modules that we are using at the moment, see Fig. 1.

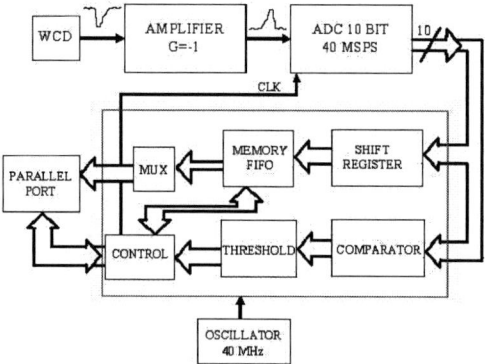

FIGURE 3. Schematic diagram of the electronics card shown in Fig. 2.

Description of the Electronics

The new electronics is meant to replace the digital oscilloscopes and the commercial NIM and CAMAC modules that we are using at the moment. Fig. 2 shows a photograph of the new electronics card and Fig. 3 shows its corresponding block diagram.

The analog signal from the PMTs is inverted by using a fast operational amplifier configured to have a gain of -1. This inverted signal is sent to an analog to digital converter (ADC) with a 10 bit resolution and a sampling speed of 40 MHz. This ADC can accept signals with amplitudes from 0 to 2 V. The 10-bit ADC output is connected to a field-programmable gate array (FPGA), Virtex XCV50 from Xilinx. This FPGA was programmed to behave like the block diagram shown inside the rectangle shown in Fig. 3.

The FPGA counts the number of events which have amplitudes above a pre-established threshold. This threshold is programmed by the PC into the *comparator* module. All events above threshold are recorded into the FIFO memory and this number is updated in the *threshold* module. Since the parallel port is not fast enough to process the output of the ADC, the 10-bit data from the ADC is saved temporarily into the *FIFO memory*. The data from the latter is downloaded into the PC in synchronous mode. The *control* module takes charge of doing this synchronization between the master clock, the ADC and the FIFO memory during the process of writing into the FIFO and also between the parallel port and the FIFO during the process of reading the FIFO memory by the PC. The *control* module also keeps track of the *empty* and *full* flags used to determine if the FIFO memory is empty of full. The *mux* module is used to pass the 10-bit data into the PC through the parallel port. This is done in two steps: in the first one the least significative 8 bits are read out by the data bus of the parallel port (connectors 2-9), and in the second step the remaining two bits are acquired by the parallel port through the state register pins (connectors 10 and 11).

CONCLUSIONS

We have described in detail the electronics cards that were designed to be the basis of the data acquisition system (DAS) of the extensive air shower detector array built in the Campus of the University of Puebla (located at 19°N, 90°W, 800 gcm^{-2}). The purpose of this observatory is to measure the energy and arrival direction of primary cosmic rays with energies around 10^{15} eV. The electronics described here uses analog to digital converters of 10 bits working at a sampling speed of 40 MS/s and a field-programmable gate array to replace the set of digital oscilloscopes, NIM and CAMAC modules presently used as the basis of the EAS-UAP data acquisition system.

ACKNOWLEDGEMENTS

We would like to thank University of Michoacan, University of Puebla and CONACyT for supporting this work.

REFERENCES

1. H. Salazar, O. Martinez, E. Moreno, J. Cotzomi, L. Villasenor, O. Saavedra, Nuclear Physics B (Proc. Suppl.) 122 (2003) 251-254.
2. J. Cotzomi, O. Martinez, E. Moreno, H. Salazar and L. Villasenor, Revista Mexicana de Fisica, 51(1)(2005) 38-46, .
3. J. Cotzomi, E. Moreno, T. Murrieta, B. Palma, E. Perez, H. Salazar and L. Villasenor, Nucl. Instrume. and Meths. in Phys. Res. A. Volume 553, Issues 1-2 (2005) 290-294.
4. M. Alarcón et al., Nucl. Instrum and Meths. in Phys. Res. A 420 (1999) 39-47.
5. H. Salazar and L. Villasenor, Nucl. Instruments and Meths. in Phys. Res. A. Volume 553, Issues 1-2 (2005) 295-298.

Proceedings Published by the Division of Particles and Fields of the Mexican Physical Society

MEXICAN SCHOOLS ON PARTICLES AND FIELDS

I. Oaxtepec, Morelos; AIP Conf. Proceedings 143, 1986; editors: J.L. Lucio, M. Moreno, A. Zepeda

II. Cuernavaca, Morelos; World Scientific, 1987; editors: J.L. Lucio, A. Zepeda

III. Oaxtepec, Morelos; World Scientific, 1989; editors: J.L. Lucio, A. Zepeda

IV. Oaxtepec, Morelos; World Scientific, 1991; editors: J.L. Lucio, A. Zepeda

V. Guanajuato, Gto.; AIP Conf. Proceedings 317, 1993; editors: J.L. Lucio, A. Zepeda

VI. Villahermosa, Tabasco; World Scietific 1995; editors: J.C. D'Olivo, M. Moreno, M.A. Pérez

VII. Mérida, Yucatán; AIP Conf. Proceedings 400, 1997; editors: J.C. D'Olivo, M. Klein-Kreisler, H. Méndez

VIII. Oaxaca, Oax.; AIP Conf. Proceedings 490, 1999; editors: J.C. D'Olivo, G. López-Castro, M. Mondragón

IX. Metepec, Puebla; AIP Conf. Proceedings 562, 2001; editors: G. Herrera, L. Nellen

X. Playa del Carmen, Quintana Roo, AIP Conf. Proceedings 670, 2003; editors: U. Cotti, M. Mondragón, G. Tavares-Velasco

XI. Xalapa, Veracruz; Journal of Physics Conf. Serv. 18, 2005; editors: A. Bashir, J. Erler, R. Hernández, M. Mondragón, L. Villaseñor

MEXICAN WORKSHOPS ON PARTICLES AND FIELDS

II. Puebla, Pue.: Rev. Mex. Fís., **36** Supl. 1, 1990; editors: S.R. Juárez, M.A. Pérez, A. Queijeiro.

IV. Mérida, Yucatán; World Scientific, 1994; editors: R. Huerta, M.A. Pérez, L. Urrutia

V. Puebla, Pue.; AIP Conf. Proceedings 359, 1996; editors: J.C. D'Olivo, A. Fernández, M.A. Pérez

VI. Morelia, Michoacán; AIP Conf. Proceedings 445, 1998; editors: J.C. D'Olivo, M. Ruiz-Altaba, L. Villaseñor

VII. Mérida, Yucatán; AIP Conf. Proceedings 531, 2000; editors: A. Ayala, G. Contreras, G. Herrera

VIII. Zacatecas, Zac.; AIP Conf. Proceedings 623, 2002; editors: L. Díaz-Cruz, J. Engelfried, M. Kirchbach, M. Mondragón

IX. Colima, Col.; Journal of Physics Conf. Serv. 37, 2006; editors: P. Amore, A. Aranda, A. Bashir, M. Mondragón, A. Raya

X. Morelia, Mich.; AIP Conf. Proceedings 857A, 2006; editors: A. Bashir, V. Villanueva, L. Villaseñor. AIP Conf. Proceedings 857B, 2006; DPF-MPS Commemorative Volume, editors: M.A. Pérez, L.F. Urrutia, L. Villaseñor

Author Index

A

Alfaro, R., 134, 218
Alvarez, C., 259
Ayala, A., 311

B

Bashir, A., 226, 279, 311, 334

C

Cabral-Rosetti, L. G., 179, 316, 321
Campos, R. G., 170
Carrillo-Ruiz, M. G., 271
Chivukula, R. S., 34
Compeán Jasso, C. B., 275
Concha-Sánchez, Y., 279
Conde, R., 338
Cruz, E., 218
Cuautle, E., 175, 283

D

de Gouvêa, A., 3
de Jesús Anguiano-Galicia, M., 334
Díaz-Cruz, L., 326
Domínguez, I., 283
Dunne, G. V., 240

E

Erler, J., 85

F

Fachini, P., 62

G

Gago Medina, A., 306
Gaitán, R., 179
García-Ferreira, Ix-B., 152
García-Herrera, J., 152
Godina, J. J., 186
Gómez-Izquierdo, J. C., 297
González Canales, F., 287
Gutierrez-Guerrero, L. X., 279
Gutiérrez-Rodríguez, A., 302

H

He, H.-J., 34
Hernández-Ruíz, M. A., 302
Herrera-Aguilar, A., 194
Herrera Corral, G., 306
Höll, A., 46

J

Jalilian-Marian, J., 76
Juárez W., S. R., 202

K

Kanakoglou, K., 194
Kielanowski, P., 202
Kirchbach, M., 275
Kurachi, M., 34

L

Li, L., 186
Lopez-Fernandez, R., 210
Luján-Peschard, C., 293

M

Martinez, M. I., 218
Martínez, O., 259, 338

Matos, T., 316
Meurice, Y., 186
Miranda, O. G., 179, 330
Mondragón, A., 287
Mondragón, M., 321
Montaño, L. M., 143, 218
Moreno, E., 113
Murrieta, T., 113, 338

N

Napsuciale, M., 271, 293
Nellen, L., 100, 321
Nieto, J. A., 249
Nieves, J. F., 234
Nunez, D., 316, 321

O

Oktay, M. B., 186

P

Paić, G., 175, 218, 283
Paschalis, J. E., 194
Pedraza-Morales, M. I., 326
Pérez, E., 338
Pérez Lara, C., 306
Pérez-Lorenzana, A., 297
Pérez Martínez, R., 330

R

Ramírez-Jiménez, F. J., 121
Ramírez-Sánchez, F., 302
Raya, A., 226, 311

Roberts, C. D., 46
Rojas, E., 311
Rosado, A., 326
Ruiz, L., 249

S

Sahu, S., 234
Salazar, H., 113, 259, 326, 338
Sandoval, A., 218
Schubert, C., 240
Silvas, J., 249
Simmons, E. H., 34
Solís R., H. G., 202
Stone, S., 18
Sussman, R. A., 316, 321

T

Tanabashi, M., 34
Tututi, E. S., 170

V

Villanueva, V. M., 249
Villaseñor, L., 113, 152, 259, 338

W

Wright, S. V., 46

Z

Zavala, J., 316, 321
Zepeda, A., 330

PARTICLES AND FIELDS

Previous Proceedings in the Series of Mexican Workshops on Particles and Fields

	Year	Held in	Publisher	ISBN
8th	2002	Zacatecas, Zacatecas	AIP Conf. Proceedings vol. 623	0-7354-0072-5
7th	1999	Mérida, Yucatán	AIP Conf. Proceedings vol. 531	1-56396-954-8
6th	1997	Morelia, Michoacán	AIP Conf. Proceedings vol. 445	1-56396-791-X
5th	1995	Puebla, Puebla	AIP Conf. Proceedings vol. 359	1-56396-548-8

Previous Proceedings in the Series of Mexican Schools on Particles and Fields

	Year	Held in	Publisher	ISBN
X	2003	Playa del Carmen, Quintana Roo	AIP Conf. Proceedings vol. 670	0-7354-0135-7
IX	2000	Metepec, Puebla	AIP Conf. Proceedings vol. 562	1-56396-998-X
VIII	1998	Oaxaca, México	AIP Conf. Proceedings vol. 490	1-56396-895-9
VII	1996	Mérida, Yucatán	AIP Conf. Proceedings vol. 400	1-56396-686-7

To learn more about these titles, or the AIP Conference Proceedings Series, please visit the webpage **http://proceedings.aip.org/proceedings**

PARTICLES AND FIELDS

Commemorative Volume of the Division of Particles and Fields of the Mexican Physical Society

Morelia Michoacán, México   6 –12 November 2005

■ **PART B**

EDITORS
Miguel A. Pérez
Cinvestav

Luis F. Urrutia
UNAM

Luis Villaseñor
UMSNH

SPONSORING ORGANIZATIONS
University of Michoacán (UMSNH)
Division of Particles and Fields of the Mexican Physical Society (DPyC-SMF)
Mexican Council on Science and Technology (CONACyT)

Melville, New York, 2006
AIP CONFERENCE PROCEEDINGS ■ VOLUME 857

Editors

Miguel A. Pérez
Departamento de Física
Cinvestav
Ap. Postal 14-740
07000 Mexico City, México

E-mail: mperez@fis.cinvestav.mx

Luis F. Urrutia
Instituto de Ciencias Nucleares
UNAM
Circuito Exterior, Ciudad Universitaria
Ap. Postal 70-543
04510 Mexico City, México

E-mail: urrutia@nucleares.unam.mx

Luis Villaseñor
Instituto de Física y Matemáticas
UMSNH
Edificio C3, Ciudad Universitaria
58040 Morelia, Michoacán, México

E-mail: villasen@ifm.umich.mx

Authorization to photocopy items for internal or personal use, beyond the free copying permitted under the 1978 U.S. Copyright Law (see statement below), is granted by the American Institute of Physics for users registered with the Copyright Clearance Center (CCC) Transactional Reporting Service, provided that the base fee of $23.00 per copy is paid directly to CCC, 222 Rosewood Drive, Danvers, MA 01923, USA. For those organizations that have been granted a photocopy license by CCC, a separate system of payment has been arranged. The fee code for users of the Transactional Reporting Services is: 978-0-7354-0354-3/06/$23.00

© 2006 American Institute of Physics

Permission is granted to quote from the AIP Conference Proceedings with the customary acknowledgment of the source. Republication of an article or portions thereof (e.g., extensive excerpts, figures, tables, etc.) in original form or in translation, as well as other types of reuse (e.g., in course packs) require formal permission from AIP and may be subject to fees. As a courtesy, the author of the original proceedings article should be informed of any request for republication/reuse. Permission may be obtained online using Rightslink. Locate the article online at http://proceedings.aip.org, then simply click on the Rightslink icon/"Permission for Reuse" link found in the article abstract. You may also address requests to: AIP Office of Rights and Permissions, Suite 1NO1, 2 Huntington Quadrangle, Melville, NY 11747-4502, USA; Fax: 516-576-2450; Tel.: 516-576-2268; E-mail: rights@aip.org.

L.C. Catalog Card No. 2006932275
ISBN 978-0-7354-0354-3
ISSN 0094-243X

Printed in the United States of America

CONTENTS

Preface ix
Acknowledgments x

HISTORICAL NOTES

High Energy Physics in Mexico: Historical Sketch and Implications 3
    A. García and M. A. Pérez
Remembrances on the Origin of the Mexican School of Particles and
Fields 11
    J. L. Lucio and A. Zepeda
Fermilab and Latin America 15
    L. M. Lederman

PHENOMENOLOGICAL HIGH ENERGY PHYSICS

Baryon Semileptonic Decays: The Mexican Contribution 27
    R. Flores-Mendieta and A. Martínez
Neutrinos 37
    J. C. D'Olivo and O. G. Miranda
Low-Energy Physics (u,d,s) 49
    J. L. Lucio and M. Napsuciale
Studies of Electroweak Symmetry Breaking at Mexican Institutions 59
    J. L. Diaz Cruz
Supersymmetry and GUTs in Mexico 66
    M. Mondragón
Applications of the Renormalization Group 79
    P. Kielanowski and S. R. Juárez W.
Deep Inelastic Scattering and Collider Physics 90
    A. Rosado
New Physics Effects in Trilinear Electroweak Gauge Boson Couplings 103
    J. J. Toscano
QCD Phenomenology 118
    P. O. Hess
Astroparticle Physics 128
    A. Zepeda
Top Quark Physics 133
    F. Larios
Relativistic Heavy-ion Physics in Mexico 144
    A. Ayala
A Brief Review on Extra Dimensions 152
    A. Pérez-Lorenzana

THEORETICAL HIGH ENERGY PHYSICS

String Theory in México 167
    H. García-Compeán and A. Güijosa
Renormalization Group Theory 181
    C. R. Stephens
Dark Energy and Dark Matter 191
    A. de la Macorra and T. Matos
Quantum Gravity Phenomenology 205
    H. A. Morales-Técotl and L. F. Urrutia
Mexican Contributions to Noncommutative Theories 218
    J. D. Vergara and H. García-Compeán
Hamiltonian Methods: BRST, BFV 228
    J. A. García
The Casimir Effect 235
    C. Villarreal and W. L. Mochán
Anomalies and Gravity 246
    E. W. Mielke
Developments in Supergravity 258
    O. Obregón and C. Ramírez
Models of Flavour with Discrete Symmetries 266
    A. Mondragón
Dynamical Mass Generation 283
    A. Bashir and A. Raya

EXPERIMENTAL HIGH ENERGY AND COSMIC-RAY PHYSICS

Reflections on a Teenage Quark 297
    J. Konigsberg
The Mexican Participation in the ALICE Experiment 300
    G. Herrera-Corral and G. Paić
Physics of the Charm Quark 311
    S. Carrillo-Moreno and E. F. Vázquez-Valencia
Λ^0 Polarization in Exclusive pp Reactions 320
    J. Felix
Towards the International Linear Collider 330
    R. López-Fernández
Ring Imaging Cherenkov Detectors 340
    J. Engelfried
Weak Decays: Early Experiments (Magnetic Moments of Hyperons) 347
    A. Morelos
Use of Silicon Detectors in Medical Physics 355
    L. M. Montaño-Zetina
The Mexican Side of the H1 Collaboration 364
    J. G. Contreras
Radiographic Images Produced by Cosmic-ray Muons 373
    R. Alfaro

Astroparticle Physics: Detectors for Cosmic Rays 382
      H. Salazar and L. Villaseñor
Application of PIN Diodes in Physics Research 395
      F. J. Ramírez-Jiménez, L. Mondragón-Contreras, and P. Cruz-Estrada
Cosmic Ray Studies at CERN 407
      A. Fernández T.
Glueballs, Hybrids and Exotics 419
      M. A. Reyes and G. Moreno

Proceedings Published by the Division of Particles and Fields of the
Mexican Physical Society 427
Author Index 429

PREFACE

The 20th anniversary of the founding of the Division of Particles and Fields of the Mexican Physical Society (DPF-MPS) has been celebrated with two scientific meetings: the X Mexican Workshop on Particles and Fields (Morelia, Michocán, November, 2005) and the XX Annual Meeting of the DPF-MPS (Mexico City, June, 2006). As part of this celebration, the goal of the present volume is precisely to put together review articles on the contributions made by the members of our division in high energy physics (HEP) over the last four decades. Even though the beginning of particle physics as a research subject has been traced to 1947 with the discovery of the π meson, the first research papers on HEP published by Mexican physicists took place in the 1950's. However, it was only in the late 1970's that activities in HEP research in Mexico became both constant and visible, with an increasing number of researchers and students working in Mexican institutions.

The DPF-MFS was founded in 1986 with the specific task of organizing academic activities that would strengthen HEP in Mexico. The wide spectrum of important scientific issues covered in this volume reflects the fact that HEP is a field cultivated seriously and with high standards in Mexico. In this respect, one can say that the main goal of DPF-MFS has been achieved. Of course, we are aware that much work still needs to be done and much more maturity remains to be reached. In the mean time, the editors of the present volume were pleased with the quick and enthusiastic response from our colleagues that were invited to review the HEP topics where the Mexican community has made important contributions.

The articles presented in this commemorative volume are organized as follows:
- Historical Notes
- Phenomenological High Energy Physics
- Theoretical High Energy Physics
- Experimental High Energy and Cosmic-Ray Physics

We hope these review articles will be useful not only as a summary of the past and present development of HEP in Mexico, but also as an introduction to the present worldwide research areas in our field.

Miguel A. Pérez, Cinvestav
Luis F. Urrutia, UNAM
Luis Villaseñor, UMSNH

ACKNOWLEDGEMENTS

We would like to thank CONACyT and the University of Michoacan for providing financial support to celebrate the X Mexican Workshop of Particles and Fields. This support made possible the publication of this special volume to commemorate the 20th anniversary of the founding of the Division of Particles and Fields of the Mexican Physical Society.

We also thank the members of the **National Organizing Committee**:
- Alfredo Aranda, Universidad de Colima
- Alejandro Ayala, Instituto de Ciencias Nucleares, UNAM
- Heriberto Castilla Valdez, Departamento de Física, Cinvestav
- Lorenzo Díaz-Cruz, FCFM, BUAP
- Jurgen Engelfried, Instituto de Física, UASLP
- Gabriel Lopez-Castro, Departamento de Física, Cinvestav
- Miguel Angel Perez Angón, Departamento de Física, Cinvestav
- Humberto Salazar, FCFM, BUAP
- Maria Elena Tejeda, Universidad de Sonora
- Luis Villaseñor, Insituto de Física y Matemáticas, UMSNH
- Arnulfo Zepeda, Departamento de Física, Cinvestav

and the members of the **Local Organizing Committee**:
- Adnan Bashir, Insituto de Física y Matemáticas, UMSNH
- Jaime Nieto, Facultad de Ciencias Físico-Matemáticas, UMSNH
- Ulises Nucamendi, Insituto de Física y Matemáticas, UMSNH
- Eduardo Tututi, Facultad de Ciencias Físico-Matemáticas, UMSNH
- Victor Villanueva, Insituto de Física y Matemáticas, UMSNH
- Axel Weber, Insituto de Física y Matemáticas, UMSNH

Finally, we would like to thank all the authors of this commemorative volume for their excellent effort to make this a valued compilation for the benefit of our present and future communities of high energy and cosmic-ray physicists.

Miguel A. Pérez, Cinvestav
Luis F. Urrutia, UNAM
Luis Villaseñor, UMSNH

HISTORICAL NOTES

High Energy Physics in Mexico: Historical Sketch and Implications

A. García and M.A. Pérez

Depto. de Física, Cinvestav, Apdo. Postal 14-740, 07000 México D.F.

Abstract. We present a personal account of the development of research groups in high energy physics and of the Division of Particles and Fields of the Mexican Physical Society. We conclude that this area qualifies as a seriously cultivated scientific speciality, with several research groups active in both theoretical and experimental high energy physics.

BUILD UP OF RESEARCH GROUPS

A detailed and precise historical account of the development of scientific research in Mexico should be done by professional historians, and since neither of us qualify as such, we shall not attempt something like that. However, a personal account which reflects our role as observers and participants and only in high energy physics (HEP) may be illustrative and, hopefully, useful to young people who envisage to embrace a long-life scientific career in developing countries. We believe it is imperative that young researchers ask themselves fundamental questions such as why should there be basic research in these countries, what is the difference between this research and the so called applied research. Of outmost importance is to answer the question: what is that really remains of their effort and work in these countries, that makes it necessary for them to develop and fulfil their careers in them and not elsewere. It is our intention to help students and young researchers to answer such questions by themselves.

Our informal account will only reflect the way we saw things and not the precise way they were. Unavoidably, we shall not be able to make justice to people who may have been involved and whose contributions escaped us then and still escape us. We can only apologize in advance and encourage those who may feel treated unfairly to give their own accounts so that a more complete picture may emerge. However, we will endeavor to keep the qualitative aspects as complete as possible.

One could observe that in the 1950's very limited but nevertheless serious efforts had taken place in mathematics, astronomy, biology, nuclear physics and a little else. Very few papers had been published in HEP on pion-nucleon interactions [1] and some extended review notes in parity violation had been written (M. Moshinsky, 1959). In the early 1960's three undergraduate thesis had been done (Colon, Chaos, Selfdovich, under J. de Oyarzabal) and one graduate student in HEP (G. Cocho, Princeton, 1962) had returned. A critical young graduating student in the mid 1960's (1965), seriously considering enterprizing a research career in physics, starting at a graduate program, could find three active researchers (that is, publishing regularly in refereed international journals): G. Cocho (HEP), L. García-Colín (statistical physics), and M. Moshinsky

CP857, *Particles and Fields, Part B: Commemorative Volume of the Division of Particles and Fields of the Mexican Physical Society*, edited by M. A. Pérez, L. F. Urrutia, and L. Villaseñor
© 2006 American Institute of Physics 978-0-7354-0354-3/06/$23.00

(nuclear physics). Of course, that student may have been aware of a few other possible thesis advisors, but could not find easily examples of their active Ph.D. students. The choice was very limited and it was quite natural that the few dozens of the finishing undergraduates, unless their interests already overlapped with the concerns of the few active researchers, would try to go abroad for their Ph.D. studies, looking for wider horizons.

Through 1971 and most of 1972, the activity in HEP was still too limited. A weekly seminar was organized to overcome isolation and to keep us updated in the very important *revolution* that was taking place then and that led to the Standard Model. This seminar was organized by G. Cocho (IF-UNAM), A. García (ESFM-IPN), and J. Pestieau (Cinvestav). People interested in the field were invited and, in order to make it accessible to more people, the seminar was rotated weekly among the three institutions. The audience was variable and rather unstable, except for the three of us. The situation may well be summarized with a little humor by saying that we had a "annual meeting on HEP" every week.

Slowly, but steadily, things were changing. The first three local Ph.D. thesis in HEP were finished by A. Zepeda (Cinvestav, 1970, under M. Zaidi), M.A. Pérez Angón (Cinvestav, 1972 under J. Pestieau) and C. Avilez, *Clic* (UNAM, 1973, under G. Cocho). Three people were encouraged and supported by Cinvestav to make a two-year post-doctoral stage (A. Zepeda, 1970-72, Rockefeller; H. Moreno, SLAC, 1971-73; and M.A. Pérez Angón, Rockefeller, 1972-74). People that graduated at the Ph.D. level abroad incorporated themselves, mainly at UNAM, ESFM, and Cinvestav. To mention a few: M. Dubovoy (Berkeley), A. Cisneros (Cal. Tech), Arturo García (Oxford), S. Hacyan (Oxford), L. Torres (Berkeley), C.A. Domínguez (Buenos Aires), M. Moreno worked his Ph.D. thesis under J. Pestieau but graduated at UNAM late 1975. *Clic* went for a theoretical post-doc at the DESY laboratory (Hamburg, 1976-78) [2].

A visible activity was taking place in the second half of the 1970's. However, by 1978 one could observe two very significant effects. The first one was that as many as 40 individuals with a Ph.D. thesis in HEP were working in Mexican institutions, but one could only count eight active researchers (as mentioned above, publishing regularly in refereed international journals). They were *Clic*, G. Cocho and M. Moreno (UNAM), C.A. Domínguez, A. García, H. Moreno, M.A. Pérez Angón and A. Zepeda (Cinvestav). One should mention that A. Mondragón had extended his research interest on nuclear physics into HEP, bringing the total to nine[1]. The survival rate was about one out of five. Of those who dropped HEP in Mexico, some moved into other fields and were active there, some moved to foreign industrial laboratories, some went to administrative positions, and some stayed in the field but found a hard time getting started on their own. A more or less systematic pattern emerged: those who left the field had not followed their Ph.D. work with a post-doctoral stage in HEP.

The second effect was that not a single one of those 40 individuals, even if most of them graduated abroad, had gone into experimental HEP. One could have expected that at least two or three of those 40, having done their graduate work at well recognized foreign institutions which offer wide horizons in physics research, would have become experimentalists. At first one would have thought that those who may have specialized in experimental HEP stayed abroad (and thus were not included in those 40) because they had no future in a country that lacked the large and expensive laboratories of

HEP. But the fact was that one could not find a single individual who had done so. It seemed as if Mexican graduate students, whether in this country or abroad, were simply "impermeable" to experimental HEP. One explanation we found for this effect was that none of those people had ever been exposed to the experience of visiting one of those laboratories. The very first visit may produce a shock that makes ones motivation and vocational determination to shake and may force one to review if, indeed, theoretical physics was the really challenging choice. Many individuals will not be motivated to move into experimental HEP, but some will.

The first effect led us to conclude that encouraging and supporting recently graduated Ph.D. students to make a two-year post-doctoral stage with a group, different from the one which they made their thesis with, was particularly helpful in allowing them to remain active in research. The explanation we had figured out for the second effect was vividly confirmed by *Clic*, after he returned from DESY in late 1978.

In mid 1979 he opened up his mind and expressed his intention of dropping out of theory to become and experimentalist in HEP. He mentioned that, even if he was already an active theoretician, he would have to start as a fresh Ph.D thesis student in experimental HEP. It took him only one year to make this conversion. Abroad he had decisive support of several senior experimentalists (at Brookhaven and Nevis laboratories) and here of J. Flores (as director of IF-UNAM). By late 1980 he was able to participate as one of the four main parties of a collaboration in an experiment in Brookhaven. The reason for this was that the mechanical workshop of IF-UNAM was first rate and better than the one at Nevis and with some money (about 100,000 U.S. dollars) and specialized training of several technicians (3-6 months at Nevis) he could provide all the mechanical parts of the detectors this collaboration had to build.

Theoretical HEP had started spontaneously and development in the mid 1960's and continued on its own to do so. No particular planning was necessary. Experimental HEP was something else. *Clic* tried to persuade others to follow his example. None of the active theoreticians converted and only a few undergraduates followed him. In the meantime an important change in the U.S HEP community took place. They deemed that experimental groups throughout Latinamerica, that could participate in the large already international collaborations which use their accelerator facilities, was in their interest. As far as we could see they did not need our people nor our little money, but the political success of CERN in Europe indicated to them that something similar in the Americas would be in their benefit. On the other hand, several of us believed that developing countries should be up-dated and active in all modern scientific activities and that our several communities should seriously participate in them. As many as possible of our young people should be encouraged to participate, without any form of censorship, in any field, be it physics, astronomy, biology or whatever else. However, it should be in basic research, whose results are published in prestigious journals that circulate freely. We also believed that other forms of research, applied science directed to attain particular goals subjected to patent policies or to confidentiality clauses, should be organized and sponsored by those who expected to obtain important non-scientific (financial or otherwise) benefits from them.

In early January 1982 a meeting was organized by the U.S. HEP community. *Clic* managed to make it take place in Cocoyoc, Mor. About 150 people participated. Among them were the directors of several U.S. laboratories and the NSF, several Nobel prize winners, many U.S experimentalists and theoreticians and about 100 scientists from most of Latin America, practically none of them a HEP experimentalist [3]. At the end of this meeting, we realized that there could be an opening to form experimentalists at the top level that could return and survive in our environment. L.M. Lederman, then the director of Fermilab, listened to our plan and accepted to back it. This plan consisted of sending groups of about five students to Fermilab for about two months every year. These students should be close to finishing the M.Sc. thesis or starting their Ph.D. thesis. Once at Fermilab, they should work in particular short projects that could give them a glimpse of what experiments are in HEP. Then, have them return and wait to see which of them would freely decide to go back to Fermilab to work for a Ph.D. thesis. We were after the second effect mentioned above. If one out of every five was converted the plan worked; if two were converted the plan would be a success. Two advisors were necessary, one at Fermilab to take care of the thesis work and a local one who would make sure the thesis met high standards and the degree could be granted by Cinvestav. Of the people who helped, three were particularly important: J. Lach and R. Rubinstein, who helped organize the short visits and find the thesis advisors and took care of administrative aspects, both at Fermilab, and D. Barnés, then director of the fellowship program of CONACyT, who authorized that the fellowships of the students were turned into foreign ones even if they were registered locally. The plan had a slow start but it gradually picked up. By the late 1980's and early 1990's as many as 10 students graduated in experimental HEP, some of them in foreign universities. A third effect emerged: students are more easily convinced by their senior mates than by their teachers.

Not everybody was happy. Some colleagues feared that the scarce resources would be eaten up by HEP, others expected that only *data analysis* would be accessible while real *hardware* would be unattainable, and others believed that this kind of research should not even be attempted. M.J. Yacaman wrote a critical letter to the editors of *Physics Today* [4] claiming that HEP should be banned not only in Mexico but in all of Latin America. There were ups and downs and students had all the reasons to be discouraged, but many of them were not, so the main credit must go to the latter.

BUILD UP OF THE DPF-MPS

The Division of Particles and Fields of the Mexican Physical Society (DPF-MPS) was founded in 1986 when Matías Moreno and M.A. Pérez Angón were members of the MPS's Executive Committee: Matías was director of the *Revista Mexicana de Física* and M.A. Pérez Angón general director of the MPS. The main goal of the new organization was clear at that time: to identify feasible and appropriate steps to strengthen HEP in Mexico. The following tasks were easily identified as the first steps needed to reach this goal.

1. *Mexican School of Particles and Fields (MSPF)*. This school was founded in 1984 by Arnulfo Zepeda, José Luis Lucio and Matías himself. Its general aim at that time was to expose our graduate students to the recent advances in HEP. For this purpose, every two years several well-known lecturers were invited to give short courses on phenomenology and theory of particle physics. This idea was well received by the Mexican HEP community since at that time Mexico's economy was in rather bad shape

due to several devaluations of the Mexican peso. This fact made prohibitive to travel abroad to most of the Mexican researchers and, of course, also to graduate students. The funds necessary to run the MSPF were collected directly by the organizers in several Mexican agencies which support science and technology activities, and the USA National Science Foundation was congenial to the organization of this school. However, the third edition of the MSPF had already problems to get enough financial support and the DPF-MPS provided a convenient framework to guarantee the continuity of the MSPF. This goal has been fulfiled: the MSPF has taken place every two years since 1984 and the XII-MSPF will be held in Puerto Vallarta in November 1-8, 2006, jointly with the VI Latin American Symposium on High Energy Physics.

2. *Mexican Workshop on Particles and Fields (MWPF)*. Just after the foundation of DPF-MPS, in Mexico we started to learn how to survive in a national economy with frequent devaluations and high inflationary rates. The MSPF reached a steady situation and since we had and increasing number of graduate students, it was natural to plan the organization of another meeting which could take place every two years in between the years in which the MSPF was held. For this pourpose, the MWPF was founded and it took place for the first time in León, Gto., in 1987. In principle, the MWPF was designed as a meeting involving more tutorial courses for our graduate students. However, we have experimented with different formats and sometimes it is difficult to distinguish between a MSPF and a MWPF. The XI-MWPF will take place in 2007 in Saltillo, Coah.

3. *Annual Meeting of the DPF-MPS*. In 1986 there were in Mexico 22 active researchers in HEP [5] and we felt that it was necessary to have an Annual Meeting along the lines of this type of meetings organized the DPF-APS: few invited talks and a lot of short presentations of graduate students on their Ph.D thesis work. It was recognized that offering graduate students a forum to expose their work at a professional level, open to the criticism of the HEP community, was an indispensable experience for them. This idea has been successful: we have had continuously an annual meeting since the summer of 1986 and almost all the times it has been held in the rather pleasant atmosphere of the Sala de Seminarios Ignacio Chávez of UNAM in Mexico City.

4. *Medal of the DPF-MPS*. In order to recognize Mexican or foreign scientists, based in Mexico or abroad, who have contributed significantly to the development of HEP in Mexico, our division created the Medal of the DPF-MPS which has been awarded since 1999 to the following colleagues:

Leon M. Lederman (Fermilab, 1999); James W. Cronin, (U. Chicago, 2000); Alfonso Mondragón (IF-UNAM, 2001); Augusto García and Arnulfo Zepeda (Cinvestav, 2002); Benjamín Grinstein (U. California, San Diego, 2003); Jacobo Konigsberg, (U. Florida, 2004); Luis F. Urrutia (ICN-UNAM, 2005); Marleigh Sheaff (U. Wisconsin, 2006).

5. *Summer Program in Foreign Laboratories*. The DPF-MPS has institutionalized the summer visits of under-graduate students to Fermilab, CERN and DESY. This program has been implemented for more than 30 years and has been instrumental in promoting experimental HEP in young Mexican students. As a result of this effort, we have now experimental HEP groups in several institutions: Cinvestav (Zacatenco, Mérida), IFUG, IF-UASLP, IFM-UMSNH, FCFM-BUAP,IF-UNAM, ICN-UNAM, UIA, UAS.

Finally, it is important to mention the colleagues who have had the responsabilty of working as Presidents of the DPF-MPS:

Rodrigo Huerta Quintanilla (Cinvestav-Mérida, 1986-1988); Manuel Torres (IF-UNAM, 1988-1990); Rebeca Juárez (ESFM-IPN, 1990-1992); Luis F. Urrutia(ICN-UNAM, 1992-1994); Miguel A. Pérez Angón (Cinvestav, 1994-1996); Juan Carlos D'Olivo (ICN-UNAM, 1996-1998); Arnulfo Zepeda (Cinvestav, 1998-2000); Gerardo Herrera (Cinvestav, 2000-2001); Lorenzo Díaz Cruz (FCFM-BUAP, 2001-2002); Myriam Mondragón (IF-UNAM, 2002-2004); Luis M. Villaseñor (IFM-UMSNH, 2004-2006); Heriberto Castilla (Cinvestav, 2006-2008).

Currently, the Mexican HEP Community is in good health: there were only 22 researchers (holding Ph.D's) in 1986 [5] and by 2005 the number already jumped to 95, 24 of them working in experimental HEP in international Collaborations such as DO, H1, ALICE, P. Auger, FOCUS, SELEX, and CDF [6]. It is also important to mention the colleagues that have received the highest academic distinction awarded in Mexico to young researchers under 40 years old, the Research Prize of the Mexican Academy of Sciences:

Germinal Cocho Gil (T; IF-UNAM, 1969); Peter O. Hess (T; ICN-UNAM, 1993); José Luis Lucio (T; IFUG, 1995); Gabriel López Castro (T, Cinvestav, 1999); Gerardo Herrera Corral (E; Cinvestav, 2001); Myriam Mondragón (T; IF-UNAM, 2003); Héctor H. García Compeán (T; Cinvestav, 2004); Guillermo Contreras (E; Cinvestav-Mérida, 2005).

To the present, one can say that HEP, both theoretical and experimental, is a field cultivated seriously and with high standards in Mexico. To our minds, the most stringent criterion to be met, so that a scientific specialty qualifies as "seriously cultivated", is that young scientists, within not too long a period, by themselves graduate younger fellows with high quality Ph.D. thesis, whose results are published in prestigious journals and who in their own turn survive so as to continue with the *reproduction* chain. There should be intellectual parents, children, grand children, etc. We may recognize already that this is the case of the HEP community in Mexico and we have high hopes that it will continue to be so. For those interested in learning more on the development of HEP in Mexico, we suggest the excellent account published recently by Fondo de Cultura Económica [7].

WHAT HAVE WE LEARNT?

The experience we gained as participants and observers can be ordered in the form of opinions and convictions, with the purpose of stimulating others, specially students and young scientists, to form their own opinions and convictions.

The distinctive characteristic of the human species is its extraordinary instinct of curiosity. The role and the only one of science is to provide satisfaction to this instinct as objectively and accurately as possible. This is done by scientists by solving the problems raised by their own curiosity. It is then indispensable to distinguish science from the use of science. This distinction helps understand the enormous difference between basic and applied research. In the former, the scientist is led by his/her curiosity to find problems by him/herself and solve them also by him/herself; otherwise he/she is fired. In the latter the scientist is provided with a problem (found by others, say industrialists, sociologists, economists, etc.) and he/she must solve it, otherwise, he/she is fired. The basic scientist is free to choose problems, the applied one is not. However, both must master the art of solving problems and this requires that their academic training be of the best quality, all the way up to the obtention of a Ph.D. degree.

The natural places for basic researchers to work are universities and technological institutes. The natural places for applied scientists to work are industrial laboratories, hospitals, etc. National laboratories and institutes may have either of both roles. For basic research, they provide the very expensive facilities required by basic science and they are shared by very many scientists. For applied research, they play the same role but their purposes are determined by non-scientific parts of society. The former are open, freely accessible institutions, the latter are subject to confidentiality. Modern science, whether basic or applied, is a team-work enterprise, carried on by many thousands of scientists scattered throughout the world. An individual scientist must share part of this work and make sure he/she is active and well placed within this huge team.

The basic and applied scientific careers split radically right after the obtention of the Ph.D. degree. The former becomes substantially harder than the latter. The greatest difficulty is to master the art of finding problems by oneself. This must be achieved within the first two to four years after graduating from a Ph.D. program. Statistics show that in *first world* environments one out of four graduates succeed. The other three go into applied research or into non-scientific activities (administrative, commercial, or else). These statistics are also confirmed in Mexico.

What has just been said covers only half of the responsibilities of a basic scientist. The other half is to organize and control the work of young people so that they learn to work well both at professional and scientific levels. This is referred to as "teaching", but this term must not be confused with giving lectures. The only way for students to learn to work is by doing their own work ant the role of the teacher is to make sure they work well, steadily and efficiently (just making them listen to beautiful lectures is tantamount to doing nothing). This second responsibility cannot be fulfilled if the scientist does not fulfill well his/her first one.

At this point it should be clear the enormous local responsibility of basic scientists. They must contribute decisively to build up a network of universities and technological institutes throughout the country and make sure it works well, in order for it to teach to work millions of young people so that they perform to serve society at professional and scientific levels. This precisely is what remains locally of the work and effort of basic scientists. The results of basic research must, independently of the authors and their countries, belong to the scientific knowledge of Mankind. But without excelling in this activity it is point-less to attempt the second activity.

Clearly, the best way to beat statistics is to support our students and young scientists. This is done by implementing high quality academic programs, post-doctoral programs and, very importantly, by providing young scientists with good working conditions (salaries, budgets, etc.) so that they can survive the still quite dissipative environment of our network of higher education institutions.

The development of HEP is similar to the development of other scientific specialities in Mexico. Much progress has been attained in the past four decades. Today a student considering embarking into a scientific career can find several hundreds of possible Ph.D. thesis advisors in physics, compared to the three he would have found in 1965. However, today, as was the case then, the development of science in Mexico relies almost exclusively on the shoulders of the scientists. If Mexican scientists loose their initiative and enthusiasm then the rest of society would do nothing about it. Much work remains to be done and much more maturity remains to be reached. Nevertheless, as long as a country has young people there is hope and, in as-much-as a country accepts and supports the initiative and enthusiasm of its young people, there is progress.

REFERENCES

1. A. Mondragón, *AIP Conference Proceedings* **623**, 117 (2002).
2. For an account of the early development of Mexican HEP, see *http://www.fis.cinvestav.mx/~ zepeda*.
3. Proceedings of the I Latinamerican Symposium on High Energy Physics and Technology, *Eds. M. Moreno et al., Notas de Física* **6** (1983).
4. L.M. Lederman/M.J. Yacamán, *Physics Today* **35**, No. 8, 9 (1982); *ibid.* **36**, No. 2, 107 (1983).
5. M. A. Pérez, G. Torres-Vega, *Bol. Soc. Mex. Fis.* **2**, *22 (1988.*
6. *Atlas de la Ciencia Mexicana*, Academia Mexicana de Ciencias, htttp://www.amc.unam.mx/atlas.htm.
7. L. Morales Luna, J. Russell, *El uso de nuevas tecnologías de información y comunicación científica en el área de las partículas elementales: el caso de la física mexicana* (FCE, México, 2006).

Remembrances on the origin of the Mexican School of Particles and Fields

¹José Luis Lucio* and ²Arnulfo Zepeda†

*Instituto de Física, Universidad de Guanajuato
P.O. Box E-143, 37150 León, México
†Centro de Investigación y de Estudios Avanzados
P.O. 14-740, 07000 México, D.F.

Abstract. The Mexican School of Particles and Fields was borne at a time when scientific communications in Mexico, specially with the first world community, were very slow and inefficient. Nowadays, after 22 years, the school is well attended every two years by students and young researchers from Mexico and Latin America. By introducing participants to the fast moving frontier of scientific developments in the area of of high energy physics and astroparticle physics, the School has played an important role in fostering the career of young researchers working in these fields.

Keywords: Particles, Fields, Mexican School
PACS: 10, 90.50.S-, 29

The purpose of this note is to present our point of view regarding the origin, development and impact of the "Mexican School of Particles and Fields" (MSPF). The next school, to be held in Puerto Vallarta this year 2006, will be the twelfth edition, after 22 years of holding it every two years. Here we will be concerned with the schools number one (1984) to five (1992), which were organized by the authors, together with Matias Moreno from UNAM, of the present text. The rest, from the sixth onward, were organized by the Particle and Fields' Division (P&FD) of the Mexican Physical Society (SMF), which was created in 1986 with the aim, among others, to promote the meetings of the community. In order to have a program defined by the institutions, and considering that the endorsement of an organization will facilitate the procurement of funds and the handling of financial resources, the authors delegated to the P&FD the responsibility of the organization of the MSPF.

Before entering into the subject it is important to set up the context. For this reason we start with a short summary of the situation in Mexico in those days. Necessarily this reflects a partial and very summarized view of the always changing reality. As poor as it may be, the view we present will allow the reader to have an idea on the reasons we had to create the School.

1982 was the last year of the administration of the Mexican President José López Portillo. López Portillo based his economical policy on the exploitation of petroleum, using to this end credits obtained abroad. Unfortunately the prices of petroleum went down

---

[1] E-mail: lucio@fisica.ugto.mx
[2] E-mail: zepeda@fis.cinvestav.mx

CP857, *Particles and Fields, Part B: Commemorative Volume of the Division of Particles and Fields of the Mexican Physical Society,* edited by M. A. Pérez, L. F. Urrutia, and L. Villaseñor
© 2006 American Institute of Physics 978-0-7354-0354-3/06/$23.00

causing serious problems, leading to a devaluation of the currency and the nationalization of the banks. All this traduced in cuts to the budget for Science and Technology and strong problems to get grants from the National Council for Science and Technology (CONACyT). From 1982 to 1988 the president of Mexico was Miguel de La Madrid. He received the country in a disastrous economical situation, worsened by the increment in the external debt and political and economic circumstances at world-wide level. The approach he used to face the problems was an austere economic policy. Towards the end of his six year period a new strong devaluation occurred affecting again all the Mexican society, in particular the academic community. The crisis still existed in 1988 when Carlos Salinas assumed the presidency of Mexico. The period 1988-1994 was characterized by NAFTA (Free Trade Agreement with USA and Canada), an improvement of the economy based on the selling of an important number of - until then - state owned enterprises and also, of course, by the uprising of the guerilla Zapatista. And again the Mexican currency was devaluated.

In summary, during the 1982-1994 period the economical and political crisis were recurrent. The academic community suffered not only shortage of economical resources, but mainly a strong uncertainty on the possibilities of development. The consequences of such a uncertainty was that important members of the community decided to leave the country producing thus further delay in the consolidation of the groups.

In the early 80's the Mexican community of particle physicist was small, both in locations as well as in the number of professionals of the field. In fact the community was mostly concentrated in Mexico City at Cinvestav, IPN and UNAM, although a few bright people, then recently graduated, worked in three other cities: Leon, Morelia and Puebla. To have an idea of what we are saying, it is enough to say that none of the groups included more than five professional researchers. In contradistinction to the 70's, the groups in the early 80's were already formed by Mexican researchers, most of them graduated abroad.

Although nowadays information appears to be a granted gift, in the 80's information was a serious problem so that subscription to journals was an absolute requirement to be aware of the developments of the field. Although we have to say that the "preprint" system worked pretty well, so that joining a place which was already in the lists of the main groups was an important advantage. Other sources of information were seminars and conferences. In northern countries it was normal to attend several week-long conferences containing talks by the most outstanding researchers. Also, graduate students had to select the schools and/or conferences that they would be able to attend each year. The normality in Mexico was quite different and thus, in the early 80's, being well informed in Mexico was still a problem to be faced.

"The Mexican School of Particles and Fields" started as an effort to remediate that situation. The main problems that its organization faced were the smallness of the community and the diversity of topics of interests to be covered. Still, the attendance increased systematically every two years. Since the very beginning the school had an international scope, which at the beginning was reflected only in the invited speakers.

The first meeting involved 45 participants, all of them Mexicans [1]. The second school was attended by 52 persons, including students and researchers from non-Mexican institutions [2]. Ever since then the participation of Latin-American physicist has been a constant. It is not our purpose to describe the statistics of every school we organized, it is enough to say that at the peak, in this set of the first five schools, we reached an attendance of 60 participants, including physicists from Argentina, Brazil, Cuba, Colombia and Puerto Rico.

To set high standards we invited leaders in the HEP community to talk about physics we considered of interest, not only fashionable in the sense of being short term problems, but fundamental ideas that should be considered as long term research topics. A good example of this is the lectures by M.A.B. Bég on Higgs physics and Dynamical symmetry breaking, and D. Wyler lectures on neutrino masses. The subjects of the remaining lectures of the First Mexican school were composite models (H. Terazawa), supersymmetry and supergravity (B. Ovrut) and Hadron collider physics presented by D. Cline, who then held a position at CERN [1]. It was particularly important for us the participation of D. Cline since we were interested in promoting experimental high energy physics among the young Mexican graduate students. In retrospective, we can safely assert that the impact of the schools we organized was not minor, as seen from the list of invited speakers, the students they form and the long term collaborations that have arisen. Certainly, the merit is totally from the students, who were able to contact the leaders and stand out in a competitive medium.

The other four schools we organized were not different in concepts, i.e. always included speakers covering theory, phenomenology and experiments. The experimental physicist we invited were:

- L. Lederman (FERMILAB)[3]
- D. Weneger (Institut fűr Physik, Universität Dortmund)
- J. Lach (FERMILAB)
- J. Bufler (FERMILAB)
- R.L. Dixon (FERMILAB)
- R. Raja (FERMILAB)

Among others, the following are theoretical topics covered in the first editions of the MSPF [2, 3, 4, 5]:

- Chiral perturbation theory
- Precision Tests of the Electroweak Theory
- Heavy Quark Physics
- Light Mesons Physics
- Chiral Perturbation Theory

---

[3] A worth recalling memory is the telegram by Leo Lederman who excused to attend the school due to a compromise to be in Stockholm.

- Electroweak Symmetry Breaking
- Neutrino Physics
- CP Violation
- Grand Unification
- Strings

At this point it should be stressed that the idea behind the MSPF was to gather together graduate students and researchers to spend a couple of weeks in an isolated but agreeable ambiance so that they could share experiences with invited speakers. At the same time faculty and graduate students had the opportunity to present advances of their work and to expose themselves both to the scrutiny of researchers of other latitudes as well as to compare their work with research with international standards.

As far as we can see, the following are areas where the MSPF had impact:

- Students came to Mexico to pursue graduated studies.
- Mexican students contacted first level researchers to pursue graduate studies abroad, both in theory and experiments. Among these, present leaders of recognized groups (Luis Villaseñor, G. Herrera, Hector Mendez Mella, Lorenzo Diaz, etc.)

ACKNOWLEDGMENTS

We thank the editors for inviting us to write this article and particularly Luis for having the initiative to publish this compendium.

REFERENCES

1. Mexican School of Particles and Fields. J.L. Lucio M, M. Moreno and A. Zepeda editors. American Institute of Physics Conference Proceedings 143, Oaxtepec, Mexico, 1984.
2. Proceedings of the Second Mexican School of Particles and Fields. J.L. Lucio M, M. Moreno and A. Zepeda editors. World Scientific, Cuernavaca, Mexico, 1986.
3. Third Mexican School of Particles and Fields. J.L. Lucio M, M. Moreno and A. Zepeda editors. World Scientific, Oaxtepec, Mexico, 1988.
4. Forth Mexican School of Particles and Fields. J.L. Lucio M, M. Moreno and A. Zepeda editors. World Scientific, Oaxtepec, Mexico, 1990.
5. Fifth Mexican School of Particles and Fields. J.L. Lucio M, M. Vargas editors. American Institute of Physics Conference Proceedings 317, Guanajuato, Mexico, 1992.
6. M.A. Perez Angon and A. Garcia. These proceedings.

Fermilab and Latin America

Leon M. Lederman

*Pritzker Professor of Science, Illinois Institute of Technology
Resident Scholar, Illinois Mathematics and Science Academy
Director Emeritus, Fermi National Accelerator Laboratory
P. O. Box 500, Batavia, IL 60510
lederman@fnal.gov*

Abstract. As Director of Fermilab, starting in 1979, I began a series of meetings with scientists in Latin America. The motivation was to stir collaboration in the field of high energy particle physics, the central focus of Fermilab. In the next 13 years, these Pan American Symposia stirred much discussion of the use of modern physics, created several groups to do collaborative research at Fermilab, and often centralized facilities and, today, still provides the possibility for much more productive North-South collaboration in research and education. In 1992, I handed these activities over to the AAAS, as President. This would, I hoped, broaden areas of collaboration. Such collaboration is unfortunately very sensitive to political events. In a rational world, it would be the rewards, cultural and economic, of collaboration that would modulate political relations. We are not there yet.

Before becoming Director of the Fermi National Accelerator Laboratory, I had the opportunity of visiting laboratories in Europe and in the U.S. in which I met many scientists from Latin America. Many of these were refugees from political instability in their home countries. I was impressed by their achievements in physics, and this resonated with my long-term interest in science for development.

Latin America interested me because the scientists came from the same European tradition that was responsible for the development of science in the U.S. Yet the U.S. (in the late 1970s) was thriving from the economic benefits of technology derived from scientific knowledge, and Latin American science was relatively impoverished, slowed by political instability, by corruption and by variable educational systems.

Strong groups did exist in Brazil, Mexico and Argentina, but progress in the construction of modern scientific instruments lagged. Another problem was the almost total absence of the use of scientists by industry. Nevertheless, to me Latin America represented a huge potential treasure of human resources which would, I was sure, eventually be devoted to scientific research to the benefit of the nations of South and Central America and, indeed, the world.

In 1978, I accepted the appointment as Director Designate of Fermilab. This was the largest laboratory (30 miles west of Chicago) devoted to research in high energy particle physics. My tenure would officially begin in July of 1979, and I used the Designate period to complete some research at Columbia University and to become familiar with the problems I would meet at Fermilab.

Early in 1979, I began to correspond with friends in Mexico and Brazil about a possible Center for collaborative research, located at Fermilab. This was in part inspired by the Center for Theoretical Physics founded in Trieste by Abdus Salam. The goal was to promote basic science and advanced technology in the developing countries of Latin America. I was encouraged by responses to this idea from UNESCO, the U.S. State Department and DoE. In 1980, I visited scientists in Colombia and Mexico. In Mexico, scientists at the Instituto de Fisica de la Universidad Nacional Autonoma de Mexico (UNAM) worked with me to plan a Pan American Symposium on Elementary Particles and Technology. This took place in the city of Cocoyoc on January 5, 6 and 7, 1982.

The organizing Committee consisted of Clicerio Avilez, Jorge Flores, Carlos Hojvat, myself and Matias Moreno. The U.S. delegation consisted of L. Lederman, J.D. Bjorken, S Glashow, R. Feynman, B. Richter, A. Tollestrup, N. Samius, R. Marshak, M. Moravcsik, W. Panofsky, Richard Taylor and R.R. Wilson. [Note the six Nobel Laureates here!]

There were about 50 attendees from the U.S. and Latin American with one European (G. Charpak). In addition to invited talks reviewing the status and prospects in High Energy Physics, there were extensive group discussions in HEP, on its sociology and also on the prospects for such a research activity in Latin America. It was emphasized that Latin American groups should feel that the extensive facilities in Europe and the U.S. were available to them. The meeting gave specific encouragement to the formation of the UNAM HEP group by led by C. Avilez.

The first symposium is widely believed to have stirred interest in science which had been suppressed by a drop in fiscal and political stability. A report of this meeting and a subsequent proposal was published in Physics Today (Guest Comment, August 1982).
In September of 1982, I visited Brazil and thanks to my host, Moyses Nussenzweig, President of the Brazilian Physical Society, was introduced to many of the activities of Brazilian physicists. This visit was followed by the Second Pan American Symposium, held in Rio de Janeiro in July of 1983, co-sponsored by Fermilab and the Brazilian Physical Society. This meeting included branches of physics outside of HEP and generated lively discussions of collaboration between physics and industry. The presence of Brazilian science funding officers was welcome and, we felt, would advance the cause of science and North-South collaboration.

Here again, the conference aided the formation of a Brazilian HEP group led by A. Santoro of CBPF and C. Escobar from Sao Paulo. Four Brazilian theoretical physicists were granted support for two years at Fermilab to train in experimental physics. They participated in a major Fermilab experiment using the new Superconducting 900 GeV Tevatron and provided a strong basis for Santoro's CBPF HEP research. The meeting generated an idea for a much more extensive U.S. – Latin American collaboration which I articulated in 1984 in a short paper: *A Latin American Caper* (Appendix I).

The third Pan American Symposium was then planned for Rio in 1987. There our host was Jose Leite Lopes, Director of CBPF. Subsequent Pan American Symposia were held in Argentina (Bariloche, host Mario Mariscotti) in 1989 and in Colombia (Cartagena, host Eduardo Posada) in 1992.

The symposia, which collected scientists from many Latin American countries and Canada, certainly established many close connections between Fermilab and its 90 associated universities and scientific leaders with centers throughout Latin America. Many examples exist of collaborative enterprises. For example, in 1984, in response to an urgent "hard currency" crisis in Brazil and other nations, in which travel and subscriptions to U.S. journals were cancelled, Fermilab, through the collaboration of the American Physical Society appealed to the U.S. NSF and to the DoE for assistance.

A grant of $300,000 was made to the APS, administered by Fermilab (Roy Rubinstein) and designed to be used to assist physics in the five nations with major physics activity: Argentina, Brazil, Chile, Mexico and Venezuela. This was to be spent in four areas: U.S. journal subscriptions, payment of publication page charges, spare parts and maintenance items for existing equipment in Latin American labs, and per diem support for urgent short trips by physicists to the U.S. Although an elaborate mechanism was set up for the distribution of funds, both Roy Rubinstein and I were awakened more than once by a Latin American physicist stranded without funds in some remote U.S. outpost like Miami or San Diego! The $330,000 was later supplemented by another $150,000 and expended over three years in grants of ~$500 or so. The feedback for these small grants was truly impressive, enabling physicists to continue their research in spite of the severe foreign currency squeeze in these countries.

In both the purchases of equipment from U.S. vendors and in Fermilab's efforts, with DoE blessings, to send surplus lathes, milling machines, etc., to Latin America, the services of Fermilab's purchasing and accounting departments, administrative and secretarial assistance, and computerized record keeping were of crucial assistance. Most important was to subvert hopelessly stagnant customs procedures. Here we found that our embassies, often with young science attachés, were very helpful in a kind of collective smuggling operation (how do you smuggle in a 1,000 kg. milling machine?).

A strong element in these Pan American Symposia was the intense discussions of the political obstacles to scientific development and collaboration. Scientists in countries with oppressive governments were often harassed or dismissed. Over the years, we were witness to a gradual replacement of military dictatorships by democratic governments. The problems of political dismissals were then replaced by other problems: low salaries, low publicity for science, low esteem of national accomplishments, poor primary and secondary education.

The bringing in a group of "gringo Nobel Laureates" and other distinguished scientists was, I thought, sure to impress the science officials of the host nations. So would active collaboration in High Energy Physics which, after all, is played on the world stage of scientific attention. This intensified my belief that collaboration of Latin American physicists at Fermilab (and even, choke!, at CERN) would bring attention to the capability for doing frontier research. Then too, the connections generated by collaboration would be useful in bringing home advanced technologies which made the accelerators and detectors go.

An excellent example: Alberto Santoro's Brazilian group participated in the early use of Fermilab-developed Advanced Computer Project (ACP) microprocessors to analyze the huge amount of proton-antiproton collision data emerging from Fermilab's Tevatron. A significant number of modules were brought back to Rio and introduced to engineers at CBPF and their use resulted in several research programs. This surely reduced the skepticism among Santoro's colleagues concerning their collaborations with the North.

Collaborations which emerged from the Symposia were especially fruitful in Mexico. At UNAM (Mexico), these were facilitated by Prof. Miguel Yacaman, Jorge Flores, Matias Moreno and especially Clicerio Avilez. Clicerio headed the high energy group which eventually moved to Leon, the city in close proximity to the old town of Cuernavaca.

The Fermilab-Latin American collaboration touched much more than the formation of teams from Brazil, Mexico and Colombia participating in research at the Fermilab (later the CERN facilities) accelerator. At any epoch in the 1980-2000 period, some 50 Latin American engineers, physicists and technicians were in residence at Fermilab, participating in accelerator and instrumentation development. Workshops on accelerator technology, on science education, and other topics in scientific research stimulated by the collaboration, were held in Mexico, Brazil, Costa Rica, Bolivia, Honduras, and Peru.

I was honored on an extended visit to Peru to receive an honorary degree from Universidad Nacional de San Antonio Abad del Cusco in 1995. A Mexican School of Particles and Fields was held in 1986 and repeated in 1992, organized by Arnulfo Zepeda.

Important Latin American scientists, Cesar Laltes of Campinas, Roberto Salmeron who commuted from France to Rio, and distinguished theoretician Marcos Moshinsky of UNAM participated in workshops and mini-conferences. For example, a Conference of the Teaching of Modern Physics, held at Fermilab, with strong Latin American attendance, was repeated several times in several Latin American nations.

The fifth Symposium, taking place at Cartagena, Colombia in 1992 now had a vastly more cheerful, more optimistic mood. By this time, the idea of science for development was no longer a new idea and local research activity was picking up. This is described in my trip report with, again, a magnificent idea which turned out to be totally impractical. It was modestly entitled: *Cartagena Manifesto* and was greeted enthusiastically by the designated beneficiaries. However, people who controlled funding were somewhat less captured (Appendix II).

Looking back and surveying science now, we see a quite more promising landscape. Nevertheless, there are still large pockets of poverty, corruption, and populations that have little or no access to the miracle of the internet. What will it take to have Latin American science on a plane with the U.S.? As an added question, we see huge scientific and technological developments in India and China. The issue is clearly a political one. Scientists in Latin America did not have the traumatic issue of World War which compelled a massive technological effort, aided in no small part by the immigration of European scientists fleeing from persecution.

Critical mass is another handicap, i.e. the failure of inter-Latin American collaboration to, for example, create a center based upon major facilities for research in biotechnology, environmental remediation or nanotechnology. In the present epoch (2000-2010), U.S. scientists do not have the influence or the energy to help. This does not mean that it won't happen.

Such a dramatic collaboration which took place after WWII in Europe was in fact stimulated and facilitated by the U.S. This created the CERN laboratory in Geneva. It can (and should!) happen again to the benefit of *all*! The presence of 300 million people, with a GDP of $500 billion and with tens of thousands of scientists and engineers, represents an incomparable potential which the U.S. tends to ignore. Western hemisphere collaboration with large investments in education and research would provide a third axis to match Asian evolution and European (East and West) renewal. All this for a peaceful, more equitable globalization.

Appendix I

A LATIN AMERICAN CAPER

The current obsession with Central America recalls an idea I have been trying to sell for three years. It derives from a deep appreciation of the talents of Latin American scientists encountered over the past thirty years. They are to be found in laboratories all over the world, usually as valued and honored emigrés or refugees from some political crisis. The idea is available to any statesman who would like to become at least as immortal as Fulbright or Rhodes. It is this: The United States should offer our Latin American neighbors a massive graduate fellowship program, in science and engineering. I estimate that we could satisfy all the reasonable requirements for graduate study in U.S. universities for the sum of less than $50 million or so, paying about $8K of subsistence and full tuition. Largesse? Not really. There are large short-term and even larger long-term benefits to be collected for this apparent generosity. It is reasonable to assume that the technological leaders of the developing countries will be oriented towards the country in which they trained, towards U.S. products, styles and democratic forms. Although we will eventually face competition, the resulting development will create a vast new population of consumers.

Application of the results of increasing technological self-sufficiency to the desperate internal social problems will bring profound benefits to the developing nations. This comes from enhanced stability, vigorous trade (yes, and competition) on the basis of increasing equality and, of primary importance from the long range point of view, an increased world scientific base for the advancement of compassionate technology. As we enter the 21st century, we'll need all the help we can get to fulfill the promise of science and to spread its benefits more generously.

This is a propitious time to make the offer. There is a growing awareness that the increasing gap between industrial nations and developing nations is inherently unstable. Teach them how to fish is an appropriate saying. The encouraging restoration of democratic governments in Brazil, Bolivia and Argentina was a necessary component to scientific vigor and the return of the disaffected scientists. The potential of many Latin American scientific infrastructures is impressive and U.S. assistance at this juncture can have a very large leverage in advancing their progress. Then too, the current economic crisis will render this assistance even more valuable and appreciated.

Of course, there are drawbacks and problems. Some, like the danger of brain drain, are, I believe, not serious. We do need to be very sensitive to the individual needs of each country. There is the problem of converting attitudes and habits of Latin American industry so that they adapt to and make use of national scientific talent. The cooperation of multinational corporations must be obtained here. The contrast of this U.S. assistance to Latin America and U.S. aid to students from Alabama and California will be made. I believe all of these objections can be countered. As the program matures, the needs will decrease. Local universities, many with distinguished traditions, will increasingly take over. One can ask for matching contributions, initially via reciprocal grants to U.S. scholars in Latin countries, and, as the economies improve, to defray U.S. costs.

It would be useful to consider the fellowship program as a tax on foreign aid in order to avoid a contrast with domestic U.S. aid to education. At the less than one percent level, it will be money extremely well-spent. Central American problems remind us that we really need to pay at least as much attention to our hemispheric neighbors as we do to Europe and Asia.

A very important byproduct of this self-interested program is the aid to U.S. research universities. The paying student represents an important strengthening of a valuable and endangered national resource. There is ample graduate school capacity and the large injection of able Latin American students (added to the already large number of able Asiatic students) will create a new kind of transient melting pot. The stimulating broth can have far-reaching positive effects.

Leon M. Lederman
Fermilab
Batavia, Illinois
January, 1984

Appendix II

CARTEGENA MANIFESTO:

PAN AMERICAN SCIENCE TECHNOLOGY FOUNDATION

The Fifth Pan American Symposium on Collaboration in Physics met in Cartagena, Colombia, August 17-21, 1992. As with the previous symposia, the goal was to foster communications among the scientists of some 20 nations, mainly in the Western hemisphere, but with important contributions from Europe. The exchange of information as to the state of science in each country is essential to the progress of each nation since this progress will increasingly depend on regional collaboration.

Many experiences in collaboration were shared at the meeting. Typically these were between small groups in Latin America and groups and facilities in the U.S. or in Europe. However, here and there, the growth of collaborations between Latin American groups were described.

As a general statement, in each country there is an increasing recognition of the importance of scientific research to sustainable development. The traditional excitement in the doing of good science, whether basic or applied, was well demonstrated in Cartagena. The new factor has to do with global and local environmental problems which not only profit from regional collaboration but absolutely require such collaboration. Issues of science education and of the public understanding of science were also discussed.

Inspired by the report of the establishment of a bilateral US-Mexico Science Foundation, this Symposium discussed the idea of the creation of a Pan American Science and Technology Foundation (PASTF).

This independent Foundation would have an initial goal of raising an endowment of $100M, the income of which would be used to support research in the Pan American hemisphere with emphasis on furthering fundamental knowledge. Potential sources of endowment funds would be the member nations, World Bank, B.I.D., large foundations and governments around the world who would benefit from the development and environmental consequences.

As an example of how one might organize PASTF, a governing Board, nominated by the member nations, could appoint a Director General who would assemble a small staff remembering that bureaucracy must be held to the minimum consistent with responsible administration. Peer review of new proposals and assessment of the progress of on-going programs would be carried out with the help of the world scientific community at large.

Criteria for approval of proposals would involve scientific and technical merit, primarily, but would also encourage regional collaboration and see to the balance between research designed for development, for environment and for fundamental knowledge.

Such a venture, specifically focussed on the support of scientific and technology in the Americas could also have the support of the world organizations, whose important work we must continue to support. It is also important to note that it is far easier to get attention if our goals are ambitions and our vision is unlimited.

The Cartagena assembly generally agreed to begin the process of creating such a unique institution by holding discussions of the idea among their colleagues, both physicists and non-physicists, and government officials. One possible mechanism for consolidating ideas is to aim for a more formal planning group to be organized at the December meeting of the Federation of Associations for the Advancement of Science in Miami. This group has the advantage of representing all disciplines. By this Manifesto, the organizers of this Fifth Symposium are encouraged to coordinate the efforts to achieve the goal of creating PASTF. It was further agreed that CLAF, under Professor Juan Jose Giambiagi would provide administrative assistance and that the Venezuelan representative to CLAF, Prof. Miguel Octavio (Instituto Venezolano de Investigaciones Cientificas, Caracas, Venezuela) would serve as temporary communications secretariat to receive ideas and comments.

Leon M. Lederman
August 1992

PHENOMENOLOGICAL HIGH ENERGY PHYSICS

Baryon semileptonic decays: the Mexican contribution

Rubén Flores-Mendieta* and Alfonso Martínez[†]

*Instituto de Física, UASLP, Álvaro Obregón 64, San Luis Potosí, S.L.P. 78000, Mexico
[†]ESFM del IPN, Apartado Postal 75-702, México, D.F. 07738, Mexico

Abstract. We give a detailed account of the techniques to compute radiative corrections in baryon semileptonic decays developed over the years by Mexican collaborations. We explain how the method works by obtaining an expression for the Dalitz plot of semileptonic decays of polarized baryons including radiative corrections to order $\mathcal{O}(\alpha q/\pi M_1)$, where q is the four-momentum transfer and M_1 is the mass of the decaying baryon. From here we compute the totally integrated spin angular asymmetry coefficient of the emitted baryon and compare its value with other results.

Keywords: Weak decays, baryons, radiative corrections
PACS: 14.20.Lq, 13.30.Ce, 13.40.Ks

INTRODUCTION

Right after Pauli formulated the neutrino postulate, Fermi introduced the field theoretical treatment of the process $n \to p + e^- + \bar{\nu}_e$ in the early 1930's [1]. This is the first known example of a V-theory. In the ensuing years other forms of nuclear β decays were observed, which prompted Gamow and Teller to formulate an A-theory. The concept of a new class of interactions, the *weak interactions*, had just emerged.

Both the V and A theories were fused into the $V - A$ theory by Sudarshan and Marshak and also independently by Feynman and Gell-Mann in the late 1950's, motivated by the discovery of parity non-conservation in weak interactions by Lee and Yang (theoretically) and Wu and Telegdi (experimentally).

In this context the weak interactions are described by an effective Lagrangian $\mathscr{L}_{\text{eff}}(x) = -(G_F/\sqrt{2}) J_\lambda^\dagger(x) J^\lambda(x) + \text{h.c.}$, where G_F is the Fermi coupling constant and the weak current $J_\lambda(x)$ has the $V - A$ structure. $J_\lambda(x)$ can be separated into weak leptonic $J_\lambda^l(x)$ and weak hadronic $J_\lambda^h(x)$ currents, namely, $J_\lambda(x) = J_\lambda^l(x) + J_\lambda^h(x)$, where the leptonic current can be written directly in terms of the lepton fields whereas the hadronic one can be decomposed into parts having definite flavor SU(3) transformation properties and can be written in terms of quark fields [2].

The Lagrangian $\mathscr{L}_{\text{eff}}$ however, faces many problems and cannot be taken as a self-consistent quantum field theory of weak interactions. Among other aspects, *i*) it is not renormalizable; *ii*) at high energies it leads to a violation of unitarity, i.e., it brings in probability non-conservation; and *iii*) it has no room for neutral currents.

With the advent of gauge theories, the cornerstone of the theory of weak interactions became the SU(2)×U(1) Weinberg-Salam theory, which currently possesses a quite impressive experimental evidence [3]. Further work, both theoretical and experimental, has finally yielded to the standard model of elementary particles, which embodies our knowledge of the strong and electroweak interactions.

The purpose of this paper is to briefly review the achievements of the past thirty years of theoretical activity in baryon semileptonic decays (BSD) from the Mexican perspective. We give a detailed account of BSD, focusing on techniques to the calculation of radiative corrections to observables which have been the major contribution of local research groups. The paper opens with a historical account of the development of the theoretical approach and continues with an application to the evaluation of radiative corrections to the spin-asymmetry coefficient of the emitted baryon. Numerical results are discussed afterwards. The paper closes with a brief summary and conclusions.

RADIATIVE CORRECTIONS IN BSD: THE EARLY YEARS

In dealing with the radiative corrections (RC) to BSD it is very important to have an organized systematic approach to deal with the complications that accompany them. These RC depend on an ultraviolet cutoff, on strong interactions, and on details of the weak interactions, other than the effective $V - A$ theory. In other words, they have a model-dependent part [4, 5]. They also depend on the charge assignments of the decaying and emitted baryons. Their final form depends on the observed kinematical and angular variables and on certain experimental conditions. Over the years, our approach to the calculation of these RC has been to advance results which can be established in as-much-as possible once and for all. This task is considerably biased by the experimental precision attained in given experiments and by the available phase space in each decay.

A systematic study of the calculation of RC in BSD was initiated back in 1980 [6] following an approach originally introduced by Sirlin [4] for the electron energy spectrum in neutron decay and later extended to the decay of polarized neutrons [7, 8]. An expression for the electron energy spectrum including radiative corrections was obtained in Ref. [6], which was accurate enough to allow experimental analyses to be performed in a model-independent fashion, provided hard bremsstrahlung photons are experimentally discriminated. At that time, those results were not directly applicable to the experiments being performed, which had lower statistics and made no provision to discriminate against hard photons. Later, in the same spirit, the RC to the differential decay rate of polarized neutral and charged baryons were presented in Ref. [9]. From that result, obtaining the RC to the decay rates and angular coefficients was straightforward. This was the first attempt to obtain RC to integrated observables.

A step further was taken in 1987, when in order to improve previous results on the RC to the electron energy spectrum of baryon β decay [9], a new theoretical expression was obtained, this time including all terms of order $\alpha q/\pi M_1$ where q is the four-momentum transfer and M_1 is the mass of the decaying baryon [10]. The result is suitable for model-independent analyses of very-high-statistics experiments, without any experimental constraint on the detection of hard inner-bremsstrahlung photons. In Ref. [10] it was shown that the bremsstrahlung contribution to BSD can be computed in a model-independent way up to terms of first order in q, by using the Low theorem [5].

In the meantime, new high-statistics experiments in BSD [11] made Dalitz plot (DP) measurements feasible and the application of RC was necessary. Much of the work had already been advanced in early calculations for the energy spectrum of the charged lepton and for the decay distribution of polarized baryons [6, 8, 9, 10], so the new task was to adapt the results for the DP. In this respect, an expression for the DP of semileptonic decays of charged and neutral baryons including RC to order α and neglecting terms of order αq was introduced in Ref. [12]. The virtual RC presented no new complications. In contrast, the bremsstrahlung RC became rather involved. The approximation implemented in that work was that the real photons are not observed directly but indirectly when the energies of the final baryon and charged lepton are found not to satisfy the final three-body overall energy-momentum conservation. In consequence, a detailed kinematical analysis for determining the integration limits over the photon variables was mandatory. Besides, the infrared divergence in the bremsstrahlung had to be identified carefully along with the finite terms that accompany it. A proper choice of variables yielded the integrations feasible. No doubt this paper marked an important path toward new results. The immediate application was to the computation of RC to the DP to order $\alpha q/\pi M_1$, both for charged [13] and neutral [14] decaying baryons.

In summary, from the early years of research we learned that following the analysis of Sirlin of the virtual RC in neutron beta decay [4] and armed with the theorem of Low for the bremsstrahlung RC [5] one can show that the model-dependent contributions to both corrections (introducing new form factors) contribute to orders $(\alpha/\pi)(q/M_1)^n$ with $n = 2, 3, \ldots$, while orders $n = 0, 1$ lead to model-independent final expressions, because their model dependence is absorbed into the already existing form factors. The RC to BSD obtained to these latter orders are then suitable for model-independent experimental analyses and are valid to an acceptable degree of precision in the near and intermediate future: they will be useful in BSD involving heavy quarks.

RADIATIVE CORRECTIONS TO BSD: THE RECENT YEARS

There are six different charge assignments in BSD, namely, $A^- \to B^0(l^-\bar{v}_l)$, $A^0 \to B^+(l^-\bar{v})$, $A^+ \to B^0(l^+v_l)$, $A^0 \to B^-(l^+v_l)$, $A^{++} \to B^+(l^+v_l)$, and $A^+ \to B^{++}(l^-\bar{v}_l)$. In Ref. [15] it was shown that it is not necessary to calculate each case separately. The final results of the last four cases can be directly obtained from the final results of the first two cases. This saves a considerable amount of effort, since only the first two cases need be calculated in detail. Such detail requires the choice of specific kinematic situations. It is the DP that is normally studied experimentally. However, energy-momentum conservation may allow to discriminate events where photons are emitted carrying away energy such that the BSD is placed outside the so called three-body region (TBR) of the DP of non-radiative BSD. The events with those photons belong to what we have referred to as the four-body region (FBR). In addition, when the initial baryon is polarized the angular correlations between that polarization and the emitted baryon and the emitted charged lepton involve different RC, and it is not possible to obtain the final results of one correlation from the final results of the other correlation. Back in the middle 1990's the systematic study of RC to BSD was complete to both orders $n = 0$ and $n = 1$ for unpolarized decaying baryons so it was time to tackle new problems by considering the polarization of either the decaying or emitted baryons. After some work, the analysis was finished to order $n = 0$ for polarized decaying baryons, covering the spin-final baryon momentum and spin-final charged lepton momentum angular correlations [16, 17]. Those results were indeed useful for obtaining theoretical expressions for the angular spin asymmetry coefficients of the emitted baryon and charged lepton, respectively. Within our approximations and after producing some numbers, our results agreed well with others already published [18].

Nowadays our goal has been extended to cover both the spin-final baryon momentum and the spin-final charged lepton angular correlations to order $n = 1$ of both neutral and negatively charged decaying baryons, restricted to the TBR. The former problem has been already solved [19] whereas the latter is in progress. A further analysis will take into account the FBR contribution to the RC [20] and also the polarization of the emitted baryon. This will be done in the near future. Let us mention that up to this order our results will be useful in high-precision experiments involving heavy quarks.

In the next sections we will describe how to apply our methods to get the RC to the TBR of the spin-final baryon momentum angular correlation to order $n = 1$. Basically we borrow some recent results of Ref. [19] and analyze the case of a negatively charged decaying baryon in order to illustrate the procedure.

VIRTUAL RADIATIVE CORRECTIONS

For definiteness, let us consider the BSD

$$A \to B + l + \overline{v}_l, \quad (1)$$

and let A denote a negatively charged baryon and B a neutral one, so that l represents a negatively charged lepton and $\overline{v}_l$ its accompanying antineutrino. The four momenta and masses of the particles involved in process (1) will be denoted by $p_1 = (E_1, \mathbf{p}_1)$, $p_2 = (E_2, \mathbf{p}_2)$, $l = (E, \mathbf{l})$, and $p_v = (E_v^0, \mathbf{p}_v)$, and by M_1, M_2, m, and m_v, respectively [19]. Further notation and conventions can be found in those references. We organize our results to explicitly exhibit the angular correlation $\hat{\mathbf{s}}_1 \cdot \hat{\mathbf{p}}_2$, where $\hat{\mathbf{s}}_1$ is the spin of A. This choice of the kinematical variables yields the angular spin-asymmetry coefficient of the emitted baryon, denoted hereafter by α_B.

The uncorrected transition amplitude M_0 for process (1) is given by

$$\mathsf{M}_0 = \frac{G_V}{\sqrt{2}}[\overline{u}_B(p_2) W_\mu(p_1, p_2) u_A(p_1)][\overline{u}_l(l) O^\mu v_v(p_v)], \quad (2)$$

where u_A, u_B, u_l, and v_v are the Dirac spinors of the corresponding particles and

$$\begin{aligned} W_\mu(p_1, p_2) = {} & f_1(q^2)\gamma_\mu + f_2(q^2)\sigma_{\mu\nu}\frac{q^\nu}{M_1} + f_3(q^2)\frac{q_\mu}{M_1} \\ & + \left[g_1(q^2)\gamma_\mu + g_2(q^2)\sigma_{\mu\nu}\frac{q^\nu}{M_1} + g_3(q^2)\frac{q_\mu}{M_1} \right]\gamma_5. \end{aligned} \quad (3)$$

Here $O_\mu = \gamma_\mu(1+\gamma_5)$, $q \equiv p_1 - p_2$ is the four-momentum transfer, and $f_i(q^2)$ and $g_i(q^2)$ are the conventional vector and axial-vector form factors, respectively, which are assumed to be real in this work.

The observable effects of spin polarization are analyzed by the replacement

$$u_A(p_1) \to \Sigma(s_1) u_A(p_1) \tag{4}$$

where $\Sigma(s_1) = (1 - \gamma_5 \slashed{s}_1)/2$ is the spin projection operator, and the polarization four-vector s_1 satisfies $s_1 \cdot s_1 = -1$ and $s_1 \cdot p_1 = 0$. In the center-of-mass frame of A, s_1 becomes the purely spatial unit vector $\hat{\mathbf{s}}_1$ which points along the spin direction.

The virtual RC can be separated into a model-independent part M_v which is finite and calculable and into a model-dependent one which contains the effects of the strong interactions and the intermediate vector boson [6, 13, 14]. This model-dependence can be absorbed into M_0 through the definition of effective form factors, hereafter referred to as f_i' and g_i'. The decay amplitude with virtual radiative corrections M_V is given by

$$M_V = M_0' + M_v, \tag{5}$$

where

$$M_v = \frac{\alpha}{2\pi} \left[M_0 \Phi + M_{p_1} \Phi' \right]. \tag{6}$$

The model-independent functions Φ and Φ' contain the terms to order $\mathcal{O}(\alpha q/\pi M_1)$ [13] and they reduce respectively to ϕ and ϕ' of Ref. [12], in the limit of vanishing $\alpha q/\pi M_1$. The second term in Eq. (6) can also be found in this reference.

At this point we can construct the DP with virtual RC by leaving E_2 and E as the relevant variables in the differential decay rate for process (1). After making the replacement (4) in (5), squaring it, summing over final spin states, we have

$$\sum_{\text{spins}} |M_V|^2 = \frac{1}{2} \sum_{\text{spins}} |M_V'|^2 - \frac{1}{2} \sum_{\text{spins}} |M_V^{(s)}|^2. \tag{7}$$

The spin-independent contribution to M_V in Eq. (7), denoted here by M_V', was obtained to order $\mathcal{O}(\alpha/\pi)$ in Ref. [12] and later improved to order $\mathcal{O}(\alpha q/\pi M_1)$ in Ref. [13], so we will borrow the latter result. We now focus here on the spin-dependent part $M_V^{(s)}$ to this order of approximation along the lines of Ref. [16], where only terms of order $\mathcal{O}(\alpha/\pi)$ were retained. The differential decay rate can be written as

$$d\Gamma_V = \frac{dE_2 dE d\Omega_2 d\varphi_l}{(2\pi)^5} M_2 m m_v \left[\frac{1}{2} \sum_{\text{spins}} |M_V'|^2 - \frac{1}{2} \sum_{\text{spins}} |M_V^{(s)}|^2 \right] = d\Gamma_V' - d\Gamma_V^{(s)}, \tag{8}$$

where $d\Gamma_V'$ corresponds to the differential decay rate with virtual RC of unpolarized baryons given by Eq. (9) of Ref. [13], except that we have chosen here, without loss of generality, a coordinate frame in the center-of-mass frame of A with the z axis along the three-momentum of the emitted baryon, whereas in Ref. [13] the z axis was chosen along the three-momentum of the emitted charged lepton. Similarly $d\Gamma_V^{(s)}$ can be obtained by standard techniques. Thus the decay rate with virtual RC is

$$d\Gamma_V = d\Omega \left\{ A_0' + \frac{\alpha}{\pi}(B_1' \Phi + B_1'' \Phi') - \hat{\mathbf{s}}_1 \cdot \hat{\mathbf{p}}_2 \left[A_0'' + \frac{\alpha}{\pi}(B_2' \Phi + B_2'' \Phi') \right] \right\}, \tag{9}$$

where $d\Omega$ is a phase space factor and A_0', B_1', and B_1'' are given in Ref. [13] whereas A_0'' can be found in Ref. [16]. They all depend on the kinematical variables and are quadratic functions of the form factors. Equation (9) is the differential decay rate with virtual RC to order $\mathcal{O}(\alpha q/\pi M_1)$. It is model-independent and contains an infrared divergent term in Φ which at any rate will be canceled when the bremsstrahlung RC are added.

When including terms of order $\mathcal{O}(\alpha q/\pi M_1)$ the model dependence of the RC shows up. For the virtual part, one can handle this by introducing effective form factors, f_i' and g_i', in the uncorrected amplitude M_0, in such a way that [6]

$$f_1'(q^2, p_+ \cdot l) = f_1(q^2) + \frac{\alpha}{\pi} a_1(p_+ \cdot l), \quad g_1'(q^2, p_+ \cdot l) = g_1(q^2) + \frac{\alpha}{\pi} b_1(p_+ \cdot l),$$

$$f_k'(q^2, p_+ \cdot l) = f_k(q^2) + \frac{\alpha}{\pi} a_k, \quad g_k'(q^2, p_+ \cdot l) = g_k(q^2) + \frac{\alpha}{\pi} b_k, \quad (k=2,3)$$

i.e., f_1' and g_1' have a new dependence in the electron and emitted baryon energies other than the ones in the q^2 dependence of the original form factors; this can be seen through a_1 and b_1, which are functions of the product $p_+ \cdot l = (p_1 + p_2) \cdot l$. For the remaining form factors, within our approximations, a_k and b_k ($k = 2, 3$) are constant.

BREMSSTRAHLUNG RADIATIVE CORRECTIONS

In this section we turn to the emission of real photons in the process

$$A \rightarrow B + \ell + \overline{v}_l + \gamma, \tag{10}$$

where A, B, ℓ, and $\overline{v}_l$ denote the same particles as in the virtual case and γ represents a real photon with four-momentum $k = (\omega, \mathbf{k})$.

The process (10) itself is strictly speaking a four-body decay whose kinematically allowed region is the joined area $\mathsf{A} + \mathsf{B}$ depicted in Fig. 1. The distinction between these two regions has important physical implications. Region A is delimited by

$$E_2^{\min} \leq E_2 \leq E_2^{\max}, \quad m \leq E \leq E_m \tag{11}$$

where $E_m = (M_1^2 - M_2^2 + m^2)/2M_1$ whereas region B is delimited by

$$M_2 \leq E_2 \leq E_2^{\min}, \quad m \leq E \leq E_b \tag{12}$$

with $E_b = [(M_1 - M_2)^2 + m^2]/2(M_1 - M_2)$. Finding an event with energies E and E_2 in region B demands the existence of a fourth particle which must carry away finite energy and momentum. In contrast, in region A this fourth particle may or may not do so. Consequently, region B is exclusively a four-body region whereas region A is both a three-body and a four-body region. We will refer to them as the four-body and three-body regions (FBR and TBR), respectively. Our analysis of the bremsstrahlung RC will consider process (10) restricted to the TBR.

The starting point will be to obtain the amplitude of process (10) with the spin effects included, retaining terms of order $\mathcal{O}(\alpha q/\pi M_1)$ by following the theorem of Low [5]. The amplitude for process (10), M_B, is given in Ref. [13] and will not be repeated here.

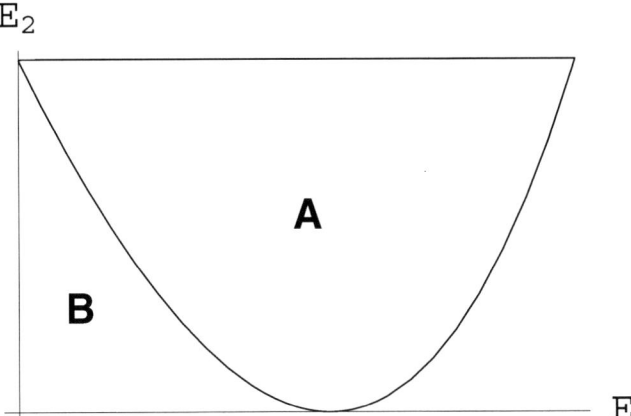

FIGURE 1. Kinematical region (A + B) as a function of E and E_2 of the four-body decay (10). The region A corresponds to what is referred to as the three-body region in this work.

The bremsstrahlung contribution to the DP is obtained from the differential decay rate

$$d\Gamma_B = \frac{M_2 m m_\nu}{(2\pi)^8} \frac{d^3 p_2}{E_2} \frac{d^3 l}{E} \frac{d^3 p_\nu}{E_\nu} \frac{d^3 k}{2\omega} \sum_{\text{spins},\varepsilon} |M_B|^2 \delta^4(p_1 - p_2 - l - p_\nu - k), \quad (13)$$

where the sum extends over the spins of the final particles and the photon polarization. In analogy to the virtual RC, the substitution (4) in $\sum |M_B|^2$ of Eq. (13) leads to

$$\sum_{\text{spins},\varepsilon} |M_B|^2 = \frac{1}{2} \sum_{\text{spins},\varepsilon} |M'_B|^2 - \frac{1}{2} \sum_{\text{spins},\varepsilon} |M_B^{(s)}|^2, \quad (14)$$

and therefore the differential decay rate $d\Gamma_B$ is

$$d\Gamma_B = d\Gamma'_B - d\Gamma_B^{(s)}. \quad (15)$$

Except for minor changes, the quantity $d\Gamma'_B$ in Eq. (15) corresponds to the bremsstrahlung differential decay rate for unpolarized baryons given by Eq. (42) of Ref. [13]. As for the spin-dependent term, $d\Gamma_B^{(s)}$, one can find further details about it in Ref. [19]. The result can be cast into the form

$$d\Gamma_B^{(s)} = \frac{\alpha}{\pi} d\Omega \hat{\mathbf{s}}_1 \cdot \hat{\mathbf{p}}_2 [B'_2 I_0(E, E_2) + C_A], \quad (16)$$

where $I_0(E, E_2)$ contains the infrared divergent term [16] which cancels the one of its virtual counterpart and C_A in infrared convergent. C_A is presented in two forms in Ref. [19]. The first one is given in terms of triple integrals over kinematical variables of the photon and the second one is fully analytic.

The full bremsstrahlung differential decay rate $d\Gamma_B$ is now constructed by subtracting $d\Gamma_B^{(s)}$ from $d\Gamma_B'$. This $d\Gamma_B$ is added to $d\Gamma_V$ to obtain

$$d\Gamma = d\Omega \left[A_0' - A_0'' \hat{\mathbf{s}}_1 \cdot \hat{\mathbf{p}}_2 + \frac{\alpha}{\pi} [\theta_I - \theta_{II} \hat{\mathbf{s}}_1 \cdot \hat{\mathbf{p}}_2] \right], \qquad (17)$$

where the functions θ_I and θ_{II} can be found in Ref. [19].

Expression (17) is model-independent and strictly only well-defined in the TBR of the kinematically allowed region. Although this equation is composed of rather lengthy expressions, it has been organized in such a manner that is easy to use.

SPIN-ASYMMETRY COEFFICIENT α_B

The DP (17) so organized allows the calculation of α_B, which is defined as

$$\alpha_B = 2 \frac{N^+ - N^-}{N^+ + N^-}. \qquad (18)$$

Here N^+ [N^-] denotes the number of emitted baryons with momenta in the forward [backward] hemisphere with respect to the polarization of the decaying baryon. Appropriate integration of Eq. (17) leads to

$$\alpha_B = -\frac{B_2 + (\alpha/\pi)A_2}{B_1 + (\alpha/\pi)A_1}, \qquad (19)$$

where

$$B_2 = \int_m^{E_m} \int_{E_2^{\min}}^{E_2^{\max}} A_0'' dE_2 dE, \qquad A_2 = \int_m^{E_m} \int_{E_2^{\min}}^{E_2^{\max}} \theta_{II} dE_2 dE,$$

$$B_1 = \int_m^{E_m} \int_{E_2^{\min}}^{E_2^{\max}} A_0' dE_2 dE, \qquad A_1 = \int_m^{E_m} \int_{E_2^{\min}}^{E_2^{\max}} \theta_I dE_2 dE,$$

and A_0'', θ_{II}, A_0', and θ_I are defined in Ref. [19].

Equation (19) is a model-independent analytic expression for α_B, including radiative corrections to order $\mathcal{O}(\alpha q/\pi M_1)$. The uncorrected asymmetry coefficient α_B^0 is obtained by dropping the terms proportional to α/π from this equation.

We can evaluate α_B in order to make a comparison with previous works [18, 16]. We use definite values of the form factors in order to compare under the same quotations, but this does not mean that our calculation is compromised to any particular values of them. Therefore we use $f_1 = 1.27$, $g_1 = 0.89$, and $f_2 = 1.20$ for the decay $\Lambda \to pe\overline{\nu}$, $f_1 = 1$, $g_1 = -0.34$, and $f_2 = -0.97$ for the decay $\Sigma^- \to ne\overline{\nu}$, and $f_1 = 0$, $g_1 = 0.60$, and $f_2 = 1.17$ for the process $\Sigma^- \to \Lambda e\overline{\nu}$.

First, we can evaluate $\alpha_B(E, E_2)$ [the same Eq. (19) without performing the double integrals] in several points of the DP. This evaluation is presented in Table 1 for the process $\Sigma^- \to ne\overline{\nu}$, where the first part corresponds to $\alpha_B(E, E_2)$ to order $\mathcal{O}(\alpha/\pi)$ from

TABLE 1. Percentage $\delta\alpha_B(E, E_2)$ with RC over the TBR in $\Sigma^- \to n e \bar{\nu}$ decay (a) to order $\mathcal{O}(\alpha/\pi)$ of Ref. [16]; (b) to order $\mathcal{O}(\alpha q/\pi M_1)$ of this work; and (c) computed in Ref. [18].

σ	(a)									
0.8067	0.5	0.1	0.0	0.0	0.0	0.0	0.0	0.0	0.0	0.1
0.8043	50.7	0.3	0.1	0.1	0.0	0.0	0.0	0.0	0.1	0.3
0.8020		1.2	0.3	0.2	0.1	0.1	0.1	0.0	0.1	
0.7997		5.4	0.7	0.3	0.2	0.1	0.1	0.1	0.1	
0.7974			1.6	0.6	0.3	0.2	0.1	0.1	0.1	
0.7951			4.4	1.1	0.5	0.3	0.2	0.1	0.1	
0.7928			19.8	2.1	0.9	0.4	0.2	0.1		
0.7904				5.2	1.5	0.7	0.3			
0.7881					3.2	1.1	0.3			
0.7858					9.8	2.2	0.2			
	(b)									
0.8067	0.6	0.1	0.0	0.0	0.0	0.0	0.0	0.0	0.0	0.1
0.8043	51.2	0.3	0.1	0.0	0.0	0.0	0.0	0.0	0.1	0.2
0.8020		1.2	0.3	0.1	0.1	0.0	0.1	0.1	0.1	
0.7997		5.5	0.6	0.2	0.1	0.1	0.1	0.1	0.2	
0.7974			1.4	0.4	0.2	0.1	0.1	0.1	0.2	
0.7951			4.1	0.8	0.3	0.2	0.1	0.1	0.1	
0.7928			18.5	1.7	0.6	0.3	0.2	0.1		
0.7904				4.4	1.1	0.4	0.2	0.1		
0.7881					2.4	0.7	0.2			
0.7858					8.3	1.4	0.1			
	(c)									
0.8067	0.6	0.1	0.0	0.0	0.0	0.0	0.0	0.0	0.0	0.1
0.8044	50.7	0.3	0.1	0.0	0.0	0.0	0.0	0.0	0.1	0.2
0.8020		1.2	0.2	0.1	0.1	0.0	0.0	0.1	0.1	
0.7997		5.4	0.6	0.2	0.1	0.1	0.1	0.1	0.2	
0.7974			1.4	0.4	0.2	0.1	0.1	0.1	0.2	
0.7951			4.0	0.8	0.3	0.2	0.1	0.1	0.1	
0.7928			18.4	1.7	0.6	0.3	0.2	0.1		
0.7904				4.4	1.0	0.4	0.2	0.1		
0.7881				11.1	2.4	0.7	0.2			
0.7858					8.2	1.4	0.1			
δ	0.05	0.15	0.25	0.35	0.45	0.55	0.65	0.75	0.85	0.95

Ref. [16], the second part is $\alpha_B(E,E_2)$ to order $\mathcal{O}(\alpha q/\pi M_1)$ from this work, and the last part was computed in Ref. [18] and reproduced here for comparison.

Next, we integrate numerically Eq. (19) to obtain α_B. This evaluation is displayed in Table 2, where the entries of the second column correspond to α_B^0, the following column is the percentage differences defined as $\delta\alpha_B = \alpha_B - \alpha_B^0$, and the last two columns are reserved to comparisons with Refs. [16, 18]. There is a very good overall agreement.

TABLE 2. Values of α_B and comparison with other works.

Decay	α_0	$\delta\alpha_B$ (this work)	$\delta\alpha_B$ Ref. [16]	$\delta\alpha_B$ Ref. [18]
$\Lambda \to pe\bar{\nu}$	-58.6	-0.09	-0.2	-0.1
$\Sigma^- \to ne\bar{\nu}$	66.7	0.05	0.1	0.0
$\Sigma^- \to \Lambda e\bar{\nu}$	7.2	0.08		

DISCUSSION

We have presented a short review of the situation (past and present) of the achievements in the computation of RC in BSD developed by Mexican research groups. We have shown how to apply them to the particular case of the spin-asymmetry coefficient of the emitted baryon and compared with other approaches.

We can claim that the advancement in this topic has been important over the past years so that our understanding on the subject is now clear. Our approach to compute RC can be used in model-independent analyses for charm-baryon semileptonic decays to a high degree of precision. Even for semileptonic decays of baryons containing two charm quarks, they provide a good first approximation. Of course the problem is still open. More work is needed but the approach can be applied straightforwardly.

Finantial support from CONACYT and COFAA-IPN (Mexico) is acknowledged.

REFERENCES

1. For an interesting historical review see for example T.-P. Cheng and L.-F. Lee, *Gauge theory of elementary particle physics* (Oxford University Press, Oxford, 1984) and references therein.
2. A. García and P. Kielanowski, *The Beta Decay of Baryons*, Lecture Notes in Physics Vol. 222 (Springer-Verlag, Berlin, 1985).
3. S. Eidelman *et al.* [Particle Data Group], Phys. Lett. B **592** 1 (2004).
4. A. Sirlin, Phys. Rev. **164**, 1767 (1967).
5. F. E. Low, Phys. Rev. **110**, 974 (1958).
6. A. Garcia and S. R. Juarez W., Phys. Rev. D **22**, 1132 (1980); **22**, 2923(E) (1980).
7. T. Shann, Nuovo Cimento **5A**, 591 (1971).
8. A. Garcia and M. Maya, Phys. Rev. D **17** 1376 (1978).
9. A. Garcia, Phys. Rev. D **25**, 1348 (1982).
10. S. R. Juarez W., A. Martinez and A. Garcia, Phys. Rev. D **35** 232 (1987).
11. M. Bourquin *et al.* Z. Phys. C **21**, 1 (1983); S. Y. Hsueh *et al.* Phys. Rev. D **38** 2056 (1988).
12. D. M. Tun, S. R. Juarez W. and A. Garcia, Phys. Rev. D **40**, 2967 (1989).
13. D. M. Tun, S. R. Juarez W., and A. Garcia, Phys. Rev. D **44**, 3589 (1991).
14. A. Martinez, A. Garcia and D. M. Tun, Phys. Rev. D **47**, 3984 (1993).
15. A. Martinez, J. J. Torres, A. Garcia and R. Flores-Mendieta, Phys. Rev. D **66**, 074014 (2002).
16. R. Flores-Mendieta, A. Garcia, A. Martinez and J. J. Torres, Phys. Rev. D **55**, 5702 (1997).
17. A. Martinez, J. J. Torres, R. Flores-Mendieta and A. Garcia, Phys. Rev. D **63**, 014025 (2001).
18. F. Gluck and K. Toth, Phys. Rev. D **46**, 2090 (1992).
19. M. Neri, A. Martinez, R. Flores-Mendieta, J. J. Torres and A. Garcia, Phys. Rev. D **72**, 057503 (2005); **72**, 079901(E) (2005).
20. S. R. Juarez W., Phys. Rev. D **53**, 3746 (1996); **55** (1997) 2889.
21. R. Flores-Mendieta, J. J. Torres, M. Neri, A. Martinez and A. Garcia, Phys. Rev. D **71**, 034023 (2005); **70**, 093012 (2004).

Neutrinos

J. C. D'Olivo* and O. G. Miranda[†]

*Departamento de Física de Altas Energías, Instituto de Ciencias Nucleares,
Universidad Nacional Autónoma de México, A. P. 70-543, 04510, México, Distrito Federal,
México.
[†]Departamento de Física
Centro de Investigación y de Estudios Avanzados del IPN,
A. P. 14-740
México 07000 D F, México

Abstract. We present a selective overview of neutrino physics with a strong emphasis in the Mexican contribution to this topic. Our scope of the subject put the emphasis on topics like the neutrino oscillations and electromagnetic properties, as well as the treatment of neutrinos in dense media.

INTRODUCTION

The existence of the neutrino was conjectured by Pauli in December 1930 [1], as a "desperate remedy" to save the law of energy conservation and explain the continuous spectrum of the electron emitted in nuclear beta decays [2]. Few years later, Enrico Fermi used the new hypothetical particle to develop his successful theory of beta decay [3]. Despite their elusive nature, in the fifties, neutrinos were detected for the first time by Cowan and Reines [4]. From that moment, they have played a key role in the understanding of weak interactions and the construction of the theory that unified them with the electromagnetic interaction, the Standard Model (SM). Just to mention some remarkable facts, we could remind the observation of parity violation in the beta decay of ^{60}Co [5, 6] and the discovery of the weak neutral currents by the Gargamelle collaboration at CERN [7], which gave the first evidence for the electroweak unification.

A characteristic feature of the SM is that weak interactions couple only to left-handed neutrinos (or right-handed antineutrinos). There is one massless neutrino associated with each charged lepton (e, μ, or τ) and lepton flavor is rigorously conserved. In the last decade this picture has suffered a dramatic change. Conclusive evidence has been accumulated in the sense that neutrinos undergo flavor oscillations [8, 9, 10], implying that they have nonzero mass and mix among themselves much like quarks.

In this work, we review the topic of neutrino physics and astrophysics with a bias on the participation of Mexican researchers. After twenty years of the Particles and Fields community in Mexico, it is a good time to review its contributions to this very active field. We will not discuss here the models underlying neutrino physics since the most important contributions in this subject will be covered in a different article inside this volume [11].

NEUTRINO PROPAGATION IN MATTER

In what constitutes the most important discovery in particle physics during the last years, several experiments with solar, atmospheric, accelerator, and reactor neutrinos have provided us with conclusive evidence that neutrinos produced in a certain flavor eigenstate ν_ℓ ($\ell = e, \mu, \tau$), after traveling a macroscopic distance, can be detected in a different flavor. Neutrino oscillations [12, 13] of mixed massive neutrinos are the simplest explanation of this phenomena: Suppose that flavor states are linear combinations of the mass eigenstates ν_j ($j = 1, 2, 3$), with masses m_j, i.e., $\nu_\ell = U_{\ell j} \nu_j$ where $U_{\ell j}$ are the elements of a unitary mixing matrix, the analogue of the CKM matrix in the quark sector. If a ν_ℓ is created at a certain point, with a 3-momentum $\mathbf{p}$, then when the state evolves, each of the mass eigenstates acquires a different phase $e^{-iE_j t}$, with $E_j = \sqrt{m_j^2 + p^2}$ ($p = |\mathbf{p}|$). As a consequence, at a certain distance $r \cong t$ from the source the neutrino beam will not correspond to pure ν_ℓ but turns partially into other flavors.

Neutrino oscillations are affected when neutrinos propagate in a material medium [14]. The basic reason is that, because of its coherent interactions with the particles in the background, the energy-momentum relation of a neutrino differs from the one in vacuum. The modifications are conveniently described in terms of a potential energy V_ℓ, which is related to the real part of the refraction index n_ℓ according to $\operatorname{Re} n_\ell = 1 - V_\ell/k$. Within the framework of Thermal Field Theory (TFT), V_ℓ is determined from the temperature- and density-contributions to the neutrino self-energy. In normal matter (electrons and nucleons) like, for example, the solar plasma, no heavier leptons are present in the background. Then, the neutral current contribution (Z-exchange) is the same for all neutrino flavors, whereas only the ν_e receive a charged-current contribution (W-exchange). As a consequence, the effective potential differs for neutrinos of different flavors and they acquire distinct phases when propagating through a medium. This fact can have impressive consequences in the case of mixed neutrinos and, under favorable conditions, flavor transformations are significantly enhanced in a medium with a varying density even when mixing in vacuum is small. This is the essence of the MSW mechanism [15, 16] whose large mixing realization provides the solution to the solar neutrino problem, i.e., the discrepancy between the predicted and the observed electron-neutrino flux from the Sun [17].

The differential equations governing the evolution of flavor amplitudes have been studied for neutrinos in various environments. They are manageable numerically and, in fact, there exist extensive numerical calculation for the Sun. Yet, the essential underlying physics of the MSW mechanism can be understood analytically, which also allows a better understanding of the effects of changes in the oscillations parameters. Useful result have been derived from the exact analytic solutions for certain density profiles [18] and also by means of different approximation methods [19], which can be applied to more general situations. In this sense, the Magnus exponential representation of the evolution operator provides an elegant approach that is independent of the number of generations involved and is specially suited to describe the sudden regime of the MSW theory [20]. A variant of the method applies to the full interval of the energy, rendering a better result for the electron-neutrino survival probability [21].

In environments like the early Universe or a supernova core, where the neutrinos represent an appreciable fraction of the total density, the neutral-current contributions to the potential energy arising from the neutrino-neutrino scattering are not in general the same for all flavors and need to be included in the analysis of the resonant flavor transformations. The calculation of these contributions are complicated because the background composition is modulated by the flavor amplitudes, whose time evolution is in turn determined by the matter oscillations. A thermal-field treatment of this (non-linear) problem, which is simpler that the standard density-matrix approach [22], was presented in Ref. [23]. The neutrino contributions to their self-energies are calculated not in terms of the free thermal propagator but instead the one for the neutrino modes in the medium. By using the effective field associated with these modes, the neutrino densities can be expressed in a precise form in terms of the dynamical flavor amplitudes. More recently, this approach has been further elaborated by implementing a nonequilibrium formulation of quantum field theory [24].

The dispersion relation of a neutrino in a medium is, in general, a complex quantity, whose real and imaginary parts correspond to the energy and the damping γ of the propagating mode. To leading order in g^2/M_W^2, the real contributions to the energy (i.e. to V_ℓ) are proportional to the particle-antiparticle asymmetry in the background. In a nearly CP-symmetric plasma, as may be the case during the early Universe or in a gamma-ray burst fireball [25], the asymmetry is small. Then the imaginary contributions and corrections of order g^2/M_W^{-4} to the effective potential may become important because they do not depend on the differences between the number of particles and antiparticles. In the formalism of the FTF, the $O(g^2/M_W^{-4})$ corrections to the real part of the neutrino dispersion relations arise from the momentum-dependent terms of the boson propagators in the self-energy diagrams [26, 27]. Although the self-energy depends on the gauge parameter, in Ref. [27] was explicitly shown that the dispersion relation (and, therefore, V_ℓ) is independent of it. On the other hand, a detailed calculation of the neutrino damping rate was carried out also using the methods of the TFT, for a medium composed of charged leptons, neutrinos, and nucleons [28]. Starting from the discontinuity of the neutrino self-energy at the two-loop level, γ was expressed as integrals over phase space of amplitudes squared, weighted with statistical factors that account for the possibility of particle absorption or emission from the medium.

The resonant production of a Z boson trough $\nu\bar{\nu}$ annihilation results in absorption dips in the spectrum of cosmic neutrinos at ultra-high energies (UHEν) traveling through the cosmological background of relic (anti)neutrinos (CνB) [29]. Their observation would provide us with evidence for the existence of the CνB and a way to determine the absolute neutrino masses [30]. The same process has also been proposed as a possible mechanism for generating UHE cosmic rays ("Z-burst mechanism") [31]. The transmission probability is determined by integrating the damping rate along the UHEν path. As explained above, γ can be obtained from the imaginary part of the neutrino self-energy calculated, in this case, in terms of the dressed Z boson propagator near the resonance [32]. This approach incorporate the thermal effects introduced by the momentum distribution of the relic neutrinos. Such effects significantly affect the shape and position of the absorption dips as soon as the ratio between the neutrino mass and the CνB temperature goes below $\approx 10^2$, i.e., even when relic neutrinos are non-relativistic [32].

NEUTRINO ELECTROMAGNETIC PROPERTIES

The electromagnetic properties of neutrinos have become a subject of increasing research efforts over the last years. In this section, we review some aspects of this subject, such as the neutrino charge radius and magnetic moment. In particular, the case of a non-zero neutrino magnetic moment is discussed both in the context of a possible solution to the solar neutrino problem as well as in the context of the limits on this property. At the end of the section we describe how the neutrino electromagnetic interaction can be modified by the presence of a medium.

Neutrino Magnetic Moments and the Solar Neutrino Problem

Amongst the neutrino electromagnetic properties, the magnetic moment μ_ν has received special attention in connection with various chirality flip processes that could have important consequences for the explanation of the solar neutrino problem [33, 34], the dynamics of stellar collapse [35, 36], and the evolution of the early universe [37].

Although a non-zero neutrino magnetic moment was already considered in the original proposal by Pauli [1], the idea was abandoned until a correlation between the solar neutrino data and the solar cycle was suggested [38]. At the same time, the first discussion about a possible explanation of the solar neutrino data through a non-zero neutrino magnetic moment, as far as we know, was proposed by A. Cisneros, when he left Mexico to make a Ph. D. in Caltech [33]. According to his proposal neutrinos with non-zero neutrino magnetic moment could be converted into right handed sterile neutrinos with a probability

$$P(\nu_{eL} \rightarrow \nu_{eR}; r) = \sin^2\left(\int_0^r \mu_\nu B_\perp(r')\,dr'\right), \qquad (1)$$

where μ_ν is the neutrino magnetic moment and $B_\perp$ the solar magnetic field perpendicular to the neutrino trajectory. Later on, this idea was developed by Okun, Voloshin and Visotsky [34] who incorporate the evolution in matter. At last both cases were ruled out.

Within this scheme, an alternative mechanism was posed in terms of a non-zero neutrino magnetic moment in the Majorana case, first in vacuum [39] and then in matter [40]. This gives rise to the so called resonant spin flavor precession (RSFP) phenomenon, in which an electron neutrino can be converted into a muon anti neutrino. For a long time this mechanism gave a good explanation to the solar neutrino data [41] and it was even possible to consider well motivated regular magnetic field profiles giving a χ^2 analysis as good as other neutrino oscillation mechanisms [42], both for small mixing angles as well as for the dark side [43] of neutrino parameters [44]. However, after KamLAND [45] first results this solution was discarded [46] remaining only as a sub-leading effect.

Limits on Neutrino Magnetic Moments

In the context of the SM, dipole moments of the neutrino vanish because its left-handed nature. If a ν_R is included, then a magnetic moment is induced by radiative corrections but is too small to be of any phenomenological significance: $\mu_\nu \lesssim 10^{-19}\mu_B$ for $m_\nu \lesssim 1$ eV, where μ_B is the Bohr magneton. Larger values for the neutrino electromagnetic couplings can be obtained in non trivial extensions of the SM like left-right symmetric models. In what follows we examine some astrophysical and laboratory limits on μ_ν [47].

A non-vanishing neutrino magnetic moment implies that left-handed neutrinos produced inside a supernova core during the collapse, could change their chirality becoming sterile with respect to the weak interaction. These sterile neutrinos would fly away from the star leaving essentially no energy to explain the observed luminosity of the supernova. The most efficient process for conversion of left-handed to right-handed neutrinos happens through scattering off electrons with the exchange of effective space-like photons. The production rate for right-handed neutrinos can be determined from the imaginary part of the ν_R self-energy, calculated in terms of the resummed photon propagator, which consistently incorporates the background effect. By applying this method to the case of the supernova core collapse, an upper limit on the neutrino magnetic moment $\mu_\nu < (0.1 - 0.4) \times 10^{-11}\mu_B$ was obtained [48]. It improves the one obtained introducing a Debye mass screening prescription to regularize infrared divergences [36] and is comparable to the limit derived from the plasmon decay in globular-cluster stars [49].

Even though the SFP has been ruled out as a solution to the solar neutrino problem, it is still interesting to consider the inverse problem, that is, to use the solar neutrino data to constrain the neutrino magnetic moment. An attempt in this direction, based on the neutrino electron scattering, used the Super-Kamiokande data [50, 51]. A different approach is to take into account that, even in the LMA regime, the RSFP mechanism could produce a sub-leading effect that could spoil the solution to the solar neutrino problem. Moreover, RSFP could lead to antineutrino production in the Sun. Although the effect would be small, there are strong limits from different experiments [52], especially from KamLAND [53], that restrict the antineutrino flux from the Sun to be less than $2.8 \times 10^{-4}\Phi(^8 B)$ with $\Phi(^8 B)$ the predicted Boron flux. This limit can be translated into a constrain on the neutrino transition magnetic moment, provided that there is a good model for the magnetic field profile inside the Sun. This was actually done [54] and it was found that the random magnetic field would produce an efficient antineutrino yield, providing us with the constraint $\mu_\nu < 5 \times 10^{-12}\mu_B$. As we see, this is competitive with the other astrophysical limits given above.

On the other hand, the rate of the decay $Z \to \nu\bar{\nu}\gamma$ is negligible small in the SM, providing a valuable tool to search for evidences of new physics [55]. By using the experimental bound on the branching ratio for such rare decay, an analysis in the framework of the effective Lagrangian formalism, yields $\mu_{\nu_\tau} \leq 2.6 \times 10^{-6}\mu_B$ [56, 57]. This bound for the tau-neutrino magnetic moment is more stringent than the one obtained from $Z \to \nu\bar{\nu}\gamma\gamma$ [57] and is compatible with the limit set by the DONUT collaboration [58]. The same process has also been used to put limits on the neutrino magnetic moment in specific models beyond the SM [59].

Neutrino Charge Radius

Within the SM there is no coupling between the photon and neutrinos at tree-level; however, an effective $\gamma\nu\nu$ vertex is generated through radiative corrections, giving rise to the neutrino electromagnetic form factor $F(q^2)$. Its static limit at zero momentum transfer, $q^2 \to 0$, corresponds to the electric charge, which is a gauge independent and finite (vanishing) quantity [60]. A similar result was explicitly shown to be true for the magnetic moment of the neutrino by a one-loop calculation, in the minimal extension of the SM with massive Dirac neutrinos [61]. On the contrary, the calculation of the neutrino charge radius (NCR) $\langle r^2 \rangle = -6 \partial F / \partial q^2|_{q^2=0}$ was faced with severe complications [62, 63], which, in turn, can be traced back to the fact that, in non-Abelian gauge theories, off-shell Green functions depend in general on the gauge-fixing parameter. To attribute a characteristic size to the neutrino, in ref. [63] the authors introduced the "electroweak radius", which is proportional to $1 - \kappa(0)$, with $\kappa(q^2)$ being a radiative correction factor in the ν-lepton scattering amplitude. This is an observable quantity and, therefore, the electroweak radius satisfies the crucial requirement of being finite and gauge invariant.

The problem of the NCR has been conclusively solved [64] by a well-defined separation of physical amplitudes into effective (as opposed to diagrammatic) self-energy, vertex, and box sub-amplitudes, implemented by means of the pinch technique. These effective Green's functions are completely independent of the gauge-fixing parameter and satisfy simple QED-like Ward identities. The neutrino charge radius obtained in this way is (i) independent of the gauge-fixing parameter, (ii) ultraviolet finite, (iii) couples electromagnetically to the target, and (iv) process (target) independent. Therefore, it can be considered as an intrinsic property of the neutrino and is given by

$$\langle r_\ell^2 \rangle = \frac{G_F}{4\sqrt{2}\pi^2} \left[3 - 2 \ln \left(\frac{m_\ell^2}{M^2} \right) \right], \quad \ell = e, \mu, \tau, \qquad (2)$$

where m_ℓ denotes the mass of the charged lepton partner of the neutrino under consideration. Numerically, $\langle r_\ell^2 \rangle \sim 10^{-33} \text{cm}^2$.

Another static quantity of the neutrinos that has been examined in the literature is the so called anapole momentum a_V [65]. In fact, this is related to the only non-vanishing form factor that a Majorana neutrino could have. For a massless Dirac neutrino, the anapole moment is not an independent quantity, but it is related to the charge radius according to $a_V = \langle r_\ell^2 \rangle / 6$. Consequently, like the charge radius, it is a physical quantity which can be calculated from the effective neutrino-photon vertex [66]. Constraints on possible contributions to the neutrino charge radius and anapole momentum arising from physics beyond the SM has been imposed by considering their effects on the anomalous magnetic moment of the muon [67].

Neutrino Electromagnetic Properties in a Medium

Most neutrinos in the universe are expected to be produced in the presence of environments characterized by high temperatures and densities. In such media neutrinos behave quite differently from the way they do in vacuum. The coherent interactions of neutrinos with the particles in the background can enhance or induce effects that otherwise would be very feeble or absent. As already discussed, the neutrino resonant oscillations represents the best known example of such kind of unusual behavior. Electromagnetic properties are other of the physical neutrino attributes that can be significantly affected when they traverse a material medium [68]. For example, a Majorana neutrino may have electric and magnetic dipole moments in a medium, whereas in the vacuum this quantities must vanish. The explicit expression of the $\nu\nu\gamma$ vertex in an electron gas was determined in Ref. [69], by a one-loop calculation performed to the leading order in G_F, using the methods of the FTF. Later on, the calculation was extended by including the contribution from the nucleon background, taken into account the anomalous magnetic moment coupling of the nucleons to the photon [70]. The form factors yield additional contributions to the neutrino charge radius and the electric and dipole form factors. They depend not only on the energy and momentum of the photon, but also on the characteristic of the medium. The formulas derived for them are valid for general conditions of the gas, degenerate or nondegenerate, whether it is relativistic or not [69, 70].

The most relevant physical consequence of the neutrino electromagnetic interaction in matter is the plasmon decay $\gamma \to \nu\bar{\nu}$ [71], which is an important cooling mechanism of white dwarfs. The rate for this decay can be deduced in a simple way from the results presented in [69]. Besides it, other possible physical process are the radiative decay [72, 73] and the Cherenkov radiation [74]. Unlike the vacuum, in a medium that contains electrons but no μ's or τ's, the decay rate of a heavy neutrino into a lighter one and a photon is not suppressed by the GIM mechanism [72]. In materials that are transparent to photons well above the optical range, at neutrino energies around 1 MeV, the effect considered in Ref. [74] can be as important as the Cherenkov emission by a intrinsic magnetic moment of order $10^{-10}\mu_B$ [75]. However, it is pertinent to notice that such effect is induced trough the weak interaction of the neutrino with the particles in the background and exist even if neutrinos were massless and do not have intrinsic dipole moments.

Due to the effective $\nu\nu\gamma$ vertex, in the presence of an external magnetic field there are additional contributions to the neutrino refraction index [69, 70]. Flavor transformations are affected, but in a way that preserves chirality [76], contrary to what happens for neutrino oscillations driven by transition magnetic moments [40, 41, 42]. The potential energy for active neutrinos can be written as $V_\ell = b_\ell - c_\ell \hat{\mathbf{k}} \cdot \mathbf{B}$, where the coefficients b_ℓ and c_ℓ depend on the properties of the thermal background. An important feature of the second term in V_ℓ is that it is anisotropic and depends on the direction between the neutrino momentum and the external magnetic field. This phenomenon can induce an asymmetric neutrino emission by a protoneutron star and is the basis of the kick mechanism proposed by Kusenko and Segrè [77, 78] to explain the large drift velocities observed in pulsars. With standard (active) neutrinos the mechanism requires an exceedingly high square mass difference, in conflict with the existing limits. A variant avoiding such a limitation can be implemented in terms of active-sterile neutrino oscillation [79, 80]. The viability of the mechanism has been shown by means of a detailed calculations implemented in terms of the distribution function for active neutrinos diffusing within their neutrinosphere [80]. When combined with the idea of a spherical resonance shell for neutrinos that move in different directions relative to $\mathbf{B}$, this scheme provides a more accurate description of the problem than the one formulated in terms of a single deformed sphere acting as an effective emission surface of the sterile neutrinos. According to estimations made for simple models of a protostar, magnetic fields of the order of $10^{16}-10^{17}$ G are required. The oscillation parameters are compatible with the allowed region for sterile neutrinos to be warm dark matter [81], leaving the mechanism as an attractive possibility to explain the proper motion of pulsars.

NON-STANDARD PHYSICS

An exciting feature about neutrinos is that most of the original ideas were only partially correct. Their history has been full of surprises and we may expect more in the future. Therefore, we consider pertinent to end this review by presenting some works along more exotic directions.

Non-Standard Interactions

Non-standard interactions (NSI) appear naturally in Standard Model extensions. Just to cite some examples, in $v_e d$ scattering we can have additional non-universal and flavor changing contributions to the cross section if we consider models with broken R parity, in which case we can have a v_τ in the final state due to a squark exchange [82]. The violation of the GIM mechanism arises naturally when considering massive neutrinos [83]. Massive neutrinos may also imply lepton flavor violation in the charged sector [84]. NSI appears also in the context of the solar neutrino analysis [85], Supersymmetry [86] and models of neutrino mass [87].

It is possible to parametrize the NSI contribution by using a typical phenomenological description so that, at the four-fermion approximation (energies $\ll M_Z$) the neutrino NSI with charged fermions can be described by the effective Lagrangian

$$\mathscr{L}_{vf}^{NSI} = -\frac{G_F}{\sqrt{2}} \sum_{\substack{f=e,u,d \\ \alpha,\beta=e,\mu,\tau}} \left[\bar{v}_\alpha \gamma^\mu(1-\gamma^5)v_\beta\right] \left(\varepsilon_{\alpha\beta}^{fL}\left[\bar{f}\gamma_\mu(1-\gamma^5)f\right] + \varepsilon_{\alpha\beta}^{fR}\left[\bar{f}\gamma_\mu(1+\gamma^5)f\right]\right), \tag{3}$$

where the parameters $\varepsilon_{\alpha\beta}^{fP}$ ($f=e,u,d$ and $P=L,R$) describe non-standard neutrino interactions, both for the non-universal terms, $\varepsilon_{\alpha\alpha}^{fP}$, as well as for the flavor-changing contributions, $\varepsilon_{\alpha\beta}^{fP}$ ($\alpha \neq \beta$). Allowed regions for these NSI parameters have been obtained both from accelerator and reactor experiments [88, 89] as well as from solar [90, 91] and atmospheric [92] neutrino data. The general conclusion of these works is that, with the exception of the case $\alpha = \mu$, constraints are rather looses with allowed values as big as one [93]. Nevertheless, there are good expectations for improving these constraints by the use of low energy neutrino experiments, both for the case of couplings with electrons, through an artificial neutrino source such as tritium [89, 94, 95], as well as for the case of quarks, where the proposal of detecting the coherent neutrino scattering off nuclei [96] for the first time opens the possibility of restringing these parameters to the level of less than one percent [97]. The same idea, but using a stopped-pion neutrino source has also been developed [98].

Gravity Effects on Neutrino Propagation

The behavior of quantum systems in curved space time is relevant both in astrophysics and cosmology. Quantum mechanical process can be affected by strong gravitational fields as is well illustrated by Hawking radiation. The possible effects of the expansion of the Universe on the oscillations of mixed massive neutrinos has been examined from the solutions of the covariant Dirac equation [99]. Although small, these effects appear in observable quantities and could be relevant for neutrinos emitted by distant objects, in much the same way as for electromagnetic radiation.

A less orthodox mechanism for neutrino oscillations, which does not need neutrino masses, is based on the idea of a flavor dependent coupling of neutrinos to gravity [100]. Consequences of such a violation of the equivalence principle (VEP) in the neutrino sector have been examined in a number of papers [101]. A generalized VEP mechanism for neutrino oscillations has been formulated within the framework of the parametrized post-Newtonian formalism [102]. This allows to incorporate effects that are of the same order as those produced by the Newtonian potential, but have a very characteristic directional dependence. When applied to the concrete situation of atmospheric and solar neutrinos, these effects provide new constraints for some VEP parameters [103]. It has also been shown that resonant VEP neutrino transitions consistent with present bounds can generate the kicks responsible for both the translational and rotational motion of pulsars [104].

The formulation of a satisfactory theory of quantum gravity may necessitate a drastic modification or our perception of space-time, by giving it a foamy structure at scales comparable to the Planck length ℓ_P. Then, it can be expected that what we consider flat space actually involves quantum gravity corrections to the dispersion relations arising from the discreteness of space-time at ℓ_P. Such tiny effects might become observable over travels through cosmological distances L by energetic photons [105] and neutrinos [106] produced in gamma ray bursts. For neutrinos the dominant correction is helicity independent and, to leading order in the momentum p ($p \approx 10^5$ GeV), leads to a time delay of order $p\ell_P L/c \approx 10^4$s with respect to neutrinos traveling at the speed of light [106]. Besides, a dependence $L_{os} \propto p^2 \ell_p$ was found for the characteristic length of two-flavor neutrino oscillations.

Acknowledgements. This work has been partially supported by CONACYT under grants 46999-F and 43649 and by DGAPA-UNAM under grant PAPIIT IN112105.

REFERENCES

1. W. Pauli, Open Letter to Radioactive Persons, 1930, for an English translation see Physics Today **31** 27 (1978).

2. J. Chadwick, Verh. dtsch. phys. Ges. **16** 383 (1914); C. D. Ellis, W. A. Wooster, Proc. Roy. Soc. **A117** 109 (1927); L. Meitner, W. Orthmann, Z. Phys. **60** 143 (1930); also in C. S. Wu *The Neutrino*, Theoretical Physics in the Twentieth Century, A memorial Volume to Wolgang Pauli, edited by M. Fierz and V. F. Weisskopf (Interscience Publishers Inc., 1960).
3. E. Fermi, Ricercha Scient. **2**, No 12 (1933); Z. Phys. **88**, 161 (1934).
4. F. Reines and C. L. Cowan, Jr., Phys. Rev. Lett. **92**, 330 (1953); C. L. Cowan, Jr., *et al.*, Science **124**, 103 (1956).
5. T. D. Lee and C. N. Yang, Phys. Rev. **104**, 254 (1956).
6. C. S. Wu *et al.*, Phys. Rev. **105**, 1423 (1957).
7. F. J. Hasert *et al.* [Gargamelle Neutrino Collaboration], Phys. Lett. B **46**, 138 (1973).
8. R. Davis, Prog. Part. Nucl. Phys. **32**, 13 (1994); B. T. Cleveland *et al.*, Astrophys. J. **496**, 505 (1998); SAGE Collaboration, J. N. Abdurashitov *et al.*, Phys. Rev. C **60**, 055801 (1999); GNO Collaboration, M. Altmann *et al.*, Phys. Lett. B **490**, 16 (2000); GNO Collaboration, Nucl. Phys. Proc. Suppl. **110**, 311 (2002); GALLEX Collaboration, W. Hampel *et al.*, Phys. Lett. **B447**, 127 (1999); Super-Kamiokande Collaboration, S. Fukuda *et al.*, Phys. Lett. B **539**, 179 (2002); SNO Collaboration, Q. R. Ahmad *et al.*, Phys. Rev. Lett. **89**, 011301 (2002); *ibid.* **89**, 011302 (2002); SNO Collaboration, S. N. Ahmed *et al.*, Phys. Rev. Lett., 041801 (2003).
9. KamLAND Collaboration, K. Eguchi *et al.*, Phys. Rev. Lett. **90**, 021802 (2003).
10. Super-Kamiokande Collaboration, Y. Fukuda *et al.*, Phys. Rev. Lett. **81**, 1562 (1998); MACRO Collaboration, A. Surdo, Nucl. Phys. Proc. Suppl. **110**, 342 (2002); G. Giacomelli and A. Margiotta, Phys. Atom. Nucl. **67**, 1139 (2004); Soudan 2 Collaboration, M. Sanchez *et al.*, Phys. Rev. D **68**, 113004 (2003); K2K Collaboration, M. H. Ahn *et al.*, Phys. Rev. Lett. **90**, 041801 (2003); M. H. Ahn [K2K Collaboration], arXiv: hep-ex/0606032.
11. A. Perez-Lorenzana, A brief review on extra dimensions, this volume; see also A. Perez-Lorenzana, AIP Conf. Proc. **809** 164 (2006).
12. B. Pontecorvo, Zh. Eksp. Teor. Fiz. **34**, 247 (1958) [Sov. Phys. JETP **7**, 172 (1958)].
13. Z. Maki, M. Nakagawa, and S. Sakata, Prog. Theor. Phys. **28**, 870 (1962).
14. L. Wolfenstein, Phys. Rev. D **17**, 2369 (1978); **20**, 2634 (1979); V. Barger, K. Whisnant, S. Pakvasa, and R. J. N. Phillips, *ibid.* D **22**, 2718 (1980).
15. S. P. Mikheev and A. Yu. Smirnov, Yad. Fis. **42**, 1441 (1985) [Sov. J. Nucl. Phys. **42**, 913 (1985)] ; Nuovo Cimento **C9**, 17 (1986).
16. R. N. Mohapatra and P. B. Pal, *Massive Neutrinos in Physics and Astrophysics*, 3rd ed. (World Scientific, Singapore, 2004).
17. J. N. Bahcall, M. C. Gonzalez-Garcia, and C. Peña-Garay, JHEP08, 016 (2004), arXiv: hep-ph/0406294.
18. See T. K. Kuo and J. Pantaleone, Rev. Mod. Phys. **61**, 937 (1989), and references therein.
19. W. Haxton, Phys. Rev. Lett. **57**, 1271 (1986); P. Pizzochero Phys. Rev. D **36**, 2293 (1987); A. B. Balantekin and J. F. Beacom, Phys. Rev. D **54**, 6323 (1996).
20. J. C. D'Olivo and J. A. Oteo, Phys. Rev. D **42**, 256 (1990); *i*bid. **54**, 1187 (1996).
21. J. C. D'Olivo, Phys. Rev. D **45**, 924 (1992).
22. A. D. Dolgov, Sov. J. Nucl. Phys. **33**, 700 (1981); Phys. Rep. **370**, 333 (2002); G. Sigl and G. Raffelt, Nucl. Phys. **B406**, 423 (1993); B. H. J. McKellar and M. J. Thompson, Phys. Rev. D **49**, 2710 (1994).
23. J. C. D'Olivo and J. F. Nieves, Int. J. Mod. Phys. A **11**, 141 (1996).
24. C. M. Ho, D. Boyanovsky, and H. J. de Vega, Phys. Rev. D **72**, 085016 (2005).
25. S. Sahu and J. C. D'Olivo, Phys. Rev. D **71**, 047303 (2005).
26. D. Nötzold and G. Raffelt, Nucl. Phys. **B307**, 924 (1988).
27. J. C. D'Olivo, J. F. Nieves, and M. Torres, Phys. Rev. D **46**, 1172 (1992).
28. E. S. Tututi, M. Torres, and J. C. D'Olivo, Phys. Rev. D **66**, 043001 (2002).
29. T. J. Weiler, Phys. Rev. Lett. **49**, 234 (1982); Astrophys. J. **285**, 495 (1984); E. Roulet, Phys. Rev. D **47**, 5247 (1993).
30. B. Eberle, A. Ringwald, L. Song, and T. J. Weiler, Phys. Rev. D **70**, 023007 (2004); G. Barenboim, O. Mena Requejo, and C. Quigg, Phys. Rev. D **71**, 083002 (2005).
31. S. Yoshida, G. Sigl, and S. J. Lee, Phys. Rev. Lett. **81**, 5505,1998; D. Fargion, B. Mele, and A. Salis, Astrophys. J. **517**, 725 (1999); T. J.Weiler, Astropart. Phys. **11**, 303 (1999).
32. J. C. D'Olivo, L. Nellen, S. Sahu, and V. Van Elewyck, Astropart. Phys. **25**, 47 (2006).
33. A. Cisneros, Astrophys. Space Sci. **10**, 87 (1971).

34. L. B. Okun, M. B. Voloshin, and M. I. Vysotsky, Zh. Eksp. Teor. Fiz. **91**, 754 (1986)[Sov. Phys. JETP **64**, 446 (1986)]; L. B. Okun, M. B. Voloshin, and M. I. Vysotsky, Yad. Fiz. **44**, 677 (1986) [Sov. J. Nucl. Phys. **44**, 440 (1986)].
35. S. Nussinov and Y. Rephaeli, Phys. Rev. D **36**, 2278 (1987); I. Goldman, Y. Aharonov, G. Alexander, and S. Nussinov, Phys. Rev. Lett. **60**, 1789 (1988); J. M. Lattimer and J. Cooperstein, Phys. Rev. Lett. **61** (1988) 23; D. Notzold, Phys. Rev. D **38**, 1658 (1988).
36. R. Barbieri and R. N. Mohapatra, Phys. Rev. Lett. **61**, 27 (1988).
37. B.D. Fields, K. Kainulainen, and K.A. Olive, Astropart. Phys. **6** 169 (1997).
38. G. A. Bazilevskaya, Y. I. Stozhkov, and T. N. Charakhchian, Pisma Zh. Eksp. Teor. Fiz. **35**, 273 (1982) [JETP Lett. **35**, 341 (1982)].
39. J. Schechter and J. W. F. Valle, Phys. Rev. D **24**, 1883 (1981); Erratum-ibid. **25**, 283 (1982).
40. E. K. Akhmedov, Phys. Lett. B **213**, 64 (1988); C. S. Lim and W. J. Marciano, Phys. Rev. D **37**, 1368 (1988).
41. H. Minakata and H. Nunokawa, Phys. Rev. Lett. **63**, 121 (1989); A. B. Balantekin, P. J. Hatchell, and F. Loreti, Phys. Rev. D **41**, 3583 (1990); M. M. Guzzo and H. Nunokawa, Astropart. Phys. **12**, 87 (1999); E. K. Akhmedov and J. Pulido, Phys. Lett. B **529**, 193 (2002); A. M. Gago *et al.*, Phys. Rev. D **65**, 073012 (2002).
42. O. G. Miranda, C. Pena-Garay, T. I. Rashba, V. B. Semikoz, and J. W. F. Valle, Nucl. Phys. **B595**, 360 (2001).
43. A. de Gouvea, A. Friedland, and H. Murayama, Phys. Lett. B **490**, 125 (2000).
44. O. G. Miranda, C. Pena-Garay, T. I. Rashba, V. B. Semikoz, and J. W. F. Valle, Phys. Lett. B **521**, 299 (2001).
45. K. Eguchi et al. [KamLAND Collaboration], Phys. Rev. Lett. **90**, 021802 (2003).
46. J. Barranco, O. G. Miranda, T. I. Rashba, V. B. Semikoz, and J. W. F. Valle, Phys. Rev. D **66**, 093009 (2002).
47. For a review on limits on neutrino electromagnetic properties, see G. Raffelt, Phys. Rep. **320**, 319 (1999).
48. A. Ayala, J. C. D'Olivo, and M. Torres, Phys. Rev. D **59**, 111901 (1999).
49. G. G. Raffelt, Phys. Rev. Lett. **64**, 2856 (1990).
50. J. F. Beacom and P. Vogel, Phys. Rev. Lett. **83**, 5222 (1999).
51. W. Grimus, M. Maltoni, T. Schwetz, M. A. Tortola, and J. W. F. Valle, Nucl. Phys. **B648**, 376 (2003).
52. M. Balata *et al.* [Borexino Collaboration], arXiv: hep-ex/0602027.
53. K. Eguchi *et al.* [KamLAND Collaboration], Phys. Rev. Lett. **92**, 071301 (2004).
54. O. G. Miranda, T. I. Rashba, A. I. Rez, and J. W. F. Valle, Phys. Rev. Lett. **93**, 051304 (2004); O. G. Miranda, T. I. Rashba, A. I. Rez, and J. W. F. Valle, Phys. Rev. D **70**, 113002 (2004).
55. J. M. Hernández, M. A. Pérez, G. Tavares-Velasco, and J. J. Toscano, Phys. Rev. D **60**, 013004 (1999).
56. M. Maya, M. A. Pérez, G. Tavares-Velasco, and B. Vega, Phys. Lett. B **434**, 354 (1998).
57. F. Larios, M. A. Pérez, and G. Tavares-Velasco, Phys. Lett. B **531**, 231 (2002).
58. R. Schwienhorst *et al.* [DONUT Collaboration], Phys. Lett. B **513**, 23 (2001).
59. A. Gutiérrez-Rodríguez, M. A. Hernández-Ruíz, B. Jayme-Valdés, and M. A. Pérez, arXiv: hep-ph/0605277; A. Gutiérrez-Rodríguez, M. A. Hernández-Ruíz, and A. Del Rio-De Santiago, Phys. Rev. D **69**, 073008 (2004).
60. J. L. Lucio, A. Rosado, and A. Zepeda, Phys. Rev. D **29**, 1539 (1984).
61. L. G. Cabral-Rosetti, J. Bernabeu, J. Vidal, and A. Zepeda, Eur. Phys. J. C **12**, 633 (2000).
62. W. A. Bardeen, R. Gastmans, and B. Lautrup, Nucl. Phys. **B46**, 319 (1972); S.Y. Lee, Phys. Rev. D **6**, 1701 (1972); B.W. Lee and R. Shrock, *ibid.* **16**, 1444 (1977); N. M. Monyonko and J. H. Reid, Prog. Theor. Phys. **73**, 734 (1985); G. Degrassi, A. Sirlin, and W. J. Marciano, Phys. Rev. D **39**, 287 (1989).
63. J. L. Lucio, A. Rosado and A. Zepeda, Phys. Rev. D **31**, 1091 (1985).
64. J. Bernabeu, L. G. Cabral-Rosetti, J. Papavassiliou, and J. Vidal, Phys. Rev. D **62**, 113012 (2000).
65. A. Rosado, Phys. Rev. D **61**, 013001 (2000); A. Rosado, Mod. Phys. Lett. A **14**, 929 (1999).
66. L. G. Cabral-Rosetti, M. Moreno, and A. Rosado, arXiv: hep-ph/0206083.
67. A. Rosado, Mod. Phys. Lett. A**14**, 929 (1999); R. M. García-Hidalgo and A. Rosado, Int. J. Mod. Phys. A **18**, 1587 (2003). See also A. Rosado and A. Zepeda, Phys. Rev. D **9**, 2517 (1982).
68. J. F. Nieves and P. B. Pal, Phys. Rev. D **40**, 1693 (1989); V. N. Oraevsky and V. B. Semikoz, Physica A **142**, 135 (1987); V. Semikoz, Yad. Fiz. **46** 1592 (1987) [Sov. J. Nucl. Phys. **46**, 946 (1987)].
69. J. C. D'Olivo, J. F. Nieves, and P. B. Pal, Phys. Rev. D **40**, 3679 (1989); V. N. Oraevsky,

A. Yu. Plakhov, V. B. Semikoz, Ya. A. Smorodinsky, Sov. Phys. JETP **66**, 890 (1987), Erratum-ibid. **68**, 1309 (1989).
70. J. C. D'Olivo and J. F. Nieves, Phys. Rev. D **57**, 3116 (1998).
71. J. B. Adams, M. A. Ruderman, and C. H. Woo, Phys. Rev. **129**, 1383 (1963); M. H. Zaidi, Nuovo Cimento **40A**, 502 (1965); D.A. Dicus, Phys. Rev. D **6**, 941 (1965).
72. J. C. D'Olivo, J. F. Nieves, and P. B. Pal, Phys. Rev. Lett. **64**, 1088 (1990).
73. D. Grasso and V. Semikoz, Phys. Rev. D **60**, 053010 (1999).
74. J. C. D'Olivo, J. F. Nieves, and P. B. Pal, Phys. Lett. B **365**, 178 (1996).
75. W. Grimus and H. Neufeld, Phys. Lett. B **315** 129 (1993).
76. J. C. D'Olivo and J. F. Nieves, Phys. Lett. B **383**, 87 (1996).
77. A. Kusenko and G. Segrè, Phys. Rev. Lett. **77**, 4872 (1996); *ibid.* **79**, 2751 (1997); Phys. Rev. D **59**, 061302 (1999).
78. Y. Z. Qian, Phys. Rev. Lett. **79**, 2750 (1997); C. W. Kim, J. D. Kim, and J. Song, Phys. Lett. B **419**, 279 (1998); M. Barkovich, J. C. D'Olivo, R. Montemayor, and J. F. Zanella, Phys. Rev. D **66**, 123005 (2002).
79. A. Kusenko and G. Segrè, Phys. Lett. B **396**, 197 (1997); G. M. Fuller, A. Kusenko, I. Mocioiu, and S. Pascoli, Phys. Rev. D **68**, 103002 (2003); A. Kusenko, Int. J. Mod. Phys. D **13**, 2065 (2004).
80. M. Barkovich, J. C. D'Olivo, and R. Montemayor, Phys. Rev. D **70**, 043005 (2004); *Neutrinospheres, resonant neutrino oscillations, and pulsar kicks*, to be published in Pulsars: New Research, Nova Science Publishers, arXiv: hep-ph/0503113.
81. A. D. Dolgov and S. H. Hansen, Astropart. Phys. **16**, 339 (2001).
82. V. D. Barger, R. J. N. Phillips, and K. Whisnant, Phys. Rev. D **44**, 1629 (1991).
83. J. Schechter and J. W. F. Valle, Phys. Rev. D **22**, 2227 (1980).
84. W. Buchmuller, D. Delepine and L. T. Handoko, Nucl. Phys. B **576**, 445 (2000); W. Buchmuller, D. Delepine and F. Vissani, Phys. Lett. B **459**, 171 (1999).
85. J. W. F. Valle, Phys. Lett. B **199** 432 (1987); E. Roulet, Phys. Rev. D **44**, 935 (1991); M. M. Guzzo, A. Masiero, and S. T. Petcov, Phys. Lett. B **260**, 154 (1991).
86. V. D. Barger, G. F. Giudice, and T. Han, Phys. Rev. D **40**, 2987 (1989).
87. A. Zee, Phys. Lett. B **93**, 389 (1980); Erratum-ibid. B **95** 461 (1980).
88. Z. Berezhiani and A. Rossi, Phys. Lett.B **535**, 207 (2002); Z. Berezhiani, R. S. Raghavan, and A. Rossi, Nucl. Phys. **B638**, 62 (2002).
89. J. Barranco, O. G. Miranda, C. A. Moura, and J. W. F. Valle, Phys. Rev. D **73**, 113001 (2006); J. Barranco, AIP Conf. Proc. **809** 193 (2006).
90. A. Friedland, C. Lunardini, and C. Pena-Garay, Phys. Lett. B **594**, 347 (2004), [hep-ph/0402266]; M. M. Guzzo, P. C. de Holanda, and O. L. G. Peres, Phys. Lett. B **591**, 1 (2004).
91. O. G. Miranda, M. A. Tortola, and J. W. F. Valle, hep-ph/0406280.
92. A. Friedland and C. Lunardini, Phys. Rev. D **72**, 053009 (2005); A. Friedland, C. Lunardini, and M. Maltoni, *ibid.***70**, 111301 (2004).
93. S. Davidson, C. Pena-Garay, N. Rius, and A. Santamaria, JHEP **03**, 011 (2003).
94. I. Giomataris and J. D. Vergados, Phys. Atom. Nucl. **67**, 1097 (2004).
95. B. S. Neganov *et al.*, Phys. Atom. Nucl. **64**, 1948 (2001).
96. S. Aune *et al.*, NOSTOS: A spherical TPC to detect low energy neutrinos, arXiv: hep-ex/0503031.
97. J. Barranco, O. G. Miranda, and T. I. Rashba, JHEP **0512**, 021 (2005).
98. K. Scholberg, Phys. Rev. D **73**, 033005 (2006).
99. J. C. D'Olivo, D. Núñez, and M. Torres, Int. J. Mod. Phys. D **1**, 193 (1992).
100. M. Gasperini, Phys. Rev. D **38**, 2635 (1988); A. Halprin and C. N. Leung, Phys. Rev. Lett. **67**, 1833 (1991).
101. A. Halprin, C. N. Leung, and J. Pantaleone, Phys. Rev. D **53**, 5365 (1995); J. N. Bahcall, P. I. Krastev, and C. N. Leung, *ibid.* **52**, 1770 (1995); J. R. Mureika and R. B. Mann, *ibid.* **54**, 2761 (1996); R. Horvat, Mod. Phys. Lett. A **13**, 2379 (1998).
102. H. Casini, J. C. D'Olivo, R. Montemayor, and L. F. Urrutia, Phys. Rev. D **59**, 062001 (1999).
103. H. Casini, J. C. D'Olivo, and R. Montemayor, Phys. Rev. D **61**, 105004 (2000).
104. M. Barkovich, H. Casini, J. C. D'Olivo, and R. Montemayor, Phys. Lett. B **506**, 20 (2001).
105. G. Amelino-Camelia, J. Ellis, N. E. Mavromatos, D. V. Nanopoulos, and S. Sarkar, Nature (London) **393**, 763 (1998).
106. J. Alfaro, H. A. Morales-Técotl, and L. F. Urrutia, Phys. Rev. Lett. **84**, 2318 (2000).

Low-energy physics (u,d,s)

J.L.Lucio, and M. Napsuciale

*Instituto de Física de la Universidad de Guanajuato
Loma del Bosque 103, Col. Lomas del Campestre, CP 37150, León, Gto. México*

Abstract. We review the contributions of Mexican particle physicists to the description of the properties of hadrons composed of u,d and s quarks. Due to the limitation of space, we focus on the description of strong and electromagnetic interactions.

Keywords: chiral symmetry
PACS: 12.39.Fe, 14.40.Aq, 14.40.Cs

THE EARLY DAYS: 70'S

Our present understanding of hadrons composed of light quarks (u,d,s) evolved from the symmetry arguments for the strong interactions developed in the 60's. Light hadrons were the first to be discovered and a description of Mexican physicists contributions to the field requires a brief historical review of these developments.

In the early seventies a revolution was taking place in HEP which eventually lead to the Standard model of particle physics. After the completion of the electroweak theory [1] in 1967-68 and the demonstration of the renormalizability of the spontaneously broken gauge theories with a Higgs mechanism [2], the basic principles for the formulation of a gauge theory of strong interactions based on quarks were given. The natural group to gauge was SU(3) and this theory, called Quantum Chromodynamics (QCD), was shown to be asymptotically free in 1973 [3]. In spite of this, the fundamental parameters of this theory, the current quark masses, were unknown and the link to phenomenology, based mainly on current algebra, the partial conservation of the axial current (PCAC) hypothesis and dispersion relations was, far from being established.

In this context a small group of Mexican theoreticians started working on HEP around 1970. An active group on nuclear physics already existed at UNAM some of whose members eventually evolved to HEP. Perhaps the oldest Mexican contribution to the field came from this group [4]. In this work, canonical quantization of finite-component, first- and second-order Largrangians in general, and a class of fourth-order Lagrangians are studied.

As for 1971-72 a lot of activity around chiral symmetry of strong interactions was taking place around the world based on phenomenological Lagrangians where the main task was to clarify the mechanism for chiral symmetry breaking and to asses the relative importance of the different proposed terms. Another approach to hadrons was the description of their properties based on quark models. Some of the oldest contributions of Mexican physicists to the field were precisely on those topics. In [5] A. Zepeda, computed the one-meson-to-vacuum matrix element of the equal-time commutator of the charge density of a vector current and a scalar density. On the other side, relativistic harmonic oscillator models of nuclei were applied to the structure of baryons by G. Cocho [6]. An important contribution to the description of the properties of light hadrons dates from 1972 [7]. Here, the leading terms to the pion radius and the isovector nucleon radii in the limit in which the pion mass approaches zero are calculated. Corrections to the Golberger-Trieman relation (GTR) were also calculated in [8].

Around 1975 activity in HEP increased considerably in the two small active groups in Mexico. The group at UNAM focused on the description of phenomenology of electro and photo-production of pseudoscalar mesons in the context of $SU(6)$ models for electromagnetic and strong interactions [9]. The group at Cinvestav continued working on the chiral symmetry of strong interactions. As for 1978 it was not still clear what the origin of chiral symmetry is, a problem related to the lack of reliable framework to estimate the current quark masses (the masses appearing in the QCD Lagrangian). From this time dates one of the most important contribution of Mexican high energy physicists. Indeed, one of the possibilities to avoid strong CP violation was a massless u quark. However a variety of arguments existed against this possibility [10]. In this context A. Zepeda showed that the possibility $m_u = 0$ is not in disagreement with measured physical quantities like pion and kaon masses and weak decay constants and that "it implies an improvement of the $SU(2) \times SU(2)$ symmetry" [11]. The price to pay was to admit that flavor $SU(3)$ was not as good as supposed to be and m_s should be much large than people conceived. This result, and a calculation of quark masses in terms of hadron properties under reasonable assumptions [12], opened the venue to our present understanding of a good $SU(2) \otimes SU(2)$ chiral symmetry as a consequence of small m_u and m_d and an approximate $SU(3) \otimes SU(3)$ chiral symmetry due to a much larger m_s.

Some other contributions are related to the "extended PCAC" hypothesis formulated by C. A. Domínguez whose initial purpose was to explain the higher corrections to the GTR [13], and the study of weak neutral currents in electron-positron annihilation into three pions with polarized beams [14].

WORK AFTER 1980

The understanding of chiral symmetry as a consequence of the structure of the quark mass matrix in QCD was the starting point for our present understanding of the lightest pseudoscalar mesons (π, K and η) as pseudo-Golstone-bosons (PGB) associated to the chiral symmetry of the QCD Lagrangian in the strict massless quark limit. This picture opened the venue to the development of a program depicted by Weinberg in 1979 [15] whose goal was the systematic expansion of PGB properties in terms of powers of quark masses. Mexican physicist contributed in the initial stage of the development of this program [16]. In this work the leading non-analytic corrections to the mass of pions were calculated and corresponding expressions for bosons and fermions of any spin were given. The power counting scheme developed by Manohar and Georgi [17] revealed the chiral symmetry breaking scale as $\Lambda_\chi = 4\pi f_\pi \approx 1\ GeV$. This scale was used in [18] to develop a systematic expansion of green functions of the light sector of QCD in terms of powers of $\frac{p}{\Lambda_\chi}, \frac{m_q}{\Lambda_\chi}$ where p denotes the momentum of PGB's, in the so called chiral perturbation theory (CHPT). In this formalism the QCD Lagrangian is expanded as an infinite series involving the effective degrees of freedom at low energy (PGB) and powers of the derivatives of the field $\frac{\partial}{\Lambda_\chi}$ and quark mass matrix $\frac{m_q}{\Lambda_\chi}$. At a given order, there exist a finite number of terms allowed by chiral symmetry and all the underlying QCD dynamics is encoded in coefficients accompanying these terms, the so-called low energy constants (LEC). These free coefficients appear in all processes involving PGB's and in practice are fixed from some processes giving definite predictions for all the remaining ones. Evenmore, when applied to strong processes parity invariance restricts the series only to even powers of derivatives of the field and the systematic expansion for the effective Lagrangian reads

$$\mathscr{L}_{eff} = \mathscr{L}_{eff}^{(2)} + \mathscr{L}_{eff}^{(4)} + \mathscr{L}_{eff}^{(6)} + ... \quad (1)$$

At leading order ($\mathcal{O}((\frac{p}{\Lambda_\chi})^2)$) the formalism is highly predictive since $\mathscr{L}_{eff}^{(2)}$ contains only two LEC's which can be fixed from the well known pion weak decay constant, f_π, and the vacuum expectation value of quark bilinears $<\bar{q}_i q_j>$, which are assumed SU(3) symmetric. This Lagrangian reproduces all the results extracted at leading order from current algebra. Corrections to these results arise from $\mathscr{L}_{eff}^{(4)}$. This Lagrangian contains 10 new terms and correspondingly there are 10 new free parameters. In the calculation of S-matrix elements to this order we have to consider one-loop diagrams from $\mathscr{L}_{eff}^{(2)}$ and tree level diagrams from $\mathscr{L}_{eff}^{(4)}$. Infinities from the loop diagrams are renormalized by the tree level terms, the theory being thus renormalizable in the Weinberg sense [15]. At this level, we need to fix 12 LEC's. In spite of this the formalism has predictive power since a large amount of data is available. At $\mathcal{O}((\frac{p}{\Lambda_\chi})^6)$, the theory contains more than one hundred LEC's and the predictive power is lost unless we find an alternative way to fix these parameters. The simplest possibility is the saturation of the underlying dynamics by the formation and exchange of resonances. This possibility was explored in [19] and although the saturation of LEC's in the vector and tensor channel was indeed satisfied, things were less clear in the scalar channel were experimental data was rather scarce. The CHPT expansion has been extensively applied to the calculation of properties of PGB and was also extended to include baryons.

Spin $\frac{1}{2}$ and spin $\frac{3}{2}$ baryons.

The seminal work concerning effective theories for baryons was the scheme proposed in [20] where the chiral properties of the spin $\frac{1}{2}$ baryon octet were studied and the corresponding effective lagrangian formulated. In this case the power counting is much more complicated than in CHPT for one reason. The baryon mass in the chiral limit is of the same order as Λ_χ and baryon loops typically yield terms counting as $\frac{m_B}{\Lambda_\chi}$. This produces terms coming from loops which are not in a one-to-one correspondence with derivatives of the fields causing an extremely complicated power counting scheme. The way out of this problem was to benefit from the heavy quark techniques and to apply them at the hadron level, a formalism developed by A. Manohar and E. Jenkins in [21]. Here, a new expansion was introduced, the heavy baryon expansión, which removes the baryon mass from the baryon field rendering a triple expansion in terms of $\frac{\partial}{\Lambda_\chi}$, $\frac{m_q}{\Lambda_\chi}$ and $\frac{\partial}{m_B}$ and taking advantage of the fact that $m_B \approx \Lambda_\chi$ to reduce it to a double expansion in powers of $\frac{\partial}{\Lambda}$, $\frac{m_q}{\Lambda}$ where now Λ stands for a mass scale of the order of $1\,GeV$. This theory known as Heavy Baryon CHPT (HBCHPT), describes interactions of the octet of PGB and the octet of light spin 1/2 baryons. States higher in mass are integrated out and their effects are supposed to be reproduced by operators of higher order in the chiral expansion built in terms of PGB's and the baryon octet. However, the mass difference between the spin $\frac{3}{2}$ decuplet (T) and the spin $\frac{1}{2}$ octet (B), is small, $\Delta m \approx 230\,MeV$ and the coupling $\mathscr{C}$ associated to $TB\Phi$, where Φ denotes the PGB octet, is too strong $\mathscr{C} \simeq 1.5$ causing operators obtained integrating the decuplet to be 6-7 orders of magnitude larger than those supposed to reproduce their effects in HBCHPT. In other words, it is necessary to introduce explicitly in the theory the spin 3/2 degrees of freedom. This formalism has been extensively used in the calculation of processes involving light baryons and improved using a simultaneous $\frac{1}{N_c}$ expansion [22] and a new power counting scheme to account for the new small scale Δm [23]. However, it is well known that Quantum Field Theory for spin 3/2 interacting fields based on the Rarita-Schwinger formalism [24], suffers of serious inconsistencies [25]. The R-S spinor ψ_μ has more degrees of freedom than required, which makes the classical Lagrangian non-unique. In fact there exist a whole family of one parameter Lagrangians $\mathscr{L}(A)$ from which the Dirac-Fierz-Pauli equation and the necessary free field constraints for ψ_μ are obtained [26]. These Lagrangians remain invariant under point transformations (R-invariance) which mix the spurious spin 1/2 fields contained in ψ_μ. This invariance ensures that the unphysical spin 1/2 degrees of freedom have no observable effects. Generalizations of this scheme to describe interacting spin 3/2 fields requires that the corresponding Lagrangian preserves R invariance, this fact introduces new ambiguities into the theory (the so-called "off-shell" parameters)[27]. These fundamental problems are not considered in the HBCHPT and are important in the unambiguous interpretation of experimental data. In this concern we must mention that at the time HBCHPT was developed there was interest in México on the description of processes involving spin $\frac{3}{2}$ light fields. This interest dates from the Ph. D. thesis by Jesús Urías, under the advise of Jean Pestieau, who addressed the problem of the $\Delta(1232)^{++}$ contribution to the radiative pion-proton scattering [28]. In this work, following [27] the involved "off-shell" parameter is fixed to $Z = \frac{1}{2}$. A more refined calculation of this process was given in [29], and a similar approach used to study isospin breaking in the $\Delta^0 - \Delta^{++}$ system, and more recently, the properties of the Δ, and different contributions to the pion-proton radiative cross section [31]. Previously, the possibility that the decay $\Omega^- \to \Lambda K^- \gamma$ can be used as an alternative experimental procedure to determine the Ω^- magnetic moment was explored in [30] using soft photon theorems.

Concerning the old problems of spin $\frac{3}{2}$ quantum field theory and the systematic expansion in HBCHPT, in [32] the authors showed that the HBCHPT expansion if free of inconsistencies and ambiguities only to the leading order in the $\frac{1}{m_B}$ expansion. The corresponding formalism is used to perform a gauge invariant calculation of the radiative decays of the decuplet ($\frac{3}{2}^+ \to \frac{1}{2}^+ \gamma$) to lowest order in $\frac{1}{m}$ and to order $\frac{\omega^2}{\Lambda_\chi^2}$ in the chiral expansion [33] (ω is the photon energy).

A different approach to the spectra of light baryons was developed by M. Kirchbach and co-workers in [34] which pointed out that the Hilbert space of three quarks, with one excited quark, can be decomposed into Lorentz group representations. Reported and "missing" resonances are shown to populate distinct representations that differ by their parity or/and charge conjugation properties. In particular, light baryon resonances reported by the Particle Data Group are accommodated by Rarita-Schwinger (RS) type representations $(k/2,k/2) \otimes [(1/2,0)+(0,1/2)]$ with $k = 1, 3$, and 5. Rarita-Schwinger fields with physical resonances as lower-spin components are given physical meaning as a whole without imposing auxiliary conditions on them and describe multispin-parity clusters.

Light mesons physics

In practice, CHPT yields a systematic expansion for the physics of PGB, π, K, η, up to the region where the first resonances appear. In particular, it works well in the $J^P = 0^-, 1^\pm, 2^\pm$ where the first resonances appear quite far from the $\pi - \pi$ threshold (the closest one is the channel $J^P = 1^-$ with $m_\rho = 770 MeV$). However, in the scalar channel, $J^P = 0^+$ the formalism does not work that well. In this channel, even the lowest energy behavior of $\pi - \pi$ scattering (threshold behavior) cannot be properly reproduced [35]. This is a serious problem because threshold energy for pion scattering is precisely where CHPT is supposed to work better. Similar problems arise in the so called "gold-plated" processes such as $\gamma\gamma \to \pi\pi$ induced at one loop level in CHPT ($\mathcal{O}(\frac{p^4}{\Lambda_\chi^4})$) but which do not receive three level contributions and are determined by the two LEC's at $\mathcal{O}(\frac{p^2}{\Lambda_\chi^2})$ [36]. These problems are related to the existence of a light scalar meson and opened an intense debate about the existence and properties of the lightest scalar mesons. This is an important topic since these states have the same quantum numbers as the vacuum and the appropriate description of their properties can shed some light on the non-trivial structure of the vacuum in QCD and ultimately on the understanding of the mechanism for confinement in this theory. The Guanajuato group made important contributions to this field which we review in the following.

The existence of the $f_0(980)$ and the $a_0(980)$ was firmly established since the early 70's due to their relatively narrow width. Both couple strongly to $\bar{K}K$ and lie so close to the $\bar{K}K$ threshold at 987 MeV, that their shapes are distorted by threshold effects. The correct interpretation of data requires a coupled channel scattering analysis, being $\pi\pi$ and $\bar{K}K$ for the f_0, and $\pi\eta$ and $\bar{K}K$ for the a_0, the relevant processess. Wealth of work has been done trying to understand this phenomena and even a strong isospin violation has been suggested due to the strong coupling of these states to the $\bar{K}K$ channel [37, 38, 39].

Relevant in this analysis is the S matrix approach, which was also strongly advocated by Mexican physicist in order to obtain the fundamental properties associated to resonances. Thus, the ρ resonance has been studied in detail [40] using this approach, including the isospin breaking *i.e.* the charged-neutral mass and width differences [41]. In the scalar sector, the same approach has been used to determine the masses, widths and coupling constants [42].

Evidence in the isoscalar and isospinor channels is not conclusive. A light isoscalar scalar, the so-called sigma meson, was predicted by unitarized models [43, 44], and effective models [45, 46, 47, 48]. It was reconsidered by the PDG classification of particles as the $f_0(400-1200)$ [49] only after its prediction by the unitarization of the results of chiral perturbation theory as a state that do not survive in the large N_c limit [50]. After an intense activity on the experimental side over the past few years, compelling evidence has been accumulated for the existence of this state and its mass has been measured [51]. We can now safely say that there is a universal acceptance for the existence of enhancements at low energy in the s-wave isoscalar meson-meson scattering but the interpretation of this phenomena is still controversial. Concerning the isospinor channel a resonance with a mass around 900 MeV has been predicted by unitarized quark models [44], effective models [45, 47], and unitarized chiral perturbation theory calculations [50]. The πK s-wave data has been reanalyzed finding a relatively strong attraction in the $I = 1/2$ channel [52, 53]. This enhancement in the isoscalar channel of πK scattering is identified with an S-matrix pole at approximately 900 MeV ($\kappa(900)$). Although some doubt has been shed on the existence of this resonance [54], recent experimental results put this particle on firm basis [55].

As a result of different analysis a general consensus emerged on the existence of a light scalar nonet composed of the $\sigma(500)$, $f_0(980)$, $\kappa(900)$ and $a_0(980)$ states [44, 45, 47, 50, 51, 52, 55, 56, 57].

The spectrum of this nonet contradicts the naive quark models expectations which requires $^3P_0$ states to be heavier than the $^3S_1$ states. Some members of the light scalar nonet have a mass smaller than the corresponding members of the vector nonet. On the other hand, the nearly degeneracy of the $f_0(980)$ and $a_0(980)$ suggest an $\omega - \rho$ like quark composition but this is incompatible with the strong coupling of the $f_0(980)$ to $\bar{K}K$ system. Furthermore, the quark structure as probed by the electromagnetic interaction is definitively not consistent with a naive $\bar{q}q$ composition [58, 59]. The latter problem was evident shortly after the discovery of the $f_0(980)$ and $a_0(980)$ mesons and alternative models for its internal structure were proposed. A $\bar{q}^2q^2$ structure was suggested by Jaffe [60] and a "molecular" (clustered four-quark) composition was put forth in [61]. The former proposal has gained renewed interest due to a striking prediction of this model which is consistent with the spectrum of the light scalar nonet: an *inverted mass spectrum* for four-quark states, when compared with a $\bar{q}q$ nonet, due to the presence of a hidden $\bar{s}s$ pair in some of the $\bar{q}^2q^2$ states [62].

Since the mass of a hadron should increase roughly linearly with the number of constituent quarks, four-quark states are naively expected around 1.5 GeV. In Jaffe's picture, a magnetic gluon exchange interaction accounts for the lowering of this scale down to around 900 MeV in such a way that the lowest lying scalar nonet can be in principle identified as four-quark states. In this model, the nearly degeneracy of the $f_0(980)$ and $a_0(980)$ and their strong coupling to $\bar{K}K$ is just a consequence of the flavor structure. The couplings to $\bar{q}q$ mesons follow from the flavor structure also. A calculation of two photon decays of neutral scalars in this formalism would be desirable since these decays provide direct evidence on the constituents of mesons. The calculation of electromagnetic transitions such as $V \to S\gamma \to P P\gamma$ [63] where P stands for a pseudoscalar are also necessary to understand data recently presented by the experimental groups at ϕ factories [64, 65]. These decays have been studied e.g. in the unitarized chiral perturbation theory framework [66, 67] or the three flavor linear sigma model [68, 69]. Finally, four-quark states are expected to mix with conventional $\bar{q}q$ states to yield physical mesons and this mixing should be accounted for in this formalism.

A different explanation to the light scalar spectrum was proposed in [47, 48, 59, 70, 71] based on phenomenological Lagrangians for QCD formulated firstly in [72]. In this formalism, the *inverted mass spectrum* of the lightest scalar nonet arises from an interplay between the spontaneous breaking of chiral symmetry (SBχS) and a trilinear interaction between fields. This interaction is a remnant at the hadron level of the six-quark interaction due to instantons in QCD [73] and is trilinear for $\bar{q}q$ states only, hence scalars are interpreted as $\bar{q}q$ in this framework. Under SBχS, two of the quarks entering in the six-quark 't Hooft interaction acquires non-vanishing v.e.v. generating flavour-dependent mass terms which mix the fields and drastically modify the mass spectrum for scalars as compared with naive quark model expectations. This mechanism yields also an inverted mass spectrum and describes this phenomena and the distorting of the pseudoscalar spectrum in a unified way [70]. In this formalism meson couplings are dictated by chiral symmetry and can be related to meson masses. The small two photon decay width of scalars is also satisfactorily described in this framework [68].

Radiative ϕ decays have been studied in this formalism [69] yielding results which accommodate the experimental data from high luminosity ϕ factories, thus rendering a simple framework where the measured properties of scalar are satisfactorily explained. However, the $\eta - \eta'$ the estimate for the pseudoscalar mixing angle turned out to be too small: $-5° \leq \theta_p \leq 2°$ in the singlet-octet basis or $49° \leq \varphi_p \leq 57°$ in the strange-nonstrange basis [70, 71]. An estimate of this angle from the two gamma decays of η and η' using the strong contribution to the singlet anomalies as predicted by the model, the electromagnetic contributions to such anomalies as calculated in QCD, and the anomaly matching arguments by 't Hooft yield $39° \leq \varphi_p \leq 41°$ ($-15° \leq \theta_p \leq -13°$ in the singlet-octet basis) [74].

The three flavor linear sigma model was also used in [75] as a "toy model" to study properties of light mesons. A convenient unitarization procedure was proven to give encouraging results for meson-meson scattering. In addition, the possible quark composition of the light scalar (and pseudoscalar) fields was discussed on the light of the chiral transformation properties of the corresponding quark structures. The conclusion is that non-standard structures such as $\bar{q}\bar{q}qq$ have the same transformation properties under $SU(3)_L \times SU(3)_R$ as the conventional $\bar{q}q$ structures and they cannot be distinguished by chiral effective Lagrangians. Based on this observation, in [75] the conclusion is reached that we cannot assign a specific quark structure to the fields used in the construction of phenomenological chiral Lagrangians realizing this symmetry, which on the other side are the appropriate tools for the long distance description of the physics of mesons composed by light quarks, whatever this composition be. However, the full chiral symmetry at the level of massless QCD is actually $U(3)_L \times U(3)_R$ and the violation of $U(1)_A$ by non-perturbative effects can distinguish at least a $\bar{q}q$ state from a four-quark state in any of its internal configurations.

The existence of two scalar nonets, one below 1 GeV and another one around 1.4 GeV, lead to the exploration of two nonet models. A first step in this direction was given in [76] where a light four-quark scalar nonet and a heavy $\bar{q}q$ -like nonet are mixed up to yield physical scalar mesons above and below 1 GeV. In the same spirit, in Ref. [75] a linear sigma model is coupled to a nonet field with well defined transformation properties under chiral rotations and trivial dynamics, except for a $U_A(1)$ (non-determinantal) violating interaction. The possibility for the pseudoscalar light mesons to have a small four-quark content was speculated on the light of this "toy" model. An alternative approach was formulated in [77] where two linear sigma models were coupled using the very same interaction as in [76] and a Higgs mechanism.

Beyond the lowest lying $\sigma(500)$, $f_0(980)$, $\kappa(900)$ and $a_0(980)$ states, the next group of scalars listed by Particle Data Group are all in the energy region around 1.4 GeV: $f_0(1370)$, $K_0^*(1430)$, $a_0(1450)$, and $f_0(1500)$ [78]. In addition, we have the $f_0(1710)$ at a slightly higher mass [78]. In [79] it was noticed that if we consider the scalar states around 1.4 GeV as the members of a nonet, then it presents a slightly inverted mass spectrum: the heavier states are the cuasi-degenerate isotriplet and isosinglet states $a_0(1450)$, and $f_0(1500)$, the lightest one is the isosinglet $f_0(1370)$ and the isovector $K_0^*(1430)$ is in between. This structure is characteristic of a four-quark nonet. Furthermore, a look onto the pseudoscalar side at the same energy yields the following states: $\eta(1295), \pi(1300)$, $\eta(1450), K(1469)$ [78]. Thus *data seems to indicate the existence of a cuasi-degenerate chiral nonet around 1.4 GeV whose scalar component has a slightly inverted mass spectrum.* On the other hand, the linear rise of the mass of a hadron with the number of constituent quarks indicates that four-quark states should lie slightly below 1.5 GeV. This lead to conjecture that this chiral nonet comes from tetraquark states mixed with conventional $\bar{q}q$ mesons to form physical mesons. The cuasi-degeneracy of this chiral nonet indicates that chiral symmetry is realized in a direct way for tetraquark states. This possibility was considered in [80] in the framework of an effective chiral Lagrangian with two chiral nonets, one around 1.4 GeV with chiral symmetry realized directly, and another one at low energy with chiral symmetry spontaneously broken. In contrast to previous studies, mesons in the "heavy" nonet are considered as four-quark states. The nature of these states is distinguished from conventional $\bar{q}q$ mesons by terms breaking the $U_A(1)$ symmetry. The parameters of the model are fixed from the masses of $\pi(137)$, $a_0(980)$, $K(495)$, $\eta(548)$, $\eta'(957)$, $\eta(1295)$ in addition to the weak decay constants f_π and f_K. The model gives definite predictions for the masses of the remaining members of the nonets. Couplings of all mesons entering the model are also predicted. The outcome of the fit allow us to identify the remaining mesons as the $\pi(1300), \eta(1450), K(1469)$ on the pseudoscalar side and $f_0(1370)$, $K_0^*(1430)$, $a_0(1450)$, $f_0(1500)$ on the scalar side. In general, isospinor and isovector pseudoscalar mesons arise as weak mixing of $\bar{q}q$ and tetraquark fields. In contrast, scalars in these isotopic sectors are strongly mixed. Concerning the isoscalar sector we obtain a strong mixing between $\bar{q}q$ and tetraquarks for both scalar and pseudoscalar fields. A calculation of the two photon decay of neutral scalars in this framework refine the $U(3) \otimes U(3)$ predictions in [68] yielding results closer to the experimental data [81]. Finally we should mention that B factories will provide large samples of scalar mesons, providing thus another arena where the properties and interactions of the scalars can be tested [82].

ACKNOWLEDGMENTS

This work was supported by U. Gto. under project DINPO-000085/05, and CONACyT, México under project CONACyT SEP-2003-C02-44644.

REFERENCES

1. S.L. Glashow, Nucl. Phys. **22**, (1961) 579; S. Weinberg, Phys. Rev. Lett. **19** (1967) 1264; A. Salam in Elementary Particle Theory, Ed. N. Swartholm, Almquist and Wissel, Ttockholm (1968).
2. G. 't Hooft Nucl. Phys. **35** (1971) 167.
3. D. Gross, F. Wilczek, Phys. Rev. Lett. **30** (1973) 1343; H. D. Politzer, Phys. Rev. lett. **30** (1973) 1346.
4. G. Cocho, C. Fronsdal and R. White, Phys. Rev. **180**, 1547 (1969) [Erratum-ibid. **187**, 2280 (1969)].
5. A. Zepeda, Phys. Rev. D **4**, 1072 (1971).
6. G. Cocho and J. Flores, Phys. Rev. D **3**, 157 (1971).
7. M. A. B. Beg and A. Zepeda, Phys. Rev. D **6**, 2912 (1972).
8. H. Pagels and A. Zepeda, Phys. Rev. D **5**, 3262 (1972).
9. C. Avilez and G. Cocho, Phys. Rev. D **10**, 3638 (1974); C. Avilez, G. Cocho and M. Dubovoy, Phys. Rev. D **11**, 555 (1975).
10. S. Weinberg, Phys. Rev. Lett. **40**, 223 (1978); F. Wilczek, Phys. Rev. Lett. 40, 279 (1978); C. A. Dominguez, Phys. Rev. Lett. **41**, 605 (1978).
11. A. Zepeda, Phys. Rev. Lett. **41**, 139 (1978).
12. P. Minkowski and A. Zepeda, Nucl. Phys. B **164**, 25 (1980).
13. C. A. Dominguez, Phys. Rev. D **15**, 1350 (1977); C. A. Dominguez, Phys. Rev. D **16**, 2313 (1977); C. A. Dominguez, Phys. Rev. D **16**, 2320 (1977); C. A. Dominguez and A. Zepeda, Phys. Rev. D **18**, 884 (1978).
14. J. L. Lucio and A. Zepeda, Phys. Rev. D **16**, 42 (1977).
15. S. Weinberg, Physica A**96**(1979) 327.
16. J. Gasser and A. Zepeda, Nucl. Phys. B **174**, 445 (1980).
17. A. Manohar and H. Georgi, Nucl Phys. B**234** (1984), 189.
18. J. Gasser, L. Leutwyler, Nucl. Phys. B**250** (1985) 465.
19. G. Ecker, J. Gasser, A. Pich, E. de Rafael, Nucl Phys. B**321**,311 (1989).
20. J. Gasser, M. E. Sainio, A. Svarc, Nucl. Phys. B**307** (1988) 779.
21. E. Jenkins, A. Manohar, Phys. Lett B**255** (1991), 558.
22. R. Flores-Mendieta, C. P. Hofmann and E. Jenkins, Phys. Rev. D **61**, 116014 (2000); R. Flores-Mendieta, C. P. Hofmann, E. Jenkins and A. V. Manohar, Phys. Rev. D **62**, 034001 (2000).
23. T. R. Hemmert, B. R. Holstein and J. Kambor, Phys. Lett. B **395**, 89 (1997); J. Phys. G **24**, 1831 (1998);
24. W. Rarita & J. Schwinger. Phys. Rev. 60, 61 (1941).
25. K. Jhonson & E.C.G. Sudharsan, Ann. Phys. 13, 126, (1961). G. Velo & D. Zwanziger, Phys. Rev. 186, 1337 (1969); C.R. Hagen, Phys. Rev. D4, 2204 (1971).
26. C. Fronsdal Nuovo Cimento (suppl) IX, 416 (1958).
27. L.M. Nath, B. Etemadi & J.D. Kimel, Phys. Rev. D3, 2153 (1971).
28. J. Urias, Ph D. tesis, Universite Católique de Louvain (1976), unpublished.
29. M. El Amiri, J. Pestieau and G. Lopez Castro, Nucl. Phys. A **543**, 673 (1992).
30. M. Hernandez, J. L. Lucio M. and G. Lopez Castro, Phys. Rev. D **44**, 794 (1991).
31. A. Bernicha, G. Lopez Castro and J. Pestieau, Nucl. Phys. A **597**, 623 (1996); A. Mariano and G. Lopez Castro, Phys. Rev. C **62**, 014604 (2000); Nucl. Phys. A **697**, 440 (2002).
32. M. Napsuciale and J. L. Lucio, Phys. Lett. B **384**, 227 (1996)
33. M. Napsuciale and J. L. Lucio, Nucl. Phys. B **494**, 260 (1997).
34. M. Kirchbach, Int. J. Mod. Phys. A **15**, 1435 (2000); Nucl. Phys. A **689**, 157 (2001); Rev. Mex. Fis. **50**, 54 (2004); M. Kirchbach, M. Moshinsky and Y. F. Smirnov, Phys. Rev. D **64**, 114005 (2001).
35. J. L. Lucio, M. Napsuciale and M. Ruiz-Altaba, arXiv:hep-ph/9903420.
36. J. F. Donoghue, B. R. Holstein and Y. C. Lin, Phys. Rev. D **37**, 2423 (1988).
37. N. N. Achasov, S. A. Devyanin and G. N. Shestakov, Phys. Lett. B **88**, 367 (1979); N. N. Achasov

and G. N. Shestakov, Phys. Rev. D **56**, 212 (1997); N. N. Achasov and A. V. Kiselev, Phys. Lett. B **534**, 83 (2002).
38. F. E. Close and A. Kirk, Phys. Lett. B **489**, 24 (2000); Phys. Lett. B **515**, 13 (2001); O. Krehl, R. Rapp and J. Speth, Phys. Lett. B **390**, 23 (1997).
39. A. Gallegos and J. L. Lucio M., Found. Phys. **33**, 855 (2003).
40. A. Bernicha, G. Lopez Castro, J. Pestieau, Phys.Rev. D**50**, 4454 (1994); Phys.Rev. D**53**, 4089 (1996).
41. M. Feuillat, J.L.Lucio M., J. Pestieau, Phys.Lett.B501:37-43,2001.
42. A. Gallegos, J.L. Lucio M., J. Pestieau, Phys.Rev.D69:074033,2004; A. Gallegos, J.L. Lucio M, G. Moreno, J. Pestieau Eur.Phys.J.C28:107-114,2003.
43. N. A. Tornqvist, Phys. Rev. Lett. **49**, 624 (1982); Z. Phys. C **68**, 647 (1995); N. A. Tornqvist and M. Roos, Phys. Rev. Lett. **76**, 1575 (1996).
44. E. Van Beveren, T. A. Rijken, K. Metzger, C. Dullemond, G. Rupp and J. E. Ribeiro, Z. Phys. C **30**, 615 (1986).
45. M. D. Scadron, Phys. Rev. D **26**, 239 (1982).
46. M. Harada, F. Sannino and J. Schechter, Phys. Rev. D **54**, 1991 (1996).
47. M. Napsuciale, arXiv:hep-ph/9803396, unpublished.
48. N. A. Tornqvist, Eur. Phys. J. C **11**, 359 (1999);
49. C. Caso et al. [Particle Data Group Collaboration], Eur. Phys. J. C **3**, 1 (1998).
50. J. A. Oller, E. Oset and J. R. Pelaez, Phys. Rev. Lett. **80**, 3452 (1998); Phys. Rev. D **59**, 074001 (1999) [Erratum-ibid. D **60**, 099906 (1999)]; J. A. Oller and E. Oset, Phys. Rev. D **60**, 074023 (1999).
51. E. M. Aitala et al. [E791 Coll.], Phys. Rev. Lett. **86**, 770 (2001); Phys. Rev. Lett. **86**, 765 (2001).
52. S. Ishida, M. Ishida, T. Ishida, K. Takamatsu and T. Tsuru, Prog. Theor. Phys. **98**, 621 (1997).
53. D. Black, A. H. Fariborz, F. Sannino and J. Schechter, Phys. Rev. D **58**, 054012 (1998).
54. S. N. Cherry and M. R. Pennington, Nucl. Phys. A **688**, 823 (2001).
55. E. M. Aitala et al. [E791 Coll.], Phys. Rev. Lett. **89**, 121801 (2002).
56. D. Black, A. H. Fariborz, F. Sannino and J. Schechter, Phys. Rev. D **59**, 074026 (1999).
57. J. A. Oller, Nucl. Phys. A **727**, 353 (2003).
58. T. Barnes, Phys. Lett. B **165**, 434 (1985).
59. M. Napsuciale, AIP Conf. Proc. **623**, 131 (2002).
60. R. Jaffe, Phys. Rev. D**15** (1977) 267.
61. J. Weinstein and N. Isgur, Phys. Rev. D**41** (1990) 2236.
62. M. G. Alford and R. L. Jaffe, Nucl. Phys. B **578**, 367 (2000).
63. J.L. Lucio Martinez, J. Pestieau. Phys.Rev.D42:3253-3254,1990; J.L. Lucio Martinez, M. Napsuciale, Phys.Lett.B331:418-424,1994.
64. R. R. Akhmetshin et al. [CMD-2 Coll.], Phys. Lett. B **462**, 380 (1999); Phys. Lett. B **462**, 371 (1999).
65. A. Aloisio et al. [KLOE Coll.], Phys. Lett. B **537**, 21 (2002), Phys. Lett. B **536**, 209 (2002).
66. J. A. Oller and E. Oset, Nucl. Phys. A **629**, 739 (1998);
67. J. E. Palomar, L. Roca, E. Oset and M. J. Vicente Vacas, Nucl. Phys. A **729**, 743 (2003).
68. J. L. Lucio Martinez and M. Napsuciale, Phys. Lett. B **454**, 365 (1999).
69. A. Bramon, R. Escribano, J. L. Lucio M., M. Napsuciale and G. Pancheri, Phys. Lett. B **494**, 221 (2000); Eur. Phys. J. C **26**, 253 (2002); A. Bramon, R. Escribano, J. L. Lucio Martinez and M. Napsuciale, Phys. Lett. B **517**, 345 (2001).
70. M. Napsuciale and S. Rodriguez, Int. J. Mod. Phys. A **16**, 3011 (2001).
71. M. Napsuciale, A. Wirzba and M. Kirchbach, Nucl. Phys. A **703**, 306 (2002).
72. J. Schechter and Y. Ueda, Phys. Rev. D3, 2874 (1971).
73. G. 't Hooft, Phys. Rev. D **14**, 3432 (1976) [Erratum-ibid. D **18**, 2199 (1978)]; G. 't Hooft, Phys. Rept. **142**, 357 (1986); G. 't Hooft, arXiv:hep-th/9903189, unpublished;
74. M. Napsuciale and S. Rodriguez, Phys. Rev. D**65**, 096013 (2002).
75. D. Black, A. H. Fariborz, S. Moussa, S. Nasri and J. Schechter, Phys. Rev. D**64**, 014031 (2001).
76. D. Black, A. H. Fariborz and J. Schechter, Phys. Rev. D**61**, 074001 (2000).
77. N. A. Tornqvist, arXiv:hep-ph/0201171; F. E. Close and N. A. Tornqvist, J. Phys. G**28**, R249 (2002).
78. K. Hagiwara et al. [Particle Data Group Collaboration], Phys. Rev. D**66**, 010001 (2002).
79. M. Napsuciale and S. Rodriguez, Phys. Lett. B **603**, 195 (2004)
80. M. Napsuciale and S. Rodriguez, Phys. Rev. D **70**, 094043 (2004).
81. S. Rodriguez and M. Napsuciale, Phys. Rev. D **71**, 074008 (2005).
82. Delepine, J.L. Lucio M., Carlos A. Ramirez, Eur.Phys.J.C45:693-700,2006.

Studies of Electroweak Symmetry Breaking at Mexican Institutions

J. Lorenzo Diaz Cruz

Facultad de Ciencias Físico-Matemáticas, BUAP. Apdo. Postal 1364, C.P. 72000 Puebla, Pue., México

Abstract. This paper is aimed to review the contribution made by scientists working at Mexican Institutions on the subject of Electroweak symmetry breaking. This review covers the period from the 80 up to the present.

Keywords: Electroweak symmetry breaking, Higgs
PACS: 13.60.Hb, 13.60.-r, 13.15.+g

INTRODUCTION: HOW DID IT START?

"When I was younger, so much younger than today...", around 1983, I decided to study a master degree in physics at CINVESTAV-IPN, after having finished studies in nuclear engineering at UAM-Azcapotzalco. At that time, the Physics Department of CINVESTAV was already known in Mexico, and surrounded by its own legends, full of color and drama. Between courses, soccer and beer, we learned that once upon a time there was a brilliant student, that got tired of climbing buildings and carrying his heavy briefcase full of stones. Suddenly, he refused to continue a research career and went north. There was another student, who was eating just grass, and died when he was left locked inside the building. We also learned that our academic life would be at risk if one were assigned some of the courses of the Pakistan-Cuba-Argentina axis. In that case, the advanced students would be eager to organize lotteries, to guess who among the newcomers would fail the exams and be expelled from paradise. Several generation proved those asserts, specially one that arrived claiming that good physicist should not do homework problems at all, "the real good ones should learn and understand everything during the exams", obviously all of them failed.

Most students at the Physics department came from the province states in Mexico, and damn it! we were nearly poor. Our fellowships were quite limited, and we had to stretch our money in order to survive till the end of the month. Some students used to sleep in the offices, take shower in the labs and would go for breakfast to the cafeteria still wearing a towel. Nevertheless, there were some luxuries that we could afford, such as accompanying visitors to "El Tenampa", in Garibaldi square; among these visitors, one can recall some famous Nobel prize winners. Other luxuries, include the occasional visits to the "non-sanct" places of the Mexico City's night life, an style that was about to disappear.

Among classes and exams, we also learned that the real challenge was in particle physics or general relativity. However, I found that the classical approach followed by the CINVESTAV GR group was not suited for me. So, I decided to write a thesis in high energy physics (HEP). At that period there was some level of excitement, because of the intermediate gauge bosons, recently discovered at CERN. Once I read in the papers that the existence of another particle had been suggested at DESY, so I asked Rodrigo Huerta and Jose Luis Lucio, the younger faculty at that time, what this new particle was about. I remember they said that it may be the Higgs boson or a superpartner?. It sound interesting to me, so when trying to decide for a thesis advisor, after I learned from Alfonso Rosado that Miguel Angel Perez Angon, was studying the Higgs particles, I chose to work with him. I believe that it was Miguel, who started the study of the Higgs sector in Mexico, mostly in collaborations with students. Their interest included the phenomenology of the Higgs sector in several extensions of the SM, such as the two-Higgs doublet and left-right models, and more lately within the effective lagragian approach.

It also happened that during the thesis process I read some papers from G.L. Kane, and also learned that the theory group at Michigan included Martin Veltman, so I told Miguel Angel Perez (MAPA from now on) that after finishing the Masters I would like to study abroad, at Michigan if possible. Once I got there, in 1985, I decided to work with Kane, and he suggested me to study the Higgs sector of the Minimal SUSY extension of the Standard model. At Michigan I also met some Spanish theorists, including Tony Mendez from Barcelona. After graduation I spent one year at CINVESTAV, and for the postdoc I spent two years at Barcelona, where I continued working on the Higgs sector. So, without really planning it, most of my career had been devoted to the study of the Higgs boson, or as it said sometimes: on the physics of electroweak symmetry breaking.

As graduate student, I heard in Gordy Kane's courses that the nature of the Higgs sector would be elucidated around 1998, as at that time nobody would ever suspected that the SSC would not see the day light. Now we say the same, but with a different machine and date: after LHC would enter in operation, with real Higgs data probably will come after 2010. In other courses, Tini Veltman was discussing the screening theorem and how precise the W, Z masses should be measured in order to get indirect information on the Higgs mass. We were also thought that essentially there were two options for the physics of electroweak symmetry breaking (EWSB): either there is a light Higgs boson, standard or beyond, that would be associated with perturbative physics, or else, there is a heavy Higgs boson (or no Higgs at all), that would point towards the presence of strong interactions associated with EWSB. It seems quite amazing in how many presentations this classification idea comes nowadays. In the first place we have the popular SUSY models, minimal or more exotic, which can arise in four or more dimensions, while the second option can be associated with the fat or little Higgs, obese or composite, etc, etc, [1, 2].

The purpose of this article is to present a short review of how the Mexican HEP community has contributed to the study of the Higgs sector or EWSB. There is a minimal reason for this association, namely that the Higgs mechanism is based on the so-called "Mexican hat", so at least the image was created by our ancestors. I tried to cover the time period from early 80's up to the present, and in trying to separate the real contributions from others, I had only covered material that was published as regular articles. At several conferences, I have heard some more exotic ideas that unfortunately have not been published. Sorry about that.

HIGGS BOSON DECAYS IN THE SM AND BEYOND

One of the first aspects of the Higgs sector studied in Mexico was the production of Higgs boson in decays of the W boson by MAPA et al. [3]. Then, limits on the Higgs mass were also obtained using the top decays into charged Higgs bosons [4, 5], as at that time there was a claim that CERN had discovered the top quark, with a mass of order 45 GeV. Since the reported decay modes were in agreement with SM, it would mean that a charged Higgs with mass of order 40 GeV would be excluded; but as it is well known now, the top signal was not there.

On the other hand, by the early 80's most SM Higgs decays were known, including the dominant modes into fermion or gauge boson pairs, as well as the rare loop-induced decays into gluon or photon pairs [1]. Similarly, the 3-body decays into one vector boson and fermion pairs, were known. Then, MAPA and collaborators, started to study the corresponding decays for both neutral and charged Higgs bosons; they considered extensions of the SM, such as the THDM-II [6], the LR models [7, 8], and the effective largrangian approach [9, 10, 11, 12]. Higgs boson decays within the context of the Minimal SUSY extension of the SM, were studied by Diaz-cruz et al. [13].

HIGGS BOSON PRODUCTION AT HADRON COLLIDERS

The subject of Higgs boson production at hadron colliders, was considered in [14], although this thesis was written at the U. of Michigan, under the advice of Porf. G.L. Kane, the resulting article was published by myself, as a faculty member of CINVESTAV-IPN [14]. This paper considered the production of the neutral Higgs bosons of the MSSM through gluon fusion, including the possible detectable signatures at LHC. The inclusions of radiative effects was considered in the subsequent paper [15].

Later on, the production of neutral and charged Higgs boson at hadron colliders was discussed by Diaz-Cruz and Sampayo [16, 17]. For the neutral Higgs boson they considered the associated production with a top-bottom pair, while for the charged Higgs they calculated the mechanism of gluon fusion, i.e. $pp \to t + \bar{b} + H^-$, at SSC energies; an attempt was also made in that paper, to estimate the heavy parton effects. In the late 90's, Diaz-Cruz et al, considered the associated production of the neutral Higgs bosons in association with a pair of $b\bar{b}$ quarks, as a mean to detect the signal in the large $tan\beta$ region [18, 19], both for the MSSM and composite Higgs scenarios, such as the so called top-color models.

More exotic models were also studied by our groups, these include the invisible Higgs decay into majorons that arises in models with R-parity breaking [20], the Higgs boson decay into photons and gravitinos which arises in models with Gauge-Mediated SUSY breaking [21]. Effects of modified top-Higgs couplings on Higgs production by gluon fusion, using effective lagrangians, were considered in [22]

HIGGS BOSON PRODUCTION AT LINEAR COLLIDERS

Several groups in Mexico studied Higgs production at linear collider, including the production in γe collisions, as well as the associated production with heavy quarks in e^+e^- collisions. The production of the SM Higgs in the reaction $\gamma e \to e + H$, was discussed in [23], while the inclusion of charged and neutral Higgs bosons of the MSSM was discussed in [24]. Production of neutral Higgs bosons in photon-photon collisions, within the effective lagrangian approach were reported in refs. [25, 26]. On the other hand, the study of Higgs boson production in association with heavy quarks had been discussed by A. Rodriguez and O.A. Sampayo in several papers, [27, 28, 29]. Pair production of Higgs bosons has been studied too [30, 31].

VACUUM ALIGNMENT AND PRECISION TESTS

The issue of vaccum alignment, which had been stressed by Tini Veltman for many years, namely whether the minimum of the potential respects the charge conservation, was studied for multi-Higgs models in ref. [32], with the conclusion that vacuum alignment does not pose a fine-tunning problem, and furthermore the constraints imposed by supersymmetry force the theory to align itself into the correct vacuum.

The possibility to modify the standard bounds on the Higgs boson mass from precision measurements, by considering the effective lagrangian approach was also studied in ref. [33]. Constraints on the presence of heavy quanta from the induced modification to Higgs boson production were studied in [34]. A definition of a rho parameter in terms of Higgs couplings was proposed in [35], as a mean to test the origin of electroweak symmetry breaking.

A MORE FLAVORED HIGGS SECTOR: THDM AND SUSY

The search for connections between the Higgs and flavor sector had been pursued in our group too. Early work on the two-Higgs doublet model with FCNC (THDM-III), and its implication for top quark decays was discussed in [36, 37]. The possibility to detect the Lepton Flavor Violating decays of Higgs bosons at current and future colliders, within the context of multi-Higgs models was first considered in [38]. Subsequently, this issue was also discussed in SUSY models in [39], in models with additional fermion in [40]. More recntly, systematic studies of the THDM-III incorporating specific texture ansatz for the Yukawa matrices were considered in [41, 42].

RECENT TRENDS IN EWSB

More recently, together with my collaborators Alfredo Aranda from the University of Colima and C. Balazs, from ANL, we have studied new ideas for electroweak symmetry breaking. These include the discussion of Higgs mechanism in the framework of extra-dimensions, and the possible unification of gauge and Higgs interactions.

Among the new approaches to discuss the problems of the SM that have been proposed, the possible existence of extra-dimensions is certainly a very attractive one. In particular, there could be extra-dimensions with sizes that ranges from microscopic up to GUT scale. Extra dimensions of microscopic scale could be required in order to offer a new solution to the hierarchy problem, where it is gravity scale which is modified, while the weak scale becomes the fundamental scale in nature. Extra-dimensions of TeV-size could also lead to new physics, close to the electroweak scale, and this could in turn provide a solution to the little hierarchy problem. Even an extra-dimension with a size of the GUT scale, offers interesting new physics, as it provides a very plausible solution to the doublet-triplet problem.

In our first approach, we have assumed the presence of a new dimension with TeV scale, and asked about the implications of locating the Higgs boson in different points of the extra-dimension. Using a two-Higgs doublet model, we have considered an scenario where one Higgs doublet is located at the fixed points, while the other Higgs can have allocation in the bulk. Implications of this scenario for the Higgs search at future linear collider are presented in our paper [43]. Another option that we have considered in the extra-dimensional setting, deals with the mixing between the scalars that participate in the generation of the flavor hierarchies, namely the flavon fields [44].

Understanding the origin of the parameters of the SM constitutes another line of research that we have considering recently [45]. This includes both the flavor parameters, as well as the ones that appear in the process of EWSB. In this regard we have also attempted to discuss possible scenarios where the Higgs slef-coupling unfies with the electroweak couplings [46], in a scenario where the $SU(2)_L$ and $U(1)_Y$ couplings unfie at an intermediate scale. This permits to use the RGE to relate the Higgs boson mass with the gauge ones, which results in the prediction $m_h \simeq 200$ GeV.

Finally, we have considered the possibility that the Higgs boson has as origin the extra-dimensional components of gauge vector fields. We have studied a six-dimensional model with gauge group $SU(3)_W$, which breaks into the SM $SU(2)_L \times U(1)_Y$ through appropriate orbifold boundary conditions. The 5th and 6th dimensions of the 6D gauge field combines to form a Higgs sector that may include one or two Higgs doublets. The minimal model is not realistic as it predicts a too large value for the weak mixing angle. In order to fix this problem we have argued that the so-called Brane kinetic terms can help to fix the correct value of this weak mixing angle [47].

CONCLUSIONS

One of the goals of this document, is to offer the possibility to the reader to appreciate the contributions of the subset of the Mexican HEP community working on the topics of Higgs physics and EWSB. Among the papers listed above there are several ones that have received a large number of citations. When asking spires for the entry "find t Higgs and cc mexico", it gives 83 total papers, which have received 1068 citations, not bad for such an small community that includes less than 10 faculty members. These papers include the participation from several scientists working at Mexican institutions in the international working groups devoted to this topic [48, 49, 50, 50]. It is also our hope that this review can motivate the younger contribution to make even more important contributions to the topic. There are great chances because, in the first place, the problem of explaining what breaks the electroweak symmetry remains as open as ever.

ACKNOWLEDGMENTS

The author thanks the *Sistema Nacional de Investigadores* and *CONACyT* (México) for financial support.

REFERENCES

1. An excellent review of Higgs phenomenology is still: V.D. Barger and R.J.N. Phillips, Collider Physics (Updated Edition),Addison-Wesley Publishing Company, Inc., Reading, Massachusetts, 1997.
2. For a review of SUSY phenomenology see: H. E. Haber and G. L. Kane, Phys. Rept. **117**, 75 (1985).
3. M. A. Perez and M. A. Soriano, Phys. Rev. D **31**, 665 (1985).
4. M. A. Perez and A. Rosado, Phys. Rev. D **30**, 228 (1984).
5. M. A. Perez, Z. Phys. C **18**, 113 (1983).
6. J. L. Diaz-Cruz and M. A. Perez, Phys. Rev. D **33**, 273 (1986).
7. R. Martinez, M. A. Perez and J. J. Toscano, Phys. Lett. B **234**, 503 (1990).
8. R. Martinez and M. A. Perez, Nucl. Phys. B **347**, 105 (1990).
9. M. A. Perez and J. J. Toscano, Phys. Lett. B **289**, 381 (1992).
10. J. M. Hernandez, M. A. Perez and J. J. Toscano, Phys. Rev. D **51**, 2044 (1995).
11. J. L. Diaz-Cruz, J. Hernandez-Sanchez and J. J. Toscano, Phys. Lett. B **512**, 339 (2001) [arXiv:hep-ph/0106001].
12. J. Hernandez-Sanchez, M. A. Perez, G. Tavares-Velasco and J. J. Toscano, Phys. Rev. D **69**, 095008 (2004) [arXiv:hep-ph/0402284].
13. E. Barradas, J. L. Diaz-Cruz, A. Gutierrez and A. Rosado, Phys. Rev. D **53**, 1678 (1996).
14. J. L. Diaz-Cruz, Nucl. Phys. B **358**, 97 (1991).
15. J. L. Diaz-Cruz and O. A. Sampayo, Int. J. Mod. Phys. A **8**, 4339 (1993).
16. J. L. Diaz-Cruz and O. A. Sampayo, Phys. Lett. B **276**, 211 (1992).
17. J. L. Diaz-Cruz and O. A. Sampayo, Phys. Rev. D **50**, 6820 (1994).
18. J. L. Diaz-Cruz, H. J. He, T. Tait and C. P. Yuan, Phys. Rev. Lett. **80**, 4641 (1998) [arXiv:hep-ph/9802294].
19. C. Balazs, J. L. Diaz-Cruz, H. J. He, T. Tait and C. P. Yuan, Phys. Rev. D **59**, 055016 (1999) [arXiv:hep-ph/9807349].
20. J. C. Romao, J. L. Diaz-Cruz, F. de Campos and J. W. F. Valle, Mod. Phys. Lett. A **9**, 817 (1994) [arXiv:hep-ph/9211258].
21. J. L. Diaz-Cruz, D. K. Ghosh and S. Moretti, Phys. Rev. D **68**, 014019 (2003) [arXiv:hep-ph/0303251].
22. J. L. Diaz-Cruz, M. A. Perez and J. J. Toscano, Phys. Lett. B **398**, 347 (1997) [arXiv:hep-ph/9702413].
23. U. Cotti, J. L. Diaz-Cruz and J. J. Toscano, Phys. Lett. B **404**, 308 (1997) [arXiv:hep-ph/9704304].
24. U. Cotti, J. L. Diaz-Cruz and J. J. Toscano, Phys. Rev. D **62**, 035009 (2000) [arXiv:hep-ph/9912406].
25. M. A. Perez, J. J. Toscano and J. Wudka, Phys. Rev. D **52**, 494 (1995) [arXiv:hep-ph/9506457].
26. J. M. Hernandez, M. A. Perez and J. J. Toscano, Phys. Lett. B **375**, 227 (1996) [arXiv:hep-ph/9601378].
27. U. Cotti, A. Gutierrez-Rodriguez, A. Rosado and O. A. Sampayo, Phys. Rev. D **59**, 095011 (1999) [arXiv:hep-ph/9902417].
28. A. Gutierrez-Rodriguez, M. A. Hernandez-Ruiz and O. A. Sampayo, J. Phys. Soc. Jap. **70**, 2300 (2001) [arXiv:hep-ph/0005050].
29. A. Gutierrez-Rodriguez and O. A. Sampayo, Phys. Rev. D **62**, 055004 (2000).
30. A. Gutierrez-Rodriguez, M. A. Hernandez-Ruiz and O. A. Sampayo, Phys. Rev. D **67**, 074018 (2003) [arXiv:hep-ph/0302120].

31. A. Gutierrez-Rodriguez, M. A. Hernandez-Ruiz and O. A. Sampayo, Mod. Phys. Lett. A **20**, 2629 (2005) [arXiv:hep-ph/0504266].
32. J. L. Diaz-Cruz and A. Mendez, Nucl. Phys. B **380**, 39 (1992).
33. J. L. Diaz-Cruz, J. M. Hernandez and J. J. Toscano, Mod. Phys. Lett. A **15**, 1377 (2000) [arXiv:hep-ph/9905335].
34. J. L. Diaz-Cruz, Mod. Phys. Lett. A **16**, 863 (2001) [arXiv:hep-ph/0105036].
35. J. L. Diaz-Cruz and D. A. Lopez-Falcon, Phys. Lett. B **568**, 245 (2003) [arXiv:hep-ph/0304212]; J. L. Diaz Cruz and D. A. Lopez Falcon, Phys. Rev. D **61**, 051701 (2000) [arXiv:hep-ph/9911407].
36. J. L. Diaz-Cruz and G. Lopez Castro, Phys. Lett. B **301**, 405 (1993).
37. J. L. Diaz Cruz, J. J. Godina Nava and G. Lopez Castro, Phys. Rev. D **51**, 5263 (1995) [arXiv:hep-ph/9509229].
38. J. L. Diaz-Cruz and J. J. Toscano, Phys. Rev. D **62**, 116005 (2000) [arXiv:hep-ph/9910233]; U. Cotti, L. Diaz-Cruz, C. Pagliarone and E. Vataga, in *Proc. of the APS/DPF/DPB Summer Study on the Future of Particle Physics (Snowmass 2001)* ed. N. Graf, eConf **C010630**, P102 (2001) [arXiv:hep-ph/0111236].
39. J. L. Diaz-Cruz, JHEP **0305**, 036 (2003) [arXiv:hep-ph/0207030].
40. U. Cotti, J. L. Diaz-Cruz, R. Gaitan, H. Gonzales and A. Hernandez-Galeana, Phys. Rev. D **66**, 015004 (2002) [arXiv:hep-ph/0205170].
41. J. L. Diaz-Cruz, R. Noriega-Papaqui and A. Rosado, Phys. Rev. D **69**, 095002 (2004) [arXiv:hep-ph/0401194].
42. J. L. Diaz-Cruz, R. Noriega-Papaqui and A. Rosado, Phys. Rev. D **71**, 015014 (2005) [arXiv:hep-ph/0410391].
43. A. Aranda, C. Balazs and J. L. Diaz-Cruz, Nucl. Phys. B **670**, 90 (2003) [arXiv:hep-ph/0212133].
44. A. Aranda and J. L. Diaz-Cruz, Mod. Phys. Lett. A **20**, 203 (2005) [arXiv:hep-ph/0207059].
45. J. L. Diaz-Cruz, Mod. Phys. Lett. A **20**, 2397 (2005) [arXiv:hep-ph/0409216].
46. A. Aranda, J. L. Diaz-Cruz and A. Rosado, arXiv:hep-ph/0507230.
47. A. Aranda and J. L. Diaz-Cruz, Phys. Lett. B **633**, 591 (2006) [arXiv:hep-ph/0510138].
48. M. Carena *et al.* [Higgs Working Group Collaboration], arXiv:hep-ph/0010338.
49. U. Baur *et al.* [The Snowmass Working Group on Precision Electroweak Measurements], in *Proc. of the APS/DPF/DPB Summer Study on the Future of Particle Physics (Snowmass 2001)* ed. N. Graf, eConf **C010630**, P1WG1 (2001) [arXiv:hep-ph/0202001].
50. B. C. Allanach *et al.* [Beyond the Standard Model Working Group], arXiv:hep-ph/0402295.
51. E. Accomando *et al.* [CLIC Physics Working Group], arXiv:hep-ph/0412251.

Supersymmetry and GUTs in Mexico

Myriam Mondragón

Instituto de Física, UNAM, Apdo. Postal 20-364, 01000 México D.F., México

Abstract. Here I review briefly some of the work done in Mexico in Supersymmetry and Grand Unified Theories. The emphasis will be mainly in work that has some relation to phenomenology.

Keywords: Supersymmetry, Grand Unified Theories
PACS: 11.30.Pb,12.10.Dm, 12.10.Kt,12.60.-i,12.60.Jv

INTRODUCTION

The last thirty years have seen a very large effort devoted to the study and understanding of supersymmetry, Grand Unified Theories and the combination of both.

In particle physics there were very few people dedicated to the study of these subjects in Mexico compared to the United States and Europe, until approximately the last ten years. In the last ten years the situation changed, with an increasing number of people working on these subjects, as well as an increasing number of students doing their thesis in related topics. Supersymmetry and Grand Unified Theories (GUTs) are now topics much more widely studied in Mexico, proof of this are the themes presented at the events organized by the Division of Particles and Fields of the Mexican Physical Society, the Latin American School of Physics ELAF XXXV [1], with the title *Supersymmetries in Physics and its Applications*, which took place in Mexico City, and the various papers published on the subjects.

In this paper I will review only the work done in supersymmetry and GUTs in particle physics in Mexico, with an emphasis on the phenomenology. Most of the information I have gathered from the internet and from personal contact with colleagues. I apologize for any work that has not been included in this review. For a proper overview on the theory and phenomenology of supersymmetry and/or GUTs, the interested reader may consult some of the excellent reviews in the literature, for instance [2–4].

SUPERSYMMETRY

The Standard Model (SM) works remarkably well at the energies reached by present accelerators. Still, there are several issues of the SM which makes us think that it must be the low-energy limit of a more fundamental theory. These features manifest itself in a large number of free parameters; free in the sense that we cannot predict their value from the theory but have to extract if from the experiment. Among these are all the quark and lepton masses, Yukawa couplings and mixing angles. We do not know, or have in the context of the SM, any hint of why there is the large variety of masses in the SM, or why there are only three families. This problem is only enhanced by the fact that neutrinos have also masses, which are extremely small compared to the ones of the other particles. Moreover, all the particles acquire masses through the Higgs mechanism (except possibly Majorana neutrinos), but the Higgs itself recieves large quadratic corrections to its mass, which means that to keep it at the elecroweak scale a very accurate fine tuning of the parameters has to be performed. This is known as the gauge hierarchy problem.

The fundamental theory of particle and interactions has to be compatible also with theories of creation and evolution of the Universe, and with astrophysical observations where the nature of the particles is involved. There are cosmological observations which cannot be explained within the context of the SM. One of them is the absence of antimatter in the observed Universe, which points to a baryon-antibaryon asymmetry. To create this asymmetry dynamically would require new particles, since it is not possible to do it in the context of the SM. Another one is the dark matter problem: there is much more matter in the Universe than we observe, which means we need much more non-baryonic matter. The SM does not include any scalar particle capable of creating the inflationary epoch needed in cosmological models. Thus again, to keep compatibility of a theory of fundamental interactions with Big Bang nucleosynthesis and inflation we need to introduce new particles.

The immediate question is: at which energies do these new particles appear? Or, what is the scale of this new physics?

Supersymmetry Breaking

Supersymmetry (SUSY) has emerged as a possible solution of the gauge hierarchy problem. Supersymmetry is indeed from the theoretical point of view, a very attractive theory. It is the most general symmetry compatible with local, relativistic quantum field theory, it is the only possible extension of the Poincaré group. It has "better behaved" or softer divergencies than a non-supersymmetric field theory. Local superysmmetry gives supergravity, which could lead to a unification of gravity with the other fundamental forces. But SUSY has to be broken at a certain scale, otherwise it would have been already observed in nature. The mechanism and scale of SUSY breaking are unknown, but if SUSY is broken close to the electroweak scale (a few TeV at most) it could solve the gauge hierarchy problem, due to the cancellation of divergencies between the known particles and its superpartners. This observation already triggered a large amount of work in low-energy SUSY, which was enhanced by the second observation that gauge coupling unification with low-energy SUSY works much better with the low energy precision data, than within ordinary GUTs.

The interest in SUSY was triggerd, among others, by the work of Wess and Zumino [5, 6], Haag, Lopuszanski and Sohnius [7], and Ferrara and Fayet [8]. SUSY phenomenology appeared shortly afterwards [9, 10].

From then on a substantial effort and work has been done in the subject, with the emergence of the Minimal Supersymmetric Standard Model (MSSM) as the simplest SUSY extension of the Standard Model consistent with known phenomenology, and later with the constrained MSSM (CMSSM), next to minimal SUSY model (NMSSM) and minimal Supergravity (MSUGRA), among others. Supersymmetric models have been studied in particle physics in the context of minimal settings, extra symmetries, extra dimensions, superstrings and supergravity, branes, etc. The strongest constraints for constructing a viable low-energy SUSY theory come from Flavour Changing Neutral Currents (FCNC) and CP violation, which are processes that are highly suppressed in the SM. Although the actual mechanism of SUSY breaking is not known, the favoured phenomenological way of breaking is by introducing the so-called soft-SUSY breaking (SSB) terms. These are terms that preserve the ultra-violet properties of SUSY quantum field theory, but allowing the quadratic divergencies not to cancel completely, thus providing a cure to the gauge hierarchy problem and giving masses to the superpartners of the order of a few TeV. These terms could arise from a more fundamental theory, where the SUSY breaking is transmitted from a hidden to a visible sector either through gauge or gravitational interactions.

SUSY in Mexico

The first papers on supersymmetry in particle physics in Mexico are due to José Luis Lucio, based in the confining hypercolor theory of composite hypercolour particles [11, 12]. But it was not until the last ten years that there has been a sustained effort in the subject.

SUSY Phenomenology

In the last ten years there has been work on different aspects of SUSY phenomenology, in particular those related to the Higgs sector and its possible detection in future colliders. This research is done on generic aspects of SUSY phenomenology, where no Grand Unification theory or more fundamental scheme has been assumed.

There are several phenomenological studies of possible detection of MSSM Higgs bosons at future colliders. One of these, by Enrique Barradas, Lorenzo Díaz-Cruz, Alejandro Gutiérrez and Alfonso Rosado [13], focuses on three body decays of the Higgs boson sector of the MSSM, including the ones that have additional b quarks in the final state.

In a similar context Umberto Cotti, Alejandro Gutiérrez, Alfonso Rosado and Alfredo Sampayo [14] studied the possible detections of the neutral Higgs bosons of the MSSM with energies avaibable at LEP-II and at the LC. Umberto Cotti, Lorenzo Díaz-Cruz and Jesús Toscano [15] made a complete study of the production of neutral $(h^0, H^0, A^0 (= \phi_i^0))$ and charged Higgs $(H^\pm)$ bosons at electron-photon colliders. In [16] the contribution of charginos and neutralinos of the MSSM to the one-loop vertices ZAA, ZZA, ZZZ, is studied in relation to the cancellation of anomalies.

There is also work on the constraints on the Higgs boson production associated to bottom quarks, when there is a large b-quark Yukawa coupling, via the processes $p\bar{p}/pp \to b\bar{h}(\to \bar{b}) + X$ and $p\bar{p}/pp \to \phi b\bar{b} \to b\bar{b}b\bar{b}$, at Tevatron Run II or the LHC, done by L. Díaz-Cruz, H-J He, C.P. Yuan, T. Tait and C. Balazs [17, 18]. The analysis is done with emphasis on the top-condensate or topcolor models with large bottom Yukawa couplings, and for SUSY models with large tan β. The conclusion is that the experiments can put bounds on these models if the $b\bar{b}$ signal is not found.

Also in the context of Higgs boson signals, the di-photon signature is proposed to be studied at the LHC as a test of the Gauge Mediated Symmetry Breaking scenarios [19].

SUSY breaking

In the context of SUSY breaking, FCNC and CP violation the following work has been done.

Axel de la Macorra and Graham Ross studied the possibility to break SUSY dynamically [20, 21], by using Nambu-Jona-Lasinio NJL techniques, and showing that a gaugino condensation is dynamically favored and it breaks SUSY. The dilaton field, which gives the gauge coupling constant, is stabilized with a single gauge group. Once SUSY is broken the moduli fields are fixed and its phenomenological consequences are studied in [22], where a possible fifth force due to light moduli fields is considered. Constraints on the models are worked out in order not to contradict the lack of observational evidence on this fifth force. A model where the dynamical breaking of SUSY is induced by the proper alignment of three gaugino condensates is presented in [23].

In [24, 25], C.P. Burgess, A. de la Macorra, I. Maksymyk and F. Quevedo derived non-perturbative superpotentials, using Affleck-Dine-Seiberg techniques, for gauge groups with $N_f > N_c$ and for models with two gauge groups [24, 25]. These works are relevant for SUSY phenomenology.

Different aspects of SUSY models with flavour have also been studied by L. Díaz-Cruz and collaborators. A class of models with a $U(1)_H$ was studied in [26, 27] in a flavour-changing-neutral current scheme in which the squark mixings arise from the non-diagonal scalar trilinear interactions, which generate realistic quark-mass matrices and could provide a solution to the SUSY μ-problem. In [28] consequences of having a flavoured Higgs boson, in which the flavour structure encoded in SUSY extensions of the SM is transmitted to the Higgs sector are studied. In [29] the authors consider a SUSY theory flavour and CP conserving, where CP violating phases are associated to the vacuum expectation values of flavour violating SUSY-breaking fields. Using a U(2) flavour model as an example they show that it is possible to generate radiatively the first and second generation of quark masses and mixings as well as the CKM CP phase. Another approach to the SUSY flavour problem was to apply a statistical analysis to estimate the size of the SUSY flavour problem, and to determine its dependence on the parameters of the MSSM [30].

Jens Erler has been active in precision tests of the SM and beyond. In [31] he and Damien Pierce point out that regions of parameter space with large supersymmetric corrections from light superpartners are associated, in general, with poor fits to the data. They contrast results of a simple (oblique) approximation with full one-loop results, and show that for the most important observables the non-oblique corrections can be larger than the oblique corrections, and must be taken into account. They find out the regions of parameter space in both gravity- and gauge-mediated models which are excluded. In

[32] supersymmetric models with R-parity violation are studied, where scalar neutrinos may be produced as s-channel resonances in e^+e^- colliders. They notice that within current constraints, the scalar neutrino may have a width of several GeV into $b\bar{b}$ and be produced with large cross section, leading to a discovery signal at LEP II. If the scalar neutrino mass approximately equals m_Z the resonance improves the fit to electroweak data.

Recently, J. Erler and M. Mondragón have been invited to participate in the Supersymmetry Parameter Analyisis (SPA) project [33]. The SPA is a joint project between theoreticians and experimentalists working on SUSY, and its possible LHC and Linear Collider phenomenology. The main aims of the project are the high-precision determination of the supersymmetry Lagrange parameters at the electroweak scale and the extrapolation to a high scale to reconstruct the fundamental parameters and the mechanism for supersymmetry breaking.

Non-minimal models and/or extra symmetries

In the context of non minimal extensions of SUSY, either with extra particles or symmetries, there has been work done recently.

Enrique Barradas, Olga Félix and Alfonso Rosado studied the decay $H^\pm \to W^\pm h_0$, in an extension of the MSSM with an additional Higgs triplet and found that it becomes an important decay mode and, in some cases, the dominant one [34].

There has also been work on the combination of SUSY and discrete flavour symmetries. Namely, in Torsten Schmidt's Diplom thesis [35] a model is proposed in which a continuous SU(3) flavour symmetry is broken down to a discrete Q_6 symmetry. The breaking of the two symmetries at different scales to give the SM reproduces correctly qualitatively the known fermion mass spectrum, and gives predictions in the neutrino sector.

SUPERSYMMETRY AND GRAND UNIFICATION

Grand Unified Theories (GUTs) are the natural extension of the unification programme, i.e. electromagnetism, and then electroweak unification. GUTs provide a unified framework for all interactions of the SM (electroweak and strong), and could provide solutions to some of its problems, like quark fractional charges and some mass ratios. Some of them also predict extra particles and/or proton decay. There has been some work done on different aspects of Grand Unified Theories without supersymmetry in Mexico, although there has been considerable more work done on SUSY GUTs.

The first papers done by mexicans in the context of unified theories are probably the ones due to Arnulfo Zepeda and William Ponce [36], although this first work is on electroweak unification and not on GUTs. Another work of the same authors together with Albino Galeana and Roberto Martínez [37], explored unification on electroweak, strong, and horizontal interactions with a $SU(6)_L \times U(1)_Y$ group, as a way of explaining masses and hierarchies of the SM. Related aspects of this model can be found on [38].

In proper Grand Unified Theories, [39, 40] Abdel Pérez Lorenzana, W. Ponce and A. Zepeda explored the possibility of achieving one-step unification of the coupling constants of certain class of non supersymmetric and supersymmetric gauge models, which at low energies have only the standard particle content, using as constraints the experimental values of α_{em}, α_s and $\sin^2\theta_W$. The same authors considered a similar possibility in the context of left-right simmetric models in [41], and in a model with unification symmetry $(SU(6))^4 \times Z(4)$ [42].

Rebeca Juárez, Piotr Kielanowski, Gabriel López Castro and H. González assumed that Yukawa couplings of up and down quarks are related by $Y_u \propto Y_d^2$ at Grand Unification scales [43]. The ansatz gives rise to a symmetrical CKM matrix at the GUT scale. They illustrate the results with three specific examples and conclude that the small asymmetry of the CKM matrix at low energies may be the effect of the renormalization group evolution only.

SUSY GUTs

The combination of supersymmetry and Grand Unification is attractive from many points of view. Since the realization that gauge coupling unification is achieved in a better way in low-energy SUSY GUTs [44–47] there has been a wide study of these kind of theories. In Mexico, different aspects of SUSY GUTs have been explored. These range from Gauge-Yukawa and Finite models, left-right models, extra-dimensional models, to supergravity and cosmological models.

Gauge-Yukawa Unification and Finiteness

Myriam Mondragón and George Zoupanos, together with Jisuke Kubo and various other collaborators, have pursued a program for looking for a more fundamental theory, possibly at the Planck scale, whose basic ingredients are GUTs and supersymmetry, but its consequences go beyond the known ones [48–53]. In their studies they considered gauge-Yukawa unification (GYU), which is based on the principles of reduction of couplings and in addition finiteness. The method consists of looking for renormalization group invariant (RGI) relations holding below the Planck scale, which in turn are preserved down to the GUT scale. This programme, called Gauge–Yukawa unification scheme (GYU), applied in the dimensionless couplings of supersymmetric GUTs, such as gauge and Yukawa couplings, had already noticeable successes by predicting correctly, among others, the top quark mass in the finite and in the minimal N=1 supersymmetric SU(5) GUTs [48, 49]. An impressive aspect of the RGI relations is that one can guarantee their validity to all-orders in perturbation theory by studying the uniqueness of the resulting relations at one-loop, as was proven in the early days of the programme of reduction of couplings [54, 55]. Even more remarkable is the fact that it is possible to find RGI relations among couplings that guarantee finiteness to all-orders in perturbation theory [56–58]. Applying these principles, one can relate the gauge and Yukawa couplings, thereby improving the predictive power of a model [59, 60].

The above described method of reducing the dimensionless couplings has been extended [61, 62] to the soft supersymmetry breaking (SSB) dimensionful parameters of $N = 1$ supersymmetric theories. Based on the work of [63], interesting progress was made [64–67] concerning the renormalization properties of the SSB parameter. Applying the supergraph method to softly broken theories using "spurion" external space-time independent superfields [68], the relations among the soft term renormalization and that of an unbroken supersymmetric theory have been derived. In particular the β-functions of the parameters of the softly broken theory are expressed in terms of partial differential operators involving the dimensionless parameters of the unbroken theory. The key point in the strategy of refs. [63, 63, 69–75] in solving the set of coupled differential equations so as to be able to express all parameters in a RGI way, was to transform the partial differential operators involved to total derivative operators [69]. In addition it was found that RGI SSB scalar masses in Gauge-Yukawa unified models satisfy a universal sum rule at one-loop [73]. This result was generalized to two-loops for finite theories [75], and then to all-loops for general Gauge-Yukawa and Finite Unified Theories [70]. Thus, the full theories can be made all-loop finite and their predictive power is extended to the Higgs sector and the supersymmetric spectrum (s-spectrum). An exhaustive search of these latter predictions and their confrontation to low-energy phenomenology (like $b \to s\gamma$, g-2, $B_s \to \mu^+\mu^-$) and cosmology is underway. Some of the results concerning the Higgs sector and phenomenological issues are found in [76, 77].

There has also been work done on Finite Unified Theories based on cross group structure [78] $SU(N)_1 \times SU(N)_2 \times \ldots \times SU(N)_k$ with matter content $(N, N^*, 1, ..., 1) + (1, N, N^*, ..., 1) + \ldots + (N^*, 1, 1, ..., N)$. A necessary condition for these theories to be finite is that there should be exactly three families. Several such models are discussed and the $SU(3)^3$ model emerges as the one with the most promising phenomenology. For a recent review on more theoretical aspects of finiteness and reduction of couplings see [79].

Left-right models

In order to improve the fermion mass relations in SUSY GUTs Lorenzo Díaz-Cruz, Hitoshi Murayama and Aaron Pierce propose different scenarios in which they consider only loop effects, to show that Yukawa coupling unification can occur in the framework of minimal GUTs [26]. The first scenario is to consider A-terms that follow the usual proportionality condition, the second one is to use a new ansatz for the tri-linear A-terms that satisfies all experimental and vacuum stability bounds, and in the last one they study general (non-proportional) A-terms, with large off-diagonal entries.

In the context of left-right supersymmetric models, A. Pérez-Lorenza and W. Ponce [80] studied again the possibility of one step unification, in this case with $SU(3)_L \otimes SU(3)_{CL} \otimes SU(3)_{CR} \otimes SU(3)_R$ as gauge group, which unifies in one single step the three gauge couplings of the standard model at a scale $M \sim 10^{18}$ GeV, and spontaneously breaks down to $SU(3)_c \otimes U(1)_{EM}$ using only fundamental representations of SU(3). In this model the proton decay is highly suppressed and the doublet-triplet problem is lessened. The see-saw mechanism for the neutrinos is readily implemented with the use of an extra tiny mass sterile neutral particle for each generation, which provides a natural explanation to the neutrino puzzle.

Also within left-right SUSY models, but now with a single gauge group, Francisco Astorga studied the possibility of two-step unification in a SO(10) model [81]. The results are compared to those obtained for SUSY SU(5) and L-R SO(10), among which similar analyses have revealed significant differences in the structure of unification. Unlike L-R SO(10), it is found that the Higgs contribution is indispensible to attain unification. It also turns out that a two-intermediate-scale scenario allows more freedom in the actual values of the scales constrained by unification.

GUTs and extra dimensions

There has been a line of work in the possible connection of SUSY GUTs with a more fundamental theory, like superstrings and/or branes. Many of these models are based on field theories with extra dimensions, either using the fact that the running of the gauge couplings in models with extra dimensions obeys a power law rather than a logarithmic one [82, 83], gauge-Higgs unification, or models that come from branes.

In [84], Tatsuo Kobayashi, Jisuke Kubo, Myriam Mondragón and George Zoupanos studied the renormalization group properties of the parameters of the MSSM including the soft breaking sector, with a certain set of Kaluza-Klein fields. The matching conditions between the effective, renormalizable and original, unrenormalizable theories are obtained with the continuous Wilson renormalizaion group technique. They address several phenomenological issues, in particular, whether the assumption on a large compactification radius in the model is consistent with the gauge coupling unification, the $b - \tau$ unification and the radiative breaking of the electroweak gauge symmetry with the universal SSB terms. They conclude that there is a chance of discriminating these kind of models from conventional MSSM, even if the compactification radius is small, or equivalently, if the Kaluza-Klein towers are heavy. CATA

The effects of extra dimensions on gauge coupling unification on a large class of models were also studied in [85]. By identifying products of evolution coefficients that dictate the profile of unification in different models, A. Pérez-Lorenzana and R. Mohapatra studied under what conditions unification of couplings can occur in both one and two step unification models. In particular, they find that the Kaluza-Klein towers can generate intermediate scale models, like the left-right models, with the seesaw mechanism at a $M_{W_R} \sim 10^{13}$ GeV intermediate scale.

In [86] Abdel Pérez-Lorenzana proposes that the construction of consistent string-GUT models could require the use of non canonical affine levels and list the most common level values related to realistic GUTs.

In [87] Lorenzo Diaz-Cruz proposes that all fundamental interactions of the elementary particles should be of gauge type, including the Yukawa and Higgs ones, and studies this possibility in the MSSM and in the SUSY LR model, where both extra-dimensions are needed to to express all the Yukawa and Higgs parameters in terms of gauge couplings.

Cosmology

There has also been a considerable effort in the connection between SUSY and cosmology. The work on supergravity and quantum cosmology will be reviewed in another article in this same volume, where also quintessence and dark energy are discussed. In here I present only work in Cosmology more directly related to SUSY phenomenology.

SUSY Guts and cosmology

The unification scale in SUSY models [88, 89] is determined as a function of the gravitino mass, which sets the SUSY breaking scale. Axel de la Macorra and Graham Ross also show that the unification scale has minimum value of 10^{16} GeV which coincides with the unification scale required by gauge coupling unification. In another work, it is shown that it is possible to break SUSY dynamically and have a vanishing cosmological constant [89].

Inflation

The possibility of getting inflationary potentials in supersymmetry from models derived from superstring theory was studied by A. de la Macorra and S. Lola. The problem arises due to the non inflationary behavior of the dilaton field, which is always present in these models [90]. They show that using S-duality it is possible to fix the dilaton field and allow for other scalar field to drive inflation [91].

In the light of the new evidence for inflation with a running scalar index $n_s \neq 1$ and with un upper bound on the number of e-folds, supersymmetric models of inflation are proposed by Gabriel Germán, A. de la Macorra and M. Mondragón [92]. These models required no fine tuning on the initial conditions and give the desired e-folds and n_s.

In [93] Z. Berezhiani, A. Mazumdar and A. Pérez-Lorenzana show that a non-trivial evolution of the classic sneutrino fields after inflation offers an interesting mechanism for generating a correct amount of lepton asymmetry, which being reprocessed by sphalerons, can explain the observed baryon asymmetry of the Universe.

Dark matter

Applying the microcanonical definition of entropy to a weakly interacting and self-gravitating neutralino gas, Luis Cabral, Xavier Hernández and Roberto Sussman evaluate the change in the local entropy per particle of the gas between the freeze-out era and present day virialized halo structures [94]. By comparing theoretical and empirical estimates of this entropy they deduce an entropy consistency criterion. They apply this criterion together with the usual abundance one, to the cases when neutralinos are either mostly B-inos or mostly Higgsinos, requiring that present neutralino relic density complies with $0.2 < \Omega_{\chi_0^1} < 0.4$ for $h \simeq 0.65$. This methodology can be applied to test other annihilation channels of the neutralino, as well as other particle candidates for thermal weakly interacting massive particle gas relics.

Dario Núñez, Roberto A. Sussman, Jesús Zavala, Lukas Nellen, Luis G. Cabral-Rosetti, and Myriam Mondragón explore the use of these two criteria to constrain the allowed parameter space in MSUGRA models in [95], both based in the calculation of the present density of neutralinos as dark matter in the Universe. As mentioned above, the first one is the usual "abundance" criterion which is used to calculate the relic density after the "freeze-out" era. The second one applies the microcanonical definition of entropy to a weakly interacting and self-gravitating gas evaluating then the change in the entropy per particle of this gas between the "freeze-out" era and present day virialized structures. It is determined for which regions of the parameter space in the MSUGRA model are both criteria consistent with the 2σ bounds according to WMAP for the relic density.

Superstrings and compactifications

Work on superstring and branes will be reviewed in another paper of this same volume. Here again, I will just mention work more directly related to superstring phenomenology. A nice review on the subject is done by Fernando Quevedo in [96].

In [97] Alejandro Gutiérrez-Rodríguez, M. A. Hernández-Ruiz, B. Jayme-Valdes and Miguel Angel Pérez obtain upper bounds on the anomalous magnetic moment and electric dipole moment of the tau-neutrino through the reaction $e^+e^- \to \nu\bar{\nu}\gamma$ at the Z_1-pole in the framework of a class of E_6 inspired models with an additional neutral vector boson Z_θ.

There was also work on superstring phenomenology by A. de la Macorra and M. Mondragón prior to their moving to Mexico. The possibility of constructing phenomenologically viable vacua based on the Z_3 and Z_7 orbifolds is analyzed in [98, 99], by A. Casas, A. de la Macorra, M. Mondragón and C. Muñoz. The acceptable ones are classified and drastically reduced by the use of equivalence operations between them. Phenomenological issues related to each type of orbifold are also discussed. In [100] A. Font, L. Ibáñez, M. Mondragón and F. Quevedo study the construction of (0, 2) heterotic string compactifications built by tensoring of N=2 superconformal theories. Phenomenological implications of these new constructions are briefly discussed.

CONCLUSIONS

SUSY and GUTs are now better established subjects of research in Mexico. It is expected that with the increasing number of people working on these topics, the depth and variety of the research involved, as well as the advent of new experimental data, we will see even more work with more impact on these subjects in the future.

ACKNOWLEDGMENTS

This work was partially supported by grants from Concayt 40162-F and UNAM PAPIIT-IN116202.

REFERENCES

1. R. Bijker, O. Castaños, F. D., H. Morales, L. Urrutia, and C. Villarreal, editors, *Latin-American School of Physics: XXXV ELAF*, 2005.
2. S. P. Martin (1997), hep-ph/9709356.
3. H. E. Haber, and G. L. Kane, *Phys. Rept.* **117**, 75 (1985).
4. H. P. Nilles, *Phys. Rept.* **110**, 1 (1984).
5. J. Wess, and B. Zumino, *Phys. Lett.* **B49**, 52 (1974).
6. J. Wess, and B. Zumino, *Nucl. Phys.* **B70**, 39–50 (1974).
7. R. Haag, J. T. Lopuszanski, and M. Sohnius, *Nucl. Phys.* **B88**, 257 (1975).
8. P. Fayet, and S. Ferrara, *Phys. Rept.* **32**, 249–334 (1977).
9. J. R. Ellis, and D. V. Nanopoulos, *Phys. Lett.* **B110**, 44 (1982).
10. R. Barbieri, and R. Gatto, *Phys. Lett.* **B110**, 211 (1982).
11. B.-R. Zhou, and J. L. Lucio, *Phys. Lett.* **B132**, 83 (1983).
12. B.-R. Zhou, and J. L. Lucio, *Phys. Lett.* **B149**, 152–156 (1984).
13. E. Barradas, J. L. Diaz-Cruz, A. Gutierrez, and A. Rosado, *Phys. Rev.* **D53**, 1678–1683 (1996).
14. U. Cotti, A. Gutierrez-Rodriguez, A. Rosado, and O. A. Sampayo, *Phys. Rev.* **D59**, 095011 (1999), hep-ph/9902417.
15. U. Cotti, J. L. Diaz-Cruz, and J. J. Toscano, *Phys. Rev.* **D62**, 035009 (2000), hep-ph/9912406.
16. J. L. Diaz-Cruz, *Phys. Rev.* **D56**, 523–526 (1997), hep-ph/9705476.
17. J. L. Diaz-Cruz, H.-J. He, T. Tait, and C. P. Yuan, *Phys. Rev. Lett.* **80**, 4641–4644 (1998), hep-ph/9802294.
18. C. Balazs, J. L. Diaz-Cruz, H. J. He, T. Tait, and C. P. Yuan, *Phys. Rev.* **D59**, 055016 (1999), hep-ph/9807349.
19. J. L. Diaz-Cruz, D. K. Ghosh, and S. Moretti, *Phys. Rev.* **D68**, 014019 (2003), hep-ph/0303251.
20. A. de la Macorra, and G. G. Ross, *Nucl. Phys.* **B404**, 321–341 (1993), hep-ph/9210219.
21. A. de la Macorra, and G. G. Ross, *Nucl. Phys.* **B443**, 127–154 (1995).
22. A. de la Macorra, *Phys. Lett.* **B335**, 35–40 (1994), hep-ph/9401240.
23. M. Mondragón, and G. Ross, *Rev. Mex. de Física* **S. 38, Supl. 1**, 7–11 (1991).
24. C. P. Burgess, A. de la Macorra, I. Maksymyk, and F. Quevedo, *Phys. Lett.* **B410**, 181–187 (1997), hep-th/9707062.
25. C. P. Burgess, A. de la Macorra, I. Maksymyk, and F. Quevedo, *JHEP* **09**, 007 (1998), hep-th/9808087.
26. J. L. Diaz-Cruz, H. Murayama, and A. Pierce, *Phys. Rev.* **D65**, 075011 (2002), hep-ph/0012275.
27. J. L. Diaz-Cruz, H.-J. He, and C. P. Yuan, *Phys. Lett.* **B530**, 179–187 (2002), hep-ph/0103178.
28. J. L. Diaz-Cruz, *JHEP* **05**, 036 (2003), hep-ph/0207030.
29. J. L. Diaz-Cruz, and J. Ferrandis, *Phys. Rev.* **D72**, 035003 (2005), hep-ph/0504094.
30. J. L. Diaz-Cruz, M. Gomez-Bock, R. Noriega-Papaqui, and A. Rosado (2005), hep-ph/0512168.
31. J. Erler, and D. M. Pierce, *Nucl. Phys.* **B526**, 53–80 (1998), hep-ph/9801238.
32. J. Erler, J. L. Feng, and N. Polonsky, *Phys. Rev. Lett.* **78**, 3063–3066 (1997), hep-ph/9612397.
33. J. A. Aguilar-Saavedra, et al., *Eur. Phys. J.* **C46**, 43–60 (2006), hep-ph/0511344.
34. E. Barradas-Guevara, O. Felix-Beltran, and A. Rosado, *Phys. Rev.* **D71**, 073004 (2005), hep-ph/0408196.
35. T. Schmidt, *A SU(3) Extension of the MSSM with Q_6 Flavour Symmetry*, Master's thesis, University of Regensburg and UNAM, Advisors: Myriam Mondragón and Andreas Schäfer (2005).
36. W. A. Ponce, and A. Zepeda, *Z. Phys.* **C39**, 377 (1988).
37. A. H. Galeana, W. A. Ponce, A. Zepeda, and R. E. Martinez, *Phys. Rev.* **D44**, 2166–2178 (1991).

38. A. Hernandez-Galeana, W. A. Ponce, and A. Zepeda, *Z. Phys.* **C55**, 423–434 (1992).
39. A. Perez-Lorenzana, W. A. Ponce, and A. Zepeda, *Europhys. Lett.* **39**, 141–146 (1997), hep-ph/9708441.
40. A. Perez-Lorenzana, W. A. Ponce, and A. Zepeda, *Mod. Phys. Lett.* **A13**, 2153–2162 (1998), hep-ph/9808357.
41. A. Perez-Lorenzana, W. A. Ponce, and A. Zepeda, *Phys. Rev.* **D59**, 116004 (1999), hep-ph/9812401.
42. A. Perez-Lorenzana, W. A. Ponce, and A. Zepeda, *Rev. Mex. Fis.* **45**, 109 (1999), hep-ph/9901301.
43. H. Gonzalez, S. R. Juarez W., P. Kielanowski, and G. Lopez Castro, *Phys. Lett.* **B440**, 94–100 (1998), hep-ph/9703292.
44. U. Amaldi, W. de Boer, and H. Furstenau, *Phys. Lett.* **B260**, 447–455 (1991).
45. J. R. Ellis, S. Kelley, and D. V. Nanopoulos, *Phys. Lett.* **B249**, 441–448 (1990).
46. C. Giunti, C. W. Kim, and U. W. Lee, *Mod. Phys. Lett.* **A6**, 1745–1755 (1991).
47. P. Langacker, and M.-x. Luo, *Phys. Rev.* **D44**, 817–822 (1991).
48. D. Kapetanakis, M. Mondragón, and G. Zoupanos, *Z. Phys.* **C60**, 181–186 (1993), hep-ph/9210218.
49. J. Kubo, M. Mondragón, and G. Zoupanos, *Nucl. Phys.* **B424**, 291–307 (1994).
50. J. Kubo, M. Mondragón, N. D. Tracas, and G. Zoupanos, *Phys. Lett.* **B342**, 155–162 (1995), hep-th/9409003.
51. J. Kubo, M. Mondragón, S. Shoda, and G. Zoupanos, *Nucl. Phys.* **B469**, 3–20 (1996), hep-ph/9512258.
52. J. Kubo, M. Mondragón, M. Olechowski, and G. Zoupanos, *Nucl. Phys.* **B479**, 25–45 (1996), hep-ph/9512435.
53. J. Kubo, M. Mondragón, and G. Zoupanos, *Acta Phys. Polon.* **B27**, 3911–3944 (1997), hep-ph/9703289.
54. W. Zimmermann, *Commun. Math. Phys.* **97**, 211 (1985).
55. R. Oehme, and W. Zimmermann, *Commun. Math. Phys.* **97**, 569 (1985).
56. C. Lucchesi, O. Piguet, and K. Sibold, *Helv. Phys. Acta* **61**, 321 (1988).
57. A. V. Ermushev, D. I. Kazakov, and O. V. Tarasov, *Nucl. Phys.* **B281**, 72–84 (1987).
58. D. I. Kazakov, *Mod. Phys. Lett.* **A2**, 663–674 (1987).
59. T. Kobayashi, J. Kubo, M. Mondragón, and G. Zoupanos, *Acta Phys. Polon.* **B30**, 2013–2027 (1999).
60. M. Mondragón, and G. Zoupanos, *Concise Encyclopaedia of Supersymmetry*, Kluwer Academic Publishers, 2004, editors J. Bagger, S. Duplij and W. Siegel.
61. I. Jack, and D. R. T. Jones, *Phys. Lett.* **B349**, 294–299 (1995), hep-ph/9501395.
62. J. Kubo, M. Mondragón, and G. Zoupanos, *Phys. Lett.* **B389**, 523–532 (1996), hep-ph/9609218.
63. Y. Yamada, *Phys. Rev.* **D50**, 3537–3545 (1994), hep-ph/9401241.
64. R. Delbourgo, *Nuovo Cim.* **A25**, 646 (1975).
65. A. Salam, and J. A. Strathdee, *Nucl. Phys.* **B86**, 142–152 (1975).
66. K. Fujikawa, and W. Lang, *Nucl. Phys.* **B88**, 61 (1975).
67. M. T. Grisaru, W. Siegel, and M. Rocek, *Nucl. Phys.* **B159**, 429 (1979).
68. L. Girardello, and M. T. Grisaru, *Nucl. Phys.* **B194**, 65 (1982).
69. L. V. Avdeev, D. I. Kazakov, and I. N. Kondrashuk, *Nucl. Phys.* **B510**, 289–312 (1998), hep-ph/9709397.
70. T. Kobayashi, J. Kubo, and G. Zoupanos, *Phys. Lett.* **B427**, 291–299 (1998), hep-ph/9802267.
71. T. Kobayashi, J. Kubo, M. Mondragón, and G. Zoupanos, *Surveys High Energ. Phys.* **16**, 87–129 (2001).
72. J. Hisano, and M. A. Shifman, *Phys. Rev.* **D56**, 5475–5482 (1997), hep-ph/9705417.
73. Y. Kawamura, T. Kobayashi, and J. Kubo, *Phys. Lett.* **B405**, 64–70 (1997), hep-ph/9703320.
74. I. Jack, D. R. T. Jones, and A. Pickering, *Phys. Lett.* **B426**, 73–77 (1998), hep-ph/9712542.
75. T. Kobayashi, J. Kubo, M. Mondragón, and G. Zoupanos, *Nucl. Phys.* **B511**, 45–68 (1998), hep-ph/9707425.
76. A. Djouadi, S. Heinemeyer, M. Mondragón, and G. Zoupanos, *Springer Proc. Phys.* **98** (-284), hep-ph/0404208.

77. M. Mondragón, and G. Zoupanos, *Acta Phys. Polon.* **B34**, 5459–5468 (2003).
78. E. Ma, M. Mondragón, and G. Zoupanos, *JHEP* **12**, 026 (2004), hep-ph/0407236.
79. M. Mondragón, and G. Zoupanos, "Classical and Quantum Reduction of Couplings in Unified Theories," in *Quantum Theory and Symmetries IV*, edited by V. Dobrev, Heron Press, Sofia, 2006.
80. A. Perez-Lorenzana, and W. A. Ponce, *Phys. Lett.* **B464**, 77–81 (1999), hep-ph/9812402.
81. F. Astorga, *J. Phys.* **G25**, 981–988 (1999).
82. K. R. Dienes, E. Dudas, and T. Gherghetta, *Phys. Lett.* **B436**, 55–65 (1998), hep-ph/9803466.
83. K. R. Dienes, E. Dudas, and T. Gherghetta, *Nucl. Phys.* **B537**, 47–108 (1999), hep-ph/9806292.
84. T. Kobayashi, J. Kubo, M. Mondragón, and G. Zoupanos, *Nucl. Phys.* **B550**, 99–122 (1999), hep-ph/9812221.
85. A. Perez-Lorenzana, and R. N. Mohapatra, *Nucl. Phys.* **B559**, 255 (1999), hep-ph/9904504.
86. A. Perez-Lorenzana, and W. A. Ponce, *Europhys. Lett.* **49**, 296–301 (2000), hep-ph/9911540.
87. J. L. Diaz-Cruz, *Mod. Phys. Lett.* **A20**, 2397–2408 (2005), hep-ph/0409216.
88. A. de la Macorra, *Phys. Lett.* **B341**, 31–37 (1994), hep-ph/9401239.
89. A. de la Macorra, *Int. J. Mod. Phys.* **D5**, 567–578 (1996), hep-ph/9501250.
90. A. de la Macorra, and S. Lola, *Phys. Lett.* **B373**, 299–305 (1996), hep-ph/9511470.
91. A. de la Macorra, and S. Lola, *Int. J. Mod. Phys.* **D5**, 541–566 (1996), hep-ph/9411443.
92. G. German, A. de la Macorra, and M. Mondragón, *Phys. Lett.* **B494**, 311–317 (2000), hep-th/0006200.
93. Z. Berezhiani, A. Mazumdar, and A. Perez-Lorenzana, *Phys. Lett.* **B518**, 282–293 (2001), hep-ph/0107239.
94. L. G. Cabral-Rosetti, X. Hernandez, and R. A. Sussman, *Phys. Rev.* **D69**, 123006 (2004).
95. D. Nunez, et al. (2006), astro-ph/0604127.
96. F. Quevedo, *Nucl. Phys. Proc. Suppl.* **62**, 134–143 (1998), hep-ph/9707434.
97. A. Gutierrez-Rodriguez, M. A. Hernandez-Ruiz, B. Jayme-Valdes, and M. A. Perez (2006), hep-ph/0605277.
98. J. A. Casas, M. Mondragón, and C. Munoz, *Phys. Lett.* **B230**, 63 (1989).
99. J. A. Casas, A. de la Macorra, M. Mondragón, and C. Munoz, *Phys. Lett.* **B247**, 50–56 (1990).
100. A. Font, L. E. Ibanez, M. Mondragón, F. Quevedo, and G. G. Ross, *Phys. Lett.* **B227**, 34 (1989).

Applications of the renormalization group

P. Kielanowski[*] and S.R. Juárez W.[†]

[*]*Departamento de Física, Centro de Investigación y Estudios Avanzados del IPN, Mexico*
[†]*Departamento de Física, Escuela Superior de Física y Matemáticas, IPN, Mexico*

Abstract. We review some applications of the renormalization group techniques in the standard model of elementary particles. We first introduce the systematic method of the perturbative treatment of the renormalization group that is based on the parameter λ of the CKM matrix. Next, at the one loop order, we find the analytic solutions of the renormalization group equations and obtain the evolution of the gauge couplings, quark masses and the CKM matrix. We show that the unitarity triangle is invariant under such an evolution. We also consider the two loop approximation where we find the corrections to the one loop running and derive the bounds for the physical Higgs mass $146.8 \leq M_H \leq 193.8$.

Keywords: standard model, renormalization group, Higgs mass, grand unification
PACS: 11.10.Hi, 12.10.Dm, 14.80.Bn

INTRODUCTION

The Standard Model (SM) in particle physics is a quantum field theory of leptons, quarks and intermediate vector bosons determined from the gauge group $SU(3)_c \otimes SU(2)_y \otimes U(1)$. The SM has 18 free parameters which have to be determined from the measured properties of the elementary particles. The parameters of the SM fall into the following categories:

1. gauge couplings constants g, g', g_s (or g_1, g_2 and g_3)
2. quark and lepton masses (9 masses)
3. Cabibbo-Kobayashi-Maskawa (CKM) matrix (4 parameters)
4. mass of the Higgs boson and the Higgs boson quartic coupling.

The range of values of some of these parameters is very widespread, what is commonly described as hierarchy. The values of the three gauge coupling constants are rather close and this was the first hint for the idea of grand unification. The lightest quarks m_u and m_d have masses smaller than 8 MeV while the heaviest has mass ~ 172 GeV, so the ratio of light and heavy quark masses is $\sim 5 \cdot 10^{-5}$. The elements of the CKM matrix also vary significantly: the diagonal elements are $V_{ii} \sim 1$, $|V_{12}| \sim 0.22$, $|V_{23}| \sim 4 \cdot 10^{-2}$ and $|V_{13}| \sim 3 \cdot 10^{-3}$. Additionally the CKM matrix is almost symmetric. The CP violating phase is $\delta_{13} \sim 60^0$.

The hierarchy, present for the values of these parameters of the SM, may be an important signal of the underlying physics and may lead to new ideas which can provide a better understanding of the fundamental properties of matter and may lead also to another formulation of the SM with a smaller number of free parameters. The reduction of the number of free parameters is only possible if new symmetries are discovered.

An important property of the gauge coupling constants, quark masses and the CKM matrix elements is that their values do depend on the scale of energy at which they are measured. This fact follows from the necessity of renormalization of the gauge theory of the SM and that the values of the SM parameters at different energies are related through the Renormalization Group Equations (RGE) [1].

The RGE for the parameters of the SM are written in the framework of perturbation theory where the ordering parameter is the number of loops [2]. The RGE for the SM are known exactly up to two loops. Additional parameters for the perturbative expansion are related to the observed hierarchy in the SM. The RGE in the SM form a set of coupled non-linear equations which are difficult or impossible to solve analytically. To simplify the perturbative analysis and to make all the calculations consistent we assume that the RGE equations can be written in the perturbative scheme in terms of powers of $\lambda = |V_{12}| \approx 0.22$, V_{12} is an element of the CKM matrix. The relevant hierarchy parameters can be expressed in the following way in terms of powers of λ: loop order parameter $1/(4\pi)^2 \sim 2.7 \cdot \lambda^4$, the ratio of the bottom to the top quark masses $(m_b/m_t)^2 \sim 1.32 \cdot \lambda^5$, the ratio of the electron to the top quark masses $(m_e/m_t)^2 \sim 1.32 \cdot \lambda^{17}$. From these relations we draw the following conclusions:

1. In the one loop approximation we can omit all the terms of order λ^4 and higher
2. The lepton degrees of freedom can be neglected relatively to the hadron degrees of freedom.

Based on these principles we first consider the RGE up to the order λ^3 (one loop). These equations can be solved analytically and we obtain some general properties for the running of the SM parameters within this approximation. We then derive the specific predictions for the Renormalization Group (RG) evolution of the SM "observables", quark masses, CKM matrix and the Vacuum Expectation Value (VEV) of the Higgs field.

Afterwards we consider the RGE in the next order [3]. The most important effect in this approximation is the appearance of the Higgs quartic coupling λ_H in the RGE for the observables. We thus also consider the running of λ_H from which one can derive the limits on the Higgs boson mass [4].

GENERAL PROPERTIES OF THE EVOLUTION

Let us first consider the one loop RGE in the SM:

$$\frac{dg_l(t)}{dt} = \frac{1}{16\pi^2} b_l g_l^3(t), \tag{1a}$$

$$\frac{d}{dt} y_{u,d}(t) = \frac{1}{(4\pi)^2} \beta_{u,d}^{(1)}(t) y_{u,d}(t), \tag{1b}$$

the g_l, $l = 1, 2, 3$ are the gauge coupling constants, $t = \ln(E/m_t)$, $b_l = (\frac{41}{10}, \frac{-19}{6}, -7)$ are constants and $\beta_{u,d}^{(1)}(t)$ are the one loop β functions for the up and down quarks Yukawa couplings y_u and y_d [5].

$$\beta_{u,d}^{(1)}(t) = \alpha_1^{u,d}(t) + \alpha_2^{u,d} H_u + \alpha_3^{u,d} \text{Tr}(H_u), \quad H_u = y_u y_u^\dagger \tag{2}$$

where $\alpha_1^u(t) = -\left(\frac{17}{20}g_1^2 + \frac{9}{4}g_2^2 + 8g_3^2\right)$, $\alpha_1^d(t) = -\left(\frac{1}{4}g_1^2 + \frac{9}{4}g_2^2 + 8g_3^2\right)$, $\alpha_2^u = \frac{3}{2}$, $\alpha_2^d(t) = -\frac{3}{2}$, $\alpha_3^u = \alpha_3^d = 3$.

The solutions of Eqs. (1a) are

$$g_l^2(t) = \frac{g_l^2(t_0)}{1 - \frac{2b_l}{(4\pi)^2} g_l^2(t_0)(t-t_0)}. \tag{3}$$

and from the values of b_l one can see that $g_1(t)$ increases with energy, while $g_2(t)$ and $g_3(t)$ are decreasing.

Eqs. (1b) are non-linear, coupled differential equations for the matrices of the Yukawa couplings y_u and y_d. The equations for y_u decouple from the equations for y_d so they can be solved independently. If one represents the y_u matrix using the bi-unitary representation

$$y_u \sim (U_u)_L^\dagger \text{ Diag}(Y_u, Y_c, Y_t)\, (U_u)_R \tag{4}$$

then one can show [5] that the matrices $(U_u)_{L,R}^\dagger$ do not depend on t and Eqs. (1b) for the eigenvalues of y_u can be solved explicitly.

Using the ansatz (4) one can also simplify and solve explicitly Eqs. (1b) for y_d. From the solutions one obtains the following theorems:

Theorem 1. *The one loop renormalization group evolution of the CKM matrix depends on energy only through one function $h(t)$ where*

$$h(t) = \left(\frac{1}{1 - \frac{9}{(4\pi)^2} Y_t^2(t_0) \int_{t_0}^t r(\tau) d\tau} \right)^{-\frac{1}{4}}, \tag{5}$$

$$r(t) = \exp\left[\frac{2}{(4\pi)^2} \int_{t_0}^t \alpha_1^u(\tau) d\tau\right] = \Pi_{k=1}^{k=3} \left[\frac{g_k^2(t_0)}{g_k^2(t)}\right]^{\frac{c_k}{b_k}}, \tag{6}$$

$c_i = (\frac{17}{20}, \frac{9}{4}, 8)$.

Theorem 2. *The ratio of the down quark masses m_d/m_s and m_s/m_b are functions of only $h(t)$.*

Theorem 3. *The ratio of the quark masses m_u/m_c is energy independent.*

One can also obtain the evolution of the elements of the CKM matrix, which according to Theorem 1, depend only on the function $h(t)$.

The main conclusions of this section are that the evolution of the CKM matrix depends on one function of energy $h(t)$ and that the evolution of the quark masses is also correlated. The form of the function $h(t)$ for the standard model is shown in Fig. 1.

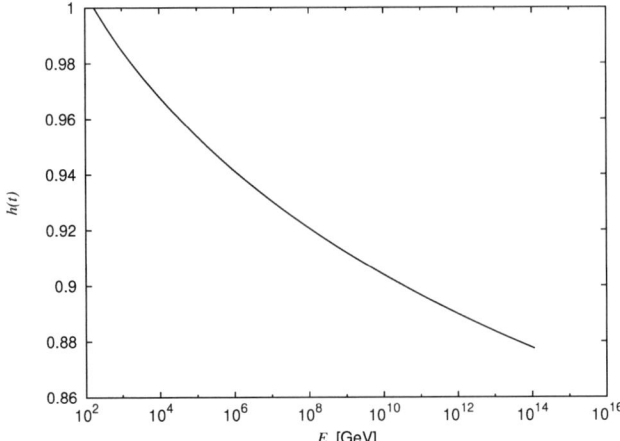

FIGURE 1. The energy dependence of the function $h(t)/h(t_0)$

From this figure one can see that $h(t)/h(m_t)$ decreases 13% between the top mass and the grand unification energy. One can expect the same order of variation for the elements of the CKM matrix.

EVOLUTION OF THE OBSERVABLES

As discussed earlier, a study of the evolution of the observables of the SM can give a deeper insight into the possible symmetries at different values of energy. We consider the RGE, neglecting terms of order λ^4 ($\lambda = |V_{us}|$) and higher. As discussed in the previous section, these equations can be solved explicitly and the evolution of the observables can be expressed in terms of known functions.

Running of the quark masses

Let us start with the evolution of the quark masses. The quark masses are expressed as the product of the VEV of the Higgs field and the eigenvalues of the quark Yukawa couplings. The equation for the evolution of the VEV is

$$\frac{dv}{dt} = \frac{1}{(4\pi)^2}\left[\alpha_1^v(t) + \alpha_3^v \text{Tr}\left(y_u y_u^\dagger\right)\right]v \qquad (7)$$

where $\alpha_1^v(t) = c_1'' g_1^2 + c_2'' g_2^2$, $\alpha_3^v(t) = -3$, $\left(c_1'', c_2''\right) = \left(-\frac{9}{20}, -\frac{9}{4}\right)$ and its solution reads

$$v(t) = v(t_0)\sqrt{r_v''(t)}\, h_m(t)^{\alpha_3^v} \qquad (8)$$

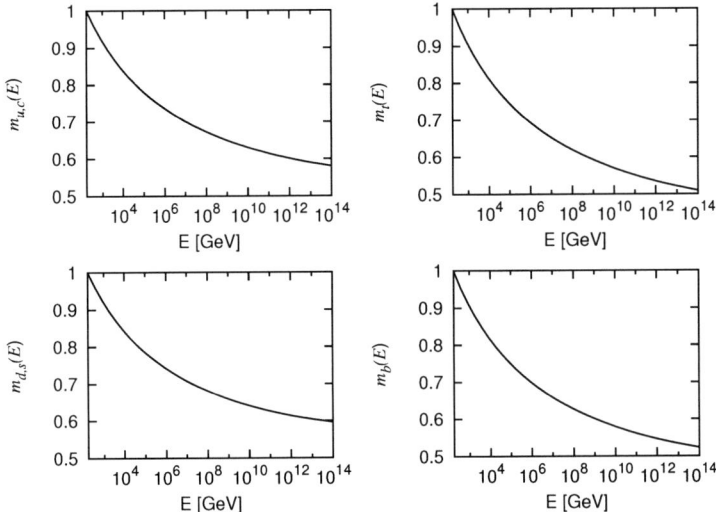

FIGURE 2. The evolution of the quark masses

where the functions $r_v''(t)$ and $h_m(t)$ are equal to

$$r_v''(t) = \exp\left[\frac{2}{(4\pi)^2}\int_{t_0}^{t}\alpha_1^v(\tau)d\tau\right],$$

$$h_m(t) = \left(\frac{1}{1-\frac{2(\alpha_2^u+\alpha_3^u)}{(4\pi)^2})\int_{t_0}^{t}r(\tau)d\tau}\right)^{\frac{1}{2(\alpha_2^u+\alpha_3^u)}}. \quad (9)$$

The eigenvalues of the quark Yukawa couplings are obtained from Eqs. (1b) and their running is the following:

$$Y_t(t) = Y_t(t_0)\sqrt{r(t)}h_m(t)^{(\alpha_2^u+\alpha_3^u)} \quad \text{for the } t \text{ quark}, \quad (10)$$

$$Y_{u,c}(t) = Y_{u,c}(t_0)\sqrt{r(t)}h_m(t)^{\alpha_3^u} \quad \text{for the } u,c \text{ quarks}, \quad (11)$$

$$Y_b(t) = Y_b(t_0)\sqrt{r'(t)}h_m(t)^{(\alpha_2^d+\alpha_3^d)} \quad \text{for the } b \text{ quark}, \quad (12)$$

$$Y_{d,s}(t) = Y_{d,s}(t_0)\sqrt{r'(t)}h_m(t)^{\alpha_3^d} \quad \text{for the } d,s \text{ quarks}, \quad (13)$$

$$r'(t) = \exp\left(\frac{2}{(4\pi)^2}\int_{t_0}^{t}\alpha_1^d(\tau)d\tau\right) = \Pi_{k=1}^{k=3}\left[\frac{g_k^2(t_0)}{g_k^2(t)}\right]^{\frac{c_k^d}{b_k}}. \quad (14)$$

The quark masses are obtained from the formulas $m_q = \frac{1}{\sqrt{2}}v(t)Y_q(t)$ so from Eqs. (8) and (10)-(13) one can obtain the evolution of the quark masses. The running of the ratio

$m_q(t)/m_q(t_0)$ is shown in Fig. 2 from which one can see that the values of the running quark masses decrease with energy, *e.g.*, the mass of the bottom quark is reduced by 50% at the scale of the grand unification. From these figures one can also see that the renormalization group running of the quark masses preserves the hierarchy, *i.e.* the quark mass hierarchy is also present at the scale of grand unification and any model of grand unification must include the quark mass hierarchy from the very beginning and the quark mass hierarchy cannot be explained as a dynamical effect.

Running of the CKM matrix

The CKM matrix in the SM is defined as

$$\widehat{V} = U_L^u \left(U_L^d\right)^\dagger \tag{15}$$

where $\widehat{V}$ is the CKM matrix and $U_L^{u,d}$ are the left diagonalizing matrices of the up and the down quarks Yukawa couplings.

To find the evolution of the CKM matrix we use the property that U_L^u does not depend on the energy and the running of U_L^d follows from the equation for y_d. After some calculations we obtain the following equations for the squares of the absolute elements of the CKM matrix [6]

$$\frac{1}{|V_{cd}(t)|^2} \frac{d}{dt} |V_{cd}(t)|^2 = \frac{2}{(4\pi)^2} \alpha_2^d Y_t^2 |V_{td}(t)|^2, \tag{16a}$$

$$\frac{1}{|V_{td}(t)|^2} \frac{d}{dt} |V_{td}(t)|^2 = -\frac{2}{(4\pi)^2} \alpha_2^d Y_t^2 (1 - |V_{td}(t)|^2), \tag{16b}$$

$$\frac{1}{|V_{ub}(t)|^2} \frac{d}{dt} |V_{ub}(t)|^2 = -\frac{2}{(4\pi)^2} \alpha_2^d Y_t^2 |V_{tb}(t)|^2, \tag{16c}$$

$$\frac{1}{|V_{cb}(t)|^2} \frac{d}{dt} |V_{cb}(t)|^2 = -\frac{2}{(4\pi)^2} \alpha_2^d Y_t^2 |V_{tb}(t)|^2. \tag{16d}$$

The solutions of Eqs. (16) read

$$|V_{td}(t)|^2 = \frac{|V_{td}^0|^2}{h^2(t) + (1 - h^2(t))|V_{td}^0|^2}, \quad |V_{cd}(t)|^2 = \frac{h^2(t)|V_{cd}^0|^2}{h^2(t) + (1 - h^2(t))|V_{td}^0|^2},$$

$$|V_{tb}(t)|^2 = \frac{h^2(t)|V_{tb}^0|^2}{1 + (h^2(t) - 1)|V_{tb}^0|^2}, \quad |V_{ub}|^2 = \frac{|V_{ub}^0|^2}{1 + (h^2 - 1)|V_{tb}^0|^2},$$

$$|V_{cb}|^2 = \frac{|V_{cb}^0|^2}{1 + |V_{tb}^0|^2(h^2 - 1)}.$$

All the remaining matrix elements can be obtained from the unitarity of the CKM matrix. From these equations one can obtain the evolution of the other important observables of the CKM matrix: Jarlskog invariant J, the Wolfenstein parameter λ and the angles of the unitarity triangle [7]–[9].

The equation for the parameter J is the following

$$\frac{d}{dt}\ln J = -\frac{1}{h^2}\frac{dh^2}{dt}\cdot|V_{tb}|^2 \tag{17}$$

and the solution

$$J(t) = \frac{J_0}{1+(h^2-1)|V_{tb}^0|^2} \approx \frac{J_0}{h^2(t)} \tag{18}$$

and $J(t)$ increases with energy.

The running of the Wolfenstein parameters is the following

$$A(t) = \frac{A}{h(t)} \text{ and } \lambda, \rho, \eta \text{ are invariant.} \tag{19}$$

From Eq. (19) one can deduce that the unitarity triangle of the CKM matrix is invariant under the renormalization group evolution.

There are two general conclusions that follow from this section. First is that the strength of the CP violation increases with energy and the Jarlskog invariant at the grand unification scale is larger by 60% than at low energy. The second is the restriction following from the unitarity triangle which requires that any grand unification model has to include such a structure of the quark Yukawa couplings that they reproduce the low energy unitarity triangle also at the grand unification energy.

TWO LOOP RGE IN THE SM AND THE HIGGS MASS

RGE in the SM are exactly known up to two loops and they enable us to study the corrections to the one loop evolution [4]. The most important correction that appears in the two loop RGE is that the equations for the observables become dependent on the Higgs quartic coupling λ_H which is defined in the Higgs potential

$$V_H = \frac{1}{2}m_H^2\phi^2 + \frac{1}{8}\lambda_H\phi^4. \tag{20}$$

The one loop equation for λ_H is the following

$$\frac{\lambda_H}{dt} = \frac{12}{4\pi^2}\{\lambda_H^2 + [\text{Tr}(H_u) - \frac{3}{4}(\frac{1}{5}g_1^2 + g_2^2)]\lambda_H$$
$$+ \frac{3}{16}(\frac{3}{25}g_1^4 + \frac{2}{5}g_1^2g_2^2 + g_2^4) - (\text{Tr}(H_u))^2)\} \tag{21}$$

Eq. (21) is non linear and is of the Riccati type. The solution of this equation can be expressed by the solutions of the second order linear equation for the auxiliary function $W(t)$ which is related to $\lambda_H(t)$ in the following way

$$\lambda_H(t) = -\frac{(4\pi)^2}{12}\frac{W'(t)}{W(t)}. \tag{22}$$

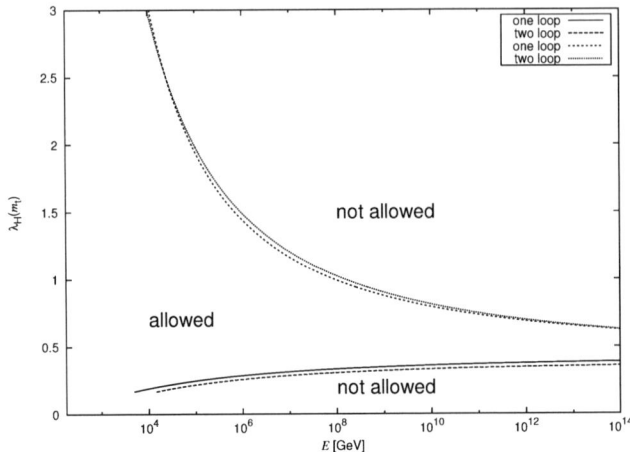

FIGURE 3. Allowed and forbidden regions in the plane $(\lambda_H(t_0), E)$

and one can find two solutions $W_1(t)$ and $W_2(t)$ with the properties

$$W_1(t_0) = 1, \quad W_1'(t_0) = 0, \\ W_2(t_0) = 0, \quad W_2'(t_0) = 1. \tag{23}$$

The solutions $W_1(t)$ and $W_2(t)$ are regular functions of t in the physical range of values of t.

With the help of the functions $W_1(t)$ and $W_2(t)$ the solution of Eq. (21) can be written as the following ratio

$$\lambda_H(t) = -\frac{(4\pi)^2}{12} \frac{W_1'(t) - \frac{12}{(4\pi)^2}\lambda_H(t_0)W_2'(t)}{W_1(t) - \frac{12}{(4\pi)^2}\lambda_H(t_0)W_2(t)}. \tag{24}$$

From Eq. (20) and the consistency of the SM one obtains that λ_H must be non-negative (stability condition) and cannot be singular (triviality condition). From Eq. (24) it turns out that for certain values of $\lambda_H(t_0)$ and t, λ_H can be negative or become singular (pole). Such values are forbidden from the physical point of view and it means that analyzing Eq. (24) we can find the regions of $\lambda_H(t_0)$ and t where the SM cannot be valid. If we impose the condition that the SM should be valid up to the energy of grand unification then we obtain the following range of the allowed values of $\lambda_H(t_0)$

$$0.389 \leq \lambda_H(t_0) \leq 0.622 \quad \text{for one loop RGE}, \tag{25}$$
$$0.360 \leq \lambda_H(t_0) \leq 0.626 \quad \text{for two loops RGE}. \tag{26}$$

The division of the $\lambda_H(t_0)$ and t plane into the allowed and forbidden regions is shown in Fig. 3.

The Higgs quartic coupling $\lambda_H(t)$ is related to the Higgs boson mass. In the one loop case this relation reads

$$m_H = v(t)\sqrt{\lambda_H(t)}. \tag{27}$$

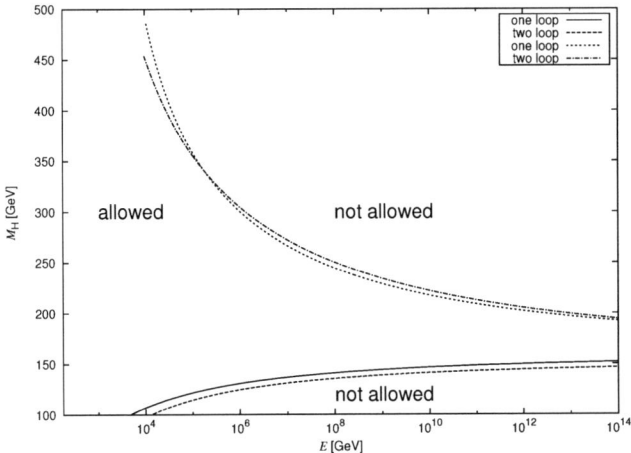

FIGURE 4. Allowed and forbidden regions in the plane (M_H, E)

In the two loop case this relation becomes more complicated and additionally we must include the Higgs field self-energy correction. Taking all these corrections into account the limits in Eqs. (25) and (26) can be expressed as bounds for the physical Higgs boson mass M_H

$$152.0 \leq M_H \leq 191.8 \quad \text{for one loop,} \tag{28}$$
$$146.8 \leq M_H \leq 193.8 \quad \text{for two loops.} \tag{29}$$

The physical meaning of Eqs. (28) and (29) is that if the Higgs boson is discovered and its mass is within the range given in these equations then the SM may be valid up to the energy of grand unification. If the Higgs mass is outside of this range then the SM breaks down above a certain value of the energy which can be estimated from Fig. 4.

CONCLUSIONS

In this paper we have presented some applications of the renormalization group technique in elementary particle physics applied to the study of the properties of the SM. The main conclusions of this study are the following

1. In the approximation of one loop we have found explicit analytic solutions of the RGE for the SM for all the observables.
2. There are many correlations in the renormalization group running of the observables of the SM, *e.g.*, in the approximation of one loop the evolution of the CKM matrix is described by only one function of energy. Such correlations can be only obtained if we know explicit solutions of the RGE.
3. We have shown that the unitarity triangle for the CKM matrix is invariant under the renormalization group running in the approximation of one loop. This is a severe

condition on the grand unified model which have to reproduce the unitary triangle at the grand unification scale.
4. The hierarchy among the parameters of the SM is also present at the energy scale of grand unification so the hierarchy must be included in any grand unified model from very beginning and cannot be dynamically generated by the group renormalization flow.
5. We have obtained numerical solutions of the two loop RGE of the SM and we have compared the one and two loop running of the observables.
6. We have shown that the one loop RGE for the quartic Higgs coupling $\lambda_H(t)$ is of the Riccati type. This makes it possible to solve this equation explicitly and to study the zeros and singularities (poles) of $\lambda_H(t)$ as the function of t. We find the position of the Landau pole and of zero of $\lambda_H(t)$ as a function of t. this information with the triviality and stability conditions lead to the bounds of the permitted values of the $\lambda_H(t_0)$ and of the physical Higgs mass M_H.

Let us summarize by saying thet the method of the renormalization group gives a very rich spectrum of physical predictions for the behavior of the SM in different ranges of energies.

ACKNOWLEDGMENTS

S.R.J.W. thanks the "Comisión de Operación y Fomento de Actividades Académicas" (COFAA) from Instituto Politécnico Nacional.

REFERENCES

1. E.C.G. Stuckelberg and A,. Peterman, Helv. Phys. Acta **26** (1953) 499. M. Gell-Mann and F. E. Low Phys. Rev. **95** (1954) 1300.
2. M.E. Machacek and M.T. Vaughn, Nucl. Phys. B **222**, 83 (1983); **236**, 221 (1984); **249**, 70 (1985); K. Sasaki, Z. Phys. C **32**, 149 (1986); K.S. Babu, Z. Phys. C **35**, 69 (1987); B. Grzadkowski and M. Lindner, Phys. Lett. B **193**, 71 (1987); B. Grzadkowski, M. Lindner and S. Theisen, Phys. Lett. B **198**, 64 (1987); M. Olechowski and S. Pokorski, Phys. Lett. B **257**, 388 (1991); H. Arason, D.J. Castano, E.J. Piard and P. Ramond, Phys. Rev. D **47**, 232 (1993); V. Barger, M.S. Berger and P. Ohmann, Phys. Rev. D **47**, 2038 (1993); D.J. Castaño, E.J. Piard and P. Ramond, Phys. Rev. D **49**, 4882 (1994). P. Binetruy and P. Ramond, Phys. Lett. B **350**, 49 (1995); R.D. Peccei and K. Wang, Phys. Rev. D **53**, 2712 (1996); K. Wang, Phys. Rev. D **54**, 5750 (1996); H. González *et al.*, Phys. Lett. B **440**, 94 (1998); H. González, S.R. Juárez W., P. Kielanowski, G. López C., AIP CP 400, 245 (1997); H. Fusaoka and Y. Koide, Phys. Rev. D **57**, 3986 (1998); H. González et al., *A New symmetry of Quark Yukawa Couplings*, page 755, International Europhysics Conference on High Energy Physics, Jerusalem 1997, eds. Daniel Lellouch, Giora Mikenberg, Eliezer Rabinovici, Springer-Verlag 1998; G. Cvetic, C.S. Kim and S.S. Hwang, Int. J. Mod. Phys. A **14**, 769 (1999); C. Balzereit, Th. Hansmann, T. Mannel and B. Plümper, Eur. Phys. J. C **9** 197 (1999); S.R. Juárez W., S.F. Herrera, P. Kielanowski and G. Mora, *Energy dependence of the quark masses and mixings* in Particles and Fields, Ninth Mexican School, Metepec, Puebla, México, AIP Conference Proceedings 562 (2001) 303-308, ISBN 1-56396-998-X, ISSN 0094-243X; S.R. Juárez W., A. Morales S. and P. Kielanowski, Revista Mexicana de Física **50**, 401 (2004).
3. Mingxing Luo and Yong Xiao, Phys. Rev. Lett. **90**, 011601 (2003). Mingxing Luo, Huawen Wang and Yong Xiao, *Two Loop Renormalization Group Equations in General Gauge Field Theories*, Phys. Rev.D **67** , 065019 (2003)

4. P. Kielanowski, S.R. Juárez W. and H.G. Solís-Rodríguez, Phys. Rev. D **72**, 96003 (2005); P. Kielanowski, S. R. Juárez W., AIP CP **670**, 81 (2003); S.R. Juárez W., H. G. Solís Rodríguez, P. Kielanowski, AIP CP **809**, 196 (2006); H. G. Solís-Rodríguez, S.R. Juárez W. and P. Kielanowski, X Mexican Workshop on Particles and Fields to be published, AIP CP (2006).
5. P. Kielanowski, S.R. Juárez W. and G. Mora, Phys. Lett. B (479) 1-3, 181-189 (2000).
6. S.R. Juárez W., S.F. Herrera H., P. Kielanowski and G. Mora, Phys. Rev. D **66**, 116007 (2002).
7. C. Jarlskog, Phys. Rev. Lett. **55**, 1039 (1985); Z. Phys. C **29**, 491 (1985).
8. L. Wolfenstein, Phys. Rev. Lett. **51**, 1945 (1983). A. J. Buras, M. E. Lautenbacher and G. Ostermaier, Phys. Rev. D 50, 3433 (1994).
9. C. Jarlskog and R. Stora, Phys. Rev. Lett.B 208, 268 (1988), L. L Chau and W.Y. Kenng, Phys. Rev. Lett. 53, 1802 (1984).

Deep Inelastic Scattering and Collider Physics

A. Rosado

Instituto de Física, BUAP. Apdo. Postal J-48, C.P. 72570 Puebla, Pue., México

Abstract. In this work we review the contribution made by scientists working at Mexican Institutions to the topics of Deep Inelastic Scattering and Collider Physics. In particular, we review the heavy vector boson production through e^-p and $\nu_l \mathcal{N}$ deep inelastic collisions. Our review is restricted to theoretical contributions.

Keywords: Deep Inelastic Scattering, Collider Physics, Heavy Boson Production
PACS: 13.60.Hb, 13.60.-r, 13.15.+g

INTRODUCTION

Our aim in this paper is to review the contribution made by scientists working at Mexican Institutions to the topics of Deep Inelastic Scattering (DIS) and Collider Physics [1]. We review in more detail the heavy vector boson production through e^-p and $\nu_l \mathcal{N}$ deep inelastic collisions. Our review takes into account only theoretical contributions, the experimental contributions to DIS and Collider Physics will be reviewed by other authors in these Proceedings.

This paper is organized as follows. In section II, we survey the contribution done on DIS by researchers working at Mexican Institutions. In section III, we present the contributions on Collider Physics made by researchers working at Mexican Institutions. In section IV, we review the production of a massive vector bosons in deep inelastic lepton-nucleon scattering and we give a summary of results and conclusions for the total cross section and differential cross section as a function of the dimensionless variables x and y for all the different reactions and reaction mechanisms which contribute to charged vector boson and neutral vector boson production in e^-p-collisions. This section also contains the results for the energy spectrum and the angular distribution of the produced Z^0. Finally, in section V, we review the production of a massive vector bosons in deep inelastic neutrino-nucleon scattering and summarize our results and conclusions.

DEEP INELASTIC SCATTERING

At low energy the electrons are deflected slightly and the protons recoiled. As the energy of the electrons was increased a threshold energy was reached beyond which the electrons were deflected through much greater angles. At the same time the protons started to fragment into a shower of particles rather than being deflected themselves. It has been started to see evidence for charged quarks inside the proton. The first real evidence came as a result of the analysis of data taken by an experiment at the Stanford Linear Accelerator Centre (SLAC) in the late 1960s. The leaders of this experiment, Friedman, Taylor and Kendal, were awarded the 1990 Nobel Prize in Physics.

It has long been realized that the nucleons are not point particles but are of finite size $\sim 10^{-13}$ cm while the evidence from symmetries has suggested that they are built from quarks. Probing the nucleons dynamically by means of DIS of leptons (electrons, muons and neutrinos) has led to an understanding of nucleon structure, to the confirmation of the quarks as the elements of the nucleon substructure and to measurement of the quark properties. The leptons, having no strong interaction, probe the nucleon structure by means of the electromagnetic and weak interactions.

The results of the SLAC experiments were a great puzzle at that time, the quark theory was not well known. The full theory that explained the results was developed by Feynman and Bjorken in 1968. Feynman named the small charged objects inside the protons "partons". However, it soon became clear that Feynman's partons were the same as Gell-Mann's quarks. Some people maintain a distinction by using parton to refer to both the quarks and the gluons inside protons and neutrons.

When the experimental results showed that the electrons are being deflected by large angles and the proton is fragmenting into a shower of hadrons, then we can deduce that the electron is interacting directly with the quarks within the proton. At the end, it was possible to establish that there are three quarks inside the proton and that their charges correspond to those suggested by Gell-Mann and Zweig. Deep Inelastic Scattering forms the cornerstone of our physical evidence for the existence of quarks.

C. Domínguez, J. Pestieau, H. Moreno and A. Zepeda at the CINVESTAV-IPN, and G. Cocho at the National University (UNAM) were pioneers in doing research in DIS at Mexican Institutions. These researchers and their collaborators contributed to the understanding of some aspects of the physics involved in the DIS. They have done research in subjects related to DIS such as Hadron Form Factors, Scaling, The Parton Model, Radiative Correction and Sum Rules. More recently, a new generation of researchers is working in the search for new physics in neutrino nucleon dispersion. To end this section, we will list in chronological order the works made in Mexican Institutions in this area.

- "Asymptotics In Inelastic Electroproduction And Photoabsorption," by J. Pestieau (1971) [2].
- "Hadron Form Factors, Deep-Inelastic Electroproduction, and Massive-Lepton Pair Production: Kinematics or Dynamics" by G. Cocho and J.J. Salazar (1971) [3].
- "Elastic And Inelastic Form-Factors Of The Nucleons And Deep-Inelastic Electroproduction," by H. Moreno, J. Pestieau (1972) [4].
- "Comment On The Possible Relation Between Prominent Resonances And Limiting Scaling Data In Inelastic Electroproduction On Nucleons," by J. Pestieau (1972) [5].
- "Comparison Of The Parton And Generalized Vector-Dominance Analyses Of Deep-Inelastic Electron-Nucleon Scattering," by J. Pestieau and J. Urias (1973) [6].

- "Diffractive Behavior Of Deep-Inelastic Electroproduction In A Resonance Model," by C. A. Dominguez and H. Moreno (1974) [7].
- "SU(6)-W Analysis Of The Electroproduction Of Pseudoscalar Mesons In The Second Resonance Region," by C. Avilez and G. Cocho (1974) [8].
- "Deep Inelastic Electroproduction And The Q**2 Behavior Of The J = 0 Fixed Pole," by C. A. Dominguez and H. Moreno (1975) [9].
- "Present Status Of The Cabibbo - Radicati Sum Rule," by C. A. Dominguez and H. Moreno (1976) [10].
- "Subtractions In The Adler Sum Rule And Charge Symmetry Violation," by C. A. Dominguez, H. Moreno and A. Zepeda (1976) [11].
- "Modified Alder Sum Rule And Violation Of Charge Symmetry," by C. A. Dominguez, H. Moreno and A. Zepeda (1976) [12].
- "Coulomb Corrections To Deep Inelastic Electron Or Muon Scattering From Nuclei," by E. Calva-Tellez and D. R. Yennie (1979) [13].
- "Broken SU(6)-W Analysis Of Electroproduction," by C. Avilez and G. Cocho (1979) [14].
- "Does Color Shine In Deep Inelastic Muon Scattering At Large Q**2?," by C. Avilez (1979) [15].
- "Probing new physics with coherent neutrino scattering off nuclei," by J. Barranco, O. G. Miranda and T. I. Rashba (2005) [16].
- "Deeply virtual neutrino scattering (DVNS)," by P. Amore, C. Coriano and M. Guzzi (2005) [17].

COLLIDER PHYSICS

In collider experiments, two beams of accelerated particles are steered towards one another and made to cross at a specified point. The beams pass through each other and the particles interact. The advantages in this kind of experiments are the following: i) It is easier to achieve higher energies as both particles can be accelerated rather than relying on just the one to carry all the energy; ii) If the beams collide head on and the particles carry the same momentum, then the total momentum in the interaction is zero. This means that the produced particles must have a total of zero momentum as well. This is a much more efficient use of energy as it is possible for all the reaction energy to go into the mass of new particles (if they are produced at rest) — no kinetic energy is needed after the reaction. These advantages over fixed target experiments make the collider the experiment of choice in modern particle physics.

Although it is gotten plainly more center mass energy and momentum for a given beam energy with colliders there are also disadvantages and many practical difficulties to overcome.

The obvious disadvantage is low luminosity, low event rate. A normal solid or liquid fixed target has a high density of nucleons and can be made many meters long if necessary, to give a better chance of observing the less probable types of collision. Another limitation, coming from the need to store and manipulate the beam particles, is that only charged and stable particles can be used. This restricts the to e^-, e^+, p, and $\bar{p}$, but fortunately it does not much restrict the physics potential. Actually, it is under consideration the construction of a $\mu^-\mu^+$-collider. After four decades of study and development, colliders have become very essential tools of high energy physics.

E. Calva-Tellez, C. Domínguez, H. Moreno and A. Zepeda at the CINVESTAV-IPN and G. Cocho at the National University (UNAM) were pioneers in doing research in Collider Physics at Mexican Institutions. These authors and their collaborators contributed to pace the way to elucidate some aspects of the physics involved in the collider phenomenology. They have done research for the different type of colliders: e^-e^+, $p\bar{p}$, pp and e^-p, working in subjects such as Hadron Form Factors, Scaling, The Parton Model, Radiative Correction, Sum Rules and Particle Production. We want to comment that most of the works done in collider physics by scientists at Mexican Institutions, have been written after 1980, working in the context of gauge models of the electroweak and strong interactions. The works include studies of several aspects of the collider phenomenology, such as Particle Production and Physics Beyond the Standard Model, in the context, besides the Standard Model, of Multiboson Electroweak Gauge Models, Left-Right Symmetric Models and the Minimal Supersymmetric Standard Model. We end this section by giving a list in chronological order the works made in Mexican Institutions in the physics associated with colliders.

- "Scaling Models Of High-Energy Multiparticle Production In Hadron - Hadron Collisions," by H. Moreno (1973) [18].
- "Bremsstrahlung Model Calculation Of High-Energy Nucleon-Nucleon Scattering," by H. Moreno (1973) [19].
- "Theoretical Analysis Of The Frascati Data On The Single-Bremsstrahlung Process E+ E- $\to$ E+ E- Gamma," by E. Calva-Tellez (1973) [20].
- "Charm, Evdm And Narrow Resonances In E+ E- Annihilation," by C. A. Dominguez and M. Greco (1975) [21].
- "On The Calculations Of Baryon Parameters In The Extended Models Of Hadrons," by G. Cocho, M. Fortes and H. Vucetich (1976) [22].
- "Weak Neutral Currents In Electron - Positron Annihilation Into Three Pions," by E. Calva-Tellez and A. Zepeda (1976) [23].
- "Photon - Nucleus Collisions And The Relative Phase Between The Gamma P $\to$ Pi0 P And Rho0 P $\to$ Pi0 P Amplitudes," by C. Avilez and G. Cocho (1976) [24].
- "Weak Neutral Currents In Electron - Positron Annihilation Into Three Pions With Polarized Beams," by J. L. Lucio and A. Zepeda (1977) [25].
- "The Quark Quark Interaction In Extended Models Of Hadrons," by G. Cocho, M. Fortes and H. Vucetich (1977) [26].
- "Polarized Nucleon-Nucleon Scattering From 3-Gev/C To 12-Gev/C: What Can Be Inferred From The Experiments?," by C. Avilez, G. Cocho and M. Moreno (1981) [27].
- "Polarized Bhabha Scattering In Multiboson Electroweak Gauge Models," by W. Hollik and A. Zepeda (1982) [28].
- "The V-A Structure Of The Charged Weak Current Of The Tau Lepton And The Reaction Gamma N $\to$ Lepton+- Lepton-Neutrino N-Prime," by J. H. Garcia and

M. A. Perez (1985) [29].
- "Phenomenology Of A Second Neutral Gauge Boson In The Drell-Yan Process," by A. Rosado, J. L. Lucio and A. Zepeda (1985) [30].
- "Three body decays of Higgs bosons in the MSSM," by E. Barradas, J. L. Diaz-Cruz, A. Gutierrez and A. Rosado (1996) [31].
- "Bounding the tau-neutrino magnetic moment at CERN LEP in a left-right symmetric model," by A. Gutierrez, M. A. Hernandez, M. Maya and A. Rosado (1998) [32].
- "Measurement of the v_τ magnetic moment at CERN LEP in a left-right symmetric model," by A. Gutierrez-Rodriguez, M. A. Hernandez, A. Rosado and M. Maya (1999) [33].
- "Analysis of t b W and t t Z couplings from CLEO and LEP/SLC data," by F. Larios, M. A. Perez and C. P. Yuan (1999) [34].
- "Detection of neutral MSSM Higgs bosons at LEP-II and NLC," by U. Cotti, A. Gutierrez-Rodriguez, A. Rosado and O. A. Sampayo (1999) [35].
- "Top-pions and single top production at HERA and THERA," by F. Larios, F. Penunuri and M. A. Perez (2005) [36].

HEAVY VECTOR BOSON PRODUCTION IN DEEP INELASTIC ep-SCATTERING

Introduction

LEP/LHC will provide us with the possibility to observe $e^- p$-collisions with an energy $E \approx 60$ GeV of the electron and $E_P \approx 7$ TeV of the proton [37]. One of the interesting experiments at LEP/LHC will be electroproduction of weak bosons. One of the goals of LEP/LHC is to look for physics beyond the standard model [38]. However, the events rates for most of the exotic processes are of the same order or smaller [39] than the expected event rates for heavy gauge boson production in $e^- p$-scattering. Hence, although the cross sections for heavy boson production are expected to be not big enough to allow for detailed investigations of the $W^\pm$, Z^0-properties, it is necessary to calculate them. Only if $W^\pm$- and Z^0-production, which are an important background for new physics, are completely known and subtracted the latter can be investigated. Therefore our aim in this section is to review in detail the heavy boson production in deep inelastic $e^- p$-scattering. We did this using the coupling between fermions and bosons as given by the standard model of the strong and electroweak interactions and the parton model [40] with the parton distribution functions reported by J. Botts *et al.* [41], which take into account scaling violations and the heavy quarks contribution.

We discussed the following three types of reactions mechanisms which contribute to $W^\pm$ and Z^0: $W^\pm$ and Z^0-production at the lepton line (Fig.1(a),(b)), at the quark line (Fig.1(c),(d)) and through the fusion diagram with the non-Abelian coupling of the gauge bosons (Fig.1(e)). Our calculations included besides γ-exchange also Z^0- and W-exchange diagram contributions and consequently are complete.

The first estimation for $W^\pm$ and Z^0-production in e^-p-scattering were given by Llewelyn-Smith and Wiik in 1977 [42]. Other authors have reported more detailed calculations [43, 44, 45, 46, 47, 48, 49, 50, 51]. It is not possible to compare directly all our results with those of former papers because different cutoffs were applied there. In [45] the total cross section for the reaction $e^-p \to \nu_e W^- X$ was calculated with a cutoff of $O(1)GeV^2$ taking into account the γ-exchange contribution only. For an energy in the center of mass of $1,296\ GeV$ we got with our cutoff $\sigma_T = 5.5 \times 10^{-37} cm^2$ compared to $\sigma_T = 5.05 \times 10^{-37} cm^2$ [45]. However, for the process $e^-p \to eZ^0 X$, E. Gabrielli [45] found for the lepton vertex contribution $\sigma_{lep} = 3.3 \times 10^{-37} cm^2$. Taking the same parameters we got a result similar; namely $\sigma_{lep} = 3.2 \times 10^{-37} cm^2$. We also have included in our calculations the hadron vertex contribution, $\sigma_{had} = 2.6 \times 10^{-37} cm^2$.

We presented also results for the differential cross section as a function of the dimensionless variables x and y with the aim to compare the production from the leptonic vertex and the hadronic vertex. We found that there are kinematical regions where either the lepton or the hadron vertex contributions can be neglected.

One of the aims of LEP/LHC is to look for physics beyond the standard model. For example, particles predicted by supersymmetry [52], subconstituent models [39] and others [53]. This can be done only if the results of the standard model are known in all details. Therefore we completed our discussion by presenting calculations for the energy spectrum and the angular distribution of the produced Z^0. We found that the Z^0 for fixed x and y is mainly produced in a small, well determined region of phase space. The notation and the kinematics for heavy boson production through lepton-nucleon collision can be found in Refs. [54] and [55]. Here, we just introduce the definition of some sets of variables used in this paper.

We write the production of a boson $B = (W^\pm, Z^0)$ through the inclusive $\ell(p') + N(P_N)$ scattering as follows:

$$\ell(p') + \mathcal{N}(P_\mathcal{N}) \to \ell'(p') + B(k) + X \qquad (1)$$

where $\ell, \mathcal{N}, \ell'$ and B stand for the initial lepton, the incoming nucleon, the final lepton and the produced massive boson. For explicit construction of the particle momenta we choose in accordance with LEP/LHC kinematics the following energies and angles (whenever possible we will neglect the nucleon and lepton masses):

$$\begin{aligned}
p^\mu &= E(1,0,0,1) \\
P_\mathcal{N}^\mu &= E_\mathcal{N}(1,0,0,-1) \\
p'^\mu &= E'(1,\sin\theta,0,\cos\theta) \\
k^\mu &= (E_k, k\sin\theta_k \cos\phi_k, k\sin\theta_k \sin\phi_k, k\cos\theta_k), \quad k = \sqrt{E_k^2 - M_B^2}.
\end{aligned} \qquad (2)$$

As usual we define the invariants:

$$\begin{aligned}
s &= (p+P_\mathcal{N})^2 &&= 4EE_\mathcal{N} \\
Q^2 &= -(p-p')^2 &&= 2EE'(1-\cos\theta) \\
v &= P_\mathcal{N}(p-p') &&= E_\mathcal{N}(2E-E'(1+\cos\theta)) \\
s' &= (p+P_\mathcal{N}-k)^2 &&= 4EE_\mathcal{N}+M_B^2-2E(E_k-k\cos\theta_k)-2E_\mathcal{N}(E_k+k\cos\theta_k) \\
Q'^2 &= -(p-p'-k)^2 &&= 2EE'(1-\cos\theta)-M_B^2+2E(E_k-k\cos\theta_k) \\
& && -2E'(E_k-k(\cos\theta\cos\theta_k+\sin\theta\sin\theta_k\cos\phi_k)) \\
v' &= P_\mathcal{N}(p-p'-k) &&= E_\mathcal{N}(2E-E'(1+\cos\theta)-E_k-k\cos\theta_k)
\end{aligned} \quad (3)$$

and the dimensionless variables:

$$x = \frac{Q^2}{2v}, \quad y = \frac{2v}{s}, \quad \tau = \frac{s'}{s}, \quad x' = \frac{Q'^2}{2v'}, \quad y' = \frac{2v'}{s}. \quad (4)$$

We also can obtain the following expressions:

$$\begin{aligned}
E' &= E_\mathcal{N}xy+E(1-y) \\
E'\cos\theta &= E(1-y)-E_\mathcal{N}xy \\
E_k &= E(y-y')+E_\mathcal{N}(1-\tau+\mu+y'-y) \\
k\cos\theta_k &= E(y-y')-E_\mathcal{N}(1-\tau+\mu+y'-y) \\
\cos\phi_k &= \frac{xy(y-y'-1)+x'y'-y(1-\tau+y'-y)+\mu(1-y)}{\sqrt{xy(1-y)}\sqrt{(y-y')(1-\tau+y'-y)-\mu(1-y+y')}}
\end{aligned} \quad (5)$$

where $\mu = M_B^2/s = M_B^2/4EE_\mathcal{N}$. The physical region of these kinematical variables is discussed in detail in Refs. [54] and [55].

In the next subsection B, we present a summary of results and conclusions for the total cross section and differential cross section as a function of the dimensionless variables x and y for all the different reactions and reaction mechanisms which contribute to charged vector boson and neutral vector boson production in e^-p-collisions reported in Ref.[56]. We also comment on the results for the energy spectrum and the angular distribution of the produced Z^0.

Results and Conclusions

In Ref.[56], we have presented the complete calculation of heavy vector boson production in unpolarized deep inelastic e^-p-scattering in the context of the standard model and using the parton model, at LEP/LHC energies.

Taking $M_W = 80.3\ GeV$, $M_{Z^0} = 91.2\ GeV$, $\sin^2\theta_W = 0.223$, $\sqrt{s} = 1,296\ GeV$ and using the parton distributions reported by J. Botts et al., we found a total production of around $290\, Z^0$, $3030\, W^+$, $3310\, W^-$ bosons for an integrated luminosity de $500\ pb^{-1}$, for LEP/LHC.

We have included in our calculations besides γ-exchange also Z-exchange and W-exchange diagrams, and found that these contributions can grow up to 6% of the photonic cross section (See Table 1).

TABLE 1. Contribution to the total cross sections from the different boson exchange diagrams: $E_e = 60\ GeV$, $E_P = 7\ TeV$.

Processes	$\sigma_\gamma(cm^2)$	$\sigma_{\gamma+Z}(cm^2)$	$\sigma_T(cm^2)$
$e^-p \to e^-W^-X$	5.7×10^{-36}	5.7×10^{-36}	6.1×10^{-36}
$e^-p \to e^-W^+X$	6.8×10^{-36}	6.8×10^{-36}	6.1×10^{-36}
$e^-p \to \nu W^-X$	4.6×10^{-37}	6.6×10^{-37}	5.5×10^{-37}
$e^-p \to e^-Z^0X$	5.5×10^{-37}	5.7×10^{-37}	5.8×10^{-37}
$e^-p \to \nu Z^0X$	0	0	1.4×10^{-37}

We have shown that for the y distribution heavy boson exchange contributions are important, particularly for large values of y ($y \approx 1$).

We have calculated separately the contribution from the different mechanisms of charged boson production. The investigation of the separate contributions to the differential cross sections shows the compensation mechanism which is typical for the non-Abelian structure of the electroweak standard model as a gauge theory. This compensation reaches at LEP/LHC energies up to two orders of magnitude. This means that even small deviations of the coupling structure of the standard model like anomalous magnetic moment terms will lead to observable deviations in the predictions for charged boson production in e^-p collisions [57, 58].

In the case of a neutral boson Z^0, its decays into lepton pairs provide a clear signal

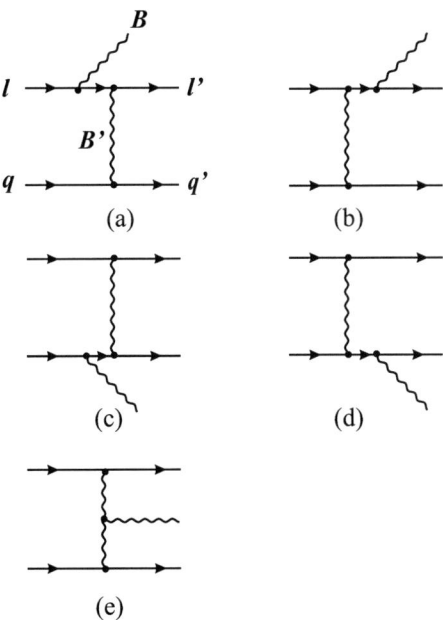

FIGURE 1. Feynman diagrams for heavy boson production from the initial (a) and final (b) lepton, from the initial (c) and final (d) quark and via the non-Abelian couplings (e).

of its production. Neutral bosons can be produced through neutral current and charged current interactions. It is expected that the production rates for neutral current processes will be bigger than those for charged current processes, because in the later case photon-exchange diagrams do not contribute in the lowest order of α.

The contribution of the Z^0-exchange diagrams to the total cross section is only a few percent, γ-exchange diagrams dominate by far. However, for Q^2, Q'^2 large (*i.e* $O(M_Z^2)$) γ-exchange as well as Z^0-exchange are important.

We compared the two production mechanisms, emission of the Z^0 at the leptonic vertex and at the hadronic vertex, and their interference. We found that there are kinematical regions where either the contribution from the leptonic vertex or that provided by the hadronic vertex can be neglected. The contribution to the cross section is mainly hadronic for $y \lesssim 0.5$ whereas for $0.8 \lesssim y < 1$ it is almost purely leptonic.

We also have analyzed in the LEP/LHC system the energy E_k and $\cos\theta_k$ distributions of the produced Z^0 for given x and y (*i.e.* for given energy and polar angle of the final electron). Working in the region where the Z^0-production is almost purely leptonic, we found that the E_k and θ_k distributions are strongly peaked. We showed that if the energy and polar angle of the scattered electron are given by scaling variables x and y, the boson Z^0 will be mostly produced with energy $E_k = Ey + E_p x(1-y) + E_p M_Z^2/(sy)$, polar angle $\cos\theta_k \approx (2Ey - E_k)/\sqrt{E_k^2 - M_Z^2}$ and azimuthal angle $\cos\phi_k \approx 0$.

Finally, we have pointed out that the resonance-like behavior of the E_k and θ_k distributions will help to discriminate between the production of a standard Z^0 and the production of other particles, such as those predicted by other theories with other reaction mechanisms, for example supersymmetry [52] and subconstituent models [39], in the experiments which will be carried at the collider LEP/LHC.

The cross section and consequently the event rate is presumable not big enough to allow for detailed investigation of the $W^\pm$-, Z^0-properties. However, it is of the same order or bigger than that of many exotic processes. Therefore it is important to have a complete and detailed calculations of $W^\pm$, Z^0-production since possible new physics can be observed only is standard background events are correctly subtracted. In this sense the results of Ref.[56] may help to pace the way to new physics at LEP/LHC or future $e^- p$-machines.

HEAVY VECTOR BOSON PRODUCTION IN DEEP INELASTIC $\nu_l \mathcal{N}$-SCATTERING

Introduction

Large-scale neutrino telescopes [59] have as a main goal the detection of ultrahigh-energy (UHE) cosmic neutrinos ($E_\nu \geq 10^{12}$ eV) produced outside the atmosphere (neutrinos produced by galactic cosmic rays interacting with interstellar gas, and extragalactic neutrinos) [60]. UHE neutrinos can be detected by observing long-range muons produced in charged-current neutrino-nucleon interactions. A very enlightening discussion on UHE neutrino interactions is given by R. Gandhi *et al.* [61].

The detection of UHE neutrinos will provide us with the possibility to observe $\nu\mathcal{N}$-collisions with a neutrino energy in the range $10^{12}\,eV \leq E_\nu \leq 10^{21}\,eV$ and a target nucleon at rest. Earlier estimations of the W production rates in $\nu\mathcal{N}$-scattering were given in 1970, by H. H. Chen [62], and by J. Reiff [63], in 1971 by R. W. Brown and J. Smith [64], and by F. A. Berends and G. B. West [65]. In those works the calculations were made using neither the standard model [38] nor the parton model [40]. One of the aims of detecting UHE neutrinos is to look for physics beyond [66] the standard model [38]. Hence, even though the cross sections for heavy boson production are expected not to be large enough to allow for detailed investigations of the heavy bosons properties, it is necessary to calculate them. Only if heavy boson production, which could be an important background for physics beyond the standard model, is completely known and subtracted, the new physics could be investigated. Further, we can expect that the cross section rates for process $\nu_l + \mathcal{N} \to l^- + W^+ + X$ could become comparable to those rates for the charged current interaction $\nu_l + \mathcal{N} \to l^- + X$. The reason is that the propagator of the virtual lepton in the diagram in which the W^+-boson is emitted by the neutrino (see Fig.2(a)) behaves like the propagator of a photon [55].

Therefore, we review in this section heavy vector boson production in the deep inelastic $\nu_l\mathcal{N}$-scattering within the frame of the standard model $SU(3)_C \times SU(2)_L \times U(1)$ of the strong and electroweak interactions and using the parton model [40] with the parton distribution functions reported by J. Pumplin *et al.* [67]. We use the CTEQ PDFs provided in a $n_f = 5$ active flavours scheme.

The cross section for the process $\nu_l + \mathcal{N} \to l^- + W^+ + X$ is expected to be the most important one because it gets contributions from photon exchange diagrams (see Fig.2). We thus present results for the total cross section for W^+ production through the processes $\nu_l + \mathcal{N} \to l^- + W^+ + X$ as a function of the neutrino energy in the range $10^{14}\,eV \leq E_\nu \leq 10^{17}\,eV$ [1] and taking $E_\mathcal{N} = m_\mathcal{N}$ for the nucleon. We also calculate $\sigma(\nu_l + \mathcal{N} \to l^- + X)$ to find out how it compares to $\sigma(\nu_l + \mathcal{N} \to l^- + W^+ + X)$.

In the next subsection B, we present a summary of results and conclusions for the total cross section of charged vector boson in $\nu_l + \mathcal{N}$-collisions, which were reported in Ref.[68].

Results and Conclusions

In Ref.[68] we have presented the general formulas for the cross section of the production of massive vector bosons in deep inelastic neutrino nucleon scattering in the framework of the standard model.

We have neglected the contribution from heavy boson exchange diagrams in our numerical calculations. Hence, we have given numerical results only for the total cross section rates of the process $\nu_l + \mathcal{N} \to l^- + W^+ + X$ ($\mathcal{N}$: P, N.) because this reaction is the only one which gets contribution from photon-exchange diagrams in the lowest order of α. Considering the nucleon at rest, taking $M_W = 80.4\,GeV$, $\sin^2\theta_W = 0.223$,

---

[1] We take $E_\nu \leq 10^{17}\,eV$, in order to use the CTEQ PDFs in the range of the (x', Q'^2) parameter space in which these have been found to be valid: $10^{-6} \leq x' \leq 1$ and $1.3\,GeV \leq \sqrt{Q'^2} \leq 10^4\,GeV$

a neutrino energy in the range $10^{14}\, eV \leq E_V \leq 10^{17}\, eV$, setting cuts of 4 GeV^2 and 10 GeV^2 for the momenta transfer square (Q^2 and Q'^2) and the invariant mass (W). We made use of the parton distribution functions of J. Pumplin et al., we used the CTEQ PDFs provided in a $n_f = 5$ active flavors scheme. In particular, we have obtained $\sigma(\nu_l + P \to l^- + W^+ + X) = 2.3 \times 10^{-35}\, cm^2$ and $\sigma(\nu_l + N \to l^- + W^+ + X) = 1.8 \times 10^{-35}\, cm^2$ for $E_V = 10^{17}\, eV$. We have also shown that the total cross section rates of the process $\nu_l + \mathcal{N} \to l^- + W^+ + X$ do not depend strongly on the choice of the cuts on the momenta transfer square (Q^2 and Q'^2), when we take them equal to a few GeV^2.

We have also presented results for $\sigma(\nu_l + \mathcal{N} \to l^- + W^+ + X)/\sigma(\nu_l + \mathcal{N} \to l^- + X)$ ($\mathcal{N}: P, N$) as a function of E_V in the range $10^{14}\, eV \leq E_V \leq 10^{17}\, eV$, with $E_\mathcal{N} = m_\mathcal{N}$. We have gotten $\sigma(\nu_l + P \to l^- + W^+ + X)/\sigma(\nu_l + P \to l^- + X) = 4.7 \times 10^{-3}$ and $\sigma(\nu_l + N \to l^- + W^+ + X)/\sigma(\nu_l + N \to l^- + X) = 3.6 \times 10^{-3}$ for $E_V = 10^{17}\, eV$.

Finally, we pointed out that the cross section rates for the process $\nu_l + \mathcal{N} \to l^- + W^+ + X$ did not become so large as one could expect, due to the strong destructive interference between the two different mechanisms which contribute at the lowest order in α (keeping only photon exchange diagrams): production at the leptonic vertex (Fig.2(a)) and through the boson self interaction (Fig.2(e)). Such destructive interference mechanism is an intrinsic attribute of the standard model as a non-Abelian gauge theory. magnitude. This means that even small deviations of the coupling structure of the standard model like anomalous magnetic moment terms could lead to observable deviations in the predictions for the process $\nu_l + \mathcal{N} \to l^- + W^+ + X$ [69].

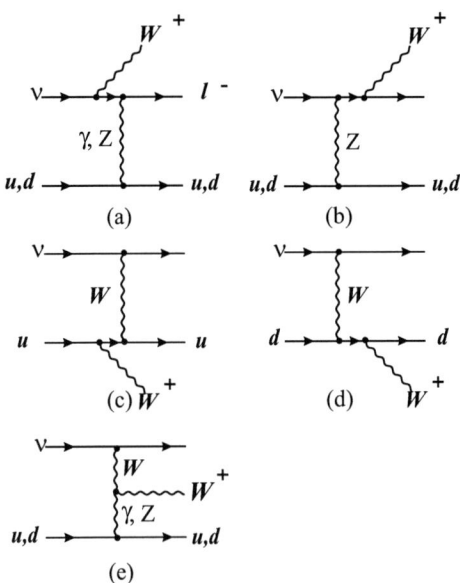

FIGURE 2. Feynman diagrams for the process $\nu_l + \mathcal{N} \to l^- + W^+ + X$: W^+-production from the incoming neutrino (a), the outgoing lepton (b), the initial (c) and final (d) quark, and through the non-Abelian couplings (e). u stands for $u, c, \bar{d}, \bar{s}, \bar{b}$; d for $d, s, b, \bar{u}, \bar{c}$.

ACKNOWLEDGMENTS

The author thanks the *Sistema Nacional de Investigadores* and *CONACyT* (México) for financial support.

REFERENCES

1. An excellent reference to Collider Physics and DIS phenomenology is: V.D. Barger and R.J.N. Phillips, Collider Physics (Updated Edition),Addison-Wesley Publishing Company, Inc., Reading, Massachusetts, 1997.
2. J. Pestieau and E. Mexico National Comm Nucl, Phys. Rev. D **4**, 1827 (1971).
3. G. Cocho and J. J. Salazar, Phys. Rev. Lett. **27**, 892 (1971).
4. H. Moreno, J. Pestieau and E. Mexico National Comm Nucl, Phys. Rev. D **5**, 1210 (1972).
5. J. Pestieau, Phys. Rev. D **6**, 1449 (1972).
6. J. Pestieau and J. Urias, Phys. Rev. D **8**, 1552 (1973).
7. C. A. Dominguez and H. Moreno, Phys. Rev. D **9**, 2584 (1974).
8. C. Avilez and G. Cocho, Phys. Rev. D **10**, 3638 (1974).
9. C. A. Dominguez and H. Moreno, Phys. Rev. D **11**, 119 (1975).
10. C. A. Dominguez and H. Moreno, Phys. Rev. D **13**, 616 (1976).
11. C. A. Dominguez, H. Moreno and A. Zepeda, Phys. Rev. D **14**, 1455 (1976).
12. C. A. Dominguez, H. Moreno and A. Zepeda,
13. E. Calva-Tellez and D. R. Yennie, Phys. Rev. D **20**, 105 (1979).
14. C. Avilez and G. Cocho, Can. J. Phys. **57**, 1141 (1979).
15. C. Avilez, Lett. Nuovo Cim. **25**, 542 (1979).
16. J. Barranco, O. G. Miranda and T. I. Rashba, JHEP **0512**, 021 (2005) [arXiv:hep-ph/0508299].
17. P. Amore, C. Coriano and M. Guzzi, JHEP **0502**, 038 (2005) [arXiv:hep-ph/0404121].
18. H. Moreno, Phys. Rev. D **8**, 268 (1973).
19. H. Moreno, Phys. Rev. D **8**, 894 (1973).
20. E. Calva-Tellez and E. Mexico National Comm Nucl, Phys. Rev. D **8**, 3856 (1973).
21. C. A. Dominguez and M. Greco, Lett. Nuovo Cim. **12**, 439 (1975).
22. G. Cocho, M. Fortes and H. Vucetich, Phys. Rev. D **13**, 1513 (1976).
23. E. Calva-Tellez and A. Zepeda, Phys. Rev. D **14**, 1867 (1976).
24. C. Avilez and G. Cocho, Nucl. Phys. B **108**, 355 (1976).
25. J. L. Lucio and A. Zepeda, Phys. Rev. D **16**, 42 (1977).
26. G. Cocho, M. Fortes and H. Vucetich, Phys. Rev. D **16**, 3339 (1977).
27. C. Avilez, G. Cocho and M. Moreno, Phys. Rev. D **24**, 634 (1981).
28. W. Hollik and A. Zepeda, Z. Phys. C **12**, 67 (1982).
29. J. H. Garcia and M. A. Perez, Phys. Rev. D **31**, 2765 (1985).
30. A. Rosado, J. L. Lucio and A. Zepeda, Z. Phys. C **29**, 197 (1985).
31. E. Barradas, J. L. Diaz-Cruz, A. Gutierrez and A. Rosado, Phys. Rev. D **53**, 1678 (1996).
32. A. Gutierrez, M. A. Hernandez, M. Maya and A. Rosado, Phys. Rev. D **58**, 117302 (1998).
33. A. Gutierrez-Rodriguez, M. A. Hernandez, A. Rosado and M. Maya, Rev. Mex. Fis. **45**, 249 (1999).
34. F. Larios, M. A. Perez and C. P. Yuan, Phys. Lett. B **457**, 334 (1999) [arXiv:hep-ph/9903394].
35. U. Cotti, A. Gutierrez-Rodriguez, A. Rosado and O. A. Sampayo, Phys. Rev. D **59**, 095011 (1999) [arXiv:hep-ph/9902417].
36. F. Larios, F. Penunuri and M. A. Perez, Phys. Lett. B **605**, 301 (2005) [arXiv:hep-ph/0409291].
37. R. Ruckl, MPI-PAE-PTH-76-90 *Plenary talk given at ECFA-LHC Workshop, Aachen, Germany, Oct 4-9, 1990*; A. Verdier, CERN-LHC-NOTE-151 *Presented at the 1991 IEEE Particle Accelerator Conference, San Francisco, CA, 6 - 9 May 1991*.
38. S. L. Glashow, Nucl. Phys. **22**, 579 (1961); S. Weinberg, Phys. Rev. Lett. **19**, 1264 (1967); A. Salam, Proc. 8th NOBEL Symposium, ed. N. Svartholm (Almqvist and Wiksell, Stockholm, 1968), p. 367.
39. R. J. Cashmore *et al.*, Phys. Rept. **122**, 275 (1985).
40. R.P. Feynman, Photon-Hadron Interactions, Benjamin, Reading, Massachusetts, 1972.
41. J. Botts, J. G. Morfin, J. F. Owens, J. Qiu, W. K. Tung and H. Weerts, preprint Fermilab-Pub-92/371.

42. C. H. Llewelyn-Smith and B. H. Wiik, DESY Report 77/36 (1977).
43. P. Salati and J. C. Wallet, Z. Phys. **C16**, 155 (1982).
44. G. Altarelli, G. Martinelli, B. Mele and R. Rückl, Nucl. Phys. **B262**, 204 (1985).
45. E. Gabrielli, Mod. Phys. Lett. **A1**, 465 (1986).
46. D. Atwood, U. Baur, G. Couture and D. Zeppenfeld, Proc. of the Snowmass DPF Summer Study 1988, p. 264.
47. H. Baer, J. Ohnemus and D. Zeppenfeld, Z. Phys. **C43**, 675 (1989).
48. D. Atwood, U. Baur, D. Goddard, S. Godfrey and B. A. Kniehl, Proc. of the Snowmass DPF Summer Study 1990, p. 557.
49. U. Baur, B. A. Kniehl, J. A. M. Vermaseren and D. Zeppenfeld, Proc. of the Aachen ECFA Workshop 1990, p. 956.
50. U. Baur, J. A. M. Vermaseren, D. Zeppenfeld, Nucl. Phys. **B375**, 3 (1992).
51. U. Baur, Proc. of the Rencontres de Moriond on QCD and High Energy Hadronic Interactions, Les Arcs, France 1992, p. 91.
52. H. E. Haber and G. L. Kane, Phys. Rept. **117**, 75 (1985).
53. M. Bohm, A. Fernandez, M. A. Perez and A. Rosado, Z. Phys. C **53**, 135 (1992).
54. M. Bohm and A. Rosado, Z. Phys. C **39**, 275 (1988).
55. M. Bohm and A. Rosado, Z. Phys. C **34**, 117 (1987).
56. A. Gutierrez and A. Rosado, Phys. Rev. D **57**, 4318 (1998).
57. M. Bohm and A. Rosado, Z. Phys. C **42**, 479 (1989).
58. A. Gutierrez-Rodriguez and A. Rosado, Rev. Mex. Fis. **46**, 440 (2000).
59. *AMANDA Collaboration*, E. Andres *et al.*, Nature **410**, 441 (2001); *ANTARES Collaboration*, Y. Becherini *et al.*, e-Print Archive: hep-ph/0211173; *AUGER Collaboration*, D. Zavrtanik *et al.*, Nucl. Phys. Proc. Suppl. **85**, 324 (2002); *NESTOR Collaboration*, P. K. F. Grieder *et al.*, Nuovo Cim. **24C**, 771 (2001); *RICE Collaboration*, I. Kravchenko *et al.*, Astropart. Phys. **19**, 15 (2003).
60. V. S. Berezinsky, Nucl. Phys. **B28A**, 352 (1992); V. S. Berezinsky and G. Zatsepin, Proceedings of the 1976 DUMAND Summer Workshop (A. Roberts and R. Donaldson, eds.), p. 215 (1977); T. Stanev, Nucl. Phys. **B14A**, 17 (1990); K. Greisen, Phys. Rev. Lett. **16**, 748 (1966); C. T. Hill and D. N. Schramm, Phys. Lett. **B131**, 247 (1983); and Phys. Rev. **D31**, 564 (1985); R. Gandhi *et al.*, Phys. Rev. **D58**, 093009 (1998) and references therein.
61. R. Gandhi *et al.*, Phys. Rev. **D58**, 093009 (1998).
62. H. H. Chen, Phys. Rev. **1**, 3197 (1970).
63. J. Reiff, Nucl. Phys. **B23**, 387 (1970); and Nucl. Phys. **B28**, 495 (1971).
64. R. W. Brown and J. Smith, Phys. Rev. **3**, 207 (1971); R. W. Brown, R. H. Hobbs and J. Smith, Phys. Rev. **4**, 794 (1971); R. W. Brown, R. H. Hobbs, J. Smith and N. Stanko, Phys. Rev. **6**, 3273 (1972).
65. F. A. Berends and G. B. West, Phys. Rev. **3**, 262 (1971).
66. Z. Fodor *et al.*, Phys. Lett. **B561**, 191 (2003); T. Han and D. Hooper, e-Print Archive: hep-ph/0307120.
67. J. Pumplin *et al.*, JHEP **207**, 12 (2002); D. Stump *et al.*, e-Print Archive: hep-ph/0303013.
68. R. M. Garcia-Hidalgo and A. Rosado, Mod. Phys. Lett. A **20**, 1465 (2005) [arXiv:hep-ph/0310107].
69. R. M. Garcia-Hidalgo and A. Rosado, Rev. Mex. Fis. **51**, 203 (2005) [arXiv:hep-ph/0310243].

New physics effects in trilinear electroweak gauge boson couplings

J. J. Toscano

Facultad de Ciencias Físico Matemáticas, Benemérita Universidad Autónoma de Puebla, Apartado Postal 1152, Puebla, Puebla, México

Abstract. Virtual effects of new physics on the trilinear electroweak couplings WWV and VVV ($V = \gamma, Z$) are reviewed, both in specific models and the effective Lagrangian approach. The impact of new particles on the static electromagnetic properties of the W boson are discussed in several contexts. In particular, the sensitivity of the CP–violating electromagnetic moments to new sources of CP violation, as general Yukawa couplings, is stressed. The one–loop contribution of new gauge bosons to the off–shell WWV vertex is analyzed in the light of nonconventional quantization methods. In particular, a covariant scheme based in the BRST symmetry endowed with a nonlinear gauge–fixing procedure is discussed. The VVV coupling is studied in the context of the effective Lagrangian approach and the role played by the Bose and Lorentz symmetries emphasized. We argue that these symmetries are so restrictive that these vertices perhaps never could be observed, unless one of these principles could not be an exact symmetry of the nature, as suggested by quantum field theories formulated in a noncommutative space–time, which violate the Lorentz symmetry and thus allow for the existence of non–vanishing on–shell VVV vertices at the level of the classical action.

Keywords: Gauge boson couplings, gauge invariance, gauge independence, effective Lagrangians.

1. INTRODUCTION

Future search for new physics will be focused on its indirect manifestation through quantum effects caused by new particles rather that in its direct observation, as it is expected that the new particles are beyond the threshold of the collider. In this context, the trilinear WWV and VVV ($V = \gamma, Z$) gauge boson couplings may serve as a window for probing new physics, as they may constitute a mechanism through which physics beyond the Fermi scale may show up. Due to this, the one–loop structure of these vertices has been considerably studied in the literature. The next generation $e^- e^+$ linear colliders (NLC) will offer very interesting possibilities for studying these couplings, since these machines will operate as W factories. Since the Ws will be produced via highly off–shell V bosons, it is of experimental interest to investigate the sensitivity of the off–shell WWV vertex to virtual effects of heavy particles. Also, these colliders will play an important role in detecting the trilinear neutral VVV ($ZZZ, ZZ\gamma, Z\gamma\gamma$) couplings, whose observation would be a clear evidence of new physics, as they are strongly suppressed in the SM.

In this article, we present a short review on the sensitivity of the trilinear electroweak gauge bosons to radiative corrections induced by physics beyond the standard model (SM). We assume that the new physics is not directly observed and can be probed only through quantum effects. Besides the WWV coupling reflects the gauge structure of the electroweak theory, its sensitivity to virtual effects of new particles can have an important impact on some rare SM processes that occur as quantum fluctuations, as for instance, the one–loop induced Higgs boson decays $H \to \gamma\gamma, \gamma Z$ [1]. On the other hand, the fact that the WWV and VVV couplings involve only fields which are subject to the gauge freedom, means that their study at the level of radiative corrections must be treated carefully. In this context, it is an interesting fact that at the one–loop level the neutral coupling VVV only can be generated by fermionic particles, in contrast with the WWV vertex which can also receive contributions from bosonic particles. A specially interesting situation arises when loop contributions from gauge particles are considered, since in this case the corresponding amplitude is not necessarily unique, as it could depend on the gauge–fixing procedure. The possibility of extracting physical information concerning the presence of new particles directly from the off–shell WWV vertex, would be very important from both theoretical and experimental point of views. This requires to critically analyze the concepts of gauge invariance and gauge independence in the quantum theory context, as they lie at the heart of gauge systems. In this work, we will stress the importance of these aspects by analyzing the one–loop structure of the off–shell WWV vertex in the context of a specific model [2], which impacts this vertex at one–loop level through new massive gauge bosons that present unusual couplings with the electroweak gauge bosons [3]. This model provides the ideal scenario to discuss the gauge structure of the WWV vertex in the light of modern quantization schemes that go beyond the conventional one based on the Faddeev–Popov method (FPM) [4], as the background field method (BFM) [5], the pinch technique [6] and the BRST [7] formalism endowed with nonlinear gauges [8, 3].

In sec.2, we will analyze the sensitivity of the WWV coupling to new physics effects. We first discuss the static properties of the W gauge boson. We will focus on the CP–odd components, which may be highly sensitive to new physics effects, as they are strongly suppressed in the SM. Next, the one–loop structure of the off–shell WWV vertex is discussed in the context of conventional and nonconventional quantization methods. In particular, we will discuss how the conventional quantization method based on the BRST symmetry endowed with a nonlinear gauge–fixing procedure, can be used to construct gauge–invariant quantum actions. In sec.3, the structure of the VVV vertex will be discussed in a model independent manner using the effective Lagrangian approach. In this category there are the standard electroweak effective Lagrangian approach [9], as well as some interesting results predicted by the noncommutative standard model [10], which naturally induces the VVV vertex at tree level. In sec.4, we conclude with some final comments.

2. NEW PHYSICS EFFECTS ON THE *WWV* VERTEX

2.1. Effective Lagrangian description of the *WWV* vertex

Assuming that the underlying physics respects the SM gauge symmetry, the *WWV* vertex can be parametrized by means of the following effective Lagrangian [11]

$$\begin{aligned}\mathscr{L} &= -\frac{1}{2}Tr[W_{\mu\nu}W^{\mu\nu}] + \frac{i\alpha_{\Phi B}}{\Lambda^2}(D^\mu\Phi)^\dagger B_{\mu\nu}(D^\nu\Phi) + \frac{i\alpha_{\Phi W}}{\Lambda^2}(D^\mu\Phi)^\dagger W_{\mu\nu}(D^\nu\Phi) \\ &+ \frac{\alpha_{WB}}{\Lambda^2}(\Phi^\dagger B_{\mu\nu}W^{\mu\nu}\Phi) + \frac{\alpha_W}{\Lambda^2}Tr[W^\mu_{\ \nu}W^\nu_{\ \rho}W^\rho_{\ \mu}] + \frac{i\tilde{\alpha}_{\Phi B}}{\tilde{\Lambda}^2}(D^\mu\Phi)^\dagger \tilde{B}_{\mu\nu}(D^\nu\Phi) \\ &+ \frac{i\tilde{\alpha}_{\Phi W}}{\tilde{\Lambda}^2}(D^\mu\Phi)^\dagger \tilde{W}_{\mu\nu}(D^\nu\Phi) + \frac{\tilde{\alpha}_{WB}}{\tilde{\Lambda}^2}(\Phi^\dagger B_{\mu\nu}\tilde{W}^{\mu\nu}\Phi) + \frac{\tilde{\alpha}_W}{\tilde{\Lambda}^2}Tr[W^\mu_{\ \nu}W^\nu_{\ \rho}\tilde{W}^\rho_{\ \mu}],\end{aligned} \quad (1)$$

where $W_{\mu\nu} = T^a W^a_{\mu\nu}$, $\tilde{W}_{\mu\nu} = \frac{1}{2}\varepsilon_{\mu\nu\alpha\beta}W^{\alpha\beta}$, with $Tr(T^a T^b) = (1/2)\delta^{ab}$, and $B_{\mu\nu}$ the Abelian field strength. After spontaneous symmetry breaking (SSB), one obtains

$$\begin{aligned}\mathscr{L}_{WWV} &= -ig_V\Big\{g_1^V\left(W^+_{\mu\nu}W^{-\mu} - W^-_{\mu\nu}W^{+\mu}\right)V^\nu + \kappa_V W^+_\mu W^-_\nu V^{\mu\nu} \\ &+ \frac{\lambda_V}{m_W^2}W^+_{\mu\nu}W^{-\nu\rho}V_\rho^{\ \mu} + \tilde{\kappa}_V W^+_\mu W^-_\nu \tilde{V}^{\mu\nu} + \frac{\tilde{\lambda}_V}{m_W^2}W^+_{\mu\nu}W^{-\nu\rho}\tilde{V}_\rho^{\ \mu}\Big\},\end{aligned} \quad (2)$$

where the overall coupling constants are $g_\gamma = gs_W$ and $g_Z = gc_W$. In addition, $g_1^\gamma = 1$, $g_1^Z = 1 + \frac{g}{8c_W^2}\varepsilon_{\Phi W}$, and

$$\kappa_\gamma = 1 - \frac{gc_W}{2s_W}\varepsilon_{BW} + \frac{g}{8s_W}(c_W\varepsilon_{\Phi B} + s_W\varepsilon_{\Phi W}), \quad \lambda_\gamma = \frac{3g}{8}\varepsilon_W, \quad (3)$$

$$\tilde{\kappa}_\gamma = 1 - \frac{gc_W}{2s_W}\tilde{\varepsilon}_{BW} + \frac{g}{8s_W}(c_W\tilde{\varepsilon}_{\Phi B} + s_W\tilde{\varepsilon}_{\Phi W}), \quad \tilde{\lambda}_\gamma = \frac{3g}{8}\tilde{\varepsilon}_W, \quad (4)$$

$$\kappa_Z = 1 + \frac{gs_W}{2c_W}\varepsilon_{BW} + \frac{g}{8c_W}(-s_W\varepsilon_{\Phi B} + c_W\varepsilon_{\Phi W}), \quad \lambda_Z = \lambda_\gamma, \quad (5)$$

$$\tilde{\kappa}_Z = 1 + \frac{gs_W}{2c_W}\tilde{\varepsilon}_{BW} + \frac{g}{8c_W}(-s_W\tilde{\varepsilon}_{\Phi B} + c_W\tilde{\varepsilon}_{\Phi W}), \quad \tilde{\lambda}_Z = \tilde{\lambda}_\gamma, \quad (6)$$

where we have introduced the definitions $\varepsilon_i = (v/\Lambda)^2 \alpha_i$ and $\tilde{\varepsilon}_i = (v/\tilde{\Lambda})^2 \tilde{\alpha}_i$.

As far as the vertex function is concerned, we will focus on the case with the two *W*s on–shell and *V* off–shell. Since the *W* pairs will be produced at NLC through the $e^+e^- \to W^+W^-$ reaction, it is only necessary to retain the transverse components of *V*. Using the notation of Figure 1, the corresponding vertex function can be written as

$$\Gamma^V_{\alpha\beta\mu} = -ig_V\Big\{g_1^V[2p_\mu g_{\alpha\beta} + 4(q_\beta g_{\alpha\mu} - q_\alpha g_{\beta\mu})] + 2\Delta\kappa_V(q_\beta g_{\alpha\mu} - q_\alpha g_{\beta\mu}) + \frac{4\Delta Q_V}{m_W^2}p_\mu\left(q_\alpha q_\beta - \frac{1}{2}q^2 g_{\alpha\beta}\right) + 2\Delta\tilde{\kappa}_V \varepsilon_{\alpha\beta\mu\lambda}q^\lambda + \frac{4\Delta\tilde{Q}_V}{m_W^2}q_\beta \varepsilon_{\alpha\mu\lambda\rho}p^\lambda q^\rho\Big\}, \quad (7)$$

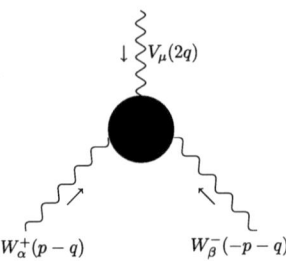

FIGURE 1. Trilinear WWV coupling. The loop denotes virtual effects of new particles.

where

$$\Delta \kappa_V = \kappa_V + \lambda_V - g_1^V, \quad \Delta \tilde{\kappa}_V = \tilde{\kappa}_V - \frac{m_W^2 - 2q^2}{m_W^2} \tilde{\lambda}_V, \quad \Delta Q_V = -2\lambda_V, \quad \Delta \tilde{Q}_V = 2\tilde{\lambda}_V. \quad (8)$$

2.2. The on–shell $WW\gamma$ vertex

The W gauge boson is the only no self–conjugate bosonic field of the SM. Since in this case the particle does not coincide with the antiparticle, this field can have static electromagnetic properties. Following angular momentum conservation, but without imposing any C, P, or T invariance, the number of electromagnetic form factors for any particle of spin s is $6s+1$ [12]. This is a general result, valid for any arbitrarily charged no self–conjugate vector field even if it is electrically neutral [13, 14, 15]. However, this number reduces to five if electromagnetic gauge invariance is invoked. On the other hand, it is a well–known fact that the electromagnetic moments of the W gauge boson, defined by the on–shell $WW\gamma$ vertex, are gauge–independent. This means that contributions from gauge bosons circulating in the loop can be calculated using the conventional method or any other quantization scheme. The result must be the same. This fact has strongly motivated the study of this vertex, as allows us to estimate the sensitivity of the Yang–Mills sector to physics beyond the SM. For this reason, the on–shell $WW\gamma$ vertex has been considerably studied in the literature, both in the context of the SM[16, 17] and in many well motivated of its extensions[18, 19, 20, 21, 22, 23, 24].

The structure of the on–shell $WW\gamma$ vertex is given through eqs. (2,3,7,8) with the change $V \to A$ and $q^2 = 0$. The $(g_1^\gamma, \kappa_\gamma, \lambda_\gamma)$ form factors are CP–conserving, whereas the $(\tilde{\kappa}_\gamma, \tilde{\lambda}_\gamma)$ ones are CP–violating. The former define the magnetic dipole (μ_W) and electric quadrupole (Q_W) moments, whereas the latter determine the electric dipole (d_W) and magnetic quadrupole ($\tilde{Q}_W$) moments. They are given by

$$\mu_W = \frac{e}{2m_V}(2 + \Delta \kappa_\gamma), \quad Q_W = -\frac{2}{m_W^2}(1 + \Delta \kappa_\gamma + \Delta Q_\gamma), \quad (9)$$

$$d_W = \frac{e}{2m_W}(\Delta \tilde{\kappa}_\gamma + \Delta \tilde{Q}_\gamma), \quad \tilde{Q}_W = -\frac{e}{m_W^2} \Delta \tilde{\kappa}_\gamma. \quad (10)$$

The CP–violating moments are interesting because they arise from quantum fluctuations in any renormalizable theory.

Before investigating the quantum effects of a given type of particles on the on–shell $WW\gamma$ vertex, notice that the presence of at least one fermionic loop involving a Dirac trace is required to generate a term proportional to the Levi–Civita tensor. This means that at the one–loop level only particles of spin $1/2$ can generate the CP–odd form factors [25, 13]. Another interesting result is that there is no contribution to $\Delta \tilde{Q}_\gamma$ at the one–loop level, which means that, at this order, the electric dipole moment of a no self–conjugate vector particle is directly proportional to the magnetic quadrupole moment [13, 25]. These features show that the CP–odd electromagnetic properties of elementary particles are quite elusive indeed. The CP-even form factors are less suppressed, as they arise always at one–loop level in a renormalizable theory. In the SM, the one–loop radiative correction of the bosonic sector and light fermions to the on–shell $WW\gamma$ vertex was calculated using the unitary gauge in Ref. [16]. Afterwards, the top quark contribution was calculated in Ref. [17]. In units of $a = g^2/96\pi^2$, it was found that $\Delta \kappa_\gamma \simeq O(10a)$ and $\Delta Q_\gamma \simeq O(a/10)$ [16, 17]. These form factors have been calculated in many renormalizable theories, such as SM with extended Higgs sectors [18, 23], supersymmetric models [19], theories with new Z bosons [20, 23], left–right symmetric models [22], and the so–called 331 models [24]. In all these weakly coupled renormalizable extensions of the SM, $\Delta \kappa_\gamma$ and ΔQ_γ are of the order of $O(a)$ and $O(a/100)$, respectively, i. e., they are about one order of magnitude lower than the values predicted by the SM. Also, their heavy–mass limit is consistent with the decoupling theorem [27]. Thus, while $\Delta \kappa_\gamma$ is sensitive to heavy–mass effects, ΔQ_γ is of decoupled nature. One exception to these results is the one obtained from composite models [21], whose prediction is comparable with those obtained from the SM.

As far as the CP–odd electromagnetic properties of the W boson are concerned, their scrutiny may provide relevant information for our knowledge of CP violation. In the SM, the only source of CP violation is the Cabbibo–Kobayashi–Maskawa (CKM) phase, which seems to be the cause of CP violation in nondiagonal processes [28, 29], as it is strongly suggested by experimental information on $B - \bar{B}$ mixing [30]. However, diverse studies [31] show that the CKM phase has a rather marginal impact on flavor–diagonal processes such as the electric dipole moments of elementary particles, which means that they could be highly sensitive to new sources of CP violation. In fact, although the magnetic quadrupole moment $\tilde{Q}_W$ receives a tiny contribution at the two–loop level in the SM [32], the electric dipole moment d_W is first generated at the three–loop level [33]. As already mentioned, the only class of models which can generate these quantities at the one–loop level are those involving both left–and right–handed currents with complex phases. This possibility has already been explored within the context of left–right symmetric models (LRSM) [25], but the corresponding contribution is strongly suppressed, as the complex phase is tiny due to experimental constraints on the $W_L - W_R$ mixing. Recently [34], a new possible source of CP violation that generate both $\Delta \tilde{\kappa}_\gamma$ and $\Delta \tilde{Q}_\gamma$ form factors has been explored in a model–independent manner using the effective Lagrangian technique. Instead of considering general couplings of W to fermions, which would be very suppressed due to a tiny complex phase, one can think in scalar–fermion couplings involving both scalar and pseudoscalar components, i.e., interactions of the way $\phi \bar{\Psi}(e + io\gamma_5)\Psi$. This Yukawa coupling induces at the one–

loop level the ϕWW coupling with both CP–even and CP–odd components, which in turns, when introduced inside a loop, induces the $WW\gamma$ vertex also with both type of components. This effect arises for instance from the Yukawa sector of the type–III two–Higgs doublet model [35]. In this model, the one–loop ϕWW vertex (here ϕ stands for one of the three neutral Higgs bosons predicted by the model) is generated by the known fermions. Another interesting possibility is when the HWW vertex is induced by new heavy fermions circulating in the loop, with H a SM–like Higgs boson. Since the new fermions are considered much heavier than the Fermi scale, they can be integrated out in the generating functional and thus, instead of focusing on a specific model, one can parametrize this class of effects in a model–independent manner via the effective Lagrangian technique [9]. Apart from the advantages of working in a model–independent fashion, this approach has the technical advantage that a two–loop calculation can be treated as an one–loop effect.

In the effective Lagrangian framework, the SM CP–even tree level HWW vertex can be modified in a $SU_L(2) \times U_Y(1)$–invariant way by adding a CP–odd dimension–six term, as follows:

$$\mathscr{L}_{eff} = gm_W HW^-_\mu W^{+\mu} + \frac{\tilde{\alpha}_{HWW}}{\tilde{\Lambda}^2}(\Phi^\dagger\Phi)W^a_{\mu\nu}\tilde{W}^{\mu\nu}_a. \quad (11)$$

After SSB, the relevant couplings which can contribute in the unitary gauge to the on–shell $WW\gamma$ vertex are given by

$$\frac{\mathscr{L}_{HW}}{gm_W} = H\left\{W^-_\mu W^{+\mu} + \frac{\tilde{\varepsilon}_{HWW}}{4m^2_W}[W^-_{\mu\nu}\tilde{W}^{+\mu\nu} + 2ie(\tilde{W}^+_{\mu\nu}A^\mu W^{-\nu} - \tilde{W}^-_{\mu\nu}A^\mu W^{+\nu})]\right\}. \quad (12)$$

These couplings contribute to both CP–even ($\Delta\kappa_\gamma, \Delta Q_\gamma$) and CP–odd ($\Delta\tilde{\kappa}_\gamma, \Delta\tilde{Q}_\gamma$) form factors, but we are interested only in the CP–odd ones. Since both HWW and $HWW\gamma$ have a nonrenormalizable structure, they yield an ultraviolet divergent amplitude for $\tilde{\kappa}_\gamma$, thereby requiring the adoption of a renormalization scheme. We have used the $\overline{MS}$ scheme with the renormalization scale $\mu = \tilde{\Lambda}$, which leads to a logarithmic dependence of the form $log(\tilde{\Lambda}^2/m^2_W)$. As far as the $\Delta\tilde{Q}_\gamma$ form factor is concerned, it cannot arise at one–loop level in any renormalizable theory [13]. Since the operator $(\Phi^\dagger\Phi)W^a_{\mu\nu}\tilde{W}^{\mu\nu}_a$ can only be generated at one loop or higher orders by the fundamental theory [36], our calculation represents a two–loop or higher order effect in this theory. Thereby, the contribution to $\Delta\tilde{Q}_\gamma$ must be necessarily finite, in accordance with renormalization theory. These form factors can be written in terms of two–point Passarino–Veltman scalar functions as follows [34]:

$$\Delta\tilde{\kappa}_\gamma = \frac{\tilde{\varepsilon}_{HWW}\alpha}{4\pi^2_W\delta}\left\{16(7+6(B_W - B_{WH})) - 8(23 + 21B_H + 15B_W - 36B_{WH})x_H + \right.$$
$$\left. 3(25 + 26B_H + 4B_W - 30B_{WH})x^2_H - 3(1 + B_H - B_{WH})x^3_H - 9\delta x_H log\left(\frac{\tilde{\Lambda}^2}{m^2_W}\right)\right\} (13)$$

$$\Delta\tilde{Q}_\gamma = \frac{\tilde{\varepsilon}_{HWW}\alpha}{4\pi^2\delta}\left\{-4 + (5 + 2(2B_H + B_W - 3B_W))x_H - 2(1 + B_H - B_{WH})x^2_H\right\}, \quad (14)$$

where $\delta = 4 - x_H$, $s_W = sin\theta_W$, $x_H = (m_H/m_W)^2$, $B_H = B_0(0, m_H^2, m_H^2)$, $B_W = B_0(0, m_W^2, m_W^2)$, and $B_{WH} = B_0(m_W^2, m_H^2, m_W^2)$.

In order to make predictions, some values for the $\tilde{\varepsilon}_{HWW}$ parameter and the $\tilde{\Lambda}$ scale must be assumed. Assuming that $(\Phi^\dagger\Phi)W_{\mu\nu}^a\tilde{W}_a^{\mu\nu}$ is generated at the one loop level by the fundamental theory, the $\tilde{\varepsilon}_{HWW}$ parameter must contain the loop factor $1/16\pi^2$ along a g coupling for each gauge field. From these considerations, it is reasonable to assume that $\tilde{\varepsilon}_{HWW} \sim (g^2/16\pi^2)f$, where $f = f(v, \tilde{\Lambda})$ is a dimensionless loop function, whose specific structure depends on the details of the underlying physics. Since the CP–violating effects are expected to be of decoupled nature, f is expected to be of the order $O(1)$ at most. We will adopt a somewhat optimistic scenario, which consists in assuming that $f \sim 1$. We will thus make predictions under the assumption that $\tilde{\varepsilon}_{HWW} = (v/\tilde{\Lambda})^2(\alpha/4\pi s_W^2)$. For m_H ranging between 120 and 200 GeV and for $\tilde{\Lambda} = 1, 2, 3$ TeV, the estimated values for the CP–odd electromagnetic moments are $d_W \sim 3 - 6 \times 10^{-20} e.cm$ and $\tilde{Q}_W \sim -10^{-36} e.cm^2$ [34].

It is interesting to compare these results with those obtained in other scenarios. In the SM, at the lowest order, d_W and $\tilde{Q}_W$ have been estimated to be about $10^{-29} e.cm$ [33] and $-10^{-51} e.cm^2$ [32], respectively. All studies in specific extensions of the SM have focused on d_W. Some of these extensions predict values for d_W that are several orders of magnitude larger than the SM one. For instance, a value of $10^{-22} e.cm$ was estimated for d_W in LRSM [25, 37], and similar results were found in supersymmetric models [37]. Explicit calculations carried out in the context of extended Higgs sectors show that $d_W \sim 10^{-20} - 10^{-21} e.cm$ [38]. A similar value was found in the context of the 331 model [39]. From these results, we can conclude that our model–independent estimation for d_W lies in the range of the predictions obtained from some specific models. The values given in Ref. [34] for d_W and $\tilde{Q}_W$ are 8 and 15 orders of magnitude above the respective SM contributions. To our knowledge, this model–independent estimation for the size of the $\tilde{Q}_W$ given in Ref.[34] is the first one obtained in theories beyond the SM.

2.3. Heavy gauge bosons contribution to the off–shell WWV vertex

Within the conventional quantization scheme, which is based on the BRST symmetry [7] with a linear gauge–fixing procedure, also called the FPM, it is not gauge invariance which is a fundamental concept, but gauge independence, which guarantees independence with respect to the gauge–fixing procedure used to quantize the system. Physical quantities are those which are gauge independent in this sense. However, this interpretation could be limited, as suggested by recent investigations based on nonconventional quantization schemes, as the BFM [5] and the PT [6], which allow one to establish a profound link between gauge invariance and gauge independence [40]. While the BFM was developed to construct gauge–invariant Green functions, the PT allows one to construct both gauge–invariant and gauge–independent Green functions. The connection between the PT and the BFM establishes that the Green functions calculated via the BFM Feynman rules coincide with those obtained through the PT for the specific value $\xi_A = 1$ [41]. With the generalization of the PT to any order [42], this link has been established too [43]. The reason for such a link remains a puzzle, though it is worth noting that the

Feynman–t'Hooft gauge yields no unphysical thresholds. This link has important practical advantages, as one can simply use the BFM Feynman–t'Hooft gauge to calculate gauge–independent off shell amplitudes, which happens to be much less cumbersome than the use of the PT.

Although in conventional quantization schemes the quantum action of the theory is not gauge–invariant, it is still possible to introduce gauge invariance with respect to a subgroup of such a theory. This scheme can be particularly useful in situations as the following: imagine that we are interested in assessing the virtual effects of the heavy physics lying beyond the Fermi scale on the SM Green functions in a $SU_L(2) \times U_Y(1)$–covariant manner, in which case it is only necessary to introduce a quantization scheme for the heavy fields since the SM fields would only appear as external legs. It should be emphasized that if new physics exists and it couples to the electroweak gauge bosons, it must appear in some representation of the electroweak group. This is indeed the philosophy behind the effective Lagrangian approach [9] widely used in the context of the electroweak theory, where it is assumed that the new physics effects must respect the electroweak symmetry. These gauges, first introduced by Fujikawa in the context of the SM[8], are called nonlinear or covariant gauges. This class of gauges has proved useful in many applications [44], simplifying both technically and conceptually the calculations of radiative corrections. Recently [3], this class of gauges was used to investigate the impact on the off–shell WWV vertex of the new gauge bosons predicted by the so–called 331 model [2], which is based on the $SU_C(3) \times SU_L(3) \times U_X(1)$ gauge group. It turns out that in the first stage of SSB, when $SU_L(3) \times U_X(1)$ break down into $SU_L(2) \times U_Y(1)$, there emerge singly and doubly charged gauge bosons in a doublet of $SU_L(2) \times U_Y(1)$ with hypercharge 3, $Y_\mu^\dagger = (Y_\mu^{--}, Y_\mu^{-})$. This gives rise to unusual couplings between the new gauge bosons and the electroweak gauge fields, which arise via the $SU_L(2) \times U_Y(1)$–covariant derivative [24, 45]. These massive gauge bosons are known as bileptons because they violate the lepton number in two units. In this stage of SSB, the Yang–Mills sector associated with the $SU_L(3) \times U_X(1)$ group can be expressed in terms of three $SU_L(2) \times U_Y(1)$–invariant pieces: $\mathscr{L}_{SM}$, $\mathscr{L}_{SMNP}$, and $\mathscr{L}_{NP}$, which represent the electroweak Yang–Mills sector, the interactions between the electroweak gauge bosons and the new ones, and the couplings among the new gauge bosons, respectively [45, 3]. Here we only display the part which represents the couplings of the electroweak gauge bosons with the bileptons:

$$\mathscr{L}_{SMNP} = -\frac{1}{2}(D_\mu Y_\nu - D_\nu Y_\mu)^\dagger (D^\mu Y^\nu - D^\nu Y^\mu) - Y^{\dagger \mu}(\mathbf{W}_{\mu\nu} + \mathbf{B}_{\mu\nu})Y^\nu + \cdots \quad (15)$$

where $\mathbf{W}_{\mu\nu} = ig\tau^a W_{\mu\nu}^a/2$, $\mathbf{B}_{\mu\nu} = ig'YB_{\mu\nu}/2$, and $D_\mu = \partial_\mu - \mathbf{W}_\mu - \mathbf{B}_\mu$. The above terms, which are invariant under the electroweak group, induce the trilinear (WYY, VYY) and quartic ($WWYY, VWYY$) couplings, which contribute at the one–loop level to the WWV vertex. The idea is to define the Y propagators within the conventional quantization method without spoiling this symmetry. In Ref. [3], gauge–fixing functions transforming as Y_μ under $SU_L(2) \times U_Y(1)$ were introduced for the bileptons. As a consequence, the gauge–fixing Lagrangian is manifestly invariant under this group. For the same reason, the Lagrangian of the ghost sector is also manifestly invariant under this group. The gauge–fixing Lagrangian not only allows one to define the bilepton propagators, but also modifies nontrivially the couplings between the bileptons and the electroweak gauge bosons. It also removes some unphysical couplings that greatly simplifies the calculations. As a consequence of $SU_L(2) \times U_Y(1)$–invariance, the trilinear couplings WYY and VYY share the same Lorentz structure. Let $\Gamma_{\alpha\mu\nu}(k,k_1,k_2)$ be the vertex function associated with these vertices, with k_α the four–momentum of the electroweak gauge boson. Then, these vertices satisfy the following simple Ward identities [3]:

$$k^\alpha \Gamma_{\alpha\mu\nu}(k,k_1,k_2) = \Pi_{\mu\nu}^{Y^\dagger Y^\dagger}(k_2) - \Pi_{\mu\nu}^{YY}(k_1), \tag{16}$$

where the Πs are two–point vertex functions. Once again, due to the $SU_L(2) \times U_Y(1)$ symmetry, the trilinear vertices $WS^\dagger S$ and $VS^\dagger S$ satisfy simple Ward identities

$$k^\alpha \Gamma_\alpha^{(W,V)S^\dagger S} = \Pi^{S^\dagger S^\dagger}(k_2) - \Pi^{SS}(k_1), \tag{17}$$

where k_α is the four–momentum of the electroweak gauge boson and S stands for a commutative (pseudo–Goldstone boson) or anticommutative (ghost) charged scalar.

The $SU_L(2) \times U_Y(1)$–covariant structure of the Feynman rules guarantees the gauge invariance of the WWV vertex, though it will be gauge dependent, which also occurs when the BFM is applied. In other words, gauge–invariant quantum actions render gauge–invariant but not gauge–independent Green functions. However, motivated by the link between the BFM and the PT, the results were obtained in the Feynman–'t Hooft gauge [3]. The amplitude can be written as

$$\Gamma_{\alpha\beta\mu}^V = -g_V I_{\alpha\beta\mu}, \tag{18}$$

with $I_{\alpha\beta\mu}$ the loop amplitude, which is the same for both $WW\gamma$ and WWZ vertices. It is important to notice that the associated Green functions differ only by the global factor g_V, just as occurs at the tree–level. This means that $SU_L(2) \times U_Y(1)$ invariance is preserved at the one–loop level. Thus $I_{\alpha\beta\mu}$ for on-shell W bosons must satisfy the simple Ward identity

$$q^\mu I_{\alpha\beta\mu} = 0, \tag{19}$$

which can be verified once the loop integrals are solved explicitly. At this order, the model only contributes to the CP–even form factors $\Delta\kappa_V$ and ΔQ_V [3], which are given by

$$\begin{aligned}\Delta\kappa_V =\ & \frac{6a}{x^3}\Big\{x_W x(8x_W+3) - 6x_W[x_W(1+x_W) - 3x_Y x]Q^2 C_0 \\ & + 4x_Y x^2[B_0(3) - B_0(1)] + [26x_W^2 - 32x_Y x_W x + x_W][B_0(1) - B_0(2)]\Big\},\end{aligned} \tag{20}$$

$$\begin{aligned}\Delta Q_V =\ & \frac{12a}{x^3}\Big\{6x_W[2x_W(2x_W^3 - 2x_W^2(1+4x_Y) + x_W(1+6x_Y) - 3x_Y) + x_Y]Q^2 C_0 \\ & + 4x_W[x_W(6x_W^2 - 5x_W + 8x_Y - 1) - 2x_Y][B_0(1) - B_0(2)] \\ & + 4x_W x_Y x^2[B_0(2) - B_0(3)] - x_W x[1 + 2x_W(6x_W - 1)]\Big\},\end{aligned} \tag{21}$$

where we have introduced the definitions $Q = 2q$, $x_W = m_W^2/Q^2$, and $x_Y = m_Y^2/Q^2$. $B_0(i)$ and C_0 stand for scalar functions, which are given in Ref. [3].

Before presenting numerical results, it is interesting to discuss the SM radiative correction. In the SM, the one–loop amplitudes were calculated using the conventional quantization scheme along the Feynman–t'Hooft gauge [46], but the result for $\Delta\kappa_V$ is not gauge–invariant. Later on, the amplitudes were recalculated using the PT [47]. The form factor $\Delta\kappa_V$ obtained for the authors disagrees from that presented in [46], though there is agreement for ΔQ_V. It was found that the radiative corrections to $\Delta\kappa_V$ are of the order of α/π, whereas ΔQ_V is one order of magnitude below. As far as the new gauge bosons contribution is concerned, it was found [3] that for a relatively light bilepton, with mass in the range $2m_W < m_Y < 6m_W$, both $\Delta\kappa_V$ and ΔQ_V take values within the range of the SM contribution. In the case of a more heavy bilepton, with mass in the range $8m_W < m_Y < 12m_W$, the form factors remain essentially uniform and are of order 10^{-6}. These results are in agreement with the expectations arising from a decoupling scenario of new physics, as argued in the light of the effective Lagrangian approach in Ref.[3].

On the experimental side, the D0 Collaboration set recently the following limits on the CP–even form factors: $-0.88 < \Delta\kappa_\gamma < 0.96$ and $-0.20 < \Delta Q_\gamma < 0.20$ for the $WW\gamma$ coupling [49], and $-2.0 < \Delta\kappa_Z < 2.4$, $-0.48 < \Delta Q_Z < 0.48$, and $-0.49 < \Delta g_1^Z < 2.4 <$ for the WWZ coupling [50].

3. THE VVV VERTEX

The s–channel reaction $e^+e^- \to VV$, if detected at NLC, would reveal the existence of nonvanishing neutral $VVV = ZZZ, ZZ\gamma, Z\gamma\gamma$ couplings, which would be a clear evidence of the presence of new physics, as these couplings are very suppressed in the SM [48]. These couplings are severally restricted by Bose and Lorentz symmetries. In this context, the Landau–Yang theorem invokes Bose symmetry and rotational invariance arguments to prove that a vector particle cannot decay into two massless vector particles [51]. This is the reason why the $Z \to \gamma\gamma$ decay is forbidden. The requirement of two massless vector bosons in the final state is essential for the theorem, as transitions of a vector boson into two massive vector bosons or into a massive vector boson and the photon can occur, as it has been verified in theories with an extra Z' boson, in which the $Z' \to ZZ$ and $Z' \to Z\gamma$ [45, 52] decays are allowed. Indeed, if both Bose symmetry and Lorentz invariance are simultaneously respected, the VVV coupling vanishes when the three bosons are real, they only can exist when at least one of the particles is off-shell. In particular, the Z boson cannot have static electromagnetic properties [48, 12, 53], which is due to fact that the particle coincides with the antiparticle. Only off–shell electromagnetic properties exist for this particle [48, 12, 53]. However, as soon as this requirement is violated, the VVV vertex with the three bosons on–shell can exist. For instance, in the noncommutative SM [10], which violates the Lorentz symmetry, but respects the Bose one, the on-shell VVV couplings, including the the $\gamma\gamma\gamma$ one, are generated at the tree–level. On the other hand, in Ref. [54] the Z decay into two photons was studied using a vertex function for the $Z\gamma\gamma$ coupling which respects the Lorentz symmetry, but is antisymmetric under the interchange of the photons, i.e. it violates the Bose symmetry.

Another interesting peculiarity of the VVV vertices is that they must be induced by loop effects in any renormalizable theory since they cannot possess a renormalizable structure. In addition, at the one–loop level, they only can be generated by fermionic triangles [55]. All these features naturally restrict the existence of these couplings.

Motivated by the experimental potential of the reaction $e^+e^- \to V^* \to VV$ at NLC (from now on, $*$ stands for off–shell particles), with the final state ZZ or $Z\gamma$ on–shell, a model–independent parametrization for the VVV^* vertex function, respecting the Bose symmetry, Lorentz invariance, and gauge invariance, was proposed [12, 55, 56]. The vertex functions for the $Z_{\alpha_1}(k_1)Z_{\alpha_2}(k_2)V_\alpha^*(k)$ and $Z_\alpha(k)A_\beta(q)V_\mu^*(p)$ couplings, with all momenta incoming, can be written as

$$\Gamma^{ZZV^*}_{\alpha_1\alpha_2\alpha} = \frac{i(k^2 - m_V^2)}{m_Z^2}\left[f_4^V(k_{\alpha_1}g_{\alpha_2\alpha} + k_{\alpha_2}g_{\alpha_1\alpha}) + f_5^V\varepsilon_{\alpha_1\alpha_2\alpha\beta}(k_1-k_2)^\beta\right], \quad (22)$$

$$\Gamma^{Z\gamma V^*}_{\alpha\beta\mu}(k,q,p) = \frac{i(p^2 - m_V^2)}{m_Z^2}\Bigg\{h_1^V(q_\alpha g_{\beta\mu} - q_\mu g_{\alpha\beta}) +$$
$$\frac{h_2^V}{m_Z^2}(q_\mu p_\beta - p\cdot q g_{\beta\mu}) + h_3^V\varepsilon_{\alpha\beta\mu\nu}q^\nu + \frac{h_4^V}{m_Z^2}\varepsilon_{\mu\beta\lambda\rho}p^\lambda q^\rho\Bigg\}. \quad (23)$$

In these expressions, the form factors f_4^V, h_1^V and h_2^V are CP violating, whereas f_5^V, h_3^V and h_4^V are CP conserving. This parametrization has been used to study the $ZZ\gamma$ and $Z\gamma\gamma$ couplings at NLC [57] and also in some CP–violating rare Z decays [58]. Recently, the D0 Collaboration has obtained the following limits on these form factors [59]: $|h^Z_{10,30}| < 0.23$, $|h^Z_{20,40}| < 0.020$, $|h^\gamma_{10,30}| < 0.23$, and $|h^\gamma_{20,40}| < 0.019$, where the following parametrization was used, $h_i^V = h_{i0}^V/(1 + s/\Lambda^2)^{n_i}$ ($i = 1,\cdot,4$). The values $\Lambda = 1000$ GeV, $n_1 = n_3 = 3$, and $n_2 = n_4 = 4$ were used.

The relative importance of the VVV couplings can be more easily appreciated when compared with the WWV ones in a $SU_L(2)\times U_Y(1)$–invariant scheme using the effective Lagrangian technique. It turns out that at the lowest order, the WWV couplings are generated by dimension–six operators in the linear realization of the electroweak symmetry and by dimension–4 operators in the nonlinear scheme. In contrast, the VVV couplings are induced by dimension–8 operators in the linear realization and by dimension-6 operators in the nonlinear one [60]. This means that the VVV couplings have always an additional $(v/\Lambda)^2$ suppression factor compared with the WWV ones. For instance, assuming that both type of couplings are generated at the same order for the new physics, that the masses of the new particles involved are of the same order of magnitude, and that the coefficients α are also similar and of order $O(1)$, we obtain that, the VVV couplings are suppressed in two orders of magnitude with respect to the WWV ones at $\Lambda = 1$ TeV. From these considerations, one can conclude that it is quite unlikely that these couplings could be substantially enhanced within a conventional perturbative field theory. In Ref. [60], the most general vertex functions were constructed for the off-shell VVV couplings, which contain as a particular case the vertices given by eqs. (22,23) [60]. As far as the off-shell electromagnetic properties of the Z boson is concerned, they are generated in the linear realization by four operators of dimension–8 and four ones of dimension–10 [60]. An equal number of operators contribute in the nonlinear scheme, but now with dimension two units lower. In both schemes, these operators are proportional to $\partial^\mu B_{\mu\nu}$, which means that this term can be replaced by the current J^ν through the use of the equations of motion. Two classes of off–shell electromagnetic properties of the Z boson can be identified by noting that $\partial^\mu B_{\mu\nu} \to (\partial^\mu F_{\mu\nu}, \partial^\mu Z_{\mu\nu})$. The properties associated with the tensor fields $\partial^\mu F_{\mu\nu}$ and $\partial^\mu \tilde{F}_{\mu\nu}$ define the so–called anapole moment. The CP–even one is induced by the SM at the one–loop level [48]. The structures proportional to the terms $\partial^\mu Z_{\mu\nu}$ and $\partial^\mu \tilde{Z}_{\mu\nu}$ allows one to define the transition electromagnetic properties of the Z boson, namely, the magnetic and electric dipole moments, as well as the magnetic and electric quadrupole moments [60, 61]. Of course, all these electromagnetic properties vanish on–shell.

In closing this section, we would like to present some general considerations concerning the origin and nature of the trilinear neutral gauge boson couplings. First, we would like to emphasize that none of the VVV couplings reflects the non–Abelian nature of the electroweak group, as it is frequently claimed in the literature. This is due to the fact that the Yang–Mills sector neither induces these couplings at the tree–level nor contributes to them at the one–loop order. This in contrast with the quartic VVVV couplings, which are generated by the Yang–Mills sector at the one–loop level. This is the reason why the $SU_L(2) \times U_Y(1)$–invariant operators that generate the trilinear vertices VVV [60] do not induce the quartic ones VVVV [62]. It should be possible to show that this happens indeed at higher dimensions. This should be contrasted with the fact that those operators that generate the trilinear WWV couplings always induce the quartic $WWVV$ ones [9], which in fact is a direct consequence of the non–Abelian nature of the electroweak group. As emphasized in Ref. [62], these differences between the set of vertices $(WWV, WWVV)$ and $(VVV, VVVV)$ have important experimental implications. In fact, while an experimental constraint on WWV can be translated into $WWVV$ and viceversa due to gauge invariance, it is necessary to use different physical processes to bound VVV and VVVV. On the other hand, as already commented, the simultaneous presence of the Bose and Lorentz symmetries implies that the VVV vertex cannot exist on the mass shell. However, as soon as one of these requirement is relaxed [10, 54], the on–shell VVV vertex is generated. This is the case of the noncommutative SM, which is a field theory defined on a noncommutative space–time $[x^\mu, x^\nu] = i\theta^{\mu\nu}$, that induces the $VVV = ZZZ, ZZ\gamma, Z\gamma\gamma, \gamma\gamma\gamma$ couplings at the tree–level [10]. The existence of these vertices is a consequence of the Lorentz symmetry violation through the $\theta^{\mu\nu}$ parameters.

4. FINAL REMARKS

The self–interactions of the electroweak gauge bosons constitute an important and sensitive probe of the SM. Due to this, their study attracts the attention of both theorists and experimentalists. On the theoretical side, it is interesting to study the one–loop structure of the trilinear vertices WWV and VVV, as they can be sensitive to virtual effects of new particles. In this work, the one–loop impact of new particles to these couplings was reviewed. First, the radiative contributions to the on–shell $WW\gamma$ vertex were analyzed in the context of many renormalizable SM extensions. Since this vertex is gauge–independent, it serves to investigate the sensitivity of the Yang–Mills sector to new physics effects. As far as the CP–odd form factors is concerned, they may represent a genuine effect of physics lying beyond the SM, as the CKM mechanism cannot generate observable effects. A possible new source of CP violation induced by general Yukawa couplings was studied in a model–independent manner using the effective Lagrangian approach, which induces both the electric dipole and magnetic quadruopole moments. The interesting problem of studying the virtual effects of new gauge bosons on the off–shell WWV^* vertex was examined. The concepts of gauge invariance and gauge independence were discussed in the light of nonconventional quantization schemes, as the BFM and the PT. An alternative method to construct gauge invariant Green functions, based on the BRST symmetry scheme endowed with covariant gauge–fixing procedures, was discussed. The equivalence of this scheme with the BFM was stressed and the link between the BFM and the PT was exploited. The method was illustrated in the context of the minimal 331 model, which predicts unusual couplings between the new gauge bosons and the electroweak ones. To our knowledge, apart from the SM radiative correction given in the context of the PT [47], this is the first time that a calculation concerning the virtual effects of new heavy gauge bosons on this vertex is presented. It is also the first time that a nonlinear gauge is used beyond the electroweak theory to construct gauge–invariant quantum actions characterizing new physics effects [3].

The structure of the trilinear neutral gauge boson couplings VVV was examined. General criterions based on the effective Lagrangian approach, allow one to conclude that these couplings are much more suppressed than the WWV ones. It is stressed that, in contrast with the WWV vertex, the VVV one does not reflect the $SU_L(2) \times U_Y(1)$ symmetry since it is neither generated at tree–level by the Yang–Mills sector nor receive contributions from gauge bosons at the one–loop level. Indeed, this vertex is governed by the simultaneous presence of the Bose and Lorentz symmetries, which restrict it to exist only off–shell. It seems that beyond the conventional weakly coupled quantum field theories there are more promising scenarios. Recent investigations in the context of field theories defined in a noncommutative space–time, in which Lorentz symmetry is not respected, seems to confirm this conjecture, as these vertices are generated at the tree–level. Also, explicit violation of the Bose symmetry could be the cause of a possible detection of these vertices. Anyway, the experimental scrutiny of these vertices is very important, as it would allow one to establish with a high degree of precision if the Bose and Lorentz symmetries are respected at very small distances.

ACKNOWLEDGMENTS

I thank M. A. Pérez and G. Tavares–Velasco by their participation in many of the material presented here. Also, support from Conacyt and VIEP–BUAP (México) is acknowledged.

REFERENCES

1. M. A. Pérez and J. J. Toscano, Phys. Lett. **B289**, 381 (1992); J. M. Hernández, M. A. Pérez, and J. J. Toscano, Phys. Rev. **D51**, R2044 (1995); M. A. Pérez, J. J. Toscano, and J. Wudka, Phys. Rev. **52**, 494

(1995); J. M. Hernández, M. A. Pérez, and J. J. Toscano, Phys. Lett. **375**, 227 (1996); J. L. Díaz–Cruz, M. A. Pérez, and J. J. Toscano, Phys. Lett. **398**, 347 (1997).
2. F. Pisano and V. Pleitez, Phys. Rev. **D46**, 410 (1992); P. H. Frampton, Phys. Rev. Lett. **69**, 2889 (1992).
3. J. Montaño, F. Ramírez–Zavaleta, G. Tavares–Velasco, and J. J. Toscano, Phys. Rev. **D72**, 055023 (2005).
4. L. D. Faddeev and V. N. Popov, Phys. Lett. **B25**, 29 (1967).
5. B. S. De Witt, Phys. Rev. **162**, 1195 (1967); *Dynamical Theory of Groups and Fields* (Gordon and Breach, New York, 1965); C. J. Isham, R. Penrose, and D. W. Sciama, *Quantum Gravity 2* (Oxford University Press, Oxford, 1981); G. t'Hooft, Acta Univ. Wratislavensis **368**, 345 (1976); M. T. Grisaru, P. van Nieuwenhuizen, and C. C. Wu, Phys. Rev. **D12**,b 3203 (1975); H. Kluberg–Stern and J. B. Zuber, Phys. Rev. **D12**, 482 (1975); D. G. Boulware, Phys. Rev. **D23**, 389 (1981); C. F. Hart, Phys. Rev. **D28**, 1993 (1983); L. F. Abbott, Nucl. Phys. **B185**, 189 (1981); Acta Phys. Pol. **B13**, 33 (1982); L. F. Abbott, M. T. Grisaru, and R. K. Schaefer, Nucl. Phys. **B229**, 372 (1983). For the BFM formulation of the electroweak theory see: A. Denner, G. Weglein, and S. Dittmaier, Nucl. Phys. **B440**, 95 (1995); Acta Phys. Pol. **B27**, 3645 (1996).
6. J. M. Cornwall, Phys. Rev. **D26**, 1453 (1982); J. M. Cornwall and J. Papavassiliou, Phys. Rev. **D40**, 3474 (1989); J. Papavassiliou, Phys. Rev. **D41**, 3179 (1990).
7. C. Becchi, A. Rouet, and A. Stora, Commun. Math. Phys. **42**, 127 (1975); Ann. Phys. (N.Y.) **98**, 287 (1976); I. V. Tyutin, Lebedev Report N0. FIAN 39, 1975 (unpublished). The BRST symmetry arises naturally form the field–antifield formalis. For a reviw, see J. Gomis, J. Paris, and S. Samuel, Phys. Rep. **259**, 1 (1995).
8. K. Fujikawa, Phys. Rev. **D7**, 393 (1973). The most general gauge–fixing functions of R_ξ type, non–linear in both the vector and scalar sectors, were introduced in U. Cotti, J. L. Díaz–Cruz, and J. J. Toscano, Phys. Lett. **B404**, 308 (1997) and their implications on the ghost sector discussed with detail in J. G. Méndez and J. J. Toscano, Rev. Mex. Fís. **50**, 346 (2004).
9. W. Buchmuller and D. Wyler, Nucl Phys. **B268**, 621 (1986); J. Wudka, Int, J. Mod. Phys. **A9**, 2301 (1994).
10. H. S. Snyder, Phys. Rev. **71**, 38 (1947); X. Calmet, B. Jurco, P. Schupp, and J. Wess, Eur. Phys. J. **C23**, 363 (2002); W. Behr, N. G. Deshpande, G. Duplancic, J. Trampetic, and J. Wess, Eur. Phys. J. **C29**, 441 (2003). See also, J.C. López–Domínguez, O. Obregón, C. Ramírez, and J. J. Toscano, Phys. Rev. **D73**, 095003 (2006).
11. A. Arhrib, J.-L. Kneur, and G. Moultaka, Phys. Lett. **B376**, 127 (1996); K. Hagiwara, T. Hatsukano, S. Ishihara, and R. Szalapski, Nucl. Phys. **B496**, 66 (1997); J. Ellison and J. Wudka, Ann. Rev. Nucl. Part. Sci. **48**, 33 (1998).
12. K. J. F. Gaemers and G. J. Gounaris, Z. Phys. **C1**, 259 (1979); K. Hagiwara, R. D. Peccei, D. Zeppenfeld, and K. Hikasa, Nucl. Phys. **B282**, 253 (1987). See also, F. Boudjema and C. Hamzaoui, Phys. Rev. **D43**, 3748 (1991).
13. G. Tavares–Velasco and J. J. Toscano, J. Phys. **G30**, 1299 (2004).
14. G. Tavares–Velasco and J. J. Toscano, Phys. Rev. **D70**, 053006 (2004).
15. J. F. Nieves and P. B. Pal, Phys. Rev. **D55**, 3118 (1991).
16. W. A. Bardeen, R. Gastmans, and B. Lautrup, Nucl. Phys. **B46**, 319 (1972).
17. G. Couture and J. N. Ng, Z. Phys. **C35**, 65 (1987).
18. G. Couture, J. N. Neg, J. L. Hewett, and T. G. Rizzo, Phys. Rev. **D36**, 859 (1987).
19. C. L. Bilachak, R. Gatsmans, and A. van Proeyen, Nucl. Phys. [**?**], 46 (1986); G. Couture, J. N. Ng, J. L. Hewett, and T. G. Rizzo, Phys. Rev. **D38**, 860 (1988); A. B. Lahanas and V. C. Spanos, Phys. Lett. **B334**, 378 (1994); T. M. Aliyev, *ibid.* **155**, 364 (1985); A. Arhrib, J. L. Kneur, and G. Moultaka, *ibid.* **376**, 127 (1996).
20. N. K. Sharma, P. Saxena, Sardar Singh, A. K. Nagawat, and R. S. Sahu, Phys. Rev. **D56**, 4152 (1997).
21. T. G. Rizzo and M. A. Samuel, Phys. Rev. **D35**, 403 (1987); A. J. Davies, G. C. Joshi, and R. R. Volkas, *ibid.* **42**, 3226 (1990).
22. F. Larios, J. A. Leyva, and R. Martínez, Phys. Rev. **D53**, 6686 (1996).
23. G. Tavares–Velasco and J. J. Toscano, Phys. Rev. **D69**, 017701 (2004).
24. G. Tavares–Velasco and J. J. Toscano, Phys. Rev. **D65**, 013005 (2001); J. L. García–Luna, G. Tavares–Velasco, and J. J. Toscano, Phys. Rev. **D69**, 093005 (2004).
25. D. Atwood, C. P. Burgess, C. Hamzaoui, B. Irwin, and J. A. Robinson, Phys. Rev. **D42**, 3770 (1990).

26. R. Foot, H. N. Long, and Tuan A. Tran, Phys. Rev. **D50**, R34 (1994); H. N. Long, Phys. Rev. **D54**, 4691 (1996); N, Anh Ky, H. N. Long, and D. V. Soa, Phys. Lett. **B486**, 140 (2000).
27. T. Appelquist and J. Carazzone, Phys. Rev. **D11**, 2856 (1975).
28. J. H. Christenson, J. W. Cronin, V. L. Fitch, and R. Turlay, Phys. Rev. Lett. **13**, 138 (1964).
29. J. Alexander *et al.* [Heavy Flavor Averaging Group (HFAG) Collaboration], hep-ex/0412073.
30. A. Soni, hep-ph/0509180.
31. For a recent review, see M. Pospelov and A. Ritz, Ann. Phys. (N.Y.) **318**, 119 (2005).
32. I. B. Khriplovich and M. E. Pospelov, Nucl. Phys. **B420**, 505 (1994).
33. M. E. Pospelov and I. B. Khriplovich, Sov. J. Nucl. Phys. **53**, 638 (1991) [Yad. Fiz. **53**, 1030 (1991)]; E. P. Shabalin, Sov. J. Nucl. Phys. **28**, 75 (1978) [Yad. Fiz. **28**, 151 (1978).
34. J. Montaño, F. Ramírez–Zavaleta, G. Tavares–Velasco, and J. J. Toscano, Phys. Rev. **D72**, 115009 (2005).
35. G. Tavares–Velasco and J. J. Toscano, work in progress.
36. C. Arzt, M. B. Einhorn, and J. Wudka, Nucl. Phys. Rev. **D49**, 1370 (1994); Nucl. Phys. **B433**, 41 (1994).
37. D. Chang, W. Y. Keung, and J. Liu, Nucl. Phys. **B355**, 295 (1991).
38. R. López–Mobila and T. H. West, Phys. Rev. **D51**, 6495 (1995).
39. C. S. Huang and T. J. Li, Phys. Rev. **D50**, 2127 (1994).
40. A. Denner, G. Weiglein, and S. Dittmaier, Phys. Lett. **B333**, 420 (1994); S. Hashimoto, J. Kodaira, Y. Yasui, and K. Sasaki, Phys. Rev. **D50**, 7066 (1994); J. Papavassiliou, Phys. Rev. Lett. **84**, 2782 (2000); Phys. Rev. **D62**, 045006 (2000); D. Binosi and J. Papavassiliou, Phys. Rev. **D66**, 076010 (2002); **D66**, 025024 (2002); **D66**, R111901 (2002); J. Phys. **G30**, 203 (2004); D. Binosi, J. Phys. **G30**, 1021 (2004).
41. A. Denner, G. WEiglein, and S. Dittmaier, Phys. Lett. **B333**, 420 (1994); S. Hashimoto, J. Kodaira, Y. Yasui, and K. Sasaki, Phys. Rev. **D50**, 7066 (1994).
42. D. Binosi and J. Papavassilio, Phys. Rev. **D66**, 025024 (2002); *ibid.* **66**, R111901 (2002).
43. J. Papavassiliou, Phys. Rev. Lett. **84**, 2782 (2000); Phys. Rev. **D62**, 045006 (2000); D. Binosi and J. Papavassiliou, Phys. Rev. **D66**, 076010 (2002). D. Binosi and J. Papavassiliou, J. Phys. **G30**, 203 (2004); D. Binosi, *ibid.* **30**, 1021 (2004).
44. M. Bace and N. D. Hari Dass, Ann. Phys. (N.Y.) **94**, 349 (1975); M. B. Gavela, G. Girardi, C. Malleville, and P. Sorba, Nucl. Phys. **B193**, 257 (1981); N. M. Monyonko, J. H. Reid, and A. Sen, Phys. Lett. **B136**, 265 (1984); N. M. Monyonko and J. H. Reid, Phys. Rev. **D32**, 962 (1985); G. Jikia and A. Tkabladze, Phys. Lett. **B323**, 453 (1994); **332**, 441 (1994); G. Jikia, Phys. Lett. **B298**, 224 (1993); Nucl. Phys. **B405**, 24 (1993); **412**, 57 (1994); D. A. Dicus and C. Kao, Phys. Rev. **D49**, 1265 (1994); A. Abbasabadi, D. Bowser–Chao, D. A. Dicus, and W. W. Repko, Phys. Rev. **D52**, 3919 (1995); U. Citti, J. L. Díaz–Cruz, and J. J. Toscano, Phys. Rev. **D62**, 035009 (2000); J. L. Díaz–Cruz, J. Hernández–Sánchez, and J. J. Toscano, Phys. Lett. **B512**, 339 (2001); J. Hernández–Sánchez, M. A. Pérez, G. Tavares–Velasco, and J. J. Toscano, Phys. Rev. **D69**, 095008 (2004).
45. M. A. Pérez, G. Tavares–Velasco, and J. J. Toscano, Phys. Rev. **D69**, 115004 (2004).
46. E. N. Argyres *et al.*, Nucl. Phys. **B391**, 23 (1993).
47. J. Papavassiliou and K. Philippides, Phys. Rev. **D48**, 4255 (1993).
48. F. M. Renard, Nucl. Phys. **B196**, 93 (1982); A. Barroso, F. Boudjema, J. Cole, and N. Dombey, Z. Phys. **C28**, 149 (1985); A. Barroso, P. Nogueira, and J. C. Romao, Z. Phys. **C33**, 243 (1986).
49. D0 Collaboration, V. M. Abazov *et al.*, Phys. Rev. **D71**, 091108 (2005).
50. D0 Collaboration, V. M. Abazov *et al.*, Phys. Rev. Lett. **95**, 141802 (2005).
51. L. D. Landau, Dokl. Akad. Nauk SSSR **60**, 242 (1948); C. N. Yang, Phys. Rev. **77**, 242 (1950).
52. D. Chang, W.-Y. Keung, and S.-C. Lee, Phys. Rev. **D38**, 850 (1988); G. A. Kozlov, Phys. Rev. **D72**, 075015 (2005).
53. J. M. Hernández, M. A. Pérez, G. Tavares–Velasco, and J. J. Toscano, Phys. Rev. **D60**, 01304 (1999).
54. A. Yu. Ignatiev, G. C. Joshi, and M. Matsuda, Mod. Phys. Lett. **A11**, 871 (1996).
55. G. J. Gounaris, J. Layssac, and F. M. Renard, Phys. Rev. **D62**, 073013 (2000).
56. G. J. Gounaris, J. Layssac, and F. M. Renard, Phys. Rev. **D61**, 073013 (2000).
57. R. Walsh and A. J. Romalho, Phys. Rev. **D57**, 5908 (1998); *ibid.* **65**, 055011 (2002); S. Atag and I. Sahin, Phys. Rev. **68**, 093014 (2003); *ibid.* **70**, 053014 (2004).
58. See, M. A. Pérez and F. Ramírez–Zavaleta, Phys. Lett. **B609**, 68 (2005) and references therein.
59. D0 Collaboration, V. M. Abazov *et al.*, Phys. Rev. Lett. **95**, 051802 (2005).
60. F. Larios, M. A. Pérez, G. Tavares–Velasco, and J. J. Toscano, Phys. Rev. **D63**, 113014 (2001).
61. J. Ellison and J. Wudka, Ann. Rev. Nucl. Part. Sci. 48, 33 (1998).
62. M. A. Pérez, G. Tavares–Velasco, and J. J. Toscano, Phys. Rev, D67, 017702 (2003).

QCD phenomenology

Peter O. Hess

Instituto de Ciencias Nucleares, UNAM, Ciudad Universitaria, Circuito Exterior S/N, A.P. 70-543, 04150 México D.F., Mexico

Abstract. A review is presented on the contributions of Mexican Scientists to QCD phenomenology. These contributions range from *Constituent Quark model's* (CQM) with a fixed number of quarks (antiquarks) to those where the number of quarks is not conserved. Also glueball spectra were treated with phenomenological models. Several other approaches are mentioned.

Keywords: QCD, Low energy, Phenomenological Models
PACS: 11.15.Pg, 12.38.-t, 12.39.-x

1. INTRODUCTION

QCD [1] is the favored theory of the strong interaction. It is well tested at high energy, where due to the running coupling constant the interaction is relatively weak and perturbative methods are allowed. However, at low energy the coupling constant gets large and non-perturbative methods have to be applied. The only first-principle and non-perturbative theory is *Lattice QCD* (LQCD) [2], which starts from the QCD Lagrangian and takes into account the running coupling constant and determines the particle spectrum without the need of perturbative approximations. The drawback of LQCD is the fact that only the lowest state (for a given set of quantum numbers, like spin, parity and charge conjugation) can be obtained, though in some occasions also the next one is calculated, i.e., the spectrum is not complete. In addition, the numerical procedure is quite involved, time consuming and the results are often difficult to understand from a physical point of view (e.g., why the ordering of levels is as it is).

Using phenomenological approaches to QCD is one alternative to LQCD, though, with the expense of having to introduce model parameters. One of the first attempt is the *MIT Bag Model* (MITBM) [3]. This model assumes that the quarks move within a bubble of the size of a hadron. It can be visualized as being submerged in a space with a particular vacuum structure. Within this bubble, the quarks move freely, simulating the asymptotic freedom. Confinement is introduced by hand through particular boundary conditions, not allowing the color electric field to leave this tiny volume of space. It is similar to the boundary conditions within a cavity in Classical Electrodynamics, with the electric and magnetic fields interchanged. The residual interaction can be treated with perturbative methods. The MITBM was very successful in describing the low energy spectrum of the hadrons and some transition ratios.

Still, several difficulties remained: First of all, the resulting coupling constant, when treating residual interactions, was still too large once fitted to experiment. Extensions of this model (like the cloudy bag model [4]) did correct for it partly. Also, these models assume a constant number of quarks (antiquarks) and gluons, which leads to the prediction of exotic states like *glue balls* (g^2), the pentaquark ($q^3(q\bar{q})$) and of exotic mesons (($q\bar{q})^2$), nothing having been observed up to now.

Especially the glue balls attracted a lot of attention. One reason is their exotic structure and another one is the easier treatment of gauge fields within the LQCD than fermion fields (in the latter one applies still the *quenched approximation*). In [5, 6] first attempts were presented to describe glue ball states with many body techniques and the boson approximation. Later, one of the authors left to Mexico where he continued to work on this topic and was the first mexican scientist in introducing phenomenological models of QCD at low energy in Mexico.

2. MODELS FOR GLUE BALLS

In [7, 8] the work started in [5, 6] was continued. The Hamiltonian deduced from the MITBM [9] was used, which assumes that the total number of gluons is constant. Group theoretical many body techniques were applied to determine the energy of a many gluon state with constant number of gluons N. It was shown that the energy is lowered with increasing N, which leads to inconsistencies, i.e., in a bubble of the size of a hadron the lowest state possible is with an infinite number of gluons. This is interpreted as a hint that interactions mixing the number of gluons are very important.

In [10] several orbital levels where introduced in order to correct for the problem, which did lead to the use of boson mapping techniques [11] for the treatment of many gluons. However, this attempt did not result in any practical application due to problems related to the appearance of unphysical states using the boson mapping technique [12].

A final breakthrough came in [13]. As a Hamiltonian the kinetic energy as derived in [14] was used, where one of the first attempts was realized to *derive* the Hamiltonian of pure gauge fields at low energy. The potential has a minimum at a given position. Unfortunately, the calculations [14] are only valid for a still too low coupling constant. In [13] the same kinetic energy was used with the mass parameter, depending on the strong coupling constant, substituted by a parameter and the potential was also parametrized, having a minimum in some place. The resulting Hamiltonian is similar to the one in [15] and the solution for the energy is analytic. Though, eight parameters are involved, the glue ball spectrum was very well reproduced comparing to lattice calculations [16, 17]. One remarkable observation is that the O^{+-} state resulted to be at too high energy, compared to [16]. However, the improved LQCD calculation of [17] did shift this state into the correct position. With [13] the level ordering can be understood simply by the assumption that gluons have an effective mass at low energy and using the result of a complete classification of the glue ball states [18], which determines the allowed spin combinations with color zero for a given number of gluons N. In Tabel 1 a list of the glue ball states, adjusted to the two lattice calculation, is given. A complete list of glue ball states is available in [13].

Furthermore, not only the lowest state of a given spin-parity-charge conjugation was given, but a *complete list* of glue ball states. Those with exotic quantum numbers, i.e., not being reproduced by pure three quark or one quark-antiquarks states, represent predictions of this model which can be tested experimentally. The exception is the sector with parity and charge conjugation positive, because it is suspected to mix strongly with

TABLE 1. Glue ball states as obtained with the schematic model of Ref. [13], compared to two lattice calculations. The J^{PC} refer to the spin (J), parity (P) and charge (C) conjugation. A cross (X) in front of the lattice data refers to an upper limit for the energy. The state 0^{+-} is highlighted. The model prediction is consistent with the improved LQCD results of Ref. [17].

J^{PC}	mass [GeV]	lattice [16]	lattice [17]
0^{++}	1.61	1.61±0.3	-
2^{++}	2.13	2.23±0.22	2.39±0.01
3^{++}	3.69	3.92±0.48	6.69±0.04
1^{++}	4.50	3.96±0.31	4.12±0.05
0^{-+}	2.19	2.23±0.37	2.59±0.03
2^{-+}	3.41	3.01±0.18	3.07±0.02
1^{-+}	4.07	X4.71±0.39	4.18±0.03
3^{-+}	5.03	X5.83±0.66	4.67±0.05
1^{+-}	3.03	2.90±0.26	2.94±0.02
2^{+-}	4.09	3.89±0.66	4.10±0.04
3^{+-}	4.09	X6.18±0.89	3.54±0.02
0^{+-}	**4.77**	**X2.99±0.75**	**4.74±0.05**
1^{--}	3.38	4.36±0.48	3.85±0.04
0^{--}	3.84	3.93±0.48	4.94±0.05
2^{--}	3.86	3.86±0.35	3.93±0.02
3^{--}	3.91	X3.91±0.89	41.3±0.08

quark and antiquark states.

Finally, in [19] a schematic consideration was applied to many gluon states in one level. Even the Gribov horizon [20] could be explicitly constructed for that sample case.

3. CONSTITUENT QUARK MODELS WITH CONSTANT NUMBER OF QUARKS

In this section we describe the contribution of mexican scientists to CQM's, assuming a constant number of quarks.

Of particular interest is an algebraic model, applied in [21] to mesons ($U(4)$ model), in [22] to non-strange baryons and extended in [23] to strange baryons. The model for baryons is based on group theoretical methods, using a $U(7)$ group for the orbital excitations [22]. The $U(7)$ originates from the fact that three particles have seven spatial degrees of freedom, having subtracted the center of mass motion. The flavor and spin degrees of freedom are added, considering the Pauli exclusion principle. The coordinates of the system are the relative distance of two quarks and the distance of a third quark to the first relative vector. The orbital excitations are described by oscillations of the distance between the quarks, bending modes of the relative vectors with respect to each other and by rotations of the whole system. Several *dynamical* symmetries where identified, such as the anharmonic oscillator (with two versions) and

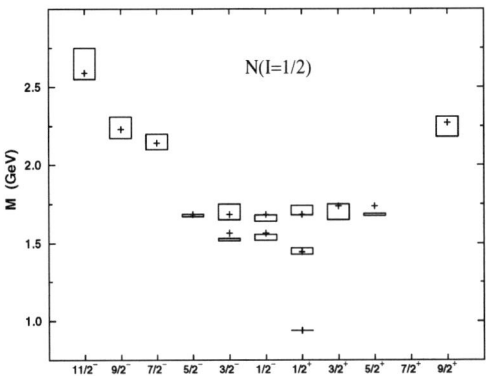

FIGURE 1. Spectrum of the nucleon resonances with spin $\frac{1}{2}$. The crosses are theoretical data [22], while the boxes represent the experimental information, including the width of the states.

the deformed oscillator. As a distinct feature, assuming the last dynamical symmetry the Roper resonance could be described very well. This is a non-trivial task, because the Roper resonance corresponds to two quanta of excitations and it has to be lower than the first negative parity state, which is a one phonon excitation. A similar good description was obtained, using the *Hypercoulomb Potential* [24]. In [25] the isoscalar form factor of the Roper resonance was investigated. The baryon spectrum at low energy can be described by that model extremely well (see Fig. 1).

Within the $U(7)$ algebraic model for baryons also decay properties were investigated, like the strong decay of a baryon into a different one plus a meson ($B \to B' + M$) [23, 26] and the emission of photons ($B^* \to B + \gamma$) [23]. With rather simple assumptions the decay properties were described with great success.

Also the electromagnetic form factors of the hadrons, in particular the one of the nucleon, where described within this model [26, 27, 28, 29].

In a two-component model with quark-like intrinsic structure and a meson cloud, the electric and the magnetic form factors where investigated [28, 29]. This topic attracted much of attention due to the observation that the ratio $\frac{\mu_p G_{E_p}}{G_{M_p}}$ is not 1 as was expected. Here, μ_p is the magnetic moment of the proton and G_{E_p}, G_{M_p} the electric and magnetic form factor, respectively, of the proton. This two-component model is able to describe the discrepancy, implying a complicated internal structure of the proton. In [29] the two-component model was applied to obtain the *strange form factors* of the nucleon without any further parameter. The agreement to experiments seems to be very good, which gives evidence of the predictive power of that model.

As a particular application, the structure of pentaquarks and their magnetic moments were investigated in [30, 31, 32]. For a more detailed information on pentaquarks, please consult Ref. [33] and references therein. Also magnetic moments were calculated and compared to other theoretical models. The pentaquarks is pictured as the combination of three quarks and an additional quark-antiquark pair. The spectrum was described and possible signatures predicted. Unfortunately, the first announcements on the observation of the pentaquark states where proven to be misleading and their existence was again put on doubt in [33].

In [34] a relativistic equation for a many particle system and a harmonic oscillator is proposed, called the *Dirac Oscillator*. The Dirac Oscillator was applied in [35, 36] to obtain a mass formula for baryons and in [37] for the mass formula of mesons. Systematic studies were performed in Refs. [38, 39, 40] concerning the structure of solutions of the Dirac Oscillator. Special attention was put on the appearance of multiplicities at low energy, presenting a serious problem to be tackled in this theory. Finally, in [41] the Dirac Oscillator was applied with success to the *bottonium*. In [42] a structural study was published on *composite particles in relativistic quantum mechanics*.

In Ref. [43] a totally new approach is pursued. The effort concentrates on classifying the hadron states not according to the standard CQM (dividing it in a flavor and spin part, plus subsequent orbital excitations) but using Rarita-Schwinger states, which follow a consistent classification according to the Lorentz group, exploiting the degeneration of its irreducible representations. In [44, 45] the details of the spectrum were nicely reproduced using a diquark-quark model obeying an $O(4)$ type of symmetry for the orbital excitations. One important conclusion is that the main degrees of freedom are determined by the relative motion between the diquark and the quark. The degrees of freedom within the diquark are apparently frozen.

One philosophical objection to CQM's with a constant number of quarks (antiquarks) has to be mentioned here: Even when the constituent quarks are assumed to have a mass, realistic interactions suggest that higher correction including $(q\bar{q})^n$ are still present. This becomes clear when pentaquarks are considered, which requires the need of an additional quark-antiquark pair. The problem of exotic states, like g^n and $(q\bar{q})^m$ ($m > 1$) is still present. Without a particle mixing interaction, the large multiplicity of states at low energy cannot be removed. This is partly solved by the two-component model [28, 29]. There, the back ground of quark-antiquark pairs and gluons is taken into account indirectly.

4. MODELS OF QCD WITH NON-FIXED PARTICLE NUMBER

In Refs. [46, 47, 48] a schematic Hamiltonian was proposed, which mixes the number of quarks and antiquarks. One basic ingredient is a two level system for the quarks, one at positive and one at negative energy. Only quarks are distributed in this system, where the lower level represents the negative energy Dirac sea. Real quarks are presented by quarks in the positive energy level and antiquarks by holes in the lower level. A meson in the CQM is presented as a particle-hole excitation. The position of the positive level is at about one third of the nucleon mass, introducing an effective mass for the quarks. This part of the model is very similar to the classical Lipkin model of nuclear physics. The fermions interact with gluon pairs of color zero in a higher energy level at about 1.6 GeV, which is the number obtained in the model for gluons of Ref. [13]. The interaction was chosen such that it can be justified in structure by the basic interactions of QCD.

The possible quark-antiquark pairs, with color zero, are listed as $[\lambda,S]$, with $\lambda=0, 1$ denoting flavor singlet and octet, while $S=0,1$ is the spin.

In a further step, the gluon and the quark-antiquark pairs are mapped to bosons. This introduces spurious states which have to be eliminated. It is done by first obtaining a complete classification of many quark-antiquark states and matching the result with the model space of the bosons. This is a useful byproduct, which can be applied also to models with a large number of quarks and antiquarks. The method used for the reduction is published in [18].

The interactions are of the type $(b^\dagger_{\lambda S} \cdot b^\dagger_{\lambda S})(b^\dagger + b)$, $(b_{\lambda S} \cdot b_{\lambda S})(b^\dagger + b)$, $(b^\dagger_{\lambda S} \cdot b_{\lambda S})(b^\dagger + b)$, for the part relevant for mesons only. The $b^\dagger_{\lambda S}$ represents a quark-antiquark pair creation operator and $b^\dagger$ the gluon pair operator. The first and the second interaction are important for the vacuum fluctuations. The Hamiltonian can be diagonalized in a seniority basis, with quite trivial matrix elements. The four parameters involved were fitted to the experimental spectrum (please, for more details see the listed references).

For the description of baryons four additional interactions are needed. Two of the parameters are adjusted to the nucleon resonances and two to the Δ resonances.

The main results are: i) With the particle mixing interaction the multiplicity problem of hadrons at low energy disappears; ii) The structure of the hadrons includes many quark-antiquarks states, gluons and the valence quarks (quark-antiquark); iii) the content of gluons in the nucleon is roughly 40% in close agreement to experiment; iv) the Roper resonance comes down to its experimental energy without the need of extra parameters; v) The position of the lowest pentaquark is correctly described, though, first investigations on its structure suggests that it is rather a molecular state consisting of a nucleon and a kaon, thus, having a large width; vi) heptaquarks are also predicted; vii) the vacuum properties were automatically well reproduced [49], comparing to results of QCD sum rules [50]; and finally viii) the model was also able to describe the production of particles in heavy ion collisions, once temperature is built into the model [47].

The model intents to describe the complicated background structure of hadrons and various aspects of QCD at low and high energy at the same time. This limits its applicability to particular structures at low energy and refinements have to be included.

Another drawback is that no decay properties were calculated within this model. For that, one needs the Clebsch-Gordan coefficients of a high rank group which contains the flavor (or color) $SU(3)$. First steps were undertaken in Ref. [51] and the Clebsch-Gordan coefficients are obtained in [52]. The hope is to calculate decay properties within this model in near future.

Now to a completely different approach to the description of meson properties at low energy: In [53] it is conjectured that the unusual properties of the low lying scalar mesons are a manifestation at the meson level of the effective interaction between quarks induced by instantons as shown by G. 't Hooft in the sixties. A $U(3) \times U(3)$ Linear Sigma Model is formulated in which the inverted mass spectrum, exhibited by the light scalar mesons, is reproduced and the whole phenomenology of scalars and pseudoscalars is predicted by using only few parameters. In this model, light scalar mesons are considered as $\bar{q}q$ states.

In [54, 55] it is proposed that scalar and pseudoscalar mesons below 1.5 GeV are actually admixtures of $\bar{q}\bar{q}qq$ and $\bar{q}q$ states. A model is presented, where the one of Ref.

[53] is coupled with a new nonet of $\bar{q}\bar{q}qq$ states with a "bare" mass consistent with the linear rising of hadron with the mass of the constituent quarks, thus lying roughly around $1.4 GeV$. This second nonet has a chiral symmetry realized directly and are described also by a Linear Sigma Model but chiral symmetry is not spontaneously broken in this case. These states get mixed yielding the observed mesons.

5. MODELS BASED ON FIELD THEORY

Here we give a short list of further important contributions based on field theoretical descriptions.

In [56] it is observed that the flavor symmetry $SU_F(3)$ is a rather bad choice and better agreement to experiment can be achieved replacing this symmetry by $SU(2)_{ud} \otimes SU(2)_{cs} \otimes U(1)$, which has the merit to be consistent from the start with the OZI rule. The new proposal divides the quarks in the natural multiplets of up-down and charm-strange. The difference between the two symmetries is revealed by the properties of the non-isotriplet part of the neutral current. The usefulness of the approach was tested on the case of the ηN coupling.

In [57] the relevance of the instanton-induced determinantal 't Hooft interaction to the ηN coupling, within the framework of a three-flavor linear sigma model, was investigated. It was shown that instantons, in combination with the spontaneous breaking of chiral symmetry, provides the major mechanism for the ideal mixing between pseudo-scalar strange and non-strange quarkonia. The unitary spin is shown to be an accidental symmetry due to the anomalous gluon dynamics rather than being a fundamental symmetry. The model allows for possible generalizations to non-ideal mixing angles and different values of the meson decay constants in the strange sectors.

In another contribution some effort was invested in understanding the structure of *bound states* at low energy. The question to address is: The fundamental particles are quarks, antiquarks and gluons. The physical states, however, are composed particles (no free quarks, antiquarks and gluons exist as a consequence of confinement). How then these bound states arise at low energy from a field theory? This question was tackled in the Refs. [58, 59, 60] with the help of the method of *environmentally friendly reonormalization* [61] and the theory of Regge trajectories, applied to a simple field theoretical model.

In [62, 63] a study is performed for the production of scalar mesons (S) at the Frascati ϕ factory in the reactions $e^+e^- \to S\gamma \to \pi^+\pi^-\gamma$ and $e^+e^- \to S\gamma \to K\bar{K}\gamma$ and all the reactions contributing to the background. This is a very phenomenological analysis where the corresponding couplings are not constrained by chiral symmetry. It is shown that there are sizable charged meson loop corrections in the mechanism for the production of scalar mesons.

In [64] is is shown that the predictions of the $U(3) \times U(3)$ Linear Sigma Model [53] for the controversial two photon decays of the $a_0(980)$ and $f_0(980)$ mesons agree with experimental data.

In [65, 66, 67] a study of the predictions of the $U(3) \times U(3)$ Linear Sigma Model [53] is performed for the more relevant processes being measured at the Daϕne ϕ factory at Frascati. All processes are properly described. It is shown that the predictions of the model reduce to those of Chiral Perturbation Theory in the heavy scalars limit. The model incorporates higher order corrections in this systematic expansion due to the scalar resonances.

In [68] the two photon decay is investigated of all neutral scalars below 1.5 GeV, in the context of the chiral model, for $\bar{q}\bar{q}qq$ and $\bar{q}q$ states. The calculations improve the previous predictions in the case of the $a_0(980)$ and $f_{(}980)$ and give definitive predictions for other not yet measured processes.

In related contributions [69, 70, 71] a mixed quark-hadron model and its phenomenological implications were studied.

Finally, in [72, 73] the feasibility was studied of Heavy Baryon Chiral Perturbation Theory in the light of the known inconsistencies-ambiguities suffered by the Quantum Field Theory of spin 3/2 (and higher) fields. It is concluded that only the leading order in the 1/m expansion is free of these problems. The radiative decay is studied of the spin 3/2 decuplet within this formalism.

6. TOWARDS THE JUSTIFICATION OF PHENOMENOLOGICAL MODELS

The motivation for the schematic model, mentioned in section 4, was the test of many body techniques on QCD. For that, we wanted a sufficient realistic model to mimic QCD with exact results (save a numerical diagonalization). As it seems, the model is more realistic than thought. Also, other CQM's seem to work rather well at low energy, proving the usefulness of phenomenological models. One can ask the question: Why do these models work?

To this end, in Ref. [74] the authors started from a QCD model, assuming $SU(2)$ for color and flavor and approximating the quark-gluon interaction by a contact interaction of the quarks. Restricting to a finite volume V and taking into account only S-orbital contributions did lead to a surprising simple Hamiltonian, with the feature that increasing the strong coupling constant leaves the states with color zero unaffected but those with color are raised in energy.

This model shows that many body methods and exploiting symmetries can probably reach out far towards the solution of QCD at low energy.

At the moment, the same authors extend the model Hamiltonian of QCD, including more orbital excitations *and* gluons, thus, starting from the real QCD. Apart from the attempt to justify the schematic models, we are interested in how chiral symmetry breaking arises.

7. CONCLUSIONS

In this contribution we tried to give a rough review on the contributions of mexican scientists on the phenomenology of QCD. The models are mainly related to the phsyics at low energy. Due to lack of space we could not mention all contributions, but rather some of them. Even those mentioned could not be explained in detail but only their main results were listed. If we omitted one or the other contribution we ask in advance for forgiveness.

Nevertheless, the contributions listed show an important participation of mexican scientists in this field, involving many subtopics, ranging from Constituent Quark Models, with and without conserving the number of particles, to field theoretical models, probing the basic structure of QCD. All this work is well recognized on the international level, involving many important collaborations.

ACKNOWLEDGMENTS

The author acknowledges financial support from CONACyT, CONACyT-CONICET and from DGAPA (IN108206).

REFERENCES

1. K. Huang, *Quarks Leptons and Gauge Fields*, 2nd edition, (World Scientific, Singapore, 1992).
2. H. J. Rothe, *Lattice Gauge Theories: An Introduction*, 2nd. edition, (World Scientific, Singapore, 1997).
3. A. Chodos, R. L. Jaffe, K. Johnson, C. B. Thorn and V. F. Weisskopf, Phys. Lett. D **9** (1974), 3471.
4. S. Théberge, G. A. Miller and A. W. Thomas, Can. J. Phys. **60** (1982), 59.
5. P.O.Hess and R.D.Viollier, Phys. Rev. D **34** (1986), 258.
6. P.O.Hess and R.D.Viollier, Nucl. Phys. A **468** (1987), 414.
7. P.O.Hess and R.López, Phys. Rev. D **36** (1987), 242.
8. P.O.Hess and R.López, Phys. Rev. D **37** (1988), 2019.
9. R. F. Buser, R. D. Viollier and P. Zimak, Int. J. Mod. Phys. **27** (1988), 925.
10. P.O.Hess and J.C.López, Nucl. Phys. B **30** (Supl.) (1993), 936.
11. P.O.Hess and J.C.López, J. of Math. Phys. **36** (1995), 1123.
12. A. Klein and E. R. Marshalek, Rev. Mod. Phys. **63** (1991), 375.
13. P.O.Hess, A.Weber, C.R. Stephens, S.A.Lerma H., J.C.López, Eur. Phys. Jour. **C9** (1999), 121.
14. J. Koller and P. Baal, Nucl. Phys. B **302** (1988), 1.
15. O.Castaños, E.Chacón, A.Frank, P.O.Hess and M.Moshinsky, J. Math. Phys. **23** (1982), 2537.
16. G. S. Bali, K. Schilling, A. Hulsebos, A. C. Irving, C. Michael and P. W. Stephenson, Phys. Lett. B **309** (1993), 378.
17. M. Peardon, Nucl. Phys. B (Proc. Suppl.) **63** (1998), 22.
18. R.Lopez, P.O.Hess, P.Rochford and J. P. Draayer, J. of Phys. A **23** (1990), L229.
19. P.O.Hess, D.Schuette, Ann. of Phys. **211** (1991), 112.
20. V. N. Gribov, Nucl. Phys. B **139** (1978), 1.
21. C. Gobbi, F. Iachello and D. Kusnezov, Phys. Rev. D **50** (1994), 2048.
22. R. Bijker, F. Iachello and A. Leviatan, Ann. Phys. (N.Y.) **236** (1994), 69.
23. R. Bijker, F. Iachello and A. Leviatan, Ann. Phys. (N.Y.) **284** (2000), 89.
24. R. Bijker and A. Leviatan, Rev. Mex. Fís. **44** (1998), 15.
25. E. Tomasi-Gustafsson, M. P. Rekalo, R. Bijker, A. Leviatan and F. Iachello, Phys. Rev. C **59** (1999), 1526.
26. R. Bijker, F. Iachello and A. Leviatan, Phys. Rev. D **55** (1997) 2862.
27. R. Bijker, F. Iachello and A. Leviatan, Phys. Rev. C **54** (1996) 1935.
28. R. Bijker and F. Iachello, Phys. Rev. C **69** (2004), 068201.
29. R. Bijker, submitted to Physical Review Letters, (2006).
30. R. Bijker, M.M. Giannini and E. Santopinto, Phys. Lett. B **595** (2004), 260.
31. R. Bijker, M.M. Giannini and E. Santopinto, Eur. Phys. J. A **22** (2004), 319.
32. R. Bijker, M.M. Giannini and E. Santopinto, Rev. Mex. Fís. **50 S2** (2004), 88.
33. A. D. Dzierba, C. A. Meyer and A. Szczepaniak, J. Phys. Conf. Ser. **9** (2005), 192.

34. A. Szczepaniak and M. Moshinsky, J. Phys. A**22** (1989), L817.
35. M. Moshinskym G. Loyola, A. Szczepaniakm, C. Villegas and N. Aquino, *Proceedings of the Relativistic Aspects of Nuclear Physics*, Ed. T. K. Odama et al., p.271, (World Scientific, Singapore, 1990).
36. G. Loyola, C. Villegas and M. Moshinsky, *Proceedings of the XIII. Symposium of Nuclear Physics*, Notas de Física, (1990), 187.
37. M. Moshinsky, A. Gonzalez and G. Loyola, Rev. Mex. Fís. **40** (1994), 12.
38. V. Kukulin, G. Loyola and M. Moshinsky, Phys. Lett. A **158** (1991), 19.
39. A. Del Sol Mesa, Can. J. Phys. **72** (1994), 453.
40. A. Del Sol Mesa, V. Riquer and M. Moshinsky, Found. Phys. **27** (1997), 1139.
41. M. Moshinsky and V. Riquer, J. Phys. A **36** (2003), 2163.
42. M. Moshinsky and E. Sadurní, *Proceedings of the 25th International Colloquium on Group Theoretical Methods in Physics*, Cocoyoc, México., IOP Publishing LtD, p. 403 (2005).
43. M. Kirchbach, Rev. Mex. Fís. **47** (S2) (2001), 123.
44. M. Kirchbach, M. Moshinsky and Yu. F. Smirnov, Phys. Rev. D **64** (2001), 114005.
45. M. Kirchbach, Rev. Mex. Fís. **50** (S2) (2004), 54.
46. S Lerma, S. Jesgarz, P. O. Hess, O. Civitarese, M. Reboiro, Phys. Rev. **C67** (2003), 055209.
47. S. Jesgarz, S Lerma, P. O. Hess, O. Civitarese, M. Reboiro, Phys. Rev. **C67** (2003), 055210.
48. M. Nuñez, S. Lerma, P. O. Hess, S. Jesgarz, O. Civitarese, M. Reboiro, Phys. Rev. C **70** (2004), 035208.
49. P. O. Hess, O. Civitarese, submitted to Phys. Rev. C (2006).
50. L. J. Reinders, H. Rubinstein and S. Yazaki, Phys. Rep. **127** (1985), 1.
51. I. Sánchez-Lima, P. O. Hess, Rev. Mex. Fis. **52** (S) (2005), 82.
52. I. Sánchez-Lima and P. O. Hess, submitted to J. Math. Phys. (2006).
53. M. Napsuciale and S. Rodriguez, Int. J. Mod. Phys. A **16** (2001), 3011.
54. M. Napsuciale and S. Rodriguez, Phys. Lett. B **603** (2004), 195.
55. M. Napsuciale and S. Rodriguez, Phys. Rev. D **70** (2004), 09404.
56. M. Kirchbach, Phys. Lett. B **455** (1999), 259.
57. M. Napsuciale, A. Wirzba and M. Kirchbach, Nucl. Phys. A **703**, (2002), 306.
58. C.R.Stephens, A.Weber, J.C.López and P.O.Hess, Phys. Lett B **414** (1997), 333.
59. C.R.Stephens, A.Weber, J.C.Lopez V. and P.O.Hess, Jour. Mod. Phys. A **15** (2000), 1773.
60. A. Weber, J. C. López Vieyra, C. R. Stephens, S. Dilcher and P. O. Hess, Int. J. Mod. Phys. A **16** (2001), 4377.
61. D. O'Connor and C. R. Stephens, Phys. Rep. **363** (2002), 425.
62. J. L. Lucio Martinez and M. Napsuciale, Phys. Lett. B **331** (1994), 418.
63. J. L. Lucio and M. Napsuciale, Nucl. Phys. B **440** (1995), 237.
64. J. L. Lucio Martinez and M. Napsuciale, Phys. Lett. B **454** (1999), 365.
65. A. Bramon, R. Escribano, J. L. Lucio M., M. Napsuciale and G. Pancheri, Phys. Lett. B **494** (2000), 221.
66. A. Bramon, R. Escribano, J. L. Lucio Martinez and M. Napsuciale, Phys. Lett. B **517** (2001), 345.
67. A. Bramon, R. Escribano, J. L. Lucio M, M. Napsuciale and G. Pancheri, Eur. Phys. J. C **26** (2002), 253.
68. S. Rodriguez and M. Napsuciale, Phys. Rev. D **71** (2005), 074008.
69. M. Napsuciale, S. Rodriguez and E. Alvarado-Anell, Phys. Rev. D **67** (2003), 036007.
70. M. Napsuciale and S. Rodriguez, Phys. Rev. D **65** (2002), 096013.
71. J. L. Lucio-Martinez, M. Napsuciale, M. D. Scadron and V. M. Villanueva, Phys. Rev. D **61** (2000), 034013.
72. M. Napsuciale and J. L. Lucio, Phys. Lett. B **384** (1996), 227.
73. M. Napsuciale and J. L. Lucio, Nucl. Phys. B **494** (1997), 260.
74. P. O. Hess, A. Szczepaniak, Phys. Rev. C **73** (2005), 025201.

Astroparticle Physics

¹Arnulfo Zepeda

Centro de Investigación y de Estudios Avanzados
P.O. 14-740, 07000 México, D.F.

Abstract. Astroparticle Physics is the union of astrophysics and elementary particle physics. Research in Mexico in this branch of physics has been carried out in several topics, mainly those associated with neutrino physics and with cosmic rays.
Keywords: Astroparticle, cosmic rays
PACS: 95.30Ky, 96.50.S-

INTRODUCTION

Astroparticle physics is a recently born branch of fundamental physics. It is the union of Elementary Particle Physics and Astrophysics. It aims at extracting from astrophysical phenomena new information about properties of the elementary components of matter and energy and their interactions. Part of this information complements that accessible at man made accelerators. There is also a strong relation between cosmology and particle physics. Graciela Gelmini considers 1905 a the initial point of modern cosmology and the early 70's as the beginning of astroparticles.

The research activities in astroparticle physics have been being developed already before the term was coined. It addresses the Big Questions such us:

- The origin and nature of ultrahigh energy cosmic rays
- The possible particle nature of dark matter
- Specific properties of the neutrino

Astroparticle physics has a deep connection with astrophysics. Together they provide models of the evolution of the universe and test the predictions of these models with precision data collected about all kind of radiations and of charged and neutral particles with observatories displayed around the Earth, placed on the ground, below the Earth surface, in the atmosphere or in the outer space. The field is well described in the review article of Carramiñana, [1].

We will not discuss here astroparticle physics topics which are already described in other contributions to this volume. For this reason we will concentrate in high energy

---

[1] E-mail: zepeda@fis.cinvestav.mx

cosmic rays. Historically, cosmic ray physics has given rise to very important discoveries of new and unexpected particles: the muon, the pion, strange particles, etc. In Mexico the study of cosmic rays started with the works of Sandoval Vallarta. In the modern times the possibility to join the Pierre Auger project triggered the interest of several groups in Mexico in the cosmic rays of the highest energies.

Nowadays experimental cosmic ray physics is carried out not only within the Pierre Auger Collaboration, which we will describe in detail below, but also developing experimental facilities in Mexico, as described in this volume by Salazar and Villaseñor, and by Fernandez. In this last case the instruments developed in Mexico are being integrated into a bigger detector, ALICE, to be placed in the LHC accelerator which will be commissioned next year, 2007. Other development in this field is the application of cosmic ray physics to search for caverns inside the Pyramid of the Sun in Teotihuacan, Mexico.

The work developed in Mexico in the field of neutrinos is summarized in the contribution of D'Olivo and Miranda to this volume.

In their contribution to this volume de la Macorra and Matos describe work done in Mexico addressing the question of the nature of the dark matter and the dark energy from the standpoint of scalar fields and of particles.

Astroparticle topics have been discussed by lecturers and participants to the Mexican School and the Mexican Workshop of Particles and Fields which are organized organized every two years on a biyearly basis. In 1998, for example, in the Eight Mexican School of Particles and Fields, G.G. Raffelt lectured on the topic "Stars as Particle Physics Laboratories". And R. Peccei lectures on Neutrino Physics include the themes of cosmological constraints on neutrino masses and the discussion on solar neutrinos. There were also several seminars presented: on a Pierre Auger Cherenkov detector prototype and on bounds on the neutrino magnetic moment from supernova data.

In the next section we describe the activities in Mexico within the Pierre Auger Project.

THE PIERRE AUGER OBSERVATORY

Fermilab played a central role in the origin of the involvement of Mexico in the Pierre Auger Project. Indeed, at the occasion of a visit of Umberto Cotti, then PhD student at Cinvestav, to Fermilab, within the program of summer stays founded by Leon Lederman, and supported by Fermilab and Cinvestav (now by Fermilab and the Mexican Academy of Sciences), he became aware of an important development taking place at Fermilab, namely the continuous workshop for the design of the Pierre Auger Observatory and he enthusiastically informed me about it. Motivated by this first approach and reinforced by recommendations from UNESCO, Professor Jim Cronin invited AZ to the final workshop for the design of the Pierre Auger Observatory that was held at Fermilab in May of 1995. These encounters lead to the formation in May of 1996 of a group of Mexican scientists, engineers and students from 5 institutions (BUAP, Cinvestav,

INAOE, UNAM, and the University of Michoacan) which joined the then recently formed Pierre Auger Collaboration.

The size of the Mexican group has been fluctuating during these years and at present it is of about 16 scientists and 12 students from 4 institutions (University of Puebla, Cinvestav, University of Michoacan and National University). More people would love to join this exciting project if sufficient funds will be provided by the Mexican agencies.

An important development worth mentioning has been the incorporation of Rotoplas, a Mexican industry, to the team in charge of finalizing the design of the container of the water Cherenkov detector, a key component of the Pierre Auger Observatory. To refine the design, several studies were made, and reported, with prototype water Cherenkov detectors installed at the University of Puebla campus. After the completion of the rotomolding machine and the mold in the installations of CTEQ in Queretaro, several test were made in Mexico and the these instruments were sent to Argentina to fabricate there hundreds of tanks and reduce in this way transportation costs.

The participation of the Mexican group in the optical design of the fluorescence detectors is also worth mentioning. Mexico has also participated in the simulation work and in the development of the front end electronics. A complete list of references to reports written by the Mexican group may be found in Ref. [2].

The design and objectives

The Pierre Auger Observatory is designed to detect with unprecedented statistics cosmic rays of the highest energy known in our Universe. The interest in these phenomena stems not only from their extreme nature but mostly from their puzzling existence. Indeed, it is believed that the change of slope in the spectrum of cosmic rays around 10^{14-15} eV, the "knee", shows that the higher energy cosmic rays are of extragalactic origin. Moreover no sources of ultra high energy cosmic rays (with E> 10^{18} eV) have been clearly identified in the neighborhood around our galaxy. Several observatories (Volcano Ranch [3], Haverah Park [4], Yakutsk [5], Fly's Eye [6], and AGASA [7]) have however detected a handful of cosmic rays with energy bigger than 10^{20} eV, well beyond the so-called GZK cut-off [8], of about 5×10^{19} eV, which is the limit for the energy of cosmic rays originating at distances greater than 50 Mpc.

Motivated by this puzzle Jim Cronin and Alan Watson steered an international collaboration, to which now belong scientists and engineers from 19 countries around the world. The aim of the collaboration is to design, construct and operate the Pierre Auger Observatory and analyze the data obtained with this fine instrument.

To improve the resolution of the observatory, it has been decided that its structure be of the hybrid type, that is composed of fluorescence and surface detectors; and to have a significant statistics of registered events in the part of the spectrum to which the Pierre Auger Observatory is aimed, the size of its surface detector component has been set up to be 60 times bigger than the largest of its predecessors (AGASA). Previous surface detector systems include Volcano Ranch, Haverah Park, Yakutsk and AGASA, while previous fluorescence detector systems were Fly's Eye and HiRes [9], this last one taking data. The Pierre Auger Observatory will be the first large hybrid detector (recently this technology has been put into operation with the HiRes/MIA experiment [10]) and will also be incorporated into the Telescope Array detector presently being deployed in Utha, USA. The Pierre Auger Observatory will be constructed in two sites, one in the northern hemisphere (Colorado, USA) and one in the southern one (Malargüe, Argentina) covering each an area of 3,000 km^2 with 1,600 water Cherenkov detectors (as in the Haverah Park experiment) and four fluorescence detector systems in the southern site and three in Colorado. The fluorescence detector systems will have a range of view large enough to cover together the whole site. At the present time, July of 2006, 1036 Cherenkov detectors and 18 fluorescence telescopes are operating in the southern site. The design of the Northern site is presently being finished. The array of surface detectors operate 100% of the time while the fluorescence detectors need, as was the case of Fly's Eye and is the case of HiRes, dark cloudless nights and have therefore only a 10% duty cycle. With this arrangement it is be possible to collect 100 events per year in the surface detector with a primary energy $> 10^{20}$ eV and 500 per year with an energy $> 4 \times 10^{19}$ eV. The designed energy resolution with the surface array alone is 15% at E = 10^{20} eV and 30% at E = 10^{19} eV. With the hybrid mode this numbers improve to 10% and 20%, respectively. A similar improvement is obtained for the angular resolution: from 1 to 0.20 deg. at E = 10^{20} eV, and from 2 to 0.35 deg. at E = 10^{19} eV. The hybrid mode is therefore a definite advantage. This follows from the complementary role that the surface array and the fluorescence system play. Ground arrays sample the ground lateral density profile of muons, electrons, and photons of the shower while atmospheric fluorescence detectors observe the evolution of the air showers, their growth and attenuation, as they move longitudinally downward.

A complete description of he first stage, the Engineering Array, of the southern site, is reported in [11]

ACKNOWLEDGMENTS

We thank the editors for inviting us to write this article and particularly Luis for having the initiative to publish this compendium. This work supported in part by Conacyt and by the HELEN project.

REFERENCES

1. Alberto Carramiñana, Astrophysics and Elementary Particles, J.Phys.Conf.Ser.18:308-337,2005
2. http://www.fis.cinvestav.mx/~auger/papers.html
3. Linsley J., *Phys. Rev. Lett.* **10**, 146 (1963).
4. Brooke, G. et al., in *Proc. 19th ICRC* **2**, 150 (1985); Lawrence M.A., Reid R.J.O., Watson A.A., *J. Phys. G.* **17**, 773 (1991).

5. Efimov, N.N. et al., in *Proc. Int. Workshop on the astrophysical aspects of the most energetic cosmic rays*, ed. Nagano M. and Takahara F., World Scientific, p.20 (1991); Glushkov A.V. et al., *Astropart. Phys.* **4**, 15 (1995).
6. Bird, D.J. et al., *Astrophys. J.*. **441**, 144 (1995).
7. Yoshida, S. et al., *Astropart. Phys.* **3**, 105 (1995).
8. Greisen, K., *Phys.Rev.Letters* **16**, 748 (1966); Zatsepin G.T. and Kuz'min V.A., *JETP Lett.* **4**, 78 (1966).
9. Corbato S.C. et al. *Proc. of the International Workshop on Techniques to Study Cosmic Rays with Energies Greater than 10^{19} eV, Nucl. Phys.* **28B**, 36 (1992).
10. Abu-Zayyad T. et al. in *26th International Cosmic Ray Conference, Contributed Papers*, ed. by Kieda D., Salamon M. & Dingus B., Vol. 3, p. 260-263, University of Utah, 1999; ibid, Vol. 5, p. 365-368.
11. Properties and performance of the prototype instrument for the Pierre Auger Observatory. By the Pierre Auger Collaboration (J. Abraham et al.). Nucl.Instrum.Meth.A523:50-95,2004

Top Quark Physics

F. Larios

Departmento de Física Aplicada, CINVESTAV-Mérida, AP 73 Cordemex, 97310 Mérida, Yucatán, Mexico

Abstract. We give an overview of the physics of the Top quark, from the experimental discovery to the studies of its properties. We review some of the work done on the Electroweak and Flavor Changing couplings associated with the Top quark in the Standard Model and beyond. We will focus on the specific contribution of phycisits working in México and Mexican physicists working abroad.

Keywords: Top quark
PACS: 14.65.Ha, 12.15.Mm, 12.60.-i, 12.60.Cn

INTRODUCTION

The fact that the Top quark has such a large mass, even close to the Electroweak Symmetry Breaking (EWSB)scale, has given this third family quark an important role in the quest for the mechanism that generates masses for all the other elementary particles known so far. From the theoretical viewpoint, even if the EWSB mechanism is originated just by a Higgs field as the SM predicts, we know that the Top quark is the main source for the higher order corrections to the electroweak observables. On the other hand, it is possible that the Top quark is directly involved in the generation of mass. From the experimental viewpoint, the Top quark is the only quark that decays through the weak interaction before it could ever hadronize. This opens a special window of opportunity to study the effects of any kind of physics beyond the Standard Model (SM).

The Top quark was discovered by the CDF [1] and the DØ [2] collaborations at Fermilab in 1995. Figure 1 depicts the typical production and decay mechanism at the Tevatron collider.

We would like to start by mentioning the active role of Jacobo Konigsberg as a member of the CDF group in the research efforts on the Top quark [3]. On the other hand, for the DØ group, our colleague H. Castilla-Valdez, along with two of his former students R. Hernández-Montoya and A. Sánchez-Hernández, have had a part in this effort as well.

There are two modes for Top quark production: $t\bar{t}$ pairs and single t. Figure 2 depicts the three processes that give rise to a single top event in a hadron collider. So far, single top events have not been identified neither at the tevatron [4], nor at HERA [5]. To end the section of the experimental discovery we mention the role of J.G. Contreras-Nuño and G. Herrera-Corral as members of the H1 group.

As is well known the outstanding feature of the Top quark is its very large mass: $m_t = 173$ GeV [6]. With a width of order 1.5 GeV, the Top quark lives very a very short time, giving no time for hadronization. It decays almost entirely into a b and W^+ pair. As with any process involving quarks, QCD corrections can be very significant.

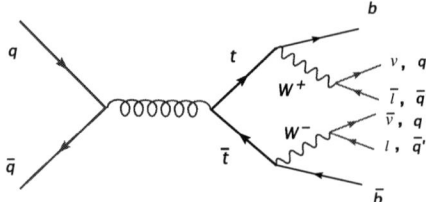

FIGURE 1. One of the processes ($q\bar{q}$ annihilation) that gives rise to $t\bar{t}$ production at the Tevatron. Three distinct decay modes: dilepton, lepton plus jets and hadronic, can identified for the detection of the Top quark.

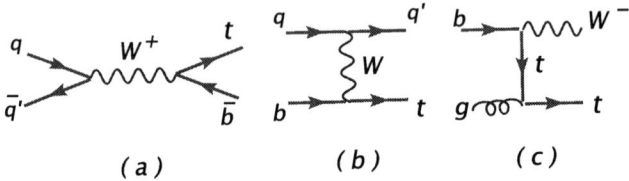

FIGURE 2. The three production modes for single top events in hadron collisions: (a) s-channel, (b) t-channel, and (c) W-associated production.

Their net effect is to reduce the width by about 10% [7]. Another correction comes from the effects due to the finite width of the W boson. In 1989 G. Sánchez-Colón and R. Huerta made a study of the finite W width effects several years before the discovery of the Top quark; however, back then m_t was not thought that could be so heavy, and they assumed $m_W > (m_t - m_b)$ [8]. Recently, G. López-Castro and G. Calderón have made a computation of these effects [9]. The net effect of the W's finite width is to reduce the Top's width about 1%. This small reduction is countered by the also $\sim 1\%$ increase from electroweak corrections.

THE TOP QUARK AND THE PRECISION TEST OF THE SM

Due to its large mass the Top quark plays a major role in the precision tests of the SM [10]. At the one loop level, electroweak corrections of the precision observables of the Z boson are driven by the contributions from the Top quark. Therefore, the LEP and the CLEO data can be used as an indirect measurement of the tbW and ttZ couplings. Figure 3 shows the contributions to the Z and W boson propagators, the Zbb vertex, and the FCNC decay $b \to s\gamma$.

It is possible to parameterize possible deviations from the SM predictions for the tbW and ttZ couplings in terms of only four coefficients $\kappa_{L,R}^{NC}$ and $\kappa_{L,R}^{CC}$ defined as follows [11]:

$$\mathcal{L} = \frac{g}{2c_W}\left(1 - \frac{4s_W^2}{3} + \kappa_L^{NC}\right)\bar{t}_L\gamma^\mu t_L Z_\mu + \frac{g}{2c_W}\left(\frac{-4s_W^2}{3} + \kappa_R^{NC}\right)\bar{t}_R\gamma^\mu t_R Z_\mu$$

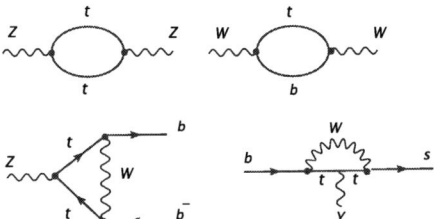

FIGURE 3. The contribution of the ttZ and tbW couplings to the electroweak precision observables and the FCNC decay $b \to s\gamma$.

$$+ \frac{g}{\sqrt{2}}\left(1+\kappa_L^{CC}\right)\bar{t}_L\gamma^\mu b_L W_\mu^+ + \frac{g}{\sqrt{2}}\left(1+\kappa_L^{CC\dagger}\right)\bar{b}_L\gamma^\mu t_L W_\mu^-$$
$$+ \frac{g}{\sqrt{2}}\kappa_R^{CC}\bar{t}_R\gamma^\mu b_R W_\mu^+ + \frac{g}{\sqrt{2}}\kappa_R^{CC\dagger}\bar{b}_R\gamma^\mu t_R W_\mu^- \qquad (1)$$

where t_L denotes a top quark with left-handed chirality, etc. While the ttZ vector and axial-vector couplings are tightly constrained by the LEP data [11], the right handed tbW coupling is severely bounded by the observed $b \to s\gamma$ rate [12] at the 2σ level,

$$|Re(\kappa_R^{CC})| \leq 0.4 \times 10^{-2}$$
$$-0.0035 \leq Re(\kappa_R^{CC}) + 20|\kappa_R^{CC}|^2 \leq 0.0039. \qquad (2)$$

On the other hand, LEP/SLC data also constrains the other top-quark couplings included in Eq.(4). Even though these data do not restrict all the anomalous κ terms, they induce the following inequalities

$$-0.019 \leq (\kappa_R^{NC} - \kappa_L^{NC}) - (\kappa_R^{NC} - \kappa_L^{NC})^2 + \kappa_L^{CC} + \kappa_L^{CC\,2} \leq 0.0013$$
$$-0.33 \leq (\kappa_R^{NC} - 4\kappa_L^{NC})(1+2\kappa_L^{CC}) \leq 0.1$$
$$\kappa_L^{CC} \sim \kappa_L^{NC} - \kappa_R^{NC} \qquad (3)$$

These relations impose in turn strong correlations on the κ couplings so that if only one coupling, κ_L^{CC} for instance, is not zero, the others are forced to be about the same order of magnitude [12].

TOP QUARK PRODUCTION AT PRESENT AND FUTURE COLLIDERS.

Currently, the production of Top quarks has only been observed at the Tevatron machine. In principle, the electrón-protón collider (HERA) runs at an C.M.S. energy of 320 GeV: enough to produce on-shell single Top events. However, the SM amplitude for single Top production in such a collision is very small, and it is not expected that HERA will be able to observe it.

The CDF and DØ groups have been able to measure the $t\bar{t}$ production cross section to about a 20% accuracy. A combination of all the current independent experimental measurements would give $\sigma^{exp}(t\bar{t}) = 7.1 \pm 0.6(stat) \pm 0.8(syst)$ pb, to be compared with the SM prediction $\sigma^{theo}(t\bar{t}) = 6.77 \pm 0.42$ pb. It is expected that with about 2 fb^{-1} of integrated luminosity the error could be reduced to about less than 10% [6].

The production of Top quarks can be divided in two modes: $t\bar{t}$ and single Top, depicted in Figures 1 and 2 respectively. However, it is also possible to produce Tops in pairs of same sign, as we shall see later.

Production of $t\bar{t}$ pairs.

The work done by the Mexican community on $t\bar{t}$ production is focused on the contribution from physics beyond the SM. In 1996 a work was done on the effects of dimension 5 ttZ and tbW couplings on $t\bar{t}$ production [13]. Due to the high energy behavior of the dimension 5 operators[1], it is possible to change significantly the SM prediction for the ZZ and the W^+W^- fusion processes $ZZ \to t\bar{t}$ and $W^+W^- \to t\bar{t}$. Several hundreds or even a few thousand events for every 100 fb^{-1} of luminosity could be generated through these modes at the LHC [13]. A more detailed study on $t\bar{t}$ production was done for the Linear Collider in Ref. [14], where some background effects from initial state radiation were also taken into account. With 200 fb^{-1} of luminosity this collider could be a good probe of the dimension 5 couplings of the Top quark. J. Wudka, et. al. have also analyzed the prospects of Top quark production at this collider in the context of the Linear effective Electroweak Lagrangian [15].

In 1997, J. L. Diaz-Cruz, M. A. Perez and J. J. Toscano considered $t\bar{t}h$ production at the LHC in a model with enhanced ttH couplings [16]. The effects of a strong ttH coupling would also modify the largest mode of Higgs production, which is gluon-gluon fusion. As it is well known, the effective ggH coupling arises at one loop level with the Top quark as the internal fermion line [16]. J. Wudka, et.al. have made an extensive study on Top quark pairs produced at a photon-photon collider, where the use of polarized beams can also distinguish CP violating effects in $tt\gamma$ and tbW couplings [17].

One more interesting mode of Top quark pairs is the same sign tt (instead of $t\bar{t}$) production mode. This mode can appear via a t-channel diagram with the FCNC effective coupling tqV (with $q = uc$ and $V = \gamma Z$) turning the incoming parton quarks into Top quarks. In Ref. [18] it was found that in some models of dynamical Electroweak Symmetry breaking theses couplings could give rise to enough same sign tt pairs to be observed at the LHC.

Production of single Top.

Single Top production is the only process that can probe directly the tbW coupling. It has not been observed at the Tevatron so far, mainly because $t\bar{t}$ production stands as an

---

[1] Dimension 5 operators arise in the non-Linear Electroweak Lagrangian. There are also many studies based on the Linear version, in which the lowest (higher than 4) dimension for the effective operators is 6.

important background process [6].

As mentioned above, there are three modes of single Top production: s-channel, t-channel and W-associated production (see Fig. 2) [19]. These three modes can be studied separately, and because of their electroweak origin, the produced Top quark has a certain degree of polarization. In the SM, NLO QCD corrections have been calculated that preserve the information of the polarized of the final state particles [20].

The polarizing effect of the SM tbW coupling can also be observed by looking at the polarization state of the decay particles. In particular, CDF and DØ have been able to measure the polarization of the W boson in the decay of the Top. There are three modes in the $t \to bW$ decay, depending on the polarization state of the W boson. Each mode is associated with a fraction, f_0, f_+ or f_-, that corresponds to the longitudinal, right-handed or left-handed polarization, respectively. By definition, we have the constraint $f_0 + f_+ + f_- = 1$. Recent reports by the DØ and CDF collaborations at Fermilab give the following (95% C.L.) results for the longitudinal and right-handed fraction of $t \to bW$ in the $t\bar{t}$ pair events [21]:

$$f_0 = 0.91 \pm 0.38 \, (\text{CDF}), \quad f_0 = 0.56 \pm 0.32 \, (\text{DØ}),$$
$$f_+ \leq 0.18 \, (\text{CDF}), \quad f_+ \leq 0.24 \, (\text{DØ}).$$

In Ref. [22], it is shown that by using the experimental information of f_0, f_+, σ_t (t-channel single Top) and σ_s (s-channel) we can determine the general tbW vertex that contains four independent coefficients $f_{1,2}^{L,R}$:

$$\mathcal{L}_{tbW} = \frac{g}{\sqrt{2}} W_\mu^- \bar{b} \gamma^\mu \left(f_1^L P_L + f_1^R P_R \right) t$$
$$- \frac{g}{\sqrt{2} M_W} \partial_\nu W_\mu^- \bar{b} \sigma^{\mu\nu} \left(f_2^L P_L + f_2^R P_R \right) t + h.c., \quad (4)$$

In the SM the values of the coefficients are $f_1^L = V_{tb} \simeq 1$, $f_1^R = f_2^L = f_2^R = 0$. With millions of Top quark events expected at every year of the LHC we should be able to measure these coefficients down to the order of 10^{-2}.

Single Top events could in principle also appear at the electron-proton collider HERA. In this case a large enough FCNC $tq\gamma$ coupling could produce an observable signal. The associated diagram would be t-channel like with the lepton line exchanging a virtual photon with the quark line. It is known that in the low Q region, when the photon gets close to the on-shell condition, the production amplitude grows indefinitely. An infrared divergence has to be avoided by using the non-zero mass of the lepton (electron for HERA). The authors of Ref. [23], calculated the effective $tq\gamma$ (also tqZ) coupling that arises in Topcolor assisted Technicolor theories (TC2) through a strong FC $tc\pi_t$ coupling that appears with the Top-pion condensate formed by the Topcolor interaction. They concluded that the single Top production cross section was high enough to give rise to an observable signal at HERA. With F. Peñuñuri and M.A. Pérez I was able to re-analyze this computation [24]. We found an unfortunate mistake in the numerical integration done in Ref. [23]. The origin of this mistake came from not knowing how to deal with the infrared divergent behavior of the amplitude. Unfortunately, a careful calculation of this process showed that the production cross section of single Top at HERA is very small, even for a model like TC2 where large FCNC Top couplings can arise.

FCNC AND RARE DECAYS OF THE TOP QUARK

In the SM, there are no flavor changing neutral couplings (FCNC) mediated by the Z, γ, g gauge bosons nor the Higgs boson H at tree level because the fermions are rotated from gauge to mass eigenstates by unitary diagonalization matrices. Furthermore, the top-quark FCNC induced by radiative effects are also highly suppressed. The higher order contributions induced by the charged currents are proportional to $(m_i^2 - m_j^2)/M_W^2$, where $m_{i,j}$ are the masses of the quarks circulating in the loop and M_W is the W gauge boson mass. As a consequence, in the SM all top-quark FCNC transitions $t \to qV, qH$, with $V = Z, \gamma, g$, which involve down-type quarks in the loops, are suppressed far below the observable level at existing or upcoming high energy colliders [25].

FCNC and rare Top decay in the SM

In 1990 J.L. Diaz-Cruz, R. Martínez, M.A. Pérez and A. Rosado performed the one loop calculation for the $t \to cV$ transitions. They found that the scale of the respective partial widths is set by the b quark mass [26]:

$$\Gamma(t \to V_i c) = |V_{bc}|^2 \alpha \alpha_i m_t \left(\frac{m_b}{M_W}\right)^4 \left(1 - \frac{m_{V_i}^2}{m_t^2}\right) \tag{5}$$

where α_i is the respective coupling for each gauge boson V_i. From the above result, it follows the approximated branching ratios $BR(t \to \gamma, Z) \sim 10^{-13}$ and $BR(t \to cg) \sim 10^{-11}$. In contrast, in the $b \to s\gamma$ transitions the leading contribution is proportional to m_t^4/M_W^4 and thus the GIM mechanism induces in this case an enhancement factor. In a similar way, it has been realized that some top-quark FCNC decay modes can be enhanced by several orders of magnitude in scenarios beyond the SM, and some of them falling within the LHC's reach. In this case, the enhancement arises either from a large virtual mass or from the couplings involved in the loop. Top-quark FCNC processes may thus serve as a window for probing effects induced by new physics.

The absence of the vertex Htc at tree-level in the SM can be traced down to the presence of only one Higgs doublet. The process involved in the diagonalization of the fermion masses induces simultaneously diagonal Yukawa couplings for the physical Higgs boson. In models with more than one Higgs doublet, additional conditions have to be imposed to ensure that no FCNC arise at tree level. In particular, a discrete symmetry that makes quarks of same charge to interact with only one of the two (or more) Higgs doublets will, by the Glashow-Weinberg mechanism, cause all the Yukawa couplings involving physical neutral Higgs boson states become diagonal. On the other hand, without any FCNC suppression mechanism these type of models may produce tqH couplings at tree level, which in turn may induce large enhancements of the FCNC tqV by radiative effects [27].

The most general effective Lagrangian describing the FCNC top-quark interactions with a light quark $q' = u, c$, containing terms up to dimension five, can be written as [28]

$$\mathscr{L} = \bar{t}\left\{\frac{ie}{2m_t}(\kappa_{tq'\gamma} + i\tilde{\kappa}_{tq'\gamma}\gamma_5)\sigma_{\mu\nu}F^{\mu\nu}\right.$$

$$+ \ \bar{t}\{\frac{ig_s}{2m_t}(\kappa_{tq'g} + i\tilde{\kappa}_{tq'g}\gamma_5)\sigma_{\mu\nu}\frac{\lambda^a}{2}G_a^{\mu\nu}$$
$$+ \ \frac{i}{2m_t}(\kappa_{tq'Z} + i\tilde{\kappa}_{tq'Z}\gamma_5)\sigma_{\mu\nu}Z^{\mu\nu}$$
$$+ \ \frac{g}{2c_w}\gamma_\mu(v_{tq'Z} + a_{tq'Z}\gamma_5)Z^\mu$$
$$+ \ \frac{g}{2\sqrt{2}}(h_{tq'H} + i\tilde{h}_{tq'H}\gamma_5)H\}q'. \tag{6}$$

where we have assumed also that the top quark and the neutral bosons are on shell or coupled effectively to massless fermions. In terms of these coupling constants, the respective partial widths for FCNC decays are given by [29]

$$\Gamma(t \to qZ)_\gamma = \frac{\alpha}{32 s_W^2 c_W^2}(|\kappa_{tqZ}|^2 + |\tilde{\kappa}_{tqZ}|^2)\frac{m_t^3}{M_Z^2}\left[1 - \frac{M_Z^2}{m_t^2}\right]^2\left[1 + 2\frac{M_Z^2}{m_t^2}\right]$$

$$\Gamma(t \to qZ)_\sigma = \frac{\alpha}{16 s_W^2 c_W^2}(|v_{tqZ}|^2 + |a_{tqZ}|^2) m_t \left[1 - \frac{M_Z^2}{m_t^2}\right]^2\left[2 + \frac{M_Z^2}{m_t^2}\right]$$

$$\Gamma(t \to q\gamma) = \frac{\alpha}{2}(|\kappa_{tq\gamma}|^2 + |\tilde{\kappa}_{tq\gamma}|^2) m_t$$

$$\Gamma(t \to qg) = \frac{2\alpha_s}{3}(|\kappa_{tqg}|^2 + |\tilde{\kappa}_{tqg}|^2) m_t$$

$$\Gamma(t \to qH) = \frac{\alpha}{32 s_W^2}(|h_{tqH}|^2 + |\tilde{h}_{tqH}|^2) m_t \left[1 - \frac{M_H^2}{m_t^2}\right]^2 \tag{7}$$

An update of the original SM calculations [26, 30, 31] for the FCNC top-quark branching ratios gives the following results [29]

$$BR(t \to q\gamma) = (4.6^{+1.2}_{-1.0} \pm 0.2 \pm 0.4^{+1.6}_{-0.5}) \times 10^{-14}$$
$$BR(t \to qg) = (4.6^{+1.1}_{-0.9} \pm 0.2 \pm 0.4^{+2.1}_{-0.7}) \times 10^{-12}$$
$$BR(t \to qZ) \approx 1 \times 10^{-14}$$
$$BR(t \to qH) \approx 3 \times 10^{-15} \tag{8}$$

where the uncertainties shown in the $t \to c\gamma, cg$ branching ratios are associated to the top and bottom quark masses, the CKM matrix elements and the renormalization scale. These updated results are about one order of magnitude smaller than the ones previously obtained [26, 30, 31]. For the decays involving the u quark, the respective BR are a factor $|Vub/Vcb|^2 \sim 0.0079$ smaller than those shown in (8).

The $BR(t \to qZ)$ refers to an on-shell Z boson in the final state. A more realistic analysis considers the subsequent Z decay into a fermion pair. Furthermore, a fermion-antifermion pair can also come from an off-shell photon or gluon. A. Cordero-Cid, J. M. Hernandez, G. Tavares-Velasco and J. J. Toscano, have calculated the rare top quark decay $t \to u(1)\bar{u}(2)u(2)$ in the SM [32]. Their conclusion was that the branching ratio for $t \to u(1)\bar{u}(2)u(2)$ is about the same as the two-body decay $t \to u(1)g$.

FCNC Top decay in models beyond the SM

In 1992, before the discovery of the Top quark, J.L. Diaz-Cruz and G. G. Lopez-Castro analized the possibility of FCNC and CP violation in a Two Higgs doublet model (THDM). The found that the presence of a charged Higgs boson can greatly enhance them in this model [33]. Many years before, in 1983 M.A. Pérez and A. Rosado had already found that contributions from this non-SM Higgs scalar would be important for the decay of a t-meson [34] (then, the Top quark was assumed to be just another heavy quark soon to be observed). In the THDM-III the heavy neutral scalar and pseudoscalar Higgs bosons H and A have non-diagonal couplings to fermions at the tree level. The FCNC decay modes $t \to cV$ and $t \to ch$ proceed at the one-loop level due to the exchange of H, A and $H^{\pm}$. In 1999 J.L. Diaz-Cruz, M.A. Pérez, G. Tavares-Velasco and J.J. Toscano obtained the branching ratios of $t \to cV$ for $V = g, \gamma, Z$ in the context of this model [35]. They found that the respective branching ratios may be enhanced by several orders of magnitude (for reasonable values of the THDM-III parameters) with respect to the SM predictions: $BR(t \to cg) \sim 10^{-4} - 10^{-8}, BR(t \to c\gamma) \sim 10^{-7} - 10^{-11}, BR(t \to cZ) \sim 10^{-6} - 10^{-8}$.

The FCNC decays $t \to cV$ have been studied extensively in the MSSM. The first studies considered one-loop SUSY-QCD and SUSY-EW contributions, which were later generalized in order to include the left-handed (LH) and right-handed (RH) squarks mixings. The SUSY-EW corrections were further generalized and included the neutralino-$q\tilde{q}$ loop, as well as the relevant SUSY mixing angles and diagrams involving a helicity flip in the gluino line. While the first calculations obtained $BR(t \to cV)$ of the order of $10^{-6} - 10^{-8}$, every new study improved these results until the range of values $BR(t \to cg) \sim 10^{-5}, BR(t \to c\gamma) \sim 10^{-6}, BR(t \to cZ) \sim 10^{-6}$ were reached. However, they are still below the estimated sensitivity at the LHC with an integrated luminosity of 100 fb^{-1}. Similar results were obtained by D. Delepine and S. Khalil in a general SUSY model with a light right-handed top-squark and a large mixing between the first or second and the third generation of up-squarks [36].

On the other hand, while the $t \to cH$ decay is the less favored channel in the SM, it is this FCNC channel which shows the most dramatic enhancements due to new physics effects. In some SUSY extensions its BR can be ten orders of magnitude larger than the SM prediction. This possibility arises not only because in some models the FCNC vertex tcH can be generated at tree level, but also because the GIM suppression does not apply in some loops. In particular, the gluino-mediated FCNC couplings $u_a \tilde{u}_b \tilde{g}$ induces a $BR(t \to ch) \sim 10^{-4}$, where h is the lightest CP-even Higgs boson predicted in the MSSM. The branching fraction for this channel has been found to be as large as $10^{-3} - 10^{-5}$ in a minimal SUSY FCNC scenario in which all the observable FCNC effects come from squark mixings $\tilde{c} - \tilde{t}$ induced by the non-diagonal scalar trilinear interactions [37]. However, it has been pointed out recently that the electroweak precision measurements may impose constraints on this squark mixing; which in turn decrease the MSSM prediction for the FCNC top quark processes [38].

The contribution of an extra neutral gauge boson Z' to the $t \to c\gamma$ decay mode has been studied also in the framework of the TC2 model and the so-called 331 model [39]. Even though the Z' boson predicted in these models couples in a non-universal way to the third generation of fermions, it was found that its contribution to the branching ratio of $t \to c\gamma$ is at most of order of 10^{-8} for $m_{Z'} \sim 500$ GeV [40].

Left-Right (LR) symmetric models are based on the gauge group $SU(2)_L \times SU(2)_R \times U(1)_{B-L}$. Their general aim is to understand the origin of parity violation in low-energy weak interactions. This gauge symmetry allows a seesaw mechanism and predicts naturally neutrino masses and mixing. FCNC top-quark decays have been studied by R. Gaitan, O.G. Miranda and L.G. Cabral-Rosetti in the alternative LR symmetric model. This is a new formulation of LR models with an enlarged fermion sector: it includes vector-like heavy fermions in order to explain the fermionic mass hierarchy. Because of the presence of extra quarks, the CKM mass matrix is not unitary and FCNC may exists at tree level. In particular, there is a top-charm mixing angle which induces the tree level couplings tcZ and tcH. Precision measurements at LEP impose rather weak constraints on this mixing angle, which in turn allows FCNC branching ratios as high as $BR(t \to cH) \sim 10^{-4}$ [41].

In models with extra quarks, the CKM matrix is no longer unitary and the tcZ and tcH couplings may arise at the tree level. When the new quarks are $SU(2)_L \ Q = 2/3$ singlets, present experimental data allow large branching ratios: $BR(t \to cZ) \sim 1.1 \times 10^{-4}$ and $BR(t \to cH) \sim 4.1 \times 10^{-5}$ [29]. The decay rates for $t \to cg, c\gamma$ are induced at the one-loop level but they receive only moderate enhancements: $BR(t \to cg) \sim 1.5 \times 10^{-7}$ and $BR(t \to c\gamma) \sim 7.5 \times 10^{-9}$. In models with $Q = -1/3$ quark singlets, the respective branching ratios are much smaller since the breaking of the CKM unitarity is very constrained by experimental data [29]. The contributions arising from a sequential fourth generation b' to the FCNC top-quark decays have been also studied [26, 42]. However, the virtual effects induced by a b' heavy quark cannot enhance the respective branching ratios to within the LHC's reach: $BR(t \to cZ) \sim 10^{-6}$, $BR(t \to cH) \sim 10^{-7} - 10^{-6}$, $BR(t \to cg) \sim 10^{-7}$, $BR(t \to c\gamma) \sim 10^{-8}$ [42].

Constraints on FCNC Top quark couplings from experimental data.

The measurement of the inclusive branching ratio for the FCNC process $b \to s\gamma$ has been used to put constraints on the $tc\gamma, tcg$ couplings [43]. These anomalous couplings modify the coefficients of the operators O_7 and O_8 of the effective Hamiltonian for the $b \to s\gamma$ transition. The known branching ratio for $t \to bW$ [44] and the CLEO bound on $b \to s\gamma$ place the limits $|\kappa_g| < 0.9$ and $|\kappa_\gamma| < 0.16$, which can be translated into the bounds $BR(t \to c\gamma) < 2.2 \times 10^{-3}$ and $BR(t \to cg) < 3.4 \times 10^{-2}$ [43].

The FCNC couplings tcZ and tcH have also been constrained by using the electroweak precision observables Γ_Z, R_c, R_b, R_l, A_c and the S/T oblique parameters [45]. The one-loop correction of these couplings to the decay modes $Z \to c\bar{c}$ and $Z \to b\bar{b}$ are shown in Fig. 4. Even though these vertices enter in the Feynman diagrams 3(b)-(d) as a second order perturbation, the known limits on the above precision observables [44] impose significant constraints on the tcH and tcZ couplings [45].

Figures show the 95% C.L. limits on the g_l/g_r and h_l/h_r FCNC top quark vertices and for several values of the intermediate-mass Higgs boson. These limits can be translated into the following bounds for the respective branching ratios: $BR(t \to cZ) < 1.6 \times 10^{-2}$ and $BR(t \to cH) < 0.9 - 29 \times 10^{-4}$ for $116\,GeV < mH < 170\,GeV$ [45]. In particular,

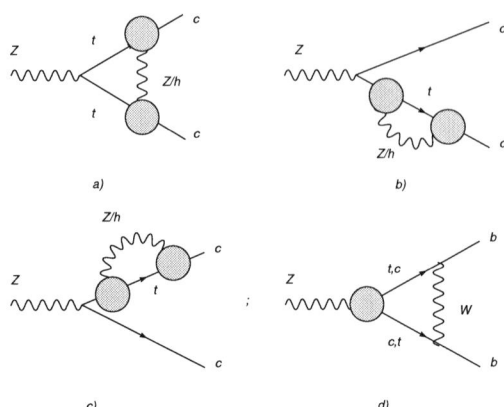

FIGURE 4. Feynman diagrams for the one-loop contribution of the FCNC tcZ/H vertices to the decay mode $Z \to c\bar{c}$

the limit on $Br(t \to cZ)$ is similar to the bound recently reported by the DELPHI Collaboration [47].

ACKNOWLEDGMENTS

I want to thank for the invitation and for Conacyt support.

REFERENCES

1. F. Abe *et al.* [CDF Collaboration], *Phys. Rev. Lett.* **73**, 225 (1994); Phys. Rev. D **50**, 2966 (1994); Phys. Rev. Lett. **74**, 2626 (1995).
2. S. Abachi *et al.* [D0 Collaboration], Phys. Rev. Lett. **74**, 2632 (1995).
3. D. Chakraborty, J. Konigsberg and D. L. Rainwater, Ann. Rev. Nucl. Part. Sci. **53**, 301 (2003) [arXiv:hep-ph/0303092].
4. V. M. Abazov *et al.* [D0 Collaboration], *Phys. Lett.* B **622**, 265 (2005) [arXiv:hep-ex/0505063].
5. A. Aktas *et al.* [H1 Collaboration], Eur. Phys. J. C **33**, 9 (2004) [arXiv:hep-ex/0310032].
6. A. Juste, *Top quark current experimental status*, PoS **TOP2006**, 007 (2006), [arXiv:hep-ex/0603007].
7. H.S. Do, S. Groote, J.G. Korner and M.C. Mauser, Phys. Rev. **D67** (2003), 091501 and references therein.
8. G. Sanchez-Colon and R. Huerta, Phys. Rev. D **40**, 1683 (1989).
9. G. Calderon and G. Lopez Castro, arXiv:hep-ph/0108088.
10. J. Erler, *Precision electroweak physics*, Talk given at 10th Mexican Workhop on Particles and Fields, Morelia, Michoacan, Mexico, 7-12 Nov 2005 [arXiv:hep-ph/0604035].
11. E. Malkawi and C.-P. Yuan, Phys. Rev. **D50**, 4462 (1994); ibid. **D52**, 472 (1995); F. Larios, E. Malkawi and C.-P. Yuan, hep-ph/9704288.
12. F. Larios, M. A. Perez and C. P. Yuan, Phys. Lett. B **457**, 334 (1999)
13. F. Larios and C. P. Yuan, Phys. Rev. D **55**, 7218 (1997) [arXiv:hep-ph/9606397].
14. F. Larios, T. Tait and C. P. Yuan, Phys. Rev. D **57**, 3106 (1998) [arXiv:hep-ph/9709316].
15. S. Bar-Shalom and J. Wudka, Phys. Rev. D **60**, 094016 (1999) [arXiv:hep-ph/9905407].
16. J. L. Diaz-Cruz, M. A. Perez and J. J. Toscano, Phys. Lett. B **398**, 347 (1997) [arXiv:hep-ph/9702413].

17. B. Grzadkowski, Z. Hioki, K. Ohkuma and J. Wudka, Nucl. Phys. B **689**, 108 (2004) [arXiv:hep-ph/0310159]; Phys. Lett. B **593**, 189 (2004) [arXiv:hep-ph/0403174]; Acta Phys. Polon. B **36**, 3531 (2005) [arXiv:hep-ph/0511038]; JHEP **0511**, 029 (2005) [arXiv:hep-ph/0508183].
18. F. Larios and F. Penunuri, J. Phys. G **30**, 895 (2004) [arXiv:hep-ph/0311056].
19. Z. Sullivan, Phys. Rev. **D70**, 114012(2004), and references therein; J. Campbell, R.K. Ellis and F. Tramontino, Phys. Rev. **D70**, 094012(2004).
20. Q. H. Cao, R. Schwienhorst, J. A. Benitez, R. Brock and C. P. Yuan, Phys. Rev. D **72**, 094027 (2005) [arXiv:hep-ph/0504230]; Phys. Rev. D **71**, 054023 (2005) [arXiv:hep-ph/0409040]; Q. H. Cao and C. P. Yuan, Phys. Rev. D **71**, 054022 (2005) [arXiv:hep-ph/0408180].
21. By the DØ collaboration, hep-ex/0404040 and references there in; by the CDF collaboration, hep-ex/0411070 and references therein.
22. C. R. Chen, F. Larios and C. P. Yuan, Phys. Lett. B **631**, 126 (2005) [AIP Conf. Proc. **792**, 591 (2005)] [arXiv:hep-ph/0503040].
23. C. X. Yue, W. Wang and H. J. Zong, Nucl. Phys. B **667**, 349 (2003).
24. F. Larios, F. Penunuri and M. A. Perez, Phys. Lett. B **605**, 301 (2005) [arXiv:hep-ph/0409291].
25. F. Larios, R. Martinez and M. A. Perez, *New physics effects in the flavor-changing neutral couplings of the top quark*, to appear in Int. J. Mod. Phys. A, [arXiv:hep-ph/0605003].
26. J.L. Díaz-Cruz, R. Martínez, M.A. Pérez, A. Rosado, *Phys. Rev.* **D41**, 891 (1990).
27. A. Cordero-Cid, M. A. Perez, G. Tavares-Velasco and J. J. Toscano, Phys. Rev. D **70**, 074003 (2004) [arXiv:hep-ph/0407127].
28. T. Han and J.L. Hewett, *Phys. Rev.* **D60**, 074015 (1999).
29. Aguilar-Saavedra, *Acta Phys. Pol.* **B35**, 2695 (2004); *Phys. Rev.* **D67**, 035003 (2003); ibid. **D69**, 099901 (2004); J.A. Aguilar-Saavedra and B.M. Nobre, *Phys. Lett.* **B553**, 251 (2003); J.A. Aguilar-Saavedra and G.C. Branco, *Phys. Lett.* **B495**, 347 (2000); J.A. Aguilar-Saavedra, *Phys. Lett.* **B502**, 115 (2001).
30. G. Eilam, J.L. Hewett, A. Soni, *Phys. Rev.* **D44**, 1473 (1991); Erratum-ibid. **D59**, 039901 (1999).
31. B. Mele, S. Petrarca, A. Soddu, *Phys. Lett.* **B435**, 401 (1998); H. Fristch, *Phys. Lett.* **B224**, 179 (1989).
32. A. Cordero-Cid, J. M. Hernandez, G. Tavares-Velasco and J. J. Toscano, arXiv:hep-ph/0411188.
33. J. L. Diaz-Cruz and G. Lopez Castro, Phys. Lett. B **301**, 405 (1993).
34. M. A. Pérez, Z. Phys. C **18**, 113 (1983); M. A. Pérez and A. Rosado, Phys. Rev. D **30**, 228 (1984).
35. J. L. Diaz-Cruz, M. A. Perez, G. Tavares-Velasco and J. J. Toscano, Phys. Rev. D **60**, 115014 (1999) [arXiv:hep-ph/9903299].
36. D. Delepine and S. Khalil, Phys. Lett. B **599**, 62 (2004) [arXiv:hep-ph/0406264].
37. J.L. Díaz-Cruz, H.-J. He and C.-P. Yuan, Phys. Lett. **B530**, 179 (2002).
38. J. Cao, G. Eilam, K.-I. Hikasa and J.-M. Yang, hep-ph/0604163.
39. F. Pisano and V. Pleitez, *Phys. Rev.* **D46**, 410 (1992); P.H. Frampton, *Phys. Rev. Lett.* **69**, 2889 (1992).
40. C.X. Yue, H.J. Zong and L.J. Liu, *Mod. Phys. Lett.* **A18**, 2187 (2003); A. Cordero-Cid, G. Tavares-Velasco and J. J. Toscano, Phys. Rev. D **72**, 057701 (2005).
41. R. Gaitan, O. G. Miranda and L. G. Cabral-Rosetti, Phys. Rev. D **72**, 034018 (2005); ibid. arXiv:hep-ph/0604170.
42. A. Arhrib and W.S. Hou, hep-ph/0602035.
43. R. Martinez, M. A. Perez and J. J. Toscano, Phys. Lett. B **340**, 91 (1994); T. Han et al., Phys. Rev. **D55**, 7241 (1997); G. Burdman, M.C. González García and S.F. Novaes, Phys. Rev. **D61**, 114016 (2000).
44. S. Eidelman et al., Particle Data Group *Phys. Lett.* **B592**, 1 (2004).
45. F. Larios, R. Martinez and M. A. Perez, Phys. Rev. D **72**, 057504 (2005) [arXiv:hep-ph/0412222].
46. J. L. Diaz-Cruz and C. E. Pagliarone, arXiv:hep-ph/0412329.
47. I. Abdalla et al., DELPHI Collaboration, Phys. Lett. **B590**, 21 (2004).

Relativistic heavy-ion physics in Mexico

Alejandro Ayala

Instituto de Ciencias Nucleares, Universidad Nacional Autónoma de México, Apartado Postal 70-543, México D.F. 04510, Mexico.

Abstract.
The field of heavy-ion physics in Mexico is a rather young but nonetheless very successful enterprise. The community of practitioners working in the field has been growing steadily both in quantity and degree of specialization; during this year, the field counted with about 18 Ph. D's and about the same number of students pursuing different degrees, coming from 4 different institutions in Mexico. The amount of people along with the quality of their research make this team one of the strongest in high energy physics in Mexico. The key to the success lies on the ability of the community to cluster around well defined goals and to work coherently to solve specific tasks. The effort has been both experimental and theoretical, which adds to the strength of the collaboration. The defining moment that made the launching of the enterprise possible was the early involvement in the ALICE collaboration and in particular the taking of the responsibility for the design and construction of two of its detectors, the V0A and ACORDE triggers. In this short review, I concentrate in describing the contributions to the field by colleagues whose research has been developed in Mexico, on theoretical, phenomenological and Monte Carlo simulation aspects.

Keywords: Relativistic heavy-ions
PACS: 25.75.-q, 01.65.+g

THE EARLY DAYS

During the early to mid 90's, the field of heavy-ion physics was one of a few in high energy physics that experienced a continuous worldwide growth. This is reflected on today's size of the world community working in the field that consists of about 1500 experimentalists and about 150 theoreticians. This growth was mainly stimulated by the existence of new and intriguing data. The results from the AGS and SPS –where heavy ions were being collided in fixed target experiments at nucleon-nucleon center of mass energies of 5 and 17 GeV, respectively– were prompting the interest in the properties of matter at high temperatures and densities and aiming to identify the so called Quark Gluon Plasma (QGP), a state of matter that, according to the current theories of the evolution of the Universe, existed 1 to 10 μs after the Big Bang. At the same time, these studies required the development of state-of-the-art detectors to distinguish tracks in the crowded environment produced in heavy-ion reactions. Moreover, the future looked promising with the upcoming Relativistic Heavy Ion Collider (RHIC) at BNL in the USA and the Large Hadron Collider (LHC) in Europe at CERN, where the study of matter would be taken to the extreme conditions reached in collisions of heavy systems at nucleon-nucleon center of mass energies of 200 GeV and 14 TeV, respectively.

In contrast, the world's excitement in this area was not being shared by the physics community in Mexico. Since the field lies at the intersection of nuclear and particle physics and both disciplines enjoyed a relatively long standing tradition, it could have been expected that efforts to take part in these developments were being made, but this was not the case. Given that the field is mainly driven by experimentalists, it was natural to foresee that also Mexican experimentalists would have taken the necessary steps to join the venture. However, by those days, the recently established, experimental high energy physics community in Mexico was mainly interested –with very few and remarkable exceptions– in continuing their participation in the analysis of data from those experiments where they were formed. Although the experimental nuclear physics community was better established, it is also fair to say that in general they always looked with suspicion at being part of a large collaboration. A change of attitude was needed in order to bridge the gap.

THE YOUNG AND THE RESTLESS

The niche was wide open to remain unoccupied for longer. What was needed was the will to use the contacts with colleagues in the world interested in the field, to present a coherent proposal and to knock at the right doors for local support. This finally happened in the late 90's. By then, RHIC was almost ready to start operations. From the logistical point of view, it could have been a good option to join one of the existing teams in one of the RHIC collaborations. However, in terms of visibility, a better option was to join the only LHC collaboration devoted to study heavy-ion physics, ALICE (an acronym whose meaning is A Large Ion Collider Experiment) that by then was in the midsts of defining specific tasks for the members of the collaboration. A proposal by a Mexican team was put forward to take part in the construction of the *wake up* trigger, the so called V0 detector [1]. Another Mexican team proposed also to construct the cosmic ray trigger, the so called ACORDE detector [2]. Details about these developments along with a review about experimental aspects in the field can be found in the contribution from G. Herrera and G. Paić in this volume.

With the proposals approved, the leaders devoted themselves to obtain local financing and to recruit colleagues to join efforts for these goals. Both calls for support were successful to the point that existing labs at CINVESTAV, BUAP, IFyM-UMSNH and IF-UNAM started to be better equipped and a new detectors lab at ICN-UNAM was built and equipped in a one year time.

At the same time, around the late 90's, the interest was also growing from the theory side, from isolated efforts to the formation of a small but active group mainly interested on phenomenological aspects of heavy-ion data. The point to remark here is the merging of the efforts. From the beginning it became clear that in order to have an impact in the worldwide developments it was mandatory to take part in all of the steps, namely the formation of a group of experts in detector building along with a group working in theory, phenomenology and simulation and data analysis. In this short review, I concentrate in describing the contributions in the latter three aspects of the field in Mexico.

THEORY

The pure theoretical aspects of the work in relativistic heavy-ion physics carried out by members of the field working in Mexico are very scarce. The work has concentrated in the early formulation of a QCD-like model to describe the parton distribution of a heavy nucleus at small values of Bjorken x and on field theoretical aspects of hadron interactions at finite temperature.

In Refs. [3] the authors use the idea that at very low x-values in a heavy nucleus, the largest of all energy scales is the parton density and thus that the strong coupling constant α_s, being a running parameter of the energy scales, should mainly depend on this density and thus become small as x becomes small enough. This behavior allows working with weak coupling techniques to compute the parton distribution function at low x in terms of fluctuations of the gluon field around the background field produced by incoherent sources taken as the valence quarks of the heavy nucleus. Though strictly speaking this work was not carried out in Mexico, I mention it in this review because it represents a seminal work that prompted the interest in the theory/phenomenology aspects of the field latter developed by the local community.

In Refs. [4] the authors compute one-loop effective vertices and propagators in the linear sigma model at finite temperature. The guideline for the construction of these objects is the constraint imposed by the chiral Ward identities. These vertices and propagators are used in Ref. [5] to compute the pion dispersion relation in a pion gas. In Ref. [6] the pion contribution to the modification of the mass and width of the ρ^0 is computed at finite temperature and finite pion density in a pion gas. In Ref. [7], these modifications are computed within a hadron gas accounting for interactions of ρ^0 with nucleons, pions, kaons and several baryonic and mesonic resonances.

PHENOMENOLOGY

The formulation of models to understand existing relativistic heavy-ion data has been one of the main activities in the field in Mexico. This line of research has touched upon several aspects in the field ranging from HBT analyses, Coulomb effects, QGP signals, thermal models, etc.

The Hanbury-Brown Twiss (HBT) interference effect for bosons is also named Bose-Einstein correlation effect. The importance of this effect in heavy-ion physics is that it can be implemented, by measuring two-meson distributions, to extract the space-time size of the region of emission of particles. It is also known that if other space-time scales are present during the process of particle emission/detection, the HBT results will also contain information about these scales [8]. An earlier example of this fact is discussed in Ref. [9] where the variation of the HBT radius is studied as a function of the mass number of the projectile nucleus. In that work, it is argued that, besides the size of the interaction region, proportional to the cube root of the projectile's mass number, an additional scale involved during particle emission is the size of the fragments originated in the collision. For very energetic collisions, this size is of the order of the size of a hadron. The extracted HBT radius is thus the square root of the sum of the squares of the average fragment size and the size of the projectile. The result is compared in Ref. [9]

to data of the HBT size as a function of the projectile's mass number with reasonably good agreement.

The large amount of charge present in collisions of heavy nuclei is known to produce distortions on particle spectra. This phenomenon is known as Coulomb effect. For AGS and SPS energies, where there is a large amount of stopping, the net positive charge after a central collision is carried by protons in the central region and thus positively/negatively charged particles are pushed/pulled towards higher/lower p_t values. This effect is more pronounced for low p_t particles. During kinetic freeze-out, it is possible to think that the phase space occupancy of the thermally produced particles, mainly mesons, can be described in terms of a collisionless transport equation, the Vlasov equation, where the driving force for charged particles is of electromagnetic type. Assuming that in the center of mass frame of the collision, the fireball is spherically symmetric, the solution of Vlasov equation contains as one of the parameters the fireball size, since the net charge is fixed by the net number of protons in the colliding system and the radial flow velocity can be taken from independent measurements. In Refs. [10], this study was implemented for kaons and pions and a fireball size of about 10 fm for Pb-Pb collisions at 158 A GeV in SPS was thus inferred.

In general, hadronization in high energy collisions and in particular in heavy-ion reactions is a subject where, despite the intense research devoted, the picture is by no means clear. In the latter kind of reactions, where at sufficiently high energies there is evidence for the existence of a state of matter where the quarks and gluons of QCD are the correct degrees of freedom for the description, one possible hadronization process is the thermal recombination of quarks into meson and baryons, due to the high density of partons produced in the reaction. However a commonly overlooked fact in the thermal models aiming to describe such thermal recombination is that the processes take place over small time scales and consequently over small volumes. The study of this finite size effects has been a subject of research for the Mexican community over some time already. Finite size effects at kinetic freeze-out can be implemented by imposing that the states describing particles are confined within the volume of the reaction. The immediate consequence of such description is a broadening of the momentum distribution due to the Heisenberg uncertainty principle whereby restricting the spatial region where particles are found, there is a larger spread in their momentum. In Refs. [11], the foundations of this approach for the computation of one and two particle distributions, have been given, including also the implementation of radial flow in such a description. A very good agreement of the model with pion and proton transverse momentum distributions for AGS and RHIC energies has been achieved.

The search for signals for the production of QGP in heavy-ion reactions has also been a subject of interest for the Mexican community. One proposed signal is to look for the change in the transverse (with respect to the reaction plane) polarization properties of Λ^0 hyperons in heavy-ion collisions compared to p-p collisions as a function of impact parameter [12]. The idea is that since the core of a heavy nucleus is denser than the outer layers, collisions with small impact parameter, where larger densities are achieved, are more likely to form QGP than peripheral ones which are more similar to p-p collisions. In central collisions, the recombination of the liberated partons do not form polarized Λ^0s, as opposed to the case of peripheral ones where, as in the case of p-p reactions, the transverse polarization is attributed to the Thomas precession phenomenon of the spin of the strange quark pulled by a u-d diquark coming from the valence nucleons in the collision beams. Since both production mechanisms are present in a heavy-ion reaction and contribute to the net number of Λ^0s but only peripheral collisions produced polarized ones, a diminishing of the net transverse Λ^0 polarization can be linked to the formation of QGP. This idea has been extended to considering the longitudinal $\bar{\Lambda}^0$ polarization in heavy-ion reactions [13]. In this case, this polarization is due to the weak decay of $\bar{\Xi}$ into $\bar{\Lambda}^0$ and π, which is the main decay channel for $\bar{\Xi}$. For central heavy-ion reactions, where the formation of QGP is more likely, the number of $\bar{\Xi}$s (with a quark composition $\bar{s}\bar{s}\bar{q}$) formed by thermal recombination is expected to be much greater than the number of antihyperons (with a quark composition $\bar{s}\bar{q}\bar{q}$) and, in particular, than the number of $\bar{\Lambda}^0$s, reflecting the enhancement of strangeness production in a QGP environment. Therefore, an increase in longitudinally polarized $\bar{\Lambda}^0$s can also be linked to the formation of QGP.

MONTE CARLO SIMULATIONS

An important aspect in relativistic heavy-ion physics to understand the experimental signals has to do with Monte Carlo simulations. This aspect has also been developed by part of the community in Mexico. Studies have concentrated on the performance of the ALICE detector and on simulations in the p-p mode that will be the first working scenario for the LHC.

Monte Carlo simulations to understand the design requirements for the resolution of the trigger detector V0 in ALICE, for the rejection of interactions of the proton beams with residual gas in the beam pipes, have been carried out in Ref. [14]. The result of this analysis was a key element to convince the ALICE collaboration for the need of having a resolution better than 1 ns for the V0 detectors. Since the resolution is inversely proportional to the length of the optical fiber needed to transport the signal to the photomultipliers (PMTs), this study showed that these latter needed to be placed near the detectors and thus inside the magnetic field of the spectrometer and consequently that the PMTs should be magnetic field shielded ones.

Proton and pion production in p-p collisions have also been simulated by Monte Carlo analyses in Ref. [15] where predictions from three of the standard particle production mechanisms in Pythia 6.3 were explored. This study concentrated on the p_t dependence of the $\bar{p}/p$ and p/π ratios at collision energies of 200 GeV and 14 TeV. The conclusion is that, although the behavior of the $\bar{p}/p$ ratio is the same, neither of the considered scenarios are in agreement with existing RHIC data and that the predictions for the p/π ratio at 14 TeV are significantly different. This study calls thus attention for the tuning of the standard event generators to have a better working tool when the LHC comes on line.

Parton energy loss effects in heavy-ion collisions have also been studied by means of Monte Carlo techniques in Ref. [16]. In that work, attention is paid to a realistic geometry for the collision by considering a Glauber model of the reaction with Wood-Saxon density profiles of the colliding nuclei, therefore accounting for the spacial distribution of (hard) parton production points and thus for the amount of (varying density) matter traversed by these partons. The conclusion of this work is that if realistic density profiles are considered, the resulting transport coefficients, that quantify the amount of energy loss by the parton, have to take on very large values, perhaps signaling that the hard partons interact with the medium much stronger than perturbatively expected. A prediction for the nuclear modification factor R_{AA} for LHC energies is given consisting of a flat p_t distribution, and as low as 0.2, up to the highest parton energies, perhaps signaling a decrease in the number of expected high-energy jets, fact that the authors use to call attention for future planning of experimental studies.

HBT analyses are a favorite tool to extract the space-time sizes of the so called *source* of emitting particles in high energy collisions. This technique is widely used in heavy-ion physics. In the case of p-p and p-p̄ reactions at very high energies, there have been claims that the large (of order 10 fm) HBT source sizes, indicate the presence of a thermalized source of particles after the collision and the possible existence of QGP even in this reactions. An alternative, more conventional scenario advocated in Ref. [17], proposes that the extracted HBT sizes are due to an extended distribution of the hadronization points of hard scattered partons in this kind of reactions. The mean distance between the hard scattering and the hadronization point of a parton is proportional to the parton's energy which is also related to the mean charged particle multiplicity in the jet. The authors of Ref. [17] implement a Monte Carlo simulation for this kind of collisions, assuming a reasonable distribution of hadronization points after a hard scattering, to compute the two-particle correlation function and compare to Tevatron data. The good agreement with these data for reasonable values of their proportionality factor between the hadronization length and the parton's energy is taken as an indication of the validity of the analysis.

The deconfined state of quark matter that consists of almost equal amounts of u,d and s quarks is called strange matter and it has been speculated that this is the absolute ground state of hadronic matter at high energies. In Ref. [18], a variational Monte Carlo method in a three-color, three-flavor, string-flip model, that reproduces various aspects of QCD phenomenology such as quark clustering at low density and color deconfinement at high density, is used to study the strangeness content of hadronic matter as a function of density. The authors identify two phase transitions; one is a transition from three-quark to multi-quark cluster configurations where the length scale for quark confinement jumps discontinuously at the transition; a second transition to strange matter was also identified at a higher density. They also found that once that the system is well into the strange matter domain, there was an additional phase transition characterized by the competition between two ground states, one with a low and one with a high strange-quark content.

Effects of color screening as a function of density on the formation of heavy quark-antiquark bound states such as the J/ψ meson have also been studied by variational Monte Carlo techniques in Ref. [19] also in the context of a string flip model. In this work, it is found that the length scale for quark confinement drops abruptly to zero at the transition density indicating how the bounding of the heavy quark-antiquark pair weakens as the system evolves into the quark-matter phase in agreement with expectations on the suppression of J/ψ in heavy-ion collisions at high energy.

Ref. [20] addresses the question of whether the p/π ratios in p-p collisions could be understood including radial flow in the analysis. The authors use a constant radial flow profile and superimpose to it a random spectra generated by the standard event generators Hijing and Pythia. The final particle spectra is obtained by adding vectorially the random and radial velocity components. The results are compared to data on p-p collisions obtained by STAR and the agreement is reasonable though the sample generated with Hijing requires a large value of the expansion velocity (0.6 c) whereas the sample generated with Pythia requires a low value of the expansion velocity (0.2-0.3 c).

In Ref. [21] a search for a signal to determine jet quenching in an event by event basis is presented. Jet quenching increases as the centrality of the collision increases, thus the desired variable should also be able to capture this feature. The author proposes the variable $\langle p_t^{trunc} \rangle$, which represents the difference between the average p_t and the minimum p_t of the jet. This variable shows a dependence with the centrality of the collision. The author reproduces PHENIX data for this variable using the event generator Hijing in the jet quenching mode corresponding to the default setting for energy loss of 2 GeV/fm when identifying jets using a minimum p_t of 1 GeV.

STAR data on back to back jet azimuthal correlations in p-p collisions at center of mass energy of 200 GeV is analyzed in Ref. [22] by Monte Carlo techniques. The authors study the effect of the parton intrinsic momentum on this correlation. They find that a single Gaussian distribution of intrinsic momentum, cannot reproduce data and thus propose that this distribution is made out of the combination of two Gaussians. With this model, they are able to obtain a good description of data.

CONCLUSIONS

The field of heavy-ion physics has become an important research activity both in high energy and nuclear physics in Mexico. Among its strengths there is the close connection between theoretical and experimental efforts. This connection can be easily appreciated by realizing that many of the phenomenological works in the field are coauthored both by theorists and experimentalists. An important venue of the work in the field is that it has also sparked the growth of GRID activities motivated by the eventual participation in the analysis of the forthcoming ALICE-LHC data. A main weakness, that threatens to stall this healthy activity, is the shortsightedness of some local and national evaluation committees that demand high productivity (measured by the number of papers) from people involved in research such as software development for data analysis and detector R&D. This kind of research is of course crucial to take part in a large international collaboration but its fruits can only be appreciated after a relatively long time. Another problem that has been difficult to overcome concerns the lack of experience by local funding agencies about the meaning of being involved in a large international collaboration, which implies taking care of manitenance and operation expenses during the many stages of an experimet's lifetime. A second change of attitude is needed and an extra coherent effort in this direction by the community should be one of the main short term objectives.

ACKNOWLEDGMENTS

The author acknowledges the valuable input of many of the colleagues whose work is referred in this short review. Support has been received in part by DGAPA under PAPIIT grant No. IN107105-3 and by CONACyT under grant No. 40025-F.

REFERENCES

1. "VOL", ALICE-PR-2003-104 **v1**, EDMS Id 376975 Ext. Ref. Annex 9.
2. "ACORDE workplan", ALICE-PR-2002-56 **v1**, EDMS Id 342019 Ext. Ref. Annex 3.
3. A. Ayala, L. McLerran, J. Jalilian-Marian and R. Venugopalan, Phys. Rev. D **52**, 2935–2943 (1995); *ibid* **53**, 458-475 (1996).
4. A. Ayala and S. Sahu, Phys. Rev. D **62**, 056007 (2000); A. Ayala, S. Sahu and M. Napsuciale, Phys. Lett. **B479**, 156–162 (2000).
5. A. Ayala, P. Amore and A. Aranda, Phys. Rev. C **66**, 045205 (2002).
6. A. Ayala and J. Magnin, Phys. Rev. C **68**, 014902 (2003).
7. A. Ayala, J.G. Contreras and J. Magnin, Phys. Lett. **B603**, 165–172 (2004); A. Ayala, "ρ^0 mass in a hot hadron gas", hep-ph/0410255.
8. A. Ayala, G. Baym and J.L. Popp, Nucl. Phys. **A660**, 101–117 (1999)
9. A. Gago and G. Herrera, Mod. Phys. Lett. A **10**, 1435–1440 (1995).
10. A. Ayala and J.I. Kapusta, Phys. Rev. C **56**, 407–411 (1997); A. Ayala, S. Jeon and J.I. Kapusta, *ibid* **59**, 3324-3328 (1999); A. Ayala, S. Jeon and J.I. Kapusta, Nucl. Phys. **A661**, 573–576 (1999).
11. A. Ayala and A. Smerzi, Phys. Lett. **B405**, 20–24 (1997); A. Ayala, J. Barreiro and L.M. Montaño, Phys. Rev. C **60**, 014904 (1999); A. Ayala and A. Sánchez, Phys. Rev. C **63**, 064901 (2001); A. Ayala, E. Cuautle, J. Magnin, L.M. Montaño and A. Raya, Phys. Lett. **B634**, 200–204 (2006); A. Ayala, E. Cuautle, J. Magnin and L.M. Montaño, "Proton and pion transverse spectra at RHIC from radial flow and finite size effects", nucl-th/0603039, submitted to Phys. Lett. **B**.
12. A. Ayala, E. Cuautle, G. Herrera and L.M. Montaño, Phys. Rev. C **65**, 024902 (2002).
13. G. Herrera, J. Magnin and L.M. Montaño, Eur. Phys. J. C **39**, 95-99 (2005).
14. R. Alfaro, E. Cuautle and G. Paić, ALICE-INT-2004-021 in *ALICE Technical Design Report: (CERN/LHCC 2004-023, Sept. 2004 Sec. 3.3.4, pp 60)*.
15. E. Cuautle and G.Paić, "Study of the pion and proton production in pp collisions at 14 TeV" ALICE-INT-2005-027, in *ALICE PPR, vol. II, part I: (CERN/LHCC 2005-030, Dec. 2005 Sec. 6.2.2.3, pp 267-270)*.
16. A. Dainese, C. Loizides and G. Paić, Eur. Phys. J. C **38**, 461–474 (2005).
17. G. Paić and P.K. Skowroński, J. Phys. G **31**, 1045–1053 (2005).
18. G. Toledo Sánchez and J. Piekarewicz, Phys. Rev. C **65**, 045208 (2002).
19. G. Toledo Sánchez and J. Piekarewicz, Phys. Rev. C **70**, 035206 (2004).
20. E. Cuantle and G. Paić, "Proton/pion ratios and radial flow in pp and peripheral heavy ion collisions", hep-ph/0604246.
21. I. Domínguez, "Jet quenching study in heavy ion collisions", *M.S. Thesis, CINVESTAV U. Mérida, in spanish, Thesis Advisor J.G. Contreras, Aug. 2003*.
22. E. Cuautle, I. Domínguez and G. Paić, "Azimuthal correlations in p-p collisions", hep-ph/0604257.

A Brief Review on Extra Dimensions

Abdel Pérez-Lorenzana

Departamento de Física, Cinvestav, Apdo. Post. 14-740, 07000, México, D.F., México

Abstract.
  Considering models with extra dimensions have introduced completely new ways of looking up on old problems in theoretical physics, and conversely, it has also introduced new interesting problems of theoretical interest. Here we present a brief review on the developments of the idea and the current trends. We discuss both theoretical and phenomenological aspects of some models where extra dimensions play a role, and which provide particular insights for the possible use of additional dimensions in particle model building as well as in some new interesting scenarios of possible cosmological applications. Given the interest of this special volume, we particularly underline some of the contributions to the field by people working in Mexico, from the dawn to the present days of extra dimensional models.

INTRODUCTION.

Possible existence of new spatial dimensions beyond the four we see have been under consideration since the early works by Kaluza and Klein around the 1920's [1], where they tried to unify electromagnetism with Einstein gravity by proposing a theory with a compact fifth dimension, where the electromagnetism was originated from the extra components of the metric. Nowadays, extra dimensions are a known fundamental ingredient of the most appealing candidate for a Unified theory of Gravity and quantum phenomena, String Theory, which is naturally and consistently formulated only in a ten dimensions (or eleven for M-theory). However, it used to be conventional to assume that the six additional dimensions were compactified into manifolds of Planck size, $\ell_P \sim 10^{-33}$ cm. Planck length and its associated energy scale, the Planck mass, $M_P = c\,\hbar/\ell_P$, are defined through the fundamental constants, such that

$$M_P c^2 = \left[\frac{\hbar c^5}{8\pi G_N}\right]^{1/2} \sim 2.4 \times 10^{18} \text{ GeV} ; \qquad (1)$$

with G_N the gravity Newton constant. In what follows we will use natural units $c = \hbar = 1$.

  Planck mass was supposed to be the relevant scale where quantum nature of gravity should become evident, and thus, the natural cut-off scale for the validity of the Standard Model of particle physics (in the lack of a Grand Unified Theory). A known physical draw back of this scenario is, however, the impossibility to directly test such a regime in the near future. Nevertheless, new developments by Witten and Horava [2] on the studies of the non-perturbative regime of the $E_8 \times E_8$ theory have suggested that some, if not all, of the extra dimensions could rather be larger than ℓ_P, thus, motivating the idea that possible physical effects associated to those large extra dimensions could rather be accessible in the close future. Then, the natural question to ask is where and how would this extra dimensions manifest themselves. Of course, a complete answer relies very much on the characteristics of the final theory of quantum gravity, but good insights can be obtained already from the more simplified low energy effective field theory point of view, which needs almost no any hard string theory calculations, since it is likely that the excited modes of the string could appear on the experiments way before any quantum gravity effect, in which case the effective field theory approach would be acceptable.

The first intriguing observation that triggered the interest of the community towards models with extra dimensions was the possibility that extra dimensions as large as millimeters [3] could exist and yet remain hidden to the experiments [4, 5, 6, 7, 8, 9]. This would be possible if our observable world is constrained to live on a four dimensional hypersurface (the brane) embedded in a higher dimensional space (the bulk), such that the extra dimensions can only be tested by gravity, a picture that clearly resembles D-brane theory constructions [2, 10].

The second point that attracted the attention was the observation that such large extra dimensions would indicate a scale of quantum gravity much smaller than M_P, even closer to m_{EW}, thus offering an alternative solution to the long standing hierarchy problem which competes with supersymmetry and compositness.

Over the years a diversity of new ideas had been explored intending to revisit old theoretical problems, and many new directions for physics beyond Standard Model and Standard Cosmology had been considered. With in this context, the Mexican scientific community had also something to contribute to the discussions, and in the spirit of the present volume, we would like to present a short review of the current status of the field, trying to underline some of the contributions by authors linked to Mexico. For more general reviews the reader can consult Ref. [11] Here, we will concentrate on some of those models that we consider more interesting, either because its generality, or due to the particular interest they have generated in the study of extra dimensions.

The first part of these review will cover some general aspects of models with flat extra dimensions. We start our discussion with the concept of Kaluza-Klein (KK) mode expansion of bulk fields, which provides the effective four dimensional theory framework. Then, we briefly resume some of the expected phenomenology for gravitons. Third section is devoted to general ideas for the use of extra dimensions in model building. We address the appearance of power law running of couplings constants and the use of extra dimensions on the breaking of symmetries, as well as some ideas for models with gauge symmetries, the origin of neutrino masses and mixings, baryon number violation; and flavor models. Section four will cover some aspects on cosmology of models with extra dimensions.

GENERAL ASPECTS OF FLAT EXTRA DIMENSIONAL MODELS

Contrary to earlier believes, the existence of more than four dimensions in nature could have visible manifestations in our effective four dimensional world. Indeed, first thing to notice is that in a higher dimensional theory, where Einstein gravity is assumed to hold, the true fundamental gravity coupling, G_*, is not the well known Newton constant G_N, which is, nevertheless, the gravity coupling we do observe. Last is rather an effective quantity obtained through dimensional reduction. For instance, in the simplest case where the space factorizes as a $\mathcal{M}_4 \times T^n$ manifold, such that the n extra space-like dimensions are compactified into a torus of radius R, Newton constant is given by a volumetric scaling of the truly fundamental gravity scale [3],

$$G_N = G_*/V_n .\qquad(2)$$

Even if G_* were a large coupling (as an absolute number), one can still understand a small G_N via the volumetric suppression, $V_n \sim R^n$. The size of the extra dimensions also marks the limit of validity for the effective four dimensional gravity. If one performs a measurement of the gravitational interaction among two test masses at distances shorter that R the true higher dimensional nature of the theory should become explicit. Current tests had gone down to 160 microns, with no signals of extra dimensions so far [4].

We should now recall that the Planck scale is defined in terms of the Newton constant, it is then clear that M_P is not fundamental anymore. The true scale for quantum gravity should rather be given in terms of G_* instead, as $M_* = (8\pi G_*)^{1/(2+n)}$, Clearly, both scales are related to each other by [3]

$$M_P^2 = M_*^{n+2} V_n .\qquad(3)$$

There is no evidence of quantum gravity (neither supersymmetry, nor stringy effects) well up to energies around few hundred GeV, which says that $M_* \geq 1$ TeV. On the other hand, if the volume were large enough, then the fundamental scale could be as low as the electroweak scale, and there would be no hierarchy in the fundamental scales of physics. Of course, one still would have to understand why the extra dimensions are so large. Using $V \sim R^n$ one can reverse above relation and get a feeling of the possible values of R for given M_*. For $n=2$ one gets $R \sim 0.2$ mm, that is just at the current limit of short distance gravity experiments [4]. One single large extra dimension is not totally ruled out. Indeed, if one imposes the condition that $R < 160 \mu m$ for $n=1$, we get $M_* > 10^8 GeV$. More than two extra dimensions are in fact expected (assuming string theory), but those dimensions may turn out to have different sizes, or even geometries. For getting an insight of the theory, however, one usually relies in toy models with a single compact extra dimension, decoupling the effects of any shorter dimension.

Since particle physics forces have certainly been accurately measured up to weak scale distances (about 10^{-18} cm), the Standard Model (SM) particles can not freely propagate in larger extra dimensions. Either they are constrained to live on a four dimensional submanifold (a 3 brane), embedded in a higher dimensional bulk only tested by gravity, or at most they probe extra dimensions of a shorter subspace. The branes we refer to are actually an effective theory description. We may think up on them as topological defects (domain walls) of almost zero width, which could have fields trapped on it. Such hypersurface would be located on an specific point on the extra space, usually, at the fixed points of the compact manifold. This description mimics String theory D-branes (Dirichlet branes), which are surfaces where open string end on. Open strings give rise to all kinds of fields localized to the brane, including gauge fields. In the supergravity approximation these D-branes will also appear as solitons of the supergravity equations of motion. In our approach we simply assume there is some consistent high-energy theory, that would give rise to these objects, which should appear at the fundamental scale M_*. Thus, the natural UV cutoff of our models would always be given by the quantum gravity scale.

Brane world picture breaks translational invariance, which is reflected in two ways, either affecting the flatness of the extra space as in the Randall-Sundrum (RS) models, or introducing a source for violations of the extra linear momentum conservations as brane-bulk couplings. For reasons of space we will consider only brane models on flat space along this short review. For references on RS see [11]

To describe field theories where fields live either on branes (as the Standard Model) or in the bulk (like gravity and perhaps SM singlets), as well as the interactions among the two sectors, one make use of an effective prescription starting from the complete action written on the whole $4+n$ dimensions. In this prescription, both the brane and the brane-bulk interactions are localized terms in the higher dimensional action by a δ function picked at brane position in bulk coordinates. For instance a pure four dimensional action S_{4D} would appear in the whole theory as $S_{4+nD} = \int d^n y \, S_{4D} \delta(\vec{y} - \vec{y}_0)$, for a brane located at $\vec{y} = \vec{y}_0$. To properly handle this expressions it is much better to integrate over the extra dimensions and work in the resulting effective four dimensional theory. This procedure is generically called dimensional reduction. Bulk fields, in general, have $d + n/2$ mass dimensions, with d their usual 4D mass dimension. Thus, usually brane-bulk interactions are non renormalizable. Thus, the effective four dimensional theory would be weaker compared to the bulk theory, just what happens to gravity. Since the extra dimensions are compact, bulk fields can be decomposed in terms of the eigenmodes of the extra momentum, as for quantum particles in a box. For instance, for a single extra dimension, compactified on a circle, a free bulk scalar field ϕ should satisfy the periodicity condition, $\phi(y) = \phi(y + 2\pi R)$, which allows for a Fourier expansion as

$$\phi(x,y) = \frac{1}{\sqrt{2\pi R}} \phi_0(x) + \sum_{n=1}^{\infty} \frac{1}{\sqrt{\pi R}} \left[\phi_n(x) \cos\left(\frac{ny}{R}\right) + \hat{\phi}_n(x) \sin\left(\frac{ny}{R}\right) \right]. \quad (4)$$

The y independent term, ϕ_0, is usually referred as the zero mode. Other Fourier modes, ϕ_n and $\hat{\phi}_n$; are called the excited or Kaluza-Klein (KK) modes. The index n is clearly associated to the fifth component of the momentum, which is discrete due to compactification. The effective 4D action obtained from this theory would be $S_{eff}[\phi] = \sum_{n=0}^{\infty} S_{odd}[\hat{\phi}_n] + S_{even}[\phi_n]$ where the action of both even and odd modes is that of a usual 4D massive field, $S[\phi_n] = \frac{1}{2}(\partial^\mu \phi_n \partial_\mu \phi_n - m_n^2 \phi_n^2)$ where the KK mass is given as $m_n^2 = m^2 + \frac{n^2}{R^2}$. Therefore, in the effective theory, the higher dimensional field appears as an infinite tower of fields with masses m_n, with degenerated massive levels, but the zero mode. Similar KK expansions would be obtained for vector and fermion fields. All excited modes for a given field have same spin and quantum numbers and differ only in the KK number n. This can be also understood in general from the higher dimensional invariant $p^A p_A = m^2$, $A = \mu, 5$, which can be rewritten as the 4D invariant $p^\mu p_\mu = m^2 + \vec{p}_\perp^{\,2}$, where $\vec{p}_\perp$ stands for the extra momentum components.

In higher dimensions the degeneracy would be larger on the torus (2^n states per KK level). Different compactifications would lead to different mode expansions. Eq. (4) would had to be chosen accordingly to the geometry of the extra space by typically using wave functions for free particles on such a space as the basis for the expansion. Extra boundary conditions associated to specific topological properties of the compact space may also help for a proper selection of the basis. A useful example is the one dimensional orbifold, $U(1)/Z_2$, which is built out of the circle, by identifying the opposite points around zero, so reducing the physical interval to $[0,\pi]$ only. Operatively, this is done by requiring the theory to be invariant under the extra parity symmetry $Z_2: y \to -y$. Under this symmetries all fields should pick up a specific parity, such that $\phi(-y) = \pm\phi(y)$. Even (odd) fields would be expanded only into cosine (sine) modes, thus, the KK spectrum would have only half of the modes. Clearly, odd fields do not have zero modes and thus do not appear at the low energy theory. This later would allow for breaking the symmetries of the theory, and one of the first known examples of this is the breaking of parity by bulk fields presented in Ref. [12].

The appreciation of the impact of KK excitations depends on the relevant energy of the experiment, and on the compactification scale $\frac{1}{R}$. For $m = 0$, it is clear that for energies below $\frac{1}{R}$ only the massless zero mode is kinematically accessible, thus the theory looks four dimensional. At energies such that $\frac{1}{R} < E < M_*$, or equivalently, as we do measurements at shorter distances, a large number of KK excitations, $\sim (ER)^n$, becomes kinematically accessible, and their contributions relevant for the physics. Therefore, right above the threshold of the first excited level, the KK modes should evidence the higher dimensional nature of the theory. At larger energies, above M_*, however, our effective approach has to be replaced by the more fundamental theory of quantum gravity.

Graviton Phenomenology.- One of the first physical examples of a brane-bulk interaction is the effective gravitational coupling of particles located at the brane. The problem has been extendedly discussed in Refs. [5, 6] assuming a flat bulk. The effective matter to graviton (h) coupling, at first order, is described by the action

$$S_{int} = \int d^4x \frac{h_{\mu\nu}(0)}{M_*^{n/2+1}} T^{\mu\nu}. \tag{5}$$

h_{MN} have different 4D Lorentz components: the true four dimensional graviton, in $h_{\mu\nu}$, graviphotons from $h_{a\mu}$, and some graviscalar fields, in h_{ab}, one of which corresponds to the partial trace $h^a{}_a$, usually called the radion field. All those effective fields would of course have a KK decomposition. Each of them comes with a volume suppression that exchanges the $M_*^{n/2+1}$ by an M_P suppression. Thus, all KK modes couple with standard gravity strength. Moreover, only 4D gravitons, $G_{\mu\nu}$, and the radion field, $b(x)$, get couple at first order level to the brane energy momentum tensor [5, 6]

$$\mathscr{L} = -\frac{1}{M_P} \sum_{\vec{n}} \left[G^{(\vec{n})\mu\nu} - \frac{1}{3}\sqrt{\frac{2}{3(n+2)}} b^{(\vec{n})} \eta^{\mu\nu} \right] T_{\mu\nu}. \tag{6}$$

The massless $G^{(0)\mu\nu}$ is the source of long range four dimensional gravity interactions. On the contrary, $b^{(0)}$ should be massive, otherwise it would violate the equivalence principle. $b^{(0)}$ should get a mass from the stabilization mechanism that keeps the extra volume finite. Clearly, at large distances only the massless graviton can be exchanged, making gravity looking weaker. At shorter distances, however, more and more massive KK modes can be exchanged, thus making gravity stronger. Each KK mode contribute to the gravitational potential among two test particles with a Yukawa potential:

$\Delta_{\vec{n}} U(r) \simeq U_N(r) e^{-m_{\vec{n}} r}$, with $U_N(r)$ the usual Newtonian potential. The sum over all KK modes, estimated in the continuum limit, becomes $U_T(r) \simeq -G_N V_n (n-1)! \frac{m_1 m_2}{r^{n+1}}$. Around threshold, only the first excited modes would be relevant, so, the potential one should see in short distance tests of Newton's law is actually of the form [13] $U(r) \simeq U_N(r) \left(1 + \alpha e^{-r/R}\right)$, with $\alpha = 8n/3$. Current bound is about 160 μm

As gravity may become comparable in strength to the gauge interactions at energies $M_* \sim$ TeV, the nature of the quantum theory of gravity would become accessible to LHC and NLC. The effects would be mostly of two types [5, 6, 7, 14]: (i) missing energy by graviton production; and (ii) corrections to the standard cross sections from graviton exchange. At $e^+ e^-$ colliders, the best signals would be the production of gravitons with Z, γ or fermion pairs $\bar{f}f$. In hadron colliders one could see graviton production in Drell-Yang processes, and there is also the so far unseen monojet production [5, 6]. LHC could be able to impose bounds up to 4 TeV for M_* for 10 fb^{-1} luminosity. Graviton exchange either leads to modifications of the SM cross sections and asymmetries, or to new processes not allowed in the SM at tree level. The amplitude for exchange of the entire tower naively diverges when $n > 1$ and has to be regularized, as already mentioned. An interesting channel is $\gamma\gamma$ scattering, which appears at tree level, and may surpasses the SM background at $s = 0.5$ TeV for $M_* = 4$ TeV. Bi-boson productions of $\gamma\gamma$, WW and ZZ may also give some competitive bounds [5, 6, 7] roughly at the TeV scale. There are some important examples of recent experimental searches for extra dimensions where Mexican groups are involved. H1 [15] collaboration at HERA, who has looked at $e^+ p \to e^+ X$ for Q^2 between 200 GeV^2 and 30,000 GeV^2 with an integrated luminosity of 35.6 pb^{-1}. They impose the bound $M_* > .48$ TeV. D0 [16, 17, 18] collaboration have looked in $p\bar{p}$ collisions for dielectron and diphoton production [16]; monojets [17] and dimuon production [18]. Imposed limits are about 1 TeV.

Another intriguing phenomena in colliders, associated to a low gravity scale, is the possible production of microscopic Black Holes [19]. LHC running at maximal energy could even be producing about 10^7 of those Black Holes per year, if $M_* \sim TeV$. These objects are unstable, they thermally evaporate within $\tau < 10^{-25}$ sec. This efficient conversion of collider energy into thermal radiation would be a clear signature of having reached the quantum gravity regime.

MODEL BUILDING IN EXTRA DIMENSIONS

There has also been quite a large interest on the community in studying more complicated constructions where other fields, besides gravity, live on more than four dimensions. Most contributions from members of Mexican community have been along this lines. The first simple extension one can think of is to assume than some other singlet fields may also propagate in the bulk. These fields can either be scalars or fermions, and can be useful for a diversity of new mechanisms. One more step on this line of thought is to also promote SM fields to propagate in the extra dimensions, with some modifications on the profile of compact space in order to control the spectrum and masses of KK excitations of SM fields. These constructions contain a series of interesting properties that may be of some use for model building.

If SM particles are zero modes of a higher dimensional theory, as it would be the case in string theory, the mass of the very first excited states has to be larger than the current collider energies, that is well up to some hundred GeV. In such a scenario, the volume of the compact space is characterized by n large and δ short dimensions of radius r and R, respectively. Thus, one can write the effective Planck scale as

$$M_P^2 = M_*^{2+\delta+n} r^\delta R^n . \tag{7}$$

Keeping M_* around few tenths of TeV, only requires that the larger compactification scale $M_c = 1/r$ be also about TeV, provided R is large enough, say in the submillimeter range. This way, short distance gravity experiments and collider physics could be complementary to test the profile and topology of the compact space.

A priori, SM particles may or may not propagate in all dimensions. When all SM fields do feel the same dimensions, the scenario is usually referred as having Universal Extra Dimensions (UED). The effects of such extra dimensions can be studied either on the direct production of KK excitations, or through the exchange of these modes [9, 14]. In non universal extra dimension single KK modes can be produced directly in high energy particle collisions, and future colliders may be able to observe KK resonances if the M_c turns out to be on the TeV range. In hadron colliders (TEVATRON, LHC) the KK excitations might be directly produced in Drell-Yang processes $pp(p\bar{p}) \to \ell^-\ell^+ X$ where the lepton pairs ($\ell = e, \mu, \tau$) are produced via the subprocess $q\bar{q} \to \ell^+\ell^+ X$. Current search for Z' on these channels (CDF) impose $M_c > 510\ GeV$. Future bounds could be raised up to 650 GeV in TEVATRON and 4.5 TeV in LHC, which with 100 fb^{-1} of luminosity can discover modes up to $M_c \approx 6\ TeV$. In UED models KK number is conserved, thus things may be more subtle since pair production of KK excitations would require more energy. On the other hand, the lighter KK modes would be stable and thus of easy identification, either as large missing energy, when neutral, or as a heavy stable particles if charged. It can also be a candidate for dark matter [20] In any case, precision test may be the very first place to look for constraints to the compactification scale [9, 14]. Fermi constant current precision implies for instance $M_c \gtrsim 1.6\ TeV$.

Power Law Running of Gauge Couplings. KK excitations affect the evolution of couplings in gauge theories and may alter the whole picture of unification of couplings. This question was first studied in Ref. [21] on the base of the effective theory approach at one loop level. A step by step analysis of the running was made in Ref. [22]. The result is a power law behavior of the gauge coupling constants first noted in Ref. [23]:

$$\alpha_i^{-1}(\mu) = \alpha_i^{-1}(M_c) - \frac{b_i - \tilde{b}_i}{2\pi} \ln\left(\frac{\mu}{M_c}\right) - \frac{\tilde{b}_i}{2\pi} \cdot \frac{X_\delta}{\delta} \left[\left(\frac{\mu}{M_c}\right)^\delta - 1\right], \tag{8}$$

which accelerates the meeting of the α_i's. Here b_i are the beta functions of the theory below M_c, and $\tilde{b}_i$ are the KK beta function contributions. In the Supersymmetric models the energy range between M_c and the crossing energy Λ –identified as M_*– is relatively small due to the steep behavior in the evolution of the couplings [21, 22]. For instance, for a single extra dimension the ratio Λ/M_c has an upper limit of the order of 30, and it substantially decreases for larger δ. This would requires the short extra dimension where SM propagates to be rather closer to the fundamental length. A common observation of this picture is, however the poor one-loop precision on the unification of couplings [22, 24]. It is still possible that high order joint to threshold corrections might improve one step unification, so one can not rule it out on the simple basis of one-loop running. Multi-step models where also analyzed in Ref. [24] where there is also an attempt to identify possible unification symmetries, other than SU(5) or SO(10). Other coupling constants also present power law running and some studies can be found in [21, 25, 26]. The influence of brane kinetic terms on the running has been also considered in Ref. [27].

Symmetry Breaking with Extra Dimensions. Old and new ideas on symmetry breaking have been revisited and further developed in the context of extra dimension models by many authors in the last few years. Standard spontaneous symmetry breaking as implemented with bulk fields was re-considered in first Ref. on [11] where it is shown that, at the absolute minimum, only the zero mode gets a vacuum expectation value (vev), which is in general enhanced respect to the standard 4D result, due to the large suppression on the bulk quartic couplings.

Symmetries can also be broken at distant branes, and the breaking be communicated by the mediation of bulk fields to some other brane, in what is called the shinning mechanism [28, 29]. The brane vacuum serves as a point like source for the vacuum of the bulk field and thus, this gets suppressed as one moves away from that brane. The result is an smaller amount of breaking with the distance due to the propagation in coordinate space. If all the fields carry a global $U(1)$ charge, the mechanism may induce the breaking of such a global symmetry on a second distant brane. This way, we get a suppressed effect with out the use of large energy scales.

Orbifolding of the extra space serves to project unwanted degrees of freedom of the bulk fields via the imposition of the extra discrete symmetries that are used in the construction of the orbifold out of the compact space. However orbifolding gives enough freedom as to choose which components of the bulk fields should remain at zero mode level. Fermions on 5D, for instance, are vector-like and the 5D theory left-right symmetric. However under Z_2 fermions transform as $\Psi \to \pm \gamma_5 \psi$, where the $\pm$ sign can be freely chosen, thus, only one chiral component would survive at zero mode level and the low energy theory would have less symmetry. This can be used to break both global and local symmetries [12, 30] and it is extendedly exploited in model building. Consider for instance a global or local $SU(2)$ 5D model with a bulk scalar doublet Φ On the $U(1)/Z_2$ orbifold, most general Z_2 transformation rule for the scalar would be $\Phi \to P_g \Phi$; where P_g satisfies $P_g^\dagger = P_g^{-1}$, thus, the simplest choices are $P_g = \pm \mathbf{1}$; $\pm \sigma_3$. Clearly the first option only means taking both fields on the doublet to be simultaneously even or odd, with no further implication for the theory. However, if one takes, for instance, $P_g = \sigma_3$, this selection explicitely means that under Z_2 one of the field components of the doublet is odd, and does not appear at zero mode level. Hence, at this level the original $SU(2)$ symmetry would not be evident. In fact, the lack of the whole symmetry would be clear by looking at the whole KK spectrum, where at each level there is not appropriate doublet pairing of fields. Actually, by choosing Z_2 parities, the theory is forced to have less symmetry at the boundaries, which results in the effective breaking of $SU(2)$ down to $U(1)$. The effect, of course, is manifested in the gauge sector, $\mathscr{A}_M = A_M^a \sigma^a / \sqrt{2}$, whose components should transform as $\mathscr{A}_\mu \to P_g \mathscr{A}_\mu P_g^{-1}$ and $\mathscr{A}_5 \to -P_g \mathscr{A}_5 P_g^{-1}$, respectively. Now, for $P_g = \sigma_3$ we get the following assignment of parities: $W_\mu^3(+)$; $W_\mu^\pm(-)$; $W_5^3(-)$ and $W_5^\pm(+)$. Thus, zero mode gauge fields are only those of $U(1)$. Extra zero mode fields (as the $W_5^\pm$ scalar) can be removed by a further orbifolding of the compact space, for instance using $U(1)/Z_2 \times Z_2'$, where the second identification of points is defined by the transformation $Z_2': y' \to -y'$, where $y' = y + \frac{\pi}{2}$.

A more general example of the use of non trivial boundary conditions to break the symmetries is the Scherk-Schwarz mechanism [31] which twist the fields under the action of the discrete symmetries used to compactify and orbifold the space. It takes advantage of a non smooth manifold which has singularities at the fixed points. The result is actually equivalent to the so called Wilson/Hosotani mechanism [32, 33]. where the A_5^3 component of the gauge vector, A_M^3, may by some dynamics acquire a non zero vev, and induce a mass term for the 4D gauge fields, A_μ^a, through the term $\langle A_5^3 \rangle^2 (A^{\mu 1} A_\mu^1 + A^{\mu 2} A_\mu^2)$, which is contained in $Tr F_{MN} F^{MN}$. An interesting use of these mechanisms could also be the breaking of supersymmetry [34].

Neutrino mass models. Experiments have provided conclusive evidence for neutrino oscillations and thus for neutrino mass. Standard see-saw does not work for a low M_*. Therefore some alternative ideas had to be proposed. Simple consideration of dimension five operators, $(LH)^2$, would give a too large neutrino mass. First ideas suggested that the neutrino may rather come in a theory with global lepton number conservation and the coupling to a bulk isosinglet neutrino [35, 36], $\frac{h}{\sqrt{M}} \bar{L} H v_{BR} \delta(y)$. Volume suppression and Higgs vacuum would naturally provide an small Dirac mass, $m = hvM_*/rM_P \sim 10^{-2}$ $eV \times hM_*/100$ TeV. General mixing with KK modes would imply that mass eigenstate v_L is actually a coherent superposition of an infinite number of massive modes. An analysis of the mixing profile in these models for solar neutrinos was presented in [37]. Implications for atmospheric neutrinos were discussed in [38], and some early phenomenological bounds were given in [38, 39]. It has also been suggested that bulk neutrino could play the role of an sterile on the brane [40]. A comprehensive analysis for three flavors is given in [41]. Overall, no effects attainable to extra dimensional oscillations are observed so far, which means that $R \lesssim 10^{-2}$ μm. Bulk neutrinos with Dirac masses are also a nice way to understand quantization of hypercharge, as noted in Ref. [42]

Some extended scenarios that consider the generation of Majorana masses from the spontaneous breaking of lepton number either on the bulk or on a distant brane have been considered in Refs. [43, 44]. Spontaneous breaking by a bulk scalar field, χ, carrying lepton number 2 may generate a mass to the neutrinos via non renormalizable couplings of the form [44] $m_v \sim \langle H \rangle^2 \langle \chi \rangle_B / M_*^{2+n/2}$. For n=2 and M_* of the order of 100 TeV, we need $\langle \chi \rangle_B \sim (10\ GeV)^2$ to get $m_v \sim 10\ eV$. In this scenario there is also a Majoron bulk field whose phenomenology depends on the details of the specific model. It may also gives an important contribution for neutrinoless double beta decay which is just right at the current experimental limits [44].

Local lepton number can be introduced through a simple $B - L$ extension of the theory to generate neutrino masses [45]. Such 5D model is based in the gauge group $SU(2) \times U(1)_I \times U(1)_{B-L}$, built on the orbifold $U(1)/Z_2$, with the matter content $\mathscr{L}(2,0,-1)$; $E(1,-1/2,-1)$. The scalar sector contains a doublet, $H(2,-1/2,0)$, and a singlet $\chi(1,1/2,1)$, which are used to break the symmetry down to the electromagnetic $U(1)_{em}$. Particularly $\langle \chi \rangle$ produce the breaking of $U(1)_I \times U(1)_{B-L}$ down to the hypercharge group $U(1)_Y$. The model uses orbifold breaking such that $L = \mathscr{L}_L$, and $e_R = E_R$, are even, and thus $\mathscr{L}_R$, and E_L are odd fields. At zero mode level one gets usual SM lepton content. With this parities, since there is no right handed zero mode neutrino, the theory has not Dirac mass terms $\bar{L}H\nu_R$. Neutrino masses come from the simplest dimension 10 operator in 5D [45] $(\mathscr{L}H)^2\chi^2$; which generates the Majorana mass term $(h/(M_*r)^2)$ $(LH\chi)^2/M_*^3)$. If one takes $\langle\chi\rangle \sim 800\ GeV$, assuming that $M_*r \sim 100$ as suggested from the running of gauge couplings, and $M_* \sim 100\ TeV$ one easily gets a neutrino mass in the desired range, $m_\nu \sim h \cdot eV$. Embedding in a 5D left-right symmetric version are also discussed in Ref. [45]

Models for Baryon number violation. Controlling fast proton decay via dimension six operators $\bar{Q}^c Q \bar{Q}^c L$ without having large scales on the theory was one challenges in extra dimensional models. The most elegant solution comes from a 6D SM theory built on a T^2/Z_2 orbifold [46]. Point is that the extra spatial dimensions provide a new $U(1)$ symmetry, associated to rotations in the plane formed by the fifth and sixth extra dimensions, under which the SM fermions are charged. Enough of this symmetry survives the process of orbifold compactification that it suppresses proton decay to a very high degree. Indeed it turns out that $\bar{Q}^c Q \bar{Q}^c L$ is non invariant under this extra symmetry (it fact it is a non 6D Lorentz invariant).

6D SM have the remarkable property of being free of local and global anomalies only for a minimum of three generations [47]. Cancellation of anomalies also implies that a isosinglet neutrino should be in the spectrum, and thus, it was natural to consider a left right extended model, which maintains all above features and may incorporate the above mentioned mechanism to generate masses, via the breaking of $B - L$. Details of this model were presented in Ref. [48]. Among some of the interesting processes allowed in these models for baryon number violation (see references [46, 48] for more explicit examples) one has the nucleon decay mode $N \to \pi \nu_e \nu_s \nu_s$ with a life time

$$\tau_p \approx 6 \times 10^{30}\ \text{yr} \cdot \left[\frac{10^{-4}}{\Phi_n}\right] \left(\frac{\pi r M_*}{10}\right)^{10} \left(\frac{M_*}{10\ \text{TeV}}\right)^{12}, \qquad (9)$$

which is large enough as to be consistent with the experiment even with a fundamental scale as low as 10 TeV. Φ_n is a kinematical phase space factor which depends in the specific process with n final states. r represents the size of the compact space. There is also an intriguing invisible decay for the neutron $n \to \nu_e \nu_s \nu_s$ [46, 48], that is still above current experimental limits for which $\tau_n > 2 \cdot 10^{29}$ yrs. [49].

Another interesting mechanism that may explain how proton decay could get suppressed at the proper level appeared in [50]. It relays on the idea that branes can be seen as domain walls of thickness $L \sim M_*^{-1}$, where the fermions may be stuck at different points. Hence, fermion-fermion couplings may get suppressed due to exponentially small overlaps of their wave functions. This provides a framework for understanding both the fermion mass hierarchy and proton stability without imposing extra symmetries, but rather in terms of a higher dimensional geometry [51].

Other gauge models.- Partial and total unification models had also been considered. Of particular interest of some people in Mexico had been the models with flavor symmetries that use extra dimensions [52]. Refs. [45, 48] are also good examples of models with an extended gauge sector.

COSMOLOGY AND ASTROPHYSICS IN EXTRA DIMENSIONS

Graviton production may also posses strong constraints on the theory when considering that the early Universe was an important resource of energy, and those may be produced in excess. The standard Universe evolution would be conserved as far as the total number density of KK gravitons, n_g, remains small when compared to photon number density, n_γ. This condition can be translated into a bound for a maximal reheating temperature [3] $T_r^{n+1} < M_*^{n+2}/M_P$, which indicates preference for a rather cold Early Universe. For instance, for $M_* = 10$ TeV and $n = 2$, one gets $T_r < 100$ MeV, just about to what is needed to have BBN working. This would be reflected in some difficulties for standard models trying to implement baryogenesis or leptogenesis.

Inflation is also problematic. First it has to work on time scales H^{-1} much grater than M_*^{-1}, and usually four dimensional inflation produces too few amounts of density perturbations as a consequence of the suppression on $U(\phi) < M_*^4$ [53]. Alternative, successful models that use bulk fields to drive inflation had been considered in Ref. [54]. As such, its effective potential energy is enhanced by the volume of the extra space, which allows to have larger densities contributing to Hubble expansion. Reheating of the brane in this scenario is Planck suppressed and thus usual T_r are just as low as required for consistency with graviton bounds. On the same footing the scenario allows for the introduction of extra bulk fields that may contribute to baryogenesis [55]. A stabilization potential has to be provided to fix the size of the extra dimension at early Universe, which is generically difficult and usually not enough to produce a stable radion, since the radion potential gets corrections from matter energy density [56]. Notice however that the inflaton to radion coupling induces an effective mass term for the radion which is proportional to Hubble scale, $m_{eff}^2 \sim \alpha^2 V_{eff}(\phi)/M_P = \alpha^2 H^2$. As a consequence, the radion gets a very steep effective potential term, which easily drives the radion towards the local minimum within a Hubble time [57], which at least justify the scenario discussed above.

Gravitons emitted by stellar objects contributes to cool down the star. Data obtained from the supernova 1987a gives $M_* \gtrsim 10^{\frac{15-4.5n}{n+2}}$, which for $n = 2$ means $M_* > 30$ TeV [8]. Graviton relics also decay back into the brane re-injecting energy in the form of gamma rays at current time. EGRET and COMPTEL observations indicate that $M_* > 500$ TeV [58, 59]. Massive KK gravitons have small kinetic energy, so that a large fraction of those produced in the inner supernovae core remain gravitationally trapped, which by decay should give particular signals of gamma rays coming from the remnant star. GLAST, for instance, could be in position of finding the KK signature, well up M_* as large as 1300 TeV for $n = 2$ [60]. Constraints from gamma ray emission from the whole population of neutron stars in the galactic bulge against EGRET observations gives limits on about $M_* > 450$ TeV for $n = 2$ [61]. KK decay may also heat the neutron star up to levels above the temperature expected from standard cooling models. Direct observation of neutron star luminosity provides the most restrictive lower bound on M_* at about 1700 TeV for $n = 2$ [60]. Nevertheless, if heavy KK gravitons decay into more stable lighter KK modes, with large kinetic energies, such bounds can be avoided [62]. Supernova cooling and BBN bounds are, on the hand, more robust.

CONCLUDING REMARKS

The study of models with extra dimensions has become a fruitful industry that has involved several areas of theoretical physics in matter of few years. Although many of the current leading directions of research obey more to speculative ideas that to well established facts, the study of the brane world is guided by the principle of physical and mathematical consistency, and inspired on the possibility of connecting, at some point, the models with a more fundamental theory, perhaps String theory from where the idea of extra dimensions and branes is natural. Further motivation also comes from the possibility of experimentally testing these ideas within the near future, something that was just unthinkable in many old models where the fundamental gravity scale was the Planck scale. The present notes have intended to give a very short overview of the field, paying more attention to those topics that have particularly interested to the people working at Mexican institutions.

Acknowledgments.- This work was supported in part by CONACyT Mexico, under grant J44596-F.

REFERENCES

1. Th. Kaluza, Sitzungober. Preuss. Akad. Wiss. Berlin (1921) 966; O. Klein, Z. Phys. **37** (1926) 895.
2. E. Witten, Nucl. Phys. **B471** (1996) 135; P. Horava and E. Witten, Nucl. Phys. **B460** (1996) 506.
3. N. Arkani-Hamed, S. Dimopoulos and G. Dvali, Phys. Lett. **B429** (1998) 263; I. Antoniadis, *et al.*, Phys. Lett. **B436** (1998) 257; I. Antoniadis, S. Dimopoulos, G. Dvali, Nucl. Phys. **B516** (1998) 70.
4. J.C. Long *et al.*, Nature **421** (2003) 27; Hoyle, *et al.*, Phys. Rev. **D70** (2004) 042004.
5. G. Giudice, R. Rattazzi and J. Wells, Nucl. Phys. **B544** (1999) 3.
6. T. Han, J. Lykken and R. J. Zhang, Phys. Rev. D **59** (1999) 105006.
7. See for instance: E. Mirabelli, M. Perelstein and M. Peskin, Phys. Rev. Lett. **82** (1999) 2236; S. Nussinov, R. Shrock, Phys. Rev. **D59** (1999) 105002; J. L. Hewett, Phys. Rev. Lett **82** (1999) 4765; T. G. Rizzo, Phys. Rev. D **59** (1999) 115010; K. Aghase and N. G. Deshpande, Phys. Lett. **B456** (1999) 60; L3 Coll. Phys. Lett. **B464** (1999) 135; G.F. Giudice, A. Strumia, Nucl. Phys. **B663** (2003) 377.
8. N. Arkani-Hamed, S. Dimopoulos and G. Dvali, Phys. Rev. **D59** (1999) 086004; V. Barger, *et al.*, Phys. Lett. **B461** (1999) 34; L.J. Hall and D. Smith. Phys. Rev. **D60** (1999) 085008; S. Cullen, M. Perelstein, Phys. Rev. Lett. **83** (1999) 268; C. Hanhart, *et al.*, Nucl. Phys. **B595** (2001) 335.
9. See for instance: P. Nath and M. Yamaguchi, Phys. Rev. **D60** (1999) 116004; P. Nath, Y. Yamada and M. Yamaguchi, Phys. Lett. **B466** (1999) 100; W. J. Marciano, Phys. Rev. **D60** (1999) 093006; I. Antoniadis, K. Benakli, M. Quiros, Phys. Lett. **B460** (1999) 176; M. L. Graesser, Phys. Rev. **D61** (2000) 074019; T. G. Rizzo, J. D. Wells, Phys. Rev. **D61** (2000) 016007.
10. For a review on D branes see J. Polchinski, *Preprint* hep-th/9611050.
11. See for instance: A. Pérez-Lorenzana, J. Phys. Conf. Ser. **18** (2005) 224; arXives: hep-ph/0406279; AIP Conf. Proc. **562** (2001) 53.
12. R.N. Mohapatra and A. Pérez-Lorenzana, Phys. Lett. **B468** (1999) 195.
13. A. Kehagias and K. Sfetsos, Phys. Lett. **B472** (2000) 39.
14. T.G. Rizzo, *Preprint* hep-ph/9910255; K. Cheung, *Preprint* hep-ph/0003306; I. Antoniadis and K. Benakli, Int. J. Mod. Phys. **A15** (2000) 4237 (*Preprint* hep-ph/0007226); and references therein.
15. C. Adloff *et al.* [H1 Collaboration] Phys. Lett. B **479**, 358 (2000).
16. B. Abbott *et al.* [D0 Collaboration] Phys. Rev. Lett. **86**, 1156 (2001);
17. V. M. Abazov *et al.* [D0 Collaboration] Phys. Rev. Lett. **90**, 251802 (2003);
18. V. M. Abazov *et al.* [D0 Collaboration] Phys. Rev. Lett. **95**, 161602 (2005)

19. S.B. Giddings, E. Katz and L. Randall, J. High Energy Phys. **03** (2000) 023; S.B. Giddings and S. Thomas, Phys. Rev. D **65** (2000) 056010.
20. H.C. Cheng, J.L. Feng and K.T. Matchev, Phys. Rev. Lett. **89** (2002) 211301.
21. K.R. Dienes, E. Dudas and T. Gherghetta, Nucl. Phys. **B537** (1999) 47.
22. A. Pérez-Lorenzana and R.N. Mohapatra, Nucl. Phys. **B559** (1999) 255.
23. T. Taylor and G. Veneziano, Phys. Lett. **B212** (1988) 147.
24. D. Ghilencea and G.G. Ross, Phys. Lett. **B442** (1998) 165.
25. T. Kobayashi, J. Kubo, M. Mondragon and G. Zoupanos, Nucl.Phys. **B550** (1999) 99.
26. S.A. Abel, S.F. King Phys. Rev. D**59** (1999) 095010.
27. A. Aranda and J. L. Diaz-Cruz Phys. Lett. B **633** (2006) 591.
28. N. Arkani-Hamed, S. Dimopoulos, Phys. Rev. D**65** (2002) 052003.
29. N. Arkani-Hamed, L. Hall, D. Smith, N. Weiner, Phys. Rev. D**61** (2000) 116003;
30. Y. Kawamura, Prog. Theor. Phys. **105** (2001) 999; A. Hebecker, J. March-Russell, Nucl. Phys. **B625** (2002) 128; L. Hall and Y. Nomura, Phys. Rev. D**64** (2001) 055003; A. Hebecker, J. March-Russell, Nucl. Phys. **B613** (2001) 3.
31. J. Scherk and J.H. Schwarz, Phys. Lett. **B82** (1979) 60; Nucl. Phys. **B153** (1979) 61; E. Cremmer, J. Scherk and J.H. Schwarz, Phys. Lett. **B84** (1979) 83.
32. Y. Hosotani, Phys. Lett. **B126** (1983) 309; Annals of Phys. **190** (1989) 233.
33. H.C. Cheng, K. T. Matchev, and M. Schmaltz, Phys. Rev. D**66** (2003) 036005.
34. A. Delgado, A. Pomarol and M. Quirós, Phys. Lett. **B438** (1998) 255.
35. K.R. Dienes, E. Dudas and T. Gherghetta, Nucl. Phys. **B557** (1999) 25; N. Arkani-Hamed, *et al.*, Phys. Rev. D**65** (2002) 024032.
36. R. N. Mohapatra, S. Nandi and A. Pérez-Lorenzana, Phys. Lett **B466** (1999) 115.
37. G. Dvali and A.Yu. Smirnov, Nucl. Phys. **B563** (1999) 63.
38. R. Barbieri, P. Creminelli and A. Strumia, Nucl. Phys. **B585** (2000) 28.
39. A. Faraggi and M. Pospelov, Phys. Lett **B458** (1999) 237; G. C. McLaughlin, J. N. Ng, Phys. Lett. **B470** (1999) 157; Phys. Rev. D**63** (2001) 053002; A. Ioannisian, A. Pilaftsis, Phys. Rev. D**62** (2000) 066001; A. Lukas *et al.*, Phys. Lett. **B495** (2000) 136.
40. R. N. Mohapatra and A. Pérez-Lorenzana, Nucl. Phys. **B576** (2000) 466.
41. R. N. Mohapatra and A. Pérez-Lorenzana, Nucl. Phys. **B593** (2001) 451.
42. A. Pérez-Lorenzana and C. A. de S. Pires Mod. Phys. Lett. A **18** (2003) 65.
43. A. Ioannisian and J. W. F Valle, *Preprint* hep-ph/9911349; E. Ma, M. Raidal and U. Sarkar, Phys. Rev. Lett. **85** (2000) 3769.
44. R. N. Mohapatra, A. Pérez-Lorenzana and C. A. de S. Pires, Phys. Lett. **B491** (2000) 143.
45. R. N. Mohapatra and A. Pérez-Lorenzana, Phys. Rev. D**66** (2002) 035005.
46. T. Appelquist, B.A. Dobrescu, E. Ponton, Ho-U. Yee, Phys. Rev. Lett. **87** (2001) 181802.
47. B. Dobrescu and E. Poppitz, Phys. Rev. Lett. **87** (2001) 031801.
48. R.N. Mohapatra and A. Perez-Lorenzana, Phys. Rev. D**67** (2003) 075015.
49. S.N. Ahmed, *et al.* [SNO Collaboration], Phys. Rev. Lett. **92** (2004) 102004.
50. N. Arkani-Hamed, M. Schmaltz, Phys. Rev. D**61** (2000) 033005.
51. N. Arkani-Hamed, Y. Grossman and M. Schmaltz, Phys. Rev. D**61** (2000) 115004; E.A. Mirabelli and M. Schmaltz, Phys. Rev. D**61** (2000) 113011.
52. A. Aranda and J. L. Diaz-Cruz, Mod. Phys. Lett. A **20** (2005) 203 (2005).
53. D. H. Lyth, Phys. Lett. **B448**, 191 (1999); N. Kaloper and A. Linde, Phys. Rev. D **59**, 101303 (1999).
54. R.N. Mohapatra, A. Pérez-Lorenzana, C.A. de S. Pires, Phys. Rev. D. **62** (2000) 105030.
55. R. Allahverdi, K. Enqvist, A. Mazumdar and A. Perez-Lorenzana Nucl. Phys. B **618** (2001) 277.
56. A. Mazumdar, R. N. Mohapatra and A. Perez-Lorenzana, JCAP **0406** (2004) 004; A. Perez-Lorenzana, AIP Conf. Proc. **758** (2005) 182.
57. A. Mazumdar and A. Pérez-Lorenzana, Phys. Rev. D **65** (2002) 107301.
58. S. Hannestad, Phys. Rev. D**64** (2001) 023515.
59. S. Hannestad and G. Raffelt Phys. Rev. Lett. **87** (2001) 051301; Phys. Rev. D**67** (2003) 125008; *Erratum*, Phys. Rev. D**69** (2004) 0229901;
60. S. Hannestad and G. Raffelt Phys. Rev. Lett. **88** (2002) 171301.
61. M. Casse, J. Paul, G. Bertone, G. Sigl, Phys. Rev. Lett. **92** (20040) 111102.
62. R.N. Mohapatra, S. Nussinov and A. Pérez-Lorenzana, Phys. Rev. D**68** (2003) 116001.

THEORETICAL HIGH ENERGY PHYSICS

String Theory in México

Hugo García-Compeán[*,†] and Alberto Güijosa[**]

[*]*Centro de Investigación y de Estudios Avanzados del IPN, Unidad Monterrey
Cerro de las Mitras 2565, Col. Obispado, Monterrey N.L. 64060, México*
[†]*Departamento de Física, Centro de Investigación y de Estudios Avanzados del IPN
Apdo. Postal 14-740, México D.F. 07000, México*
[**]*Departamento de Física de Altas Energías, Instituto de Ciencias Nucleares
Universidad Nacional Autónoma de México, Apdo. Postal 70-543, México D.F. 04510*

Abstract. We review the work related to string theory that has been carried out in Mexico in the past two decades.

Keywords: strings, branes, AdS/CFT, black hole entropy, NCOS
PACS: 11.25.-w, 11.25.Tq, 11.25.Uv, 11.25.Mj, 11.25.Wx

INTRODUCTION

The interest in string theory [1] has increased in Mexico over the years and nowadays there is a number of people working on the subject. In the present text we attempt to provide an overview of the string-oriented research that has been carried out in Mexico in the past two decades. We have organized the material roughly along historical lines, beginning with the work done in the wake of the first superstring revolution, and proceeding on to the third. Due to space constraints and our focus on research carried out in Mexico, we have only been able to include a limited number of references to the huge body of international literature on the various topics covered.

AFTER THE FIRST SUPERSTRING REVOLUTION

In the early days people attempted to understand the basic formalism of string theory, giving seminars, courses and lectures. A good set of lecture notes was published in [2].

One of the first works related to string theory was dedicated to the study of relativistic extended objects, in particular, the so called *terron* [3]. This work analyzed the theory of a free relativistic object extended in 3 dimensions (a threebrane, in modern parlance). A preliminary quantization of these objects was performed from the viewpoint of Dirac's method and the functional integral.

As is well-known, the Polyakov action for the string is Weyl-invariant, and consequently conformally-invariant after covariant gauge-fixing. On the two-dimensional worldsheet theory, Weyl/conformal invariance is associated with an infinite-dimensional Lie algebra known as the Virasoro algebra, and plays a crucial role in many features of string theory, including the critical dimension and the description of scattering amplitudes.

On the mathematical side, a topological and geometric description of the Virasoro algebra regarded as the Lie algebra of smooth vector fields on the circle was given in [4]. On the physical side, some attempts to incorporate Weyl invariance in the description of higher-dimensional objects, i.e., p-branes, were summarized in [5]. Soon after the extension to supersymmetric p-branes was given in [6]. Much later, in the post-second-revolution era, this line of work was revisited in [7], which derived in particular a Weyl-invariant action for D-branes, and also in [8].

Part of the hope for the early work in this direction was that Weyl invariance would play as central a role for the quantization of higher-dimensional branes as it did for the string. Unfortunately, this goal remained elusive and it was never possible to use this symmetry to determine, for instance, the critical dimension for p-branes with $p \neq 1$.

At that time it seemed natural for various people to study p-branes as candidate elementary objects, generalizing the step that have been taken previously in going from point particles to strings. Nowadays we understand that there is no renormalizable theory for these extended objects, and their worldvolume actions are generally regarded as effective field theories, as befits objects that are solitonic rather than elementary. However in those early days some alternatives were explored. One of them was proposed in [9] based on the relativistic top, with Weyl invariance still playing an important role. A relativistic stringtop consists of the ordinary relativistic string with additional degrees of freedom. In this case a relativistic spherical top is attached to each point along of the string. The quantization of the superstringtop was performed in [10]. In this reference, the critical dimension is indeed computed.

There were also some proposals at the level of the worldvolume description of the string itself, including the codification of the worldsheet metric $h_{\alpha\beta}$ and the target space metric $G_{\mu\nu}$ into a single unified metric $G_{\mu\nu}^{\alpha\beta} \equiv \sqrt{-h} h^{\alpha\beta} G_{\mu\nu}$ [11], such that the Polyakov action is written as $S = \frac{1}{2} \int d^2\sigma G_{\mu\nu}^{\alpha\beta} \frac{\partial X^\mu}{\partial \sigma^\alpha} \frac{\partial X^\nu}{\partial \sigma^\beta}$. Originally, the motivation was of mathematical nature, i.e., one would like to have a unified framework of the worldsheet and the target space which facilitates, for instance, the computations of the effective spacetime actions.

It seems appropriate to also mention here (even if it was conducted much later) a Hamiltonian analysis of the Polyakov action [12], which led to a reformulation of the bosonic string with constraints linear in the momenta, as well as a study of closely related systems including one associated with tensionless strings. Another Hamiltonian analysis is that of the supergravity in $2D$ in a covariant and gauge independent way [13]. From this one can obtain bosonic gravity following the square root method and the diffeomorphism superalgebra can be explicitly computed.

Another contribution to string theory in the pre-second revolution era was the derivation of the string action with $SO(d,1)$ target space starting from an action describing a special kind of relativistic point particle, the *quatl* [14]. This is an example of the well-known fact that string theory can also be defined on target spaces which have the structure of a Lie group. The worldsheet theory corresponds to the celebrated Wess-Zumino-Witten (WZW) theory. Its corresponding conformal field theory at he quantum level is characterized by a integer level k and a finite dimensional Hilbert space (the finite basis of vectors are the conformal blocks). Also, the existence of a quantum algebra structure $su_q(2)$ was found in this context. On the other hand, one of the known topological field theories is the so-called *intrinsic theories* whose Lagrangian is manifestly metric independent. A prototype of this kind of theories is the Chern-Simons action. It is also known that the dimensional reduction of this theory on the boundary (on a Riemann surface) is precisely the WZW model. One of the natural questions is to find realizations of the quantum group, in the context of the three-dimensional theory. This problem was studied in [15]. An early work explaining in detail the contour representation of finite quantum groups which appears naturally in the Coulomb gas representation of conformal field theories, appeared in Ref. [16].

Just before the second superstring revolution, the soliton solutions of the supergravity coming from string theory and the effects over them of the diverse dualities became important. Specially, under some compactifications the equations of motion of the effective field theory associated to the heterotic string were carried over to the familiar form of some solitonic integrable equations. This procedure was widely explored in [17, 18, 19, 20]. Also, a new approach for generating solutions in heterotic string theory compactified down to three dimensions on a torus was proposed [21]. Some extensions of the Einstein-Maxwell in the context of string theory is performed, focusing in the stationary case. In [22], a similar discussion for compactifications to four dimensions and for the static case was discussed.

In the context of toy models of string theory we have the $\mathcal{N}=2$ strings, i.e., string theories with two local supersymmetries on the worldsheet, critical dimension $D=4$ and $2+2$ signature. One of the surprising results found by Ooguri and Vafa [23] is that the effective field theory of the closed string on the target space is described by self-dual gravity in 4 dimensions and the only scalar degree of freedom (the Kahler potential) satisfies the Plebański heavenly equations. Open strings give rise to Self-dual Yang-Mills theory in various representations. Heterotic $\mathcal{N}=2$ strings describe Self-dual gravity coupled to Self-dual Yang-Mills theory. Some of these results were surveyed in [24].

AFTER THE SECOND SUPERSTRING REVOLUTION

The study of superstring phenomenology (for some reviews see [25, 26]) began during the period when Fernando Quevedo worked in Mexico. He and coworkers studied a way of fixing the moduli, especially the dilaton, at weak coupling by using a non-vanishing effective superpotential which is not asymptotically free [27]. At the same time supersymmetric gauge theories with the gauge group being the product of groups were also characterized, in particular, the dependence of the dilaton of the effective action for $\mathcal{N}=1$ $SU(N_1) \times SU(N_2)$ models with a generation of matter transforming in the bi-fundamental representation of this group product was described. The Coulomb phase was studied in detail and it was found that a U(1) factor is responsible for writing the gauge coupling in terms of the modulus of the elliptic curve. The behavior of the dilaton was studied also in the confining phase [28].

A crucial ingredient of the second revolution was the notion of duality, which culminated in the discovery of M-theory as the structure that underlies and unifies all of the superstring theories. S-duality, in particular, relates theories with inversely-related couplings and has played a central role in elucidating the nature of strongly-coupled string (and field) theory. The existence of S-duality is a very delicate property which can be spoiled when supersymmetry is absent. However in non-Abelian gauge theories it is still possible to define a kind of S-duality even when there is no supersymmetry, following the algorithm of Buscher (for a review see [29]). It is interesting to explore if this kind of procedure can be carried over to gravitational theories. For theories of gravity quadratic in the Riemann tensor, this procedure seems to work out. In particular, this was performed for topological gravity and MacDowell-Mansouri theories of gravity including their supersymmetric versions [30]. This, in part, motivated Hull to introduce the bold proposal of a new stringy symmetry, responsible for this gravitational S-duality, leading him to speculate about some conformal limit of string theory in [31, 32]. Other work involving duality was extended to supersymmetric theories [33]. To be more precise, the algorithm of Buscher's duality was extended to superspace, taking advantage that Poincaré as well as supersymmetry are global symmetries, that can subsequently be gauged. This was applied to theories in two dimensions involving linear and non-linear sigma models of the type (1,1) and (2,2). It was shown that the duality works interchanging chiral and twisted-chiral multiplets, like in the case of mirror symmetry.

Another important development in the mid-nineties was the formulation of M(atrix) theory [34] as an attempt to describe the fundamental degrees of freedom of M-theory in a restricted kinematic setup. In the M(atrix) context, compactifications of M-theory on tori are described by supersymmetric Yang-Mills theory on the dual torus with sixteen supercharges. In [35] a description of the Yang-Mills theory on the dual noncommutative torus is provided by using the Fedosov deformation quantization description [36]. Other methods of deformation quantization have emerged in the context of string theory. For instance, in [37] the Kontsevich star product on a dual Lie algebra was derived from the system of a D-brane defined in the string theory propagating on a Lie group. Thus the Kontsevich star product is encoded in the quantum correlation functions of the open string effective action and it is defined on the dual Lie algebra of this group. Deformation quantization has also been employed in order to give an alternative quantization procedure for the bosonic string [38]. Additionally, some work was done in the context of noncommutative solitons in string theory. In general terms it was found that the configurations of D-branes and anti-D-branes with a NS B-field, carries a noncommutative GMS soliton which in some circumstances, behaves as a superconducting wire [39].

Matroid theory is a generalization of graph theory and matrices which might provide a framework to define M-theory. This possibility in the context of string theory, p-branes and M-theory was explored recently in [40].

One of the most useful lessons gleaned from the second superstring revolution was the fact that the rich dynamics of supersymmetric gauge theories could be unravelled through a study of configurations of D-branes and NS5-branes. This opened a window into the nonperturbative sector of these theories, which allowed many important developments, including the elucidation of the structure of the moduli space of magnetic monopoles and instantons. In this context, perturbative and nonperturbative properties of field theories at the classical and, mainly, quantum level such as chirality, confinement and mass gap, symmetry breaking and chiral symmetry breaking etc., have a geometric counterpart (for a review, see [41]).

It has been found in [42, 43] that interesting four-dimensional chiral theories can be obtained from a configuration of periodic squares formed by NS5 and NS5' branes containing D4 branes (of the Type IIA theory) limited in two directions inside the boxes.

The spectrum and interactions could be obtained directly from this brane configuration. In this context brane box configurations were constructed with an additional orthogonal NS5" brane, such that three-dimensional boxes are formed. These boxes contain D3-branes (with 2 directions limited). It was shown that these periodic configurations can give rise to chiral theories in two dimensions with diverse degrees of supersymmetry. They are chiral theories with enhanced supersymmetry (0,2), (0,4), (0,6) and (0,8). This latter configuration is equivalent, via T-duality in 3 compact directions where the D3 is limited, to D1-branes living in orbifold singularities. All this was explored in [44].

In the late nineties work spearheaded by Sen made it possible to go beyond the study of BPS D-branes and the supersymmetric gauge theories that live on their worldvolume. In particular, Sen proposed the construction of nonsupersymmetric configurations by considering both D-branes and anti-D-branes [45]. These configurations break all supersymmetry and the spectrum of open strings stretching between D-branes and anti-D-branes includes a tachyon which survives the GSO projection. The presence of this tachyon implies that the brane configuration is unstable and will decay into the vacuum configuration. In the case where there is some topologically nontrivial flux on the D-branes (or anti-D-branes), the system does not decay into the vacuum but to a lower dimensional D-brane. The classification of stable and unstable D-branes is done according to their RR charge. In this case RR charge takes values not in the ordinary cohomology buy in K-theory, and as a result D-branes are classified by K-theory [46]. In Type IIA or IIB, BPS D-branes are classified by the group $K^0(X)$ or $K^1(X)$. Non-BPS states are classified by $K^1(X)$ in Type IIA and $K^0(X)$ in the type IIB. Bott periodicity for complex K-theory is 2 and it implies that there will only be Dp-branes with p even for the Type IIA and p odd for the Type IIB. Dp-branes in orbifold singularities are described by equivariant K-theory [46, 47]. In Type I theory Dp-branes are classified by $KO(X)$ theory and orientifolds by $KR(X)$ (for a review, see [48]). In this context, the non-BPS D7-brane of the Type I theory was studied [49]. It was found that it decays into topologically non-trivial gauge field configurations on the background D9-branes. In the uncompactified theory the decay proceeds to infinity, while with a transverse torus the decay reaches a final state, a toron gauge configuration with vanishing Chern classes but non-trivial $\mathbf{Z}_2$ charge. Similar results were obtained for the type I non-BPS D8-brane, and other related systems. Type IIA theory with O6-planes is lifted to M-theory and it was elucidated the relation between string theory K-theory and M-theory cohomology, and its interplay with NS-NS charged objects. Moreover, all stable BPS and non-BPS D-branes in the Gimon-Polchinski and Dabholkar-Park-Blum-Zaffaroni orientifolds were constructed, and their stability regions in moduli space were determined together with their decay products [50]. They found a certain kind of non-BPS branes which have integer and torsion charge. Some of them have projective representations of the orientifold × GSO group on the Chan-Paton factors. It was found that the Gimon-Polchinski orientifold is not described by equivariant orthogonal K-groups $KO_G(X)$ as may have been at first expected. Instead a twisted version of this K-theory is expected to be relevant. In a recent paper [51], RR fields in string backgrounds including orientifold planes and branes on top of them are classified by K-theory. Here such fluxes are also classified by cohomology, and both classifications are compared through the Atiyah-Hirzebruch Spectral Sequence. Some new correlations between branes on orientifold planes $Op^{\pm}$ and obstructions to the existence of some branes were found.

171

There have also been several works related to cosmology in string theory. For instance in Ref. [52] some potentials of the slow roll type were studied in low energy actions from superstring theory. In this paper it was observed that there is no unnatural fine tuning due the fact that there is only one free parameter in the theory. All moduli fields and the dilaton are stabilized and therefore the inflaton field must be identified with chiral matter. Some other important problems as the vanishing of the cosmological constant, and the consequences of the S-duality in inflation were discussed in Refs. [53, 54].

The problem of inflation in the context of the pre-big-bang scenario was studied in [55]. In particular it was shown that the hypothesis of inflation in this framework can solve the horizon problem. In this same scenario, which separates the problem of the initial conditions for the universe from that of high curvatures, it was shown that contrary to the expectation that radiation should have been the primary content of the universe before the big-bang, the use of 'string matter' seems to fit better with the fact that for all pre-big-bang cosmologies there is an universal attractor, which is represented by a Milne open universe [56].

In the context of configurations of D-branes and anti-D-branes the existing tachyon of open strings can be described through an effective field action, which can be coupled to Type IIA supergravity. This coupling was analyzed and the Hamiltonian formulation for it given in [57]. In the limit of late-time of the rolling tachyon, an exact solution of the Wheeler-deWitt equation was found. In the process the effects of electromagnetic fields were also incorporated and it was shown that in this case the interpretation of tachyon regarded as 'matter clock' is modified.

During this period there also appeared in the literature several overall surveys about string theory [58].

AFTER THE THIRD SUPERSTRING REVOLUTION

Without a doubt, the most important development in string theory in the past decade was the discovery of the gauge/gravity duality [59, 60, 61], a remarkable statement of equivalence between an ordinary gauge theory and a higher-dimensional gravitational theory. Through its study one obtains, in one direction, a potential resolution to some of the longstanding mysteries of quantum gravity, and in the other, a beautiful portrait of the strong-coupling behavior of gauge theories.

Various incarnations of this duality have been found which involve gauge theories with physics similar to that of QCD. A notable example is the deformed conifold background [62], dual to a four-dimensional $\mathcal{N}=1$ gauge theory that displays confinement and chiral symmetry breaking. A number of interesting quantities in the strongly-coupled gauge theory have been computed by means of this duality. In particular, it is possible to determine the spectrum of glueballs by solving a Schrödinger-type equation for the linearized fluctuations about the deformed conifold background [63, 64, 65]. The results are similar to those found by lattice gauge theorists [66]. A particularly striking result is that the 0^{++}, 1^{--} and 2^{++} states lie on a straight line in the J vs. M^2 plane, which is strongly reminiscent of a Regge trajectory. This analysis was generalized in [67] to the Maldacena-Nuñez background [68], which is dual to a five-dimensional theory that reduces to $\mathcal{N}=1$ SYM in the infrared.

The path that led to the discovery of the gauge/gravity duality [59, 60, 61] began with the realization that the physics of D-branes can be captured from two quite distinct perspectives: as extended objects in supergravity, or as localized objects with intrinsic worldvolume dynamics [69]. The direct comparison of various quantities in the two pictures, most notably by Klebanov and collaborators [70], supported the idea that these two perspectives can operate concurrently, in which case one would be dealing with a *duality* at the string theory level.

By adopting this point of view, and considering a low-energy decoupling limit, Maldacena was able to derive his celebrated correspondence [59]. Given the impressive body of evidence that has accumulated in support of this gauge/gravity duality [71], one is compelled to take the starting point of Maldacena's argument seriously. It then becomes natural to inquire about the precise nature and origin of the duality that operates at the level of the full-fledged string theory, *before* taking any low energy limits [72]. A related, but more modest, goal is to study this correspondence in a regime of energies that are low enough for the massive string modes to be negligible, but not low enough for the branes to decouple from the bulk [73]. To this aim, a prescription was developed in [72, 74] that generalizes the recipe of [60, 61] to the full asymptotically flat (or Ricci-flat) spacetime, and allows one to compute correlation functions in the worldvolume theory in terms of the dual geometry, with results that were successfully subjected to various checks.

The best-understood example of gauge/gravity duality, the so-called AdS/CFT correspondence, involves anti-de Sitter (AdS) spacetimes, which have negative cosmological constant. A generalization to the de Sitter (dS) case, where the cosmological constant is positive, has been proposed in [75], but there are important differences with the AdS case and as a result many open questions remain (for a review, see [76]). In [77] it was emphasized that this dS/CFT correspondence should be formulated using principal series representations of the isometry/conformal group, rather than the standard highest-weight representations as originally proposed. It was then argued that the finite entropy of dS space points to the need of q-deforming this group, to obtain representations that are at the same time unitary and finite-dimensional, and moreover reduce to the principal series representations in the classical limit $q \to 1$. In [78] the dS/CFT correspondence was examined for the case of 2-dimensional dS space, where the isometry algebra appears as a subalgebra of a much larger asymptotic symmetry algebra, the Virasoro algebra. Unitary representations for this full algebra were constructed which provide a prototype for a new class of conformal field theories dual to de Sitter backgrounds.

A different generalization of AdS/CFT replaces the AdS background with a domain-wall (DW) spacetime, and goes under the name of DW/QFT correspondence [79]. Domain walls also play important roles in world-brane scenarios and in cosmology. In the search for new domain-wall solutions, it is important to identify the various supergravity theories that may be obtained by deforming and/or dimensionally reducing the known ten- and eleven-dimensional supergravities. A number of gauged supergravities in nine dimensions were obtained in [80]. Eight-dimensional supergravities and their corresponding domain-wall solutions were constructed in [81, 82]. The phenomenon of string creation was discussed in the context of domain walls in [83]. Other novel reductions of gravity theories were considered in [84].

Over the past decade string theory has also made remarkable headway on the problem of finding the statistical-mechanical underpinnings of black hole thermodynamics, a development that is closely related to the discovery of the gauge/gravity duality. Complete success has been attained in the calculation of the entropy and the emission and absorption rates for a variety of black holes (or branes) that are either extremal or near-extremal [85]. The far-from-extremal regime, which includes in particular the Schwarzschild black hole, is conceptually more challenging and has thus far resisted fully quantitative analysis, but recent developments have brought us closer to this goal. In particular, a study of brane-antibrane systems at finite temperature led the authors of [86] to construct a microscopic model for the black threebrane of Type IIB supergravity and the black twobrane and fivebrane of eleven-dimensional supergravity. The model is based on decoupled stacks of branes and antibranes, with a gas of massless particles on each stack, and was shown to successfully reproduce the corresponding entropies arbitrarily far from extremality, up to a factor of 2. It also correctly accounts for various other properties of the black branes; in particular, their negative specific heat and pressure find a natural explanation in terms of brane-antibrane annihilation.

In the past few years, these results have been generalized in various directions. It has been shown that the brane-antibrane model also predicts the correct entropy for other singly- and multiply-charged black branes [87, 88] (see also the older works [89]). The model has been found to reproduce even the highly non-trivial entropy formulas for branes rotating with arbitrary amounts of angular momentum [90, 87]. Remarkably, it is also successful in yielding (again up to a factor of 2) the correct emission and absorption rates for arbitrary partial waves, at lowest order in the radiation frequency [91]. In spite of this already quite significant body of evidence, there are aspects of the brane-antibrane model that are still not fully understood (for a critical discussion, see [91]).

In the context of black holes in string theory, we should also mention the older work [92], where the stretched horizons for non-extremal black holes in four dimensions were studied. It was found that only for small masses of the order of the Veneziano wavelength is the stretched horizon bigger than the event horizon.

In recent times, interesting limits of string/M theory have been discovered which give rise to decoupled open brane theories exhibiting some form of noncommutativity. Best understood among these are the $(p+1)$-dimensional Noncommutative Open String (NCOS) theories [93], defined as a low-energy limit of a stack of Dp-branes in the presence of a near-critical worldvolume electric field. While decoupling the usual closed string modes, the limit in question remarkably manages to retain the whole tower of open string excitations; the result is a non-gravitational open string theory which displays noncommutativity between space and time.

The nature of NCOS theories came to be better understood through the work of [94], who discovered that, upon compactification of the electric field direction, the NCOS spectrum contains not only open strings but also closed strings with strictly positive winding number. After that, it was shown in [95, 96] that the NCOS limit may be taken with or without D-branes, and consequently defines a fully D-dimensional string theory ($D = 10$ for the superstring), known as Wound [95] or Non-relativistic [96] string theory. Much like its parent string theory, Wound string theory contains strings and branes of various kinds. Gravity also turns out to be present, but in a vastly simplified form: it is Newtonian when the theory is formulated on a flat background [96, 97], and 'asymptotically Newtonian' in a more general background [97, 98].

Shortly after the formulation of NCOS theories, generalizations based on other types of open branes were found [99, 100]. Foremost among these is Open Membrane (OM) theory [99], defined as a low-energy limit of a stack of M5-branes with a near-critical worldvolume 'electric' field strength. Just like NCOS describes a specific sector of Wound string theory, OM theory is part of an eleven-dimensional theory known as Wrapped [95] or Galilean [96] M2-brane theory. A membrane action for this theory was constructed in [101].

In the exploration of noncommutativity in string/M theory, boundary conditions play a central role. Following ideas in [102], the work [103] gave a prescription for treating boundary conditions as second-class constraints in a manner that is fully compatible with Dirac's method.

Another important development that has taken place in the past few years is the work of Berkovits on the quantization of the superstring in a manifestly super-Poincaré-covariant way (for a review, see [104]). In [105] a connection between this new 'pure spinor' method and the standard Green-Schwarz formalism was established by working out for the latter the Batalin-Fradkin-Tyutin program for converting second-class constraints into first-class ones (see also [106]).

On the more mathematical side, it is well known that twistor methods are suitable to describe Yang-Mills instantons which receive corrections from string theory in powers of α'. The description of full-fledged Yang-Mills fields can be also addressed through twistor methods, which is known as the Witten-Isenberg-Yasskin-Green (WIYG) correspondence . In [107] a stable brane configuration was proposed that naturally describes the setup involved in the WIYG correspondence. This seems to be the first step towards a description of this correspondence in string theory.

Recently, compactifications with fluxes have been an active area of research. One of the most important results is that in this context some or all of the moduli fields can be stabilized. This has been studied in a particular cosmological model [108] derived from type IIB superstring theory with fluxes [109], where usually the dilaton is interpreted as a quintessence field. Instead of that, in [110] the dilaton was interpreted as the dark matter of the universe. With this alternative interpretation it was found that in this supergravity model the dilaton dominates the universe before recombination. More surprisingly, it was found that the model gives a similar evolution and structure formation of the universe compared with the ΛCDM model in the linear regime of fluctuations of the structure formation. Some free parameters of the theory were fixed using the present cosmological observations. In the non-linear regime there are some differences between the type IIB supergravity theory and the traditional CDM paradigm. The supergravity theory predicts the formation of galaxies earlier than the CDM and there is no density cusp in the center of galaxies. These differences can distinguish both models and thus constitute an interesting feature of the cosmology obtained from superstring theory with fluxes.

REFERENCES

1. M. Green, J.H. Schwarz and E. Witten, *Superstring Theory*, Two volumes, Cambridge Univerity Press, Cambridge (1987); J. Polchinski, *String Theory*, Two volumes, Cambridge University Press, Cambridge (1998).

2. J. M. López R., M. A. Rodríguez S., M. Socolovsky and J. L. Vázquez B., "Introducción a la teoría de cuerdas: caso bósonico," Rev. Mex. Fis. **34** (1988) 452
3. J. A. Nieto, "A Relativistic Three-Dimensional Extended Object: The Terron," Rev. Mex. Fis. **34** (1988) 597.
4. M. Socolovsky, "Virasoro algebra and geometry," Rev. Mex. Fis. **36** (1990) 157.
5. J. A. Nieto, "Weyl Invariant Null P-Branes," Rev. Mex. Fis. **36** (1990) S204.
6. L. Cendejas and J. A. Nieto, "Comments On Super P-Branes," Nuovo Cim. B **105**, 1159 (1990).
7. J. A. Garcia, R. Linares and J. D. Vergara, "Weyl invariant p-brane and Dp-brane actions," Phys. Lett. B **503** (2001) 154 [arXiv:hep-th/0011085].
8. J. A. Nieto, "Remarks on Weyl invariant p-branes and Dp-branes," Mod. Phys. Lett. A **16** (2001) 2567 [arXiv:hep-th/0110227].
9. J. A. Nieto and S. A. Tomas, "Superstring Top," Phys. Lett. B **232**, 307 (1989).
10. J. A. Nieto, "Quantum superstring top," Nuovo Cim. B **109**, 411 (1994).
11. J. A. Nieto, "Comments on the string theory," Nuovo Cim. B **110**, 225 (1995).
12. M. Montesinos and J. D. Vergara, "Bosonic string theory with constraints linear in the momenta," Rev. Mex. Fis. **49S1** (2003) 53 [arXiv:hep-th/0105026].
13. J. Gamboa and C. Ramirez, Phys. Lett. B **301**, 20 (1993) [arXiv:hep-th/9211020].
14. J. A. Nieto, "From a relativistic point particle to string theory," Mod. Phys. Lett. A **10**, 3087 (1995) [arXiv:gr-qc/9508006].
15. C. Ramirez and L. F. Urrutia, "The Quantum group realizations of 3-D Chern-Simons theories," Mod. Phys. Lett. A **10**, 39 (1995).
16. C. Ramirez, H. Ruegg and M. Ruiz-Altaba, Nucl. Phys. B **364**, 195 (1991).
17. A. Herrera-Aguilar and O. V. Kechkin, Mod. Phys. Lett. A **16**, 29 (2001) [arXiv:gr-qc/0101007].
18. A. Herrera-Aguilar and O. V. Kechkin, Gen. Rel. Grav. **34**, 1331 (2002) [arXiv:gr-qc/0110096].
19. A. Herrera-Aguilar, Mod. Phys. Lett. A **19**, 2299 (2004) [arXiv:hep-th/0201126].
20. N. Barbosa-Cendejas and A. Herrera-Aguilar, Gen. Rel. Grav. **35**, 449 (2003) [arXiv:hep-th/0202006].
21. A. Herrera-Aguilar and O. V. Kechkin, J. Math. Phys. **45**, 216 (2004) [arXiv:hep-th/0203001].
22. A. Herrera-Aguilar and O. V. Kechkin, Int. J. Mod. Phys. A **17**, 2485 (2002) [arXiv:hep-th/0203002].
23. H. Ooguri and C. Vafa, "Geometry of N=2 strings," Nucl. Phys. B **361**, 469 (1991).
24. H. Garcia-Compean, "N = 2 string geometry and the heavenly equations," arXiv:hep-th/0405197.
25. A. de la Macorra, arXiv:hep-ph/9501322.
26. F. Quevedo, "Superstring phenomenology: An overview," Nucl. Phys. Proc. Suppl. **62**, 134 (1998) [arXiv:hep-ph/9707434].
27. C. P. Burgess, A. de la Macorra, I. Maksymyk and F. Quevedo, "Fixing the dilaton with asymptotically-expensive physics?," Phys. Lett. B **410**, 181 (1997) [arXiv:hep-th/9707062].
28. C. P. Burgess, A. de la Macorra, I. Maksymyk and F. Quevedo, "Supersymmetric models with product groups and field dependent gauge couplings," JHEP **9809**, 007 (1998) [arXiv:hep-th/9808087].
29. F. Quevedo, "Duality and global symmetries," Nucl. Phys. Proc. Suppl. **61A**, 23 (1998) [arXiv:hep-th/9706210].
30. H. Garcia-Compean, O. Obregon, J. F. Plebanski and C. Ramirez, "Towards a gravitational analog to S-duality in non-Abelian gauge theories," Phys. Rev. D **57**, 7501 (1998) [arXiv:hep-th/9711115]; H. Garcia-Compean, O. Obregon and C. Ramirez, "Gravitational duality in MacDowell-Mansouri gauge theory," Phys. Rev. D **58**, 104012 (1998) [arXiv:hep-th/9802063]; H. Garcia-Compean, J. A. Nieto, O. Obregon and C. Ramirez, "Dual description of supergravity MacDowell-Mansouri theory," Phys. Rev. D **59**, 124003 (1999) [arXiv:hep-th/9812175]; J. A. Nieto, "S-duality for linearized gravity," Phys. Lett. A **262**, 274 (1999) [arXiv:hep-th/9910049].
31. C. M. Hull, "Strongly coupled gravity and duality," Nucl. Phys. B **583**, 237 (2000) [arXiv:hep-th/0004195]; "Symmetries and compactifications of (4,0) conformal gravity," JHEP **0012**, 007 (2000) [arXiv:hep-th/0011215]; "Duality in gravity and higher spin gauge fields," JHEP **0109**, 027 (2001) [arXiv:hep-th/0107149].
32. J. H. Schwarz, "Does superstring theory have a conformally invariant limit?," arXiv:hep-th/0008009.
33. C. P. Burgess, M. T. Grisaru, M. Kamela, M. E. Knutt-Wehlau, P. Page, F. Quevedo and M. Zebarjad, "Spacetime duality and superduality," Nucl. Phys. B **542**, 195 (1999) [arXiv:hep-th/9809085].
34. T. Banks, W. Fischler, S. H. Shenker and L. Susskind, "M theory as a matrix model: A conjecture," Phys. Rev. D **55**, 5112 (1997) [arXiv:hep-th/9610043].

35. H. Garcia-Compean, "On the deformation quantization description of matrix compactifications," Nucl. Phys. B **541**, 651 (1999) [arXiv:hep-th/9804188].
36. B. Fedosov, J. Diff. Geom. **40**, 213 (1994); *Deformation Quantization and Index Theory* (Akademie Verlag, Berlin, 1996).
37. H. Garcia-Compean and J. F. Plebanski, "D-branes on group manifolds and deformation quantization," Nucl. Phys. B **618**, 81 (2001) [arXiv:hep-th/9907183].
38. H. Garcia-Compean, J. F. Plebanski, M. Przanowski and F. J. Turrubiates, "Deformation quantization of bosonic strings," J. Phys. A **33**, 7935 (2000) [arXiv:hep-th/0002212].
39. H. Garcia-Compean and J. Moreno, "Remarks on noncommutative solitons," Rev. Mex. Fis. **49**, 28 (2003) [arXiv:hep-th/0110119].
40. J. A. Nieto, "Searching for a connection between matroid theory and string theory," J. Math. Phys. **45**, 285 (2004) [arXiv:hep-th/0212100]; "Matroids and p-branes," Adv. Theor. Math. Phys. **8**, 177 (2004) [arXiv:hep-th/0310071]; "Oriented matroid theory as a mathematical framework for the M-theory," arXiv:hep-th/0506106.
41. A. Giveon and D. Kutasov, "Brane Dynamics and Gauge Theory", Rev. Mod. Phys. **71** (1999) 983, hep-th/9802067.
42. A. Hanany and A. Zaffaroni, "On the Realization of Chiral Four-dimensional Gauge Theories Using Branes", JHEP **05** (1998) 001, hep-th/9801134.
43. A. Hanany, M. J. Strassler and A. M. Uranga, "Finite theories and marginal operators on the brane," JHEP **9806**, 011 (1998) [arXiv:hep-th/9803086].
44. H. Garcia-Compean and A. M. Uranga, "Brane box realization of chiral gauge theories in two dimensions," Nucl. Phys. B **539**, 329 (1999) [arXiv:hep-th/9806177].
45. A. Sen, "Non-BPS states and branes in string theory," arXiv:hep-th/9904207.
46. E. Witten, "D-branes and K-theory," JHEP **9812**, 019 (1998) [arXiv:hep-th/9810188].
47. H. Garcia-Compean, "D-branes in orbifold singularities and equivariant K-theory," Nucl. Phys. B **557**, 480 (1999) [arXiv:hep-th/9812226].
48. K. Olsen and R. J. Szabo, "Brane descent relations in K-theory," Nucl. Phys. B **566**, 562 (2000) [arXiv:hep-th/9904153].
49. O. Loaiza-Brito and A. M. Uranga, Nucl. Phys. B **619**, 211 (2001) [arXiv:hep-th/0104173].
50. N. Quiroz and B. . J. Stefanski, "Dirichlet branes on orientifolds," Phys. Rev. D **66**, 026002 (2002) [arXiv:hep-th/0110041].
51. H. Garcia-Compean and O. Loaiza-Brito, "Branes and fluxes in orientifolds and K-theory," Nucl. Phys. B **694**, 405 (2004) [arXiv:hep-th/0206183].
52. A. de la Macorra and S. Lola, Int. J. Mod. Phys. D **5**, 541 (1996) [arXiv:hep-ph/9411443].
53. A. de la Macorra, Int. J. Mod. Phys. D **5**, 567 (1996) [arXiv:hep-ph/9501250].
54. A. de la Macorra and S. Lola, Phys. Lett. B **373**, 299 (1996) [arXiv:hep-ph/9511470].
55. M. Borunda and M. Ruiz-Altaba, "Pre-big-bang in string cosmology," arXiv:hep-th/9803016.
56. M. Borunda and M. Ruiz-Altaba, "On the initial conditions for pre-big-bang cosmology," arXiv:hep-th/9804082.
57. H. Garcia-Compean, G. Garcia-Jimenez, O. Obregon and C. Ramirez, "Tachyon driven quantum cosmology in string theory," Phys. Rev. D **71**, 063517 (2005).
58. H. Garcia-Compean and O. Loaiza-Brito, "Lectures on strings, D-branes and gauge theories," arXiv:hep-th/0003019; H. Garcia-Compean and O. Loaiza-Brito, "Topics on strings, branes and Calabi-Yau compactifications," AIP Conf. Proc. **562**, 86 (2001) [arXiv:hep-th/0010046]; H. Garcia-Compean "Introduction to string theory and string compactifications," J. Phys. Conf. Ser. **18**, 270 (2005).
59. J. M. Maldacena, "The large N limit of superconformal field theories and supergravity," Adv. Theor. Math. Phys. **2**, 231 (1998) [Int. J. Theor. Phys. **38**, 1113 (1999)] [arXiv:hep-th/9711200].
60. S. S. Gubser, I. R. Klebanov and A. M. Polyakov, "Gauge theory correlators from non-critical string theory," Phys. Lett. B **428**, 105 (1998) [arXiv:hep-th/9802109].
61. E. Witten, "Anti-de Sitter space and holography," Adv. Theor. Math. Phys. **2**, 253 (1998) [arXiv:hep-th/9802150].
62. I. R. Klebanov and M. J. Strassler, "Supergravity and a confining gauge theory: Duality cascades and chiSB-resolution of naked singularities," JHEP **0008** (2000) 052 [arXiv:hep-th/0007191].
63. E. Caceres and R. Hernandez, "Glueball masses for the deformed conifold theory," Phys. Lett. B **504** (2001) 64 [arXiv:hep-th/0011204].

64. M. Krasnitz, "A two point function in a cascading N = 1 gauge theory from supergravity," arXiv:hep-th/0011179.
65. X. Amador and E. Caceres, "Spin two glueball mass and glueball Regge trajectory from supergravity," JHEP **0411** (2004) 022 [arXiv:hep-th/0402061].
66. M. J. Teper, "Glueball masses and other physical properties of SU(N) gauge theories in D = 3+1: A review of lattice results for theorists," arXiv:hep-th/9812187.
67. E. Caceres and C. Nunez, "Glueballs of super Yang-Mills from wrapped branes," JHEP **0509** (2005) 027 [arXiv:hep-th/0506051].
68. J. M. Maldacena and C. Nunez, "Towards the large N limit of pure N = 1 super Yang Mills," Phys. Rev. Lett. **86**, 588 (2001) [arXiv:hep-th/0008001].
69. J. Polchinski, "Dirichlet-Branes and Ramond-Ramond Charges," Phys. Rev. Lett. **75**, 4724 (1995) [arXiv:hep-th/9510017].
70. S. S. Gubser, I. R. Klebanov and A. W. Peet, "Entropy and Temperature of Black 3-Branes," Phys. Rev. D **54**, 3915 (1996) [arXiv:hep-th/9602135]; I. R. Klebanov, "World-volume approach to absorption by non-dilatonic branes," Nucl. Phys. B **496**, 231 (1997) [arXiv:hep-th/9702076]; S. S. Gubser, I. R. Klebanov and A. A. Tseytlin, "String theory and classical absorption by three-branes," Nucl. Phys. B **499**, 217 (1997) [arXiv:hep-th/9703040]; S. S. Gubser and I. R. Klebanov, "Absorption by branes and Schwinger terms in the world volume theory," Phys. Lett. B **413**, 41 (1997) [arXiv:hep-th/9708005].
71. O. Aharony, S. S. Gubser, J. M. Maldacena, H. Ooguri and Y. Oz, "Large N field theories, string theory and gravity," Phys. Rept. **323**, 183 (2000) [arXiv:hep-th/9905111].
72. U. H. Danielsson, A. Güijosa, M. Kruczenski and B. Sundborg, "D3-brane holography," JHEP **0005**, 028 (2000) [arXiv:hep-th/0004187].
73. S. S. Gubser, A. Hashimoto, I. R. Klebanov and M. Krasnitz, "Scalar absorption and the breaking of the world volume conformal invariance," Nucl. Phys. B **526**, 393 (1998) [arXiv:hep-th/9803023]; S. P. de Alwis, "Supergravity, the DBI action and black hole physics," Phys. Lett. B **435** (1998) 31 [arXiv:hep-th/9804019]; S. S. Gubser and A. Hashimoto, "Exact absorption probabilities for the D3-brane," Commun. Math. Phys. **203**, 325 (1999) [arXiv:hep-th/9805140]; K. A. Intriligator, "Maximally supersymmetric RG flows and AdS duality," Nucl. Phys. B **580**, 99 (2000) [arXiv:hep-th/9909082].
74. X. Amador, E. Caceres, H. Garcia-Compean and A. Güijosa, "Conifold holography," JHEP **0306** (2003) 049 [arXiv:hep-th/0305257].
75. A. Strominger, "The dS/CFT correspondence," JHEP **0110** (2001) 034 [arXiv:hep-th/0106113].
76. D. Klemm and L. Vanzo, "Aspects of quantum gravity in de Sitter spaces," JCAP **0411**, 006 (2004) [arXiv:hep-th/0407255].
77. A. Güijosa and D. A. Lowe, "A new twist on dS/CFT," Phys. Rev. D **69** (2004) 106008 [arXiv:hep-th/0312282].
78. A. Güijosa, D. A. Lowe and J. Murugan, "A prototype for dS/CFT," Phys. Rev. D **72** (2005) 046001 [arXiv:hep-th/0505145].
79. H. J. Boonstra, K. Skenderis and P. K. Townsend, "The domain wall/QFT correspondence," JHEP **9901** (1999) 003 [arXiv:hep-th/9807137]; K. Behrndt, E. Bergshoeff, R. Halbersma and J. P. van der Schaar, "On domain-wall/QFT dualities in various dimensions," Class. Quant. Grav. **16** (1999) 3517 [arXiv:hep-th/9907006].
80. E. Bergshoeff, T. de Wit, U. Gran, R. Linares and D. Roest, "(Non-)Abelian gauged supergravities in nine dimensions," JHEP **0210** (2002) 061 [arXiv:hep-th/0209205].
81. N. Alonso Alberca, E. Bergshoeff, U. Gran, R. Linares, T. Ortin and D. Roest, "Domain walls of D = 8 gauged supergravities and their D = 11 origin," JHEP **0306** (2003) 038 [arXiv:hep-th/0303113].
82. E. Bergshoeff, U. Gran, R. Linares, M. Nielsen, T. Ortin and D. Roest, "The Bianchi classification of maximal D = 8 gauged supergravities," Class. Quant. Grav. **20** (2003) 3997 [arXiv:hep-th/0306179].
83. E. Bergshoeff, U. Gran, R. Linares, M. Nielsen and D. Roest, "Domain walls and the creation of strings," Class. Quant. Grav. **20** (2003) 3465 [arXiv:hep-th/0303253].
84. R. Linares, "S**3 dimensional reduction of Einstein gravity," JHEP **0507** (2005) 042 [arXiv:hep-th/0405217]; R. Linares, "Bianchi IX group-manifold reductions of gravity," Mod. Phys. Lett. A **20** (2005) 3115.
85. S. R. Das and S. D. Mathur, "The Quantum Physics Of Black Holes: Results From String Theory," Ann. Rev. Nucl. Part. Sci. **50** (2000) 153 [arXiv:gr-qc/0105063].
86. U. Danielsson, A. Güijosa and M. Kruczenski, "Brane-Antibrane systems at finite temperature and the

entropy of black holes," JHEP **0109** (2001) 011, [arXiv:hep-th/0106201]; U. H. Danielsson, A. Güijosa and M. Kruczenski, "Black brane entropy from brane-antibrane systems," Rev. Mex. Fís. **49S2** (2003) 61 [arXiv:gr-qc/0204010].
87. O. Saremi and A. W. Peet, "Brane-antibrane systems and the thermal life of neutral black holes," arXiv:hep-th/0403170.
88. O. Bergman and G. Lifschytz, "Schwarzschild black branes from unstable D-branes," JHEP **0404**, 060 (2004) [arXiv:hep-th/0403189]; S. Kalyana Rama, "A description of Schwarzschild black holes in terms of intersecting M-branes and antibranes," arXiv:hep-th/0404026; G. Lifschytz, "Charged black holes from near extremal black holes," arXiv:hep-th/0405042. S. K. Rama and S. Siwach, "A description of multicharged black holes in terms of branes and antibranes," arXiv:hep-th/0405084; E. Halyo, "Black hole entropy and superconformal field theories on brane-antibrane systems," arXiv:hep-th/0406082; K. I. Ohshima, "Comments on the entropy and the temperature of non-extremal black p-brane," arXiv:hep-th/0508100; S. Siwach, "Note on brane-antibrane description of non-extremal black holes," arXiv:hep-th/0503164.
89. G. T. Horowitz, J. M. Maldacena and A. Strominger, "Nonextremal Black Hole Microstates and U-duality," Phys. Lett. B **383** (1996) 151, [arXiv:hep-th/hep-th/9603109]; M. Cvetič and D. Youm, "Entropy of Non-Extreme Charged Rotating Black Holes in String Theory," Phys. Rev. D **54** (1996) 2612, [arXiv:hep-th/9603147]; G. T. Horowitz, D. A. Lowe and J. M. Maldacena, "Statistical Entropy of Nonextremal Four-Dimensional Black Holes and U-Duality," Phys. Rev. Lett. **77** (1996) 430, [arXiv:hep-th/9603195]; R. Kallosh and A. Rajaraman, "Brane-anti-brane Democracy," Phys. Rev. D **54** (1996) 6381, [arXiv:hep-th/9604193]; J. Zhou, H. J. Müller-Kirsten, J. Q. Liang and F. Zimmerschied, "M-branes, anti-M-branes and non-extremal black holes," Nucl. Phys. B **487** (1997) 155 [arXiv:hep-th/9611146]; M. S. Costa and M. J. Perry, "Landau degeneracy and black hole entropy," Nucl. Phys. B **520** (1998) 205, [arXiv:hep-th/9712026].
90. A. Güijosa, H. H. Hernández Hernández and H. A. Morales Técotl, "The entropy of the rotating charged black threebrane from a brane-antibrane system," JHEP **0403** (2004) 069 [arXiv:hep-th/0402158].
91. J. A. García and A. Güijosa, "Threebrane absorption and emission from a brane-antibrane system," JHEP **0409** (2004) 027 [arXiv:hep-th/0407075].
92. C. Espinoza and M. Ruiz-Altaba, "Stretched horizon for non-supersymmetric black holes," AIP Conf. Proc. **490**, 335 (1999) [arXiv:gr-qc/9904065].
93. R. Gopakumar, J. Maldacena, S. Minwalla and A. Strominger, "S-duality and noncommutative gauge theory," JHEP **0006** (2000) 036 [arXiv:hep-th/0005048]; N. Seiberg, L. Susskind and N. Toumbas, "Strings in background electric field, space/time noncommutativity and a new noncritical string theory," JHEP **0006** (2000) 021 [arXiv:hep-th/0005040].
94. I. R. Klebanov and J. Maldacena, "1+1 dimensional NCOS and its U(N) gauge theory dual," Int. J. Mod. Phys. A **16** (2001) 922 [Adv. Theor. Math. Phys. **4** (2001) 283] [arXiv:hep-th/0006085].
95. U. H. Danielsson, A. Güijosa and M. Kruczenski, "IIA/B, wound and wrapped," JHEP **0010** (2000) 020 [arXiv:hep-th/0009182].
96. J. Gomis and H. Ooguri, "Non-relativistic closed string theory," arXiv:hep-th/0009181.
97. U. H. Danielsson, A. Güijosa and M. Kruczenski, "Newtonian gravitons and D-brane collective coordinates in Wound string theory," JHEP **0103** (2001) 041 [arXiv:hep-th/0012183].
98. V. Sahakian, "The large M limit of non-commutative open strings at strong coupling," arXiv:hep-th/0107180.
99. R. Gopakumar, S. Minwalla, N. Seiberg and A. Strominger, "OM theory in diverse dimensions," JHEP **0008** (2000) 008 [arXiv:hep-th/0006062]; E. Bergshoeff, D. S. Berman, J. P. van der Schaar and P. Sundell, "Critical fields on the M5-brane and noncommutative open strings," Phys. Lett. B **492** (2000) 193 [arXiv:hep-th/0006112].
100. T. Harmark, "Open branes in space-time non-commutative little string theory," Nucl. Phys. B **593** (2001) 76 [arXiv:hep-th/0007147]; J. X. Lu, "(1+p)-dimensional open D(p-2) brane theories," JHEP **0108** (2001) 049 [arXiv:hep-th/0102056]; H. Larsson and P. Sundell, "Open string/open D-brane dualities: Old and new," JHEP **0106** (2001) 008 [arXiv:hep-th/0103188]; U. Gran and M. Nielsen, "Non-commutative open (p,q) string theories," JHEP **0111** (2001) 022 [arXiv:hep-th/0104168].
101. J. A. García, A. Güijosa and J. D. Vergara, "A membrane action for OM theory," Nucl. Phys. B **630**, 178 (2002) [arXiv:hep-th/0201140].
102. C. S. Chu and P. M. Ho, "Noncommutative open string and D-brane," Nucl. Phys. B **550** (1999) 151

[arXiv:hep-th/9812219]; F. Ardalan, H. Arfaei and M. M. Sheikh-Jabbari, "Noncommutative geometry from strings and branes," JHEP **9902** (1999) 016 [arXiv:hep-th/9810072]; F. Ardalan, H. Arfaei and M. M. Sheikh-Jabbari, "Dirac quantization of open strings and noncommutativity in branes," Nucl. Phys. B **576** (2000) 578 [arXiv:hep-th/9906161]; M. M. Sheikh-Jabbari and A. Shirzad, "Boundary conditions as Dirac constraints," Eur. Phys. J. C **19**, 383 (2001) [arXiv:hep-th/9907055].
103. J. M. Romero and J. D. Vergara, "Boundary conditions as constraints," arXiv:hep-th/0212035.
104. N. Berkovits, "ICTP lectures on covariant quantization of the superstring," arXiv:hep-th/0209059.
105. A. Gaona and J. A. Garcia, "BFT embedding of the Green-Schwarz superstring and the pure spinor formalism," JHEP **0509** (2005) 083 [arXiv:hep-th/0507076].
106. N. Berkovits and D. Z. Marchioro, "Relating the Green-Schwarz and pure spinor formalisms for the superstring," JHEP **0501**, 018 (2005) [arXiv:hep-th/0412198].
107. H. Garcia-Compean, O. Loaiza-Brito and N. Quiroz, "Towards the Witten-Isenberg-Yasskin-Green correspondence in string theory," AIP Conf. Proc. **758**, 143 (2005).
108. A. R. Frey and J. Polchinski, "N = 3 warped compactifications," Phys. Rev. D **65**, 126009 (2002) [arXiv:hep-th/0201029].
109. A. R. Frey and A. Mazumdar, "3-form induced potentials, dilaton stabilization, and running moduli," Phys. Rev. D **67**, 046006 (2003) [arXiv:hep-th/0210254].
110. T. Matos, J. R. Luevano and H. Garcia-Compean, "An alternative interpretation for the moduli fields of the cosmology associated to type IIB supergravity with fluxes," arXiv:hep-th/0511098.

Renormalization Group Theory

C. R. Stephens

Instituto de Ciencias Nucleares
Universidad Nacional Autonoma de México
Circuito Exterior, A. Postal 70-543, México D.F. 04510

Abstract. In this article I give a brief account of the development of research in the Renormalization Group in Mexico, paying particular attention to novel conceptual and technical developments associated with the tool itself, rather than applications of standard Renormalization Group techniques. Some highlights include the development of new methods for understanding and analysing two extreme regimes of great interest in quantum field theory - the "high temperature" regime and the Regge regime.

Keywords: Renormalization Group, field theory, scaling, Regge theory
PACS: 11.10.Gh,11.10.Wx,11.55.Jy

1. INTRODUCTION

Works that involve the Renormalization Group (RG) in high energy physics can be divided into two types - those that use a well known and studied version to attack a new problem (see the contribution of [1, 2] for this perspective); and those that develop new RG concepts and techniques to study either an old or new problem. This article will delineate some of the advances made by Mexican scientists to the latter. RG research in this sense was a relative newcomer to Mexico, there being little activity before the mid-90s when a group formed at the Instituto de Ciencias Nucleares of the UNAM and slightly later at the CINVESTAV. Since then further groups have sprung up at the UMSNH and the Universidad Autónoma del Estado de Hidalgo.

Before talking about the work done in Mexico it's worthwhile putting it in context by considering a bit of the history of renormalization and the RG. There have been, in fact, two routes to the RG: one with its origin in the notion of renormalization (reparametrization) in high energy physics and the other in that of coarse graining in statistical physics. The reparametrization approach had a strange early history, being associated with what might be viewed as somewhat of an embarrassment - the existence of divergences in the perturbative expansion of energy levels, scattering amplitudes and other quantities in quantum field theories such as QED. It was noticed that these divergences could be absorbed into a redefinition of the parameters that characterised the theory, such as mass parameters, coupling constants etc., passing from the underlying "bare" parameters, which were deemed to be unobservable, to "renormalized" parameters in terms of which physical observables were described. This was the "sweeping things under the rug" approach to renormalization, where a recipe was followed for a given field theory, and if the divergences fitted "under the rug" then the theory was said to be *renormalizable*. Those theories where the divergences kept sticking out were called *non-renormalizable*. As the sophistication of the formal mathematical framework of renormalization theory increased, it became possible to rigorously prove in some field theories that all of the divergences could be swept under the rug, without the labour of having to show it order by order in perturbation theory.

Early on, an observation was made by Stückelberg and Petermann [3] that there was an ambiguity associated with just how much was actually under the rug and how much was outside it, as long as the larger (infinite in the case where the ultraviolet cutoff goes to infinity) part was under. This ambiguity manifested itself in the fact that the renormalized parameters depended on an arbitrary scale - the renormalization scale. Changing this scale led to new values for the different parameters - coupling constants, mass parameters etc. Physical quantities however, were independent of these changes. In other words, one was writing the same physics, but in terms of a different set of parameters, associated with different renormalization scales. Changing from one scale to another was in this sense just a change of parametrization. What is more, one could take a theory, sweep the divergences under the rug with one bigger sweep, or two separate smaller sweeps, and still leave what was outside the rug the same. The sweepings thus had a group structure (you could also lift the rug and sweep some back out again - inverse transformation - or not sweep anything at all - the identity). Thus was born the RG.

More formally: In terms of the operators associated with a field theory, there are two basic types of "sweeping" (or more correctly - renormalization/reparametrization) - additive and multiplicative. The former are associated with tuning of parameters, such as masses/critical temperature shifts, the cosmological constant, the specific heat and composite operators in general, wherein a parameter or operator $O \to O + O_c$. Multiplicative renormalization on the other hand is associated with the scaling of parameters/operators, wherein a parameter or operator $O \to Z_O O$, and, in particular, with the correlation functions of different operators. It is precisely these multiplicative factors that determine scaling exponents for observables derived from the underlying correlation functions, such as the susceptibility, coupling constants, scattering amplitudes etc. A typical scenario for the perturbative expansion of some Greens function as a function of a parameter, such as momentum, in a scaling limit $p/m \gg 1$ is described by

$$G(p) = A p^a + B \lambda p^a \ln p + C \lambda^2 p^a (\ln p)^2 + D \lambda^2 p^a \ln p + O(\lambda^3 (\ln p)^3) \tag{1}$$

where λ is a coupling constant taken as the perturbative expansion parameter. For large p the perturbative expansion breaks down. A multiplicative renormalization of $G(p)$ is introduced and because the constants A, B, C etc. are not all independent, a multiplicative renormalization $G(p) \to Z G(p)$ can be introduced such that the large logarithms are eliminated order by order in perturbation theory. One effect of the RG is to resum these logarithms so as to be able to write for example to $O(\lambda)$

$$G(p) = A p^{(a+(B/A)\lambda)} \tag{2}$$

Upon expanding the exponent in λ one recovers the first three terms of (1) with $C = B^2/2A$. Thus, the scaling exponent characteristic of $G(p)$ is found to be $(a+(B/A)\lambda)$ rather than just a. One could make a fairly convincing argument that in high energy physics the vast amount of research has been associated with this simple abstract idea implemented at a perturbative, diagrammatic level. Thus, although it was felt that a better, still to be found, description of reality would not need renormalization, it was also felt that at least a self-consistent framework had been developed in which the whole process could be carried out. In fact, this approach was so successful that in the "Diagrammar" program of 't Hooft and Veltman a field theory was *defined* by its perturbative expansion in terms of Feynman diagrams and their associated counterterms. As far as predictions in QED and, more generally, the standard model are concerned, the entire program has worked, in pragmatic terms, fabulously well.

In the 1960s the problem of how to theoretically describe the scaling phenomena seen in second order phase transitions was very much on people's minds. Although a phenomenological framework had been developed, nothing much was known from first principles. Leo Kadanoff's notion of "block spins" - effective degrees of freedom representing sums of the underlying microscopic spins - captured very intuitively the idea that coherent fluctuations near the phase transition strongly coupled together large numbers of degrees of freedom. It was left to Ken Wilson to turn this idea into a quantitative tool, whereby such "coarse grainings", which form a semi-group (there are no inverse transformations to undo a coarse graining), could be implemented in a mathematically controllable fashion, such as with the famous ε-expansion of Wilson and Fisher [4].

One of the deep insights of Wilson was to realize that the natural arena in which to consider coarse grainings was the space of parameters, $\mathcal{M}$, or space of Hamiltonians as it is more often termed. This, actually, is equally true irrespective of whether one adopts the reparametrization point of view or the coarse graining approach, although this does not seem to have been realized in QFT before the advent of Wilson's work. Coarse graining (RG by an abuse of language) transformations yield a flow in the (potentially) infinite dimensional space of parameters $\mathcal{M}$ (or the space of all Hamiltonians). The transformations are associated with a one-parameter flow with respect to a quantity which can be interpreted as a change in lattice spacing or some UV or IR cutoff. Of particular interest are the fixed points of the coarse graining, as they represent points of scale invariance. The RG transformation can be linearized around the fixed points, the resulting eigenvalues yielding the scaling exponents associated with that fixed point. From a calculational point of view it throws the emphasis onto calculating the parameter flows, the idea being that one may wish to calculate the physics of a system at certain parameter values, where an approximate calculation is extremely difficult, such as near a second order phase transition, by relating it to the system at some other values where the calculation is more reliable, the two systems being connected by an RG flow. Thus a coarse graining relates the same system in two different physical states.

So, how are the two RGs - reparametrization and coarse graining - related?. Both have one parameter flows and fixed points for those flows. For coarse graining, two points on a flow trajectory represent the same system in two different physical states, whereas in the reparametrization approach they represent two different sets of parameters (coordinate systems) associated with a fiducial, or reference, system with which to describe a potentially different system of interest. In this case, there is also an arbitrary renormalization scale of which the physics is independent. By making a choice for this scale that is related to the physics of the system of interest a "gauge" choice has been made and the coordinate invariance is lost. If the reference system is the same as the system of interest then this "gauge-fixed" reparametrization RG will be equivalent to the coarse-grained one.

Both RG "schools of thought" - reparametrization and coarse graining - have had a profound impact on late 20th century physics, in many different fields. Each has its own particular adherents, though there are some who are equally comfortable with both paradigms. However, a difficulty of the RG is that it is a tool, and what unifies people is the application area not the tool per se. The reparametrization approach still resonates more in QFT and the coarse graining approach more in statistical physics, and hence the two communities of RG users have not mixed as much as might be wished. An exception to this rule has been the quadriennial RG conferences, organized originally by one of the most important pioneers of the reparametrization RG - Dimitri Shirkov. The fourth conference in the series, the first to be held outside the Soviet Union/Russia, was held in Taxco, México in 1999 and had as honoured speakers, among others, Dimitri Shirkov and Michael Fisher - it was the first time that these two great pioneers of the (two different) RGs had ever met!. The proceedings of this important meeting, which brought together many of the most important contributors to RG theory, both from high energy and statistical physics, were published in four special volumes of Physics Reports [8, 7, 6, 5] reviewing the state of the art in RG theory at the turn of the millenium - an important contribution from Mexico to the field. I will now turn to some specific areas of RG research that have been particularly active in Mexico.

2. CROSSOVER BEHAVIOUR

The region where it is essential to use the RG is when there is a scaling regime, wherein many microscopic degrees of freedom are strongly coupled together. This is most obviously the case in the vicinity of a second order phase transition and also, analogously, in quantum field theory, and manifests itself perturbatively by the existence of divergences as a function of, for example, a ratio of length scales, such as p/m, Λ/m or Λ/p, where p is a momentum scale, m a mass scale and Λ a cutoff. As mentioned, the reparametrization RG is a tool for relating systems at different renormalization "scales". A renormalization scale in high energy physics is often taken to be a *single* momentum scale, associated with a scattering amplitude chosen at a *symmetric* point where different momentum invariants are equal. Generally there are many other choices, such as masses, UV or IR cutoffs or, less studied, asymmetric momentum invariants.

The first triumphs of the RG were associated with the description of a single scaling regime associated with a single fixed point of the RG transformation. Generically, however, there is more than one scaling regime as a function of the different length scales inherent in a problem. A particularly relevant example of this, is finite temperature field theory, where the existence of the temperature scale, T, is equivalent to having a field theory on a cylindrical space, where Euclidean time is compactified with radius proportional to the inverse temperature. In this case, at small scales relative to $1/T$ the system is d-dimensional, whereas at large scales it looks $(d-1)$-dimensional. Hence, there are two distinct scaling regimes associated with d and $(d-1)$-dimensional physics and a corresponding *crossover* between them as a function of T/m. Such crossovers are ubiquitous, being intrinsically "infrared" phenomena involving a characteristic "environmental" scale, such as temperature, magnetic/electric field, "size" of the universe etc. Momentum scales far removed from this scale in the UV are insensitive to it and therefore insensitive to the crossover.

In terms of the RG, as a function of T and the renormalization scale, κ, these two different scaling regimes can be reached in the scaling limits where $\kappa \to \infty$, and either $T/\kappa \to 0$ or, in the "high temperature" limit, where $T/\kappa \to \infty$. As the physically relevant effective degrees of freedom of a system depend on its "environment", in this case the temperature, if one wishes to implement an RG that captures both scaling regimes, i.e., both d- and $(d-1)$-dimensional physics, then that RG had better be temperature dependent or, more generally, "environmentally friendly" [9, 10].

Such environmentally friendly RGs have been a topic of intense study by Mexican researchers, both directly, in the context of critical phenomena [11, 12, 13, 14], and, as mentioned, in high energy physics, in the context of finite temperature field theory [15, 16, 17, 18, 19]. In the latter a particular innovation was to use as renormalization scale a fiducial temperature. The corresponding RG then runs a temperature and not a mass or momentum scale. By so doing it becomes much more transparent what is meant by a "high-temperature" regime as high has to be referred to some other scale. Thus, the vicinity of a second-order or weakly first-order phase transition is definitively high temperature as the finite temperature mass in this regime is very small. In such a regime procedures such as the hard-thermal loop resummation cannot work, not least because, generically, there are important thermal corrections to the coupling constant.

3. MORE THAN ONE SINGULARITY PER GREENS FUNCTION

In the introduction I pointed out how a multiplicative renormalization, in conjunction with the RG, is able to "exponentiate" a series of potentially divergent terms in perturbation theory, thus giving rise to an asymptotic scaling behaviour. The corresponding singularity is just the *leading*, i.e., dominant, singularity. However, we know, for example by considerations of the transfer matrix, that there are many singularities inherent in a single Greens function of a theory. How does one access more than one singularity? An important example of this phenomenon in high energy physics is the small-x limit associated with Regge theory and Regge trajectories. In this extreme kinematic limit the scaling form of the four-point scattering amplitude shows an apparent dimensional reduction, or perhaps better said, a dimensional "factorization, manifest for ϕ^3 theory for example, in the dimensionally reduced form of the dominant ladder diagrams in this limit, the dimensional reduction being $d \to (d-2)$. A similar phenomenon occurs in gravity and could well play an important role in quantum gravity as has been emphasized by 't Hooft [20]. In work related to the above, 't Hooft [21] also showed that certain Planck scale processes can be calculated using known laws of physics. Specifically, Regge limit graviton-graviton scattering amplitudes with $s \gg t$ can be calculated, the resultant amplitudes possessing certain features in common with string theory amplitudes. This work was extended by the Verlindes [22] who showed that in the "Regge" regime quantum gravity separates into strongly coupled (longitudinal) and weakly coupled (transverse) sectors. This separation into weak and strong coupling sectors is very characteristic of a kinematic crossover.

Work by Mexican researchers has shown how RG techniques can be adapted to deal with such kinematic dimensional reductions [23, 24, 25]. I will briefly illustrate the phenomenon within the context of a cubic scalar field theory. Rather than deal with the connected Greens functions directly it is convenient to use the quantities $\tilde{\Gamma}_B^{i_1...i_n}(p_1,...,p_n)$, which are closely related to the S-matrix elements and are obtained from the connected Greens functions by removing the external legs. The superscript notation $i_1...i_n$ denotes the external legs, i.e. $i_1,...,i_n$, which can take different values depending on the field content of the theory. The relation to fully one-particle irreducible Greens functions is found using the standard "tree theorem".

For the pure cubic scalar theory in four dimensions with coupling g and mass m, to one loop the only UV divergence is a logarithmically divergent correction to m. This UV divergence can be removed in the standard fashion. After this renormalization, although UV finite the theory is not perturbatively reliable in the extreme, asymmetric scaling limits as can be illustrated using the two-particle scattering amplitude in the asymmetric limit $t \to \infty$ for fixed s. The one-loop diagrams of the on-shell amplitude can be classified according to whether they contain a factor of $\ln t$, a factor of $\ln(-t)$, or are "finite", a separation which can in principle be carried out to all orders leading to the following decomposition for $\tilde{\Gamma}_B^{ijkl}$ [23]:

$$\tilde{\Gamma}_B^{ijkl}(s,t,u) = B_{B,t}^{ijkl}(s,t) + B_{B,u}^{ijkl}(s,u) + B_{B,s}^{ijkl}(t,u). \tag{3}$$

The function $B_{B,s}^{ijkl}$ contains no large logarithms and does not play an important role in the asymptotic t behaviour. However, in the large-t limit, $B_{B,t}^{ijkl}$ contains logarithms of t and $B_{B,u}^{ijkl}$ logarithms of $-t$. In some simple cases these logarithms can be summed by hand. However, Mexican researchers have shown how this can be done systematically using the RG [23, 24]. In distinction to the standard case, as there are two sets of large logarithms, one associated with $B_{B,t}$ and another set with $B_{B,u}$, an overall multiplicative renormalization of the connected four-point function will not be sufficient. Rather, one needs to get inside the Greens functions and identify the parts, $B_{B,t}$ and $B_{B,u}$, that will renormalize "naturally". For example, $B_t(s,t,g(\kappa),m(\kappa),\kappa) = Z_t B_{B,t}(s,t,g_B,m_B,\Lambda)$, where Z_t depends on the specific normalization condition chosen, satisfies the RG equation

$$\kappa \frac{d}{d\kappa} B_t(s,t,g(\kappa),m(\kappa),\kappa) = \gamma_t B_t(s,t,g(\kappa),m(\kappa),\kappa), \tag{4}$$

where $\gamma_t = d\ln Z_t/d\ln\kappa$ is the anomalous dimension of B_t. This equation can be integrated and a fiducial value of t chosen as the RG scale κ. In the limit of large t there exists a one parameter family of "fixed points" wherein $g(t) \to g(\infty)$ and γ_t is purely a function of s.

To obtain the correct ratio of t- and u-contributions we have to consider the large-u limit for the u-contributions. Proceeding in exactly the same way as for the t-contributions we can now replace u by $-t$, to be interpreted as $e^{i\pi}t$, and add the t- and u-contributions to find

$$\tilde{\Gamma}_B^{\phi\phi\phi\phi}(s,t) = \frac{g^2(t)}{-\kappa} \frac{G(s)(1+e^{i\pi\alpha(s)})}{Z(s,g(\kappa),m(\kappa),\kappa)} \left(\frac{t}{\kappa}\right)^{\alpha(s)} + B_s(t,u), \tag{5}$$

where $\alpha(s) = -1 - \gamma_t(s)$ and $B_s(t,u)$ contains the finite (in the large-t limit) s-contributions and Z is a non-universal amplitude that should be determined via an experimental result on the two-point scattering amplitude at some value of t for fixed s. Thus we see it is possible using the RG to produce signatured amplitudes. The renormalization here is completely crossing symmetric yielding analagous expressions in the large s or large u limits. Naturally, in the different asymptotic limits the diagrams that contribute to the renormalization are different.

An explicit one-loop calculation (5) yields [23, 24]

$$\tilde{\Gamma}_B^{\phi\phi\phi\phi}(s,t) = \frac{g_B^2}{-\kappa} \frac{1+e^{i\pi\alpha(s)}}{Z(s,\kappa)} \left(\frac{t}{\kappa}\right)^{\alpha(s)} + \frac{g_B^2}{m^2-s}. \tag{6}$$

where the one-loop Regge trajectory is

$$\alpha(s) = \frac{g_B^2}{16\pi^2} \int_0^1 \frac{d\beta}{m^2 - \beta(1-\beta)s} - 1 \tag{7}$$

and, as mentioned, $Z(s,\kappa)$ must be determined via an appropriate experiment.

The above methodology is generalizable to much more complicated cases. Interestingly, even the simple case of an interaction $\phi^\dagger \phi \psi$ leads to quite complicated renormalization scenario where a matrix renormalization is necessary [24, 25] which leads to an extremely rich set of Regge trajectories. As Regge trajectories also give important information about the bound states of a theory the results of this section show that it is possible to use an environmentally friendly renormalization to access some aspects of the crossover between boundstates and unbound states.

3.1. Bound States

By developing an RG methodology for calculating Regge trajectories then, given the relationship between such trajectories and the existence of bound states in a field theory, it has been possible to use the RG in this extreme asymmetric limit to calculate properties of bound states [28, 29]. In [28] different charge sectors of a scalar theory with interaction $\phi^\dagger \phi \psi$ were considered and found to have a surprisingly rich bound state spectrum. The results were compared and contrasted with known results of the Bethe-Salpeter equation in the ladder approximation and, in the non-relativistic limit, with the corresponding Schrödinger equation. An advantage of using this methodology is that it preserves all relevant symmetries such as crossing symmetry and gauge invariance as well as being readily extendable to higher loop orders.

4. OTHER AREAS OF RESEARCH

A hallmark of research on the RG in Mexico has been its willingness to see the RG, in both its reparametrization and coarse graining guises, as a versatile and quite universally applicable tool for describing systems where collectivity is an important phenomenon.

Some areas outside of high energy physics that have been studied using the RG for the first time in Mexico are:

- Causal sets: Causal sets [26] offer an alternative potential model for quantum gravity based on an underlying finite set of space-time points as opposed to a continuum. In [27] it was shown how the "coupling constants" that parameterize the dynamics of a causal set are renormalized through cycles of expansion and contraction of the causal set universe. The consequent RG flows were analysed.
- Genetic Dynamics: In the dynamics of genetic systems composed of genes, the natural effective degrees of freedom are not individual genes on a single chromosome, but rather, due to the action of recombination in meiosis, are subsets of genes defined in a population that form "building blocks". These building blocks in their turn are composed of even more coarse grained blocks, which in their turn ... The associated coarse grainings form a semigroup - a RG [30, 31]. The formal solution of the associated RG equations allows for a much more intuitive and quantitatively feasible analysis of the dynamics.

5. CONCLUSIONS AND FUTURE PROSPECTS

Although RG theory is a relative newcomer to Mexico, there has been a great deal of progress in the last 10-15 years, both in terms of theoretical developments and applications. In the area of high energy physics, the main developments have been in finite temperature field theory and asymmetric momentum scaling limits in quantum field theory, i.e. the Regge limit. This work is novel in that the physics of systems as a function of momenta, both in critical phenomena and in quantum field theory, have traditionally been associated with RGs that depend on only one momentum invariant, e.g. momentum associated with the symmetric point. However, there are many phenomena that cannot be easily accessed by RG methods with this restriction due to the fact that they occur as a function of very asymmetric ratios of momenta. The paradigmatic example considered was that of physics in the Regge limit. An environmentally friendly renormalization capable of accessing this limit introduces several novel features, such as having to renormalize subparts of Greens or vertex functions rather than the functions themselves. This extreme kinematic limit also appears naturally in stellar collapse and black hole formation and, importantly, in turbulence where it has been identified as a possible mechanism originating anomalous scaling.

Another area of strong interest has been that of using more than one RG. One of the key elements of the RG methodology is to use it to map to a region of parameter space where a reliable calculation may be carried out starting from a region where standard approximation techniques are invalid. Physically, this often entails mapping from a region with a diverging length scale, such as the correlation length near a critical point, to a region where the length scale is small in an appropriate sense. A particular choice of sliding scale, or number of iterations of the RG map for a Wilsonian RG, is made to achieve this. In the case where a system may exhibit more than one diverging length scale however this artifice may become somewhat problematical as a matching can be made to one of the scales but the others are left as potentially dangerous. In the context of a reparametrization RG such problems can in principle be attacked by implementing more than one RG. The principal advantages of using more than one RG are: access to complementary information, as shown in the example of comparing finite temperature field theory with a running finite temperature mass RG and with a running temperature RG; less physical input and therefore more predictive power. This can be illustrated in the case of finite temperature QCD where with two RG's specification of the coupling at one momentum and one temperature was sufficient for the RG to be able to calculate the coupling at any other momentum or temperature. In the case of one RG it would be neccessary to have a line of initial conditions.

A description of QCD in the IR starting with the QCD Lagrangian in terms of quarks and gluons remains an unsolved problem. This is principally due to the fact that the effective degrees of freedom in the IR are baryons, mesons and glueballs — bound states of gluons and quarks. This type of crossover in the effective degrees of freedom also appears in other important problems such as superconductivity. In terms of a coarse graining RG what is required is an RG that coarse grains bound states at low energies and their constituents at high energies. As of yet such an RG has not been developed. For a reparametrization RG some progress can and has been made by exploiting the information on bound states contained in the Regge limit [28]. A direct access, however, remains very much an open problem.

ACKNOWLEDGEMENTS

I am grateful to many collaborators through the years, but especially Denjoe O'Connor.

REFERENCES

1. M. Mondragon, "SUSY and GUT's ", this volume.
2. R. Juarez and P. Kielanowski, "Applications of the renormalization group", this volume.
3. E.C.G. Stückelberg and A. Peterman, Helv. Phys. Acta **26**, 499 (1953).
4. K.G. Wilson and M.E. Fisher, Phys. Rev. Lett. **28**, 248, (1972).
5. Renormalization group theory in the new millennium Edited by D. O'Connor, C.R. Stephens Volume 344, Issue 4-6 (2001).
6. Renormalization group theory in the new millennium. II Edited by D. O'Connor, C.R. Stephens Volume 348, Issue 1-2 (2001).
7. Renormalization group theory in the new millennium. III Edited by D. O'Connor, C.R. Stephens Volume 352, Issue 4-6 (2001).
8. Renormalization group theory in the new millennium. IV Edited by D. O'Connor, C.R. Stephens Volume 363, Issue 4-6 (2002).
9. D. O'Connor and C.R. Stephens, Phys. Rev. Lett. **72**, 506, (1994).
10. D. O'Connor and C.R. Stephens, Int. J. Mod. Phys. **A9**, 2805 (1994).
11. F. Freire, D.J. O'Connor and C.R. Stephens, Phys. Rev. **E53**, 189 (1996).
12. D.J. O'Connor, C.R. Stephens and A.J. Bray, J. Stat. Phys. **87**, 273 (1997).
13. Denjoe O'Connor, C.R. Stephens and J.A. Santiago, Revista Mexicana de Física **48**, 300-306 (2002).
14. D.J. O'Connor, J.A. Santiago and C.R. Stephens, submitted to Phys. Rev. **B**, (2006).
15. M.A. van Eijck, D.J. O'Connor and C.R. Stephens, Int. J. Mod. Phys. **A10**, 3343 (1995).
16. F. Freire, D.J. O'Connor, C.R. Stephens y M.A. van Eijck, "Finite Temperature Renormalization Group Predictions: the critical temperature, exponents and amplitude ratios", Proceedings of Thermal Fields '95, 383-390, Edited by Y.X. Gui, F.C. Khanna, Z.B. Su. World Scientific (1996).

17. C.R. Stephens, "Environmentally Friendly Renormalization in Finite Temperature QCD", First Latin American Symposium on High Energy Physics and VII Mexican School of Particles and Fields, 559-563, AIP (1997).
18. F. Astorga, Environmentally Friendly Renormalization Group and Phase Transitions, arXiv preprint hep-th/0202004 (2004).
19. C.R. Stephens, A. Weber, P.O. Hess and F. Astorga, Phys. Rev. **D70**, 045024 (2004).
20. G. 't Hooft, in *"Salamfestschrift"*, eds. A. Ali, J. Ellis and S. Randjbar-Daemi, World Scientific (1993).
21. G. 't Hooft, Phys. Lett., **B198**, 61 (1987).
22. E. Verlinde and H. Verlinde, Nucl. Phys. **B371**, 246 (1992).
23. C.R. Stephens, A. Weber, J.C. López Vieyra and P.O. Hess, Phys. Lett. **B414**, 333 (1997).
24. C.R. Stephens, A. Weber, J.C. López Vieyra and P.O. Hess, Int. J. Mod. Phys. **A15**, 1773-1816 (2000).
25. A. Weber, J.C. López Vieyra, C.R. Stephens, S. Dilcher and P.O. Hess, Int. J. Mod. Phys. **A16**, 4377-4400 (2001).
26. L. Bombelli, J. Lee, D. Meyer and R.D. Sorkin, Phys. Rev. Lett. **59**, 521-524 (1987).
27. Xavier Martín, Denjoe O'Connor, David Rideout and Rafael D. Sorkin, On the "renormalization" transformations induced by cycles of expansion and contraction in causal set cosmology, arXiv:gr-qc/0009063 (2000).
28. A. Weber, J.C. López Vieyra, C.R. Stephens, S. Dilcher and P.O. Hess, Int. J. Mod. Phys. **A16**, 4377-4400 (2001).
29. A. Weber, Relativistic Bound States, AIP Conf. Proc. 670, 210-218 (2003).
30. C.R. Stephens, Acta Phys. Slov. **52**, 515-524 (2002).
31. C.R. Stephens, C. Chryssomalakos and A. Zamora, Rev. Mex. Fís. **50**, 388-396 (2004).

Dark Energy and Dark Matter

A. de la Macorra* and T. Matos[†]

Instituto de Física, UNAM
Apdo. Postal 20-364, 01000 México D.F., México [1]
[†]*Departamento de Física, Centro de Investigación y de Estudios Avanzados del IPN,*
A.P. 14-740, 07000 México D.F., México.

Abstract. We present the research carried out in México in the area of cosmology. In particular the contributions towards elucidating the nature and dynamics of dark energy and dark matter.

Keywords: Dark matter, dark energy, cosmology.
PACS: 98.20.Cq, 98.80.Es, 98.80.Jk, 04.20.-q, 04.50.+h

1. INTRODUCTION

In the past few years different observations have lead to conclude that the universe is flat and filled with an energy density with negative pressure generically called Dark Energy "DE". This energy density gives at present time 73% of the energy density of our universe. Besides DE the universe contains Dark Matter "DM" required by matter clustering. Neither DE nor DM are contained in the well established standard model of particle physics "SM" which represents only $(4-5)\%$ of the energy density of our universe.

The physical process that gives rise to dark matter and dark energy is yet unclear. It is therefore very interesting that cosmology rises today some of the most relevant questions in particle physics, namely, what is the nature of dark energy and dark matter, i.e. of 95% of the energy density of our universe.

This paper is organized as follows. In section 2 we present the works done in trying to understand the dynamics of dark matter from a phenomenologically and numerical point of view. In section 2.2 a new proposal in determining the Hubble constant using gamma ray burst is described. In section 3 we present research carried out in parameterizing dark energy and dark matter as scalar fields. We give in section 3.1 a general analysis on the cosmological evolution of scalar fields as dark energy, in section 3.2 a model of dark energy and dark matter is derived from particle physics (gauge theory), while in section 3.3 explicit examples of dark energy and dark matter as scalar fields are presented. In section 4 works on scalar tensor theories applied to cosmology are considered. We present in section 4.1 a proposal to explain an apparent galactic periodicity, in section 4.2 we discuss galaxy formation and dynamics, and finally in section 4.3 models of dark matter in the Newtonian limit and models of dark energy with inhomogeneities are presented.

---

[1] macorra@fisica.unam.mx

2. DARK MATTER AND DARK ENERGY PHENOMENOLOGY

2.1. DARK MATTER

One of the goals of the UNAM astrophysics group was to connect the predictions of the current cosmological paradigm of cosmic structure formation, the so-called Cold Dark Matter (CDM) model, with the properties of present-day disk galaxies -the most abundant in the Universe. By modelling in detail the formation and evolution of disk galaxies inside hierarchically growing CDM halos, several properties and correlations of this population of galaxies were predicted (e.g.,[1, 3],[13]). It was evidenced that the CDM model provides initial and boundary conditions to disk galaxy evolution, which allow to explain important observed properties and correlations of galaxies, for instance the nearly exponential mass surface density distribution and the nearly flat rotation curves of disk galaxies, as well as the observed tight luminosity-circular velocity correlation (Tully-Fisher relation) and its scatter. It was shown that even local properties, as the ones of the solar neighborhood in our Galaxy, can be well explained by the conditions provided by the CDM model [4].

An extensive work has been also done in the direction of predicting the properties of CDM halos as a function of their environment [2]. The results obtained in these works carried out with cosmological N-body simulations, suggest that the CDM model could also be at the basis of the well known dependence of galaxy properties on environment.

The UNAM astrophysics group has also explored the consequences of modifying the CDM model at small scales, where apparently some potential difficulties arise. They were among the first groups in proposing weak self-interacting CDM particles in order to reduce the high inner mass concentration predicted for the CDM halos [9, 10] . First empirically, and then by means of high-resolution cosmological N-body simulations [7], they concluded that the cross-section per unit of dark particle mass should be inversely proportional to the interaction velocity.

Other alternative studied by this group was Warm Dark Matter (WDM). By means of high-resolution N body simulation, they found that the amount of substructure (sub-halos) inside Milky-Way-size halos decrease from a very high number, when CDM is used, to a low number, comparable with the observed abundance of satellites galaxies, when WDM, with particle masses of 1-2 KeV, is used [6]. The inner structure of the halos remains roughly the same in both cases, thought the halos are less concentrated for WDM particles [4]. These works prompted the intensive research of WDM particles, like sterile neutralinos, from the point of view of particle physics.

2.2. DARK ENERGY

The UNAM astrophysics group, in collaboration with INAF-OAB (Italy) researchers, has introduced a new kind of standard candle able to extend the Hubble diagram to redshifts much higher than those of type Ia supernovae. With this new cosmological probe, tight constraints to the dark energy (DE) parameters can be obtained, because some model degeneracies are broken due to the high redshift range. The new standard candles are the Gamma Ray-Bursts (GRBs) with measured redshift, which obey tight correlations among their energetics and some observable properties. The group has developed a Bayesian approach to calibrate these correlations and at the same time to constrain cosmological parameters using the Hubble diagram [11]. Besides, a new improved correlation has been discovered by Firmani et al. [12]. First cosmological constraints based on this correlation were reported for GRBs alone [13] and for the combination of GRBs and SNIa [14]. The results are in excellent agreement with the CDM concordance cosmological model (minimal case), i.e. the cosmological constant as DE. Models that imply an evolving equation-of-state parameter for the DE are not favored.

3. SCALAR FIELDS IN COSMOLOGY

3.1. General Analysis of Scalar Fields

At the Instituto de Física (UNAM), there has been considerable effort to work on problems in the interface between particle physics and cosmology [17]-[31]. In particular on the nature of dark energy. Perhaps, DE is best described in terms of scalar fields. The evolution of scalar fields has been widely studied and some general approaches can be found in [17]. It is shown that a scalar field with potential V only leads to a late time acceleration (i.e. dark energy) only if $V'/V \to const < \sqrt{3}$ or $V'/V \to 0$ with a non oscillating behavior [17].

3.2. Dark Group: Dark Energy and Dark Matter

Applications of scalar fields in cosmology derived form gauge field theory can be found in [17]-[31]. A very interesting model of dark energy and dark matter derived from particle physics (field and gauge theory) was proposed in [20, 22, 26, 31]. The starting point of this model is a dark gauge group "DG" whose particles interact with the standard model "SM" only via gravity. The dark energy model is a $SU(N_c = 3)$ gauge group with $N_f = 6$ elementary particles in the fundamental representation and with only gravitational interaction with the standard model of particle physics. Besides the choice of $N_c = 3$ and $N_f = 6$ there are no other free parameters (notice that it is at the same footing as the SM since there is no explanation in the SM of the choice of gauge groups nor why there are three families).

At high energies the dark elementary fields are massless and the energy density redshifts as $\rho_{DG} \propto a^{-4}$. The ratio ρ_{DG}/ρ_r is therefore constant and it is given only in terms of the number of particles. The gauge coupling constant becomes strong at lower energies, i.e. the gauge group is asymptotically free (as the strong force given by QCD). At low energies a phase transition takes place due to a strong gauge coupling constant. At this scale the dark elementary fields are bound together producing gauge invariant states. The relevant scale for this process is the condensation scale Λ_c and for a gauge group $SU(N_c)$ with N_f matter fields it is given by the one-loop renormalization group equation [20] $\Lambda_c = \Lambda_{gut} e^{-8\pi^2/b_o g_{gut}^2}$ where $b_o = 3N_c - N_f$ is the one-loop beta function and $\Lambda_{gut} \simeq 10^{16} GeV, g_{gut}^2 \simeq 4\pi/25.7$ are the unification energy scale and coupling constant, respectively. Motivated by grand unification theories and by string theory we have constrained the dark gauge group and Λ_{DE} to be unified with the standard model gauge groups at the unification scale. Our dark energy model has $N_c = 3, N_f = 6$ giving $b_o = 3$ and $\Lambda_c = 42\,eV$. Strong gauge interactions produce a non-perturbative scalar potential V below Λ_{DE}. This potential can be calculated using Affleck-Dine-Seiberg potential. The superpotential for a non-abelian $SU(N_c)$ gauge group with N_f massless fields can be calculated and the resulting scalar potential in SUSY for one dynamical meson field $\phi^2 \equiv \langle Q\tilde{Q} \rangle$ (Q are the fundamental fields of the original gauge group) which represents a pseudo-Goldstone boson, is $V = \Lambda_c^{4+n} \phi^{-n}$ with the exponent of ϕ given by $n = 2[1 + 2/(N_c - N_f)] = 2/3$ [20, 22, 31].

In order to study the cosmological evolution of ϕ with potential V in a Friedman-Robertson-Walker metric the initial value on ϕ must be chosen. Since the elementary fields are bounded at the condensation scale Λ_{DE}, when the gauge coupling constant becomes strong, it is this scale the relevant physical scale and we therefore take $\phi_i = \Lambda_c$. With this choice the initial mass of ϕ is $m^2 = \partial^2 V/\partial \phi^2 \simeq \Lambda_{DE}^2$ which sets the correct mass scale for the process. Notice that in QCD the mass of the pion and proton is of the same order of magnitude as the QCD scale $\Lambda_{QCD} \simeq 200 MeV$. Global symmetries and SUSY protect the mass of the quintessence field ϕ. In fact the ADS superpotential is exact and receives no corrections. The energy density ρ_{DG} tracks radiation for a long period of time, including nucleosynthesis "NS" epoch, since all the particles are massless. The onset of the quintessence field is at a very late time ($a_c \simeq 10^{-6}$) and close to matter-radiation equality ($a_{eq} \simeq 10^{-4}$). The fact that Λ_{DE} is so small and the appearance of the quintessence field is at such a late time solves the coincidence problem since it implies that an accelerating universe will necessarily be at a scale factor larger than a_c and a_{eq}, i.e. close to present time.

If we have $N_c < N_f$ then on top of the gauge singlet meson fields we can have gauge singlet dark baryons $B^{i,...,N_c} = \prod_i^{N_c} Q^i$ and anti baryons. These particles get a non-vanishing mass due to non-perturbative effects (like protons and neutrons in QCD). These baryons could give the dark matter of the universe [26]. The order of magnitude of the mass of the DM particle can be estimated by the condensation $m = c\Lambda_{DE}$ with $c = O(1-10)$ a constant (in the case of QCD $m_{proton}/\Lambda_{QCD} = 4.5$). From cosmology it is know that $\Omega_{DMo} = 0.25 \pm 0.05$. From entropy conservation Ω_{DM} can be determined giving $\Omega_{DMo} = 0.25(0.74/h_o)^2 (m/g_{dec}\,eV)$ [26]. So a dark matter mass of $m \simeq g_{dec}\,eV$ is of the same order of magnitude ($g_{dec} = 106, 228$ for the SM and SUSY-SM, respectively) as the mass given by gauge theory dynamics $m = c\Lambda_{DE} = c\,42\,eV$. Comparing the model with the data (SN1a Golden set and WMAP1) it has an excellent fit and equivalent to ΛCDM giving a $\chi^2/d.o.f. = 1.08$ [31].

3.3. Scalar Fields, Dark Energy and Dark Matter

There is a group at Centro de Investigación y de Estudios Avanzados del IPN, in México, which is working in problems of dark matter and dark energy. This group has produced research in dark matter and dark energy since 1999. They have worked on different research lines, specially on the nature of the cosmological and galactic dark matter and on the nature of dark energy. The first work was a proposition about the nature of the dark matter in galaxies [32]. In this work they argue that all unification theories, from the standard model of physics to the superstrings theory, need to postulate scalar fields somehow, in order to have certain consistency. Therefore they propose to investigate the hypothesis that the dark matter is of scalar field nature, calling it the Scalar Field Dark Matter (SFDM) model. In [32] they show that using a scalar field with an exponential potential, the rotation curves of galaxies can be fitted well. In a further paper [33] they investigate the hypothesis that not only the dark matter, but also the dark energy are provided by the same scalar field. They came to the conclusion that this is not possible; later it was shown that a complex scalar field as dark matter in cosmos is also ruled out [35]. Therefore, in [34] they investigate a model with two scalar fields, the first one contains a cosh potential, it is interpreted as the dark matter, and the second one contains a sinh potential and it is interpreted as the dark energy, a quintessence field. This last hypothesis, consisting in studying a quintessence field with a sinh potential, was introduced first by Luis Ureña and Tonatiuh Matos in [36]. This potential is one of the most popular potentials for quintessence in the literature. The study of the perturbations and the structure formation in the linear regime of the two scalar fields model was investigated in [37]. They conclude that the Λ Cold Dark Matter (ΛCDM) model and the SFDM model give exactly the same predictions for the structure formation of the universe at cosmological level. They concluded that if there is a difference between the ΛCDM model and the SFDM model it must be at galactic level. One interesting feature of this model, is that the free constants of the model, all of them, were fixed using cosmological observations. In [38] the group of researchers was extended to UNAM, and together they investigated the behavior of the collapse of a scalar field, in order to understand it in the non-linear regime of fluctuations. The result was a big surprise, they found that using the values of the fixed free parameters of this model, the scalar field will collapse with a critical mass $M \sim 10^{12} M_\odot$, which implies that this model could be able to explain the structure formation in cosmos and also the galaxy formation. In 1990's it was proved that the cosh potential is a renormalizable potential, with a renormalization scale $\Lambda \sim 2 M_{Planck}$. This fact gives the possibility that the scalar field has a fundamental origin or that the scalar field goes beyond the Planck era. In [40] two members of the group have summarized the results and the main ideas about the hypothesis of the SFDM model at cosmological level.

Other branch of research of this group is the study of the dark matter in galaxies. In [41], they studied the general conditions of a metric in general relativity, in order to reproduce the rotation curves of galaxies. They gave a general metric in terms of the rotations velocity, using the geodesics of the metric. This study was applied to many systems and proposals of dark matter in [42]. One of the fields which was able to reproduce the behavior of the galaxies was just the scalar field, as proposed in [32]. In [44] it was proposed that the scalar field could have other kind of scalar field potentials. So, in [45] general statements and conditions for the formation of a scalar field halo were established. Some galaxies were fitted using this hypothesis (see [43]); it was then clear that for big galaxies, it was possible to explain their rotation curves, starting with the hypothesis of a scalar field as dark matter in galaxies. Using n-body simulations it was also possible to reproduce the baryonic part of a galaxy formation, using as a background the gravitational potential of a scalar field [46]. The question about the nature of this scalar field is still open, nevertheless, in [47] the hypothesis that the dark matter and the quintessence fields could be the same field was investigated and discarded in [48]. Within this hypothesis, the first investigations on the collapse of the scalar field as dark matter in galaxies were done in [38], and a further investigation on the cooling conditions [49] and stability [50], have shown that scalar fields collapse very early in the universe and are very stable. The authors have shown in these papers that the scalar field virialize very fast and remains oscillating for a very long time. This is the reason of the name of these objects, they are called oscillatons. These oscillatons have very interesting features, but the most important one is that they have a flat density profile, resembling the real halo of a galaxy. In [51], they show, using some approximations, that the central density profile of an oscillaton is flat. This fact is confirmed in [52], using many quantum states of the Bose-Einstein Condensate. This feature is maybe one the most important ones of the SFDM hypothesis. Observations show that dwarf galaxies have a flat density profile, and the traditional ΛCDM paradigm contains a cusp density profile, in contradictions with observations. Here, the SFDM model show an advantage over the ΛCDM paradigm. Other advantage of the SFDM model over the ΛCDM paradigm is that the SFDM model contains a natural cut of the mass power spectrum, causing a suppression of the formation of small galaxies. This prediction of the SFDM model is confirmed by observations in satellite galaxies. Observations show that there are very few satellite galaxies around the Milky way, in contradiction with the ΛCDM paradigm, which predicts more than 10 times in excess [37]. Other interesting feature of the SFDM model is that in the presence of a supermassive black hole, these oscillatons are swallow by the black hole as expected, but in [53] the authors show that the rate of scalar field matter that is swallowed by the black hole is of order of a planet per year. Too small to be significant. That means that scalar field can live together with a supermassive black hole in the galaxy.

Other line of research of the groups in Cinvestav and UNAM is the study of general properties of the halos of galaxies, using a more powerful tool, i.e., general relativity. Using the fact that the Navarro-Frenck-White (NFW) is the profile derived from n-body simulations for the halo of a galaxy, this group gives a general relativistic analysis of this profile. First they found the space-time metric arising from this profile [54]. They argue that we do not know the nature of the dark matter, thus it is necessary to use the best tool for analyzing the gravitational fields. Second, they study the thermodynamic of this space-time [55] and obtain some conclusions about the thermodynamical nature of dark matter, if the density profile is of the NFW (see also [56]). It is also possible to carry out the thermodynamical analysis of the halo of a galaxy [57] (see also [58]), without supposing any special profile. With this analysis it is possible to constrain the mass of the dark matter particles. These criteria are able to give some constrains on the central temperature of the halo. There are some other alternative candidates studied by the Mexican groups. Some examples are stellar mini-Machos [59], which consists in small scalar field stars which behave like dark stars. This hypothesis do not violate the Big Bang Nucleosyntheses constrains, because the minimachos are not baryonic. So, the minimachos can be stars around the galaxy. Nevertheless, microlensing Macho observations constrain a lot this hypothesis. Other hypothesis under research by the

Mexican groups is a thermodynamic study of galaxies, but using the alternative entropy introduced by Tsallis [60],[61], [62], where the non-local character of it enables to fit the rotation curves of galaxies. A further alternative is a non-linear theory of relativity: in [63] the authors show that with this theory it is possible to recover the results of SFDM in some cases. A review on Structure Formation in spanish can be found in [64].

In the context of the Dark Energy, the group at Cinvestav proposed a model of quintessence using a sinh scalar field potential. This is convenient because the sinh behaves as power law potential for large scalar field values and this permits to fit well the Big Bang Nucleosynthesis constrains. On the other hand, for small scalar field values, the sinh potential is an exponential and this behavior permits to fit very well the SNIa constrains. This potential has received a good acceptance in the cosmology community. There is also the study of the dynamics of a scalar field with a negative kinetic term, the so called phantom field [65].

Other line of research is the hypothesis that the inflaton is at the same time the dark matter or the dark energy or both. This is possible in the context of Brane cosmology. Using an exponential potential in the Brane cosmology paradigm, in [66] the authors show that an inflaton field can remain as dark matter after inflation. One problem they have to face is that the scalar field can not reheat the universe, as in the traditional inflation paradigm. Therefore they reheat the universe supposing that a fraction of 10^{-16} of the scalar field energy density becomes primordial black holes. These black holes can evaporate by Hawking radiation producing a huge ammount of heat, enough to reheat the universe. The primordial black holes burn before BBN. Other interesting feature is that the branes provoke a quadratic term of the density in the Friedmann equation. This term dominates the evolution and inflates the universe. After inflation the density becomes very small and the linear term in the Friedmann equation dominates. If one starts with a normal non-inflationary potential, this causes a natural gracefull exit, an exit of the inflationary epoch, when the universe passes from the quadratic to the linear behavior of the density. The value for the free constants used in this model are the ones used in the SFDM model, in such a way that the SFDM can inflate the universe in the presence of branes. In [67], solutions for different potentials in this paradigm were found, in all of them the gracefull exit was confirmed. This show that the inflaton in the brane paradigm can exit from inflation easily and can become dark matter in an easy way. In [68], another way to obtain reheating was studied in the framework of brane cosmology. A further investigation starts from string theory. Using the Landscape string theory and a quadratic potential plus a constant, in [69] it was shown that this potential can produce the inflaton together with the dark matter and that the constant can be the cosmological constant. Thus, with this potential in the Landscape theory it is possible to have it all: inflation, dark matter and dark energy. In the same way, starting with the IIB superstring theory, compactifying on an orbifold with 32 fluxes, the dilaton acquires an effective potential, which has a cosh term and a cosmological constant. With this model, in [70] (see also [71]) the authors recover the ΛCDM model, but because of the presence of the branes, the dilaton inflates the universe as well. This is the first time that a string theory makes contact with reality, the first time that this theory has a phenomenology. The result is that there are some differences between string theory and the ΛCDM model. The epoch of recombination is very different in both models, in the ΛCDM model the density rates evolve very smoothly, while in the string theory models they oscillate.

Other difference is that the density profile of galaxies is cusp in the ΛCDM model and flat, as observed, in the string theory models. This differences can be used to ruled out one of these models.

Some reviews of the SFDM model can be found in [72], [73], [74], [75], [76], [77] and [78], in different contexts. Some reviews for no-specialists (in spanish) can be found in [79] and [80].

4. COSMOLOGY WITH SCALAR-TENSOR MODELS

4.1. Galactic Periodicity

In the work of [81]-[86] they propose a scalar-tensor system, by including a scalar field non minimal coupled to the curvature. In this model the resulting gravitational constant varies with time and allows for a possible explanation the apparent galactic periodicity of 128Mpc detected at the early 1990. This model is compatible with the bound on primordial nucleosynthesis as well as with solar system bounds on time variation of coupling constants. Furthermore, the model allows for an accelerating universe at late times without the need to introduce a cosmological constant.

4.2. Galaxy formation and dynamics

Within the context of galaxy formation and galaxy dynamics several works have been presented [87]-[91] and make use of scalar tensor theories. These works can allow for a better understanding of the nature and dynamics of dark matter.

In [87], simulations within the framework of scalar-tensor theories, in the Newtonian limit, to investigate the influence of massive scalar fields on the dynamics of the collision of two equal spherical clouds are presented. They employ a SPH code modified to include the scalar field to simulate two initially non-rotating protogalaxies that approach each other, and as a result of the tidal interaction, an intrinsic angular momentum is generated. They have obtained sufficient large values of J/M to suggest that intrinsic angular momentum can be the result of tidal interactions. In [88], a family of potential density pairs has been found for spherical halos and bulges of galaxies within the Newtonian limit of scalar tensor theories of gravity. The scalar field is described by a Klein Gordon equation with a source, that is coupled to the standard Poisson equation of Newtonian gravity. The net gravitational force is given by two contributions: the standard Newtonian potential plus a term stemming from massive scalar fields. General solutions have been found for spherical systems. In particular, the authors compute potential–density pairs of spherical, galactic systems, and some other astrophysical quantities that are relevant to generate initial conditions for spherical galaxy simulations. In [89], they present a formulation for potential–density pairs to describe axi–symmetric galaxies in the Newtonian limit of scalar–tensor theories of gravity. The scalar field is described by a modified Helmholtz equation with a source that is coupled to the standard Poisson equation of Newtonian gravity. The net gravitational force is given by two contributions: the standard Newtonian potential plus a term stemming from massive scalar fields. General solutions have been found for axisymmetric systems and the multipole expansion of the Yukawa potential is given. In particular, the authors have computed potential density pairs of galactic disks for an exponential profile and their rotation curves. In [90], they study the dynamics of spherical galaxies with the Navarro-Frenk-White (NFW) density profile within the Newtonian limit of scalar-tensor theories of gravity. The scalar field is described by a modified Helmholtz equation with a source and it is coupled to the Poisson equation of standard Newtonian gravity. The net gravitational force is given by two contributions: one coming from the standard Newtonian potential and other coming from the massive scalar fields. The authors found general solutions for spherical systems, and in particular, they obtain results for the potential density pairs and other relevant quantities of galactic spherical systems with the NFW density profile. Finally, in [91], the joint influence of numerical parameters such as the number of particles N, the gravitational softening length $\dot{a}$ and the time-step t is investigated in the context of galaxy simulations. For isolated galaxy models the authors have performed a convergence study and estimated the numerical parameters ranges for which the relaxed models do not deviate significantly from its initial configuration. By fixing N, they calculate the range of the mean interparticle separation $\ddot{e}(r)$ along the disc radius. Uniformly spaced values of $\ddot{e}$ are used as $\dot{a}$ in numerical tests of disc heating. They have found that in the simulations with N = 1 310 720 particles $\ddot{e}$ varies by a factor of 6, and the corresponding final Toomre's parameters Q change by only about 5 per cent. By decreasing N, the $\ddot{e}$ and Q ranges broaden. Large $\dot{a}$ and small N cause an earlier bar formation. In addition, the numerical experiments indicate, that for a given set of parameters the disc heating is smaller with the Plummer softening than with the spline softening. For galaxy collision models we have studied the influence of the selected numerical parameters on the formation of tidally triggered bars in galactic discs and their properties, such as their dimensions, shape, amplitude and rotational velocity. Numerical simulations indicate that the properties of the formed bars strongly depend upon the selection of N and $\dot{a}$. Large values of the gravitational softening parameter and a small number of particles results in the rapid formation of a well defined, slowly rotating bar. On the other hand, small values of $\dot{a}$ produce a small, rapidly rotating disc with tightly wound spiral arms, and subsequently a weak bar emerges. The authors have found that by increasing N, the bar properties converge and the effect of the softening parameter diminishes. Finally, in some cases short spiral arms are observed at the ends of the bar that change periodically from trailing to leading and vice-versa the wiggle.

4.3. Newtonian Limit and Inhomogeneities

At the Instituto de Ciencias Nucleares, UNAM, research on dark matter has also been carried out [92]-[101]. In particular the studies on galactic dark matter can be placed in the context of incorporating models of galactic CDM halos in the framework of the Newtonian limit of General Relativity, as opposed to a purely Newtonian construction. They have studied the effects of the cosmological constant in the hydrostatic equilibrium of a CDM gas and they have also proposed a Relativistic Kinetic Theory approach to a collisionless CDM gas in which the particles are subjected to a generalized vector force interaction (as opposed to be free falling). In either case, the Newtonian limit obtained from the strict general relativistic approach yields small (but significant) differences with a purely Newtonian framework. In other work they propose a CDM model in which the "particles" are not the usual supersymmetric neutralinos but asteroid sized "mini-MACHO's" formed by condensations of primordial scalar fields (larger MACHO's would have been detected by lensing). Current N-body numerical simulations cannot distinguish this change of granularity of CDM.

On the issue of dark energy, they have studied simple semi-analytical models of self-gravitating dark matter and dark energy in inhomogeneous conditions. These models allow for minimal and non-minimal coupling between DM and DE. The non-minimal coupling can be understood within the framework of scalar-tensor theories. These models illustrate how cosmological observational parameters, associated with a cosmic scale (> 300 Mpc), are affected by inhomogeneity and anisotropy in scales smaller than the homogeneity scale [92]-[101].

REFERENCES

1. V. Avila-Reese, C. Firmani and X. Hernandez, "On formation and evolution of disk galaxies: cosmological initial conditions and the gravitational collapse," *Astrophys. J.* **505**, 37 (1998), astro-ph/9710201.
2. V. Avila-Reese, C. Firmani, A. Klypin and A.V. Kravtsov, "Density profiles of dark matter haloes: diversity and dependence on environment," *Mon. Not. Roy. Astron. Soc.* **309**, 507 (1999), astro-ph/9906260.
3. V. Avila-Reese and C. Firmani, *Rev. Mex. AA.* **36**, 23 (2000).
4. V. Avila-Reese et al., *Ap. J.* **559**, 516 (2001).
5. V. Avila-Reese, P. Colin, S. Gottlober, C. Firmani and C. Maulbetsch, "The Dependence on environment of cold dark matter halo properties," *Astrophys. J.* **634**, 51-69 (2005), astro-ph/0508053.
6. P. Colin, V. Avila-Reese and O. Valenzuela, "Substructure and halo density profiles in a warm dark matter cosmology," *Astrophys. J.* **542**, 622-630 (2000), astro-ph/0004115.
7. P. Colin, V. Avila-Reese, O. Valenzuela and C. Firmani, "Structure and subhalo population of halos in a selfinteracting dark matter cosmology," *Astrophys. J.* **581**, 777-793 (2002), astro-ph/0205322.
8. C. Firmani and V. Avila-Reese, "Disc galaxy evolution models in a hierarchical formation scenario: structure and dynamics," *Mon. Not. Roy. Astron. Soc.* **315**, 457 (2000), astro-ph/0001219.
9. C. Firmani, E. D'Onghia, V. Avila-Reese, G. Chincarini and X. Hernandez, "Evidence of self-interacting cold dark matter from galactic to galaxy cluster scales," *Mon. Not. Roy. Astron. Soc.* **315**, L29 (2000), astro-ph/0002376.
10. C. Firmani, E. D'Onghia, G. Chincarini, X. Hernandez and V. Avila-Reese, "Constraints on dark matter physics from dwarf galaxies through galaxy cluster haloes," *Mon. Not. Roy. Astron. Soc.* **321**, 713 (2001), astro-ph/0005001.
11. C. Firmani, G. Ghisellini, G. Ghirlanda and V. Avila-Reese, "A New method optimized to use Gamma-Ray Bursts as cosmic rulers," *Mon. Not. Roy. Astron. Soc.* **360**, L1 (2005), astro-ph/0501395.
12. C. Firmani et al., *Mon. Not. Roy. Astron. Soc.* (2006), in press, astro-ph/0605073.
13. C. Firmani et al. (2006), astro-ph/0605267.
14. C. Firmani et al., *Mon. Not. Roy. Astron. Soc.* (2006), in press, astro-ph/0605430.
15. X. Hernandez, V. Avila-Reese, and C. Firmani, "A cosmological study of the star formation history in the solar neighbourhood," *Mon. Not. Roy. Astron. Soc.* **327**, 329 (2001), astro-ph/0105092.
16. J. Zavala, V. Avila-Reese, H. Hernández-Toledo and C. Firmani, " The luminous and dark matter content of disk galaxies," *Astron. Astrophys.* **412**, 633 (2003), astro-ph/ 0305516.
17. A. de la Macorra and G. Piccinelli, "General scalar fields as quintessence," *Phys. Rev. D* **61**, 123503 (2000), hep-ph/9909459.
18. A. de la Macorra, "Can moduli fields parameterize the cosmological constant?," *Int. J. Mod. Phys. D* **11**, 653-668 (2002), hep-ph/9910330.

19. A. de la Macorra, "Model independent accelerating universe and the cosmological coincidence problem," *Int. J. Mod. Phys. D* **9**, 661-668 (2000), astro-ph/9911079.
20. A. de la Macorra and C. Stephan-Otto, "Natural quintessence with gauge coupling unification," *Phys. Rev. Lett.* **87**, 271301 (2001), astro-ph/0106316.
21. A. de la Macorra and C. Stephan-Otto, "Quintessence restrictions on negative power and condensate potentials," *Phys. Rev. D* **65**, 083520 (2002), astro-ph/0110460.
22. A. De la Macorra, "Quintessence unification models from nonAbelian gauge dynamics," *JHEP* **0301**, 033 (2003), hep-ph/0111292.
23. A. de la Macorra and G. German, "Acceptable IPL quintessence with n = 18/7," *Phys. Lett. B* **549**, 1-6 (2002), astro-ph/0203094.
24. A. de la Macorra, "Implications to CMB from model independent evolution of W and late time phase transition," *Phys. Rev. D***67**,103511 (2003), astro-ph/0211519.
25. A. de la Macorra and H. Vucetich, "Causality, Stability And Sound Speed In Scalar Field Models," *JCAP* **09**, 012 (2004).
26. A. de la Macorra, "Dark group: dark energy and dark matter," *Phys. Lett. B* **585**,17-23 (2004), astro-ph/0212275.
27. A. de la Macorra, "Late time phase transition as dark energy," AIP Conf. Proc. 670, 272-280 (2003), *Pramana* **62**, 779-783 (2004).
28. A. de la Macorra, "Quintessence: A natural model to parameterize the cosmological constant," *Rev. Mex. Fis* **49S2**, 80-84 (2003).
29. A. de la Macorra and G. German, "Cosmology with negative potentials with w(phi) < -1," *Int. J. Mod. Phys. D* **13**, 1939-1953 (2004).
30. A. de la Macorra, "Quintessence and dark energy," *Lect. Notes Phys.* **646**, 225-257 (2004).
31. A. de la Macorra, "A Realistic particle physics dark energy model," *Phys. Rev. D* **72**, 043508 (2005), astro-ph/0409523.
32. F. S. Guzmán and T. Matos, "Scalar Fields as Dark Matter in Spiral Galaxies," *Class. Quant. Grav.* **17**, L9-L16 (2000), gr-qc /9810028. Highlight paper at the period 1999-2000, nominated by the Editorial Board of the Journal Classical and Quantum Gravity.
33. T. Matos and L. A. Ureña. "Quintessence and Scalar Dark Matter in the Universe," *Class. Quant. Grav.* **17**, L75-L81 (2000), astro-ph/ 0004332.
34. T. Matos, F. S. Guzmán and L. A. Ureña, "Scalar Fields as Dark Matter in the Universe," *Class. Quant. Grav.* **17**, 1707-1712 (2000), astro-ph/ 9908152.
35. L. A. Ureña and T. Matos, "Complex Scalar Field Dark Matter," In Proceedings of the IX Marcel Grossman Meeting, Roma, Italia, Ed. by R. Ruffini et. al., Word Scientific Singapore, (2002), p. 2049.
36. L. A. Ureña and T. Matos, "New Cosmological Tracker Solution for Quintessence," *Phys. Rev. D* **62**, 081302(R) (2000), astro-ph/0003364.
37. T. Matos and L. A. Ureña, "Further Analysis of a Model with Quintessence and Scalar Dark Matter," *Phys. Rev. D* **63**, 063506 (2001), astro-ph/ 0006024.
38. M. Alcubierre, F. S. Guzmán, T. Matos, D. Nuñez, L. A. Ureña and P. Wiederhold, "Galactic Collapse of Scalar Field Dark Matter," *Class. Quant. Grav.* **19**, 5017-5024 (2002), gr-qc/0110102. Highlight paper at the period 2001-2002, nominated by the Editorial Board of the Journal Classical and Quantum Gravity.
39. T. Matos and L. A. Ureña. "Scalar Field Dark Matter, Cross Section and Planck-Scale Physics," *Phys. Lett. B* **538**, 246-250 (2002), astro-ph/ 0010226.
40. T. Matos and L. A. Ureña-Lopez. "On the Nature of Dark Matter," *Int. J. Mod. Phys. D* **13**, 2287-2291 (2004), astro-ph/0406194. Essay selected to receive Honorable Mention from the Gravity Research Foundation, USA.
41. T. Matos, F. S. Guzmán and D. Nuñez. "Spherical Scalar Field Halo in Galaxies," *Phys. Rev. D* **62**, 061301(R) (2000), astro-ph/ 0003398.
42. T. Matos, F. S. Guzmán, D. Nuñez and E. Ramirez, "Geometric Conditions on the Type of Matter Determining the Flat Behavior of the Rotational Curves in Galaxies," *Gen. Rel. Grav.* **34**, 283-306 (2002), astro-ph/ 0005528.
43. F. S. Guzmán, H. Villegas-Brena and T. Matos, "Scalar Fields as Dark Matter in Spiral Galaxies. Comparison with Experiments," *Astron. Nachr.* **320**, 97-104 (1999).
44. T. Matos and F. S. Guzmán, "Scalar Dark Matter in Spiral Galaxies," *Rev. Mex. A. A.* **37**, 63 -72 (2001), astro-ph/ 9811143.

45. T. Matos and F. S. Guzmán, "On the Space-time of a Galaxy," *Class. Quant. Grav.* **18**, 5055-5064 (2001), gr-qc / 0108027.
46. T. Matos and G. Torres, "Galaxies Formation Simulations with Scalar Field Dark Matter Haloes," *Rev. Mex. A. A.* **39**, 113-118 (2003).
47. T. Matos and F. S. Guzmán, "Quintessence at Galactic Level?," *Ann. Phys. (Leipzig)* **9**, S133-S136 (2000), astro-ph/ 0002126.
48. T. Matos, D. Nuñez, F. S. Guzmán and E. Ramrez, "Quintessence-like Dark Matter in Spiral Galaxies," *Rev. Mex. Fis.* **49**, 203-206 (2003), astro-ph/ 0003105.
49. F.S. Guzmán, L. A. Ureña-Lopez, "Newtonian collapse of scalar field dark matter," *Phys. Rev. D* **68**, 024023 (2003), astro-ph/0303440.
50. F. S. Guzmán, L. A. Ureña-Lopez, "Gravitational cooling of self-gravitating Bose-Condensates," *The Astrophysical Journal* (2006), in press, astro-ph/0603613.
51. A. Bernal, T. Matos and D. Nuñez, "Flat Central Density Profiles from Scalar Field Dark Matter Halo," astro-ph/ 0303455.
52. T. Matos and L. A. Ureña-Lopez, "Flat Rotation Curves in Scalar Field Galaxy Halos," *Gen. Rel. Grav.* **38** (2006), in press.
53. L. A. Ureña-Lopez and A. R. Liddle, "Supermassive black holes in scalar field galaxy halos," *Phys. Rev. D* **66**, 083005 (2002), astro-ph/ 0207493.
54. T. Matos and D. Nuñez, "The general relativistic geometry of the Navarro-Frenk-White model," *Rev. Mex. Fis.* **50**, 71-75 (2005), astro-ph/ 0303594.
55. T. Matos, R. A. Sussman and D. Nuñez, "A general relativistic approach to the Navarro-Frenk-White (NFW) galactic halos," *Class. Quant. Grav.* **21**, 5275-5293 (2004), astro-ph/ 0410215.
56. T. Matos, D. Nuñez and R. A. Sussman "The Spacetime associated with Galactic Dark Matter Halos," *Gen. Rel. Grav.* **37**, 769-779 (2005), astro-ph/ 0402157.
57. L. G. Cabral-Rosetti, T. Matos, D. Nuñez and R. A. Sussman, "Hydrodynamics of Galactic Dark Matter," *Class. Quant. Grav.* **19**, 3603-3615(2002), gr-qc/ 0112044.
58. L. G. Cabral-Rosetti, D. Nuñez, R. A. Sussman and T. Matos, "Hydrodynamical description of Galactic Dark Matter," *Rev. Mex. Fis.* **49**, S91-S97 (2003), hep-th/ 0206082.
59. X. Hernández, T. Matos, R. A. Sussman and Y. Verbin, "Scalar Field MACHOS: a new approach to Galactic Dark Matter," *Phys. Rev. D* **68**, 043537 (2004), astro-ph/ 0407245.
60. L. G. Cabral-Rosetti, T. Matos, D. Nuñez, R. A. Sussman and J. Zavala, "Fitting Stelar Polytropes to Navarro-Frenk-White Dark Matter halos: a connection to Tsallis Entropy," astro-ph/0405242.
61. J. Zavala, D. Nuñez, R. A. Sussman, L. G. Cabral-Rosetti and T. Matos, "Stellar polytropes and NFW halo model: Stellar polytropes and Navarro-Frenk-White halo models: comparison with observations," *JCAP* **0606**, 008 (2006), astro-ph/ 0605665.
62. D. Nuñez, R. A. Sussman, J. Zavala, L. G. Cabral-Rosetti and T. Matos, "Empirical testing of Tsallis' Thermodynamics as a model for dark matter halos." In Proceedings of X Mexican Workshop on Particles and Fields, Morelia Michoacan, Mexico, November 7-12, 2005, astro-ph/0604126.
63. O. Obregón, L. A. Ureña-Lopez and F. E. Schunck, "Oscillatons formed by nonlinear gravity," *Phys. Rev. D* **72**, 024004 (2005), gr-qc/ 0404012
64. T. Matos, "Formación de Estructura en el Universo," *Rev. Mex. Fis.* **49**, S16-S25 (2003).
65. L. A. Ureña-Lopez, "Scalar phantom energy as a cosmological dynamical system," *J. Cosmology and Astrophysics* **0509**, 013 (2005), astro-ph/ 0507350.
66. J. Lidsey, T. Matos and L. A. Ureña, "The Inflaton Field as Self-Interacting Dark Matter in the Braneworld Scenario," *Phys. Rev. D* **66**, 023514 (2002), astro-ph/ 0111292.
67. A. Gonzalez, T. Matos and I. Quiros, "Unified Models of Inflation and Quintessence," *Phys. Rev. D* **69**, 084029 (2005), hep-th/ 0410069.
68. A. R. Liddle and L. A. Ureña-Lopez, "Curvaton reheating: An Application to brane world inflation," *Phys. Rev. D* **68**, 043517 (2003), astro-ph/ 0302054.
69. A. R. Liddle and L. A. Ureña-Lopez, "Inflation, dark matter and dark energy in the string landscape," astro-ph/ 0605205.
70. T. Matos, J. R. Luebano and H. García-Compeán, "An Alternative Interpretation for the Moduli Fields of the Cosmology Associated to Type IIB Supergravity with Fluxes," hep-th/ 0511098.
71. T. Matos. "From the Strings Theory to the Dark Matter in Galaxies." In Proceedings of the VIII Mexican School of Particles and Fields, Ed. by M. Mondragon et. al., American Institut of Physics (AIM), (1999), p 382.

72. L. A. Ureña-Lopez, T. Matos and F. S. Guzmn. "Cosmic Scalar Field," In Proceedings of the Third Mexican School on Gravitation and Mathematical Physics, (2000). Ed. by Nora Breton et. al., Universidad de Guanajuato.
73. Tonatiuh Matos, Dario Nuñez, Luis Ureña and F. S. Guzmán, "The Scalar Field Dark Matter Model." In Exact Solutions and Scalar Fields in Gravity, edited by A. Macías, J. Cervantes and C. Lmerzhal., Ed. Kluwer Academic (2001), p. 166-184. 2.
74. M. Alcubierre, F. S. Guzmán, T. Matos, D. Nuñez, L. A. Ureña-López and Petra Wiederhold, "Scalar Field Dark Matter and Galaxy Formation." In Dark Matter in Astro-and Particle Physics, Edited by H. V. Klapdor-Kleingrothuas and R. D. Viollier. Springer Verlag (2002), p. 356-364, astro-ph/ 0204307.
75. T. Matos, "Galaxy Formation from the Scalar Field Dark Matter Model." In Developments in Mathematical and Experimental Physics. Ed. by A. Macías, F. Uribe and E. Díaz. Kluwer Academic (2002), p. 111.
76. F. S. Guzmán, T. Matos and D. Nuñez, "Relativistic Dark Matter in Spiral Galaxies." In Proceedings of the IX Marcel Grossman Meeting, Roma, Italia, Ed. by D. Ruffini et. al., Word Scientific Singapore, (2002), p. 2151.
77. T. Matos, L. A. Ureña, M. Alcubierre, R. Becerril, F. S. Guzmán and D. Nuñez, "The Scalar Field Dark Matter Model: A Braneword Connection." In Lecture Notes in Physics, 646, Springer-Verlag, Berlin Heidelberg, (2004), p 401-420.
78. F. S. Guzmán and L. A. Ureña-Lopez, "The hypothesis of scalar field dark matter: the cosmological evolution challenge." In Trends in Dark Matter Research, chapter 2, p. 39.
79. E. de la Cruz Burelo, F. S. Guzmán and T. Matos," Materia Obscura en el Universo: el nuevo ter, " *Avance y Perspectiva* **18**, 139-147 (1999).
80. H. García Compeán and Tonatiuh Matos, "La Influencia de Einstein en la Fisica Moderna," *Avance y Perspectiva* **23**, 7-18 (2004).
81. M. Salgado, D. Sudarsky and Hernando Quevedo,"Galactic periodicity and the oscillating G model," *Phys. Rev. D* **53**, 6771-6783 (1996).
82. D. Sudarsky, H. Quevedo and M. Salgado, "Cosmological dark matter and galactic periodicity," *Rev. Mex. Fis.* **43**, 699-710 (1997).
83. M. Salgado, D. Sudarsky and H. Quevedo, "Has cosmological dark matter been observed?," *Phys. Lett. B* **408**, 69-74 (1997).
84. H. Quevedo, M. Salgado and D. Sudarsky , "The oscillating G model: a possible explanation for the nature of the cosmological nonbaryonic matter," *Astrophys. J.* **488**, 14-26 (1997).
85. H. Quevedo, M. Salgado and D. Sudarsky, "A scalar field as a candidate for the cosmological nonbaryonic dark matter," *Gen. Rel. Grav.* **31**, 767-774 (1999).
86. J. A. González, H. Quevedo, M. Salgado and D. Sudarsky,"Local constraints on the oscillating G model," *Phys. Rev. D* **64**, 047504-1 (2001).
87. M. A. Rodriguez-Meza, J. Klapp and J. L. Cervantes-Cota, "The influence of scalar fields in protogalatic interactions," in *Exact solutions and scalar fields in gravity: Recent developments*, Kluwer Academic/Plenum Publishers (2001), New York, pp. 213-221.
88. M. A. Rodríguez-Meza and J. L. Cervantes-Cota, "Potential-density pairs for spherical galaxies and bulges: the influence of scalar fields," *Mon. Not. Roy. Astron. Soc.* **350**, 671-678 (2004).
89. M. A. Rodríguez-Meza, J. L. Cervantes-Cota, M.I. Pedraza, J.F. Tlapanco and E.M. De la Calleja, "Potential-density pairs for axisymmetric galaxies: the influence of scalar fields," *Gen. Rel. Grav.* **37**, 823-829 (2005).
90. M. A. Rodríguez-Meza and J. L. Cervantes-Cota, "NFW galactic profiles in the Newtonian limit of scalar-tensor theories," Proceedings of the X Marcel Grossman Meeting, edited by M. Novello, S. Perez-Bergliaffa and R. Ruffini, World Scientific, Singapore, 2006.
91. R. Gabbasov, M. A. Rodríguez-Meza, J. Klapp, and J. L. Cervantes-Cota, "The influence of numerical parameters on tidally triggered bar formation," *Astronomy and Astrophysics* **449**, 1043-1060 (2006).
92. R. A. Sussman, I. Quirós and O. Martín-González, "Inhomogeneous models of interacting dark matter and dark energy," *Gen. Rel. Grav.* **37**, 2117-2143 (2005), astro-ph/ 0503609.
93. D. Núñez, T. Matos and R. A. Sussman, "The spacetime associated with galactic dark matter halos," *Gen. Rel. Grav.* **37**, 769-779 (2005), astro-ph/ 0402157.
94. D. Núñez, T. Matos and R. A. Sussman, "A general relativistic approach to the Navarro–Frenk–White galactic halos," *Class. Quant. Grav.* **21**, 1-19 (2004), astro-ph/ 0410215.

95. A. Balakin, R. A. Sussman and W. Zimdahl, "The Maxwell–Boltzmann gas with non-standard self-interactions: a novel approach to galactic dark matter," *Phys. Rev. D* **70**, 064027 (2004), astro-ph/ 0406535.
96. X. Hernández, T. Matos, R. A. Sussman and Y. Verbin, "Scalar field 'mini–MACHOs': a new explanation for galactic dark matter," *Phys. Rev. D* **70**, 043537 (2004), astro-ph/ 0407245.
97. L. G. Cabral-Rosetti, X. Hernández and R. A. Sussman, "Using Entropy to Discriminate Annihilation Channels in Neutralinos Making up Galactic Halos," *Phys. Rev. D* **69**, 123006 (2004), hep-ph/ 0208131.
98. R. A. Sussman and X. Hernández, "On the newtonian limit and cut–off scales of isothermal dark matter halos with cosmological constant," *Mon. Not. Roy. Astron. Soc.* **345**, 871 (2003), astro-ph/ 0304385.
99. A. Coley, C. A. Clarkson, E. S. D. O'Neill, R. A. Sussman and R. K. Barret, "Inhomogeneous cosmologies, the Copernican principle and the cosmic microwave background: More on the EGS theorem," *Gen. Rel. Grav.* **35**, 969-990 (2003), gr-qc/ 0302068.
100. A. Coley, A. Sarmiento and R. A. Sussman, "Qualitative and numerical analysis of matter-radiation interaction in Kantowski-Sachs cosmologies," *Phys. Rev. D* **66**, 124001 (2002), astro-ph/0210100.
101. L. G. Cabral-Rosetti, D. Núñez, T. Matos and R. A. Sussman, "Hydrodynamics of galactic dark matter," *Class. Quant. Grav.* **19**, 3603-3616 (2002), gr-qc/0112044.

Quantum Gravity Phenomenology

Hugo A. Morales-Técotl* and Luis F. Urrutia†

*Departamento de Física, Universidad Autónoma Metropolitana Iztapalapa,
San Rafael Atlixco 186, CP 09340, México D.F., México.
†Departamento de Física de Altas Energías, Instituto de Ciencias Nucleares, Universidad
Nacional Autónoma de México, Apartado Postal 70-543, 04510 México D.F.

Abstract. The old order of magnitude argument stating that physics from Planck scale is out of experimental reach has been reconsidered in recent years in the light of current astrophysical and high precision tests indicating possible enhancement of the otherwise tiny quantum gravity effects. Nowadays it seems feasible to probe and discard effects associated with some simplified models and situations describing the space time structure at Planck length by these means. In this work we intend to survey contributions that have been developed by at least one collaborator researcher working in México, but showing them within the whole framework of the field. The corresponding list of works start from Ref.[57] on. As in any paper of this kind an inevitable or even not adverted biased point of view can appear in spite of our efforts to avoid it. Clearly our list of references is incomplete and we apologize in advance for omissions due mainly to lack of space.

Keywords: Quantum Gravity and Experimental Tests, Effective Field Theories, Deformed Lorentz Covariance.
PACS: 01.30.Rr, 04.60.-m, 04.80.Cc, 11.30.Cp, 04.90.+e

1. INTRODUCTION

Nowhere is the appreciation of the microstructure of spacetime more deeply underlined than in the 1967 paper of J.A. Wheeler [1] stating:

... *"Quantum fluctuations in the geometry of space is the other pressing field of application of quantum geometrodynamics". What a strange combinations of words! Fluctuations are well known. The term "quantum fluctuations" carries a deeper meaning. It stands for a movement that can never be frozen out, however low temperature. Such fluctuations are universal. In the Hydrogen molecule both the separation of the two atoms and their relative momentum continually fluctuate. Fixity of both would violate the uncertainty principle. In the frozen vacuum of quantum electrodynamics the electric and magnetic fields both fluctuate. Were both of these dynamically conjugated variables to vanish, the uncertainty principle would likewise fail. The same is true in quantum geometrodynamics ... The space of quantum geometrodynamics can be compared to a carpet of foam spread over a slowly undulating landscape . The undulations symbolize deterministic classical geometrodynamics. The continual microscopic changes in the carpet of foam as new bubbles appear and old ones disappear symbolize the quantum fluctuations in the geometry. The fluctuations change the microscopic connectivity of space itself. No longer is one entitled to take it for granted that space is Euclidean in the small. ...*

Accordingly, most quantum gravity proposals suggest that our notion of space-time as a continuum needs to be revised at short distances (high energies) of the order of the Planck length $\ell_P \approx 10^{-33}$ cm (Planck mass, $M_P \approx 10^{19}$ GeV). The notion of space-time as a continuum has been successfully probed up to energies even much lower than those of the standard model of particle physic, namely $\approx 10^3$ GeV. Clearly there is plenty of room for modifications of our generally accepted ideas of space-time to take place. Following Wheeler, the microscopic features of space-time have been referred to as foam space-time generically. Describing these features should be intimately related to the time honored problem of finding a consistent unification of gravity and quantum mechanics. Two of the efforts actively pursued are loop quantum gravity (LQG) [2] and string theory [3], among others lines of research (e.g. [4, 5]). In any case, a complete quantum gravity theory must contain, as a limit, the standard notion of space-time at macroscopical length scales, thus being consistent with the tests of classical Einstein gravity well established by now.

A physically crucial question is to investigate whether or not such short length (high energy) foamy features leave any imprint in the dynamics of particles at standard model energies, which we could detect with present day technology and observational sensibilities. This question has been answered in the positive, thus opening the possibility to provide observational guidance to the construction of quantum theories of gravity.

Phenomenologically, just as matter propagating in a medium probes its properties, the microstructure of spacetime may become evident for travelling matter in what we would classically recognize as smooth spacetime. As opposed to a material medium, spacetime is not embedded in something else but is instead the arena of any physical process. Thus this idea already reveals potential non trivial difficulties both technical and conceptual that at the same time make the problem rather appealing.

While traditionally it has been taken for granted that Planck scale phenomena are simply out of reach from present experiments and/or observations, astrophysical say, recent investigations point out this not necessarily the case. Indeed a whole line of research based on it has developed under the name of Quantum Gravity Phenomenology (QGP). It is meant to use current and future experiments/observations to investigate different quantum gravity proposals [6, 57, 7, 58]. Such efforts are partially given in [8, 9]. Progress in this direction lies along the lines of posing physical quantitative questions that were formerly considered uninteresting. Hence QGP has made advances in restricting the different parameters encoding some simplified situations capturing the modifications induced at standard model energies. In some quantum gravity approaches like LQG there still remains open the problem of calculating such modifications to matter dynamics in a rigorous semiclassical approximation. Once this gap is filled one could really test theories with observations. In the present work we adopt a heuristical approach as a starting point in order to convey the idea of what a semiclassical limit in LQG might yield as what we perceive as flat spacetime macroscopically. This will allow us to describe the evolution of the subject and in particular to identify contributions from researchers in Mexico.

As in any field under construction there are alternative points of view and QGP is not an exception. There is the question of whether or not the spacetime foam will yield a modified matter dynamics at standard model energies or even lower enough energies. If any modifications arise, the effects should not be too big, otherwise we would already have noticed them. Also we expect them not to be very small so that they would we be out of the capability of really constraining possible theories through their detection.

Now, most of the modifications that have been considered can be characterized by not preserving the standard local Lorentz covariance (LC): one class alluding to preferred reference frames and another to deformed/extended Lorentz Covariance.

Nevertheless the possibility is not automatically excluded that standard local Lorentz Covariance still holds in spite of whatever foam structure of spacetime we have at Planck scale. For instance this has been argued to be the case within LQG in regard to the discreteness of geometric quantities like areas or volumes just like the discrete spectrum of angular momentum is compatible with invariance under the continuous rotation group [10] or as in the case of causal sets approach to quantum gravity where discreteness is the starting point to approximate a smooth manifold at large scales [4].

The possibility that matter dynamics corrections do not respect local LC have been considered as follows:

(i) A phenomenological parameterization within the standard model was put forward in [11]. A more comprehensive analysis of all possible corrections terms, according to the dimensionality of the corresponding Lorentz invariance violating (LIV) operators, which are assumed to arise via a spontaneous Lorentz symmetry breaking of string theory, has been put forward [12]. These vacuum expectation values define a set of preferred frames, called concordant frames [13], in which the LIV terms can be maintained appropriately small when going from one frame to another via a passive (observer) Lorentz transformation. This is analogous to the description of atomic phenomena in the presence of an external magnetic field, where rotational invariance is broken by active (particle) transformation. Nevertheless, one can rotate the apparatus and perform the experiment in the presence of the rotated external field. The Standard Model Extension (SME)[14] belongs to this class and it has succeeded in accommodating experimental data gathered since 1960 [15] regarding isotropy of space, transformations among inertial frames and the validity of the discrete transformations C, P and T and combinations thereof [18]. The SME has been recently generalized to incorporate gravity [19]. More recently LIV models based upon dimension five operators have been constructed [20] and thoroughly analyzed in [21, 22]. Within the approach (i) we find also direct extensions of Dirac and Maxwell equations incorporating modifications which go beyond effective field theories [70]. An alternative approach to LIV via spontaneous symmetry breaking can be found in Ref.[23, 24].

(ii) Dynamical corrections have also been argued to admit a description in terms of an extension or deformation of the standard Lorentz relativity principle [25, 26, 27, 28, 29, 30], however fundamentally connected to quantum aspects of gravity for example as an effective field theory description of quantum general relativity [31, 32], non-critical string theory [33, 34, 35] or LQG [36, 61, 62, 63, 113, 117, 37]. Yet alternative views to account for dynamical modifications can be found in Ref. [38, 89, 99, 100, 101, 102, 103].

Corrected dynamics could be easily recognized through modifications to particle dispersion relations, say for fermions and photons,

$$\omega_\pm^2 = |\mathbf{k}|^2 \pm \xi \frac{|\mathbf{k}|^3}{E_{QG}}, \qquad (1)$$

$$E_{R,L}^2 = |\mathbf{p}|^2 + m^2 + \eta_{R,L}\frac{|\mathbf{p}|^3}{E_{QG}}. \qquad (2)$$

Here E_{QG} denotes the scale where quantum gravity effects become relevant, which is usually taken as the Planck energy and $\xi, \eta_{R,L}$ are dimensionless parameters for different particles to be determined from specific models and contrasted with observations which provide bounds for them. The connection of these modifications with gravity and possible astrophysical observations in Gamma Ray Bursts was advanced in [39]. Such corrected dispersion relations yield a time delay Δt between two photons having an energy difference ΔE that have travelled a distance L

$$\Delta t \approx \xi \frac{\Delta E}{E_{QG}} \frac{L}{c}. \qquad (3)$$

The suppressive effect of E_{QG} here can be compensated by selecting large energy differences or, even better, cosmological distances L [39]. The orders of magnitude $L \approx 10^{10} l.y., \Delta E \approx 20 MeV, E_{QG} = 10^{19} GeV, \xi \approx 1$, lead to $\Delta t \approx 10^{-3} s$, which is within the range of sensitivities δt in actual and forthcoming gamma ray bursts (GRB) observations. In order to measure such effect it is necessary that $\delta t < \Delta t$. Future planned observations of GRB at cosmological distances having well determined red shifts z, together with greater sensitivities ranging form $10^{-6} s$ (RHESSI: Reuben Ramaty High Energy Solar Spectrometer) to $10^{-7} s$ (GLAST: Gamma Ray Large Area Telescope) will allow a substantial increase of such bounds. The fireball model of GRB emission [40, 41] predicts also the generation of $10^5 - 10^{10} GeV$ neutrino bursts which will be detected by observatories like NUBE (Neutrino Burst Experiment), OWL (Orbiting Wide-angle Light collector experiment) and EUSO (Extreme Universe Space Observatory), for example. This will open up the possibility of using such particles, some times detected in coincidence with the respective photons, to set further bounds upon the quantum gravity scale E_{QG}.

Soon after the proposal of Ref. [39], an effective electrodynamics was put forward within LQG leading to dispersion relations (1) and having the form [36]

$$\nabla \cdot \mathbf{E} = 0, \quad \partial_t \mathbf{E} = -\nabla \times \mathbf{B} + 2\xi \ell_P \nabla^2 \mathbf{B}, \qquad (4)$$
$$\nabla \cdot \mathbf{B} = 0, \quad \partial_t \mathbf{B} = \nabla \times \mathbf{E} - 2\xi \ell_P \nabla^2 \mathbf{E}. \qquad (5)$$

Additional bounds upon the parameters describing the quantum gravity induced modifications have been obtained by incorporating the dynamics through the SME or similar constructions. In particular, the topics of Lorentz and CPT violations have been thoroughly studied in low energy physics via theory and experiments related to: Penning traps, clock comparison measurements, hydrogen-antihydrogen studies, spin polarized dispersion and muon experiments, among others subjects [42].

Finally we mention the use of astrophysical phenomena to discuss such modified theories. Distinguished examples are the bounds imposed by polarization measurements from astrophysical sources [43], the study of ultra high energy physics processes, among them cosmic rays, [44, 45, 46, 48, 49] and the consequences of the detected synchrotron radiation from the Crab nebulae, as well as that from other objects [21, 22, 69]. For recent reviews about such topics see Ref.[50].

2. HEURISTICAL APPROACHES AND SEMICLASSICAL STATES.

The possibility of exploring the fuzziness of space-time via either terrestrial experiments or astrophysical observations was added to the list of experimental tests related to the interface of quantum and gravitational realms in Refs. [39, 6] and further emphasized in [57]. One of the proposed manifestations of such space-time fluctuations was a modification or deformation of the standard particle dispersion relations yielding, for example, an energy dependent photon velocity. Soon after these works, the first heuristic derivation of a quantum gravity induced modified Hamiltonian for Maxwell equations in a Loop Quantum Gravity inspired description of Einstein-Maxwell theory was reported in [36]. This sparked a great deal of interest and it was subsequently followed by generalizations to the fermionic case [63, 61], further elaborations in the Maxwell case [62, 59] and the first steps towards the Yang-Mills case [60]. The idea behind all these heuristic derivations starts from the corresponding well defined operators in the quantum theory [51] and relies in the construction of quantum gravity semiclassical states which approximate a given geometry at macroscopical distances while retaining their quantum granular structure at scales of the order of the Planck length. Since those states were not known, their expected properties were postulated and estimations of the required expectation values were performed. Accordingly, this procedure left overall dimensionless coefficients undetermined. Furthermore, extracting the dynamics is a precess less clear to develop due to the known difficulties with time evolution, see for instance [94].

The construction of such semiclassical states is still an open problem in canonical quantum gravity [37, 64] and the main difficulty resides in going from the kinematical to the physical Hilbert space. Recent progress in such construction has been made using the group averaging technique in the case of quantum Bianchi I cosmology [67], in systems with linear and quadratic constraints[65] and in statistics of combinatorial Riemann geometries [66].

3. MODIFIED DIRAC EQUATION

The Hamiltonian found in Ref. [61] adapted for two components fermions was obtained under the assumption of flat space isotropy and it was assumed to account for the dynamics in a preferred reference frame, identified as the one in which the Cosmic Microwave Background looks isotropic. The earth velocity $\mathbf{w}$ with respect to that frame has already been determined to be $w/c \approx 1.23 \times 10^{-3}$. Thus, in the earth reference frame one expects the appearance of signals indicating minute violations of space isotropy encoded in the diurnal variation of the $\mathbf{w}$-dependent terms appearing in the transformed Hamiltonian or Lagrangian. On the other hand, many high precision experimental test of such variations, using atomic and nuclear systems for example, have been already reported in the literature [15] and already analyzed in terms of the Standard Model Extension. Amazingly enough, such precision is already adequate to set very strong bounds on some of the parameters arising from the quantum gravity corrections. The case of a non-relativistic valence nucleon was considered in clock comparison experiments with pairs like $(^{21}\text{Ne},^3\text{He})$ [16] and $(^{129}\text{Xe},^3\text{He})$ [17]. Stringent bounds of the order 10^{-9} for the correction terms which involve the coupling of the spin to the CMB velocity and of the order 10^{-5} for those describing a quadrupolar anisotropy of the inertial mass [72, 73, 74] were obtained. Similar analysis was applied to some string theory inspired quantum gravity models described in terms of an induced metric originating from the recoil velocity of topological defects called D-particles, which fill our universe, colliding with standard particles [34]. The bounds $M_D \geq 10^5 M_P$ and $v_D \leq 10^{-27} c$ were found for the mass and recoil velocity of the D-particle [73]. A proposal for modifying the Dirac equation by adding a second-order time derivative term, which predicts a modified Larmor frequency when the particle is minimally coupled to electromagnetism, was presented in [71]. Experiments related to the spreading of wave packets, scattering of electrons from an spherical symmetric potential and measurements of energy differences in silver atoms were also proposed to bound such correction.

4. MODIFIED ELECTRODYNAMICS

The study of radiation in LIV electrodynamics constitutes an interesting problem on its own whose resolution will subsequently allow the use of additional astrophysical information to put bounds upon the correction parameters. The observation of $100 MeV$ synchrotron radiation from the Crab Nebula has been used to impose stringent limits upon the parameters describing a quantum gravity inspired modified electrodynamics together with the corresponding coupled equations for the charges [21, 22]. Such bounds were based on a set of very reasonable assumptions on how some of the standard results of synchrotron radiation extend to the Lorentz non-invariant situation. This caused some controversy in the literature [52]. Moreover, an assessment of such assumptions requires the introduction of specific dynamical models. One of them is the Myers-Pospelov effective theory at the classical level, which parameterizes LIV using dimension five operators Ref. [20]. A detailed description of synchrotron radiation in this model has been presented in Ref.[69]. Subsequently, an unifying point of view was adopted by describing models with linear and quadratic correction terms in the Lagrangian via constitutive relations in Maxwell equations. The effective description corresponds to an electrodynamics in a birefringent dispersive and absorptive non-local medium [68]. Also, a general phenomenological framework for extending electrodynamics, based solely upon the equations of motion, which includes LIV as well as charge non-conservation has been reported in [70]. These additional parameters cannot be probed by optical experiments but can be subjected to experimental constraints by exploring the electromagnetic field created by a point charge or a magnetic dipole.

5. RADIATIVE CORRECTIONS

Most of the quantum gravity inspired correction to the dynamics have been studied in the non-interacting theory and arise through factors of the type $(E/M_P)^\Upsilon$, which are directly relevant at unaccessible energies $E \approx M_P$. An alternative possibility of probing such high energies is through the inclusion of radiative corrections (particle's self energies, for example), because the internal momenta are integrated up to the maximum allowed in a given reference frame. The standard folklore with respect to any new physics entering at high energy scales (here Planck scale) via an effective low energy field theory is that such scales should have negligible effects upon the leading-order low-energy physics (here free particle corrections). Contrary to this belief, the first indications that this is not necessarily true in the LIV case appeared in [20], where a specific form of the coupling was proposed in order to avoid such contributions. Subsequently it was shown that the proposed cure did not survive higher orders in perturbation theory [98, 96]. The fact that LIV induced by modelling space granularity via the introduction of a physical cutoff, which defines a preferred reference frame, adds a new item to the list of fine-tuning problems in high energy physics, leading to unsuppressed dimension four LIV contributions, was finally recognized in [97]. To this end the calculation of one-loop radiative corrections in the Yukawa model

$$L = 1/2(\partial \phi)^2 - 1/2\mu^2\phi^2 + \bar{\psi}(i\gamma^\mu \partial_\mu - m)\psi + g\phi\bar{\psi}\psi + (LVT) \tag{6}$$

was considered. Here LVT refers to the highly suppressed zeroth order Lorentz violating terms that take into account the previously discussed free particle dynamical modifications. The space granularity was modelled by introducing a physical cut-off Λ in such a way that the magnitude of the three-momentum in any loop is bounded by this quantity. The parameter Λ defines the onset of the scale at which the granularity of space becomes manifest. A convenient way of incorporating this requirement is to introduce the physical cutoff function $f(|\mathbf{k}|/\Lambda)$ with $f(0) = 1$, $f(\infty) = 0$, which suppresses the internal momenta having $|\mathbf{k}| \geq \Lambda$. The one-loop calculation of the boson self-energy $\Pi(E, \mathbf{p}, \Lambda)$ produces a finite contribution

$$\delta\Pi(E, \mathbf{p}, \Lambda) = \eta_{RC} p_\mu p_\nu W^\mu W^\nu, \quad \eta_{RC} = \frac{g^2}{12\pi^2}\left[1 + 2\int_0^\infty dx\, x f'(x)^2\right], \tag{7}$$

where W^μ is the four-velocity of the frame in which the momentum cut-off Λ is defined. Using standard model couplings the coefficient is estimated in the range $\eta_{RC} \geq 10^{-3}$. On the other hand, η_{RC} can be interpreted as a correction δc to the boson (photon) speed, for which extremely tight bounds exists: $\eta_{RC}/2 = \delta c/c = \leq 10^{-20}$. Similar results are found in the calculation of the fermion self-energy. These expected sizes are in extreme contrast to the measurements limits and pose serious problems for the preferred frame interpretation of the possible quantum gravity induced dynamical modifications [97, 92]. On the other hand, some mechanisms to evade the naturalness problem for LIV have been proposed. Supersymmetry, with only the unbroken translation subalgebra, has been considered as a custodial symmetry to protect Lorentz symmetry, which works at the one loop order [53]. Also, regularization alternatives which continuously interpolate between LIV and LI, depending only on one parameter, and which produce correct order of magnitude predictions for maximal attainable particle velocities has been proposed in Ref.[47]. This parameter can be estimated using data from the Ultra High Energy Cosmic Ray spectrum, which flux has been also successfully explained in the region of the current AGASA data using LIV ideas [48].

6. EXTENDED RELATIVITY

An alternative way of incorporating Planck scale corrections to standard particle dynamics has been proposed under the name of special relativity with two or three invariant scales (called Double or Triple special relativity in the literature), which are expected to replace Special Relativity at ultra-high energies and which should be regarded as the flat space semiclassical limit of gravity. The basic idea is to incorporate additional observer-independent scales, besides the velocity of light, into a relativity principle relating the physics observed in different inertial frames, by extending (deforming) the Lie algebra of the Lorentz group. This has the advantage of avoiding preferred frames and the situation is similar to the transition from Galilean to Lorentz relativity: the number of generators is the same, but the Lie algebra gets modified, the transformation laws pick up the additional parameters and the standard kinematical relations get modified accordingly [25, 26, 27, 28, 29]. Initially the deformation of the Poincare-Heisenberg algebra was performed in terms of non-linear realizations, which poses many unanswered questions regarding the operational interpretation of the involved operators in the light of their modified commutation relations, mainly in relation with the phase space interpretation of the theory, as well as with the many particle sector [87, 86, 85, 88]. A change in perspective has arisen motivated by the work of Ref.[90] showing that an appropriate redefinition of the generators together with a deformation of the central charges allow to cast the non-linear algebra of Triple Special relativity [27] in a linear (*i. e.* Lie) form. The resulting Lie algebra is a stable form proposed previously in the literature [30]. This has further motivated the use of Lie algebra deformation theory to the problem of identifying the stable form of the quantum relativistic kinematical algebra. The most general unique stable extension of such algebra allows the identification of two additionally invariant length scales [91, 89]. Some interpretational issues still remains and the physical consequences of this approach deserve to be explored.

7. DISPERSION RELATIONS AND MODIFIED HEISENBERG ALGEBRA

Possible manifestations of the modified dispersion relations associated to quantum gravity effects has been explored mainly in connection with photons in different interferometric experiments [75, 76, 77] and also in the context of finite temperature by determining the induced corrections to Planck, Boltzmann and Wein radiation laws, together with the modifications to the equation of state of the black body radiation [78].

The idea that quantum measurements in the Planck realm necessarily alter the local space-time metric in a manner that destroys the commutativity of the position measurements of two different particles was presented in Ref. [54]. This noncommutativity easily extends to measurements of different components of the position vector of a single particle and modifies the fundamental commutators of the Heisenberg algebra [82]. Alternative arguments motivating such modifications arise in string theory [55] and also from the existence of an upper bound for proper acceleration [56], for example.

Consequences of these gravitationally modified expressions for the wave particle duality have been further studied in relation to the early universe [81, 83]; in the case of quantum electrodynamics [79] and in the propagation of a particle in an external gravitational field [80].

In closing, and due to lack of space, we only mention some interesting related problems such as: alternative proposals to explore quantum gravity phenomenology [99, 100, 101, 104], the issue of time in quantum gravity [105, 106, 107], the quantum ambiguities arising in symplectic quantization [108] and quantum measurements in the context of the κ-Poincare group [109]. Finally some review articles can be found in [110, 111, 112, 113, 114, 115, 116, 117, 118, 119].

ACKNOWLEDGMENTS

L.F.U is partially supported by projects CONACYT-40745F and DGAPA-UNAM-IN104503-3. H.A.M.T. has been partially supported by Mexico's National Council of Science and Technology (CONACyT), under grant CONACyT 51132-F

REFERENCES

1. J.A. Wheeler, "Superspace and the nature of quantum geometrodynamics", in Batelle Rencontres, 1967 Lectures in Mathematical Physics, edited by C.M. DeWitt and J.A. Wheeler, W.A. Benjamin, New York, 1968.
2. See for example: C. Rovelli: Loop quantum gravity, *Livings Reviews*, **1**, 1 (1998), URL http://www.livingreviews.org/Articles; C. Rovelli: *Quantum Gravity*, Cambridge University Press, Cambridge 2004 ; R. Gambini and J. Pullin: *Loops, Knots, Gauge Theories and Quantum Gravity*, Cambridge University Press, Cambridge 1996; C. Beetle and A. Corichi: *Bibliography of publications related to classical and quantum gravity in terms of connections and loop variables*,e-Print Archive: gr-qc/9703044.
3. See for example: L. Dolan: "TASI Lectures on perturbative string theory and Ramond-Ramond flux", in *String, branes and extra dimensions: TASI 2001*, edited by S.S. Gubser and J.D. Lykken, World Scientific, River Edge, N. J. 2004 pp. 161; R.J. Szabo: *BUSSTEPP Lectures on string theory: an introduction to string theory and D-brane dynamics*, arXiv: hep-th/0207142; J. Polchinski: *String Theory, Vols I and II*, Cambridge University Press, Cambridge, 2000.
4. F. Dowker, J. Henson, and R.D. Sorkin: *Mod. Phys. Lett.* **A19**, 1829 (2004).
5. R. Gambini and J. Pullin, *Phys. Rev. Lett.* **90**, 021301 (2003).
6. G. Amelino-Camelia, *Nature (London)* **398**, 216 (1999).
7. G. Amelino-Camelia, *Lect. Notes Phys.* **541**, 1 (2000); G. Amelino-Camelia, *Int. Jour. Mod. Phys.* **D10**, 1 (2001); N.E. Mavromatos, "The quest for quantum gravity: testing times for theories?", in *From Particles to the Universe*, edited by A. Astbury et al., World Scientific, Singapore, 2001, pp. 335; J. Ellis, "Perspectives in High-Energy Physics", in *Proceedings of the 3rd Latin American Symposium on High-Energy Physics (SILAFAE III)*, edited by E. Nardi, Institute of Physics, Bristol, 2000; J. Ellis, *Nuovo Cim.* **24C**, 483 (2001); G. Amelino-Camelia, *Mod. Phys. Lett.* **A17**, 899 (2002); S. Sarkar, *Mod. Phys. Lett.* **A17**, 1025 (2002).
8. S.D. Biller et. al., *Phys. Rev. Lett.* **83**, 2108 (1999).
9. O. Betolami and C.S. Carvalho, *Phys. Rev.* **D61**, 103002 (2000); J.M. Carmona and J.L. Cortés, *Phys. Lett. B* **494**, 75 (2000); J.M. Carmona and J.L. Cortés, *Phys. Rev.* **D65**, 025006 (2002); C. Lämmerzhal and C. Bordé, "Testing the Dirac equation", in *Lecture Notes in Physics* **562**, edited by C. Lämmerzhal, C.W.F. Everitt and F.W. Hehl, Springer, Heidelberg, 2001, pp. 463; M.P. Haugan and C. Läemmerzhal, "Principles of equivalence: their role in gravitation physics and experiments that test them", in *Lecture Notes in Physics* **562**, edited by C. Lämmerzhal, C.W.F. Everitt and F.W. Hehl, Springer, Heidelberg, 2001, pp. 195; G. Lambiase, *Gen. Rel. Grav* **33**, 2151 (2001); G. Lambiase, *Eur. Phys. J.* **C19**, 553 (2001); G. Amelino-Camelia, *Phys. Letts.* **B528**, 181 (2002); G. Lambiase, *Class. Quant. Grav.* **20**, 4213 (2003); G. Lambiase and P. Singh, *Phys. Letts.* **B565**, 27 (2003); G.

Amelino-Camelia and C. Lämmerzhal, *Class. Quant. Grav.* **21**, 899 (2004); F.W. Stecker, *Astropart. Phys.* **20**, 85 (2003); F.W. Stecker, *J. Phys.* **G29**, R47 (2003).
10. C. Rovelli and S. Speziale, *Phys. Rev.* **D67**, 064019 (2003).
11. S. Coleman and S.L. Glashow, *Phys. Rev.* **D59**, 116008 (1999).
12. V.A. Kostelecký and S. Samuel, *Phys. Rev.* **D39**, 683 (1989), V.A. Kostelecký and S. Samuel, *Phys. Rev.* **D40**, 1886 (1989); V.A. Kostelecký and R. Potting,*Nucl. Phys.* **B359**, 545 (1991); V.A. Kostelecký and R. Potting, *Phys. Lett.* **B381**, 89 (1996).
13. V.A. Kostelecký and R. Lehnert, *Phys. Rev.* **D63**, 065008 (2001).
14. D. Colladay and V.A. Kostelecký, *Phys. Rev.* **D55**, 6760 (1997); D. Colladay and V.A. Kostelecký, *Phys. Rev.* **D58**, 116002 (1998). For reviews see for example *Proceedings of the Meeting on CPT and Lorentz Symmetry*, edited by V.A. Kostelecký, World Scientific, Singapore, 1999; C.M. Will, *The Confrontation between General Relativity and Experiment*, Living Reviews in Relativity 4, 2001; *Proceedings of the Second Meeting on CPT and Lorentz Symmetry*, edited by V. A. Kostelecký, World Scientific, Singapore, 2002.
15. V.W. Hughes, H.G. Robinson and V. Beltrán-López, *Phys. Rev. Lett.* **4**, 342 (1960); R.W.P. Drever, *Phil. Mag.* **6**, 683 (1961); J. D. Prestage et al.,*Phys. Rev. Lett.* **54**, 2387 (1985); S. K. Lamoreaux et al., *Phys. Rev. Lett.* **57**, 3125 (1986); S. K. Lamoreaux et al., *Phys. Rev.* **A 39**, 1082 (1989); C. J. Berglund et al., *Phys. Rev. Lett.* **75**, 1879 (1995); D. F. Phillips et al., *Phys. Rev* **D63**, 111101 (2001).
16. T.E. Chupp et al., *Phys. Rev. Lett.* **63**, 1541 (1989).
17. D. Bear et al., *Phys. Rev. Lett.* **85** 5038 (2000).
18. J.R. Ellis, J.L. Lopez, N.E. Mavromatos and D.V. Nanopoulos, *Phys. Rev.* **D53**, 3846 (1996).
19. V. A. Kostelecký, *Phys. Rev.* **D69**, 105009 (2004).
20. R.C. Myers and M. Pospelov, *Phys. Rev. Lett.* **90**, 211601 (2003); R.C. Myers and M. Pospelov, "Experimental Challenges of Quantum Gravity", in *Quantum Theory and Symmetries*, edited by P.C. Argyres, T.J. Hodges, F. Mansouri, J.J. Scanio, P. Suranyi and L.C.R. Wijewardhana, World Scientific, New Jersey, 2004, pp. 732.
21. T. Jacobson, S. Liberati and D. Mattingly, *Nature (London)* **424**, 1019 (2003).
22. T. Jacobson, S. Liberati and D. Mattingly, *Phys. Rev.* **D67**, 124011 (2003); T. Jacobson, S. Liberati and D. Mattingly, *Phys. Rev.* **D66**, 081302 (2002); T. Jacobson, S. Liberati, D. Mattingly and F. Stecker, *Phys. Rev. Lett.* **93**, 021101 (2004). For a review see for example T. Jacobson, S. Liberati and D. Mattingly, *Annals Phys.* **321**, 150 (2006).
23. A.A. Andrianov and R. Soldati, *Phys. Rev.* **D51**, 5961 (1995); A.A. Andrianov and R. Soldati, *Phys. Letts.* **B435**, 449 (1998); A.A. Andrianov, P. Giacconi and R. Soldati, *JHEP* **0202**, 030 (2002).
24. J. Alfaro, A.A. Andrianov, M. Cambiaso, P. Giacconi, R. Soldati, "On the consistency of Lorentz invariance violation in QED induced by fermions in constant axial-vector background", arXiv: hep-th/0604164; J. Alfaro, *Phys. Rev.* **D72**, 024027 (2005); J.Alfaro, *Phys. Rev. Lett.* **94**, 221302 (2005).
25. G. Amelino-Camelia, *Nature (London)* **418**, 34 (2002); G. Amelino-Camelia, *Int. J. Mod. Phys.* **D11**, 35 (2002); for a recent review see for example "Doubly special relativity as a limit of gravity", K. Imilkowska and J. Kowaslki-Glikman, in *Proceedings of the 339th WE Heraeus Seminar on Special Relativity: Will It Survive the Next 100 Years?*, edited by J. Ehlers and C. Lämmerzhal, Springer, Berlin, 2006, pp. 279.
26. J. Magueijo and L. Smolin, *Phys. Rev. Lett.* **88**, 190403 (2002); J. Magueijo and L. Smolin, *Phys. Rev.* **D67**, 044017 (2003).
27. J. Kowaslki-Glikman and L. Smolin, *Phys. Rev.* **D70**, 065020 (2004).
28. S. Liberati, S. Sonego and M. Visser, *Phys. Rev.* **D71**, 045001 (2005).
29. L. Smolin, "Falsifiable predictions from semiclassical quantum gravity", arXiv: hep-th/0501091.
30. C. N. Yang, *Phys. Rev.* **72**, 874 (1947); R. Vilela Mendes, *J. Phys.* **A27**, 8091 (1994).
31. J. F. Donoghue, *Phys. Rev. Lett.* **72**, 2996 (1994); J. F. Donoghue, *Phys. Rev.* **D50**, 3874 (1994).
32. D.A.R. Dalvit, F.D. Mazzitelli and C. Molina-Paris, *Phys. Rev.* **D63**, 084023 (2001).
33. G. Amelino-Camelia, J.R. Ellis, N.E. Mavromatos and D.V. Nanopoulos, *Int. J. Mod. Phys.* **A12**, 607 (1997); J.R. Ellis, N.E. Mavromatos and D.V. Nanopoulos, *Gen. Rel. Grav.* **31**, 1257 (1999); J.R. Ellis, N.E. Mavromatos and D.V. Nanopoulos, *Phys. Rev.* **D62**, 084019 (2000); J.R. Ellis, K. Farakos, N.E. Mavromatos, V.A. Mitsou and D.V. Nanopoulos, *Astrophys. J.* **535**, 139 (2000); J.R. Ellis, N.E. Mavromatos, D.V. Nanopoulos and G. Volkov, *Gen. Rel. Grav.* **32**, 1777 (2000); E. Gravanis and N.E. Mavromatos: *JHEP* **0206**, 019 (2002).
34. J.R. Ellis, N.E. Mavromatos and D.V. Nanopoulos, *Gen. Rel. Grav.* **32**, 127 (2000); J. Ellis, E.

34. Gravanis, N. E. Mavromatos and D. V. Nanopoulos:, *Phys. Rev.* **D61**, 027503 (2000); J.R. Ellis, E. Gravanis, N.E. Mavromatos and D.V. Nanopoulos,"Impact of low-energy constraints on Lorentz violation ", arXiv: gr-qc/0209108.
35. J.R. Ellis, N.E. Mavromatos, D.V. Nanopoulos and A.S. Sakharov, *Nature (London)* **428**, 386 (2004).
36. R. Gambini and J. Pullin, *Phys. Rev.* **D59**, 124021 (1999).
37. H. Sahlmann, T. Thiemann and O. Winkler, *Nucl. Phys.* **B606**, 401 (2001); H. Sahlmann and T. Thiemann, "Towards the QFT on curved spacetime limit of QRG. I: A general scheme", arXiv: gr-qc/ 0207030; H. Sahlmann and T. Thiemann, "Towards the QFT on curved spacetime limit of QRG.II: A concrete implementation",arXiv: gr-qc/ 0207031.
38. T. Padmanabhan, *Phys. Rev.* **D57**, 6206 (1998); K. Srinivasan, L. Sriramkumar and T. Padmanabhan, *Phys. Rev.* **D58**, 044009 (1998); S. Shankaranarayanan and T. Padmanabhan, *Int. J. Mod. Phys.* **D10**, 351 (2001).
39. G. Amelino-Camelia, J.R. Ellis, N.E. Mavromatos, D.V. Nanopoulos and S. Sarkar, *Nature (London)* **393**, 763 (1998).
40. E. Waxman and J. Bahcall, *Phys. Rev. Lett.* **78**, 2292 (1997); E. Waxman, *Nucl. Phys. (Proc. Supl)* **91**, 494 (2000); E. Waxman, *Nucl. Phys. (Proc. Supl)* **87**, 345 (2000).
41. M. Vietri, *Phys. Rev. Lett.* **80**, 3690 (1998).
42. R. Bluhm, "QED Test of Lorentz Symmetry", in *Proceedings of the 339th WE Heraeus Seminar on Special Relativity: Will It Survive the Next 100 Years?*, edited by J. Ehlers and C. Lämmerzhal, Springer, Berlin, 2006, pp. 191.; R. Bluhm, V.A. Kostelecký, C.D. Lane and N. Russell, *Phys. Rev. Lett.* **88**, 090801 (2002); R. Bluhm and V.A. Kostelecký, *Phys. Rev. Lett.* **84**, 1381 (2000).
43. R.J. Gleiser and C. N. Kozameh, *Phys. Rev.* **D64**, 083007 (2001).
44. T. Kifune, *Astrophys. J. Lett.* **518**, L21 (1999).
45. G. Amelino-Camelia and T. Piran, *Phys. Rev.* **D64**, 036005 (2001).
46. G. Amelino-Camelia, *Phys. Lett.* **B 528**, 181 (2002).
47. J. Alfaro, *Phys. Rev. Lett.* **94**, 221302 (2005), *Phys. Rev. D* **72**, 024027 (2005).
48. J. Alfaro and G. Palma, *Phys. Rev.* **D65**, 103516 (2002); J. Alfaro and G. Palma, *Phys. Rev.* **D67**, 083003 (2003).
49. T.J. Konopka and S.A. Major, *New J. Phys.* **4**, 57 (2002).
50. H. Vucetich,"Testing Lorentz invariance violation in quantum gravity theories", arXiv: gr-qc/ 0502093; D. Mattingly, "Modern test of Lorentz invariance", *Living Rev. Rel.* **8**, 5 (2005) ; T. Jacobson, S. Liberati and D. Mattingly, *Annals Phys.* **321**, 150 (2006).
51. T. Thiemann, *Phys. Lett.* **B380**, 257 (1996).
52. G. Amelino-Camelia, *New J. Phys.***6**, 188 (2004); T. Jacobson, S. Liberati and D. Mattingly, "Comments on "Improved limit on quantum-spacetime modifications of Lorentz symmetry from observations on gamma-ray blazers"", arXiv: gr-qc/0212002 v2.
53. S. G. Nibbelink and M. Pospelov, *Phys. Rev. Lett.* **90**, 081601 (2005); P. Jain and J. P. Ralston, *Phys. Lett.* **B621**, 213 (2005).
54. D. V. Ahluwalia, *Phys. Lett.* **B339**, 301 (1994).
55. S. de Haro, *Class. Quant. Grav.* **15**, 519 (1998); G. Veneziano, *Eur. Phys. Lett.* **2**, 199 (1986).
56. S. Capozziello, G. Lambiase and G. Scarpetta, *Int. J. Theor. Phys.* **39**, 15 (2000).
57. D.V. Ahluwalia, *Nature (London)* **398** 199 (1999).
58. G.Z. Adunas, E. Rodriguez-Milla and D.V. Ahluwalia, *Phys. Lett.* **B485**, 215 (2000).
59. J. Alfaro, M. Reyes , H.A. Morales-Tecotl, L.F. Urrutia, *Phys. Rev.* **D70** 084002 (2004).
60. J. Alfaro, H.A. Morales-Tecotl, M. Reyes, L.F. Urrutia, *J. Phys.* **A36** 12097-12107 (2003).
61. J. Alfaro, H.A. Morales-Tecotl, L.F. Urrutia, *Phys.Rev.* **D66** 124006 (2002).
62. J. Alfaro, H.A. Morales-Tecotl, L.F. Urrutia, *Phys.Rev.* **D65** 103509 (2002).
63. J. Alfaro, H.A. Morales-Tecotl, L.F. Urrutia, *Phys. Rev. Lett.* **84** 2318-2321 (2000).
64. M. Varadarajan and J. A. Zapata, *Class. Quant. Grav.* **17**, 4085 (2000).
65. A. Ashtekar, L. Bombelli and A. Corichi. *Phys.Rev.* **D72** 025008 (2005).
66. L. Bombelli, A. Corichi and O. Winkler, *Annalen Phys.* **14** 499 (2005).
67. L, Bombelli and A. Corichi, *Class. Quant. Grav.* **21**, 4087 (2004).
68. S.A. Martinez, R. Montemayor, L.F. Urrutia, "Reality and causality in quantum gravity modified electrodynamics ", arXiv: gr-qc/0511117.
69. B. Gonzalez, S.A. Martinez, R. Montemayor, L.F. Urrutia, *J. Phys. Conf. Ser.* **24**, 58 (2005); R. Montemayor, L.F. Urrutia. *Phys.Rev.* **D72** 045018 (2005); R. Montemayor, L.F. Urrutia, AIP Con-

ference Proceedings. **758**, American Institute of Physics, New York, 2005, pp. 81; R. Montemayor, L.F. Urrutia, *Phys.Lett.* **B606** 86 (2005).
70. C. Lammerzahl, A. Macias and H. Mueller, *Phys.Rev.* **D71**, 025007 (2005).
71. A. Camacho, A. Macias, *Phys. Lett.* **B582**, 229 (2004); A. Macias and A. Camacho, in *Boston 2004, Particles, strings and cosmology*, pp. 539.
72. L.F. Urrutia, "Loop quantum gravity induced corrections to fermion dynamics in flat space",in *Quantum Theory and Symmetries*, edited by P.C. Argyres, T.J. Hodges, F. Mansouri, J.J. Scanio, P. Suranyi and L.C.R. Wijewardhana, World Scientific, New Jersey, 2004, pp. 683.
73. D. Sudarsky, L.F. Urrutia, H. Vucetich *Phys. Rev.* **D68**, 024010 (2003).
74. D. Sudarsky, L. Urrutia, H. Vucetich, *Phys. Rev. Lett.* **89**, 231301 (2002.
75. A. Camacho, E. Castellanos," Deformed dispersion relations and Sagnac interferometry" in *Gravitation and Cosmology*, edited by A. Macías, C. Lämmerzahl and D. Nuñez, AIP Conference Proceedings**758**, American Institute of Physics, New York, 2005, pp. 90.
76. A. Camacho, *Class. Quant. Grav.* **22**, 2101 (2005).
77. A. Camacho, *Gen. Rel. Grav.* **37**, 1405 (2005).
78. H. Yepez, J.M. Romero, A. Zamora, "Corrections to the Planck's radiation law from loop quantum gravity", arXiv: hep-th/0407072.
79. A. Camacho, *Int. J. Mod. Phys.* **D12**, 1687 (2003); A. Camacho, *Gen. Rel. Grav.* **35**, 1153 (2003); A. Camacho, *Gen. Rel. Grav.* **34**, 1839 (2002).
80. A. Camacho, *Rel. Grav. Cosmol.* **1**, 89 (2004.
81. D.V. Ahluwalia and M. Kirchbach, *Int. J. Mod. Phys.* **D10**, 811 (2001).
82. D.V. Ahluwalia, *Phys.Lett.* **A275**, 31 (2000).
83. D.V. Ahluwalia, "Principle of equivalence and wave particle duality in quantum gravity", arXiv: gr-qc/0009033.
84. A. Camacho and A. Camacho-Galvan, *Gen. Rel.Grav.* **37**, 651 (2005).
85. D.V. Ahluwalia-Khalilova, *Int. J. Mod. Phys.* **D13**, 335 (2004).
86. . D.V. Ahluwalia-Khalilova, "Fermions, bosons, and locality in special relativity with two invariant scales", *in Tegernsee 2003, Beyond the desert 2003*, pp. 503.
87. D.V. Ahluwalia-Khalilova, " Operational indistinguishability of doubly special relativities from special relativity", arXiv: gr-qc/0212128.
88. D.V. Ahluwalia-Khalilova, *Class. Quant. Grav.* **22**,1433 (2005).
89. C. Chryssomalakos, E. Okon, *Int. J. Mod. Phys.* **D13**, 2003 (2004).
90. C. Chryssomalakos, E. Okon, *Int. J. Mod. Phys.* **D13**, 1817 (2004)
91. C. Chryssomalakos, *Mod. Phys. Lett.* **A16**, 197 (2001).
92. J. Collins, A. Perez and D. Sudarsky, " Lorentz invariance violation and its role in quantum gravity phenomenology", arXiv: hep-th/0603002.
93. D. Sudarsky and J.A. Caicedo, *J. Phys. Conf. Ser.* **24**, 69 (2005).
94. M. Bojowald, H.A. Morales-Tecotl and H. Sahlmann, *Phys.Rev.* **D71**, 084012 (2005).
95. A. Perez and D. Sudarsky, *Gen. Rel. Grav.* **37**, 813 (2005).
96. D. Sudarsky, "Quantum gravity induced granularity of spacetime and Lorentz invariance violation", in *Bloomington 2004, CPT and Lorentz symmetry*, edited by V.A. Kostelecký, World Scientific, Singapore, 2005, pp. 201.
97. J. Collins, A. Perez, D. Sudarsky, L. Urrutia and H. Vucetich *Phys. Rev. Lett.* **93**, 191301 (2004).
98. A. Perez and D. Sudarsky, *Phys. Rev. Lett.* **91**, 179101 (2003).
99. A. Perez, H. Sahlmann and D. Sudarsky *Class. Quant. Grav.* **23**, 2317 (2006).
100. A. Corichi and D. Sudarsky, *Int. J. Mod. Phys.* **D14**, 1685 (2005).
101. A. Corichi and D. Sudarsky," New quantum gravity phenomenology" , arXiv: gr-qc/0503078.
102. A. Camacho,*Gen. Rel. Grav.* **35**, 319 (2003).
103. , D. Ahluwalia, *Int. J. Mod. Phys.* **D8**, 651 (1999).
104. C. Chryssomalakos and D. Sudarsky, *Gen. Rel. Grav.* **35**, 605 (2003).
105. A. Camacho, A. Camacho-Galván, *Nuovo Cim.* **B114** 923 (1999).
106. M. Montesinos, *Gen. Rel. Grav.* **33**, 1 (2001).
107. M. Montesinos and C. Rovelli, *Class. Quant. Grav.* **18**, 555 (2001).
108. M. Montesinos, G.F. Torres del Castillo, *Phys. Rev.* **A 70**, 032104 (2004)
109. A. Camacho and A. Camacho-Galván, *Gen. Rel. Grav.* **37**, 651 (2005).
110. D. Sudarsky, *Int. J. Mod. Phys.* **D14**, 2069 (2005).

111. L.F. Urrutia, "Corrections to flat-space particle dynamics arising from space granularity", in *Proceedings of the 339th WE Heraeus Seminar on Special Relativity: Will It Survive the Next 100 Years?*, edited by J. Ehlers and C. Lämmerzhal, Springer, Berlin, 2006, pp. 299.
112. G. Amelino-Camelia, C. Lammerzahl, A. Macias, H. Muller, "The Search for quantum gravity signals", in *Gravitation and Cosmology*, edited by A. Macías, C. Lämmerzahl and D. Nuñez, AIP Conference Proceedings **758**, American Institute of Physics, New York, 2005, pp. 30; L.F. Urrutia, "Flat space modified particle dynamics induced by loop quantum gravity", in *Rio de Janeiro 2003, Recent developments in theoretical and experimental general relativity, gravitation, and relativistic field theories, pt. C*, pp. 2105.
113. L.F. Urrutia, *Mod. Phys. Lett.* **A17**, 943 (2002).
114. D.V. Ahluwalia, *Mod. Phys. Lett.* **A17**, 1135 (2002).
115. D.V. Ahluwalia, "At the interface of quantum and gravitational realms", arXiv: gr-qc/0202098.
116. *The interface of gravitational and quantum realms. Proceedings, 1st Inter-University Centre for Astronomy and Astrophysics Meeting*, edited by D.V. Ahluwalia and N.K. Dadhich, *Mod. Phys. Lett.* **A17**, 899 (2002).
117. L.F. Urrutia, "Loop quantum gravity induced modifications to particle dynamics", in *Playa del Carmen 2002, Particles and Fields*, edited by U. Cotti, M. Mondragon and G. Tavares-Velasco, AIP Conference Proceedings **670**, American Institute of Physics, New York, 2003, pp. 289.
118. L.F. Urrutia, "Quantum gravity corrections to neutrino propagation", in *JHEP Proceedings: Cartagena de Indias 2000, High energy physics*, edited by E. Nardi, Institute of Physics, Bristol, 2000.
119. J. Alfaro, H.A. Morales-Tecotl and L.F. Urrutia, "Loop quantum gravity effective matter dynamics" in *Rome 2000, Recent developments in theoretical and experimental general relativity, gravitation and relativistic field theories, Pt. B*, edited by R.T Jantzen, V. Gurzadyan and R. Rufini, World Scientific, Singapore, 2002, pp. 1265.

Mexican contributions to Noncommutative Theories

J. David Vergara* and H. García-Compeán[†,**]

*Instituto de Ciencias Nucleares, UNAM, A. Postal 70-543, México, D.F. México
[†]Centro de Investigación y de Estudios Avanzados del IPN, Unidad Monterrey
Cerro de las Mitras 2565, Col. Obispado, Monterrey N.L. 64060, México
[**]Departamento de Física, Centro de Investigación y de Estudios Avanzados del IPN
Apdo. Postal 14-740, México D.F. 07000, México

Abstract. In this paper we summarize the Mexican contributions to the subject of Noncommutative theories. These contributions span several areas: Quantum Groups, Noncommutative Field Theories, Hopf algebra of renormalization, Deformation Quantization, Noncommutative Gravity, and Noncommutative Quantum Mechanics.

Keywords: Noncommutative Field Theories, Deformation Quantization, Noncommutative Quantum Mechanics
PACS: 11.10.Nx, 02.20.Uw, 02.40.Gh

1. INTRODUCTION

One of the most important open problems in theoretical physics is to understand the microscopic structure of the space-time. Because of the interplay of Quantum Mechanics and General Relativity at the Planck length scale, there are good reasons for believing that at short distances the structure of space-time may no longer be considered as a 4-dimensional continuum. Furthermore, under these premises the very concept of manifold as an underlying mathematical structure in the construction of unified physical theories, applicable to distances of the order of the Planck length, becomes questionable and some people have been convinced that a new paradigm of geometric space is needed that would allow us to incorporate into our theoretical formalisms small-scale structures completely different from those to which we are usually accustomed. Among mathematicians mainly, one such outstanding paradigm is the noncommutative geometry invented by Connes, which considers a new calculus, the so called spectral calculus, based on operators in Hilbert space and the use of the tools of spectral analysis [1]. This geometry has among its features that it includes ordinary Riemannian space; discrete spaces are treated on the same footing as the continuum, thus allowing for a mixture of the two. Furthermore, it allows the possibility of noncommuting coordinates.

On the other hand, in Physics it would appear reasonable the need to extend the phase-space noncommutativity of quantum mechanics to a noncommutativity of space-time in order to study the microstructure of the space-time. It then becomes a major issue the study of the noncommutative spaces. These spaces are characterized by the commutation relations satisfied by their coordinate operators,

$$[\hat{x}^i, \hat{x}^j] = i\hbar \Theta^{ij}. \tag{1}$$

Here the constant parameters of the non-commutativity are given by the real and antisymmetric matrix elements $\hbar\Theta^{ij}$, having dimensions of area. Noncommutativity of space-time has been in the literature for many years [2] and was studied in some systems. However, recently this subject has received a renewed interest, and now there are several reasons why in physics we are interested in this subject. For example, the quantum Hall effect, one of most studied phenomenon in condensed matter, presents noncommutativity in the canonical coordinates and momenta [3]. Further evidence along this line of thought has been provided by recent developments in string theory [4]. In their proposal, Seiberg and Witten have found signals of noncommutativity in the description of the low energy excitations of open strings (possibly attached to D-branes) in the presence of a Neveu-Schwarz constant background B-field. In the decoupling limit, the low energy effective field theory is described in terms of a supersymmetric gauge theory. By means of two different regularization schemes, two versions of the same effective theory are obtained in this limit, corresponding to the commutative and the noncommutative gauge theories. The independence on the regularization scheme leads to an equivalence of these theories which gives rise to the so called *Seiberg-Witten map*. This map relates the noncommutative and the commutative gauge theories, both of them with the same number of physical degrees of freedom. It has also been shown recently that in noncommutative field theories the Seiberg-Witten map can be interpreted as a field dependent gravitational background [5]. In fact, it is not difficult to show that a similar interpretation can be carried out even at the level of quantum mechanics on noncommutative phase-space. These recent results, as well as others (*c.f.* examples of noncommutative geometry in field theory listed in [6]), have generated a considerable interest to understand the role played by noncommutative geometry in different theoretical sectors of physics.

One way to deal with these noncommutative spaces is to construct a new kind of field theory, changing the standard product of the fields by the star product (Weyl-Moyal):

$$(f*g)(x) = \exp(\frac{i}{2}\Theta^{ij}\partial_i\partial_j)f(x)g(y)|_{x=y}, \quad (2)$$

where f and g are infinitely differentiable functions. In this theory some interesting results have been found [7]; for instance, it was shown that there is a relation between the infrared and ultraviolet divergences [8]. Several interactions have been analyzed in this context, for example in the case of the Standard Model see the references [9] and [10].

If, on the other hand, one assumes the commutation rules:

$$[\hat{x}^i, \hat{x}^j] = i\hbar\Theta^{ij}, \quad [\hat{x}^i, \hat{p}_j] = i\hbar\delta^i_j, \quad [\hat{p}_i, \hat{p}_j] = 0. \quad (3)$$

a noncommutative quantum mechanics can be formulated, of which some relevant results have already been obtained [11], [12].

Although none of the above mentioned apparently conceptually different approaches and their variants are anywhere close to a final theory of grand unification, and probably no single one of this directions will succeed in producing it, there appears to be emerging a common denominator of noncommutativity in some of their ingredients which points to the fact that when considering the problem of coordinates below the Planck length, there is no good reason to presume that the texture of space-time will still have a 4-dimensional continuum structure.

The purpose of this paper is to review the Mexican contributions to the subject of noncommutative theories. The subject includes contributions no only from physicists, and mathematical physicists, also there are important contributions from mathematicians, and we have tried to incorporate most of the references including topics related to particles and fields. We divide these contributions in several areas that we order chronologically. In Section 2 we review the contributions to Quantum Groups and Braid Groups. In Section 3 we present the contributions to Deformation Quantization. In Section 4 we present the Mexican contributions to Noncommutative Field theories, Noncommutative Gravity and Noncommutative Quantum Cosmology. In Section 5 we summarize the area of Noncommutative Geometry and Hopf algebras of renormalization. Finally, in Sec. 6 we present the contributions to Noncommutative Quantum Mechanics.

2. QUANTUM GROUPS

Quantum groups were first introduced independently by Drinfeld [13], Jimbo [14], and Woronowicz [15]. The motivation that leads to the study of these mathematical structures was the consideration of the symmetries of the Yang-Baxter equations, in the case of Drinfeld and Jimbo. On the other hand, starting from a deformation of the universal enveloping algebra of $SU(2)$ Woronowicz got the same structure. From the mathematical point of view Quantum groups are not groups, are Quasitriangular Hopf algebras [16]. But in some sense, these structures represent to Noncommutative Geometry the equivalent of a Lie Group to Differential Geometry. In this area of study there were very important contributions to the subject by Mexican researchers. The main contribution is the book *Quantum Groups in two-dimensional Physics* by M. Ruiz-Altaba, et al [17]. This book was the first serious treatment of the subject by physicists and is a fundamental reference for this area. In the case of papers, an interesting contribution to the area is [18]. In this article the authors found an explicit quantum group structure of WZNW models, in their Coulomb gas representation. They construct the quantum group generators in the Chevalley basis as operators creating, destroying or counting screening charges, and show that they satisfy all the required properties. In another paper [19], the same authors describe in detail the contour representation of finite quantum groups which appears naturally in the Coulomb gas representation of conformal field theories. Another important contribution to this area is the construction of Quantum Clifford algebras from spinor representations [20]. Here using the quantum group formalism of bicovariant bimodules it was possible to built the quantum analogues of Dirac and Laplace operators and to construct quantum Spin(n) groups [21]. This construction was one of the first fully consistent Quantum Clifford algebras. One interesting extension of this subject was the construction of Involutive braided Spin groups [22]. These mathematical structures are generalizations of quantum groups, that include a braid operator. Other contributions to this area are [23], [24], [25] and [26].

3. DEFORMATION QUANTIZATION

The goal of classical and quantum mechanics is to study the evolution of a system. In classical mechanics the set of possible states of a system are points in the phase space and the observables are functions of a commutative algebra. In quantum mechanics the physical states form a projective Hilbert space and the set of observables are self-adjoint operators that form a noncommutative algebra. The aim of Deformation Quantization is to obtain the quantum theory of a system by deforming the product that is used to multiply the observables of the theory.

In mathematics, the deformation of an object is a family of objects depending on a parameter. In the case of Deformation Quantization it is possible to define a family of associative products, star products, that depend on the deformation parameter $\hbar$ (or θ). In this context, research in Mexico has produced several interesting proposals. For example in [27] was shown how the reduced Self-dual Yang-Mills theory described by the Nahm equations can be carried over to the Weyl-Wigner-Moyal formalism employed in Self-dual gravity. This shows the existence of a correspondence between BPS magnetic monopoles and space-time hyper-Kähler metrics. On the other hand, the Deformation quantization of the bosonic string was constructed in [28]. In this paper it was shown that the light-cone gauge is the most convenient classical description to perform the quantization of bosonic strings in the deformation quantization formalism. The mass spectrum, propagators and the Virasoro algebra were finally described within this deformation quantization scheme. This work was extended in [29]. Here, the deformation quantization of scalar and abelian gauge classical free fields was studied and the Stratonovich-Weyl quantizer, together with star-products and Wigner functionals were obtained. Another interesting contributions to this area are: [30] and [31]. For Kahler manifolds it is known that an alternative procedure to quantize these spaces is that given by Berezin in Ref. [32]. In this procedure it is also defined an star product called the Berezin star. This was used in geometric quantum mechanics in [33]. It is known that several field theories have as moduli spaces which are compact Kahler manifolds. This idea was precisely implemented in [34] for the study of a quantization of Chern-Simons gauge theories and in [35] for WZW and coset models.

The more general associative star product is due to Kontsevich [36]. This construction allows us to define star products on general phase spaces, with coordinate dependent and degenerated symplectic structures. In physics this construction has some applications, for example in [37] it was shown that the Hopf algebra of renormalization and the universal formula of Kontsevich for quantum deformation are equivalent. This result could be useful to study in more depth the perturbative structure of quantum field theory.

4. NONCOMMUTATIVE FIELD THEORIES

Noncommutative gravity has been formulated in the literature by using different approaches (for instance, see, Refs. [38, 39, 40]). More recently Chamseddine has made several proposals for noncommutative formulations of Einstein's gravity [41, 42, 43], where a Moyal deformation is done. A more recent proposal of a noncommutative deformation of Einstein-Hilbert Lagrangian in four dimensions is given in [44]. The study of models of noncommutative gravity, trying to imitate Yang-Mills gauge theory, arising from string or M-theory, is very important. Such models could be obtained starting from those formulations of gravitation which are based on a gauge principle. Two of these formulations are the topological gravity and the Plebański formulation of self-dual gravity [45], from which the hamiltonian Ashtekar's formulation [46] can be obtained.

In México, different proposals of noncommutative gravity in four dimensions have been formulated recently. A proposal for gauge topological gravity is given in terms of a noncommutative gauge invariant action [47]. Here the possibility of noncommutative topological gravity arising in the same manner as a Yang-Mills theory is explored. The Seiberg-Witten map is used to construct such a theory based on a $SL(2,C)$ complex connection, from which the Euler characteristic and the signature invariant are obtained. This gives a way towards the description of noncommutative gravitational instantons as well as noncommutative local gravitational anomalies.

Moreover, using a noncommutative formulation of Plebański's self-dual gravity a noncommutative theory of pure Einstein theory in four dimensions was obtained [48]. In order to do that, the Seiberg-Witten map was used. It is shown that the noncommutative torsion constraint is solved by the vanishing of commutative torsion. Also, the noncommutative corrections to the action are computed up to second order. In the case of linearized gravity an interesting approach was presented in [49].

On the other hand a formulation of noncommutative quantum cosmology starting from a noncommutative parametrization of the minisuperspace has been also given [50]. Here it was proposed a model for noncommutative quantum cosmology by means of a deformation of minisuperspace. For the Kantowski-Sachs metric the exact wave function was found. Thus, wave packets are constructed and it is shown that new quantum states appear that "compete" to be the most probable state, in clear contrast with the commutative case. A tunnelling process could be possible among these states. In this same direction, the CFT construction of S-branes describing the rolling and bouncing tachyons is analyzed in the context of a θ-noncommutative deformation of minisuperspace. Half s-brane and s-brane in the noncommutative minisuperspace are studied and exact analytic solutions, involving the noncommutative parameter θ and compatible with the boundary conditions at infinity, are found. Also it was performed a comparison with the usual commutative minisuperspace. This was described in Ref. [51]. Another different proposal to noncommutative quantum cosmology is given in [52].

Going back to topic of noncommutative gravity, a noncommutative description of topological half-flat gravity in four dimensions was formulated. BRST symmetry of this topological gravity is deformed through a twisting of the usual BRST quantization of noncommutative gauge theories. It is argued also that the resulting moduli space of instantons is characterized by the solutions of a noncommutative version of the Plebański's heavenly equation [53].

The analogous calculation in Yang-Mills theories can also be addressed, for instance the cohomological Yang-Mills theory is formulated on a noncommutative differentiable four manifold through the θ-deformation of its corresponding BRST algebra. The resulting noncommutative field theory is a natural setting to define the θ-deformation of Donaldson invariants and they are interpreted as a mapping between the Chevalley-Eilenberg homology of noncommutative spacetime and the Chevalley-Eilenberg cohomology of noncommutative moduli of instantons. In the process one can find that in the weak coupling limit the quantum theory is localized at the moduli space of noncommutative instantons [54].

Gravitational axial and chiral anomalies in a noncommutative space are examined through the explicit perturbative computation of one-loop diagrams in various dimensions [55]. The analysis depend on how gravity is coupled to noncommutative matter fields. The Delbourgo-Salam computation of the gravitational axial anomaly contribution to the pion decay into two photons, is studied in detail in this context. In the process one can see that the two-dimensional chiral pure gravitational anomaly does not receive noncommutative corrections. Pure gravitational chiral anomaly in $4k+2$ dimensions with matter fields being chiral fermions of spin-1/2 and spin-3/2, is discussed and a noncommutative correction is found. In this paper, the mixed anomalies are considered in both cases. Finally in [56], the influence of higher dimensions in noncommutative field theories was studied. For this purpose, the bosonic sector of a recently proposed 6 dimensional $SU(3)$ orbifold model for the electroweak interactions was analyzed. The corresponding noncommutative theory is constructed by means of the Seiberg-Witten map in 6D. In the reduced bosonic interactions for a 4D theory, new couplings (with respect to those known in others 4D noncommutative formulations of the Standard Model) were found using the Seiberg-Witten map.

5. NONCOMMUTATIVE GEOMETRY AND HOPF ALGEBRAS OF RENORMALIZATION

In mathematics, Gaussian and Riemannian geometric spaces are usually defined as manifolds where the metric is given by the geodesic distance

$$d_\gamma(x,y) = \inf \gamma \{ \text{lenght of paths } \gamma \text{ from } x \text{ to } y \}. \tag{4}$$

However, in order to reduce the geometry to algebraic form and to arrive at a formulation which can be extended to noncommutative spaces, Connes [1] has proposed, as starting point, the following dual form of (4):

$$d(x,y) = \sup\{|f(x) - f(y)| f \in \mathscr{A}, \|\frac{df}{ds}\|\}, \tag{5}$$

where $\mathscr{A}$ is the algebra of $C^\infty(M)$ functions over M, and ds is the line element in Riemannian geometry. To measure distances in a possible noncommutative space X, equation (5) is generalized by first introducing a Fredholm module $(\mathscr{H}, F)$ over the involutive algebra $\mathscr{A}$. Here F is a selfadjoint involutive operator acting on the Hilbert space $\mathscr{H} = L^2(M,S)$ of square integrable sections of the irreducible spinor bundle over M. The differential calculus is quantized by using the operator quantum theoretic notion for the differential

$$df = [F, f], \tag{6}$$

where $f \in \mathscr{A}$. One further specifies a metric structure on X by defining a unit length via an operator of the form

$$G = (dx^\mu)^* g_{\mu\nu}(dx^\nu), \tag{7}$$

with $x^\mu \in \mathscr{A}$, then $dx = [F,x]$, and $g_{\mu\nu}$ is a positive element of the matrix algebra $M_q(\mathscr{A})$. Thus, G is a positive compact operator, and we can think of its positive square root as the line element of Riemannian geometry, $\sqrt{G} = ds$. Connes identifies this operator with the inverse of the Dirac operator. A more interesting idea is to think instead that the geometry of the space-time is dictated by Quantum Field Theory. According to Connes and Kreimer [57], this idea stresses the fact that space-time ought to be regarded as a derived concept. A remarkable result that gives support to this idea is the equivalence between the Hopf algebra of Connes-Moscovici [58], found in the context of Noncommutative Geometry, and the Hopf algebra of Kreimer [59], found in the context of Quantum Field Theory. Using this identification it is possible to associate the renormalization process in quantum field theory with a general mathematical procedure of extraction of finite values based on the Riemann-Hilbert problem [60].

In this area there were some interesting Mexican contributions. For example, the introduction of normal coordinates on the infinite dimensional group G studied by Connes and Kreimer in their analysis of the Hopf algebra of rooted trees was proposed in [61]. Furthermore, the primitive elements of the algebra were studied and it was shown that they are generated by a simple application of the inverse Poincaré lemma, starting from a closed left invariant 1-form on G. For the special case of the ladder primitives, a second description was found that relates them to the Hopf algebra of functionals on a power series with the usual product. Either approach shows that the ladder primitives are given by the Schur polynomials. This analysis reduce considerably the renormalization procedure of primitive ladder Feynman diagrams. On the other hand in [62], the combinatorics resulting from the perturbative expansion of the transition amplitude in quantum field theories was analyzed, and its relation to the Hausdorff series. It was shown that in the context of these structures the power sum symmetric functionals of the perturbative expansion are Hopf primitives and that they are given by linear combinations of Hall polynomials, or diagrammatically by Hall trees. Furthermore, it was shown that each Hall tree corresponds to sums of Feynman diagrams each with the same number of vertices, external legs and loops. In addition, since the Lie subalgebra admits a derivation endomorphism, it was also shown that, with respect to it, these primitives are cyclic vectors generated by the free propagator, and thus provide a recursion relation by means of which the $(n+1)$-vertex connected Green functions can be derived systematically from the n-vertex ones. Another contributions to this area are: [63], [64], [65], [66], [67], [68].

6. NONCOMMUTATIVE QUANTUM MECHANICS

From the intrinsically noncommutative operator point of view, the development of a formulation for noncommutative quantum mechanics requires: (1) a specification of a representation for the phase-space algebra, (2) a specification of the Hamiltonian which governs the time evolution of the system and (3) a specification of the Hilbert space on which these operators and the other observables of the theory act. Regarding the choice of a representation for the intrinsic Heisenberg noncommutative phase-space algebra, several recent works in the literature have suggested to use a quantum mechanical equivalent to the Seiberg-Witten map [4], whereby the noncommutative Heisenberg algebra is mapped into a commutative one [69], [70], [71]. Since in all generality this map admits many possible realizations, one could have in principle also many possible resulting self-consistent quantum mechanics of which the proper one could only be discerned by experiment.

Moreover, since single particle quantum mechanics can be seen, in the free field or weak coupling limit, as a mini-superspace sector of quantum field theory where most degrees of freedom have been frozen. It is interesting to continue a more detailed study of exactly solvable models in noncommutative quantum mechanics and perhaps this results will be helpful both for the understanding of the effects of noncommutativity in field theory, as well as of its possible phenomenological consequences.

In this area, an interesting work produced in México is [72], where it was shown that corrections to the Newton's second law appear if it is assumed that the phase space has a symplectic structure consistent with the rules of commutation of noncommutative quantum mechanics. In the central field case was found that the correction term breaks the rotational symmetry. In particular, for the Kepler problem, this term takes the form of a Coriolis force produced by the weak gravitational field far from a rotating massive object. Following this line of thinking the Kepler problem was studied in more depth in [73] and it was shown that a noncommutative parameter of the order of $10^{-58} m^2$ gives observable corrections to the movement of the solar system. In this way, modifications in the physics at smaller scales imply modifications at large scales, something similar to the UV/IR mixing. Another interesting contribution is [74], where the authors formulate non-relativistic classical and quantum mechanics in the noncommutative two dimensional plane. The approach used is based on the Galilei group, where the noncommutativity is seen as a central extension upon identification of the boost generators with the position operator. Furthermore, they perform a systematic study of the free particle, defined by the symmetries of the space-time, which include the noncommutativity. The symmetries at the classical level are analyzed in terms of Noether's theorem. Canonical quantization is presented and the representation of the corresponding Heisenberg algebra is obtained. The path integral representation and Wigner distribution function in phase space are also discussed. Finally, they use Einstein's model for a solid to corroborate that, according with intuition, the entropy is a growing function of θ in the low temperature regime, as a consequence of the space fuzziness. On the other hand in [75], following the idea that noncommutative quantum mechanics is related to noncommutative field theory, it was shown that introducing an extended Heisenberg algebra in the context of the Weyl-Wigner-Groenewold-Moyal formalism leads to a deformed product of the classical dynamical variables that is inherited at the level of quantum field theory. This allows to relate the operator space noncommutativity in quantum mechanics to the quantum group inspired algebra deformation noncommutativity in field theory.

ACKNOWLEDGMENTS

The authors acknowledge partial support from CONACyT projects SEP-2004-C01-47211(J.D.V.), 40745-F(J.D.V.), 45713-F(H.G.-C.), as well as by DGAPA-UNAM grant IN104503.

REFERENCES

1. A. Connes, *Noncommutative Geometry*, Academic Press, San Diego, 1994.
2. H. Snyder, *Phys. Rev.* **71**, 38 (1947).
3. Z. F. Ezawa, *Quantum Hall Effects: Field Theoretical Approach and Related Topics*, World Scientific, Singapore, 2000.
4. N. Seiberg and E. Witten, *J. High Energy Phys.* **09**, 032 (1999).
5. V. O. Rivelles, *Phys. Lett. B* **558**, 191 (2003).
6. L. Alvarez-Gaume, and S. R. Wadia, *Phys. Lett. B* **501**, 319 (2001).
7. R. J. Szabo, *Phys. Rep.* **378**, 207–299 (2003).
8. S. Minwalla, M. V. Raamsdonk and N. Seiberg, *JHEP* **0002**, 020 (2000).
9. M. Chaichian, P. Presnajder, M. M. Sheikh-Jabbari and A. Tureanu, *Eur. Phys. J. C* **29**, 413–432 (2003).
10. X. Calmet, B. Jurco, P. Schupp, J. Wess and M. Wohlgenannt, *Eur. Phys. J. C* **23**, 363 (2002).
11. M. Chaichian, M. M. Sheikh-Jabbari and A. Tureanu, *Phys. Rev. Lett.* **86**, 2716 (2001).
12. S. Bellucci, A. Nersessian and C. Sochichiu, *Phys. Lett. B* **522**, 345 (2001).
13. V. G. Drinfeld, *Sov. Math. Doklady* **36**, 212 (1988).
14. M. Jimbo, *Commun. Math. Phys.* **102**, 537 (1986).
15. S. L. Woronowicz, *Commun. Math. Phys.* **111**, 613 (1987).
16. S. Majid, *Int. J. Mod. Phys. A* **5**, 1 (1990).
17. C. Gómez, M. Ruiz-Altaba, and G. Sierra, *Quantum Groups in Two-dimensional Physics*, Cambridge University Press, Cambridge, 1996.
18. C. Ramírez, H. Ruegg and M. Ruiz-Altaba, *Phys. Lett. B* **247**, 499 (1990).
19. C. Ramírez, H. Ruegg and M. Ruiz-Altaba, *Nucl. Phys. B* **364**, 195 (1991).
20. R. Bautista, A. Criscuolo, M. Durdevic, M. Rosenbaum and J. D. Vergara, *J. Math. Phys.* **37**, 5747 (1996).
21. A. Criscuolo, M. Durdevic, M. Rosenbaum and J. D. Vergara, *J. Phys. A* **30**, 6451 (1997).
22. A. Criscuolo, M. Rosenbaum and J. D. Vergara, *J. Geom. and Phys.* **25**, 46 (1998).
23. M. Ruiz-Altaba, *Phys. Lett. B* **279**, 326 (1992).
24. C. Gómez, M. Ruiz-Altaba and G. Sierra, *Phys. Lett. B* **265**, 95 (1991).
25. N. Fleury and A. Turbiner, *J. Math. Phys.* **35**, 6144 (1994).
26. M. Durdevic, *J. Phys. A* **30**, 326 (1997).
27. H. García-Compean and J. F. Plebanski, *Phys. Lett. A* **234**, 5 (1997).
28. H. García-Compeán, J. F. Plebanski, M. Przanowski and F. J. Turrubiates, *J. Phys. A* **33**, 7935 (2000).
29. H. García-Compeán, J. F. Plebanski, M. Przanowski and F. J. Turrubiates, *Int. J. Mod. Phys. A* **16**, 2533 (2001).
30. H. García-Compeán, *Nucl. Phys. B* **541**, 651 (1999).
31. M. Rosenbaum and J. D. Vergara, *Gen. Relativ. Gravit.* (2006), in press.
32. F.A. Berezin, *Commun. Math. Phys.* **40**, 153 (1975).
33. H. García-Compeán, J. F. Plebanski, M. Przanowski and F. J. Turrubiates, *J. Phys. A* **35**, 4301 (2002).
34. I. Carrillo-Ibarra and H. García-Compeán, "On the Berezin description of Kaehler quotients," arXiv:hep-th/ 0202015.
35. I. Carrillo-Ibarra, H. García-Compeán and W. Herrera-Suarez, "Berezin quantization of gauged WZW and coset models," arXiv:hep-th/ 0408207.
36. M. Kontsevich, *Lett. Math. Phys.* **66** 157 (2003).
37. M. Rosenbaum and J. D. Vergara, *J. Phys. A* **37**, 7939 (2004).
38. A.H. Chamseddine, *Commun. Math. Phys.* **155**, 205 (1993).
39. W. Kalau and M. Walze, *J. Geom. Phys.* **16**, 327 (1995).
40. A. Connes, *Commun. Math. Phys.* **182**, 155 (1996); A.H. Chamseddine and A. Connes, *Phys. Rev. Lett.* **77**, 4868 (1996).
41. A.H. Chamseddine, *Commun. Math. Phys.* **218**, 283 (2001).
42. A.H. Chamseddine, *Phys. Lett.* **B504**, 33 (2001).
43. A.H. Chamseddine, "Invariant Actions for Noncommutative Gravity," arXiv: hep-th/ 0202137.
44. M.A. Cardella and D. Zanon, *Class. Quant. Grav.* **20**, L95 (2003).
45. J. Plebański, *J. Math. Phys.* **18**, 2511 (1977).

46. A. Ashtekar, *Phys. Rev. Lett.* **77**, 3228 (1986); *Phys. Rev. D* **36**, 1587 (1987); *Lectures on Nonperturbative Canonical Gravity*, World Scientific, Singapore (1991).
47. H. García-Compeán, O. Obregón, C. Ramírez and M. Sabido, *Phys. Rev. D* **68**, 045010 (2003).
48. H. García-Compeán, O. Obregón, C. Ramírez and M. Sabido, *Phys. Rev. D* **68**, 044015 (2003).
49. H. Quevedo and J. G. Tafoya, *Gen. Rel. Grav.* **37**, 2083 (2005).
50. H. García-Compeán, O. Obregón and C. Ramírez, *Phys. Rev. Lett.* **88**, 161301 (2002).
51. H. García-Compeán and J. González-Beltrán, arXiv:hep-th/ 0404164.
52. L. O. Pimentel and C. Mora, *Gen. Rel. Grav.* **37**, 817 (2005).
53. H. García-Compeán, O. Obregón and C. Ramírez, *Gen. Rel. Grav.* **37**, 713 (2005).
54. H. García-Compeán and P. Paniagua, *Gen. Rel. Grav.* **37**, 723 (2005).
55. S. Estrada-Jiménez, H. García-Compeán and C. Soto-Campos, arXiv:hep-th/ 0404095.
56. J. C. López-Domínguez, O. Obregón and C. Ramírez, *Phys. Rev. D* **73** 095003 (2006).
57. A. Connes and D. Kreimer, *Lett. Math. Phys.* **48**, 85 (1999).
58. A. Connes and H. Moscovici, *Commun. Math. Phys.* **198**, 198 (1998).
59. D. Kreimer, *Adv. Theor. Math. Phys.* **2**, 303 (1998).
60. A. Connes and D. Kreimer, *Commun. Math. Phys.* **210**, 249 (2000).
61. C. Chryssomalakos, H. Quevedo, M. Rosenbaum and J. D. Vergara, *Commun. Math. Phys.* **225**, 465 (2002).
62. M. Rosenbaum, J. D. Vergara and H. Quevedo, *J. Geom. Phys.* **49**, 206 (2004).
63. H. García-Compeán, L. E. Morales and J. F. Plebanski, *Rev. Mex. Fis.* **42**, 695 (1996).
64. M. Durdevich, *Commun. Math. Phys.* **175**, 457 (1996).
65. M. Durdevich, *Rev. Math. Phys.* **9**, 531 (1997).
66. M. Durdevich, *Rep. Math. Phys.* **41**, 91 (1998).
67. C. Chryssomalakos and M. Durdevich, *Mod. Phys. Lett. A* **19**, 197 (2004).
68. M. Rosenbaum and J. D. Vergara, "Dirac Operator, Hopf Algebra of Renormalization, and Structure of Spacetime," in *Clifford Algebras and their Applications in Mathematical Physics*, edited by R. Ablamowicz, B. Fauser, Birkhäuser, Boston, 2000, pp. 283–302.
69. V. P. Nair and A. P. Polychronakos, *Phys. Lett. B* **505**, 267 (2001).
70. C. Sochichiu, "A Note on noncommutative and false noncommutative spaces," arXiv:hep-th/ 0010149.
71. A. Smailagic and E. Spallucci, *Phys. Rev. D* **65**, 107701 (2002)
72. J. M. Romero, J. A. Santiago and J. D. Vergara, *Phys. Lett. A* **310**, 9 (2003).
73. J. M. Romero and J. D. Vergara, *Mod. Phys. Lett. A* **18**, 1673 (2003).
74. C. A. Vaquera-Araujo and M. J. L. Lucio, "Non-commutative mechanics as a modification of spacetime," arXiv:math-ph/ 0512064.
75. M. Rosenbaum, J. D. Vergara and L. R. Juárez, *Phys. Lett. A* **354**, 389 (2006).

Hamiltonian methods: BRST, BFV

J. Antonio García

Instituto de Ciencias Nucleares
Universidad Nacional Autónoma de México
Apartado Postal 70-543, 04510
México, D.F.

Abstract. The range of applicability of Hamiltonian methods to gauge theories is very diverse and cover areas of research from phenomenology to mathematical physics. We review some of the areas developed in México in the last decades. They cover the study of symplectic methods, BRST-BFV and BV approaches, Klauder projector program, and non perturbative technics used in the study of bound states in relativistic field theories.
Keywords: Gauge field theories, BRST cohomology, BFV, BV, covariant methods.
PACS: 11.10.Ef, 11.15.-q, 11.30-j

1. QUANTIZATION OF GAUGE SYSTEMS, DIRAC METHOD, BRST-BFV AND BV

The theories that describe the fundamental interactions –electromagnetic, electroweak, strong and gravitational– are gauge theories. Recently another gauge theory, string theory, has attracted a lot of attention because it could be used to construct a consistent theory of quantum gravity[1] It is important to mention here that classical general relativity as a field theory can be quantized using the methods that we will describe here. This approach has been pursued by Ashtekar and collaborators in the last two decades. It is know as Loop Quantum Gravity. A basic property of such theories is that they have constraints among the fields and its conjugated momenta in phase space or among fields and its "velocities" in configuration space. This imply, in particular, that the physical degrees of freedom are not the same as the ones used to construct the theory from first principles. This type of theories are based on variational principles and symmetries that are cornerstones in the theoretical construction of any physical acceptable interacting theory. They have peculiar symmetries called gauge symmetries that are deeply connected with the fact that these theories have constraints. The interactions are constructed following the principle of gauge invariance. The systematic research of these type of theories was initiated by Dirac [1] whose aim was to construct a general procedure to quantize the general theory of relativity. From this seminal work a wide trend of research was opened: the study of the classical constrained dynamics, its consistency conditions and quantization.

Nowadays we have at our disposal some methods to analyze the dynamical consistency, the symmetries, physical content and quantization of a given constrained field

---

[1] String theory will be reviewed in another entry of this volume.

theory. They are the Dirac method and its extensions (see below), loop quantization, geometric quantization, Klauder projector program and symplectic covariant quantization.

Gauge theories can not be quantized using the standard canonical quantization in the operator or path integral approaches. One of the reasons is that not all the degrees of freedom of the theory are physical. Some of them must be eliminated because they are not gauge invariant observables, i.e., they transform under the gauge transformations. At the end when theory is compared with experimental results these degrees of freedom must be eliminated using a covariant or noncovariant method. The power of covariant methods is that the gauge symmetry (and some other global symmetries like Poincaré symmetry) can be used to prove the renormalizability and unitarity of the given theory.

With the aim to try to understand in better grounds the structure and physical content of a gauge theory we can analyze the intrinsic properties (independently from the constraint algebra) of the gauge theory and the gauge fixing method. This perspective helps also in the understanding of anomalies, renormalizability, consistent interactions, structure of the path integral measure, among others. A systematic Hamiltonian approach based on the BRST symmetry [2] developed by E. S. Fradkin and his collaborators [3], the BFV (Batalin, Fradkin, Vilkoviskii) method take full advantage of this perspective. Some of the properties of this powerful approach based on the Dirac method are

- It can be applied to open algebras (algebras that closes when the equations of motion are taken into account)
- The BRST transformation is based on intrinsic properties of the constraint surface. The Noether generator of the transformation is a classical object that can be quantized using the standard canonical approach. It is nilpotent to implement the constraint surface and the Poisson structure in the dynamics.
- It is fully based on the Hamiltonian formalism and allows the use of Hamiltonian technics like Liouville measure, canonical transformations, definition of the kinematic Hilbert space, among others.

As the Dirac method, the BFV approach can be very difficult to implement in systems with general covariance or in systems where the first and second class constraints can not be separated in a covariant way. This last point is crucial in the Green-Schwarz approach to the superstring.

An alternative Lagrangian approach know as BV (Batalin, Vilkoviskii) [4], is covariant and implement the dynamics on cohomology through the Kozul-Tate resolution. This deep property allow this method to be used for the study of anomalies, consistent deformations of a given theory, and renormalizability using covariant technics.

In this context a wide range of applications and intrinsic studies of these methods was developed in the last decades. On one hand, the study of the classical and quantum properties of constrained systems, and intrinsic properties of the Dirac method [5, 6, 7, 8, 9, 10]. On the other, applications of Dirac method to Ashtekar formulation of general relativity [11, 12, 13] and relational dynamics [15]. The BFV method has been extended to systems with time dependent constraints [16], and its formulation in terms of the Schwinger quantum principle was studied in [17]. An application of the BV approach to the study of the general form of the strict gauge invariant observables of exotic gauge theories (theories with field tensors with mixed type symmetry) is [18]. This type of theories are relevant in recent studies about duality. In particular, they are dual to the Fierz-Pauli Lagrangian. From this perspective, the structure and coupling of this theories and the corresponding Dirac analysis of them in 4D was studied in [19].

2. KLAUDER PROYECTOR METHOD

As an alternative approach to Hamiltonian Quantization, the physical Projector Operator Approach initiated and implemented by Klauder using coherent state techniques [20], has been applied to some simple gauge invariant quantum mechanical models. In this approach gauge fixing is not necessary and thus, it could avoid [21] potential Gribov ambiguities [22] which arise in the quantization of gauge invariant systems. As is well known, some gauges may suffer Gribov ambiguities. In fact it is only for an admissible gauge fixing that we can define the correct dynamical description of the system in reduced space. This admissible gauges must be well defined globally and this property is crucial in the description of non perturbative phenomena. The aim of the Projector Operator Approach is to construct a systematic method to project the dynamics in the space defined by the solutions to a given field theory. In this way the gauge fixing procedure is avoided.

These aspects of gauge invariant systems were explicitly analyzed within a solvable U(1) gauge invariant quantum mechanical model [23] related to the dimensional reduction of Yang-Mills theory. In this model, even at the classical level, one can parameterize the space of gauge orbits in terms of a classical parameter called the Teichmüller parameter [24]. It is through this parameter that all the gauge orbits are included in the quantization of the system, in agreement with the Friedberg *et al.* [23] point of view that all such gauge copies should be included in a correct quantization of gauge systems. These points were discussed and analyzed in Refs.[25, 26].

During the process of quantization of physical systems through Hamiltonian procedures, it was necessary to investigate the general construction of self-adjoint configuration space representations of the Heisenberg algebra over arbitrary manifolds not necessarily cartesian or parameterized with cartesian coordinates. All such inequivalent representations are parameterized in terms of the topology classes of flat U(1) bundles over the configuration space manifold. In the case of Riemannian manifolds, these representations are also manifestly diffeomorphic covariant. The general discussion, illustrated by some simple examples in non relativistic quantum mechanics, is of particular relevance to systems whose configuration space is parameterized by curvilinear coordinates or is not simply connected, which thus include for instance the modular spaces of theories of non abelian gauge fields and gravity. This was the main motivation of Ref. [27].

Finally, in order to study Hamiltonian gauge invariance, a Hamiltonian version of the Noether theorem for constrained systems is formulated in [28]. In particular, a novel method is presented to show that the gauge transformations are generated by the conserved quantities associated with the first class constraints. These results are applied to the relativistic point particle, to the Friedberg *et al.* model and, with special emphasis, to two time physics.

3. SYMPLECTIC GEOMETRY IN GAUGE THEORY

The symplectic geometry constitutes a modern Hamiltonian scheme in the study of symmetries and quantization of gauge theories from a geometrical point of view. The basic idea of this scheme is the construction of a Hamiltonian structure on the phase space of the theory which contains all physically relevant information, and does not require the choice of phase space coordinates p's and q's as in the traditional approach. Geometrically the Hamiltonian structure plays the role of a field strength, obtained by (*exterior*) derivative from a symplectic potential, which can be considered as a gauge field on the phase space.

The symplectic scheme was originally introduced by Witten *et al.* with applications to Yang-Mills theory, General Relativity, and string field theory [29]. The applications to string/brane theory have been given recently revealing a rich underlying geometrical structure of the theory [30]. Additionally we have undertaken the study of topological gauge theories, which are relevant in the context of formulating background-independent theories, and in the construction of topological invariants of four-manifolds. Specifically the symplectic scheme reveals that the topological action related with the Euler characteristic of the world-sheet in string theory mimics the geometrical structure of a two-dimensional gauge theory [31]. Moreover, the corresponding lower bound state for that topological string action is given by a loop state, described by a Wilson loop along the spatial configuration of the string [32].

On the other hand, the symplectic scheme allows us to prove that some properties of instantons in Yang-Mills theory traditionally associated with their self-duality, actually come from their topological nature. Specifically the only solution for the Schrödinger equation for quantum Yang-Mills theory known as the Chern-Simons wave functional and associated with instantons, exists actually for the (four-dimensional) topological Yang-Mills theory, reducing the self-dual property to a spurious condition [33]. This result can be generalized for topological actions and Chern-Simons functionals in spaces of even dimension.

Classically all topological action associated with curvatures coming from a gauge connection can be expressed as the (*exterior*) derivative of a Chern-Simons form; the quantum Hamiltonian of the topological action has as its lower state a functional of the corresponding Chern-Simons form [34], in such a way that the results previously described for topological Yang-Mills theory (and consequently for the absolute minimum of conventional Yang-Mills theory) corresponds only to a particular case. Furthermore, starting from the topological Yang-Mills theory, it can be proved that a moduli space of finite dimension can be obtained without invoking self-duality, leading to the idea of *fluctons* [35].

4. HAMILTONIAN NON PERTURBATIVE METHODS IN GAUGE THEORIES

A given Hamiltonian H is split into a "free" (solvable) part H_0 and an "interacting" part H_1, $H = H_0 + H_1$. Then a generalization of the Gell-Mann–Low Theorem [36] provides a similarity transform U_{BW} from any H_0-invariant subspace Ω_0 of the Hilbert space to an exactly H-invariant subspace Ω. Consequently, the diagonalization of H in Ω is equivalent to the diagonalization of an "effective" Hamiltonian H_{BW} in Ω_0. The map U_{BW}, and hence the Hamiltonian H_{BW}, are given in terms of a perturbative series. However, the results of the application of the generalized Gell-Mann–Low theorem will typically be nonperturbative.

This framework has so far been applied to the bound state problem in quantum field theory, by taking Ω_0 as the subspace of Fock space consisting of all H_0- (and momentum-) eigenstates of the would-be constituents as free particles. The effective Hamiltonian then consists of the relativistic kinetic energies of the constituents and an effective potential for their interaction generated as a perturbative series. The solution of the corresponding Schrödinger equation yields (to any finite order of the perturbative series an approximation to) the bound state energies of the full theory and the wave functions of the constituents (for N-particle bound states, the N-particle components of the full states in Fock space).

Applications to two-particle bound states in the Wick-Cutkosky model, Yukawa theory and Coulomb gauge QED, including a determination of the lowest-order fine and hyperfine structures in these theories, can be found in Ref. [37]. The most important results, particularly in comparison with the Bethe-Salpeter approach, are

- UV divergencies (as far as they have appeared in the lowest-order calculations) can be absorbed in a renormalization of the parameters. The renormalization procedure can be set up entirely in the Hamiltonian framework.
- The nonrelativistic and one-body limits are particularly transparent and lead to the correct results.
- Even in Yukawa theory, the effective Schrödinger equation is a well-defined eigenvalue equation.
- No abnormal solutions have been found. All solutions are consistent with physical expectations (symmetry properties).

ACKNOWLEDGMENTS

It is a pleasure to thanks V. M. Villanueva, R. Cartas-Fuentevilla and A. Weber for their contribution to sections 2, 3, and 4 respectively.

REFERENCES

1. P. A. M. Dirac, "Generalized Hamiltonian Dynamics," *Can. J. Math.* **2** (1950) 129-148; " Lectures on Quantum Mechanics" (Yeshiva Univ. Press, New York, 1964). P. A. M. Dirac,
2. C. Becchi, A. Rouet and R. Stora, "Renormalization Of Gauge Theories," Annals Phys. **98**, 287 (1976). C. Becchi, A. Rouet and R. Stora, "The Abelian Higgs-Kibble Model. Unitarity Of The S Operator," Phys. Lett. B **52**, 344 (1974). I. A. Batalin and G. A. Vilkovisky, "Relativistic S Matrix Of Dynamical Systems With Boson And Fermion Constraints," Phys. Lett. B **69**, 309 (1977). I. A. Batalin and I. V. Tyutin, "Existence theorem for the effective gauge algebra in the generalized canonical formalism with Abelian conversion of second class constraints," Int. J. Mod. Phys. A **6**, 3255 (1991).

3. E. S. Fradkin and T. E. Fradkina, "Quantization Of Relativistic Systems With Boson And Fermion First And Second Class Constraints," Phys. Lett. B **72**, 343 (1978). I. A. Batalin and E. s. Fradkin, "A Generalized Canonical Formalism And Quantization Of Reducible Gauge Theories," Phys. Lett. B **122**, 157 (1983). I. A. Batalin and E. S. Fradkin, "Operator Quantization Of Relativistic Dynamical Systems Subject To First Class Constraints," Phys. Lett. B **128**, 303 (1983) [Sov. J. Nucl. Phys. **39**, 145.1984 YAFIA,39,231 (1984 YAFIA,39,231-239.1984 YAFIA,41,278-281.1985)]. J. Math. Phys. **25**, 2426 (19yy).
4. I. B. Batalin and G.A.Vilkovisky, *Phys. Lett.* **B102** (1981), 27; *Phys. Rev.* **D28** (1983), 2567.
5. M. Henneaux, C. Teitelboim and J. D. Vergara, "Gauge invariance for generally covariant systems," Nucl. Phys. B **387**, 391 (1992) [arXiv:hep-th/9205092].
6. M. Montesinos and J. D. Vergara, "Linear constraints from generally covariant systems with quadratic constraints," Phys. Rev. D **65**, 064002 (2002) [arXiv:gr-qc/0111006].
7. J. A. Garcia and J. M. Pons, "Lagrangian Noether symmetries as canonical transformations," Int. J. Mod. Phys. A **16**, 3897 (2001) [arXiv:hep-th/0012094].
8. J. M. Pons and J. A. Garcia, "Rigid and gauge Noether symmetries for constrained systems," Int. J. Mod. Phys. A **15**, 4681 (2000) [arXiv:hep-th/9908151].
9. J. A. Garcia and J. M. Pons, "Equivalence of Faddeev-Jackiw and Dirac approaches for gauge theories," Int. J. Mod. Phys. A **12**, 451 (1997) [arXiv:hep-th/9610067].
10. J. A. Garcia and J. M. Pons, "Faddeev-Jackiw approach to gauge theories and ineffective constraints," Int. J. Mod. Phys. A **13**, 3691 (1998) [arXiv:hep-th/9803222].
11. H. A. Morales-Tecotl, L. F. Urrutia and J. D. Vergara, "Reality conditions for Ashtekar variables as Dirac constraints," Class. Quant. Grav. **13**, 2933 (1996) [arXiv:gr-qc/9607044].
12. M. Montesinos and J. D. Vergara, "Gauge invariance of complex general relativity," Gen. Rel. Grav. **33**, 921 (2001) [arXiv:gr-qc/0010113].
13. M. Montesinos, H. A. Morales-Tecotl, L. F. Urrutia and J. D. Vergara, "Real sector of the non-minimally coupled scalar field to self-dual gravity," J. Math. Phys. **40**, 1504 (1999) [arXiv:gr-qc/9903043].
14. A. Ashtekar, L. Bombelli and A. Corichi, "Semiclassical states for constrained systems," Phys. Rev. D **72**, 025008 (2005) [arXiv:gr-qc/0504052].
15. M. Montesinos, C. Rovelli and T. Thiemann, "An SL(2,R) model of constrained systems with two Hamiltonian constraints," Phys. Rev. D **60**, 044009 (1999) [arXiv:gr-qc/9901073].
16. J. A. Garcia, J. D. Vergara and L. F. Urrutia, "BRST-BFV method for nonstationary systems," Phys. Rev. D **51**, 5806 (1995) [arXiv:hep-th/9608178].
17. J. A. Garcia, J. D. Vergara and L. F. Urrutia, "BRST-BFV quantization and the Schwinger action principle," Int. J. Mod. Phys. A **11**, 2689 (1996) [arXiv:hep-th/9511092].
18. J. A. Garcia and B. Knaepen, "Couplings between generalized gauge fields," Phys. Lett. B **441**, 198 (1998) [arXiv:hep-th/9807016].
19. H. Casini, R. Montemayor and L. F. Urrutia, "Dual theories for mixed symmetry fields. Spin-two case: (1,1) versus (2,1) Young symmetry type fields," Phys. Lett. B **507**, 336 (2001) [arXiv:hep-th/0102104]. H. Casini, R. Montemayor and L. F. Urrutia, "Duality for symmetric second rank tensors. I: The massive case," Phys. Rev. D **66**, 085018 (2002) [arXiv:hep-th/0206129]. H. Casini, R. Montemayor and L. F. Urrutia, "Duality for symmetric second rank tensors. II: The linearized gravitational field," Phys. Rev. D **68**, 065011 (2003) [arXiv:hep-th/0304228].
20. Coherent states quantization of constraint systems, By J.R. Klauder, Published in Ann. Phys.254:419-453, 1997; Universal procedure for enforcing quantum constarints, By J. Klauder, Published in Nucl. Phys. B547:38-47, 1999; Quantization of constrained systems, By J.R. Klauder, Published in Lect. Notes Phys. 572:143-182, 2001.
21. Proyector operator approach to constrained systems, By J. Govaerts, Published in J. Phys. A30:603-617, 1997.
22. Quantization of non-abelian gauge theories, By V.N. Gribov, Published in Nucl. Phys. B139:1-19,1978.
23. A soluble gauge model with gribov type copies, By R. Friedberg, T.D. Lee, Y. Pang and H.C. Ren, Published in Ann. Phys. 246: 381-445,1996.
24. Hamiltonian quantization and constrained dynamics, J. Govaerts, (Leuven University Press, Leuven, 1991).

25. Quantizing gauge theories without gauge fixing: The physical proyector. V.M. Villanueva (IFM-UMSNH, Michoacan), J. Govaerts (Louvain U.), J.L. Lucio (Guanajuato U.),. Jun 2000. 7pp. Published in *Moscow 2000, Quantization, gauge theory, and strings, vol. 2* 270-276
26. Quantization without gauge fixing: avoiding Gribov ambiguities through the physical proyector. By Victor M. Villanueva (IFM-UMSNH, Michoacan), Jan Govaerts (Louvain U.), Jose-Luis Lucio-Martinez (Guanajuato U.),. Sep 1999. 21pp. Published in J.Phys.A33:4183-4202,2000
27. *Topology classes of flat U(1) bundles and diffeomorphic covariant representations of the Heisenberg algebra.* By Jan Govaerts (Louvain U.), Victor M. Villanueva (IFM-UMSNH, Michoacan),. Aug 1999. 21pp. Published in Int.J.Mod.Phys.A15:4903-4932,2000
28. Hamiltonian Noether theorem for gauge systems and two time physics. By V.M. Villanueva (IFM-UMSNH, Michoacan), J.A. Nieto, L. Ruiz, J. Silvas (Sinaloa U.),. Mar 2005. 19pp. Published in J.Phys.A38:7183-7196,2005
29. E. Witten, Nucl. Phys. **B276**, 291 (1986); C. Crncović and E. Witten, in *Three Hundred Years of Gravitation*, edited by S. W. Hawking and W. Israel (Cambridge University Press. Cambridge, 1987). C. Crnkovic, Nucl. Phys., **B288**, 419 (1987);
30. R. Cartas-Fuentevilla, Phys. Lett. B., **536**, 283 (2002); **536** 289 (2002); **563**, 107, 2003; Class. Quant. Grav. **19**, 3571 (2002); B. Carter, Int. J. Theor. Phys. **42**, 1317 (2003).
31. R. Cartas-Fuentevilla, math-ph/0404011.
32. R. Cartas-Fuentevilla, hep-th/0411251, to be published, Int. J. Theor. Phys. (2006).
33. R. Cartas-Fuentevilla, and F. Tlapanco-Limón, Phys. Lett. B., **623**, 165 (2005).
34. R. Cartas-Fuentevilla, *Quantization of the Chern invariant polynomial and its topological quantum ground state*; to be published, Ann. Phys. (2006).
35. R. Cartas-Fuentevilla, and J.M. Solano-Altamirano, *Fluctons*; in preparation (2006).
36. A. Weber, in *Particles and Fields — Seventh Mexican Workshop*, Eds. A. Ayala, G. Contreras und G. Herrera, AIP Conf. Proc. 531 (AIP, New York, 2000), p. 305, hep-th/9911198.
37. A. Weber und N.E. Ligterink, Phys. Rev. D **65**, 025009 (2002); preprint hep-ph/0506123; A. Weber, hep-ph/0509019.

The Casimir Effect

C. Villarreal[*] and W. L. Mochán[†]

[*]Instituto de Física, Universidad Nacional Autónoma de México, Apartado Postal 20-362, 01000, México, D.F., México
[†]Centro de Ciencias Físicas, Universidad Nacional Autónoma de México, Apartado Postal 48-3, 62251 Cuernavaca, Morelos, México.

Abstract. A review of relevant contributions of Mexican researchers to the understanding of the Casimir effect is presented.

Keywords: Casimir forces, Lifshitz formula
PACS: 12.20.Ds,42.50.Ct, 42.50.Lc, 42.50.Nn,

1. INTRODUCTION

In 1948, H. B. G. Casimir [1] predicted that the fluctuations of the electromagnetic field within two perfectly conducting plates separated by a distance a would produce an attractive force between them with magnitude $F/A = -\pi^2 \hbar c/240 a^4$, where A is the area. However, it was not until half a century later that experimental studies finally reached the necessary accuracy to test in detail the theoretical predictions, and experimental precisions of the order of 1% have now become common [2], boosting theoretical investigations on the Casimir forces among different research groups. In this work, we briefly review investigations performed in Mexico devoted to the understanding of the quantum vacuum structure, specially, the Casimir effect. We apologize with our foreign colleagues whose work is not properly cited here. In the following review, we sketch the formalism for calculating the energy-momentum tensor of quantum fields subject to external constraints, based on a Green's function approach. We then discuss the use of this formalism to study the Casimir forces and energy density in perfectly conducting cavities, both at zero and finite temperatures. A similar formalism is employed in the next section, however introducing supplementary conditions, to study the Casimir forces for materials with arbitrary dielectric response and composition. Finally, Casimir-related effects associated to space-time deformations, non-inertial movement, or the action of strong external fields, are presented in the last section.

2. GENERAL FORMALISM

Consider an arbitrary classical field $\Phi_i(x)$ described by a Lagrangian density $\mathscr{L} \equiv \mathscr{L}(\Phi_i, \partial_\mu \Phi_i)$. According to the Stationary Action Principle, $\Phi_i(x)$ satisfies the Euler-Lagrange equations. In general, the free Lagrangian density is a quadratic function of the quantum fields and their space-time derivatives, thus yielding a wave equation $(\partial_\mu \partial^\mu - m^2)\Phi_i = 0$, with eigenfunctions determined by the boundary conditions of the system under consideration. The energy-momentum tensor of the field is calculated as $T^{\mu\nu}(x) = (\partial \mathscr{L}/\partial \Phi_{i,\nu})\Phi_{i,\mu} - \mathscr{L}g^{\mu\nu}$, with $,\nu \equiv \partial_\nu$. Quantization of the theory is achieved by replacing the classical fields by quantum field operators $\phi_i(x) \to \hat{\phi}_i(x)$, which admit a linear expansion in terms of orthogonal eigenfunctions of a wave equation $\hat{\phi}_i(x) = \sum_n \left(\hat{a}_n \phi_{n,i}(x) + \hat{a}_n^\dagger \phi_{n,i}^*(x) \right)$, where the creation, $\hat{a}_n^\dagger$, and annihilation, $\hat{a}_n$, operators satisfy either bosonic $[\hat{a}_n, \hat{a}_{n'}^\dagger] = \delta_{n,n'}$ or fermionic commutation relations $\{\hat{a}_n, \hat{a}_{n'}^\dagger\} = \delta_{n,n'}$. From the structure of the above relations it follows that the vacuum expectation value (VEV) of the energy-momentum tensor, is straightforwardly expressed in terms of a Green's function of the wave equation satisfied by the quantum field $< T^{\mu\nu}(x) >= \lim_{x \to x'} \Delta^{\mu\nu}(x,x') G(x,x')$, where $\Delta^{\mu\nu}(x,x')$ represents a differential operator, and $G(x,x')$ is a solution of either a homogeneous or inhomogeneous Klein-Gordon or wave equation. The spectral density of the physical variables is obtained by means of the Fourier transform of the VEV of fields separated by a temporal interval $t - t' = \sigma$. The spectrum of the energy momentum tensor is $< T^{\mu\nu}(\mathbf{r}; \omega) >= \int d\sigma e^{i\omega\sigma} < T^{\mu\nu}(\mathbf{r}; \sigma) >$. We discuss the use of these relationships in the next sections. Unless otherwise stated, we use units with $\hbar = c = k_b = 1$.

3. CASIMIR FORCES IN PERFECTLY CONDUCTING CAVITIES.

The Casimir forces for perfectly conducting cavities have been studied for a long time by a variety of approaches [3, 4, 5, 6, 7]. In [3] it was shown that they may be regarded as a consequence of the radiation pressure associated to the zero-point energy $\hbar\omega/2$. On the other hand, the spectrum of quantum fluctuations and the energy-momentum tensor of the electromagnetic field between parallel plates at zero were studied in [4].

In reference [4], the Maxwell equations $\partial_\nu F^{\mu\nu} = 0$, and $\partial_\rho F_{\mu\nu} + \partial_\nu F_{\rho\mu} + \partial_\mu F_{\nu\rho} = 0$, are solved in the Lorentz gauge, $\partial^\mu A_\mu = 0$, and subject to the boundary conditions that the tangential component of the electric field $F_{0i}(x)$, and the normal component of the magnetic field $F_{ij}(x)$, vanish at the inner surfaces of the conductors, where the index notation $a, b = 0, 1$, and $i, j = 2, 3$, is used. The planar symmetry of the system permits to impose additional constraints $\partial_a A^a = 0$, and $\partial_i A^i = 0$, thus indicating the existence of scalar Hertz potentials, ϕ and ψ, such that the electromagnetic four-potential $A_\mu = (\partial_1 \psi, -\partial_0 \psi, \partial_3 \phi, -\partial_2 \phi)$. These potentials represent the two independent degrees of freedom of the field, yielding transverse electric (TE) and transverse magnetic (TM) electromagnetic modes, respectively. They satisfy the wave equations $\partial_\mu \partial^\mu \phi = 0 = \partial_\mu \partial^\mu \psi$, and Dirichlet, $\phi = 0$, or Neumann, $\partial_1 \psi = 0$, boundary conditions. Quantization is achieved by writing the solutions of Maxwell's equations as a linear superposition of primitive modes, and the bosonic commutation relations. This procedure allows the calculation of scalar Wightman correlation functions $D(x,x') =< \hat{\phi}(x)\hat{\phi}(x') >$ and $N(x,x') =< \hat{\psi}(x)\hat{\psi}(x') >$, whose linear combinations yields an image-like result:

$$N \pm D = \frac{1}{2\pi^2} \sum_n \frac{1}{(x \mp x' - 2an)^2 + (y-y')^2 + (z-z')^2 - (t-t'+i\varepsilon)^2}. \quad (1)$$

Notice that the term with $n = 0$ leads to the usual free-space correlation $\sim 1/(x-x')^\mu (x-x')_\mu$ which diverges in the limit $x \to x'$. Therefore, field correlations are regularized by subtracting this contribution. The electromagnetic correlation function $D_{\mu\nu}(x,x') = <A_\mu(x)A_\nu(x')>$ is then expressed as a block diagonal matrix with non-null elements $D_{ab} = \varepsilon(a,b)\partial_a\partial_{b'}N(x,x')$, and $D_{ij} = \varepsilon(i,j)\partial_i\partial_{j'}D(x,x')$, with $\varepsilon(r,s) = \pm 1$, and the upper sign holds for $r = s$, and the lower for $r \neq s$. As a result, the energy-momentum tensor can be obtained as the limit $x \to x'$ of $T^{\mu\nu}(x,x') = 2\partial^\mu\partial^\nu D(x,x')$. Considering instead the limit $\mathbf{x} \to \mathbf{x}'$, but $t - t' = \sigma$, we get

$$T^{\mu\nu}(\sigma) = \frac{1}{\pi^2}\sum_n{}' \frac{\text{diag}(3\sigma^2 + (2an)^2, \sigma^2 + 3(2an)^2, \sigma^2 - (2an)^2, \sigma^2 - (2an)^2)}{[(\sigma - i\varepsilon)^2 - (2an)^2]^3}, \quad (2)$$

where the prime in the summation means that the $n = 0$ term is omitted. In the limit $\sigma \to 0$ the usual expressions for the energy per unit volume $\mathcal{E} = T^{00} = -\pi^2/720$ and $P_x = T^{11} = \pi^2/240$ are recovered. Notice, however, that the pressure in the y and z directions is positive, $P_y = P_z = \pi^2/720$, in accordance with the null trace property of the energy-momentum tensor: $T^{00} = T^{11} + T^{22} + T^{33}$.

By performing a Fourier transformation, the spectrum of the energy density is expressed as $\mathcal{E}(\omega) = \rho(\omega)\hbar\omega/2$, where the mode distribution per unit volume is

$$\rho(\omega) = \rho(\omega)^\infty + \rho(\omega)^f = \frac{\omega^2}{\pi^2} + \frac{\omega}{\pi^2 a}\frac{\pi - 2\omega a}{2}, \quad (3)$$

where last term is periodic, with period 2π. Here, ρ^∞, arises from the $n = 0$ term, and represents Weyl's asymptotic mode density for a large cavity, yielding an infinite result when integrations over all frequencies are performed; on the other hand, $\rho^f(\omega)$ has the form of a saw-tooth, and leads upon integration to the finite value of the Casimir energy density.

The saw-tooth structure of the mode density of vacuum electromagnetic fields in the presence of conducting plates induces modifications in the spontaneous emission rates of atomic radiators placed in the inner region. The interaction of matter and radiation is described by a term $V(x) = -ej^\mu(x)A_\mu(x)$, where e is the electric charge, and j^μ the matter current satisfying the continuity equation $\partial_\mu j^\mu = 0$. First order perturbation theory then yields the transition rates between states i and f for an atomic system with a dipole moment $\mathbf{p}_{fi}$ either parallel or perpendicular to the plates. Denoting by ω_0 the frequency of a dipole atomic transition, and A the Einstein coefficient for spontaneous emission, they are expressed as

$$w^\parallel = \frac{3\pi A}{2\omega_0 a}\sum_{l=1}^{\omega_0 a/\pi}\left(1 - \frac{l^2\pi^2}{\omega_0^2 a^2}\right)\sin^2(l\pi x/a), \quad (4)$$

$$w^\perp = \frac{3\pi A}{\omega_0 a}\left[\frac{1}{2} + \sum_{l=1}^{\omega_0 a/\pi}\left(1 - \frac{l^2\pi^{o2}}{\omega_0^2 a^2}\right)\cos^2(l\pi x/a)\right], \quad (5)$$

where x stands for the position of the atomic radiator relative to one of the plates. Therefore, the spontaneous emission of an atomic system may be either enhanced or suppressed depending on the separation between the cavity plates and its relative position with respect to them.

More complicated geometries like rectangular waveguides and cavities were studied in [5, 6]. In Ref.[5] the spectrum of quantum noise was studied, concluding that the eigenfrequencies in a waveguide show a very slow convergence to a Poissonian distribution, proper of integrable systems. On the other hand, the Casimir energy-momentum tensor inside single rectangular cavities with edge sizes a_i and volume $V = a_1 a_2 a_3$ was calculated in [6]. In that case, an expression for the energy density was derived

$$\mathcal{E}(\sigma) = -\frac{1}{\pi^2} \sum_{l,m,n} \frac{3\sigma^2 + u_{lmn}^2}{[u_{lmn}^2 - \sigma^2]^3} + \frac{1}{4\pi V} \sum_{i=1}^{3} a_i \sum_{n_i} \frac{\sigma^2 + (2a_i n_i)^2}{[(2a_i n_i)^2 - \sigma^2]^2}, \tag{6}$$

with $u_{lmn}^2 = (2a_1 l)^2 + (2a_2 m)^2 + (2a_3 n)^2$, while the pressure acting at the wall, with normal directed along the direction a_i, is:

$$P_i(\sigma) = -\frac{1}{\pi^2} \sum_{l,m,n} \frac{4(2a_i l)^2 - u_{lmn}^2 + \sigma^2}{[u_{lmn}^2 - \sigma^2]^3} + \frac{V}{4\pi a_i} \sum_{l} \frac{\sigma^2 + (2a_i l)^2}{[(2a_i l)^2 - \sigma^2]^2}. \tag{7}$$

The terms with $l = m = n = 0$ and $n_i = 0$ must be excluded since they lead to divergent quantities as $\sigma \to 0$ irrespective of the box size. For example, for the energy density we have: $\mathcal{E}(\sigma) = 3\pi^{-2}\sigma^{-4} + (4\pi V)^{-1}(a_1 + a_2 + a_3)\sigma^{-2} + E^f(\sigma)$, where E^f is finite as $\sigma \to 0$. As before, the energy density spectrum can be written in the form $\mathcal{E}(\omega) = \rho(\omega)\hbar\omega/2$, but with a mode density now given by

$$\rho(\omega) = \frac{\omega^2}{\pi^2} \sum_{l,m,n} j_0(\omega u_{lmn}) - \sum_{i=1}^{3} \frac{a_i}{2\pi V} \sum_{n_i} \cos(2a_i n_i \omega), \tag{8}$$

which, when isolating the terms with $l = m = n = 0$ and $n_i = 0$ can be reexpressed as $\rho(\omega) = \rho^\infty(\omega) + \rho^f(\omega) \equiv \omega^2/\pi^2 - (a_1 + a_2 + a_3)/(2\pi V) + \rho^f(\omega)$. In this case, $\rho^\infty(\omega)$ represents Weyl's asymptotic mode density for large cavities corrected by edge effects. Since this expansion is valid in the limit $V^{1/3}\omega \to \infty$, the terms associated with ρ^∞ must be discarded in the calculations. A relevant prediction of this study is that Casimir energies and pressures on the different walls of the cavity can be either positive or negative, depending on the specific geometry. For instance, all of these quantities are positive in a cube, confirming the results obtained by other authors [8] using alternative approaches and regularization methods.

The finite temperature Casimir effect in closed cavities was recently discussed by in Ref.[9]. In that paper, fully regularized expressions for the thermodynamic variables associated to the electromagnetic field inside a conducting box were obtained. The regularization procedure used is based on the condition that in the zero temperature limit, the entropy should vanish: $S(T = 0^+) = -(\partial F/\partial T)|_{0^+} = 0$, which is equivalent to imposing the uniqueness of the electromagnetic vacuum. The thermodynamic quantities can be evaluated in terms of the mode density employed in the zero-temperature calculations $\rho(\omega) = \rho^\infty(\omega) + \rho^f(\omega)$. The free energy density is given by

$$\mathcal{F} = \int_0^\infty d\omega \left(\frac{\omega}{2} - \beta^{-1} \ln(1 + e^{-\beta\omega}) \right) \rho(\omega), \tag{9}$$

where $\beta = 1/T$. Now, the temperature dependent part of the free energy associated to ρ^∞ is finite: $\mathscr{F}^\infty = -(\pi^2 T^4)/45 - [\pi^2 T^2(a_1 + a_2 + a_3)]/12$, corresponding to the (large volume) Stefan-Boltzmann formula corrected by edge effects. The free energy arising from ρ^f is

$$F^f = \mathscr{E}_0 + \frac{\pi^2}{2}T^4 \sum_{\mathbf{n}}{}' h_V(v_\mathbf{n}) + \frac{\pi}{4V}T^2 \sum_{n_i}{}' \left[a_i f_E(2\pi T n_i a_i) \right], \quad (10)$$

where $\mathscr{E}_0$ is the Casimir energy density at zero temperature, and $f_{V,E}(v) = [g(v) - K_{V,E}(v)]/v$, $h_{V,E}(v) = h(v) + K_{V,E}(v)/v^3$, $g(v) = \coth(v) - 1/v$, and $h(v) = v^{-1}d(v^{-1}g(v))/dv$ are functions of the dimensionless variable $v_\mathbf{n} = \pi T u_\mathbf{n}$. Here, $K_{V,E}(v) = \Theta(v - v_{V,E})$, where $v_{V,E}$ is chosen so to guarantee that the entropy derived from (10) vanishes at zero temperature. Other thermodynamic variables may be derived from well known thermodynamic relations. A striking prediction of this approach is that, except for the internal energy, the other thermodynamic variables: free energy, entropy, and pressures at the cavity walls, show discontinuities as functions of the temperature. These predictions should be tested experimentally.

4. CASIMIR FORCES IN ARBITRARY MATERIALS

In 1956, Lifshitz proposed a macroscopic theory for two semi-infinite homogeneous dielectric slabs [10] characterized by complex frequency dependent dielectric functions $\varepsilon(\omega)$ and separated by a distance L. However, Lifshitz derivation is directly applicable only to semiinfinite homogeneous local media, and could not cope with more complex systems, such as thin films, layered systems, superlattices, photonic crystals and metamaterials. Based on a surface-plasmon mechanism, short-distance corrections to the Lifshitz formula were obtained in [11]. In particular, a finite value of the force was predicted for $L \to 0$, by assuming that for wavenumbers greater than K_c, the surface excitations decay into electron-hole pairs. An alternative derivation of Lifshitz force was proposed in [12, 13] using both a scattering approach and a dissipationless ancillary system. In those works, it is argued that in thermal equilibrium, all of the properties of the radiation field within a cavity are completely determined by the optical reflection amplitudes of the walls. Thus, the Casimir force may be obtained from the stress tensor of a fictitious system, as long as its reflection amplitudes are chosen to be identical to those of the real system. By choosing a fictitious system that has infinitesimal thin walls, and selecting their transmission amplitudes in such a way that their scattering matrix obeys the unitarity condition, it is ensured that the electromagnetic energy is conserved in the fictitious system and that no degrees of freedom beyond those of the field itself are excited. This permits a full quantum mechanical calculation of the fields, even when the real system is dissipative.

Consider a setup consisting of two planar slabs (termed 1 and 2), a distance L apart. The force per unit area of a slab can be calculated by subtracting the momentum flux of the electromagnetic field evaluated at both sides of the slab. Denoting by $f(k) = N_k + 1/2$ to the photon occupation number of state k, the momentum flux from the vacuum gap into a slab is [12, 13]: $T_{zz} = \sum_\alpha \int Q dQ dk^2 k^3 / q f(k) \rho^\alpha_{k^2}$, where $\mathbf{Q}$ and k represent the transverse and longitudinal momenta, $q = \omega/c$, and $\rho_{k^2}^\alpha$ is the density of states of α-polarized photons. A similar contribution exists related to the semi-infinite vacuum on the other side of the slab, obtained by substituting $r_2^\alpha \to 0$ above and reversing the flux direction $z \to -z$. The density of states is derived from the Green function of the electromagnetic field. For each $\mathbf{Q}$, it has the form

$$\rho_{k^2}^s(z) = -\frac{1}{2\pi}\mathrm{Im}\left(G_{\tilde{k}^2}^E(z,z) + G_{\tilde{k}^2}^B(z,z)\right), \tag{11}$$

with ($\tilde{k} \equiv k + i0^+$), where the E and B refer to electric and magnetic contributions. After explicit evaluation of the Green functions [12, 13] the total force per unit area is finally obtained:

$$T_{zz} = P_z = \frac{\hbar c}{2\pi^2}\int_0^\infty dQ Q \int_{q \geq 0} dk \frac{\tilde{k}^3}{q}\mathrm{Re}\frac{1}{\tilde{k}}\left[\frac{1}{\xi^s - 1} + \frac{1}{\xi^p - 1}\right], \tag{12}$$

with $\xi^\alpha = (r_1^\alpha r_2^\alpha \exp(2ikL))^{-1}$. The integral over k runs from iQ to 0 and then to ∞, so that q remains real and positive. This expression depends only on the reflection coefficients or, equivalently, on the surface impedances of the system and the slab separation. Lifshitz formula is recovered upon substitution of the *local* Fresnel amplitudes, whereas for perfect mirrors ($r_a^\alpha = \pm 1$) Eq.(12) yields the expected Casimir force. The main result is that if Lifshitz formula is written in terms of the reflection coefficients of the walls of the cavity, or equivalently, in terms of their exact surface impedance [14], it becomes applicable to any system with translational invariance along the surfaces and isotropy around their normal. Thus, it may be employed to calculate the Casimir force between semiinfinite or finite, homogeneous or layered, local or spatially dispersive, transparent or opaque, finite or semi-infinite systems. For example, through a simple substitution of the appropriate optical coefficients, the formalism has allowed the calculation of the Casimir force between photonic structures [15, 16], between non-local excitonic semiconductors [17] and between non-local plasmon-supporting metals with sharp boundaries [13, 19] and between realistic spatially dispersive metals with a smooth self-consistent electronic density profile [12, 18].

The relative simplicity of the formalism allowed its generalization to non-isotropic systems and the calculation of Casimir torques [20] in one dimensional systems. In that work, a formalism for the calculation of the flow of angular momentum carried by the fluctuating electromagnetic field within a cavity bounded by two flat anisotropic materials was performed. The formalism yields an expression for the torque between anisotropic plates in terms of their reflection amplitude coefficients. The torque between two opposite slabs forming an angle γ between their respective principal axes is

$$\tau_z = -\frac{1}{2\pi}\int dK \frac{\Delta r_1 \Delta r_2 \sin(2\gamma) e^{-2KL}}{\Delta r_1 \Delta r_2 \sin^2(\gamma) e^{-2KL}(1 - r_{1x}r_{2x}e^{-2KL})(1 - r_{1y}r_{2y}e^{-2KL})}. \tag{13}$$

where γ is the angle between the principal axes of the material. In the case of two perfectly reflecting mirrors covered by ideal perfectly absorbing polarizers, *i.e.*, $r_{x1} = r_{x2} = \pm 1$, $r_{y1} = r_{y2} = 0$, Eq.(13) can be integrated analytically, yielding a $1/L$

decay: $\tau_z = \tan\gamma\log[\sin^2\gamma]/2\pi L$. The same formula was also employed to study the torques between dichroic materials with absorptive and dissipative response functions, as well as the torque between dissimilar materials, suggesting procedures to manipulate it.

With a few modifications, a similar formalism has been employed for the calculation of other macroscopic Casimir-like forces, such as those due to electronic tunnelling across an insulating gap separating two conductors [21]. The expression for these forces is derived by calculating the flux of momentum arising from the overlap of evanescent fields, solutions of Schrödinger's equation. The tunnelling force per unit area is expressed in terms of the reflection amplitude coefficients of the materials, and it has the same structure as Lifshitz's formula [10] for electromagnetically mediated Casimir forces. For a configuration consisting of two conducting regions separated by a thin insulating gap $\mathscr{V}$, we have:

$$\mathscr{T}_z^z = \frac{1}{\pi^2}\mathrm{Re}\int dk \left(K_F - \frac{\hbar^2}{2m}k^2\right) k^2 \frac{1+r_1 r_2 e^{2ikL}}{1-r_1 r_2 e^{2ikL}}, \tag{14}$$

where the integration region includes all states below the Fermi level with kinetic energy K_F within $\mathscr{V}$ and for which the zero-temperature Fermi-Dirac distribution $f(E) = 2$, has been assumed, including the spin degeneracy. The formalism has been employed to calculate the electronic pressure within a conductor, the force between two semiinfinite conductors, and between two thin films at very small distances, of the order of the Fermi wavelength. The resulting forces in these latter cases is extremely small, but it could important to study the forces acting, for example, between the tip and the substrate of a scanning tunnelling microscope, or an atomic force microscope.

Spectral representation techniques have been also applied to calculate the nonretarded Casimir force between a spherical particle of radius R placed a distance z above a substrate with arbitrary dielectric properties [22]. In the dipole approximation, it is assumed that the quantum electromagnetic fluctuations induce a dipole moment $\mathbf{p}_{sph}$ at the sphere, inducing an image distribution on the substrate, which reacts back by modifying the sphere's dipole moment by a local field. The equation for the induced dipole moment is rewritten as an eigenvalue equation which provides the eigenfrequencies ω_s of the proper electromagnetic modes of the sphere-substrate system, and the zero-point energy can be easily calculated as $\mathscr{E} = \sum_s 1/2\hbar\omega_s$. The Casimir interaction $V(z)$ is obtained as the difference $V(z) = \mathscr{E}(z) - \mathscr{E}(\infty)$. It turns out that $V(z) \sim R^3/z^3$. The generalization to include higher n-pole moments is straightforward, and leads to interactions $V(z) \sim R^n/z^n$. The same formalism has been used to study setups involving dissimilar materials [23].

5. CASIMIR EFFECT IN NON-INERTIAL SYSTEMS AND EXTERNAL FIELDS.

Casimir-related effects arise when the spectral density of free vacuum fields $\mathscr{E}(\omega) = \omega^3/2\pi^2$ is distorted, not by the presence of physical boundaries, but by the action of strong external fields, or by deformations of the geometry of space-time, or in non-inertial systems. This kind of effects was studied in a series of works [24, 25] using a formalism similar to that presented in section 2, however generalized to arbitrary space-time metrics. Their results may be summarized in a formula for the VEV of the energy-momentum tensor for a massless field with spin s:

$$<T_\nu^\mu> = -\frac{32h(s)}{\pi^6}\int d\omega \frac{\omega(\omega^2+4\pi s^2 A^2)}{e^{\omega/A}-1} diag[-1,1/3,1/3,1/3], \qquad (15)$$

where ω is the frequency, $h(s) = 1, 2$ for $s = 0, 1$, and the parameter $A = 1/(8\pi GM)$ for a Schwarzchild black hole of mass M, or $A = a$, the uniform acceleration of a non-inertial observer, or $A = 1/R$ for an Einstein universe. The same techniques were applied to study the vacuum spectral density detected by observers in rotating motion [26], and uniformly accelerated observers inside a waveguide [27]. In both cases, a turbulent-like spectrum was gotten.

The energy-momentum tensor of the Dirac vacuum in presence of a magnetic field was calculated in Ref.[28]. In that work the energy density of the Dirac sea is considered $\mathscr{E}_0 \sim -\int d\mathbf{p}(\mathbf{p}^2+m^2)^2$, where $\mathbf{p}$ is the momentum, and m the electron mass. Under the action of a uniform magnetic field of strength B, the transverse component of the momentum is discretized $\mathbf{p}_\perp^2 = 2neB$ ($n = 0, 1, ...$), and the energy of the electron corresponds to the Landau levels: $\mathscr{E}_L^2 = m^2 + \mathbf{p}_\parallel^2 + 2neB$. The Casimir energy is then obtained as the difference $\Delta\mathscr{E} = \mathscr{E}_L - \mathscr{E}_0$. From the calculations it follows that in the presence of magnetic fields, the Dirac vacuum exhibits parallel and perpendicular pressures $P_\parallel = -\Delta E + (eB/m)^2/12\pi^2$, and $P_\perp = 2\Delta E - (eB)^4/90\pi^2$.

The properties of accelerated quantum systems were further discussed in [29], where the elementary quantum electrodynamic vertex was evaluated in Rindler coordinates. In that paper, general features of the emitted radiation were discussed. In particular, spin depolarization was predicted as a consequence of acceleration, even in absence of an external magnetic field.

The quantization of the electromagnetic field between moving uniformly boundaries in four-dimensional space-time was studied in [30]. The natural coordinates of the system are the Milne coordinates $t = \tau\cosh\zeta$, and $z = \tau\sinh\zeta$, where the world line of each plate can be taken as $\zeta = \pm\zeta_0$, with $\tanh\zeta_0$ the speed as seen in the center of velocity frame. Using the formalism depicted above, the tetradial components of the energy-momentum tensor $T_{(\mu\nu)}$ are calculated on a given hypersurface $\tau = $const.: $T_{(\mu\nu)} = \mathscr{E}_\mathscr{C} diag(1,-1,-1,3)$, with the Casimir energy density $\mathscr{E}_\mathscr{C}$ given by

$$\mathscr{E}_\mathscr{C} = \frac{-1}{8\pi^2\tau^4}\sum_{n=1}^{\infty}\sinh^{-4}(2\zeta_0 n) \approx -\frac{\pi^2}{720 L_t^4} + \frac{\zeta_0^2}{18 L_t^4}, \qquad (16)$$

where the last relationship holds in the nonrelativistic limit $\zeta_0 \ll 1$. The first term in the right hand side of the expansion represents the usual Casimir energy density between fixed plates, while the second term is a positive second-order correction associated to the motion of the plates.

Using the Bogoliubov transformation mechanism it is shown in Ref. [30] that the vacuum expectation value of the particle number in mode i is $P_i = <\hat{c}_i^\dagger \hat{c}_i + \hat{d}_i^\dagger \hat{d}_i>_{dyn} = \sum_j(|\beta_{ij}^D|^2 + |\beta_{ij}^N|^2)$, where the Bogoliubov coefficients β_{ij} are obtained from the inner product of operators pertaining to the dynamic and static Fock spaces. The crucial point is that the dynamical field with only positive frequency modes is a sum of both positive and negative frequency static modes, and thus the dynamical vacuum and static vacua are not equivalent. It turns out that the total energy per unit volume $\mathscr{E}$ produced by the motion of the plates is

$$\mathscr{E} = \frac{1}{2\pi^2 L_t} \sum_{n=0}^{\infty} \int d^2k \, \omega_t P_{n,\mathbf{k}_\perp} = \frac{\zeta_0^2}{18 L_t^4}, \qquad (17)$$

where a Riemann ζ function regularization was employed to evaluate the second equality. This expression coincides with the correction to the Casimir energy in (16) and it may be directly associated to photon production associated to the motion of the plates. Furthermore, another consequence of the Bogoliubov approach is that as the plates continue to move, the originally coherent vacuum state that becomes squeezed, *i.e.* the variances of the "position" and "momentum" operators, defined in terms of creation and annihilation operators, oscillate around the value $\hbar/2$.

In a subsequent work, the transition probabilities of an atomic detector placed between the moving plates were evaluated in the dipole approximation [31]. The interaction with the vacuum fluctuations of the quantum electromagnetic field induces modifications on the spontaneous emission probabilities. In the nonrelativistic limit, either enhancement or inhibition of the spontaneous emission is achieved, depending on the relative orientation of the detectort's dipole moment and the plates. In the ultrarelativistic limit, the standard expressions for free atomic transitions are recovered, since the time for signal exchange between detector and plates becomes exceedingly large.

The vacuum fluctuations and symmetry breaking mechanism in spaces with nontrivial topology were investigated in [32]. Feynman rules were obtained for a scalar field ϕ defined on a four-dimensional torus, and with a self-interaction $\lambda \phi^4$. This procedure permits to calculate the effective potential Ω up to the two-loop level in perturbation theory. The periodicities in space generate Casimir forces that tend to compactify the associated spatial dimensions, while the periodicity in the time allows to derive the finite temperature potential, according to Matsubara's formalism. It is found that for a critical value λ_c of the coupling constant a phase transition occurs, giving rise to a vacuum structure with stable compactified dimensions, thus suggesting to investigate the action of this Casimir compactification mechanism within the framework of n-dimensional gauge theories.

ACKNOWLEDGMENTS

This work was partially supported by DGAPA-UNAM under grants IN111306, and IN118605.

REFERENCES

1. H. B. G. Casimir, *Proc. Kon. Ned. Akad. Wet.* **51**, pp. 793–795 (1948).
2. S. K. Lamoreaux, *Phys. Rev. Lett.* **78**, pp. 5–8 (1997); H. B. Chan, V. A. Aksyuk, R. N. Kliman, D. J. Bishop and F. Capasso, *Science* **291**, pp. 1942–1945 (2001); U. Mohideen and Anushree Roy, *Phys. Rev. Lett.* **81**, pp. 4549-4552 (1998); R. S. Decca, E. Fischbach, G. L. Klimchitskaya, D. E. Krause, D. López, and V. M. Mostepanenko, *Phys. Rev. D* **68**, pp. 116003-1–10 (2003); F. Chen, G. L. Klimchitskaya, U. Mohideen, and V. M. Mostepanenko, *Phys. Rev. A* **69**, pp. 022117-1–8 (2004); H. B. Chan, V. A. Aksyuk, R,. N. Kleiman, D. J. Bishop, and Federico Capsso, *Phys. Rev. Lett.* **87**, pp. 211801–211805 (2001).
3. A. González, *Physica A* **131**, pp. 228–250 (1985).
4. S. Hacyan, R. Jáuregui, F. Soto, and C. Villarreal, *J. Phys. A* **23**, pp. 2401–2412 (1990).
5. C. Villarreal, R. Jáuregui, S. Hacyan, and G. Cocho, *Mod. Phys. Lett. A* **7**, pp. 2957–2960 (1992).
6. S. Hacyan, R. Jáuregui, F. Soto, and C. Villarreal, *Phys. Rev. A* **47**, pp. 4204–4211 (1993).
7. A. M. Cetto and L. de la Peña, *Il Nuovo Cim. B* **108**, pp 447–458 (1993).
8. W. Lukosz, *Physica* **56**, pp. 109–120 (1971); J. Ambjorn and S. Wolfram, *Ann. Phys. (N.Y.)* **147**, pp. 1–23 (1983).
9. R. Jáuregui, C. Villarreal, and S. Hacyan, *Ann. Phys. (N.Y.)* (2006). In press.
10. E. M. Lifshitz, *Soviet Physics*, **2**, pp. 73–108 (1956) [*J. Exper. Theoret. Phys.* USSR **29**, 94–128 (1955)].
11. R. G. Barrera and E. Gerlach, *Chem. Phys. Lett.* **25**, pp. 443-444 (1974).
12. W. Luis Mochán, Ana M. Contreras-Reyes, Raul Esquivel-Sirvent, and Carlos Villarreal in *Statistical Physics and Beyond: Proceedings of the Second Mexican Meeting on Mathematical and Experimental Physics (El Colegio Nacional, México, DF, septiembre 6-10, 2004)*, edited by Francisco Uribe, Leopoldo García Colín S., and Enrique Díaz-Herrera, AIP Conference Proceedings **757**, pp. 66-76 (2005).
13. R. Esquivel, C. Villarreal, W. Luis Mochán, *Phys. Rev. A* **68**, 052103 (2003); W. Luis Mochán, C. Villarreal, and R. Esquivel-Sirvent, *Rev. Mex. Fis.* **48**, pp. 339-342 (2002); Raul Esquivel, Carlos Villarreal, and W. Luis Mochán, *Phys. Rev. A* **71**, 029904 (2005); R. Esquivel-Sirvent, C. Villarreal, W. L. Mochán, and G. H. Cocoletzi, *Phys. Status Solidi B* **230**, pp. 409–413 (2002).
14. J. A. Stratton, *Electromagnetic Theory* (McGraw-Hill, N.Y., 1941).
15. R. Esquivel-Sirvent, C. Villarreal, and G. H. Cocoletzi, *Phys. Rev. A* **64**, pp. 052108-052111 (2001).
16. C. Villarreal, R. Esquivel-Sirvent and G. H. Cocoletzi, *Int. J. of Modern Phys. A* **17**, pp. 798–803 (2002).
17. A. D. H. de la Luz, A. F. Alvarado-Garcia, G. H. Cocoletzi et al., *Solid State Commun.* **132**, pp. 623–627 (2004).
18. Ana M. Contreras-Reyes and W. Luis Mochán, *Physical Review A* **72**, pp. 034102-1–6 (2005).
19. R Esquivel-Sirvent, C Villarreal, W L Mochán, A M Contreras-Reyes and V B Svetovoy, *J. of Phys. A: Math. Gen.* **39**, pp. 6323–6333 (2006).
20. José C. Torres-Guzmán and W. Luis Mochán, *J. of Phys. A: Math. Gen.* **39**, pp. 6791–6798 (2006).
21. LM Procopio, C Villarreal, and WL Mochán, *J. of Phys. A: Math. Gen.* **39**, pp. 6679–6686 (2006).
22. C. E. Roman-Velazquez, C. Noguez, C. Villarreal, and R. Esquivel-Sirvent, *Phys. Rev. A* **69**, 042109-1-5 (2004); C. Noguez, C. E. Roman-Velazquez, R. Esquivel-Sirvent, and C. Villarreal, *Europhys. Lett.* **67**, pp 191–194 (2004).
23. C. Noguez and C. E. Roman-Velazquez, *Phys. Rev. B* **70**, pp. 195412-1–8 (2004).
24. S. Hacyan, A. Sarmiento, G. Cocho, and F. Soto, *Phys. Rev. D* **32** pp. 914–922 (1985); F. Soto, G. Cocho, C. Villarreal, S. Hacyan, and A. Sarmiento, *Rev. Mex. Fis.* **33**, pp. 389-403 (1987); S. Hacyan, *Rev Mex. Fis.* **39**, pp. S154–S162 (1993).
25. G. Cocho, S. Hacyan, A. Sarmiento, and F. Soto, *Int. J. Theo. Phys.* **28**, pp. 699–705 (1985).

26. S. Hacyan and A. Sarmiento, *Phys. Rev. D* **40**, pp. 2641–2651 (1989).
27. A. Sarmiento, S. Hacyan, G. Cocho, F. Soto, and C. Villarreal, *Phys. Lett. A* **142** pp. 194–199 (1989).
28. S. Hacyan, R. Jáuregui, and M. Torres, *Phys. Lett. A* **150**, pp. 345–348 (1990).
29. R. Jáuregui, *Int. J. Mod. Phys. A* **10**, pp. 1483–1493 (1995).
30. C. Villarreal, S. Hacyan, and R. Jáuregui, *Phys. Rev. A* **52**, pp. 594–601 (1995).
31. R. Jáuregui and C. Villarreal, *Phys. Rev. A* **54**, pp. 3480–3488 (1996).
32. C. Villarreal, *Phys. Rev. D* **51**, pp. 2959–2967 (1995).

Anomalies and gravity

Eckehard W. Mielke[1]

*Universidad Autónoma Metropolitana–Iztapalapa,
Apartado Postal 55-534, C.P. 09340, México, D.F., MEXICO*

Abstract. Anomalies in Yang-Mills type gauge theories of gravity are reviewed. Particular attention is paid to the relation between the Dirac spin, the axial current j_5 and the non-covariant gauge spin C. Using diagrammatic techniques, we show that only generalizations of the $U(1)$- Pontrjagin four–form $F \wedge F = dC$ arise in the *chiral anomaly*, even when coupled to gravity. Implications for Ashtekar's canonical approach to quantum gravity are discussed.

Keywords: Gauge theories, anomalies, gravity, torsion, Pontrjagin terms
PACS: PACS no.: 04.50.+h; 04.20.Jb; 03.50.Kk

INTRODUCTION: ANOMALIES FOR PEDESTRIANS

Anomalies can be viewed as a breaking of some Noether symmetry through the effects of the vacuum. In relativistic quantum field theory (QFT), such a (classical) symmetry is *broken* by field quantization, cf. Refs [20, 7, 51] for recent reviews. This has important implications on such physical processes as the decay of the neutral π-meson [6], induced instanton effects [43], or underlies the postulation of the *axion* in quantum chromodynamics (QCD), cf. Ref. [41].

In quantum electrodynamics (QED), Schwinger [46] demonstrated that the charge current j can be retained conserved, i.e. $\langle dj \rangle = 0$, whereas the conservation of the axial current j_5 is broken, $\langle dj_5 \rangle \neq 0$.

In the Fujikawa approach [14], the right-hand side can obtain from considering the point-splitted current $j_5(x;\varepsilon) := \overline{\psi}(x)\gamma_5{}^*\gamma\psi(x+\varepsilon)$, where ε is an infinitesimal four-vector in spacetime. Such an expression can be rendered invariant by dressing it with a path-ordered exponential

$$\overline{\psi}(x)\gamma_5{}^*\gamma\psi(x+\varepsilon) \to \overline{\psi}(x)\gamma_5{}^*\gamma\psi(x+\varepsilon)\mathbf{P}\exp\left\{i\int_x^{x+\varepsilon} A\right\}. \tag{1}$$

The variation $\delta/\delta A$ of the current $j_5(x;\varepsilon)$ is compensated by the variation of the exponential. As the parallel transport from $x^i \to x^i + \varepsilon^i$ along the infinitesimal line element can be expanded perturbatively, it is clear that the net effect of this approach is just the standard result $\langle dj_5(x)\rangle = 2im\langle P\rangle - (1/96\pi^2)F \wedge F$ for massive fermions, where $F := dA$ is the gauge field strength. Further details of the path integral formulation were developed, e.g., in Refs. [2, 3, 50] with extension of the regularized Jacobian, as well as in the light-cone gauge [15] of the Schwinger model.

---

[1] E-mail: ekke@xanum.uam.mx

CP857, *Particles and Fields, Part B: Commemorative Volume of the Division of Particles and Fields of the Mexican Physical Society,* edited by M. A. Pérez, L. F. Urrutia, and L. Villaseñor
© 2006 American Institute of Physics 978-0-7354-0354-3/06/$23.00

There is an intuitive physical interpretation of this result: The additional Chern-Simons (CS) term $C := A \wedge dA$ corresponds to the spin or helicity of the photon, with its spacelike part $\vec{A} \cdot \vec{B}$ known as magnetic helicity [22]. Since the axial current j_5 is proportional to the spin of a fermion, the deformed current $\tilde{j}_5 := j_5 + (1/96\pi^2)A \wedge dA$, includes the spin of the photon, lacking, however, gauge invariance. The chiral anomaly can then be understood as the 'conservation law'

$$\langle d\tilde{j}_5 \rangle = 0, \qquad (2)$$

such that in QFT "...the flow of electronic spin drags some photon spin and vice versa" [53].

Anomalies were studied also in Yang-Mills type gauge models of gravity [19] with Einsteinian instanton solutions [40]. Then, the equivalence principle not only requires a coupling of gravity to the energy-momentum current of matter, but also to the spin current. Here we will focus on the intricate interaction between the *chiral anomaly* and the spin or helicity of the gravitational gauge field and extend it to post-Riemannian spacetimes with torsion.

DIRAC FIELDS IN RIEMANN–CARTAN SPACETIME

In our notation [30], a Dirac field is a bispinor–valued zero–form ψ for which $\overline{\psi} := \psi^\dagger \gamma_0$ denotes the Dirac adjoint. The minimal coupling to the gauge (electromagnetic) potential $A = A_i dx^i$ is accounted for via $\mathscr{D} := D + iA\wedge$, where $D\psi := d\psi + \Gamma \wedge \psi$ is the exterior covariant derivative with respect to the Riemann-Cartan (RC) connection one-form $\Gamma^{\alpha\beta} = \Gamma_i{}^{\alpha\beta} dx^i$.

The Dirac Lagrangian is given by the manifestly *Hermitian* four–form

$$L_D = L(\gamma, \psi, \mathscr{D}\psi) = \frac{i}{2}\{\overline{\psi}{}^*\gamma \wedge \mathscr{D}\psi + \overline{\mathscr{D}\psi} \wedge {}^*\gamma\psi\} + m\overline{\psi}\psi\eta, \qquad (3)$$

where $\gamma := \gamma_\alpha \vartheta^\alpha$ is the Clifford algebra-valued coframe, see the Appendix.

The Dirac equation and its adjoint can be obtained by varying L_D independently with respect to $\overline{\psi}$ and ψ. Making use of the torsion $\Theta := D\gamma$ and of the properties of the Hodge dual, the Dirac equation assumes the form

$$i^*\gamma \wedge \left(\mathscr{D} + \frac{i}{4}m\gamma - \frac{1}{2}T\right)\psi = 0, \qquad (4)$$

where $T := \frac{1}{4}Tr(\check{\gamma}\rfloor\Theta) = e_\alpha \rfloor T^\alpha$ is the one–form of the trace (or vector) torsion. However, the covariant derivative D also contains torsion.

In order to separate out the purely Riemannian piece from torsion terms, let us decompose the Riemann–Cartan connection $\Gamma = \Gamma^{\{\}} - K$ into the Riemannian (or Christoffel) connection $\Gamma^{\{\}}$ and the *contortion* one–form $K = \frac{1}{4}K^{\alpha\beta}\sigma_{\alpha\beta}$, obeying $D\gamma = [\gamma, K] = \gamma_\alpha T^\alpha$. Accordingly, the Dirac Lagrangian (3) splits [32] into a Riemannian and a spin–contortion piece:

$$L_D = L(\gamma, \psi, D^{\{\}}\psi) - \frac{i}{2}\overline{\psi}(^*\gamma \wedge K - K \wedge {}^*\gamma)\psi + A \wedge j$$

247

$$\begin{aligned} &= L(\gamma, \psi, D^{\{\}}\psi) + \frac{1}{4}\mathscr{A} \wedge j_5 + A \wedge j \\ &= L(\gamma, \psi, D^{\{\}}\psi) - T^\alpha \wedge \mu_\alpha + A \wedge j. \end{aligned} \quad (5)$$

The covariant derivative with respect to the Riemannian connection $\Gamma^{\{\}}$ satisfies $D^{\{\}}\gamma = 0$. Hence, in a RC spacetime a Dirac spinor only feels the *axial torsion* one–form

$$\mathscr{A} := \frac{1}{4}{}^*Tr(\gamma \wedge D\gamma) = {}^*(\vartheta^\alpha \wedge T_\alpha) = \frac{1}{2}T^{[\alpha\beta\gamma]}\eta_{\alpha\beta\gamma} = \mathscr{A}_i dx^i, \quad (6)$$

which is invariant under Weyl rescalings and *chiral transformations* $\gamma \to \gamma^\beta = e^{i\gamma^5\beta}\gamma e^{-i\gamma^5\beta}$ of the coframe, but odd under parity $P: \vartheta^B \to -\vartheta^B$, where $B = 1,2,3$, cf. Ref. [38].

CLASSICAL AXIAL ANOMALY AND SPIN

Similarly as in QED, the gravitational coupled Dirac Lagrangian $L_D = \overline{L}_D = L_D^\dagger$ is *Hermitian* as required, even in an anholonomic frame. Then minimal coupling prescribes us automatically with the following *charge and axial currents*, respectively,

$$j = \overline{\psi}{}^*\gamma\psi = j^\mu \eta_\mu, \qquad j_5 := \overline{\psi}\gamma_5{}^*\gamma\psi = \frac{1}{3}\overline{\psi}\sigma \wedge \gamma\psi = j_5^\mu \eta_\mu. \quad (7)$$

From the Dirac equation (4) and its adjoint one can readily deduce that $dj \simeq 0$ 'on shell', whereas for the axial current we find the well–known "classical axial anomaly"

$$dj_5 = 2imP = 2im\overline{\psi}\gamma_5\psi\eta \quad (8)$$

for *massive* Dirac fields [21]. The same holds in a RC spacetime. If we restore chiral symmetry in the limit $m \to 0$, this leads to classical conservation law $dj_5 = 0$ of the axial current for massless Weyl spinors, or since $dj \simeq 0$, equivalently, for the *chiral current* $j_\pm := \frac{1}{2}\overline{\psi}(1 \pm \gamma_5){}^*\gamma\psi = \overline{\psi}_{L,R}{}^*\gamma\psi_{L,R}$.

As mentioned in the Introduction, the axial current has an intriguing relation to the (dynamical) *spin current* of the Dirac field canonically defined by the Hermitian three–form

$$\begin{aligned} \tau_{\alpha\beta} &:= \frac{\partial L_D}{\partial \Gamma^{\alpha\beta}} = \frac{1}{8}\overline{\psi}\left({}^*\gamma\sigma_{\alpha\beta} + \sigma_{\alpha\beta}{}^*\gamma\right)\psi \\ &= \frac{1}{4}\eta_{\alpha\beta\gamma\delta}\overline{\psi}\gamma^\delta\gamma_5\psi\eta^\gamma = \tau_{\alpha\beta\gamma}\eta^\gamma = \vartheta_{[\alpha} \wedge \mu_{\beta]}. \end{aligned} \quad (9)$$

Its components $\tau_{\alpha\beta\gamma} = \tau_{[\alpha\beta\gamma]}$ are *totally antisymmetric*. Equivalently, in Eq. (5) torsion merely couples to the two-form

$$\mu_\alpha = \frac{1}{4}\vartheta_\alpha \wedge {}^*j_5, \quad (10)$$

commonly referred to as the *spin-energy potential* [18, 32]. Consequently, we obtain the remarkable result that he vector one-form

$$\mu := e_\alpha \rfloor \mu^\alpha = \frac{3}{4} {}^* j_5, \qquad (11)$$

of the *Dirac spin* is dual to the axial current j_5.

There is also a relation to the axion: In Ref. [41] it is tentatively assumed that the dimensionless pseudo-scalar θ serves as a potential for the axial torsion via $\mathscr{A} = 2d\theta$. Then, there arises in (5) a derivative coupling of the would-be axion $a = \theta f_a$ to two fermions via the CPT-invariant term

$$L_{a\psi\psi} = \frac{1}{2} d\theta \wedge j_5 = \frac{1}{2f_a} da \wedge \overline{\psi} {}^* \gamma\gamma_5 \psi, \qquad (12)$$

exactly as in the usual formulation, where the axial current j_5 is the Noether current associated with a spontaneously broken Peccei-Quinn symmetry $U(1)_{PQ}$.

AXIAL CURRENT IN THE EINSTEIN–CARTAN THEORY

The Einstein–Cartan (EC) theory of a gravitationally coupled spin 1/2 Dirac field provides a *dynamical* understanding of the axial anomaly on a semi-classical level: The Lagrangian reads:

$$L = \frac{i}{2\ell^2} Tr(\Omega \wedge {}^*\sigma) + L_D = \frac{1}{2\ell^2} R^{\alpha\beta} \wedge \eta_{\alpha\beta} + L_D, \qquad (13)$$

where $\eta^{\alpha\beta} := {}^*(\vartheta^\alpha \wedge \vartheta^\beta)$ is dual to the unit two–form.

In EC–theory, Cartan's algebraic relation between torsion and spin implies the following relation [37] between the *axial current* j_5 of the Dirac field and the translational Chern-Simons (CS) term (27), or equivalent, for the axial torsion one-form:

$$C_{TT} \cong \frac{1}{4} j_5, \qquad \mathscr{A} = 2\ell^2 {}^* C_{TT} = (\ell^2/2)\overline{\psi}\gamma_5\gamma\psi. \qquad (14)$$

Thus in EC-theory, the net axial current production

$$dj_5 \cong 4dC_{TT} = \frac{2}{\ell^2}\left(T^\alpha \wedge T_\alpha + R_{\alpha\beta} \wedge \vartheta^\alpha \wedge \vartheta^\beta\right) \qquad (15)$$

establishes a link to the NY four form [44] for *massive* fields.

This result, cf. [33, 37], holds on the level of first quantization. Since the Hamiltonian of the semi-classical Dirac field is not bounded from below, one has to go over to second quantization, where the vacuum expectation value $\langle dj_5 \rangle$ of the axial current picks up anomalous terms.

Restoring chiral invariance for the Dirac fields, the limit $m \to 0$, implies that the NY four–form tends to zero "on shell", i.e. $dC_{TT} \cong (1/4)dj_5 \to 0$. This is consistent with the fact that a Weyl spinor does not couple to torsion at all, because then the axial torsion $\mathscr{A}$

becomes a *lightlike* covector, i.e. $\mathscr{A}_\alpha \mathscr{A}^\alpha \eta = \mathscr{A} \wedge {}^*\mathscr{A} \cong (\ell^4/4) {}^*j_5 \wedge j_5 = 0$. Here we implicitly assume that the light-cone structure of the axial covector ${}^*j_5$ is not spoiled by quantum corrections, i.e. that no "Lorentz anomaly" occurs as in $n = 4k+2$ dimensions [26].

CHIRAL ANOMALY IN QUANTUM FIELD THEORY

Let us recall a couple of distinguished features of the axial anomaly: Most prominent is its relation with the Atiyah–Singer index theorem [5]. But also from the viewpoint of perturbative QFT, the chiral anomaly has some features which signal its conceptual importance. For all topological field theories and topological effects like the anomaly, there is the remarkable fact that it does not renormalize — higher order loop corrections do not alter its one-loop value. This very fact guarantees that the anomaly can be given a topological interpretation. For the anomaly, this is the Adler–Bardeen theorem [1], while other topological field theories are carefully designed to have vanishing beta functions, for example. Another feature is its finiteness: in any approach, the chiral anomaly as a topological invariant is a finite quantity.

Now, to approach the anomaly in the context of spacetime with torsion, let us first switch off the Riemannian curvature and concentrate on the last but one term in the decomposed Dirac Lagrangian (5).

Then, this term can be regarded as an *external* axial covector $\mathscr{A}$ coupled to the axial current j_5 of the Dirac field in an *initially flat* spacetime. By applying the result (11–225) of Itzykson and Zuber [21], we find that only the term $d\mathscr{A} \wedge d\mathscr{A}$ arises in the axial anomaly, but *not* the NY type term $d^*\mathscr{A} \sim dC_{TT}$ as was recently claimed [10]. After switching on the Yang-Mills field G as well as the curved spacetime of Riemannian geometry, we finally obtain for the vacuum expectation value of the *axial anomaly*

$$\langle dj_5 \rangle = 2im \langle \overline{\psi}\gamma_5\psi \rangle \eta - \frac{1}{4\pi^2} Tr(G \wedge G) - \frac{1}{96\pi^2}\left[2R^{\{\}}_{\alpha\beta} \wedge R^{\{\}\alpha\beta} + d\mathscr{A} \wedge d\mathscr{A} \right]. \quad (16)$$

This result [25] is based on diagrammatic techniques and the Pauli–Villars regularization scheme. In this respect, it is a typical perturbative result, and in agreement with [16, 54, 52] no NY term arises in the anomaly. Thus only the Weyl invariant term $d\mathscr{A} \wedge d\mathscr{A} = -2\vec{\mathscr{E}} \cdot \vec{\mathscr{B}}\eta$ for the axial torsion contributes to the axial anomaly, resembling the $U(1)$ part $F \wedge F = dA \wedge dA$ of the Pontrjagin term (26). Torsion terms like $d\mathscr{A} \wedge d\mathscr{A}$ and $d^*\mathscr{A} \wedge {}^*(d^*\mathscr{A}) = 4\ell^4 V_{NY} \wedge {}^*V_{NY}$ have been considered previously, as part of the Lagrangian, in order to make the axial torsion propagating. Due to the geometric identity (28) for the NY term $d^*\mathscr{A} = 2\ell^2 dC_{TT} = 2\ell^2 V_{NY}$, the second term is really quartic in torsion and not scale invariant.

A rescaling of the tetrad has been proposed, however, one should not ignore the presence of renormalization conditions and the generation of a scale upon renormalization. Rescaling the tetrad would ultimately change the wave function renormalization Z-factor which would creep into the definition of the NY term, in sharp contrast to proper topological invariants at the quantum level, which remain unchanged under renormalization.

With no renormalization condition available for the NY term, and other methods obtaining it as zero, we can only conclude that the response function of QFT to a gauge variation (this is the anomaly) delivers no NY term. Or, saying it differently, its finite value is zero after renormalization.

Chiral anomaly in SUGRA

Simple supergravity consists in a consistent coupling of the EC to the Rarita–Schwinger spinor-valued one-form $\Psi = \Psi_i dx^i$, cf. Refs. [49, 36] for more details.

The anomaly for the corresponding *axial current* $J_5 := i\overline{\Psi} \wedge \gamma \wedge \Psi$ is $-21\times$ the anomaly for Dirac fields, whereas for the corresponding supersymmetric Yang–Mills anomaly one finds $3\times$ the Dirac result:

Spin	Gravitational	YM anomaly
1/2	1	1
3/2	−21	3

Depending on the asymptotic helicity states, there occur contributions of topological origin of the Riemannian Pontrjagin or Euler type, respectively. The role of spinors for the index theorem and in the $4D$ Donaldson invariants via Seiberg–Witten equation has recently been reviewed by Atiyah [5]. Six dimensional supergravity free of gauge and gravitational anomalies is studied in Ref. [12].

COMPARISON WITH THE HEAT KERNEL METHOD

In the heat kernel approach, there exists for small $t \to +0$ the asymptotic expansion

$$K(t,x,\slashed{D}^2) = (4\pi)^{-n/2} \sum_{k=0}^{\infty} t^{(k-n)/2} K_k(x,\slashed{D}^2) \tag{17}$$

of the kernel in n dimensions, where the usual Feynman "dagger" convention $\slashed{A} := \check{\gamma}\rfloor A = \gamma^\alpha e_\alpha \rfloor A = (-1)^{s+1} * [*\gamma \wedge A]$ for one–forms is used.

The squared Dirac operator

$$\begin{aligned}
\slashed{D}^2 &= -\frac{1}{2}\gamma^\alpha\gamma^\beta\left(\{D_\alpha^{\{\}}, D_\beta^{\{\}}\} + [D_\alpha^{\{\}}, D_\beta^{\{\}}]\right) - 2im\slashed{D}^{\{\}} \\
&\quad - \frac{i}{4}\gamma_5(\slashed{D}^{\{\}}\slashed{\mathscr{A}}) + \frac{1}{2}\gamma_5\sigma^{\alpha\beta}\mathscr{A}_\alpha D_\beta^{\{\}} + m^2 - \frac{1}{2}m\gamma_5\slashed{\mathscr{A}} - \frac{1}{16}\slashed{\mathscr{A}}\slashed{\mathscr{A}} \\
&\cong -\Box - \frac{1}{8}\sigma^{\alpha\beta}R^{\{\}}_{\alpha\beta\mu\nu}\sigma^{\mu\nu} \\
&\quad - \frac{i}{4}\gamma_5(\slashed{D}^{\{\}}\slashed{\mathscr{A}}) + \frac{1}{2}\gamma_5\sigma^{\alpha\beta}\mathscr{A}_\alpha D_\beta^{\{\}} - \frac{1}{16}\mathscr{A}_\alpha\mathscr{A}^\alpha - m^2,
\end{aligned} \tag{18}$$

has been explicitly calculated in Refs.[54, 45], and the terms additional to the generally covariant Riemannian d'Alembertian operator $\Box := \partial_\mu\left(\sqrt{|g|}g^{\mu\nu}\partial_\nu\right)/\sqrt{|g|}$ are identified [34]. Not unexpectedly, besides the familiar Riemannian curvature scalar, only the axial torsion (6) contributes to the squared Dirac operator for *massive* spinor fields.

The coefficients $K_k(x,\slashed{D}^2)$ are completely determined by the form of the second-order differential operator $\slashed{D}^2$, which is positive for Euclidean signature diag $o_{\alpha\beta} = (-1,\cdots,-1)$. For odd $k = 1,3,\ldots$ these coefficients are zero, while the first nontrivial terms [54], which potentially could contribute to the axial anomaly, read

$$Tr(\gamma_5 K_2) = -d^*\mathscr{A},$$
$$Tr(\gamma_5 K_4) = \frac{1}{6}\left[Tr\left(R^{\{\}} \wedge R^{\{\}}\right) - \frac{1}{4}d\mathscr{A} \wedge d\mathscr{A} + d\mathscr{K}\right], \quad (19)$$

where the higher order term $d\mathscr{K} = d^*\breve{D} \wedge^* \breve{D}\,^*\mathscr{A}$ involves the covariant derivative $\breve{D} = D^{\{\}} + i\mathscr{A}\gamma_5/4$ modified by the axial torsion.

However, there is an essential difference in the physical dimensionality of the terms K_2 and K_4. Whereas in $n = 4$ dimensions the Pontrjagin type term K_4 is dimensionless and thus, for $k = 4$, multiplied by $t^{(k-4)/2} = 1$, the term $K_2 \sim d^*\mathscr{A} = 2\ell^2 dC_{TT}$ carries dimensions. Since a *massive* Dirac spinor has canonical dimension $[length]^{-3/2}$, it scales as $\psi \sim m^{3/2}$. Moreover, the term $t = 1/M^2$ is related to the regulator mass $M \to \infty$ in Fujikawa method [14]. Then the second order term in the heat kernel expansion scales as $-K_2/t = (2\ell^2/t)dC_{TT} \cong (\ell^2/2t)dj_5 = (im\ell^2/t)\overline{\psi}\gamma_5\psi \sim \ell^2 M^2 m^4 \to 0$. If we assume in the renormalization procedure, that the fundamental length ℓ does not scale (no running coupling constant), the second order term in the heat kernel expansion will tend to zero in the chiral limit $m \to 0$. In the case $m \neq 0$, this term diverges and the Fujikawa regulator method $M \to \infty$ cannot be applied. To rescale the coframe by $\vartheta^\alpha \to \tilde{\vartheta}^\alpha = M\vartheta^\alpha$ does not help, since this would change also the dimension of the Dirac field, in order to retain the physical dimension $[\hbar]$ of the Dirac action.

Thus the NY term dC_{TT} does NOT contribute to the *chiral anomaly* in four dimensions, neither classically nor in QFT. On would surmise that in $n = 2$ dimensional models only the term $d^*\mathscr{A}$ survives in the heat kernel expansion, since it then has the correct dimensions. However, it is well-known [19] that in 2D the axial torsion $\mathscr{A}$ vanishes identically. Moreover, gravitational anomalies [26], specifically the Einstein anomaly and the Weyl anomaly, are fully determined by means of dispersion relations [8].

Let us stress the interrelation between the scale and chiral invariance: The renormalized conformal (or trace) anomaly [11]

$$\langle \vartheta^\alpha \wedge \sigma_\alpha \rangle = -\frac{1}{3\pi^2}\left[Tr(G \wedge {}^*G) + \frac{1}{24}\left(2R^{\alpha\beta\{\}} \wedge R^{\{\}\star}_{\alpha\beta} + d\mathscr{A} \wedge {}^*d\mathscr{A}\right)\right] \quad (20)$$

for the energy-momentum current $\sigma_\alpha := \Sigma_\alpha - D\mu_\alpha$ Belinfante symmetrized via the spin energy (10) receives, in addition to the Riemannian Euler term, a kinetic contribution of the Maxwell type from the axial torsion $\mathscr{A}$. The coefficients are similar to those in Eq. (16), due to the fact that chiral and trace anomalies constitute a supermultiplet [55].

HAMILTONIAN INTERPRETATION OF ANOMALIES

In the canonical formulation á la Ashtekar [4], the translational NY term dC_{TT} plays via

$$\overset{(\pm)}{V}_{EC} := V_{EC} \pm idC_{TT} = \pm\frac{1}{2\ell^2}Tr\left\{(1\mp\gamma_5)\Omega\wedge\sigma\right\} = -\frac{1}{2\ell^2}\overset{(\pm)}{R}{}^{\alpha\beta}\wedge\eta_{\alpha\beta} + \frac{\Lambda}{\ell^2}\eta \quad (21)$$

the role of the *generating functional* [28] for chiral, i.e. self– or antiselfdual variables $\overset{(\pm)}{\Gamma}$ in EC theory as well as in simple supergravity [33, 36].

The appearance of the Riemannian Pontrjagin term $dC_{RR}^{\{\}}$ in the anomaly (16) could pose problems for the canonical approach to gravity, since the anomaly does not renormalize. In the presence of gravitational instantons, which due to the necessary condition $\Lambda \neq 0$ could even be the dominating configurations, one gets a net production of chiral zero modes and a global symmetry is broken.

One could argue that this is a perturbative effect. In the Wilson type loop approach to gravity [9, 17, 31], the tangential complexified CS term $\underline{C}_{RR}^{\{\}}$ is known [24] to solve the Hamiltonian constraint $\mathcal{H}_\Lambda \Psi(\overset{(\pm)}{\underline{\Gamma}}) = 0$ of gravity, where the complex Ashtekar variable $\overset{(\pm)}{\underline{\Gamma}}$ is the tangential part of the self– or antiselfdual spin connection one–form. Since this solution is intrinsically non–perturbative, no anomaly should occur. In the lattice gauge approach this is indeed the case, but the problem of fermion doubling [47] appears to be another manifestation of the anomaly.

It is instructive to look at the problem from an Hamiltonian point of view since the canonical formalism of chiral gravity is closely related to the $SU(2)$ CS gauge theory on the three–dimensional hypersurface, with $\mathscr{C} := \Gamma/\mathscr{G}$ of non-equivalent gauge connections as configuration space.

Gauge anomalies are related to the global topology and have the common feature [42] that the *Gauss constraint* $\mathscr{G}^A \cong 0$ *cannot* anymore be implemented on the physical states [13]. The reason is that the anomalous Ward identity

$$\text{Ł}_n \mathscr{G}^A \cong n\rfloor (D\underline{\tau}^A), \quad (22)$$

where $\text{Ł}_n := n\rfloor D + Dn\rfloor$ is the gauge–covariant Lie derivative along the normal direction, relates the time evolution of the Gauss constraint to the conservation law for the matter current τ^A on the spacelike hypersurface [23]. Only when the individual contributions to the anomaly cancel each other, a gauge theory can be consistently quantized. In the EC formulation of the gravitationally coupled Dirac field, it is the canonical spin $\tau^A := (1/2)\eta^{0A\beta\gamma}\tau_{\beta\gamma}$ which appears on the right-hand side of the Gauss constraint. Since spin is via (9,11) related to the axial current j_5, it is precisely the *chiral anomaly* which prevents the Gauss constraint to remain a proper constraint under time evolution. This result confronts the Ashtekar approach based on loop variables, and thus on notions of parallel transport, with the chiral anomaly [34, 35]. The teleparallelism equivalent [28] of chiral gravity, where Wilson loops are replaced by Cartan circuits [29, 31], may avoid some of these obstacles.

APPENDIX A: GRAVITATIONAL CHERN–SIMONS AND PONTRJAGIN TERMS

When the constant Dirac matrices γ_α obeying $\gamma_\alpha \gamma_\beta + \gamma_\beta \gamma_\alpha = 2 o_{\alpha\beta}$ are saturating the index of the orthonormal coframe one–form $\vartheta^\alpha = e_j{}^\alpha dx^j$ and its Hodge dual $\eta^\alpha := {}^*\vartheta^\alpha$, we obtain a basis of Clifford–algebra valued exterior forms [33, 30] via:

$$\gamma := \gamma_\alpha \vartheta^\alpha, \qquad {}^*\gamma = \gamma^\alpha \eta_\alpha. \tag{23}$$

In terms of the Clifford algebra–valued *connection* $\Gamma := \frac{i}{4}\Gamma_i{}^{\alpha\beta}\sigma_{\alpha\beta}dx^i$, the $SL(2,C)$–covariant exterior derivative is given by $D = d + \Gamma \wedge$, where $\sigma_{\alpha\beta} = \frac{i}{2}(\gamma_\alpha\gamma_\beta - \gamma_\beta\gamma_\alpha)$ are the Lorentz generators entering in the two-form $\sigma := \frac{i}{2}\gamma \wedge \gamma = \frac{1}{2}\sigma_{\alpha\beta}\vartheta^\alpha \wedge \vartheta^\beta$.

Differentiation of these basic variables leads to the Clifford algebra–valued *torsion* and *curvature* two–forms:

$$\Theta := D\gamma = T^\alpha \gamma_\alpha, \qquad \Omega := d\Gamma + \Gamma \wedge \Gamma = \frac{i}{4}R^{\alpha\beta}\sigma_{\alpha\beta} \tag{24}$$

in RC geometry. The *Chern–Simons term* for the Lorentz connection reads

$$C_{RR} := -Tr\left(\Gamma \wedge \Omega - \frac{1}{3}\Gamma \wedge \Gamma \wedge \Gamma\right). \tag{25}$$

The corresponding *Pontrjagin topological term* can be obtained by exterior differentiation

$$\begin{aligned}
dC_{RR} &= -Tr(\Omega \wedge \Omega) \\
&= \frac{1}{2}R^{\{\}}_{\alpha\beta} \wedge R^{\{\}\alpha\beta} \\
&\quad + \frac{1}{12}d\left[{}^*\mathcal{A} \wedge R^{\{\}} - \frac{1}{3}\mathcal{A} \wedge d\mathcal{A} + \frac{1}{9}{}^*\mathcal{A} \wedge {}^*(\mathcal{A} \wedge {}^*\mathcal{A})\right].
\end{aligned} \tag{26}$$

The latter contains [41], amongst others, a term proportional to the curvature scalar $R := {}^*(R^{\alpha\beta} \wedge \eta_{\beta\alpha})$ and the axial torsion piece $d\mathcal{A} \wedge d\mathcal{A}$ of the axial anomaly with a relative factor 9 as required by the supersymmetric path integral [27].

Since the coframe is the 'soldered' translational part [39, 48] of the Cartan connection, a related *translational* CS term arises

$$C_{TT} := \frac{1}{8\ell^2}Tr(\gamma \wedge \Theta) = \frac{1}{2\ell^2}\vartheta^\alpha \wedge T_\alpha. \tag{27}$$

By exterior differentiation we obtain the NY four–form [44]:

$$dC_{TT} = \frac{1}{8\ell^2}Tr(\Theta \wedge \Theta - 4i\Omega \wedge \sigma) = \frac{1}{2\ell^2}\left(T^\alpha \wedge T_\alpha + R_{\alpha\beta} \wedge \vartheta^\alpha \wedge \vartheta^\beta\right). \tag{28}$$

It is crucial to note that a fundamental length ℓ necessarily occurs here for dimensional reasons. This can be also understood by a de Sitter type gauge approach, in which the $sl(5,R)$–valued connection $\hat{\Gamma} = \Gamma + (\vartheta^\alpha L^4{}_\alpha + \vartheta_\beta L^\beta{}_4)/\ell$ is expanded into the dimensionless linear connection Γ plus the coframe $\vartheta^\alpha = e_i{}^\alpha dx^i$ which carries canonical dimension [*length*]. The corresponding Pontrjagin term $\hat{C}_{RR}$ splits via

$$\hat{C}_{RR} = C_{RR} - 2C_{TT} \tag{29}$$

into the linear one and the translational CS term, see the footnote 31 of Ref. [19] for details.

ACKNOWLEDGMENTS

We would like to thank Hugo Morales–Técotl and Luis Urrutia for useful hints and comments. This article is dedicated to the memory of Yuval Ne'eman.

REFERENCES

1. S. L. Adler and W. A. Bardeen: "Absence of higher order corrections in the anomalous axial vector divergence equation," Phys. Rev. **182**, 1517 (1969).
2. J. Alfaro, L. F. Urrutia and J. D. Vergara, "Extended definition of the regulated Jacobian in the path integral calculation of anomalies," Phys. Lett. B **202**, 121 (1988).
3. J. Alfaro, L.F. Urrutia y J.D. Vergara: "Anomalous Jacobians and the vector anomaly", Proceedings of the Second Meeting of Quantum Mechanics of Fundamental Systems, Santiago, Ed. C. Teitelboim, (Plenum Press, 1989), p. 1.
4. A. Ashtekar, Phys. Rev. Lett. **57**, 2244 (1986); *New Perspectives in Canonical Gravity* (Bibliopolis, Napoli 1988).
5. M.F. Atiyah: "The Dirac equation and geometry", in A. Pais, et al.: *Paul Dirac* (Cambridge University Press, 1998.), p. 108-124.
6. J. S. Bell and R. Jackiw: "A PCAC puzzle: $\pi^0 \to 2\gamma$ in the σ model," Nuovo Cim. **60A**, 47 (1969).
7. R.A. Bertlmann: *Anomalies in quantum field theory* (Oxford University Press, Oxford UK: 2000)
8. R. A. Bertlmann and E. Kohlprath: "Gravitational anomalies in a dispersive approach," Nucl. Phys. Proc. Suppl. **96**, 293 (2001).
9. B. Brügmann, R. Gambini, and J. Pullin, Nucl. Phys. **B385**, 587 (1992).
10. O. Chandia and J. Zanelli, Phys. Rev. **D55**, 7580 (1997).
11. S. Deser and A. Schwimmer: "Geometric classification of conformal anomalies in arbitrary dimensions," Phys. Lett. B **309**, 279 (1993).
12. J. Erler: "Anomaly cancellation in six-dimensions," J. Math. Phys. **35**, 1819 (1994).
13. L. D. Faddeev: "Operator anomaly for the Gauss law," Phys. Lett. B **145**, 81 (1984).
14. K. Fujikawa: "Path integral measure for gauge invariant Fermion theories," Phys. Rev. Lett. **42**, 1195 (1979).
15. J. Gamboa, I. Schmidt and L. Vergara: "Anomaly and condensate in the light-cone Schwinger model," Phys. Lett. B **412**, 111 (1997).
16. G. Grensing, Phys. Lett. **B169**, 333 (1986).
17. J. Griego, Phys. Rev. **D53**, 6966 (1996).
18. F.W. Hehl, A. Macías, E.W. Mielke, and Yu. N., Obukhov: "On the structure of the energy–momentum and the spin currents in Dirac's electron theory", in: *On Einstein's Path*. Festschrift for E. Schucking on the occasion of his 70th birthday, A. Harvey, ed. (Springer, New York, 1999) pp. 257–274.
19. F.W. Hehl, J.D. McCrea, E.W. Mielke, and Y. Ne'eman: "Metric–affine gauge theory of gravity: Field equations, Noether identities, world spinors, and breaking of dilation invariance", Phys. Rept. **258** (1995) 1–171.
20. B.R. Holstein: "Anomalies for pedestrians", Am. J. Phys. **61**, 142 (1993).

21. C. Itzykson and J.-B. Zuber: *Quantum Field Theory* (McGraw Hill, New York 1980).
22. R. Jackiw and S. Y. Pi: "Creation and evolution of magnetic helicity," Phys. Rev. D **61**, 105015 (2000).
23. W. Jiang, J. Math. Phys. **32**, 3409 (1991).
24. H. Kodama, Phys. Rev. **D 42**, 2548 (1990).
25. D. Kreimer and E.W. Mielke: "Comment on: Topological invariants, instantons, and the chiral anomaly on spaces with torsion", Phys. Rev. **D63**, 048501 (2001).
26. H. Leutwyler, Helvetia Physica Acta **59**, 201 (1986).
27. N.E. Mavromatos, J. Phys. A: Math. Gen. **21**, 2279 (1988).
28. E.W. Mielke: "Ashtekar's complex variables in general relativity and its teleparallelism equivalent", Ann. Phys. (N.Y.) **219**, 78– 108 (1992).
29. E.W. Mielke: "Anomaly–free solution of the Ashtekar constraints for the teleparallelism equivalent of gravity", Phys. Lett. **A 251**, 349 – 353 (1999).
30. E.W. Mielke: "Beautiful gauge field equations in Clifforms", Int. J. Theor. Phys. **40**, 171 – 190 (2001). (Proceedings of Ixtapa Conference on Clifford Algebra, June 27 –July 1999).
31. E.W. Mielke: "Chern–Simons solution of the chiral teleparallelism constraints of gravity" Nucl. Phys. **B 622**, 457–471 (2002).
32. E.W. Mielke: "Consistent coupling to Dirac fields in teleparallelism: Comment on "Metric-affine approach to teleparallel gravity", Phys. Rev. **D 69**, 128501 (2004).
33. E.W. Mielke, P. Baekler, F.W. Hehl, A. Macías, and H.A. Morales-Técotl, in: *Gravity, Particles and Space–Time*, ed. by P. Pronin and G. Sardanashvily (World Scientific, Singapore, 1996), pp. 217–254.
34. E.W. Mielke and D. Kreimer: "Chiral anomaly in Ashtekar's approach to canonical gravity", Int. J. Mod. Phys. **D7**, 535 – 548 (1998).
35. E.W. Mielke and D. Kreimer: "Chiral anomaly in contorted spacetimes", J. Gen. Rel. Gravitation **31**, 701–712 (1999).(Hehl Special Issue: Proccedings of *Mexican Meeting on Gauge Theories of Gravity* UAM-I, México).
36. E.W. Mielke and A. Macías: "Chiral supergravity and anomalies ", Annalen der Physik (Leipzig) **8**, 301- 317 (1999).
37. E.W. Mielke, A. Macías, and H.A. Morales–Técotl: "Chiral fermions coupled to chiral gravity", Phys. Lett. **A 215**, 14 – 20 (1996).
38. E.W. Mielke, A. Macías, and Y. Ne'eman: "CP–symmetry in chiral gravity", *Proc. of the Eigth Marcel Grossman Meeting on General Relativity*, Jerusalem, 1997, T. Piran and R. Ruffini, eds. (World Scientific, Singapore, 1999) pp. 901–903.
39. E.W. Mielke, J.D. McCrea, Y. Ne'eman, and F.W. Hehl: "Avoiding degenerate coframes in an affine gauge approach to quantum gravity", Phys. Rev. **D48**, 673 – 679 (1993).
40. E.W. Mielke and A. A. Rincón Maggiolo: "Duality in Yang's theory of gravity", Gen. Rel. Grav. **37**, 997-1007 (2005).
41. E.W. Mielke and E. S. Romero: "Cosmological evolution of a torsion-induced quintaxion," Phys. Rev. D **73**, 043521 (2006).
42. Ph. Nelson and L. Alvarez–Gaumé, Commun. Math. Phys. **99**, 103 (1985).
43. M. Napsuciale, A. Wirzba and M. Kirchbach: "Instantons as unitary spin maker," Nucl. Phys. A **703**, 306 (2002).
44. H.T. Nieh and M.L. Yan, J. Math. Phys. **23**, 373–374 (1982).
45. Obukhov, Yu. N., E.W. Mielke, J. Budczies, and F.W. Hehl: "On the chiral anomaly in non-Riemannian spacetimes", Found. Phys. **27**, 1221 (1997). (Biedenharn Festschrift).
46. J. S. Schwinger: "On gauge invariance and vacuum polarization," Phys. Rev. **82**, 664 (1951).
47. L.L. Smalley, Nuovo Cimento **92 A**, 25 (1986).
48. R. Tresguerres and E.W. Mielke: "Gravitational Goldstone fields from affine gauge theory" Phys. Rev. **D62** 44004 (2000).
49. L. F. Urrutia and J. D. Vergara: "Consistent coupling of the gravitino field to a gravitational background with torsion," Phys. Rev. D **44**, 3882 (1991).
50. L. F. Urrutia and J. D. Vergara: "Anomalies in the Fujikawa method using parameter dependent regulators," Phys. Rev. D **45**, 1365 (1992).
51. J. W. Van Holten: "Aspects of BRST quantization," Lect. Notes Phys. **659**, 99 (2005).
52. C. Wiesendanger, Class. Quant. Grav. **13**, 681-699 (1996).

53. A. Widom and Y. Srivastava:"A simple physical view of the quantum electrodynamic chiral anomaly", Am. J. Phys. **56** (9), 824 (1988).
54. S. Yajima, Class. Quantum Grav. **13**, 2423-2435 (1996).
55. J. F. Yang: "Trace anomalies and chiral Ward identities," Chin. Phys. Lett. **21**, 792 (2004).

Developments in Supergravity

O. Obregón* and C. Ramírez[†]

*Instituto de Física de la Universidad de Guanajuato
P.O. Box E-143, 37150 León Gto., México
[†]Facultad de Ciencias Físico Matemáticas,
Universidad Autónoma de Puebla, P.O. Box 1364, 72000 Puebla, México

Abstract. We present an account of the development of supergravity in México.

Keywords: Supergravity
PACS: 04.65.+e,12.10.-g,11.30.Pb,11.10.Kk

INTRODUCTION

A nice account of the discovery of Supersymmetry and of its salient aspects has been recently written by Bruno Zumino [79]. Apart from his well known formulation together with Julius Wess, two groups in the former Soviet Union have also, in an independent and different form, proposed Supersymmetry. Supergravity, the supersymmetric extension of general relativity, can be seen as the gauge theory of Supersymmetry [2] and was first formulated by Daniel Freedman, Peter van Nieuwenhuizen and Sergio Ferrara [1]. Supergravity N=1 was soon extended in a consistent way to include other fields and it was then shown that it could be consistently formulated only until N=8. It was also realized that there is an interplay between the number of dimensions and Supersymmetry in only four dimensions. This was a way in which Supergravity in eleven dimensions was found. As we know today, D=11 Supergravity plays a central role in connection with the well known string formulations and is related to them through the interesting web of dualities that connect these theories. Of course, many supergravities can be obtained as the low energy limit of String Theory.

Here we make an account of the work realized in México, in which Mexican institutions appear. We refer only to work published in Journals. There are contributions that lie at the border between Supergravity and other topics, for instance Supersymmetry and Superstrings. In those cases our colleagues in charge of these other topics included some of the work sent to us and vice versa. Some of our colleagues kindly sent us a list and a description of their work. In other cases we have searched for the information, particularly in the Internet. So we apologize if we have not been able to find some of the relevant contributions. Many of our colleagues have produced scientific work together. So we have made a list of the publications and have enumerated them according with the year of appearance and then refer each of the author's contribution to the articles in the list. Finally, we want to thank the Editors for inviting us to contribute to this report, with the wonderful world of Supergravity.

Supersymmetry has attracted the attention of Mexican researchers from the beginning of the 80's. The first International School of Supersymmetry in México took place in

México City during the period December 14-18, 1981 and served to spark the local research in this field. The event was hosted by the Instituto de Ciencias Nucleares, UNAM and took place in the Unidad de Seminarios Ignacio Chávez. The lecturers were D.Z. Freedman, M.T. Grisaru, E. Witten, S. Ferrara and I. Bars. The lectures were published in [6].

CLASSICAL SOLUTIONS OF SUPERGRAVITY

The first works on supergravity have been related to classical solutions. Such solutions have been studied in four dimensions in Refs. [3, 4]. The latter is an exact plane-wave type solution that was subsequently generalized by C.M. Hull to SUGRA in eleven dimensions and also to N=2 extended supergravity. [7]. Also solutions to Kasner type cosmological models [11], together with colliding plane waves [9] were also found in N=1 SUGRA. In the latter case, contrary to the situation in ordinary gravity, the solution is non-singular everywhere due to the contributions of the underlying Grassmann algebra.

Further, there has been work on linearized supergravity. The usual equations in curved spate-time for massless fields with spin bigger than one possesses quite restrictive integrability conditions. However the supergravity field equations, linearized with respect to the massless 3/2-spin field, in the background of a vacuum solution of the Einstein equations, do not have this problem. In [12, 13, 14], in the case of the $O(2)$-supergravity, the equations of a $O(2)$-doublet of a 3/2-spin field on a background solution of the Einstein-Maxwell equations are deduced. In [15, 17], the background metrics are the ones of a neutral or charged rotating black hole, as well as a family which contains them. In [39], these linearized equations are analyzed on a flat Minkowski background.

Later, in [26], the author considers N=2 supergravity in five dimensions in the most general spherically symmetric space-time. In [34], exact solutions of N=2 supergravity in five dimensions corresponding to the exterior of a cosmic string are found. In [27, 33] the Bianchi V and VI(0) models in $N = 2$, $d = 5$ supergravity are studied.

Returning to the framework of linearized supergravity and expressing the complete solution of the Rarita-Schwinger equation in terms of Debye potentials, the spin 3/2 perturbations of certain background solutions of the Einstein and Einstein-Maxwell equations have been analyzed in [30, 31, 32, 40, 41].

FORMULATIONS OF SUPERGRAVITY

Other field of early work has been the study of formulations of supergravity, attempting to better understand its mathematical structure, in particular if it is of the type of a gauge theory.

The first order formalism of supergravity is formulated by means of the tetrads, the Ricci rotation coefficients and the gravitino field. In the second order formalism only the tetrads and the gravitino fields are considered independently. In [5] a different formulation of supergravity was given, where the basic fields in the supergravity action are the metric, the torsion and the gravitino field. So supergravity could be seen as a theory of Gravitation with torsion.

In [8] a formulation for supergravity based on supertwistor fiber bundles is given.

The techniques of QFT have been applied to supergravity in [16], where the consistency conditions for the supergravity coupling of the gravitino to a Riemann-Cartan background were obtained. The torsion needs not be zero and a particular solution of the consistency conditions allows to calculate the gravitino contribution to the axial anomaly, with the result that the torsion does not contribute, in accordance with the index theorem.

In [20] $2D$ supergravity is studied in a covariant and gauge independent way. The theory is obtained from $2D$ bosonic gravity following the square root method and the diffeomorphism superalgebra is explicitly computed.

In [24] the authors were able to show that the square root of the dynamical equations of linearized gravity are the Rarita-Schwinger equations. This is similar, but not the same, as what happens among the constraints in canonical supergravity.

In [80] the authors discuss the local Lorentz invariance in the context of N=1 supergravity. In [38], the authors give a formulation of the chiral, i.e. (anti) self-dual gravity or $N = 1$ supergravity, starting from the observation that the corresponding lagrangian can be obtained from the usual one, by the addition of a boundary term corresponding to the 'translational' Chern-Simons term. In the second case they consider a spinorial representation.

It seems that the first attempt to formulate of Loop Quantum Supergravity, which extends to supergravity the corresponding procedure to quantize gravity, has been done in [35].

Further, well known formulations of gravity and supergravity as gauge theories are due to Mac-Dowell and Mansouri. In [36] the authors were able to extend this formulation of Supergravity when the spin connection is self-dual, a particular and relevant complex form. In [78], a superfield description of self-dual Macdowell-Mansouri supergravity is given, considering the superfield formulation of a $OSp(1|4)$ super Yang-Mills theory. Further, its generalization to eight dimensions is discussed.

Regarding duality, a crucial property of string theory, S-duality allows us to go from a system that interacts strongly to one that interacts weakly like in electromagnetism where the charge of the monopole is the inverse of the electric charge. We were the first to propose in the literature to explore S-duality in connection with gravity and supergravity. This is the topic of [45] and in (2+1) dimensions of [55] .

Also in connection to duality, in [78] a relation between an old concept in mathematics, which arises from a subset of the axioms of a matrix, called matroid and which has interesting duality properties has been studied. The relation of this object to octonions lead the author to establish a connection with eleven dimensional supergravity and further with M-theory.

SUPERSYMMETRIC QUANTUM COSMOLOGY

One important contribution in México to quantum supergravity has been through the study of the quantization of homogeneous supergravity models, called Supersymmetric Quantum Cosmology. The first publication on the subject in the literature was [10]. Already in standard Quantum Cosmology (the "quantum mechanics" of the early Universe) attempts were made to find a square root. In a certain sense to search for something similar to the Dirac Equation. All methods were applied, in particular to cosmological models, but a general principle was lacking. The discovery by C. Teitelboim and later by S. Deser that, in the canonical formulation of Supergravity a constraint appears (a supercharge), that is the square root of the Hamiltonian, allowed us to apply this idea to Quantum Cosmology. Instead of having the Wheeler-DeWitt Equation, a kind of Klein-Gordon equation, one has its square root from a general principle. The corresponding wave function has now various entries that represent the degrees of freedom of the gravitation field.

These publications can be divided in three different groups according with the approach used. In the first group, that we called the matrix approach, the algebra among the gravitino field components is represented by matrices [10, 18, 19, 21, 23, 72] and [43]. In [22] and [25], a different method is applied where the algebra between the gravitino fields components is realized through differential operators. In [37, 42, 46] and [56], the bosonic gravitational field components of the cosmological metric are generalized to superfields, they depend then on the time and two associated Grassman variables.

In [28, 29] the Bianchi type A supergravity models are studied. In the first work it was shown that there are nontrivial physical states only when $\Lambda = 0$. In the second work, the authors consider the Jacobson formulation of N=1 canonical supergravity in terms of Ashtekar's new variables for spatially homogenous solutions of the mentioned type.

In [47, 48, 49] the authors consider a formulation of supergravity for cosmological models, i.e. with homogeneous metrics. These models have a finite number of degrees of freedom, and their gravity versions are described by mechanical models. Supersymmetry is introduced by means of a superfield formulation, whose bosonic limit is the classical mechanics of the homogeneous theory. These models are coupled to various configurations of matter and their quantization is studied.

COSMOLOGICAL MODELS IN SUPERGRAVITY

Another field of work is the study of supergravity inspired models with well behaved inflationary potentials. In [51], the authors study supergravity low scale inflation, on scales in the range between supersymmetry and electroweak breaking scales. In [59] supergravity inspired models are considered, with scales lower than the electroweak scale. In [66] inflation along the angular direction of a chiral superfield is considered. In [76] a supergravity inspired two-stage inflationary model is constructed and in [70] supergravity inflation with one parameter is studied. This model gives a bounded total number of e-folds and the authors argue that this number is close to the required one for observable inflation, being also independent of the initial conditions for the inflaton.

In other works, supergravity is considered in connection to dark energy models. In [63] the dynamics of the moduli T of compactified dimensions is studied in order to see if they can generate the cosmological constant. It is shown that in general it would require fine tuning. In the works [74, 75, 77, 69], the authors propose dark energy models based on condensates of gauge symmetries in supergravity. These models depend only on the gauge group and on their matter content. The scale of the condensates is very small, which would explain why the dark energy manifests at very low cosmological energies. These models have a natural explanation in terms of elementary particles physics and shares the best fit of the cosmological data.

AdS/CFT IN SUPERGRAVITY

In [62, 57, 58, 53, 54], the authors consider various ansatzes of supergravity on the brane, by use of the effective action formulation and exploiting the AdS/CFT correspondence.

In [68] a new static partially twisted solution of N=4, SO(4) gauged supergravity in D=11 is obtained from the embedding of four dimensional supergravity. Further two more solutions in four dimensions are obtained, as well as an exact solution interpolating them, which are also uplifted to 11 dimensions: an asymptotic one corresponding to AdS_4 and a near horizon fixed point solution of the form $AdS_2 \times H_2$.

EXPERIMENTAL SEARCH OF SUPERGRAVITY

The experimental search of signatures of supergravity has been done in the frame of the collaborations D0 and HERA.

In HERA, with collaboration of G. Herrera of Cinvestav [50, 52, 60, 61, 71], the minimal supergravity model is tested, at different center of mass energies, by the search for associated production of supersymmetric particles, or missing energy or momentum events. In this way, limits on the corresponding parameters are being set.

In D0, with collaboration of H. Castilla of Cinvestav, [64, 65, 67, 73], using data from the Fermilab Tevatron $p\bar{p}$ Collider, at different center of mass energies, the minimal supergravity model is tested, by the search for associated production of supersymmetric particles, or missing energy or momentum events. In this way, limits on the corresponding parameters are being set.

ACKNOWLEDGMENTS

We thank Prof. Luis Urrutia for his invitation to contribute to this proceedings.

REFERENCES

1. D. Z. Freedman, P. van Nieuwenhuizen and S. Ferrara, *Phys. Rev. D* **13**, 3214 (1976).
2. P. van Nieuwenhuizen, *Phys. Rep.* **68**, 189 (1981).
3. L. F. Urrutia, "Further Discussion of the Ansatz $\psi_\mu = \delta_{\mu t}\xi$ for an Exact Solution of the Helicity 3/2 Field in Classical Supergravity, " *Phys. Lett. B* **98**, 427 (1981).
4. L. F. Urrutia, "A New Exact Solution of Classical Supergravity," *Phys. Lett. B* **102**, 393 (1981).
5. O. Obregón and M. P. Ryan, Jr., "An Einstein-Cartan Formulation of Supergravity," *Kinam* **5**, 3-19 (1983).
6. "Introduction to Supersymmetry in Particle and Nuclear Physics", edited by O. Castaños, A. Frank and L. Urrutia, *Plenum Press, New York*, 1984.

7. C. M. Hull, "Exact pp wave solutions of eleven dimensional supergravity," *Phys. Lett. B* **139**, 39 (1984), C. M. Hull, "Killing spinors and exact plane wave solutions of extended supergravity, " *Phys. Rev. D* **30**, 334 (1984).
8. C. P. Luehr, M. Rosenbaum, "Supertwistor fiber bundles as a formalism for supergravities," *J. Math. Phys.* **26**, 1834-1846 (1985).
9. M. Rosenbaum, M. Ryan, L. F. Urrutia and R. Matzner, "Colliding Plane Waves in Supergravity, " *Phys. Rev. D* **34**, 409 (1986).
10. A. Macías, O. Obregón and M. P. Ryan Jr., "Quantum Cosmology: The Supersymmetric Square Root," *Class. Quantum Grav.* **4**, 1477-1486 (1987).
11. M. Rosenbaum, M. Ryan and L. F. Urrutia, "Kasner-Type Cosmological Models in N=1 Supergravity," *Class. Quantum Grav.* **4**, 79 (1987).
12. G. F. Torres del Castillo, "Rarita–Schwinger fields in algebraically special vacuum space-times," *J. Math. Phys.* **30**, 446 (1989).
13. G. F. Torres del Castillo, "Spin-3/2 perturbations of algebraically special solutions of the Einstein–Maxwell equations," *J. Math. Phys.* **30**, 2114 (1989).
14. G. F. Torres del Castillo, "Debye potentials for Rarita–Schwinger fields in curved space-times," *J. Math. Phys.* **30**, 1323 (1989).
15. G. F. Torres del Castillo and G. Silva Ortigoza, "Rarita–Schwinger fields in the Kerr geometry," *Phys. Rev. D* **42**, 4082 (1990).
16. L. F. Urrutia and J. D. Vergara, "Consistent coupling of the Gravitino Field to a Gravitational Background with Torsion," *Phys. Rev. D* **44**, 3882 (1991).
17. G. F. Torres del Castillo and G. Silva Ortigoza, "Spin-3/2 perturbations of the Kerr–Newman solution," *Phys. Rev. D* **46**, 5395 (1992).
18. J. Benítez, O. Obregón and J. Socorro, "Supersymmetric Taub Model, the Micro-Superspace Sector," *Astrophys. and Sp. Sci.* **193**, 61-68 (1992).
19. J. Socorro, O. Obregón and A. Macías, "Supersymmetric Microsuperspace Quantization for the Taub Model," *Phys. Rev. D* **45**, 2026-2032 (1992).
20. J. Gamboa and C. Ramirez, "Hamiltonian approach to 2-D supergravity," *Phys. Lett. B* **301**, 20 (1993).
21. A. Macías, O. Obregón and J. Socorro, "Supersymmetric Quantum Cosmology," *Int. J. Mod. Phys. A.* **8**, 4291-4317 (1993).
22. P. D. D'Eath, S. W. Hawking and O. Obregón, "Supersymmetric Bianchi Models and the Square Root of The Wheeler-DeWitt Equation," *Phys. Lett. B* **300**, 44-48 (1993).
23. O. Obregón, J. Socorro and J. Benítez, "Supersymmetric Quantum Cosmology Proposals and the Bianchi Type-II Model," *Phys. Rev. D* **47**, 4471-4475 (1993).
24. J. A. Nieto and O. Obregón, "Classical Supersymmetric Spin-3/2 Particle," *Phys. Lett. A*, **175**, 11-13 (1993).
25. O. Obregón, J. Pullin and M. P. Ryan Jr., "Bianchi Cosmologies: New Variables and a Hidden Supersymmetry," *Phys. Rev. D* **48**, 5642-5647 (1993).
26. L. O. Pimentel, "Birkhoff theorem in N=2, D = 5 supergravity," *Phys. Rev. D* **47**, 4780 (1993).
27. L. O. Pimentel and J. Socorro, "Bianchi VI(0) models in N=2, D = 5 supergravity," *Gen. Rel. Grav.*, **25**, 1159 (1993).
28. R. Capovilla and O. Obregón, "No quantum superminisuperspace with Lambda not = 0," *Phys. Rev. D* **49**, 6562 (1994).
29. R. Capovilla and J. Guven, "Superminisuperspace and new variables," *Class. Quant. Grav.* **11**, 1961 (1994).
30. G. Silva-Ortigoza, "The Starobinsky constant and the algebraically special perturbations of Carter's A solution," *Rev. Mex. Fís.* **40**, 730 (1994).
31. G. Silva-Ortigoza, "Perturbations of Carter's A solution," *Rev. Mex. Fís.* **41**, 728 (1995).
32. G. Silva-Ortigoza, "Killing spinors and separability of Rarita-Schwinger's equation in type {2,2} backgrounds," *J. Math. Phys.* **36**, 6929 (1995).
33. L. O. Pimentel and J. Socorro, "Bianchi V models in N=2, D = 5 supergravity," *Int. J. Theor. Phys.* **34**, 701 (1995).
34. L. O. Pimentel, "Cosmic strings from N=2, 5=D supergravity," *Mod. Phys. Lett. A* **10**, 1997 (1995).
35. D. Armand-Ugon, R. Gambini, O. Obregón and J. Pullin, "Towards a Loop Representation for Quantum Canonical Supergravity," *Nucl. Phys. B* **460**, 615-631 (1996).

36. J. A. Nieto, J. Socorro and O. Obregón, "Gauge Theory of Supergravity Based Only on Self-Dual Spin Connection," *Phys. Rev. Lett.* **6**, 3482-3485 (1996).
37. V. I. Tkach, J. J. Rosales and O. Obregón, "Supersymmetric Action for Bianchi Type Models," *Class. Quantum Grav.* **13**, 2349-2356 (1996).
38. A. Macías, "Chiral (N=1) supergravity," *Class. Quant. Grav.* **13**, 3163 (1996).
39. G. F. Torres del Castillo and A. Herrera Morales, "Spin-3/2 fields in flat space-time," *Int. J. Theor. Phys.* **35**, 569 (1996).
40. G. Silva-Ortigoza, "Covariant characterization of the separation constants for spin-3/2 and gravitational perturbations of the Carter A solution," *Class. Quantum Grav.* **13**, 3107-3113 (1996).
41. G. Silva-Ortigoza, "Killing spinors and spin-3/2 fields in type-D space-times," *Class. Quantum Grav.* **14**, 795-804 (1997).
42. V. I. Tkach, O. Obregón and J. J. Rosales, "FRW Model and Spontaneous Breaking of Supersymmetry," *Class. Quantum Grav.* **14**, 339-350 (1997).
43. O. Obregón and C. Ramírez, "Dirac-Like Formulation of Quantum Supersymmetric Cosmology", *Phys. Rev. D.* **57**, 1015-1026 (1998).
44. J. A. Nieto, "Matroid theory and supergravity," *Rev. Mex. Fis.* **44**, 358 (1998).
45. H. García-Compeán, J. A. Nieto, O. Obregón and C. Ramírez, "Dual Description of Supergravity MacDowell-Mansouri Theory", *Phys. Rev. D* **59**, 124003 (1999).
46. O. Obregón, J. J. Rosales, J. Socorro and V. I. Tkach, "Supersymmetry Breaking and a Normalizable Wavefunction for the FRW (k=0) Cosmological Model", *Class. Quantum Grav.* **16**, 2861-2870 (1999).
47. V. I. Tkach, J. J. Rosales and J. Socorro, "Supersymmetric FRW model and the ground state of supergravity," *Class. Quant. Grav.* **16**, 797 (1999).
48. V. I. Tkach, J. J. Rosales and J. Socorro, "Spontaneous breaking of supersymmetry in cosmological models and supergravity theories," *Mod. Phys. Lett. A* **14**, 1209 (1999).
49. V. I. Tkach, J. J. Rosales and J. Torres, "On the relation of the gravitino mass and the GUT parameters," *Mod. Phys. Lett. A* **14**, 169 (1999).
50. B. Abbott et al. [D0 Collaboration], "Search for squarks and gluinos in events containing jets and a large imbalance in transverse energy," *Phys. Rev. Lett.* **83**, 4937 (1999).
51. G. Germán, A. de la Macorra and M. Mondragón, "Low scale supergravity inflation with R-symmetry," *Phys. Lett. B* **494**, 311 (2000).
52. B. Abbott et al. [D0 Collaboration], "Search for R-parity violation in multilepton final states in p anti-p collisions at s**(1/2) = 1.8-TeV," *Phys. Rev. D* **62**, 071701 (2000).
53. S. Nojiri, S. D. Odintsov and S. Ogushi, "Scheme-dependence of holographic conformal anomaly in d5 gauged supergravity with non-trivial bulk potential," *Phys. Lett. B* **494**, 318 (2000).
54. S. Nojiri, S. D. Odintsov and S. Ogushi, "Holographic conformal anomaly with bulk scalars potential from d3 and d5 gauged supergravity," *Prog. Theor. Phys.* **104**, 867 (2000).
55. H. García-Compean, O. Obregón, C. Ramírez and M. Sabido, "On S-duality in (2+1)-dimensional Chern-Simons Supergravity," *Physical Review D* **64**, 024002 (2001).
56. O. Obregón, H. Quevedo and M.P. Ryan, "Kerr metric as an exact solution for unpolarized S1S2 Gowdy Models," *Phys. Rev. D* **65**, 024022 (2001).
57. S. Nojiri, S. D. Odintsov and S. Ogushi, "Quantum stabilization of thermal brane-worlds in M-theory," *Phys. Lett. B* **506**, 200 (2001).
58. S. Nojiri and S. D. Odintsov, "Supersymmetric new brane world," *Phys. Rev. D* **64**, 023502 (2001).
59. G. Germán, G. G. Ross and S. Sarkar, "Low-scale inflation," *Nucl. Phys. B* **608**, 423 (2001).
60. B. Abbott et al. [D0 Collaboration], "A search for dilepton signatures from minimal low-energy supergravity in p anti-p collisions at s**(1/2) = 1.8-TeV," *Phys. Rev. D* **63**, 091102 (2001).
61. C. Adloff et al. [H1 Collaboration], "Searches at HERA for squarks in R-parity violating supersymmetry," *Eur. Phys. J. C* **20**, 639 (2001).
62. S. Nojiri and S. D. Odintsov, "On the way to brane new world," *Grav. Cosmol.* **8**, 73 (2002).
63. A. de la Macorra, "Can moduli fields parameterize the cosmological constant?," *Int. J. Mod. Phys. D* **11**, 653-668 (2002).
64. V. M. Abazov et al. [D0 Collaboration], "Search for the production of single sleptons through R-parity violation in p anti-p collisions at s**(1/2) = 1.8-TeV," *Phys. Rev. Lett.* **89**, 261801 (2002).
65. V. M. Abazov et al. [D0 Collaboration], "Search for mSUGRA in single electron events with jets and large missing transverse energy in p anti-p collisions at s**(1/2) = 1.8-TeV," *Phys. Rev. D* **66**, 112001 (2002).

66. G. Germán, A. Mazumdar and A. Perez-Lorenzana, "Natural inflation from supergravity," *Mod. Phys. Lett. A* **17**, 1627 (2002).
67. V. M. Abazov *et al.* [D0 Collaboration], "Search for R-parity violating supersymmetry in dimuon and four-jets channel," *Phys. Rev. Lett.* **89**, 171801 (2002).
68. H. H. Hernández Hernández and H. A. Morales-Técotl, "New solution of D = 11 supergravity on S**7 from D = 4," *JHEP* **0309**, 039 (2003).
69. A. De la Macorra, "Quintessence unification models from non-abelian gauge dynamics", JHEP **0301** 033, 2003.
70. G. Germán and A. de la Macorra, "A model of inflation independent of the initial conditions, with bounded number of e-folds and n(s) larger or smaller than one," *Phys. Rev. D* **70**, 103521 (2004).
71. A. Aktas *et al.* [H1 Collaboration], "Search for squark production in R-parity violating supersymmetry at HERA," *Eur. Phys. J. C* **36**, 425 (2004).
72. O. Obregón, L. A. Ureña-López and F. E. Schunck, "Oscillatons formed by non-linear gravity," *Phys. Rev. D* **72**, 024004 (2005).
73. V. M. Abazov *et al.* [D0 Collaboration], "Search for supersymmetry via associated production of charginos and neutralinos in final states with three leptons," *Phys. Rev. Lett.* **95**, 151805 (2005).
74. A. de la Macorra and C. Stephan-Otto, "Natural quintessence with gauge coupling unification," *Phys. Rev. Lett.* **87**, 271301 (2001).
75. A. de la Macorra, "Dark group: dark energy and dark matter," *Phys. Lett. B* **585**, 17-23 (2004).
76. G. Germán and A. de la Macorra, "Inflation at the maxima of the potential inspired by brane models," *Int. J. Mod. Phys. A* **20**, 6451 (2005).
77. A. de la Macorra, "A Realistic particle physics dark energy model," *Phys. Rev. D* **72**, 043508 (2005).
78. J. A. Nieto, "Superfield description of a self-dual supergravity a la Macdowell-Mansouri," to appear in *Class. Quant. Grav*, e-Print Archive: hep-th/0509169.
79. B. Zumino, *Fortsch. Phys.* **54**, 199 (2006).
80. A. Macías, H. Quevedo and A. Sánchez, "On the local Lorentz invariance in N=1 supergravity," *Phys. Rev. D* **73**, 027501 (2006).

Models of flavour with discrete symmetries

A. Mondragón

Instituto de Física, UNAM, Apdo. Postal 20-364, 01000 México D.F., México

Abstract. We briefly review some recent developments in theoretical models of fermion masses, mixings and CP violation with discrete non-Abelian symmetries. Then, we explain the main ideas of a recently proposed Minimal S_3−invariant Extension of the Standard Model and its application to a unified analysis of masses, mixings and CP violation in the leptonic and quark sectors as well as the explicit computation of the V_{PMNS} and V_{CKM} mixing matrices.

Keywords: Flavor symmetries; Quark and lepton masses and mixings; Neutrino masses and mixings
PACS: 11.30.Hv, 12.15.Ff,14.60.Pq

INTRODUCTION

In the last six or seven years, great advances have been made in the experimental knowledge of flavour physics, fermion masses and mixings and CP violation. These advances initiated a huge upsurge of theoretical activity aimed at uncovering the nature of this new physics. In the next section of this paper I will very briefly outline some recent theoretical developments on models of flavour with discrete non-Abelian symmetries in which the participation of the mexican community of particles and fields is visible. Section 3 is devoted to a brief explanation of the recently proposed Minimal S_3−invariant Extension of the Standard Model [33]. The paper ends with a short summary and some conclusions.

MODELS OF FLAVOUR WITH DISCRETE SYMMETRIES

The history of models for the quark mass matrices may possibly be traced back to Weinberg's[1] observation that the Gatto, Sartori, Tonin[2] relation for the Cabbibo angle may be expressed as a relation between the Cabbibo angle and the quark masses of the first two generations,

$$V_{us} \approx \sqrt{\frac{m_d}{m_s}}, \qquad (1)$$

and that mass matrices of the form

$$M_u = \begin{pmatrix} m_u & 0 \\ 0 & m_c \end{pmatrix}, \qquad M_d = \begin{pmatrix} 0 & p \\ p & q \end{pmatrix} \qquad (2)$$

can account for the approximate equality (1). As a consequence, in the early approaches to the problem of quark masses and mixings, it was natural to postulate that some entries in the Yukawa matrix were equal to zero, the so called "texture zeroes" [3, 4], thereby reducing the number of free parameters of the theory. Since then many approaches have been developed in the context of different theoretical and phenomenological models.

In the Standard $[SU_C(3) \times SU_L(2) \times U_Y(1)]$ Model of the strong, weak and electromagnetic interactions, it is the Higgs mechanism that provides a theoretically consistent framework to generate masses for gauge bosons and fermions - the latter acquire masses after spontaneous breaking of the $SU(2)$ gauge symmetry, through the Yukawa couplings and the vacuum expectation value of the neutral Higgs field. However, this framework can neither predict the values of fermion masses nor interpret the observed hierarchy of their spectra. Hence, the three charged lepton masses, the six quark masses as well as the four parameters in the quark mixing matrix are free parameters of the Standard Model. As a straightforward consequence of the symmetry structure of the Standard Model, the renormalizable Yukawa couplings do not allow neutrino masses, although they can be introduced through the addition of non-renormalizable, higher-dimensional operators, presumably originating in physics beyond the Standard Model.

In the late 70's and early 80's, there already were a few promising indications for theoretical structures beyond the Standard Model which addressed the fermion mass problem. For simple symmetry breaking schemes, grand unification can relate quark and lepton masses. The most promising of such relations is the equality $m_b = m_\tau$, a result which applies at the Grand Unified Theory (GUT) scale. Radiative corrections are dominated by QCD interactions which increase the bottom quark mass in fair agrement with experiment. A fairly straightforward supersymmetric generalization of such GUT relations, which involves a new family symmetry, also provides the Georgi-Jarlskog [5] relations between the down quarks and charged leptons of the first two generations. Supersymmetry also enables the gauge couplings to meet at the GUT scale to give a self-consistent unification picture [6, 7, 8, 9].

The past ten years have seen great advances in the experimental and theoretical knowledge of flavour physics, CP-violation and fermion masses. In 1999, direct CP-violation in the Kaon system was established through the NA48 (CERN) and KTeV(FNAL) collaborations. In this decade, huge experimental efforts have been made to further explore CP-violation and the quark-flavour sector of the Standard Model. The main actor in these studies has been the B-meson system. In 2001, CP-violating effects were discovered and measured in the B-meson system by the BaBar [10] and Belle [11] Colaboration. A detailed investigation was also made of some benchmark, rare decay modes such as $B_d^o \to J/\psi K_s$, $B_d^o \to \phi K_s$ and $B_d^o \to \pi^+\pi^-$ and many others, for a recent review see R. Fleischer [12]. As of May 2006, it can be said that all existing data on CP-violation and rare decays in the quark sector can be described by the Standard Model within the theoretical and experimental uncertainties. The recent discovery and measurement of flavour conversion of solar [13, 14, 15, 16] atmospheric [17, 18], reactor [19, 20] and accelerator [21, 22] neutrinos have conclusively established that neutrinos have non-vanishing mass and they mix among themselves much like the quarks, thereby providing the first evidence of new physics beyond the Standard Model. The difference of the squared neutrino masses and the mixing angles in the lepton mixing matrix, U_{PMNS}, were determined, but neutrino oscillation data are insensitive to the absolute value of neutrino masses and also to the fundamental issue of wether neutrinos are Dirac or Majorana particles. Upper bounds on neutrino masses were provided by the searches that probe the neutrino mass values at rest: beta decay experiments [23], neutrinoless double beta decay [24] and precision cosmology [25].

These recent experimental advances triggered an enormous theoretical activity attempting to uncover the nature of this new physics. This includes further developments of the already existing mechanisms and theories such as GUT's and Supersymmetric Grand Unified Theories (SUSY GUTs) and the appearance of new ideas and approaches implemented at a variety of different energy scales.

Regardless of the energy scales at which those theoretical models are built, the mechanisms for fermion mass generation and flavour mixing can roughly be classified into four different types:

1. Texture zeroes,
2. Family or flavour symmetries
3. Radiative mechanisms and
4. Seesaw mechanisms

These mechanisms are not disjoint but rather they are related and in many cases they complement and support each other. In the last six or seven years, important theoretical advances have been made in the understanding of these four mechanisms. The following points should be stressed.

1. Phenomenologically, some striking progress has been made with the help of texture zeroes and flavour symmetries in specifying the quantitative relationship between flavour mixing angles and quark or lepton mass ratios [26, 27].

2. After all the recent developments, the seesaw mechanism with large scale of the B-L violations still looks as the most appealing and natural mechanism of neutrino mass generation. At the same time, it is not excluded that some more complicated version of this mechanism is realized [28].

3. Gran Unification plus supersymmetry in some form still looks like the most plausible scenario of physics which naturally embeds the seesaw mechanism.

4. At the same time, it seems now clear that the "seesaw GUT" scenario does not provide a complete understanding of the neutrino masses and mixings as well as the quark masses and mixings or, in other words, the flavour structure of the mass matrices. Some new physics on top of this scenario seems essential. In this connection, two important questions arise:

- the possible existence of new symmetries that show up mainly or only in the lepton sector
- the need to understand the relation between quarks and leptons and the picture of flavour physics and CP violation in a unified way. The corresponding phenomenology is very rich.

These two issues point to the need of simpler models.

5. The search for simpler models starts by first constructing a low energy theory with the Standard Model and a discrete non-Abelian flavour symmetry group $\tilde{G}_F$ and then showing the possible embeddings of this theory into a GUT, like $SO(10)$ or $SU(5)$. The discrete symmetry will therefore be a subgroup of $SO(3)_f$ or $SU(3)_f$. Models in which the discrete non-Abelian flavour symmetry is only broken at low energies became very

popular in the last few years [29, 30, 31, 32, 33, 34, 35, 36, 37, 38]. The search for an adequate discrete group has concentrated on the smallest subgroups of $SO(3)$ or $SU(3)$ that have at least one singlet and one doublet irreducible representations to accommodate the fermions in each family [39].

To end this section, I will very briefly outline some recent developments on these questions in which the participation of the mexican community of particles and fields is visible.

Flavour permutational symmetry and Fritzsch textures

The non-Abelian flavour permutational symmetry $S_{3L} \otimes S_{3R}$ and its explicit sequential breaking according to $S_{3L} \otimes S_{3R} \supset S_{3diag} \supset S_{2diag}$ was used by A. Mondragón and E. Rodríguez-Jáuregui[40, 41] to characterize the quark mass matrices, M_u and M_d, with a texture of the same modified Fritzsch type. In a symmetry adapted basis, different patterns for breaking the permutational symmetry give rise to quark mass matrices which differ in the ratio $Z^{1/2} = M_{23}/M_{22}$ and are labelled in terms of the irreducible representations of an auxiliary S_2 group. After analytically diagonalizing the mass matrices, these authors derive explicit, exact expressions for the elements of the quark mixing matrix, V_{CKM}^{th}, the Jarlskog invariant, J, and the three inner angles, α, β and γ of the unitarity triangle as functions of the quark mass ratios and only two free parameters, the symmetry breaking parameter $Z^{1/2}$ and one CP-violating phase Φ. The numerical values of these parameters which characterize the experimentally preferred symmetry breaking pattern $Z^{1/2} = 9/2\sqrt{2}$ and $\Phi = 90°$, were extracted from a χ^2 fit of the theoretical expressions for the moduli, $|V^{th}|$, to the experimentally determined values of the moduli of the elements of the quark mixing matrix $|V_{CKM}^{exp}|$. The agreement between theory and experiment, which initially was fairly good, improved as the experimental determination of the elements of the mixing matrix and the inner angles of the unitarity triangle improved [42]. The phase equivalence of V_{CKM}^{th} and the mixing matrix V_{CKM}^{PDG} in the standard parameterization advocated by the Particle Data Group allowed to translate those results into explicit exact expressions for the three mixing angles $\theta_{12}, \theta_{13}, \theta_{23}$ and the CP-violating phase δ_{13} in terms of the four quark mass ratios and the symmetry breaking parameters $Z^{1/2}$ and Φ [41].

The main point in these results is simply that the hierarchy of quark masses and the texture of quark mass matrices are enough to determine, at least partly, some important features of the quark flavour mixing. In this sense, it was established that a scheme in which the two quark mass matrices, M_u and M_d, have the same modified Fritzsch texture with the same value of the symmetry breaking parameter has some predictive power for the flavour mixing angles and CP-violating phase.

Models of flavour with continuous symmetry

There is a large variety of possible candidates for supersymmetric models of new physics beyond the Standard Model based in $N = 1$ SUSY with commuting GUT's and family symmetry groups, $G_{GUT} \otimes G_f$. This is so because there are many possible candidate GUT's and family symmetry groups G_f. The model dependence does not end there since the details of the symmetry breaking vacuum plays a crucial role in specifying the model and determining the masses and mixing angles, for a recent review see S.F. King [43]. G. G. Ross and L. Velasco-Sevilla chose the largest family symmetry group, $SU(3)$, consistent with $SO(10)$ GUT's and with additional Abelian family symmetries chosen to restrict the allowed Yukawa couplings. In a series of interesting papers[44, 45, 46] they explored the phenomenological implications of their model and were able to find a symmetry breaking scheme in which the observed hierarchical quark masses and mixings are described together with the hierarchy of charged lepton masses and a hierarchical structure for the neutrino masses. The significant differences between quark and lepton mixings are explained as due to the seesaw mechanism. Given the very large underlying symmetry, the fermion masses are heavily constrained. This $SO(10) \otimes SU(3)_f$ model provides a consistent description of the known masses and mixings of quarks and leptons. In the quark sector, the presence of CP violating phases is necessary, not only to reproduce CP violating processes, but also to reproduce the observed masses and mixings. In this model, the spontaneous breaking of CP in the flavour sector naturally solves the supersymmetric CP problem and the SUSY flavour problem, although flavour changing processes must occur at a level close to current experimental bounds. Motivated by the fact that leptogenesis is a very attractive candidate for explaining the large baryon asymmetry observed in the universe L. Velasco-Sevilla [46] also explored the very interesting possible connections between low energy CP violating phases appearing in the lepton mixing matrix and those phases relevant for leptogenesis.

SUSYGUT Models with discrete flavour symmetry

As noted above, in supersymmetric Grand Unified models of flavour with a non-Abelian continuous family group, such as $SO(10) \otimes SU(3)_f$ or $SU(5) \otimes SU(2)_f$, the phenomenological success depends crucially on the details of the symmetry breaking vacuum and its alignment.

For instance, neutrino mixing angle relations such as the bimaximal mixings of the left handed neutrinos is achieved only if the Yukawa couplings involving different families are related in some special way. The condition for the required equalities of Yukawa couplings to emerge is that the several scalar fields which break the family symmetry, called flavons, have their vacuum expectation values carefully aligned (or misaligned) along special directions in family space. Then, if these flavons appear in the effective operators responsible for the Yukawa couplings, the relations between the Yukawa couplings may be due to the particular alignment of the flavons responsible for that particular operator.

In an interesting series of papers, A. Aranda, C.D. Carone and R.F. Lebed [47, 48, 49] showed that the physics of vacuum alignment simplifies if the continuous family symmetry $SU(2)_f$ is replaced by the discrete non-Abelian family symmetry $T' \otimes Z_3$ in the SUSYGUT model of flavour $SU(5) \otimes SU(2)_f$ proposed by Romanino, Barbieri and Hall [50, 51, 52]. The group T' is the group of proper rotations that leave a regular tetrahedron invariant in the $SU(2)$ double covering of $SO(3)$. It has singlet, doublet and triplet irreducible representations with the multiplication rule $2 \otimes 2 = 1 \oplus 3$, which is a requisite to reproduce the phenomenologically successful mass textures derived from GUT $SU(5) \otimes SU(2)_f$. The extra Abelian Z_3 factor in $G_f = T' \otimes Z_3$ is included in order to obtain the minimal extension needed to reproduce the $SU(2)$ model textures and satisfy discrete anomaly cancellation conditions. The flavons have non-trivial transformation properties under the GUT SU(5) symmetry and the up-type and down-type quark mass textures are accordingly modified. Additionally, in the lepton sector, the rich representation structure of T' allows for the neutrinos to be placed in different reps than the charged leptons, which, in this model is the origin of different hierarchies in the two sectors. The symmetry breaking pattern is $T' \otimes Z_3 \to Z_3^{diag} \to$ nothing. The light neutrino masses are generated through the seesaw mechanism. Three generations of right-handed neutrinos are introduced with the assignments $2^{0-} \oplus 1^{-+}$. This assignment leads to Dirac and Majorana mass matrices that allow the introduction of flavons that do not contribute at all to the charged fermion mass matrices. In this way, mass matrices with a modified Fritzsch texture are generated for the u and d−type quarks, and for the charged leptons while the light Majorana neutrino mass matrix has a texture that naturally leads to the bimaximal U_{PMNS} lepton mixing matrix.

Some further advantages of using a finite, discrete family symmetry are the following:

1. The breaking of a discrete symmetry does not lead to unwanted massless Goldstone bosons, unlike continuous symmetries
2. If this breaking is only spontaneous, it might produce domain walls [53] which can be a serious problem. However, it can be solved by either invoking low scale inflation or embedding the discrete symmetry group into a continous group [54] as is the case for $T' \otimes Z_3 \subset SU(2)$.
3. In the context of SUSY, discrete gauge symmetries do not give rise to excessive flavour changing neutral currents (FCNC) as is the case for continous symmetries.

Type II seesaw and $S_3 \otimes U(1)_{e-\mu-\tau}$ symmetry

One of the first phenomenologically successful models for reproducing the bimaximal mixing among the neutrinos was presented by R.N. Mohapatra, A. Pérez-Lorenzana and C. Pires [55]. This model is an extension of the Standard Model where the bimaximal mixing pattern among the neutrinos naturally arises via the type II seesaw mechanism. The model does not include right handed neutrinos, the lepton content of the SM is left unaltered but the Higgs sector is modified. The $SU_L(2)$ content of the Higgs sector consists of three doublets, two triplets with $Y = 2$ and a charged isosinglet with $Y = +2$. The model has a global $S_3 \otimes U(1)_{e-\mu-\tau}$ flavour symmetry. The charged μ and τ fields are in doublet representations, while the e field is in a singlet representation of S_3. The pattern of $SU_L(2)$ Higgs doublet vacuum expectation values leads to a diagonal mass matrix for the charged leptons while the additional Higgs triplet acquires naturally small vacuum expectation values due to the type II see saw mechanism. At tree level, the v_μ and v_e masses are degenerate, but the presence of the global $L_e - L_\mu - L_\tau$ and S_3 symmetry leads naturally to the desired mass splittings among neutrinos at the one loop level. The resulting neutrino masses have an inverted hierarchy $|m_1| \geq |m_2| \gg |m_3|$. There is a well known difficulty of this very interesting model to fit the large angle solution of the solar neutrino problem. Indeed, barring cancellations between the perturbations, these must be very small in order to obtain a Δm_{sun}^2 close to the best fit value, but then, the value of $\sin^2 2\theta$ comes out to close to unity in disagreement with the best global fits of solar data [56].

A minimal S_3-invariant extension of the Standard Model

The discovery of neutrino masses and mixings added ten new parameters to the already long list of free parameters in the Standard Model and made evident the urgent need of a systematic and unified treatment of all fermions in the theory. These two facts, taken together, pointed to the necessity and convenience of eliminating parameters and systematizing the observed hierarchies of masses and mixings as well as the presence or absence of CP violating phases by means of a flavour or family symmetry under which the families transform in a non-trivial fashion. As explained above, such a flavour symmetry might be a continuous or, more economically, a finite group.

In a recent paper, J. Kubo, A. Mondragón, M. Mondragón and E. Rodríguez-Jáuregui[33] argued that such a flavour symmetry, unbroken at the Fermi scale, is the permutational symmetry of three objects, S_3, and introduced a Minimal S_3-invariant Extension of the Standard Model. In this model, S_3 is imposed as a fundamental symmetry in the matter sector which is only spontaneously broken together with the electroweak gauge symmetry. This assumption leads to extend the concept of flavour and generations to the Higgs sector. Hence, going to the irreducible representations of S_3, the model has one Higgs $SU(2)_L$ doublet in the S_3-singlet representation plus two more Higgs $SU(2)_L$ doublets which can only belong to the two components of the S_3-doublet representation. The fermion content of the Standard Model is left unaltered. In this way, all the matter fields - Higgs, quarks and lepton fields including the right-handed neutrino fields - belong to the three dimensional representation $1_s \oplus 2$ of the permutation group S_3. The leptonic sector is further constrained by an Abelian Z_2 symmetry. A defined structure of the Yukawa couplings is obtained which permits the calculation of mass and mixing matrices for quarks and leptons in a unified way. The Majorana neutrinos acquire mass via the type I seesaw mechanism. In a recent paper, O. Felix, A. Mondragón, M. Mondragón and E. Peinado [57] reparametrized the mass matrices of charged leptons and neutrinos in terms of the respective mass eigenvalues and derived explicit analytic and exact expressions in closed form for the mixing angles appearing in the U_{PMNS} matrix as functions of the masses of charged leptons and neutrinos and one Majorana phase Φ_ν. The U_{PMNS} matrix has also one Dirac phase which has its origin in the charged lepton mass matrix. The numerical values of the mixing angles θ_{13} and θ_{23} are determined by the mass of charged leptons only in very good agreement with the best fit experimental values. The solar mixing angle θ_{12} is almost insensitive to the values of the masses of the charged leptons, but its experimental value allows the determination of the neutrino mass spectrum which has an inverted hierarchy with the values $|m_{\nu_2}| = 0.0507$ eV, $|m_{\nu_1}| = 0.0499$ eV and $|m_{\nu_3}| =$

0.0193 eV. A complete and detailed discussion of the Majorana phases of the neutrino mixing matrix in this model is given in J. Kubo [58]. A numerical analysis of the quark mass matrices and the V_{CKM} matrix gives one set of parameters that are consistent with the experimental values given by the Particle Data Group [59]. A slightly less sketchy explanation of this model is given in the next section.

A MINIMAL S_3- INVARIANT EXTENSION OF THE STANDARD MODEL

Recently, a minimal S_3-invariant extension of the Standard Model was suggested in [33], in this section I will explain in a slightly more detailed fashion the main ideas of this model and some recent results on neutrino masses and mixings.

S_3-**symmetric Lagrangian and fermions masses** . In the Standard Model analogous fermions in different generations have completely identical couplings to all gauge bosons of the strong, weak and electromagnetic interactions. Prior to the introduction of the Higgs boson and mass terms, the Lagrangian is chiral and invariant with respect to permutations of the left and right fermionic fields.

The six possible permutations of three objects (f_1, f_2, f_3) are elements of the permutational group S_3. This is the discrete, non-Abelian group with the smallest number of elements. The three-dimensional real representation is not an irreducible representation of S_3, it can be decomposed into the direct sum of a doublet and a singlet, $\mathbf{1}_s \oplus \mathbf{2}$. The direct product of two doublets may be decomposed into the direct sum of two singlets and one doublet, $\mathbf{2} \otimes \mathbf{2} = \mathbf{1}_s \oplus \mathbf{1}_A + \mathbf{2}$. The antisymmetric singlet is not invariant under S_3.

Since the Standard Model has only one Higgs $SU(2)_L$ doublet, which can only be an S_3 singlet, it can only give mass to the quark or charged lepton in the S_3 singlet representation, one in each family, without breaking the S_3 symmetry. Therefore, in order to impose S_3 as a fundamental symmetry, unbroken at the Fermi scale, we are led to extend the concept of flavour and generations to the Higgs sector of the theory. Hence, going to the irreducible representations of S_3, we add to the Higgs $SU(2)_L$ doublet in the S_3- singlet representation, two more $SU(2)_L$ doublet in the S_3-doublet representation. In this way, all the quark, lepton and Higgs fields, $Q^T = (u_L, d_L)$, u_R , d_R , $L^T = (\nu_L, e_L)$, e_R , ν_R and H, are in reducible representations $\mathbf{1}_s \oplus \mathbf{2}$. The most general renormalizable Yukawa interactions are given by

$$\mathscr{L}_Y = \mathscr{L}_{Y_D} + \mathscr{L}_{Y_U} + \mathscr{L}_{Y_E} + \mathscr{L}_{Y_\nu}, \qquad (3)$$

where

$$\begin{aligned}\mathscr{L}_{Y_{D,E}} &= -Y_1^d \overline{Q}_I H_S d_{IR} - Y_3^d \overline{Q}_3 H_S d_{sR} \\ &\quad -Y_2^d [\,\overline{Q}_I(\sigma_1)_{IJ} H_1 d_{JR} - \overline{Q}_I(\sigma_3)_{IJ} H_2 d_{JR}\,] \\ &\quad -Y_4^d \overline{Q}_s H_I d_{IR} - Y_5^d \overline{Q}_I H_I d_{sR} + \text{h.c.},\end{aligned} \qquad (4)$$

$$\mathscr{L}_{Y_{U,\nu}} = -Y_1^u \overline{Q}_I (i\sigma_2) H_S^* u_{IR} - Y_3^u \overline{Q}_3 (i\sigma_2) H_S^* u_{sR}$$

$$-Y_2^u [\overline{Q}_I(\sigma_1)_{IJ}(i\sigma_2)H_1^* u_{JR} - \overline{Q}_I(\sigma_3)_{IJ}(i\sigma_2)H_2^* u_{JR}]$$
$$-Y_4^u \overline{Q}_s(i\sigma_2)H_I^* u_{IR} - Y_5^u \overline{Q}_I(i\sigma_2)H_I^* u_{sR} + \text{h.c.}, \qquad (5)$$

The fields in the S_3–doublets carry capital indices I and J, which run from 1 to 2 and the singlets are denoted by the subscript s.

Furthermore, we add to the Lagrangian the Majorana mass terms for the right-handed neutrinos

$$\mathscr{L}_M = -M_1 v_{IR}^T C v_{IR} - M_3 v_{3R}^T C v_{3R}. \qquad (6)$$

Due to the presence of three Higgs fields, the Higgs potential $V_H(H_S, H_D)$ is more complicated than that of the Standard Model. This potential was analyzed by Pakvasa and Sugawara [60], see also Kubo[61], who found that in addition to the S_3 symmetry, it has a permutational symmetry S_2: $H_1 \leftrightarrow H_2$, which is not a subgroup of the flavour group S_3 and an Abelian discrete symmetry that will be used for selection rules of the Yukawa couplings in the leptonic sector. Here, we will assume that the vacuum respects the accidental S_2 symmetry of the Higgs potential and $\langle H_1 \rangle = \langle H_2 \rangle$.

With these assumptions, the Yukawa interactions, eqs. (4)- (5) yield mass matrices, for all fermions in the theory, of the general form

$$\mathbf{M} = \begin{pmatrix} \mu_1 + \mu_2 & \mu_2 & \mu_5 \\ \mu_2 & \mu_1 - \mu_2 & \mu_5 \\ \mu_4 & \mu_4 & \mu_3 \end{pmatrix}. \qquad (7)$$

The Majorana masses for the left neutrinos v_L will be obtained from the see-saw mechanism. The corresponding mass matrix is given by

$$\mathbf{M}_v = \mathbf{M}_{v_D} \tilde{\mathbf{M}}^{-1} (\mathbf{M}_{v_D})^T \qquad (8)$$

where $\tilde{\mathbf{M}} = \text{diag}(M_1, M_1, M_3)$.

In principle, all entries in the mass matrices can be complex since there is no restriction coming from the flavour symmetry S_3.

The mass matrices are diagonalized by bi-unitary transformations as

$$\begin{aligned} U_{d(u,e)L}^\dagger \mathbf{M}_{d(u,e)} U_{d(u,e)R} &= \text{diag}(m_{d(u,e)}, m_{s(c,\mu)}, m_{b(t,\tau)}), \\ U_v^T \mathbf{M}_v U_v &= \text{diag}(m_{v_1}, m_{v_2}, m_{v_3}). \end{aligned} \qquad (9)$$

The entries in the diagonal matrices may be complex, so the physical masses are their absolute values.

The mixing matrices are, by definition,

$$V_{CKM} = U_{uL}^\dagger U_{dL}, \quad V_{PMNS} = U_{eL}^\dagger U_v. \qquad (10)$$

Leptonic sector and Z_2 symmetry.

A further reduction of the number of parameters in the leptonic sector may be achieved by means of an Abelian Z_2 symmetry. A set of charge assignments of Z_2, compatible with the experimental data on masses and mixings in the leptonic sector is given in Table 1

−	+
H_S, v_{3R}	H_I, L_3, L_I, e_{3R}, e_{IR}, v_{IR}

Table I. Z_2 assignment in the leptonic sector.

These Z_2 assignments forbid certain Yukawa couplings,

$$Y_1^e = Y_3^e = Y_1^v = Y_5^v = 0. \tag{11}$$

Therefore, the corresponding entries in the mass matrices vanish, i.e., $\mu_1^e = \mu_3^e = 0$ and $\mu_1^v = \mu_5^v = 0$.

The mass matrix of the charged leptons

The mass matrix of the charged leptons takes the form

$$M_e = m_\tau \begin{pmatrix} \tilde{\mu}_2 & \tilde{\mu}_2 & \tilde{\mu}_5 \\ \tilde{\mu}_2 & -\tilde{\mu}_2 & \tilde{\mu}_5 \\ \tilde{\mu}_4 & \tilde{\mu}_4 & 0 \end{pmatrix}. \tag{12}$$

The unitary matrix U_{eL} that enters in the definition of the mixing matrix, U_{PMNS}, is calculated from

$$U_{eL}^\dagger M_e M_e^\dagger U_{eL} = \text{diag}(m_e^2, m_\mu^2, m_\tau^2), \tag{13}$$

The entries in the mass matrix squared, $M_e M_e^\dagger$, may readily be expressed in terms of the mass eigenvalues $(m_e^2, m_\mu^2, m_\tau^2)$. Then, the matrix U_{eL} may be expressed in terms of the charged lepton masses and one Dirac phase,

$$U_{eL} \approx \begin{pmatrix} \dfrac{1}{\sqrt{2}} \dfrac{\frac{m_e}{m_\mu}}{\sqrt{1-\left(\frac{m_e}{m_\mu}\right)^2}} & \dfrac{1}{\sqrt{2}} \dfrac{1}{\sqrt{1+\left(\frac{m_e}{m_\mu}\right)^2}} & \dfrac{1}{\sqrt{2}} \dfrac{e^{i\delta_e}}{\sqrt{1+\frac{m_e m_\mu}{m_\tau^2}}} \\ -\dfrac{1}{\sqrt{2}} \dfrac{\frac{m_e}{m_\mu}}{\sqrt{1-\left(\frac{m_e}{m_\mu}\right)^2}} & -\dfrac{1}{\sqrt{2}} \dfrac{1}{\sqrt{1+\left(\frac{m_e}{m_\mu}\right)^2}} & \dfrac{1}{\sqrt{2}} \dfrac{e^{i\delta_e}}{\sqrt{1+\frac{m_e m_\mu}{m_\tau^2}}} \\ \dfrac{\sqrt{1-2\left(\frac{m_e}{m_\mu}\right)^2}}{\sqrt{1-\left(\frac{m_e}{m_\mu}\right)^2}} & \dfrac{\frac{m_e}{m_\mu}}{\sqrt{1+\left(\frac{m_e}{m_\mu}\right)^2}} & \dfrac{\frac{m_e m_\mu}{m_\tau^2} e^{i\delta_e}}{\sqrt{1+\frac{m_e m_\mu}{m_\tau^2}}} \end{pmatrix} \tag{14}$$

The mass matrix of the neutrinos

According with the Z_2 selection rule eq. (11), $\mu_1^{\nu_D} = \mu_5^{\nu} = 0$ in (12). Then, the mass matrix for the left-handed Majorana neutrinos obtained from the see-saw mechanism takes the form

$$\mathbf{M}_\nu = \mathbf{M}_{\nu_D}\tilde{\mathbf{M}}^{-1}(\mathbf{M}_{\nu_D})^T$$
$$= \begin{pmatrix} m_{\nu_3} & 0 & \sqrt{(m_{\nu_3}-m_{\nu_1})(m_{\nu_2}-m_{\nu_3})} \\ 0 & m_{\nu_3} & 0 \\ \sqrt{(m_{\nu_3}-m_{\nu_1})(m_{\nu_2}-m_{\nu_3})} & 0 & m_{\nu_1}+m_{\nu_2}-m_{\nu_3} \end{pmatrix} \quad (15)$$

as in the case of the charged leptons, the matrix M_{ν_D} has been reparametrized in terms of its eigenvalues, the complex neutrino masses.

The unitary matrix U_ν that brings M_{ν_D} to diagonal form is

$$U_\nu = \begin{pmatrix} \sqrt{\frac{m_{\nu_2}-m_{\nu_3}}{m_{\nu_2}-m_{\nu_1}}} & \sqrt{\frac{m_{\nu_3}-m_{\nu_1}}{m_{\nu_2}-m_{\nu_1}}} & 0 \\ 0 & 0 & 1 \\ -\sqrt{\frac{m_{\nu_3}-m_{\nu_1}}{m_{\nu_2}-m_{\nu_1}}} & \sqrt{\frac{m_{\nu_2}-m_{\nu_3}}{m_{\nu_2}-m_{\nu_1}}} & 0 \end{pmatrix}. \quad (16)$$

The unitarity of U_ν constrains its entries to be real. This condition fixes the phases ϕ_1 and ϕ_2 as

$$|m_{\nu_1}|\sin\phi_1 = |m_{\nu_2}|\sin\phi_2 = |m_{\nu_3}|\sin\phi_\nu = 0 \quad (17)$$

The only free parameter in M_ν and U_ν, other than the real neutrino masses $|m_{\nu_1}|$, $|m_{\nu_2}|$ and $|m_{\nu_3}|$, is the phase ϕ_ν.

The neutrino mixing matrix

The neutrino mixing matrix V_{PMNS}, in the standard form advocated by the *PDG*, is obtained by taking the product $U_{eL}^\dagger U_\nu$ and making an appropriate transformation of phases, U_{PMNS} is, then equal to

$$\begin{pmatrix} \frac{1}{\sqrt{2}}\frac{x}{\sqrt{1-x^2}}\sin\eta + \frac{\sqrt{1-2x^2}}{\sqrt{1-x^2}}\cos\eta & \frac{1}{\sqrt{2}}\frac{x}{\sqrt{1-x^2}}\cos\eta - \frac{\sqrt{1-2x^2}}{\sqrt{1-x^2}}\sin\eta & -\frac{1}{\sqrt{2}}\frac{x}{\sqrt{1-x^2}}e^{-i\delta_e} \\ \frac{1}{\sqrt{2}}\frac{1}{\sqrt{1+x^2}}\sin\eta - \frac{x}{\sqrt{1-x^2}}\cos\eta\, e^{i\delta_e} & \frac{1}{\sqrt{2}}\frac{1}{\sqrt{1+x^2}}\cos\eta + \frac{x}{\sqrt{1-x^2}}\sin\eta\, e^{i\delta_e} & -\frac{1}{\sqrt{2}}\frac{1+2\frac{z}{y(1-y)}}{\sqrt{1+x^2}} \\ \frac{1}{\sqrt{2}}\frac{1}{\sqrt{1+\sqrt{z}}}\sin\eta - \frac{\sqrt{z}}{\sqrt{1+\sqrt{z}}}\cos\eta\, e^{i\delta_e} & \frac{1}{\sqrt{2}}\frac{1}{\sqrt{1+\sqrt{z}}}\cos\eta + \frac{\sqrt{z}}{\sqrt{1+\sqrt{z}}}\sin\eta\, e^{i\delta_e} & \frac{1}{\sqrt{2}}\frac{1}{\sqrt{1+\sqrt{z}}} \end{pmatrix}$$
$$(18)$$

where

$$\sin\eta = \sqrt{\frac{m_{\nu_2}-m_{\nu_3}}{m_{\nu_2}-m_{\nu_1}}} \quad x = \frac{m_e}{m_\mu}, \quad y = \frac{m_e^2+m_\mu^2}{m_\tau^2}, \quad \text{and} \quad z = \left(\frac{m_e m_\mu}{m_\tau^2}\right)^2 \quad (19)$$

and $K = \text{diag}(1, e^{i\alpha}, e^{i\beta})$ is the diagonal matrix of the Majorana phases.

Explicit expressions for the mixing angles in terms of the lepton masses are obtained from a comparison of U_{PMNS}^{th}, eq.(18) with the standard parameterization advocated by the PDG[59].

$$\tan\theta_{12} \approx \sqrt{\frac{(|m_{\nu_2}|^2 - |m_{\nu_3}|^2 \sin^2\phi_\nu)^{1/2} - |m_{\nu_3}||\cos\phi_\nu|}{(|m_{\nu_2}|^2 - |m_{\nu_3}|^2 \sin^2\phi_\nu)^{1/2} + |m_{\nu_3}||\cos\phi_\nu|}}, \qquad (20)$$

$$\sin\theta_{13} \approx \frac{\frac{1}{\sqrt{2}}\frac{m_e}{m_\mu}}{\sqrt{1-\left(\frac{m_e}{m_\mu}\right)^2}}, \quad \text{and} \quad \sin\theta_{23} \approx -\frac{1}{\sqrt{2}}\frac{\sqrt{1-\left(\frac{m_e}{m_\mu}\right)^2}}{\sqrt{1-\frac{1}{2}\left(\frac{m_e}{m_\mu}\right)^2}} \qquad (21)$$

Similarly, the Majorana phases are given by

$$\sin 2\alpha = \sin(\phi_1 - \phi_2) = \frac{|m_{\nu_3}|\sin\phi_\nu}{|m_{\nu_1}||m_{\nu_2}|} \times$$
$$\left(\sqrt{|m_{\nu_2}|^2 - |m_{\nu_3}|^2 \sin^2\phi_\nu} + \sqrt{|m_{\nu_1}|^2 - |m_{\nu_3}|^2 \sin^2\phi_\nu}\right) \qquad (22)$$
$$\sin 2\beta = \sin(\phi_1 - \phi_\nu) =$$
$$\frac{\sin\phi_\nu}{|m_{\nu_1}|}\left(|m_{\nu_3}|\sqrt{1-\sin^2\phi_\nu} + \sqrt{|m_{\nu_1}|^2 - |m_{\nu_3}|^2\sin^2\phi_\nu}\right)$$

A detailed discussion of the Majorana phases in the neutrino mixing matrix U_{PMNS} obtained in our model is given in J. Kubo [58].

Neutrino masses and mixings

In this model, $\sin^2\theta_{13}$ and $\sin^2\theta_{23}$ are determined by the masses of the charged leptons in very good agreement with the experimental values [62, 63, 64, 65],

$$(\sin^2\theta_{13})^{th} = 1.1 \times 10^{-5}, \quad (\sin^2\theta_{13})^{exp} \leq 0.046, \qquad (23)$$

and

$$(\sin^2\theta_{23})^{th} = 0.49, \quad (\sin^2\theta_{23})^{exp} = 0.5^{+0.06}_{-0.05}. \qquad (24)$$

In the present model, the experimental restriction $|\Delta m_{12}^2| < |\Delta m_{13}^2|$ implies an inverted neutrino mass spectrum, $|m_{\nu_3}| < |m_{\nu_1}| < |m_{\nu_2}|$ [33].

The mass $|m_{\nu_2}|$ assumes its minimal value when $\sin\phi_\nu$ vanishes, then

$$|m_{\nu_2}| \approx \frac{\sqrt{\Delta m_{13}^2}}{\sin 2\theta_{12}}. \qquad (25)$$

Hence, we find

$$|m_{\nu_2}| \approx 0.0507 eV, \quad |m_{\nu_1}| \approx 0.0499 eV, \quad |m_{\nu_3}| \approx 0.0193 eV \qquad (26)$$

where we used the values $\Delta m_{13}^2 = 2.2^{+0.37}_{-0.27} \times 10^{-3} eV^2$ and $\sin^2 \theta_{12} = 0.31^{+0.02}_{-0.03}$ taken from M. Maltoni et al. [64, 65] and G. L. Fogli et al. [63].

With those values for the neutrino masses we compute the effective electron neutrino mass m_β,

$$m_\beta = \left[\sum_i |U_{ei}|^2 m_{\nu_i}^2\right]^{\frac{1}{2}} = 0.0502 eV, \qquad (27)$$

well below the upper bound $m_\beta < 1.8 eV$ coming from the tritium β-decay experiments [23, 63, 66].

The hadronic sector

The Z_2 assignments in the hadronic sector are independent of those in the leptonic sector. Hence, in principle, it can be assumed that Z_2 is a good symmetry at a more fundamental level and verify that Z_2 is free from any quantum anomaly. However, if we give all quarks even parity as we did to the charged leptons, the Yukawa couplings $Y_1^{u,d}$ and $Y_3^{u,d}$ will be forbidden. Consequently, the squared mass matrix, of the u–quarks, $M_u M_u^\dagger$ would have a texture similar to the texture of the mass matrix of the Majorana neutrinos, given in (12), which would lead to a U_u that produces large values of the quark mixing angles in disagreement with the small experimental values.

Therefore, to give one set of parameters that are consistent with the experimental values given by the Particle Data Group[59], and show that the model is phenomenologically viable, we proceeded under the assumption that Z_2 is explicitly broken in the hadronic sector. Since all the S_3 invariant Yukawa couplings are now allowed, the mass matrices for the quarks take the general form (7), where all the entries can be complex. One can easily see that all the phases, except for those of $\mu_1^{u,d}$ and $\mu_3^{u,d}$, can be removed through an appropriate redefinition of the quark fields. Of course, only one of the four phases of $\mu_1^{u,d}$ and $\mu_3^{u,d}$ is observable in V_{CKM}. So, we assume that only m_3^d is a complex number.

The gross structure of realistic mass matrices can be obtained, if $\mu_3^{u,d} \sim O(m_{t,b})$ and $\mu_{1,2}^{u,d} \sim O(m_{c,s})$ (to achieve realistic mass hierarchies), and the non-diagonal elements $\mu_4^{u,d}$ and $\mu_5^{u,d}$ along with $\mu_{1,2}^{u,d}$ can produce a realistic mixing among the quarks. There are 10 real parameters and one phase to produce six quark masses, three mixing angles and one CP-violating phase. The set of dimensionless parameters

$$\begin{aligned}
m_1^u/m_0^u &= -0.000293 \,,\ m_2^u/m_0^u = -0.00028 \,,\ m_3^u/m_0^u = 1 \,, \\
m_4^u/m_0^u &= 0.031 \,,\ m_5^u/m_0^u = 0.0386, \\
m_1^d/m_0^d &= 0.0004 \,,\ m_2^d/m_0^d = 0.00275 \,,\ m_3^d/m_0^d = 1 + 1.2I \,, \\
m_4^d/m_0^d &= 0.283 \,,\ m_5^d/m_0^d = 0.058
\end{aligned} \qquad (28)$$

yields the mass hierarchies

$$m_u/m_t = 1.33 \times 10^{-5} \,,\ m_c/m_t = 2.99 \times 10^{-3},$$

$$m_d/m_b = 1.31 \times 10^{-3}, \; m_s/m_b = 1.17 \times 10^{-2}, \tag{29}$$

where $m_0^u = \mu_3^u$ and $m_0^d = \text{Re}(\mu_3^d)$, and the mixing matrix becomes

$$V_{CKM} = U_{uL}^\dagger U_{dL}$$
$$= \begin{pmatrix} 0.968 + 0.117I & 0.198 + 0.0974I & -0.00253 - 0.00354I \\ -0.198 + 0.0969I & 0.968 - 0.115I & -0.0222 - 0.0376I \\ 0.00211 + 0.00648I & 0.0179 - 0.0395I & 0.999 - 0.00206I \end{pmatrix} \tag{30}$$

The magnitudes of the elements are given by

$$|V_{CKM}| = \begin{pmatrix} 0.975 & 0.221 & 0.00435 \\ 0.221 & 0.974 & 0.0437 \\ 0.00682 & 0.0434 & 0.999 \end{pmatrix}, \tag{31}$$

which should be compared with the experimental values[59]

$$|V_{CKM}^{exp}| = \begin{pmatrix} 0.9741 \text{ to } 0.9756 & 0.219 \text{ to } 0.226 & 0.0025 \text{ to } 0.0048 \\ 0.219 \text{ to } 0.226 & 0.9732 \text{ to } 0.9748 & 0.038 \text{ to } 0.044 \\ 0.004 \text{ to } 0.014 & 0.037 \text{ to } 0.044 & 0.9990 \text{ to } 0.9993 \end{pmatrix}. \tag{32}$$

Note that the mixing matrix (30) is NOT in the standard parameterization. So, we give the invariant measure of CP-violations[67]

$$J = \text{Im}\left[(V_{CKM})_{11}(V_{CKM})_{22}(V_{CKM}^*)_{12}(V_{CKM}^*)_{21}\right] = 2.5 \times 10^{-5} \tag{33}$$

for the choice (28), which is slightly larger than the experimental value $(3.0 \pm 0.3) \times 10^{-5}$ (see [59] and also [68]). The angles of the unitarity triangle for V_{CKM} (30) are given by

$$\phi_1 \simeq 22°, \; \phi_3 \simeq 38°, \tag{34}$$

where the experimental values are: $\phi_1 = 24° \pm 4°$ and $\phi_3 = 59° \pm 13°$ [59]. The normalization masses m_0^u and m_0^d are fixed at

$$m_0^u = 174 \text{ GeV}, \; m_0^d = 1.8 \text{ GeV} \tag{35}$$

for $m_t = 174$ GeV and $m_b = 3$ GeV, yielding that $m_u \simeq 2.3$ MeV, $m_c \simeq 0.52$ GeV, $m_d \simeq 3.9$ MeV and $m_s = 0.035$ GeV. Although these values cannot be directly compared with the running masses, because our calculation is of the tree level, it is nevertheless worthwhile to observe how close they are to [26]

$$\begin{aligned} m_u(M_Z) &= 0.9 - 2.9 \text{ MeV}, \; m_d(M_Z) = 1.8 - 5.3 \text{ MeV}, \\ m_c(M_Z) &= 0.53 - 0.68 \text{ GeV}, \; m_s(M_Z) = 0.035 - 0.100 \text{ GeV}, \\ m_t(M_Z) &= 168 - 180 \text{ GeV}, \; m_b(M_Z) = 2.8 - 3.0 \text{ GeV}. \end{aligned} \tag{36}$$

CONCLUSIONS

The recent advances in the experimental knowledge of flavour physics, CP-violation and fermion masses and mixings triggered an enormous theoretical activity aimed to uncover the nature of this new physics. Important advances have been made in the further development of the already existing mechanisms and theories and the proposal of ingenious new ideas.

As an instance of the first approach, here I discussed the unified SUSY SO(10) theory with an additional $SU(3)_f$ flavour symmetry explored by G.G. Ross and L. Velasco-Sevilla [44, 45, 46]. The phenomenological success of this kind of theories with a continuous gauged family symmetry is achieved through the details of the symmetry breaking vacuum and elaborate mechanisms for its alignment. In this class of models, the physics of vacuum alignment simplifies if the continuum family symmetry is replaced by a discrete non-Abelian family symmetry as shown by A. Aranda, C.D. Carone and R.F. Lebed [47, 48, 49] who replaced the discrete non-Abelian group $T' \otimes Z_3$ for $SU(2)_f$ in the SUSY $SU(5) \times SU(2)$ unified theory of Barbieri, Romanino et al [50, 51, 52], and found that the reduction of the underlying continuous family symmetry to a discrete subgroup renders the desired vacuum alignment a generic property of such models.

As an example of the second approach, I discussed two extensions of the Standard Model in which the Higgs sector is modified and have an additional S_3 non-Abelian symmetry. They have this symmetry in common with the phenomenologically successful efforts of A. Mondragón and E. Rodríguez-Jáuregui [40, 41] to uncover a flavour S_3 symmetry in the Fritzsch texture zeroes of the quark mass matrices and the V_{CKM} phenomenology.

In the model proposed and discussed by R. N. Mohapatra, A. Pérez-Lorenzana and C.A. de S. Pires [55], there are no right handed neutrinos but additional Higgs triplets which acquire naturally small vacuum expectation values due to the type II see-saw mechanism. The presence of a global $S_3 \otimes U(1)_{e-\mu-\tau}$ symmetry leads naturally to the desired neutrino mass textures and generates the desired small splittings among neutrinos in fair agreement with experiment.

In the Minimal S_3-invariant Extension of the Standard Model proposed by J. Kubo, A. Mondragón, M. Mondragón and E. Rodríguez-Jáuregui [33], the concept of flavour and generations is extended to the Higgs sector by introducing three Higgs fields that are $SU(2)_L$ doublets in such away that all matter fields - lepton, quark and Higgs fields - belong to the three dimensional reducible $1 \oplus 2$ representation of the permutation group S_3. A well defined structure of the Yukawa couplings is obtained which permits the calculation of mass and mixing matrices for quarks and leptons in a unfied way. A further reduction of redundant parameters is achieved in the leptonic sector by introducing a Z_2 symmetry. In this model, the Majorana neutrinos acquire mass via the type I see-saw mechanism. The flavour symmetry group $S_3 \otimes Z_2$ relates the mass spectrum and mixings. This allows the computation of the neutrino mixing matrix explicitly in terms of the masses of the charged leptons and neutrinos [57]. The magnitudes of the three neutrino mixing angles are determined by the interplay of the flavour $S_3 \times Z_2$ symmetry, the see-saw mechanism and the charged lepton mass hierarchy. It is also found that the lepton mixing matrix V_{PMNS} has one Dirac CP-violating phase and two Majorana phases. The numerical values of the θ_{13} and θ_{23} mixing angles are determined by the charged leptons only in very good agreement with experiment. The solar mixing angle θ_{12} is almost insensitive to the values of the masses of the charged leptons but its experimental value allows to fix the scale and origin of the neutrino mass spectrum which has an inverted hierarchy with the values $|m_{\nu_2}|$ =0.0507 eV $|m_{\nu_1}|$ = 0.0499 eV and $|m_{\nu_3}|$ =0.0193 eV.

In conclusion, a discernible trend is perceptible in the formulation of symmetry based models of flavour,fermion masses and mixings and CP violation. In a bottom up approach, the search for simpler models starts with the formulation of a phenomenologically successful low energy theory with a minimal extension of the Standard Model and a discrete, non-Abelian flavour of family group $\tilde{G}_f$, and then, showing the possible embeddings of this theory into an SO(10) or SU(5) GUT and a continuous flavour group $G_f \subset \tilde{G}_f$.

ACKNOWLEDGMENTS

This work was partially supported by CONACYT México under contract No 40162-F and by DGAPA-UNAM under contract PAPIIT-IN116202-3.

REFERENCES

1. S. Weinberg, *Transactions of the New York Academy of Sciences II*, **38**, 185- (1977).
2. R. Gatto, G. Sartori and M. Tonin, *Phys. Lett. B*, **28**, 128- (1968).
3. H. Fritzsch *Phys. Lett.* B**70**, 436- (1977).
4. H. Fritzsch *Nucl. Phys.* B **155**,189- (1979).
5. H. Georgi and C. Jarlskog, *Phys. Lett.* B **86**, 297- (1979).
6. J. Ellis, S. Kelley and D.V. Nonoupoulos, *Phys. Lett.* B **249**, 441- (1990); *Phys. Lett.*B **260**, 131- (1991).
7. U. Amaldi, W. de Boer and H. Fürstenau, *Phys. Lett.* B **260**, 447- (1991).
8. C. Giunti, C.W. Kim and U.W. Lee,*Mod. Phys. Lett.* A**6**, 1745- (1991).
9. P. Langacker and M.-X Luo, *Phys. Rev.* D **44**, 817- (1991).
10. B. Aubert et al. *Phys. Rev. Lett.* **87**, 091801, (2001).
11. K. Abe et al. *Phys. Rev. Lett.* **87**, 091802, (2001).
12. R. Fleischer, *J. Phys. G: Nucl. Part. Phys.* **32**, R71-R125, (2006).
13. M. Altmann et al., (GNO collaboration), *Phys. Lett.* B **616**, 174- (2005).
14. M. B. Smy et al., (SK collaboration), *Phys. Rev.* D **69**, 011104, (2004).
15. Q.R. Ahmad et al., (SNO collaboration), *Phys. Rev. Lett.* **89**, 011301, (2002).
16. B. Aharmin et al., (SNO Collaboration), *nucl-ex/0502021*.
17. S. Fukuda et al., (SK Collaboration) *Phys. Lett.* B **539**, 179- (2002).
18. Y. Ashie et al., *Phys. Rev. Lett.* **93**,101801, (2004); [hep-ex/0501064]; *Phys. Rev. D*, **71**, 112005, (2005).
19. C. Bemporad, G. Gratta and P. Vogel, *Rev. Mod. Phys.*, **74**, 297- (2002).
20. T. Araki et al., (KamLAND Collaboration), *Phys. Rev. Lett.*, **94**, 081801 (2005).
21. M. Apollonio et al., (CHOOZ Collaboration), *Eur. Phys. J. C* **27**, 331- (2003).
22. E. Aliu et al., (K2K Collaboration), *Phys. Rev. Lett.* **94**, 081802 (2005).
23. K. Eitel, "Neutrino 2004", in *Nucl. Phys.* B (*Proc Suppl.*) **143**, (2005), edited by J. Dumarchey, Th. Patyak and F. Vanuucci, 21st International Conference on Neutrino Physics and Astrophysics, Paris, France 2004, pp 3-
24. S. R. Eliot and J. Engel *J. Phys.* G **30**, 3- (2005).
25. O. Elgaroy, and O. Lahav, *New J Phys.* **7**, 61- (2005).
26. H. Fritzsch and Z.-Z.Xing, "Mass and Flavour Mixing Schemes of Quarks and Leptons", in *Progress in Particle and Nuclear Physics*, Elsevier Science, The Netherlands 2000, pp 1-

27. Z. -Z. Xing, *Int. J. of Modern Physics* A **19**, 1-79 (2004).
28. R. N. Mohapatra and A. Yu. Smirnov, "Neutrino Mass and New Physics" in *Annual Review of Nuclear and Particle Science* **56**, Annuals Reviews Inc. (2006); *hep-ph/0603118*.
29. H.Harari, H.Haut, J.Weyers *Phys. Lett.* **B78**, 459- (1978).
30. E. Derman, *Phys. Rev.* D **19**, 317- (1979).
31. L.J. Hall and H. Murayama, *Phys. Rev. Lett.* **75**, 3985- (1995).
32. E. Ma, *Phys. Rev.* D **61**, 033012, (2000).
33. Kubo J, Mondragon A, Mondragon M, Rodriguez-Jauregui E. *Prog. Theor. Phys.* **109**, 795- (2003); Erratum - ibid, **114**, 287- (2005).
34. S.-L.Chen, M. Frigerio, and E. Ma, *Phys. Rev.* **D70**, 073008 (2004); Erratum: ibid D**70**, 079905, (2004).
35. W. Grimus and Lavoure *J. High Energy Phys.* **08**, 013- (2005).
36. P.H. Frampton and T.W. Kephart, *Int. J. Mod. Phys.* A **10**, 4689- (1995).
37. A. Aranda, C.D. Carone and R.F. Lebed, *Phys. Lett.* B **474**, 170- (2000).
38. A. Aranda, C.D. Carone and R.F. Lebed, *Phys. Rev.* D **62**, 016009, (2000).
39. I. de Medeiros Varzielas, S.F. King and G.G. Ross, *arXiv:hep-ph*/0512313
40. A. Mondragón and E. Rodríguez-Jáuregui, *Phys. Rev.* D **59**, 093009, (1999).
41. A. Mondragón and E. Rodríguez-Jáuregui, *Phys. Rev.* D **61**, 113002, (2000).
42. Z. -Z. Xing and H. Zhang, *J. Phys. G: Nucl. Part. Phys.***30**, 129- (2004).
43. S.F. King, *Rep. Prog. Phys.* **67**, 107- (2004).
44. G.G. Ross and L. Velasco-Sevilla, *Nucl. Phys.* B **653**, 3- (2003).
45. G.G. Ross, L. Velasco-Sevilla and O. Vives, *Nucl. Phys.* B **692**, 90- (2004).
46. L. Velasco-Sevilla, *J. H. P.* **10**, 035- (2003).
47. A. Aranda, C.D. Carone and R.F. Lebed, *Int. J. Mod. Phys.* A **16S1C**, 896-, (2001).
48. A. Aranda, C.D. Carone and R.F. Lebed, *Phys. Rev.* D **62**, 016009 (2000).
49. A. Aranda, C.D. Carone and R.F. Lebed, *Phys. Lett.* B **474**, 170-, (2000)
50. R. Barbieri, G. Dvali and L.J. Hall, *Phys. Lett.* B **377**, 76-, (2000).
51. R. Barbieri, L.J. Hall and A. Romanino, *Phys. Lett.* B **401**, 47- (1997).
52. R. Barbieri, L.J. Hall, S. Raby and A. Romanino, *Nucl. Phys.* B **493**, 3-, (1997)
53. Y.B. Zel'dovich, I.Y. Kobzarev and L.B. Okun, *Zk. Eks. Theor. Fiz.* **67**, 3- (1974); *Sov. Phys.JETP* **40**, 1- (1974).
54. G. Lazarides and Q. Shafy, *Phys. Lett.* B **115**, 2- (1987).
55. R.N. Mohapatra, A. Pérez-Lorenzana and C. Pires *Phys. Lett.* B **474**, 355- (2000).
56. G. Altarelli and F. Feruglio, *New J. of Physics* **6**,106- (2004).
57. O. Felix, A. Mondragon, M. Mondragon and E. Peinado, *Rev. Mex. Fis.* **52**, (2006) in press.
58. J. Kubo *Phys. Lett.* B**578**, 156- (2004) *Erratum: Ibid.* B **619**, 387- (2005).
59. Particle Data Group (F.J. Gilman, K. Kleinknect and B. Renk), *Phys. Rev.*, **D66**, 010001, (2002).
60. S. Pakvasa and H. Sugawara, *Phys. Lett.* **73B**, 61- (1978).
61. J. Kubo, *Phys. Rev.* D **70**, 036007 (2004).
62. C. K. Jung, C. Mc Grew, T. Kajita and T. Mann, *Annu. Rev. Nucl. Part. Sci* **51**, 451- (2001).
63. G.L. Fogli, et al., *hep-ph/0506083*;
64. M. Maltoni, T. Schwetz, M.A. Tórtola and J.W.F. Valle, *New J. Phys.* **6**, 122- (2004).
65. T. Schwetz, "Neutrino oscillations: Current status and prospects", *hep-ph/0510331*.
66. G.L. Fogli, et al., *Phys. Rev.* D **70**, 113003 (2004).
67. C. Jarlskog, *Phy. Rev. Lett.* **55**, 1039- (1985).
68. B. Aubert et al., *Phys. Rev. Lett.* **89**, 201802, (2002); K. Abe et al., *Phys. Rev.* D **66**, 071102, (2002).

Dynamical Mass Generation

A. Bashir*,† and A. Raya*

Instituto de Física y Matemáticas, Universidad Michoacana de San Nicolás de Hidalgo, Apartado Postal 2-82, Morelia, Michoacán 58040, México
†*Institute for Particle Physics Phenomenology, University of Durham, Durham DH1 3LE, U.K.*

Abstract.
Understanding the origin of mass, in particular that of the fermions, is one of the most uncanny problems which lie at the very frontiers of particle physics. Although the celebrated Standard Model accommodates these masses in a gauge invariant fashion, it fails to predict their values. Moreover, the mass thus generated accounts for only a very small percentage of the mass which permeates the visible universe. Most of the observed mass is accounted for by the strong interactions which bind quarks into protons and neutrons. How does that exactly happen in its quantitative details is still an unsolved mystery. Lattice formulation of quantum chromodynamics (QCD) or continuum studies of its Schwinger-Dyson equations (SDEs) are two of the non-perturbative means to try to unravel how quarks, starting from negligible current masses can acquire enormously large constituent masses to account for the observed proton and neutron masses. Analytical studies of SDEs in this context are extremely hard as one has to resort to truncation schemes whose quantitative reliability can be established only after a very careful analysis. Let alone the far more complicated realm of QCD, arriving at reliable truncation schemes in simpler scenarios such as quantum electrodynamics (QED) has also proved to be a hard nut to crack. In the last years, there has been an increasing group of physicists in Mexico which is taking up the challenge of understanding how the dynamical generation of mass can be understood in a reliable way through SDEs of gauge theories in various contexts such as (i) in arbitrary space-time dimensions d as well as $d \leq 4$, (ii) finite temperatures and (ii) in the presence of magnetic fields. In this article, we summarise some of this work.

Keywords: Dynamical mass generation, Schwinger-Dyson equations, Non-perturbative phenomena.
PACS: 11.15.Tk,12.20.-m

PRESENTATION

In the words of Salam [1], *"scientific thought and its creation is the common and shared heritage of mankind. The history of science, like the history of all civilization, has gone through cycles"*. He goes on to mention the *"scientific creativity among the pre-spanish Mayas and Aztecs, with their invention of the zero, of the calendars of the moon and Venus and of their diverse pharmacological discoveries, including quinine."* A healthy unrest after a long period of relative scientific silence is clearly visible in Mexico, augmenting our belief that science just as *"literature is the expression of a feeling of deprivation, a recourse against a sense of something missing"* [1]. The Nobel Prize winner of 1995 in Chemistry, Mario Molina, is a prime example. Although a long and arduous path may still have to be traversed, science in general and particle physics in particular

---

[1] A quote from Octavio Paz - Mexican poet, critic and diplomat. Born 1914. Died 1998. Nobel Prize winner of 1990 in Literature.

have made considerable advances in this country in the last few decades. Building upon this work, we hope that the scientific future of Mexico will see several collections of 55,000 words of quality [2].

Whereas the reader of this volume will have the chance to have an access to several important topics of research undertaken by Mexican scientific community in the field of particle physics, we shall restrict ourselves to the study of the origin of mass through the dynamics of forces that fundamental particles experience. This field of research is paving its way up the ladder of interest here. As we all know, the celebrated gauge theories of fundamental interactions have been highly successful in collating experimental results in a regime where they are only weakly felt. However, not all interesting phenomena can be understood in this approximation scheme. Confinement of *colourful* quarks and dynamical origin of mass are two well known examples. There has been a host of problems related to dynamical mass generation, ranging from QCD phenomenology [3, 4, 5, 6, 7, 8, 9, 10, 11, 12, 13], to more modern problems of strong dynamics beyond the standard model, [14], and theoretical issues in simpler gauge theory settings, see for example [15] and references therein, which have attracted Mexican scientific community over the years. In particular, pioneering work has been carried out by A. Zepeda in 70s and 80s, [3, 4, 5, 6, 7, 8, 9] in low energy QCD in collaboration with some of the leading names on the international stage such as H. Pagels, M.A.B. Beg, A. Ali, J. Bernstein, J. Gasser and P. Minkowski, attracting hundreds of citations. In his article [7], Zepeda argues that the zero light quark mass could not be ruled out with the available theoretical and experimental input at the time. This goes in keeping with modern lattice and Schwinger-Dyson calculations which suggest that QCD with zero up and down quark masses gives almost the right answer for the observed hadronic spectrum.

The interest related with theoretical issues concerning SDEs is relatively newer in Mexico though the activity associated with these problems has gradually filtered into various institutions such as BUAP, ICN-UNAM, FC-UCOL and IFM-UMSNH, with a varying degree of intensity. As a consequence, the topic of dynamical mass generation and/or SDEs is not only becoming more popular in the national and international conferences organized within Mexico [16, 17, 18, 19, 20, 21], but has also made its way to international events held abroad.

Phenomenon of dynamical mass generation is of direct physical relevance in QCD. Although lattice calculations yield several important physical observables in hadronic physics *"but simply having a computer spit out the answer, after gigantic and totally opaque calculations does not satisfy our hunger for understanding"* [2]. Moreover, massless particles are difficult to fit on a lattice of finite size. In practice, one deals with finite masses and extrapolates the results to zero masses. The process of extrapolation is likely to lie outside the domain of lattice calculations. In contrast, Schwinger-Dyson approach naturally encompasses physics at all momentum scales. Therefore, it is the most obvious candidate to make connections with lattice on one hand and with experimental reality on the other. The relation between SDE and the Wilsonian *exact* renormalization group has been discussed in [22]. SDEs are the non-perturbative field equations of a given quantum field theory, which form an infinite tower of coupled integral equations between

---

[2] Frank Wilczek.

various n-point Green's functions. Dynamical mass generation is studied through the non-perturbative structure of the fermion propagator. When fundamental fermions are initially massless, their self-interactions at strong coupling can account for the formation of chiral condensates and mass for fermions.

Given the complexity and vastness of the material related to dynamical mass generation on one hand and our limited domain of expertise on another, we would only attempt to discuss a part consisting of an ongoing work in the non-perturbatibve truncations of SDEs. Owing to its non-abelian structure, corresponding equations in QCD are very complicated. Existence of Gribov copies [23] (a topic which has also been explored in Mexico, [24]) can also potentially cast a shadow of ambiguities. Therefore, QED serves as a simple testing ground for the truncation schemes employed to study SDEs. We present a summary of these studies in different contexts such as (i) in arbitrary space-time dimensions d as well as $d \leq 4$, (ii) finite temperatures and (ii) in the presence of magnetic fields.

DYNAMICAL MASS GENERATION IN QED

Forming an infinite tower of prohibitively complicated coupled integral equations, SDEs must be truncated in order to extract predictions. Fermion propagator is the Green's function of our interest :

Figure 1. The SDE for the (inverse) fermion propagator.

Of course, the full Green functions involved in the self energy obey their own SDE. The one for the boson propagator is shown in Figure 2

Figure 2. The SDE for the boson propagator.

and the one for the fermion-boson vertex is depicted in Figure 3.

Figure 3. The SDE for the fermion-boson vertex.

A favourite starting point to evaluate non-perturbative fermion propagator is to make an *ansatz* for the remaining Green's functions coupled to it so that it could be solved for self-consistently. We shall consider quenched approximation alone in which photon propagator remains bare. Any choice of the fermion-photon interaction employed must ensure that the following fundamental requirements of a gauge theory are fulfilled :

- Ward and Ward-Green-Takahashi identities, i.e., $\Gamma^\mu(p,p) = \partial S_F^{-1}(p)/\partial p_\mu$ and $(k-p)_\mu \Gamma^\mu(k,p) = S_F^{-1}(k) - S_F^{-1}(p)$ which relate the fermion propagator to the three-point vertex should be satisfied.
- Physical observables related with the fermion propagator such as the physical mass of the fermion, the condensate, etc. should be independent of the gauge parameter.
- A gauge independent regularization scheme should be employed.
- It should be independent of any kinematic singularities such as the ones when $k^2 \to p^2$.
- Under the operations of charge conjugation, parity and time reversal, it should transform in the same way as the bare vertex γ^μ.
- It should reduce to the perturbative Feynman expansion for the vertex in the limit when the coupling is small.
- Under a variation of gauge, fermion propagator and fermion-boson vertex should change in accordance with the Landau-Khalatnikov-Fradkin transformations (LKFT), [25].

Reference [15] (see also references therein) provides a brief review on the hunt for non-perturbative truncations of SDEs compatible with the above mentioned criteria. The hope is then to achieve gauge independence of physical observables. Here we shall only summarise the attempts relevant to the purpose of this article.

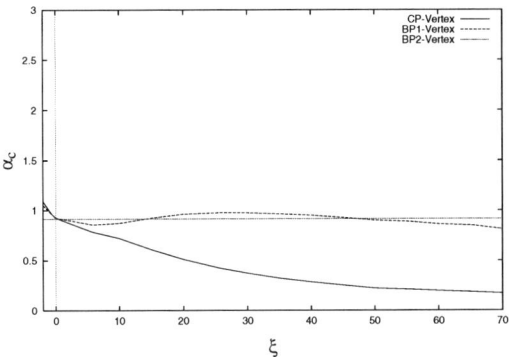

Figure 4. α_c versus ξ for different truncation schemes.

An important physical observable of interest in QED4 has been the critical value, α_c, of the electromagnetic coupling above which masses are dynamically generated. Perturbatively inspired bare vertex truncation of the SDEs yields a highly gauge dependent α_c. A breakthrough was achieved when Curtis and Pennington, [26], invoked multiplicative renormalizability of the fermion propagator (which is also a statement of the LKFT in 4 dimensions) to introduce what is now generally referred to as the Curtis-Pennington vertex (CP-vertex) and which reduces the above mentioned gauge dependence appreciably in the neighbourhood of the Landau gauge. Following the same line of thought, in a later work, Bashir and Pennington, (BP1-vertex) [27], were able to discover another truncation with which the improvement not only exists in the neighbourhood of the Landau gauge but that the curve continues to remain flat around $\alpha_c = 0.93$ even up to quite large values of the gauge parameter. Going from $\xi = 0$ to $\xi = 10$ reduces the gauge dependence by about 15%. The improvement becomes more significant when we are further away from the Landau gauge, Figure 4. For example, in going up to $\xi = 70$, the change in α_c is improved by more than 60% as compared to the CP-vertex. These results are encouraging in the sense that one finds a vertex which serves the aim of reducing the gauge dependence better than the ones constructed before, in particular the CP-vertex, without introducing any significant complication. But, however weak the variation of α_c with ξ may be, any gauge dependence shows that this vertex cannot be the exact choice. Even if it achieves a lot, it does not do it all. The list of requirements outlined previously led to various subsequent works, [28, 29, 30], constraining the vertex in such a fashion that a *complete gauge independence* of α_c could be guaranteed as shown in Figure 4, (BP2 -vertex). The next question to address is naturally if one could generalize the findings in the context of QED4 to other observables, dimensions and more realistic gauge theories such as QCD.

PLANAR QED

In the absence of ultraviolet divergences, planar QED (QED3) can provide a neat testing ground to verify the ideas presented in the previous section. In addition to its role as a useful toy model, it has many useful applications. It has growing application in condensed matter systems and it displays the finite temperature behaviour of QED4. Planar physics has additional intrinsically interesting features not seen in 3+1 dimensional space-time. It includes arbitrary statistics, giving rise to anyons, arbitrary spin for massless particles and novel features regarding parity and chiral symmetry. Some of these issues have been studied in Mexico, [31].

In the context of SDEs studies, perturbation theory constraints on the electron-photon interaction were studied in [32, 33, 34, 35] and corresponding numerical results were reported in [36]. As mentioned before, multiplicative renormalizability of the fermion propagator in QED4 is a statement of LKFT. Therefore, it is natural to explore the role played by LKFT in dimensions other than 4. Considering a perturbative fermion propagator in the Landau gauge, LKFT give a non-perturbative expression for the two-point function in an arbitrary covariant gauge, [37, 38, 39]. In the weak coupling regime, this expression contains valuable information of the Feynman expansion of the propagator at all orders. Owing to the fact that dynamical mass generation is a non-perturbative phenomenon, there must exist a way to incorporate LKFT in a quest to obtain gauge invariance of physical observables. This was the basis of a program that has been carried out in the past few years in a series of articles [40, 41]. The reasoning was based on the gauge covariance properties of the fermion propagator and the LKFT. Since these transformations leave the SDEs and the WGTI invariant, they provide an ideal framework to study the gauge dependence of the fermion propagator. The idea is to require the reliability of a truncation scheme to obtain non-perturbative solutions to the SDE in one particular gauge (here the Landau gauge). In other words, instead of looking for a reliable truncation scheme, i.e., a good vertex ansatz, in all gauges, one should only focus in achieving the same in one particular gauge. LKFT will take it from there, leading us along the path of the varying gauge.

As an example, we recall the study of dynamical mass generation in unquenched QED3 in [42]. Despite the fact that they employ a gauge invariant truncation scheme, their results show a severe gauge dependence for the condensate. However, in [41], we propose LKFT to generate solutions in gauges other than Landau. In Figure 5, we plot the condensate value for $N_f = 2$ as a function of ξ for the two methods. Along the (almost) flat line, both the WGTI and the LKFT are satisfied, whereas, in the sharply rising graph, LKFT are violated although the WGTI continues to hold true. Just like in QED4, we confirm the importance of LKFT in addition to WGTI in obtaining gauge invariant physical observables also in QED3. A natural temptation stemming from these studies is to explore generalised LKFT in QCD. This is for future.

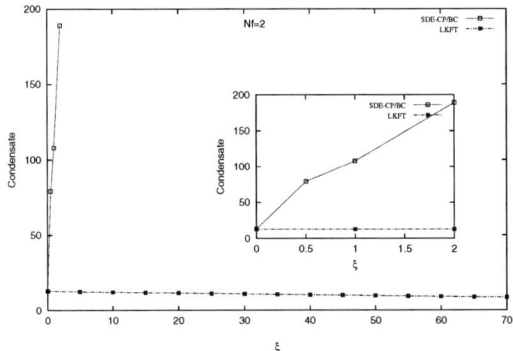

Figure 5. $-\langle\bar{\psi}\psi\rangle$ versus ξ in unquenched QED3 using LKFT.

FINITE TEMPERATURE STUDIES

Finite temperature field theories are of great importance, specially in the studies of symmetry restoration at non-zero temperature and density. Following the work done at zero temperature in the study of Schwinger-Dyson equations, corresponding studies at finite temperature have also taken a start in Mexico. In this scenario, complexity of these equations and the number of unknown functions to be evaluated grows enormously. For example in QED at zero temperature, the three-point function can be written in terms of twelve spin amplitudes while at finite temperature this number becomes 32, almost a 3-fold increase in the number of unknowns involved. Therefore, the hunt for reliable truncation schemes is even harder. Adopting a systematic approach, we invoke WGTI at finite temperature to constrain the longitudinal part of the fermion-photon vertex, [43]. The transverse vertex can then be expanded out in the remaining 24 basis vectors. Following the singularity-free scheme provided by Ball and Chiu at zero temperature, we propose a special basis of 24 vectors at finite temperature. They are chosen in such a way that their coefficients are free of any kinematic singularities. We check this by an explicit one loop calculation of these coefficients in the hard thermal loop (HTL) approximation. This work was followed by a numerical study of dynamical mass generation at finite temperature in [44] in the quenched approximation with the bare vertex ansatz. At fixed temperature, masses were found to be dynamically generated above a critical value of the effective coupling. Expectedly, when temperature increases, chiral symmetry is found to be restored, [44]. It would be useful to explore the persistence of these qualitative and quantitative predictions with a more sophisticated form of the fermion-photon interaction advocated in [43]. This work is planned.

DYNAMICAL MASS GENERATION IN MAGNETIC FIELDS

In the presence of an external magnetic field, propagation properties of particles get modified. Back in the 50s this fact was shown by Schwinger in his seminal work [45] which established his famous proper time method. There are, however, other methods which can be applied to study the dynamics of fermions in the presence of magnetic fields, like the eigenfunction method by Ritus [46]. Dynamical mass generation in the context of SDEs has been largely studied in the presence of strong magnetic fields. Such intense field catalyses the generation of masses for any weak value of electromagnetic coupling, a phenomenon that has been dubbed as "magnetic catalysis" [47]. The case of the weak fields has attracted little attention. In QED4, the only prior work to our knowledge is that of Kikuchi and Ng, [48]. However, they concentrate mainly on the behaviour of the critical coupling α_c when magnetic fields are weakly turned on. They do not evaluate the magnetic field dependence of the dynamically generated fermion mass or the chiral condensate. In Mexico, dynamical mass generation in the presence of a weak magnetic field was only recently studied [49]. This regime is defined when the coupling in the absence of the field exceeds a critical value, in such a way that a mass can be generated. The strength of the field is then taken to be be much smaller than the square of this dynamical mass. Being the field of weak strength, all Landau levels contribute and the calculational playground becomes crowded and messy. The resulting SDE can, however, be simplified by the use of eigenfunctions method [46] under plausible assumptions. In the lowest Landau level, the dynamical mass and the chiral condensate are found to increase quadratically with the strength of the external field as shown in Figure (6), Λ being the ultraviolet cut-off. These findings are in sharp contrast with the corresponding behaviour in the case of strong magnetic field which is proportional to $\sqrt{eB}$. The region which is left to be explored is that of intermediate field strengths, currently under scrutiny.

For the sake of completeness, it is worth mentioning that the dynamical mass generation for fermions has also been studied for the Wess-Zumino model of supersymmetry, [50]. It was observed that under simple but still rather general assumptions, supersymmetry prevents the generation of fermion masses. However, it is unclear, whether or not these conclusions will persist for physically more relevant models of supersymmetry.

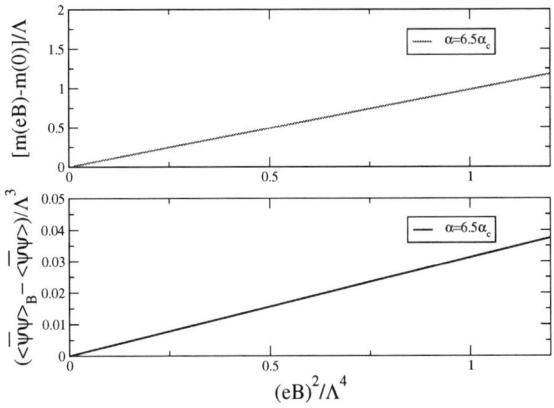

Figure 6. Weak magnetic field dependence of mass and condensate in QED4.

SUMMARY AND PERSPECTIVES

In summary, although the field of dynamical mass generation through SDEs is still a fledgling enterprise in Mexico, its efforts are being echoed nationally and internationally. This is reflected in the increasing number of publications, events where this subject is presented by host and guest researchers and citations it has attracted. The small but enthusiastic community involved in this research is growing both in number and scientific production. Bridges are being built at personal and institutional levels within the country and abroad. Several students have carried out (or are currently in the process of carrying out) their undergraduate, masters or Ph.D. theses in this field of research. So far, attention has been devoted to discovering reliable non-perturbative truncation schemes of SDEs in QED4 and QED3 especially as regards rendering physical observables gauge independent, Figures 4 and 5. We have recently started to expand these studies to include the presence of background magnetic fields and finite temperatures. Studying physically more realistic gauge theories such as QCD and the models of strong dynamics beyond the Standard Model of particle physics can be an attractive next step.

ACKNOWLEDGMENTS

AB is grateful for the invitation to present this contribution on the special occasion of the XX Anniversary of the Division of Particles and Fields of the Mexican Society of Physics. AB also wishes to acknowledge a short term visitor grant by a joint scheme of Royal Society and Mexican Academy of Sciences. We thank CIC and CONACyT (grants 4.10 and 46614-I) for financial support. We acowledge the input of our colleagues whose work has facilitated the compilation of this brief review.

REFERENCES

1. A. Salam, *"Gauge Unification of Fundamental Forces"*, Nobel Lectures, Physics 1971-1980, Editor Stig Lundqvist, World Scientific Publishing, Co., Singapore, (1992).
2. "55,000 words of some of the most savagely eloquent prose in the history of Mexico.", a quote by Alexander Cockburn, an English journalist, commenting on a collection of Mexican communiqués reprinted in California's Anderson Valley Advertiser.
3. H. Pagels and A. Zepeda, Phys. Rev. **D5** 3262 (1972).
4. M.A.B. Beg, A. Zepeda, Phys. Rev. **D6** 2912 (1972).
5. A. Ali, J. Bernstein and A. Zepeda, Phys. Rev. **D12** 503 (1975).
6. C.A. Dominguez and A. Zepeda, Phys. Rev. **D18** 884 (1978).
7. A. Zepeda, Phys. Rev. Lett. **41** 139 (1978).
8. J. Gasser and A. Zepeda, Nucl. Phys. **B174** 445 (1980).
9. P. Minkowski and A. Zepeda, Nucl. Phys. **B164** 25 (1980).
10. M. Socolovsky, Phys. Lett. **B168** 110 (1986).
11. D.P. Page and M. Socolovsky, Phys. Rev. **D31** 923 (1985).
12. O. Civitarese, P.O. Hess, J.G. Hirsch and M. Reboiro, Phys. Rev. **C61** 064303 (2000).
13. P. Maris, A. Raya, C.D. Roberts and S.M. Schmidt, Eur. Phys. J. **A18** 231 (2003).
14. A. Aranda and C.D. Carone, Phys. Lett. **B488** 351 (2000).
15. A Bashir and A. Raya, *"Gauge Symmetry and its implications for the Schwinger-Dyson equations"* In "Trends in Boson Research". Nova Science Publishers Inc. N. Y., ISBN 1-59454-521-9, (2006). Preprint hep-ph/0411310.
16. A. Bashir, *"Dynamical Symmetry Breaking"*, invited talk given at the XIII Annual Meeting of the Division of Particles and Fields of the Mexican Society of Physics, June (1999).
17. A. Bashir, *"On the Dynamical Origin of Fermion Masses"*, invited talk given at the XIX Annual Meeting of the Division of Particles and Fields of the Mexican Society of Physics, June (2005).
18. A. Bashir, *"Non-Perturbative Aspects of Schwinger-Dyson Equations"*, invited talk given at the X Mexican School of Particles and Fields, Playa del Carmen, November (2002), published in AIP Conference Proceedings, **670** 145 (2003).
19. M.R. Pennington, *"Swimming with Quarks"*, A series of six invited lectures given at the XI Mexican School of Particles and Fields, Xalapa (2004), published in J. Phys. Conf. Ser. **18** 1 (2005).
20. A. Höll, C.D. Roberts and S.V. Wright, *"A perspective on hadron physics"*, invited talk given at the X Mexican Workshop on Particles and Fields, Morelia (2005), to be published in the proceedings. Preprint nucl-th/0604029.
21. E.H. Simmons, *"Strong Dynamics and Electroweak Symmetry Breaking"*, invited talk given at the X Mexican Workshop on Particles and Fields, Morelia (2005), to be published in the proceedings.
22. U. Ellwanger, M. Hirsch and A. Weber, Eur. Phys. J. **C1** 563 (1998).
23. V.N. Gribov, Nucl. Phys. **B149** 1 (1978).
24. V.M. Villanueva, J. Govaerts and J.L. Lucio-Martinez, J. Phys. **A33** 4183 (2000); *ibid*, talk given at the 7th Mexican School Of Particles And Fields And 1st Latin American Symposium On High-Energy Physics, Merida, 1996, published in AIP Conference Proceedings **400** 578 (1997); *ibid*, talk given at the International Conference On Quantization, Gauge Theory, And Strings, Moscow (2000), published by World Scientific **2** 270 (2000).
25. L.D. Landau and I.M. Khalatnikov, Zh. Eksp. Teor. Fiz. **29** 89 (1956); L.D. Landau and I.M. Khalatnikov, Sov. Phys. JETP **2** 69 (1956); E.S. Fradkin, Sov. Phys. JETP **2** 361 (1956).
26. D.C. Curtis and M.R. Pennington, Phys. Rev. **D42** 4165 (1990).
27. A. Bashir, *Constructing Vertices in QED*, University of Durham, England, (1995).
28. A. Bashir and M.R. Pennington, Phys. Rev. **D50** 7679 (1994).
29. A. Bashir and M.R. Pennington, Phys. Rev. **D53** 4694 (1996).
30. A. Bashir, A. Kızılersü and M.R. Pennington, Phys. Rev. **D57** 1242 (1998).
31. A. Bashir and Ma. De Jesús Anguiano, Few Body Syst. **37** 71 (2005); *ibid*, J. Phys. Conf. Ser. **37** 141 (2006); M. Napsuciale and M.G. Carrillo-Ruiz, *"Planar electrons from first principles"* in Proceedings of the Mini-Workshop "Simetrías" (2006); M. Klein-Kreisler and M. Torres, Phys. Lett. **B347** 361 (1995).
32. A. Bashir, A. Kızılersü and M.R. Pennington, Preprint hep-ph/9907418; A. Bashir, *"Perturbation theory constraints on the 3-point vertex in massless QED3"* Proceedings of "Adelaide 1999, Light-

cone QCD and Non-Perturbative hadron physics" 227 (2000).
33. A. Bashir, A. Kızılersü and M.R. Pennington, Phys. Rev. **D62** 085002 (2002).
34. A. Bashir and A. Raya, Phys. Rev. **D64** 105001 (2001); *ibid* AIP Conf. Proc. **623** 313 (2002).
35. A. Raya, Rev. Mex. Fís. **50** 104 (2004).
36. A. Bashir, A. Huet and A. Raya, Phys. Rev. **D66** 025029 (2002); *ibid*, AIP Conf. Proc. **623** 355 (2002); *ibid*, J. Phys. Conf. Ser. **37** 144 (2006).
37. A. Bashir, Phys. Lett. **B491** 280 (2000).
38. A. Bashir and A. Raya, Phys. Rev. **D66** 105005 (2002); *ibid*, J. Phys. Conf. Ser. **37** 79 (2006).
39. A. Bashir and R. Delbourgo J. Phys. **A37** 6587 (2004); *ibid*, J. Phys. Conf. Ser. **37** 73 (2006).
40. A. Bashir and A. Raya, Nucl. Phys. **B709** 307 (2005); *ibid*, Nucl. Phys. Proc. Suppl. **141** 259 (2005); *ibid*, J. Phys. Conf. Ser. **37** 90 (2006).
41. A. Bashir and A. Raya, *"On the gauge independence of the chiral condensate in QED3"*, Preprint hep-ph/0511291; *ibid*, *"Schwinger-Dyson equations and the problem of gauge invariance"* Proceedings of "Nagoya 2004, Dynamical Symmetry Breaking" 257 (2005); *ibid*, *"On gauge independent chiral symmetry breaking"* to be published in the Proceedings of the X Mexican Workshop on Particles and Fields, Morelia, (2005).
42. C.S. Fischer, R. Alkofer, T. Dahm and P. Maris, Phys. Rev. **D70** 073007 (2004).
43. A. Ayala and A. Bashir Phys. Rev. **D64** 025015 (2001).
44. A. Ayala and A. Bashir Phys. Rev. **D67** 076005 (2003).
45. J. Schwinger, Phys. Rev. **82** 664 (1951).
46. V.I. Ritus in *"Issues in Intense-Field Quantum Electrodynamics"*, Ed. V.L. Ginzburg Nova Science, Commack, New York (1987).
47. V.P. Gusynin, V.A. Miransky and I.A. Shovkovy, Phys. Lett. B **349** 477 (1995); D.-S. Lee, C.N. Leung and Y.J. Ng, Phys. Rev. D **55** 6504 (1997); E.J. Ferrer and V. de la Incera, Phys. Lett. B **481** 287 (2000).
48. Y. Kikuchi and Y.J. Ng, Phys. Rev. **D38** 3578 (1988).
49. A. Ayala, A. Bashir, A. Raya and E. Rojas, Phys. Rev. **D73** 105009 (2006); *ibid*, to be published in the Proceedings of the X Mexican Workshop on Particles and Fields, Morelia, (2005).
50. A. Bashir and J.L. Díaz-Cruz, J. Phys. **G25** 1797 (1999); *ibid*, AIP Conf. Proc. **490** 323 (1999).

EXPERIMENTAL HIGH ENERGY
AND COSMIC-RAY PHYSICS

Reflections on a Teenage Quark

Jacobo Konigsberg

University of Florida

Abstract. I present a brief, personal, perspective of the first years of life of the top quark and the prospects for surprises that detailed studies of this young particle may bring about.

Keywords: top quark
PACS: 14.65.Ha

On October 21st, 2005, we celebrated at Fermilab the 10th anniversary of the discovery of the top quark by the CDF and D0 experiments. We called it the "Top Turns Ten" symposium. It was a wonderful celebration of a wonderful achievement. We invited speakers from both experiments to share their experiences, and those of their colleagues, during the intense and exciting discovery period. We made sure that not only senior researchers, at the time, spoke at the symposium but that also young student physicists, who participated in the discovery and did much of the work, were represented appropriately. We also heard from accelerator physicists about the Tevatron's birth, growing pains, and its slow march to success. Fermilab's director, at the time, also gave us his unique perspective. We also created a very interesting collection of posters, all related to the top quark discovery, showing the international flavor of the teams, the intricate works of the accelerators and detectors critical to this feat and personal perspectives and accounts of the events of the period -including e-mail excerpts that offered windows into the sociology and the human side of the enterprise.[1]

Speaking of sociology, it was not entirely easy to organize such a successful, collaborative, event. It is no secret that the CDF and D0 collaborations were engaged in a fierce competition to discover the top quark. CDF published the "evidence" paper in April 2004 [1], where a top quark signal with a significance of 2.8-sigma was announced and the top mass was measured at a whopping 174 GeV! The production cross section of the dozen or so identified top-antitop pairs candidate events was, within the large statistical uncertainty, consistent with theory. However this wasn't enough to claim unequivocally that the top quark was discovered. A few months later, with more data analyzed, both collaborations had enough evidence to finally publish (simultaneously, by accord) their discovery papers [2].

Those first papers generated a tremendous production of theoretical work and, by now, have been cited near an infinite number of times. The top quark discovery ignited a new era of exploration in particle physics. Several studies with top quarks were done with the full datasets of the so-called "Run 1" of the Tevatron (the run that spanned from 1992 to

---

[1] At http://www.fnal.gov/pub/news05/TopTurnsTen.html one can view the presentations and posters of the Top Turns Ten symposium

CP857, *Particles and Fields, Part B: Commemorative Volume of the Division of Particles and Fields of the Mexican Physical Society,* edited by M. A. Pérez, L. F. Urrutia, and L. Villaseñor
© 2006 American Institute of Physics 978-0-7354-0354-3/06/$23.00

1996). However not until today, more than ten years later, we can say that we are really deepening our understanding of this wonderfully rare particle that presumably inhabited the Universe only near its very birth.

At the Tevatron energies, it takes about 10^{10} proton-antiproton collisions to produce a single top-antitop pair. The experiments need to trigger on their decay products[2], which include leptons, jets and missing transverse energy in various mixes, depending on the decay channel. This, together with event selection criteria designed to reduce backgrounds and end up with a relatively pure sample of top quark events, results in only a small fraction of the produced events being used in the various analysis (of order 10%, depending on the decay channel). Clearly a most important element in this business is the ability of Fermilab's accelerator complex (and dedicated staff!) to produce proton-antiproton collision for the experiments with a rate as high as possible. The so-called "luminosity" is a measure of this rate.

Run 1 was terminated so that Fermilab could implement a series of important upgrades to its accelerators so that they can deliver much more luminosity than previously achieved. The CDF and D0 experiments also performed major upgrades to their detectors and data acquisition systems so they can in turn cope with the higher luminosity. After a relatively slow start in 2001, Run 2 is now going at full steam with luminosity records being achieved on a regular basis and a strong and exciting physics program with datasets already more than ten times larger than what was achieved in Run 1. In these datasets we now find hundreds of top-antitop events with which we perform every possible measurement we can think of. These measurements evolve with time and become more and more precise as more data is added to them and, very importantly, as experimenters become more comfortable with their understanding of their detectors and more creative with their analysis techniques.

At the time of this writing the top quark mass has been measured with the amazing precision of 1.3% ($m_{top} = 172.5 \pm 2.3$ GeV), already exceeding the expectations for Run 2. This measurement helps constrain the mass of the Standard Model Higgs towards light values. The production cross section of top-antitop pairs is now measured at the 10% level, which is approximately the level of the theoretical uncertainties in the QCD calculation. For the first time attempts to determine the charge and the lifetime of the top quark are being made, with preliminary indications that exclude a possible charge of 4/3 that of the electron and a lifetime larger than about 2×10^{-13} sec. The V-A nature of the top-W-b vertex is being tested via measurements of the W-helicity with results consistent so far with what is expected from the Standard Model. Top decay branching ratios also seem consistent with the Standard Model. The electroweak production of a single top quark is expected to occur with a cross section of about half of that of top pair-production. However the much more copious backgrounds make the observation of this production process extremely difficult. Still, both CDF and D0 are very close to having enough data to measure the cross section for this process and I suspect that in the very near future they'll announce strong evidence for such a "discovery". The most recent results on top quark physics from the CDF and D0 collaborations can always be found in their public web pages [3] [4].

---

[2] The top quark lifetime is only about 10^{-25} sec

Even though the top quark so far is behaving according to what is expected from the Standard Model it is very important to continue testing for possible connections to new physics through it. Theoretical models on how these connections might happen abound. However, we, as experimentalists, must be guided by detailed tests of the data. Top-enriched samples are being tested in every possible way to try to find evidence for "contaminations" from new physics processes. For example, we've tested for possible decays of the top quark to heavy charged particles, for a heavy t' 4th-generation-like quark with similar decays as those of the top quark and for possible resonances that decay to top-antitop pairs. This last search has created some excitement with a possible peak at around 500 GeV in the invariant mass distribution of the top-antitop pair coming and going as more data was looked at.

As Run 2 progresses towards its inevitable end in a few years, we expect to gather datasets several times larger that the ones currently being analyzed. Some of us harbor much hope that, as many youngsters do, the top quark will misbehave in its teenage years and we'll find evidence for new physics in the top quark sector!

REFERENCES

1. Abe F. et al. (CDF Collab) Phys. Rev. D 50:2966 (1994).
2. Abe F. et al. (CDF Collab) Phys. Rev. Lett 74:2626 (1995); Abachi S, Et el. (D0 Collab) Phys.Rev. Lett. 74:2632 (1995).
3. http://www-cdf.fnal.gov/physics/new/top/top.html
4. http://www-d0.fnal.gov/Run2Physics/top/top_public_web_pages/top_public.html

The Mexican Participation in the ALICE Experiment

G. Herrera Corral[1] and G. Paic[2]

1) Departamento de Física, CINVESTAV
2) Instituto de Ciencias Nucleares, UNAM
México

Abstract. A large portion of the Mexican community of experimental high energy and nuclear physicists has joined the ALICE collaboration with the aim to contribute effectively to the design and construction of the experiment at the Large Hadron Collider at CERN. This decision has long term consequences on the development of the physics in Mexico. We will review the main features of this commitment and the results obtained so far.

Keywords: LHC, ALICE, V0A
PACS: 29.40.Mc

INTRODUCTION

In the nineties it became very clear that the ultra-relativistic heavy ion collisions represent a unique interface between the underlying simplicity of the high energy particle physics and the immense complexity of the nuclear systems relying on the concepts of Quantum Chromodynamics (QCD) in a typical many body system. The main guiding idea is that nuclear matter once brought to extreme conditions of temperature or pressure will transit from a system consisting of confined hadrons to a more elementary system of quarks and gluons in a new state of matter called Quark Gluon Plasma (QGP). The analysis of the experiment has somewhat changed our perception of this state but we do believe that the evidence for such a state is very strong.

At the end of the nineties one of us (G.H) initiated the move of the Mexican experimental community toward this promising field joining the ALICE experiment in construction at the CERN Large Hadron Collider to be put in operation in 2007.

ALICE is the one of the four existing experiments dedicated to the heavy ions collisions at energies unprecedented (5.5 TeV in the nucleon nucleon centre of mass).

The guiding idea was to initiate a collaboration of a new kind for the mexican physicists. Namely, to try to participate in the design, and construction phase as well as in the later data analysis. Considering the ever greater complexity of high energy physics experiments they offer enormous technological challenges that have to be solved by the participants themselves – be it in electronics, computation, detector performances etc. For that reason high energy physics is an excellent conduct for the development of technological skills for young people in a very competitive international atmosphere.

In the present work we will shortly overview the main contributions of the Mexican groups (listed at the end of the article) in two main areas: The so called V0A detector, which purpose is to give together with the V0C detector a trigger signal to all the other detectors of ALICE. This signal is of crucial importance for the experiment, since it has to cover a number of strict requirements on efficiency and timing. The second detector is the detector ACORDE whose function is to emit a trigger signal for cosmic rays allowing thus calibrations in the tracking detectors of ALICE, and to continue the studies of cosmic ray phenomena like the "muon bundles" studied earlier in the COSMOLEP experiments.

THE ALICE EXPERIMENT

The Large Hadron Collider (LHC) is now in construction at the European Center for Nuclear Research (CERN). In year 2007 it will provide beams of protons and heavy ions at a very high energy in the center of mass. Four experiments will be performed at the LHC. The ALICE experiment (**A** **L**arge **I**on **C**ollider **E**xperiment) will be dedicated to the study of heavy ion collisions.
The accelerator will run in proton-proton mode 90% of the time and will deliver heavy ions the remaining 10% of the run time. In lead – lead collisions the bunch crossing will be 125 ns while in the proton-proton mode it will be 25 ns.
At the energies achieved by the LHC, the density, the size and the lifetime of the excited quark matter will be high enough as to allow a careful investigation of the properties of this new state of matter. The temperature will exceed by much the critical value predicted for the transition to take place. The particle production will be dominated by hard processes providing a powerful tool to study the evolution of the system from the earliest stage. The ALICE detector will have a tracking system over a wide range of transverse momentum which goes from 100 MeV/c to 100 GeV/c as well as particle identification able to separate pions, kaons, protons electron and photons. Topics like parton energy loss in a dense medium, heavy quark and jet production, prompt photons etc. will be addressed providing important information in that energy regime.

We do not give a description of the physics that will be studied with the ALICE detector. We refer the interested reader to ref. [1].

A longitudinal view of the ALICE detector is shown in Fig. 1. A detailed description of the ALICE detector can be found in ref. [2].

The central part of the detector is located in a magnetic field of 0.5 T. The track measurement is performed with a set of six barrels of silicon detectors a large Time Projection Chamber (TPC), and the Transition Radiation Detector. The TPC has an effective volume of 88 m^3. It is the largest TPC ever built. The photons will be measured in a high resolution calorimeter 5 m below from the interaction point in the central part. The photon spectrometer (PHOS) is built from a novel scintillating compound $PBWO_4$ which has a high light output. In addition to this, the TRD will provide electron identification and a Time of Flight system will provide excellent particle separation at intermediate momentum . The Time of Flight system which uses Multi-gap Resistive Plate Chambers (MRPC) with a total of 160,000 readout channels represents also a complete novel approach to this kind of systems, built so far with expensive photomultipliers.. A Ring Imaging Cherenkov using a CsI photocathode will extend the particle identification capability to higher momentum particles. It covers 15% of the acceptance in the central area and will separate pions from kaons with momenta up to 3 GeV/c and kaons from protons with momenta up to 5 GeV/c. A Forward Multiplicity Detector (FMD) consisting of silicon strip detectors, and a Zero Degree Calorimeter will cover the very forward region providing information on the charge multiplicity and energy flow. The V0 system consisting of two scintillation counters on each side of the interaction point will be used as the main interaction trigger. In the top of the magnet a Cosmic Ray Detector will signal cosmic muons arrival.

In the forward direction a set of tracking chambers inside a dipole magnet will measure muons while electrons and photons are measured in the central region.

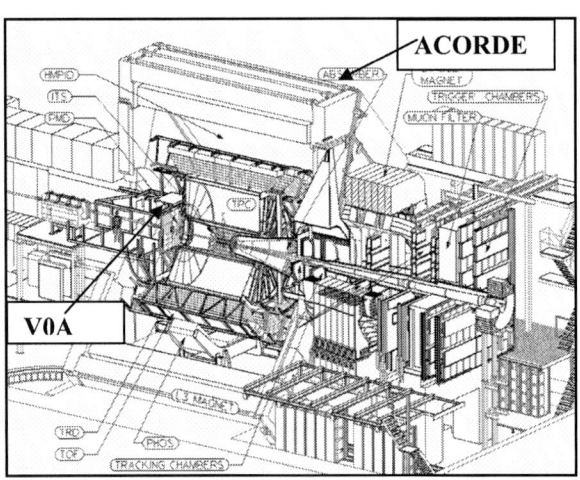

FIGURE 1. The ALICE detector and its components. ACORDE (A Cosmic Ray Detector) on the top of the magnet and the V0A location is shown here.

THE V0 DETECTOR

The V0 system consists of two detectors (V0A and V0C) which will be located in the central part of ALICE. The V0A will be installed at a distance of 340 cm from the interaction point as shown in Fig. 2, mounted in two rigid half boxes around the beam pipe. Each detector is an array of 32 cells of scintillator plastic, distributed in 4 rings forming a disc with 8 sectors as shown in Fig. 2. For the V0C the cells of rings 3 and 4 are divided into two identical pieces that will be read with a single photo-multiplier. This is done to achieve uniformity of detection and a small time fluctuation.

The light produced in the scintillator plastic is collected by wavelength shifting fibers and transported to photomultipliers. This setup is based on experimental tests and simulations.

The V0 detector system has several functions:
- On line vertex determination. With the use of the timing information one should be able to locate the main vertex.
- reduction of beam gas events (namely of events caused by the interaction of either beam with molecules of gas) due to discrimination according to the time distribution of signals in one and the other detector.
- On-line centrality measurement. With a rough multiplicity measurement one may discriminate central from non-central collisions.
- It will provide a wake up signal to the electronics of the Transition Radiation Detector within 100 ns after the interaction.
- Luminosity measurement by counting triggered events.

In the pp mode the mean number of charged particles within 0.5 units of rapidity is about 3. Each ring covers approximately 0.5 units of rapidity. The particles coming from the main vertex will interact with other components of the detector generating secondary particles. In general, each cell of the V0 detector will on the average register one hit. For this reason the detector should have a very high efficiency. In Pb-Pb collisions the number of particles in a similar pseudo-rapidity range could be up to 4000 once secondary particles are included. Comparing the number of hits in the detector in the pp and Pb-Pb mode we can see that the required dynamic range will be 1 – 500 minimum ionizing particles.

The segments of the V0A detector will be obtained with a megatile construction (see ref. [3]). This technique consists of machining the scintillator plastic and filling the grooves with TiO_2 loaded epoxy in order to separate one sector from the other. Since this is done all the way through the plastic the epoxy restores mechanical strength. The surface of the slices are then wrapped with teflon tape.

Wavelength shifting fibers are then embedded in the plastic in 3 *mm* deep grooves on both sides of the cell. These fibers bring the light to the edge of the disc where they are coupled to clear fibers with optical connectors.

The construction is made of 25 *mm* thick scintillator plastic from Bicron (BC404). The wavelength shifting fibers WLS (BC9929AMC) are 1 *mm* diameter. At the far end of the WLS, aluminum coating is used as a reflector. The fibers are located every 12 *mm* from each other on each face. The fibers from one face are displaced with respect to those on the other face by 6 *mm*.

The photomultiplier tubes (PMT) will be installed inside the magnet not far from the detector. In order to tolerate the magnetic field, fine mesh tubes have been chosen: H6153-70 from Hamamatsu.

In order to fulfill its functions the V0 electronics should be compatible with the proton-proton and heavy ion modes. The system will provide the following information:

- Minimum-bias trigger: this signal is generated if at least one channel of V0A and one of V0C is fired in the right time window.
- Beam-gas trigger: a signal is generated if the time difference between the signals in V0A and V0C do not correspond to a beam-beam interactions.
- Centrality triggers: these are generated if one of the following two conditions or the two of them are fulfilled:
  - the charged seen in V0A and V0C after a collisions is larger than a threshold
  - the number of channels after the collision is larger than an expected value.
- A measure of the multiplicity. This is obtained measuring the total charge by analyzing the anode pulse from the PMTs.
- A measure of the time difference between particles detected and the beam crossing signal.
- A fast wake-up signal for the Transition Radiation Detector.

A more detailed description of the V0 system can be found in ref. [4].

Simulation of the set up

We have undertaken two types of simulations related to the functioning of the V0 detectors – more specifically the V0A. The first is connected to the efficiency and homogeneity of the detector.

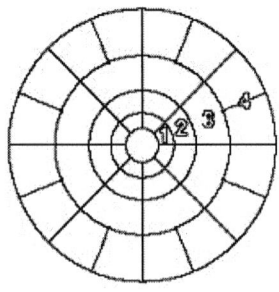

FIGURE 2. Segmentation of the V0 detectors. On the left, V0A in its final position.

These simulations were made using the LITRANI code.
LITRANI stands for LIght TRansmission for ANIsotropic Media [5]. It is a general purpose Monte-Carlo for the simulation of propagation of optical photons. It is written in C++/Root and was originally developed for the calibration of the CMS calorimeter. The code was developed using LITRANI v3.0 in a Root v 4.08f framework. With this we analyzed the homogeneity in each cell of the detector the homogeneity in the hole detector and the propagation of photons through the wavelength shifters. From Fig. 3 we see that a satisfactory uniformity and efficiency is achieved with the adopted fiber arrangement.

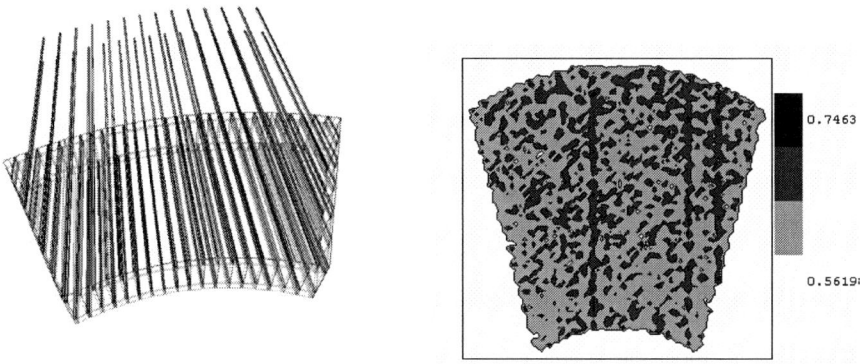

FIGURE 3. One of the cells of the V0A detector (left) with WLS fibers visible and the simulated geometric efficiency (right). The geometric efficiency is the number of photons seen at the PMT divided by the number of photons produced when a particle goes trough the cell.

The other simulation had to do with the evaluation of the time resolution of the detector necessary to efficiently discriminate beam-gas events from beam-beam events rejection of beam gas interactions in proton-proton collisions have been studied with a Monte Carlo simulation, using the ALICE geometry and transporting the produced particles through the detector. In Fig. 4 we see the distribution of the arrival times of primary and secondary particles for the V0A and the V0C detector.

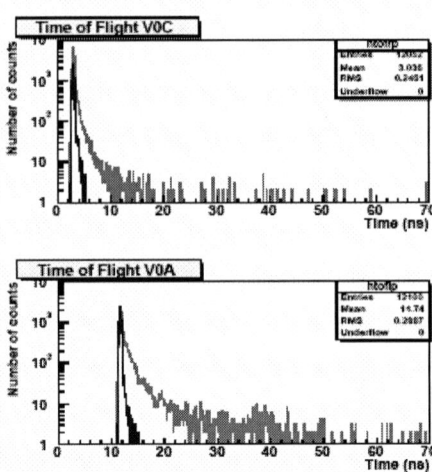

FIGURE 4. In the upper plot, the time of flight of primary (lower curve) and secondary particles (upper curve) arriving at V0C is shown. In the bottom a similar timing distributions for primary and secondary particles arriving at V0A.

Fig.5 shows the time-difference and time-sum distributions of primary and secondary particles. As one can see, it is possible to separate signal from background by measuring these distributions.

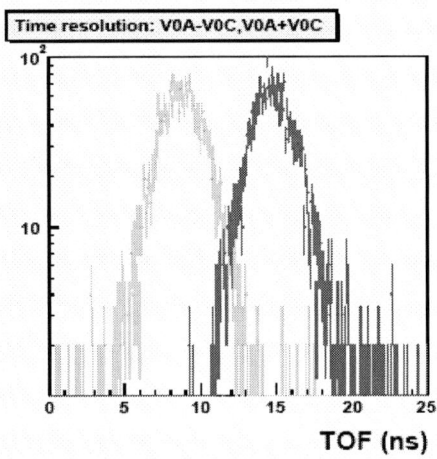

FIGURE 5. Time distribution of the signal from beam-beam events (curve on the left) and signal from beam-gas interactions (curve at the right) as a function of the time of flight. The beam-beam events are chosen imposing a window on the time distribution. The plots are for a time resolution of ~800 ps.

The conclusion of this study was that a time resolution better that 1 ns is imperative to achieve a reasonable separation between beam-gas and real proton-proton events.

Irradiation test

We have performed radiation tolerance tests on the BCF-99-29MC wavelength shifting fibers and the BC404 plastic scintillator from Bicron as well as on silicon rubber optical couplers ref [6]. We used the Co60 gamma source at the Instituto de Ciencias Nucleares facility to irradiate 30-cm fiber samples with doses from 50 Krad to 1 Mrad. We also irradiated a 10x10 cm^2 scintillator detector with the WLS fibers embedded on it with a 200 krad dose and the optical connectors between the scintillator and the PMT with doses from 100 to 300 krad. We measured the radiation damage on the materials by comparing the pre- and post-irradiation optical transparency as a function of time.
The results obtained confirmed that up to doses of 300 krad, which is the expected lifetime doses of the detector, no significant deterioration occurs.

COSMIC RAY DETECTOR

The ALICE cosmic ray trigger is based on the ACORDE system, which is an array of 60 plastic scintillator slabs 188×20 cm^2, placed on the top sides of the central magnet (see Fig.1).

The cosmic array will serve two purposes:
1) Performing as a cosmic ray trigger for the TRD and TPC detectors in the calibration phases with single muon tracks.
2) Study of rare cosmic ray events in conjunction with the ALICE tracking detectors.

Together with other ALICE devices it will provide interesting information about cosmic rays originated by primaries with an energy of 10^{15-17} eV. In particular multi-muon events can be studied with the use of the Time Projection Chamber.
The Cosmic Ray Detector consists of an array of 60 scintillator counters located in the upper part of the ALICE magnet. Fig. 6 shows the distribution of the panels in their final position.

The plastic used for the construction of the detector was part of the DELPHI detector [8]. The material was carefully studied and the design of the detector was done according to the capabilities of the plastic available. The material was transported to Mexico where the assembly of modules has taken place.
Each module has a sensitive area of $1.9 \times .195 m^2$ and is built with two superimposed plastics. Fig. 7 shows the production process of the panels. The doublet has an efficiency of 90 % or more along the module.

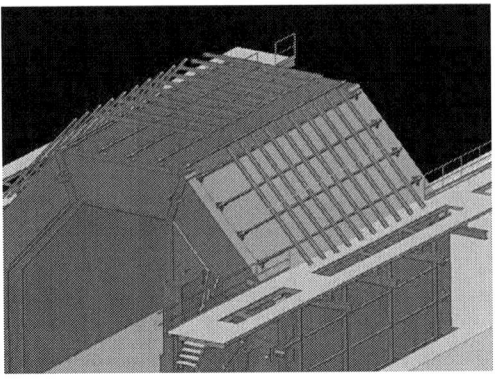

FIGURE 6. Array of panels on the top of the ALICE magnet.

The electronics design is also made in Mexico [9] using the latest FPGA designs and fiber optics to link the signal to the central trigger processor and the Data acquisition of ALICE. The requirements are as follows:

- It should generate a single muon trigger at the level0 to calibrate the Time Projection Chamber and other components of ALICE.
- It should generate a multi-muon trigger to study cosmic rays.
- It will provide a wake-up signal for the Transition Radiation Detector.
- It should include a calibration system for periodical testing.
- It will scan 120 channels (but it is designed to scan up to 200 channels for possible enlargement of the surface covered) in synchronization with the LHC clock.
  - It will produce a coincidence signal when two overlapped plastics are fired.

The electronics should generate the signal in less than 100 ns and will provide spatial information of the scintillator fired. It will also store the information of that spatial location.

Fig. 8 shows the array of cosmic ray detectors during test and calibration of the Time Projection Chamber at CERN. The electronic card is also shown here with the labels describing the components.

FIGURE 7. Panel production process and testing of the panels.

 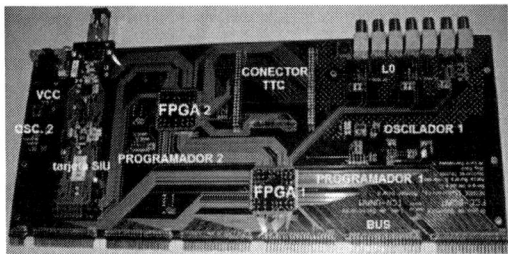

FIGURE 8. ACORDE modules on the top of the Time Projection Chamber during test and calibration. On the right side an ACORDE electronics card is shown.

SUMMARY

In a relatively short time a competent team of scientists, students and engineers has been created generating a momentum towards close collaboration, pooling the human and material resources in a project that should tomorrow lead us to one of the frontier of human research.

ACKNOWLEDGMENTS

Supported by the World Bank through the Iniciativa Cientifica del Milenio 2001 and by CONACYT through several research projects submitted along the years.

The Mexican group is formed by researchers from several institutions:

G. Contreras (Departamento de Física Aplicada, CINVESTAV), L. Montaño, G. Herrera Corral, A. Zepeda (Departamento de Física, CINVESTAV),
A. Ayala, E. Cuautle, J.C. D Olivo, M. I. Martínez, L. Nellen, G. Paic (Instituto de Ciencias Nucleares, UNAM), R. Alfaro, E. Belmont, V. Grabski, A. Martínez, A. Menchaca, A. Sandoval (Instituto de Física, UNAM),
L. Villaseñor (Universidad de Michoacán), A. Fernández, R. López, M. A. Vargas, S. Vergara (Universidad de Puebla).

REFERENCES

1. F.Carminati et al. (ALICE collaboration) Journal Phys. G: Nucl. Part. Phys 30 (2004)1517.
2. ALICE Technical Proposal, Ahmad et al. CERN/LHCC/ 95-71, 1995
3. S. Kim, et al. Nucl. Inst. Meth. Phys. Res. A360(1995)206
4. ALICE Technical Design Report, Forward Detectors FMD, T0, V0 CERN/LHCC-2004-025, 2004.
5. LITRANI, Light Transmission for Anisotropic Media, F. Gentit, DAPNIA-SPP,CEA-Saclay, http://gentit.home.cern.ch/gentit/ Francois-Xavier Gentit, Nucl. Instr. and Meth. **A486** (2002)35
6. R. Alfaro, E. Cruz, M. I. Martínez, G. Paic and A. Sandoval proceedings of 10h Mexican Workshop on Particles and Fields, Morelia, Michoacan, Mexico, 7-12 Nov. 2005.
7. A. Fernández et al, Czechoslovak Journal of Physics,Vol. 55 (2005) B801 - B807
8. R.I. Dzhelyadin, et al., DELPHI Internal Note 86-108, TRACK 42 , CERN 1986
9. S. Vergara (private communication). ALICE Internal Note in preparation.

Physics of the Charm Quark

Salvador Carrillo Moreno* and Elsa Fabiola Vázquez Valencia[†,*]

*Universidad Iberoamericana, México.
[†]CINVESTAV, México

Abstract. This is a brief summary about the development of the charm quark physics in the area of experimental physics. The summary is centered in what is done by mexican physicists, particularly in the E791 and the FOCUS Experiment at FERMILAB. FOCUS (or E831) was designed to detect states of matter combining one or more charm quarks with light quarks (strange, up, down). The experiment created 10 times as many such particles as in previous experiments and investigated several topics on charm physics including high precision studies of charm semileptonic decays, studies of hadronic charm decays (branching ratios and Daltiz analyses), lifetime measurements of all charm particles, searches for mixing, CP/CPT violation, rare and forbidden decays, spectroscopy of excited charm mesons and baryons, charm production asymmetry measurements, light quark diffractive studies, QCD studies using charm pair events and searches for and upper limits on: charm pentaquarks, double charm baryons, $D_{SJ}(2632)$.

Keywords: charm quark
PACS: 14.65.Dw

E791 EXPERIMENT

By the beginning of the last decade a Mexican group joined Fermilab-E791 Collaboration, a charm hadroproduction experiment at Fermilab. Gerardo Herrera (Cinvestav) and Arturo Fernández(first Cinvestav/Fermilab posdoc, later BUAP collaborator) started a new branch, pioneering the experimental charm physics research in México.

E791 [1] recorded 2×10^{10} events with a loose transverse energy trigger. These events were produced by a 500 GeV/cπ^- beam interacting in a target consisting of five thin foils that had 15 mm center-to-center separation along the beamline. The most upstream foil was 0.5 mm thick platinum. It was followed by four foils consisting of 1.6 mm thick diamond. Momentum analysis was provided by two dipole magnets that bent particles in the horizontal (x-z) plane. Position information for track and vertex reconstruction was provided by 23 silicon microstrip detectors (6 upstream and 17 downstream of the target) along with 10 planes of proportional wire chambers (8 upstream and 2 downstream of the target), and 35 drift chamber planes. The experiment also included electromagnetic and hadronic calorimeters, a muon detector, and two multicell Čerenkov counters that provided π/K separation in the momentum range 6- 60 GeV/c. The kaon identification criteria varied by search decay mode. We typically required that the momentum dependent light yield in the Čerenkov counters be consistent with that of a kaon track measured in the spectrometer. Electrons were identified by an electromagnetic calorimeter that consisted of lead sheets and liquid scintillator located 19 m downstream of the target. See Figure 1.

E791 experiment has published 34 papers in refereed journals up to now, with eight

E-791 Spectrometer

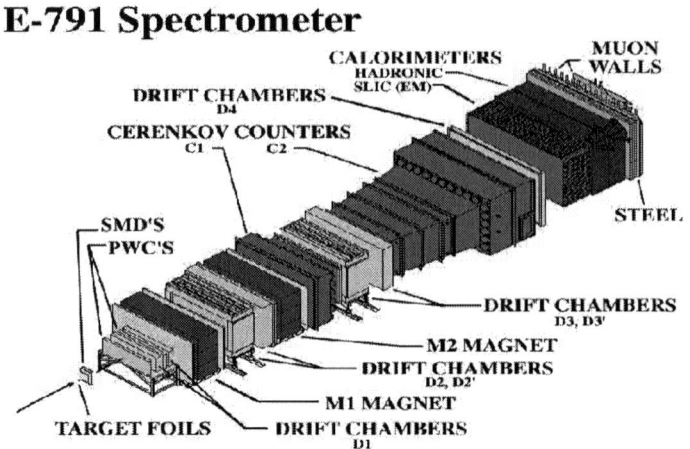

FIGURE 1. E791 spectrometer. The Mexican group worked on the offline alignment, characterization and parametrización of the proportional wire chambers and drift chamber planes.

of those publications being cited more than 100 times each. Most of this production is related with studies on the charm hadroproduction mechanism (differential cross-sections, charged production asymmetries), charmed baryon and meson lifetime, search for doubly Cabibbo suppressed decays, rare and forbidden decays of D mesons, search for the pentaquark $P_{\bar{c}s}^0$, hyperon studies and evidences of light resonances in D meson decays.

In 1992 the Mexican group joined this experiment, few months after the conclusion of the E791 data taking period. For the detector comprehension, the Mexicans contributed mostly on the offline alignment, characterization and parametrization of the efficiency for the wire chambers. This turn out to be crucial in the understanding of the detector that allowed a number of analysis later on. They participated in the elaboration of the general data reconstruction program, the complete data and MC event reconstruction using a multiprocessor management for event-parallel computing system, (for those days, it was a technical challenge to use parallel computing code programming for physics analysis) as well as the x_F dependence of the D meson width, the study of several properties of charmed baryons[2], and asymmetries in the production of the Λ^0, Ξ^- and Ω^- hyperons. It is important to mention that the strong evidence of the $\sigma(500)$ resonance (a scalar resonance of light mass and narrow width) found by E791 using 1200 $D^+ \to \pi^-\pi^+\pi^+$ decays[3], is one of the most cited papers on charm physics nowadays, with more than 250 cites. Our group participated very closely in the internal discussion and final approval of this controversial result. In the discussion process there were many possible explanations. At some point the idea of having identical bosons in the decay final state was put forward as a possible source of distortion of the phase space. The study was finally published in Reference [4].

Analyzing E791 data, several Mexican physicist began their scientific carrier: Pablo Medina, Raul Hernández, Luis Montaño and Edgar Linares from Cinvestav, and Eleazar Cuautle from BUAP worked on Master thesis. Alberto Gago, from the Pontificia Universidad Catolica de Perú obtained his Master degree under the advice of G. Herrera. Another Peruvian physicist, Javier Solano, obtained his PhD working on a hyperon physics, topic proposed by G. Herrera and the Brazilian CBPF group.

E831 (FOCUS) EXPERIMENT

In order to be in the last frontier of the High Energy Physics Experiments it is necessary to join experiments of international size. There are many of them around the world where phycists from many countries work together to give their contributions to find new discoveries in their areas. In our case, our little contribution began 10 years ago, in the Fermi National Accelarator Laboratory (FERMILAB) in Batavia, IL. in a fixed target experiment: E831 or FOCUS *Fotoproduction (sic) of Charm with an Upgraded Spectrometer* (It is an upgraded version of its predecessor, E687). FOCUS is an International Collaboration of the following Institutions: Univ. of California at Davis; CBPF, Brazil; CINVESTAV, México; Univ. of Colorado, Boulder; Fermilab, Batavia; LNF and INFN Frascati, Italy; Univ. of Guanajuato, México; Univ. of Illinois, Urbana-Champaign; Indiana Univ., Bloomington; Korea Univ., Korea; Kyungpook National Univ., Korea; INFN and Univ. of Milano, Italy; Univ. of North Carolina, Asheville; Univ. and INFN Pavia, Italy; Pontificia Univ. Catolica, Brazil; Univ. of Puerto Rico, Mayaguez; Univ. of South Carolina, Columbia; Univ. of Tennessee, Knoxville; Vanderbilt Univ., Nashville; Univ. of Wisconsin, Madison.

The experiment accumulated data during 1996-1997 fixed target run and has fully reconstructed more than one million charm particles using the 'golden mode' decays, $D^0 \to K^-\pi^+$, $D^0 \to K^-\pi^+\pi^-\pi^+$, and $D^+ \to K^-\pi^+\pi^+$.

In those days, all the experiments in Fermilab were running, the fixed target experiments and the collision experiments (CDF & D0). Fermilab was a place where physicists all around the world met. We were several students then from CINVESTAV, and the experience to be part of FOCUS served to obtain our PhD. and to make us return to FERMILAB in many occasions.

We were 5 physicists from CINVESTAV, México: S. Carrillo, E. Casimiro, E. Cuautle, A. Sánchez-Hernández, F. Vázquez; one from IFUAP, Puebla, México: C. Uribe; and one that later participated as a member of University of Guanajuato, Leon, Guanajuato, México: M. Reyes.

Four of us made our thesis with data collected in FOCUS:

- Asymmetry in the production of baryons and anti-baryons Λ_c in high energy interactions between fotons and nucleons. Casimiro Linares, Edgar.
- Confirmation of the observed decay that violates isospin: $D_s^{*+} \to D_s^+ + \pi^0$ Carrillo Moreno, Salvador.
- Charm-anticharm baryon production asymmetries in photon-nucleon interactions Cecilia Uribe [12].
- Asymmetric photoproduction study of charm mesons, using data from FOCUS experiment. Vázquez Valencia, Elsa Fabiola.

Up to a point each of us apport something in the many articles, results and discoveries in the area of the Charm Physics.

We will present a brief summary of the experiment, some important results and how, by these, the Charm Quark Physics have evolved in what it is today.

FOCUS Spectrometer

The FOCUS detector is a large aperture fixed target multiparticle spectrometer which features excellent particle identification and vertexing for charged hadrons and leptons. A photon beam is derived from the bremmstrahlung of secondary electrons ($E_{max} = 250$ GeV) produced from the Tevatron proton beam. This beam impinges on a beryllium target; charged particles which emerge from the target are tracked by two systems of silicon microvertex detectors. The first system is comingled with the experimental target, the second is just downstream of the target and consists of twelve planes of microstrips arranged in three views. These detectors provide high resolution separation of primary (production) and secondary (decay) vertices. The momentum of charged particles is determined by measuring their deflections in two analysis magnets of opposite polarity with five stations of multiwire proportional chambers. Straw tubes are used to supplement tracking in the central pair region. Three threshold multicell Cerenkov counters are used to identify electrons, pions, kaons and protons. There are two electromagetic calorimeters. The inner calorimeter, a lead glass block array, covers the central solid angle and detects particles which pass through the apertures of both magnets. The outer calorimeter covers the outer angular annulus described by particles that pass through the first magnet but not the second. Muons are identified in either a fine grained scintillator hodoscope with an iron filter (covering the inner region) or in an outer system that uses resistive plate chambers and the iron yoke of the second magnet as a filter. A hadron calorimeter consisting of iron and scintillating tile is used primarily in the experiment trigger but is also used to reconstruct neutral hadrons.

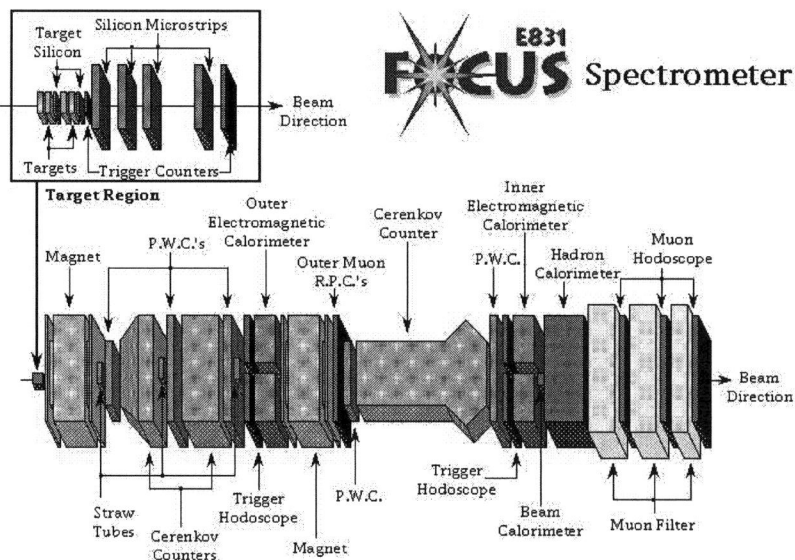

FIGURE 2. The E831 Spectrometer

Production of Charm Particles

In the photon-gluon fusion production mechanism the photon interacts with a gluon from the target nucleon that has fluctuated into a charm-anticharm quark pair. The quarks are forced out of the nucleon and are dressed to produce the charm mesons and baryons. Charm particles live aproximately 1 ps., then decay weakly to strange and other particles. E831 FOCUS spectrometer was designed to detect these decay products. The finite lifetime of charm particles is the principal property exploited to isolate signals from copious non-charm backgrounds.

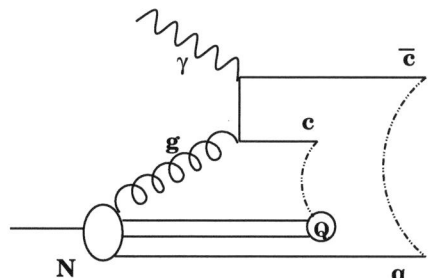

FIGURE 3. Photon-gluon fusion production mechanism.

Charm Particle Lifetimes

Precise measurements of all the singly charmed particles lifetimes have been made by FOCUS [5, 6, 7, 8, 9, 10]. The latest is for the D_s^+ using $\phi\,\pi$ and K* K decay modes.

We use the reduced proper time, $t' = (L - 6\sigma_L)/\beta\gamma c$, to minimize acceptance corrections. Where $L > 6\sigma_L$ is the decay vertex detachment cut used. The reduced proper time distributions for signal and sideband regions are binned in two histograms. The expected number in each time bin in the signal region, and the likelihood used in the fit are:

$$n_i = S \frac{f(t_i')exp\left[-\frac{t_i'}{\tau}\right]}{\sum_i f(t_i')exp\left[-\frac{t_i'}{\tau}\right]} + B\frac{b_i}{\sum_i b_i} \quad (1)$$

$$L = \prod_i \frac{n_i^{s_i} exp[-n_i]}{s_i!} \times \frac{(\alpha B)^{N_b} exp[-\alpha B]}{N_b!} \quad (2)$$

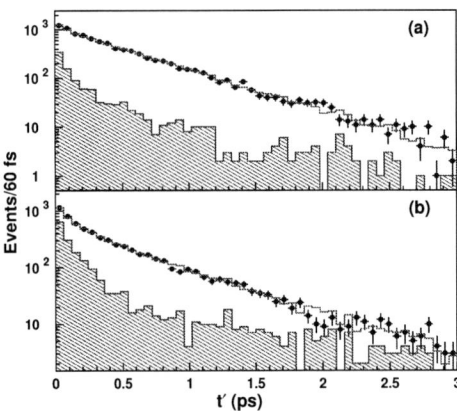

FIGURE 4. a) $\phi\,\pi^+$ decay. b) $\bar{K}^{*0}K^+$ decay. Red: Sideband region data. Blue: Signal region data. Violet: Lifetime fit.

Lifetime and background level, B, are the fit parameters. Results for the lifetimes:

- $\phi\pi^+$: 507.6 ±6.5(stat.) ±4.3(syst.) fs.
- $\bar{K}^{*0}K^+$: 506.9 ±10.6(stat.) ±8.8(syst.) fs.
- $\tau(D_s^+)$: 507.4 ±5.5(stat.) ±5.1(syst.) fs.

Many systematic studies were done, see hep-ex/0504056 for details.
Comparing with other measurements: See Figure 6.

Search for the Strongly Decaying Neutral Charmed Pentaquark

Theoretically, the existance of multiquark states, like the $\theta_c^0(\bar{c}uudd)$, as an observable bound state is inconclusive [11]. The H1 Collaboration reported evidence of a S=0

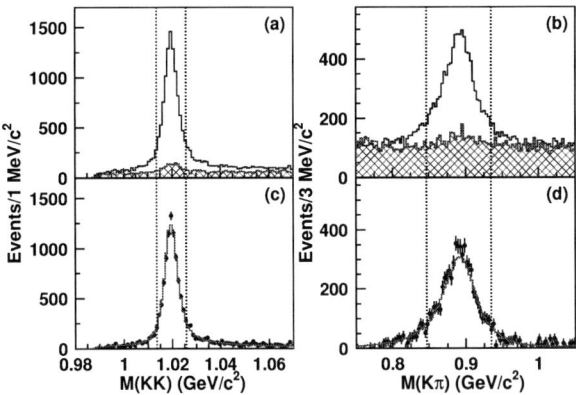

FIGURE 5. Left: $\phi \pi^+$ decay. Right: $\bar{K}^{*0}K^+$ decay. Red: Sideband region data. Blue (top): Signal region data. Blue (bottom): SR - SBR. Violet: MC.

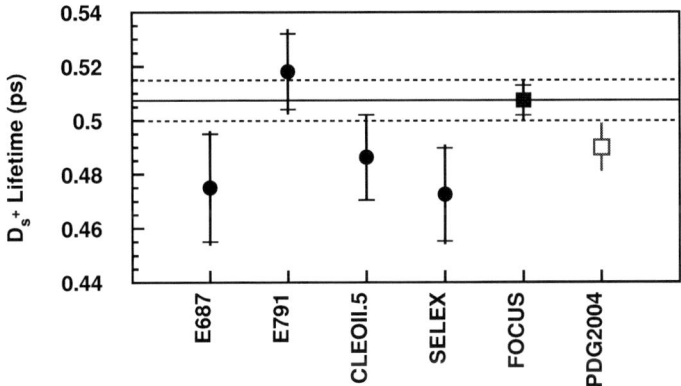

FIGURE 6. Comparison of D_s^+ lifetime with other experiments

charmed pentaquark state decaying to $D^{*-}p$ at a mass of $3099 \pm 3 \pm 5$ MeV/c^2. We search for the $\theta_c^0(\bar{c}uudd)$ pentaquark candidate in the $D^{*-}p$ D^-p decay modes. Since the $D^{*+}p$ statistics and data quality of the FOCUS experiment is better than seen in H1 and the production mechanism is similar we should be able to confirm or refute the existence of the purposed state.

A decay vertex detachment requirement gives clean samples of the decays $D^0 \to K^-\pi^+, K^-\pi^+\pi^-\pi^+$, and $D^+ \to K^-\pi^+\pi^+$. The D^{*+} are obtained from $D^{*+} \to D^0\pi_s$. The selection criteria were chosen to maximise $S/\sqrt{B}$. The signal, S, comes from Monte Carlo Simulation. The background, B, is obtained from wrong sign data ($D^{*+}p$) over the entire mass ragne (threshold to 4 GeV/c^2). This is an unbiased method of determining the selection criteria.

No evidence for a pentaquark at 3.1 GeV/c^2 or at any mass less than 4 GeV/c^2 is observed in the $D^{*-}p$ and D^-p invariant mass plots.

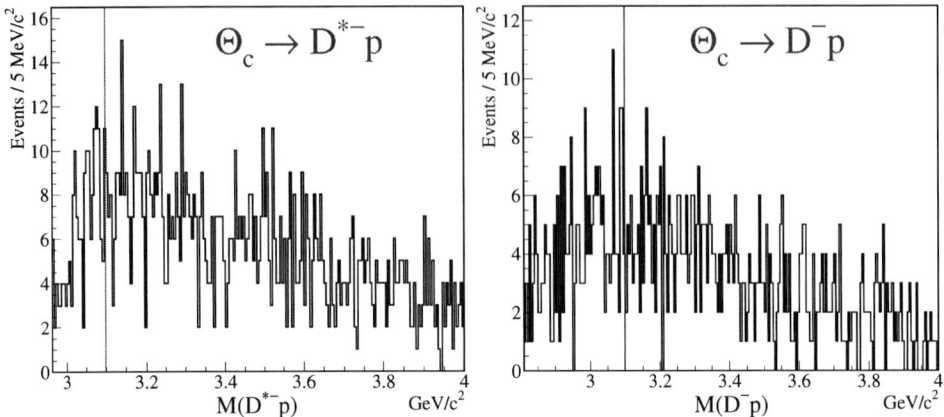

FIGURE 7. No conclusive signal.

The sample of D^{*-} events is more than 30 times larger as well as cleaner than the sampler used by H1 in the paper which reported evidence for the charm pentaquark. While H1 find 1% of the $D^{*-}p$ from θ_c^0, FOCUS sets an upper limit of 0.075% at 95 %CL. The production is similar between the two experiments; virtual (real) photons on protons (nucleons) for H1 (FOCUS). Thus, the H1 result is either a statistical fluctuation or the result of an unusual production mechanism which increases the charm pentaquark to charm cross section by a factor of least 10 in H1 relative to FOCUS. More details of this analysis can be found at hep-ex/0506013.

With more than a million of charm particles reconstructed and more than 50 publications, FOCUS has provided a much better understanding of the charm quark physics.

ACKNOWLEDGMENTS

Thanks to the pioneer effort that CINVESTAV have made in experimental high energy physics in México for more than 10 years, providing opportunities to many students, it has accomplish a base group of experimental high energy mexican physicist that today continue to grow and collaborate in many other experiments around the world. We are glad to be part and continue the effort of such group of scientists.

We wish to acknowledge the assistance of the staffs of Fermi National Accelerator Laboratory, the INFN of Italy, and the physics departments of the collaborating institutions. This research was supported in part by the U.S. National Science Foundation, the U.S. Department of Energy, the Italian Instituto Nazionale di Fisica Nucleare and Ministero dell'Istruzione dell'Università e della Ricerca, the Brazilian Conselho Nacional de Desenvolvimento Científico e Tecnológico, CONACyT-México, the Korean Ministry of Education, and the Korean Science and Engineering Foundation.

REFERENCES

1. See http://ppd.fnal.gov/experiments/e791/welcome.html for a complete list of E791 scientific publications.
2. A. Fernandez, et. al. "Λ_c studies at Fermilab E791", *Proc. of the IV Mexican Workshop of Particles and Field, World Scientific(1995)*; A. Fernandez, "Results on charm physiscs", *Canadian, American and Mexican Meeting, AIP Proc. 342, p. 409-342(1995)*; A. Fernandez, et. al. "Hadroproduction of charm at Fermilab E791", *Proc. of the V Mexican School of Particles and Fields, World Scientific(1992)*; J. dos Anjos, G. Herrera, J. Magnin, F.R.A. Simao , "The Charm of the proton and the Lambda(c) + production", *Proc. of the 7th Mexican School of Particles and Fields and 1st Latin American Symposium on High-Energy Physics, Merida, Yucatan, Mexico (1996)*. e-Print Archive: hep-ph/9702257.
3. E791 Collaboration (E. M. Aitala et.al), *Phys. Rev. Lett.* **86**,(2001), 770.
4. E. Cuautle and G. Herrera, *Phys. Lett.* **B434**,(1998), 153.
5. A Measurement of the Ds+ Lifetime, The FOCUS Collaboration, *Phys. Rev. Lett.* vol. **95** pg. 052003
6. A new measurement of the Xi_c^0 lifetime, The FOCUS Collaboration, *Phys. Lett. B* vol. **541** pg. 211.
7. Measurements of the D^0 and D^+ lifetimes, The FOCUS Collaboration, *Phys. Lett. B* vol. **537** pg. 192.
8. Lifetime Measurement of the Charmed Strange Baryon Cascade-c+, Eduardo Ramírez thesis, E831-ths-2002/1
9. A high statistics measurement of the Lambda-c+ lifetime, The FOCUS Collaboration, *Phys. Rev. Lett.* vol. **88** pg. 161801.
10. A new measurement of the Xi_c^+ lifetime, The FOCUS Collaboration, *Phys. Lett. B* vol. **523** pg. 53.
11. Search for a strongly decaying neutral charmed pentaquark, The FOCUS Collaboration, *Phys. Lett. B* vol. **622** pg. 229.
12. Charm-anticharm baryon production asymmetries in photon nucleon interactions, The FOCUS Collaboration, *Phys. Lett. B* vol. **581** pg. 39 (2004)

Λ^0 Polarization in Exclusive pp Reactions

J. Felix

Instituto de Física, Universidad de Guanajuato
Lomas del Bosque 103, Frac. Lomas del Campestre, León GTO., México
felix@fisica.ugto.mx

Abstract. Among all properties of baryons, the polarization they acquire when created from unpolarized **p-nucleus** collisions is the most recent discovered one; so far, the origin of this polarization remains unexplained in spite of the experimental evidences accumulated in the past thirty years. Up to these days, Λ^0 is the most studied baryon for polarization, due to it is very easy to produce Λ^0's at the energies of the principal high energy physics accelerators of the world. This article is a review of the experimental experience accumulated on the polarization of Λ^0 in unpolarized exclusive pp collisions as function of x_F, P_T, and $M(\Lambda^0 K^+)$ in the past fifteen years here at the Instituto de Física, Universidad de Guanajuato, inside Fermilab e690 and Brookhaven National Laboratory e766 collaborations.

Keywords: Hyperon, Λ^0, Polarization, Exclusive Reactions, Baryon
PACS: 13.88.+e, 13.90.+i, 14.20.Gk

INTRODUCTION

It is an experimental evidence that baryons in general are created polarized, when they are produced from unpolarized *particle-nucleus* collisions. For instance Ξ, Σ, Λ^0, etc., are produced polarized in exclusive as well as in inclusive reactions at different energies[1, 2, 3]. In the same experimental circumstances, it appears that Ω^-(K.B. Luk et al)[1] and $\overline{\Lambda}^0$(K. Heller et al)[1] [18] do not appear polarized. However, it appears that $\overline{\Lambda}^0$ is created polarized in $K^- p$ reactions[19]. The origin of baryon polarization is not known up to these days.

From the above searched baryons, Λ^0 in the most studied one for polarization, for it is very easy to produce and analyze for polarization. Many experiments have reveal that Λ^0's from unpolarized pp inclusive and exclusive collisions, at different energies, are produced polarized[2]; and that this polarization depends on x_F, P_T, and $\Lambda^0 K^+$ invariant mass[3].

Based on the above experimental evidences, some authors have proposed many theoretical ideas -in the context of the Lund model, parton recombination, multiple scattering of the strange quark, gluon fusion, Regge theory, coherent scattering, low and high order QCD calculations, valence quark effects, and quark condensates- trying to understand Λ^0 polarization[4]. See reference [5] for a review of the theoretical ideas proposed to explain Λ^0 polarization. To date, there is no a satisfactory explanation. These models lack of predictive power, and the problem of Λ^0 polarization, and in general of baryon polarization, remains as an open problem. Therefore many more measurements for polarization, in all possible experimental circumstances, are needed to formulate a solution of Λ^0 polarization problem.

CP857, *Particles and Fields, Part B: Commemorative Volume of the Division of Particles and Fields of the Mexican Physical Society,* edited by M. A. Pérez, L. F. Urrutia, and L. Villaseñor
© 2006 American Institute of Physics 978-0-7354-0354-3/06/$23.00

This paper reviews the results of the studies of Λ^0 polarization in exclusive unpolarized pp collisions carried out at the Instituto de Física, Universidad de Guanajuato, in collaboration with the Fermilab e690 and Brookhaven National Laboratory e766, during the last fifteen years.

Inside these collaborations a research program has been conducted to measure Λ^0 polarization, in exclusive pp collisions, trying to unveil Λ^0 polarization origin studying specific final states where Λ^0 is produced.

This paper resumes the results of this research program on the Λ^0 polarization, as function of X_F, P_T $M_{\Lambda^0 K^+}$, in the specific final states

$$pp \to p\Lambda^0 K^+ (\pi^+\pi^-)^N, N = 1,2,3,4,5, \quad (1)$$

at 27.5 GeV, inside the BNL e766 collaboration. And in the specific final state

$$pp \to p\Lambda^0 K^+, \quad (2)$$

at 800 GeV, inside the FNAL e690 collaboration.

In those reactions Λ^0 is produced polarized. And this polarization is function of x_F, P_T, and $M_{\Lambda^0 K^+}$. This paper reports on these results.

The FNAL e690 and BNL e766 collaborations are described in detail elsewhere[6].

TECHNIQUE TO MEASURE Λ^0 POLARIZATION

There are many methods to measure Λ^0 polarization, all of them, in principle, very straightforward. Here is the technique that I have used mainly to measure Λ^0 polarization:

For Λ^0 decays weakly into $p\pi^-$ and the angular distribution of those protons in the reference system where Λ^0 is at rest is very well known[20], the measurement of Λ^0 polarization consists in determining the angular distribution of the proton from the Λ^0 and fitting it to a straight line, using the least square technique; for an ample description of this technique, see reference[13].

The polarization of Λ^0 is measured with respect to the vector normal to the Λ^0 production plane defined as

$$\hat{n} \equiv \frac{\vec{P}_{beam} \times \vec{P}_{\Lambda^0}}{|\vec{P}_{beam} \times \vec{P}_{\Lambda^0}|}, \quad (3)$$

where $\vec{P}_{beam}$ and $\vec{P}_{\Lambda^0}$ are the momentum vectors of the incident beam proton and of the Λ^0, respectively. This normal to the production plane is calculated event by event.

The angular distribution of the proton in the Λ^0 rest frame is described by this expression, which is related with Λ^0 polarization[20]:

$$\frac{dN}{d\Omega} = N_0(1 + \alpha \mathscr{P} \cos\theta), \quad (4)$$

where α is the Λ^0 decay asymmetry parameter -its most recent value is 0.642 ± 0.013[7]- and $\mathscr{P}$ is the Λ^0 polarization. The polarization is determined by a linear fit of a function of the above form to the $cos\theta$ distribution of the proton -after Monte Carlo detector acceptance corrections- for two free parameters: A normalization constant, N_0, and polarization, $\mathscr{P}$. From these parameters the Λ^0 polarization and its statistical error are measured. The systematic errors are obtained simulating the detector acceptance and efficiency using the Monte Carlo technique.

Λ^0 POLARIZATION AS FUNCTION OF x_F AND P_T

The discovery of Λ^0 polarization in inclusive pp reactions showed that it depends linearly on P_T and that it is negative with respect the Λ^0 production plane[8]: At $P_T = 0.0$, $\mathscr{P} = 0.0$, and it decreases as P_T increases up to close $\mathscr{P} = -0.25$ at $P_T = 1.2$ GeV. Also it was determined in inclusive pp reactions that Λ^0 polarization depends on x_F. See Reference [9] for a review of the experimental results on Λ^0 polarization.

In the reaction $pp \to p\Lambda^0 K^+(\pi^+\pi^-)^2$ at 27.5 GeV, Λ^0 polarization in a function of x_F and P_T[10]. This is the first report on Λ^0 polarization in exclusive reactions at high energies.

The linear dependence of Λ^0 polarization on P_T is given by $\mathscr{P} = (-0.250 \pm 0.067)P_T + (0.063 \pm 0.041)$ -$\frac{\chi^2}{N_{DOF}} = 1.04$, for $x_F > 0.0$-; by $\mathscr{P} = (0.147 \pm 0.056)P_T + (-0.007 \pm 0.033)$ -$\frac{\chi^2}{N_{DOF}} = 1.04$, for $x_F < 0.0$-; and by $\mathscr{P} = (-0.189 \pm 0.042)P_T + (0.029 \pm 0.025)$ -$\frac{\chi^2}{N_{DOF}} = 1.68$, for $x_F > 0.0$ and $x_F < 0.0$- combined.

The average polarization is roughly linear in x_F; its polarization is given by $\mathscr{P} = (-0.20 \pm 0.06)x_F + (-0.032 \pm 0.018)$ -$\frac{\chi^2}{N_{DOF}} = 0.247$, for $x_F > 0.0$ and $x_F < 0.0$ combined-. The average polarization changes sign at $x_F \simeq 0.0$.

The polarization data were examined for possible correlations with respect to the directions of other particles in the final state, the p and the K^+, for some models correlate the *s-quark* direction and the *s̄-quark* direction with the polarization of Λ^0; for those models the polarization of the $s\bar{s}$ system spin is the origin of Λ^0 polarization. These correlations can be investigated by looking for correlations between the polarization and the motions of the $\Lambda^0(s\text{-}quark)$ and the K^+ ($\bar{s}\text{-}quark$), where it was assumed that the motion direction of the particle is the motion direction of the corresponding *quark*. The Λ^0 polarization was the same even the Λ^0 and K^+ were in the same or in opposite hemisphere and these coincide with the Λ^0 polarization measured without cuts in the Λ^0-K^+ direction. Statistically, no correlations between Λ^0 polarization and Λ^0-K^+ relative motion direction were detected. These results differ from the above model propositions.

Λ^0 polarization also was determined in the subset in which the Λ^0 and the p have opposite signs in their respective x_F. For both the $x_F < 0.0$ Λ^0 sample and the $x_F > 0.0$ Λ^0 sample Λ^0 polarization agree, inside statistics, with that Λ^0 polarization without these cuts; the result is as follows: $\mathscr{P} = (-0.291 \pm 0.068)P_T + (0.082 \pm 0.041)$ -$\frac{\chi^2}{N_{DOF}} = 1.03$, for $x_F > 0.0$-; and $\mathscr{P} = (0.185 \pm 0.070)P_T + (-0.013 \pm 0.039)$ -$\frac{\chi^2}{N_{DOF}} = 1.39$, for $x_F < 0.0$-. These results are within two standard deviations of the results in which no conditions on the relative $\Lambda^0 - p$ relative directions were imposed.

When p and Λ^0 have the same sign of their respective x_F the polarization is consistent

TABLE 1. Values of parameter a in Eq. 5 as found by the maximum likelihood analysis for each reaction (N=1,2,3,4) and for the combined sample.

N reaction	$a\ (\frac{1}{GeV/c})$
1	0.390 ± 0.079
2	0.490 ± 0.051
3	0.366 ± 0.064
4	0.515 ± 0.143
all combined	0.443 ± 0.037

with zero, this is $\mathscr{P} = (0.041 \pm 0.096)P_T + (0.026 \pm 0.060)$ -$\frac{\chi^2}{N_{DOF}} = 0.38$-. This result is dominated by the fact that $\sim 75\%$ of these Λ^0's has $|x_F| < 0.3$; for the Λ^0 polarization is linear in x_F of Λ^0 a small Λ^0 polarization is expected.

A complete analysis of Λ^0 polarization in the particular reactions described by Eq. 1, with $N = 1, 2, 3, 4$, was performed at high statistics, searching for Λ^0 polarization as function of x_F and P_T[11]. To analyze the data, $\mathscr{P}$ is parameterized as a function of x_F and P_T simultaneously: $\mathscr{P} = \mathscr{P}(x_F, P_T)$. The parameters of this function are determined using the maximum likelihood method[12] with Eq. 4 as the probability distribution for having dN protons in a solid angle of $d\Omega$. For the maximum likelihood analysis, the chosen function that represents the simplest bilinear combination of x_F and P_T is as follows:

$$\mathscr{P} = -ax_F P_T. \quad (5)$$

Other functions, with different P_T dependencies, were also investigated expressing $\mathscr{P}(x_F, P_T)$ as a power series expansion in P_T; the best solution was given by the above equation. The used method was as follows:

Using Eq. 4, the probability for the extended likelihood as a function of $\cos\theta$ is defined as

$$\mathscr{P}(\cos\theta)_{exp} = C_0 A (1 + \alpha \mathscr{P} \cos\theta), \quad (6)$$

C_0 is a normalization constant and A is the detector acceptance determined by the Monte Carlo technique. Since the acceptance corrections are symmetric in $\cos\theta$ these results are without acceptance corrections.

The parameter a from Eq. 5 was determined for each $N = 1, 2, 3, 4$ of the Eq. 1 and for the combined sample -for this parameter, inside statistics, is the same for all samples- minimizing the negative ln of the extended likelihood. The results from this analysis are in Table 1.

This table shows that, within errors, the dependence of Λ^0 polarization of x_F and P_T is independent of the reaction. In other words, this function is independent of the number of pion's in the final state reactions. Furthermore, the polarization for the combined sample is consistent with the results for the individual reactions. Using the value for the combined sample from Table 1, the Λ^0 polarization function has the following form:

$$\mathscr{P}(x_F, P_T) = (-0.443 \pm 0.037) x_F P_T, \quad (7)$$

TABLE 2. Λ^0 polarization for bins of x_F and P_T. The number of events is indicated in each case.

P_T bin (GeV/c)	x_F bin $-1.0\cdots-0.8$	$-0.8\cdots-0.7$	$-0.7\cdots-0.6$	$-0.6\cdots-0.5$
0.0 – 0.2	+0.27±0.05 3098	+0.24±0.04 3560	+0.09±0.07 1765	··· ···
0.2 – 0.4	+0.39±0.05 3569	+0.39±0.03 11645	+0.16±0.04 5181	−0.07±0.18 314
0.4 – 0.6	+0.47±0.06 487	+0.57±0.04 4099	+0.26±0.06 2481	+0.20±0.18 262
0.6 – 0.8	··· ···	+0.17±0.09 1055	−0.22±0.09 959	−0.04±0.21 208
0.8 – 1.0	··· ···	−0.24±0.13 448	−0.57±0.13 475	··· ···
1.0 – 1.2	··· ···	−0.38±0.19 203	−0.42±0.17 217	··· ···

for $-1.0 \leq x_F \leq 1.0$ and $0.0 \leq P_T \leq 1.8\ GeV/c$-.

The conclusion is as follows: The mechanism responsible for Λ^0 polarization is independent of the Λ^0 production mechanism, at least for the final state reactions examined.

Furthermore, in general, equation

$$\mathscr{P}(x_F, P_T) = (-0.443 \pm 0.037) x_F P_T, \quad (8)$$

for $-1.0 \leq x_F \leq 1.0$ and $0.0 \leq P_T \leq 1.8\ GeV/c$, is valid for Λ^0 polarization in exclusive and inclusive *p-nucleus* reactions[14] with the proper a value.

Both Λ^0 polarization from inclusive and exclusive $p-nucleus$ reactions follow the same trend as function of x_F and P_T: $\mathscr{P}(x_F, P_T) = (a) x_F P_T$. This equation is valid for $-1.0 < x_F < +1.0$ and $0.0 < P_T < 2.0\ GeV/c$ and for *nucleus* $= p, Be, Cu, W$ and Pb. The value of a depends on the target nature: It is $-(0.443 \pm 0.037)$, in exclusive pp reactions; $-(0.376 \pm 0.024)$, in inclusive pp reactions; -there is no statistical difference between inclusive and exclusive reactions-; $-(0.232 \pm 0.012)$, in inclusive pBe reactions; $-(0.124 \pm 0.015)$, in inclusive pCu reactions; $-(0.274 \pm 0.023)$, in inclusive pW reactions; $-(0.147 \pm 0.023)$, in inclusive pPb reactions. This equation is independent of the beam energy. This equation is the most simple bilinear form that describes Λ^0 polarization in $-1.0 < x_F < +1.0$ and $0.0 < P_T < 2.0\ GeV/c$.

For the Eq. 2, Λ^0 polarization at 800 GeV as function of x_F and P_T is shown in Table 2 (J. Félix et al)[3]. This functionality of Λ^0 polarization in terms of x_F and P_T is not observed in the reactions described by Eq. 1. This is the first time that large Λ^0 polarization is reported at low P_T. At large P_T -greater than 0.6 GeV/c- Λ^0 polarization agrees with previous reported measurements.

From Eq. 1, with $N = 5$, the polarization of 1973 Λ^0's from the specific reaction $pp \rightarrow p\Lambda^0 K^+(\pi^+\pi^-)^5$ created from 27.5 GeV incident protons on a liquid Hydrogen target, as function of x_F, P_T, and $M_{\Lambda^0 K^+}$, is, inside statistics, consistent with the polarization of Λ^0's from $pp \rightarrow p_{fast}\Lambda^0 K^+$ at 800 GeV [15, 16].

TABLE 3. Λ^0 polarization for bins of $M(\Lambda^0 K^+)$ (J. Félix *et al*)[3].

$M_X(GeV/c^2)$:	1.630	1.675	1.715	1.745	1.775
$\mathscr{P}$:	0.71 ± 0.05	0.60 ± 0.04	0.34 ± 0.04	0.26 ± 0.05	0.28 ± 0.05
$M_X(GeV/c^2)$:	1.810	1.850	1.895	1.940	2.000
$\mathscr{P}$:	0.21 ± 0.05	0.17 ± 0.05	0.14 ± 0.05	0.40 ± 0.07	0.37 ± 0.06
$M_X(GeV/c^2)$:	2.090	2.195	2.325	2.500	2.750
$\mathscr{P}$:	0.20 ± 0.06	0.03 ± 0.06	-0.11 ± 0.07	-0.14 ± 0.07	-0.43 ± 0.08

Λ^0 POLARIZATION AS FUNCTION OF M_X

Λ^0 polarization depends on x_F, P_T -parameters associated with Λ^0- as well as on $M(\Lambda^0)K^+$ -the diffractive mass of the system $\Lambda^0 K^+$-. In general Λ^0 polarization depends on the diffractive mass of the system $\{\Lambda^0 K^+(\pi^+\pi^-)^N\}$, with $N = 0,1,2,3,4,5$.

In Eq. 1, for $N = 2$, Λ^0 polarization depends on $(\Lambda^0 K^+)$, $(\Lambda^0 K^+\pi^+\pi^-)$, $(\Lambda^0 K^+\pi^+\pi^-\pi^+\pi^-)$ diffracted mass systems[10]. This is a novel results indicating that Λ^0 polarization could be related with the polarization of some resonances, from were Λ^0's are created. These results are as follows:

Λ^0 polarization is roughly linear with $M(\Lambda^0 K^+)$, decreasing from zero at $M(\Lambda^0 K^+) = 1.63\ GeV$ to -0.16 at $M(\Lambda^0 K^+) = 2.47\ GeV$. The linear fit is $\mathscr{P} = (-0.185 \pm 0.068)M(\Lambda^0 K^+) + (0.30 \pm 0.14)$, with $\chi^2/N_{dof} = 0.002$. These results agree with those reported in Reference[3](T. Henkes *et al*) in the reaction $pp \rightarrow p\Lambda^0 K^+$ at higher energies and for masses from 1.6 to 2.4 GeV/c^2. To boot, Λ^0 polarization is roughly linear with $(\Lambda^0 K^+\pi^+\pi^-)$, decreasing from zero at $(\Lambda^0 K^+\pi^+\pi^-) = 2.75\ GeV$ to -0.16 at $(\Lambda^0 K^+\pi^+\pi^-) = 3.75\ GeV$; the linear fit is $\mathscr{P} = (-0.170 \pm 0.057)M(\Lambda^0 K^+\pi^+\pi^-) + (0.475 \pm 0.187)$, with $\chi^2/N_{dof} = 0.073$. And additionally, Λ^0 polarization is roughly linear with $(\Lambda^0 K^+\pi^+\pi^-\pi^+\pi^-)$, decreasing from zero at $(\Lambda^0 K^+\pi^+\pi^-\pi^+\pi^-) = 3.9\ GeV$ to -0.14 at $(\Lambda^0 K^+\pi^+\pi^-\pi^+\pi^-) = 5.1\ GeV$; the linear fit is $\mathscr{P} = (-0.154 \pm 0.048)M(\Lambda^0 K^+\pi^+\pi^-\pi^+\pi^-) + (0.63 \pm 0.22)$, with $\chi^2/N_{dof} = 0.55$.

Furthermore, in Eq. 1, for $N = 1,2,3,4,5$, Λ^0 polarization depends on $(\Lambda^0 K^+)$. The results are plotted in Fig. 1; there, the results are compared with the data from both papers in Reference[3]. The behavior of the Λ^0 polarizations is different. As function of pair of pions $(\pi^+\pi^-)$, in each final state, the Λ^0 polarization is still under investigation.

In Eq. 2, the Λ^0 polarization results are in Table 3 as function of $M(\Lambda^0 K^+)$. Also they are plotted in Fig. 1 and compared with Λ^0 polarization from Eq. 1 for $N = 1,2,3,4$ and 5.

There is some relation between $\Lambda^0 K^+$ mass distribution and Λ^0 polarization. That relation evidences the relationship between Λ^0 polarization and $\Lambda^0 K^+$ resonance structure -where the variations in $M(\Lambda^0 K^+)$, from Eq. 2, are bigger, the variations of Λ^0 polarization are bigger; see, and compare, Fig. 6 and Fig. 4 from (J. Félix *et al*) [3]. Maybe both polarizations, the Λ^0 polarization and the $\Lambda^0 K^+$ system polarization, could be related. In this case the origin of Λ^0 polarization -and in general of all baryons- could be parity

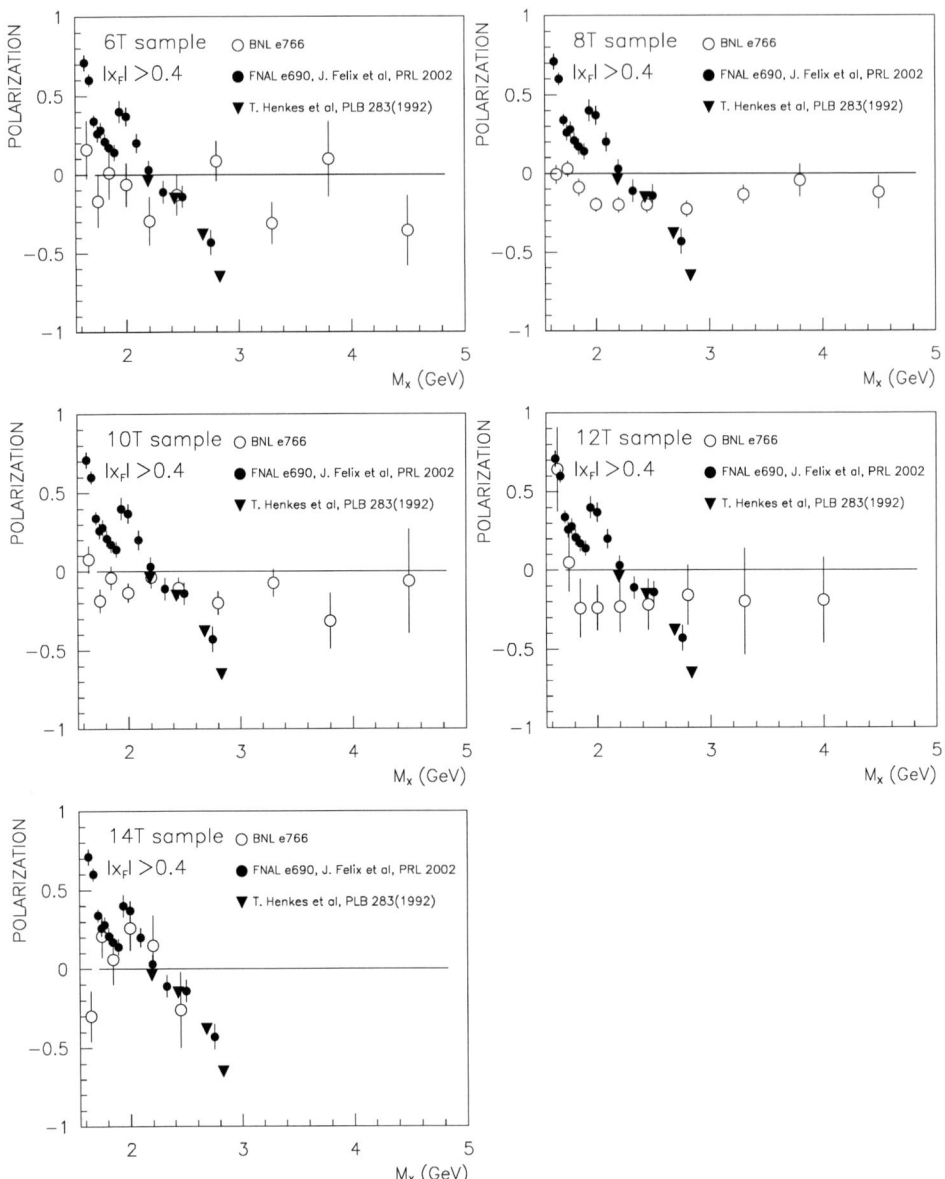

FIGURE 1. Λ^0 Polarization as function of the $M(\Lambda^0 K^+)$ invariant mass, at 27.5 GeV, for $N = 1,2,3,4,5$ according to Eq. 1. From Reference[15]. In each case Λ^0 polarization from Eq. 2, also as function of $\Lambda^0 K^+$, is shown; the Λ^0 polarization from (T. Henkes *et al*)[3] additionally is superimposed.

Λ^0 POLARIZATION AS FUNCTION OF TARGET DENSITY

The general characteristics of Λ^0 polarization from $p - Nucleus$ are very well known: It strongly depends on x_F and P_T; weakly, on target nature. Λ^0 polarization does not depend on the beam energy, but depends on the effective energy that generates Λ^0 and some associated particles -like the K^+-, for instance the diffractive mass of the system $\Lambda^0 K^+$.

The way Λ^0 depends on target nature was investigated to give that the important parameter is the target material density[17]. And that this dependence dilutes, in some sense, the Λ^0 polarization measured.

The above dependence is given by the following equation:

$$\mathcal{P}(x_F, P_T, \rho) = \frac{-\kappa_0}{\lambda_0 + \frac{\rho}{\rho_w}} x_F P_T, \qquad (9)$$

where $\frac{\rho}{\rho_w}$ is the material density divided by Tungsten's density, $\kappa_0 = 0.423 \pm 0.065 \ (GeV/c)^{-1}$, and $\lambda_0 = 1.191 \pm 0.200$, inside the kinematical region defined by $0.0 < x_F < 1.0$ and $0.0 < P_T < 1.2 \ GeV/c$ variables.

The above equation shows that the dependence of Λ^0 polarization on target nature is very well described by the target density: The absolute value of Λ^0 polarization decreases as the density of the target increases; this is compatible with that after Λ^0 is created inside the target, it goes through a multiple scattering processes reducing its polarization; the greater the target material density the more likely Λ^0 collides with some nucleus.

The dependence of Λ^0 polarization on target nature, through its density is a secondary effect that models on Λ^0 polarization must take into account. This is for that the measured Λ^0 polarization is smaller than the polarization that Λ^0 has when it is produced.

CONCLUSIONS

From the above described Λ^0 polarization measurements -inclusive and exclusive-, in different final states, in different beam energies, in different beams and targets, it is followed these conclusions:

Λ^0 polarization is a function of x_F, P_T, and the diffractive mass of $\Lambda^0 K^+$ system. This functionality is not unique; it seems that it depends on the production mechanism.

For exclusive $pp \to p\Lambda^0 K^+(\pi^+\pi^-)^N$, with $N = 1,2,3,4$ at 27.5 GeV and for inclusive p-nucleus reactions, Λ^0 polarization is described very well, inside statistics, by $\mathcal{P}(x_F, P_T) = a x_F P_T$, with a a constant, negative, that depends on the target nature; a negative implies that Λ^0 polarization from pp reactions is opposite to the normal of the production plane; that Λ^0 polarization from inclusive and from exclusive reactions are, inside statistics, the same means that Λ^0 polarization is independent if the rest of the particles in the final state reaction is observed or not and that Λ^0 polarization is independent of the number of pairs of pions, $(\pi^+\pi^-)^N - N = 1,2,3,4-$, that the exclusive pp reaction has in the final state. The dependence of Λ^0 polarization on the target nature is via the density of the target, this dependence is described very well by $\mathscr{P}(x_F, P_T, \rho) = \frac{-\kappa_0}{\lambda_0 + \frac{\rho}{\rho_w}} x_F P_T$, where $\frac{\rho}{\rho_w}$ is the material density divided by Tungsten's density, $\kappa_0 = 0.423 \pm 0.065 \ (GeV/c)^{-1}$, and $\lambda_0 = 1.191 \pm 0.200$, inside the kinematical region defined by $0.0 < x_F < 1.0$ and $0.0 < P_T < 1.2 \ GeV/c$ variables. This means that the produced Λ^0 polarization -at the $p - nucleus$ reaction- is bigger than the Λ^0 polarization measured outside the target.

All models to explain Λ^0 polarization must take this fact into account.

The dependence that Λ^0 polarization has, in exclusive $pp \to p\Lambda^0 K^+(\pi^+\pi^-)^N$, with $N = 1,2,3,4$ at 27.5 GeV, is very different from that the Λ^0 polarization has from exclusive $pp \to p\Lambda^0 K^+(\pi^+\pi^-)^N$, with $N = 5$ at 27.5 GeV and from $pp \to p\Lambda^0 K^+$, at 800 GeV; it seems that the production mechanism are different and plays a fundamental role in Λ^0 polarization, due to this, the Λ^0 polarization is different -and its dependence on x_F, P_T, and diffractive mass of $\Lambda^0 K^+$ is different- in each case. In this last case, Λ^0 polarization is very big and positive at P_T close to zero and decreases to big negative values at approximately $P_T = 2.0 \ GeV$; as function of the diffractive mass of the object $\Lambda^0 K^+$, we have discovered that Λ^0 polarization is very big at the threshold of Λ^0 mass, it decreases to big negative values, having zero value at approximately $(\Lambda^0 K^+) = 2.2 \ GeV$, at approximately $(\Lambda^0 K^+) = 3.5 \ GeV$, in the exclusive reaction $pp \to p\Lambda^0 K^+$ at 800 GeV and in $pp \to p\Lambda^0 K^+(\pi^+\pi^-)^5$ at 27.5 GeV, even thought this dependence is no exactly the same.

For exclusive $pp \to p\Lambda^0 K^+(\pi^+\pi^-)^N$, with $N = 1,2,3,4$ at 27.5 GeV, Λ^0 polarization depends on the diffractive mass of the $\Lambda^0 K^+$ system and on the diffractive mass of $(\Lambda^0 K^+(\pi^+\pi^-)^N, N = 1,2,3,4)$, differing from that from $pp \to p\Lambda^0 K^+$ at 800 GeV and from $pp \to p\Lambda^0 K^+(\pi^+\pi^-)^5$ at 27.5 GeV.

The production mechanism of Λ^0 polarization is very important and plays a fundamental role in the Λ^0 polarization.

It appears that Λ^0 polarization does not depend on the *beam* energy but on the nature -quark content- of the beam, on the energy of the effective beam -this is, the beam that really produces Λ^0- and on the isospin channel through Λ^0 is produced.

To reach a complete knowledge about Λ^0 polarization many more studies and measurements must be performed in all possible experimental circumstances. For instance, a very important example is the measurement of Λ^0 polarization in exclusive $pp \to pp\Lambda^0\overline{\Lambda}^0$ at different energies and $pp \to p\Lambda^0 K_s^0 \pi^+$ at different energies. These studies are running at the Instituto de Física, Universidad de Guanajuato.

Additionally, to acquire a complete knowledge on *baryon* polarization all the members of 20-plet with an SU(3) octet baryons and all the members of 20-plet with an SU(3) decuplet baryons[6] must be searched for polarizations. To understand what is the origin of baryon polarization and what is the role that spin of particle plays during the creation of the particle it is mandatory to study all baryons for polarization.

In spite of all the above experimental efforts, it seems that the explanation of Λ^0 polarization is still faraway.

ACKNOWLEDGMENTS

Thanks to the BNL e766 and FNAL e690 collaborations (http://www-e690.fnal.gov/). Thanks to this volumen Editors for their kind invitation to write this paper. This research was supported in part by CONACYT, México, under grant No. 2002-C01-39941.

REFERENCES

1. J. Duryea et al, Phys. Rev. Lett. 67, 1193(1991). R. Rameika et al, Phys. Rev. **D33**, 3172(1986). C. Wilkinson et al, Phys. Rev. Lett. 58, 855(1987). B. Lundberg et al, Phys. Rev. **D40**, 39(1989). F. Lomanno et al, Phys. Rev. Lett. 43, 1905(1979). S. Erhan et al, Phys. Lett. **B82**, 301(1979). F. Abe et al, Phys. Rev. Lett. 50, 1102(1983). K. Raychaudhuri et al, Phys. Lett. **B90**, 319(1980). K. Heller et al, Phys. Rev. Lett. 41, 607(1978). F. Abe et al, J. of the Phys. S. of Japan. 52, 4107(1983). P. Aahlin et al, Lettere al Nuovo Cimento 21, 236(1978). A. M. Smith et al, Phys. Lett. **B185**, 209(1987). V. Blobel et al, Nuclear Physics **B122**, 429(1977). G. Zapalac et al, Phys. Rev. Lett. 57, 1526(1986). A. Morelos et al, Phys. Rev. Lett. 71 2172(1993). Y. W. Wah et al, Phys. Rev. Lett. 55 2551(1985). L. Deck et al, Phys. Rev. **D28**, 1(1983). E. C. Dukes et al, Phys. Lett. **B193** 135(1987). K. B. Luk et al, Phys. Rev. Lett. 70, 900(1993).
2. K. Heller et al, Phys. Lett. **B68**, 480(1977). G. Bunce et al, Phys. Lett. **B86**, 386(1979).
3. T. Henkes et al, Phys. Lett. B283, 155(1992). J. Félix et al, Phys. Rev. Lett. 88, 061801-4(2002).
4. T. A. DeGrand et al, Phys. Rev. **D24**, 2419(1981). B. Andersson et al, Phys. Lett. **B85**, 417(1979). J. Szweed et al, Phys. Lett. **B105**, 403(1981). K. J. M. Moriarty et al, Lett. Nuovo Cimento 17, 366(1976). S. M. Troshin and N. E. Tyurin, Sov. J. Nucl. Phys. 38, 4(1983). J. Soffer and N.E. Törnqvist, Phys. Rev. Lett. 68, 907(1992). Y. Hama and T. Kodama, Phys. Rev. **D48**, 3116(1993). R. Barni et al, Phys. Lett. **B296**, 251(1992). W. G. D. Dharmaratna and G. R. Goldstein, Phys. Rev. **D53**, 1073(1996). W. G. D. Dharmaratna and G. R. Goldstein, Phys. Rev. **D41**, 1731(1990). S. M. Troshin and N. E. Tyurin, Phys. Rev. **D55**, 1265(1997). L. Zuo-Tang and C. Boros, Phys. Rev. Lett. 79, 3608(1997).
5. J. Félix, Mod. Phys. Lett. **A14**, 827(1999).
6. J. Uribe et al, Phys. Rev. D49, 4373(1994). E. P. Hartouni et al, Nucl. Inst. Meth. A317, 161(1992). E. P. Hartouni et al, Phys. Rev. Lett. 72, 1322(1994). E. E. Gottschalk et al, Phys. Rev. D53, 4756(1996). D. C. Christian et al, Nucl. Instr. and Meth. A345, 62(1994). B. C. Knapp and W. Sippach, IEEE Trans. on Nucl. Sci. NS 27, 578(1980). E. P. Hartouni et al, IEEE Trans. on Nucl. Sci. NS 36, 1480(1989). B. C. Knapp, Nucl. Instrum. Methods **A289**, 561(1980).
7. S. Eidelman et al, Phys. Lett. **B592**, 1(2004).
8. G. Bunce et al, Phys. Rev. Lett. 36, 1113(1976).
9. J. Félix, Mod. Phys. Lett. **A12**, 363(1997).
10. J. Félix et al, Phys. Rev. Lett. 76, 22(1996).
11. J. Félix et al, Phys. Rev. Lett. 82, 5213(1999).
12. L. Montanet et al, Phys. Rev. D **50**, 1173(1994).
13. J. Félix, Rev. Mex. de Fís. 45(4) 421(1999).
14. J. Félix, Mod. Phys. Lett. **A16**, 1741(2001).
15. J. Félix et al, Λ^0 Polarization in $pp \to p\Lambda^0 K^+(\pi^+\pi^-)^5$ at 27.5 GeV. SPIN2004, proceedings. The 16th International Spin Physics Symposium. Trieste, IT. 10-16 October 2004.
16. J. Félix et al, Λ^0 Polarization in $pp \to p\Lambda^0 K^+(\pi^+\pi^-)^5$ at 27.5 GeV. HEPP-EPS 2005, proceedings. International Europhysics Conference on High Energy Physics. Lisboa, Portugal July 21-27, 2005.
17. V. M. Castillo, V. Gupta, J. Félix, Int. J. of Mod. Phys. **A20**, 2047(2005).
18. E. J. Ramberg et al, Phys. Lett. B 338, 403(1994).
19. S. A. Gourlay et al, Phys. Rev. Lett. 56, 2244(1986).
20. J. Félix, Ph.D. Thesis, Universidad de Guanajuato-Universidad de Massachusetts, México (1994).

Towards the International Linear Collider

Ricardo López-Fernández

Departamento de Física, CINVESTAV, AP 14-740, 07000 México DF, Mexico

Abstract. The broad physics potential of e^+e^- linear colliders was recognized by the high energy physics community right after the end of LEP in 2000. In 2007, the Large Hadron Collider (LHC) now under construction at CERN will obtain its first collisions. The LHC, colliding protons with protons at 14 TeV, will discover a standard model Higgs boson over the full potential mass range, and should be sensitive to new physics into the several TeV range. The program for the Linear Collider (LC) will be set in the context of the discoveries made at the LHC. All the proposals for a Linear Collider will extend the discoveries and provide a wealth of measurements that are essential for giving deeper understanding of their meaning, and pointing the way to further evolution of particle physics in the future. For the mexican groups is the right time to join such an effort.

Keywords: Linear Collider, Higgs, Supersymmetry, Top, Electroweak Physics

FOREWORD

I had the fortune to work at the Large Electron-Positron Collider LEP. At the very end of the LEP operations there was an euphorical activity around all the reachable physics subjects. The four detectors were hungry of luminosity and each increase of the centre of mass enery was a new hope for discovery. Now is the turn of the Large Hadron Collider (LHC). There is already an historical example of how well a lepton collider (LEP) can improve the knowledge on a discovery made by a hadron collider (SPS). From all the physics perspectives and expectations coming from a new electron-positron collider it was found a long consensus about the necessity of this next step and some the facility, machine and detectors general requirements.

PHYSICS WITH e^+e^- COLLIDERS

In this section I just remark the main subjects already stated in the review by E. Accomando *et al.* [1]. This review was published three years before the end of LEP operations and shows the strong case since then, from the scientific point of view.

Two strategies can be followed in future experiments to explore the area beyond the Standard Model and to reveal the signals of new physical phenomena. First, the properties of the particles and forces in the Standard Model may be affected by new energy scales. Precision studies of the top quark and the electroweak gauge bosons can thus reveal clues to the physics beyond the Standard Model. Second, if the machine energies are high enough to cross the relevant thresholds, new phenomena can be searched for directly and studied thoroughly. This is of course the prime raison d'etre for any new accelerator. While the presently operating collider facilities, the e^+e^- collider LEP2, the ep collider HERA and the pp collider Tevatron, cover the energy range up to a scale of 200 to 300 GeV, the pp collider LHC and e^+e^- linear colliders will enable us to explore the energy range up to the TeV scale.

While new vector bosons and particles carrying color quantum numbers can be searched for very efficiently at the hadron collider LHC, e^+e^- colliders provide in many ways unique opportunities to discover and explore the non-colored particles. This is most obvious in supersymmetric theories. Combining LEP2 analyses with future searches at the LHC, the individual light and heavy Higgs bosons can be found only in part of the supersymmetry parameter space; even if all channels are combined, the coverage of the entire parameter space is guaranteed only if non-supersymmetric decay modes of the Higgs bosons prevail. Squarks and gluinos can be searched for very efficiently at the LHC. Yet, the detailed experimental study of their properties is very difficult at this machine. Likewise, cascade decays proceeding in several steps, will allow the search for other, non-colored supersymmetric particles, yet a general model independent analysis of gauginos/higgsinos and scalar sleptons can only be carried out at e^+e^- colliders with well-determined kinematics at the level of the individual subprocesses. They will allow to perform high-precision studies which are impossible or very difficult to carry out at hadron colliders. Only the detailed knowledge of all the properties of the colored and non-colored supersymmetric states, gathered both at the LHC and e^+e^- experiments, will finally enable us to reveal the structure of the underlying theory.

The Top Quark

The large mass renders the top quark a very interesting object, the properties of which should be studied with high precision. Being the leading particle in the fermion spectrum of the Standard Model, it likely plays a key role in any theory of flavor dynamics. Moreover, due to the large mass, its properties are most strongly affected by Higgs particles and nearby new physics scales. High-precision measurements of the properties of top quarks are therefore mandatory at any future collider.

Since the lifetime of the t quark is much shorter than the time scale Λ_{QCD}^{-1} the strong interactions, the impact of non-perturbative aspects on the production and decay of top quarks can be neglected to a high level of accuracy. The short lifetime provides a cut-off $k > \sqrt{2m_t\Gamma_t}$ for any soft non-perturbative and infrared perturbative interactions. The t quark sector can therefore be analyzed within perturbative Quantum Choromodynamics (QCD). Unlike light quarks, the properties of t quarks are reflected directly in the distributions of the decay jets and W bosons, and they are not affected by the obscuring confinement and fragmentation effects.

e^+e^- colliders are the most suitable instruments to study the properties of top quarks. Operating the machine at the tt threshold, the mass of the top quark can be determined with an accuracy that is an order of magnitude superior to measurements at hadron colliders. The static properties of top quarks, magnetic and electric dipole moments, can be measured very accurately in continuum top-pair production at high e^+e^- colliders. Likewise, the chirality of the charged top-bottom current can be measured accurately in the decay of the top quark. In extensions of the Standard Model, supersymmetric extensions for example, top decays into novel particles, charged Higgs bosons and/or stop/sbottom particles, may be observed.

QCD Physics

The annihilation of e^+e^- into hadrons provides a high-energy source of clean quark and gluon jets: $e^+e^- \to \gamma \to q\,q,\,q\,qg$ ... This has offered unrivaled opportunities for QCD tests at machines such as PETRA and LEP. The program will be continued at a linear collider, although separation from new backgrounds such as top and W/Z pair production will require more delicate analyses of multijet events. Conversely, the study of these other processes, as well as the new particle searches, require a good understanding of the annihilation events. Topics of interest for QCD per se include the study of multijet topologies, the energy increase of charged multiplicity, particle momentum spectra and their scaling violations, angular ordering effects, hadronization phenomenology (power corrections), and so on.

One of the key elements of quantum chromodynamics is asymptotic freedom [2], a consequence of the non-abelian nature of the color gauge symmetry. This fundamental aspect has been tested in many observables measured at e^+e^- colliders and other accelerators between a minimum Q^2 of order 4 GeV^2 up to 4×10^4 GeV^2, ranging from the τ lifetime to multi-jet distributions in Z decays. The range of Q^2 can be extended at e^+e^- linear colliders by as much as two orders of magnitude to a value $Q^2 \approx 4 \times 10^6$ GeV^2. The most sensitive observable in this energy range is the fraction of events with 2, 3, 4, ... jets in the final state of $e^+e^- \to$ hadrons results of the simulations can be nicely illustrated by presenting the evolution of the three-jet fraction in the variable 1/log Q^2. Asymptotic freedom predicts this dependence to be linear, modified only slightly by higher order corrections. Based on the present theoretical accuracy of the perturbative jet calculations, the error with which the QCD coupling at $\sqrt{s} = 500$ GeV can be measured, is expected to be $\delta\alpha(s)(M_Z^2) \approx 0.005$ matching the error which can be expected from the analysis of the top excitation curve at threshold. If the theoretical analysis of the jet rates can be improved, the error on α_s can be reduced significantly.

$\gamma\gamma$ Events

$\gamma\gamma$ interactions provide a complementary way to study many aspects of new physics. The objective of a $\gamma\gamma$ physics program is to bring our understanding of the photon to the same level as HERA is achieving for the proton. Since the photon is the more complex of the two, as described below, this will offer new insights in QCD [3]. Linear e^+e^- colliders offer three sources of photons: (i) bremsstrahlung [4], (ii) beamstrahlung [5] and (iii) potentially, from laser backscattering [6]. The bremsstrahlung source provides a spectrum of different photon energies and virtualities, but distributions are peaked at the lower end so that the more interesting studies at higher $\gamma\gamma$ energies are limited by statistics. Since beamstrahlung is a drawback for the normal e^+e^- physics program, current machine designs attempt to reduce the beamstrahlung energy to a minimum, so that it may not be interesting for $\gamma\gamma$ physics.

The nature of the photon is complex. A photon can fluctuate into a virtual $q\bar{q}$ pair. The low-end part of the spectrum of virtualities is in a non-perturbative regime, where the Vector Meson Dominance (VMD) model can be used to approximate the photon properties by those of mesons with the same quantum numbers as the photon mainly the ρ^0. In the direct interactions the full photon energy is used to produce (high-p_T) jets, whereas the resolved photon leaves behind a beam remnant that does not participate in the primary interaction.

Electroweak Gauge Bosons

The fundamental electroweak and strong forces appear to be of gauge theoretical origin. This is one of the outstanding theoretical and experimental results in the past three decades. While the non-abelian symmetry of QCD, manifest in the self-coupling of the gluons, has been successfully demonstrated in the distribution of hadronic jets in Z decays, only indirect evidence has been accumulated so far for the electroweak $W^\pm$, Z, γ sector, based on loop corrections to electroweak low-energy parameters and Z observables. The direct evidence from recent Tevatron and LEP2 analyses is still feeble. Deviations from the prescriptions of gauge symmetry manifest themselves in the cross sections, destroying fine-tuned unitarity cancelations [7] at high energies. Since the deviations of the static parameters from the SM values are expected to be of order $[M_W/\Lambda]^j$, Λ denoting the energy scale at which the Standard Model breaks down, only the very high energies at the LHC and linear colliders will allow stringent direct tests of the self-couplings of the electroweak gauge bosons. The gauge symmetries of the Standard Model determine the form and the strength of the self-interactions of the electroweak bosons: the triple couplings WWγ, WWZ and the quartic couplings. Deviations from the form and the strength of these vertices predicted by the gauge symmetry, as well as novel couplings like ZZZ and ZZZZ in addition to the canonical SM couplings, could however be expected in more general scenarios, in models with composite W, Z bosons, for instance. Other examples are provided by models in which the W, Z bosons are generated dynamically or interact strongly with each other.

Pair production of W bosons in e^+e^- collisions, $e^+e^- \to W^+W^-$ is the best-suited process to study the electroweak gauge symmetries. The high efficiency for reconstructing W, Z bosons from hadronic and leptonic decays in the clean environment of e^+e^- collisions makes a 500 GeV collider superior to the LHC. Deviations from the predictions of the Standard Model for the total cross section [8], would signal non-standard self-couplings of the electroweak gauge bosons.

The Higgs Mechanism

The Higgs mechanism is the third building block in the electroweak sector of the Standard Model. The fundamental particles, leptons, quarks and weak gauge bosons, acquire masses through the interaction with a scalar field of non-zero field strength in the ground state[9].

To accommodate the well-established electromagnetic and weak phenomena, the Higgs mechanism requires the existence of at least one weak iso doublet scalar field. After absorbing three Goldstone modes to build up the longitudinal polarization states of the $W^\pm/Z$ bosons, one degree of freedom is left over, corresponding to a real scalar particle. The discovery of this Higgs boson and the verification of its characteristic properties is crucial for the theory of the electroweak interactions. The physical implications reach far beyond the canonical formulation of the Standard Model.

The only unknown parameter in the Higgs sector of the Standard Model is the mass of the Higgs particle. Stringent constraints however can be derived from the scale Λ up to which the Standard Model is assumed to be valid before the gauge and Higgs particles become strongly interacting and new physics phenomena may emerge[10]. The strength of the Higgs self-interaction is determined by the Higgs mass itself at the scale $v = 246$ GeV, the value of the Higgs field in the ground state which characterizes the spontaneous breaking of the electroweak gauge symmetries. Increasing the energy scale, the quartic self-coupling of the Higgs field increases logarithmically, in a similar way to the electromagnetic coupling in QED. If the Higgs mass is small, the energy cut-off Λ at which the coupling grows beyond any given bound, is large; conversely, if the Higgs mass is large, the cut-off Λ is small. The condition $M_H < \Lambda$ sets an upper limit on the Higgs mass in the Standard Model. It has been shown in lattice analyses, which account properly for the onset of the strong interactions in the Higgs sector, that this condition leads to an estimate of about 700 GeV for the upper limit on M_H[11]. [These analyses are based on the orthodox Φ^4 formulation of the Standard Model. Therefore, they do not exclude higher values for Higgs masses in any extension of the Standard Model.]

The SM Higgs boson can be discovered at the LHC in the region above LEP2, including a firm overlap of the two machines, up to the canonical upper limit of $M_H \approx 800$ GeV. In the theoretically preferred intermediate mass range below the ZZ decay threshold, the experimental search is difficult.

A large variety of channels can be exploited to search for Higgs particles in the Higgs-strahlung [12] and fusion processes [13] at e^+e^- colliders. The signature is very clear and the background almost negligible so that the properties of the Higgs boson can easily be reconstructed, in particular in the preferred intermediate mass region. In the Higgs-strahlung process $e^+e^- \to ZH$, recoil-mass techniques can be used in final states with leptonic Z decays, or the Higgs particle may be reconstructed in $H \to b\bar{b}, WW$ directly. The WW fusion process $e^+e^- \to e\nu e\nu H$ requires the reconstruction of the Higgs particle.

Once the Higgs boson is found, it will be very important to explore the properties which reveal the physical nature of the particle. The zero-spin of the Higgs particle is reflected in the angular distribution of the Higgs-strahlung process which asymptotically must approach the $sin^2\theta$ law. Of paramount importance is the measurement of the couplings to gauge bosons and matter particles. The strength of the couplings to Z and W bosons is reflected in the size of the e^+e^- production cross sections. The strength of the couplings to fermions can be measured through the decay branching ratios and Higgs bremsstrahlung of top quarks. These measurements are important instrumentaria to establish the Higgs mechanism experimentally. Finally, the Higgs potential itself, which provides the physical basis of the Higgs phenomenon, must be reconstructed by measuring the triple and quartic Higgs self-couplings [14]. This appears possible only by exploiting multi-Higgs production in the fusion mechanism at TeV energies and maximum possible luminosity.

To sum up, we conclude from the preceding discussion that e^+e^- linear colliders with energies in the range of 300 to 500 GeV are the ideal instruments to search for Higgs particles in the intermediate mass range, a priori the theoretically most attractive range, and to establish its characteristic properties experimentally. In the high energy phase of the colliders, important parameters of the Higgs potential can be reconstructed which are necessary for generating the spontaneous breaking of the electroweak symmetries.

Beyond the Standard Model (BSM) Physics

Supersymmetry

Even though no direct experimental evidence has emerged yet for the existence of supersymmetry[15] in Nature, the concept has so many attractive features that it may be considered as a prime target of present and future experimental particle research. Arguments in favor of supersymmetry are deeply rooted in particle physics. Supersymmetry unifies matter and forces, and if realized lo cally, it plays a crucial role in a quantum theory of gravity. In relating particles of diferent spins to each other, fermions and bosons, low-energy supersymmetry stabilizes the masses of fundamental Higgs scalars in the context of very high energy scales associated with grand unification[?]. Besides solving this part of the hierarchy problem, supersymmetry may even be closely related to the physical origin of the Higgs mechanism itself : In supergravity inspired realizations of supersymmetric theories[17] , incorporating universal scalar masses at the scale of the grand unification, one of the scalar masses squared can evolve down to negative values and thus induce spontaneous symmetry breaking in the electroweak sector. This is possible if the top mass has a value between 150 and 200 GeV; all other masses squared of squarks and sleptons remain positive so that $U(1)_{EM}$ and $SU(3)_C$ are unbroken.

It has been established that at least one of the Higgs particles in the SUSY spectrum could be discovered at the LHC. It has been assumed, however, in these analyses that the Higgs particles decay only into SM particles. The general consequences of decays to SUSY particles, partly invisible, have not yet been studied experimentally. A problem arises from the dificulties of establishing the heavy neutral Higgs bosons, above the top threshold, in the interesting parameter range of small to moderate $\tan\beta$. This is a problem for most of the Higgs mass estimates in supergravity inspired parameterizations. The detection of charged Higgs bosons is guaranteed, at this point, only in top decays, restricting the mass range accessible for this particle to rather low values. Thus, there are large areas in the SUSY Higgs parameter space where the ensemble of individual Higgs particles are not accessible at the same time.

It is evident from the table of masses derived for various SGUT scenarios that a large number of particles can be expected which are accessible in the first phase of the e^+e^- linear colliders. In particular, the color-neutral charginos/neutralinos and sleptons, lighter than the colored squarks and gluinos, can be produced in e^+e^- collisions and studied thoroughly in the clean environment of these machines. Moreover, stop particles could be light as well, partly a result of mixing effects induced by the large Yukawa coupling between L and R states in this sector. The experiments at e^+e^- colliders will not only allow high-precision measurements of masses and couplings, but also of such subtle effects as mixings.

This experimental scenario at the LHC is easy to compare with the scenario at e^+e^- colliders. While the discovery limits at the LHC can be extended to values above 1 TeV, the experiments at the e^+e^- machines provide high-precision information on all the supersymmetric particles which are kinematically accessible, in the high energy range for masses up to ≈ 1 TeV. In particular, the analysis of the color-neutral states, charginos/neutralinos and sleptons which are generally (much) lighter than the colored states, can be carried out with high accuracy, independently of specific assumptions and solely based on kinematics. Masses, for instance, can be measured by exploiting threshold effects or simple decay kinematics. Thus, a systematic and high-precision study can be performed which will resolve the complexities of the supersymmetric phenomena.

Compositeness

In supersymmetric extensions of the Standard Model, the fundamental particles are pointlike down to distances close to the Planck length. However, if light Higgs particles do not exist, the electroweak bosons become strongly interacting at energies of order 1 TeV. As one among other physical scenarios, this could be interpreted as a signal of composite substructures of these particles at a scale of 10^{-17} cm. Moreover, the proliferation of quarks and leptons could be taken as evidence for possible substructures of the matter particles[18]. In this picture, masses and mixing angles are a consequence of the interactions between a small number of elementary constituents in perfect analogy to the quark/gluon picture of hadrons. No theoretical formalism has been set up so far which would reconcile, in a satisfactory manner, the small masses in the Standard Model with the tiny radii of these particles which imply very large kinetic energies of these constituents. However, the lack of theoretical formalism does not invalidate the physical picture or its motivation.

THE LINEAR COLLIDER FACILITY

Research and development on the linear collider has been conducted continuously for over a decade, leading to very well-understood proposals. The linear collider complex will contain two accelerators, one for electrons and one for positrons, bringing beams into head-on collisions at the location of a large particle detector. To achieve the necessary collision energy, the overall complex will need to be about 30 km long. The challenges in the project are exemplified by the transverse beam sizes at collision (a few nanometers), by the requirement of providing very high electric field gradients to achieve the large energies, and by the exceptional control of the beams needed during the acceleration process. The baseline collision energy dictated by the physics program should be 500 GeV, with the capability to lower the energy to about 90 GeV for some measurements and calibrations. Several physics studies demand that the energy be adjustable so as to scan across particle production thresholds. The machine luminosity dictates the collision rate; to meet the physics goals it should be more than $10^{34} \text{cm}^{-2}\text{s}^{-1}$ at 500 GeV. The ability to distinguish many interesting processes from each other, and from backgrounds due to known reactions, is enhanced by providing electrons whose spins are aligned along their direction in a polarized beam. The baseline electron beam polarization should be 80% or greater. This should be achievable.

It is almost certain that the new discoveries made during the initial operation at 500 GeV will lead to the need for subsequent measurements at higher energy, so the ability to upgrade the energy to around 1000 GeV is essential. Two basic technologies exist today that could be chosen for building a linear collider: the TESLA design[19] pioneered at DESY in Hamburg Germany and the JLC/NLC design that emerged from a joint R&D program between SLAC in Stanford California and KEK in Tsukuba Japan. The main difference between the two technologies lies in the frequency of the electric fields used for acceleration. The lower frequency TESLA design employs superconducting radio frequency cavities, while the higher frequency JLC[20] and NLC[21] designs use room temperature accelerating structures. A comparison of the technical parameters, remaining R&D and risks has been made by a Technical Review Committee formed by the International Committee on Future Accelerators (ICFA) in 2001[22]. Both technologies were judged to be viable, and the costs should be comparable. The choices of technology and site will be addressed over the coming year or so. Meanwhile R&D continues on methods to use a low-energy high-intensity electron beam as the accelerating power source that may allow an even higher energy collider in the future. The new information gained from the initial LC operation in its baseline configuration, and the results from the LHC and current accelerators, will influence the need for an extended capability of the LC facility. The options discussed in the following three paragraphs will have a different priority depending on what the initial investigations reveal, but the potential for adding them should be retained in the machine design. The addition of positron beam polarization may be needed to perform precision studies of Z bosons, especially if the electroweak symmetry is found to be broken in a non-supersymmetric way. Positron polarization also offers a valuable tool for disentangling supersymmetric states, and can be used to enhance the rates for rare processes such as two Higgs boson production. Although the initial operation of the LC will bring electrons and positrons into collision, there are possible scenarios for which considerable benefit would derive from studying collisions of two photons, an electron and photon, or two electrons. Reasonably intense and monochromatic beams of polarized photons could be formed by backscattering bright laser beams from accelerated electrons. The use of $\gamma\gamma$ collisions with tunable energy and polarization can give unique opportunities. For example, in supersymmetry, the two heaviest Higgs bosons should be nearly degenerate in mass, and overlapped when produced in e^+e^- collisions. Choosing the appropriate photon beam polarization states, one can selectively produce one or the other. Use of $\gamma\gamma$ collisions gives a greater discovery reach for heavy Higgs bosons than is possible in e^+e^- collisions. Science has always demanded independent confirmations of new discoveries, and this will continue to be true for the physics to be explored at the LC. Besides the general advantage of having independent crosschecks of new results, providing two collision points could allow the optimization of two detectors for different studies. Two-detector operation can improve the LC efficiency, since one detector could take data while maintenance is done on the other. Finally, there are sociological reasons for two detectors. Spreading the people who work on the linear collider over two experiments allows more workable collaborations in which the inventiveness of young physicists can flower more effectively, thus providing better opportunities for developing the pool of talented scientists that is a major societal benefit of the linear collider program. For all of these reasons it is desirable to plan for two interaction regions into which two well-optimized detectors can be placed.

THE DETECTOR(S)

The detectors at the ILC have to have capability of efficient identification and precise measurement of four-momenta of the fundamental particles. In order to satisfy these requirements, the detector must have the following performances:

- good jet-energy resolution to separate W and Z in their hadronic decay mode,
- efficient jet-flavor identification capability,
- excellent charged-particle momentum resolution, and
- hermetic coverage which gives high veto efficiency against 2-photon background.

General purpose detectors for ILC experiments will be composed of a vertex detector, a central tracker, an intermediate tracker (if necessary), a calorimeter system, a solenoid coil, an iron flux return yoke interleaved with muon detector, and forward (small angle) calorimeters. The world-wide consensus of the performance goal for the detector system corresponding to the items listed above are[23];

- jet energy resolution of $\delta E_j/E_j = 30\%/\sqrt{E_j(GeV)}$
- impact parameter resolution of $\delta_b \leq 5 \frac{10}{p\beta sin^{3/2}\Theta}$ for jet flavor tagging,
- transverse momentum resolution of $\delta p_t/p_t^2 \leq 5 \times 10^{-5} (GeV/c)^{-1}$ for charged track at high momentum limit, and
- hermeticity down to 5 mrad from the beam line.

MEXICAN PERSPECTIVE

The participation of mexican physicist in Experimental High Energy Physics (EHEP) is resumed in this volume. One can extract several conclusions but one clear is that EHEP in Mexico is reaching maturity. There are mexican groups in several and diverse Collaborations (Fixed Target Experiments, Tevatron, HERA, LHC), and in most of them their participation has played, or will have a relevant role. This, together with the old tradition in Theoretical High Energy Physics could trigger a very productive participation in future experiments. The International Linear Collider is the big chance and we have to be prepared for it and start to be involved from now, from the very realisation.

ACKNOWLEDGMENTS

During my academic formation I have collaborated with most of the EHEP mexican groups: SELEX, H1, D0 and ALICE. I would like to thank all of them.

REFERENCES

1. E. Accomando *et al. Phys.Rept.***299**(1998)1-78.
2. D. Gross and F. Wilczek, *Phys.Rev.Lett.* **30** (1973) 1343; H.D. Politzer, *Phys.Rev.Lett.* **30** (1973) 1346.
3. Proceedings of the Workshop on Gamma-Gamma Colliders, Berkeley 1994 [*Nucl.Instr. Meth.* **A355** (1995) 19].
4. C.F. von Weizsäcker, *Z. Phys.* **88** (1934) 612; E.J. Williams, *Phys. Rev.* **45** (1934)729.
5. D. Schulte, in: DESY-ECFA Conceptual LC Design Report (1997).
6. I.F. Ginzburg, G.L. Kotkin, S.L. Panfil, V.G. Serbo and V.I. Telnov, *Nucl.Instr.Meth.* **219** (1984) 5.
7. C.H. Llewellyn Smith, *Phys.Lett.* **B46** (1973) 233.
8. W. Alles, C. Boyer and A. Buras, *Nucl.Phys.* **B119** (1977) 125.
9. P.W. Higgs, *Phys.Rev.Lett.* **12** (1964) 132; *Phys.Rev.* **145** (1966) 1156; F. Englert and R. Brout, *Phys.Rev.Lett.* **13** (1964) 321; G.S. Guralnik, C.R. Hagen and T.W. Kibble, *Phys.Rev.Lett.* **13** (1964) 585.
10. N. Cabibbo, L. Maiani, G. Parisi and R. Petronzio, *Nucl.Phys.* **B158** (1979) 295; G. Altarelli and G. Isidori,*Phys.Lett.* **B337** (1994) 141.
11. A. Hasenfratz, T. Neuhaus, K. Jansen, H. Yoneyama and C.B. Lang, *Phys.Lett.* **B199** (1987) 531; M. Lüscher and P. Weisz, *Phys.Lett.* **B212** (1988) 472.
12. J. Ellis, M.K. Gaillard and D.V. Nanopoulos, *Nucl.Phys.* **B106** (1976) 292; B.W. Lee, C. Quigg and H.B. Thacker, *Phys.Rev.* **D16** (1977) 1519.
13. D.R.T. Jones and S.T. Petcov, *Phys.Lett.* **B84** (1979) 440.
14. G.J. Gounaris, F. Renard and D. Schildknecht,*Phys.Lett.* **B83** (1979) 191; V. Barger, T. Han and R.J.N. Phillips,*Phys.Rev.* **D38** (1988) 2766.
15. J. Wess and B. Zumino, *Nucl.Phys.* **B70** (1974) 39.
16. E. Witten, *Nucl.Phys.* **B188** (1981) 513; J. Polchinski and L. Susskind, *Phys. Rev.* **D26** (1982) 3661; S. Dimopoulos and H. Georgi, *Nucl. Phys.* **B193** (1981) 150; N. Sakai, *Z. Phys.* **C11** (1981) 153.
17. J.-P. Derendinger, L.E. Ibanez and H.P. Nilles, *Phys.Lett.* **B155** (1985) 65; M. Dine, R. Rohm, N. Seiberg and E. Witten, *Phys.Lett.* **B156** (1985) 55.
18. H. Harari, Proceedings, 1984 Scott. Summer School (St. Andrews); M.E. Peskin, Proceedings, Int. Symp. on Lepton and Photon Interactions at High Energies, eds. M. Konuma and K. Takahashi, (Kyoto 1985).
19. TESLA Technical Design Report (March 2001), http://tesla.desy.de/new_pages/TDR_CD/start.html
20. "JLC Project: Linear Collider for TeV Physics" (February 2003), http://lcdev.kek.jp/RMdraft/
21. For information on the NLC, see http://www-project.slac.stanford.edu/lc/nlc.html
22. Report of the Linear Collider Technical Review Committee report (February 2003) http://www.slac.stanford.edu/xorg/ilc-trc/2002/2002/report/03rep.htm
23. J. Brau, et al., "Linear Collider Detector R&D", Proceedings of LCWS2002 at Jeju, Korea, p.787.

Ring Imaging Cherenkov Detectors

Jürgen Engelfried

Instituto de Física, Universidad Autónoma de San Luis Potosí, Mexico

Abstract. Ring Imaging Cherenkov Counter have become a standard detector in High Energy and Nuclear Physics in the last 15 years. We review the basic physics of these detectors, give a short historical overview on their development, and describe the involvement of Mexican groups in experiments using this type of detectors.

Introduction

The basic idea of determining the velocity of charged particles via measuring the Cherenkov angle was proposed in 1960 [1]. A first prototype was successfully operated [2] in 1977. But only during the last decade Ring Imaging Cherenkov (RICH) Detectors were successfully used in experiments. A very useful collection of review articles and detailed descriptions can be found in the proceedings of five international workshops on this type of detectors, which were held in 1993 (Bari, Italy) [3], 1995 (Uppsala, Sweden) [4], 1998 (Ein Gedi, Israel) [5], 2002 (Pylos, Greece) [6], and 2004 (Playa del Carmen, Mexico) [7], respectively.

A charged particle with a velocity v larger than the velocity of light in a medium emits light [8]. The threshold velocity v_{thres} is given by

$$\beta_{\text{thres}} = \frac{v_{\text{thres}}}{c} \geq \frac{1}{n}, \quad \text{or} \quad \gamma_{\text{thres}} = \frac{n}{\sqrt{n^2 - 1}} \tag{1}$$

n being the (wavelength dependent) refractive index of the medium. The angle of emission (relative to the particle trajectory) is given by [8]

$$\cos\theta_c = \frac{1}{\beta n} = \frac{1}{\frac{v}{c} n} \tag{2}$$

with a maximum angle of $\theta_c^{\max} = \arccos 1/n$. The number of photons (N) emitted by Cherenkov radiation (differentially per unit of photon energy E or wavelength λ, and path length in the medium l) is [9]

$$\frac{d^2 N}{dEdl} = \frac{\alpha z^2}{\hbar c}\left(1 - \frac{1}{(\beta n)^2}\right) = \frac{\alpha z^2}{\hbar c} \sin^2\theta_c \tag{3}$$

$$\frac{d^2 N}{d\lambda dl} = \frac{2\pi\alpha z^2}{\lambda^2} \sin^2\theta_c \tag{4}$$

Here z is the electric charge (in units of e).

By measuring the Cherenkov angle θ_c one can determine the velocity of the particle, which will, together with the momentum p obtained via a magnetic spectrometer, lead to the determination of the mass and therefore to the identification of the particle.

Neglecting multiple scattering and energy loss in the medium, all Cherenkov light (in one plane) is parallel, and can therefore be focused (for small θ_c) with a spherical mirror (radius R) onto a point. Since the emission is symmetrical in the azimuthal angle around the particle trajectory, this leads to a ring of radius r in the focus, which is itself a sphere with radius $R/2$. The radius r is given by

$$r = \frac{R}{2}\tan\theta_c \approx \frac{R}{2}\sqrt{2 - \frac{2}{n}\sqrt{1 + \frac{m^2 c^2}{p^2}}} \tag{5}$$

For media where the Cherenkov angle is not "small" some other surface form (elliptical) of the mirror is necessary. For high refractive index media the path length can be very short, and no focusing elements are necessary; this is called *proximity focusing*[1].

A Short History of RICHes

After the first successful laboratory tests in the 1970's [2], at the beginning of the 1980's the first generation of Ring Imaging Cherenkov Detectors were built, with mixed results. Examples are the CERN Omega RICH, used by WA69 and WA82 [10], the UA2 RICH at CERN, and the E653 RICH at Fermilab.

A second generation RICH detectors were developed and employed at the end of the 80's and beginning of the 90's, with the positive and negative experiences from the first generation incorporated. Examples are an upgraded Omega RICH, used by WA89 and WA94 [10, 16], Delphi, CERES (all at CERN), and the SLD–GRID at SLAC. All these detectors had significant startup problems to overcome, but eventually they all worked very well and contributed to the physics analysis of the experiments.

The third generation, build and used mid-end 90's, finally worked without too many problems from the beginning and all of them played an important role in the physics analysis. Examples are the SELEX RICH at Fermilab, and Hermes and Hera-B at DESY.

Since the mid-90's, the RICHes are an established detector type, and are currently employed in recent experiments: BaBar–DIRC (SLAC), PHENIX and STAR (Brookhaven-RHIC), CLEO–III (Cornell), COMPASS (CERN).

There is a long list of future experiments, who plan to or are already building RICHes: ALICE, LHCB (CERN), BTeV, CKM (Fermilab, both terminated before construction started), AMS (International Space Station), and many more. Also in other fields like nuclear physics (GSI) and in astroparticle physics experiments technology developed for RICHes is used.

---

[1] Historical node: In this mode Pavel Cherenkov actually observed Cherenkov radiation for the first time.

General Details for RICHes

Because the number of photons is $\propto \lambda^{-2}$ (eq. 4), most of the light is emitted in the VUV range. To fulfill equation 2, the refractive index has to be $n > 1$, so there will be no Cherenkov radiation in the x-ray region. Also it is very important to remember that n is a function of the wavelength ($n = n(\lambda)$, chromatic dispersion) and most materials have an absorption line in the VUV region, where $n \rightarrow \infty$. Since usually the wavelength of the emitted photon is not measured, this leads to a smearing of the measured ring radius, and one has to match carefully the wavelength ranges which one wishes to use: Lower wavelengths gives more photons, but larger chromatic dispersion.

A very useful formula is obtained by integrating eq. 4 over λ (or E), taking into account all efficiencies etc., obtaining a formula for the number of detected photons N_{ph} [2]:

$$N_{ph} = N_0 L \sin^2 \theta_c \qquad (6)$$

where N_0 is an overall performance measure (quality factor) of the detector, containing all the details (sensitive wavelength range, efficiencies), and L is the path length of the particle within the radiator. Typically around 10 to 15 photons (N_{ph}) are detected on a $\beta = 1$ ring.

The usual construction of a RICH detector is to use a radiator length of $L = R/2$, e.g. the path length is equal to the focal length; but any other configurations, like folding the light path with additional (flat) mirrors, or the before mentioned "proximity focusing" mode, are possible.

All the presented arguments only work for small θ_c, which is always fulfilled in gases, since $n \approx 1$. Should the particles not pass through the common center of curvature of mirror(s) and focal spheres, the ring gets deformed to an ellipse or, in more extreme cases, to a hyperbola. If the photon detector is able to resolve this, and the resolution is needed for the measurement, these deviations from a perfect circle have to be taken into account in determining the velocity β. In general this effect can be neglected, and all parallel particles (with the same β) will give the same ring in the focal surface, due to the fact that all emitted Cherenkov light is parallel. The position of the ring center is determined by the angle of the tracks, and not by their positions.

Applications for RICHes

The typical application for RICHes in high energy physics experiments is for the identification of particles. As mentioned before, the determination of the ring radius (r) leads to a measurement of the velocity v of the particle, which, if the momentum p is measured by other means (for example by deflection in a magnetic field) yields a determination of the rest mass of the particle.

The number of detected photons (N), together with the measured radius, gives a determination of the electric charge of the particle, as shown in equation 4. This is used in nuclear physics applications, were as in particle physics a charge of $z = 1$ is assumed.

A new old concept, call Velocity Spectrometer, was proposed recently [23, 27] to use the basic measurement of velocity in RICH detectors as a replacement or complement to a magnetic spectrometers. The RICH detector has to be operated close to threshold for the spectrometer to work.

Mexican groups have involvements in all three types of application described here.

Mexican Groups working on RICH detectors

In Mexico three groups are working currently on RICH detectors. In the *Universidad Nacional Autónoma de México* in Mexico City two groups, one at the *Instituto de Ciencias Nucleares* (ICN-UNAM) participating in STAR (Brookhaven) and ALICE (CERN); and one at the *Instituto de Física* (IF-UNAM) participating in AMS. In the *Universidad Autónoma de San Luis Potosí* (UASLP) the group is working in SELEX, CKM (both Fermilab) and on P326 (CERN).

In the following we give short descriptions for the different experimental applications with Mexican involvement.

The SELEX Phototube RICH Detector

Members of the UASLP group participated in the design, construction, and operation of the SELEX RICH detector [12, 13, 14] even before they joined UASLP. The SELEX [35] experiment studies production, spectroscopy and decay properties of charmed and non-charmed hadrons. In the last years the group worked mostly in performance measures of this detector [22], on improving the reconstruction software, and on applications of the particle identification to the physics analysis.

The AMS Proximity Focusing RICH detector

A group in IF-UNAM under the leadership of Arturo Menchaca-Rocha participates [17] in the AMS (Antimatter Magnetic Spectrometer) [36] project. The full detector should be installed on the International Space Station, but a prototype flew on a Shuttle mission. The group is mostly involved [15, 18, 19, 25, 26, 30, 33] in testing the radiator medium, but also in the physics analysis of the experiment.

A RICH for ALICE

At the ICN-UNAM a group under the leadership of Guy Paic participating in development and construction of a RICH detector for the ALICE [34] experiment at CERN [24, 31, 32]. The groups is involved in hardware and electronics designs, tests, and software for this detector. This is part of a larger Mexican involvement in ALICE. The same group participates at the moment in STAR, where a prototype of the ALICE RICH detector is used in a real experimental environment.

Velocity Spectrometers for CKM and P326

The UASLP group started to work in a project at Fermilab to measure the branching ratio the decay $K^+ \to \pi^+ \nu \bar{\nu}$ called CKM [37]. The experience in building and operating the SELEX RICH was to be transferred to a new concept called a Velocity Spectrometer [23, 27] where the basic velocity measurement is used directly as an observable in the experiment. The UASLP group worked on basic designs and simulations [20, 21, 29], the construction (in Mexico) and testing of mirrors [11, 28], and on testing photon detectors.

Unfortunately, the CKM project was terminated in 2004 after receiving Stage 1 (physics) approval due to budget constrains in the USA. Currently the group is participating in a new proposal (P326 [38]) with the same physics goals at CERN, a continuation of the NA48 experiment; the design from CKM are being transferred with a few changes to the new location.

The RICH 2004 Conference

The International Advisory Committee for the RICH Workshop Series decided in 2003 that the 2004 edition of this workshop should be held in Mexico, with the organizing committee lead by Jürgen Engelfried (UASLP) and Guy Paic (ICN-UNAM). The conference was held at the end of 2004 in Playa del Carmen [39]. About 100 participants from all over the world interchanged experiences in design, construction, operation, and applications of RICH detectors. The conference also commemorated the centenary of Pavel Cherenkov's birth, with talks by Elena Cherenkova and Boris Govorkov.

Summary

Ring Imaging Cherenkov Detectors are common detectors in many high energy and nuclear physics experiments. Mexican groups participate actively in various experiments.

REFERENCES

1. A. Roberts: *A new type of Cherenkov detector for the accurate measurement of particle velocity and direction*. Nucl. Instrum. and Meth. **9** (1960) 55 (1960).
2. J. Séguinot and T. Ypsilantis: *Photo-ionization and Cherenkov Ring Imaging*. Nucl. Instrum. and Meth. **142** 377 (1977).
3. E. Nappi, T. Ypsilantis (Eds.): *Proceedings of the First Workshop on Ring Imaging Cherenkov Detectors*. Nucl. Instrum. and Meth. A **343** (1994).
4. T. Ekelöf (Ed.): *Proceedings of the Second International Workshop on Ring Imaging Cherenkov Detectors*. Nucl. Instrum. and Meth. A**371** (1996).
5. A. Breskin, R. Chechik, T. Ypsilantis (Eds.): *Proceedings of the Third International Workshop on Ring Imaging Cherenkov Detectors*. Nucl. Instrum. and Meth. A**433** (1999).

6. *Proceedings of the IV. International Workshop on Ring Imaging Cherenkov Detectors.* Nucl. Instrum. and Meth. A **502** (2003).
7. J. Engelfried and G. Paic (Eds.): *Experimental techniques of Cherenkov light imaging. Proceedings, 5th International Workshop, RICH 2004, Playa del Carmen, Mexico, November 30-December 5, 2004.* Nucl. Instrum. Meth. A **553**.
8. P.A. Cherenkov: *Visible radiation produced by electrons moving in a medium with velocities exceeding that of light.* Phys. Rev. **52** (1937) 378.
   Pavel A. Cherenkov, Il'ja M. Frank, Igor Y. Tamm, Nobel Price 1958.
9. I. Frank, I. Tamm: *Coherent visible radiation of fast electrons passing through matter.* C. R. Acad. Sci. URSS **14** (1937) 109.
10. H.-W. Siebert et al.: *The Omega-RICH.* Nucl. Instrum. and Meth. A**343** (1994), 60.
11. L. Stutte, J. Engelfried and J. Kilmer: *A Method to evaluate mirrors for Cherenkov counters.* Nucl. Instrum. Meth. A **369**, 69 (1996).
12. J. Engelfried et al.: *The SELEX phototube RICH detector.* Nucl. Instrum. Meth. A **431**, 53 (1999) [arXiv:hep-ex/9811001].
13. J. Engelfried et al.: *The RICH detector of the SELEX experiment.* Nucl. Instrum. Meth. A **433**, 149 (1999).
14. R. Lopez-Fernandez: *Identificación de partículas producidas en interacción p-N mediante el E781 RICH.* Master Thesis, Instituto de Física, Universidad Autónoma de San Luis Potosí, unpublished (1997).
15. Z. Ren et al.: *A Ring Imaging Cherenkov Detector for the AMS experiment: Simulation and prototype.* Nucl. Instrum. Meth. A **433**, 172 (1999).
16. U. Müller et al.: *The Omega RICH in the CERN hyperon beam experiment.* Nucl. Instrum. Meth. A **433**, 71 (1999) [Nucl. Instrum. Meth. A **440**, 260 (2000)].
17. A. Menchaca-Rocha et al.: *Mexican participation in the AMS project.* AIP Conf. Proc. **566**, 172 (2000).
18. T. Thuillier et al.: *Prototype study of a proximity focusing RICH for the AMS experiment.* Nucl. Instrum. Meth. A **442**, 74 (2000).
19. T. Thuillier et al.: *Experimental study of a proximity focusing Cerenkov counter prototype for the AMS experiment.* Nucl. Instrum. Meth. A **491**, 83 (2002) [arXiv:astro-ph/0201051].
20. I. Torres, J. Engelfried and A. Morelos: *Simulation of a RICH detector for the CKM experiment.* arXiv:hep-ex/0202002.
21. A. Morelos, J. Engelfried, J. Mata, I. Torres and E. Vazquez-Jauregui: *Fundamental measurements and instrumentation 'CKM'.* AIP Conf. Proc. **623**, 369 (2002).
22. J. Engelfried et al.: *SELEX RICH performance and physics results.* Nucl. Instrum. Meth. A **502**, 285 (2003) [arXiv:hep-ex/0208046].
23. J. Engelfried et al.: *Two RICH detectors as velocity spectrometers in the CKM Experiment.* Nucl. Instrum. Meth. A **502**, 62 (2003) [arXiv:hep-ex/0209020].
24. D. Cozza et al.: *The CSI-based RICH detector array for the identification of high momentum particles in ALICE.* Nucl. Instrum. Meth. A **502**, 101 (2003).
25. M. Buenerd et al.: *The AMS-02 RICH Imager Prototype - In-Beam Tests with 20 GeV/c per Nucleon Ions.* arXiv:astro-ph/0306224.
26. B. Baret et al.: *In-beam tests of the AMS RICH prototype with 20-A-GeV/c secondary ions.* Nucl. Instrum. Meth. A **525**, 126 (2004).
27. P. S. Cooper and J. Engelfried: *Redesign of the CKM RICH velocity spectrometers for use in a 1/4-GHz unseparated beam.* Nucl. Instrum. Meth. A **553**, 220 (2005).
28. N. Estrada, J. Engelfried and A. Morelos: *Ronchi test for flat mirrors.* Nucl. Instrum. Meth. A **553**, 172 (2005).
29. A. Morelos, J. Mata, P. S. Cooper, J. Engelfried and J. L. Aguilera-Servin: *Radial tail resolution in the SELEX RICH.* Nucl. Instrum. Meth. A **553**, 237 (2005).
30. A. Martinez-Davalos, E. Belmont-Moreno and A. Menchaca-Rocha: *Optical and ageing studies of aerogel samples for RICH applications in space.* Nucl. Instrum. Meth. A **553**, 177 (2005).
31. E. Cuautle et al.: *Aerogel Cherenkov counters for high momentum proton identification.* Nucl. Instrum. Meth. A **553**, 25 (2005).
32. A. Di Mauro et al.: *Status of the HMPID CsI-RICH project for ALICE at the CERN/LHC.* IEEE Trans. Nucl. Sci. **52** (2005) 972.

33. F. Barao et al.: *The ring imaging Cherenkov detector (RICH) of the AMS experiment.* arXiv:astro-ph/0603852.
34. http://aliceinfo.cern.ch/
35. http://www-selex.fnal.gov/
36. http://ams.cern.ch/
37. http://www.fnal.gov/projects/ckm/
38. http://na48.web.cern.ch/NA48/NA48-3/
39. http://www.ifisica.uaslp.mx/rich2004/

Weak decays: early experiments (Magnetic Moments of Hyperons)

A. Morelos

Instituto de Fisica,
Universidad Autonoma de San Luis Potosi,
San Luis Potosi, S.L.P., Mexico 78000

Abstract. A summary of present hyperon magnetic moment measurement techniques is shown. The spin precession technique for high precision Σ^+ and $\bar{\Sigma}^-$ magnetic moment measurements is described along the E761 experiment. It includes a description of the E761 Σ^+ spin precession measurement by crystal channeling. Alternative techniques using radiative πp scattering or radiative decays are briefly described.

Keywords: magnetic moment; crystal channeling ; hyperon
PACS: 13.40.Em, 14.20.Jn, 29.30.-h

INTRODUCTION

The magnetic moments are a good test for a variety of phenomenological models, different authors have used chiral quark model, lattice theory, 1/Nc expansion, and variants of the non-relativistic quark model. Just to name a couple of works I am citing [1] and [2], a complete set of works is found in SPIRES data base [3].

TABLE 1. Baryon Measured Magnetic Moments, data from [9].

Particle	Magnetic Moment	± error	Precision (%)
p	2.792847351	0.000000028	0.0
n	-1.913042700	0.000000500	0.0
Λ^0	-0.6130	0.0040	0.7
Σ^+	2.4580	0.0100	0.4
Σ^-	-1.1600	0.0250	2.2
Ξ^0	-1.2500	0.0140	1.1
Ξ^-	-0.6507	0.0025	0.4
Ω^-	-2.02	0.05	2.5
Δ^{++} [6]	5.55	0.41	7.3
$\Sigma\Lambda$	-1.61	0.08	5.0

The magnetic moments of nucleons are measured to a precision of a few ppm, the long lived hyperons, including the anti-hyperons $\bar{\Sigma}^-$ [4] and $\bar{\Xi}^+$ [5], are measured on the precision range of 2% and a fraction of a %. Also the Δ^{++} magnetic moment and the transition $\Sigma\Lambda$ magnetic moment have been measured. The measurements are summarized in table 1.

The preferred way for high precision measurements of the magnetic moment of long lived hyperons has been the spin precession technique. The spin precession technique requires a polarized hyperon, a magnet for spin precession, and a particle spectrometer. In a straight forward way the magnetic moment ($\mu_{hyperon}$) is obtain from:

$$\phi = \frac{q}{m_{hyperon}c^2}\frac{1}{\beta}\frac{g-2}{2}\int Bdl \quad \text{and} \quad \mu_{hyperon} = \frac{gm_p}{2m_{hyperon}}\mu_N$$

where: ϕ is the spin precession angle; q is the *hyperon* charge; $m_{hyperon}$ is the *hyperon* mass; c is the speed of light; β is the *hyperon* speed over c, $\beta = 1$; g is the gyromagnetic ratio; $\int Bdl$ is the integral of the magnetic field, over the path the *hyperon* travels; m_p is the proton mass; and $\mu_N = e\hbar/2m_pc$ is the nuclear magneton.

To exemplify the measurement of the magnetic moment as well the measurement of the polarization I describe them around a fix target experiment, E761.

EXPERIMENT E761

Experiment E761, *An Electroweak Enigma: Hyperon Radiative Decays*, was designed to make a systematic study of the branching ratio and asymmetry parameter in hyperon weak radiative decays, $\Sigma^+ \to p\gamma$ [10], $\Xi^- \to \Sigma^-\gamma$ [11], $\Omega^- \to \Xi^-\gamma$ [12].

E761, also, measured to a high precision the Σ^+ and $\bar{\Sigma}^-$ magnetic moments [4], as well their polarization [13, 14] required to measure the angle ϕ. The study was made through the weak decay $\Sigma^+ \to p\pi^0$ and its c.c. decay.

Polarized Hyperon

It is known that a beam of unpolarized protons over an unpolarized nucleon target produce hyperon polarization. This experimental fact was born with the discovery of the polarization in inclusive reactions [15]. From parity conservation the hyperon polarization vector should be perpendicular to the plane formed by the incident proton momentum and the hyperon momentum. The angle between the incident proton and produced hyperon is named as targeting angle.

The hyperon production polarization is an experimental fact, although there are a few phenomenological models which in a partial way explain the phenomena, see this proceedings under the author J. Felix.

In E761 the Σ was produced by the interaction of the 800 GeV/c incident proton over a Cu target. The incident proton impinged on the left side of the 7.3 m long and 3.4 *Tesla* Hyperon Magnet and struck a target located at the upstream entrance of the

Hyperon Magnet, see figure 1. A thin curved channel, immerse in the Hyperon Magnet, set the Σ charge and momentum at 375 GeV/c. Figure 1 shows only the relevant parts for the Σ polarization and magnetic moment analysis. The coordinate system shown in figure 1 corresponded to: the $\hat{z}$ axis pointed downstream the particle beam; the $\hat{y}$ axis pointed to the sky; and the $\hat{x}$ axis completed a right handed system.

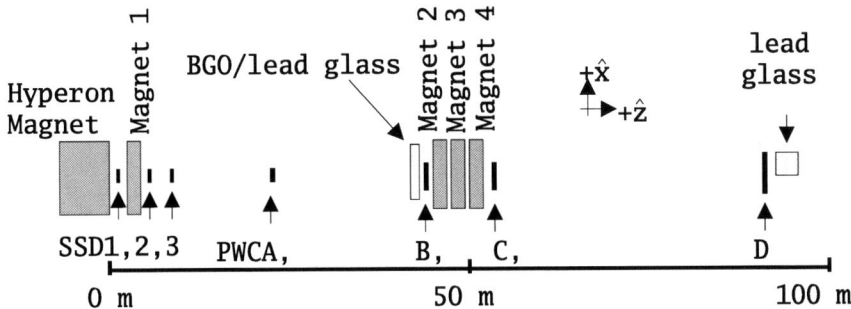

FIGURE 1. Plan view schematic of E761 spectrometer.

The proton beam line was instrumented with dipole magnets upstream the Hyperon Magnet, we tilted gradually the incident proton beam to set a variety of targeting angles.

Experiment E761 had two ways to produce the polarization vector. One mode we called horizontal, the polarization vector pointed perpendicular to the plan view, both the incident proton and the hyperon were on the horizontal plane, corresponding to the plan view in figure 1. This polarization vector pointed along the $\hat{y}$ axis.

In the horizontal mode the Σ spin did not precess since the magnetic field of all Magnets was parallel to the polarization direction.

In the relevant mode for the μ_Σ measurement the Σ spin precessed while the Σ traveled through the Hyperon Magnet and Magnet 1. This mode we called vertical production. In vertical mode the produced polarization at the interaction point was along the $\hat{x}$ axis.

In the horizontal mode we produced data over a variety of pairs of targeting angles. In the vertical mode we produced data over one pair of targeting angles, only.

Particle Spectrometer

Dedicated experiments for high precision hyperon measurements have been done in fixed target mode, E761 was one of them. At the high energy proton beam at Fermilab, 800 GeV/c, the spectrometers tend to be large, see figure 1. The E761 spectrometer consisted on silicon strip detectors SSD1-3, proportional wire chambers PWCA-D, and a photon calorimeter built with BGO & lead glass.

The E761 trigger included the existence of the Σ, proton and a photon from the decay $\pi^0 \to \gamma\gamma$. The trigger signals were generated with scintillators and the photon calorimeter. From all the produced Σ's, we recorded only the ones who decayed between SSD3 and PWCA. The spectrometer recorded the Σ hyperon track with the aid of SSD1-3 and the Hyperon Magnet and Magnet 1. Independently, the spectrometer recorded the proton track with the aid of PWCA-D and Magnets 2 - 4. We identified the decay $\Sigma^+ \to p\pi^0$ by the π^0 mass square reconstructed peak.

Polarization and spin rotation analysis

The measurement of the polarization was made with the proton angular distribution in the Σ Center of Mass system (CM), the expression is:

$$\frac{dN}{d\Omega} = a(\Omega) N_0 \frac{1}{4\pi} (1 + \vec{A} \cdot \hat{p}_b),$$

where: $\vec{A}$ is the asymmetry defined as $\vec{A} = \alpha \vec{P}$, $\vec{P}$ is the hyperon polarization vector, $\alpha = -0.980 \pm 0.016$ for $\Sigma^+ \to p\pi^0$ [9] and $\alpha = +0.980 \pm 0.016$ for its c.c.; $\hat{p}_b$ is a unit vector along the proton momentum in the hyperon CM; N_0 is the total number of events in the sample; $a(\Omega)$ is the acceptance of the apparatus, trigger, and analysis, which depends upon the proton direction cosines ($\cos\theta_x$, $\cos\theta_y$, $\cos\theta_z$) in the hyperon CM and the hyperon's laboratory angles in the decay volume.

In order to cancel the acceptance we took data on opposite production polarization directions. And to reduce systematic effects on hyperon phase space we divided the data on bins of the Σ track directions.

We plotted the horizontal production data as a function of p_t and x_f finding different behavior than the most studied Λ hyperon. We reported also the polarization for the $\bar{\Sigma}^-$. The horizontal data also let us measure false asymmetries in the other axis, where we measured zero [13, 14].

TABLE 2. Σ^+ and $\bar{\Sigma}^-$ vertical targeting angle (t.a.), events, Asymmetry, and precession angle (ϕ), only statistical errors are shown.

particle	Σ^+	$\bar{\Sigma}^-$
t.a.(*mrad*)	± 2.9	± 2.9
events	249 863	11 806
A_x	0.0094 ± 0.0036	-0.0191 ± 0.0166
A_y	-0.0055 ± 0.0036	-0.0227 ± 0.0167
A_z	0.1598 ± 0.0037	-0.0678 ± 0.0171
$\phi(rad)$	-10.937 ± 0.022	-10.72 ± 0.24

In the vertical mode the polarization direction at the production point was toward the $+\hat{x}$ axis. After the Σ left Magnet 1 the measured final direction of the polarization is approximately toward the $-\hat{z}$ axis. In table 2 is shown the measured asymmetries [4].

The precession angle ϕ is the angle between the polarization at production point and the one measured just previous to the decay point. Using two previous experiments [16, 17] we adjusted the angle ϕ for the total number of full rotation as: $\phi = tan^{-1}(|A_x/A_z|) - 3.5\pi$, where the 3.5 factor included also the quadrant correction. The angle ϕ is shown at the bottom of table 2.

Magnetic field for spin precession and magnetic moment

Another crucial part for the precision μ_Σ measurement was the accurate knowledge of the magnetic field along the Hyperon Magnet and Magnet 1. The aperture of Magnet 1 was large in such a way that it was relatively straight forward to measure the magnetic field with high accuracy. The Hyperon Magnet field measurement was more delicate, it required an NMR probe which was pulled through the thin channel immersed in the Hyperon Magnet. The value of $\int Bdl$ was $-25.215 \pm 0.031\ Tm$ for the Hyperon Magnet and $+4.750 \pm 0.007\ Tm$ for the Magnet 1.

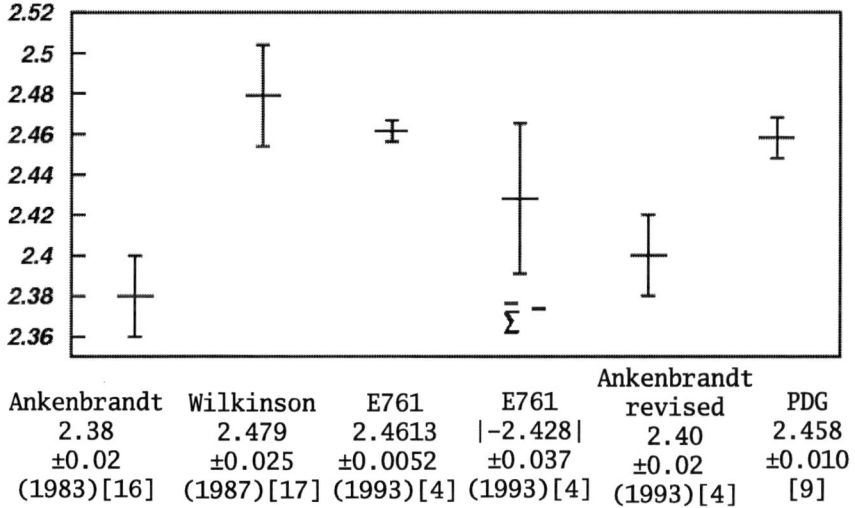

FIGURE 2. Σ^+ and $\bar{\Sigma}^-$ Magnetic Moment Measurements.

The Σ^+ and $\bar{\Sigma}^-$ measured magnetic moments are shown in figure 2. The shown error is the square sum of the statistical and systematic errors. Two previous measurements are shown, also, Ankenbrandt [16] and Wilkinson [17]. Ankenbrandt [16] measurement was revised with the high accuracy E761 Hyperon Magnet field mapping.

Actually, one of the motivations for making the μ_{Σ^+} precision measurement with E761 data was the big discrepancy between the previous two μ_{Σ^+} measurements.

CRYSTAL CHANNELING AND SPIN PRECESSION

In an attempt to reduce the magnet size an alternative technique, also spin precession, was successfully tested in the E761 experiment [18]. Crystal Channeling has previously been used for bending high energy beams. As the beam bends the particle's spin also precesses. The relation between the gyromagnetic ratio and the angle of spin precession (ϕ) is :

$$\phi = \gamma\omega(g-2)/2, \text{ for } \gamma >> 1,$$

where: γ is the lorentz factor; and ω is the deflection angle of the channeled particle.

Two Si crystals, not shown on figure 1, were installed just downstream of SSD3. The crystals were cut along their (111) plane, which were oriented in the $x-z$ plane. The crystal size was $2.50 \times 0.04 \times 4.50$ cm^3 and was bent using a multipoint bending jig. The Σ^+ hyperon beam was incident on the 2.50×0.04 cm^2 face, the hyperon traveled 4.50 cm along the crystal. One crystal bent the Σ^+ beam in the upward, $+\hat{y}$, direction and the other crystal in the opposite direction.

We used a pair of targeting angles in horizontal mode. We used the acceptance canceling technique to measure the polarization in the decay $\Sigma^+ \rightarrow p\pi^0$. At production point the polarization pointed along the $\hat{y}$ axis, the polarization direction rotated over the $y-z$ plane until the Σ decayed.

The measured deflection angles were 1.649 ± 0.043 and -1.649 ± 0.030 $mrad$. The spin precessed $60^o \pm 17^o$. The Σ^+ magnetic moment resulted in $2.40 \pm 0.46 \pm 0.40$ μ_N. The bent effect of the crystal was equivalent to a 45 $Tesla$ magnetic field.

This magnetic moment measurement was completely independent of the high precision measurement described above. The data sample was in one case from the vertical mode and the other from horizontal mode. Those sets of data were captured at completely independent times during the experiment.

Both measurements agreed and proved the crystal channeling spin precession.

OTHER MAGNETIC MOMENT MEASUREMENTS TECHNIQUES

The spin precession hyperon magnetic moment measurement technique requires a long lived polarized hyperon and a very intense hyperon beam, for both cases normal magnetic field and crystal channeling. This kind of experiments are highly specialized, in contrast to present state of the art multi-task and huge detector facilities, like CMS, see subjects around Physics at the LHC/CERN on this proceedings.

It has already been proved an alternative technique for measuring the Δ^{++} magnetic moment. The technique uses radiative πp scattering data. This analysis technique uses models which are sensible to the magnetic moment [7, 8].

There is an alternative experimental approach for the Ω^- magnetic moment using the analysis of its radiative decay $\Omega^- \to \Lambda K^- \gamma$ [19]. This procedure has not been tested with data, also the radiative decay $\Omega^- \to \Lambda K^- \gamma$ has not been recorded in the PDG [9].

The analysis of radiative decays has been extended to look into the magnetic moment of charged vector mesons (V^+) through the decay $V^+ \to P^+ P^0 \gamma$, where $P^{+/0}$ is a charge/neutral pseudoscalar [20].

The same authors have extended their approach to consider the influence of the ρ magnetic moment in the τ radiative decays. In one report they studied the decay $\tau^- \to V^- \pi^0 \nu_\tau \gamma$, were V^- can be ρ^- or K^{*-} [21]. They also consider the ρ as an intermediate particle in the decay $\tau^- \to \pi^- \pi^0 \nu_\tau \gamma$ [22]. The authors report that the observables in the case of the intermediate vector meson present no sensitivity to the magnetic moment.

The branching ratio of a particle radiative decay is suppressed by a factor of $\approx 10^4$ respect its most common hadronic decay [9]. The radiative decays are produced in quite low statistics in general experiments. If high precision is required in the measurement of the magnetic moment then the radiative decay ideas fall back into high statistics dedicated experiments.

COMMENTS

With the spin precession technique, the long lived hyperon magnetic moments have been measured to high precision, as required by the present phenomenological models. There has not been risen a demand for dedicated facilities with high intensity flux of hyperons and better precision.

The crystal channeling technique has been proved to work for measuring magnetic moments. Until now it has not been instrumented in experiments looking at short living particles, as charm baryons.

For a long time there may be new measurements on more baryon magnetic moments using sophisticated decays and models, depending on the theoretical ideas describing the decays and our ability to use the existing experimental data and facilities.

ACKNOWLEDGMENTS

The author thanks his E761 collaborators for executing with expertise a high precision experiment. Also thanks to G.Lopez-Castro for pointing me to his extensive work on looking into the magnetic moment sensitivity of radiative decays in different modes.

Special "felicitaciones" to Division de Particulas y Campos de la Sociedad Mexicana de Fisica (DPC-SMF) for its 20th. Anniversary.

REFERENCES

1. J.G.Contreras, R.Huerta, and L.R.Quintero, *Revista Mexicana de Fisica* **50** (5), 490–494 (2004).
2. M.Bohm, R.Huerta, and A.Zepeda, *Phys. Rev.* **D 25**, 223, (1982).
3. SPIRES, www.slac.stanford.edu/spires/ .
4. A.Morelos et al., *Phys. Rev. Lett.* **71**, 3417–3420, (1993).
5. P.M. Ho et al., *Phys. Rev. Lett.* **65**, 1713–1716, (1990).
6. Lopez Castro [7] and Bosshard [8] use different πp scattering data, also different models to obtain the magnetic moment. In this article I choose to present the weighted average of those measurements.
7. G. Lopez Castro and A. Mariano, *Phys.Lett.* **B 517**, 339–344 (2001).
    G. Lopez Castro and A. Mariano, *Nucl.Phys.* **A 697**, 440–468 (2002).
8. A.Bosshard et al., *Phys. Rev.* **D 44**, 1962–1974 (1991).
9. Particle Data Group, S. Eidelman et al., *Phys.Lett.* **B 592**, 1 (2004).
10. S. Timm et. al., *Phys. Rev.* **D 51**, 4638, (1995).
    M.Foucher et al., *Phys. Rev. Lett.* **68**, 3004 (1992).
11. T.Dubbs et al., *Phys. Rev. Lett.*, **72**, 808, (1994).
12. I.F. Albuquerque et al., *Phys. Rev.* **D 50**, 18-20, (1994).
13. A.Morelos et al., *Phys. Rev. Lett.* **71**, 2172–2175, (1993).
14. A.Morelos et al., *Phys. Rev.* **D 52**, 3777–3780, (1995).
15. G.Bunce et al., *Phys. Rev. Lett.* **36**, 1113, (1976).
16. C.Ankenbrandt et al., *Phys. Rev. Lett.* **51**, 863, (1983).
17. C.Wilkinson et al., *Phys. Rev. Lett.* **58**, 855, (1987).
18. D.Chen et al., *Phys. Rev. Lett.* **69**, 3286, (1992).
19. M.Hernandez, G.Lopez Castro, J.L.Lucio M., *Phys. Rev.* **D 44**, 794, (1991).
20. G. Lopez Castro, G. Toledo Sanchez, *Phys.Rev.* **D 56**, 4408-4411, (1997).
    G. Lopez Castro, G. Toledo Sanchez, *J.Phys.* **G 27**, 2203-2210, (2001).
21. G. Lopez Castro, G. Toledo Sanchez, *Phys.Rev.* **D 60**, 053004, (1999).
22. G. Lopez Castro, G. Toledo Sanchez, *Phys.Rev.* , **D 61**, 033007, (2000).
    A. Flores-Tlalpa, G. Lopez Castro and G. Toledo Sanchez, *Phys.Rev.* **D 72**, 113003, (2005).

Use of Silicon Detectors in Medical Physics

Luis Manuel Montaño Zetina

Physics Department, Cinvestav, Mexico City, Mexico

Abstract. In this document I will review the characteristics and applications of silicon detectors in Medical Physics. I will cover the activities done by some research mexican groups working with silicon detectors (Silicon Strip and PIN detectors) as devices for digital imaging supported by some Monte Carlo simulations and X-ray units parameters valuation devices for quality control. In the end I will give some perspectives on the future of these scientific activities as important contributions in the development of the area of Medical Physics around the world.

INTRODUCTION

As we know radiation is invisible to us, it cannot be smelled, touched, seen and nevertheless it can be harmful. It was necessary the study of the properties of the radiation to, in some way, have control on it and indeed apply it for our own benefit. The way radiation can be identified is through devices which can react when it passes through them. Some processes can take place when radiation interacts with matter: ionization if a charged radiation hit the material, or the well known effect as photoelectric effect, pair production, compton scattering, bremsstrahlung, if the raadiation is uncharged.

The development of radiation detectors has passed for several periods, from identifying a single sound (Geiger-Muller), to the ability of localizing the position incidence (wire chambers, silicon detectors). Before 1950 Geiger counters, photographic emulsions and Wilson's cloud chamber were the major detection instruments used; after that the bubble chamber took over much of the task. In the early sixties the spark chamber entered and evolved to the proportional wire chamber. Nowadays there exist several kind of detectors whose development depended on the application. There are ionizing chambers, proportional counters, scintillation material coupled with photomultiplier tubes, cerenkov and semiconductor detectors.

The principle of many detectors is the detection of a track left by the passage of a charged particle. When it passes through matter, it knocks out electrons from the atoms, thereby disturbing the structure of the material and also creating loose electrons. Thus a charged particle passed through matter leaves a trace of disturbed matter and move electrons from their positions that can be further collected. They can also detect uncharged radiation causing the effects mentioned above. We will see now more details of semiconductor detectors.

TABLE 1. Properties of silicon and germanium.

	Si	Ge
Z	14	32
A	28.09	72.60
Density (300 K), g/cm^3	2.33	5.33
Atoms/cm^3	4.96×10^{22}	4.41×10^{22}
ε_r	12	16
E_g (300 K), eV	1.115	0.665
Intrinsic carriers density (300 K), cm^{-3}	1.5×10^{10}	2.4×10^{13}
Intrinsic resistivity (300 K), Ωcm	2.3×10^5	47
μ_e (300 K), cm^2/Vs	1350	3900
μ_h (300 K), cm^2/Vs	480	1900

SEMICONDUCTORS

Nowadays it is difficult to find a field where semiconductors have not been applied to improve systems of any kind. They are the core of computing development as integrated circuits have more capabilities. It is known that one of the greatest development in technology which influences life in our society was the invention of transistor, made of semiconductor materials as silicon and germanium (see table 1). The transistor (transfer resistor), was invented by a research team at Bell Laboratories in 1947. It has had an unprecedented impact on the electronic industry in general and on solid-state research in particular. Prior to 1947 semiconductors were only used as thermistors, photodiodes, and rectifiers. In 1948 John Bardeen and Walter Brattain announced the development of the point-contact transistor [1]. In the following year William Shockley's classic paper on junction diodes and transistors was published [2].

The basic idea of a diode or transistor is the following: When p^+ and n^+ materials are put together there will be a junction where a depletion zone is created as well as an electric field. Then, the electrons and holes present in both materials undergo a net migration. This effect is called the drift velocity of the carriers. This migration or velocity depends on the electric field. Applying an inverse voltage this depletion region expands, decreasing the production of charges called leakage current. This low leakage current will be a basic feature for the application of semiconductor detectors in many areas.

Another application of semiconductors detectors is in high energy physic experiments, more particular in the detection of charged particles, products of the collisions of nucleons or heavy ions in the accelerators of international laboratories. Semiconductor detectors were mainly developed to identify the path of charged particles. The dominant advantage of semiconductor detectors lies in the smallness of the ionization energy. In particular detectors made of silicon, which it will be referred as silicon detectors, require only a charged particle energy of 3.6 eV to create an electron-hole pair while for a gas-filled detectors a 30 eV is needed. Thus it is expected that silicon detectors offer better energy resolution and, depending on the design, good spatial resolution as well.

SILICON DETECTORS

In general the principle of particle detectors is the identification of the tracks left by the passage of a charged particle by ionization. This interaction leaves a trace of disturbed matter and move the electrons from the original atoms to some terminals (wires, anodes) where they can be collected. This collected charge will be the signal that the rest of the electronic chain in the system will identify as the electronic pulse, it means, the signal.

As it was indicated the spatial resolution of the silicon detector depends on the design. Due to the great energy and spatial resolution and the small amount of radiation energy to generated an interaction in them, these detectors have been successfully applied in tracking and identification systems. There are pixel or strip detectors, two and one dimensional respectively. There are others which also identify the kind of particle through the energy deposited in it. Their energy resolution depends on the number of interactions on the detector. So in order to have a good energy resolution it is necessary to increase the number of information carriers creating during the interaction. The amount of charge deposited in the typical 300 μm of thickness of a silicon detector is around 25,000 electrons. Moreover the time required to collect the charge could be of nanoseconds, which makes them one of the fastest-responding of all radiation detector types. One can see in Fig. 1 a charged particle or photon hitting the detector ionizing it. Therefore, they are generally situated in the closest position of the particle interaction point of a complete detector experiment on elementary particle.

The main characteristics of the silicon detectors that make them useful devices are:

- Speed of reaction when radiation cross the surface of 10 ns.
- Spatial resolution $\sim 10\mu$m.
- Flexibility of design.
- Small amount of material (0.003 X_0 for 300 μm thick detector).
- Linearity of the response vs. the deposited energy.

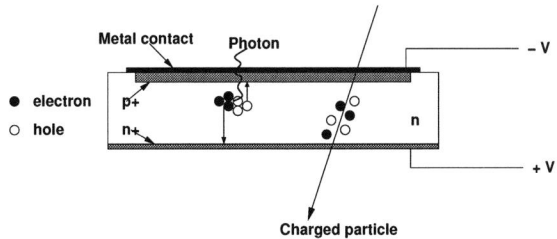

FIGURE 1. Charge particle and a photon hitting a silicon detector.

- Good resolution in the deposited energy.
- Tolerance to high radiation doses.

A silicon detector can be visualized as a diode. Therefore it needs an inverse voltage to have a big depletion region, in general, as big as the detector is wide. This let the region ready for detecting radiation that will originate charges when crossing the wafer surface. The electric field created guides the generated charge to the cathodes. These cathodes are the p^+ material which collect the charge that will be transmitted to the electronics. For our detector these cathodes are the microstrips. Above each cathode there is a metallic cover to permit the connection between the detector and readout electronics via microboundings.

SILICON DETECTORS IN MEDICAL PHYSICS

In the last decades silicon detectors have entered to help in the development of the area of medical physics, in particular in digital imaging for diagnostic. Some of the most important characteristics of silicon detectors for Medical Physics are its good energy and spatial resolution and great signal-to-noise ratio.

A large part of medical images falls in the X-ray energy range of about 20-30 keV, where tissues and bones or calcifications can absorb enough radiation to be noticed in a radiological image. Silicon detectors have the possibility of detecting X-ray of this order of energy (see Fig. 2). As the size of these objects is fairly large (of the order of tenths of microns), silicon detector seem to be the proper candidate to be used for medical imaging.

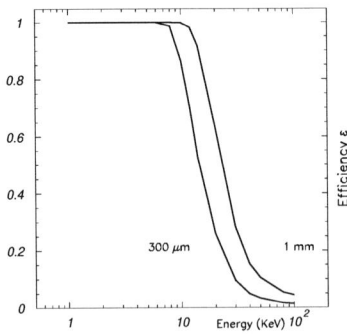

FIGURE 2. Efficiency of a Silicon wafer for different thicknesses

Silicon Strips Detectors (SSD)

As a brief historical tale one can say that, although their applications have flourished in initially unimagined ways, SSD were born of a specific scientific need in elementary particle physic experiments: to detect and study particles with "charm", it means, particles containing a charm quark. The precision needed to extrapolate the tracks in order to keep the different vertices in focus should be much less than the particle lifetime multiplied by teh speed of light. So for charmed particles the required precision is a few tens of microns, a value well within the capabilities of SSD [3].

Microstrip detectors provide therefore the measurement of one coordinate of the particle's crossing point with high precision. Using very low noise readout electronics, the measurement of the centroid of the signal over more than one strip further improves the precision. Clearly the precision of this procedure depends on the noise of the readout chain (including the quantization error introduced by the analog-to-digital converter, which is important when using small signals). If digital readout is used (strip hit or not hit), the resolution is simply $\sigma = \frac{pitch}{\sqrt{12}}$ [4]. Figure 3 shows a detail of a SSD.

As was mentioned one of the fields that SSD are being applied is in the area of imaging, which means, using these devices to obtain digital images with a great spatial resolution. Digital images present some advantages when comparing to the conventional ones. Digital images have higher range of contrast efficiency with respect to conventional screen-film, also digital technology has the advantage to process the images in order to improve them, they can be transferred in a digital way, there is no more film development, avoiding retakes if the develop process is not well done, as others. In this sense it is promising that SSD be useful in mammography where the earlier the diagnostic of the presence of cancer is identified, the chances of saving lives increase.

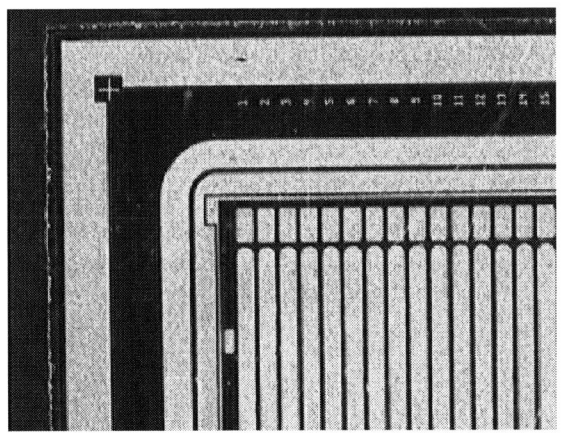

FIGURE 3. Detail of a corner of a SDD

It is known that mammography is the best method for breast cancer diagnosis. One signal of the presence of cancer is the identification of microcalcifications in breasts. Mammography refers to obtaining radiografic images and identifying ill tissues in the zone. One of the issues that can help to get a better diagnosis of cancer is having good spatial resolution radiographic images. It is clear the interest of some research groups in the world to study the possibilities to apply SSD to get better digital images for helping doctors to give a better diagnosis. We will not enter into many details on breast cancer because it is not the scope of this document, but it is fair to say that these kind of researches are of great importance. Breast cancer is the second cause of death by cancer in the world. So the efforts to get better and earlier diagnosis to prevent this cancer will be in benefit for all.

At this point I will mention the contribution that my research group is doing in the Physic Department at Cinvestav. We are working with a SSD for imaging purposes. This SSD was part of a set of prototypes for the tracking system for ALICE detector at CERN [5]. The group responsible for this system was using the SSD also for imaging and, as we were participating with them for ALICE, we were given a SSD prototype to develop our contribution in Mexico. Thus, since 2001 we have made improvements in the idea of obtaining better images using the SSD with X-rays [6].

During the development of this research we have found some limitations to get good images and every time we go further solving them. For instance one of the limitations in using silicon strip detectors is that in the interesting energy range for diagnostic (10-50keV), the absorption length of silicon is of the order of mm, then only a fraction of the X-rays are converted in the commonly used 300 microns thick detectors. Therefore there are two limits in applying these devices to low energy range: the upper energy limit caused by the reduction of efficiency and the degradation of the intrinsic spatial resolution due to Compton scattering. The last limit is comparable or larger than the effect of diffusion during the transport of the signal charge to the read-out chain.

To avoid this limit, one idea is to use a SSD in the edge-on configuration, it means, letting the X-rays passing through the sample and hit the detector by one of its edge perpendicular to the strips. This can be seen as a very narrow and long detector, but the radiation has to travel centimeters in the silicon bulk before leaving it, increasing the probability of interaction.

Some techniques have been developed to improve the process of getting better contrast images, one of them is using dual energy substraction, as it was applied by this research group [7]. It is shown in Fig. 4 a picture of the SSD coupled with two readout chip.

There are other efforts for applying this prototype in angyography. So far there have been computing simulations to test the expected resolution to be reached. In reference [8] it is explained how these tests were performed using two different Monte Carlo softwares.

PIN detectors

During the last decade much effort has been dedicated to the research of amorphous silicon PIN diodes for application as radiation detectors, imaging generation in medicine as others. These detectors offer some advantages as low cost, high stability to radiation damage and the possibility to grow large active areas. These devices can also be integrated together with read-out circuits [9].

Among possible medical applications, amorphous PIN diodes are studied as X-ray detectors, to substitute normal screens with multiple advantages. Indirect detection, previously converting through scintillate screens the X-rays to visible radiation which is detected by the PIN diodes acting as photodetectors, has already been achieved [10]. Direct detection requires much thicker layers, which is difficult to get with the required stability and quality.

One group created and used PIN diodes developed at Electric Engineering Department in Cinvestav to valuate some parameters for quality control of X-ray units. The exposed a PIN diode to X-ray single shot to measure the exposure time, high voltage applied, and dose [11]. Nowadays there are still some efforts for applying these PIN detectors in imaging.

PERSPECTIVES

High contrast digital X-ray radiography needs advances in readout electronics, solid-state coupled to it and direct image capture. Great improvements in both the quality of the X-ray systems and in the photographic films have resulted in better image quality,

FIGURE 4. SSD with two readout chips.

which is essential for the doctors to assess and to diagnose the problems of the patients. However there is still room for improvements, particulary with regard to the dose needed to produce an image and to the reduction to the inevitable noise, which is the main limiting factor in revealing low contrast objects. The application of SSD can accomplish these necessities.

For SSD there are some international collaborations to continue the application of these detectors to imaging. There have been bilateral projects between this mexican group with some other international groups for instance cuban, italian, polish, colombian. Moreover there was an ALFA project which put together several research group to develop a digital system using SSD. This ALFA project was supported by the european community and the participant institutions. In this sense Mexico is participating and will continue being part of this research. We know in our country as well as in the world death for breast cancer is increasing. Any research which is addressed to find methods to discover earlier the possible presence of cancer in breast will be very well appreciated.

However, although digital mammographic units are available in the market nowadays and the need of having a direct digital imaging system, there are some other things to be improved. The things are for instance protocols for the physicians to valuated those digital images to get a diagnosis. Also it seems to be problems in the storage of big amount of data, cause every image requires at least tenths of Mb. The quality control of the X-ray units that use digital systems are not well established too. So, for our interests, we are working in solving and contribute to the improvement of the images taking into account the advantages that digital technology offers. The other details mentioned have to be solve in the near future because the step in changing from conventional to digital mammography is inevitable. We are contributing directly to this scope.

ACKNOWLEDGMENTS

I would like to thank the organizers of this project for having invited me to write this document.

REFERENCES

1. J. Bardeen, W.H. Brattain "The transistor, A Semiconductor Triode", Phys.Rev. 74, 230 (1948).
2. W. Shockley, "The theory of p-n junctions in semiconductors and p-n junction transistors", Bell Sys. Tech., J.,28,435 (1949)
3. Litke M. and Schwarz S. "The silicon Microstrip Detector" Sci. Am. May 1995 p56-61
4. Duerdoth I. 1982, *Nucl. Instr. and Meth.* **203** 291.
5. ALICE Collaboration 1999, *CERN/LHCC, 99/12*, (Swi/CERN)
6. Montano L. et al.,"Digital Mammography: improvements in breast cancer diagnostic", AIP 809 (2005)283-289
7. Montano L. et al., "Contrast cancellation technique applied to digital X-ray imaging using silicon strip detectors", Med. Phys. 32 (2005) pp. 3755-3766

8. Montano L. et al.,"Comparison between two Monte Carlo simulation of Angiography Phantom coupled to silicon strips detector", AIP 724 (2004)221-225
9. Cerdeira A. et al. 2000, "Response of amorphous silicon PIN detectors to X-ray from a Mammograph" Proceedings of the RADECS 2000 Workshop Belgium.
10. R. L. Weisfield, "Amorphous silicon TFT X-ray imaging sensors" IEDM Technical Digest, Dic, 6-9, 1998, San Francisco, USA
11. Ramirez F. et al. 2000, *Proc. IV Mexican Symposium on Medical Physics*, (AIP538) p196-200.
    photon (ISBN
    ASIC
    (2003)185-191

The Mexican side of the H1 Collaboration

J. G. Contreras

Departamento de Física Aplicada
CINVESTAV–IPN, Unidad Mérida,
A. P. 73 Cordemex, 97310 Mérida, Yucatán, México

Abstract. This article reviews the Mexican contributions to the H1 Collaboration and describes the grow of the Mexican involvement in the collaboration, from an isolated effort to a consolidate group.

Keywords: HERA,H1,Mexico
PACS: 13.60.Hb

1. INTRODUCTION

HERA is the first, and up to now only, collider of electrons on protons. It was design to study the structure of the proton through the scattering of electrons or positrons (e) off protons (p) in a new kinematic regime. It started operation in 1992 and since then provides ep collisions to the H1 and Zeus detectors. Both were built and are maintained by huge international collaborations and have similar physics programs.

There are some 80 institutions from around the world working in either of the H1 or Zeus experiments, but only one of them is Latin American: The Department of Physics and the Department of Applied Physics of the Mexican research center Cinvestav are members of the H1 Collaboration.

Here I will review the Mexican participation in the activities of the H1 collaboration and how the relationship between Cinvestav and H1 came about. First, I will give a brief description of HERA and H1. Then, to set the appropriate context, I will explain how the work to produce a publication is distributed inside the collaboration. Finally I will discuss the areas where Mexican students and scientists have worked in H1.

2. A BRIEF DESCRIPTION OF HERA AND H1

In the last 14 years HERA has been producing ep collisions, which have been measured by the H1 detector. The analysis of these data has produced many exciting and far reaching results as evidenced by the more than 150 publications of H1 in the most prestigious journals in the area. One signal of the importance of these results is the several thousand citations that they have received.

2.1. The HERA collider

HERA has two accelerators, one for protons and one for electrons or positrons. Both are in a tunnel several tens of meter underground. The tunnel has a circumference of 6.3 km. HERA and the complex of pre-accelerators are part of the DESY laboratory located in the city of Hamburg in Germany.

The proposal to built HERA was approved in April 1984 and its construction lasted until 1990 where the commissioning of the new machine started. The first ep–collisions were observed on October 19th, 1991 at a center-of-mass energy of $\sqrt{s}$ = 152 GeV. The first HERA data used for physics analysis were produced in the Spring of 1992 at an electron beam energy of 26.6 GeV and a proton beam energy of 820 GeV for $\sqrt{s}=295$ GeV.

The protons are injected into HERA already in a bunched structure, where the distance between bunches is 96 ns and each bunch has the approximated shape of a truncated Gaussian in all three dimensions with a transversal size of less than a millimeter and longitudinal size of 11 cm, where the size given corresponds to the RMS of the Gaussian shape. The protons are injected at 40 GeV and are accelerated, using superconducting technology, to 920 GeV in HERA (820 GeV before 1998).

The second HERA accelerator can work either with electrons or positrons. These are injected, also already bunched and separated by 96 ns, at 12 GeV and are accelerated in HERA to 27.5 GeV (26.6 GeV since 1994). HERA has accelerated electrons from 1992 to 1994, during the 1998–1999 period and from 2004 to date. The rest of the time HERA has worked with positrons.

The performance of HERA has been improving continuously since the start of operations going from a few inverse pico-barns (pb^{-1}) of data in the first few years to 70 pb^{-1} in 2000 and more than 150 pb^{-1} produced in each of the last two years. HERA will stop operations in Summer of 2007

2.2. The H1 detector

The H1 Collaboration is an international group formed by approximately 350 scientists from some 40 institutions and around 15 different countries. This group designed, built and maintain the H1 detector, which is located in the hall north of HERA.

The H1 detector is an assembly of several sub-detectors arranged concentrically around the point where the electron collides with the proton. Seen from this interaction point the produced particles meet first a silicon tracker, then a set of drift chambers. Surrounding the trackers there is a liquid argon calorimeter (LAr) and a lead–scintillator spaghetti type of calorimeter (SpaCal). All these detectors are inside a solenoid creating a magnetic field parallel to the beam pipe of 1.15 T. Outside of it there is the return yoke for the field lines which is instrumented in order to measure muons and the energy not contained in the liquid argon calorimeter. The detector is completed by a luminosity detector, several tens of meters backwards from the main detector, consisting on an electron and a photon calorimeters, and forward detectors of protons and neutrons some hundred meters after the interaction point.

Given that the detector has several hundred thousand channels, that there is a collision every 96 ns and that the experiment is running almost continuously, it is clear that not all data can be stored to be analyzed later. In order to trigger the data acquisition system in the few most interesting events, a complex electronic system has been developed. The H1 trigger system has four different levels, the first three running synchronous to the HERA clock. They have the task of reducing the event rate from the original 10 MHz to the level of few tens of Hz.

A detailed description of the whole detector and all its subsystems can be found in reference [1]. A figure of one event as measured with the central part of the H1 detector is shown in Figure 1

3. THE BIRTH OF A PAPER: FROM DATA TAKING TO PUBLICATION

Due to the complexity of the detector and of the physics we want to extract from the data, the complete process from the occurrence of a collision to the publication of data takes in general from several months to a few years. A full publication list of the H1 Collaboration, can be found in [2] The first stage, called data taking, comprises the actual selection of data to be recorded in a permanent media for its ulterior study. Then there is a general set of task to be done in order to analyze the data and finally the analysis itself takes place. This process will be discussed in some detail in the following.

3.1. Detector maintenance and data taking

The H1 detector is a set of sub-detectors working together. Each of these sub-detectors has to be continuously monitored and calibrated. Furthermore the relationship and synchronization among the different sub-detectors has also to be monitored and adjusted. The same is true for the data acquisition and triggering systems as well as for the whole infrastructure to record and store in permanent media the potentially interesting events.

Each institute has a responsibility regarding some subsystem to ensure its correct function. It is customary that each Ph. D. student works at least some time in a technical issue in this area and that the bigger institutes have some of their personal permanently assigned to maintain a given sub-detector in working order.

Each member of the collaboration has to participate actively in the data taking process. This happens through so called shift blocks. The detector is in ready to take data virtually all the time. This means that at all times there has to be someone supervising the functioning of the detector. To ensure this there are each day three eight-hour shifts covering the 24 hours. During the first years of operation there were four persons assigned to a shift, but the supervision of the detector has been so streamlined that currently only two persons are needed. The shifts are organized in blocks of six shifts in one week and every member of the collaboration has to be in a shift block at least once for each data taking period.

In summary every institute contributes to the detector maintenance and each single member of the collaboration participates in the data taking itself. Every paper needs

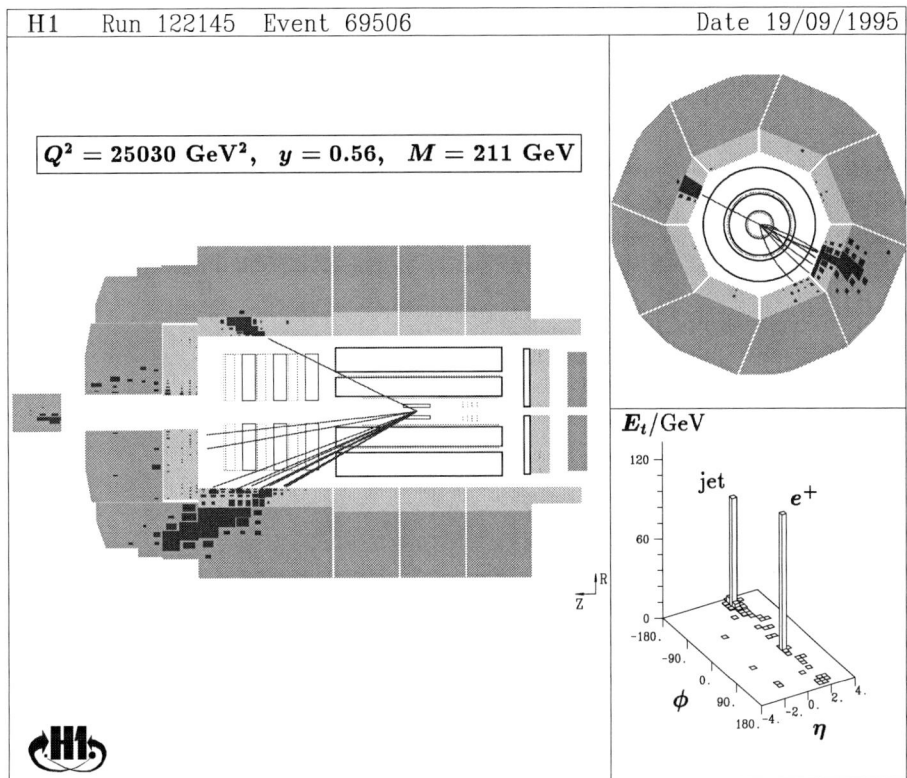

FIGURE 1. A typical deep inelastic event as seen by the H1 Detector. The darker blobs represent energy deposition in the active elements of the detector, while the lines are the result of the algorithms which associate to these energy deposition, tracks and clusters representing the produced particles. The Central and Forward Trackers as well as the Liquid Argon and Spacal calorimeters are shown in a longitudinal (left) as well as in a transversal (upper right) view. The so called lego plot of the event is also shown (lower right). The event shows clearly one jet recoiling from the scattered electron and in the forward direction (to the left) the remnants of the proton.

data and within H1 every person who has contributed in the data taking is considered an author in the corresponding H1 publications.

3.2. The role of physics working groups

The combination of HERA and H1, more than an experiment is an experimental facility in the sense that each collision is an experiment in itself. There are many types of physics that can take place in a collision so that the full set of data can be separated in different subsets according to the underlying physics which took place during the scattering. Taking this into account the H1 Collaboration has different working groups each specialized in a specific area. For example one studies the production of heavy quarks, other concentrates on inclusive cross sections, another on diffraction. There are also a group to study the physics beyond the standard model and one looking into QCD topics in exclusive final states.

These physics working groups gather the different tools, techniques and procedures developed during all these years in specific analysis and organize the tasks needed for the new analysis in the different subfields studied within the H1 Collaboration. Some of these tasks are for example: data quality checks, production of simulated collisions with Monte Carlo programs, and the development and testing of general software tools.

These groups also supervise the advance of each individual analysis, providing a forum to present the progress in a given investigation, the current status and problems found. The meetings of these groups serve to discuss ideas to solve the problems, to identify new directions for an analysis and to keep the work focused and progressing.

3.3. Data analysis

Each Ph. D. student works in a specific analysis. Several of the post-docs have also their own analysis and senior physicist normally supervise the analysis of their students. Within a physics working group each analysis is assigned a couple of referees who get involved in all the steps of the work until the publication.

Each analysis is quite thorough and normally takes one and a half to two years to be in a state to be discussed within the collaboration. The people doing the analysis write the draft and answer all the questions of the collaboration regarding the soundness of the work. This stage also takes some months, so normally one Ph. D. student works mainly in one analysis, but the technical work done for one analysis is normally very useful to other analysis.

In summary, the involvement of a member of the H1 Collaboration in the analysis of data is at different levels. As a general member of the collaboration participates in the final writing of each paper, as a member of a physics working group participates directly and indirectly in the different analysis produced within the group and finally it is strongly involved in a given analysis.

4. THE MEXICAN SIDE OF H1

Since 1993 there has been Mexicans working in the H1 Collaboration. The initial steps were through the Dortmund group and the Summer Students program at DESY. Later, in 2003, Cinvestav, thanks to the support of the Dortmund group and DESY, was accepted as an independent member of the H1 Collaboration. The next subsections give a brief account of these developments.

4.1. The H1–Dortmund–Mexico connection

The Dortmund group, under Prof. Dr. Dietrich Wegener, has been instrumental in the Mexican relation with the H1 Collaboration. It has received three Mexican Ph. D., a professor in a sabbatical stay and many Summer students. It also supported strongly the acceptance of Cinvestav as an independent Institute of H1.

The contact between the Dortmund group and Mexico, started before the work in H1. Gerardo Herrera was a Ph. D. student of Prof. Wegener working in the ARGUS experiment, also at DESY. In part, thanks to this experience, the first Mexican Ph. D. student (myself) started working for the H1 Collaboration as a part of Prof. Wegener's group. In the Autumn of 1993 I joined the Dortmund group and started work in technical issues related to the Liquid Argon Calorimeter of H1. The Dortmund group had contributed in building this detector and has been traditionally quite active in its maintenance. As a matter of fact the first Mexican working in H1 was Fabiola Vazquez who worked, during the Summer of 1993, in the reweighting of the hadronic energy measured with the Liquid Argon Calorimeter.

As physics topic I studied the dynamics of the partons in the proton in the new low x, or high energy, domain open up by HERA. This was to become a very hot topic in the early years of HERA and the results of my thesis work showed the need to go beyond the traditional DGLAP approach and supported the notion of BFKL contributions to the dynamics of the proton structure in this new kinematic region.

The second Mexicans Ph. D. student working for the Dortmund group in H1 was Miguel Mondragon, who in 1996 was a Summer Student working, for the Dortmund group, in the charge collection efficiency of the Liquid Argon Calorimeter. His technical obligations included the data quality checks for the Heavy Flavors physics working group and he studied the semileptonic decays of charm in deep inelastic scattering.

The third Mexican Ph. D. student student working for the Dortmund group in H1 is Andrea Vargas who will defend her thesis work in a few months. She was also a Summer Student in H1 working for the Dortmund group in the installation of an upgrade of the Time of Flight system. Some of her technical work has been to be the responsible for the production of simulated Monte Carlo collisions for her physics working group. Her thesis work deals with the measurement of F_2 structure function of the proton at very low values of both variables x and Q_2. This is, she studies the structure of the proton at very high energies but when seeing the proton at low resolution. The results of her analysis are already written up and are under discussion within her physics working group and about to be given to the whole collaboration for their approval to submit them to publication.

Gerardo Herrera had a sabbatical stay working for H1, financed by the Alexander von Humbold foundation as a guest of prof. Wegener. During his stay in H1 he proposed to use data from deep inelastic scattering to study hadronization models in the forward region.

4.2. Mexican Summer students in DESY

There has been at least ten Mexican Summer students working for H1, most of them within the Dortmund group, and several other Mexicans Summer students working in other experiments at DESY.

Besides the students mentioned in the previous section, there were Javier Espinoza working in the evaluation of noise for jet studies in the forward part of the Liquid Argon Calorimeter, Oscar Rodriguez working, as Fabila Vazquez did, in techniques to reweight the hadronic energy measured in the calorimeter, Gabriela Murguia who developed a WWW interface to monitor online the performance of the calorimeter, Manuel Rendon who, as Miguel Mondragon, worked in the monitoring of the charge collection efficiency of the Liquid Argon Calorimeter, J. L. Gamboa who did data quality monitoring for the SpaCal calorimeter and Joaquin Miranda who studied systematics of the hadronic final state. Most of these Summer students have finished their Ph. D studies or are working in a Ph. D. program.

4.3. Cinvestav, DESY and H1

In 1998 I returned to Mexico, to the Applied Physics Department of Cinvestav. I remained connected to the H1 Collaboration through the Dortmund group of Prof. Wegener who took over the financial and technical responsibilities tied to belonging to the collaboration. This arrangement gave time to the group of Cinvestav to be formed and to start to work.

In 2002 the former Spokeperson of the H1 Collaboration, Dr. Elsen from DESY, and Prof. Wegener suggested that it was the time for Cinvestav to be an independent member of the collaboration. This proposal was enthusiastically supported by the then spokeperson , Dr. Klein from DESY. By that time there were several Mexicans working in H1. Both Gerardo Herrera and myself based in Cinvestav, Andrea Vargas as Ph. D. student in the Dortmund group, Ricardo Lopez as a DESY postdoc and the first Cinvestav Ph. D. student, Daniel Perez. In July of 2003 the governing body of the collaboration accepted unanimously Cinvestav as a new member of H1.

Ricardo Lopez worked for the group in charge of central trigger of H1. This is a vital system of the detector. If it does not work properly the collisions can not be correctly selected and then no physics analysis is possible. Ricardo was more than two years expert on call for the group. He also was active in the Heavy Flavor physics working group studying the production of J/Ψ particles. After finishing his postdoctoral stay at DESY he got a position in the Department of Physics of Cinvestav where he continues his involvement in H1.

The technical work of Daniel Perez was in the group in charge of the Central Silicon Tracker (CST) of H1. He worked with Dr. List, from the ETH Zurich, in the general maintenance of the hardware and software of the CST. In particular he helped in the development of software to control the operation of the detector, in the first steps of the alignment procedure, studied the efficiency of the CST to distinguish true hits from noise and participated in the estimation of the correction factors to account for energy losses in the dead material of the detector. His analysis work was the study of D^* production in deep inelastic scattering where he reproduced in an independent analysis previous results obtained by other members of the collaboration and where he investigates the possibility to measure the production in the kinematic range of high Q^2, which has not been published by H1 before. He will defend his thesis in a few months.

4.4. The present and the future

Currently there are five Mexicans working in H1 and four of them are working from Cinvestav, and only one from Dortmund. From the five, two are Ph. D. students about to finish their work, but there are also two new students in Cinvestav, Karla Cantun and Julia Ruiz, starting their work in H1 and some potential candidates to do thesis work at the master level.

The whole Mexican group is working in a new field for H1. We are interested in measuring the Λ_c baryon in deep inelastic scatterings. To do it, we first have to understand the production of other particles appearing in the decay channels and which are interesting on their own, like the K_s^o and the Λ particles.

The Mexican group now is quite solid, contributes to technical tasks and to physics analysis. As HERA will stop operations in Summer 2007, we will concentrate then in the analysis side. We are currently in the second generation of Cinvestav students working in a Ph.D. and the next generation is on the way, so that we will keep contributing to the analysis of H1 Data up to the forseen end of the Collaboration around 2010 or 2011.

5. SUMMARY AND OUTLOOK

The Mexican group in H1 has grown step by step and has benefited from the strong support from the Dortmund and DESY groups in particular and the help and support of the whole Collaboration in general.

In the technical side, to date Mexico, through Summer Students Ph. D. students in the Dortmund group, a postdoc and DESY, and recently through our own Cinvestav students, has contribute to a large array of technical work needed for the proper working of the H1 detector. The main three areas have been the Liquid Argon Calorimeter, the Central Trigger system and the Central Silicon Tracker.

In the side of physics analysis, Mexicans have been involved in the studies of QCD in exclusive final states, in studies of inclusive cross sections at low resolutions and large energies and in the activities of the Heavy Flavor group where we have contributed to the search of J/Ψ and D^* particles and started the new analysis to measure Λ_c baryons.

Currently the group is stable and growing. New students are joining and this attracts still more students. The group plans to keep contributing to the H1 effort up to the end of the Collaboration around 2010 or 2011.

REFERENCES

1. H1 Collaboration, Nucl. Instr. and Meth. A386 (1997) 310-347; 348-396.
2. http://www-h1.desy.de/publications/H1_sci_results.shtml

Radiographic Images Produced by Cosmic-Ray Muons

Rubén Alfaro

Instituto de Física, Universidad Nacional Autónoma de México,
Apartado Postal 20-364, 01000, México D.F., México

Abstract. An application of high energy physics instrumentation is to look for structure or different densities (materials) hidden in a matrix (tons) of material. By tracing muons produced by primary Cosmic Rays, it has been possible to generate a kind of radiographs which shows the inner structure of dense containers, monuments or mountains. In this paper I review the basics principles of such techniques with emphasis in the Sun Pyramid project, carried out by IFUNAM in collaboration with Instituto Nacioanal de Antropologia e Historia.

INTRODUCTION

It is well known that cosmic rays muons are illuminating every substance on the earth. Since they are very penetrating they can reach any object at the surface or underground and the attenuation (or dispersion) that they suffered has been used to probe the inner-structure of very dense matrix of materials. From cross-border surveillance to Archeological or geophysical studies; radiography with muons is a today application of the instrumentation original developed for high energy experiments. Depending of the size and density of the objects two different procedures can be used to obtain an image, in the next section the physics of each approach will be discussed. In the last section, there is a detail discussion about how these techniques are helping to solve a very important question around the Mexican archeologists: Is the Pyramid of the Sun at Teotihuacan a ceremonial monument or could it be a Mausoleum?

DISPERSION AND ATTENUATION

In X-ray radiography the intensity of an image pixel is determined mainly by the attenuation of the incident beam caused by the absorption, the maximum mean free path expected for photons is around 25 g cm^{-2} for all materials, so two objects (3cm thick) of high density (steel and tungsten for example) will be almost undistinguishable. The second limitation is the sizes of the object. Sampling objects of meters required large sources of radiation which usually are not available. In these cases natural background muons are a suitable alternative to obtain radiographic imaging.

Muons are the most numerous cosmic-ray particles at sea level, having a rate of about 10,000 part m^{-2} min^{-2} in horizontal detectors [1]. These particles are highly penetrating, a typical muon of energy 3 GeV will penetrate more than 1000 g cm^{-2} (10 m of water for example). They interact with the materials mostly by electromagnetic interactions, as a result when a muon of energy E pass trough a piece of material it suffered a deviation from its original trajectory, if the piece is thick enough the muon can be stopped on it, otherwise the muon probably will go through.

In the first case, the many small coulomb interactions add up and yield an angular deviation that roughly follows a Gaussian distribution,

$$\frac{dN}{d\theta_x} \approx \frac{1}{\sqrt{2\pi}\theta_0} e^{-\theta_x^2/2\theta_0^2}$$

with the width, θ_0 related to the scattering material through its radiation length, L_0, as follows

$$\theta_0 = \frac{13.6}{\beta cp}\sqrt{\frac{L}{L_0}}[1 + 0.038 \ln(L/L_0)]$$

where p the particle's momentum in MeV c^{-1} and βc is its velocity [2]. The radiation length decreases rapidly as the atomic number of a material increases, and θ_0 increases accordingly; in a layer 10cm thick, a 3 GeV muon will scatter with and angle of 2.3 milliradians in water, 11 milliradians in iron and 20 milliradians in lead. By tracking the scattering angles of individual particles the scattering material can be mapped. Since it is necessary to know the originally direction of the muons, two tracking devices are necessary, one on top of the sampling object and one under it, besides that, high energy muons deviates less, muons with more that 20 GeV became a noise-background that is necessary to eliminate.

In the second case, only muons with enough energy will go through the sampling object, so a fraction of the muons is absorbed. The range can be obtained by integrating this expression [3]

$$\frac{dE}{dX} = [1.88 + 0.077 \ln(E/M) + 3.9E] \times 10^{-6}$$

with E in TeV and dE/dX in TeV g^{-1} cm^2.

The two first terms represents the ionization loss and the third term represents various stochastic processes, mainly due to bremsstralung. By taking a relatively small contribution of the *ln* term, at high energies, the mean range X can be approx by

$$X = 2.5 \times 10^5 \ln(1.56E + 1)$$

where E is giving in TeV and X in gr cm^{-1}. If it is known the muon angular and energy distribution then it is possible to map the object.

Both methods have their own advantages and disadvantages, the first one relies less on simulation, because the energy distribution is dominated by low energy muons, corrections by the high energy noise-background can be neglected. For the same reason is not as penetrating as the second case. Besides, it requires two tracking devices and their sizes have to been of the order magnitude of the sampling object, otherwise the method is little effective. In second case it is possible to probe large objects (like mountains, dams or monuments like Pyramids), however simulations about angular and energy distributions as long as the geometrical shape and material's density play an important role. The quality of the image depends strongly in these parameters. As a practical case, in the next section it will be described an experiment, which is carried out in Teotihuacan Mexico, to obtain a radiographic image of the Pyramid of the Sun.

THE SUN PYRAMID EXPERIMENT

24 Years Ago

In 1968 Luis Alvarez carried out his famous muon detection at Chephren Pyramid, in Giza [4], the mail goal was to search for possible hidden chambers above the Belzoni Chamber. 24 years latter the Group of Experimental High Energy Physics from IF-UNAM is working on apply the Alvarez technique to the Mexican Pyramid of the Sun at Teotihuacan. In spite of its fame, little is known about it. Some excavations, early in the 20 century showed no identifiable internal structures, however the Pyramid of the Moon has them and therefore the purpose of the Sun and Moon Pyramid is unclear. A revealing discovery made in the early 1970's was the existence of a natural tunnel running 8 meters under the Pyramid of the Sun, ending beneath its symmetry axis figure1. Besides the archeological implications, it represents the unique

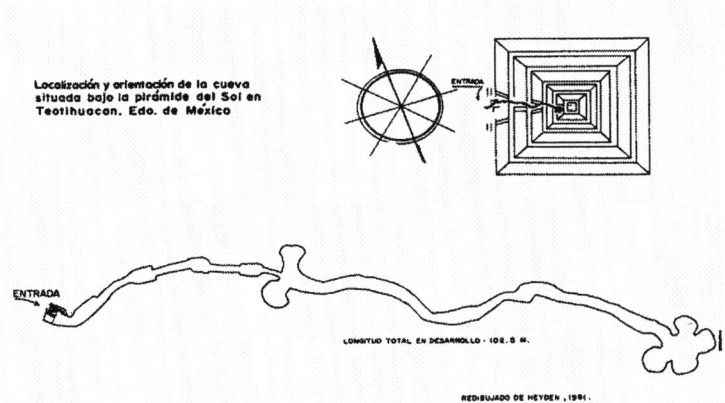

FIGURE 1. The natural tunnel runs 8m below the Pyramid of the Sun base. It almost follows a straight line to the center of the pyramid.

advantage of providing a site to install a cosmic ray muon detector to search for possible (>1m³) cavities in the body of the pyramid. In the next sections, the experimental set up and the relevant experimental parameters will be discussed

Method

As a first approach the experimental setup considered is similar to the one used by Alvarez [4] as it satisfies the basic requirements of being simple and low cost. The design consists on a three plane multiwire proportional chambers (MWPC) tracker and two scintillators counter for triggering purposes figure 2. To determine the important parameters (sensitivity, resolution and efficiency) it is necessary to take into a count what is know about the local muon spectrum, the pyramid's external geometry and the internal structure, estimations of sensitivity and expected resolution are very helpful for understanding the technique and optimizing it.

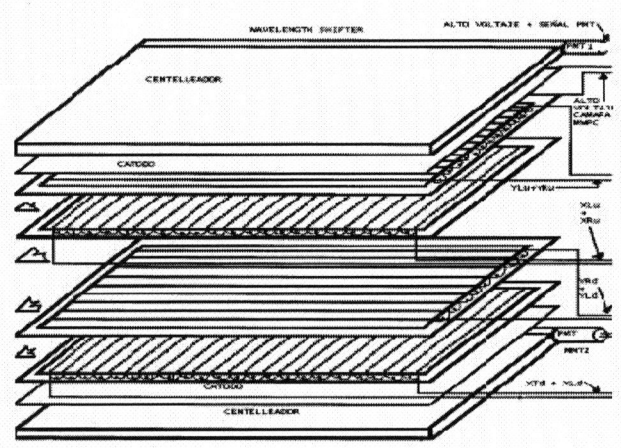

FIGURE 2. Proposed detector. .

The Alvarez method [4] consists of two important aspects, simulation and experiment. The first part requires the best possible knowledge of the pyramid's external dimensions, its detailed geometrical shape, and a guess of its internal materials (elemental composition and density distribution), all of this assuming that the pyramid is solid (cavity free). Then a simulated muon distribution reproduced by a hypothetical detector, having the same structure as the real one, is subtracted from the experimental observations to search for possible differences. Significant deviations in a given direction indicate and appreciable mater density difference in the corresponding subtended volume. About the mean composition and density distribution it was assumed that are similar to those found in early excavations by Nogera and Gamio, who excavated a 2m high, 1mwide tunnel running in a near straight line across the 226m pyramid base and 8 m above the natural observation tunnel. The Noguera-Gamio tunnel is also valuable, as it represents a well located cavity to be used for calibration purposes. Another important problem is that the density ρ of mass inside the pyramid is heterogeneous and its mean value is smaller than the density of rock. Hence if a hidden chamber has rocky walls, this could compensate the effect of a gas filled cavity on ΔM, the difference of mass with or without cavity along the muon path on which the method is based [4]. To quantify this, the detected cavity size L_d is define as

$$L_d = L_r - L_w \frac{\rho_w - \rho_p}{\rho_p} \quad (1)$$

where L_r represents the real cavity size, L_w is the total wall thickness, ρ_w and ρ_p are the wall and pyramid mean densities. Then total wall-cavity compensation for a rocky wall ($\rho_w = 2.65$ g/cm^3) results when $L_r / L_w = (\rho_w - \rho_p) / \rho_p = 0.4$.

Sensitivity

The amount of matter M traversed by a muon along its trajectory of length L, may be estimated trough the underground muon count N. The *"sensitivity"* ξ of this method may be defined as the ratio

$$\xi = \frac{\Delta N}{\sqrt{2N}} \quad (2)$$

where $\Delta N = N_1 - N_2$, using N_1, N_2 to denote the muon count in the cases where there is and there is not cavity respectively. The N value to be used in the denominator may approximated by the average number of counts $(N_1 + N_2)/2$ so that $(2N)^{1/2}$ represents the uncertainty of ΔN. For the muon flux we used the distribution proposed by Bedewi [5]

$$F(E, \theta) = kE^m \cos^n(\theta) \quad (3)$$

where E is the muon energy in GeV, θ is the polar angle referred to the vertical direction, n and m are slowly varying functions of the muon energy [5]. The proportionality constant k is mostly a function of the location altitude and affects the observation time. Since the pyramid is an energy filter inside the tunnel the important muon energy is in the region from 18 to 35 GeV, where the *m*- value rages from -1.5 to -1.72, and n from 0.3 to 0.4. By numerical integration of eq.3 and using eq.2 the statistics necessary to obtain a given sensitivity may be estimated by

$$N \approx 0.9 \left(\xi \frac{L}{X} \right)^2 \quad (4)$$

where L is the muon path inside the pyramid (maximum 80m) and X the size of the cavity. The above estimation ignores multiple scattering, a process which will be discussed in latter section. Also the above argument are based on a constant energy loss hypothesis, which not exactly because for high energy (>100 GeV) the muons loose energy through more complex mechanism due to radiative processes, but their corresponding flux is sufficiently small to make their contribution insignificant.

Resolution

Another important aspect of the experiment is the space resolution, which depends on the detector ability to reconstruct particle tracks. The resolution is quantified as the uncertainty associated to the reconstructed "entry point" of muons on the external boundary of the pyramid. Multiple scattering and detector reconstruction limitations transform this point into a finite size radial distribution known as the "point spread function". Roughly speaking this function resembles a gaussian having a standard deviation σ_r which can be taken as an estimate of the resolution. There is a compromise between resolution and sensibility, because higher the energy of the muons better the resolution (less multiple scattering) however the sensitivity decrease (less attenuation). Besides this fact, the density and shape of the pyramid are critical parameters in the simulations. The estimated mean material density originates in a description given by Millon [6] from 45 photographs taken in some regions along the Nogera-gamio tunnel (since this tunnel was walled 30 years ago Millon report is best reference). From the photographs, each showing $\approx 3m^2$ tunnel section, Millon published drawings with a code describing the space distribution of the most important filling materials such as sand, loam of various colors, adobe, volcanic tuff, scattered stones etc. From this information is expected to be fairly uniform, having an overall mean value of $\rho = 1.9$ g/cm^3. About the geometrical shape has been extracted from a level diagram obtained 30 years ago using an aero-photographic techniques. This provides information every 15-30cm in the horizontal plain, and every 100cm in the vertical axis. The topological diagram has been digitized obtaining a 750 000 point description of the pyramid surface figure 3. In particular the location of natural tunnel is only known to within an accuracy of ≈ 0.5m. Since the typical length is Ł ≈ 50m then typically $\Delta L / L \approx 1\%$. Under this conditions and combining geometrical, density and muon distribution uncertainties the minimal cavity size able to be detected is ≈ 0.7m^3 (supposing $\sigma_r \approx 50$ cm).

 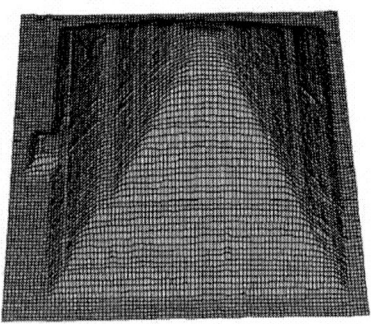

FIGURE 3. Topological diagram and the 3D reconstruction of the pyramid.

Simulations

The simulations has been carried out using the simulation package Geant4 which allows reproducing the relevant processes and helped to optimize the detector design. Since σ_r is not zero, with the simulations was possible to determine the minimal experimental σ_r as function of the wire spacing in the MWPC figure 4. As it is showed a 5mm pitch is enough to get a 50cm σ_r resolution. The impossibility to decrease σ_r mainly is due to multiple scattering interactions.

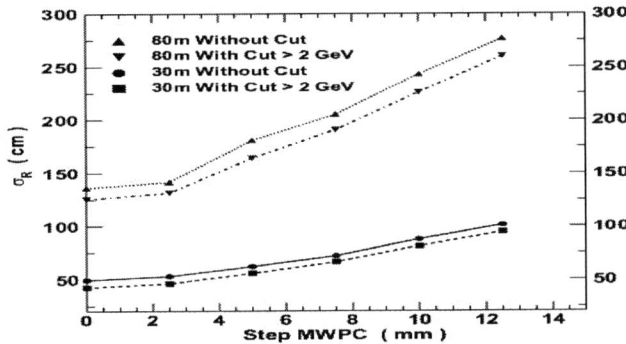

FIGURE 4. Resolution vs wire spacing in the MWPC.

Detector construction

The most complicated part of the tracker is the MPWC, as schematically is shown in figure 5, the anode consist of 200, 1 m long 25µm diameter gold-plated tungsten parallel wires having a 5mm pitch, forming a sensitive area of 1m x 1m. The wires are soldered to two, 1m long, circuit board strips, glued to (and insulated from) stress-free square aluminum frame. Two single-side G10 sheets fixed to a 1.1m x 1.1m acrylic support plates acting as cathodes, and as part of the gas container system. The cathodes are kept parallel to the anode, with a 5mm separation from it by an acrylic

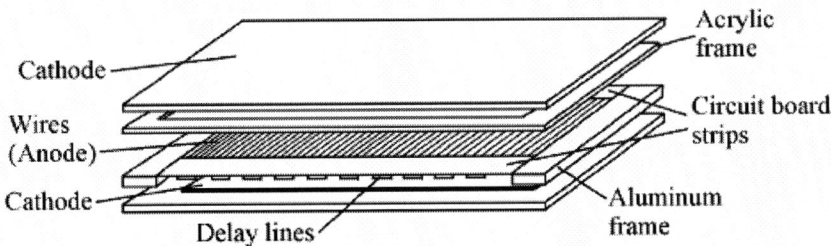

FIGURE 5. MWPC schematics.

frame. A low cost CO2 Ar (20%+80%) was used. The nominal detector bias was -1900 V. The read-out is carried out by the delay line technique. The fast trigger is generated trough the fast coincidence between 1m x 1m x 0.1cm plastic scintillator. (EJ-208) having a 1cm x 1cm x 1m wavelength shifting bar (BCF-92) optically coupled to a R1450 Hamamatsu photomultiplier. Although an interpolation read-out has been used, the position resolution achieved is very good for the detector size (FWHM ≈ 0.7mm) figure 6. The full tracker is formed by 8 modules 6 MWPC (3X and 3Y) and 2 Plastics scintillators (figure 7), this modular design is obligated by the natural tunnel dimension because the narrower part is only 1.2 m by 0.49cm. The approximate weight of each module is 60Kg, so the full tracker weight (including the mechanical support frame) is about 500Kg (figure 7).

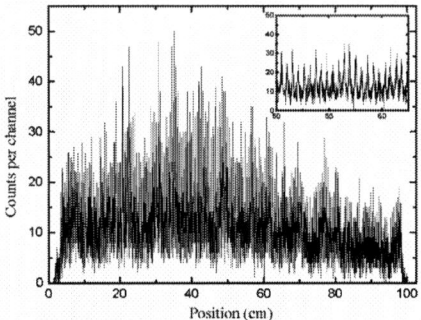

FIGURE 6. Position resolution at the IFUNAM laboratory.

FIGURE 7. The muon tracker, 6 MWPC and 2 scintillators, with electronics and gas handling system.

At the present time the full tracker is under testing and calibration, first in the lab, latter in a shack on the gardens of the IFUNAM. A similar shack has been installed at the end of the tunnel and in October of 2006 (figure 8) is expected to move the tracker in to the tunnel. Preliminary counts rates carried out at the tunnel (figure 8) implies that at least the tracker will take data for about one year in order to get enough statistic to compare with the simulations.

FIGURE 8. Pictures of the shack installed at the end of the tunnel and first muons measurements.

CONCLUSIONS

The search for possible hidden chamber in the Pyramid of Sun is possible using a detector as the one described in last section. A modified version of such tracker could be also used as monitor of gigantic geophysical structures [3], or dam walls, looking for possible cracks, obviously the simulations and resolution has to be adjusted to each situation. It can be also used in surveillance for cross-border transport checking, looking for forbidden dense objects (as the nuclear materials) hidden in big containers, which are difficult to sample by the normal X-rays monitor, actually in the United States some laboratories has been founded to do research in this directions [7].

ACKNOWLEDGMENTS

I thank to my colleagues from IFUNAM, who participate in The Pyramid of the Sun project. E. Belmont, V. Grabski, A. Martinez, A. Menchaca, M. Moreno and A. Sandoval.

REFERENCES

1. Grieder P.K.F.., *Cosmic Rays at Earth*, Amsterdam: Elsevier Science, 2001, pp. 40.
2. Hagiwara Y. et al., *Phys. Rev.* **D66**, 01001 (2002).
3. Nagamine K. et al., *Nuc. Inst. Meth.,* **A356**, 585 (1995).
4. Alvarez L.W. et al., *Science* **167**, 832 (1970).
5. Bedewi F. E.et al., *J. Phys A.* **5**, 292 (1972).
6. Millon R. et al., *Trans. Amer. Phil. Soc.* **55**, 6 (1965).
7. Priviate comunication during a meeting wtih Dr. Rich Brey from Idaho Acceletardor Center, Idaho State University.

Astroparticle Physics: Detectors for Cosmic Rays

Humberto Salazar* and Luis Villaseñor[†]

*Facultad de Ciencias Fisico-Matematicas, BUAP, Puebla Pue., 72570, Mexico.
[†]Institute of Physics and Mathematics, Universidad Michoacana, Edificio C3, Ciudad Universitaria, Morelia, Mich., 58040, Mexico.

Abstract. We describe the work that we have done over the last decade to design and construct instruments to measure properties of cosmic rays in Mexico. We describe the measurement of the muon lifetime and the ratio of positive to negative muons in the natural background of cosmic ray muons at 2000 m.a.s.l. Next we describe the detection of decaying and crossing muons in a water Cherenkov detector as well as a technique to separate isolated particles. We also describe the detection of isolated muons and electrons in a liquid scintillator detector and their separation. Next we describe the detection of extensive air showers (EAS) with a hybrid detector array consisting of water Cherenkov and liquid scintillator detectors, located at the campus of the University of Puebla. Finally we describe work in progress to detect EAS at 4600 m.a.s.l. with a water Cherenkov detector array and a fluorescence telescope at the Sierra Negra mountain.

Keywords: Cosmic rays; Extensive air showers; Water Cherenkov detectors; Liquid scintillator detectors.

DETECTION OF ISOLATED MUONS AND ELECTRONS

Muon Decay in a Liquid Scintillator Detector

Primary cosmic rays are composed of protons (85%), alpha particles (12%), electrons (2%) and heavier nuclei (1%). The primary cosmic rays that reach the earth atmosphere collide with nuclei of nitrogen and oxygen of the stratosphere to produce numerous secondary particles collectively called extensive air showers (EAS). The measurement of the muon lifetime is one of the simplest experiments that can be used to illustrate the basic elements of modern high energy physics experiments. We have used these experiments in the past to motivate students to appreciate the beauty of experimental high energy physics [1, 2, 3].

The muons used for this first experiment were cosmic ray muons resulting from the decay in flight of secondary pions and kaons. At sea level the muon flux is about 180 $m^2 s^{-1}$ and it corresponds to approximately 75% of the flux of charged cosmic rays. The muon flux has a mean energy of 2 GeV and a differential spectrum falling as E^{-2} up to a few TeV. The muon lifetime at rest is of tens of kilometers reach the ground before decaying is a consequence of the relativistic time dilation.

Fig. 1 shows 1 340 845 data events corresponding to a data taking period of 2 572 h with a 35 l detector. On this plot a four-parameter fit of these data to the function $P_1 e^{-t/P_2} + P_3 e^{-t/P_4} + 156.8$. Where 156.8 events per bin corresponds to the previously measured noise due to accidental coincidences; P_2 corresponds to the lifetime of μ^+, P_4 corresponds to the lifetime of μ^- which is reduced due to its inverse beta decay

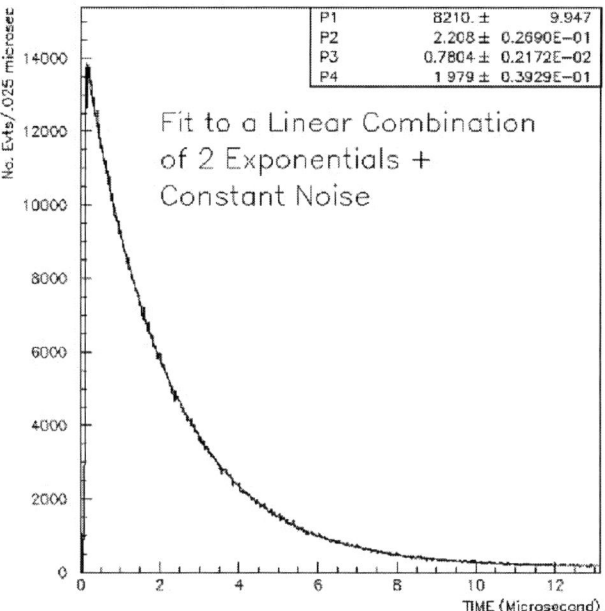

FIGURE 1. Raw muon data with a fit superimposed. The solid curve is a fit to the data using the function $P_1 e^{-t/P_2} + P_3 e^{-t/P_4} + 156.8$. The muon lifetime from this fit is $\tau = 2.208 \pm 0.027 \mu s$ and the measured plus to minus charge ratio of cosmic rays muons is $P_3^{-1} = 1.28 \pm 0.06$, i.e., in good agreement with the literature.

interaction with the protons of the nuclei of the liquid scintillator. In turn, P_2, gives a measurement of the intensity ratio μ^- to μ^+.

The result obtained for the muon lifetime was $\tau = 2.208 \pm 0.027 \mu s$, in good agreement with the literature. From the fit we also measured the plus to minus charge ratio of cosmic rays muons at a latitude of 20° North; the result obtained was 1.28 ± 0.06, also in good agreement with the literature.

Muon Decay in a Water Cherenkov Detector

Fig. 2 shows the experimental setup and data acquisition system used to study muon dacay in a water Cherenkov detector (WCD); the latter consists of a cylindrical aluminum container of 26 cm of diameter and 36 cm of height filled with 35 l of liquid scintillator. The sealed container has a 2 inch photomultiplier tube (PMT) looking downwards from the top of the liquid. We also used this setup to measure the absorption length of UV light in water [4]

This experiment confirmed the viability of the use of muons stopping and decaying inside the detector to calibrate and monitor the Auger Observatory WCDs [5] in a remote way [6, 7, 8]. Three clear peaks of PMT charge distributions were identified. All of them

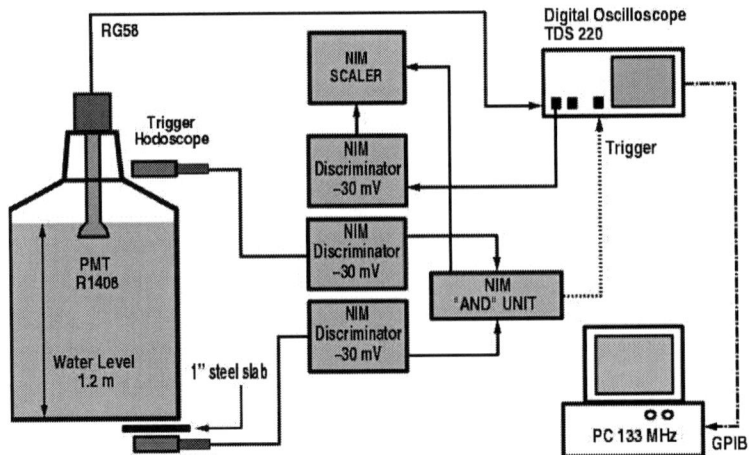

FIGURE 2. Experimental setup and data acquisition system for the muon decay experiment with a water Cherenkov detector.

are useful for calibration and monitoring of WCDs: one for stopping muons, one for decay electrons and one for crossing muons.

Fig. 3 summarizes our results for the study of muon decays in a water Cherenkov detector, for additional details see [6]. Fig. 3.a shows the charge distribution for the first pulse obtained by requiring $Q_2 > Q_1$ and time difference between consecutive pulses $< 8\mu s$. The upper plot corresponds to data and the lower to simulations. The solid line is a gaussian fit. Fig. 3.b shows the charge distribution for the second pulse obtained by requiring $Q_2 > Q_1$ and time difference between consecutive pulses $< 8\mu s$. The upper plot corresponds to data and the lower to simulations. The solid line is a gaussian fit. Fig. 3.c shows the charge distribution for the first pulse obtained by requiring $Q_2 < Q_1$. The upper plot corresponds to data and the lower to simulations. The solid line is a gaussian fit. Fig. 3.d shows the distribution of the time difference between the first and second pulses obtained by requiring $Q_2 > Q_1$. The solid line corresponds to an exponential curve with a decay constant of 2.09 μs. We see good agreement between data and simulation.

If we assume that a VEM corresponds to 2 MeV/cm times the 120 cm tank height we obtain as a first approximation that 1 VEM = 240 MeV. Assuming that this conversion factor applies for decay electrons, which are relativistic for most of their paths as long as their energies are above a few MeV, we obtain a mean energy for the decay electrons of 240 MeV times 0.17 = 41 ±11 MeV, i.e., in agreement with their maximum possible energy of 53 MeV.

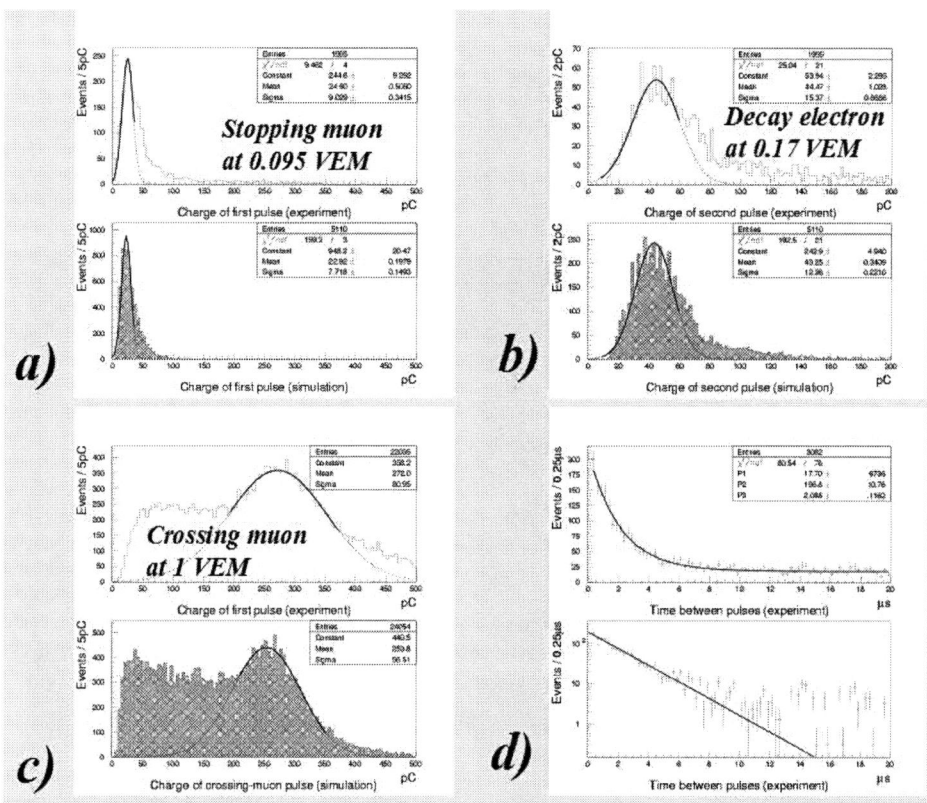

FIGURE 3. **a:** Charge distribution for the first pulse obtained by requiring $Q_2 > Q_1$ and time difference between consecutive pulses $< 8\ \mu s$. The upper plot corresponds to data and the lower to simulations. The solid line is a gaussian fit. **b:** Charge distribution for the second pulse obtained by requiring $Q_2 > Q_1$ and time difference between consecutive pulses $< 8\ \mu s$. The upper plot corresponds to data and the lower to simulations. The solid line is a gaussian fit. **c:** Charge distribution for the first pulse obtained by requiring $Q_2 < Q_1$. The upper plot corresponds to data and the lower to simulations. The solid line is a gaussian fit. **d:** Time distribution for the time difference between the first and second pulses obtained by requiring $Q_2 > Q_1$. The solid line is a linear fit to the data; the muon lifetime measured this way is 2.09 μs, i.e., in agreement with our expectation given the inverse beta decay interaction of negative muons with the oxygen nuclei.

Identification of Isolated Muons and Electrons in Water Cherenkov Detectors

The experimental setup used to identify and statistically separate isolated electrons from isolated muons is shown in Fig. 2. Fig. 4 shows the amplitude vs. rise time from 10% to 90% for a water Cherenkov detector **located inside a building**. The events labeled as Electrons, with rise times around 10 ns, include mostly knock-on and muon-decay electrons produced in the concrete ceiling and walls of our laboratory housing our

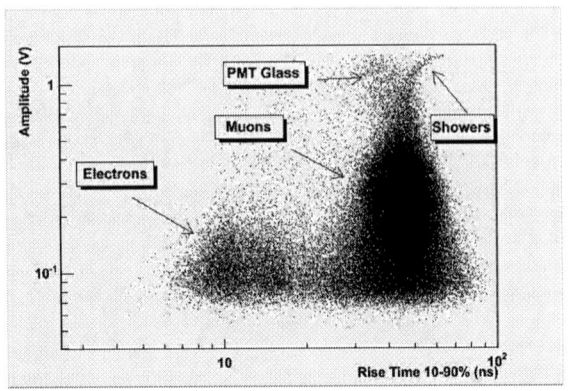

FIGURE 4. Amplitude vs. rise time from 10% to 90% for signals from the water Cherenkov detector shown in Fig. 2 **located indoors**. The different labeled components are clearly separated in this 2-parameter.

detector [9].

Fig. 5 shows the PMT charge distribution for the same data. The dashed-line data labeled as Electrons correspond to events with rise time 10-90% < 20 ns, while the solid-line data labeled as Muons correspond to events with rise time 10-90% > 20 ns. The shaded curve corresponds to vertical non-central muons. The horizontal scale has been chosen so that the MPV of the latter curve equals 1 VEM: it corresponds 74 photoelectrons.

In contrast, Fig. 6 shows data taken with an identical setup but **this time located outdoors**, i.e., outside the building. The events labeled as Muons, with rise times around 45 ns, include vertical, inclined and corner clipping muons; they occur at a measured rate of around 870 Hz. For comparison, the events labeled as Electrons, selected by requiring Q/A < 0.5 and rise time 10%-90% < 0.5 occur at a rate of 80 Hz.

Note that the ratio for the MPV of the electron/muon peaks of 0.045, i.e., in good agreement with our expectations for electrons with energies around 10 MeV which would deposit all of their energy in about 5 cm of liquid compared to the about 240 MeV energy deposition for vertical muons crossing the whole tank height of 120 cm. This value is considerably lower than the same ratio (0.12) for the tank located inside the building, see Fig. 5, in which the electron peak is dominated by knock-on and decay electrons with higher energies, i.e., up to 53 MeV for decay electrons.

We have also used these data to attempt to measure the muon contents of EAS by means of neural networks by exploiting the different temporal structures of EM showers with different muon contents [10].

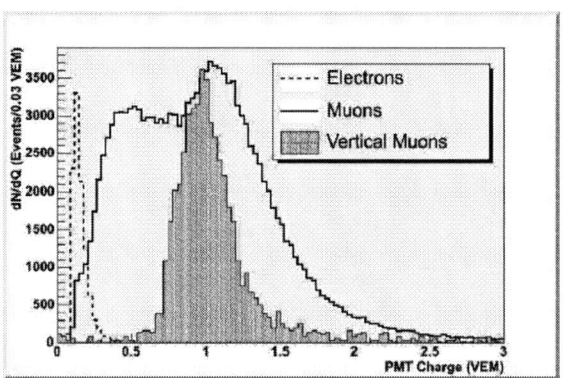

FIGURE 5. PMT charge distribution for the arbitrary muon trigger events. The dashed-line data labeled as Electrons correspond to events with rise time 10-90% < 20 ns, while the solid-line data labeled as Muons correspond to events with rise time 10-90% > 20 ns. The shaded curve corresponds to vertical non-central muons. The horizontal scale has been chosen so that the MPV of the latter curve equals 1 VEM: it corresponds to 107.5 pC, or, equivalently, 74 photoelectrons.

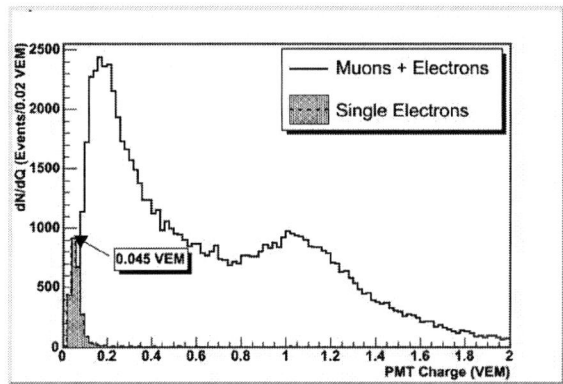

FIGURE 6. Spectrum of charge depositions scaled in VEMs; the shaded histogram corresponds to electrons selected by requiring Q/A < 0.5 and rise time from 10% to 90% in amplitude < 0.5. The first peak of the non-shaded histogram is dominated by corner-clipping muons, low-energy muons and electrons from muon decays; the second peak is dominated by muons crossing the whole depth of the the detector in all directions.

Identification of Isolated Muons and Electrons in Liquid Scintillator Detectors

Fig. 7 shows data taken with a setup similar to that shown in Fig. 2, but for a liquid scintillator detector. The shaded histogram gives the distribution of the PMT charge for penetrating muons triggered by a scintillator paddle placed below the scintillation detector and shielded with a 2 cm steel slab. The first peak of the non-shaded histogram corresponds to electrons and corner-clipping muons while the second peak is dominated

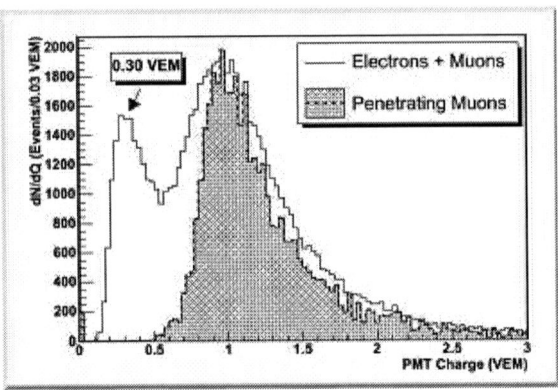

FIGURE 7. Spectrum of PMT charge deposition taken with a setup similar to that shown in Fig. 2, but for a cylindrical detector filled with liquid scintillator up to a height of 13 cm. The shaded histogram gives the distribution of the PMT charge for penetrating muons. The first peak of the non-shaded histogram corresponds to electrons and corner-clipping muons while the second peak is dominated by crossing muons.

by crossing muons. The ratio of the MPVs of these two peaks is about 3.3, i.e., in rough agreement with the fact that crossing muons deposit around 26 Mev of energy in 13 cm of liquid while low energy electrons deposit all of their energy, i..e., around 10 MeV.

DETECTION OF EXTENSIVE AIR SHOWERS

Introduction

The collisions of primary cosmic rays with nitrogen and oxygen nuclei high in the Earth atmosphere give rise to extensive air showers (EAS. The four components of extensive air showers are: 1) The hadronic component including hadrons such as protons, neutrons, pions and kaons; this component is very attenuated at sea level. 2) The electromagnetic component composed of electrons, positrons and photons originates either from primary cosmic rays or from the decay photons of neutral pions; 3) The muon component is composed of the decay muons of pions and kaons; at sea level 80% of all charged particles are muons and 20% are electrons; 4) The neutrino component is made up of neutrinos coming from decays of pions, kaons and muons; this is the non-interacting component of extensive air showers.

EASs can be studied by measuring their particle densities as they arrive at the ground by means of ground detectors or their particle densities as they traverse the atmosphere by means of fluorescence or Cherenkov light telescopes on the ground.

It has been found that the energy spectrum of primary cosmic rays is well described by a power law, i.e., $dE/dx \sim E^{-\gamma}$, over many decades of energy with the spectral index γ approximately equal to 2.7, and steepening to $\gamma = 3$ at $E = 3 \times 10^{15}$ eV [11]. This structural feature is known as the "knee" of the cosmic ray spectrum.

The nature of the knee is still a puzzle despite the fact that it was discovered more than 46 years ago [12]. Most theories consider its origin as astrophysical and relate it to the breakdown of the acceleration mechanisms of possible sources within our galaxy or to a leakage during propagation of cosmic rays in the magnetic fields within our galaxy; in particular, these theories lead to the prediction of a primary composition richer in heavy elements around the knee due to the decrease of galactic confinement of cosmic rays with increasing energy of the primary cosmic rays. Alternatively, there are scenarios where a change in the hadronic interaction at the knee energy gives rise to new heavy particles [13] which produce, upon decay, muons of higher energies than those produced by normal hadrons.

The best handle to study the composition of primary cosmic rays by using ground detector arrays is the measurement of the ratio of the muonic to the electromagnetic component of EAS; in fact, Monte Carlo simulations show that heavier primaries give rise to a bigger muon/EM ratio compared to lighter primaries of the same energy [14]. In fact, evidence for such variations has been reported recently [15].

Extensive Air Shower Detector Array at University of Puebla

The extensive air shower detector array at University of Puebla (EAS-UAP) was designed to measure the lateral distribution and arrival direction of secondary particles for EAS in the energy region of $10^{14} - 10^{16}$ eV. The special location of the EAS-UAP array; 2200 m above sea level; and all the facilities coming from the Campus of the University of Puebla make it a valuable apparatus for the long term study of cosmic rays and at the same time an important training center for new physics students interested in getting a first class education in the field of cosmic rays in Mexico.

The EAS-UAP array is located in the campus of the University of Puebla in Mexico (UAP) at 19° N, 89° W and 2100 m.a.s.l., i.e., 800 gcm^{-2}; it consists of 12 liquid scintillator detectors distributed uniformly on a square grid with spacing of 20 m, and six water Cherenkov detectors (one of 10 m^2 cross section and five smaller ones of 1.86 m^2), as shown in Fig. 8, where liquid scintillator detectors are represented by black cylinders and water Cherenkov detectors by stars. We make use of the natural flux of background muons and electrons to monitor and calibrate our detectors. For this purpose, single particle triggers are used simultaneously with EAS triggers.

For the location of the EAS-UAP, muons are the dominant contribution to the flux of secondary cosmic rays for energies above 100 MeV with a flux of about 85 $m^{-2}s^{-1}sr^{-1}$ and a mean energy of 4 GeV; at lower energies, up to 100 MeV, electrons, positrons and photons are also an important component; electrons coming from muon decays with energies up to 53 MeV dominate the flux of EM particles for detectors placed outside, while knock-on electrons are another important contribution for detectors placed inside or close to buildings.

We have reported on the performance of the EAS-UAP array elsewhere [16, 17, 18, 19]. The direction of the primary cosmic ray is inferred directly from the relative arrival times of the shower front at the different detectors. The core position, lateral distribution function and total number of shower particles on the ground are reconstructed from a fit of the measured electron-positron densities to the NKG [20] formula. A discussion of how this detector array is being used to measure the muon contents of EAS is given elsewhere [21].

Fig. 9 shows the measured particle densities and the fitted lateral distribution function

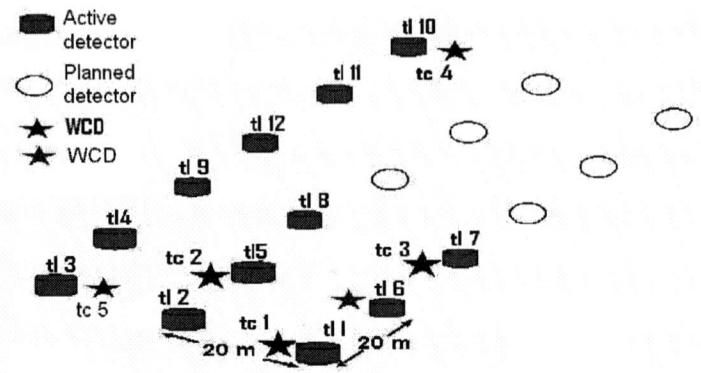

FIGURE 8. EAS-UAP array located on the Campus of the University of Puebla. Stars represent Cherenkov detectors filled with 2230 l of purified water and cylinders represent detectors filled with 130 l of liquid scintillator.

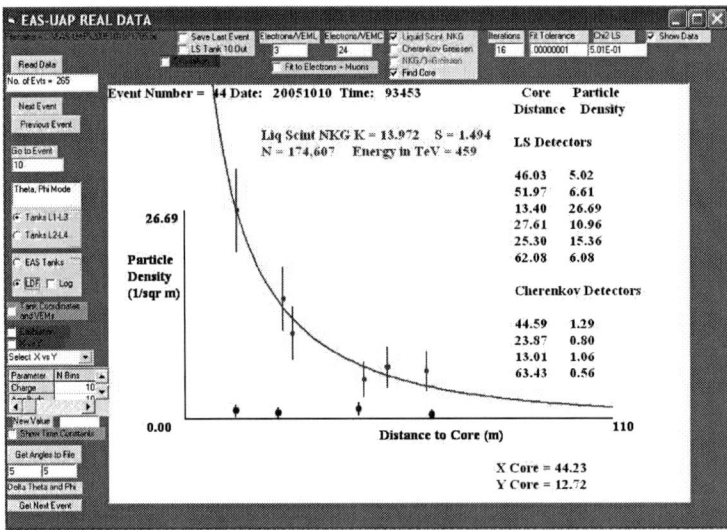

FIGURE 9. Typical event shown with the event display program of the EAS-UAP. The solid curve is a fit of the NKG formula to the measured lateral distribution particle densities using only the liquid scintillator detectors.

FIGURE 10. Fluorescence detector and electronics presently under construction to detect extensive air showers at Sierra Negra.

for a near-vertical shower. For this particular event the fitted energy of the primary cosmic ray was 459 TeV.

High Altitude Extensive Air Shower Detector Array at Sierra Negra

The Sierra Negra Observatory is a high mountain facility located at 19° N, 97.3° W and 4600 m.a.s.l., i.e., 590 gcm^{-2}, near the city of Puebla. An EAS detector is under construction at that location; this experiment has been designed to combine two detection techniques based on two different and independent detectors. For this purpose an array of water Cherenkov detectors is under construction at Pico de Orizaba at 4200 m.a.s.l. (620 gcm^{-2}) spread over 2.25 km^2 to measure the particle density at ground level, and a fluorescence telescope [22], also under construction at Sierra Negra, (590 gcm^{-2}), with a field of view overlooking the ground array.

The WCDs will have a continuum duty cycle that will allow us to use them to measure the arrival direction and energy distribution with a good statistics. In hybrid mode, the arrival direction and the lateral distribution function of the EAS will also be measured by the fluorescence telescope [23].

The TUS project [24, 25, 26] is managed by an international collaboration that aims to detect ultra high energy cosmic rays, i.e., cosmic rays with energies above 10^{19} eV, starting operation in the near future. The main innovation is that TUS will use one or more fluorescence telescopes located on a satellite. In this way the aperture of the

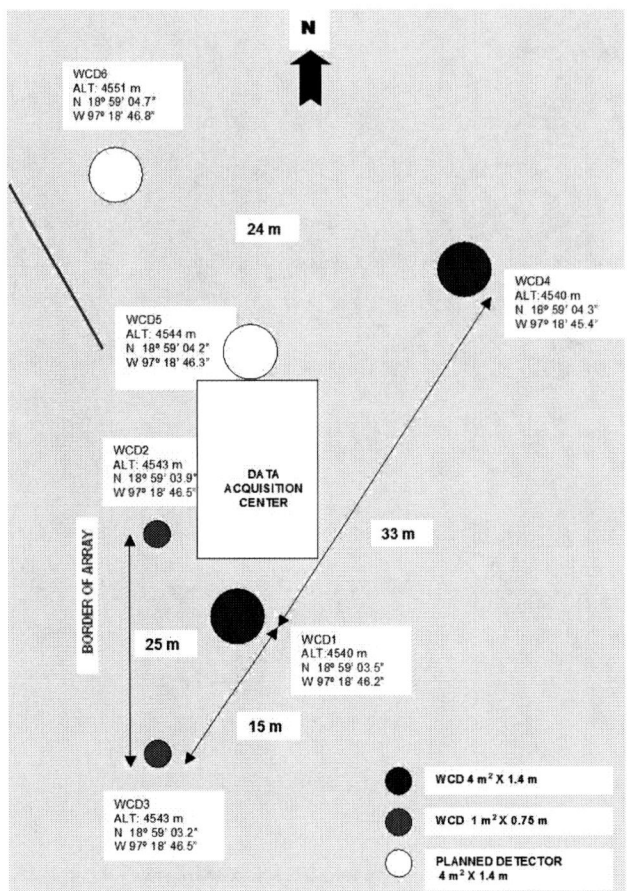

FIGURE 11. Water Cherenkov detector array located at Sierra Negra at an altitude of 4550 m.a.s.l.

instrument is greatly increased with respect to ground-based observatories. A prototype of the fluorescence telescope of TUS is being built in Puebla with the intention to test it at Sierra Negra.

Fig. 10 shows the fluorescence detector that is being built to detect EAS at Sierra Negra. The light produced by nitrogen molecules excited by the passage of charged particles is collected by the mirrors of the telescope which focuses it into a PMT camera where the signal is digitized [27].

An absolute calibration provides the conversion factor from FADC counts to calculate the number of photons arriving at the front of the mirror. The digitizing electronics required for this purpose, shown in Fig. 10, is also being developed at the University of Puebla. In addition, multi-anode PMTs are being used to develop a simple prototype of a ring imaging Cherenkov detector [28].

Search for Gamma Ray Bursts at Sierra Negra

Detection of GRBs from ground-based experiments has enormous potential for revealing new information about the energy spectrum, origin and propagation of γ-rays through space. This potential is greatly increased if GRBs are detected by two or more ground-based experiments without ambiguity. The possibility of continuously monitoring GRBs by observing GeV γ-rays by the Milagro (USA)-Sierra Negra (Mexico)-Chacaltaya (Bolivia)- Auger Argentina) network is discussed in [29].

Fig. 11 shows the array of water Cherenkov detectors located at Sierra Negra, at an altitude of 4550 m.a.s.l. At present, the array consists of 4 cylindrical light-tight water Cherenkov detectors; two of cross section equal to 4 m^2 and two of 1 m^2, filled with 750 l of ultra-pure water. The interior of each tank is covered with tyvek to reflect the Cherenvov light in a diffusive way. These detectors have a PMT each, located on the cylinder axis and looking downwards to the water volume to collect the Cherenkov light produced in the water. The PMT signals are read out by a DAQ system that measures the rates of secondary particles each tenth of second [30, 31].

Conclusions

We discussed several past and present detectors built in Mexico to measure properties of cosmic rays. First we described a simple measurement of the muon lifetime and the ratio of positive to negative muons, next we described the detection of decaying and crossing muons in water Cherenkov detectors and liquid scintillator detector. We also discussed a technique to separate isolated particles based on the temporal structure of their PMT pulses. Next we described the detection of extensive air showers (EAS) with a hybrid detector array consisting of water Cherenkov and liquid scintillator detectors, located at the campus of the University of Puebla. Finally we described work in progress to detect EAS at 4600 m.a.s.l. with a water Cherenkov detector array and a fluorescence telescope at the Sierra Negra mountain.

ACKNOWLEDGEMENTS

We would like to thank University of Michoacan, University of Puebla and CONACyT for supporting this work.

REFERENCES

1. J. Estevez, L. Villaseñor, A. Gonzalez y G. Moreno, Rev. Mex. Fis. 42 [4] (1996) 649-662.
2. L. Villaseñor, Proc. of the VII ICFA school on instrumentation in elementary particle physics, Julio 1997, AIP conference Proceedings 422. American Institute of Physics, eds. G. Herrera y M. Sosa (1998) 333-346.
3. L. Villasenor, Proc. of the First ICFA Instrumentation School at the ICFA Instrumentation Center in Morelia, eds. V. Villanueva and L. Villaseñor, AIP Conf. Proceedings Volumen 674 (2003) 237-245.

4. M. Alarcón, M. Medina, L. Villaseñor, E. Cantoral, A. Fernández, r. López, M. Rubín, S. Román, H. Salazar, M. Vargas, L. Nellen, J.C. D'Olivo, J. Valdés-Galicia, and A.Zepeda, Rev. Mex. Fis. 44 [5] (1998) 479-483.
5. Auger Collaboration, Nucl. Instr. and Meth. in Phys. Res. A. 523 issues1-2 (2004) 50-95.
6. M. Alarcon et al., Nucl. Instrum and Meths. in Phys. Res. A 420 (1999) 39-47.
7. J.C. D'Olivo, A. Fernandez, M. Medina, L. Nellen, S. Roman, H. Salazar, J. Valdes-Galicia, M. Vargas, L. Villasenor and A. Zepeda, Nucl. Phys. B(Proc.Suppl) 75A (1999) 389-391.
8. H. Salazar, L. Nellen and L. Villasenor, Proc. 27th. Intl. Cosmic Ray Conf., edited by K.H. Kampert, G. Heinzelmann and C. Spiering. Hamburg Germany, vol. 2 (2001) 752-755.
9. H. Salazar and L. Villasenor, Nucl. Instruments and Meths. in Phys. Res. A. Volume 553, Issues 1-2 (2005) 295-298.
10. L. Villasenor, Y. Jeronimo, and H. Salazar, Proc. 28th Intl. Cosmic Ray Conf., T. Kajita et al. (eds.), Universal Academy Press, Inc., Tokyo, Japan Vol. 1 (2003) 93-96.
11. S.P. Swordy et al., Astroparticle Physics, Volume 18 (2002) 129-150.
12. G.V. Kulikov and G.V. Khristiansen, Sov. Phys. JETP, 41 (1959) 8.
13. A.A. Petrukhin, Proc. XIth Rencontres de Blois "Frontiers of Matter" (The Gioi Publ., Vietnam, (2001) 401.
14. B. Alessandro, et al., Proc.27th Intl. Cosmic Ray Conf., 1 (2001) 124-127.
15. KASCADE Collaboration (Klages H. O. et al.), Nucl. Phys. B (Proc. Suppl.), 52 (1997) 92.
16. H. Salazar, O. Martinez, E. Moreno, J. Cotzomi, L. Villasenor, O. Saavedra, Nuclear Physics B (Proc. Suppl.) 122 (2003) 251-254.
17. J. Cotzomi, O. Martinez, E. Moreno, H. Salazar and L. Villasenor, Revista Mexicana de Fisica, 51(1)(2005) 38-46, .
18. J. Cotzomi, E. Moreno, T. Murrieta, B. Palma, E. Perez, H. Salazar and L. Villasenor, Nucl. Instrume. and Meths. in Phys. Res. A. Volume 553, Issues 1-2 (2005) 290-294.
19. H. Salazar, O. Martinez, J. Cotzomi, E. Moreno, S. Aguilar and L. Villasenor, Proc. of the First ICFA Instrumentation School at the ICFA Instrumentation Center in Morelia, eds. V. Villanueva and L. Villaseñor, AIP Conf. Proceedings Volumen 674 (2003) 385-388.
20. J. Nishimura, Handbuch der Physik XLVI/2, (1967) 1.
21. O. Martínez, E. Pérez, H. Salazar and Luis Villaseñor, Hybrid Extensive Air Shower Detector Array at the University of Puebla to Study Cosmic Rays, Proc. of the Mexican School on Astrophys 2005, Springer Verlag (2006) in press.
22. H. Salazar, M. Cuautle, J. Cotzomi, E. Moreno, S. Aguilar, E. Ponce, O. Martinez and L. Villasenor, Proc. of the First ICFA Instrumentation School at the ICFA Instrumentation Center in Morelia, eds. V. Villanueva and L. Villaseñor, AIP Conf. Proceedings Volumen 674 (2003) 296-304.
23. E. Ponce, H. Salazar, O. Martinez, E. Moreno, I. Pedraza, J. Cotzomi, E.Perez, L. Villasenor, B. Khrenov and G. Garipov, Proc. of the First ICFA Instrumentation School at the ICFA Instrumentation Center in Morelia, eds. V. Villanueva and L. Villaseñor, AIP Conf. Proceedings Volumen 674 (2003) 372-376.
24. V.I. Abrashkin et al., Int.J.Mod.Phys.A20 (2005) 6865-6868.
25. B.A. Khrenov et al., Phys.Atom.Nucl. 67 (2004) 2058-2061. Also in Yad. Fiz. 67 (2004) 2079-2082
26. V. Abrashkin et al., Adv. in Space Res. 37 Iss. 10 (2006) 1876-1883.
27. M. Cuautle, E. Moreno, I. Pedraza, T. Murrieta, G. Garipov, B. Khrenov, H. Salazar, O. MartInez, and L. Villaseñor, AIP Conf. Proc., Vol. 623 (2002) 403-406.
28. H. Salazar, E. Moreno, T. Murrieta and L. Villaseñor, Ring Imaging Cherenkov Detectors, Proc. of the X Mexican Workshop on Particle Phys., A. Bashir, V. Villanueva and L. Villasenor (eds.), AIP Conf. Series (2006) in press.
29. L. Villasenor, O. Martinez, H. Salazar, O. Saavedra, A. Velarde and A. Zepeda, Proc. of the 27th. Intl. Cosmic Ray Conf., edited by K.H. Kampert, G. Heinzelmann and C. Spiering. Hamburg Germany, vol. 2 (2001) 896-899.
30. H. Salazar, C. Alvarez, O. Martínez and L. Villaseñor, Proc. of the Mexican School on Astrophys 2005, Springer Verlag (2006) in press.
31. H. Salazar, L. Villasenor, C. Alvarez and O. Martinez, Search for Gamma Ray Bursts at Sierra Negra, Mexico, Proc. of the X Mexican Workshop on Particle Phys., A. Bashir, V. Villanueva and L. Villasenor (eds.), AIP Conf. Series (2006) in press.

Application of PIN diodes in Physics Research

F. J. Ramírez-Jiménez, L. Mondragón-Contreras, P. Cruz-Estrada

Instituto Nacional de Investigaciones Nucleares
Carretera México-Toluca S/N, La Marquesa, Ocoyoacac, 57150, MEXICO
e-mail: fjrj@nuclear.inin.mx

Abstract. A review of the application of PIN diodes as radiation detectors in different fields of Physics research is presented. The development and research in semiconductor technology, the use of PIN diodes in particle counting, X-and γ-ray spectroscopy, medical applications and charged particle spectroscopy are considered. Emphasis is made in the activities realized in the different research and development Mexican institutions dealing with this kind of radiation detectors.

Keywords: PIN diodes, X-rays, spectroscopy, charged particles.
PACS: 29.40.-n, 29.30.Kv

INTRODUCTION

Whereas the concept of PIN diodes is very old, it was mentioned in a paper [1] by Kleinman in 1956, its application to radiation detection is more or less recent; for example, Nowotny and Reiter [2] reported low energy photon measurements in 1977. Since the proposal of the basic silicon point contact diode by Pickard in 1946 and its application in radar technology [3], the need to have faster devices surely lead the researchers to the PIN diode concept in which less capacitance and less reverse current provided better characteristics. As any semiconductor diode, PIN diodes are very sensitive to light, thus they are widely used for light detection mainly in communications through optical fiber lines, taking advantage of its fast response, PIN diodes are used at frequencies of 50 MHz or more [4].

In this review article we describe the main applications of PIN diodes as radiation detectors in different fields of Physic research in Mexico, covering some aspects of the work realized in semiconductors technology for producing PIN diodes, nuclear counting experiments, measurement of intensity of X-ray beams, energy spectroscopy of γ-and X-rays and charged particles. These activities have led to useful practical applications in medicine, industry and basic research. Beside this description, the main applications worldwide are also mentioned including the more relevant bibliographic references.

PIN DIODE DETECTORS

The basic structure of a PIN diode is seen in Figure 1(a); the three regions, n^+, Intrinsic and p^+, are shown. Figure 1(b) shows the picture of a commercial PIN diode.

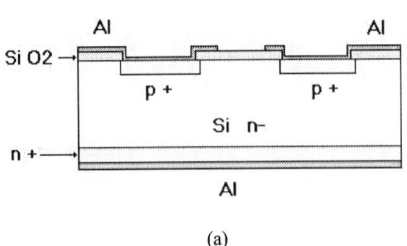 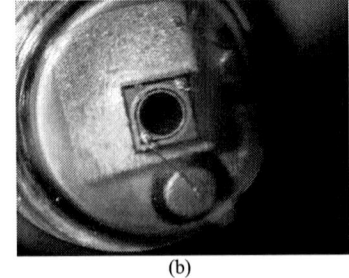

(a) (b)

FIGURE 1. a) Basic structure of a PIN diode. b) Picture of an OPF420 PIN diode as seen through an optical microscope.

Measurement of the radiation is based in the production of electron-hole pairs in the interaction of radiation with the detector material and the further collection of charges. There are two possibilities to operate the PIN diodes [5]: in photovoltaic mode without any bias voltage applied and in photoconductive mode with a reverse bias voltage applied to create a depletion zone inside the intrinsic region.

In Figure 2.a, the diode is in photovoltaic mode, the radiation is measured as an integral effect, a current is generated from the interaction of radiation with the detector, and the response is the average of the events inside the detector. This configuration is used in applications of photometry and to measure the intensity [6] of X-rays in radio diagnostic [7], the preamplifier converts the current to voltage, and the output signal has the waveform of the X-rays that reaches the detector, therefore it can be displayed in an oscilloscope.

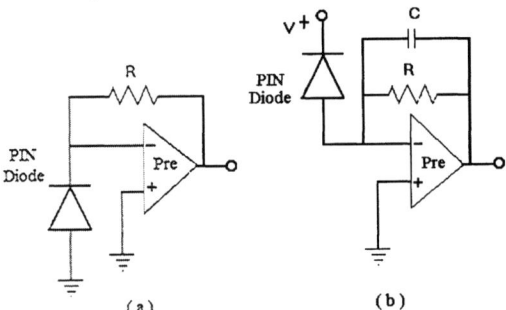

(a) (b)

FIGURE 2. Connections of PIN diode with a preamplifier to operate in: a) photovoltaic mode, b) photoconductive mode.

In Figure 2.b, the diode is connected in photoconductive mode, the radiation is measured in every single interaction, a charge is generated due to the interaction of radiation with the detector and it is directly proportional to the energy of the radiation, the preamplifier converts the input charge to voltage. With this configuration, spectrometric measurements can be done [8]. PIN diode detectors need to be operated with enough reverse voltage in order to reach the maximum depletion zone and therefore to get the best detection efficiency.

The intrinsic detection efficiency is around 100% for photons of a few keV, and decreases to around 2 % at 60 keV for a 300 μm thick PIN diode. The efficiency is around 100% for charged particles of moderate energy.

The electrical characteristics that are more related with the performance of PIN diodes as radiation detectors are the leakage current and the diode capacitance. Thanks to the good quality of silicon, the leakage current can be less than 100 pA at depletion voltage and the capacitance less than 2 pF for small diodes, therefore the PIN diodes can be used at room temperature with good performance for X-ray spectroscopy, but require a low input capacitance and low noise preamplifier as read-out circuit. Due to the very small signal generated in the detector for X-rays, the noise of the measuring system has to be considered with great care in this case.

CHARGE SENSITIVE PREAMPLIFIERS

The matching between the detector and preamplifier defines the low noise of the spectroscopy system, therefore for these applications; the best noise performance of the associated preamplifier is required. Generally, a field effect transistor, FET, is used as the input device. Special low noise preamplifiers have been developed to fulfill this requirement like the charge sensitive preamplifier that uses a feedback resistor, see Figure 2.b, in this case, the feedback capacitor is charged by the injected signal from the detector and discharged immediately through the feedback resistor. The feedback resistor has the disadvantage that it is an additional and undesired source of noise. The charge sensitive preamplifier with optical feedback [9] is another possibility that do not include an additional source of noise by a feedback resistor. The feedback capacitor is discharged through the input FET, when an optically coupled LED sends a reset light pulse to the FET every time the output voltage, reaches a defined level.

ENERGY SPECTROSCOPY OF RADIATION

In a basic energy spectroscopy system, the signal of charge generated in the detector by the radiation is conditioned in a preamplifier, a spectroscopy amplifier gives a further amplification and sets the optimal bandwidth of the system in order to get the best signal to noise ratio, the output from the amplifier is a series of analog pulses with a height directly proportional to the energy of the detected radiation, these pulses are analyzed considering its pulse height, classified an the result of this classification is shown in the screen of a multichannel analyzer system.

The total noise of the system is reflected on the spectrum in the width of the peaks, the energy resolution of the spectroscopy system is defined [10] as the Full Width at Half Maximum of a peak, FWHM, for example, the resolution for X-ray detectors is measured in the peak of 5.89 keV of an Fe-55 radiation source.

The total noise is the resultant of two uncorrelated components: the electronic noise associated mainly with the preamplifier and the electrical characteristics of the detector; and the statistical component due to the detection process inside the detector [9].

RESEARCH AND APPLICATION OF PIN DIODES

Semiconductor Technology of PIN Diodes

The research on semiconductor technology to produce PIN diodes started in Mexico in 1996 with the project: "Radiation Semiconductor Detectors" [11][12][13][14][15][16][17][18][19][20] financed by CONACYT[1] with the collaboration of several institutions as CINVESTAV[2], INAOE[3], ININ[4] and CEADEN[5]. A second project started in 1999: "Semiconductor Structures for Radiation Detectors" [21][22][23][24] with the participation of the same institutions that continued with research on crystalline and amorphous silicon PIN diodes and special preamplifiers [25][26][27][28][29][30]. Research continued to incorporate the amplifying device in the same chip with the PIN detector on high resistivity silicon [31][32][33][34][35][36][37].

Since 1996, several Italian groups have made a big effort in research on semiconductor technology to optimize PIN diodes as radiation detectors, also they have worked in the integration of detector and the amplification device to produce arrays for the application of PIN diodes in image generation [38][39][40][41][42][43].

Applications in Medicine

PIN diodes have been used to measure the intensity of X-ray beams provided by Radiographic X-ray units in order to evaluate the main operational parameters related with the quality of the final photographic image in the film [44][45][46][47][48][49][50], such parameters are: the applied high voltage; the waveform of the high voltage, see Figure 3; the beam current; the exposure time and the applied dose, see Figure 4.

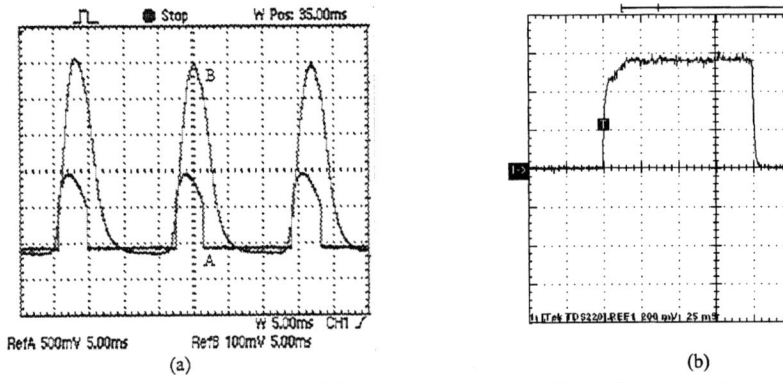

(a)   (b)

FIGURE 3. Oscilloscope screens of X-ray beams measured with an X-ray waveform tester made with a PIN diode. a) Half wave X-ray unit, b) High frequency X-ray unit.

---

[1] CONACYT, National Council of Science and Technology of México.
[2] CINVESTAV, Research and Advanced Studies Center, México.
[3] INAOE, National Institute for Astrophysics, Optics and Electronics. México.
[4] ININ, National Institute for Nuclear Research. México.
[5] CEADEN, Center for Applied Studies in the Development of Nuclear Energy, Cuba.

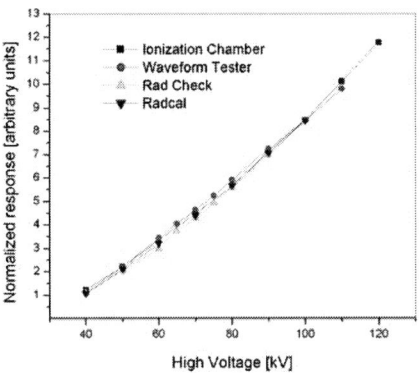

FIGURE 4. Normalized response of a waveform tester made with a PIN diode to an X-ray beam with different accelerating voltages, 100 mA, and comparison with the response of other instruments [48].

Thickness measurements of thin biological samples have been done with PIN diodes and an X-ray beam for medical applications with success [51].

The good light sensitivity of PIN diodes has been combined with scintillation radiation detectors to make position detectors in PET applications as proposed by Derenzo and Moses in 1989 [52], in this way digital images can be generated. The direct detection of radiation in image applications also has been studied [53][54], in this case the PIN diodes are used as pixel elements.

Personal monitors for radiation protection and dosimetry have been done with PIN diodes and an appropriate build–up material that is selected according with the nature of the radiation to be detected, X-and γ-rays or neutrons [55][56][57].

Photon Spectroscopy

Si-Li detectors are normally used to get the best results in X-ray analysis due to its good energy resolution, typically 180 eV, and good efficiency in the energy range from 1 kev to 100 keV, nevertheless there are special PIN diodes [58] commercially available and suitable for X-ray detection, these detectors have some nice advantages over the Si-Li detectors mainly regarding the size and compactness. We demonstrated the detection capability of an OPF420 PIN photo diode [8], in this application. The employed preamplifier configuration was the forward biased FET charge amplifier, FBFA [59][60], in order to get the minimum noise. The FBFA configuration is shown in the Figure 5, C_f is discharged through the input field effect transistor biased with the gate in forward mode [9]. The obtained characteristics of the proposed system at room temperature are: conversion gain 22 mV/fC, with a feedback capacitance of 0.045 pF; the equivalent noise referred to the input is 20 electrons (rms) with an amplifier shaping time of 1.5 μs; the resolution of the system for an Fe-55 source is 1.02 keV.

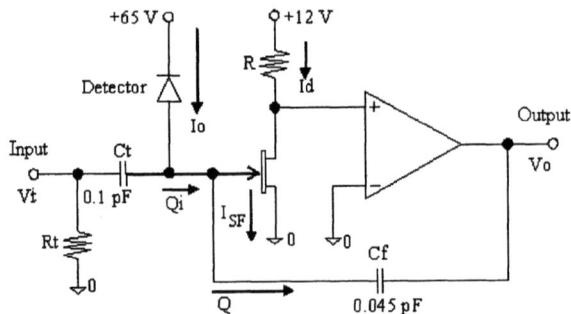

FIGURE 5. Basic idea of the preamplifier with input FET in the condition of forward biased gate.

The spectrum for an Am-241 source obtained with the PIN diode is compared with the one obtained with a Si-Li detector, as shown in Figure 6. In this figure a good correspondence of the peaks position in both detectors is observed, which guaranties the good linearity of the PIN diode experimental results. Similar results are reported in the literature [61][62]. This system has been used with success in different experiments at ININ [63].

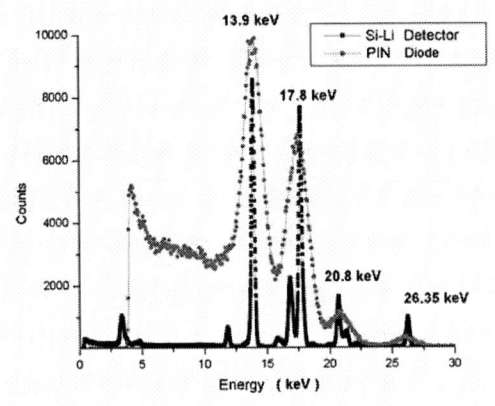

FIGURE 6. Comparison of the energy spectra for an Am-241 source, obtained with the PIN diode at room temperature and a Si-Li cooled detector [9].

In the Plasma Focus experiment at UNAM[6], PIN diodes are used to measure soft X-rays produced at the experiment and to make diagnostic of the conditions in the generated plasma [64].

Generally, γ-ray spectroscopy with PIN diodes is performed by indirect measurement of radiation employing different scintillation detectors [65].

---

[6] National Autonomous University of México

Charged Particle Spectroscopy

The application of different planar silicon detectors to the detection of charged particles has been reported widely in the literature [66], [67], [68]. Specially devised PIN photodiodes have been reported for charged particle spectroscopy [69] and also for use in heavy ion collision experiments [70].

The research for PIN diode preamplifiers [71][72] started in Mexico in 1995, later on, we proposed [73] a PIN diode–preamplifier set that could substitute a silicon surface barrier detector, SSB, in some applications; for example, a spectrum from alpha sources such as ^{239}Pu, ^{241}Am and ^{244}Cm is shown in Figure 7.a, it also has been applied in the measurement of charged particles in experiments at accelerators and in Rutherford backscattering analysis in the energy range of 4 to 13 MeV, see Figure 7.b. The electronic noise of the preamplifier, measured with a pulser, was less than 2.14 keV.

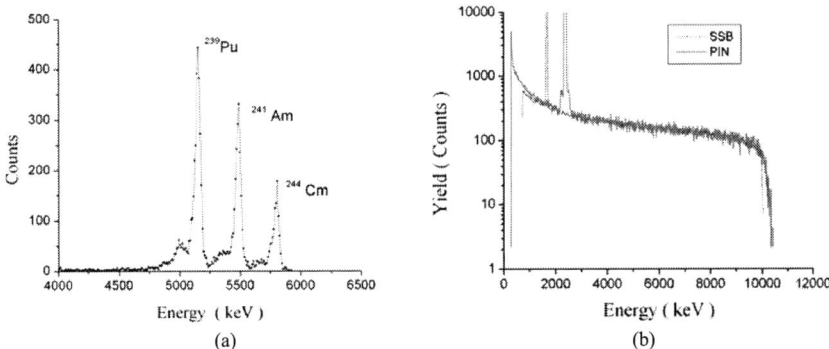

FIGURE 7. Spectra obtained with the proposed PIN diode-preamplifier set: a) spectrum of calibration alpha sources; b) Rutherford backscattering spectrum for a thin layer.

Energy loss measurements for different ions have been done by using PIN diodes and a stopping medium, thus, stopping forces have been measured accurately [74] which provided more information to ion beam analysis.

Radiation Damage on Semiconductor Structures

Damage from radiation has been observed in PIN diodes due to particle bombardment [73], the measurement of the increase in the leakage current and forward voltage, see Figure 8, which evidence the damage of the detector, can be used as a tool to evaluate the unknown total number of incident particles reaching the detector. The increase of the leakage current in the detector worsens drastically the noise behavior of the system [10].

Reinhard proved [75] that a PIN diode can be a good radiation damage monitor for mixed fields of electrons, fast and epithermal neutrons.

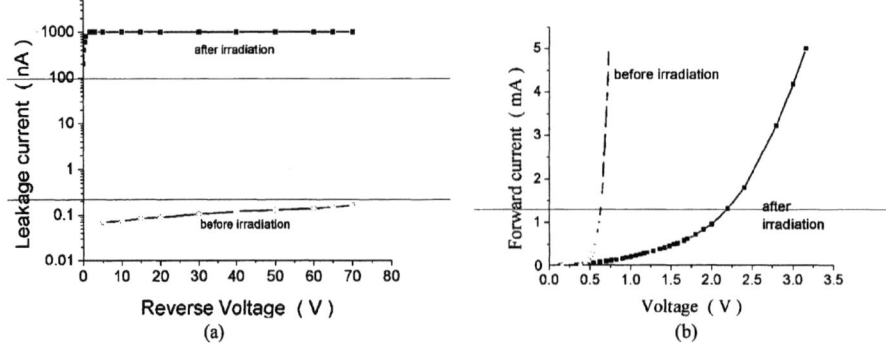

FIGURE 8. Modification of the electric characteristics of PIN diodes after bombardment with charged particles: a) leakage current; b) forward voltage.

Spatial Applications

In spatial research the PIN diodes are been widely used as environmental monitors for mixed fields of high energy particles and gamma rays [76], only some examples are mentioned as follows: the Cosmic-Ray Effects and Activation Monitor (CREAM) experiment; Cosmic Particle Experiment (CPE); Total Dose Experiment (TDE); Cosmic-Ray Effects and Dosimetry (CREDO) payload. Under UoSAT-3 experiment in 1990 and in CEDEX[7] at Surrey Space Center in 2000, PIN diodes have been used for the measurement of protons and heavy ions fluxes inside the spacecraft.

In 1997, under the Pathfinder mission, spectra of Martian rocks analyzed by X-ray fluorescence were sent to earth, the detector in this case was a cooled PIN diode [77], cooling was made by applying thermoelectric effect.

High Energy Physics, HEP, Applications

The use of PIN diodes in HEP big experiments is related mainly with the detection of charged particles. PIN diode Beam Loss Monitors (BLM) are employed commonly in: the European Synchrotron Radiation Facility[78], in HERA[8] at DESY[9] since 1987 [79]; at Tevatron of FNAL[10] to measure the loss of protons and antiprotons in the beam since 1997 [80]; the Beta meson And Anti Beta meson experiment (Babar collaboration) to measure the background produced by the beam since 1997 [81]; and the RHIC[11] [70] in the STAR[12] experiment in 1999.

The PIN diodes were used as light detectors in digital optical links for fast data transmission in the tracker control system of the Compact Muon Solenoid (CMS) at CERN [82], and in the SCT[13] and pixel detectors [83] at the ATLAS[14] detector.

---

[7] CEDEX, Cosmic Ray Energy Deposition Experiment
[8] HERA, Hadron Electron Ring Accelerator
[9] DESY, Deutsches Elektronen Synchrotron
[10] FNAL, Fermi National Accelerator Laboratory, USA
[11] RHIC, Relativistic Heavy Ion Collider at Brookhaven National Laboratory, U. S. A.
[12] STAR, Semiconductor Tracker at RHIC.
[13] SCT, Semiconductor Tracker.
[14] ATLAS, A Toroidal Large Hadron Collider (LHC) Apparatus

REFERENCES

1. D. A. Kleinman "Forward Characteristics of the PIN Diode" Bell Sys. Tech. J. Vol. 35, May 1956, pp. 685-706.
2. R. Nowotny, W. L. Reiter "The Use of Silicon PIN-Photodiodes as Low-Energy Photon Spectrometer" Nucl. Instr. and Meth., 147, 1977, pp. 477.
3. A. B. Fowler "The Long-Reaching Influence of Arthur von Hippel: Interdisciplinarity and Semiconductors" MRS Bulletin, Vol. 30, Nov. 2005, pp 854-857.
4. Optek Technology, Inc. "Optek's Product Catalog" Carrollton, TX, USA, 1997.
5. F. J. Ramírez Jimenez "Medición de Rayos X con Detectores de Semiconductor Tipo PIN" IX Congreso Técnico Científico ININ-SUTIN, Centro Nuclear, Mexico, Dec. 1999, pp. 66-69.
6. F. J. Ramírez-Jiménez "Radiation Detectors of PIN Type for X-Rays" American Institute of Physics, 674, 2003, ISBN: 0-7354-0141-1, ISSN 0094-243X, pp. 313-335,
7. F. J. Ramírez-Jiménez "Waveform Measurements in X-Ray Units" American Institute of Physics, 682, 2003, ISBN: 0-7354-0151-9, ISSN 0094-243X, pp. 86-91.
8. F. J. Ramírez-Jiménez, R. López-Callejas, A. Cerdeira-Altuzarra, J. S. Benítez-Read, M. Estrada-Cueto, J. O. Pacheco-Sotelo "PIN Diode-Preamplifier Set for the Measurement of Low Energy γ and X-Rays" Nucl. Instr. and Meth. A497, 2003, pp. 557-583.
9. F. J. Ramírez-Jimenez "X-Ray Spectroscopy with PIN Diodes" X Mexican Workshop on Particles and Fields, Morelia, México, Nov. 2005.
10. G. F. Knoll "Radiation Detection and Measurement" John Willey and S., New York, 2000.
11. A. Díaz, A. E. Cabal., F. J. Ramírez Jimenez, J. Osorio "Investigación y Selección de Parámetros de Preamplificadores de Nuevo Tipo para Detectores Semiconductores de Bajo Ruido" III Workshop on Nuclear Physics. La Habana, Cuba, Oct. 1997.
12. A. Cerdeira, M. Estrada, M. Aceves and B. S. Soto "New Aspects on the Characterization of PIN Structures Fabricated on Very High Resistivity Sustrate" Journal of Solid-State Devices and Circuits, 5 (2), 1997, pp. 1-4.
13. A. Diaz, L. M. Montaño, F. J. Ramírez-Jiménez, A. Cerdeira and M. Estrada "Simulation and Characterization of a DC Coupled Preamplifier for the Measurement of Ionizing Radiation in Crystalline PIN Diodes" Second Workshop on Simulation and Characterization Techniques in Semiconductors. IEEE-CINVESTAV, Mexico, Sept. 1998.
14. A. E. Cabal, F. J. Ramírez-Jimenez, M. Estrada, A. Cerdeira A., M. Aceves, P. Rosales "Caracterización de Diodos PIN de Silicio Cristalino Utilizando Radiaciones Nucleares" Cuarta conferencia de Ingeniería Eléctrica, CINVESTAV, Sep. 1998, pp. 206-210.
15. P. Rosales, M. Aceves, A. Cerdeira, M. Estrada, A. E. Cabal, F. J. Ramírez-Jimenez "Sensores de Radiación Utilizando Diodos PIN de Silicio" XX Congreso Internacional de Ingeniería Electrónica. Chih., Mexico, Oct. 1998. pp. 281-286.
16. M. Aceves, P. Rosales, A. Cerdeira, M. Estrada, A. E. Cabal, F. J. Ramírez-Jimenez "Investigation on the Reduction of the Dark Current for PIN Photodiodes Using Statistical Methods" 1998 IEEE International Integrated Reliability Workshop Final Report, Stanford, Cal., Oct. 1998. pp. 107-108.
17. F. J. Ramírez Jiménez, A. Cerdeira A., M. Aceves M., A. Díaz, M. Estrada, P. Rosales, A. E. Cabal, L. M. Montaño, A. Leyva "Avances en el Proyecto sobre Detectores de Radiación de Silicio Tipo PIN" VIII Congreso Técnico Científico ININ- SUTIN, Cetro Nuclear, Salazar, Estado de México, Dec. 1998, pp. 24-29.
18. S. Soto, M. Estrada, A. Merkulov, R. Asomoza "High Deposition Rates of Amorphous Silicon Thick Layers Using a Gas Mixture" Thin Solid Films 330, 1998, 83-89.
19. P. Rosales-Quintero "Fabricación de Sensores de Radiación de Alta Energía" Tesis de Maestría, INAOE, México. 1999.
20. F. J. Ramírez-Jimenez, A. Diaz, A. Cerdeira A., A. Leyva, L. M. Montaño, M. Estrada "Low Noise Charge Sensitive Preamplifier for the Measurement of Low energy γ and X rays with PIN Diodes" XXII Congreso Nacional de Ingeniería Biomédica, Ixtapa Zihuatanejo, México, Nov. 1999.
21. M. Estrada, I. Pereyra "PIN Diodes on High Deposition Rate Thick-Si:H Layers from Pure SiH4" Thin Solid Films, 346, 1999, pp.255-260.

22. A. Leyva, A. E. Cabal, F. J. Ramírez-Jiménez, M. Estrada, A. Cerdeira, A. Díaz "Estudio de la Respuesta a las Radiaciones Ionizantes de Diodos tipo PIN de A-Si:H" V Workshop on Nuclear Physics. La Habana, Cuba, Oct. 1999.
23. A. Leyva, F. J. Ramírez-Jimenez, Y. Ortega, M. Estrada, A. E. Cabal, A. Cerdeira, A. Díaz "Spectrometric Characterization of Amorphous Silicon PIN Detectors" CP 538, Am. Inst. of Phy., Conference Proceedings of the Fourth Mexican Symposium on Medical Physics, Mérida, Yucatán, march 2000, pp. 201-205.
24. K. Monfil Leiva "Efectos de los Tratamientos Térmicos en las Corrientes de Fuga de Diodos PIN" Tesis de Maestría, INAOE, México, 2002.
25. U. Macías-Aguilar "Efectos de la Oxidación Durante los Tratamientos Térmicos de Alta Temperatura en Diodos PIN de Silicio" Tesis de Maestría, INAOE, México, 2003.
26. A. Cerdeira A., A. Leyva A., F. J. Ramírez-Jimenez, M. Estrada, I. Pereyra, M. Paez "Response of Amorhous Silicon PIN Detectors to X–Rays from a Mammograph" RADECS 2000 Workshop, Sep. 11-13, 2000, Louvain la Neuve, Belgium, pp. 183-186.
27. A. Cerdeira, M. Estrada "Modeling of Reverse Current Behavior in Amorphous Thin and Thick p-i-n Diodes" IEEE Trans. on ED, 41, 11, 2000, pp. 2338.
28. F. J. Ramírez Jiménez, M. Aceves, A. Cerdeira., M. Estrada "Ion Implantation in Semiconductor Radiation Detectors" International Materials Research Congress. Can-Cun, México, Oct. 2000.
29. M. Estrada, A. Cerdeira, A. Leyva, M.N.P. Carreno, I. Pereyra "Optimization of the i-Layer Width of Cr-a-Si:H PIN X-Ray Detectors" Thin Solid Films 396, 2001, 235.
30. M. Estrada, A. Cerdeira, A. Ortiz-Conde and F. García "Determination of Trap Cross Section in a-Si:H p-i-n Diodes Parameters Using Simulation and Parameter Extraction" Microelectronics Reliability 41, 2001, pp. 605.
31. Shtejer K, Leyva A., Ramírez-Jiménez F. J. "Modelling Optimal Characteristics of a-Si:H Semiconductor Detectors For X-Ray Detection" IEEE Nuclear Science Symposium-Medical Imaging Conference, 2004, R11-16, Roma, Italia, 2004, ISSN 1082-3654.
32. A. T. Medel de Gante, M. Aceves, A. Cerdeira "Diseño de un JFET Compatible con el Proceso de Fabricación del PIN, Construido por Triple Difusión en un Substrato de Alta Resistividad" 9ª Conferencia de Ingeniería Eléctrica, CIE 2003, México, D. F., pp. 187- 192.
33. A. T. Medel de Gante, M. Aceves, A. Cerdeira "Disminución de la Capacitancia de Entrada de un JFET Compatible con el Proceso de Fabricación del PIN" Cuarto Encuentro de Investigación, 2003, INAOE, México, pp.137-140.
34. U. Macias A., M. Aceves M., K. Monfil L. "Efectos de la Oxidación Durante los Tratamientos Térmicos de Alta Temperatura en Diodos PIN de Silicio" 9ª Conferencia de Ingeniería Eléctrica, México DF, 2003, pp. 123-126.
35. K. Monfil, M. Aceves, A. Cerdiera, A. Medel "Efectos de Tratamientos Térmicos en las Corrientes de Fuga de Diodos PIN de Silicio" Revista Información Tecnológica. Vol. 14, No 3, 2003, pp. 123-130.
36. A. T. Medel de Gante, M. Aceves, A. Cerdeira, L. Sánchez "Design of a JFET and Radiation PIN Detector Integrated on a High Resistivity Silicon Substrate Using a High Temperature Process" CCCT 2004, Austin, Texas.
37. M. Aceves-Mijares, M. Estrada, A. Cerdeira, A. Cerdeira-Estrada "Silicon PIN Diodes as Radiation Detectors" Encyclopedia of Sensors, Edited by C. A. Grimes, E. C. Dickey, and M. V. Pishko, The Pennsylvania State University, University Park, USA, 2005.
38. G. F. Dalla Betta, M. Boscardin, G. Verzellesi, G. U. Pignatel, A. Fazzi, G. Soncini "A test chip for the development of PIN-type silicon radiation detectors" Atti del convegno "Proceedings of ICTMS_96, the IEEE International Conference on Microelectronic Test Structures", Trento, 26-28 March, 1996, pp. 231-235.
39. G. F. Dalla Betta, G. U. Pignatel, G. Verzellesi, M. Boscardin "Si-PIN X-ray detector technology" Nucl. Instr. & Meth. in Physics Research. Section A, 1997, v. 395, n. 3, pp. 344-348.
40. G. F. Dalla Betta, G. Verzellesi, M. Boscardin, L. Bosisio, G. U. Pignatel, L. Ferrario, M. Zen, G. Soncini "Silicon PIN radiation detectors with on-chip front-end junction field effect transistor" Nucl. Instr. & Meth. in Physics Research. Section A, 1998, v. 417, n. 2-3, pp. 325-331.

41. A. Fazzi, G. U. Pignatel, G. F. Dalla Betta, M. Boscardin, V. Varoli, G. Verzellesi "Charge preamplifier for hole collecting PIN diode and integrated tetrode N-JFET" IEEE Trans. on Nucl. Sci., 2000, v. 47, n. 3, pp. 829-833.
42. A. Fazzi, G. F. Dalla Betta, G. U. Pignatel, M. Boscardin, P. Gregori, N. Zorzi "PIN diode and integrated JFET on high resistivity silicon: a new test structure" Atti del convegno "IEEE Nuclear Science Symposium and Medical Imaging Conference-NSS-MIC 2001", San Diego (USA), 4-10 Novembre, 2001
43. S. Ronchin, M. Boscardin, G. F. Dalla Betta, P. Gregori, V. Guarnieri, C. Piemonte, N. Zorzi "Fabrication of PIN diode detectors on thinned silicon wafers" Nucl. Instr. & Meth. in Physics Research. Section A, 2004, v. 530, pp. 134-138.
44. F. J. Ramírez-Jimenez "Detectores de Radiación de Silicio tipo PIN y sus Aplicaciones en Física Medica" 3er. Simposium Mexicano sobre Física Medica. León Guanajuato, Mexico, 25-26 Feb. 1999.
45. I. Mercado, F. J. Ramírez-Jimenez, V. M. Tovar, A. Becerril "Prototipo para la Medición de Parámetros en una Unidad de Mamografía Utilizando Fotodiodos" II Conferencia Internacional y XII Congreso Nacional sobre Dosimetría de Estado Sólido. México, D. F., 239-247, Sep. 1999.
46. F. J. Ramírez-Jimenez, E. Gaytán G., I. Mercado, M. Estrada, A. Cerdeira "Measurement of Parameters for the Quality Control of X-ray Units by Using PIN Diodes and a Personal Computer" CP 538, Am. Inst. of Phy., Conference Proceedings of the Fourth Mexican Symposium on Medical Physics, Mérida, Yucatán, March 2000, pp. 196-200.
47. F. J. Ramírez-Jiménez "Measurements in X-Ray Units Used in Radio-diagnostic" American Institute of Physics, 630, 2002, ISBN: 0-7354-0084-9, ISSN 0094-243X, pp. 104-115.
48. E. Gaytan, F. J. Ramírez-Jimenez, V. M. Tovar "Sistema de Medición de los Parámetros de Operación para Máquinas de Rayos X" Contacto Nuclear, Instituto Nacional de Investigaciones Nucleares, N° 36, Mexico, 2004.
49. F. J. Ramírez-Jiménez., R. López-Callejas, J. S. Benítez-Read, J. O. Pacheco-Sotelo "Considerations on the Measurement of the Practical Peak Voltage in Diagnostic Radiology" The British Journal of Radiology, 77, 2004, ISBN: 0-7803-8257-9, ISSN 1082-3654, pp.745-750.
50. F. J. Ramírez-Jiménez "A New Instrument for the Measurement of the Waveform in X-Ray Units" American Institute of Physics, 724, issue 1, (2004), ISBN: 0-7354-0205-1, ISSN 0094-243X, pp. 249-253.
51. F. J Ramírez-Jiménez, S. Galindo "An Instrument for Measuring the Thickness of Gamma Irradiated Xenografts" IEEE Nuclear Science Symposium-Medical Imaging Conference (2003), N36-130, Portland Oregon, U.S.A., Oct. 2003. ISBN: 0-7803-8257-9, ISSN 1082-3654.
52. S. E. Derenzo, W. W. Moses, H. G. Jackson, et al. "Initial Characterization of a Position Sensitive Photodiode/BGO Detector for PET" IEEE Trans. on Nucl. Sci., NS-36, 1989, pp. 1084-1089.
53. M. Novelli, S. Amendolia, M. Bisogni, M. Boscardin, G. F. Dalla Betta, P. Delogu, M. Fantacci, M. Quattrocchi, V. Rosso, A. Stefanini, L. Venturelli, S. Zucca "Semiconductor Pixel Detectors for Digital Mammography" Nucl. Instr. & Meth. in Pysics Research, Section A, 2003, v. 509, n. 1-3, pp. 283-289.
54. M. Bisogni, M. Boscardin, G. F. Dalla Betta, P. Delogu, M. Fantacci, P. Gregori, S. Linsalata, M. Novelli, C. Piemonte, M. Quattrocchi, V. Rosso, A. Stefanini, N. Zorzi, S. Zucca "Characterization of Si Pixel Detectors of Different Thickness" Nucl. Instr. & Meth. in Physics Research, Section A, 2004, v. 518, n. 1-2, pp. 418-420.
55. K. Kovačevič, N. Stipčič, G. Paič, I. Šlaus, B. Eman, V. Pečar, M. Antič "Use of Photodiodes for Neutron Dosimetry" Nucl. Instr. & Meth. in Physics Research, V. 148, 1978, pp. 291-298.
56. R. Nowotny "A Silicon-Diode Pocket Radiation Chirper" Health Physics Vol. 44, N° 2, 1983, pp. 158-160.
57. H. J. Khoury, F. Almeida de Melo, C. A. Brayner de O. "Utilização de fotodiodos Para dosimetria das Radiações Ionizantes" 2° Congresso Geral de Energia Nuclear, Vol . II, Brazil, 1988.
58. Hamamatsu, available web site: http://usa.hamamatsu.com/cmp-detectors/xray.htm.
59. G. Bertuccio, P. Rehak and D. Xi "A Novel Charge Sensitive Preamplifier without the Feedback Resistor" Nucl. Instr. and Meth., A326, 1993, pp. 71-76.
60. G. Bertuccio, A. Pullia "Room Temperature X-Ray Spectroscopy with a Silicon Diode Detector and an Ultra Low Noise Preamplifier" IEEE Trans. Nucl. Sci., 41-4, 1994, pp. 1704-1709.

61. N. Markevich, I. Gertner, J. Felsteiner "Low Energy X-Ray and γ Spectroscopy using Silicon PIN Photodiodes" Nucl. Instr. and Meth. in Physics Research A, 269, 1988, pp. 219-221.
62. G. F. Dalla, G. U. Pignatel, G. Verzellesi, M. Boscardin, A. Fazzi, L. Bosisio "Monolithic integration of Si-Pin diodes and N-chanel and N–channel double–gate JFET's for room temperature X-ray spectroscopy" Nucl. Instr. and Meth. in Physics Research, A 458, 2001, pp. 275-280.
63. F. J Ramírez-Jimenez "Uso de Detectores PIN en la Identificación de Radioisótopos a Partir de su Espectro de Rayos X" Contacto Nuclear, Instituto Nacional de Investigaciones Nucleares, N° 37, México, 2004.
64. J. Herrera, F. Castillo "Generation of Protons, Neutrons and X-rays in the Plasma Focus FN-II" Instituto de Ciencia Nucleares, UNAM, Conference at the Instituto Nacional de Investigaciones Nucleares, México, April, 2006.
65. E. Gramsch, K. G. Lynn, M. Weber, B. DeChillo, R. R. McWilliams "Silicon PIN Photodetectors in High Resolution Nuclear Spectroscopy" Nucl. Instr. and Meth. in Physics Research, A 311, 1992 pp. 529-538.
66. L. Lavergne-Gosselin, L. Stab, M. O. Lampert, H. A. Gustafsson, B. Jakobsson, A. Kristiansson, A. Oskarsson, M. Westenius, A. J. Kordyasz, K. Aleklett, L. Westerberg, M. Ridehell and O. Tengblad "On the Use of Thin Ion Implanted Si Detectors in Heavy Ion Experiments" Nucl. Instr. & Meth., A 276, 1989, pp. 210-215.
67. L. Stab "Thin Epitaxial Silicon Detectors" Nucl. Instr. & Meth., A 288, 1990, pp. 24-30.
68. F. Foulon, L. Rousseau, L. Babadjian, S. Spirkovitch, A. Brambilla, P. Bergonzo "A New Technique for the Fabrication of Thin Silicon Radiation Detectors" IEEE Trans. Nucl. Sci., 46-3, 1999, pp. 218-220.
69. K. W. Wenzel, C. K. Li, and D. A. Pappas "Soft X-ray Silicon Photodiodes with 100% Quantum Efficiency" IEEE Trans. Nucl. Sci., 41-4, 1994, pp. 979-983.
70. R. M. Willson, S. U. Pandey, R. Bellwied, R. Beuttenmuller, A. Drees, P. Kuczewski, W. Leonhart, D. Lynn, R. Soja, J. Takahashi "Novel Applications of PIN-Photodiodes in Relativistic Heavy Ion Collisions" IEEE Trans. Nucl. Sci., 47-3, 2000, pp. 851-853.
71. A. Yot, F. J. Ramírez-Jimenez "Elaboración de un Conjunto preamplificador-Fotodiodo para Espectroscopía alfa" Informe Técnico SR001-95, ININ, México, July 1995.
72. F. J. Ramírez-Jiménez "Preamplificador Sensible a Carga para Espectrometría Alfa" XVIII Congreso Internacional Académico de Ingeniería Electrónica. Chih., Chih., Oct. 1996.
73. F. J. Ramírez-Jiménez, R. López-Callejas, J. S. Benítez-Read, J. O. Pacheco-Sotelo "A novel application of a PIN diode-preamplifier set for the measurement of charged particles" Nucl. Instr. and Meth. in Physics Research, A 545/3, 2005, DOI: 10.1016/j.nima2005.02.017, pp. 721-726.
74. H. Timmers, K. Stenström, M. Graczyk, H. J. Whitlow "Energy Loss Measurements for Mass-14 Ions Using a patterned Stopping Medium on a PIN Diode" Nucl. Instr. and Meth. in Physics Research, B 219-220, 2004, pp. 263-267.
75. M. Reihard "Radiation Hardness Study of High purity Silicon and the development of a radiation Damage Monitor System for Silicon Devices in Mixed Radiation Fields" Ph. D. thesis, University of Wollongong, Australia, 2003.
76. Available web site: http://www.amsat.org/amsat/sats/dl1fdt/CEDEX.ger.html
77. AMPTEK Inc, available web site: http://www.amptek.com/
78. Available web site: www.esrf.fr/UsersAndScience/Experiments/HRRS/ID28/BeamlineLayout/EHI/spectrometer/Detectors/.
79. Available web site: www.desy.de/~ahluwali/herareports/2000/00-03.pdf.
80. V. Shiltser "Fast PIN Diode Beam Loss Monitors at Tevatron" available web site: lss.fnal.gov/archive/1997/tm/TM-2012.pdf.
81. Available web site: hep.Stanford.edu/babar/commissioning/index.htm.
82. Available web site: www.proj-rx40.web.cern.ch.
83. Available web site: www.physics.ohio-state.edu/~gan/atlas.html

Cosmic ray studies at CERN

Arturo Fernández T.

Facultad de Ciencias Físico Matemáticas
Benemérita Universidad Autónoma de Puebla
Apartado Postal 1152, Puebla, Pue., México.

Abstract. The use of the sophisticated and large underground detectors at CERN for cosmic ray studies has been considered by several groups, e.g. UA1, LEP and LHC detectors. They offer the opportunity to provide large sensitivity area with magnetic analysis which allow a precise determination of the direction of cosmic ray muons as well as their momentum up to the order of some TeV. The aim of this article is to review the observation of high energy cosmic ray muons using precise spectrometers at CERN, mainly LEP detectors as well as the possibility of improve those measurements with LHC apparatus, giving special emphasis to the ACORDE-ALICE cosmic ray physics program.

Keywords: Cosmic rays, Muon, Astroparticle, Composition.
PACS: 96.50.sb; 96.50.Vg.

INTRODUCTION

The identification of cosmic ray primary particles with detectors on the ground level critically depends on the detailed understanding of the interaction mechanism of those primary particles with air[1]. The particle production, both at large energies and at very forward cones, can today only be estimated by model based extrapolation of accelerator data. Indeed, there are no accelerator data for particle production in the very small forward angles and in the relevant energy region around the "knee"(10^{15} to 10^{17} eV)[2], see Figure 1. In fact, exotic phenomena in very forward high-energy hadronic interactions, like e.g. coherent pion production, disoriented chiral condensate states, quark gluon plasma or heavy flavour production would significantly influence the hadronic cascade and hence the observable variables at the ground level[3, 4, 5]. This might be the cause for conflicting results from various experiments about the particle composition which have not been satisfactorily resolved.

The interpretation of the cosmic ray data depends crucially on models extrapolated well beyond the range in which they have been tuned or tested. While it may be that some of the models will converge on a common interpretation, they may still be incorrect. Hence, the greater the diversity of the measurements which the models must confront, the greater the likelihood of converging to the correct answer. We therefore believe that as many complementary measurements as possible in different kinematic regions should be made to understand more about the forward particle production and hence about the cosmic spectrum, and to provide cross-checks on the interpretation of the results. Cosmic ray air showers are characterized by the distribution of the electromagnetic and muonic components at ground level and by muon distributions at different underground levels[6]. Recent results from LEP experiments [7] have shown valuable implications of

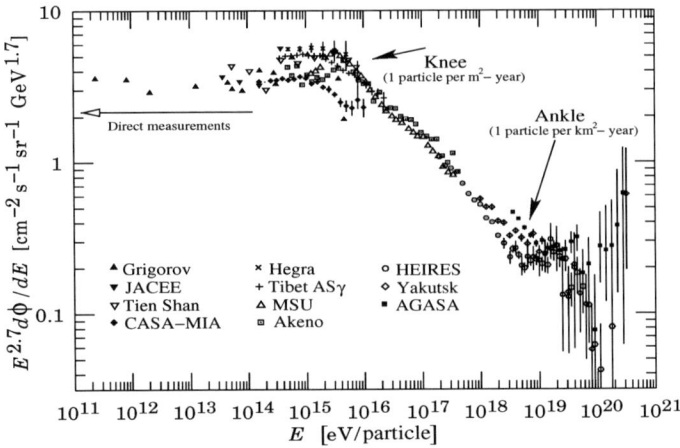

FIGURE 1. Cosmic Ray spectrum, taken from [2]. The change of the spectral index around 10^{15}eV ("knee") and $10^{18.5}$eV ("ankle") are clearly seen.

detecting atmospheric muons at CERN underground levels using high precision particle detectors.

In what follows we give a short review of resent results from LEP experiments on cosmic rays, we describe the status of the ACORDE (A cosmic ray detector for ALICE), and finally we conclude giving an outlook of the possible contribution of LHC experiments to cosmic ray physics.

RESULTS FROM LEP DETECTORS

ALEPH[8], DELPHI[9], and L3[10] are still analyzing cosmic ray data. The advantages of these LEP detectors with respect the traditional cosmic ray experimental devices are the precision of the muon momentum measurement and the excellent understanding of the detector properties trough very accurate Monte Carlo simulation. For instance L3+C obtained their detection efficiency and momentum resolution using L3 $e^-e^+ \rightarrow Z \rightarrow \mu^-\mu^+$ event information. Table 1 gives the main features of these LEP detectors, including underground location, detectors that used as well as their physics objectives.

Nowadays several results from this experimental effort have been published. Most of the results have been related with atmospheric muon momentum spectrum[10, 8, 4], studies on muon bundles where it has been demonstrated that low multiplicities favor protons as primaries, while median multiplicities show a trend to heavier elements. Very high muon bundles have been observed, without a clear explanation of their origen[4, 9]. Very interested studies on Moon shadow for different muon momentum ranges have been published recently. A limit of 11% on the antiproton/proton ratio flux could be set around 1TeV[11]. A search for Dark Matter (SUSY particles) has also been performed. As a preliminary result, an upper limit of $7.1 \times 10^{-13} cm^2 s^{-1} sr^{-1}$ has obtained[12].

To give an example of the high quality and competentness of this LEP observations

TABLE 1. LEP cosmic ray physics program.

Detector	Cosmo-ALEPH	DELPHI	L3+C
Deep underground	130 m $E_{th}^{\mu} > 70 GeV$	100 m $E_{th}^{\mu} > 50 GeV$	40 m $E_{th}^{\mu} > 17 GeV$
Detector used	Hadron calorimeter, TPC + 5 scintillators stations	Hadron calorimeter, TPC, TOF, muon chambers	Drift chambers, Timing scintillator, surface array, dedicated trigger
Physics topics	Muon energy spectrum, charge ratio, Muon multiplicity, lateral distribution, point sources	Muon multiplicity, point sources	Muon spectrum, charge ratio, antiproton flux limit, point sources, flares, muon multiplicity

we present a discussion on few key results. Figure 2 shows the measurement of the atmospheric muon momentum spectrum between 20 and 3000 GeV, with a total error of 2.4% at 150GeV [10]. A comparison is made with the most recent measurements. The combined fit of BESS[13], CAPRiCE[14] and L3+C with the phenomenological functions used by Bugaev[15] and Hebbeker[16] gives a $\chi^2/Ndf = 1.2$, after the systematic momentum scale and normalization uncertainties quoted by those experiments. The same figure shows a comparison between L3+C vertical spectrum and results from air showers calculations using different interaction models[17]. The agreement with these models highly depend of the primary spectra selected, showing that we need more information from accelerator data in order to tune the interaction models. Hopefully, LHC experimental results will reduce the inconsistencies seen in the mass composition at the knee energy region. L3+C has measured the charge ratio $R=\mu^+/\mu^-$ between 20 and 500GeV. For vertical muons they obtained a value of $1.285 \pm 0.003(stat.) \pm 0.019(syst.)$. They have also obtained the zenith angular dependence of this ratio for eight different zenith angles, from 0^0 to 58^0. Currently these calculations are of great interest, as they predict the absolute atmospheric neutrino fluxes which are needed to interpret the observed muon neutrino flux deficit and to evaluate the backgrounds for neutrino astronomy[18].

Muons observation underground with energies $E_\mu > 100GeV$ originate from the first interaction at the top of the atmosphere are very sensitive probes of the primary composition and interaction at high energies. ALEPH[8] and DELPHI[9] have obtained a multi-muon spectrum for primary spectrum runs from 10^{14} to 10^{16}eV. See Figure 3.

ACORDE, A COSMIC RAY DETECTOR FOR ALICE

ALICE is an experiment designed to detect mainly the products of nucleus-nucleus collisions at the LHC at CERN in order to study the physics of strongly interacting matter and the possible formation of quark-gluon plasma [19, 20]. It will be located in the same place in which the L3 experiment was installed. The underground site of the ALICE apparatus with 30 m of overburden composed of subalpine molasses is ideal for

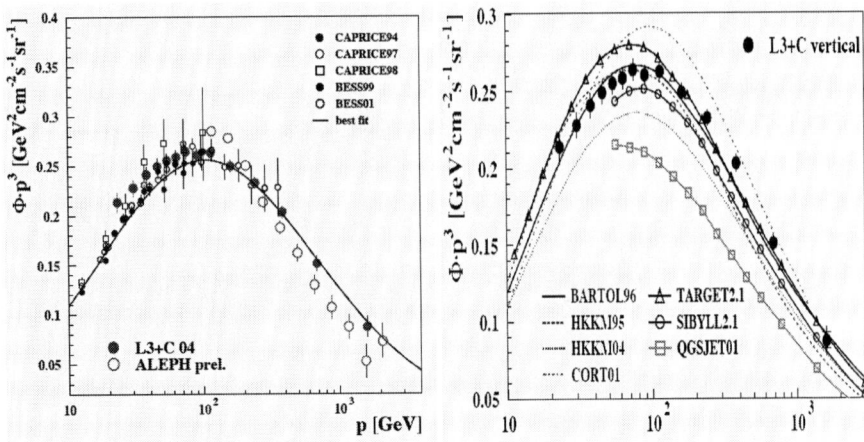

FIGURE 2. Left: L3+C absolute atmospheric muon spectrum[7, 10] compared to previous measurements providing absolute results. CosmoALEPH data is excluded from the fit. Right: Comparison of the L3+C vertical muon spectrum[17] with cascade calculations using different interaction models and primary spectra.

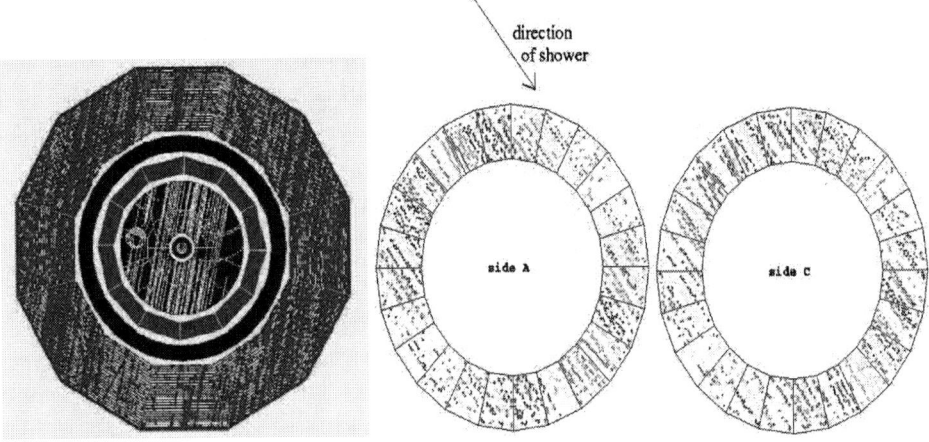

FIGURE 3. Multi-muon events from ALEPH TPC (left) and DELPHI HCAL two-sides view(right)

muon based experiments.

The ALICE cosmic ray trigger, shown in Figure 4, is based on the ACORDE detector, an array of 60 plastic scintillator modules placed on the top sides of the central magnet. The central part of the ALICE apparatus consists of an Inner Tracking System (ITS), a large volume Time Projection Chamber (TPC), a high granularity Transition Radiation Detector (TRD) and a Time Of Flight (TOF) detector with a high resolution array. Additional detectors do not cover the whole azimuth angle. The ACORDE array and the

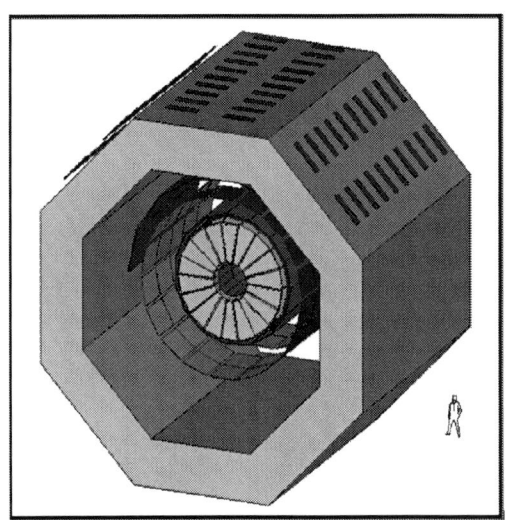

FIGURE 4. General schematic view of the ALICE setup. ACORDE array, TPC, TOF and TRD detectors, used to detect atmospheric muons.

TPC are the main detectors for atmospheric muons, while the TRD and TOF can improve the tracking performance. The molasses above imposes an average energy threshold for atmospheric muons reaching the experiment of $E_{\mu_{th}} \simeq 15\,\text{GeV}$ cutting all the hadronic and electromagnetic components, while the upper energy limit for reconstructed muons will be less than 2 TeV, depending of the magnetic field intensity (up to 0.5 T).

ACORDE will provide a fast L0 trigger signal to the central trigger processor, when atmospheric muons impinge upon the ALICE detector. The signal will be useful for calibration, alignment and performance of several ALICE tracking detectors, mainly the TPC, TRD, and the ITS. The cosmic ray trigger will be capable to deliver signals before and during the operation of the LHC beam. The typical rate for single atmospheric muons crossing the ALICE cavern will be less than 4 Hz/m^2 [21]. The rate for multi-muon events will be lower (less than 0.04 Hz/m^2 [10]) but sufficient for the study of these events provided that one can trigger and store tracking information from cosmic muons in parallel to ALICE normal data taking while beams collide.

ACORDE scintillator counters

An ACORDE module consists of two plastic scintillator paddles with 188×20 cm^2 effective area, arranged in a doublet configuration. Each doublet consists of two superimposed scintillator counters with their corresponding photomultipliers active faces looking back to back, see Figure 5. A coincidence signal (in a time window of 40 ns) from this two scintillator paddles will be the useful information from each module. We developed an PCI BUS electronics card to measure the plateau and efficiency of the

FIGURE 5. ACORDE module. This scintillator counter module consist of two superimposed scintillator counters with their corresponding photomultipliers active faces looking back to back. The active area is 188×20 cm^2

ACORDE counters[22]. The PMT signals were discriminated with a threshold of 15 mV. In average, a plateau voltage is reached at 1450-1550 V with a measured rate of about 60. Hz (at 2,200 m above the sea). The complete ACORDE module array, constructed at Cinvestav-México, in consists of 60 scintillator counter module. The performance studies (plateau, efficiency, attenuation length) of these modules were made at BUAP, Cinvestav, ICN-UNAM and IF-Umich laboratories.

ACORDE Electronics

The ACORDE Electronics is made of several parts: i) it has 60 FEE cards, one per ACORDE module, which converts the analog PMT signal to LVDS signal. For each FEE card we have three outputs: PMTA, PMTB and the two scintillator counter coincidence signal. ii) "ACORDE OR" card to generate the TRD wake up signal. This card receives the 60 coincidence LVDs signals coming from the FEE cards, it create an OR between those signal to create a single muon trigger signal for the TRD. iii) main card, which contains the electronics to receive the 120 LVDS signals coming from the same number of PMTs. This card will produce the single and the multi-coincidence trigger signals, providing also the connectivity to the ALICE trigger and DAQ systems. This electronics has been designed and implemented at BUAP and ICN-UNAM. At CERN, we have done a complete test procedure, in coordination with the Trigger and DAQ ALICE groups. See Figure 6.

FIGURE 6. Left: FEE Card. It is shown two differential LEMO input for each PMT and three output connectors (PMTA, PMTB, and signal coincidence LVDS signal) . Right: Main ACORDE electronis card. This card manage 120 LVDS signals from 60 modules to produce the single and the multi-coincidence trigger signals, providing also the synchronization with LHC clock, the connectivity to the ALICE trigger and communication with DAQ system.

ACORDE Detector Control System

We have been developed a Detector Control System (DCS) for ACORDE. The main tasks of this system is the monitoring and control of the necessary electrical and physical parameters to operate our detector. These include the high voltage (HV) and low voltage (LV) power supplies, the front-end electronics (FEE) configuration parameters, alarm handling, and parameter archiving[?]. A general view of our DCS is shown in Figure 7. We plan to use the SY1527 and the EASY [24] (*Embedded Assembly SYstem*) power supply systems from CAEN as the HV and LV supplies respectively. The latter is a radiation- and high-magnetic-field tolerant power supply system to be installed in the cavern racks to supply the LV for ACORDE. The SY1527 crate can hold up to 16 boards. Ten of its slots will be filled with 12-channel HV boards (A1733N) to feed the 120 PMTs. We will also install the EASY Branch Controller (A1676A) board that can control up to 6 EASY crates. We will need two full EASY 3000 crates each with five 12-channel 2–8 V boards (A3009) for the discriminators. The remaining four channels for the ACORDE Electronics Card and the shoebox producing the TRD wake up signal will be hosted by a third EASY 3000 crate with an additional A3009 board.

ACORDE simulation work

We have obtained a detail geometrical representation of the ALICE cavern, the calculation of the energy cut-off of (atmospheric) muons reaching the ALICE magnet, and of the corresponding angular distribution flux. We also calculate the geometric accep-

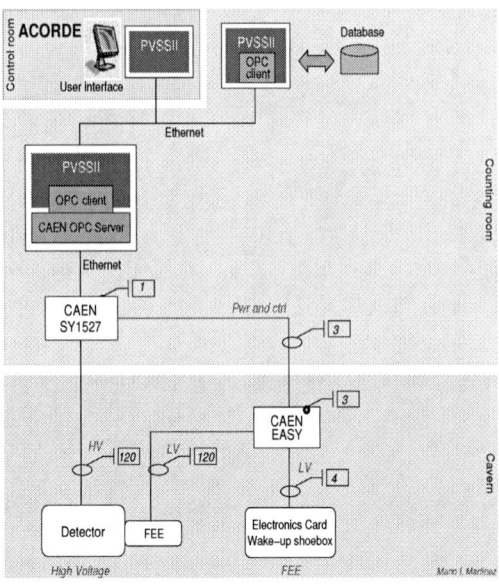

FIGURE 7. General view of the ACORDE-DCS.

tance of the proposed ACORDE scintillator counter array to atmospheric muons. We generate muons at the surface level, pointing to the IP, with a uniform azimuthal angular distribution and parametric distribution functions[16] for their initial momentum and zenithal angle. We have used GEANT3, under the AliRoot [25] framework, to transport the muons through 30m of molasse material above ALICE. See Figure 8

ACORDE status

At present, we have 20 ACORDE modules already installed and the related electronics working. The ALICE-TPC characterization is proceeding collecting data based on ACORDE cosmic ray trigger signals using 10 modules placed on top and 10 underneath on the ALICE-TPC. See Figure 9.

We have 40 modules more at CERN, under their final test before installing at the ALICE cavern. The DCS is completed and ready to be integrated in the whole ALICE DCS system. Main ACORDE electronics card, with a SIU and TTcrx installed have been tested, sending and receiving messages from the Central Trigger processor and the ALICE DAQ system.

ACORDE-ALICE PHYSICS PROGRAM

An accurate measurement of the energy spectrum and charge ratio of atmospheric muons in the energy range 20–2000 GeV can be performed in one year of data taking by

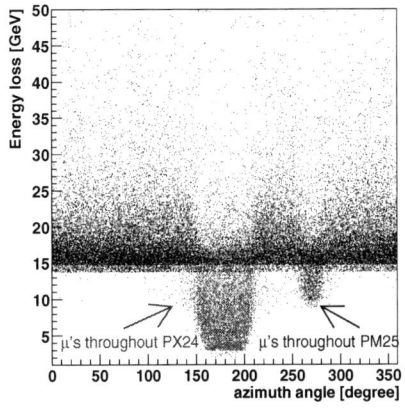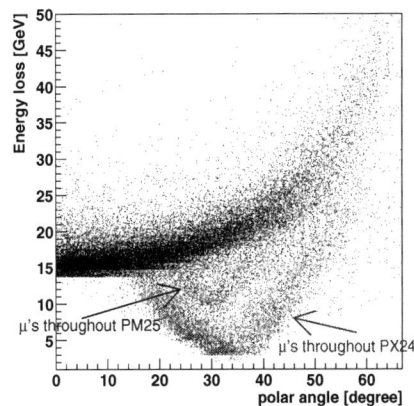

FIGURE 8. Left: Angle ϕ vs. Energy loss distribution of atmospheric muons reaching the ALICE cavern. The distribution of generated muons at surface level follows an exponential function taken from reference [16]. Right: Energy loss of (atmospheric) muons vs (initial) zenithal and azimuthal angles.

FIGURE 9. Left:ACORDE modules on top of the and ALICE-TPC. Rigth: ACORDE electronics and related DAQ system for the ALICE-TPC cosmic ray characterization.

ALICE, provided that a dedicated scintillator trigger system (ACORDE) is added to the apparatus. A precise measurement of muon spectrum constrains the absolute flux of muon neutrino, and this check the consistency of the calculations used to infer the muon neutrino oscillations. The muon flux and charge ratio analysis at different zenith angles give information on hadrons created in the air shower constraining the hadronic interaction models. As we have discussed previously all these measurements have been carried out and written in a conclusive paper by the L3 experiment[10].

As we mentioned before, ALEPH collaboration[4] shown the possibility to investigate the cosmic ray composition also with a small underground apparatus analyzing the muon content of the events. In particular their high spatial resolution TPC was able to resolve muon bundles with densities up to 20 muons/m^2. The multiplicity distribution obtained favors a chemical composition which change from light to heavier elements around the knee when the energy increases. Five highest multiplicity events occur with a frequency almost an order of magnitude greater than the expected. This point can be investigated and solved by ALICE in few months of run. This argument offers the best potentiality considering that we can improve the ALEPH result increasing the statistic and analyzing the momentum and the charge of each muon of the bundle[20].

The observation of antiprotons in cosmic rays is of paramount importance, since it can give clues on several fundamental aspects of astrophysics and cosmology. Space-based experiments, like PAMELA[26] and AMS [28], will extend the energy range for direct antiproton measurements significantly (up to several hundred GeV) in the next years. However, a different approach is also possible for ALICE and other ground-based experiments with a good tracking capability, permitting analysis at higher energies. The Earth-Moon system may be used as a long-baseline charge analyzer. Observing the shadowing effects on opposite sides of the moon due to protons and possible antiprotons, inferences can be made on the ratio of antiprotons to protons. The capability to carry out such measurements depends on the precision by which point sources may be localized and on the separation between proton and antiproton shadows. The statistic necessary is very high making difficult this analysis.

In ALICE we will have a good opportunity to study muon bundles due to the capability of ACORDE to trigger atmospheric muons, the TPC and the ALICE magnet to track very high muon densities, measure the charge and the energy of each muon at least up to 1–2 TeV. In this section we try to understand which information can be extracted from these quantities. To simplify the problem we have simulated the two extreme elements of primaries : proton (p) and iron nucleus (Fe) representing the lighter and the heavier component. The simulations have been developed using the Corsika code [27] with QGSJET as hadron interaction model. Only vertical primaries with energy $E = 10^{14}$ eV and $E = 10^{15}$ eV have been studied to get information on cosmic rays near the knee. The average lateral distributions of muons reaching the ALICE cavern, for p and Fe at $E = 10^{14}$ eV and $E = 10^{15}$ eV shown in Figure 10 give the density of muons expected in ALICE for different distances between the detected bundle and the core, and the maximum number of muons crossing the detector (located always in the core). From the lateral distribution we can also extract the lower number of muons crossing the TPC necessary to select energies near the knee, in such a way to have a low trigger rate that doesn't affect the standard ALICE data taking. Requiring at least three muons crossing the experiment we select energies near $E = 10^{15}$ eV cutting all the primaries lower than $E = 10^{14}$ eV.

The measurement of the average energy of the muons detected in the core (a circle of 2 m of radius around the core) for different number of muons in the core is shown in Figure 10 (right plot) for p and Fe with $E = 10^{15}$ eV. As we can see this correlation has a strong potentiality to distinguish among the various components of cosmic rays, permitting a new approach to study the nature of the primary. Especially at low number of muons in the core the difference on the average energy of muons due to p ($< E_\mu > \simeq$

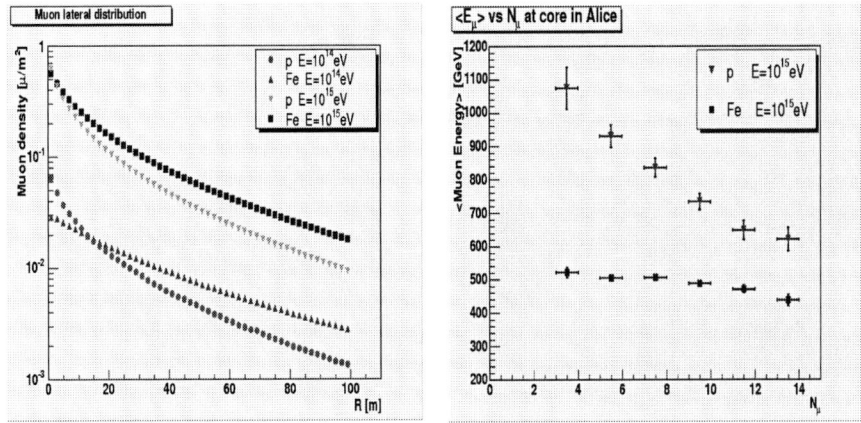

FIGURE 10. Rigth: Average lateral distribution of muons reaching the cavern of ALICE for different primaries and energies. Left: Average energy of muons in the core as a function of the number of muons in the core for p and Fe primaries with $E = 10^{15}$ eV

1.1 TeV) and Fe ($< E_\mu > \simeq 0.5$ TeV) is very large, slightly decreasing with larger number of muons.

CONCLUSIONS

We have discussed the possibilities to study cosmic ray physics using the ALICE experiment, confronting the prospects of this physics program with some recent results from LEP experiments. The ACORDE detector activities were presented with some detail. The most interesting topic seems to be a characterization of single events using the strong capability of the TPC in detecting a high density of muons, their charge and energies with a good resolution up to 1–2 TeV. In particular the correlation between the number of muons and their energies appears a powerful tool to distinguish the various components of primary cosmic rays, while triggering on the number of muons crossing the apparatus allows to select primaries near the energy of the knee. Other methods of analysis, actually not yet developed, exploiting all the possible correlations among the various observables measured in ALICE, can lead to a new approach to study the cosmic ray composition near the knee.

ACKNOWLEDGMENTS

The author would like to thank A. Cerna, J. Arteaga, J. García, E. Gámez, G. Tejeda, S. Román, R. Pelayo, A. Ortíz, R. López, G. Herrera, M. I. Martínez, G. Paic, I. León, M. A. Vargas, Sergio Vergára, L. Villaseñor and A. Zepeda for their collaboration and support to create the Mexican ACORDE-ALICE group. This work was supported by CONACyT

(Project No. 47318 / A-1) and by ALFA-EC funds in the framework of Program HELEN (High Energy Physics Latinoamerican-European Network).

REFERENCES

1. P. Lipari, *Nucl.Phys.Proc.Suppl*,**122**(2003)133.
2. P. Travnicek, PhD Thesis, Charles Univesity, Prage(2004),and references therein.
3. M.R. Krishnaswamy, et. al.,*Phys. Lett.* **B57**,(1975), 1005.
4. V. Avati, et. al., *Astroparticle Physics* **19**(2003), 513.
5. J. Ridki,*Astroparticle Physics* **17**(2002), 355.
6. V. Avati, et. al.*CERN/2001-003, SPSC/P321*
7. P. Le Coultre, *Proceedings of the 29th International Cosmic Ray Conference*, Pune(2005)00, 1001-106.
8. H. Wachsmuth, CERN CosmoLEP note (94.000); C. Grupen et.al.*Nucl. Instrum. Meth. A* **510**(2003)190.
9. P. Abreu et. al., *Nucl. Instrum. Meth.* **A 378**(1996)57; J. Ridky, P. Travnicek, DELPHI Collaboration, *Nucl. Phys. Proc. Suppl.***138**295-298,2005.
10. L3+C Collaboration, O. Adriani, et. al., *Nucl.Instr. Meth.* **A 488**(2002) 209.
11. L3+C Collaboration, O. Adriani, et. al., *Astroparticle Physics*,**23** (2005) 411Ű434.
12. L3+C Collaboration, Y. Ma, et. al., *29th International Cosmic Ray Conference Pune* (2005) 00, 1006-1010.
13. BESS Collaboration, S. Haino, et al., *Phys. Lett.* **B 594** (2004) 35.
14. CAPRICE Collaboration, M. Boezio,et al., *Phys.Rev.* **D 67** (2003) 072003.
15. E.V.Bugaev, et al., *Phys.Rev.* D **58** (1998) 054001.
16. T. Hebbeker and Ch. Timmermans, *Astropart.Phys.* **18** (2002) 107.
17. L3+C Collaboration, M. Unger, et al., *Proc. 19th ECRS 2004*.
18. Super-Kamiokande Collaboration, Y. Fukuda, et al., *Phys. Rev. Lett.* **81** (1998) 1562.
19. ALICE Physics Performance Report, Vol II, *J. Part. Nucl. Phys.* **30**(2004),1517.
20. ALICE Physics Performance Report, Vol II, Section 6.11(CERN/LHC 2005-030), to be publised in *J. Part. Nucl. Phys.*, and references therein.
21. J. Arteaga, et. al. *Proceedings of the 29th International Cosmic Ray Conference*, Pune(2005)03, 101-106; B. Alessandro, et. al., *Proceedings of the 28th International Cosmic Ray Conference*, 2003, Tokyo, Japan, pp. 1203-1206, Universal Academic Press.
22. S. Vergara, et. al. "Implementation of an automatic system to characterize the ALICE Cosmic Ray Detector", submitted for publication.
23. A. Fernández et al, *Czechoslovak J. Phys.*, **55** (2005) B801 - B807.
24. CAEN S.p.A., http://www.caen.it/nuclear/easy_info.html
25. See http://AliSoft.cern.ch/offline/
26. V. Bonvicini et al., *Nucl. Instr. Meth.*,**A 461** (1991) 262.
27. D. Heck and J. Knapp, *Extensive Air Shower Simulation with CORSIKA : 1998, Forschungszentrum Karlsruhe Report FZKA 6019*.
28. S. Ahlen et al., *Nucl. Instr. Meth.*, **A 350** (1994) 351.

Glueballs, Hybrids and Exotics

M.A. Reyes and G. Moreno

Instituto de Física, Universidad de Guanajuato,
Apartado Postal E143, 37150 León, Guanajuato, México

Abstract. We comment on the physics analysis carried out by the Experimental High Energy Physics (EHEP) group of the Instituto de Fisica of the University of Guanajuato (IFUG), Mexico. In particular, this group has been involved in analysis carried out to search for glueball, hybrid and exotic candidates.

Keywords: Spectroscopy, hybrids, glueballs
PACS: 14.40.Cs, 11.80.Et, 12.39.Mk

INTRODUCTION

The EHEP group at IFUG has been involved in studies of non-conventional states: glueballs, hybrids, and exotics. These studies have been carried out mainly using data from the Fermilab E690 experiment. Experiment E690 [1] studies proton diffraction using 800 GeV/c protons, scattering from a liquid hydrogen target (LH_2), in the reaction

$$pp \to pX .\qquad(1)$$

The diffracted forward proton was measured in a forward beam spectrometer, and the recoil system X was measured in a magnetic spectrometer. In three months of the Fermilab's 1991 fixed target run, E690 recorded more than 4.5 billion events. The trigger required the coincidence of an incoming beam particle, an outgoing beam particle within the acceptance of the forward spectrometer but scattered out of the small beam envelope, and at least one particle in the magnetic spectrometer. The E690 apparatus is described in greater detail elsewhere [2].

For studies of double Pomeron production, $pp \to p(M)p$, there are large, clean signals for meson resonances that have been considered candidates for non-$q\bar{q}$ mesons [3]. So far, the search for non-$q\bar{q}$ mesons in the data has been done in the following reactions:

$$pp \to p_{slow}(K_s K_s)p_{fast} \qquad K_s \to \pi^+\pi^- \qquad(2)$$
$$pp \to p_{slow}(\phi\phi)p_{fast} \qquad \phi \to K^+K^- \qquad(3)$$
$$pp \to p_{slow}(K_s K^\pm \pi^\mp)p_{fast} \qquad K_s \to \pi^+\pi^- \qquad(4)$$

Reactions (2-4) were analyzed as two step processes: the production of an M system formed by the collision of the Pomerons emitted by the scattered protons, and the decay step where M decays into either two particles (2,3), or a particle and an isobar (4), with the subsequent decay of the isobar into two particles.

The production coordinate system was defined in the center of mass of the (M) system, with the y-axis perpendicular to the plane of the two Pomerons in the pp c.m.

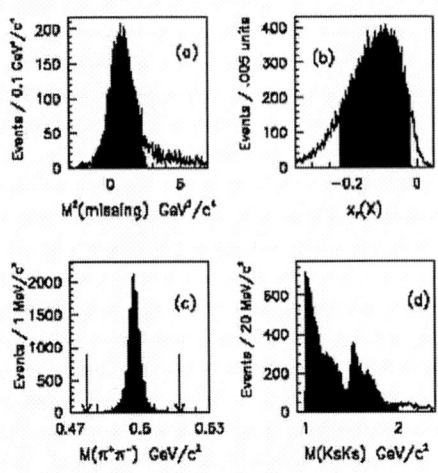

FIGURE 1. (a) Missing mass squared, (b) x_F, (c) $\pi^+\pi^-$, and (d) $K_s K_s$ invariant mass distributions for reaction (2).

system, and the z-axis in the direction of the beam Pomeron in the (M) rest system. This definition of the production coordinate system is not unique, since for central production the (opposite) directions of the Pomerons in the (M) rest system is uniquely defined, but rotations around this axis are not. Therefore, if the assumption that the (M) system is formed by the collision of two objects emitted by the scattered protons is accurate, then our selection of the y-axis is natural.

The five variables used to specify the production process were the x_F and invariant mass of the (M) system, the transverse momenta of the slow and fast protons ($p_{t,s}^2$, $p_{t,f}^2$), and δ, the angle between the planes of the scattered protons in the (M) rest system. The analyses that we comment on here were done in bins of the (M) invariant mass, for the selected region of x_F, integrating over $p_{t,s}^2$, $p_{t,f}^2$ and δ.

THE GLUEBALL CANDIDATE $f_0(1520)$

By 1995, one of the candidates for the lowest lying glueball was the $f_0(1520)$. This resonance had been observed by the Crystall Ball Collaboration [4], but had not been observed in central production [5]. For the analysis of events in reaction (2), we used 10% of the E690 data [6], requiring a primary vertex in the LH_2 target with only two K_s's, and the fast proton, assigned to it. No direct particle identification was used, and no direct measurement of the slow proton was made, but the target veto system was used to reject events with more than one missing track, the missing proton.

In Figure (1) we can see (a) a clear proton peak in the missing mass squared distribution for events in reaction (2), with little background, (b) the uncorrected x_F distribution for the (M) system, (c) the $\pi^+\pi^-$ invariant mass distribution, and (d) the $K_s K_s$ invariant

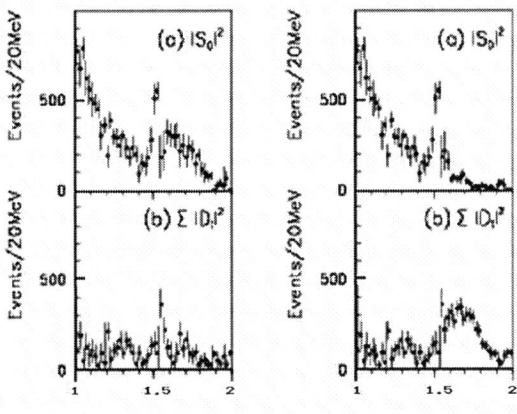

FIGURE 2. The two combinations of waves from a PWA of the $K_s K_s$ system.

mass distribution with cuts defined by the shaded events in (a-c). There is no evidence for the narrow $f_J(2220)$ state seen by the BES Collaboration [7], which had been considered a tensor glueball candidate.

Since only waves with $J^{PC} = (even)^{++}$ are produced in the $K_s K_s$ system, the lowest lying glueball 0^{++} may be seen in this channel. In order to fully understand the wave composition in the mass spectrum observed, we performed a thorough PWA, whose results can be seen in Fig.(2). Only four waves were considered dominant in the spectrum, S_0^-, D_0^-, D_1^- and D_1^+ [6]. Their amplitudes were directly determined by maximizing the extended likelihood with respect to the four wave moduli and two relative phases, $\phi(D_{0,1}^-) - \phi(S_0^-)$. Two solutions were found for each mass bin, one with mostly S-wave, and another with mostly D-wave. A bifurcation point was found at 1.55 GeV/c^2, where both solutions become identical. However, since at threshold the $K_s K_s$ cross section is dominated by the $f_0(980)$, only two combinations remained. Beyond the bifurcation point, there is no way to discriminate between both solutions, and, hence, our final results gave two combinations of amplitude waves observed in this system. Both combinations show a clear S-wave peak at 1.50 GeV/c^2 [4]. Beyond 1.55 GeV/c^2 the ambiguity above prevents a unique determination of the spin of what by that time was called the $f_J(1710)$ meson. Nowadays, it is believed that this state is also spin zero, and that the admixtures of a $q\bar{q}$ meson and the 0^{++}-glueball is what gives rise to these two states. We were the first experiment to observe the $f_0(1520)$ in central production, and this observation was later confirmed by the WA102 Collaboration [8].

A TENSOR GLUEBALL CANDIDATE

The first observation of production of the $\phi\phi$ system was made using the BNL–MPS Spectrometer in the OZI suppressed reaction $\pi^- p \rightarrow \phi\phi n$ at 22.6 GeV/c [9]. The unusual properties of this system made it a clear candidate for a bound gluon state. A PWA

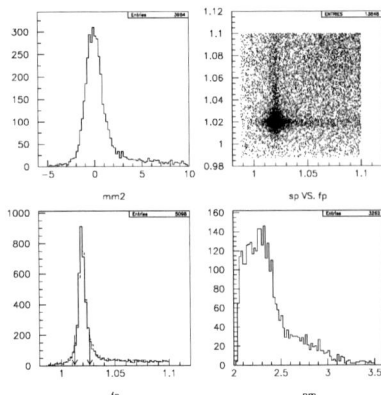

FIGURE 3. Missing mass squared minus proton mass squared, and K^+K^- invariant mass, first vs. second pair for events in reaction (3) (upper left). K^+K^- invariant mass, when the other pair lies in the ϕ–mass band, and $\phi\phi$ invariant mass (lower left).

performed by this experiment showed that three 2^{++} waves were necessary to comprise all of the cross section[10]. Due to low statistics, experiment WA76 was not able to perform a PWA for the central production of this system [5].

Using 90% of the E690 data, at IFUG we studied final state (3) requiring a primary vertex in the LH_2 target with two positive and two negative tracks assigned to it, in addition to a fast proton. At least one of the four charged tracks was required to be identified by the Cherenkov counter as an ambiguous K/p, the other were required to be compatible with a K identity. No direct measurement was made of the slow proton, but a kinematical cut of $p_z < 250$ MeV/c or $\arctan(p_t/p_z) > 45°$ for the missing momentum was used to require that it was outside the acceptance of the detector. Fig.3 shows the missing mass squared minus proton mass squared, and K^+K^- invariant mass for events in reaction (3), and the first versus second pair mass when $-2 < MM^2 - m_p^2 < 2$ GeV^2/c^4 (upper left). The lower left plots shows the K^+K^- invariant mass when the other pair lies in the ϕ–mass band of $1.0124 < m(K^+K^-) < 1.0264$ GeV/c^2, and with $-2 < MM^2 - m_p^2 < 2$ GeV^2/c^4, and the $\phi\phi$ invariant mass with all cuts.

Six angles are used to specify the spin and angular momentum of the $\phi\phi$ system, each ϕ decaying into a K^+K^- pair. Three of them were defined as the GJ angles of one of the ϕ mesons, in the rest frame of the $\phi\phi$ system, with the z-axis in the direction of $\vec{p}_{fast} - \vec{p}_{beam}$, and the y-axis perpendicular to the plane formed by $\vec{p}_{fast} - \vec{p}_{beam}$ and $\vec{p}_{slow} - \vec{p}_{tgt}$, in this frame. The rest were defined as the GJ angles for the K^+ of one ϕ, in the ϕ's rest frame, with the z'-axis in the direction of $\vec{p}_\phi$, and the y'-axis in the $\hat{z}$-$\hat{z}'$ plane.

The basis vectors for a given allowable combination of the total angular momentum of the $\phi\phi$ system J, orbital angular momentum L, $M = |J_z|$, Parity P, and exchange

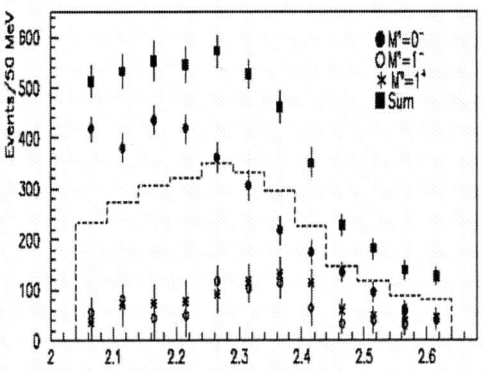

FIGURE 4. Preliminary PWA results.

naturality η, are given by [11]

$$G^{J^P LSM\eta}(\gamma,\beta,\alpha_1,\alpha_2,\theta_1,\theta_2) = \text{Real}\left[\frac{(1-i)-\eta(1+i)}{2}\sum_{\mu,\lambda}C(1,1,S|\mu,-\lambda)\times\right.$$
$$\left.C(L,S,J|0,\mu-\lambda)e^{-iM\gamma}e^{i\mu\alpha_1}e^{i\lambda\alpha_2}d^J_{M,\mu-\lambda}(\beta)\,d^1_{\mu,0}(\theta_1)\,d^1_{\lambda,0}(\theta_2)\right] \quad (5)$$

$I=0$ and $C=+$ for the $\phi\phi$ system, and Bose statistics require that $L+S=0,2,....$

The preliminary results of a PWA [12] of this system using all 54 waves with J spin up to 4, and projections $M = 0,1,2$, show production of one single 2^{++} wave ($S2$), with two projections, $M = 0$ and $M = 1$, as can be seen in Fig.(4). An amplitude analysis of the system is underway at IFUG.

THE E/ι PUZZLE

The study of mesons decaying to $K\bar{K}\pi$ is an interesting subject since in this system it is possible to observe the hybrid meson candidate $f_1(1420)$ [3], and a possible answer may be given to the "E/ι puzzle" [13]. In order to study this state, we used 10% of the E690 data.

Reaction (4) was selected by requiring a primary vertex in the LH_2 target, with two oppositely charged tracks, a K_s, and a fast proton assigned to it. At least one of the charged tracks had to be identified by the Cerenkov counter as either a π or an ambiguous K/p, the other track having an identity compatible with the final state. No direct measurement of the slow proton was made. A minimum gap of 1.8 units of rapidity between the slow proton and each individual meson was required to avoid Δ^{++} or $\Lambda(1520)$ contamination in the final state.

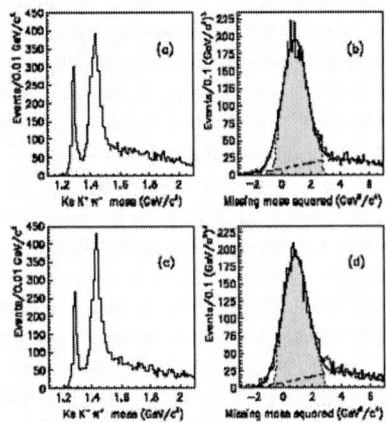

FIGURE 5. $K_s K \pi$ invariant mass distributions and missing mas squared distributions for both charges of reaction (4).

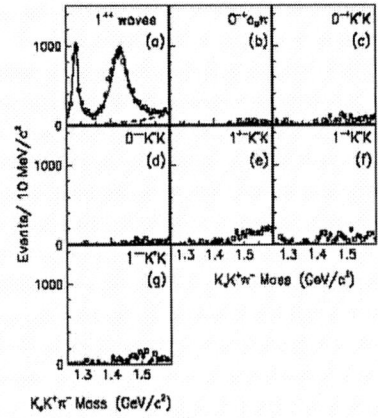

FIGURE 6. PWA results for the $K_s K \pi$ system.

The Dalitz distributions for the system showed $1^{++} K^* K$ dominance in the $f_1(1420)$ region, and a PWA of the system was performed to confirm this conclusion [14]. The decay of the system was characterized by its isobar mass, the Gottfried-Jackson (GJ) angles (θ, ϕ) of the bachelor particle in the $K\bar{K}\pi$ c.m. frame, and the GJ angles (α, γ) of the $K^{\pm}$ in the isobar rest frame. The PWA results are shown in Fig.(6) for the final state $K_s K^+ \pi^-$. As expected, the $f_1(1285)$ region was dominated by $1^{++} a_0 \pi$ wave, while the $f_1(1420)$ region was clearly dominated by $1^{++} K^* K$ wave. In both cases, each wave had similar contributions of $|J_z|^\eta = 1^{\pm}$.

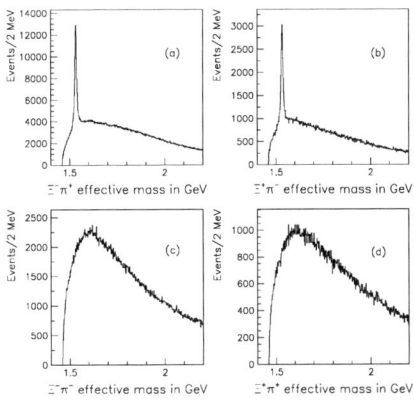

FIGURE 7. The $\Xi\pi$ system: (a) $\Xi^-\pi^+$ system, (b) $\overline{\Xi}^+\pi^-$, (c) $\Xi^-\pi^-$, and (d) $\overline{\Xi}^+\pi^+$.

SEARCH FOR AN EXOTIC PENTAQUARK

Beginning in 2003, a number of different experiments reported the observation of the exotic pentaquark candidates $\Theta^+(1540)$ [15, 16, 17, 18, 19, 20, 21, 22], and Ξ^{++} [23]. The case for the existence of these states has been contradictory from the beginning, and many experiments have failed to confirm the positive observations.

In order to look for the $\Xi^{++}(1860)$, we conducted a high-statistics, sensitive search for new resonances decaying to $\Xi^-\pi^+$, $\overline{\Xi}^+\pi^-$, $\Xi^-\pi^-$, and $\overline{\Xi}^+\pi^+$, using E690 data. The invariant mass spectrum for each of these systems can be seen in Fig. (7).

We observed strong signals for both the $\Xi^0(1530)$ and the $\overline{\Xi}^0(1530)$, but no other peak was observed in any of the four $\Xi\pi$ effective mass distributions. If a Gaussian line shape with $\sigma = 18$ MeV was assumed, the number of $\Xi^-\pi^-$ produced at 1862 MeV was found to be less than 0.6% of the observed number of $\Xi^0(1530) \to \Xi^-\pi^+$. The limit for the $\Xi^-\pi^+$ final state was 1.6% of the $\Xi^0(1530)$ yield. The limit for the $\overline{\Xi}^+\pi^-$ and $\overline{\Xi}^+\pi^+$ final states were 1.4% and 2.5% of the observed number of $\overline{\Xi}^0(1530) \to \overline{\Xi}^+\pi^-$. No evidence was found for a state near 1862 MeV [24].

ACKNOWLEDGMENTS

This work was funded in part CONACyT de México under Grant No. SEP-2003-C02-45364.

REFERENCES

1. The E690 Collaboration: D.C.Christian, J.Felix, E.E.Gottschalk, G.Gutiérrez, E.P.Hartouni, B.C.Knapp, M.N.Kreisler, G.Moreno, M.A.Reyes, M.Sosa, M.H.L.S.Wang, A.Wehmann.

2. The E690 spectrometer previously used in BNL E766 is described in J.Uribe *et al., Phys.Rev.* **D49**, 4373 (1994). The beam chambers are described in D.C.Christian *et al., Nucl.Instum.Methods Phys.Res.*, Sect **A345**, 62 (1994).
3. C.Caso *et al., Euro.Phys.Jour.* **C3**, 1 (1998).
4. T.A.Armstrong *et al., Phys.Lett. B* **227**, 186 (1989).
5. T.A.Armstrong *et al., Phys.Lett.* **B166**, 245 (1986); **B221**, 221 (1989). D.Barberis *et al., Phys.Lett.* **B432**, 436 (1998).
6. M.A.Reyes *et al., Phys.Rev.Lett.* **81**, 4079 (1998).
7. J.Z.Bai *et al., Phys.Rev.Lett.* **76**, 3502 (1996).
8. D.Barberis *et al., Phys.Lett.* **B453**, 305 (1999).
9. A.Etkin *et al., Phys.Rev.Lett.* **40**, 422 (1978).
10. A.Etkin *et al., Phys.Rev.Lett.* **49**, 1620 (1982); A.Etkin *et al., Phys.Lett.* **B165**, 217 (1985); **B201**, 568 (1988).
11. R.S.Longacre, *AIP Conf. Proc.* **113**, 0051 (1984). T.A.Armstrong *et al., Nucl.Phys. B* **196**, 176 (1982).
12. M.A.Reyes *et al., AIP Conf. Proc.* **717**, 135 (2001).
13. F.Nichitiu, in *Proceedings of Hadron'95*, edited by M.C.Birse, G.D.Laferty, and J.A.McGovern (World Sci., Singapore, 1996), p. 164; A.Lanaro, *Nucl.Phys. B, Proc. Suppl.* **56A**, 136 (1997).
14. M.Sosa *et al., Phys.Rev.Lett.* **83**, 913 (1999).
15. T. Nakano *et al.* (LEPS Collaboration), Phys. Rev. Lett **91**, 012002 (2003).
16. V. V. Barmin *et al.* (DIANA Collaboration), Phys. Atom. Nucl. **66**, 1715 (2003).
17. S. Stepanyan *et al.* (CLAS Collaboration), Phys. Rev. Lett **91**, 252001 (2003).
18. J. Barth *et al.* (SAPHIR Collaboration), Phys. Lett. B **572**, 127 (2003).
19. A. Airapetian, *et al.* (HERMES Collaboration), Phys. Lett. **B585**, 213 (2004).
20. S. Chekanov, *et al.* (ZEUS Collaboration), Phys. Lett. **B591**, 7 (2004).
21. M. Abdel-Bary, *et al.* (COSY-TOF Collaboration), Phys. Lett. **B595**, 127 (2004).
22. V. Kubarovsky, *et al.* (CLAS Collaboration), Phys. Rev. Lett. **92**, 032001 (2004).
23. C. Alt *et al.* (NA49 Collaboration), Phys. Rev. Lett.**92**, 042003 (2004).
24. D.C. Christian *et al.* (E690 Collaboration), Phys. Rev. Lett.**95**, 152001 (2005).

Proceedings Published by the Division of Particles and Fields of the Mexican Physical Society

MEXICAN SCHOOLS ON PARTICLES AND FIELDS

I. Oaxtepec, Morelos; AIP Conf. Proceedings 143, 1986; editors: J.L. Lucio, M. Moreno, A. Zepeda

II. Cuernavaca, Morelos; World Scientific, 1987; editors: J.L. Lucio, A. Zepeda

III. Oaxtepec, Morelos; World Scientific, 1989; editors: J.L. Lucio, A. Zepeda

IV. Oaxtepec, Morelos; World Scientific, 1991; editors: J.L. Lucio, A. Zepeda

V. Guanajuato, Gto.; AIP Conf. Proceedings 317, 1993; editors: J.L. Lucio, A. Zepeda

VI. Villahermosa, Tabasco; World Scietific 1995; editors: J.C. D'Olivo, M. Moreno, M.A. Pérez

VII. Mérida, Yucatán; AIP Conf. Proceedings 400, 1997; editors: J.C. D'Olivo, M. Klein-Kreisler, H. Méndez

VIII. Oaxaca, Oax.; AIP Conf. Proceedings 490, 1999; editors: J.C. D'Olivo, G. López-Castro, M. Mondragón

IX. Metepec, Puebla; AIP Conf. Proceedings 562, 2001; editors: G. Herrera, L. Nellen

X. Playa del Carmen, Quintana Roo, AIP Conf. Proceedings 670, 2003; editors: U. Cotti, M. Mondragón, G. Tavares-Velasco

XI. Xalapa, Veracruz; Journal of Physics Conf. Serv. 18, 2005; editors: A. Bashir, J. Erler, R. Hernández, M. Mondragón, L. Villaseñor

MEXICAN WORKSHOPS ON PARTICLES AND FIELDS

II. Puebla, Pue.: Rev. Mex. Fís., **36** Supl. 1, 1990; editors: S.R. Juárez, M.A. Pérez, A. Queijeiro.

IV. Mérida, Yucatán; World Scientific, 1994; editors: R. Huerta, M.A. Pérez, L. Urrutia

V. Puebla, Pue.; AIP Conf. Proceedings 359, 1996; editors: J.C. D'Olivo, A. Fernández, M.A. Pérez

VI. Morelia, Michoacán; AIP Conf. Proceedings 445, 1998; editors: J.C. D'Olivo, M. Ruiz-Altaba, L. Villaseñor

VII. Mérida, Yucatán; AIP Conf. Proceedings 531, 2000; editors: A. Ayala, G. Contreras, G. Herrera

VIII. Zacatecas, Zac.; AIP Conf. Proceedings 623, 2002; editors: L. Díaz-Cruz, J. Engelfried, M. Kirchbach, M. Mondragón

IX. Colima, Col.; Journal of Physics Conf. Serv. 37, 2006; editors: P. Amore, A. Aranda, A. Bashir, M. Mondragón, A. Raya

X. Morelia, Mich.; AIP Conf. Proceedings 857A, 2006; editors: A. Bashir, V. Villanueva, L. Villaseñor. AIP Conf. Proceedings 857B, 2006; DPF-MPS Commemorative Volume, editors: M.A. Pérez, L.F. Urrutia, L. Villaseñor

AUTHOR INDEX

A
Alfaro, R., 373
Ayala, A., 144

B
Bashir, A., 283

C
Carrillo-Moreno, S., 311
Contreras, J. G., 364
Cruz-Estrada, P., 395

D
de la Macorra, A., 191
Diaz Cruz, J. L., 59
D'Olivo, J. C., 37

E
Engelfried, J., 340

F
Felix, J., 320
Fernández T., A., 407
Flores-Mendieta, R., 27

G
García, A., 3
García, J. A., 228
García-Compeán, H., 167, 218
Güijosa, A., 167

H
Herrera-Corral, G., 300
Hess, P. O., 118

J
Juárez W., S. R., 79

K
Kielanowski, P., 79
Konigsberg, J., 297

L
Larios, F., 133
Lederman, L. M., 15
López-Fernández, R., 330
Lucio, J. L., 11, 49

M
Martínez, A., 27
Matos, T., 191
Mielke, E. W., 246
Miranda, O. G., 37
Mochán, W. L., 235
Mondragón, A., 266
Mondragón, M., 66
Mondragón-Contreras, L., 395
Montaño-Zetina, L. M., 355
Morales-Técotl, H. A., 205
Morelos, A., 347
Moreno, G., 419

N
Napsuciale, M., 49

O

Obregón, O., 258

P

Paić, G., 300
Pérez, M. A., 3
Pérez-Lorenzana, A., 152

R

Ramírez, C., 258
Ramírez-Jiménez, F. J., 395
Raya, A., 283
Reyes, M. A., 419
Rosado, A., 90

S

Salazar, H., 382
Stephens, C. R., 181

T

Toscano, J. J., 103

U

Urrutia, L. F., 205

V

Vázquez-Valencia, E. F., 311
Vergara, J. D., 218
Villarreal, C., 235
Villaseñor, L., 382

Z

Zepeda, A., 11, 128